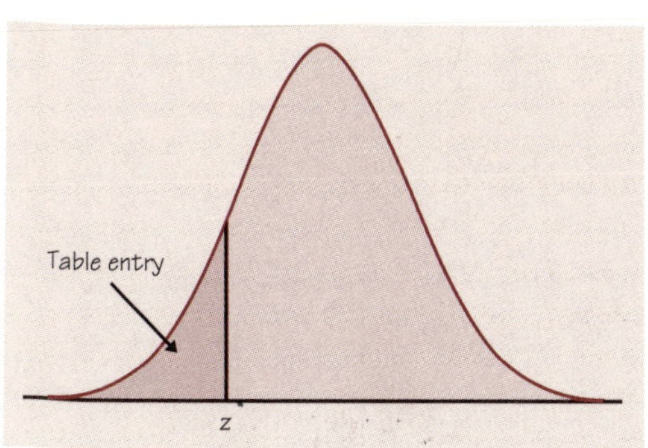

Table entry for *z* is the area under the standard normal curve to the left of *z*.

TABLE A Standard normal probabilities

z	.00	.01	.02	.03	.04	.05	.06	.07	.08	.09
−3.4	.0003	.0003	.0003	.0003	.0003	.0003	.0003	.0003	.0003	.0002
−3.3	.0005	.0005	.0005	.0004	.0004	.0004	.0004	.0004	.0004	.0003
−3.2	.0007	.0007	.0006	.0006	.0006	.0006	.0006	.0005	.0005	.0005
−3.1	.0010	.0009	.0009	.0009	.0008	.0008	.0008	.0008	.0007	.0007
−3.0	.0013	.0013	.0013	.0012	.0012	.0011	.0011	.0011	.0010	.0010
−2.9	.0019	.0018	.0018	.0017	.0016	.0016	.0015	.0015	.0014	.0014
−2.8	.0026	.0025	.0024	.0023	.0023	.0022	.0021	.0021	.0020	.0019
−2.7	.0035	.0034	.0033	.0032	.0031	.0030	.0029	.0028	.0027	.0026
−2.6	.0047	.0045	.0044	.0043	.0041	.0040	.0039	.0038	.0037	.0036
−2.5	.0062	.0060	.0059	.0057	.0055	.0054	.0052	.0051	.0049	.0048
−2.4	.0082	.0080	.0078	.0075	.0073	.0071	.0069	.0068	.0066	.0064
−2.3	.0107	.0104	.0102	.0099	.0096	.0094	.0091	.0089	.0087	.0084
−2.2	.0139	.0136	.0132	.0129	.0125	.0122	.0119	.0116	.0113	.0110
−2.1	.0179	.0174	.0170	.0166	.0162	.0158	.0154	.0150	.0146	.0143
−2.0	.0228	.0222	.0217	.0212	.0207	.0202	.0197	.0192	.0188	.0183
−1.9	.0287	.0281	.0274	.0268	.0262	.0256	.0250	.0244	.0239	.0233
−1.8	.0359	.0351	.0344	.0336	.0329	.0322	.0314	.0307	.0301	.0294
−1.7	.0446	.0436	.0427	.0418	.0409	.0401	.0392	.0384	.0375	.0367
−1.6	.0548	.0537	.0526	.0516	.0505	.0495	.0485	.0475	.0465	.0455
−1.5	.0668	.0655	.0643	.0630	.0618	.0606	.0594	.0582	.0571	.0559
−1.4	.0808	.0793	.0778	.0764	.0749	.0735	.0721	.0708	.0694	.0681
−1.3	.0968	.0951	.0934	.0918	.0901	.0885	.0869	.0853	.0838	.0823
−1.2	.1151	.1131	.1112	.1093	.1075	.1056	.1038	.1020	.1003	.0985
−1.1	.1357	.1335	.1314	.1292	.1271	.1251	.1230	.1210	.1190	.1170
−1.0	.1587	.1562	.1539	.1515	.1492	.1469	.1446	.1423	.1401	.1379
−0.9	.1841	.1814	.1788	.1762	.1736	.1711	.1685	.1660	.1635	.1611
−0.8	.2119	.2090	.2061	.2033	.2005	.1977	.1949	.1922	.1894	.1867
−0.7	.2420	.2389	.2358	.2327	.2296	.2266	.2236	.2206	.2177	.2148
−0.6	.2743	.2709	.2676	.2643	.2611	.2578	.2546	.2514	.2483	.2451
−0.5	.3085	.3050	.3015	.2981	.2946	.2912	.2877	.2843	.2810	.2776
−0.4	.3446	.3409	.3372	.3336	.3300	.3264	.3228	.3192	.3156	.3121
−0.3	.3821	.3783	.3745	.3707	.3669	.3632	.3594	.3557	.3520	.3483
−0.2	.4207	.4168	.4129	.4090	.4052	.4013	.3974	.3936	.3897	.3859
−0.1	.4602	.4562	.4522	.4483	.4443	.4404	.4364	.4325	.4286	.4247
−0.0	.5000	.4960	.4920	.4880	.4840	.4801	.4761	.4721	.4681	.4641

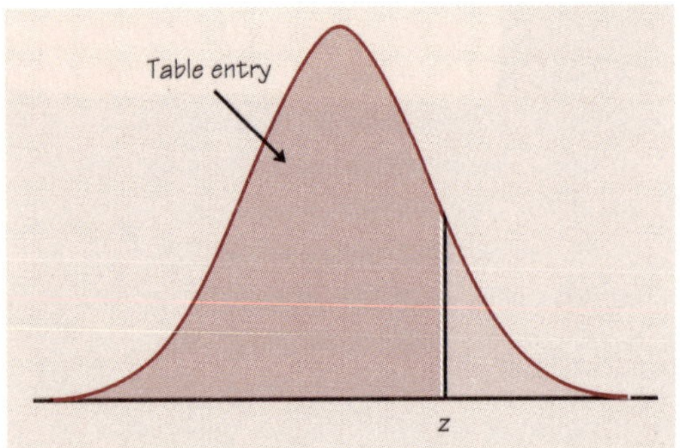

Table entry for *z* is the area under the standard normal curve to the left of *z*.

TABLE A Standard normal probabilities *(continued)*

z	.00	.01	.02	.03	.04	.05	.06	.07	.08	.09
0.0	.5000	.5040	.5080	.5120	.5160	.5199	.5239	.5279	.5319	.5359
0.1	.5398	.5438	.5478	.5517	.5557	.5596	.5636	.5675	.5714	.5753
0.2	.5793	.5832	.5871	.5910	.5948	.5987	.6026	.6064	.6103	.6141
0.3	.6179	.6217	.6255	.6293	.6331	.6368	.6406	.6443	.6480	.6517
0.4	.6554	.6591	.6628	.6664	.6700	.6736	.6772	.6808	.6844	.6879
0.5	.6915	.6950	.6985	.7019	.7054	.7088	.7123	.7157	.7190	.7224
0.6	.7257	.7291	.7324	.7357	.7389	.7422	.7454	.7486	.7517	.7549
0.7	.7580	.7611	.7642	.7673	.7704	.7734	.7764	.7794	.7823	.7852
0.8	.7881	.7910	.7939	.7967	.7995	.8023	.8051	.8078	.8106	.8133
0.9	.8159	.8186	.8212	.8238	.8264	.8289	.8315	.8340	.8365	.8389
1.0	.8413	.8438	.8461	.8485	.8508	.8531	.8554	.8577	.8599	.8621
1.1	.8643	.8665	.8686	.8708	.8729	.8749	.8770	.8790	.8810	.8830
1.2	.8849	.8869	.8888	.8907	.8925	.8944	.8962	.8980	.8997	.9015
1.3	.9032	.9049	.9066	.9082	.9099	.9115	.9131	.9147	.9162	.9177
1.4	.9192	.9207	.9222	.9236	.9251	.9265	.9279	.9292	.9306	.9319
1.5	.9332	.9345	.9357	.9370	.9382	.9394	.9406	.9418	.9429	.9441
1.6	.9452	.9463	.9474	.9484	.9495	.9505	.9515	.9525	.9535	.9545
1.7	.9554	.9564	.9573	.9582	.9591	.9599	.9608	.9616	.9625	.9633
1.8	.9641	.9649	.9656	.9664	.9671	.9678	.9686	.9693	.9699	.9706
1.9	.9713	.9719	.9726	.9732	.9738	.9744	.9750	.9756	.9761	.9767
2.0	.9772	.9778	.9783	.9788	.9793	.9798	.9803	.9808	.9812	.9817
2.1	.9821	.9826	.9830	.9834	.9838	.9842	.9846	.9850	.9854	.9857
2.2	.9861	.9864	.9868	.9871	.9875	.9878	.9881	.9884	.9887	.9890
2.3	.9893	.9896	.9898	.9901	.9904	.9906	.9909	.9911	.9913	.9916
2.4	.9918	.9920	.9922	.9925	.9927	.9929	.9931	.9932	.9934	.9936
2.5	.9938	.9940	.9941	.9943	.9945	.9946	.9948	.9949	.9951	.9952
2.6	.9953	.9955	.9956	.9957	.9959	.9960	.9961	.9962	.9963	.9964
2.7	.9965	.9966	.9967	.9968	.9969	.9970	.9971	.9972	.9973	.9974
2.8	.9974	.9975	.9976	.9977	.9977	.9978	.9979	.9979	.9980	.9981
2.9	.9981	.9982	.9982	.9983	.9984	.9984	.9985	.9985	.9986	.9986
3.0	.9987	.9987	.9987	.9988	.9988	.9989	.9989	.9989	.9990	.9990
3.1	.9990	.9991	.9991	.9991	.9992	.9992	.9992	.9992	.9993	.9993
3.2	.9993	.9993	.9994	.9994	.9994	.9994	.9994	.9995	.9995	.9995
3.3	.9995	.9995	.9995	.9996	.9996	.9996	.9996	.9996	.9996	.9997
3.4	.9997	.9997	.9997	.9997	.9997	.9997	.9997	.9997	.9997	.9998

THE PRACTICE OF STATISTICS second edition

TI-83/89 Graphing Calculator Enhanced

DANIEL S. YATES
Statistics consultant

DAVID S. MOORE
Purdue University

DAREN S. STARNES
The Webb Schools

W. H. Freeman and Company
New York

Senior Acquisitions Editor: Patrick Farace
Development Editor: Mary Johenk
Associate Editor: Danielle Swearengin
Publisher: Michelle Russel Julet
Director of High School Sales and Marketing: Mike Saltzman
New Media Editor: Brian Donnellan
Project Editing and Composition: Publication Services
Cover and Text Designer: Blake Logan
Illustrations: Publication Services
Illustration Coordinator: Bill Page
Photo Editor: Meg Kuhta
Production Coordinator: Susan Wein
Manufacturing: R R Donnelly & Sons Company

Library of Congress Cataloging-in-Publication Data

Yates, Daniel S.
The practice of statistics : TI-83/89 graphing calculator enhanced.–2nd ed. / Daniel S. Yates, David S. Moore, Daren S. Starnes.
p. cm.
Includes bibliographical references and index.
ISBN 0-7167-4773-1
1. Mathematical statistics. I. Moore, David S. II. Starnes, Daren S. III. Title.

QA276.12.Y37 2003
519.5–dc21 2001055610

Printed in the United States of America

First printing 2002

contents

preface

The Practice of Statistics, Second Edition: *TI-83/89 Graphing Calculator Enhanced* is an introductory text that focuses on data and statistical reasoning. It is intended for high school, college, and university students whose primary technological tool is the TI-83 or TI-89 graphing calculator. This book is based on the successful college texts *The Basic Practice of Statistics* (BPS) by David Moore and *Introduction to the Practice of Statistics* (IPS) by David Moore and George McCabe.

Statisticians have reached general consensus about the nature of a modern introductory statistics course. A joint committee of the American Statistical Association and the Mathematical Association of America summarized this consensus as follows:

- Emphasize statistical thinking
- Present more data and concepts with less theory and fewer recipes
- Foster active learning

The College Board* developed its Advanced Placement Statistics syllabus around these guidelines. As a result, this text follows the AP syllabus quite closely.

The Practice of Statistics takes advantage of the simulation, graphing, and computation capabilities of the TI-83/89 to promote active learning. By using the TI-83/89, students can focus on statistical concepts rather than on calculations. Although the book is elementary in the level of mathematics required and in the statistical procedures presented, it aims to give students an understanding of the main ideas of statistics and useful skills for working with data. Examples and exercises, though intended for beginners, use real data and give enough background to allow students to consider the meaning of their calculations.

THE SECOND EDITION

In revising *The Practice of Statistics*, we have attempted to build on the elements that made the first edition successful:

- Clear, concise explanations
- Interesting examples and exercises
- Essential content for the AP exam
- Activities and simulations to motivate statistical concepts
- Integration of the TI-83 graphing calculator

On the basis of classroom experiences and a careful examination of AP exam content, we have made some minor content-related changes. In addition to rewriting some explanations for greater clarity, we:

Added Ogives, linear transformations, normal approximation to the binomial distribution, and combining normal random variables, derivations of formulas using rules for random variables

Revised/expanded Comparing distributions, normal probability plots, r^2, transforming to achieve linearity, probability rules, geometric settings, Type I and Type II errors and power, tests of association/independence versus homogeneity of populations

NEW FEATURES

- **TI-89 graphing calculator** has been integrated alongside the TI-83. Screen shots and keystroke instructions are provided.

- **Historical vignettes** at the beginning of each chapter provide interesting context for the material that follows.

- **Technology toolboxes** show students how to use the statistical capabilities of the TI-83/89 and provide tips for interpreting standard computer output.

- **Highlighted application titles** at the beginning of examples and exercises emphasize the variety and abundance of real-life applications.

- **Simulations** help students investigate and learn statistical principles.

The icon marks examples and exercises in the text that involve simulations.

- **Inference Toolbox** Introduced in Chapter 10 as a way for students to organize their inference procedures. Examples that use this structure are marked with a .

- **Examples and Exercises** Every chapter includes an increased number of examples and exercises, with more real data and source references. As in the first edition, each main idea is followed by a short set of exercises for immediate reinforcement.

SUPPLEMENTS

A full range of supplements is available to help teachers and students use *The Practice of Statistics*. **For students:**

- **Student CD-ROM/Formula Card** Contains more than 80 statistical case studies ("EESEE"), most of the data sets from the text, Q & A self-quizzes, supplemental chapters in PDF format, a collection of interactive java applets, and a formula card. ISBN: 0-7167-93997.

- **Prep for Exam Supplement** By Larry J. Peterson, Bonneville High School, Ogden, Utah, provides sample exam questions that will help students refresh their skills and prepare for the actual AP exam. ISBN: 0-7167-9615-5.

- **Web site** www.whfreeman.com/tps featuring interactive Java applets, data sets from the text, extra case studies, online quizzes with instant feedback, and Prep for Exam exercises.

For teachers:

- **Golden Resource Binder (GRB)** (ISBN: 0-7167-8345-2) An indispensable resource for the first edition, the "School Bus Yellow Binder" helped teachers manage their courses, assess their students' progress on a frequent basis, and prepare their students for the AP exam. For the second edition, **the yellow binder has gone gold**. The new GRB, written by Dan Yates and Daren Starnes, offers:

    - Two or more sample quizzes for every section of every chapter
    - At least two sample chapter tests for each chapter and two sample exams for each semester
    - Extensive teaching suggestions, including additional examples, for each chapter
    - Additional activities and simulations
    - "Special Problems," which are more in-depth investigations designed to help students improve their written communication and statistical analysis skills
    - An extensive list of references to text and journal articles, video series, sources of data in both print and electronic formats, and interesting statistics-related Web sites
    - Prep for Exam questions—AP-style free response problems for use after every chapter.
    - **Instructor's Solutions Manual** by Christopher Barat, Virginia State University, contains completely worked solutions to all problems in the text.
    ISBN: 0-7167-8344-4.

- **Instructor's CD** (ISBN: 07167-9760-7) contains all the material from the student CD, plus:

    - Supplemental chapters in pdf format
        - Chapter 16: Multiple Linear Regression
        - Chapter 17: Logistic Regression
    - Word versions of the tests and quizzes from the Golden Resource Binder.
    - Added features to the Encyclopedia of Statistical Examples and Exercises (EESEE) which provide solutions to the student exercises.
    - All text figures in an exportable presentation format, JPEG for Windows and PICT for Macintosh users.
    - The Instructor's guide in Adobe PDF format.
    - Presentation Manager Pro, which creates presentation for all figures and selected tables within the text.
    - PowerPoint Slides that can be used directly or customized to fit your needs. Each slide offers an outlined text to match the textbook's table of contents along with a wide range of figures and charts, culled from the book and other sources.
    - Applets that allow students to manipulate data and see the corresponding results graphically.
    - Data sets formatted for the graphing calculator.

• **Printed Test Bank,** revised by Daren Starnes, and by Michael Fligner and William Notz, both of Ohio State University, offers over 750 multiple-choice questions in the style of the AP exam. ISBN: 0-7167-8342-8.

• **Test Bank CD-ROM.** This easy-to-use version of the printed test bank includes Windows and Mac versions on a single disc, in a format that lets you add, edit, and resequence questions to suit your needs. ISBN:0-7167-8343-6.

• **Diploma Online Testing,** from the Brownstone Research Group, allows instructors to easily create and administer secure exams over a network and over the Internet, with questions that incorporate multimedia and interactive exercises. The program lets you restrict tests to specific computers or time blocks, and includes an impressive suite of gradebook and result-analysis features. For more information, visit http://www.brownstone.net.

• **Online quizzing** at the book's Web site allows instructors to easily and securely quiz students online using prewritten, multiple choice questions from each text chapter (not from the test bank). Students receive instant feedback and can take quizzes multiple times. Instructors can enter a protected site to view results, or can get results by e-mail.

ACKNOWLEDGMENTS

First and foremost, we extend our heartfelt thanks to Patrick Farace, acquisitions editor at W. H. Freeman and Company, for his unwavering support and enthusiasm during the preparation of the second edition. Special thanks go to development editor, Mary Johenk, for patiently trying to keep us on schedule and to copy editor, Pamela Bruton, for wading deliberately but swiftly through each and every manuscript page. We would also like to thank Foti Kutil and his staff at Publication Services for turning rough manuscripts into finished product.

We both owe a large debt of gratitude to Professor David Moore for shaping the way we think about and teach statistics. Likewise, we appreciate his many contributions, in both substance and spirit, to *The Practice of Statistics*.

To our colleagues who reviewed chapters of the manuscript and offered many constructive suggestions, we give our thanks.

Jared Derksen, Rancho Cucamonga High School, Rancho Cucamonga, CA
Ron Dirkse, American School in Japan, Tokyo, Japan
Glenn Gabanski, Oak Park and River Forest High School, Oak Park, IL
Duane C. Hinders, Henry M. Gunn High School, Palo Alto, CA
Beverly A. Johnson, Fort Worth Country Day School, Fort Worth, TX
Lee E. Kucera, Capistrano Valley High School, Mission Viejo, CA
Philip Mallinson, Phillips Exeter Academy, Exeter, NH
Stoney Pryor, A&M Consolidated High School, College Station, TX
Tom Robinson, Kentridge Senior High School, Kent, WA
Albert Roos, Lexington High School, Lexington, MA
David Stein, Paint Branch High School, Burtonsville, MD
David Thiel, Math/Science Institute, Las Vegas, NV
Joseph Robert Vignolini, Glen Cove High School, Glen Cove, NY

We wish to thank the many new and experienced AP Statistics teachers across the country who have crossed paths with us. Your ideas, questions, and enthusiasm for teaching statistics were invaluable in shaping the second edition. To our students past and present who have shown us how exciting teaching statistics can be, thank you for inspiring us.

Daniel Yates and Daren Starnes

I am grateful to my friend and colleague, Daren Starnes, for joining the team to revise this book. Daren has been teaching AP Statistics from the first edition of TPS for the past four years and had great insights for the enhancements to the second edition. His work on the TI-89 component and the technology toolboxes, in particular, has been invaluable. Daren's expertise and easy-to-read writing style have helped a good book become significantly better.

I would also like to acknowledge the major assistance provided by my wife, Betty Jo, who, along with many other tasks, has proofread every word in every chapter, some multiple times, and without whose continual encouragement and support this project might still be in progress.

Daniel Yates

When I started using *The Practice of Statistics*, I never imagined that I would have a hand in writing the sequel. My sincere thanks go to Dan Yates, my friend and fellow author, for giving me a once-in-a-lifetime opportunity. His ongoing support, encouragement, and direction kept me going on more occasions than I can remember. I also want to recognize two former colleagues: Wright Robinson, who showed me first-hand the power of statistics in environmental science and spurred me to become a writer, and Sheila McGrail, who learned statistics with me in our early years teaching the AP course.

To my students, past and present, thank you for inspiring me. I owe special gratitude to my 2001-02 AP Statistics class for field testing much of the second edition.

Finally, I want to thank three people who gave me the strength and courage to tackle this project. To my mother, Hulene Hill, thank you for teaching me the value of hard work. To my grandfather, Hulon Hill, thank you for believing unconditionally in me. And to my wife Judy, thank you for lifting me up and helping me follow my dreams.

Daren Starnes

STATISTICAL THINKING

HOW TO TELL THE FACTS FROM THE ARTIFACTS

Statistics is about data. Data are numbers, but they are not "just numbers." *Data are numbers with a context.* The number 10.5, for example, carries no information by itself. But if we hear that a friend's new baby weighed 10.5 pounds at birth, we congratulate her on the healthy size of the child. The context engages our background knowledge and allows us to make judgments. We know that a baby weighing 10.5 pounds is quite large, and that a human baby can't weigh 10.5 ounces or 10.5 kilograms. The context makes the number informative.

Statistics uses data to gain insight and to draw conclusions. Our tools are graphs and calculations, but the tools are guided by ways of thinking that amount to educated common sense. Let's begin our study of statistics with an informal look at some principles of statistical thinking.

Data illuminate

What percent of the American population do you think is black? What percent do you think is white? When white Americans were asked these questions, their average answers were 23.8% black and 49.9% white. In fact, the Census Bureau tells us that 12.3% of Americans are black and 75.1% are white.[1]

Race remains a central social issue in the United States. It is illuminating to see that whites think (wrongly) that they are a minority. The census data—what really is true about the U.S. population—are also illuminating. We wonder if knowing the facts might help change attitudes.

Data beat anecdotes

An anecdote is a striking story that sticks in our minds exactly because it is striking. Anecdotes humanize an issue, but they can be misleading.

Does living near power lines cause leukemia in children? The National Cancer Institute spent 5 years and $5 million gathering data on the question. Result: no connection between leukemia and exposure to magnetic fields of the kind produced by power lines. The editorial that accompanied the study report in the *New England Journal of Medicine* thundered, "It is time to stop wasting our research resources" on the question.[2]

Now compare the effectiveness of a television news report of a 5-year, $5 million investigation against a televised interview with an articulate mother whose child has leukemia and who happens to live near a power line. In the public mind, the anecdote wins every time. A statistically literate person, however, knows that data are more reliable than anecdotes because they systematically describe an overall picture rather than focusing on a few incidents.

Beware the lurking variable

Air travelers would like their flights to arrive on time. Airlines collect data about on-time arrivals and report them to the Department of Transportation. Here are one month's data for flights from several western cities for two airlines:

	On time	Delayed
Alaska Airlines	3274	501
America West	6438	787

You can see that the percentages of late flights were:

$$\text{Alaska Airlines} \quad \frac{501}{3775} = 13.3\%$$

$$\text{America West} \quad \frac{787}{7225} = 10.9\%$$

It appears that America West does better.

This isn't the whole story, however. Almost all relationships between two variables are influenced by other variables lurking in the background. We have data on two variables, the airline and whether or not the flight was late. Let's add data on a third variable, which city the flight left from.[3]

	Alaska Airlines		America West	
	On time	Delayed	On time	Delayed
Los Angeles	497	62	694	117
Phoenix	221	12	4840	415
San Diego	212	20	383	65
San Francisco	503	102	320	129
Seattle	1841	305	201	61
Total	3274	501	6438	787

The "Total" row shows that the new table describes the same flights as the earlier table. Look again at the percentages of late flights, first for Los Angeles:

$$\text{Alaska Airlines} \quad \frac{62}{559} = 11.1\%$$

$$\text{AmericaWest} \quad \frac{117}{811} = 14.4\%$$

Alaska Airlines wins. The percentage delayed for Phoenix are:

$$\text{Alaska Airlines} \quad \frac{12}{233} = 5.2\%$$

$$\text{AmericaWest} \quad \frac{415}{5255} = 7.9\%$$

Alaska Airlines wins again. In fact, as Figure 1 shows, Alaska Airlines has a lower percentage of late flights at *every one* of these cities.

How can it happen that Alaska Airlines wins at every city but America West wins when we combine all the cities? Look at the data: America West flies most often from sunny Phoenix, where there are few delays. Alaska Airlines flies most often from Seattle, where fog and rain cause frequent delays. What city we fly from has a major influence on the chance of a delay, so including the city data reverses our conclusion. The message is worth repeating: almost all relationships between two variables are influenced by other variables lurking in the background.

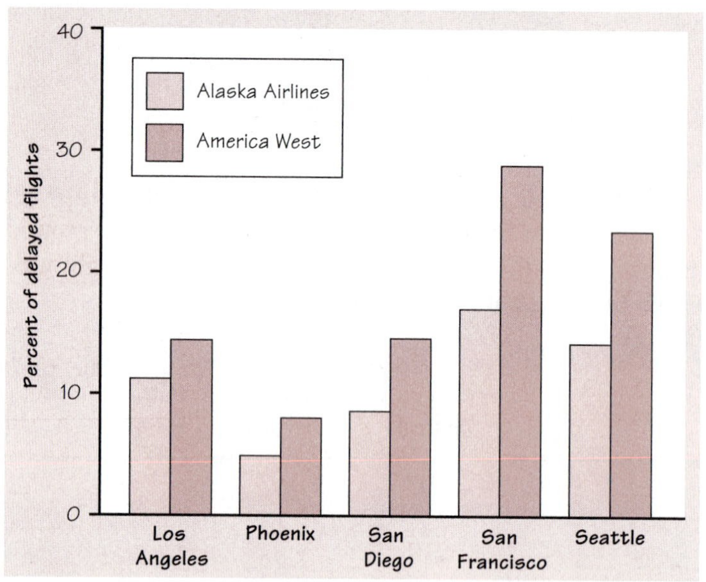

FIGURE 1 Comparing the percents of delayed flights for two airlines at five airports.

Where the data come from is important

The advice columnist Ann Landers once asked her readers, "If you had it to do over again, would you have children?" A few weeks later, her column was headlined "70% OF PARENTS SAY KIDS NOT WORTH IT." Indeed, 70% of the nearly 10,000 parents who wrote in said they would not have children if they could make the choice again. Do you believe that 70% of all parents regret having children?

You shouldn't. The people who took the trouble to write Ann Landers are not representative of all parents. Their letters showed that many of them were angry at their children. All we know from these data is that there are some unhappy parents out there. A statistically designed poll, unlike Ann Landers's appeal, targets specific people chosen in a way that gives all parents the same chance to be asked. Such a poll showed that 91% of parents *would* have children again.

The lesson: if you are careless about how you get your data, you may announce 70% "No" when the truth is close to 90% "Yes."

Variation is everywhere

The company's sales reps file into their monthly meeting. The sales manager rises. "Congratulations! Our sales were up 2% last month, so we're all drinking champagne this morning. You remember that when sales were down 1% last month I fired half of our reps." This picture is only slightly exaggerated. Many managers overreact to small short-term variations in key figures. Here is Arthur Nielsen, head of the country's largest market research firm, describing his experience:

> *Too many business people assign equal validity to all numbers printed on paper. They accept numbers as representing Truth and find it difficult to work with the concept of probability. They do not see a number as a kind of shorthand for a range that describes our actual knowledge of the underlying condition.*[4]

Business data such as sales and prices vary from month to month for reasons ranging from the weather to a customer's financial difficulties to the inevitable errors in gathering the data. The manager's challenge is to say when there is a real pattern behind the variation. Statistical tools can help. Sometimes it is enough to simply plot the data. Figure 2 plots the average price of a gallon of unleaded gasoline each month for a decade. Behind the month-to-month variation you can see a gradual upward trend, the higher prices during each summer driving season, the upward spike during the 1990 Gulf War, and the sudden drop in the spring of 1998.[5]

Variation is everywhere. Individuals vary; repeated measurements on the same individual vary; almost everything varies over time. One reason we need to know some statistics is that statistics helps us deal with variation.

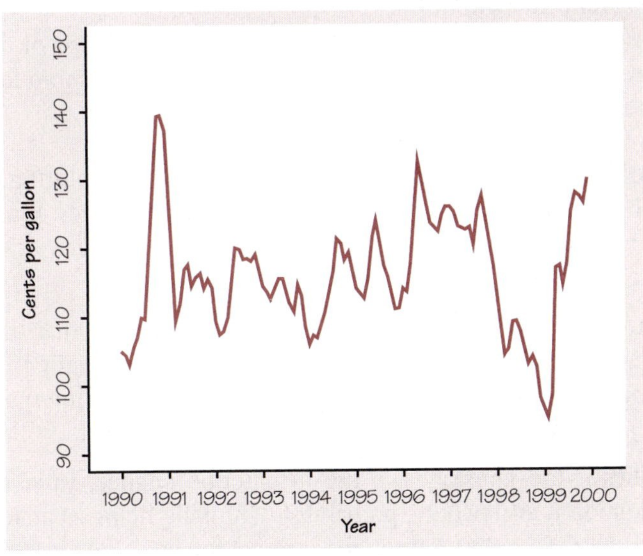

FIGURE 2 Variation is everywhere: the price at the pump of unleaded gasoline, January 1990 to December 1999.

Conclusions are not certain

Most women who reach middle age have regular mammograms to detect breast cancer. Do mammograms really reduce the risk of dying of breast cancer? To seek answers, doctors rely on "randomized clinical trials" that compare different ways of screening for breast cancer. We will see later that data from randomized comparative experiments are as good as it gets. The conclusion from 13 such trials is that mammograms reduce the risk of death in women aged 50 to 64 years by 26%.[6]

On the average, then, women who have regular mammograms are less likely to die of breast cancer. But because variation is everywhere, the results are different for different women. Some women who have mammograms every year die of breast cancer, and some who never have mammograms live to 100 and die when they crash their motorcycles. Can we be sure that mammograms reduce risk on the average? No, we can't be sure. *Because variation is everywhere, conclusions are uncertain.*

Statistics gives us a language for talking about uncertainty that is used and understood by statistically literate people everywhere. In the case of mammograms, the doctors use that language to tell us that "mammography reduces the risk of dying of breast cancer by 26 percent (95 percent confidence interval, 17 to 34 percent)." That 26% is, in Arthur Nielsen's words, "a shorthand for a range that describes our actual knowledge of the underlying condition." The range is 17% to 34%, and we are 95 percent confident that the truth lies in that range. We will soon learn to understand this language. We can't escape variation and uncertainty. Learning statistics enables us to live more comfortably with these realities.

STATISTICAL THINKING AND YOU

WHAT LIES AHEAD IN THIS BOOK

The purpose of this book is to give you a working knowledge of the ideas and tools of practical statistics. We will divide practical statistics into four parts that reflect our short introduction to statistical thinking:

1. **Data analysis** concerns methods and strategies for exploring, organizing, and describing data using graphs and numerical summaries. Only organized data can illuminate. Only thoughtful exploration of data can defeat the lurking variable. Chapters 1 through 4 discuss data analysis.

2. **Data production** provides methods for producing data that can give clear answers to specific questions. Where the data come from really is important—basic concepts about how to select samples and design experiments are the most influential ideas in statistics. These concepts are the subject of Chapter 5.

3. **Probability** is the language we use to describe chance, variation, and risk. Because variation is everywhere, probabilistic thinking helps separate reality from background noise. Chapters 6 through 9 present the essential ideas of probability.

4. **Statistical inference** moves beyond the data in hand to draw conclusions about a wider universe, taking into account that variation is everywhere and that conclusions are uncertain. Chapter 10 discusses the reasoning of statistical inference, and Chapters 11 through 15 present inference as used in practice in several settings.

Because data are numbers with a context, doing statistics means more than manipulating numbers. *The Practice of Statistics* is full of data, and each set of data has some brief background to help you understand what the data say. Examples and exercises usually express briefly some understanding gained from the data. In practice, you would know much more about the background of the data you work with and about the questions you hope the data will answer. No textbook can be fully realistic. But it is important to form the habit of asking "What do the data tell me?" rather than just concentrating on making graphs and doing calculations. This book tries to encourage good habits.

Nonetheless, statistics involves lots of calculating and graphing. The text presents the techniques you need, but you should use a calculator or computer software to automate calculations and graphs as much as possible.

Ideas and judgment can't (at least yet) be automated. They guide you in telling the computer what to do and in interpreting its output. This book tries to explain the most important ideas of statistics, not just teach methods.

You learn statistics by doing statistical problems. This book offers three types of exercises, arranged to help you learn. Short problem sets appear after each major idea. These are straightforward exercises that help you solidify the main points before going on. The Section Exercises at the end of each numbered section help you combine all the ideas of the section. Finally, the Chapter Review Exercises look back over the entire chapter. At each step you are given less advance knowledge of exactly what statistical ideas and skills the problems will require, so each step requires more understanding. Each chapter ends with a Summary section that includes a detailed list of specific things you should now be able to do. Go through that list, and be sure you can say "I can do that" to each item. Then try some chapter exercises.

The basic principle of learning is persistence. The main ideas of statistics, like the main ideas of any important subject, took a long time to discover and take some time to master. The gain will be worth the pain.

NOTES AND DATA SOURCES

PREFACE NOTES

1. D. S. Moore and discussants, "New pedagogy and new content: the case of statistics" *International Statistics Review*, 65 (1997), pp. 123–165. Richard Scheaffer's comment appears on page 156.
2. This summary of the committee's report was unanimously endorsed by the Board of Directors of the American Statistical Association. The full report is George Cobb, "Teaching statistics," in L. A. Steen (ed.), *Heeding the Call for Change: Suggestions for Curricular Action*, Mathematical Association of America, Washington, D.C., 1990, pp. 3–43.
3. David S. Moore and George P. McCabe, *Introduction to the Practice of Statistics*, 3rd ed., W. H. Freeman, New York, 1999.
4. A. C. Nielsen, Jr., "Statistics in Marketing," in *Making Statistics More Effective in Schools of Business*, Graduate School of Business, University of Chicago, 1986.
5. The data in Figure 2 are based on a component of the Consumer Price Index, from the Bureau of Labor Statistics on-line data center at http://stats.bls.gov:80/datahome.htm. To convert the index numbers into dollars, I used the national average price of $1.131 per gallon for January 1998, from the Energy Information Agency Web site, http://www.eia.doe.gov/.
6. H. C. Sox, "Editorial: benefit and harm associated with screening for breast cancer," *New England Journal of Medicine*, 338, No. 16 (1998).

PART I

Organizing Data:

Looking for Patterns and Departures from Patterns

1. Exploring Data
2. The Normal Distributions
3. Examining Relationships
4. More on Two-Variable Data

The Granger Collection, New York

FLORENCE NIGHTINGALE

Using Statistics to Save Lives

Florence Nightingale (1820–1910) won fame as a founder of the nursing profession and as a reformer of health care. As chief nurse for the British army during the Crimean War, from 1854 to 1856, she found that lack of sanitation and disease killed large numbers of soldiers hospitalized by wounds. Her reforms reduced the death rate at her military hospital from 42.7% to 2.2%, and she returned from the war famous. She at once began a fight to reform the entire military health care system, with considerable success.

One of the chief weapons Florence Nightingale used in her efforts was data. She had the facts, because she reformed record keeping as well as medical care. She was a pioneer in using graphs to present data in a vivid form that even generals and members of Parliament could understand. Her inventive graphs are a landmark in the growth of the new science of statistics. She considered statistics essential to understanding any social issue and tried to introduce the study of statistics into higher education.

In beginning our study of statistics, we will follow Florence Nightingale's lead. This chapter and the next will stress the analysis of data as a path to understanding. Like her, we will start with graphs to see what data can teach us. Along with the graphs we will present numerical summaries, just as Florence Nightingale calculated detailed death rates and other summaries. Data for Florence Nightingale were not dry or abstract, because they showed her, and helped her show others, how to save lives. That remains true today.

One of the chief weapons Florence Nightingale used in her efforts was data.

chapter 1

Exploring Data

ACTIVITY 1 How Fast Is Your Heart Beating?

Materials: Clock or watch with second hand

A person's pulse rate provides information about the health of his or her heart. Would you expect to find a difference between male and female pulse rates? In this activity, you and your classmates will collect some data to try to answer this question.

1. To determine your pulse rate, hold the *fingers* of one hand on the artery in your neck or on the inside of the wrist. (The thumb should not be used, because there is a pulse in the thumb.) Count the number of pulse beats in one minute. Do this three times, and calculate your *average* individual pulse rate (add your three pulse rates and divide by 3.) Why is doing this three times better than doing it once?
2. Record the pulse rates for the class in a table, with one column for males and a second column for females. Are there any unusual pulse rates?
3. For now, simply calculate the average pulse rate for the males and the average pulse rate for the females, and compare.

INTRODUCTION

Statistics is the science of data. We begin our study of statistics by mastering the art of examining data. Any set of data contains information about some group of *individuals*. The information is organized in *variables*.

INDIVIDUALS AND VARIABLES

Individuals are the objects described by a set of data. Individuals may be people, but they may also be animals or things.

A **variable** is any characteristic of an individual. A variable can take different values for different individuals.

A college's student data base, for example, includes data about every currently enrolled student. The students are the *individuals* described by the data set. For each individual, the data contain the values of *variables* such as age, gender (female or male), choice of major, and grade point average. In practice, any set of data is accompanied by background information that helps us understand the data.

When you meet a new set of data, ask yourself the following questions:

1. **Who?** What **individuals** do the data describe? **How many** individuals appear in the data?

2. **What?** How many **variables** are there? What are the **exact definitions** of these variables? In what **units** is each variable recorded? Weights, for example, might be recorded in pounds, in thousands of pounds, or in kilograms. Is there any reason to mistrust the values of any variable?

3. **Why?** What is the reason the data were gathered? Do we hope to answer some specific questions? Do we want to draw conclusions about individuals other than the ones we actually have data for?

Some variables, like gender and college major, simply place individuals into categories. Others, like age and grade point average (GPA), take numerical values for which we can do arithmetic. It makes sense to give an average GPA for a college's students, but it does not make sense to give an "average" gender. We can, however, count the numbers of female and male students and do arithmetic with these counts.

CATEGORICAL AND QUANTITATIVE VARIABLES

A **categorical variable** places an individual into one of several groups or categories.

A **quantitative variable** takes numerical values for which arithmetic operations such as adding and averaging make sense.

EXAMPLE 1.1 EDUCATION IN THE UNITED STATES

Here is a small part of a data set that describes public education in the United States:

State	Region	Population (1000)	SAT Verbal	SAT Math	Percent taking	Percent no HS	Teachers' pay ($1000)
⋮							
CA	PAC	33,871	497	514	49	23.8	43.7
CO	MTN	4,301	536	540	32	15.6	37.1
CT	NE	3,406	510	509	80	20.8	50.7
⋮							

Let's answer the three "W" questions about these data.

1. Who? The *individuals* described are the states. There are 51 of them, the 50 states and the District of Columbia, but we give data for only 3. Each row in the table describes one individual. You will often see each row of data called a ***case.***

case

2. What? Each column contains the values of one variable for all the individuals. This is the usual arrangement in data tables. Seven variables are recorded for each state. The first column identifies the state by its two-letter post office code. We give data for California, Colorado, and Connecticut. The second column says which region of the country the state is in. The Census Bureau divides the nation into nine regions. These three are Pacific, Mountain, and New England. The third column contains state populations, in thousands of people. Be sure to notice that the *units* are thousands of people. California's 33,871 stands for 33,871,000 people. The population data come from the 2000 census. They are therefore quite accurate as of April 1, 2000, but don't show later changes in population.

The remaining five variables are the average scores of the states' high school seniors on the SAT verbal and mathematics exams, the percent of seniors who take the SAT, the percent of students who did not complete high school, and average teachers' salaries in thousands of dollars. Each of these variables needs more explanation before we can fully understand the data.

3. Why? Some people will use these data to evaluate the quality of individual states' educational programs. Others may compare states on one or more of the variables. Future teachers might want to know how much they can expect to earn.

A variable generally takes values that vary. One variable may take values that are very close together while another variable takes values that are quite spread out. We say that the *pattern of variation* of a variable is its *distribution.*

DISTRIBUTION

The **distribution** of a variable tells us what values the variable takes and how often it takes these values.

exploratory data analysis

Statistical tools and ideas can help you examine data in order to describe their main features. This examination is called ***exploratory data analysis.*** Like an explorer crossing unknown lands, we first simply describe what we see. Each example we meet will have some background information to help us, but our emphasis is on examining the data. Here are two basic strategies that help us organize our exploration of a set of data:

- Begin by examining each variable by itself. Then move on to study relationships among the variables.
- Begin with a graph or graphs. Then add numerical summaries of specific aspects of the data.

We will organize our learning the same way. Chapters 1 and 2 examine single-variable data, and Chapters 3 and 4 look at relationships among variables. In both settings, we begin with graphs and then move on to numerical summaries.

EXERCISES

1.1 FUEL-EFFICIENT CARS Here is a small part of a data set that describes the fuel economy (in miles per gallon) of 1998 model motor vehicles:

Make and Model	Vehicle type	Transmission type	Number of cylinders	City MPG	Highway MPG
⋮					
BMW 318I	Subcompact	Automatic	4	22	31
BMW 318I	Subcompact	Manual	4	23	32
Buick Century	Midsize	Automatic	6	20	29
Chevrolet Blazer	Four-wheel drive	Automatic	6	16	20
⋮					

(a) What are the individuals in this data set?

(b) For each individual, what variables are given? Which of these variables are categorical and which are quantitative?

1.2 MEDICAL STUDY VARIABLES Data from a medical study contain values of many variables for each of the people who were the subjects of the study. Which of the following variables are categorical and which are quantitative?

(a) Gender (female or male)

(b) Age (years)

(c) Race (Asian, black, white, or other)

(d) Smoker (yes or no)

(e) Systolic blood pressure (millimeters of mercury)

(f) Level of calcium in the blood (micrograms per milliliter)

1.3 You want to compare the "size" of several statistics textbooks. Describe at least three possible numerical variables that describe the "size" of a book. In what *units* would you measure each variable?

1.4 Popular magazines often rank cities in terms of how desirable it is to live and work in each city. Describe five variables that you would measure for each city if you were designing such a study. Give reasons for each of your choices.

1.1 DISPLAYING DISTRIBUTIONS WITH GRAPHS

Displaying categorical variables: bar graphs and pie charts

The values of a categorical variable are labels for the categories, such as "male" and "female." The distribution of a categorical variable lists the categories and gives either the **count** or the **percent** of individuals who fall in each category.

EXAMPLE 1.2 THE MOST POPULAR SOFT DRINK

The following table displays the sales figures and market share (percent of total sales) achieved by several major soft drink companies in 1999. That year, a total of 9930 million cases of soft drink were sold.[1]

Company	Cases sold (millions)	Market share (percent)
Coca-Cola Co.	4377.5	44.1
Pepsi-Cola Co.	3119.5	31.4
Dr. Pepper/7-Up (Cadbury)	1455.1	14.7
Cott Corp.	310.0	3.1
National Beverage	205.0	2.1
Royal Crown	115.4	1.2
Other	347.5	3.4

How to construct a bar graph:

Step 1: Label your axes and title your graph. Draw a set of axes. Label the horizontal axis "Company" and the vertical axis "Cases sold." Title your graph.

Step 2: Scale your axes. Use the counts in each category to help you scale your vertical axis. Write the category names at equally spaced intervals beneath the horizontal axis.

Step 3: Draw a vertical bar above each category name to a height that corresponds to the count in that category. For example, the height of the "Pepsi-Cola Co." bar should be at 3119.5 on the vertical scale. *Leave a space between the bars in a bar graph.*

Figure 1.1(a) displays the completed bar graph.

How to construct a pie chart: Use a computer! Any statistical software package and many spreadsheet programs will construct these plots for you. Figure 1.1(b) is a pie chart for the soft drink sales data.

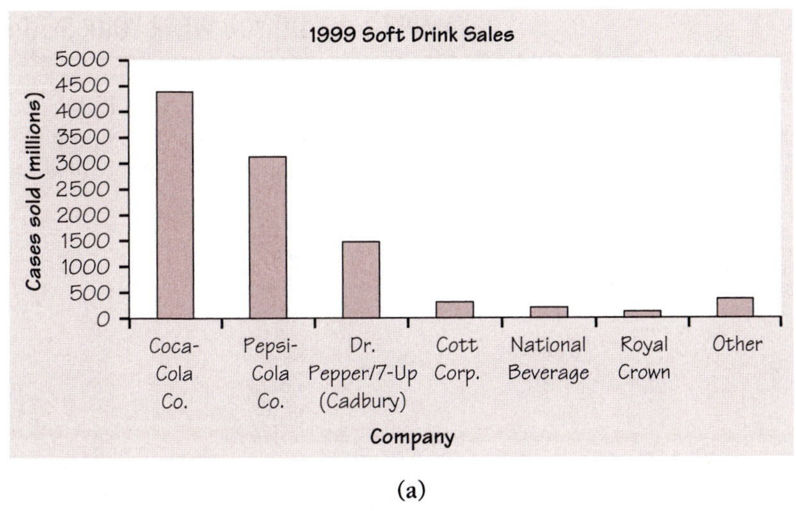

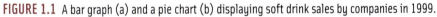

(a)

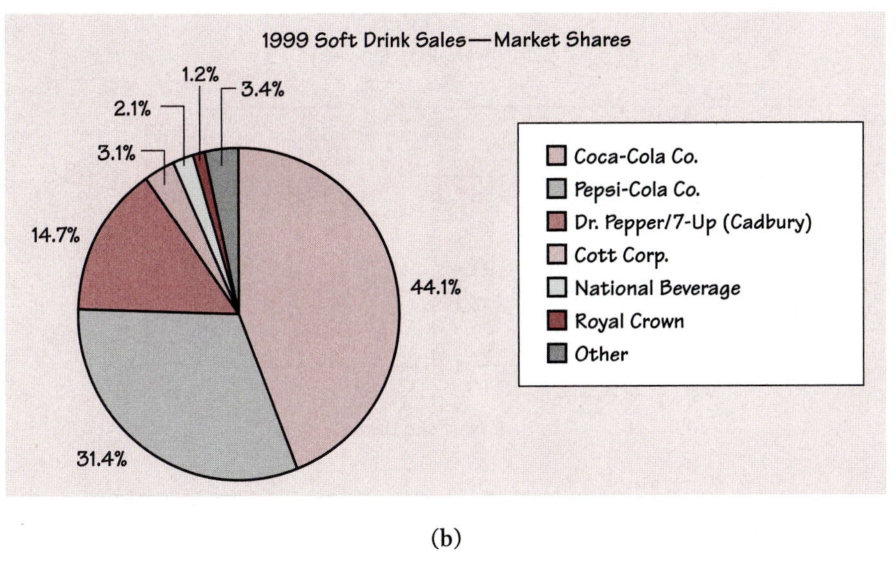

(b)

FIGURE 1.1 A bar graph (a) and a pie chart (b) displaying soft drink sales by companies in 1999.

The **bar graph** in Figure 1.1(a) quickly compares the soft drink sales of the companies. The heights of the bars show the counts in the seven categories. The **pie chart** in Figure 1.1(b) helps us see what part of the whole each group forms. For example, the Coca-Cola "slice" makes up 44.1% of the pie because the Coca-Cola Company sold 44.1% of all soft drinks in 1999.

Bar graphs and pie charts help an audience grasp the distribution quickly. To make a pie chart, you must include all the categories that make up a whole. Bar graphs are more flexible.

EXAMPLE 1.3 DO YOU WEAR YOUR SEAT BELT?

In 1998, the National Highway and Traffic Safety Administration (NHTSA) conducted a study on seat belt use. The table below shows the percentage of automobile drivers who were observed to be wearing their seat belts in each region of the United States.[2]

Region	Percent wearing seat belts
Northeast	66.4
Midwest	63.6
South	78.9
West	80.8

Figure 1.2 shows a bar graph for these data. Notice that the vertical scale is measured in percents.

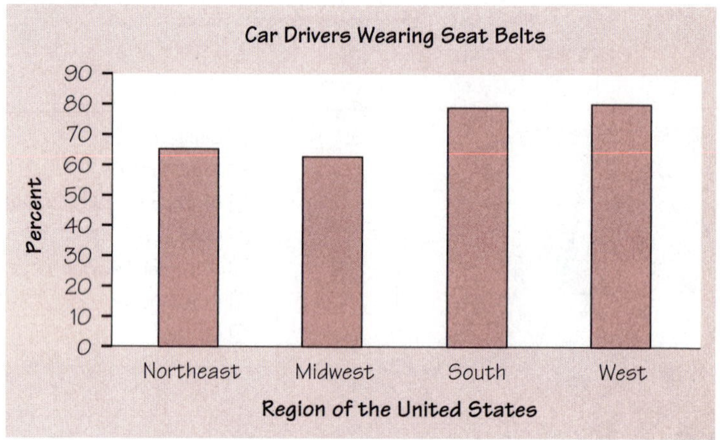

FIGURE 1.2 A bar graph showing the percentage of drivers who wear their seat belts in each of four U.S. regions.

Drivers in the South and West seem to be more concerned about wearing seat belts than those in the Northeast and Midwest. It is not possible to display these data in a single pie chart, because the four percentages cannot be combined to yield a whole (their sum is well over 100%).

EXERCISES

1.5 FEMALE DOCTORATES Here are data on the percent of females among people earning doctorates in 1994 in several fields of study:[3]

Computer science	15.4%	Life sciences	40.7%
Education	60.8%	Physical sciences	21.7%
Engineering	11.1%	Psychology	62.2%

(a) Present these data in a well-labeled bar graph.

(b) Would it also be correct to use a pie chart to display these data? If so, construct the pie chart. If not, explain why not.

1.6 ACCIDENTAL DEATHS In 1997 there were 92,353 deaths from accidents in the United States. Among these were 42,340 deaths from motor vehicle accidents, 11,858 from falls, 10,163 from poisoning, 4051 from drowning, and 3601 from fires.[4]

(a) Find the percent of accidental deaths from each of these causes, rounded to the nearest percent. What percent of accidental deaths were due to other causes?

(b) Make a well-labeled bar graph of the distribution of causes of accidental deaths. Be sure to include an "other causes" bar.

(c) Would it also be correct to use a pie chart to display these data? If so, construct the pie chart. If not, explain why not.

Displaying quantitative variables: dotplots and stemplots

Several types of graphs can be used to display quantitative data. One of the simplest to construct is a **dotplot.**

EXAMPLE 1.4 GOOOOOOOOAAAAALLLLLLLLL!!!

The number of goals scored by each team in the first round of the California Southern Section Division V high school soccer playoffs is shown in the following table.[5]

5	0	1	0	7	2	1	0	4	0	3	0	2	0
3	1	5	0	3	0	1	0	1	0	2	0	3	1

How to construct a dotplot:

Step 1: Label your axis and title your graph. Draw a horizontal line and label it with the variable (in this case, number of goals scored). Title your graph.

Step 2: Scale the axis based on the values of the variable.

Step 3: Mark a dot above the number on the horizontal axis corresponding to each data value. Figure 1.3 displays the completed dotplot.

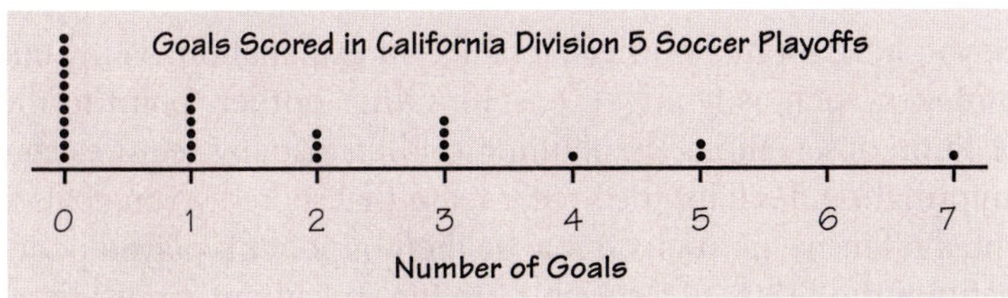

FIGURE 1.3 Goals scored by teams in the California Southern Section Division V high school soccer playoffs.

Making a statistical graph is not an end in itself. After all, a computer or graphing calculator can make graphs faster than we can. The purpose of the graph is to help us understand the data. After you (or your calculator) make a graph, always ask, "What do I see?" Here is a general tactic for looking at graphs: *Look for an overall pattern and also for striking deviations from that pattern.*

OVERALL PATTERN OF A DISTRIBUTION

To describe the overall pattern of a distribution:

- Give the **center** and the **spread**.
- See if the distribution has a simple **shape** that you can describe in a few words.

Section 1.2 tells in detail how to measure center and spread. For now, describe the *center* by finding a value that divides the observations so that about half take larger values and about half have smaller values. In Figure 1.3, the center is 1. That is, a typical team scored about 1 goal in its playoff soccer game. You can describe the *spread* by giving the smallest and largest values. The spread in Figure 1.3 is from 0 goals to 7 goals scored.

The dotplot in Figure 1.3 shows that in most of the playoff games, Division V soccer teams scored very few goals. There were only four teams that scored 4 or more goals. We can say that the distribution has a "long tail" to the right, or that its *shape* is "skewed right." You will learn more about describing shape shortly.

Is the one team that scored 7 goals an *outlier*? This value certainly differs from the overall pattern. To some extent, deciding whether an observation is an outlier is a matter of judgment. We will introduce an objective criterion for determining outliers in Section 1.2.

OUTLIERS

An **outlier** in any graph of data is an individual observation that falls outside the overall pattern of the graph.

Once you have spotted outliers, look for an explanation. Many outliers are due to mistakes, such as typing 4.0 as 40. Other outliers point to the special nature of some observations. Explaining outliers usually requires some background information. Perhaps the soccer team that scored seven goals has some very talented offensive players. Or maybe their opponents played poor defense.

Sometimes the values of a variable are too spread out for us to make a reasonable dotplot. In these cases, we can consider another simple graphical display: a **stemplot.**

EXAMPLE 1.5 WATCH THAT CAFFEINE!

The U.S. Food and Drug Administration limits the amount of caffeine in a 12-ounce can of carbonated beverage to 72 milligrams (mg). Data on the caffeine content of popular soft drinks are provided in Table 1.1. How does the caffeine content of these drinks compare to the USFDA's limit?

TABLE 1.1 Caffeine content (in milligrams) for an 8-ounce serving of popular soft drinks

Brand	Caffeine (mg per 8-oz. serving)	Brand	Caffeine (mg per 8-oz. serving)
A&W Cream Soda	20	IBC Cherry Cola	16
Barq's root beer	15	Kick	38
Cherry Coca-Cola	23	KMX	36
Cherry RC Cola	29	Mello Yello	35
Coca-Cola Classic	23	Mountain Dew	37
Diet A&W Cream Soda	15	Mr. Pibb	27
Diet Cherry Coca-Cola	23	Nehi Wild Red Soda	33
Diet Coke	31	Pepsi One	37
Diet Dr. Pepper	28	Pepsi-Cola	25
Diet Mello Yello	35	RC Edge	47
Diet Mountain Dew	37	Red Flash	27
Diet Mr. Pibb	27	Royal Crown Cola	29
Diet Pepsi-Cola	24	Ruby Red Squirt	26
Diet Ruby Red Squirt	26	Sun Drop Cherry	43
Diet Sun Drop	47	Sun Drop Regular	43
Diet Sunkist Orange Soda	28	Sunkist Orange Soda	28
Diet Wild Cherry Pepsi	24	Surge	35
Dr. Nehi	28	TAB	31
Dr. Pepper	28	Wild Cherry Pepsi	25

Source: National Soft Drink Association, 1999.

The caffeine levels spread from 15 to 47 milligrams for these soft drinks. You could make a dotplot for these data, but a stemplot might be preferable due to the large spread.

How to construct a stemplot:

Step 1: Separate each observation into a *stem* consisting of all but the rightmost digit and a *leaf*, the final digit. A&W Cream Soda has 20 milligrams of caffeine per 8-ounce serving. The number 2 is the stem and 0 is the leaf.

Step 2: Write the stems vertically in increasing order from top to bottom, and draw a vertical line to the right of the stems. Go through the data, writing each leaf to the right of its stem and spacing the leaves equally.

1 | 5 5 6
2 | 0 3 9 3 3 8 7 4 6 8 4 8 8 7 5 7 9 6 8 5
3 | 1 5 7 8 6 5 7 3 7 5 1
4 | 7 7 3 3

Step 3: Write the stems again, and rearrange the leaves in increasing order out from the stem.

Step 4: Title your graph and add a key describing what the stems and leaves represent. Figure 1.4(a) shows the completed stemplot.

What *shape* does this distribution have? It is difficult to tell with so few stems. We can get a better picture of the caffeine content in soft drinks by "splitting stems." In Figure 1.4(a), the values from 10 to 19 milligrams are placed on the "1" stem. Figure 1.4(b) shows another stemplot of the same data. This time, values having leaves 0 through 4 are placed on one stem, while values ending in 5 through 9 are placed on another stem.

Now the bimodal (two-peaked) *shape* of the distribution is clear. Most soft drinks seem to have between 25 and 29 milligrams or between 35 and 38 milligrams of caffeine per 8-ounce serving. The center of the distribution is 28 milligrams per 8-ounce serving. At first glance, it looks like none of these soft drinks even comes close to the USFDA's caffeine limit of 72 milligrams per 12-ounce serving. Be careful! The values in the stemplot are given in milligrams per 8-ounce serving. Two soft drinks have caffeine levels of 47 milligrams per 8-ounce serving. A 12-ounce serving of these beverages would have 1.5(47) = 70.5 milligrams of caffeine. Always check the units of measurement!

CAFFEINE CONTENT (MG) PER 8-OUNCE SERVING OF VARIOUS SOFT DRINKS

1 | 5 5 6
2 | 0 3 3 3 4 4 5 5 6 6 7 7 7 8 8 8 8 8 9 9
3 | 1 1 3 5 5 5 6 7 7 7 8
4 | 3 3 7 7

(a)

Key:
3|5 means the soft drink contains 35 mg of caffeine per 8-ounce serving.

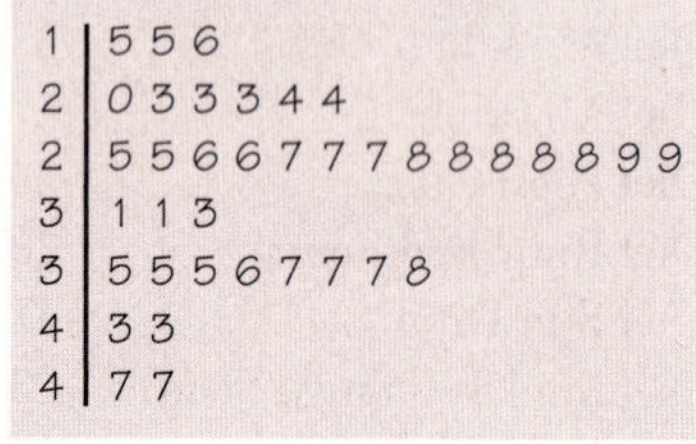

(b)

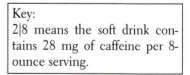

Key:
2|8 means the soft drink contains 28 mg of caffeine per 8-ounce serving.

FIGURE 1.4 Two stemplots showing the caffeine content (mg) of various soft drinks. Figure 1.4(b) improves on the stemplot of Figure 1.4(a) by splitting stems.

Here are a few tips for you to consider when you want to construct a stemplot:

- Whenever you split stems, be sure that each stem is assigned an equal number of possible leaf digits.
- There is no magic number of stems to use. Too few stems will result in a skyscraper-shaped plot, while too many stems will yield a very flat "pancake" graph.

- Five stems is a good minimum.
- You can get more flexibility by *rounding* the data so that the final digit after rounding is suitable as a leaf. Do this when the data have too many digits.

The chief advantages of dotplots and stemplots are that they are easy to construct and they display the actual data values (unless we round). Neither will work well with large data sets. Most statistical software packages will make dotplots and stemplots for you. That will allow you to spend more time making sense of the data.

TECHNOLOGY TOOLBOX *Interpreting computer output*

As cheddar cheese matures, a variety of chemical processes take place. The taste of mature cheese is related to the concentration of several chemicals in the final product. In a study of cheddar cheese from the Latrobe Valley of Victoria, Australia, samples of cheese were analyzed for their chemical composition. The final concentrations of lactic acid in the 30 samples, as a multiple of their initial concentrations, are given below.[6]

A dotplot and a stemplot from the Minitab statistical software package are shown in Figure 1.5. The dots in the dotplot are so spread out that the distribution seems to have no distinct shape. The stemplot does a better job of summarizing the data.

0.86	1.53	1.57	1.81	0.99	1.09	1.29	1.78	1.29	1.58
1.68	1.90	1.06	1.30	1.52	1.74	1.16	1.49	1.63	1.99
1.15	1.33	1.44	2.01	1.31	1.46	1.72	1.25	1.08	1.25

Dotplot for Lactic

1.0 1.5 2.0

Lactic

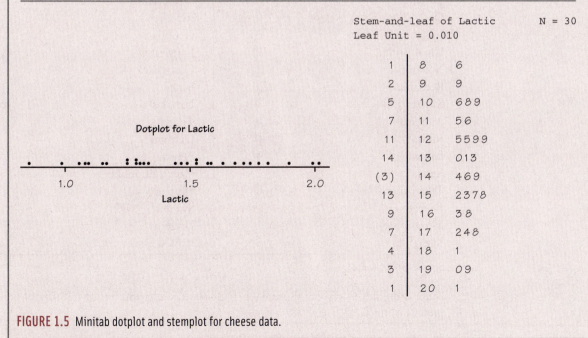

```
Stem-and-leaf of Lactic          N = 30
Leaf Unit = 0.010

   1    8  6
   2    9  9
   5   10  689
   7   11  56
  11   12  5599
  14   13  013
  (3)  14  469
  13   15  2378
   9   16  38
   7   17  248
   4   18  1
   3   19  09
   1   20  1
```

FIGURE 1.5 Minitab dotplot and stemplot for cheese data.

TECHNOLOGY TOOLBOX *Interpreting computer output (continued)*

Notice how the data are recorded in the stemplot. The "leaf unit" is 0.01, which tells us that the stems are given in tenths and the leaves are given in hundredths. We can see that the *spread* of the lactic acid concentrations is from 0.86 to 2.01. Where is the *center* of the distribution? Minitab counts the number of observations from the bottom up and from the top down and lists those counts to the left of the stemplot. Since there are 30 observations, the "middle value" would fall between the 15th and 16th data values from either end—at 1.45. The (3) to the far left of this stem is Minitab's way of marking the location of the "middle value." So a typical sample of mature cheese has 1.45 times as much lactic acid as it did initially. The distribution is roughly symmetrical in *shape*. There appear to be no *outliers*.

EXERCISES

1.7 OLYMPIC GOLD Athletes like Cathy Freeman, Rulon Gardner, Ian Thorpe, Marion Jones, and Jenny Thompson captured public attention by winning gold medals in the 2000 Summer Olympic Games in Sydney, Australia. Table 1.2 displays the total number of gold medals won by several countries in the 2000 Summer Olympics.

TABLE 1.2 Gold medals won by selected countries in the 2000 Summer Olympics

Country	Gold medals	Country	Gold medals
Sri Lanka	0	Netherlands	12
Qatar	0	India	0
Vietnam	0	Georgia	0
Great Britain	28	Kyrgyzstan	0
Norway	10	Costa Rica	0
Romania	26	Brazil	0
Switzerland	9	Uzbekistan	1
Armenia	0	Thailand	1
Kuwait	0	Denmark	2
Bahamas	1	Latvia	1
Kenya	2	Czech Republic	2
Trinidad and Tobago	0	Hungary	8
Greece	13	Sweden	4
Mozambique	1	Uruguay	0
Kazakhstan	3	United States	39

Source: BBC Olympics Web site.

Make a dotplot to display these data. Describe the distribution of number of gold medals won.

1.8 ARE YOU DRIVING A GAS GUZZLER? Table 1.3 displays the highway gas mileage for 32 model year 2000 midsize cars.

TABLE 1.3 Highway gas mileage for model year 2000 midsize cars

Model	MPG	Model	MPG
Acura 3.5RL	24	Lexus GS300	24
Audi A6 Quattro	24	Lexus LS400	25
BMW 740I Sport M	21	Lincoln-Mercury LS	25
Buick Regal	29	Lincoln-Mercury Sable	28
Cadillac Catera	24	Mazda 626	28
Cadillac Eldorado	28	Mercedes-Benz E320	30
Chevrolet Lumina	30	Mercedes-Benz E430	24
Chrysler Cirrus	28	Mitsubishi Diamante	25
Dodge Stratus	28	Mitsubishi Galant	28
Honda Accord	29	Nissan Maxima	28
Hyundai Sonata	28	Oldsmobile Intrigue	28
Infiniti I30	28	Saab 9-3	26
Infiniti Q45	23	Saturn LS	32
Jaguar Vanden Plas	24	Toyota Camry	30
Jaguar S/C	21	Volkswagon Passat	29
Jaguar X200	26	Volvo S70	27

(a) Make a dotplot of these data.

(b) Describe the shape, center, and spread of the distribution of gas mileages. Are there any potential outliers?

1.9 MICHIGAN COLLEGE TUITIONS There are 81 colleges and universities in Michigan. Their tuition and fees for the 1999 to 2000 school year run from $1260 at Kalamazoo Valley Community College to $19,258 at Kalamazoo College. Figure 1.6 (next page) shows a stemplot of the tuition charges.

(a) What do the stems and leaves represent in the stemplot? Have the data been rounded?

(b) Describe the shape, center, and spread of the tuition distribution. Are there any outliers?

1.10 DRP TEST SCORES There are many ways to measure the reading ability of children. One frequently used test is the Degree of Reading Power (DRP). In a research study on third-grade students, the DRP was administered to 44 students.[7] Their scores were:

40	26	39	14	42	18	25	43	46	27	19
47	19	26	35	34	15	44	40	38	31	46
52	25	35	35	33	29	34	41	49	28	52
47	35	48	22	33	41	51	27	14	54	45

Display these data graphically. Write a paragraph describing the distribution of DRP scores.

1	355555555666666777778888889999
2	8
3	155666899
4	0125
5	01
6	133466
7	014567
8	169
9	7
10	139
11	13678
12	39
13	24457
14	39
15	16
16	0
17	
18	2
19	3

FIGURE 1.6 Stemplot of the Michigan tuition and fee data, for Exercise 1.9.

1.11 **SHOPPING SPREE!** A marketing consultant observed 50 consecutive shoppers at a supermarket. One variable of interest was how much each shopper spent in the store. Here are the data (in dollars), arranged in increasing order:

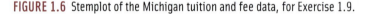

3.11	8.88	9.26	10.81	12.69	13.78	15.23	15.62	17.00	17.39
18.36	18.43	19.27	19.50	19.54	20.16	20.59	22.22	23.04	24.47
24.58	25.13	26.24	26.26	27.65	28.06	28.08	28.38	32.03	34.98
36.37	38.64	39.16	41.02	42.97	44.08	44.67	45.40	46.69	48.65
50.39	52.75	54.80	59.07	61.22	70.32	82.70	85.76	86.37	93.34

(a) Round each amount to the nearest dollar. Then make a stemplot using tens of dollars as the stem and dollars as the leaves.

(b) Make another stemplot of the data by splitting stems. Which of the plots shows the shape of the distribution better?

(c) Describe the shape, center, and spread of the distribution. Write a few sentences describing the amount of money spent by shoppers at this supermarket.

Displaying quantitative variables: histograms

Quantitative variables often take many values. A graph of the distribution is clearer if nearby values are grouped together. The most common graph of the distribution of one quantitative variable is a **histogram.**

EXAMPLE 1.6 PRESIDENTIAL AGES AT INAUGURATION

How old are presidents at their inaugurations? Was Bill Clinton, at age 46, unusually young? Table 1.4 gives the data, the ages of all U.S presidents when they took office.

TABLE 1.4 Ages of the presidents at inauguration

President	Age	President	Age	President	Age
Washington	57	Lincoln	52	Hoover	54
J. Adams	61	A. Johnson	56	F. D. Roosevelt	51
Jefferson	57	Grant	46	Truman	60
Madison	57	Hayes	54	Eisenhower	61
Monroe	58	Garfield	49	Kennedy	43
J. Q. Adams	57	Arthur	51	L. B. Johnson	55
Jackson	61	Cleveland	47	Nixon	56
Van Buren	54	B. Harrison	55	Ford	61
W. H. Harrison	68	Cleveland	55	Carter	52
Tyler	51	McKinley	54	Reagan	69
Polk	49	T. Roosevelt	42	G. Bush	64
Taylor	64	Taft	51	Clinton	46
Fillmore	50	Wilson	56	G. W. Bush	54
Pierce	48	Harding	55		
Buchanan	65	Coolidge	51		

How to make a histogram:

Step 1: Divide the range of the data into classes of equal width. Count the number of observations in each class. The data in Table 1.4 range from 42 to 69, so we choose as our classes

$$40 \leq \text{president's age at inauguration} < 45$$
$$45 \leq \text{president's age at inauguration} < 50$$
$$\vdots$$
$$65 \leq \text{president's age at inauguration} < 70$$

Be sure to specify the classes precisely so that each observation falls into exactly one class. Martin Van Buren, who was age 54 at the time of his inauguration, would fall into the third class interval. Grover Cleveland, who was age 55, would be placed in the fourth class interval.

Here are the counts:

Class	Count
40–44	2
45–49	6
50–54	13
55–59	12
60–64	7
65–69	3

Step 2: Label and scale your axes and title your graph. Label the horizontal axis "Age at inauguration" and the vertical axis "Number of presidents." For the classes we chose, we should scale the horizontal axis from 40 to 70, with tick marks 5 units apart. The vertical axis contains the scale of counts and should range from 0 to at least 13.

Step 3: Draw a bar that represents the count in each class. The base of a bar should cover its class, and the bar height is the class count. Leave no horizontal space between the bars (unless a class is empty, so that its bar has height 0). Figure 1.7 shows the completed histogram.

Graphing note: It is common to add a "break-in-scale" symbol (//) on an axis that does not start at 0, like the horizontal axis in this example.

Interpretation:

Center: It appears that the typical age of a new president is about 55 years, because 55 is near the center of the histogram.

Spread: As the histogram in Figure 1.7 shows, there is a good deal of variation in the ages at which presidents take office. Teddy Roosevelt was the youngest, at age 42, and Ronald Reagan, at age 69, was the oldest.

Shape: The distribution is roughly symmetric and has a single peak (unimodal).

Outliers: There appear to be no outliers.

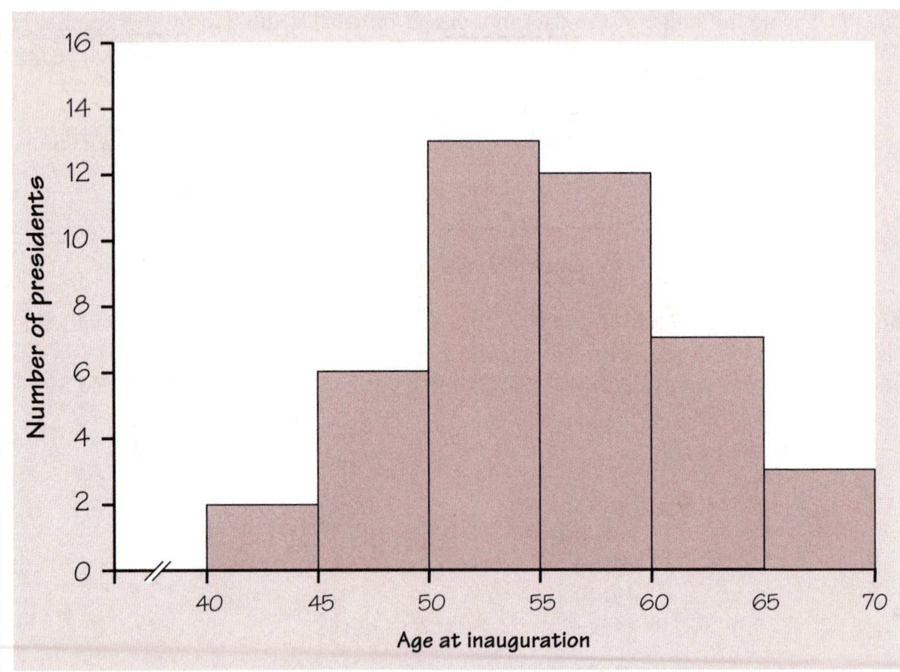

FIGURE 1.7 The distribution of the ages of presidents at their inaugurations, from Table 1.4.

You can also use computer software or a calculator to construct histograms.

TECHNOLOGY TOOLBOX *Making calculator histograms*

1. Enter the presidential age data from Example 1.6 in your statistics list editor.

TI-83

- Press STAT and choose 1:Edit....
- Type the values into list L_1.

TI-89

- Press APPS, choose 1:FlashApps, then select Stats/List Editor and press ENTER.
- Type the values into list1.

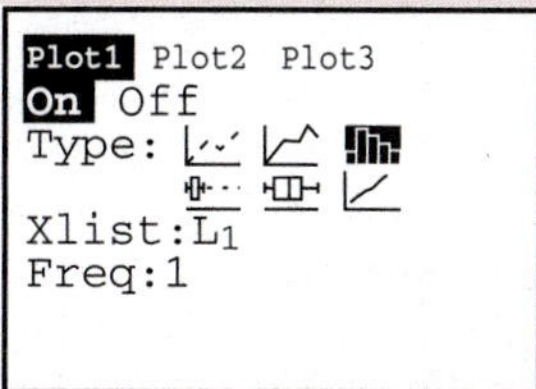

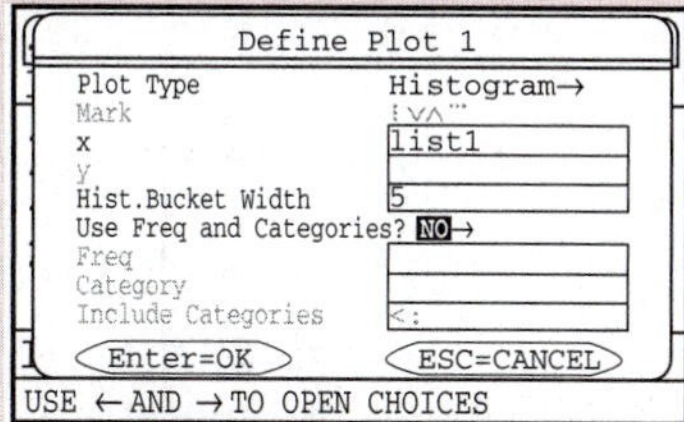

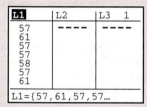

2. Set up a histogram in the statistics plots menu.

TI-83

- Press 2nd Y= (STAT PLOT).
- Press ENTER to go into Plot1.
- Adjust your settings as shown.

TI-89

- Press F2 and choose 1:Plot Setup....
- With Plot 1 highlighted, press F1 to define.
- Change Hist. Bucket Width to 5, as shown.

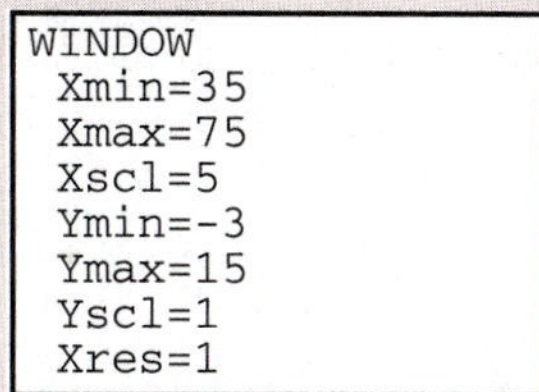

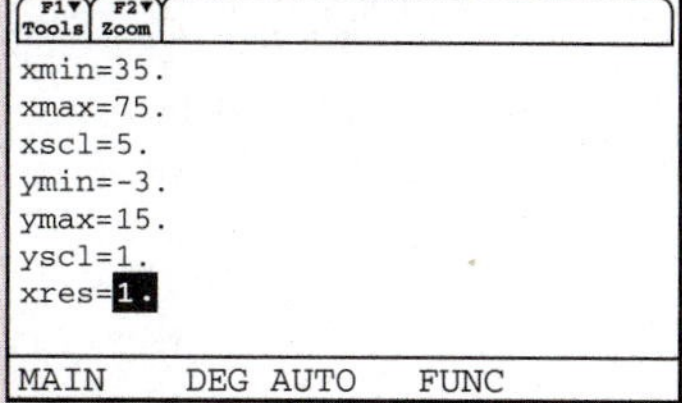

3. Set the window to match the class intervals chosen in Example 1.6.

TI-83

- Press WINDOW.
- Enter the values shown.

TI-89

- Press ◆ F2 (WINDOW).
- Enter the values shown.

```
WINDOW
 Xmin=35
 Xmax=75
 Xscl=5
 Ymin=-3
 Ymax=15
 Yscl=1
 Xres=1
```

```
F1 Tools  F2 Zoom
xmin=35.
xmax=75.
xscl=5.
ymin=-3.
ymax=15.
yscl=1.
xres=1.
MAIN DEG AUTO FUNC
```

4. Graph the histogram. Compare with Figure 1.7.

TI-83

- Press GRAPH.

TI-89

- Press ◆ F3 (GRAPH).

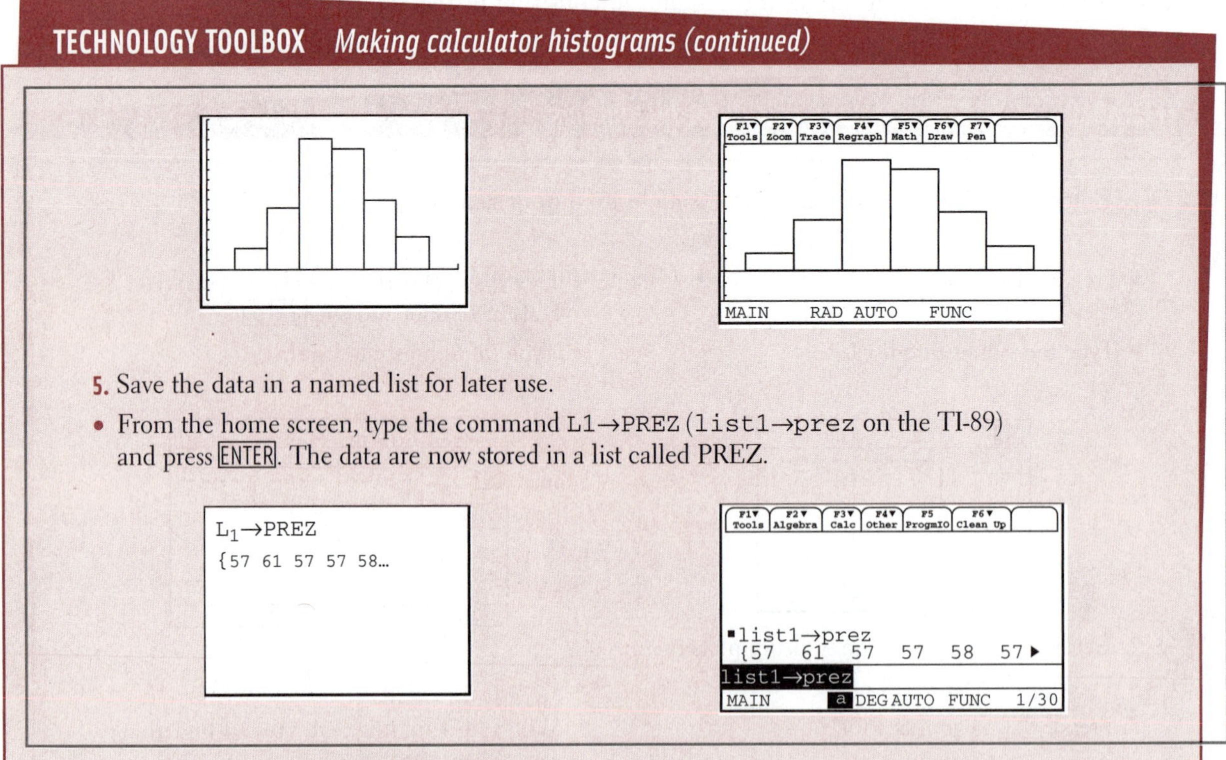

TECHNOLOGY TOOLBOX *Making calculator histograms (continued)*

5. Save the data in a named list for later use.

- From the home screen, type the command `L1→PREZ` (`list1→prez` on the TI-89) and press ENTER. The data are now stored in a list called PREZ.

Histogram tips:

- There is no one right choice of the classes in a histogram. Too few classes will give a "skyscraper" graph, with all values in a few classes with tall bars. Too many will produce a "pancake" graph, with most classes having one or no observations. Neither choice will give a good picture of the shape of the distribution.
- Five classes is a good minimum.
- Our eyes respond to the *area* of the bars in a histogram, so be sure to choose classes that are all the same width. Then area is determined by height and all classes are fairly represented.
- If you use a computer or graphing calculator, beware of letting the device choose the classes.

EXERCISES

1.12 WHERE DO OLDER FOLKS LIVE? Table 1.5 gives the percentage of residents aged 65 or older in each of the 50 states.

TABLE 1.5 Percent of the population in each state aged 65 or older

State	Percent	State	Percent	State	Percent
Alabama	13.1	Louisiana	11.5	Ohio	13.4
Alaska	5.5	Maine	14.1	Oklahoma	13.4
Arizona	13.2	Maryland	11.5	Oregon	13.2
Arkansas	14.3	Massachusetts	14.0	Pennsylvania	15.9
California	11.1	Michigan	12.5	Rhode Island	15.6
Colorado	10.1	Minnesota	12.3	South Carolina	12.2
Connecticut	14.3	Mississippi	12.2	South Dakota	14.3
Delaware	13.0	Missouri	13.7	Tennessee	12.5
Florida	18.3	Montana	13.3	Texas	10.1
Georgia	9.9	Nebraska	13.8	Utah	8.8
Hawaii	13.3	Nevada	11.5	Vermont	12.3
Idaho	11.3	New Hampshire	12.0	Virginia	11.3
Illinois	12.4	New Jersey	13.6	Washington	11.5
Indiana	12.5	New Mexico	11.4	West Virginia	15.2
Iowa	15.1	New York	13.3	Wisconsin	13.2
Kansas	13.5	North Carolina	12.5	Wyoming	11.5
Kentucky	12.5	North Dakota	14.4		

Source: U.S. Census Bureau, 1998.

(a) Construct a histogram to display these data. Record your class intervals and counts.

(b) Describe the distribution of people aged 65 and over in the states.

(c) Enter the data into your calculator's statistics list editor. Make a histogram using a window that matches your histogram from part (a). Copy the calculator histogram and mark the scales on your paper.

(d) Use the calculator's zoom feature to generate a histogram. Copy this histogram onto your paper and mark the scales.

(e) Store the data into the named list ELDER for later use.

1.13 DRP SCORES REVISITED Refer to Exercise 1.10 (page 17). Make a histogram of the DRP test scores for the sample of 44 children. Be sure to show your frequency table. Which do you prefer: the stemplot from Exercise 1.10 or the histogram that you just constructed? Why?

1.14 CEO SALARIES In 1993, *Forbes* magazine reported the age and salary of the chief executive officer (CEO) of each of the top 59 small businesses.[8] Here are the salary data, rounded to the nearest thousand dollars:

145	621	262	208	362	424	339	736	291	58	498	643	390	332
750	368	659	234	396	300	343	536	543	217	298	1103	406	254
862	204	206	250	21	298	350	800	726	370	536	291	808	543
149	350	242	198	213	296	317	482	155	802	200	282	573	388
250	396	572											

Construct a histogram for these data. Describe the shape, center, and spread of the distribution of CEO salaries. Are there any apparent outliers?

1.15 CHEST OUT, SOLDIER! In 1846, a published paper provided chest measurements (in inches) of 5738 Scottish militiamen. Table 1.6 displays the data in summary form.

TABLE 1.6 Chest measurements (inches) of 5738 Scottish militiamen

Chest size	Count	Chest size	Count
33	3	41	934
34	18	42	658
35	81	43	370
36	185	44	92
37	420	45	50
38	749	46	21
39	1073	47	4
40	1079	48	1

Source: Data and Story Library (DASL), http://lib.stat.cmu.edu/DASL/.

(a) You can use your graphing calculator to make a histogram of data presented in summary form like the chest measurements of Scottish militiamen.

- Type the chest measurements into L_1/list1 and the corresponding counts into L_2/list2.
- Set up a statistics plot to make a histogram with *x*-values from L_1/list1 and *y*-values (bar heights) from L_2/list2.

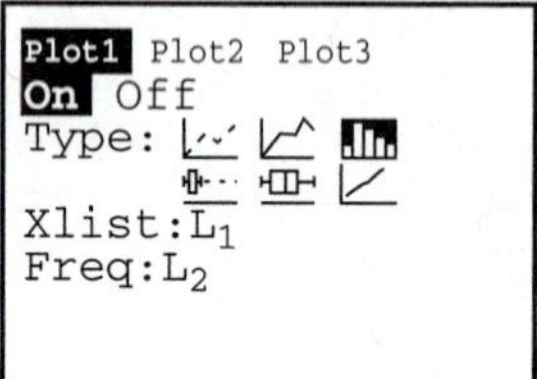

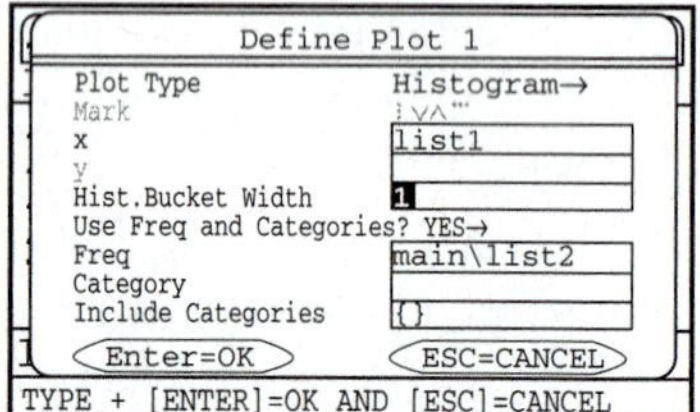

- Adjust your viewing window settings as follows: xmin = 32, xmax = 49, xscl = 1, ymin = −300, ymax = 1100, yscl = 100. From now on, we will abbreviate in this form: $X[32,49]_1$ by $Y[-300,1100]_{100}$. Try using the calculator's built-in ZoomStat/ZoomData command. What happens?
- Graph.

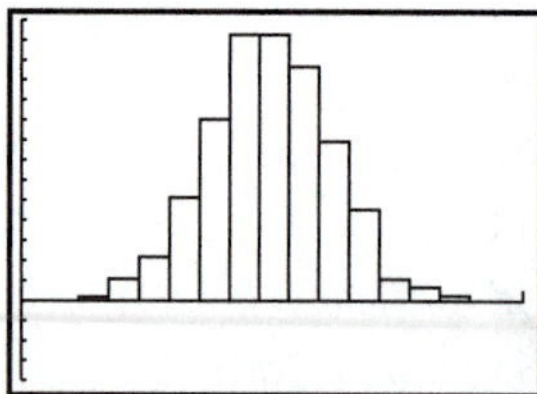

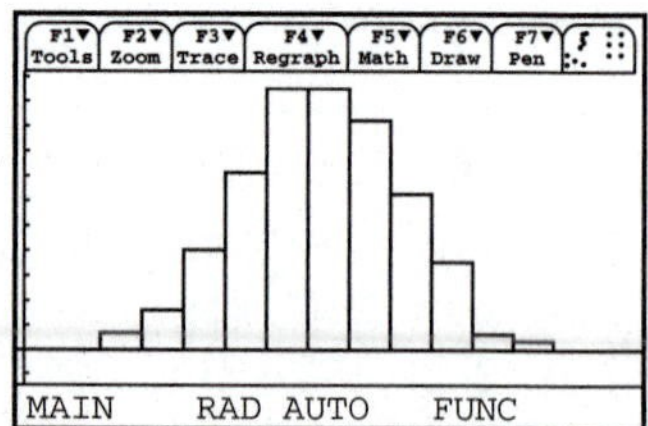

(b) Describe the shape, center, and spread of the chest measurements distribution. Why might this information be useful?

More about shape

When you describe a distribution, concentrate on the main features. Look for major peaks, not for minor ups and downs in the bars of the histogram. Look for clear outliers, not just for the smallest and largest observations. Look for rough *symmetry* or clear *skewness.*

SYMMETRIC AND SKEWED DISTRIBUTIONS

A distribution is **symmetric** if the right and left sides of the histogram are approximately mirror images of each other.

A distribution is **skewed to the right** if the right side of the histogram (containing the half of the observations with larger values) extends much farther out than the left side. It is **skewed to the left** if the left side of the histogram extends much farther out than the right side.

In mathematics, symmetry means that the two sides of a figure like a histogram are exact mirror images of each other. Data are almost never exactly symmetric, so we are willing to call histograms like that in Exercise 1.15 approximately symmetric as an overall description. Here are more examples.

EXAMPLE 1.7 LIGHTNING FLASHES AND SHAKESPEARE

Figure 1.8 comes from a study of lightning storms in Colorado. It shows the distribution of the hour of the day during which the first lightning flash for that day occurred. The distribution has a single peak at noon and falls off on either side of this peak. The two sides of the histogram are roughly the same shape, so we call the distribution symmetric.

Figure 1.9 shows the distribution of lengths of words used in Shakespeare's plays.[9] This distribution also has a single peak but is skewed to the right. That is, there are many short words (3 and 4 letters) and few very long words (10, 11, or 12 letters), so that the right tail of the histogram extends out much farther than the left tail.

Notice that the vertical scale in Figure 1.9 is not the *count* of words but the *percent* of all of Shakespeare's words that have each length. A histogram of percents rather than counts is convenient when the counts are very large or when we want to compare several distributions. Different kinds of writing have different distributions of word lengths, but all are right-skewed because short words are common and very long words are rare.

The overall shape of a distribution is important information about a variable. Some types of data regularly produce distributions that are symmetric or skewed. For example, the sizes of living things of the same species (like lengths of cockroaches) tend to be symmetric. Data on incomes (whether of individuals, companies, or nations) are usually strongly skewed to the right. There are many moderate incomes, some large incomes, and a few very large incomes. Do remember that

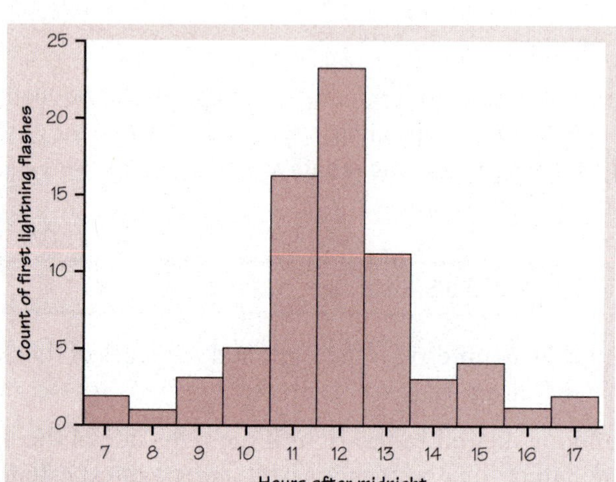

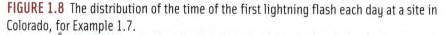

FIGURE 1.8 The distribution of the time of the first lightning flash each day at a site in Colorado, for Example 1.7.

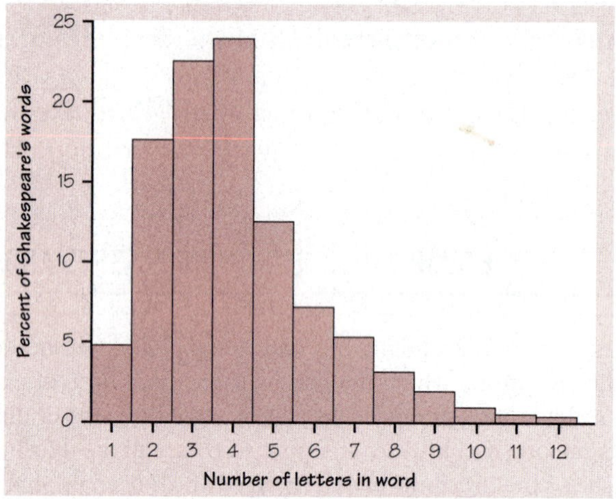

FIGURE 1.9 The distribution of lengths of words used in Shakespeare's plays, for Example 1.7.

many distributions have shapes that are neither symmetric nor skewed. Some data show other patterns. Scores on an exam, for example, may have a cluster near the top of the scale if many students did well. Or they may show two distinct peaks if a tough problem divided the class into those who did and didn't solve it. Use your eyes and describe what you see.

EXERCISES

1.16 STOCK RETURNS The total return on a stock is the change in its market price plus any dividend payments made. Total return is usually expressed as a percent of the beginning price. Figure 1.10 is a histogram of the distribution of total returns for all 1528 stocks listed on the New York Stock Exchange in one year.[10] Like

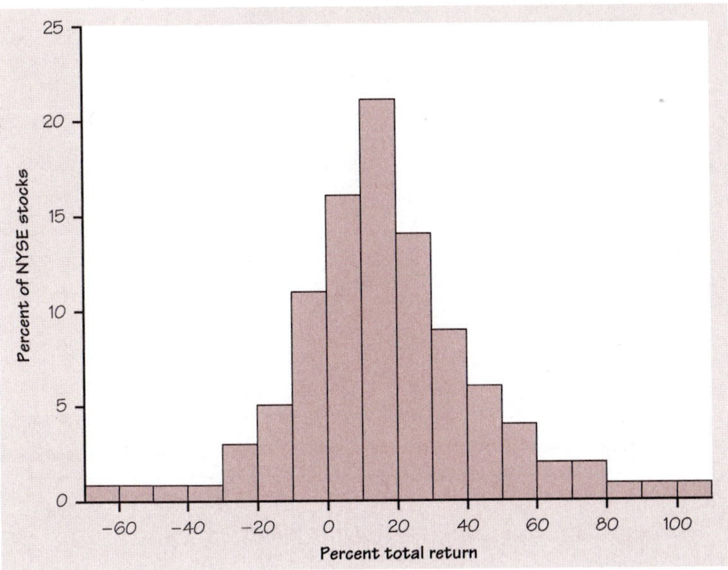

FIGURE 1.10 The distribution of percent total return for all New York Stock Exchange common stocks in one year.

Figure 1.9, it is a histogram of the percents in each class rather than a histogram of counts.

(a) Describe the overall shape of the distribution of total returns.

(b) What is the approximate center of this distribution? (For now, take the center to be the value with roughly half the stocks having lower returns and half having higher returns.)

(c) Approximately what were the smallest and largest total returns? (This describes the spread of the distribution.)

(d) A return less than zero means that an owner of the stock lost money. About what percent of all stocks lost money?

1.17 FREEZING IN GREENWICH, ENGLAND Figure 1.11 is a histogram of the number of days in the month of April on which the temperature fell below freezing at Greenwich, England.[11] The data cover a period of 65 years.

(a) Describe the shape, center, and spread of this distribution. Are there any outliers?

(b) In what percent of these 65 years did the temperature never fall below freezing in April?

1.18 How would you describe the center and spread of the distribution of first lightning flash times in Figure 1.8? Of the distribution of Shakespeare's word lengths in Figure 1.9?

Relative frequency, cumulative frequency, percentiles, and ogives

Sometimes we are interested in describing the relative position of an individual within a distribution. You may have received a standardized test score report that said you were in the 80th percentile. What does this mean? Put simply,

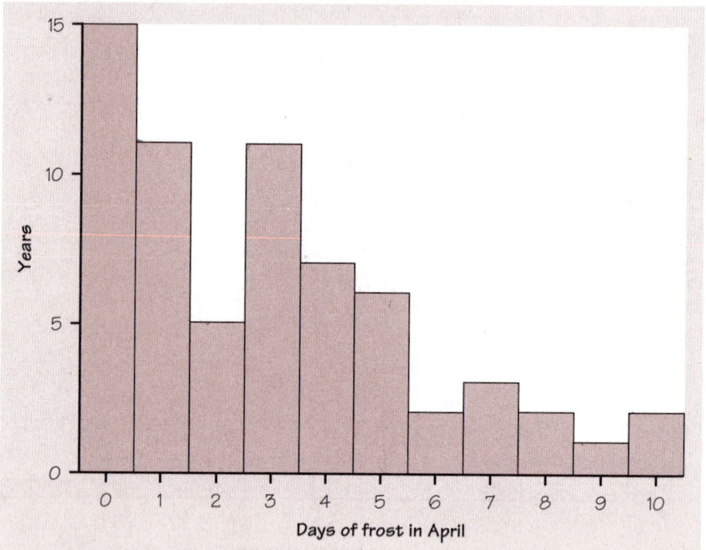

FIGURE 1.11 The distribution of the number of frost days during April at Greenwich, England, over a 65-year period, for Exercise 1.17.

80% of the people who took the test earned scores that were less than or equal to your score. The other 20% of students taking the test earned higher scores than you did.

PERCENTILE

The ***p*th percentile** of a distribution is the value such that p percent of the observations fall at or below it.

A histogram does a good job of displaying the distribution of values of a variable. But it tells us little about the relative standing of an individual observation. If we want this type of information, we should construct a **relative cumulative frequency graph,** often called an **ogive** (pronounced O-JIVE).

EXAMPLE 1.8 WAS BILL CLINTON A YOUNG PRESIDENT?

In Example 1.6, we made a histogram of the ages of U.S. presidents when they were inaugurated. Now we will examine where some specific presidents fall within the age distribution.

How to construct an ogive (relative cumulative frequency graph):

Step 1: Decide on class intervals and make a frequency table, just as in making a histogram. Add three columns to your frequency table: relative frequency, cumulative frequency, and relative cumulative frequency.

• To get the values in the *relative frequency* column, divide the count in each class interval by 43, the total number of presidents. Multiply by 100 to convert to a percentage.

• To fill in the *cumulative frequency* column, add the counts in the frequency column that fall in or below the current class interval.

• For the *relative cumulative frequency* column, divide the entries in the cumulative frequency column by 43, the total number of individuals.

Here is the frequency table from Example 1.6 with the relative frequency, cumulative frequency, and relative cumulative frequency columns added.

Class	Frequency	Relative frequency	Cumulative frequency	Relative cumulative frequency
40–44	2	$\frac{2}{43}$ = 0.047, or 4.7%	2	$\frac{2}{43}$ = 0.047, or 4.7%
45–49	6	$\frac{6}{43}$ = 0.140, or 14.0%	8	$\frac{8}{43}$ = 0.186, or 18.6%
50–54	13	$\frac{13}{43}$ = 0.302, or 30.2%	21	$\frac{21}{43}$ = 0.488, or 48.8%
55–59	12	$\frac{12}{43}$ = 0.279, or 27.9%	33	$\frac{33}{43}$ = 0.767, or 76.7%
60–64	7	$\frac{7}{43}$ = 0.163, or 16.3%	40	$\frac{40}{43}$ = 0.930, or 93.0%
65–69	3	$\frac{3}{43}$ = 0.070, or 7.0%	43	$\frac{43}{43}$ = 1.000, or 100%
TOTAL	43			

Step 2: Label and scale your axes and title your graph. Label the horizontal axis "Age at inauguration" and the vertical axis "Relative cumulative frequency." Scale the horizontal axis according to your choice of class intervals and the vertical axis from 0% to 100%.

Step 3: Plot a point corresponding to the relative cumulative frequency in each class interval at the *left endpoint* of the *next* class interval. For example, for the 40–44 interval, plot a point at a height of 4.7% above the age value of 45. This means that 4.7% of presidents were inaugurated before they were 45 years old. Begin your ogive with a point at a height of 0% at the left endpoint of the lowest class interval. Connect consecutive points with a line segment to form the ogive. The last point you plot should be at a height of 100%. Figure 1.12 shows the completed ogive.

How to locate an individual within the distribution:

What about Bill Clinton? He was age 46 when he took office. To find his relative standing, draw a vertical line up from his age (46) on the horizontal axis until it meets the ogive. Then draw a horizontal line from this point of intersection to the vertical axis. Based on Figure 1.13(a), we would estimate that Bill Clinton's age places him at the 10% *relative cumulative frequency* mark. That tells us that about 10% of all U.S. presidents were the same age as or younger than Bill Clinton when they were inaugurated. Put another way, President Clinton was younger than about 90% of all U.S. presidents based on his inauguration age. His age places him at the *10th percentile* of the distribution.

How to locate a value corresponding to a percentile:

• What inauguration age corresponds to the 60th percentile? To answer this question, draw a horizontal line across from the vertical axis at a height of 60% until it meets the ogive. From the point of intersection, draw a vertical line down to the horizontal axis.

In Figure 1.13(b), the value on the horizontal axis is about 57. So about 60% of all presidents were 57 years old or younger when they took office.

• Find the center of the distribution. Since we use the value that has half of the observations above it and half below it as our estimate of center, we simply need to find the 50th percentile of the distribution. Estimating as for the previous question, confirm that 55 is the center.

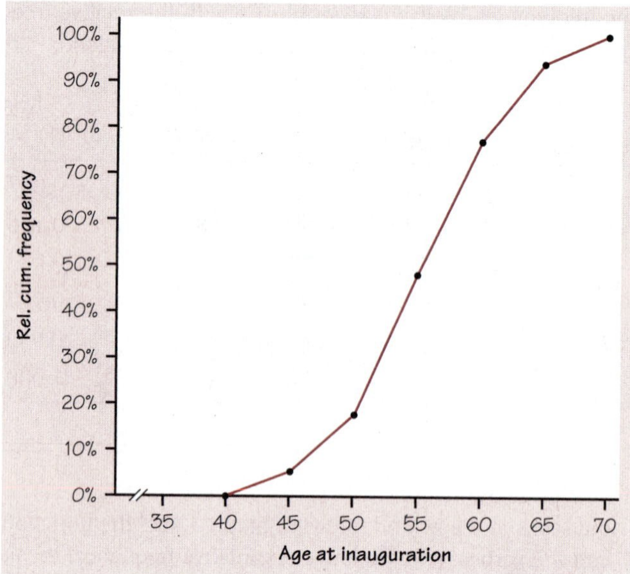

FIGURE 1.12 Relative cumulative frequency plot (ogive) for the ages of U.S. presidents at inauguration.

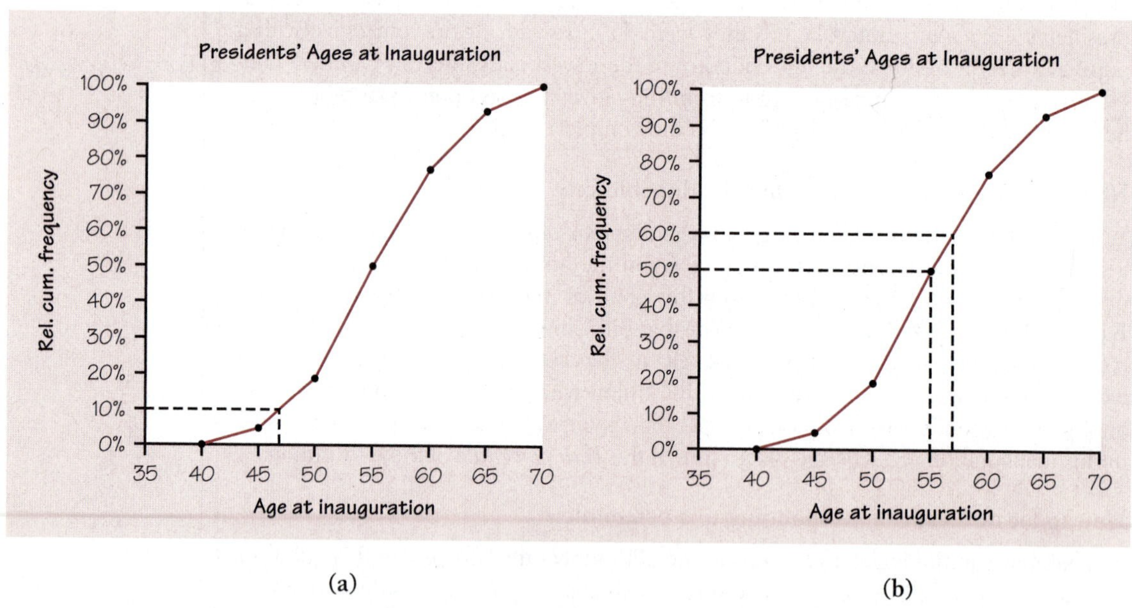

FIGURE 1.13 Ogives of presidents' ages at inauguration are used to (a) locate Bill Clinton within the distribution and (b) determine the 60th percentile and center of the distribution.

EXERCISES

1.19 OLDER FOLKS, II In Exercise 1.12 (page 22), you constructed a histogram of the percentage of people aged 65 or older in each state.

(a) Construct a relative cumulative frequency graph (ogive) for these data.

(b) Use your ogive from part (a) to answer the following questions:

- In what percentage of states was the percentage of "65 and older" less than 15%?
- What is the 40th percentile of this distribution, and what does it tell us?
- What percentile is associated with your state?

1.20 SHOPPING SPREE, II Figure 1.14 is an ogive of the amount spent by grocery shoppers in Exercise 1.11 (page 18).

(a) Estimate the center of this distribution. Explain your method.

(b) At what percentile would the shopper who spent $17.00 fall?

(c) Draw the histogram that corresponds to the ogive.

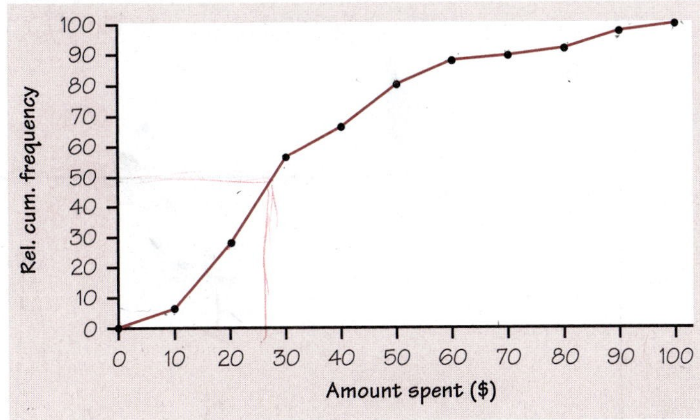

FIGURE 1.14 Amount spent by grocery shoppers in Exercise 1.11.

Time plots

Many variables are measured at intervals over time. We might, for example, measure the height of a growing child or the price of a stock at the end of each month. In these examples, our main interest is change over time. To display change over time, make a time plot.

TIME PLOT

A **time plot** of a variable plots each observation against the time at which it was measured. Always mark the time scale on the horizontal axis and the variable of interest on the vertical axis. If there are not too many points, connecting the points by lines helps show the pattern of changes over time.

trend

seasonal variation

When you examine a time plot, look once again for an overall pattern and for strong deviations from the pattern. One common overall pattern is a ***trend,*** a long-term upward or downward movement over time. A pattern that repeats itself at regular time intervals is known as ***seasonal variation.*** The next example illustrates both these patterns.

EXAMPLE 1.9 ORANGE PRICES MAKE ME SOUR!

Figure 1.15 is a time plot of the average price of fresh oranges over the period from January 1990 to January 2000. This information is collected each month as part of the government's reporting of retail prices. The vertical scale on the graph is the orange price index. This represents the price as a percentage of the average price of oranges in the years 1982 to 1984. The first value is 150 for January 1990, so at that time oranges cost about 150% of their 1982 to 1984 average price.

Figure 1.15 shows a clear *trend* of increasing price. In addition to this trend, we can see a strong *seasonal variation,* a regular rise and fall that occurs each year. Orange prices are usually highest in August or September, when the supply is lowest. Prices then fall in anticipation of the harvest and are lowest in January or February, when the harvest is complete and oranges are plentiful. The unusually large jump in orange prices in 1991 resulted from a freeze in Florida. Can you discover what happened in 1999?

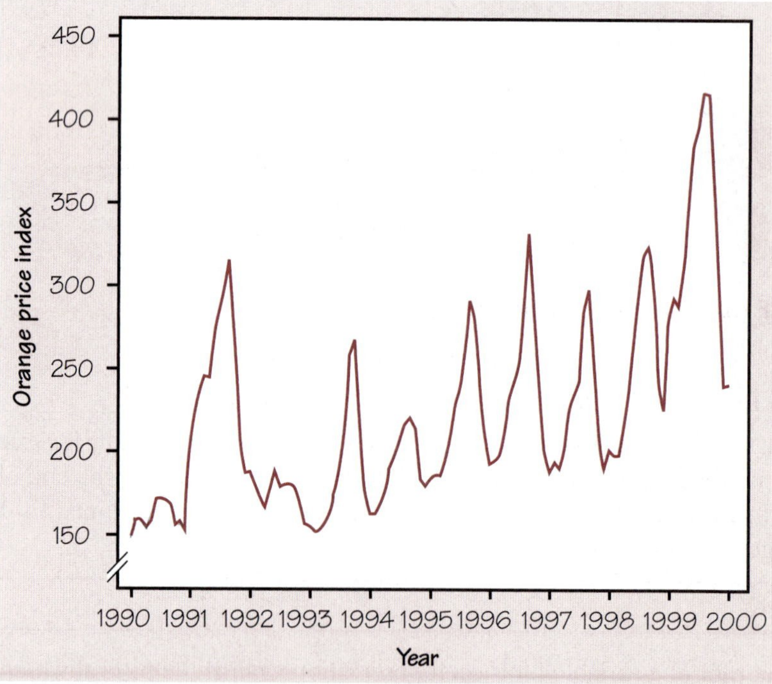

FIGURE 1.15 The price of fresh oranges, January 1990 to January 2000.

EXERCISES

1.21 CANCER DEATHS Here are data on the rate of deaths from cancer (deaths per 100,000 people) in the United States over the 50-year period from 1945 to 1995:

Year:	1945	1950	1955	1960	1965	1970	1975	1980	1985	1990	1995
Deaths:	134.0	139.8	146.5	149.2	153.5	162.8	169.7	183.9	193.3	203.2	204.7

(a) Construct a time plot for these data. Describe what you see in a few sentences.

(b) Do these data suggest that we have made no progress in treating cancer? Explain.

1.22 CIVIL UNREST The years around 1970 brought unrest to many U.S. cities. Here are data on the number of civil disturbances in each three month period during the years 1968 to 1972:

Period		Count	Period		Count
1968	Jan.–Mar.	6	1970	July–Sept.	20
	Apr.–June	46		Oct.–Dec.	6
	July–Sept.	25	1971	Jan.–Mar.	12
	Oct.–Dec.	3		Apr.–June	21
1969	Jan.–Mar.	5		July–Sept.	5
	Apr.–June	27		Oct.–Dec.	1
	July–Sept.	19	1972	Jan.–Mar.	3
	Oct.–Dec.	6		Apr.–June	8
1970	Jan.–Mar.	26		July–Sept.	5
	Apr.–June	24		Oct.–Dec.	5

(a) Make a time plot of these counts. Connect the points in your plot by straight-line segments to make the pattern clearer.

(b) Describe the trend and the seasonal variation in this time series. Can you suggest an explanation for the seasonal variation in civil disorders?

SUMMARY

A data set contains information on a number of **individuals.** Individuals may be people, animals, or things. For each individual, the data give values for one or more **variables.** A variable describes some characteristic of an individual, such as a person's height, gender, or salary.

Exploratory data analysis uses graphs and numerical summaries to describe the variables in a data set and the relations among them.

Some variables are **categorical** and others are **quantitative.** A categorical variable places each individual into a category, like male or female. A quantitative variable has numerical values that measure some characteristic of each individual, like height in centimeters or annual salary in dollars.

The **distribution** of a variable describes what values the variable takes and how often it takes these values.

To describe a distribution, begin with a graph. Use **bar graphs** and **pie charts** to display categorical variables. **Dotplots, stemplots,** and **histograms** graph the distributions of quantitative variables. An **ogive** can help you determine relative standing within a quantitative distribution.

When examining any graph, look for an **overall pattern** and for notable **deviations** from the pattern.

The **center, spread,** and **shape** describe the overall pattern of a distribution. Some distributions have simple shapes, such as **symmetric** and **skewed.** Not all distributions have a simple overall shape, especially when there are few observations.

Outliers are observations that lie outside the overall pattern of a distribution. Always look for outliers and try to explain them.

When observations on a variable are taken over time, make a **time plot** that graphs time horizontally and the values of the variable vertically. A time plot can reveal **trends, seasonal variations,** or other changes over time.

SECTION 1.1 EXERCISES

1.23 **GENDER EFFECTS IN VOTING** Political party preference in the United States depends in part on the age, income, and gender of the voter. A political scientist selects a large sample of registered voters. For each voter, she records gender, age, household income, and whether they voted for the Democratic or for the Republican candidate in the last congressional election. Which of these variables are categorical and which are quantitative?

1.24 What type of graph or graphs would you plan to make in a study of each of the following issues?

(a) What makes of cars do students drive? How old are their cars?

(b) How many hours per week do students study? How does the number of study hours change during a semester?

(c) Which radio stations are most popular with students?

1.25 **MURDER WEAPONS** The 1999 *Statistical Abstract of the United States* reports FBI data on murders for 1997. In that year, 53.3% of all murders were committed with handguns, 14.5% with other firearms, 13.0% with knives, 6.3% with a part of the body (usually the hands or feet), and 4.6% with blunt objects. Make a graph to display these data. Do you need an "other methods" category?

1.26 **WHAT'S A DOLLAR WORTH THESE DAYS?** The buying power of a dollar changes over time. The Bureau of Labor Statistics measures the cost of a "market basket" of goods and services to compile its Consumer Price Index (CPI). If the CPI is 120, goods and services that cost \$100 in the base period now cost \$120. Here are the yearly average values of the CPI for the years between 1970 and 1999. The base period is the years 1982 to 1984.

Year	CPI	Year	CPI	Year	CPI	Year	CPI
1970	38.8	1978	65.2	1986	109.6	1994	148.2
1972	41.8	1980	82.4	1988	118.3	1996	156.9
1974	49.3	1982	96.5	1990	130.7	1998	163.0
1976	56.9	1984	103.9	1992	140.3	1999	166.6

(a) Construct a graph that shows how the CPI has changed over time.

(b) Check your graph by doing the plot on your calculator.

• Enter the years (the last two digits will suffice) into L_1/list1 and enter the CPI into L_2/list2.

• Then set up a statistics plot, choosing the plot type "xyline" (the second type on the TI-83). Use L_1/list1 as X and L_2/list2 as Y. In this graph, the data points are plotted and connected in order of appearance in L_1/list1 and L_2/list2.

• Use the zoom command to see the graph.

(c) What was the overall trend in prices during this period? Were there any years in which this trend was reversed?

(d) In what period during these decades were prices rising fastest? In what period were they rising slowest?

1.27 THE STATISTICS OF WRITING STYLE Numerical data can distinguish different types of writing, and sometimes even individual authors. Here are data on the percent of words of 1 to 15 letters used in articles in Popular Science magazine:[12]

Length:	1	2	3	4	5	6	7	8	9	10	11	12	13	14	15
Percent:	3.6	14.8	18.7	16.0	12.5	8.2	8.1	5.9	4.4	3.6	2.1	0.9	0.6	0.4	0.2

(a) Make a histogram of this distribution. Describe its shape, center, and spread.

(b) How does the distribution of lengths of words used in *Popular Science* compare with the similar distribution in Figure 1.9 (page 26) for Shakespeare's plays? Look in particular at short words (2, 3, and 4 letters) and very long words (more than 10 letters).

1.28 DENSITY OF THE EARTH In 1798 the English scientist Henry Cavendish measured the density of the earth by careful work with a torsion balance. The variable recorded was the density of the earth as a multiple of the density of water. Here are Cavendish's 29 measurements:[13]

5.50	5.61	4.88	5.07	5.26	5.55	5.36	5.29	5.58	5.65
5.57	5.53	5.62	5.29	5.44	5.34	5.79	5.10	5.27	5.39
5.42	5.47	5.63	5.34	5.46	5.30	5.75	5.68	5.85	

Present these measurements graphically in a stemplot. Discuss the shape, center, and spread of the distribution. Are there any outliers? What is your estimate of the density of the earth based on these measurements?

1.29 DRIVE TIME Professor Moore, who lives a few miles outside a college town, records the time he takes to drive to the college each morning. Here are the times (in minutes) for 42 consecutive weekdays, with the dates in order along the rows:

8.25	7.83	8.30	8.42	8.50	8.67	8.17	9.00	9.00	8.17	7.92
9.00	8.50	9.00	7.75	7.92	8.00	8.08	8.42	8.75	8.08	9.75
8.33	7.83	7.92	8.58	7.83	8.42	7.75	7.42	6.75	7.42	8.50
8.67	10.17	8.75	8.58	8.67	9.17	9.08	8.83	8.67		

(a) Make a histogram of these drive times. Is the distribution roughly symmetric, clearly skewed, or neither? Are there any clear outliers?

(b) Construct an ogive for Professor Moore's drive times.

(c) Use your ogive from (b) to estimate the center and 90th percentile for the distribution.

(d) Use your ogive to estimate the percentile corresponding to a drive time of 8.00 minutes.

1.30 THE SPEED OF LIGHT Light travels fast, but it is not transmitted instantaneously. Light takes over a second to reach us from the moon and over 10 billion years to reach us from the most distant objects observed so far in the expanding universe. Because radio and radar also travel at the speed of light, an accurate value for that speed is important in communicating with astronauts and orbiting satellites. An accurate value for the speed of light is also important to computer designers because electrical signals travel at light speed. The first reasonably accurate measurements of the speed of light were made over 100 years ago by A. A. Michelson and Simon Newcomb. Table 1.7 contains 66 measurements made by Newcomb between July and September 1882.

Newcomb measured the time in seconds that a light signal took to pass from his laboratory on the Potomac River to a mirror at the base of the Washington Monument and back, a total distance of about 7400 meters. Just as you can compute the speed of a car from the time required to drive a mile, Newcomb could compute the speed of light from the passage time. Newcomb's first measurement of the passage time of light was 0.000024828 second, or 24,828 nanoseconds. (There are 10^9 nanoseconds in a second.) The entries in Table 1.7 record only the deviation from 24,800 nanoseconds.

TABLE 1.7 **Newcomb's measurements of the passage time of light**

28	26	33	24	34	−44	27	16	40	−2	29	22	24	21
25	30	23	29	31	19	24	20	36	32	36	28	25	21
28	29	37	25	28	26	30	32	36	26	30	22	36	23
27	27	28	27	31	27	26	33	26	32	32	24	39	28
24	25	32	25	29	27	28	29	16	23				

Source: S. M. Stigler, "Do robust estimators work with real data?" *Annals of Statistics*, 5 (1977), pp. 1055–1078.

(a) Construct an appropriate graphical display for these data. Justify your choice of graph.

(b) Describe the distribution of Newcomb's speed of light measurements.

(c) Make a time plot of Newcomb's values. They are listed in order from left to right, starting with the top row.

(d) What does the time plot tell you that the display you made in part (a) does not?

Lesson: Sometimes you need to make more than one graphical display to uncover all of the important features of a distribution.

1.2 DESCRIBING DISTRIBUTIONS WITH NUMBERS

Who is baseball's greatest home run hitter? In the summer of 1998, Mark McGwire and Sammy Sosa captured the public's imagination with their pursuit of baseball's single-season home run record (held by Roger Maris). McGwire eventually set a new standard with 70 home runs. Barry Bonds broke Mark McGwire's record when he hit 73 home runs in the 2001 season. How does this accomplishment fit Bonds's career? Here are Bonds's home run counts for the years 1986 (his rookie year) to 2001 (the year he broke McGwire's record):

1986	1987	1988	1989	1990	1991	1992	1993	1994	1995	1996	1997	1998	1999	2000	2001
16	25	24	19	33	25	34	46	37	33	42	40	37	34	49	73

The stemplot in Figure 1.16 shows us the *shape, center,* and *spread* of these data. The distribution is roughly symmetric with a single peak and a possible high outlier. The center is about 34 home runs, and the spread runs from 16 to the record 73. Shape, center, and spread provide a good description of the overall pattern of any distribution for a quantitative variable. Now we will learn specific ways to use numbers to measure the center and spread of a distribution.

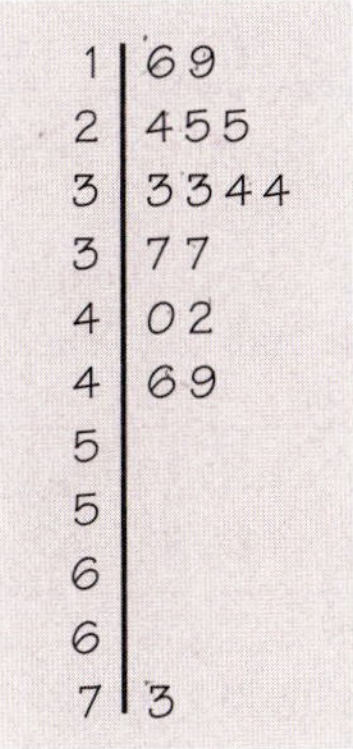

FIGURE 1.16 Number of home runs hit by Barry Bonds in each of his 16 major league seasons.

Measuring center: the mean

A description of a distribution almost always includes a measure of its center or average. The most common measure of center is the ordinary arithmetic average, or *mean*.

THE MEAN $\bar{x}$

To find the **mean** of a set of observations, add their values and divide by the number of observations. If the n observations are $x_1, x_2, ..., x_n$, their mean is

$$\bar{x} = \frac{x_1 + x_2 + \cdots + x_n}{n}$$

or in more compact notation,

$$\bar{x} = \frac{1}{n}\sum x_i$$

The • (capital Greek sigma) in the formula for the mean is short for "add them all up." The subscripts on the observations x_i are just a way of keeping the n observations distinct. They do not necessarily indicate order or any other special facts about the data. The bar over the x indicates the mean of all the x-values. Pronounce the mean $\bar{x}$ as "x-bar." This notation is very common. When writers who are discussing data use $\bar{x}$ or $\bar{y}$, they are talking about a mean.

EXAMPLE 1.10 BARRY BONDS VERSUS HANK AARON

The mean number of home runs Barry Bonds hit in his first 16 major league seasons is

$$\bar{x} = \frac{x_1 + x_2 + \cdots + x_n}{n} = \frac{16 + 25 + \cdots + 73}{16} = \frac{567}{16} = 35.4375$$

We might compare Bonds to Hank Aaron, the all-time home run leader. Here are the numbers of home runs hit by Hank Aaron in each of his major league seasons:

13	27	26	44	30	39	40	34	45	44	24
32	44	39	29	44	38	47	34	40	20	

Aaron's mean number of home runs hit in a year is

$$\bar{x} = \frac{1}{21}(13 + 27 + \cdots + 20) = \frac{733}{21} = 34.9$$

Barry Bonds's exceptional performance in 2001 stands out from his home run production in the previous 15 seasons. Use your calculator to check that his mean home run production in his first 15 seasons is $\bar{x} = 32.93$. One outstanding season increased Bonds's mean home run count by 2.5 home runs per year.

Example 1.10 illustrates an important fact about the mean as a measure of center: it is sensitive to the influence of a few extreme observations. These may be outliers, but a skewed distribution that has no outliers will also pull the mean toward its long tail. Because the mean cannot resist the influence of extreme observations, we say that it is not a ***resistant measure*** of center.

resistant measure

Measuring center: the median

In Section 1.1, we used the midpoint of a distribution as an informal measure of center. The *median* is the formal version of the midpoint, with a specific rule for calculation.

THE MEDIAN *M*

The **median M** is the midpoint of a distribution, the number such that half the observations are smaller and the other half are larger. To find the median of a distribution:

1. Arrange all observations in order of size, from smallest to largest.
2. If the number of observations n is odd, the median M is the center observation in the ordered list.
3. If the number of observations n is even, the median M is the mean of the two center observations in the ordered list.

Medians require little arithmetic, so they are easy to find by hand for small sets of data. Arranging even a moderate number of observations in order is very tedious, however, so that finding the median by hand for larger sets of data is unpleasant. You will need computer software or a graphing calculator to automate finding the median.

EXAMPLE 1.11 FINDING MEDIANS

To find the median number of home runs Barry Bonds hit in his first 16 seasons, first arrange the data in increasing order:

16	19	24	25	25	33	33	**34**	**34**	37	37	40	42	46	49	73

The count of observations $n = 16$ is even. There is no center observation, but there is a center pair. These are the two bold 34s in the list, which have 7 observations to their left in the list and 7 to their right. The median is midway between these two observations. Because both of the middle pair are 34, $M = 34$.

How much does the apparent outlier affect the median? Drop the 73 from the list and find the median for the remaining $n = 15$ years. It is the 8th observation in the edited list, $M = 34$.

How does Bonds's median compare with Hank Aaron's? Here, arranged in increasing order, are Aaron's home run counts:

13	20	24	26	27	29	30
32	34	34	**38**	39	39	40
40	44	44	44	44	45	47

The number of observations is odd, so there is one center observation. This is the median. It is the bold 38, which has 10 observations to its left in the list and 10 observations to its right. Bonds now holds the single-season record, but he has hit fewer home runs in a typical season than Aaron. Barry Bonds also has a long way to go to catch Aaron's career total of 733 home runs.

Comparing the mean and the median

Examples 1.10 and 1.11 illustrate an important difference between the mean and the median. The one high value pulls Bonds's mean home run count up from 32.93 to 35.4375. The median is not affected at all. The median, unlike the mean, is *resistant*. If Bonds's record 73 had been 703, his median would not change at all. The 703 just counts as one observation above the center, no matter how far above the center it lies. The mean uses the actual value of each observation and so will chase a single large observation upward.

The mean and median of a symmetric distribution are close together. If the distribution is exactly symmetric, the mean and median are exactly the same. In a skewed distribution, the mean is farther out in the long tail than is the median. For example, the distribution of house prices is strongly skewed to the right. There are many moderately priced houses and a few very expensive mansions. The few expensive houses pull the mean up but do not affect the median. The mean price of new houses sold in 1997 was $176,000, but the median price for these same houses was only $146,000. Reports about house prices, incomes, and other strongly skewed distributions usually give the median ("midpoint") rather than the mean ("arithmetic average"). However, if you are a tax assessor interested in the total value of houses in your area, use the mean. The total value is the mean times the number of houses; it has no connection with the median. The mean and median measure center in different ways, and both are useful.

EXERCISES

1.31 Joey's first 14 quiz grades in a marking period were

86	84	91	75	78	80	74	87	76	96	82	90	98	93

(a) Use the formula to calculate the mean. Check using "one-variable statistics" on your calculator.

(b) Suppose Joey has an unexcused absence for the fifteenth quiz and he receives a score of zero. Determine his final quiz average. What property of the mean does this situation illustrate? Write a sentence about the effect of the zero on Joey's quiz average that mentions this property.

(c) What kind of plot would best show Joey's distribution of grades? Assume an 8-point grading scale (A: 93 to 100, B: 85 to 92, etc.). Make an appropriate plot, and be prepared to justify your choice.

1.32 SSHA SCORES The Survey of Study Habits and Attitudes (SSHA) is a psychological test that evaluates college students' motivation, study habits, and attitudes toward school. A private college gives the SSHA to a sample of 18 of its incoming first-year women students. Their scores are

154	109	137	115	152	140	154	178	101
103	126	126	137	165	165	129	200	148

(a) Make a stemplot of these data. The overall shape of the distribution is irregular, as often happens when only a few observations are available. Are there any potential outliers? About where is the center of the distribution (the score with half the scores above it and half below)? What is the spread of the scores (ignoring any outliers)?

(b) Find the mean score from the formula for the mean. Then enter the data into your calculator. You can find the mean from the home screen as follows:

TI-83	TI-89
• Press 2nd STAT (LIST) ▶▶ (MATH).	• Press CATALOG then 5 (M).
• Choose 3:mean(, enter list name, press ENTER.	• Choose mean(, type list name, press ENTER.

(c) Find the median of these scores. Which is larger: the median or the mean? Explain why.

1.33 Suppose a major league baseball team's mean yearly salary for a player is \$1.2 million, and that the team has 25 players on its active roster. What is the team's annual payroll for players? If you knew only the median salary, would you be able to answer the question? Why or why not?

1.34 Last year a small accounting firm paid each of its five clerks \$22,000, two junior accountants \$50,000 each, and the firm's owner \$270,000. What is the mean salary paid at this firm? How many of the employees earn less than the mean? What is the median salary? Write a sentence to describe how an unethical recruiter could use statistics to mislead prospective employees.

1.35 U.S. INCOMES The distribution of individual incomes in the United States is strongly skewed to the right. In 1997, the mean and median incomes of the top 1% of Americans were \$330,000 and \$675,000. Which of these numbers is the mean and which is the median? Explain your reasoning.

Measuring spread: the quartiles

The mean and median provide two different measures of the center of a distribution. But a measure of center alone can be misleading. The Census Bureau reports that in 2000 the median income of American households was \$41,345. Half of all households had incomes below \$41,345, and half had higher incomes. But these figures do not tell the whole story. Two nations with the same median household income are very different if one has extremes of wealth and poverty and the other has little variation among households. A drug with the correct mean concentration of active ingredient is dangerous if some batches are much too high and others much too low. We are interested in the *spread* or *variability* of incomes and drug potencies as well as their centers. The simplest useful numerical description of a distribution consists of both a measure of center and a measure of spread.

range

One way to measure spread is to calculate the ***range,*** which is the difference between the largest and smallest observations. For example, the number of home runs Barry Bonds has hit in a season has a *range* of 73 – 16 = 57. The range shows the full spread of the data. But it depends on only the smallest observation and the largest observation, which may be outliers. We can improve our description of spread by also looking at the spread of the middle half of the data. The *quartiles* mark out the middle half. Count up the ordered list of observations, starting from the smallest. The *first quartile* lies one-quarter of the way up the list. The *third quartile* lies three-quarters of the way up the list. In other words, the first quartile is larger than 25% of the observations, and the third quartile is larger than 75% of the observations. The second quartile is the median, which is larger than 50% of the observations. That is the idea of quartiles. We need a rule to make the idea exact. The rule for calculating the quartiles uses the rule for the median.

THE QUARTILES Q_1 and Q_3

To calculate the *quartiles*

1. Arrange the observations in increasing order and locate the median M in the ordered list of observations.

2. The **first quartile** Q_1 is the median of the observations whose position in the ordered list is to the left of the location of the overall median.

3. The **third quartile** Q_3 is the median of the observations whose position in the ordered list is to the right of the location of the overall median.

Here is an example that shows how the rules for the quartiles work for both odd and even numbers of observations.

EXAMPLE 1.12 FINDING QUARTILES

Barry Bonds's home run counts (arranged in order) are

```
16  19  24  25  25  33  33  34  34  37  37  40  42  46  49  73
              ↑                  ↑                  ↑
              Q1                 M                  Q3
```

There is an even number of observations, so the median lies midway between the middle pair, the 8th and 9th in the list. The first quartile is the median of the 8 observations to the left of $M = 34$. So $Q_1 = 25$. The third quartile is the median of the 8 observations to the right of M. $Q_3 = 41$. Note that we don't include M when we're computing the quartiles.

The quartiles are *resistant*. For example, Q_3 would have the same value if Bonds's record 73 were 703.

Hank Aaron's data, again arranged in increasing order, are

```
                 Q1                                 M
                 ↓                                  ↓
13  20  24  26  27  29  30  32  34  34  38  39  39
40  40  44  44  44  44  45  47
           ↑
           Q3
```

In Example 1.11, we determined that the median is the bold 38 in the list. The first quartile is the median of the 10 observations to the left of $M = 38$. This is the mean of the 5th and 6th of these 10 observations, so $Q_1 = 28$. $Q_3 = 44$. The overall median is left out of the calculation of the quartiles.

Be careful when, as in these examples, several observations take the same numerical value. Write down all of the observations and apply the rules just as if they all had distinct values. Some software packages use a slightly different rule to find the quartiles, so computer results may be a bit different from your own work. Don't worry about this. The differences will always be too small to be important.

The distance between the first and third quartiles is a simple measure of spread that gives the range covered by the middle half of the data. This distance is called the *interquartile range*.

THE INTERQUARTILE RANGE (IQR)

The **interquartile range (IQR)** is the distance between the first and third quartiles,

$$IQR = Q_3 - Q_1$$

If an observation falls between Q_1 and Q_3, then you know it's neither unusually high (upper 25%) or unusually low (lower 25%). The IQR is the basis of a rule of thumb for identifying suspected outliers.

OUTLIERS: THE 1.5 × *IQR* CRITERION

Call an observation an outlier if it falls more than $1.5 \times IQR$ above the third quartile or below the first quartile.

EXAMPLE 1.13 DETERMINING OUTLIERS

We suspect that Barry Bonds's 73 home run season is an outlier. Let's test.

$$IQR = Q_3 - Q_1 = 41 - 25 = 16$$
$$Q_3 + 1.5 \times IQR = 41 + (1.5 \times 16) = 65 \text{ (upper cutoff)}$$
$$Q_1 - 1.5 \times IQR = 25 - (1.5 \times 16) = 1 \text{ (lower cutoff)}$$

Since 73 is above the upper cutoff, Bonds's record-setting year was an outlier.

The five-number summary and boxplots

The smallest and largest observations tell us little about the distribution as a whole, but they give information about the tails of the distribution that is missing if we know only Q_1, M, and Q_3. To get a quick summary of both center and spread, combine all five numbers.

THE FIVE-NUMBER SUMMARY

The **five-number summary** of a data set consists of the smallest observation, the first quartile, the median, the third quartile, and the largest observation, written in order from smallest to largest.

In symbols, the five-number summary is

$$\text{Minimum} \quad Q_1 \quad M \quad Q_3 \quad \text{Maximum}$$

These five numbers offer a reasonably complete description of center and spread. The five-number summaries from Example 1.12 are

16 25 34 41 73

for Bonds and

13 28 38 44 47

for Aaron. The five-number summary of a distribution leads to a new graph, the ***boxplot.*** Figure 1.17 shows boxplots for the home run comparison.

boxplot

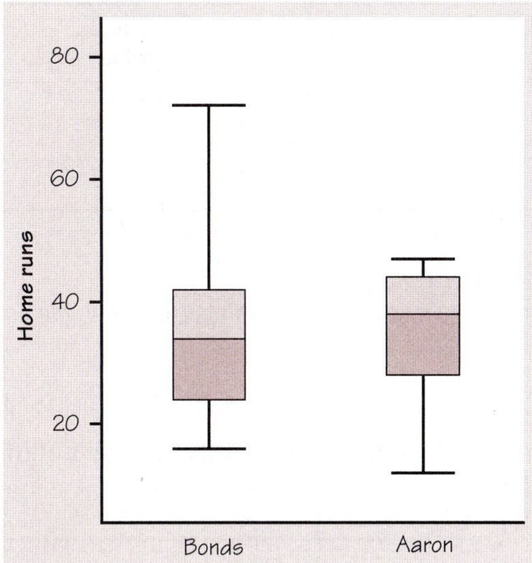

FIGURE 1.17 Side-by-side boxplots comparing the numbers of home runs per year by Barry Bonds and Hank Aaron.

Because boxplots show less detail than histograms or stemplots, they are best used for side-by-side comparison of more than one distribution, as in Figure 1.17. You can draw boxplots either horizontally or vertically. Be sure to include a numerical scale in the graph. When you look at a boxplot, first locate the median, which marks the center of the distribution. Then look at the spread. The quartiles show the spread of the middle half of the data, and the extremes (the smallest and largest observations) show the spread of the entire data set. We see from Figure 1.17 that Aaron and Bonds are about equally consistent when we look at the middle 50% of their home run distributions.

A boxplot also gives an indication of the symmetry or skewness of a distribution. In a symmetric distribution, the first and third quartiles are equally distant from the median. In most distributions that are skewed to the right, however, the third quartile will be farther above the median than the first quartile is below it. The extremes behave the same way, but remember that they are just single observations and may say little about the distribution as a whole. In Figure 1.17, we can see that Aaron's home run distribution is skewed to the left. Barry Bonds's distribution is more difficult to describe.

Outliers usually deserve special attention. Because the regular boxplot conceals outliers, we will adopt the ***modified boxplot,*** which plots outliers as isolated points. Figures 1.18(a) and (b) show regular and modified boxplots for the home runs hit by Bonds and Aaron. The regular boxplot suggests a very large spread in the upper 25% of Bonds's distribution. The modified boxplot shows that if not for the outlier, the distribution would show much less variability. Because the modified boxplot shows more detail, when we say "boxplot" from now on, we will mean "modified boxplot." Both the TI-83 and the TI-89 give you a choice of regular or modified boxplot. When you construct a (modified) boxplot by hand, extend the "whiskers"

modified boxplot

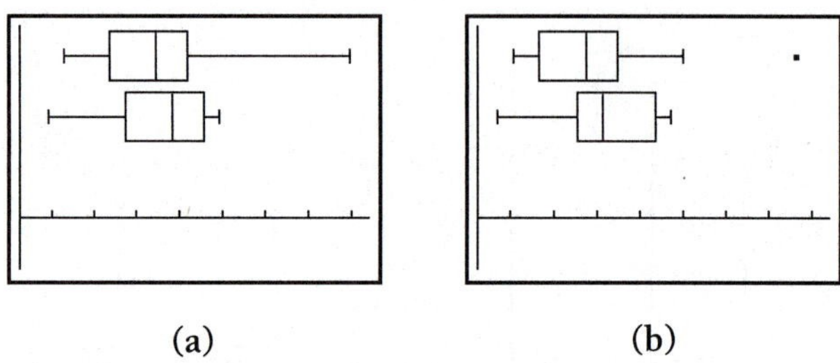

FIGURE 1.18 Regular (a) and modified (b) boxplots comparing the home run production of Barry Bonds and Hank Aaron.

out to the largest and the smallest data points that are not outliers. Then plot outliers as isolated points.

BOXPLOT (MODIFIED)

A **modified boxplot** is a graph of the five-number summary, with outliers plotted individually.

- A central box spans the quartiles.
- A line in the box marks the median.
- Observations more than $1.5 \times IQR$ outside the central box are plotted individually.
- Lines extend from the box out to the smallest and largest observations that are not outliers.

TECHNOLOGY TOOLBOX *Calculator boxplots and numerical summaries*

The TI-83 and TI-89 can plot up to three boxplots in the same viewing window. Both calculators can also calculate the mean, median, quartiles, and other one-variable statistics for data stored in lists. In this example, we compare Barry Bonds to Babe Ruth, the "Sultan of Swat." Here are the numbers of home runs hit by Ruth in each of his seasons as a New York Yankee (1920 to 1934):

54	59	35	41	46	25	47	60	54	46	49	46	41	34	22

1. Enter Bonds's home run data in L_1/list1 and Ruth's in L_2/list2.
2. Set up two statistics plots: Plot 1 to show a modified boxplot of Bonds's data and Plot 2 to show a modified boxplot of Ruth's data.

TECHNOLOGY TOOLBOX *Calculator boxplots and numerical summaries (continued)*

TI-83

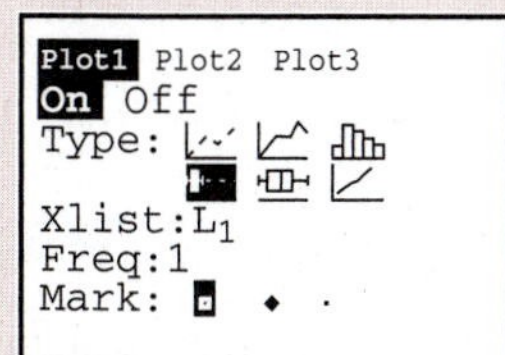

TI-89

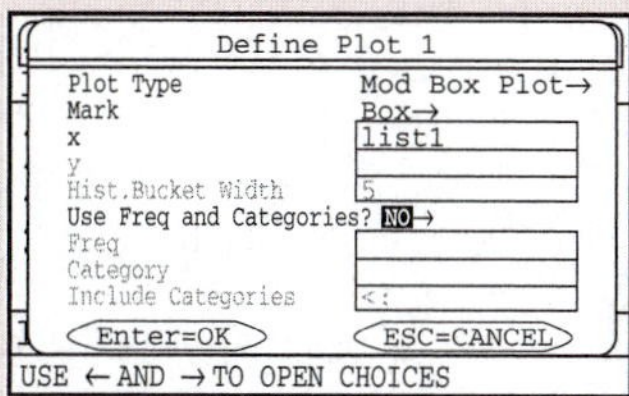

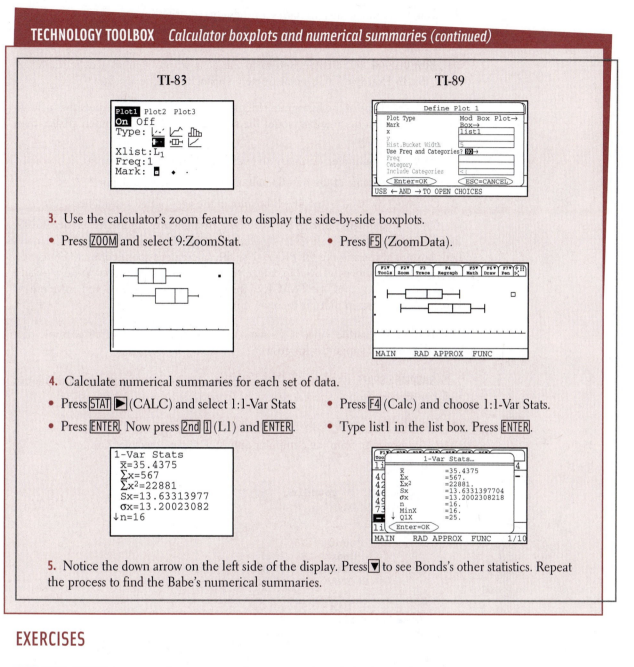

3. Use the calculator's zoom feature to display the side-by-side boxplots.

- Press ZOOM and select 9:ZoomStat.
- Press F5 (ZoomData).

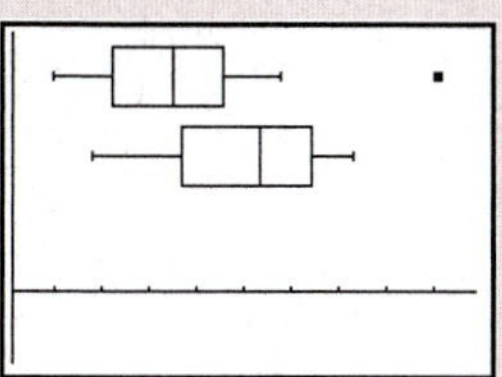

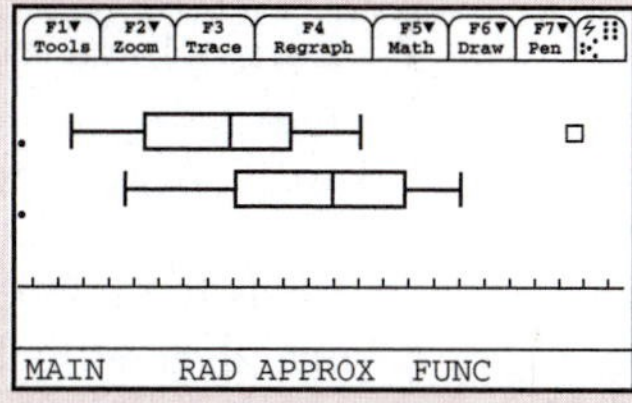

4. Calculate numerical summaries for each set of data.

- Press STAT ▶ (CALC) and select 1:1-Var Stats
- Press ENTER. Now press 2nd 1 (L1) and ENTER.
- Press F4 (Calc) and choose 1:1-Var Stats.
- Type list1 in the list box. Press ENTER.

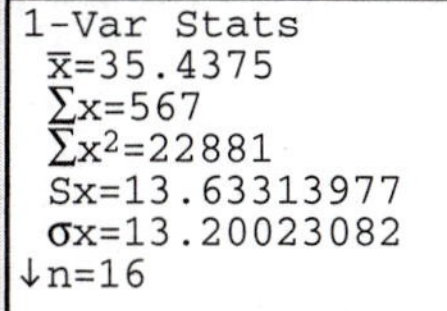

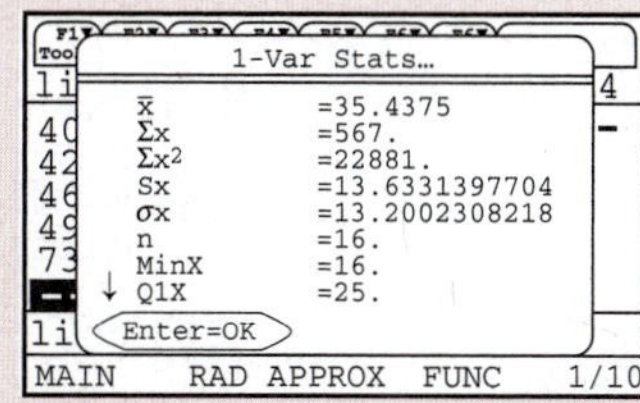

5. Notice the down arrow on the left side of the display. Press ▼ to see Bonds's other statistics. Repeat the process to find the Babe's numerical summaries.

EXERCISES

1.36 SSHA SCORES Here are the scores on the Survey of Study Habits and Attitudes (SSHA) for 18 first-year college women:

154 109 137 115 152 140 154 178 101 103 126 126 137 165 165 129 200 148

and for 20 first-year college men:

108 140 114 91 180 115 126 92 169 146 109 132 75 88 113 151 70 115 187 104

(a) Make side-by-side boxplots to compare the distributions.

(b) Compute numerical summaries for these two distributions.

(c) Write a paragraph comparing the SSHA scores for men and women.

1.37 HOW OLD ARE PRESIDENTS? Return to the data on presidential ages in Table 1.4 (page 19). In Example 1.6, we constructed a histogram of the age data.

(a) From the shape of the histogram (Figure 1.7, page 20), do you expect the mean to be much less than the median, about the same as the median, or much greater than the median? Explain.

(b) Find the five-number summary and verify your expectation from (a).

(c) What is the range of the middle half of the ages of new presidents?

(d) Construct by hand a (modified) boxplot of the ages of new presidents.

(e) On your calculator, define Plot 1 to be a histogram using the list named PREZ that you created in the Technology Toolbox on page 22. Define Plot 2 to be a (modified) boxplot also using the list PREZ. Use the calculator's zoom command to generate a graph. To remove the overlap, adjust your viewing window so that Ymin = –6 and Ymax = 22. Then graph. Use TRACE to inspect values. Press the up and down cursor keys to toggle between plots. Is there an outlier? If so, who was it?

1.38 Is the interquartile range a resistant measure of spread? Give an example of a small data set that supports your answer.

1.39 SHOPPING SPREE, III Figure 1.19 displays computer output for the data on amount spent by grocery shoppers in Exercise 1.11 (page 18).

(a) Find the total amount spent by the shoppers.

(b) Make a boxplot from the computer output. Did you check for outliers?

DataDesk

```
Summaryof     spending
No Selector

Percentile   25

         Count    50
          Mean    34.7022
        Median    27.8550
        StdDev    21.6974
           Min     3.11000
           Max    93.3400
Lower ith %tile   19.2700
Upper ith %tile   45.4000
```

Minitab

```
Descriptive Statistics

Variable      N     Mean   Median   TrMean    StDev   SEMean
spending     50    34.70    27.85    32.92    21.70     3.07

Variable    Min      Max       Q1       Q3
spending   3.11    93.34    19.06    45.72
```

FIGURE 1.19 Numerical descriptions of the unrounded shopping data from the Data Desk and Minitab software.

Measuring spread: the standard deviation

The five-number summary is not the most common numerical description of a distribution. That distinction belongs to the combination of the mean to measure center and the *standard deviation* to measure spread. The standard deviation measures spread by looking at how far the observations are from their mean.

THE STANDARD DEVIATION s

The **variance s^2** of a set of observations is the average of the squares of the deviations of the observations from their mean. In symbols, the variance of n observations $x_1, x_2, \ldots, x_n$ is

$$s^2 = \frac{(x_1 - \bar{x})^2 + (x_2 - \bar{x})^2 + \cdots + (x_n - \bar{x})^2}{n-1}$$

or, more compactly,

$$s^2 = \frac{1}{n-1}\sum(x_i - \bar{x})^2$$

The **standard deviation s** is the square root of the variance s^2:

$$s = \sqrt{\frac{1}{n-1}\sum(x_i - \bar{x})^2}$$

In practice, use software or your calculator to obtain the standard deviation from keyed-in data. Doing a few examples step-by-step will help you understand how the variance and standard deviation work, however. Here is such an example.

EXAMPLE 1.14 METABOLIC RATE

A person's metabolic rate is the rate at which the body consumes energy. Metabolic rate is important in studies of weight gain, dieting, and exercise. Here are the metabolic rates of 7 men who took part in a study of dieting. (The units are calories per 24 hours. These are the same calories used to describe the energy content of foods.)

1792	1666	1362	1614	1460	1867	1439

The researchers reported $\bar{x}$ and s for these men.

First find the mean:

$$\bar{x} = \frac{1792 + 1666 + 1362 + 1614 + 1460 + 1867 + 1439}{7} = \frac{11,200}{7} = 1600 \text{ calories}$$

To see clearly the nature of the variance, start with a table of the deviations of the observations from this mean.

Observations x_i	Deviations $x_i - \bar{x}$	Squared deviations $(x_i - \bar{x})^2$
1792	1792 − 1600 = 192	192^2 = 36,864
1666	1666 − 1600 = 66	66^2 = 4,356
1362	1362 − 1600 = −238	$(-238)^2$ = 56,644
1614	1614 − 1600 = 14	14^2 = 196
1460	1460 − 1600 = −140	$(-140)^2$ = 36,864
1867	1867 − 1600 = 267	267^2 = 71,289
1439	1439 − 1600 = −161	$(-161)^2$ = 25,921
	sum = 0	sum = 214,870

The variance is the sum of the squared deviations divided by one less than the number of observations:

$$s^2 = \frac{214{,}870}{6} = 35{,}811.67$$

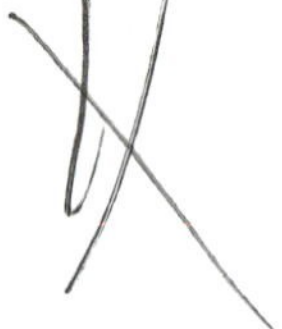

The standard deviation is the square root of the variance:

$$s = \sqrt{35{,}811.67} = 189.24 \text{ calories}$$

Compare these results for s^2 and s with those generated by your calculator or computer.

Figure 1.20 displays the data of Example 1.14 as points above the number line, with their mean marked by an asterisk (*). The arrows show two of the deviations from the mean. These deviations show how spread out the data are about their mean. Some of the deviations will be positive and some negative because observations fall on each side of the mean. In fact, *the sum of the deviations of the observations from their mean will always be zero.* Check that this is true in Example 1.14. So we cannot simply add the deviations to get an overall measure of spread. Squaring the deviations makes them all nonnegative, so that observations far from the mean in either direction will have large positive squared deviations. The variance s^2 is the average squared deviation. The variance is large if the observations are widely spread about their mean; it is small if the observations are all close to the mean.

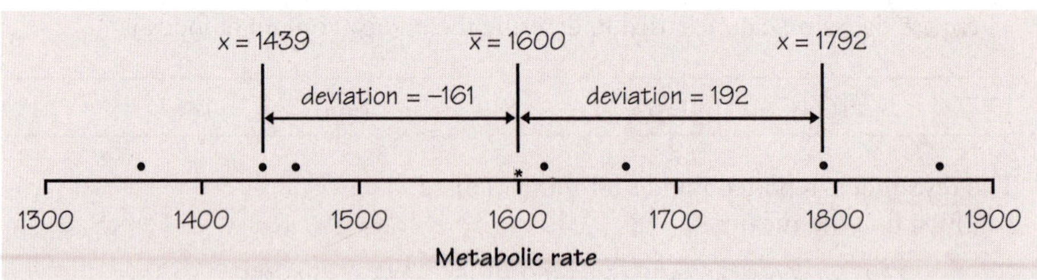

FIGURE 1.20 Metabolic rates for seven men, with their mean (*) and the deviations of two observations from the mean.

Because the variance involves squaring the deviations, it does not have the same unit of measurement as the original observations. Lengths measured in centimeters, for example, have a variance measured in squared centimeters. Taking the square root remedies this. The standard deviation s measures spread about the mean in the original scale.

If the variance is the average of the squares of the deviations of the observations from their mean, why do we average by dividing by $n - 1$ rather than n? Because the sum of the deviations is always zero, the last deviation can be found once we know the other $n - 1$ deviations. So we are not averaging n unrelated numbers. Only $n - 1$ of the squared deviations can vary freely, and we average by dividing the total by $n - 1$. The number $n - 1$ is called the ***degrees of freedom*** of the variance or of the standard deviation. Many calculators offer a choice between dividing by n and dividing by $n - 1$, so be sure to use $n - 1$.

degrees of freedom

Leaving the arithmetic to a calculator allows us to concentrate on what we are doing and why. What we are doing is measuring spread. Here are the basic properties of the standard deviation s as a measure of spread.

PROPERTIES OF THE STANDARD DEVIATION

- s measures spread about the mean and should be used only when the mean is chosen as the measure of center.
- $s = 0$ only when there is *no spread*. This happens only when all observations have the same value. Otherwise, $s > 0$. As the observations become more spread out about their mean, s gets larger.
- s, like the mean $\bar{x}$, is not resistant. Strong skewness or a few outliers can make s very large. For example, the standard deviation of Barry Bonds's home run counts is 13.633. (Use your calculator to verify this.) If we omit the outlier, the standard deviation drops to 9.573.

You may rightly feel that the importance of the standard deviation is not yet clear. We will see in the next chapter that the standard deviation is the natural measure of spread for an important class of symmetric distributions, the normal distributions. The usefulness of many statistical procedures is tied to distributions of particular shapes. This is certainly true of the standard deviation.

Choosing measures of center and spread

How do we choose between the five-number summary and $\bar{x}$ and s to describe the center and spread of a distribution? Because the two sides of a strongly skewed distribution have different spreads, no single number such as s describes the spread well. The five-number summary, with its two quartiles and two extremes, does a better job.

CHOOSING A SUMMARY

The five-number summary is usually better than the mean and standard deviation for describing a skewed distribution or a distribution with strong outliers. Use $\bar{x}$ and s only for reasonably symmetric distributions that are free of outliers.

Do remember that a graph gives the best overall picture of a distribution. Numerical measures of center and spread report specific facts about a distribution, but they do not describe its entire shape. Numerical summaries do not disclose the presence of multiple peaks or gaps, for example. **Always plot your data.**

EXERCISES

1.40 PHOSPHATE LEVELS The level of various substances in the blood influences our health. Here are measurements of the level of phosphate in the blood of a patient, in milligrams of phosphate per deciliter of blood, made on 6 consecutive visits to a clinic:

5.6	5.2	4.6	4.9	5.7	6.4

A graph of only 6 observations gives little information, so we proceed to compute the mean and standard deviation.

(a) Find the mean from its definition. That is, find the sum of the 6 observations and divide by 6.

(b) Find the standard deviation from its definition. That is, find the deviations of each observation from the mean, square the deviations, then obtain the variance and the standard deviation. Example 1.14 shows the method.

(c) Now enter the data into your calculator to obtain $\bar{x}$ and s. Do the results agree with your hand calculations? Can you find a way to compute the standard deviation without using one-variable statistics?

1.41 ROGER MARIS New York Yankee Roger Maris held the single-season home run record from 1961 until 1998. Here are Maris's home run counts for his 10 years in the American League:

14	28	16	39	61	33	23	26	8	13

(a) Maris's mean number of home runs is $\bar{x} = 26.1$. Find the standard deviation s from its definition. Follow the model of Example 1.14.

(b) Use your calculator to verify your results. Then use your calculator to find $\bar{x}$ and s for the 9 observations that remain when you leave out the outlier. How does the outlier affect the values of $\bar{x}$ and s? Is s a resistant measure of spread?

1.42 OLDER FOLKS, III In Exercise 1.12 (page 22), you made a histogram displaying the percentage of residents aged 65 or older in each of the 50 U.S. states. Do you prefer the five-number summary or $\bar{x}$ and s as a brief numerical description? Why? Calculate your preferred description.

1.43 This is a standard deviation contest. You must choose four numbers from the whole numbers 0 to 10, with repeats allowed.

(a) Choose four numbers that have the smallest possible standard deviation.

(b) Choose four numbers that have the largest possible standard deviation.

(c) Is more than one choice possible in either (a) or (b)? Explain.

Changing the unit of measurement

The same variable can be recorded in different units of measurement. Americans commonly record distances in miles and temperatures in degrees Fahrenheit. Most of the rest of the world measures distances in kilometers and temperatures in degrees Celsius. Fortunately, it is easy to convert from one unit of measurement to another. In doing so, we perform a *linear transformation.*

LINEAR TRANSFORMATION

A linear transformation changes the original variable x into the new variable x_{new} given by an equation of the form

$$x_{\text{new}} = a + bx$$

Adding the constant a shifts all values of x upward or downward by the same amount.

Multiplying by the positive constant b changes the size of the unit of measurement.

EXAMPLE 1.15 LOS ANGELES LAKERS' SALARIES

Table 1.8 gives the approximate base salaries of the 14 members of the Los Angeles Lakers basketball team for the year 2000. You can calculate that the mean is $\bar{x}$ = \$4.14 million and that the median is M = \$2.6 million. No wonder professional basketball players have big houses!

TABLE 1.8 Year 2000 salaries for the Los Angeles Lakers

Player	Salary	Player	Salary
Shaquille O'Neal	\$17.1 million	Ron Harper	\$2.1 million
Kobe Bryant	\$11.8 million	A. C. Green	\$2.0 million
Robert Horry	\$5.0 million	Devean George	\$1.0 million
Glen Rice	\$4.5 million	Brian Shaw	\$1.0 million
Derek Fisher	\$4.3 million	John Salley	\$0.8 million
Rick Fox	\$4.2 million	Tyronne Lue	\$0.7 million
Travis Knight	\$3.1 million	John Celestand	\$0.3 million

Figure 1.21(a) is a stemplot of the salaries, with millions as stems. The distribution is skewed to the right and there are two high outliers. The very high salaries of Kobe Bryant and Shaquille O'Neal pull up the mean. Use your calculator to check that $s =$ \$4.76 million, and that the five-number summary is

\$0.3 million	\$1.0 million	\$2.6 million	\$4.5 million	\$17.1 million

(a) Suppose that each member of the team receives a \$100,000 bonus for winning the NBA Championship (which the Lakers did in 2000). How will this affect the shape, center, and spread of the distribution?

```
(a)            (b)            (c)
 0 | 378        0 | 489        0 | 389
 1 | 00         1 | 11         1 | 11
 2 | 01         2 | 12         2 | 23
 3 | 1          3 | 2          3 | 4
 4 | 235        4 | 346        4 | 67
 5 | 0          5 | 1          5 | 05
 6 |            6 |            6 |
 7 |            7 |            7 |
 8 |            8 |            8 |
 9 |            9 |            9 |
10 |           10 |           10 |
11 | 8         11 | 9         11 |
12 |           12 |           12 |
13 |                          13 | 0
14 |
15 |
16 |                          16 |
                              17 |
17 | 1         17 | 2         18 | 9
```

FIGURE 1.21 Stemplots of the salaries of Los Angeles Lakers players, from Table 1.8.

Since \$100,000 = \$0.1 million, each player's salary will increase by \$0.1 million. This linear transformation can be represented by $x_{new} = 0.1 + 1x$, where x_{new} is the salary after the bonus and x is the player's base salary. Increasing each value in Table 1.8 by 0.1 will also increase the mean by 0.1. That is, $\bar{x}_{new} =$ \$4.24 million. Likewise, the median salary will increase by 0.1 and become $M =$ \$2.7 million.

What will happen to the spread of the distribution? The standard deviation of the Lakers' salaries after the bonus is still $s =$ \$4.76 million. With the bonus, the five-number summary becomes

$0.4 million	$1.1 million	$2.7 million	$4.6 million	$17.2 million

Both before and after the salary bonus, the *IQR* for this distribution is $3.5 million. *Adding a constant amount to each observation does not change the spread.* The shape of the distribution remains unchanged, as shown in Figure 1.21(b).

(b) Suppose that, instead of receiving a $100,000 bonus, each player is offered a 10% increase in his base salary. John Celestand, who is making a base salary of $0.3 million, would receive an additional (0.10)($0.3 million) = $0.03 million. To obtain his new salary, we could have used the linear transformation $x_{new} = 0 + 1.10x$, since multiplying the current salary (x) by 1.10 increases it by 10%. Increasing all 14 players' salaries in the same way results in the following list of values (in millions):

$0.33	$0.77	$0.88	$1.10	$1.10	$2.20	$2.31
$3.41	$4.62	$4.73	$4.95	$5.50	$12.98	$18.81

Use your calculator to check that $\bar{x}_{new}$ = $4.55 million, s_{new} = $5.24 million, M_{new} = $2.86 million, and the five-number summary for x_{new} is

$0.33	$1.10	$2.86	$4.95	$18.81

Since $4.14(1.10) = $4.55 and $2.6(1.10) = $2.86, you can see that both measures of center (the mean and median) have increased by 10%. This time, the spread of the distribution has increased, too. Check for yourself that the standard deviation and the *IQR* have also increased by 10%. The stemplot in Figure 1.21(c) shows that the distribution of salaries is still right-skewed.

Linear transformations do not change the shape of a distribution. As you saw in the previous example, changing the units of measurement can affect the center and spread of the distribution. Fortunately, the effects of such changes follow a simple pattern.

EFFECT OF A LINEAR TRANSFORMATION

To see the effect of a linear transformation on measures of center and spread, apply these rules:

- Multiplying each observation by a positive number b multiplies both measures of center (mean and median) and measures of spread (standard deviation and *IQR*) by b.
- Adding the same number a (either positive or negative) to each observation adds a to measures of center and to quartiles but does not change measures of spread.

EXERCISES

1.44 COCKROACHES! Maria measures the lengths of 5 cockroaches that she finds at school. Here are her results (in inches):

1.4	2.2	1.1	1.6	1.2

(a) Find the mean and standard deviation of Maria's measurements.

(b) Maria's science teacher is furious to discover that she has measured the cockroach lengths in inches rather than centimeters. (There are 2.54 cm in 1 inch.) She gives Maria two minutes to report the mean and standard deviation of the 5 cockroaches in centimeters. Maria succeeded. Will you?

(c) Considering the 5 cockroaches that Maria found as a small sample from the population of all cockroaches at her school, what would you estimate as the average length of the population of cockroaches? How sure of your estimate are you?

1.45 RAISING TEACHERS' PAY A school system employs teachers at salaries between \$30,000 and \$60,000. The teachers' union and the school board are negotiating the form of next year's increase in the salary schedule. Suppose that every teacher is given a flat \$1000 raise.

(a) How much will the mean salary increase? The median salary?

(b) Will a flat \$1000 raise increase the spread as measured by the distance between the quartiles?

(c) Will a flat \$1000 raise increase the spread as measured by the standard deviation of the salaries?

1.46 RAISING TEACHERS' PAY, II Suppose that the teachers in the previous exercise each receive a 5% raise. The amount of the raise will vary from \$1500 to \$3000, depending on present salary. Will a 5% across-the-board raise increase the spread of the distribution as measured by the distance between the quartiles? Do you think it will increase the standard deviation?

Comparing distributions

An experiment is carried out to compare the effectiveness of a new cholesterol-reducing drug with the one that is currently prescribed by most doctors. A survey is conducted to determine whether the proportion of males who are likely to vote for a political candidate is higher than the proportion of females who are likely to vote for the candidate. Students taking AP Calculus AB and AP Statistics are curious about which exam is harder. They have information on the distribution of scores earned on each exam from the year 2000. In each of these situations, we are interested in comparing distributions. This section presents some of the more common methods for making statistical comparisons.

EXAMPLE 1.16 COOL CAR COLORS

Table 1.9 gives information about the color preferences of vehicle purchasers in 1998.

TABLE 1.9 Colors of cars and trucks purchased in 1998

Color	Full-sized or intermediate-sized car	Light truck or van
Medium or dark green	16.4%	15.5%
White	15.6%	22.5%
Light brown	14.1%	6.1%
Silver	11.0%	6.2%
Black	8.9%	11.5%

Source: The World Almanac and Book of Facts, 2000.

Figure 1.22 is a graph that can be used to compare the color distributions for cars and trucks. By placing the bars side-by-side, we can easily observe the similarities and differences within each of the color categories. White seems to be the favorite color of most truck buyers, while car purchasers favor medium or dark green. What other similarities and differences do you see?

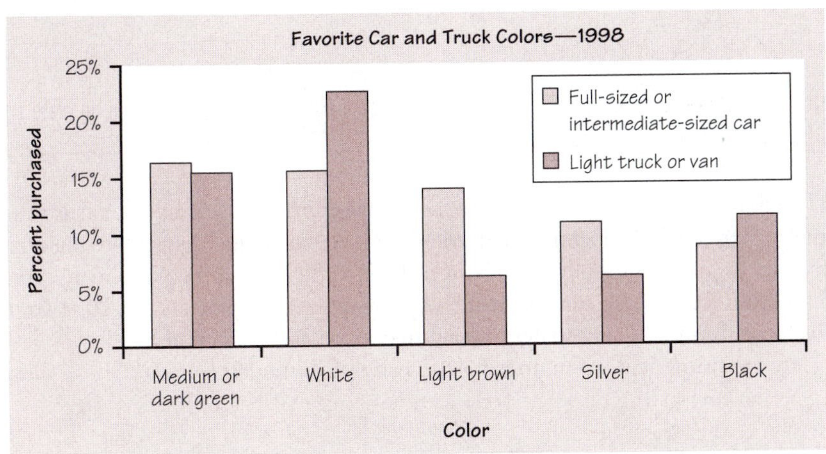

FIGURE 1.22 Side-by-side bar graph of most-popular car and truck colors from 1998.

An effective graphical display for comparing two fairly small quantitative data sets is a *back-to-back stemplot*. Example 1.17 shows you how.

EXAMPLE 1.17 SWISS DOCTORS

A study in Switzerland examined the number of cesarean sections (surgical deliveries of babies) performed in a year by doctors. Here are the data for 15 male doctors:

27	50	33	25	86	25	85	31	37	44	20	36	59	34	28

The study also looked at 10 female doctors. The number of cesareans performed by these doctors (arranged in order) were

5	7	10	14	18	19	25	29	31	33

We can compare the number of cesarean sections performed by male and female doctors using a back-to-back stemplot. Figure 1.23 shows the completed graph. As you can see, the stems are listed in the middle and leaves are placed on the left for male doctors and on the right for female doctors. It is usual to have the leaves increase in value as they move away from the stem.

NUMBER OF CESAREAN SECTIONS PERFORMED BY MALE AND FEMALE DOCTORS

Male		Female
	0	57
	1	0489
87550	2	59
76431	3	13
4	4	
90	5	
	6	
	7	
65	8	

Key:
|2| 5 means that a female doctor performed 25 cesarean sections that year
0 |5| means that a male doctor performed 50 cesarean sections that year

FIGURE 1.23 Back-to-back stemplot of the number of cesarean sections performed by male and female Swiss doctors.

The distribution of the number of cesareans performed by female doctors is roughly symmetric. For the male doctors, the distribution is skewed to the right. More than half of the female doctors in the study performed fewer than 20 cesarean sections in a year. The minimum number of cesareans performed by any of the male doctors was 20. Two male physicians performed an unusually high number of cesareans, 85 and 86.

Here are numerical summaries for the two distributions:

	$\bar{x}$	s	Min.	Q_1	M	Q_3	Max.	IQR
Male doctors	41.333	20.607	20	27	34	50	86	23
Female doctors	19.1	10.126	5	10	18.5	29	33	19

The mean and median numbers of cesarean sections performed are higher for the male doctors. Both the standard deviation and the *IQR* for the male doctors are much larger than the corresponding statistics for the female doctors. So there is much greater variability in the number of cesarean sections performed by male physicians. Due to the apparent outliers in the male doctor data and the lack of symmetry of their distribution of cesareans, we should use the medians and *IQR*s in our numerical comparisons.

We have already seen that boxplots can be useful for comparing distributions of quantitative variables. Side-by-side boxplots, like those in the Technology Toolbox on page 47, help us quickly compare shape, center, and spread.

EXERCISES

1.47 GET YOUR HOT DOGS HERE! "Face it. A hot dog isn't a carrot stick." So said *Consumer Reports*, commenting on the low nutritional quality of the all-American frank. Table 1.10 shows the magazine's laboratory test results for calories and milligrams of sodium (mostly due to salt) in a number of major brands of hot dogs. There are three types: beef, "meat" (mainly pork and beef, but government regulations allow up to 15% poultry meat), and poultry. Because people concerned about their health may prefer low-calorie, low-sodium hot dogs, we ask: "Are there any systematic differences among the three types of hot dogs in these two variables?" Use side-by-side boxplots and numerical summaries to help you answer this question. Write a paragraph explaining your findings.

TABLE 1.10 Calories and sodium in three types of hot dogs

Beef hot dogs		Meat hot dogs		Poultry hot dogs	
Calories	Sodium	Calories	Sodium	Calories	Sodium
186	495	173	458	129	430
181	477	191	506	132	375
176	425	182	473	102	396
149	322	190	545	106	383
184	482	172	496	94	387
190	587	147	360	102	542
158	370	146	387	87	359
139	322	139	386	99	357
175	479	175	507	170	528
148	375	136	393	113	513
152	330	179	405	135	426
111	300	153	372	142	513
141	386	107	144	86	358
153	401	195	511	143	581
190	645	135	405	152	588
157	440	140	428	146	522
131	317	138	339	144	545
149	319				
135	298				
132	253				

Source: Consumer Reports, June 1986, pp.366–367

1.48 WHICH AP EXAM IS EASIER: CALCULUS AB OR STATISTICS? The table below gives the distribution of grades earned by students taking the Calculus AB and Statistics exams in 2000.[14]

	5	4	3	2	1
Calculus AB	16.8%	23.2%	23.5%	19.6%	16.8%
Statistics	9.8%	21.5%	22.4%	20.5%	25.8%

(a) Make a graphical display to compare the AP exam grades for Calculus AB and Statistics.

(b) Write a few sentences comparing the two distributions of exam grades. Do you now know which exam is easier? Why or why not?

1.49 WHO MAKES MORE? A manufacturing company is reviewing the salaries of its full-time employees below the executive level at a large plant. The clerical staff is almost entirely female, while a majority of the production workers and technical staff are male. As a result, the distributions of salaries for male and female employees may be quite different. Table 1.11 gives the frequencies and relative frequencies for women and men.

(a) Make histograms for these data, choosing a vertical scale that is most appropriate for comparing the two distributions.

(b) Describe the shape of the overall salary distributions and the chief differences between them.

(c) Explain why the total for women is greater than 100%.

TABLE 1.11 Salary distributions of female and male workers in a large factory

Salary ($1000)	Women		Men	
	Number	%	Number	%
10–15	89	11.8	26	1.1
15–20	192	25.4	221	9.0
20–25	236	31.2	677	27.9
25–30	111	14.7	823	33.6
30–35	86	11.4	365	14.9
35–40	25	3.3	182	7.4
40–45	11	1.5	91	3.7
45–50	3	0.4	33	1.4
50–55	2	0.3	19	0.8
55–60	0	0.0	11	0.4
60–65	0	0.0	0	0.0
65–70	1	0.1	3	0.1
Total	756	100.1	2451	100.0

1.50 BASKETBALL PLAYOFF SCORES Here are the scores of games played in the California Division I-AAA high school basketball playoffs:[15]

71–38 52–47 55–53 76–65 77–63 65–63 68–54 64–62
87–47 64–56 78–64 58–51 91–74 71–41 67–62 106–46

On the same day, the final scores of games in Division V-AA were

98–45 67–44 74–60 96–54 92–72 93–46
98–67 62–37 37–36 69–44 86–66 66–58

(a) Construct a back-to-back stemplot to compare the number of points scored by Division I-AAA and Division V-AA basketball teams.

(b) Compare the shape, center, and spread of the two distributions. Which numerical summaries are most appropriate in this case? Why?

(c) Is there a difference in "margin of victory" in Division I-AAA and Division V-AA playoff games? Provide appropriate graphical and numerical support for your answer.

SUMMARY

A numerical summary of a distribution should report its **center** and its **spread**, or **variability.**

The **mean** $\bar{x}$ and the **median** M describe the center of a distribution in different ways. The mean is the arithmetic average of the observations, and the median is the midpoint of the values.

When you use the median to indicate the center of a distribution, describe its spread by giving the **quartiles.** The **first quartile** Q_1 has one-fourth of the observations below it, and the **third quartile** Q_3 has three-fourths of the observations below it. An extreme observation is an **outlier** if it is smaller than $Q_1 - (1.5 \times IQR)$ or larger than $Q_3 + (1.5 \times IQR)$.

The **five-number summary** consists of the median, the quartiles, and the high and low extremes and provides a quick overall description of a distribution. The median describes the center, and the quartiles and extremes show the spread.

Boxplots based on the five-number summary are useful for comparing two or more distributions. The box spans the quartiles and shows the spread of the central half of the distribution. The median is marked within the box. Lines extend from the box to the smallest and the largest observations that are not outliers. Outliers are plotted as isolated points.

The **variance** s^2 and especially its square root, **the standard deviation** s, are common measures of spread about the mean as center. The standard deviation s is zero when there is no spread and gets larger as the spread increases.

The mean and standard deviation are strongly influenced by outliers or skewness in a distribution. They are good descriptions for symmetric distributions and are most useful for the normal distributions, which will be introduced in the next chapter.

The median and quartiles are not affected by outliers, and the two quartiles and two extremes describe the two sides of a distribution separately. The five-number summary is the preferred numerical summary for skewed distributions.

When you add a constant a to all the values in a data set, the mean and median increase by a. Measures of spread do not change. When you multiply all the values in a data set by a constant b, the mean, median, IQR, and standard deviation are multiplied by b. These **linear transformations** are quite useful for changing units of measurement.

Back-to-back stemplots and **side-by-side boxplots** are useful for comparing quantitative distributions.

SECTION 1.2 EXERCISES

1.51 MEAT HOT DOGS Make a stemplot of the calories in meat hot dogs from Exercise 1.47 (page 59). What does this graph reveal that the boxplot of these data did not? *Lesson:* Be aware of the limitations of each graphical display.

1.52 EDUCATIONAL ATTAINMENT Table 1.12 shows the educational level achieved by U.S. adults aged 25 to 34 and by those aged 65 to 74. Compare the distributions of educational attainment graphically. Write a few sentences explaining what your display shows.

TABLE 1.12 Educational attainment by U.S. adults aged 25 to 34 and 65 to 74

	Number of people (thousands)	
	Ages 25–34	Ages 65–74
Less than high school	4474	4695
High school graduate	11,546	6649
Some college	7376	2528
Bachelor's degree	8563	1849
Advanced degree	3374	1266
Total	35,333	16,987

Source: Census Bureau, *Educational Attainment in the United States,* March 2000.

1.53 CASSETTE VERSUS CD SALES Has the increasing popularity of the compact disc (CD) affected sales of cassette tapes? Table 1.13 shows the number of cassettes and CDs sold from 1990 to 1999.

TABLE 1.13 Sales (in millions) of full-length cassettes and CDs, 1990–1999

	1990	1991	1992	1993	1994	1995	1996	1997	1998	1999
Full-length cassettes	54.7	49.8	43.6	38.0	32.1	25.1	19.3	18.2	14.8	8.0
Full-length CDs	31.1	38.9	46.5	51.1	58.4	65.0	68.4	70.2	74.8	83.2

Source: The Recording Industry Association of America, *1999 Consumer Profile.*

Make a graphical display to compare cassette and CD sales. Write a few sentences describing what your graph tells you.

1.54 $\bar{x}$ AND s ARE NOT ENOUGH The mean $\bar{x}$ and standard deviation s measure center and spread but are not a complete description of a distribution. Data sets with different shapes can have the same mean and standard deviation. To demonstrate this fact, use your calculator to find $\bar{x}$ and s for the following two small data sets. Then make a stemplot of each and comment on the shape of each distribution.

Data A:	9.14	8.14	8.74	8.77	9.26	8.10	6.13	3.10	9.13	7.26	4.74
Data B:	6.58	5.76	7.71	8.84	8.47	7.04	5.25	5.56	7.91	6.89	12.50

1.55 In each of the following settings, give the values of a and b for the linear transformation $x_{new} = a + bx$ that expresses the change in measurement units. Then explain how the transformation will affect the mean, the *IQR*, the median, and the standard deviation of the original distribution.

(a) You collect data on the power of car engines, measured in horsepower. Your teacher requires you to convert the power to watts. One horsepower is 746 watts.

(b) You measure the temperature (in degrees Fahrenheit) of your school's swimming pool at 20 different locations within the pool. Your swim team coach wants the summary statistics in degrees Celsius (° F = (9/5)° C + 32).

(c) Dr. Data has given a very difficult statistics test and is thinking about "curving" the grades. She decides to add 10 points to each student's score.

1.56 A change of units that multiplies each unit by b, such as the change $x_{new} = 0 + 2.54x$ from inches x to centimeters x_{new}, multiplies our usual measures of spread by b. This is true of the *IQR* and standard deviation. What happens to the variance when we change units in this way?

1.57 BETTER CORN Corn is an important animal food. Normal corn lacks certain amino acids, which are building blocks for protein. Plant scientists have developed new corn varieties that have more of these amino acids. To test a new corn as an animal food, a group of 20 one-day-old male chicks was fed a ration containing the new corn. A control group of another 20 chicks was fed a ration that was identical except that it contained normal corn. Here are the weight gains (in grams) after 21 days:[16]

Normal corn				New corn			
380	321	366	356	361	447	401	375
283	349	402	462	434	403	393	426
356	410	329	399	406	318	467	407
350	384	316	272	427	420	477	392
345	455	360	431	430	339	410	326

(a) Compute five-number summaries for the weight gains of the two groups of chicks. Then make boxplots to compare the two distributions. What do the data show about the effect of the new corn?

(b) The researchers actually reported means and standard deviations for the two groups of chicks. What are they? How much larger is the mean weight gain of chicks fed the new corn?

(c) The weights are given in grams. There are 28.35 grams in an ounce. Use the results of part (b) to compute the means and standard deviations of the weight gains measured in ounces.

1.58 Which measure of center, the mean or the median, should you use in each of the following situations?

(a) Middletown is considering imposing an income tax on citizens. The city government wants to know the average income of citizens so that it can estimate the total tax base.

(b) In a study of the standard of living of typical families in Middletown, a sociologist estimates the average family income in that city.

CHAPTER REVIEW

Data analysis is the art of describing data using graphs and numerical summaries. The purpose of data analysis is to describe the most important features of a set of data. This chapter introduces data analysis by presenting statistical ideas and tools for describing the distribution of a single variable. The figure below will help you organize the big ideas.

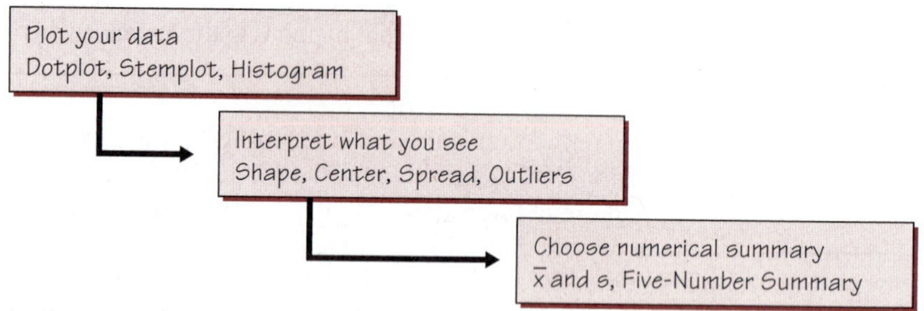

Here is a review list of the most important skills you should have acquired from your study of this chapter.

A. DATA

1. Identify the individuals and variables in a set of data.
2. Identify each variable as categorical or quantitative. Identify the units in which each quantitative variable is measured.

B. DISPLAYING DISTRIBUTIONS

1. Make a bar graph and a pie chart of the distribution of a categorical variable. Interpret bar graphs and pie charts.
2. Make a dotplot of the distribution of a small set of observations.
3. Make a stemplot of the distribution of a quantitative variable. Round leaves or split stems as needed to make an effective stemplot.
4. Make a histogram of the distribution of a quantitative variable.
5. Construct and interpret an ogive of a set of quantitative data.

C. INSPECTING DISTRIBUTIONS (QUANTITATIVE VARIABLES)

1. Look for the overall pattern and for major deviations from the pattern.

2. Assess from a dotplot, stemplot, or histogram whether the shape of a distribution is roughly symmetric, distinctly skewed, or neither. Assess whether the distribution has one or more major peaks.

3. Describe the overall pattern by giving numerical measures of center and spread in addition to a verbal description of shape.

4. Decide which measures of center and spread are more appropriate: the mean and standard deviation (especially for symmetric distributions) or the five-number summary (especially for skewed distributions).

5. Recognize outliers.

D. TIME PLOTS

1. Make a time plot of data, with the time of each observation on the horizontal axis and the value of the observed variable on the vertical axis.

2. Recognize strong trends or other patterns in a time plot.

E. MEASURING CENTER

1. Find the mean $\bar{x}$ of a set of observations.

2. Find the median M of a set of observations.

3. Understand that the median is more resistant (less affected by extreme observations) than the mean. Recognize that skewness in a distribution moves the mean away from the median toward the long tail.

F. MEASURING SPREAD

1. Find the quartiles Q_1 and Q_3 for a set of observations.

2. Give the five-number summary and draw a boxplot; assess center, spread, symmetry, and skewness from a boxplot. Determine outliers.

3. Using a calculator, find the standard deviation s for a set of observations.

4. Know the basic properties of s: $s \geq 0$ always; $s = 0$ only when all observations are identical; s increases as the spread increases; s has the same units as the original measurements; s is increased by outliers or skewness.

G. CHANGING UNITS OF MEASUREMENT (LINEAR TRANSFORMATIONS)

1. Determine the effect of a linear transformation on measures of center and spread.

2. Describe a change in units of measurement in terms of a linear transformation of the form $x_{new} = a + bx$.

H. COMPARING DISTRIBUTIONS

1. Use side-by-side bar graphs to compare distributions of categorical data.

2. Make back-to-back stemplots and side-by-side boxplots to compare distributions of quantitative variables.

3. Write narrative comparisons of the shape, center, spread, and outliers for two or more quantitative distributions.

CHAPTER 1 REVIEW EXERCISES

1.59 Each year *Fortune* magazine lists the top 500 companies in the United States, ranked according to their total annual sales in dollars. Describe three other variables that could reasonably be used to measure the "size" of a company.

1.60 ATHLETES' SALARIES Here is a small part of a data set that describes major league baseball players as of opening day of the 1998 season:

Player	Team	Position	Age	Salary
⋮				
Perez, Eduardo	Reds	First base	28	300
Perez, Neifi	Rockies	Shortstop	23	210
Pettitte, Andy	Yankees	Pitcher	25	3750
Piazza, Mike	Dodgers	Catcher	29	8000
⋮				

(a) What individuals does this data set describe?

(b) In addition to the player's name, how many variables does the data set contain? Which of these variables are categorical and which are quantitative?

(c) Based on the data in the table, what do you think are the units of measurement for each of the quantitative variables?

1.61 HOW YOUNG PEOPLE DIE The number of deaths among persons aged 15 to 24 years in the United States in 1997 due to the seven leading causes of death for this age group were accidents, 12,958; homicide, 5793; suicide, 4146; cancer, 1583; heart disease, 1013; congenital defects, 383; AIDS, 276.[17]

(a) Make a bar graph to display these data.

(b) What additional information do you need to make a pie chart?

1.62 NEVER ON SUNDAY? The Canadian Province of Ontario carries out statistical studies of the working of Canada's national health care system in the province. The bar graphs in Figure 1.24 come from a study of admissions and discharges from community hospitals in Ontario.[18] They show the number of heart attack patients admitted and discharged on each day of the week during a 2-year period.

(a) Explain why you expect the number of patients admitted with heart attacks to be roughly the same for all days of the week. Do the data show that this is true?

(b) Describe how the distribution of the day on which patients are discharged from the hospital differs from that of the day on which they are admitted. What do you think explains the difference?

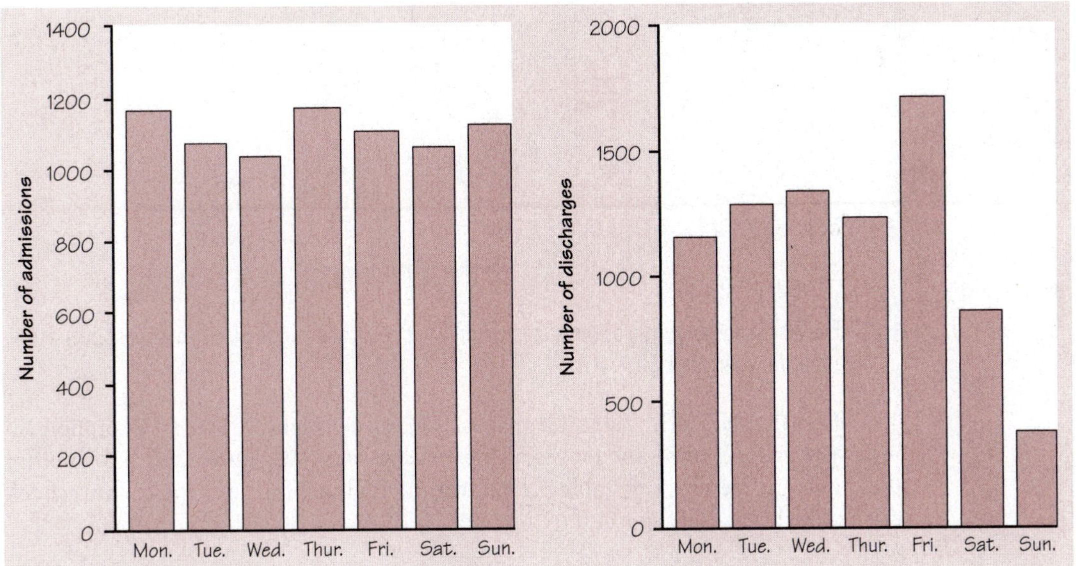

FIGURE 1.24 Bar graphs of the number of heart attack victims admitted and discharged on each day of the week by hospitals in Ontario, Canada.

1.63 PRESIDENTIAL ELECTIONS Here are the percents of the popular vote won by the successful candidate in each of the presidential elections from 1948 to 2000:

Year:	1948	1952	1956	1960	1964	1968	1972	1976	1980	1984	1988	1992	1996	2000
Percent:	49.6	55.1	57.4	49.7	61.1	43.4	60.7	50.1	50.7	58.8	53.9	43.2	49.2	47.9

(a) Make a stemplot of the winners' percents. (Round to whole numbers and use split stems.)

(b) What is the median percent of the vote won by the successful candidate in presidential elections? (Work with the unrounded data.)

(c) Call an election a landslide if the winner's percent falls at or above the third quartile. Find the third quartile. Which elections were landslides?

1.64 HURRICANES The histogram in Figure 1.25 (next page) shows the number of hurricanes reaching the east coast of the United States each year over a 70-year period.[19] Give a brief description of the overall shape of this distribution. About where does the center of the distribution lie?

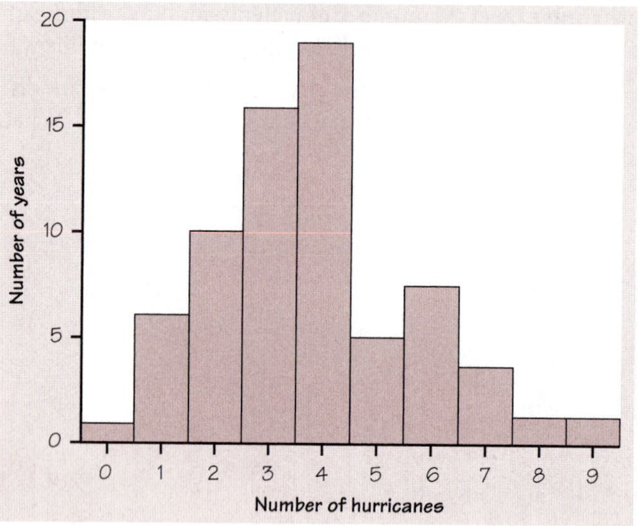

FIGURE 1.25 The distribution of the annual number of hurricanes on the U.S. east coast over a 70-year period, for Exercise 1.64.

1.65 DO SUVS WASTE GAS? Table 1.3 (page 17) gives the highway fuel consumption (in miles per gallon) for 32 model year 2000 midsize cars. We constructed a dotplot for these data in Exercise 1.8. Table 1.14 shows the highway mileages for 26 four-wheel-drive model year 2000 sport utility vehicles.

(a) Give a graphical and numerical description of highway fuel consumption for SUVs. What are the main features of the distribution?

(b) Make boxplots to compare the highway fuel consumption of midsize cars and SUVs. What are the most important differences between the two distributions?

TABLE 1.14 Highway gas mileages for model year 2000 four-wheel-drive SUVs

Model	MPG	Model	MPG
BMW X5	17	Kia Sportage	22
Chevrolet Blazer	20	Land Rover	17
Chevrolet Tahoe	18	Lexus LX470	16
Dodge Durango	18	Lincoln Navigator	17
Ford Expedition	18	Mazda MPV	19
Ford Explorer	20	Mercedes-Benz ML320	20
Honda Passport	20	Mitsubishi Montero	20
Infinity QX4	18	Nissan Pathfinder	19
Isuzu Amigo	19	Nissan Xterra	19
Isuzu Trooper	19	Subaru Forester	27
Jeep Cherokee	20	Suzuki Grand Vitara	20
Jeep Grand Cherokee	18	Toyota RAV4	26
Jeep Wrangler	19	Toyota 4Runner	21

1.66 DR. DATA RETURNS! Dr. Data asked her students how much time they spent using a computer during the previous week. Figure 1.26 is an ogive of her students' responses.

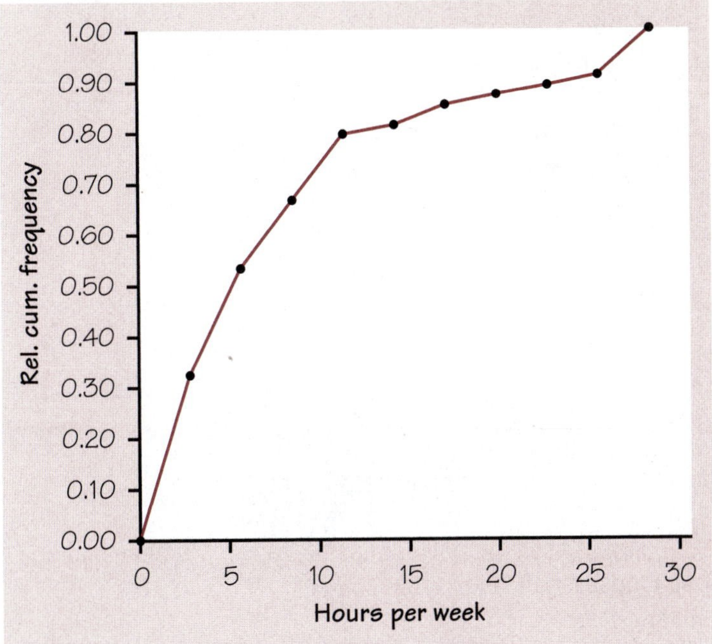

FIGURE 1.26 Ogive of weekly computer use by Dr. Data's statistics students.

(a) Construct a relative frequency table based on the ogive. Then make a histogram.

(b) Estimate the median, Q_1, and Q_3 from the ogive. Then make a boxplot. Are there any outliers?

(c) At what percentile does a student who used her computer for 10 hours last week fall?

1.67 WAL-MART STOCK The rate of return on a stock is its change in price plus any dividends paid. Rate of return is usually measured in percent of the starting value. We have data on the monthly rates of return for the stock of Wal-Mart stores for the years 1973 to 1991, the first 19 years Wal-Mart was listed on the New York Stock Exchange. There are 228 observations.

Figure 1.27 (next page) displays output from statistical software that describes the distribution of these data. The stems in the stemplot are the tens digits of the percent returns. The leaves are the ones digits. The stemplot uses split stems to give a better display. The software gives high and low outliers separately from the stemplot rather than spreading out the stemplot to include them.

(a) Give the five-number summary for monthly returns on Wal-Mart stock.

(b) Describe in words the main features of the distribution.

(c) If you had \$1000 worth of Wal-Mart stock at the beginning of the best month during these 19 years, how much would your stock be worth at the end of the month? If you had \$1000 worth of stock at the beginning of the worst month, how much would your stock be worth at the end of the month?

(d) Find the interquartile range (IQR) for the Wal-Mart data. Are there any outliers according to the $1.5 \times IQR$ criterion? Does it appear to you that the software uses this criterion in choosing which observations to report separately as outliers?

```
Mean  =  3.064
Standard deviation  =  11.49

N  = 228    Median  =  3.4691
Quartiles  =  -2.950258,   8.4511

Decimal point is 1 place to the right of the colon

Low:   -34.04255  -31.25000  -27.06271  -26.61290

 -1 : 985
 -1 : 444443322222110000
 -0 : 999988777666666665555
 -0 : 44444444333333322222222222221111111100
  0 : 0000011111111111112222222333333334444444444
  0 : 55555555555555555555556666666666667777778888888888899999
  1 : 0000000001111111112223333334444
  1 : 55566667889
  2 : 011334

High:  32.01923  41.80531  42.05607  57.89474  58.67769
```

FIGURE 1.27 Output from software describing the distribution of monthly returns from Wal-Mart stock.

1.68 A study of the size of jury awards in civil cases (such as injury, product liability, and medical malpractice) in Chicago showed that the median award was about $8000. But the mean award was about $69,000. Explain how this great difference between the two measures of center can occur.

1.69 You want to measure the average speed of vehicles on the interstate highway on which you are driving. You adjust your speed until the number of vehicles passing you equals the number you are passing. Have you found the mean speed or the median speed of vehicles on the highway?

TABLE 1.15 **Data on education in the United States for Exercises 1.70 to 1.73**

State	Region	Population (1000)	SAT Verbal	SAT Math	Percent taking	Percent no HS diploma	Teachers' pay ($1000)
AL	ESC	4,447	561	555	9	33.1	32.8
AK	PAC	627	516	514	50	13.4	51.7
AZ	MTN	5,131	524	525	34	21.3	34.4
AR	WSC	2,673	563	556	6	33.7	30.6
CA	PAC	33,871	497	514	49	23.8	43.7

TABLE 1.15 Data on education in the United States, for Exercises 1.70 to 1.73 (*continued*)

State	Region	Population (1000)	SAT Verbal	SAT Math	Percent taking	Percent no HS diploma	Teachers' pay ($1000)
CO	MTN	4,301	536	540	32	15.6	37.1
CT	NE	3,406	510	509	80	20.8	50.7
DE	SA	784	503	497	67	22.5	42.4
DC	SA	572	494	478	77	26.9	46.4
FL	SA	15,982	499	498	53	25.6	34.5
GA	SA	8,186	487	482	63	29.1	37.4
HI	PAC	1,212	482	513	52	19.9	38.4
ID	MTN	1,294	542	540	16	20.3	32.8
IL	ENC	12,419	569	585	12	23.8	43.9
IN	ENC	6,080	496	498	60	24.4	39.7
IA	WNC	2,926	594	598	5	19.9	34.0
KS	WNC	2,688	578	576	9	18.7	36.8
KY	ESC	4,042	547	547	12	35.4	34.5
LA	WSC	4,469	561	558	8	31.7	29.7
ME	NE	1,275	507	503	68	21.2	34.3
MD	SA	5,296	507	507	65	21.6	41.7
MA	NE	6,349	511	511	78	20.0	43.9
MI	ENC	9,938	557	565	11	23.2	49.3
MN	WNC	4,919	586	598	9	17.6	39.1
MS	ESC	2,845	563	548	4	35.7	29.5
MO	WNC	5,595	572	572	8	26.1	34.0
MT	MTN	902	545	546	21	19.0	30.6
NE	WNC	1,711	568	571	8	18.2	32.7
NV	MTN	1,998	512	517	34	21.2	37.1
NH	NE	1,236	520	518	72	17.8	36.6
NJ	MA	8,414	498	510	80	23.3	50.4
NM	MTN	1,819	549	542	12	24.9	30.2
NY	MA	18,976	495	502	76	25.2	49.0
NC	SA	8,049	493	493	61	30.0	33.3
ND	WNC	642	594	605	5	23.3	28.2
OH	ENC	11,353	534	568	25	24.3	39.0
OK	WSC	3,451	567	560	8	25.4	30.6
OR	PAC	3,421	525	525	53	18.5	42.2
PA	MA	12,281	498	495	70	25.3	47.7
RI	NE	1,048	504	499	70	28.0	44.3
SC	SA	4,012	479	475	61	31.7	33.6
SD	WNC	755	585	588	4	22.9	27.3
TN	ESC	5,689	559	553	13	32.9	35.3
TX	WSC	20,852	494	499	50	27.9	33.6
UT	MTN	2,233	570	568	5	14.9	33.0
VT	NE	609	514	506	70	19.2	36.3
VA	SA	7,079	508	499	65	24.8	36.7
WA	PAC	5,894	525	526	52	16.2	38.8
WV	SA	1,808	527	512	18	34.0	33.4
WI	ENC	5,364	584	595	7	21.4	39.9
WY	MTN	494	546	551	10	17.0	32.0

Source: U.S. Census Bureau Web site, http://www.census.gov, 2001.

Table 1.15 presents data about the individual states that relate to education. Study of a data set with many variables begins by examining each variable by itself. Exercises 1.70 to 1.73 concern the data in Table 1.15.

1.70 POPULATION OF THE STATES Make a graphical display of the population of the states. Briefly describe the shape, center, and spread of the distribution of population. Explain why the shape of the distribution is not surprising. Are there any states that you consider outliers?

1.71 HOW MANY STUDENTS TAKE THE SAT? Make a stemplot of the distribution of the percent of high school seniors who take the SAT in the various states. Briefly describe the overall shape of the distribution. Find the midpoint of the data and mark this value on your stemplot. Explain why describing the center is not very useful for a distribution with this shape.

1.72 HOW MUCH ARE TEACHERS PAID? Make a graph to display the distribution of average teachers' salaries for the states. Is there a clear overall pattern? Are there any outliers or other notable deviations from the pattern?

1.73 PEOPLE WITHOUT HIGH SCHOOL EDUCATIONS The "Percent no HS" column gives the percent of the adult population in each state who did not graduate from high school. We want to compare the percents of people without a high school education in the northeastern and the southern states. Take the northeastern states to be those in the MA (Mid-Atlantic) and NE (New England) regions. The southern states are those in the SA (South Atlantic) and ESC (East South Central) regions. Leave out the District of Columbia, which is a city rather than a state.

(a) List the percents without high school for the northeastern and for the southern states from Table 1.15. These are the two data sets we want to compare.

(b) Make numerical summaries and graphs to compare the two distributions. Write a brief statement of what you find.

NOTES AND DATA SOURCES

1. Data from *Beverage Digest*, February 18, 2000.
2. Seat-belt data from the National Highway and Traffic Safety Administration, *NOPUS Survey*, 1998.
3. Data from the 1997 *Statistical Abstract of the United States.*
4. Data on accidental deaths from the Centers for Disease Control Web site, www.cdc.gov.
5. Data from the *Los Angeles Times*, February 16, 2001.
6. Based on experiments performed by G. T. Lloyd and E. H. Ramshaw of the CSIRO Division of Food Research, Victoria, Australia, 1982–83.
7. Maribeth Cassidy Schmitt, from her Ph.D. dissertation, "The effects of an elaborated directed reading activity on the metacomprehension skills of third graders," Purdue University, 1987.
8. Data from "America's best small companies," *Forbes*, November 8, 1993.
9. The Shakespeare data appear in C. B. Williams, *Style and Vocabulary: Numerological Studies*, Griffin, London, 1970.

10. Data from John K. Ford, "Diversification: how many stocks will suffice?" *American Association of Individual Investors Journal*, January 1990, pp. 14–16.
11. Data on frosts from C. E. Brooks and N. Carruthers, *Handbook of Statistical Methods in Meteorology*, Her Majesty's Stationery Office, London, 1953.
12. These data were collected by students as a class project.
13. Data from S. M. Stigler, "Do robust estimators work with real data?" *Annals of Statistics*, 5 (1977), pp. 1055–1078.
14. Data obtained from The College Board.
15. Basketball scores from the *Los Angeles Times*, February 16, 2001.
16. Based on summaries in G. L. Cromwell et al., "A comparison of the nutritive value of *opaque-2*, *floury-2*, and normal corn for the chick," *Poultry Science*, 57 (1968), pp. 840–847.
17. Centers for Disease Control and Prevention, *Births and Deaths: Preliminary Data for 1997*, Monthly Vital Statistics Reports, 47, No. 4, 1998.
18. Based on Antoni Basinski, "Almost never on Sunday: implications of the patterns of admission and discharge for common conditions," Institute for Clinical Evaluative Sciences in Ontario, October 18, 1993.
19. Hurricane data from H. C. S. Thom, *Some Methods of Climatological Analysis*, World Meteorological Organization, Geneva, Switzerland, 1966.

AT&T Archives

JOHN W. TUKEY

The Philosopher of Data Analysis

He started as a chemist, became a mathematician, and was converted to statistics by what he called "the real problems experience and the real data experience" of war work during the Second World War. *John W. Tukey (1915–2000)* came to Princeton University in 1937 to study chemistry but took a doctorate in mathematics in 1939. During the war, he worked on the accuracy of range finders and of gunfire from bombers, among other problems. After the war he divided his time between Princeton and nearby Bell Labs, at that time the world's leading industrial research group.

Tukey devoted much of his attention to the statistical study of messy problems with complex data: the safety of anesthetics used by many doctors in many hospitals on many patients, the Kinsey studies of human sexual behavior, monitoring compliance with a nuclear test ban, and air quality and environmental pollution.

From this "real problems experience and real data experience," John Tukey developed exploratory data analysis. He invented some of the tools we have met, such as boxplots and stemplots. More important, he developed a philosophy for data analysis that changed the way statisticians think. In this chapter, as in Chapter 1, the approach we take in examining data follows Tukey's path.

Tukey was converted to statistics by "the real problems experience and the real data experience" during the Second World War.

chapter 2

The Normal Distributions

ACTIVITY 2A A Fine-Grained Distribution

Materials: Sheet of grid paper; salt; can of spray paint; paint easel; newspapers

1. Place the grid paper on the easel with a horizontal fold as shown, at about a 45° angle to the horizontal. Provide a "lip" at the bottom to catch the salt. Place newspaper behind the grid and extending out on all sides so you will not get paint on the easel.

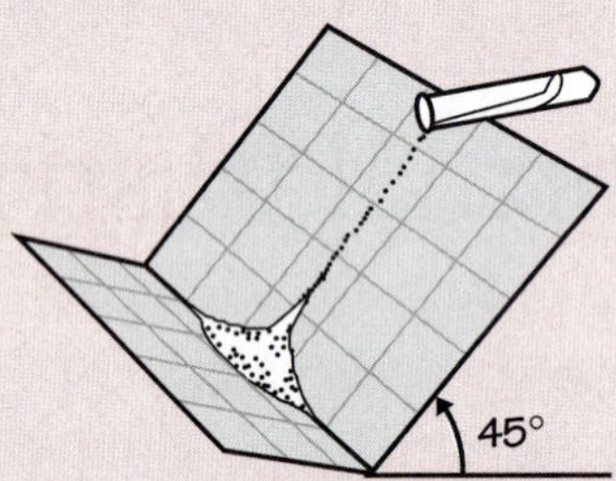

2. Pour a stream of salt slowly from a point near the middle of the top edge of the grid. The grains of salt will hop and skip their way down the grid as they collide with one another and bounce left and right. They will accumulate at the bottom, piled against the grid, with the smooth profile of a bell-shaped curve, known as a normal distribution. We will learn about the normal distribution in this chapter.

3. Now carefully spray the grid—salt and all—with paint. Then discard the salt. You should be able to easily measure the height of the curve at different places by simply counting lines on the grid, or you could approximate areas by counting small squares or portions of squares on the grid.

How could you get a tall, narrow curve? How could you get a short, broad curve? What factors might affect the height and breadth of the curve? From the members of the class, collect a set of normal curves that differ from one another.

ACTIVITY 2B Roll a Normal Distribution

Materials: Several marbles, all the same size; two metersticks for a "ramp"; a ruled sheet of paper; a flat table about 4 feet long; carbon paper; Scotch Tape or masking tape

ACTIVITY 2B Roll a Normal Distribution *(continued)*

1. At one end of the table prop up the two metersticks in a "V" shape to provide a ramp for the marbles to roll down. The marble will roll down the chute, continue across the table, and fall off the table to the floor below. Make sure that the ramp is secure and that the tabletop does not have any grooves or obstructions.

2. Roll the marble down the ramp several times to get a good idea of the area of the floor where the marble will fall.

3. Center the ruled sheet of paper (see Figure 2.1) over this area, face up, with the bottom edge toward the table and parallel to the edge of the table. The ruled lines should go in the same direction as the marble's path. Tape the sheet securely to the floor. Place the sheet of carbon paper, carbon side down, over the ruled sheet.

4. Roll the marble for a class total of 200 times. The spots where it hits the floor will be recorded on the ruled paper as black dots. When the marble hits the floor, it will probably bounce, so try to catch it in midair after the impact so that you don't get any extra marks. After the first 100 rolls, replace the sheet of paper. This will make it easier for you to count the spots. Make sure that the second sheet is in exactly the same position as the first one.

5. When the marble has been rolled 200 times, make a histogram of the distribution of the points as follows. First, count the number of dots in each column. Then graph this number by drawing horizontal lines in the columns at the appropriate level. Use the scale on the left-hand side of the sheet.

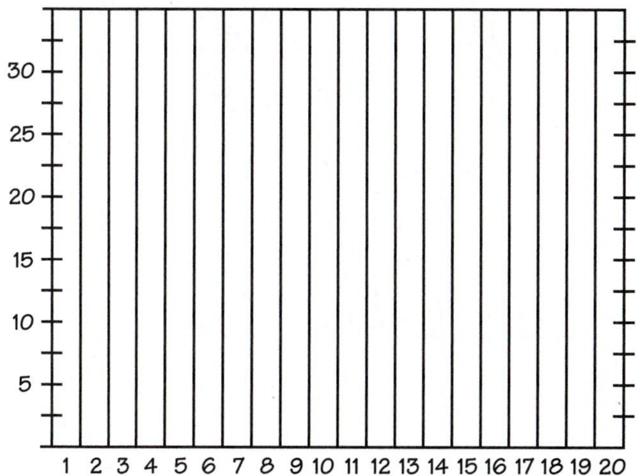

FIGURE 2.1 Example of ruled sheet for Activity 2B.

2.1 DENSITY CURVES AND THE NORMAL DISTRIBUTIONS

We now have a kit of graphical and numerical tools for describing distributions. What is more, we have a clear strategy for exploring data on a single quantitative variable:

- Always plot your data: make a graph, usually a histogram or a stemplot.
- Look for the overall pattern (shape, center, spread) and for striking deviations such as outliers.
- Calculate a numerical summary to briefly describe the center and spread.

Here is one more step to add to the strategy:

- Sometimes the overall pattern of a large number of observations is so regular that we can describe it by a smooth curve.

Density curves

Figure 2.2 is a histogram of the scores of all 947 seventh-grade students in Gary, Indiana, on the vocabulary part of the Iowa Test of Basic Skills.[1] Scores of many students on this national test have a quite regular distribution. The histogram is symmetric, and both tails fall off quite smoothly from a single center peak. There are no large gaps or obvious outliers. The smooth curve drawn through the tops of the histogram bars is a good description of the overall pattern of the data.

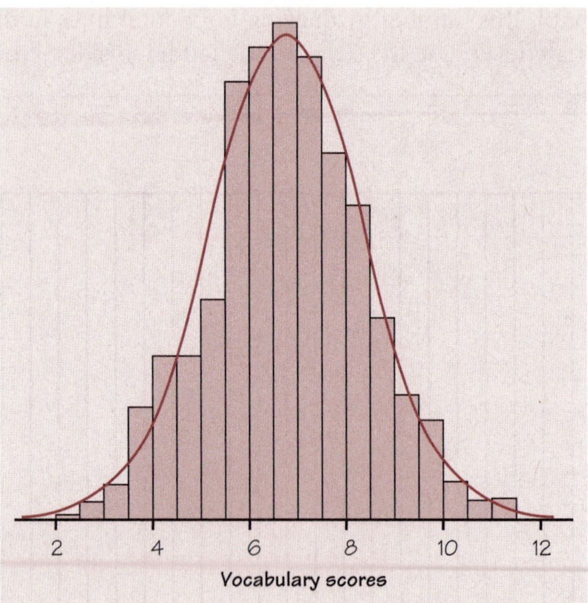

FIGURE 2.2 Histogram of the vocabulary scores of all seventh-grade students in Gary, Indiana. The smooth curve shows the overall shape of the distribution.

The curve is a ***mathematical model*** for the distribution. A mathematical model is an idealized description. It gives a compact picture of the overall pattern of the data but ignores minor irregularities as well as any outliers.

mathematical model

We will see that it is easier to work with the smooth curve in Figure 2.2 than with the histogram. The reason is that the histogram depends on our choice of classes, while with a little care we can use a curve that does not depend on any choices we make. Here's how we do it.

EXAMPLE 2.1 FROM HISTOGRAM TO DENSITY CURVE

Our eyes respond to the *areas* of the bars in a histogram. The bar areas represent proportions of the observations. Figure 2.3(a) is a copy of Figure 2.2 with the leftmost bars shaded. The area of the shaded bars in Figure 2.3(a) represents the students with vocabulary scores 6.0 or lower. There are 287 such students, who make up the proportion 287/947 = 0.303 of all Gary seventh graders.

Now concentrate on the curve drawn through the bars. In Figure 2.3(b), the area under the curve to the left of 6.0 is shaded. Adjust the scale of the graph so that *the total area under the curve is exactly 1*. This area represents the proportion 1, that is, all the observations. Areas under the curve then represent proportions of the observations. The curve is now a *density curve*. The shaded area under the density curve in Figure 2.3(b) represents the proportion of students with score 6.0 or lower. This area is 0.293, only 0.010 away from the histogram result. You can see that areas under the density curve give quite good approximations of areas given by the histogram.

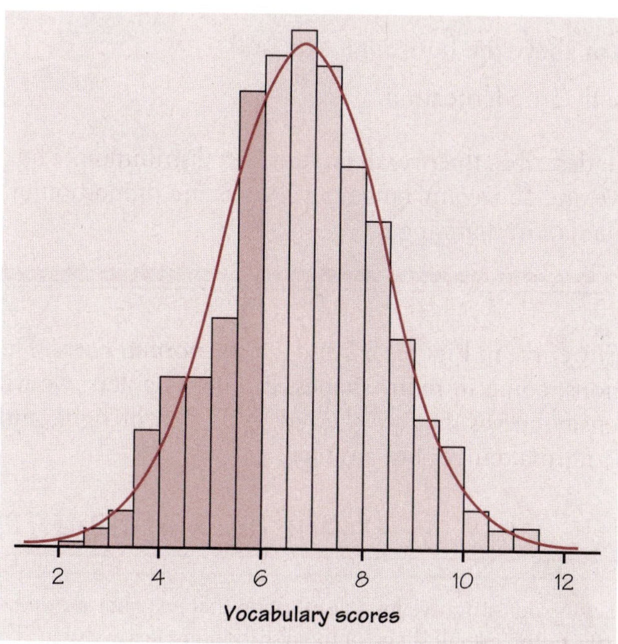

FIGURE 2.3(a) The proportion of scores less than or equal to 6.0 from the histogram is 0.303.

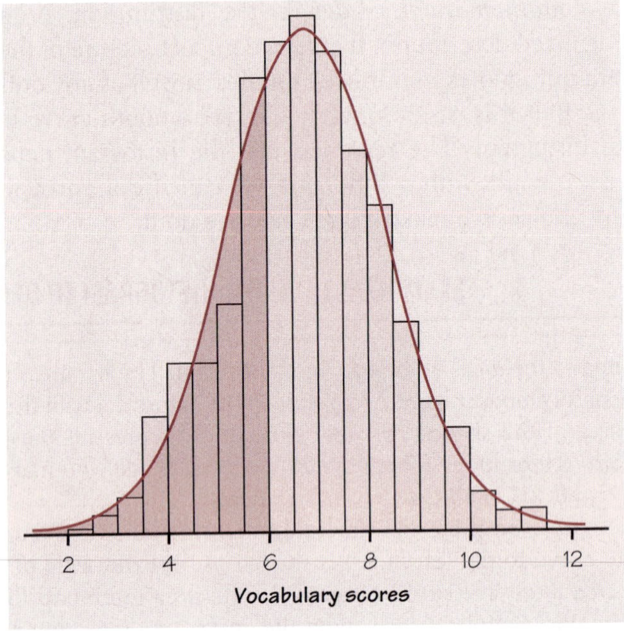

FIGURE 2.3(b) The proportion of scores less than or equal to 6.0 from the density curve is 0.293.

DENSITY CURVE

A **density curve** is a curve that

- is always on or above the horizontal axis, and
- has area exactly 1 underneath it.

A density curve describes the overall pattern of a distribution. The area under the curve and above any range of values is the proportion of all observations that fall in that range.

normal curve

The density curve in Figures 2.2 and 2.3 is a ***normal curve.*** Density curves, like distributions, come in many shapes. In later chapters, we will encounter important density curves that are skewed to the left or right, and curves that may look like normal curves but are not.

EXAMPLE 2.2 A SKEWED-LEFT DISTRIBUTION

Figure 2.4 shows the density curve for a distribution that is slightly skewed to the left. The smooth curve makes the overall shape of the distribution clearly visible. The shaded area under the curve covers the range of values from 7 to 8. This area is 0.12. This means that the proportion 0.12 of all observations from this distribution have values between 7 and 8.

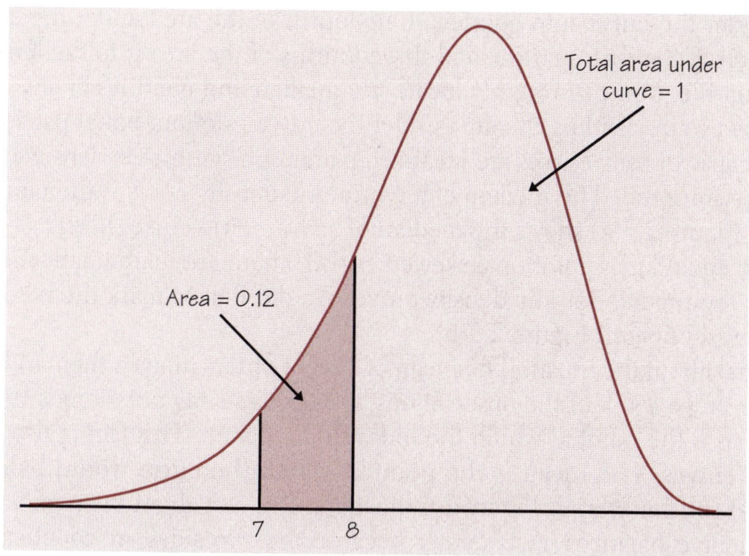

FIGURE 2.4 The shaded area under this density curve is the proportion of observations taking values between 7 and 8.

Figure 2.5 shows two density curves: a symmetric normal density curve and a right-skewed curve. A density curve of the appropriate shape is often an adequate description of the overall pattern of a distribution. Outliers, which are deviations from the overall pattern, are not described by the curve. Of course, no set of real data is exactly described by a density curve. The curve is an approximation that is easy to use and accurate enough for practical use.

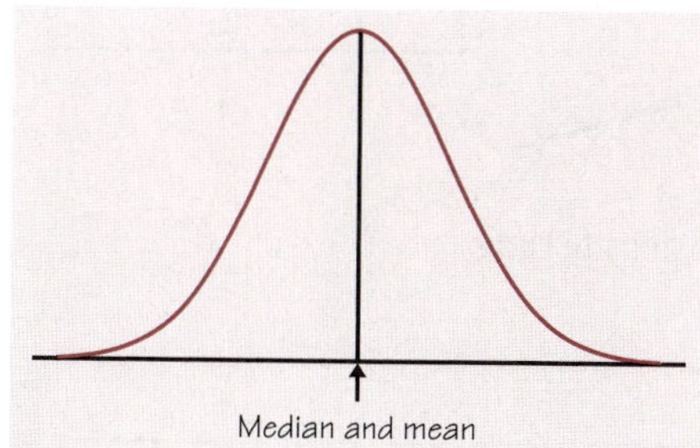

FIGURE 2.5(a) The median and mean of a symmetric density curve.

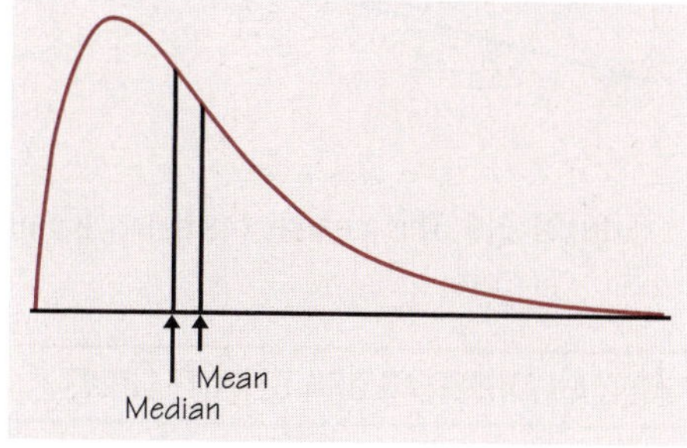

FIGURE 2.5(b) The median and mean of a right-skewed density curve.

The median and mean of a density curve

Our measures of center and spread apply to density curves as well as to actual sets of observations. The median and quartiles are easy. Areas under a density curve represent proportions of the total number of observations. The median is the point with half the observations on either side. So **the median of a density curve is the equal-areas point,** the point with half the area under the curve to its left and the remaining half of the area to its right. The quartiles divide the area under the curve into quarters. One-fourth of the area under the curve is to the left of the first quartile, and three-fourths of the area is to the left of the third quartile. You can roughly locate the median and quartiles of any density curve by eye by dividing the area under the curve into four equal parts.

Because density curves are idealized patterns, a symmetric density curve is exactly symmetric. The median of a symmetric density curve is therefore at its center. Figure 2.5(a) shows the median of a symmetric curve. It isn't so easy to spot the equal-areas point on a skewed curve. There are mathematical ways of finding the median for any density curve. We did that to mark the median on the skewed curve in Figure 2.5(b).

What about the mean? The mean of a set of observations is their arithmetic average. If we think of the observations as weights strung out along a thin rod, the mean is the point at which the rod would balance. This fact is also true of density curves. **The mean is the point at which the curve would balance if made of solid material.** Figure 2.6 illustrates this fact about the mean. A symmetric curve balances at its center because the two sides are identical. **The mean and median of a symmetric density curve are equal,** as in Figure 2.5(a). We know that the mean of a skewed distribution is pulled toward the long tail. Figure 2.5(b) shows how the mean of a skewed density curve is pulled toward the long tail more than is the median. It's hard to locate the balance point by eye on a skewed curve. There are mathematical ways of calculating the mean for any density curve, so we are able to mark the mean as well as the median in Figure 2.5(b).

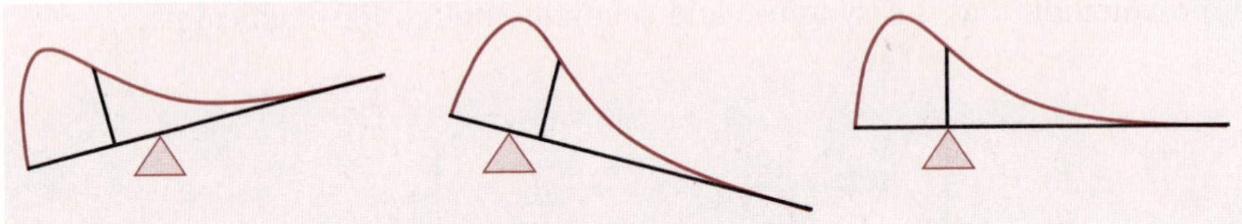

FIGURE 2.6 The mean is the balance point of a density curve.

MEDIAN AND MEAN OF A DENSITY CURVE

The **median** of a density curve is the equal-areas point, the point that divides the area under the curve in half.

The **mean** of a density curve is the balance point, at which the curve would balance if made of solid material.

The median and mean are the same for a symmetric density curve. They both lie at the center of the curve. The mean of a skewed curve is pulled away from the median in the direction of the long tail.

We can roughly locate the mean, median, and quartiles of any density curve by eye. This is not true of the standard deviation. When necessary, we can once again call on more advanced mathematics to learn the value of the standard deviation. The study of mathematical methods for doing calculations with density curves is part of theoretical statistics. Though we are concentrating on statistical practice, we often make use of the results of mathematical study.

Because a density curve is an idealized description of the distribution of data, we need to distinguish between the mean and standard deviation of the density curve and the mean $\bar{x}$ and standard deviation s computed from the actual observations. The usual notation for the mean of an idealized distribution is μ (the Greek letter mu). We write the standard deviation of a density curve as σ (the Greek letter sigma).

mean μ
standard deviation σ

EXERCISES

2.1 DENSITY CURVES

(a) Sketch a density curve that is symmetric but has a shape different from that of the curve in Figure 2.5(a).

(b) Sketch a density curve that is strongly skewed to the left.

2.2 A UNIFORM DISTRIBUTION Figure 2.7 displays the density curve of a *uniform distribution.* The curve takes the constant value 1 over the interval from 0 to 1 and is zero outside the range of values. This means that data described by this distribution take values that are uniformly spread between 0 and 1. Use areas under this density curve to answer the following questions.

(a) Why is the total area under this curve equal to 1?

(b) What percent of the observations lie above 0.8?

(c) What percent of the observations lie below 0.6?

(d) What percent of the observations lie between 0.25 and 0.75?

(e) What is the mean μ of this distribution?

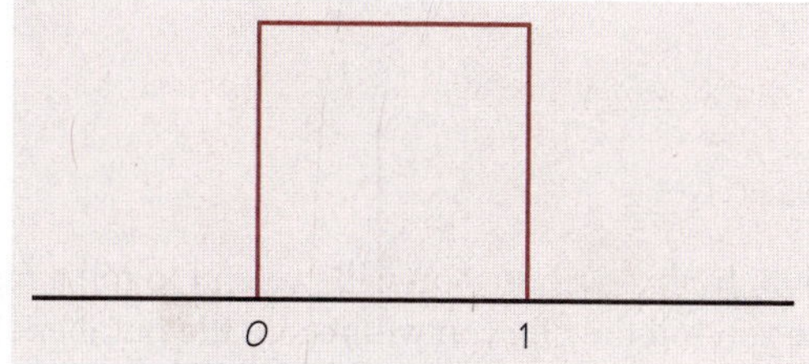

FIGURE 2.7 The density curve of a uniform distribution.

2.3 A WEIRD DENSITY CURVE A line segment can be considered a density "curve," as shown in Exercise 2.2. A "broken line" graph can also be considered a density curve. Figure 2.8 shows such a density curve.

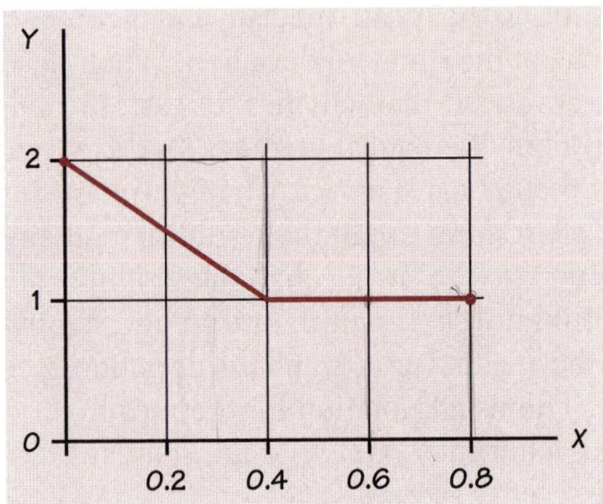

FIGURE 2.8 An unusual "broken line" density curve.

(a) Verify that the graph in Figure 2.8 is a valid density curve.

For each of the following, use areas under this density curve to find the proportion of observations within the given interval:

(b) $0.6 \leq X \leq 0.8$

(c) $0 \leq X \leq 0.4$

(d) $0 \leq X \leq 0.2$

(e) The median of this density curve is a point between $X = 0.2$ and $X = 0.4$. Explain why.

2.4 FINDING MEANS AND MEDIANS Figure 2.9 displays three density curves, each with three points indicated. At which of these points on each curve do the mean and the median fall?

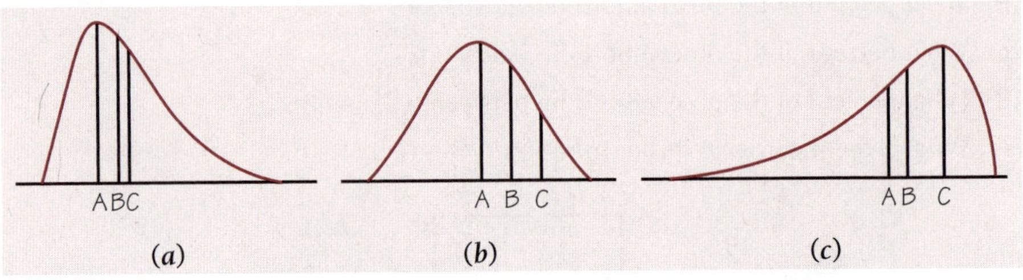

FIGURE 2.9 Three density curves.

2.5 ROLL A DISTRIBUTION In this exercise you will pretend to roll a regular, six-sided die 120 times. Each time you roll the die, you will record the number on the up-face. The numbers 1, 2, 3, 4, 5, and 6 are called the ***outcomes*** of this chance experiment.

outcomes

In 120 rolls, how many of each number would you expect to roll? The TI-83 and TI-89 are useful devices for conducting chance experiments, especially ones like this that involve performing many repetitions. Because you are only pretending to roll the die repeatedly, we call this chance experiment a ***simulation.*** There will be a more formal treatment of simulations in Chapter 5.

simulation

- Begin by clearing L_1 or list1 on your calculator.
- Use your calculator's random integer generator to generate 120 random whole numbers between 1 and 6 (inclusive), and then store these numbers in L_1 or list1.

TI-83

- Press MATH, choose PRB, then `5:RandInt.`
- Complete the command `RandInt(1,6,120)` STO➡ L_1.

TI-89

- Press CATALOG F3 and choose `randInt...`
- Complete the command `tistat.randint(1,6,120)` STO➡ `list1.`

- Set the viewing window parameters: $X[1, 7]_1$ by $Y[-5, 25]_5$.
- Specify a histogram using the data in L_1/list1.
- Then graph. Are you surprised? This is called a **frequency histogram** because it plots the **frequency** of each outcome (number of times each outcome occurred).
- Repeat the simulation several times. You can recall and reuse the previous command by pressing 2nd ENTER. It's a good habit to clear L_1/list1 before you roll the die again.

In theory, of course, each number should come up 20 times. But in practice, there is chance variation, so the bars in the histogram will probably have different heights. Theoretically, what should the distribution look like?

Normal distributions

One particularly important class of density curves has already appeared in Figures 2.2, 2.3, and 2.5(a) and the "fine-grained distribution" of Activity 2A. These density curves are symmetric, single-peaked, and bell-shaped. They are called *normal curves*, and they describe ***normal distributions.*** All normal distributions have the same overall shape. The exact density curve for a particular normal distribution is described by giving its mean μ and its standard deviation σ. The mean is located at the center of the symmetric curve, and is the same as the median. Changing μ without changing σ moves the normal curve along the horizontal axis without changing its spread. The standard deviation σ controls the spread of a normal curve. Figure 2.10 shows two normal curves

normal distributions

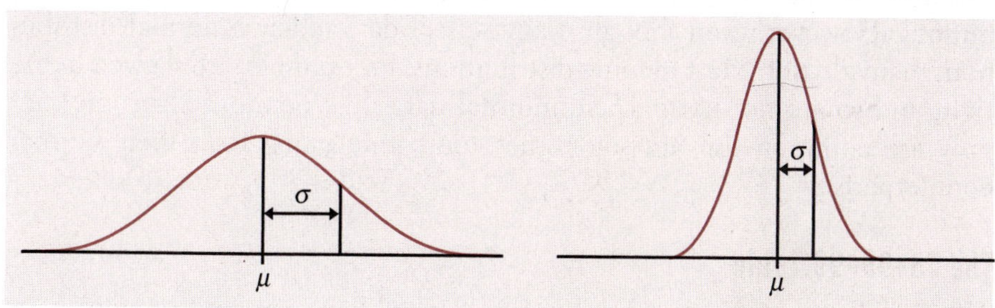

FIGURE 2.10 Two normal curves, showing the mean μ and standard deviation σ.

with different values of σ. The curve with the larger standard deviation is more spread out.

The standard deviation σ is the natural measure of spread for normal distributions. Not only do μ and σ completely determine the shape of a normal curve, but we can locate σ by eye on the curve. Here's how. As we move out in either direction from the center μ, the curve changes from falling ever more steeply

to falling ever less steeply.

inflection points

The points at which this change of curvature takes place are called ***inflection points*** *and are located at distance σ on either side of the mean μ.* Figure 2.10 shows σ for two different normal curves. You can feel the change as you run a pencil along a normal curve, and so find the standard deviation. Remember that μ and σ alone do not specify the shape of most distributions, and that the shape of density curves in general does not reveal σ. These are special properties of normal distributions.

Why are the normal distributions important in statistics? Here are three reasons. First, normal distributions are good descriptions for some distributions of *real data.* Distributions that are often close to normal include scores on tests taken by many people (such as SAT exams and many psychological tests), repeated careful measurements of the same quantity, and characteristics of biological populations (such as lengths of cockroaches and yields of corn). Second, normal distributions are good approximations to the results of many kinds of *chance outcomes,* such as tossing a coin many times. Third, and most important, we will see that many *statistical inference* procedures based on normal distributions work well for other roughly symmetric distributions. However, even though many sets of data follow a normal distribution, many do not. Most income distributions, for example, are skewed to the right and so are not normal. Nonnormal data, like nonnormal people, not only are common but are sometimes more interesting than their normal counterparts.

The 68–95–99.7 rule

Although there are many normal curves, they all have common properties. In particular, all normal distributions obey the following rule.

THE 68–95–99.7 RULE

In the normal distribution with mean μ and standard deviation σ:

- 68% of the observations fall within σ of the mean μ.
- 95% of the observations fall within 2σ of μ.
- 99.7% of the observations fall within 3σ of μ.

Figure 2.11 illustrates the 68–95–99.7 rule. Some authors refer to it as the "empirical rule." By remembering these three numbers, you can think about normal distributions without constantly making detailed calculations, and when rough approximations will suffice.

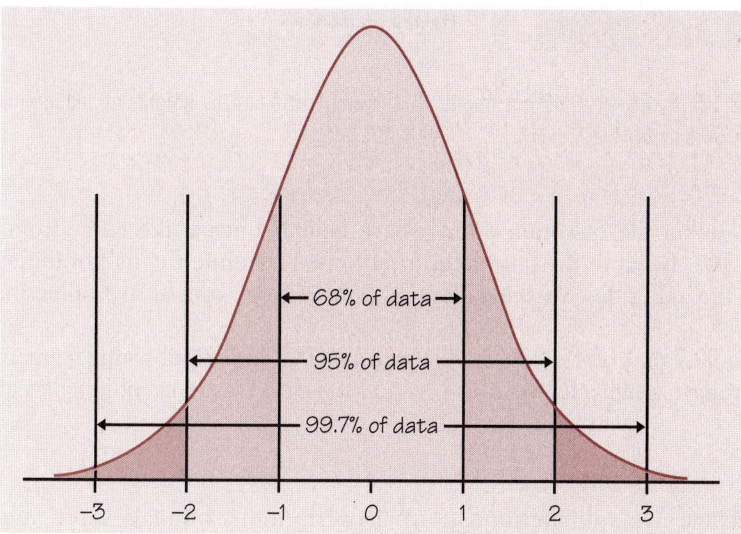

FIGURE 2.11 The 68–95–99.7 rule for normal distributions.

EXAMPLE 2.3 YOUNG WOMEN'S HEIGHTS

The distribution of heights of young women aged 18 to 24 is approximately normal with mean μ = 64.5 inches and standard deviation σ = 2.5 inches. Figure 2.12 shows the application of the 68–95–99.7 rule in this example.

Two standard deviations is 5 inches for this distribution. The 95 part of the 68–95–99.7 rule says that the middle 95% of young women are between 64.5 – 5 and 64.5 + 5 inches tall, that is, between 59.5 and 69.5 inches. This fact is exactly true for an exactly normal distribution. It is approximately true for the heights of young women because the distribution of heights is approximately normal.

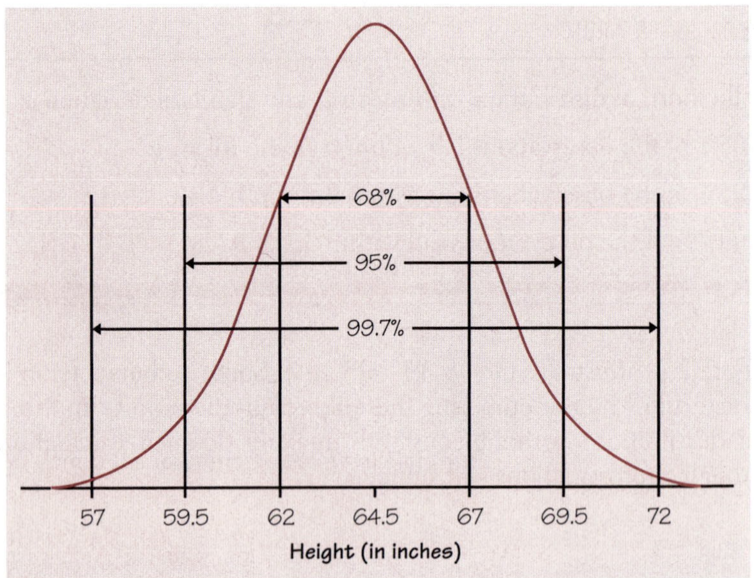

FIGURE 2.12 The 68–95–99.7 rule applied to the distribution of the heights of young women. Here, μ = 64.5 and σ = 2.5.

The other 5% of young women have heights outside the range from 59.5 to 69.5 inches. Because the normal distributions are symmetric, half of these women are on the tall side. So the tallest 2.5% of young women are taller than 69.5 inches.

The 99.7 part of the 68–95–99.7 rule says that almost all young women (99.7% of them) have heights between $\mu - 3\sigma$ and $\mu + 3\sigma$. This range of heights is 57 to 72 inches.

Because we will mention normal distributions often, a short notation is helpful. We abbreviate the normal distribution with mean μ and standard deviation σ as $N(\mu, \sigma)$. For example, the distribution of young women's heights is N(64.5, 2.5).

National test scores are frequently reported in terms of percentiles, rather than raw scores. If your score on the math portion of such a test was reported as the 90th percentile, then 90% of the students who took the math test scored *lower than or equal to* your score. Percentiles are used when we are most interested in seeing where an individual observation stands relative to the other individuals in the distribution. Typically, in practice, the number of observations is quite large so that it makes sense to talk about the distribution as a density curve. The median score would be the 50th percentile because half the scores are to the left of (i.e., lower than) the median. The first quartile is the 25th percentile and the third quartile is the 75th percentile.

EXERCISES

2.6 MEN'S HEIGHTS The distribution of heights of adult American men is approximately normal with mean 69 inches and standard deviation 2.5 inches. Draw a normal curve on which this mean and standard deviation are correctly located. (Hint: Draw the curve first, locate the points where the curvature changes, then mark the horizontal axis.)

2.7 MORE ON MEN'S HEIGHTS The distribution of heights of adult American men is approximately normal with mean 69 inches and standard deviation 2.5 inches. Use the 68–95–99.7 rule to answer the following questions.

(a) What percent of men are taller than 74 inches?

(b) Between what heights do the middle 95% of men fall?

(c) What percent of men are shorter than 66.5 inches?

(d) A height of 71.5 inches corresponds to what percentile of adult male American heights?

2.8 IQ SCORES Scores on the Wechsler Adult Intelligence Scale (WAIS, a standard "IQ test") for the 20 to 34 age group are approximately normally distributed with $\mu = 110$ and $\sigma = 25$. Use the 68–95–99.7 rule to answer these questions.

(a) About what percent of people in this age group have scores above 110?

(b) About what percent have scores above 160?

(c) In what range do the middle 95% of all IQ scores lie?

2.9 WOMEN'S HEIGHTS The distribution of heights of young women aged 18 to 24 is discussed in Example 2.3. Find the percentiles for the following heights.

(a) 64.5 inches

(b) 59.5 inches

(c) 67 inches

(d) 72 inches

2.10 FINE-GRAINED DISTRIBUTION You can do this exercise if you spray-painted a normal distribution in Activity 2A. On your "fine-grained distribution," first count the number of whole squares and parts of squares under the curve. Approximate as best you can. This represents the total area under the curve.

(a) Mark vertical lines at $\mu - 1\sigma$ and $\mu + 1\sigma$. Count the number of squares or parts of squares between these two vertical lines. Now divide the number of squares within one standard deviation of μ by the total number of squares under the curve and express your answer as a percent. How does this compare with 68%? Why would you expect your answer to differ somewhat from 68%?

(b) Count squares to determine the percent of area within 2σ of μ. How does your answer compare with 95%?

(c) Count squares to determine the percent of area within 3σ of μ. How does your answer compare with 99.7%?

SUMMARY

We can sometimes describe the overall pattern of a distribution by a **density curve.** A density curve always remains on or above the horizontal axis and has total area 1 underneath it. An area under a density curve gives the proportion of observations that fall in a range of values.

A density curve is an idealized description of the overall pattern of a distribution that smooths out the irregularities in the actual data. Write the mean of a density curve as μ and the standard deviation of a density curve as σ to distinguish them from the mean $\bar{x}$ and the standard deviation s of the actual data.

The **mean,** the **median,** and the **quartiles** of a density curve can be located by eye. The mean μ is the balance point of the curve. The median divides the area under the curve in half. The quartiles with the median divide the area under the curve into quarters. The **standard deviation** σ cannot be located by eye on most density curves.

The mean and median are equal for symmetric density curves. The mean of a skewed curve is located farther toward the long tail than is the median.

The **normal distributions** are described by a special family of bell-shaped symmetric density curves, called **normal curves.** The mean μ and standard deviation σ completely specify a normal distribution $N(\mu, \sigma)$. The mean is the center of the curve, and σ is the distance from μ to the inflection points on either side.

In particular, all normal distributions satisfy the **68–95–99.7 rule,** which describes what percent of observations lie within one, two, and three standard deviations of the mean.

An observation's percentile is the percent of the distribution that is at or to the left of the observation.

SECTION 2.1 EXERCISES

2.11 ESTIMATING STANDARD DEVIATIONS Figure 2.13 shows two normal curves, both with mean 0. Approximately what is the standard deviation of each of these curves?

2.12 HELMET SIZES The army reports that the distribution of head circumference among male soldiers is approximately normal with mean 22.8 inches and standard deviation 1.1 inches. Use the 68–95–99.7 rule to answer the following questions.

(a) What percent of soldiers have head circumference greater than 23.9 inches?

(b) A head circumference of 23.9 inches would be what percentile?

(c) What percent of soldiers have head circumference between 21.7 inches and 23.9 inches?

2.13 GESTATION PERIOD The length of human pregnancies from conception to birth varies according to a distribution that is approximately normal with mean 266 days and standard deviation 16 days. Use the 68–95–99.7 rule to answer the following questions.

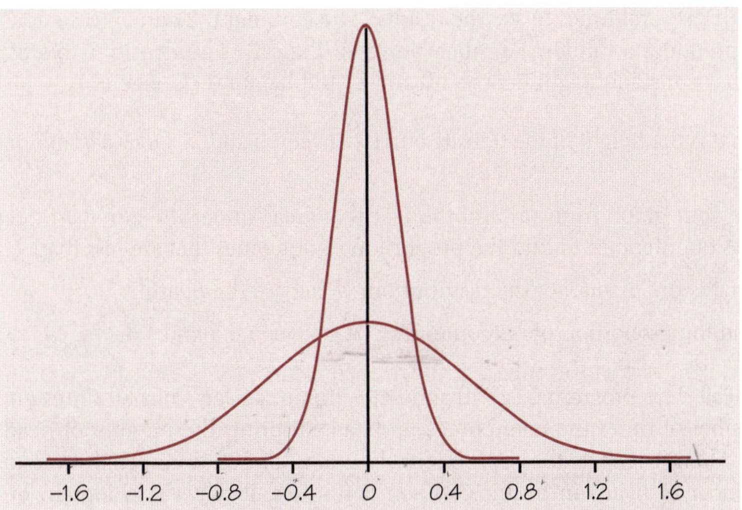

FIGURE 2.13 Two normal curves with the same mean but different standard deviations, for Exercise 2.11.

(a) Between what values do the lengths of the middle 95% of all pregnancies fall?

(b) How short are the shortest 2.5% of all pregnancies?

(c) How long are the longest 2.5% of all pregnancies?

2.14 IQ SCORES FOR ADULTS Wechsler Adult Intelligence Scale (WAIS) scores for young adults are N(110, 25).

(a) If someone's score were reported as the 16th percentile, about what score would that individual have?

(b) Answer the same question for the 84th percentile and the 97.5th percentile.

2.15 WEIGHTS OF DISTANCE RUNNERS A study of elite distance runners found a mean body weight of 63.1 kilograms (kg), with a standard deviation of 4.8 kg.

(a) Assuming that the distribution of weights is normal, sketch the density curve of the weight distribution with the horizontal axis marked in kilograms.

(b) Use the 68–95–99.7 rule to find intervals centered at the mean that will include 68%, 95%, and 99.7% of the weights of the runners.

2.16 CALCULATOR GENERATED DENSITY CURVE Like Minitab and similar computer utilities, the TI-83/TI-89 has a "random number generator" that produces decimal numbers between 0 and 1.

- On the TI-83, press MATH, then choose PRB and `1:Rand`.
- On the TI-89, press 2nd 5 (MATH), then choose `7:Probability` and `4:Rand(`. Be sure to close the parentheses.

Press ENTER several times to see the results. The command `2rand` (`2rand()` on the TI-89) produces a random number between 0 and 2. The density curve of the outcomes has constant height between 0 and 2, and height 0 elsewhere.

(a) What is the height of the density curve between 0 and 2? Draw a graph of the density curve.

(b) Use your graph from (a) and the fact that areas under the curve are relative frequencies of outcomes to find the proportion of outcomes that are less than 1.

(c) What is the median of the distribution? What are the quartiles?

(d) Find the proportion of outcomes that lie between 0.5 and 1.3.

2.17 FLIP50 The program FLIP50 simulates flipping a fair coin 50 times and counts the number of times the coin comes up heads. It prints the number of heads on the screen. Then it repeats the experiment for a total of 100 times, each time displaying the number of heads in 50 flips. When it finishes, it draws a histogram of the 100 results. (You have to set up the plot first on the TI-89.)

(a) What outcomes are likely? What outcome(s) are the most likely? If you made a histogram of the results of the 100 replications, what shape distribution would you expect?

(b) The program is listed below. Enter the program carefully, or link it from a classmate or your teacher. Run the program and observe the variations in the results of the 100 replications.

(c) When the histogram appears, TRACE to see the classes and frequencies. Record the results in a frequency table.

(d) Describe the distribution: symmetric versus nonsymmetric; center; spread; number of peaks; gaps; suspected outliers. What shape density curve would best fit your distribution?

TI-83

```
prgm:FLIP50
100→DIM(L1)
For(I,1,100)
0→H
For(J,1,50)
randInt(0,1)→N
If N=1:H+1→H
End
Disp H
H→L1(I)
End
PlotsOff
10→Xmin
40→Xmax
2→Xscl
-6→Ymin
```

TI-89

```
flip50()
Prgm
tistat.clrlist(list1)
For i,1,100
0→h
For j,1,50
tistat.randint(0,1)→n
If n=1
h+1→h
EndFor
Disp h
h→list1[i]
EndFor
PlotsOff
10→xmin
40→xmax
```

```
25→Ymax                     2→xscl
5→Yscl                      -6→ymin
Plot1(Histogram,L1)         25→ymax
DispGraph                   5→yscl
                            EndPrgm
```

Set up Plot 1 to be a histogram of list1 with a bucket width of 2. Then press ◆ F3 (GRAPH).

2.18 NORMAL DISTRIBUTION ON THE CALCULATOR The normal density curves are defined by a particular equation:

$$y = \frac{1}{\sigma\sqrt{2\pi}} e^{-\frac{1}{2}\left(\frac{x-\mu}{\sigma}\right)^2}$$

We can obtain individual members of this family of curves by specifying particular values for the mean μ and the standard deviation σ. If we specify the values $\mu = 0$ and $\sigma = 1$, then we have the equation for the *standard* normal distribution. This exercise will explore two functions.

- Enter as Y_1 the following equation for the standard normal distribution:

```
Y1=(1/√(2π))( e^(-.5x²)))
```

- For Y_2, position your cursor after Y_2=. On the TI-83, press 2nd VARS (DISTR) and choose `1:normalpdf(`. On the TI-89, press CATALOG F3 (Flash Apps) and choose `normpdf(`. Finish defining Y_2 as `normPdf(x)` (`tistat.normPdf(x)` on the TI-89).
- Turn off all plots and any functions other than Y_1 and Y_2. Change the graph style for Y_2 to a thick line by highlighting the slash \ to the left of Y_2 and pressing ENTER once. (On the TI-89, press 2nd F1 ([F6]) and choose 4:Thick.)
- Specify a viewing window $X[-3,3]_1$ and $Y[-0.1,0.5]_{0.1}$.
- Press GRAPH (◆ F3 on the TI-89.)

Write a sentence that describes the connection between these two functions. *Note:* normalpdf stands for "normal probability density function." We'll learn more about pdf's in Chapter 8.

2.2 STANDARD NORMAL CALCULATIONS

The standard normal distribution

As the 68–95–99.7 rule suggests, all normal distributions share many common properties. In fact, all normal distributions are the same if we measure in units of size σ about the mean μ as center. Changing to these units is called *standardizing*. To standardize a value, subtract the mean of the distribution and then divide by the standard deviation.

STANDARDIZING AND Z-SCORES

If x is an observation from a distribution that has mean μ and standard deviation σ, the **standardized value** of x is

$$z = \frac{x - \mu}{\sigma}$$

A standardized value is often called a **z-score.**

A z-score tells us how many standard deviations the original observation falls away from the mean, and in which direction. Observations larger than the mean are positive when standardized, and observations smaller than the mean are negative.

EXAMPLE 2.4 STANDARDIZING WOMEN'S HEIGHTS

The heights of young women are approximately normal with $\mu = 64.5$ inches and $\sigma =$ 2.5 inches. The standardized height is

$$z = \frac{\text{height} - 64.5}{2.5}$$

A woman's standardized height is the number of standard deviations by which her height differs from the mean height of all young women. A woman 68 inches tall, for example, has standardized height

$$z = \frac{68 - 64.5}{2.5} = 1.4$$

or 1.4 standard deviations above the mean. Similarly, a woman 5 feet (60 inches) tall has standardized height

$$z = \frac{60 - 64.5}{2.5} = -1.8$$

or 1.8 standard deviations less than the mean height.

If the variable we standardize has a normal distribution, standardizing does more than give a common scale. It makes all normal distributions into a single distribution, and this distribution is still normal. Standardizing a variable that has any normal distribution produces a new variable that has the *standard normal distribution.*

STANDARD NORMAL DISTRIBUTION

The **standard normal distribution** is the normal distribution N(0, 1) with mean 0 and standard deviation 1 (Figure 2.14).

If a variable x has any normal distribution $N(\mu, \sigma)$ with mean μ and standard deviation σ, then the standardized variable

$$z = \frac{x - \mu}{\sigma}$$

has the standard normal distribution.

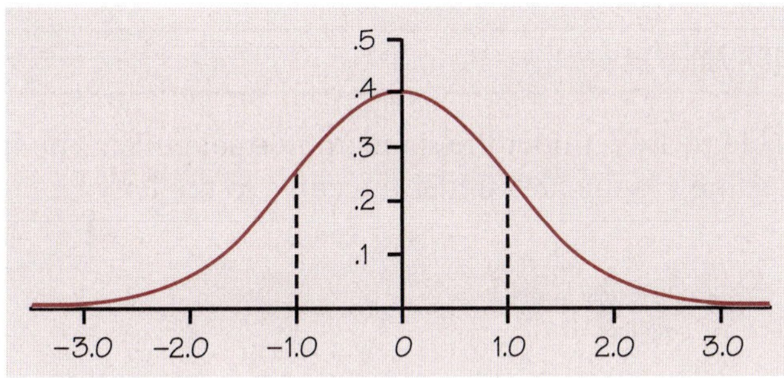

FIGURE 2.14 Standard normal distribution.

EXERCISES

2.19 SAT VERSUS ACT Eleanor scores 680 on the mathematics part of the SAT. The distribution of SAT scores in a reference population is normal, with mean 500 and standard deviation 100. Gerald takes the American College Testing (ACT) mathematics test and scores 27. ACT scores are normally distributed with mean 18 and standard deviation 6. Find the standardized scores for both students. Assuming that both tests measure the same kind of ability, who has the higher score?

2.20 COMPARING BATTING AVERAGES Three landmarks of baseball achievement are Ty Cobb's batting average of .420 in 1911, Ted Williams's .406 in 1941, and George Brett's .390 in 1980. These batting averages cannot be compared directly because the distribution of major league batting averages has changed over the years. The distributions are quite symmetric and (except for outliers such as Cobb, Williams, and Brett) reasonably normal. While the mean batting average has been held roughly constant by rule changes and the balance between hitting and pitching, the standard deviation has dropped over time. Here are the facts:

Decade	Mean	Std. dev.
1910s	.266	.0371
1940s	.267	.0326
1970s	.261	.0317

Compute the standardized batting averages for Cobb, Williams, and Brett to compare how far each stood above his peers.[2]

Normal distribution calculations

An area under a density curve is a proportion of the observations in a distribution. Any question about what proportion of observations lie in some range of values can be answered by finding an area under the curve. Because all normal distributions are the same when we standardize, we can find areas under any normal curve from a single table, a table that gives areas under the curve for the standard normal distribution.

THE STANDARD NORMAL TABLE

Table A is a table of areas under the standard normal curve. The table entry for each value z is the area under the curve to the left of z.

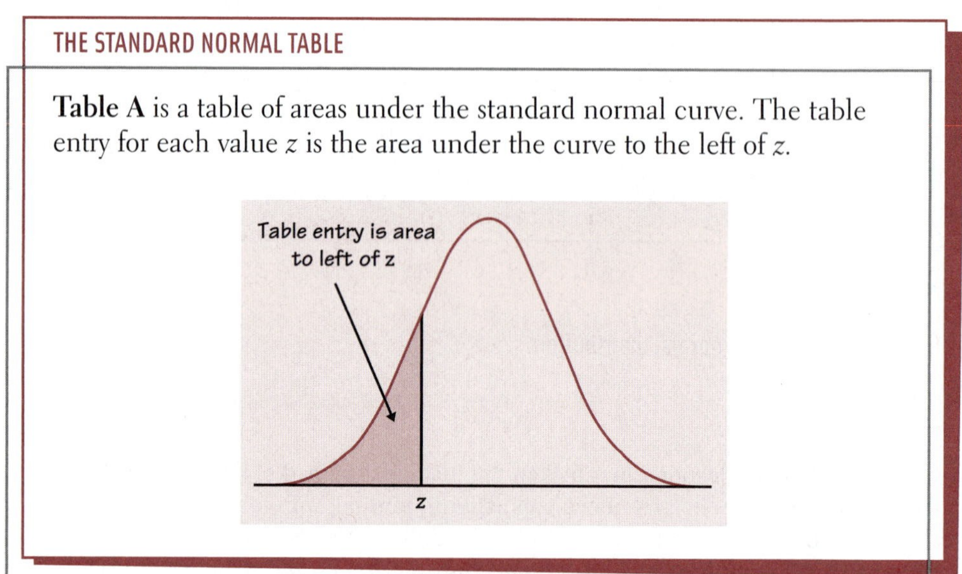

Table A, inside the front cover, gives areas under the standard normal curve. The next two examples show how to use the table.

EXAMPLE 2.5 USING THE z TABLE

Problem: Find the proportion of observations from the standard normal distribution that are less than 1.4.

Solution: To find the area to the left of 1.40, locate 1.4 in the left-hand column of Table A, then locate the remaining digit 0 as .00 in the top row. The entry opposite 1.4 and under .00 is 0.9192. This is the area we seek. Figure 2.15 illustrates the relationship between the value $z = 1.40$ and the area 0.9192.

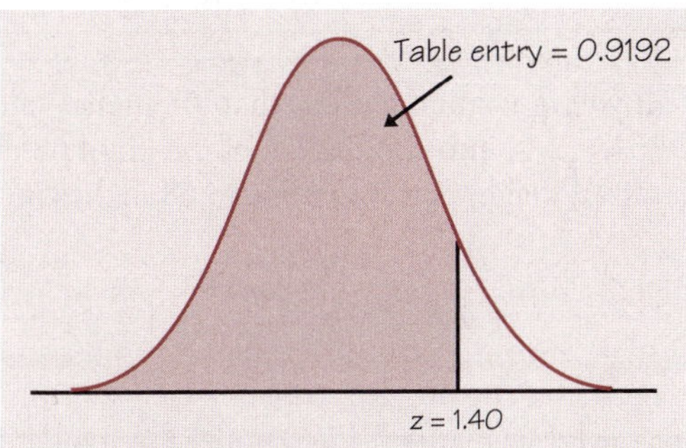

FIGURE 2.15 The area under a standard normal curve to the left of the point $z = 1.4$ is 0.9192. Table A gives areas under the standard normal curve.

EXAMPLE 2.6 MORE ON USING THE *z* TABLE

Problem: Find the proportion of observations from the standard normal distribution that are greater than –2.15.

Solution: Enter Table A under $z = -2.15$. That is, find –2.1 in the left-hand column and .05 in the top row. The table entry is 0.0158. This is the area to the *left* of –2.15. Because the total area under the curve is 1, the area lying to the *right* of –2.15 is 1 – 0.0158 = 0.9842. Figure 2.16 illustrates these areas.

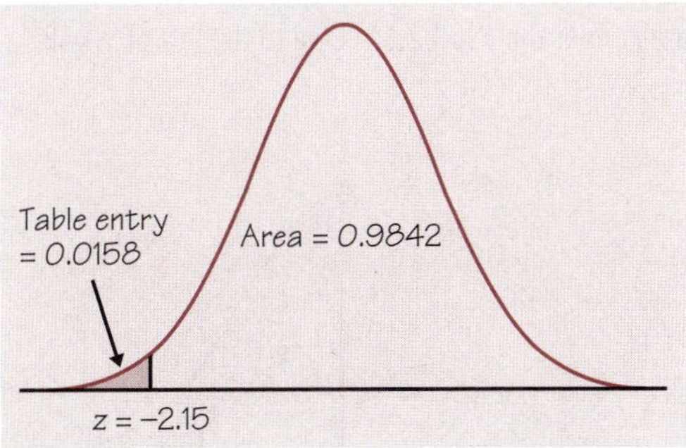

FIGURE 2.16 Areas under the standard normal curve to the right and left of $z = -2.15$. Table A gives only areas to the left.

Caution! A common student mistake is to look up a *z*-value in Table A and report the entry corresponding to that *z*-value, regardless of whether the problem asks for the area to the left or to the right of that *z*-value. Always sketch the standard normal curve, mark the *z*-value, and shade the area of interest. And before you finish, make sure your answer is reasonable in the context of the problem.

The value of the z table is that we can use it to answer any question about proportions of observations in a normal distribution by standardizing and then using the standard normal table.

EXAMPLE 2.7 USING THE STANDARD NORMAL DISTRIBUTION

What proportion of all young women are less than 68 inches tall? This proportion is the area under the N(64.5, 2.5) curve to the left of the point 68. Figure 2.17(a) shows this area. The standardized height corresponding to 68 inches is

$$z = \frac{x - \mu}{\sigma} = \frac{68 - 64.5}{2.5} = 1.4$$

The area to the left of $z = 1.4$ in Figure 2.17(b) under the standard normal curve is the same as the area to the left of $x = 68$ in Figure 2.17(a).

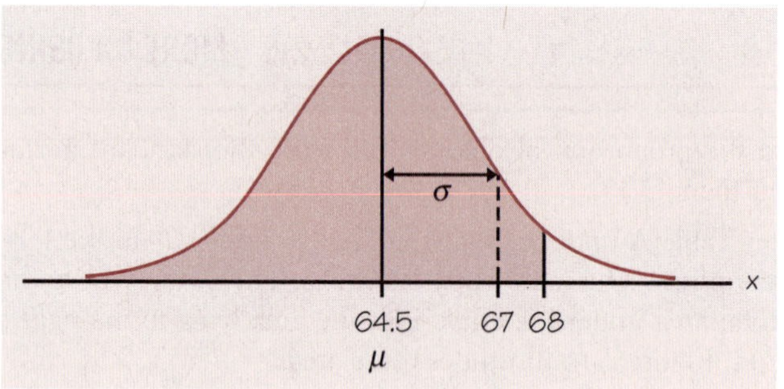

FIGURE 2.17(a) The area under the $N(68, 2.5)$ curve to the left of $x = 68$.

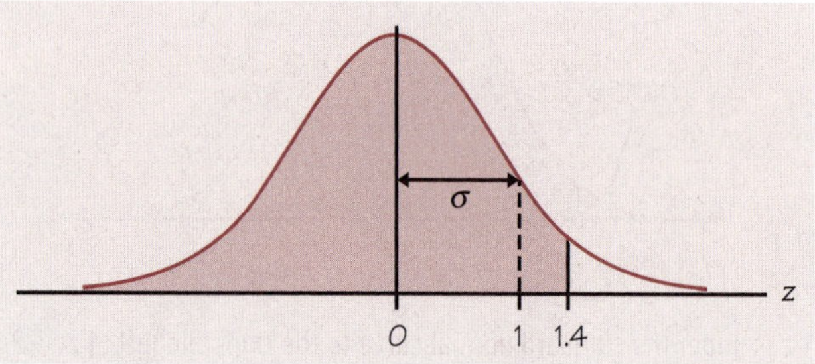

FIGURE 2.17(b) The area to the left of $z = 1.4$ under the standard normal curve $N(0, 1)$. This area is the same as the shaded area in Figure 2.17(a).

In Example 2.5, we found this area to be 0.9192. Our conclusion is that 91.92% of all young women are less than 68 inches tall.

Here is an outline of the method for finding the proportion of the distribution in any region.

FINDING NORMAL PROPORTIONS

Step 1: State the problem in terms of the observed variable x. Draw a picture of the distribution and shade the area of interest under the curve.

Step 2: Standardize x to restate the problem in terms of a standard normal variable z. Draw a picture to show the area of interest under the standard normal curve.

Step 3: Find the required area under the standard normal curve, using Table A and the fact that the total area under the curve is 1.

Step 4: Write your conclusion in the context of the problem.

EXAMPLE 2.8 IS CHOLESTEROL A PROBLEM FOR YOUNG BOYS?

The level of cholesterol in the blood is important because high cholesterol levels may increase the risk of heart disease. The distribution of blood cholesterol levels in a large population of people of the same age and sex is roughly normal. For 14-year-old boys, the mean is $\mu = 170$ milligrams of cholesterol per deciliter of blood (mg/dl) and the standard deviation is $\sigma = 30$ mg/dl.[3] Levels above 240 mg/dl may require medical attention. What percent of 14-year-old boys have more than 240 mg/dl of cholesterol?

Step 1: State the problem. Call the level of cholesterol in the blood x. The variable x has the N(170,30) distribution. We want the proportion of boys with cholesterol level $x > 240$. Sketch the distribution, mark the important points on the horizontal axis, and shade the area of interest. See Figure 2.18(a).

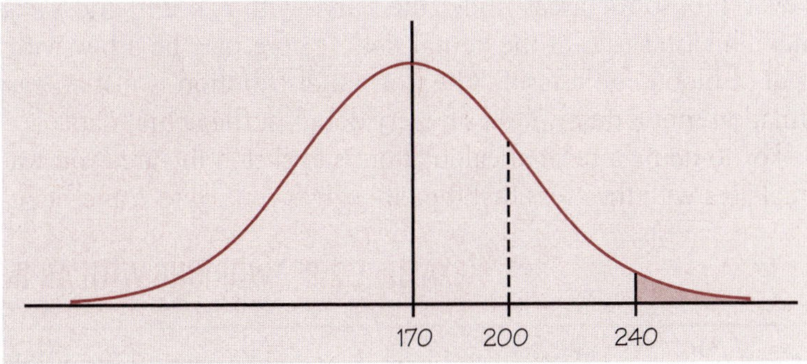

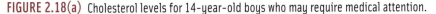

FIGURE 2.18(a) Cholesterol levels for 14-year-old boys who may require medical attention.

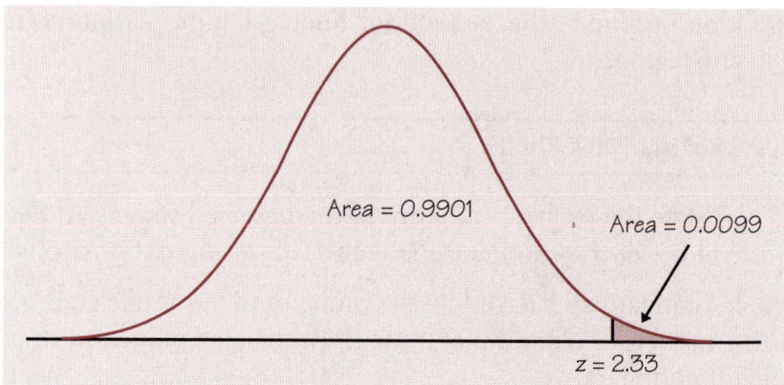

FIGURE 2.18(b) Areas under the standard normal curve.

Step 2: Standardize x and draw a picture. On both sides of the inequality, subtract the mean, then divide by the standard deviation, to turn x into a standard normal z:

$$x > 240$$
$$\frac{x - 170}{30} > \frac{240 - 170}{30}$$
$$z > 2.33$$

Sketch a standard normal curve, and shade the area of interest. See Figure 2.18(b).

Step 3: Use the table. From Table A, we see that the proportion of observations less than 2.33 is 0.9901. About 99% of boys have cholesterol levels less than 240. The area to the right of 2.33 is therefore 1 – 0.9901 = 0.0099. This is about 0.01, or 1%.

Step 4: Write your conclusion in the context of the problem. Only about 1% of boys have high cholesterol.

In a normal distribution, the proportion of observations with $x > 240$ is the same as the proportion with $x \geq 240$. There is no area under the curve and exactly over 240, so the areas under the curve with $x > 240$ and $x \geq 240$ are the same. This isn't true of the actual data. There may be a boy with exactly 240 mg/dl of blood cholesterol. The normal distribution is just an easy-to-use approximation, not a description of every detail in the actual data.

The key to doing a normal calculation is to sketch the area you want, then match that area with the areas that the table gives you. Here is another example.

EXAMPLE 2.9 WORKING WITH AN INTERVAL

What percent of 14-year-old boys have blood cholesterol between 170 and 240 mg/dl?

Step 1: State the problem. We want the proportion of boys with $170 \leq x \leq 240$.

Step 2: Standardize and draw a picture.

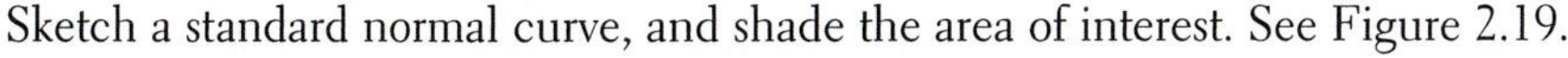

Sketch a standard normal curve, and shade the area of interest. See Figure 2.19.

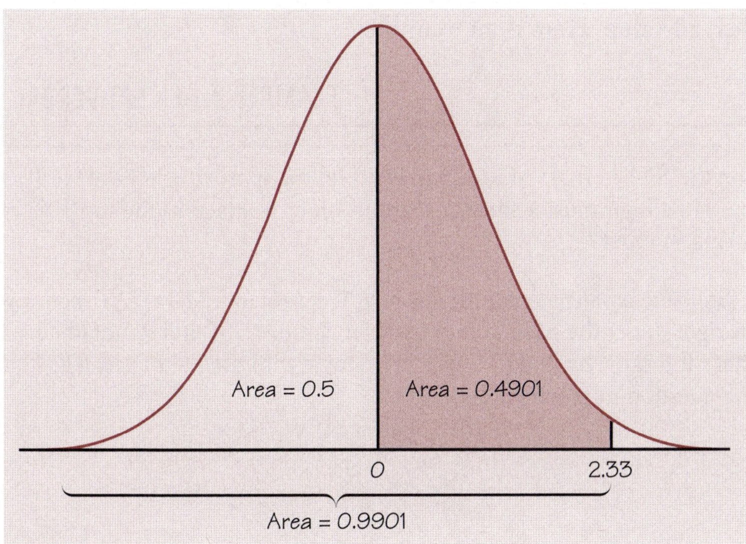

FIGURE 2.19 Areas under the standard normal curve.

$$170 \le x \le 240$$
$$\frac{170-170}{30} \le \frac{x-170}{30} \le \frac{240-17}{30}$$
$$0 \le z \le 2.33$$

Step 3: Use the table. The area between 2.33 and 0 is the area below 2.33 minus the area below 0. Look at Figure 2.19 to check this. From Table A,

$$\begin{aligned}\text{area between 0 and 2.33} &= \text{area below 2.33} - \text{area below 0.00}\\ &= 0.9901 - 0.5000 = 0.4901\end{aligned}$$

Step 4: State your conclusion in context. About 49% of boys have cholesterol levels between 170 and 240 mg/dl.

What if we meet a z that falls outside the range covered by Table A? For example, the area to the left of $z = -4$ does not appear in the table. But since -4 is less than -3.4, this area is smaller than the entry for $z = -3.40$, which is 0.0003. There is very little area under the standard normal curve outside the range covered by Table A. You can take this area to be zero with little loss of accuracy.

Finding a value given a proportion

Examples 2.8 and 2.9 illustrate the use of Table A to find what proportion of the observations satisfies some condition, such as "blood cholesterol between 170 mg/dl and 240 mg/dl." We may instead want to find the observed value with a given proportion of the observations above or below it. To do this, use

Table A backward. Find the given proportion in the body of the table, read the corresponding z from the left column and top row, then "unstandardize" to get the observed value. Here is an example.

EXAMPLE 2.10 SAT VERBAL SCORES

Scores on the SAT Verbal test in recent years follow approximately the N(505,110) distribution. How high must a student score in order to place in the top 10% of all students taking the SAT?

Step 1: State the problem and draw a sketch. We want to find the SAT score x with area 0.1 to its *right* under the normal curve with mean $\mu = 505$ and standard deviation $\sigma = 110$. That's the same as finding the SAT score x with area 0.9 to its *left*. Figure 2.20 poses the question in graphical form.

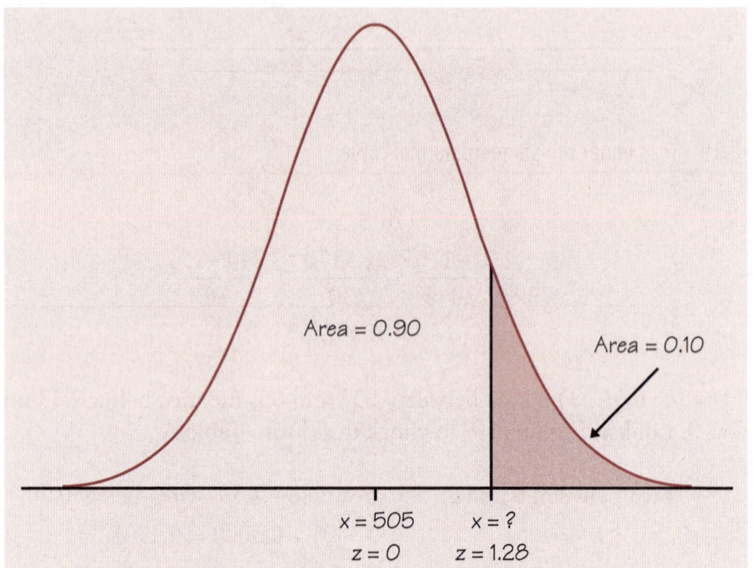

FIGURE 2.20 Locating the point on a normal curve with area 0.10 to its right.

Because Table A gives the areas to the left of z-values, always state the problem in terms of the area to the left of x.

Step 2: Use the table. Look in the body of Table A for the entry closest to 0.9. It is 0.8997. This is the entry corresponding to $z = 1.28$. So $z = 1.28$ is the standardized value with area 0.9 to its left.

Step 3: Unstandardize to transform the solution from the z back to the original x scale. We know that the standardized value of the unknown x is $z = 1.28$. So x itself satisfies

$$\frac{x-505}{110}=1.28$$

Solving this equation for x gives

$$x = 505 + (1.28)(110) = 645.8$$

This equation should make sense: it finds the x that lies 1.28 standard deviations above the mean on this particular normal curve. That is the "unstandardized" meaning of $z = 1.28$. We see that a student must score at least 646 to place in the highest 10%.

EXERCISES

2.21 TABLE A PRACTICE Use Table A to find the proportion of observations from a standard normal distribution that satisfies each of the following statements. In each case, sketch a standard normal curve and shade the area under the curve that is the answer to the question.

(a) $z < 2.85$

(b) $z > 2.85$

(c) $z > -1.66$

(d) $-1.66 < z < 2.85$

2.22 MORE TABLE A PRACTICE Use Table A to find the value z of a standard normal variable that satisfies each of the following conditions. (Use the value of z from Table A that comes closest to satisfying the condition.) In each case, sketch a standard normal curve with your value of z marked on the axis.

(a) The point z with 25% of the observations falling below it.

(b) The point z with 40% of the observations falling above it.

2.23 HEIGHTS OF AMERICAN MEN The distribution of heights of adult American men is approximately normal with mean 69 inches and standard deviation 2.5 inches.

(a) What percent of men are at least 6 feet (72 inches) tall?

(b) What percent of men are between 5 feet (60 inches) and 6 feet tall?

(c) How tall must a man be to be in the tallest 10% of all adult men?

2.24 IQ TEST SCORES Scores on the Wechsler Adult Intelligence Scale (a standard "IQ test") for the 20 to 34 age group are approximately normally distributed with $\mu = 110$ and $\sigma = 25$.

(a) What percent of people age 20 to 34 have IQ scores above 100?

(b) What percent have scores above 150?

(c) How high an IQ score is needed to be in the highest 25%?

2.25 HOW HARD DO LOCOMOTIVES PULL? An important measure of the performance of a locomotive is its "adhesion," which is the locomotive's pulling force as a multiple of its weight. The adhesion of one 4400-horsepower diesel locomotive model varies in actual use according to a normal distribution with mean $\mu = 0.37$ and standard deviation $\sigma = 0.04$.

(a) What proportion of adhesions measured in use are higher than 0.40?

(b) What proportion of adhesions are between 0.40 and 0.50?

(c) Improvements in the locomotive's computer controls change the distribution of adhesion to a normal distribution with mean $\mu = 0.41$ and standard deviation $\sigma = 0.02$. Find the proportions in (a) and (b) after this improvement.

Assessing normality

In the latter part of this course we will want to invoke various tests of significance to try to answer questions that are important to us. These tests involve sampling people or objects and inspecting them carefully to gain insights into the populations from which they come. Many of these procedures are based on the assumption that the host population is approximately normally distributed. Consequently, we need to develop methods for assessing normality.

Method 1 Construct a frequency histogram or a stemplot. See if the graph is approximately bell-shaped and symmetric about the mean.

A histogram or stemplot can reveal distinctly nonnormal features of a distribution, such as outliers, pronounced skewness, or gaps and clusters. You can improve the effectiveness of these plots for assessing whether a distribution is normal by marking the points $\bar{x}$, $\bar{x} \pm s$, and $\bar{x} \pm 2s$ on the x axis. This gives the scale natural to normal distributions. Then compare the count of observations in each interval with the 68–95–99.7 rule.

EXAMPLE 2.11 ASSESSING NORMALITY OF THE GARY VOCABULARY SCORES

The histogram in Figure 2.2 (page 78) suggests that the distribution of the 947 Gary vocabulary scores is close to normal. It is hard to assess by eye how close to normal a histogram is. Let's use the 68–95–99.7 rule to check more closely. We enter the scores into a statistical computing system and ask for the mean and standard deviation. The computer replies,

```
MEAN = 6.8585
STDEV = 1.5952
```

Now that we know that $\bar{x} = 6.8585$ and $s = 1.5952$, we check the 68–95–99.7 rule by finding the actual counts of Gary vocabulary scores in intervals of length s about the mean $\bar{x}$. The computer will also do this for us. Here are the counts:

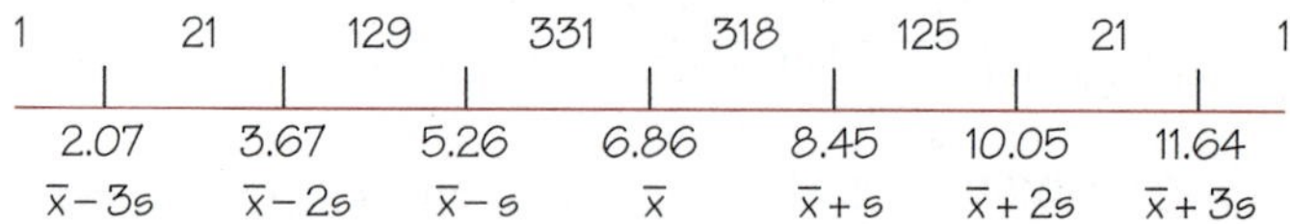

The distribution is very close to symmetric. It also follows the 68–95–99.7 rule closely: there are 68.5% of the scores (649 out of 947) within one standard deviation of the mean, 95.4% (903 of 947) within two standard deviations, and 99.8% (945 of the 947 scores) within three. These counts confirm that the normal distribution with $\mu = 6.86$ and $\sigma = 1.595$ fits these data well.

Smaller data sets rarely fit the 68–95–99.7 rule as well as the Gary vocabulary scores. This is true even of observations taken from a larger population that really has a normal distribution. There is more chance variation in small data sets.

Method 2 Construct a ***normal probability plot.*** A normal probability plot provides a good assessment of the adequacy of the normal model for a set of data. Most statistics utilities, including Minitab and Data Desk, can construct normal probability plots from entered data. The TI-83/89 will also do normal probability plots. You will need to be able to produce a normal probability plot (either with a calculator or with computer software) and interpret it. We will do this part first, and then we will describe the steps the calculator goes through to produce the plot.

normal probability plot

TECHNOLOGY TOOLBOX *Normal probability plots on the TI-83/89*

If you ran the program FLIP50 in Exercise 2.17, and you still have the data (100 numbers mostly in the 20s) in L_1/list1, then use these data. If you have not entered the program and run it, take a few minutes to do that now. Duplicate this example with *your* data. Here is the histogram that was generated at the end of one run of this simulation on each calculator.

TI-83

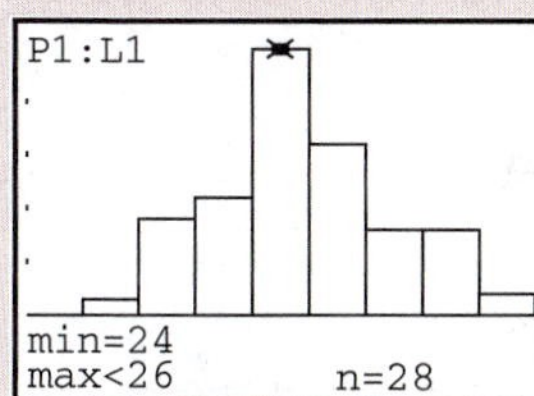

Ask for one-variable statistics:

- Press STAT, choose CALC, then `1:1-Var Stats` and 2nd 1 (L_1)

This will give us the following:

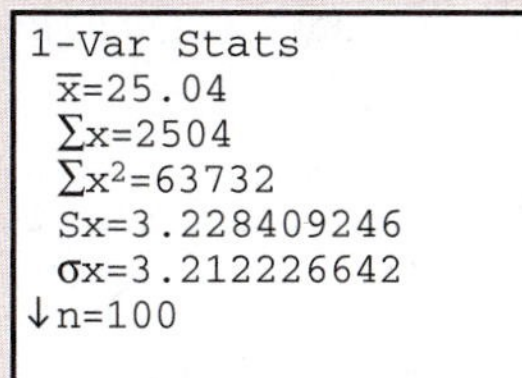


```
1-Var Stats
↑n=100
 minX=18
 Q1=23
 Med=25
 Q3=27
 maxX=33
```

TI-89

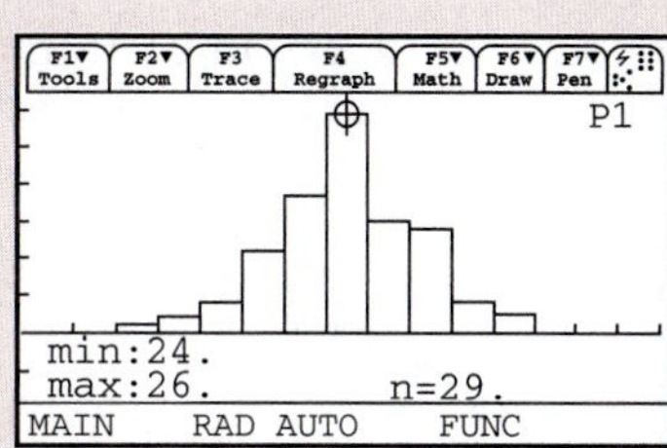

- In the Statistics/List Editor, press F4 (Calc) and choose `1:1-Var Stats` for list1.

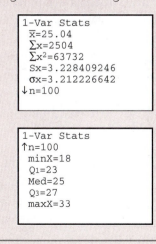

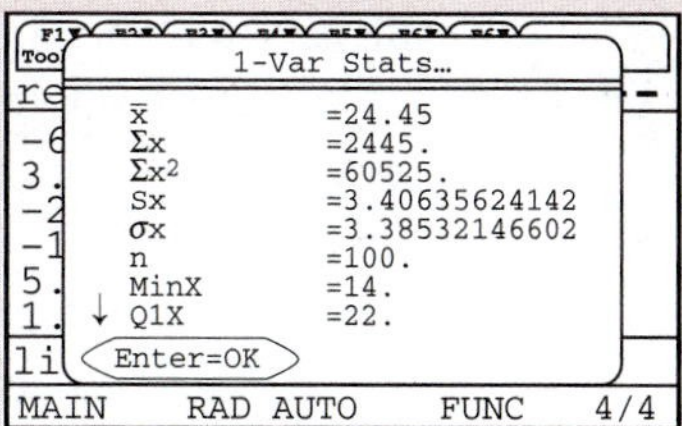

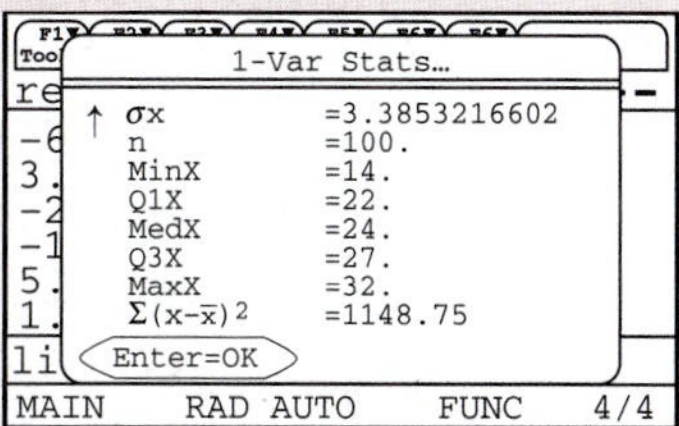

TECHNOLOGY TOOLBOX *Normal probability plots (continued)*

- Comparing the means and medians ($\bar{x} = 25.04$ vs. $M = 25$ on the TI-83 and $\bar{x} = 24.45$ vs. $M = 24$ on the TI-89) suggests that the distributions are fairly symmetric. Boxplots confirm the roughly symmetric shape.

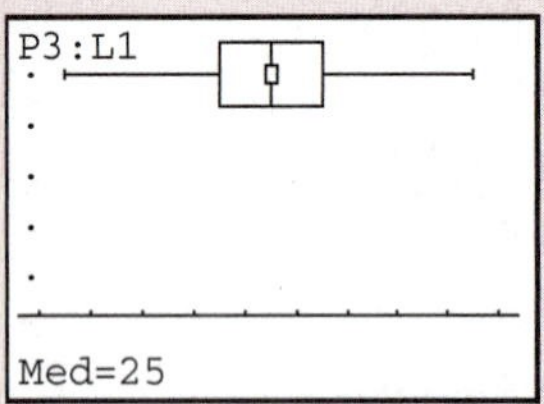

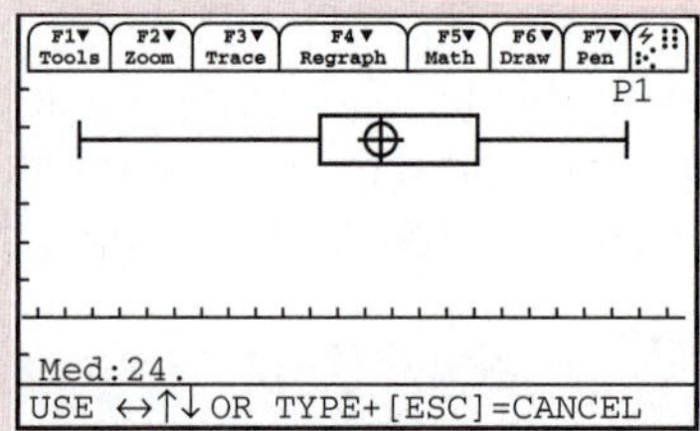

- To construct a normal probability plot of the data, define Plot 1 like this:

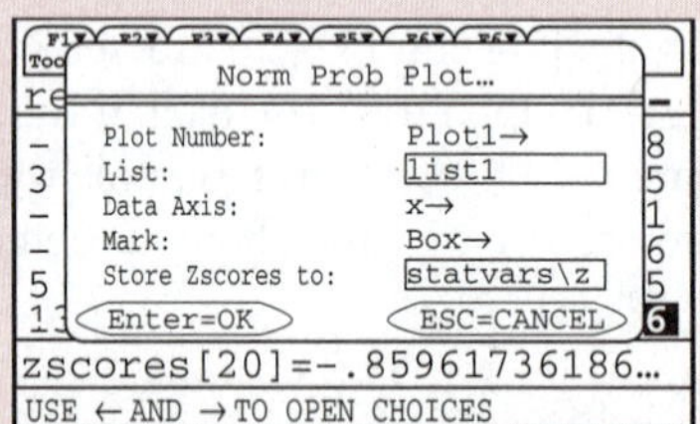

- Use ZoomStat (ZoomData on the TI-89) to see the finished graph.

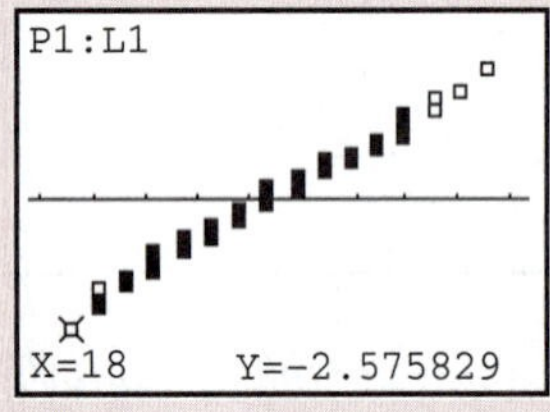

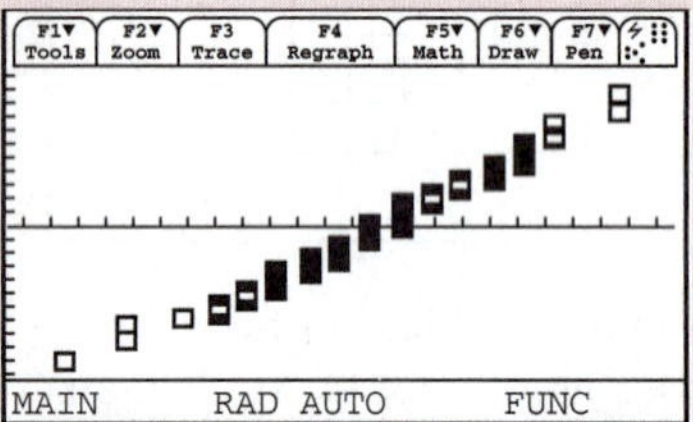

Interpretation: If the data distribution is close to a normal distribution, the plotted points will lie close to a straight line. Conversely, nonnormal data will show a nonlinear trend. Outliers appear as points that are far away from the overall pattern of the plot. Since the above plot is quite linear, our conclusion is that it is reasonable to believe that the data are from a normal distribution.

The next example uses a very simple data set to illustrate how a normal probability plot is constructed.

EXAMPLE 2.12 HOW NORMAL PROBABILITY PLOTS ARE CONSTRUCTED

To show how the calculator constructs a normal probability plot, let's look at a very simple data set: {1,2,2,3}. Here, $n = 4$ and a dotplot shows that the distribution is perfectly symmetric if not exactly bell-shaped.

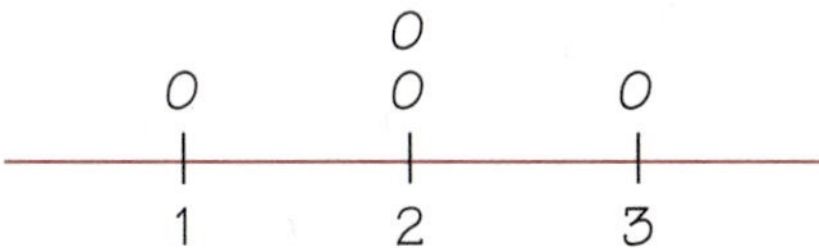

Step 1: Order the observations from smallest to largest. In this case, the points are already ordered. Since $n = 4$, divide the interval [0,1] on the horizontal axis into four subintervals.

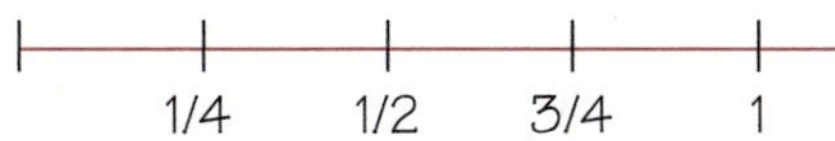

Mark the midpoint of each subinterval: 1/8, 3/8, 5/8, and 7/8. In the general case, we would mark the points corresponding to $1/2n$, $3/2n$, $5/2n$, . . ., $(2n - 1)/2n$.

Step 2: For the first midpoint, 1/8, find the z-value that has area 1/8 = 0.125 lying to the left of it. The closest value in the body of the table is 0.1251, and the corresponding z-value is –1.15. Do the same for the other midpoints. Here is a table of our results:

x	Midpoint	y
1	1/8 = 0.1250	–1.15
2	3/8 = 0.3750	–0.319
3	5/8 = 0.6250	0.319
4	7/8 = 0.8750	1.15

Step 3: Plot the points (x,y). This is the normal probability plot for our simple data set.

P1:L1
X=1 Y=-1.150349

If an outlier were added, say 10, then the table would look like this:

x	Midpoint	y
1	0.1	–1.28
2	0.3	–0.52
2	0.5	0
3	0.7	0.52
10	0.9	1.28

and the normal probability plot becomes

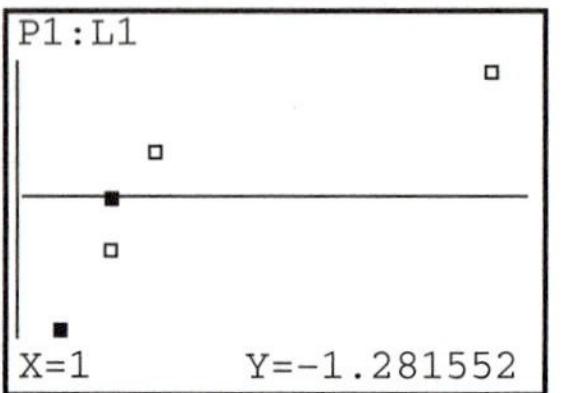

This last picture shows a normal probability plot for a data set that is clearly not approximately normally distributed.

Any normal distribution produces a straight line on the plot because standardizing is a transformation that can change the slope and intercept of the line in our plot but cannot change a line into a curved pattern.

EXERCISES

2.26 CAVENDISH AND THE DENSITY OF THE EARTH Repeated careful measurements of the same physical quantity often have a distribution that is close to normal. Here are Henry Cavendish's 29 measurements of the density of the earth, made in 1798. (The data give the density of the earth as a multiple of the density of water.)

5.50	5.61	4.88	5.07	5.26	5.55	5.36	5.29	5.58	5.65
5.57	5.53	5.62	5.29	5.44	5.34	5.79	5.10	5.27	5.39
5.42	5.47	5.63	5.34	5.46	5.30	5.75	5.68	5.85	

(a) Construct a stemplot to show that the data are reasonably symmetric.

(b) Now check how closely they follow the 68–95–99.7 rule. Find $\bar{x}$ and s, then count the number of observations that fall between $\bar{x} - s$ and $\bar{x} + s$, between $\bar{x} - 2s$ and $\bar{x} + 2s$, and between $\bar{x} - 3s$ and $\bar{x} + 3s$. Compare the percents of the 29 observations in each of these intervals with the 68–95–99.7 rule.

(c) Use your calculator to construct a normal probability plot for Cavendish's density of the earth data, and write a brief statement about the normality of the data. Does the normal probability plot reinforce your findings in (a)?

We expect that when we have only a few observations from a normal distribution, the percents will show some deviation from 68, 95, and 99.7. Cavendish's measurements are in fact close to normal.

2.27 GREAT WHITE SHARKS Here are the lengths in feet of 44 great white sharks:

18.7	12.3	18.6	16.4	15.7	18.3	14.6	15.8	14.9	17.6	12.1
16.4	16.7	17.8	16.2	12.6	17.8	13.8	12.2	15.2	14.7	12.4
13.2	15.8	14.3	16.6	9.4	18.2	13.2	13.6	15.3	16.1	13.5
19.1	16.2	22.8	16.8	13.6	13.2	15.7	19.7	18.7	13.2	16.8

(a) Use the methods of Chapter 1 to describe the distribution of these lengths.

(b) Compare the mean with the median. Does this comparison support your assessment of the shape of the distribution in (a)? Explain.

(c) Is the distribution approximately normal? If you haven't done this already, enter the data into your calculator, and reorder them from smallest to largest. Then calculate the percent of the data that lies within one standard deviation of the mean. Within two standard deviations of the mean. Within three standard deviations of the mean.

(d) Use your calculator to construct a normal probability plot. Interpret this plot.

(e) Having inspected the data from several different perspectives, do you think these data are approximately normal? Write a brief summary of your assessment that combines your findings from (a) to (d).

SUMMARY

To **standardize** any observation x, subtract the mean of the distribution and then divide by the standard deviation. The resulting ***z*-score**

$$z = \frac{x - \mu}{\sigma}$$

says how many standard deviations x lies from the distribution mean.

All normal distributions are the same when measurements are transformed to the standardized scale. If x has the $N(\mu, \sigma)$ distribution, then the **standardized variable** $z = (x - \mu)/\sigma$ has the **standard normal distribution** $N(0, 1)$ with mean 0 and standard deviation 1. Table A gives the proportions of standard normal observations that are less than z for many values of z. By standardizing, we can use Table A for any normal distribution.

In order to perform certain inference procedures in later chapters, we will need to know that the data come from populations that are approximately normally distributed. To assess normality, one can observe the shape of histograms, stemplots, and boxplots and see how well the data fit the 68–95–99.7 rule for normal distributions. Another good method for assessing normality is to construct a **normal probability plot**.

SECTION 2.2 EXERCISES

2.28 TABLE A PRACTICE Use Table A to find the proportion of observations from a standard normal distribution that falls in each of the following regions. In each case, sketch a standard normal curve and shade the area representing the region.

(a) $z \leq -2.25$

(b) $z \geq -2.25$

(c) $z > 1.77$

(d) $-2.25 < z < 1.77$

2.29 MORE TABLE A PRACTICE Use Table A to find the value z of a standard normal variable that satisfies each of the following conditions. (Use the value of z from Table A that comes closest to satisfying the condition.) In each case, sketch a standard normal curve with your value of z marked on the axis.

(a) The point z with 70% of the observations falling below it.

(b) The point z with 85% of the observations falling above it.

(c) Find the number z such that the proportion of observations that are less than z is 0.8.

(d) Find the number z such that 90% of all observations are greater than z.

2.30 THE STOCK MARKET The annual rate of return on stock indexes (which combine many individual stocks) is approximately normal. Since 1945, the Standard & Poor's 500 Index has had a mean yearly return of 12%, with a standard deviation of 16.5%. Take this normal distribution to be the distribution of yearly returns over a long period.

(a) In what range do the middle 95% of all yearly returns lie?

(b) The market is down for the year if the return on the index is less than zero. In what proportion of years is the market down?

(c) In what proportion of years does the index gain 25% or more?

2.31 GESTATION PERIOD The length of human pregnancies from conception to birth varies according to a distribution that is approximately normal with mean 266 days and standard deviation 16 days.

(a) What percent of pregnancies last less than 240 days (that's about 8 months)?

(b) What percent of pregnancies last between 240 and 270 days (roughly between 8 months and 9 months)?

(c) How long do the longest 20% of pregnancies last?

2.32 ARE WE GETTING SMARTER? When the Stanford-Binet "IQ test" came into use in 1932, it was adjusted so that scores for each age group of children followed roughly the normal distribution with mean $\mu = 100$ and standard deviation $\sigma = 15$. The test is readjusted from time to time to keep the mean at 100. If present-day American children took the 1932 Stanford-Binet test, their mean score would be about 120. The reasons for the increase in IQ over time are not known but probably include better childhood nutrition and more experience in taking tests.[4]

(a) IQ scores above 130 are often called "very superior." What percent of children had very superior scores in 1932?

(b) If present-day children took the 1932 test, what percent would have very superior scores? (Assume that the standard deviation $\sigma = 15$ does not change.)

2.33 QUARTILES The quartiles of any density curve are the points with area 0.25 and 0.75 to their left under the curve.

(a) What are the quartiles of a standard normal distribution?

(b) How many standard deviations away from the mean do the quartiles lie in any normal distribution? What are the quartiles for the lengths of human pregnancies? (Use the distribution in Exercise 2.31.)

2.34 DECILES The *deciles* of any distribution are the points that mark off the lowest 10% and the highest 10%. The deciles of a density curve are therefore the points with area 0.1 and 0.9 to their left under the curve.

(a) What are the deciles of the standard normal distribution?

(b) The heights of young women are approximately normal with mean 64.5 inches and standard deviation 2.5 inches. What are the deciles of this distribution?

2.35 LACTIC ACID IN CHEESE The taste of mature cheese is related to the concentration of lactic acid in the cheese. The concentrations of lactic acid in 30 samples of cheddar cheese are given in the Technology Toolbox on page 15.

(a) Enter the data into your calculator. Make a histogram and overlay a boxplot. Sketch the results on your paper. Compare the mean with the median. Describe the distribution of these data in a sentence.

(b) Calculate the percent of the data that lies within one, two, and three standard deviations of the mean.

(c) Use your calculator to construct a normal probability plot. Sketch this plot on your paper.

(d) Having inspected the data from several different perspectives, do you think these data are approximately normal? Write a brief statement of your assessment that combines your findings from (a) to (c).

2.36 ARE THE PRESIDENTS' AGES NORMAL? The histogram for the ages of the 43 presidents was very symmetric (see Figure 1.7, page 20). Use the list that we named PREZ to construct a normal probability plot for this data set, and confirm the linear trend. Write a statement about your assessment of normality of the presidents' ages.

2.37 STANDARDIZED VALUES BY CALCULATOR This exercise uses the TI-83/89 to calculate standardized values for a familiar data set and then calculates the mean and standard deviation for these transformed values. Without knowing the data set, can you guess the mean and standard deviation?

Set up your Statistics/List Editor so that the list PREZ (the presidents' ages from Exercise 2.36) is the first list:

- TI-83: Press STAT, choose `5:SetUpEditor`, then press 2nd STAT (LIST), choose PREZ, and press ENTER.
- TI-89: Press CATALOG F3 (Flash Apps), choose `setupEd(`, then type `prez)` and press ENTER.

In the Statistics/List Editor, move your cursor to the header of the next (blank) list and name it STDSC (for standardized scores). With the name of this list highlighted, define the list by carefully entering (PREZ-mean(PREZ))/stdDev(PREZ). The mean and stdDev commands are found under the LIST/MATH menu.

Scroll through the list STDSC to verify that the values range from about −3 to 3. Then construct a histogram of STDSC, and calculate one-variable statistics for STDSC. What are the mean and standard deviation?

CHAPTER REVIEW

Here is a review list of the most important skills you should have acquired from your study of this chapter.

A. DENSITY CURVES

1. Know that areas under a density curve represent proportions of all observations and that the total area under a density curve is 1.

2. Approximately locate the median (equal-areas point) and the mean (balance point) on a density curve.

3. Know that the mean and median both lie at the center of a symmetric density curve and that the mean moves farther toward the long tail of a skewed curve.

B. NORMAL DISTRIBUTIONS

1. Recognize the shape of normal curves and be able to estimate both the mean and standard deviation from such a curve.

2. Use the 68–95–99.7 rule and symmetry to state what percent of the observations from a normal distribution fall between two points when both points lie at the mean or one, two, or three standard deviations on either side of the mean.

3. Find the standardized value (z-score) of an observation. Interpret z-scores and understand that any normal distribution becomes standard normal $N(0, 1)$ when standardized.

4. Given that a variable has the normal distribution with a stated mean μ and standard deviation σ, use Table A and your calculator to calculate the proportion of values above a stated number, below a stated number, or between two stated numbers.

5. Given that a variable has the normal distribution with a stated mean μ and standard deviation σ, calculate the point having a stated proportion of all values above it. Also calculate the point having a stated proportion of all values below it.

C. ASSESSING NORMALITY

1. Plot a histogram, stemplot, and/or boxplot to determine if a distribution is bell-shaped.

2. Determine the proportion of observations within one, two, and three standard deviations of the mean, and compare with the 68–95–99.7 rule for normal distributions.

3. Construct and interpret normal probability plots.

CHAPTER 2 REVIEW EXERCISES

2.38 A certain density curve consists of a straight-line segment that begins at the origin, (0, 0), and has slope 1.

(a) Sketch the density curve. What are the coordinates of the right endpoint of the segment? (*Note*: The right endpoint should be fixed so that the total area under the curve is 1. This is required for a valid density curve.)

(b) Determine the median, the first quartile (Q_1), and the third quartile (Q_3).

(c) Relative to the median, where would you expect the mean of the distribution?

(d) What percent of the observations lie below 0.5? Above 1.5?

2.39 A certain density curve looks like an inverted letter "V." The first segment goes from the point (0, 0.6) to the point (0.5, 1.4). The second segment goes from (0.5, 1.4) to (1, 0.6).

(a) Sketch the curve. Verify that the area under the curve is 1, so that it is a valid density curve.

(b) Determine the median. Mark the median and the approximate locations of the quartiles Q_1 and Q_3 on your sketch.

(c) What percent of the observations lie below 0.3?

(d) What percent of the observations lie between 0.3 and 0.7?

2.40 STANDARDIZED TEST SCORES AS PERCENTILES Joey received a report that he scored in the 97th percentile on a national standardized reading test but in the 72nd percentile on the math portion of the test. Explain to Joey's grandmother, who knows no statistics, what these numbers mean.

2.41 TABLE A PRACTICE Use Table A to find the proportion of observations from a standard normal distribution that falls in each of the following regions. In each case, sketch a standard normal curve and shade the area representing the region.

(a) $z < 1.28$

(b) $z > -0.42$

(c) $-0.42 < z < 1.28$

(d) $z < 0.42$

2.42 WORKING BACKWARD, FINDING *z*-VALUES

(a) Find the number z such that the proportion of observations that are less than z in a standard normal distribution is 0.98.

(b) Find the number z such that 22% of all observations from a standard normal distribution are greater than z.

2.43 QUARTILES FROM A NORMAL DISTRIBUTION Find the quartiles for the distribution of blood cholesterol levels for 14-year-old boys (see Example 2.8, page 99). This distribution is N(170 mg/dl, 30 mg/dl).

2.44 ARE YOU A GOOD JUDGE OF PEOPLE? The Chapin Social Insight Test evaluates how accurately the subject appraises other people. In the reference population used to develop the test, scores are approximately normally distributed with mean 25 and standard deviation 5. The range of possible scores is 0 to 41.

(a) What proportion of the population has scores below 20 on the Chapin test?

(b) What proportion has scores below 10?

(c) What proportion has scores above 35?

(d) How high a score must you have in order to be in the top quarter of the population in social insight?

2.45 IQ SCORES FOR CHILDREN The scores of a reference population on the Wechsler Intelligence Scale for Children (WISC) are normally distributed with $\mu = 100$ and $\sigma = 15$. A school district classified children as "gifted" if their WISC score exceeds 135. There are 1300 sixth-graders in the school district. About how many of them are gifted?

2.46 CULTURE SHOCK The Acculturation Rating Scale for Mexican Americans (ARSMA) is a psychological test that measures the degree to which Mexican Americans are adapted to Mexican/Spanish versus Anglo/English culture. The range of possible scores is 1.0 to 5.0, with higher scores showing more Anglo/English acculturation. The distribution of ARSMA scores in a population used to develop the test is approximately normal with mean 3.0 and standard deviation 0.8. A researcher believes that Mexicans will have an average score near 1.7 and that first-generation Mexican Americans will average about 2.1 on the ARSMA scale. What proportion of the population used to develop the test has scores below 1.7? Between 1.7 and 2.1?

2.47 HELMET SIZES The army reports that the distribution of head circumference among soldiers is approximately normal with mean 22.8 inches and standard deviation 1.1 inches. Helmets are mass-produced for all except the smallest 5% and the largest 5% of head sizes. Soldiers in the smallest or largest 5% get custom-made helmets. What head sizes get custom-made helmets?

2.48 ADAPTING CULTURALLY The ARSMA test is described in Exercise 2.46. How high a score on this test must a Mexican American obtain to be among the 30% of the population used to develop the test who are most Anglo/English in cultural orientation? What scores make up the 30% who are most Mexican/Spanish in their acculturation?

2.49 PROFESSOR MOORE'S DRIVING TIMES Exercise 1.29 (page 36) shows driving times between home and college for Professor Moore.

(a) Make a histogram of these drive times. Is the distribution roughly symmetric, clearly skewed, or neither? Are there any clear outliers?

(b) The data show three unusual situations: the day after Thanksgiving (no traffic on campus); a delay due to an accident; and a day with icy roads. Identify and remove these three observations. Are the remaining observations reasonably close to having a normal distribution? Write a short statement that describes your analyses and your conclusions.

2.50 CORN-FED CHICKS Exercise 1.57 (page 63) presents data on the weight gains of chicks fed two types of corn. The researchers use $\bar{x}$ and s to summarize each of the two distributions. Make a normal probability plot for each group and report your findings. Is the use of $\bar{x}$ and s justified?

TECHNOLOGY TOOLBOX *Finding areas with ShadeNorm*

The TI-83/89 can be used to find the area to the left or right of a point or above an interval without referring to a standard normal table. Consider the WISC scores for children of Exercise 2.45. This distribution is N(100, 15). Suppose we want to find the percent of children whose WISC scores are above 125. Begin by specifying a viewing window as follows: $X[55, 145]_{15}$ and $Y[-0.008, 0.028]_{.01}$. You will generally need to experiment with the y settings to get a good graph.

TI-83

- Press 2nd VARS (DISTR), then choose DRAW and `1:ShadeNorm(`.
- Complete the command `ShadeNorm(125,1E99,100,15)` and press ENTER.

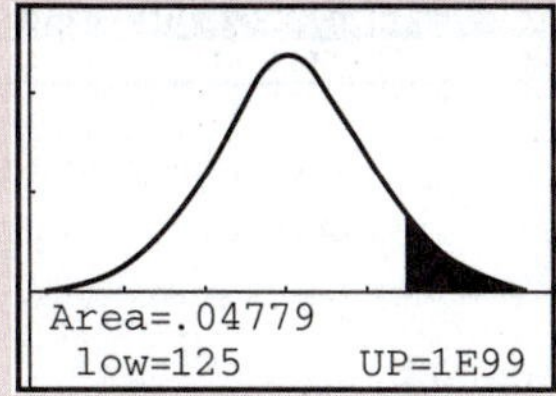

TI-89

- Press CATALOG F3 (Flash Apps) and choose `shadNorm(`.
- Complete the command `tistat.shadNorm(125,1E99,100,15)` and press ENTER.

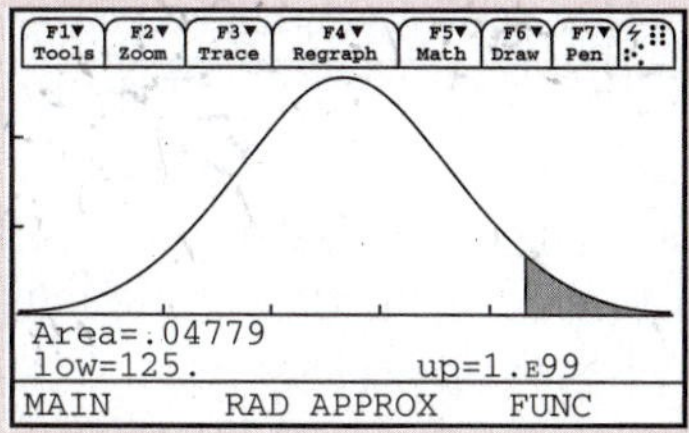

You must always specify an interval. An area in the right tail of the distribution would theoretically be the interval (125, ∞). The calculator limitation dictates that we use a *number* that is at least 5 or 10 standard deviations to the right of the mean. To find the area to the left of 85, you would specify ShadeNorm(–1E99,85,100,15). Or we could specify ShadeNorm(0,85,100,15) since WISC scores can't be negative. Both yield at least four-decimal-place accuracy. If you're using standard normal values, then you need only specify the endpoints of the interval; the mean 0 and standard deviation1 will be understood. For example, use ShadeNorm(1,2) to find the area above the interval z = 1 to z = 2.

2.51 MADE IN THE SHADE Use the calculator's ShadeNorm feature to find the following areas correct to four-decimal-place accuracy. Then write your findings in a sentence.

(a) The relative frequency of scores greater than 110.

(b) The relative frequency of scores lower than 85.

(c) Show two ways to find the relative frequency of scores within two standard deviations of the mean.

TECHNOLOGY TOOLBOX *Finding areas with normalcdf*

The **normalcdf** command on the TI-83/89 can be used to find the area under a normal distribution and above an interval. This method has the advantage over ShadeNorm of being quicker to do, and the disadvantage of not providing a picture of the area it is finding. Here are the keystrokes for the WISC scores of Exercise 2.45:

TI-83

- Press 2nd VARS (DISTR) and choose `2:normalcdf(`.
- Complete the command `normalcdf(125, 1E99,100,15)` and press ENTER.

```
normalcdf(125,1E99,
100,15)
            .0477903304
```

TI-89

- Press CATALOG F3 (Flash Apps) and choose `normCdf(`.
- Complete the command `tistat.normCdf(125,1E99,100,15)` and press ENTER.

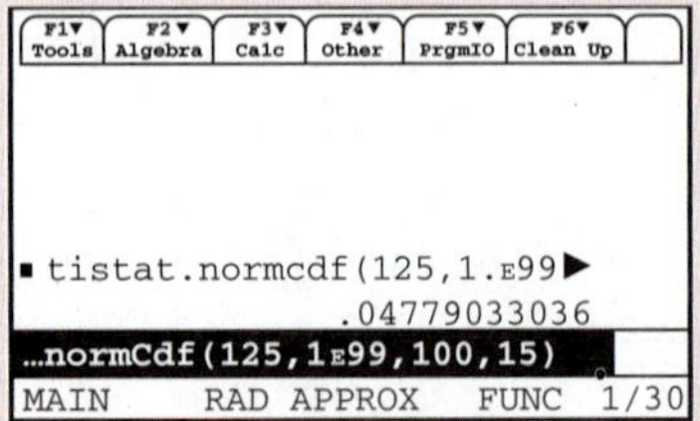

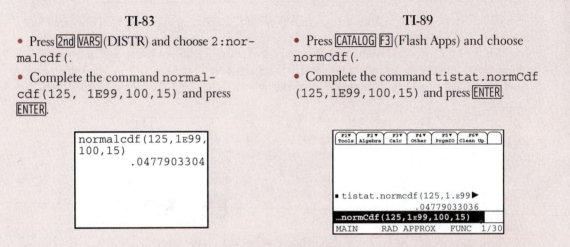

We can say that about 5% of the WISC scores are above 125. If the normal values have already been standardized, then you need only specify the left and right endpoints of the interval. For example, `normalcdf(-1,1)` returns 0.6827, meaning that the area from $z = -1$ to $z = 1$ is approximately 0.6827, correct to four decimal places.

2.52 AREAS BY CALCULATOR Use the calculator's `normalcdf` function to verify your answers to Exercises 2.41 (page 113), and 2.46 (page 114).

2.53 IQ SCORES FOR ADULTS Wechsler Adult Intelligence Scale (WAIS) scores for young adults are N(110, 25). Use your calculator to show that the area under the entire curve is equal to 1. Note that you can't specify the interval $(-\infty, +\infty)$, so you'll have to decide on some endpoints that are far enough from the center (110) of the distribution to give at least four-decimal-place accuracy. Record the interval that you use and the area that the calculator reports. Will it suffice to go out four standard deviations on either side of the center? Five standard deviations?

2.54 Use the calculator's **invNorm** function to verify your answers to Exercises 2.42 (page 113) and 2.47 (page 114). Use the method described in the Technology Toolbox on page 117.

TECHNOLOGY TOOLBOX *Finding z-values with invNorm*

The TI-83/89 **invNorm** function calculates the raw or standardized normal value corresponding to a known area under a normal distribution or a relative frequency. The following example uses the WISC scores, which have a $N(100, 15)$ distribution. Here are the keystrokes:

TI-83

- Press [2nd] [VARS] (DISTR), then choose `3:invNorm(`.
- Complete the command `invNorm(.9, 100,15)` and press [ENTER]. Compare this with the command `invNorm(.9)`.

```
invNorm(.9,100,15)
        119.2232735
invNorm(.9)
        1.281551567
```

TI-89

- Press [CATALOG] [F3] (Flash Apps) and choose `invNorm(`.
- Complete the command `tistat.invNorm (.9,100,15)` and press [ENTER]. Compare this with the command `invNorm(.9)`.

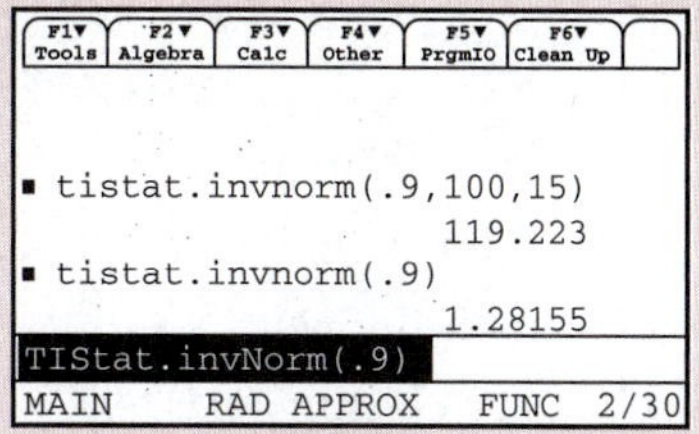

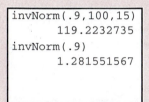

The first command finds that the *raw* WISC score that has 90% of the scores below it from the $N(100, 15)$ distribution is $x = 119$. The second command says that the *standardized* WISC score that has 90% of the scores below it is $z = 1.28$.

NOTES AND DATA SOURCES

1. Data from Gary Community School Corporation, courtesy of Celeste Foster, Department of Education, Purdue University.
2. Data from Stephen Jay Gould, "Entropic homogeneity isn't why no one hits 400 anymore," *Discover*, August 1986, pp. 60–66. Gould does not standardize but gives a speculative discussion instead.
3. Detailed data appear in P. S. Levy et al., "Total serum cholesterol values for youths 12–17 years," *Vital and Health Statistics Series 11*, No. 150 (1975), U.S. National Center for Health Statistics.
4. Ulric Neisser, "Rising scores on intelligence tests," *American Scientist*, September–October 1997, online edition.

The Granger Collection, New York

SIR FRANCIS GALTON

Correlation, Regression, and Heredity

The least-squares method will happily fit a straight line to any two-variable data. It is an old method, going back to the French mathematician Legendre in about 1805. Legendre invented least squares for use on data from astronomy and surveying. It was *Sir Francis Galton (1822–1911)*, however, who turned "regression" into a general method for understanding relationships. He even invented the word. While he was at it, he also invented "correlation," both the word and the definition of r.

Galton was one of the last gentleman scientists, an upper-class Englishman who studied medicine at Cambridge and explored Africa before turning to the study of heredity. He was well connected here also: Charles Darwin, who published *The Origin of Species* in 1859, was his cousin.

Galton was full of ideas but was no mathematician. He didn't even use least squares, preferring to avoid unpleasant computations. But Galton was the first to apply regression ideas to biological and psychological data. He asked: If people's heights are distributed normally in every generation, and height is inherited, what is the relationship between generations? He discovered a straight-line relationship between the heights of parent and child and found that tall parents tended to have children who were taller than average but less tall than their parents. He called this "regression toward mediocrity." The name "regression" came to be applied to the statistical method.

Galton was full of ideas but was no mathematician. He didn't even use least squares, preferring to avoid unpleasant computations.

chapter 3

Examining Relationships

ACTIVITY 3 SAT/ACT Scores

Materials: Pencil, grid paper

Is there an association between SAT Math scores and SAT Verbal scores? If a student performs well on the Math part of the SAT exam, will he or she do well on the Verbal part, too? If a student performs well on one part, does that suggest that the student will not do as well on the other? Is it rare or fairly common for students to score about the same on both parts of the SAT? In this activity you will collect, anonymously of course, the SAT Math and SAT Verbal scores for each member of the class who has taken the SAT exam. You will then plot these data and inspect the graph to see if a pattern is evident. If your school is in a state where the ACT exam is the principal college placement test, then use ACT scores.

1. Begin by writing your Math score and Verbal score on an index card or similar uniform "ballot." Label your Math score M, and your Verbal score V. A selected student should collect the folded index cards in a box or other container. When all of the index cards have been placed in the box, mix them without looking, so that each student's privacy is protected.

   If the size of your class is "small," then you may need to supplement your data with the scores of students in other classes. Perhaps your teacher can request that scores from other AP classes be provided to make a larger data set. Try to obtain data from at least 25 or 30 students.

2. The scores should be called out by the student who collects the data and recorded on the blackboard as ordered pairs in the form (Math, Verbal).

3. Each student should construct a plot of the data with pencil and paper. Since the Math scores appear first in the ordered pairs, label your horizontal axis "Math" and label the vertical axis "Verbal." Determine the range of the Math scores and the range of the Verbal scores, and then construct scales for both axes. Note that axes don't have to intersect at the point (0,0), but the scales on both axes should be uniform.

4. When you finish constructing your graph, look to see if there is any discernible pattern. If so, can you describe the pattern? Does the graph provide any insight into a possible association between SAT Math and SAT Verbal scores?

We will return to analyze these data in more detail after we develop some methodology.

INTRODUCTION

Most statistical studies involve more than one variable. Sometimes we want to compare the distributions of the same variable for several groups. For example, we might compare the distributions of SAT scores among students at several colleges. Side-by-side boxplots, stemplots, or histograms make the comparison visible. In this chapter, however, we concentrate on relationships among several variables for the same group of individuals. For example, Table 1.15 (page 71) records seven variables that describe education in the United States. We have already examined some of these variables one at a time. Now we might ask how SAT Mathematics scores are related to SAT Verbal scores or to the percent of a state's high school seniors who take the SAT or to what region a state is in.

When you examine the relationship between two or more variables, first ask the preliminary questions that are familiar from Chapters 1 and 2.

- What *individuals* do the data describe?
- What exactly are the *variables?* How are they measured?
- Are all the variables *quantitative* or is at least one a *categorical* variable?

We have concentrated on quantitative variables until now. When we have data on several variables, however, categorical variables are often present and help organize the data. Categorical variables will play a larger role in the next chapter. There is one more question you should ask when you are interested in relations among several variables:

- Do you want simply to explore the nature of the relationship, or do you think that some of the variables explain or even cause changes in others? That is, are some of the variables *response variables* and others *explanatory variables?*

RESPONSE VARIABLE, EXPLANATORY VARIABLE

A **response variable** measures an outcome of a study. An **explanatory variable** attempts to explain the observed outcomes.

independent variable
dependent variable

You will often find explanatory variables called ***independent variables,*** and response variables called ***dependent variables.*** The idea behind this language is that the response variable depends on the explanatory variable. Because the words "independent" and "dependent" have other, unrelated meanings in statistics, we won't use them here.

It is easiest to identify explanatory and response variables when we actually set values of one variable in order to see how it affects another variable.

EXAMPLE 3.1 EFFECT OF ALCOHOL ON BODY TEMPERATURE

Alcohol has many effects on the body. One effect is a drop in body temperature. To study this effect, researchers give several different amounts of alcohol to mice, then measure the change in each mouse's body temperature in the 15 minutes after taking the alcohol. Amount of alcohol is the explanatory variable, and change in body temperature is the response variable.

When you don't set the values of either variable but just observe both variables, there may or may not be explanatory and response variables. Whether there are depends on how you plan to use the data.

EXAMPLE 3.2 ARE SAT MATH AND VERBAL SCORES LINKED?

Jim wants to know how the median SAT Math and Verbal scores in the 51 states (including the District of Columbia) are related to each other. He doesn't think that either score explains or causes the other. Jim has two related variables, and neither is an explanatory variable.

Julie looks at some data. She asks, "Can I predict a state's median SAT Math score if I know its median SAT Verbal score?" Julie is treating the Verbal score as the explanatory variable and the Math score as the response variable.

In Example 3.1 alcohol actually *causes* a change in body temperature. There is no cause-and-effect relationship between SAT Math and Verbal scores in Example 3.2. Because the scores are closely related, we can nonetheless use a state's SAT Verbal score to predict its Math score. We will learn how to do the prediction in Section 3.3. Prediction requires that we identify an explanatory variable and a response variable. Some other statistical techniques ignore this distinction. Do remember that calling one variable explanatory and the other response doesn't necessarily mean that changes in one *cause* changes in the other.

The statistical techniques used to study relations among variables are more complex than the one-variable methods in Chapters 1 and 2. Fortunately, analysis of several-variable data builds on the tools used for examining individual variables. The principles that guide examination of data are also the same:

- First plot the data, then add numerical summaries.
- Look for overall patterns and deviations from those patterns.
- When the overall pattern is quite regular, use a compact mathematical model to describe it.

EXERCISES

3.1 EXPLANATORY AND RESPONSE VARIABLES In each of the following situations, is it more reasonable to simply explore the relationship between the two variables or to view one of the variables as an explanatory variable and the other as a response variable? In the latter case, which is the explanatory variable and which is the response variable?

(a) The amount of time a student spends studying for a statistics exam and the grade on the exam

(b) The weight and height of a person

(c) The amount of yearly rainfall and the yield of a crop

(d) A student's grades in statistics and in French

(e) The occupational class of a father and of a son

3.2 **QUANTITATIVE AND CATEGORICAL VARIABLES** How well does a child's height at age 6 predict height at age 16? To find out, measure the heights of a large group of children at age 6, wait until they reach age 16, then measure their heights again. What are the explanatory and response variables here? Are these variables categorical or quantitative?

3.3 **GENDER GAP** There may be a "gender gap" in political party preference in the United States, with women more likely than men to prefer Democratic candidates. A political scientist selects a large sample of registered voters, both men and women. She asks each voter whether they voted for the Democratic or for the Republican candidate in the last congressional election. What are the explanatory and response variables in this study? Are they categorical or quantitative variables?

3.4 **TREATING BREAST CANCER** The most common treatment for breast cancer was once removal of the breast. It is now usual to remove only the tumor and nearby lymph nodes, followed by radiation. The change in policy was due to a large medical experiment that compared the two treatments. Some breast cancer patients, chosen at random, were given each treatment. The patients were closely followed to see how long they lived following surgery. What are the explanatory and response variables? Are they categorical or quantitative?

3.5 What are the variables in Activity 3 (page 120)? Is there an explanatory/response relationship? If so, which is the explanatory variable and which is the response variable? Are the variables quantitative or categorical?

3.1 SCATTERPLOTS

The most effective way to display the relation between two quantitative variables is a *scatterplot*. Here is an example of a scatterplot.

EXAMPLE 3.3 STATE SAT SCORES

Some people use average SAT scores to rank state or local school systems. This is not proper, because the percent of high school students who take the SAT varies from place to place. Let us examine the relationship between the percent of a state's high school graduates who take the exam and the state average SAT Mathematics score, using data from Table 1.15 on page 70.

We think that "percent taking" will help explain "average score." Therefore, "percent taking" is the explanatory variable and "average score" is the response variable.

We want to see how average score changes when percent taking changes, so we put percent taking (the explanatory variable) on the horizontal axis. Figure 3.1 is the scatterplot. Each point represents a single state. In Alabama, for example, 9% take the SAT, and the average SAT Math score is 555. Find 9 on the *x* (horizontal) axis and 555 on the *y* (vertical) axis. Alabama appears as the point (9, 555) above 9 and to the right of 555. Figure 3.1 shows how to locate Alabama's point on the plot.

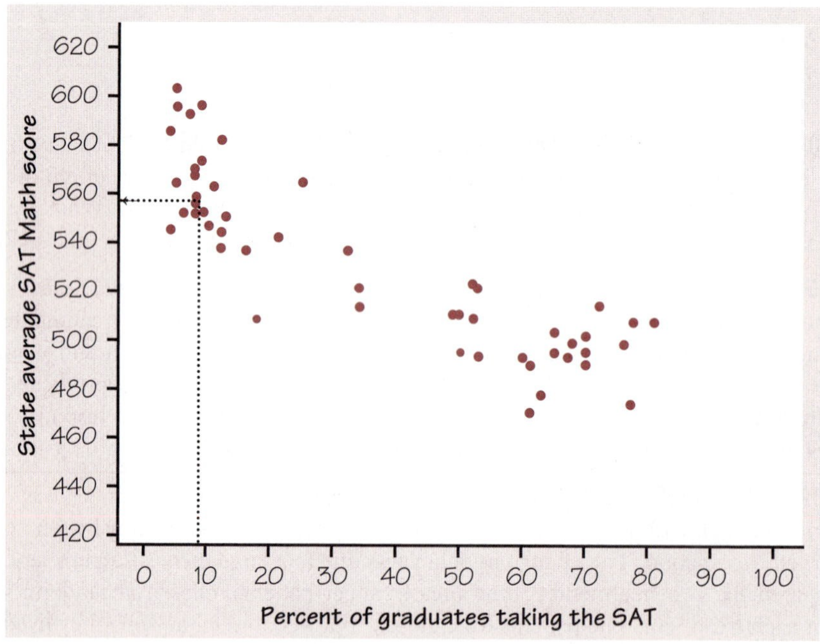

FIGURE 3.1 Scatterplot of the average SAT Math score in each state against the percent of that state's high school graduates who take the SAT, from Table 1.15. The dotted lines intersect at the point (9, 555), the data for Alabama.

SCATTERPLOT

A **scatterplot** shows the relationship between two quantitative variables measured on the same individuals. The values of one variable appear on the horizontal axis, and the values of the other variable appear on the vertical axis. Each individual in the data appears as the point in the plot fixed by the values of both variables for that individual.

Always plot the explanatory variable, if there is one, on the horizontal axis (the *x* axis) of a scatterplot. As a reminder, we usually call the explanatory variable *x* and the response variable *y*. If there is no explanatory-response distinction, either variable can go on the horizontal axis.

EXERCISES

3.6 THE ENDANGERED MANATEE Manatees are large, gentle sea creatures that live along the Florida coast. Many manatees are killed or injured by powerboats. Here are data on powerboat registrations (in thousands) and the number of manatees killed by boats in Florida in the years 1977 to 1990:

Year	Powerboat registrations (1000)	Manatees killed	Year	Powerboat registrations (1000)	Manatees killed
1977	447	13	1984	559	34
1978	460	21	1985	585	33
1979	481	24	1986	614	33
1980	498	16	1987	645	39
1981	513	24	1988	675	43
1982	512	20	1989	711	50
1983	526	15	1990	719	47

(a) We want to examine the relationship between number of powerboats and number of manatees killed by boats. Which is the explanatory variable?

(b) Make a scatterplot of these data. (Be sure to label the axes with the variable names, not just x and y.) What does the scatterplot show about the relationship between these variables?

3.7 ARE JET SKIS DANGEROUS? Propelled by a stream of pressurized water, jet skis and other so-called wet bikes carry from one to three people, retail for an average price of $5,700, and have become one of the most popular types of recreational vehicle sold today. But critics say that they're noisy, dangerous, and damaging to the environment. An article in the August 1997 issue of the *Journal of the American Medical Association* reported on a survey that tracked emergency room visits at randomly selected hospitals nationwide. Here are data on the number of jet skis in use, the number of accidents, and the number of fatalities for the years 1987–1996:[1]

Year	Number in use	Accidents	Fatalities
1987	92,756	376	5
1988	126,881	650	20
1989	178,510	844	20
1990	241,376	1,162	28
1991	305,915	1,513	26
1992	372,283	1,650	34
1993	454,545	2,236	35
1994	600,000	3,002	56
1995	760,000	4,028	68
1996	900,000	4,010	55

(a) We want to examine the relationship between the number of jet skis in use and the number of accidents. Which is the explanatory variable?

(b) Make a scatterplot of these data. (Be sure to label the axes with the variable names, not just *x* and *y*.) What does the scatterplot show about the relationship between these variables?

3.8 Make a scatterplot of the (Math SAT/ACT score, Verbal SAT/ACT score) data from Activity 3, if you haven't done so already. Does the scatterplot describe a strong association, a moderate association, a weak association, or no association between these variables?

Interpreting scatterplots

To interpret a scatterplot, apply the strategies of data analysis learned in Chapters 1 and 2.

> **EXAMINING A SCATTERPLOT**
>
> In any graph of data, look for the **overall pattern** and for striking **deviations** from that pattern.
>
> You can describe the overall pattern of a scatterplot by the **form, direction,** and **strength** of the relationship.
>
> An important kind of deviation is an **outlier,** an individual value that falls outside the overall pattern of the relationship.

clusters

Figure 3.1 shows a clear *form:* there are two distinct ***clusters*** of states with a gap between them. In the cluster at the right of the plot, 45% or more of high school graduates take the SAT, and the average scores are low. The states in the cluster at the left have higher SAT scores and lower percents of graduates taking the test. There are no clear outliers. That is, no points fall clearly outside the clusters.

What explains the clusters? There are two widely used college entrance exams, the SAT and the American College Testing (ACT) exam. Each state favors one or the other. The left cluster in Figure 3.1 contains the ACT states, and the SAT states make up the right cluster. In ACT states, most students who take the SAT are applying to a selective college that requires SAT scores. This select group of students has a higher average score than the much larger group of students who take the SAT in SAT states.

The relationship in Figure 3.1 also has a clear *direction:* states in which a higher percent of students take the SAT tend to have lower average scores. This is a *negative association* between the two variables.

POSITIVE ASSOCIATION, NEGATIVE ASSOCIATION

Two variables are **positively associated** when above-average values of one tend to accompany above-average values of the other and below-average values also tend to occur together.

Two variables are **negatively associated** when above-average values of one tend to accompany below-average values of the other, and vice versa.

The *strength* of a relationship in a scatterplot is determined by how closely the points follow a clear form. The overall relationship in Figure 3.1 is not strong—states with similar percents taking the SAT show quite a bit of scatter in their average scores. Here is an example of a stronger relationship with a clearer form.

EXAMPLE 3.4 HEATING DEGREE-DAYS

The Sanchez household is about to install solar panels to reduce the cost of heating their house. In order to know how much the solar panels help, they record their consumption of natural gas before the panels are installed. Gas consumption is higher in cold weather, so the relationship between outside temperature and gas consumption is important.

Table 3.1 gives data for 16 months. The response variable y is the average amount of natural gas consumed each day during the month, in hundreds of cubic feet. The explanatory variable x is the average number of heating degree-days each day during the month. (Heating degree-days are the usual measure of demand for heating. One degree-day is accumulated for each degree a day's average temperature falls below 65° F. An average temperature of 20° F, for example, corresponds to 45 degree-days.)

TABLE 3.1 Average degree-days and natural gas consumption for the Sanchez household

Month	Degree-days	Gas (100 cu. ft.)	Month	Degree-days	Gas (100 cu. ft.)
Nov.	24	6.3	July	0	1.2
Dec.	51	10.9	Aug.	1	1.2
Jan.	43	8.9	Sept.	6	2.1
Feb.	33	7.5	Oct.	12	3.1
Mar.	26	5.3	Nov.	30	6.4
Apr.	13	4.0	Dec.	32	7.2
May	4	1.7	Jan.	52	11.0
June	0	1.2	Feb.	30	6.9

Source: Data provided by Robert Dale, Purdue University.

The scatterplot in Figure 3.2 shows a strong positive association. More degree-days means colder weather and so more gas consumed. The form of the relationship is ***linear.*** That is, the points lie in a straight-line pattern. It is a strong relationship because the points

linear

lie close to a line, with little scatter. If we know how cold a month is, we can predict gas consumption quite accurately from the scatterplot. That strong relationships make accurate predictions possible is an important point that we will soon discuss in more detail.

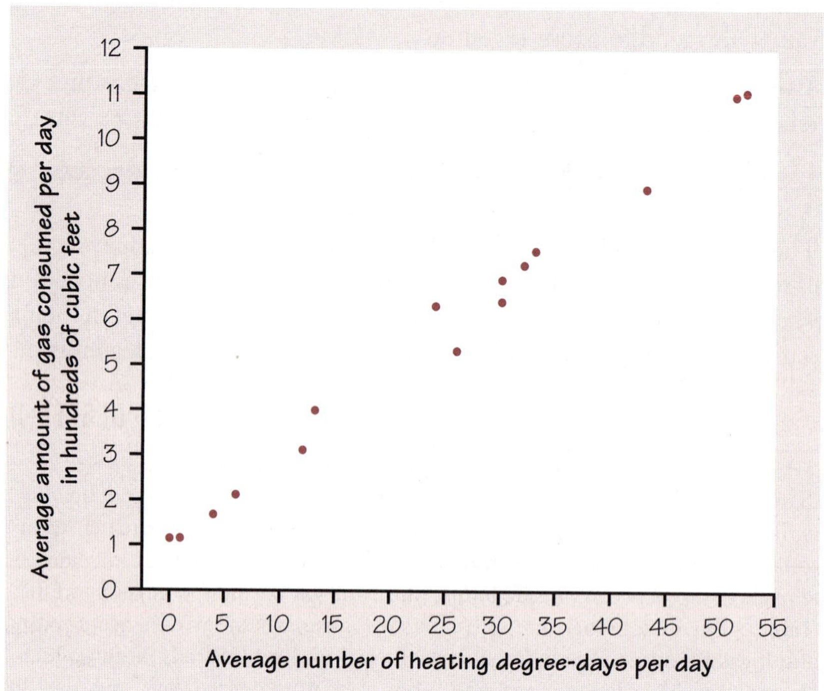

FIGURE 3.2 Scatterplot of the average amount of natural gas used per day by the Sanchez household in 16 months against the average number of heating degree-days per day in those months, from Table 3.1.

Of course, not all relationships are linear in form. What is more, not all relationships have a clear direction that we can describe as positive association or negative association. Exercise 3.11 gives an example that is not linear and has no clear direction.

Tips for drawing scatterplots

1. Scale the horizontal and vertical axes. The intervals must be uniform; that is, the distance between tick marks must be the same. If the scale does not begin at zero at the origin, then use the symbol shown to indicate a break.

2. Label both axes.

3. If you are given a grid, try to adopt a scale so that your plot uses the whole grid. Make your plot large enough so that the details can be easily seen. Don't compress the plot into one corner of the grid.

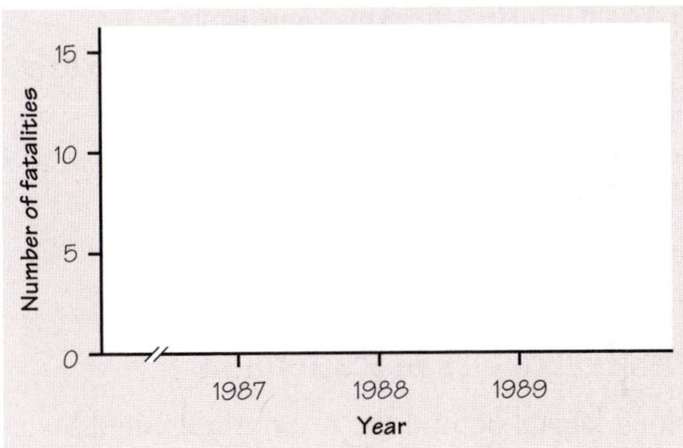

EXERCISES

3.9 MORE ON THE ENDANGERED MANATEE In Exercise 3.6 (page 125) you made a scatterplot of powerboats registered in Florida and manatees killed by boats.

(a) Describe the direction of the relationship. Are the variables positively or negatively associated?

(b) Describe the form of the relationship. Is it linear?

(c) Describe the strength of the relationship. Can the number of manatees killed be predicted accurately from powerboat registrations? If powerboat registrations remained constant at 719,000, about how many manatees would be killed by boats each year?

3.10 MORE JET SKIS In Exercise 3.7 (page 125) you made a scatterplot of jet skis in use and number of accidents.

(a) Describe the direction of the relationship. Are the variables positively or negatively associated?

(b) Describe the form of the association. Is it linear?

3.11 DOES FAST DRIVING WASTE FUEL? How does the fuel consumption of a car change as its speed increases? Here are data for a British Ford Escort. Speed is measured in kilometers per hour, and fuel consumption is measured in liters of gasoline used per 100 kilometers traveled.[2]

Speed (km/h)	Fuel used (liters/100 km)	Speed (km/h)	Fuel used (liters/100 km)
10	21.00	90	7.57
20	13.00	100	8.27
30	10.00	110	9.03
40	8.00	120	9.87
50	7.00	130	10.79
60	5.90	140	11.77
70	6.30	150	12.83
80	6.95		

(a) Make a scatterplot. (Which is the explanatory variable?)

(b) Describe the form of the relationship. Why is it not linear? Explain why the form of the relationship makes sense.

(c) It does not make sense to describe the variables as either positively associated or negatively associated. Why?

(d) Is the relationship reasonably strong or quite weak? Explain your answer.

Adding categorical variables to scatterplots

The South has long lagged behind the rest of the United States in the performance of its schools. Efforts to improve education have reduced the gap. We wonder if the South stands out in our study of state average SAT scores.

EXAMPLE 3.5 IS THE SOUTH DIFFERENT?

Figure 3.3 enhances the scatterplot in Figure 3.1 by plotting the southern states with plus signs. (We took the South to be the states in the East South Central and South Atlantic regions.) Most of the southern states blend in with the rest of the country. Several southern states do lie at the lower edges of their clusters, along with the District of Columbia, which is a city rather than a state. Georgia, South Carolina, and West Virginia have lower SAT scores than we would expect from the percent of their high school graduates who take the examination.

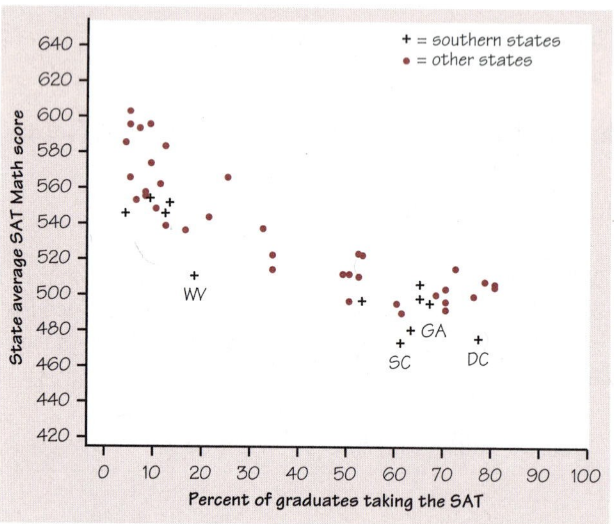

FIGURE 3.3 Average SAT Math score and percent of high school graduates who take the test, by state, with the southern states highlighted.

Dividing the states into "southern" and "nonsouthern" introduces a third variable into the scatterplot. This is a categorical variable that has only two values. The two values are displayed by the two different plotting symbols. Use different colors or symbols to plot points when you want to add a categorical variable to a scatterplot.[3]

EXAMPLE 3.6 DO SOLAR PANELS REDUCE GAS USAGE?

After the Sanchez household gathered the information recorded in Table 3.1 and Figure 3.2 (pages 127 and 128), they added solar panels to their house. They then measured their natural gas consumption for 23 more months. To see how the solar panels affected gas consumption, add the degree-days and gas consumption for these months to the scatterplot. Figure 3.4 is the result. We use different symbols to distinguish before from after. The "after" data form a linear pattern that is close to the "before" pattern in warm months (few degree-days). In colder months, with more degree-days, gas consumption after installing the solar panels is less than in similar months before the panels were added. The scatterplot shows the energy savings from the panels.

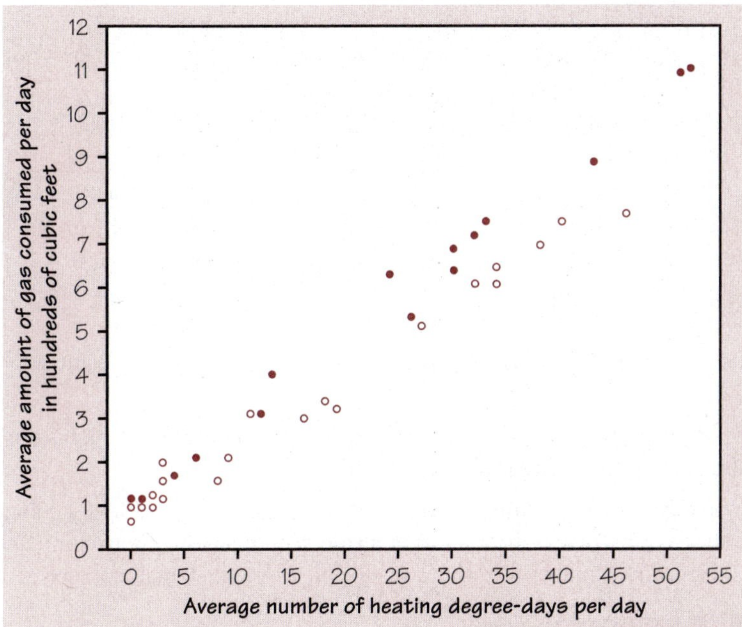

FIGURE 3.4 Natural gas consumption against degree-days for the Sanchez household. The observations indicated by filled circles are for 16 months before installing solar panels. The observations indicated by open circles are for 23 months with the panels in use.

Our gas consumption example suffers from a common problem in drawing scatterplots that you may not notice when a computer does the work. When several individuals have exactly the same data, they occupy the same point on the scatterplot. Look at June and July in Table 3.1. Table 3.1 contains data for 16 months, but there are only 15 points in Figure 3.2. June and July both occupy the same point. You can use a different plotting symbol to call attention to points that stand for more than one individual. Some computer software does this automatically, but some does not. We recommend that you do use a different symbol for repeated observations when you plot a small number of observations by hand.

EXERCISES

3.12 DO HEAVIER PEOPLE BURN MORE ENERGY? Metabolic rate, the rate at which the body consumes energy, is important in studies of weight gain, dieting, and exercise. Table 3.2 gives data on the lean body mass and resting metabolic rate for 12 women and 7 men who are subjects in a study of dieting. Lean body mass, given in kilograms, is a person's weight leaving out all fat. Metabolic rate is measured in calories burned per 24 hours, the same calories used to describe the energy content of foods. The researchers believe that lean body mass is an important influence on metabolic rate.

TABLE 3.2 Lean body mass and metabolic rate

Subject	Sex	Mass (kg)	Rate (cal)	Subject	Sex	Mass (kg)	Rate (cal)
1	M	62.0	1792	11	F	40.3	1189
2	M	62.9	1666	12	F	33.1	913
3	F	36.1	995	13	M	51.9	1460
4	F	54.6	1425	14	F	42.4	1124
5	F	48.5	1396	15	F	34.5	1052
6	F	42.0	1418	16	F	51.1	1347
7	M	47.4	1362	17	F	41.2	1204
8	F	50.6	1502	18	M	51.9	1867
9	F	42.0	1256	19	M	46.9	1439
10	M	48.7	1614				

(a) Make a scatterplot of the data for the female subjects. Which is the explanatory variable?

(b) Is the association between these variables positive or negative? What is the form of the relationship? How strong is the relationship?

(c) Now add the data for the male subjects to your graph, using a different color or a different plotting symbol. Does the pattern of relationship that you observed in (b) hold for men also? How do the male subjects as a group differ from the female subjects as a group?

TECHNOLOGY TOOLBOX *Making a calculator scatterplot*

We will use the gas consumption data from Example 3.4 to show how to construct a scatterplot on the TI-83/89.

• Begin by entering the degree-days data and assigning the values to a list named DEGDA, as shown. Then press ENTER.

TI-83

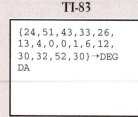

```
{24,51,43,33,26,
13,4,0,0,1,6,12,
30,32,52,30}→DEG
DA
```

TI-89

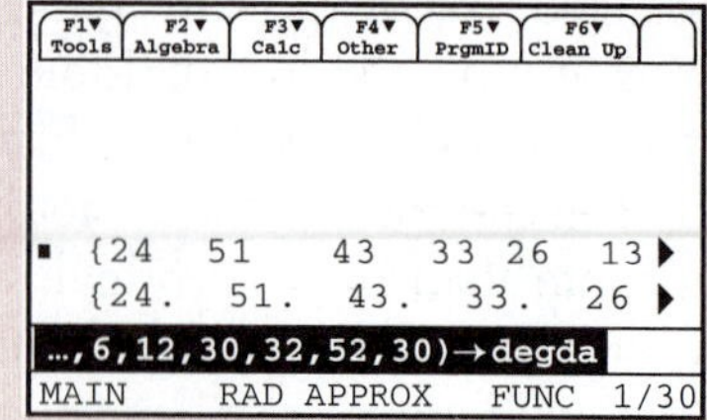

TECHNOLOGY TOOLBOX *Making a calculator scatterplot (continued)*

- Then enter the gas consumption data and assign them to the list GAS. Press ENTER.

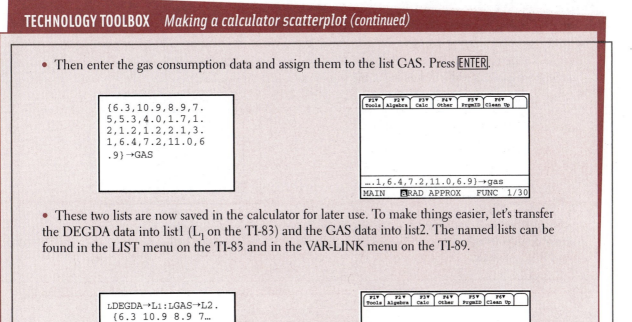

```
{6.3,10.9,8.9,7.
5,5.3,4.0,1.7,1.
2,1.2,1.2,2.1,3.
1,6.4,7.2,11.0,6
.9}→GAS
```

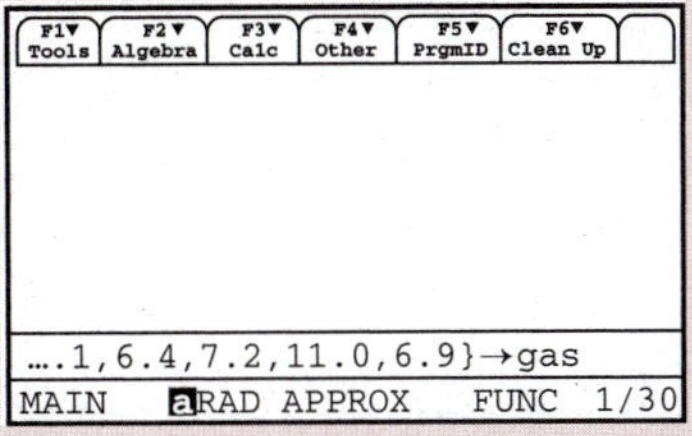

- These two lists are now saved in the calculator for later use. To make things easier, let's transfer the DEGDA data into list1 (L_1 on the TI-83) and the GAS data into list2. The named lists can be found in the LIST menu on the TI-83 and in the VAR-LINK menu on the TI-89.

```
LDEGDA→L1:LGAS→L2.
 {6.3 10.9 8.9 7...
```

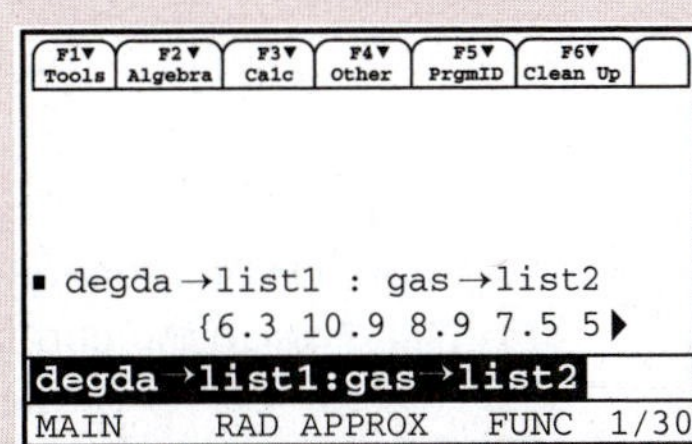

- You can verify that the two lists of data are now in L_1/list1 and L_2/list2 in the Statistics/List Editor.

L1	L2	L3 1
24	6.3	-----
51	10.9	
43	8.9	
33	7.5	
26	5.3	
13	4	
4	1.7	

L1(1)=24

F1 Tools | F2 Plots | F3 List | F4 Calc | F5 Distr | F6 Tests | F7 Ints

list1	list2	----	----
24.	6.3		
51.	10.9		
43.	8.9		
33.	7.5		
26.	5.3		
13.	4.		

list2[1]=6.3
MAIN RAD APPROX FUNC 2/2

- Next, define a scatterplot in the statistics plot menu (press F2 on the TI-89). Specify the settings shown.

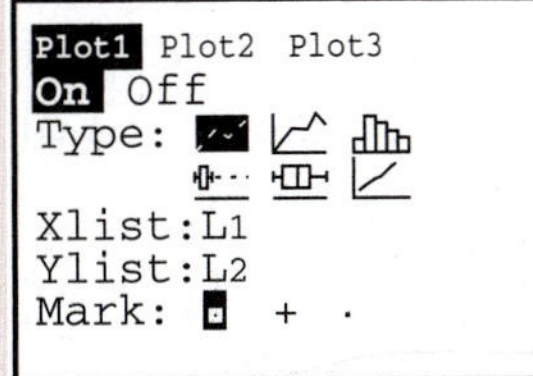

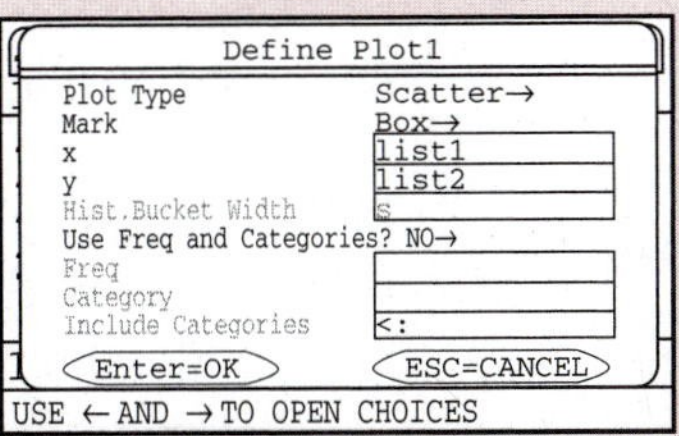

TECHNOLOGY TOOLBOX *Making a calculator scatterplot (continued)*

- Use ZoomStat (ZoomData on the TI-89) to obtain the graph. The calculator will set the window dimensions automatically by looking at the values in L_1/list1 and L_2/list2.

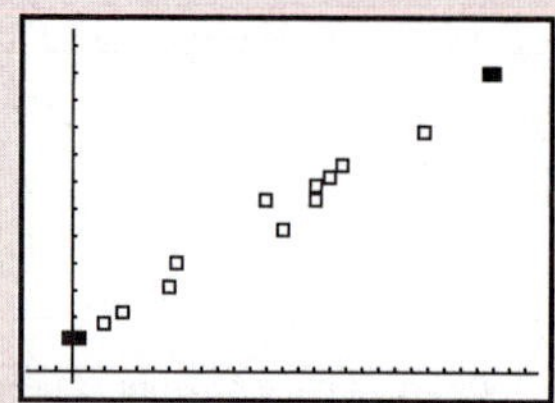

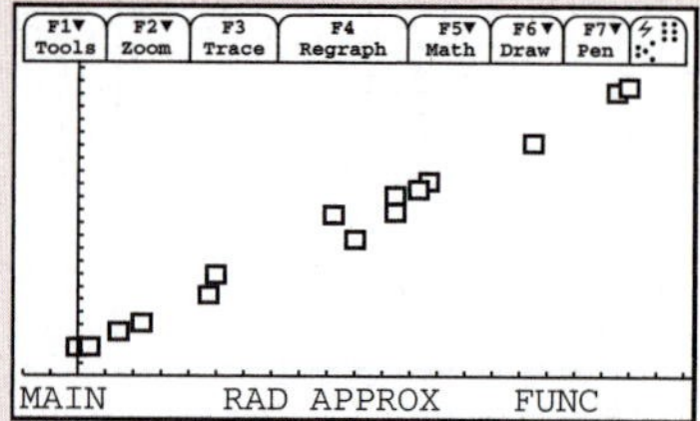

- Notice that there are no scales on the axes, and that the axes are not labeled. If you copy a scatterplot from your calculator onto your paper, make sure that you scale and label the axes. You can use TRACE to help you get started.

3.13 SCATTERPLOT BY CALCULATOR, I Rework Exercise 3.11 (page 129) using your calculator. The command `seq(10X,X,1,15)→SPEED` will create a list named SPEED and assign the numbers 10, 20, . . ., 150 to the list. (Note that `seq` is found under 2nd / LIST / OPS on the TI-83 and under CATALOG on the TI-89). Then assign the fuel data to the list FUEL, and copy the list SPEED to L_1/list1 and the list FUEL to L_2/list2. Define Plot 1 to be a scatterplot, and then ZOOM / 9:ZoomStat (ZoomData on the TI-89) to graph it. Verify your answers to Exercise 3.11.

3.14 SCATTERPLOT BY CALCULATOR, II Rework Exercise 3.12 (page 132) using your calculator. Verify your answers to Exercise 3.12.

SUMMARY

To study relationships between variables, we must measure the variables on the same group of individuals.

If we think that a variable *x* may explain or even cause changes in another variable *y*, we call *x* an **explanatory variable** and *y* a **response variable.**

A **scatterplot** displays the relationship between two quantitative variables measured on the same individuals. Mark values of one variable on the horizontal axis (*x* axis) and values of the other variable on the vertical axis (*y* axis). Plot each individual's data as a point on the graph.

Always plot the explanatory variable, if there is one, on the *x* axis of a scatterplot. Plot the response variable on the *y* axis.

Plot points with different colors or symbols to see the effect of a categorical variable in a scatterplot.

In examining a scatterplot, look for an overall pattern showing the **form, direction,** and **strength** of the relationship, and then for **outliers** or other deviations from this pattern.

Form: Linear relationships, where the points show a straight-line pattern, are an important form of relationship between two variables. Curved relationships and **clusters** are other forms to watch for.

Direction: If the relationship has a clear direction, we speak of either **positive association** (high values of the two variables tend to occur together) or **negative association** (high values of one variable tend to occur with low values of the other variable).

Strength: The **strength** of a relationship is determined by how close the points in the scatterplot lie to a simple form such as a line.

SECTION 3.1 EXERCISES

3.15 IQ AND SCHOOL GRADES Do students with higher IQ test scores tend to do better in school? Figure 3.5 is a scatterplot of IQ and school grade point average (GPA) for all 78 seventh-grade students in a rural Midwest school.[4]

(a) Say in words what a positive association between IQ and GPA would mean. Does the plot show a positive association?

(b) What is the form of the relationship? Is it roughly linear? Is it very strong? Explain your answers.

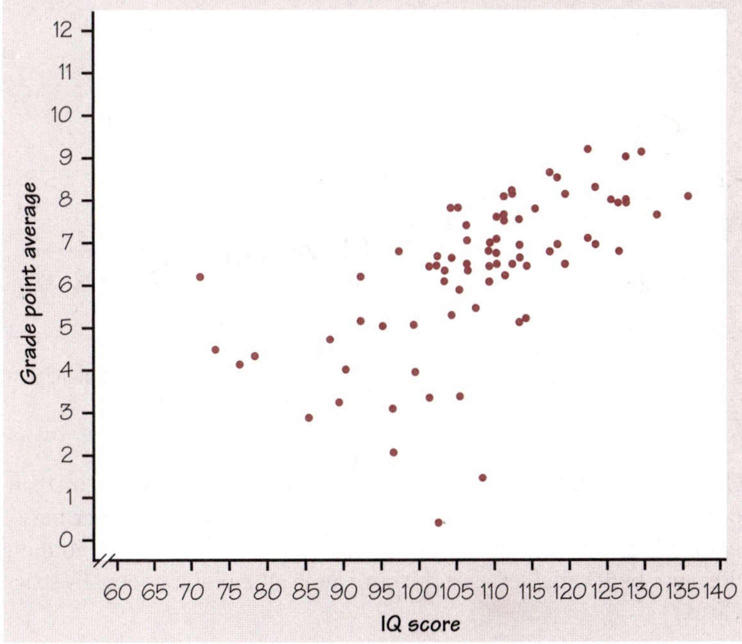

FIGURE 3.5 Scatterplot of school grade point average versus IQ test score for seventh-grade students.

(c) At the bottom of the plot are several points that we might call outliers. One student in particular has a very low GPA despite an average IQ score. What are the approximate IQ and GPA for this student?

3.16 CALORIES AND SALT IN HOT DOGS Are hot dogs that are high in calories also high in salt? Figure 3.6 is a scatterplot of the calories and salt content (measured as milligrams of sodium) in 17 brands of meat hot dogs.[5]

(a) Roughly what are the lowest and highest calorie counts among these brands? Roughly what is the sodium level in the brands with the fewest and with the most calories?

(b) Does the scatterplot show a clear positive or negative association? Say in words what this association means about calories and salt in hot dogs.

(c) Are there any outliers? Is the relationship (ignoring any outliers) roughly linear in form? Still ignoring outliers, how strong would you say the relationship between calories and sodium is?

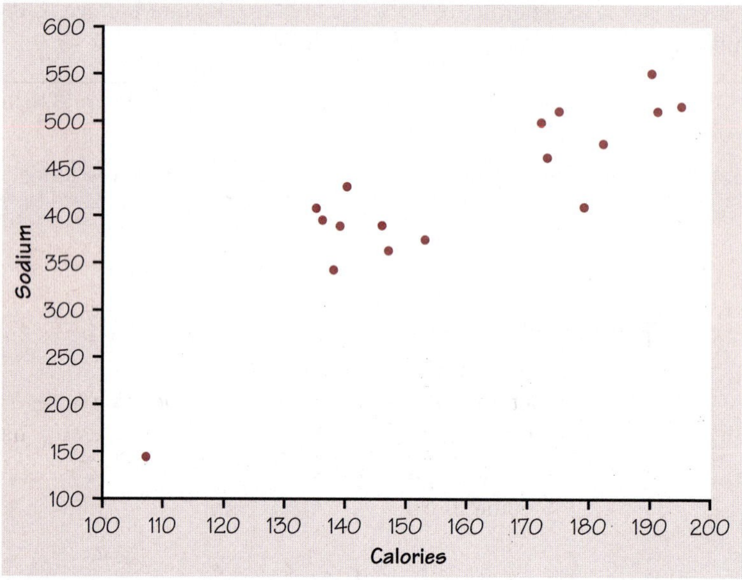

FIGURE 3.6 Scatterplot of milligrams of sodium and calories in each of 17 brands of meat hot dogs.

3.17 RICH STATES, POOR STATES One measure of a state's prosperity is the median income of its households. Another measure is the mean personal income per person in the state. Figure 3.7 is a scatterplot of these two variables, both measured in thousands of dollars. Because both variables have the same units, the plot uses equally spaced scales on both axes.[6]

(a) We have labeled the point for New York on the scatterplot. What are the approximate values of New York's median household income and mean income per person?

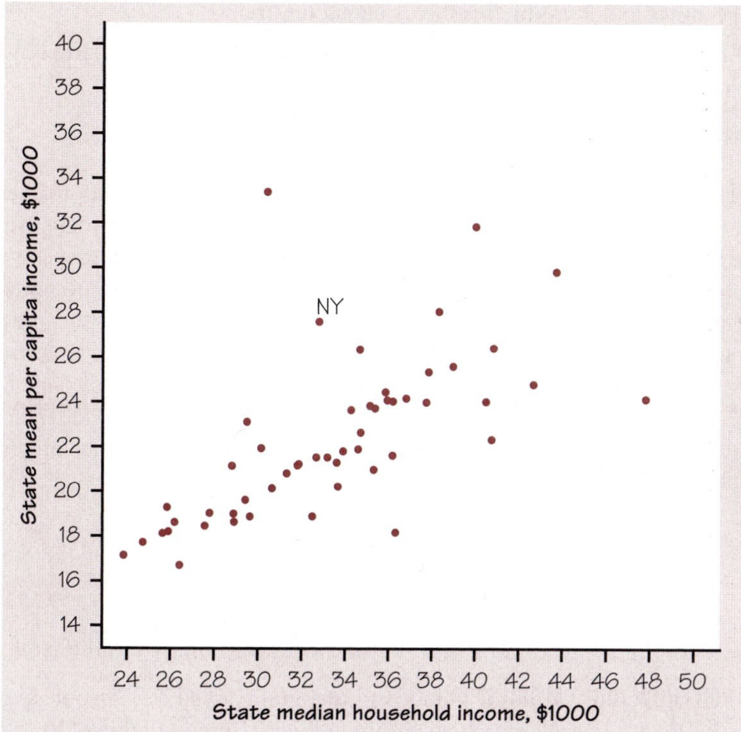

FIGURE 3.7 Scatterplot of mean income per person versus median household income for the states.

(b) Explain why you expect a positive association between these variables. Also explain why you expect household income to be generally higher than income per person.

(c) Nonetheless, the mean income per person in a state can be higher than the median household income. In fact, the District of Columbia has median income $30,748 per household and mean income $33,435 per person. Explain why this can happen.

(d) Alaska is the state with the highest median household income. What is the approximate median household income in Alaska? We might call Alaska and the District of Columbia outliers in the scatterplot.

(e) Describe the form, direction, and strength of the relationship, ignoring the outliers.

3.18 THE PROFESSOR SWIMS Professor Moore swims 2000 yards regularly in a vain attempt to undo middle age. Here are his times (in minutes) and his pulse rate after swimming (in beats per minute) for 23 sessions in the pool:

Time:	34.12	35.72	34.72	34.05	34.13	35.72	36.17	35.57	35.37
Pulse:	152	124	140	152	146	128	136	144	148
Time:	35.57	35.43	36.05	34.85	34.70	34.75	33.93	34.60	34.00
Pulse:	144	136	124	148	144	140	156	136	148
Time:	34.35	35.62	35.68	35.28	35.97				
Pulse:	148	132	124	132	139				

(a) Make a scatterplot. (Which is the explanatory variable?)

(b) Is the association between these variables positive or negative? Explain why you expect the relationship to have this direction.

(c) Describe the form and strength of the relationship.

3.19 MEET THE *ARCHAEOPTERYX* *Archaeopteryx* is an extinct beast having feathers like a bird but teeth and a long bony tail like a reptile. Only six fossil specimens are known. Because these specimens differ greatly in size, some scientists think they are different species rather than individuals from the same species. We will examine some data. If the specimens belong to the same species and differ in size because some are younger than others, there should be a positive linear relationship between the lengths of a pair of bones from all individuals. An outlier from this relationship would suggest a different species. Here are data on the lengths in centimeters of the femur (a leg bone) and the humerus (a bone in the upper arm) for the five specimens that preserve both bones:[7]

Femur:	38	56	59	64	74
Humerus:	41	63	70	72	84

Make a scatterplot. Do you think that all five specimens come from the same species?

3.20 DO YOU KNOW YOUR CALORIES? A food industry group asked 3368 people to guess the number of calories in each of several common foods. Here is a table of the average of their guesses and the correct number of calories:[8]

Food	Guessed calories	Correct calories
8 oz. whole milk	196	159
5 oz. spaghetti with tomato sauce	394	163
5 oz. macaroni with cheese	350	269
One slice wheat bread	117	61
One slice white bread	136	76
2-oz. candy bar	364	260
Saltine cracker	74	12
Medium-size apple	107	80
Medium-size potato	160	88
Cream-filled snack cake	419	160

(a) We think that how many calories a food actually has helps explain people's guesses of how many calories it has. With this in mind, make a scatterplot of these data. (Because both variables are measured in calories, you should use the same scale on both axes. Your plot will be square.)

(b) Describe the relationship. Is there a positive or negative association? Is the relationship approximately linear? Are there any outliers?

3.21 MAXIMIZING CORN YIELDS How much corn per acre should a farmer plant to obtain the highest yield? Too few plants will give a low yield. On the other hand, if there are too many plants, they will compete with each other for moisture and nutrients, and yields will fall. To find the best planting rate, plant at different rates on several plots of ground and measure the harvest. (Be sure to treat all the plots the same except for the planting rate.) Here are the data from such an experiment:[9]

Plants per acre	Yield (bushels per acre)			
12,000	150.1	113.0	118.4	142.6
16,000	166.9	120.7	135.2	149.8
20,000	165.3	130.1	139.6	149.9
24,000	134.7	138.4	156.1	
28,000	119.0	150.5		

(a) Is yield or planting rate the explanatory variable?

(b) Make a scatterplot of yield and planting rate.

(c) Describe the overall pattern of the relationship. Is it linear? Is there a positive or negative association, or neither?

(d) Find the mean yield for each of the five planting rates. Plot each mean yield against its planting rate on your scatterplot and connect these five points with lines. This combination of numerical description and graphing makes the relationship clearer. What planting rate would you recommend to a farmer whose conditions were similar to those in the experiment?

3.22 TEACHERS' PAY Table 1.15 (page 70) gives data for the states. We might expect that states with less educated populations would pay their teachers less, perhaps because these states are poorer.

(a) Make a scatterplot of average teachers' pay against the percent of state residents who are not high school graduates. Take the percent with no high school degree as the explanatory variable.

(b) The plot shows a weak negative association between the two variables. Why do we say that the association is negative? Why do we say that it is weak?

(c) Circle on the plot the point for the state your school is in.

(d) There is an outlier at the upper left of the plot. Which state is this?

(e) We wonder about regional patterns. There is a relatively clear cluster of nine states at the lower right of the plot. These states have many residents who are not high school graduates and pay low salaries to teachers. Which states are these? Are they mainly from one part of the country?

3.23 CATEGORICAL EXPLANATORY VARIABLE A scatterplot shows the relationship between two quantitative variables. Here is a similar plot to study the relationship between a categorical explanatory variable and a quantitative response variable.

The presence of harmful insects in farm fields is detected by putting up boards covered with a sticky material and then examining the insects trapped on the board. Which colors attract insects best? Experimenters placed six boards of each of four colors in a field of oats and measured the number of cereal leaf beetles trapped.[10]

Board color	Insects trapped					
Lemon yellow	45	59	48	46	38	47
White	21	12	14	17	13	17
Green	37	32	15	25	39	41
Blue	16	11	20	21	14	07

(a) Make a plot of the counts of insects trapped against board color (space the four colors equally on the horizontal axis). Compute the mean count for each color, add the means to your plot, and connect the means with line segments.

(b) Based on the data, what do you conclude about the attractiveness of these colors to the beetles?

(c) Does it make sense to speak of a positive or negative association between board color and insect count?

3.2 CORRELATION

A scatterplot displays the direction, form, and strength of the relationship between two quantitative variables. Linear relations are particularly important because a straight line is a simple pattern that is quite common. We say a linear relation is strong if the points lie close to a straight line, and weak if they are widely scattered about a line. Our eyes are not good judges of how strong a linear relationship is. The two scatterplots in Figure 3.8 depict exactly the same data, but the lower plot is drawn smaller in a large field. The lower plot seems to show a stronger linear relationship. Our eyes can be fooled by changing the plotting scales or the amount of white space around the cloud of points in a scatterplot.[11] We need to follow our strategy for data analysis by using a numerical measure to supplement the graph. *Correlation* is the measure we use.

CORRELATION r

The **correlation** measures the direction and strength of the linear relationship between two quantitative variables. Correlation is usually written as r.

Suppose that we have data on variables x and y for n individuals. The values for the first individual are x_1 and y_1, the values for the second individual are x_2 and y_2, and so on. The means and standard deviations of the two variables are $\bar{x}$ and s_x for the x-values, and $\bar{y}$ and s_y for the y-values. The correlation r between x and y is

$$r = \frac{1}{n-1}\sum\left(\frac{x_i - \bar{x}}{s_x}\right)\left(\frac{y_i - \bar{y}}{s_y}\right)$$

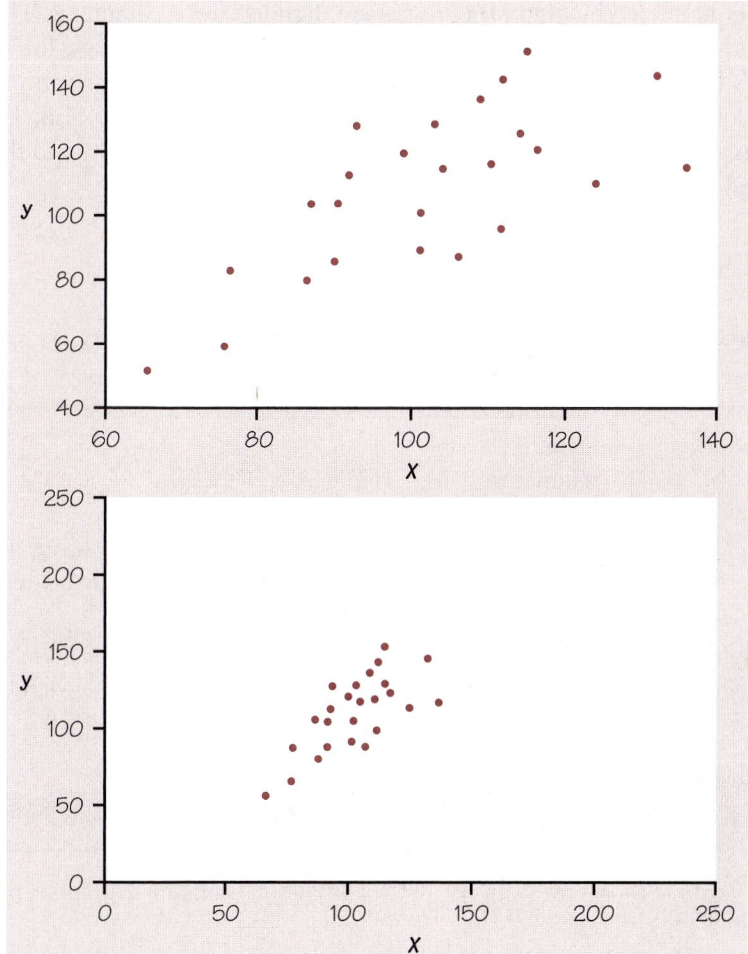

FIGURE 3.8 Two scatterplots of the same data; the straight-line pattern in the lower plot appears stronger because of the surrounding white space.

As always, the summation sign Σ means "add these terms for all the individuals." The formula for the correlation r is a bit complex. It helps us see what correlation is, but in practice you should use software or a calculator that finds r from keyed-in values of two variables x and y. Exercise 3.24 asks you to calculate a correlation step-by-step from the definition to solidify its meaning.

The formula for r begins by standardizing the observations. Suppose, for example, that x is height in centimeters and y is weight in kilograms and that we have height and weight measurements for n people. Then $\bar{x}$ and s_x are the mean and standard deviation of the n heights, both in centimeters. The value

$$\frac{x_i - \bar{x}}{s_x}$$

is the standardized height of the ith person, familiar from Chapter 2. The standardized height says how many standard deviations above or below the mean a person's height lies. Standardized values have no units—in this example, they are no longer measured in centimeters. Standardize the weights also. The correlation r is an average of the products of the standardized height and the standardized weight for the n people.

EXERCISE

3.24 CLASSIFYING FOSSILS Exercise 3.19 (page 138) gives the lengths of two bones in five fossil specimens of the extinct beast *Archaeopteryx*:

Femur:	38	56	59	64	74
Humerus:	41	63	70	72	84

(a) Find the correlation r step-by-step. That is, find the mean and standard deviation of the femur lengths and of the humerus lengths. Then find the five standardized values for each variable and use the formula for r.

(b) Duplicate the steps in the Technology Toolbox below to obtain the correlation for the *Archaeopteryx* data, and compare your result with that calculated by hand in (a).

TECHNOLOGY TOOLBOX *Using the definition to calculate correlation*

We will use the *Archaeopteryx* data to show how to calculate the correlation using the definition and the list features of the TI-83/89.

- Begin by entering the femur lengths (x-values) in L_1/list1 and the humerus lengths (y-values) in L_2/list2. Then calculate two-variable statistics for the x- and y-values. The calculator will remember all of the computed statistics until the next time you calculate one- or two-variable statistics.

TI-83

- Press STAT, choose CALC, then `2:2-Var Stats`.
- Complete the command `2-Var Stats` L_1, L_2, and press ENTER.

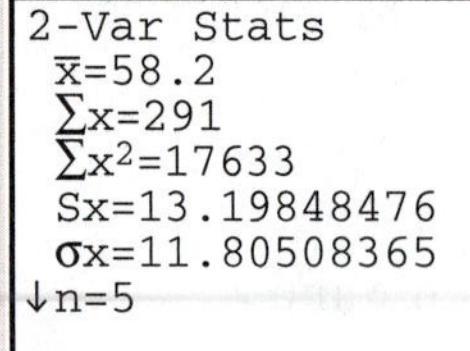

TI-89

- In the Statistics/List Editor, press F4 and choose `2:2-Var Stats`.
- In the new window, enter list1 as the Xlist and list2 as the Ylist, then press ENTER.

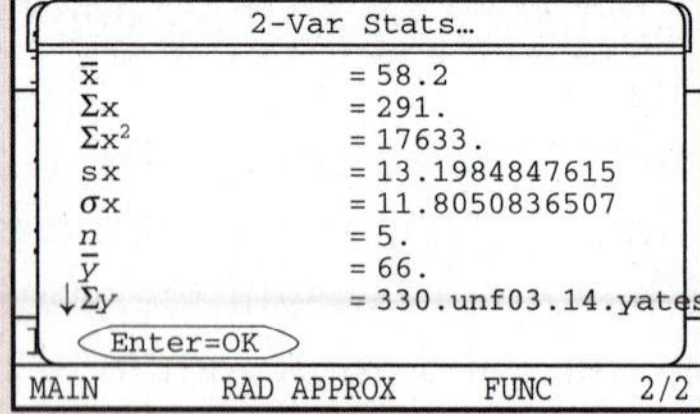

TECHNOLOGY TOOLBOX *Using the definition to calculate correlation (continued)*

- Next, define L_3/list3 = ((list1 − $\bar{x}$)/s_x)((list2 − $\bar{y}$)/s_y) from the home screen as shown. Note that $\bar{x}$, $\bar{y}$, s_x, and s_y can be found under VARS/5:Statistics (in the VAR-LINK menu on the TI-89).

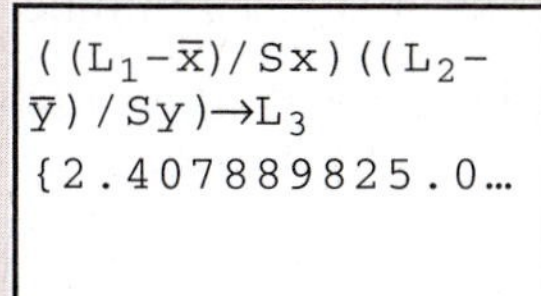

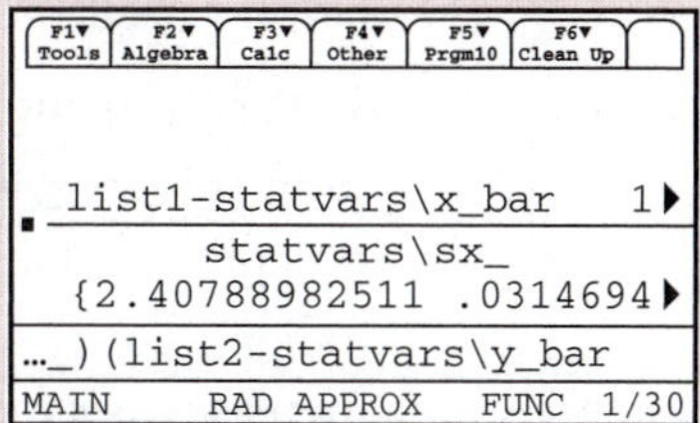

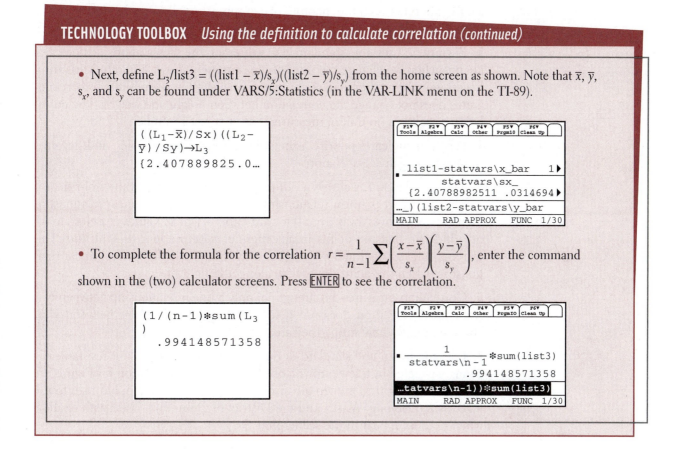

- To complete the formula for the correlation $r = \frac{1}{n-1}\sum\left(\frac{x-\bar{x}}{s_x}\right)\left(\frac{y-\bar{y}}{s_y}\right)$, enter the command shown in the (two) calculator screens. Press ENTER to see the correlation.

```
(1/(n-1)*sum(L3
)
   .994148571358
```

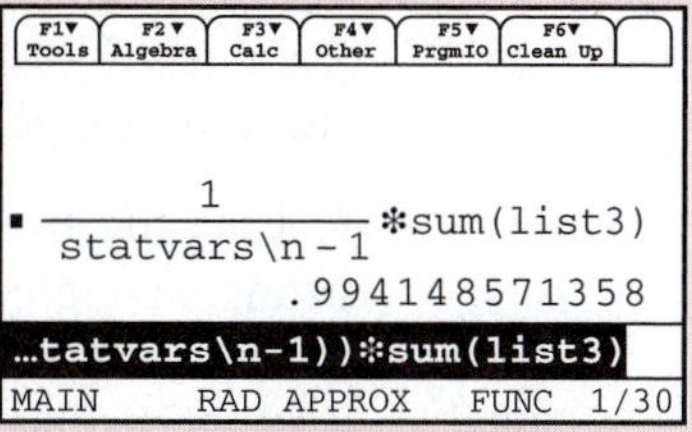

Facts about correlation

The formula for correlation helps us see that r is positive when there is a positive association between the variables. Height and weight, for example, have a positive association. People who are above average in height tend to also be above average in weight. Both the standardized height and the standardized weight are positive. People who are below average in height tend to also have below-average weight. Then both standardized height and standardized weight are negative. In both cases, the products in the formula for r are mostly positive and so r is positive. In the same way, we can see that r is negative when the association between x and y is negative. More detailed study of the formula gives more detailed properties of r. Here is what you need to know in order to interpret correlation.

1. Correlation makes no distinction between explanatory and response variables. It makes no difference which variable you call x and which you call y in calculating the correlation.

2. Correlation requires that both variables be quantitative, so that it makes sense to do the arithmetic indicated by the formula for r. We cannot calculate a correlation between the incomes of a group of people and what city they live in, because city is a categorical variable.

3. Because r uses the standardized values of the observations, r does not change when we change the units of measurement of x, y, or both. Measuring height in inches rather than centimeters and weight in pounds rather than kilograms does not change the correlation between height and weight. The correlation r itself has no unit of measurement; it is just a number.

4. Positive r indicates positive association between the variables, and negative r indicates negative association.

5. The correlation r is always a number between –1 and 1. Values of r near 0 indicate a very weak linear relationship. The strength of the linear relationship increases as r moves away from 0 toward either –1 or 1. Values of r close to –1 or 1 indicate that the points in a scatterplot lie close to a straight line. The extreme values $r = -1$ and $r = 1$ occur only in the case of a perfect linear relationship, when the points lie exactly along a straight line.

6. Correlation measures the strength of only a linear relationship between two variables. Correlation does not describe curved relationships between variables, no matter how strong they are.

7. Like the mean and standard deviation, the correlation is not resistant: r is strongly affected by a few outlying observations. The correlation for Figure 3.7 (page 137) is $r = 0.634$ when all 51 observations are included, but rises to $r = 0.783$ when we omit Alaska and the District of Columbia. Use r with caution when outliers appear in the scatterplot.

The scatterplots in Figure 3.9 illustrate how values of r closer to 1 or –1 correspond to stronger linear relationships. To make the meaning of r clearer, the standard deviations of both variables in these plots are equal and the horizontal and vertical scales are the same. In general, it is not so easy to guess the value of r from the appearance of a scatterplot. Remember that changing the plotting scales in a scatterplot may mislead our eyes, but it does not change the correlation.

The real data we have examined also illustrate how correlation measures the strength and direction of linear relationships. Figure 3.2 (page 128) shows a very strong positive linear relationship between degree-days and natural gas consumption. The correlation is $r = 0.9953$. Check this on your calculator using the data in Table 3.1. Figure 3.1 (page 124) shows a clear but weaker negative association between percent of students taking the SAT and the median SAT Math score in a state. The correlation is $r = -0.868$.

Do remember that **correlation is not a complete description of two-variable data,** even when the relationship between the variables is linear. You should give the means and standard deviations of both x and y along with the correlation. (Because the formula for correlation uses the means and standard deviations, these measures are the proper choice to accompany a correlation.) Conclusions based on correlations alone may require rethinking in the light of a more complete description of the data.

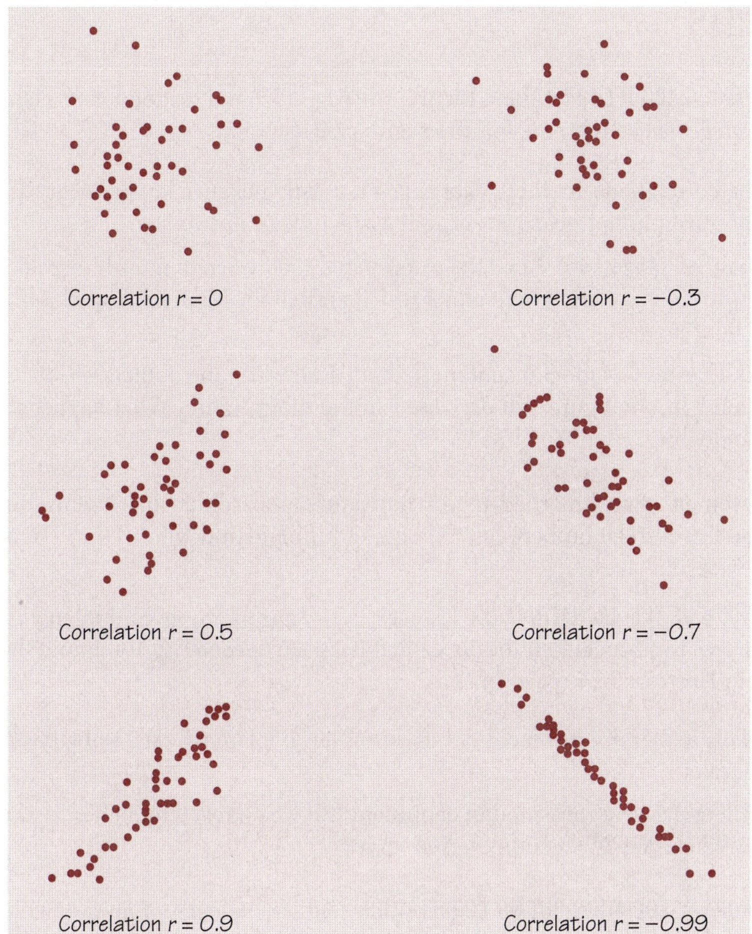

FIGURE 3.9 How correlation measures the strength of a linear relationship. Patterns closer to a straight line have correlations closer to 1 or −1.

EXAMPLE 3.7 SCORING DIVERS

Competitive divers are scored on their form by a panel of judges who use a scale from 1 to 10. The subjective nature of the scoring often results in controversy. We have the scores awarded by two judges, Ivan and George, on a large number of dives. How well do they agree? We do some calculation and find that the correlation between their scores is $r = 0.9$. But the mean of Ivan's scores is 3 points lower than George's mean.

These facts do not contradict each other. They are simply different kinds of information. The mean scores show that Ivan awards much lower scores than George. But because Ivan gives *every* dive a score about 3 points lower than George, the correlation remains high. Adding or subtracting the same number to all values of either x or y does not change the correlation. If Ivan and George both rate several divers, the contest is fairly scored because Ivan and George agree on which dives are better than others. The high r shows their agreement. But if Ivan scores one diver and George another, we must add 3 points to Ivan's scores to arrive at a fair comparison.

EXERCISES

3.25 THINKING ABOUT CORRELATION Figure 3.5 (page 135) is a scatterplot of school grade point average versus IQ score for 78 seventh-grade students.

(a) Is the correlation r for these data near -1, clearly negative but not near -1, near 0, clearly positive but not near 1, or near 1? Explain your answer.

(b) Figure 3.6 (page 136) shows the calories and sodium content in 17 brands of meat hot dogs. Is the correlation here closer to 1 than that for Figure 3.5, or closer to zero? Explain your answer.

(c) Both Figures 3.5 and 3.6 contain outliers. Removing the outliers will *increase* the correlation r in one figure and *decrease* r in the other figure. What happens in each figure, and why?

3.26 If women always married men who were 2 years older than themselves, what would be the correlation between the ages of husband and wife? (*Hint:* Draw a scatterplot for several ages.)

3.27 RETURN OF THE *ARCHAEOPTERYX* Exercise 3.19 (page 138) gives the lengths of two bones in five fossil specimens of the extinct beast *Archaeopteryx*. You found the correlation r in Exercise 3.24 (page 142).

(a) Make a scatterplot if you did not do so earlier. Explain why the value of r matches the scatterplot.

(b) The lengths were measured in centimeters. If we changed to inches, how would r change? (There are 2.54 centimeters in an inch.)

3.28 STRONG ASSOCIATION BUT NO CORRELATION The gas mileage of an automobile first increases and then decreases as the speed increases. Suppose that this relationship is very regular, as shown by the following data on speed (miles per hour) and mileage (miles per gallon):

Speed:	20	30	40	50	60
MPG:	24	28	30	28	24

Make a scatterplot of mileage versus speed. Show that the correlation between speed and mileage is $r = 0$. Explain why the correlation is 0 even though there is a strong relationship between speed and mileage.

SUMMARY

The **correlation** r measures the strength and direction of the linear association between two quantitative variables x and y. Although you can calculate a correlation for any scatterplot, r measures only straight-line relationships.

Correlation indicates the direction of a linear relationship by its sign: $r > 0$ for a positive association and $r < 0$ for a negative association.

Correlation always satisfies $-1 \le r \le 1$ and indicates the strength of a relationship by how close it is to -1 or 1. Perfect correlation, $r = \pm 1$, occurs only when the points on a scatterplot lie exactly on a straight line.

Correlation ignores the distinction between explanatory and response variables. The value of r is not affected by changes in the unit of measurement of either variable. Correlation is not resistant, so outliers can greatly change the value of r.

SECTION 3.2 EXERCISES

3.29 THE PROFESSOR SWIMS Exercise 3.18 (page 137) gives data on the time to swim 2000 yards and the pulse rate after swimming for a middle-aged professor.

(a) If you did not do Exercise 3.18, do it now. Find the correlation r. Explain from looking at the scatterplot why this value of r is reasonable.

(b) Suppose that the times had been recorded in seconds. For example, the time 34.12 minutes would be 2047 seconds. How would the value of r change?

3.30 BODY MASS AND METABOLIC RATE Exercise 3.12 (page 132) gives data on the lean body mass and metabolic rate for 12 women and 7 men.

(a) Make a scatterplot if you did not do so in Exercise 3.12. Use different symbols or colors for women and men. Do you think the correlation will be about the same for men and women or quite different for the two groups? Why?

(b) Calculate r for women alone and also for men alone. (Use your calculator.)

(c) Calculate the mean body mass for the women and for the men. Does the fact that the men are heavier than the women on the average influence the correlations? If so, in what way?

(d) Lean body mass was measured in kilograms. How would the correlations change if we measured body mass in pounds? (There are about 2.2 pounds in a kilogram.)

3.31 HOW MANY CALORIES? Exercise 3.20 (page 138) gives data on the true calorie counts in ten foods and the average guesses made by a large group of people.

(a) Make a scatterplot if you did not do so in Exercise 3.20. Then calculate the correlation r (use your calculator). Explain why your r is reasonable based on the scatterplot.

(b) The guesses are all higher than the true calorie counts. Does this fact influence the correlation in any way? How would r change if every guess were 100 calories higher?

(c) The guesses are much too high for spaghetti and snack cake. Circle these points on your scatterplot. Calculate r for the other eight foods, leaving out these two points. Explain why r changed in the direction that it did.

3.32 BRAIN SIZE AND IQ SCORE Do people with larger brains have higher IQ scores? A study looked at 40 volunteer subjects, 20 men and 20 women. Brain size was measured by magnetic resonance imaging. Table 3.3 gives the data. The MRI count is the number of "pixels" the brain covered in the image. IQ was measured by the Wechsler test.[13]

TABLE 3.3 Brain size (MRI count) and IQ score

Men				Women			
MRI	IQ	MRI	IQ	MRI	IQ	MRI	IQ
1,001,121	140	1,038,437	139	816,932	133	951,545	137
965,353	133	904,858	89	928,799	99	991,305	138
955,466	133	1,079,549	141	854,258	92	833,868	132
924,059	135	945,088	100	856,472	140	878,897	96
889,083	80	892,420	83	865,363	83	852,244	132
905,940	97	955,003	139	808,020	101	790,619	135
935,494	141	1,062,462	103	831,772	91	798,612	85
949,589	144	997,925	103	793,549	77	866,662	130
879,987	90	949,395	140	857,782	133	834,344	83
930,016	81	935,863	89	948,066	133	893,983	88

Source: There are some of the data from the EESEE story "Brain Size and Intelligence." The study is described in L. Willerman, R. Schultz, J. N. Rutledge, and E. Bigler, "In vivo brain size and intelligence," *Intelligence*, 15 (1991), pp. 223–228.

(a) Make a scatterplot of IQ score versus MRI count, using distinct symbols for men and women. In addition, find the correlation between IQ and MRI for all 40 subjects, for the men alone, and for the women alone.

(b) Men are larger than women on the average, so they have larger brains. How is this size effect visible in your plot? Find the mean MRI count for men and women to verify the difference.

(c) Your result in (b) suggests separating men and women in looking at the relationship between brain size and IQ. Use your work in (a) to comment on the nature and strength of this relationship for women and for men.

3.33 Changing the units of measurement can dramatically alter the appearance of a scatterplot. Consider the following data:

x	−4	−4	−3	3	4	4
y	0.5	−0.6	−0.5	0.5	0.5	−0.6

(a) Enter the data into L_1/list1 and L_2/list2. Then use Plot1 to define and plot the scatterplot. Use the box (□) as your plotting symbol.

(b) Use L_3/list3 and the technique described in the Technology Toolbox on page 142 to calculate the correlation.

(c) Define new variables $x^* = x/10$ and $y^* = 10y$, and enter these into L_4/list4 and L_5/list5 as follows: list4 = list1/10 and list5 = 10 × list2. Define Plot2 to be a scatterplot with Xlist: list4 and Ylist: list5, and Mark: +. Plot both scatterplots at the same time, and on the same axes, using ZoomStat/ZoomData. The two plots are very different in appearance.

(d) Use L_6/list6 and the technique described in the Technology Toolbox to calculate the correlation between x^* and y^*. How are the two correlations related? Explain why this isn't surprising.

3.34 TEACHING AND RESEARCH A college newspaper interviews a psychologist about student ratings of the teaching of faculty members. The psychologist says, "The evidence indicates that the correlation between the research productivity and teaching rating of faculty members is close to zero." The paper reports this as "Professor McDaniel said that good researchers tend to be poor teachers, and vice versa." Explain why the paper's report is wrong. Write a statement in plain language (don't use the word "correlation") to explain the psychologist's meaning.

3.35 INVESTMENT DIVERSIFICATION A mutual fund company's newsletter says, "A well-diversified portfolio includes assets with low correlations." The newsletter includes a table of correlations between the returns on various classes of investments. For example, the correlation between municipal bonds and large-cap stocks is 0.50 and the correlation between municipal bonds and small-cap stocks is 0.21.[12]

(a) Rachel invests heavily in municipal bonds. She wants to diversify by adding an investment whose returns do not closely follow the returns on her bonds. Should she choose large-cap stocks or small-cap stocks for this purpose? Explain your answer.

(b) If Rachel wants an investment that tends to increase when the return on her bonds drops, what kind of correlation should she look for?

3.36 DRIVING SPEED AND FUEL CONSUMPTION The data in Exercise 3.28 were made up to create an example of a strong curved relationship for which, nonetheless, $r = 0$. Exercise 3.11 (page 129) gives actual data on gas used versus speed for a small car. Make a scatterplot if you did not do so in Exercise 3.11. Calculate the correlation, and explain why r is close to 0 despite a strong relationship between speed and gas used.

3.37 SLOPPY WRITING ABOUT CORRELATION Each of the following statements contains a blunder. Explain in each case what is wrong.

(a) "There is a high correlation between the gender of American workers and their income."

(b) "We found a high correlation ($r = 1.09$) between students' ratings of faculty teaching and ratings made by other faculty members."

(c) "The correlation between planting rate and yield of corn was found to be $r = 0.23$ bushel."

3.3 LEAST-SQUARES REGRESSION

Correlation measures the strength and direction of the linear relationship between any two quantitative variables. If a scatterplot shows a linear relationship, we would like to summarize this overall pattern by drawing a line through the scatterplot. *Least-squares regression* is a method for finding a line that summarizes the relationship between two variables, but only in a specific setting.

REGRESSION LINE

A **regression line** is a straight line that describes how a response variable y changes as an explanatory variable x changes. We often use a regression line to predict the value of y for a given value of x. Regression, unlike correlation, requires that we have an explanatory variable and a response variable.

model

The least-squares regression line, which we will occasionally abbreviate LSRL, is a ***model***—or more formally, a ***mathematical model***—for the data. If we believe that the data show a linear trend, then it would be appropriate to try to fit an LSRL to the data. In the next chapter, we will explore data that are not linear and for which a curve is a more appropriate model. At the beginning, though, we will focus our discussion on linear trends.

EXAMPLE 3.8 PREDICTING NATURAL GAS CONSUMPTION

A scatterplot shows that there is a strong linear relationship between the average outside temperature (measured by heating degree-days) in a month and the average amount of natural gas that the Sanchez household uses per day during the month. The Sanchez household wants to use this relationship to predict their natural gas consumption. "If a month averages 20 degree-days per day (that's 45° F), how much gas will we use?

prediction

In Figure 3.10 we have drawn a regression line on the scatterplot. To use this line to ***predict*** gas consumption at 20 degree-days, first locate 20 on the x axis. Then go "up

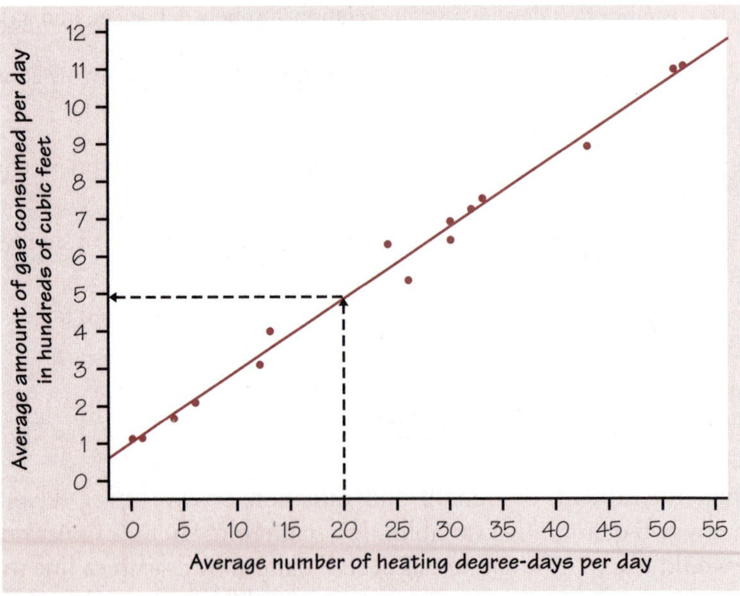

FIGURE 3.10 The Sanchez household gas consumption data, with a regression line for predicting gas consumption from degree-days. The dashed lines illustrate how to use the regression line to predict gas consumption for a month averaging 20 degree-days per day.

and over" as in the figure to find the gas consumption y that corresponds to $x = 20$. We predict that the Sanchez household will use about 4.9 hundreds of cubic feet of gas each day in such a month.

The least-squares regression line

Different people might draw different lines by eye on a scatterplot. This is especially true when the points are more widely scattered than those in Figure 3.10. We need a way to draw a regression line that doesn't depend on our guess as to where the line should go. No line will pass exactly through all the points, so we want one that is as close as possible. We will use the line to predict y from x, so we want a line that is as close as possible to the points in the *vertical* direction. That's because the prediction errors we make are errors in y, which is the vertical direction in the scatterplot. If we predict 4.9 hundreds of cubic feet for a month with 20 degree-days and the actual usage turns out to be 5.1 hundreds of cubic feet, our error is

$$\begin{aligned} \text{error} &= \text{observed} - \text{predicted} \\ &= 5.1 - 4.9 = 0.2 \end{aligned}$$

We want a regression line that makes the vertical distances of the points in a scatterplot from the line as small as possible. Figure 3.11(a) illustrates the idea. For clarity, the plot shows only three of the points from Figure 3.10, along with the line, on an expanded scale. The line passes above two of the points and below one of them. The vertical distances of the data points from the line appear as vertical line segments. A "good" regression line makes these distances as small as possible. There are many ways to make "as small as possible" precise. The most common is the *least-squares* idea.

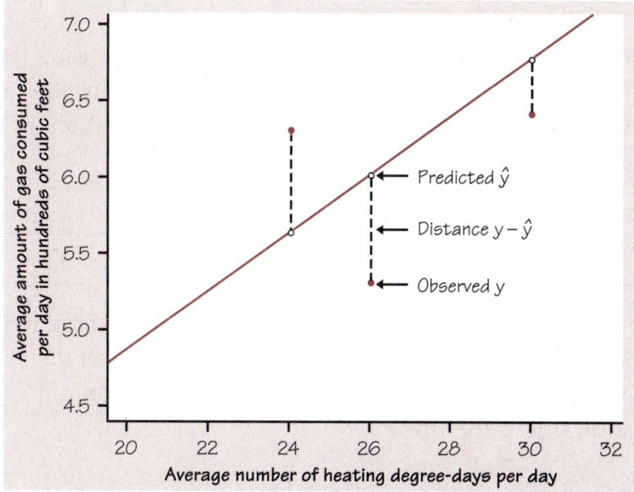

FIGURE 3.11(a) The least-squares idea. For each observation, find the vertical distance of each point on the scatterplot from a regression line. The least-squares regression line makes the sum of the squares of these distances as small as possible.

LEAST-SQUARES REGRESSION LINE

The **least-squares regression line** of y on x is the line that makes the sum of the squares of the vertical distances of the data points from the line as small as possible.

Figure 3.11(b) gives a geometric interpretation to the phrase "sum of the squares of the vertical distances of the data points from the line."

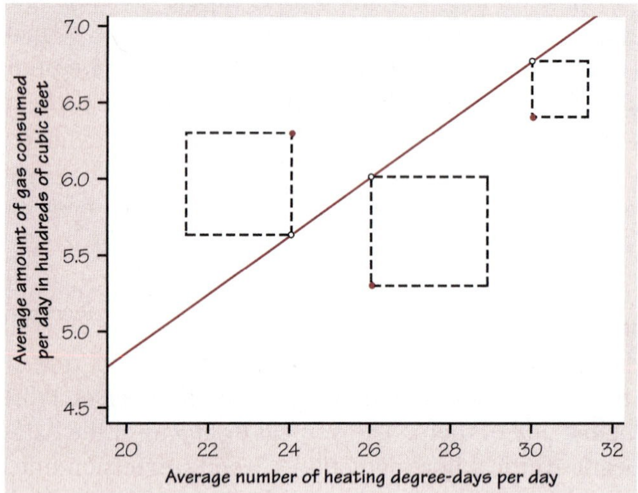

FIGURE 3.11(b) Equivalently, the least-squares regression line is the line that minimizes the total *area* in the squares.

One reason for the popularity of the least-squares regression line is that the problem of finding the line has a simple answer. We can give the recipe for the least-squares line in terms of the means and standard deviations of the two variables and their correlation.

EQUATION OF THE LEAST-SQUARES REGRESSION LINE

We have data on an explanatory variable x and a response variable y for n individuals. From the data, calculate the means $\bar{x}$, and $\bar{y}$ and the standard deviations s_x and s_y of the two variables, and their correlation r. The least-squares regression line is the line

$$\hat{y} = a + bx$$

EQUATION OF THE LEAST-SQUARES REGRESSION LINE *(continued)*

with **slope**

$$b = r\frac{s_y}{s_x}$$

and **intercept**

$$a = \overline{y} - b\overline{x}$$

Although you are probably used to the form $y = mx + b$ for the equation of a line from your study of algebra, statisticians have adopted $\hat{y} = a + bx$ as the form for the equation of the least-squares line. We will adopt this form, too, in the interest of good communication. The variable y denotes the *observed* value of y, and the term $\hat{y}$ means the *predicted* value of y. We write $\hat{y}$ (read "y hat") in the equation of the regression line to emphasize that the line gives a predicted response $\hat{y}$ for any x. When you are solving regression problems, make sure you are careful to distinguish between y and $\hat{y}$.

To determine the equation of a least-squares line, we need to solve for the intercept a and the slope b. Since there are two unknowns, we need two conditions in order to solve for the two unknowns. It can be shown that *every* least-squares regression line passes through the point $(\overline{x}, \overline{y})$. This is one important piece of information about the least-squares line. The other fact that is known is that the slope of the least-squares line is equal to the product of the correlation and the quotient of the standard deviations:

$$b = r\frac{s_y}{s_x}$$

Commit these two facts to memory, and you will be able to find equations of least-squares lines.

EXAMPLE 3.9 CONSTRUCTING THE LEAST-SQUARES EQUATION

Suppose we have explanatory and response variables and we know that $\overline{x} = 17.222$, $\overline{y} = 161.111$, $s_x = 19.696$, $s_y = 33.479$, and the correlation $r = 0.997$. Even though we don't know the actual data, we can still construct the equation for the least-squares line and use it to make predictions. The slope and intercept can be calculated as

$$b = r\frac{s_y}{s_x} = 0.997\frac{33.479}{19.696} = 1.695$$

$$a = \overline{y} - b\overline{x} = 161.111 - (1.695)(17.222) = 131.920$$

so that the least-squares line has equation $\hat{y} = 131.920 + 1.695x$

In practice, you don't need to calculate the means, standard deviations, and correlation first. Statistical software or your calculator will give the slope b and intercept a of the least-squares line from keyed-in values of the variables x and y. You can then concentrate on understanding and using the regression line.

TECHNOLOGY TOOLBOX *Least-squares lines on the calculator*

We will use the gas consumption and degree-days data from Example 3.8 to show how to use the TI-83/89 to determine the equation of the least-squares line.

- Enter the degree-days data into L_1/list1 and the gas consumption data into L_2/list2. (Recall that you saved these lists as DEGDA and GAS, respectively.) Refer to the Technology Toolbox on page 132 for details on copying these lists of data into L_1/list1 and L_2/list2.
- Define a scatterplot using L_1/list1 and L_2/list2, and then use ZoomStat (ZoomData) to plot the scatterplot.

TI-83 **TI-89**

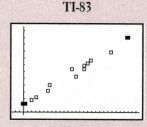

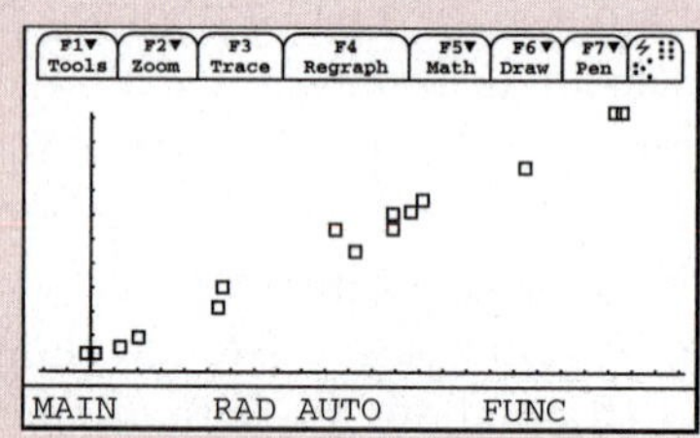

To determine the LSRL:

- Press STAT, choose CALC, then `8:LinReg (a+bx)`. Finish the command to read `LinReg (a+bx) L1, L2, Y1`. (Y_1 is found under VARS/Y-VARS/1:Function.)
- In the Statistics/ListEditor, press F4 (CALC), choose `3: Regressions`, then `1:LinReg(a+bx)`.
- Enter list1 for the Xlist, list2 for the Ylist, choose to store the RegEqn to y1(x) and press ENTER.

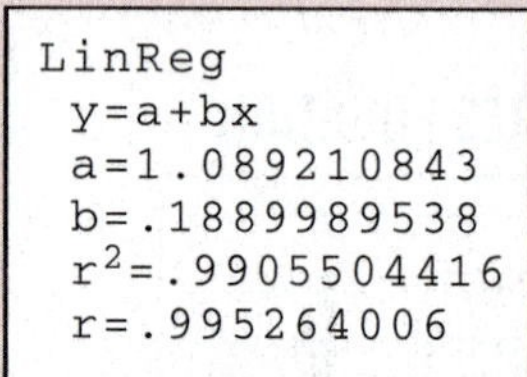

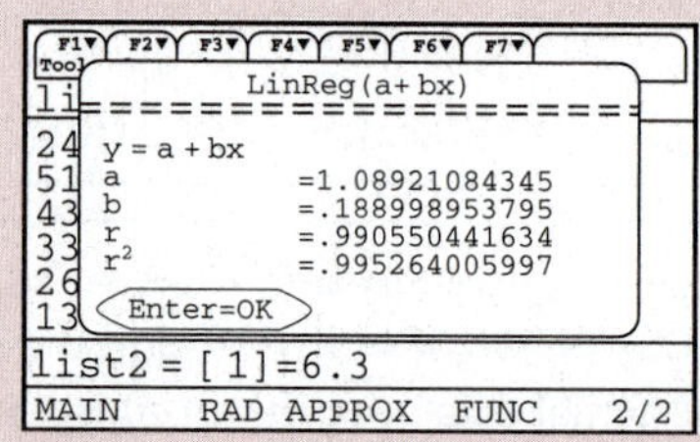

Note: If r^2 and r do not appear on your TI-83 screen, then do this one-time series of keystrokes: Press 2nd 0 (CATALOG), scroll down to `DiagnosticOn` and press ENTER. Press ENTER again to execute the command. The screen should say "Done." Then press 2nd ENTER (ENTRY) to recall the regression command and ENTER again to calculate the LSRL. The r^2- and r-values should now appear.

TECHNOLOGY TOOLBOX *Least-squares lines on the calculator (continued)*

- Deselect all other equations in the Y=screen and press GRAPH (◆ F3 on the TI-89) to overlay the LSRL on the scatterplot.

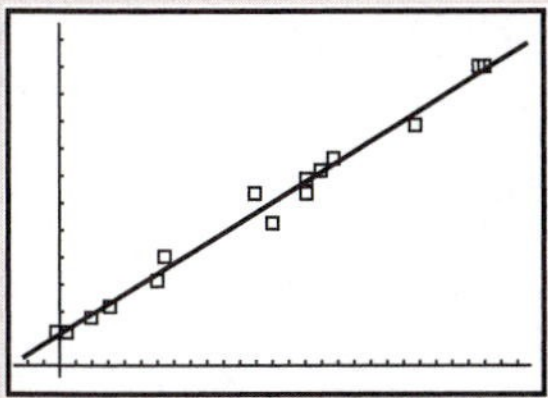

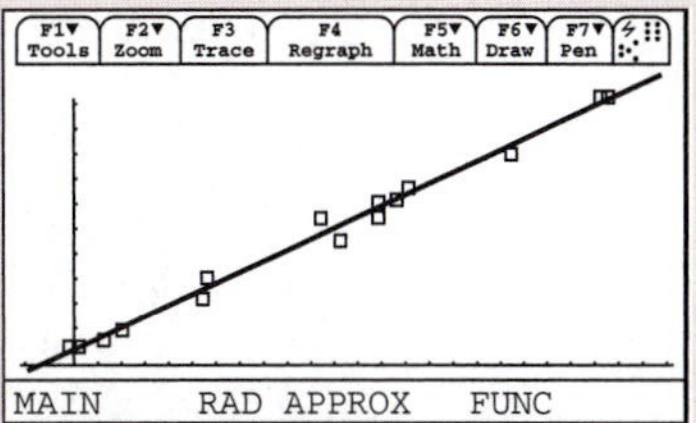

Although the calculator will report the values for *a* and *b* to nine decimal places, we usually round off to four decimal places. You would write the LSRL equation as

$$\hat{y} = 1.0892 + 0.1890x$$

When you write the equation, don't forget the hat symbol over the *y*; this means *predicted value*.

Figure 3.12 displays the regression output for the gas consumption data from two statistical software packages. Each output records the slope and intercept of the least-squares line, calculated to more decimal places than we need. The software also provides information that we do not yet need—part of the art of using software is to ignore the extra information that is almost always present. We will make use of other parts of the output in Chapters 14 and 15.

slope

The ***slope*** of a regression line is usually important for the interpretation of the data. The slope is the rate of change, the amount of change in $\hat{y}$ when x increases by 1. The slope $b = 0.1890$ in this example says that, on the average, each additional degree-day predicts consumption of 0.1890 more hundreds of cubic feet of natural gas per day.

intercept

The ***intercept*** of the regression line is the value of $\hat{y}$ when $x = 0$. Although we need the value of the intercept to draw the line, it is statistically meaningful only when x can actually take values close to zero. In our example, $x = 0$ occurs when the average outdoor temperature is at least 65° F. We predict that the Sanchez household will use an average of $a = 1.0892$ hundreds of cubic feet of gas per day when there are no degree-days. They use this gas for cooking and heating water, which continue in warm weather.

The equation of the regression line makes prediction easy. Just substitute an x-value into the equation. To predict gas consumption at 20 degree-days, substitute $x = 20$.

$$\begin{aligned}\hat{y} &= 1.0892 + (0.1890)(20) \\ &= 1.0892 + 3.78 = 4.869\end{aligned}$$

The regression equation is
Gas Used = 1.09 + 0.189 D-days

Predictor	Coef	Stdev	t-ratio	p
Constant	1.0892	0.1389	7.84	0.000
D-days	0.188999	0.004934	38.31	0.000

s = 0.3389 R-sq = 99.1% R-sq(adj) = 99.0%

Analysis of Variance

SOURCE	DF	SS	MS	F	p
Regression	1	168.58	168.58	1467.55	0.000
Error	14	1.61	0.11		
Total	15	170.19			

(a)

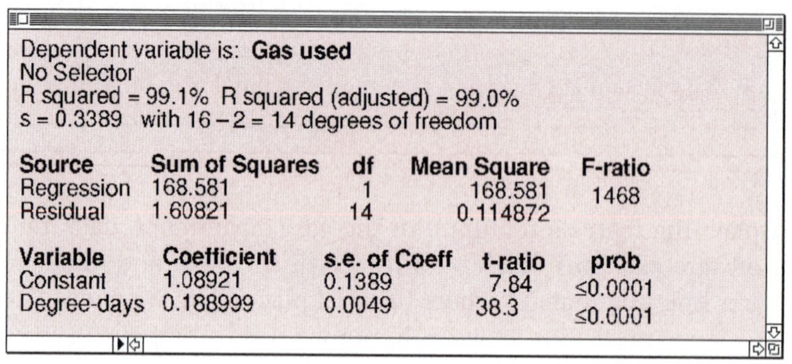

Dependent variable is: **Gas used**
No Selector
R squared = 99.1% R squared (adjusted) = 99.0%
s = 0.3389 with 16 − 2 = 14 degrees of freedom

Source	**Sum of Squares**	**df**	**Mean Square**	**F-ratio**
Regression	168.581	1	168.581	1468
Residual	1.60821	14	0.114872	

Variable	**Coefficient**	**s.e. of Coeff**	**t-ratio**	**prob**
Constant	1.08921	0.1389	7.84	≤0.0001
Degree-days	0.188999	0.0049	38.3	≤0.0001

(b)

FIGURE 3.12 Least-squares regression output for the gas consumption data from two statistical software packages: (a) Minitab and (b) Data Desk.

plot the line

To ***plot the line*** on the scatterplot by hand, use the equation to find $\hat{y}$ for two values of x, one near each end of the range of x in the data. Plot each $\hat{y}$ above its x and draw the line through the two points.

EXERCISES

3.38 GAS CONSUMPTION The Technology Toolbox (page 154) gives the equation of the regression line of gas consumption y on degree-days x for the data in Table 3.1 as

$$\hat{y} = 1.0892 + 0.1890x$$

Use your calculator to find the mean and standard deviation of both x and y and their correlation r. Find the slope b and the intercept a of the regression line from these, using the facts in the box *Equation of the least-squares regression line.* (page 152) Verify that you get the equation above. (Results may differ slightly because of rounding off.)

3.39 ARE SAT SCORES CORRELATED? If you previously plotted a scatterplot for the ordered-pairs (Math SAT scores, Verbal SAT scores) data collected by the class in Activity 3, then ask yourself, "Do these data describe a linear trend?" If so, then use your calculator to determine the LSRL equation and correlation coefficient. Overlay this regression line on your scatterplot. Considering the appearance of the scatterplot, the regression line, and the correlation, write a brief statement about the appropriateness of this regression line to model the data. Is the line useful?

3.40 ACID RAIN Researchers studying acid rain measured the acidity of precipitation in a Colorado wilderness area for 150 consecutive weeks. Acidity is measured by pH. Lower pH values show higher acidity. The acid rain researchers observed a linear pattern over time. They reported that the least-squares regression line

$$\text{pH} = 5.43 - (0.0053 \times \text{weeks})$$

fit the data well.[13]

(a) Draw a graph of this line. Is the association positive or negative? Explain in plain language what this association means.

(b) According to the regression line, what was the pH at the beginning of the study (weeks = 1)? At the end (weeks = 150)?

(c) What is the slope of the regression line? Explain clearly what this slope says about the change in the pH of the precipitation in this wilderness area.

3.41 THE ENDANGERED MANATEE Exercise 3.6 (page 125) gives data on the number of powerboats registered in Florida and the number of manatees killed by boats in the years from 1977 to 1990.

(a) Use your calculator to make a scatterplot of these data.

(b) Find the equation of the least-squares line and overlay that line on your scatterplot.

(c) Predict the number of manatees that will be killed by boats in a year when 716,000 powerboats are registered.

(d) Here are four more years of manatee data, in the same form as in Exercise 3.6:

1991	716	53	1993	716	35
1992	716	38	1994	735	49

Add these points to your scatterplot. Florida took stronger measures to protect manatees during these years. Do you see any evidence that these measures succeeded?

(e) In part (c) you predicted manatee deaths in a year with 716,000 powerboat registrations. In fact, powerboat registrations were 716,000 for three years. Compare the mean manatee deaths in these three years with your prediction from part (c). How accurate was your prediction?

The role of r^2 in regression

Calculator and computer output for regression report a quantity called r^2. Some computer packages call it "R-sq." For examples, look at the calculator screen shots in the Technology Toolbox on page 154 and the computer output in Figure 3.12(a) on page 156. Although it is true that this quantity is equal to the square of r, there is much more to this story.

To illustrate the meaning of r^2 in regression, the next two examples use two simple data sets and in each case calculate the quantity r^2. In the first example, a line would be a poor model, and the r^2-value turns out to be small (closer to 0). In the second example, a straight line would fit the data fairly well, and the r^2 value is larger (closer to 1).

EXAMPLE 3.10 SMALL r^2

One way to determine the usefulness of the least-squares regression model is to measure the contribution of x in predicting y. A simple example will help clarify the reasoning. Consider data set A:

x	0	3	6
y	0	10	2

and its scatterplot in Figure 3.13(a). The association between x and y appears to be positive but weak. The sample means are easily calculated to be $\bar{x} = 3$ and $\bar{y} = 4$. Knowing that x is 0 or 3 or 6 gives us very little information to predict y, and so we have to fall back to $\bar{y}$ as a predictor of y. The deviations of the three points about the mean $\bar{y}$ are shown in Figure 3.13(b). The horizontal line in Figure 3.13(b) is at height $\bar{y} = 4$. The sum of the squares of the deviations for the prediction equation $\hat{y} = \bar{y}$ is

$$\text{SST} = \sum (y - \bar{y})^2$$

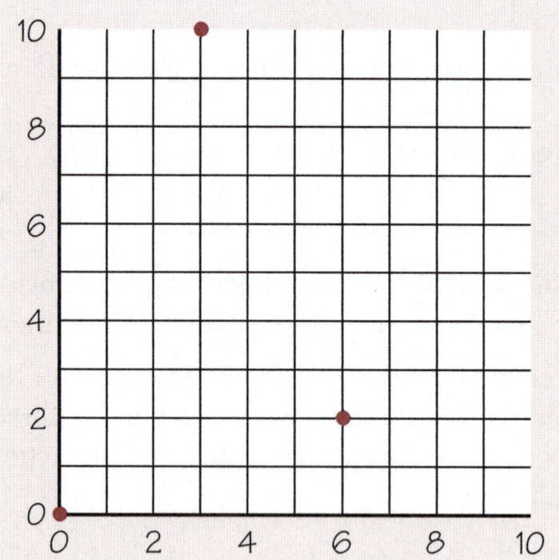

FIGURE 3.13(a) Scatterplot for data set A.

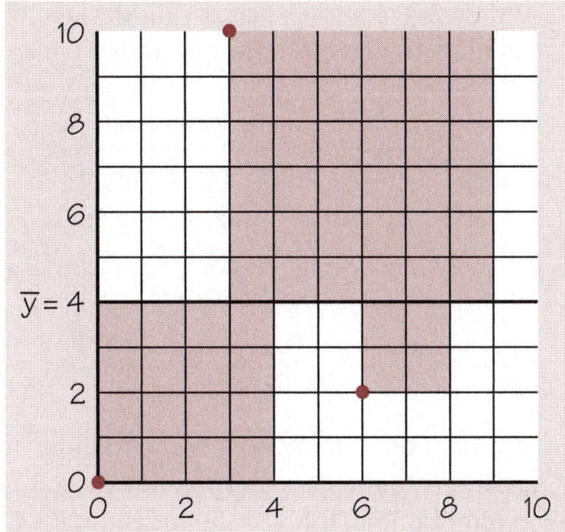

FIGURE 3.13(b) Squares of deviations about $\bar{y}$.

Geometric squares have been constructed on the graph with the deviations from the mean as one side. The total area of these three squares is a measure of the total sample variability. So we call this quantity SST for "total sum of squares about the mean $\overline{y}$."

The LSRL has equation $\hat{y} = 3 + (1/3)x$; see Figure 3.13(c). It has y intercept 3 and passes through the point $(\overline{x}, \overline{y}) = (3, 4)$.

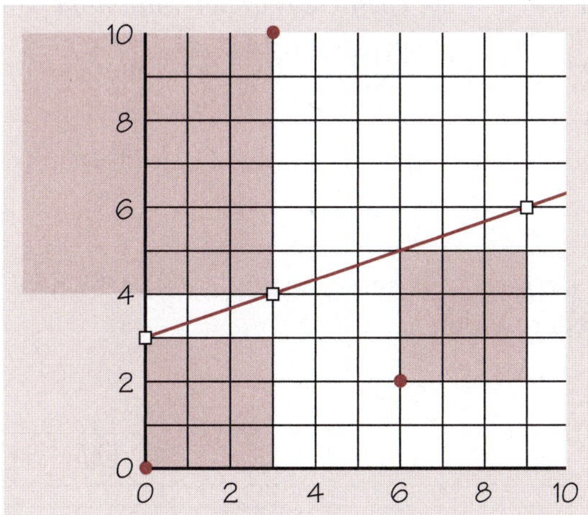

FIGURE 3.13(c) Squares of deviations about $\hat{y}$.

Now we want to consider the sum of the squares of the deviations of the points about this regression line. We call this SSE for "sum of squares for error."

$$\text{SSE} = \sum (y - \hat{y})^2$$

Figure 3.13(c) also shows geometric squares with deviations from the regression line as one side. The calculations can be summarized in a table:

x	y	$(y - \bar{y})^2$	$(y - \hat{y})^2$
0	0	16	9
3	10	36	36
6	2	4	9
		56	54
		SST	SSE

If x is a poor predictor of y, then the sum of squares of deviations about the mean $\bar{y}$ and the sum of squares of deviations about the regression line $\hat{y}$ would be approximately the same. This is the case in our example. If SST = 56 measures the total sample variation of the observations about the mean $\bar{y}$, then SSE = 54 is the remaining "unexplained sample variability" after fitting the regression line. The difference, SST – SSE, measures the amount of variation of y that can be explained by the regression line of y on x. The ratio of these two quantities

$$\frac{\text{SST} - \text{SSE}}{\text{SST}}$$

is interpreted as *the proportion of the total sample variability that is explained by the least-squares regression of y on x*. It can be shown algebraically that this fraction is equal to the square of the correlation coefficient. For this reason, we call this fraction r^2 and refer to it as the ***coefficient of determination.*** For data set A,

coefficient of determination

$$r^2 = \frac{\text{SST} - \text{SSE}}{\text{SST}} = \frac{56 - 54}{56} = 0.0357$$

We say that 3.57% of the variation in y is explained by least-squares regression of y on x.

For contrast, the next example shows a simple data set where the least-squares line is a much better model.

EXAMPLE 3.11 LARGE r^2

Consider data set B and its accompanying scatterplot in Figure 3.14(a):

x	0	5	10
y	0	7	8

The association between x and y appears to be positive and strong. The sample means are $\bar{x} = 5$ and $\bar{y} = 5$. The squares of the deviations about the mean $\bar{y}$ are shown in Figure 3.14(b), and the squares of the deviations about the regression line $\hat{y}$ are shown in Figure 3.14(c).

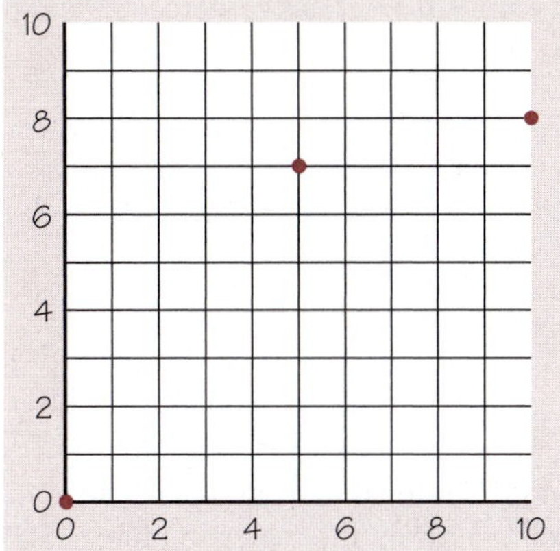

FIGURE 3.14(a) Scatterplot for data set B.

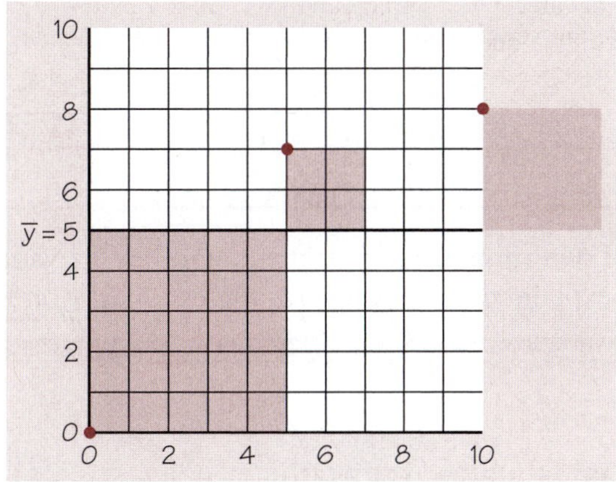

FIGURE 3.14(b) Squares of deviations about $\bar{y}$.

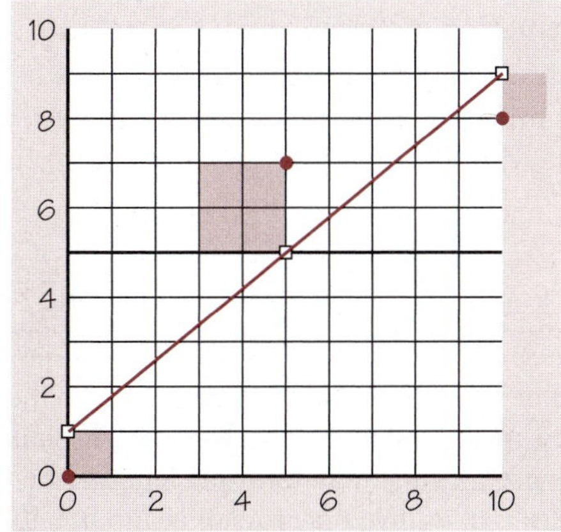

FIGURE 3.14(c) Squares of deviations about $\hat{y}$.

The LSRL has equation $\hat{y} = 1 + 0.8x$. It has y intercept 1 and passes through the points $(\bar{x}, \bar{y}) = (5,5)$ and (10,9). Here are the calculations:

x	y	$(y - \bar{y})^2$	$(y - \hat{y})^2$
0	0	25	1
5	7	4	4
10	8	9	1
		38	6
		SST	SSE

If x is a good predictor of y, then the deviations and hence the SSE would be small; in fact, if all of the points fell exactly on the regression line, SSE would be 0. For data set B, we have

$$r^2 = \frac{\text{SST} - \text{SSE}}{\text{SST}} = \frac{38 - 6}{38} = 0.842$$

We say that 84% of the variation in y is explained by least-squares regression of y on x.

r^2 IN REGRESSION

The **coefficient of determination, r^2**, is the fraction of the variation in the values of y that is explained by least-squares regression of y on x.

Facts about least-squares regression

Regression is one of the most common statistical settings, and least-squares is the most common method for fitting a regression line to data. Here are some facts about least-squares regression lines.

Fact 1. The distinction between explanatory and response variables is essential in regression. Least-squares regression looks at the distances of the data points from the line only in the y direction. If we reverse the roles of the two variables, we get a different least-squares regression line.

EXAMPLE 3.12 THE EXPANDING UNIVERSE

Figure 3.15 is a scatterplot of data that played a central role in the discovery that the universe is expanding. They are the distances from earth of 24 spiral galaxies and the speed at which these galaxies are moving away from us, reported by the astronomer Edwin Hubble in 1929.[14] There is a positive linear relationship, $r = 0.7842$, so that more distant galaxies are moving away more rapidly. Astronomers believe that there is in fact a perfect linear relationship, and that the scatter is caused by imperfect measurements.

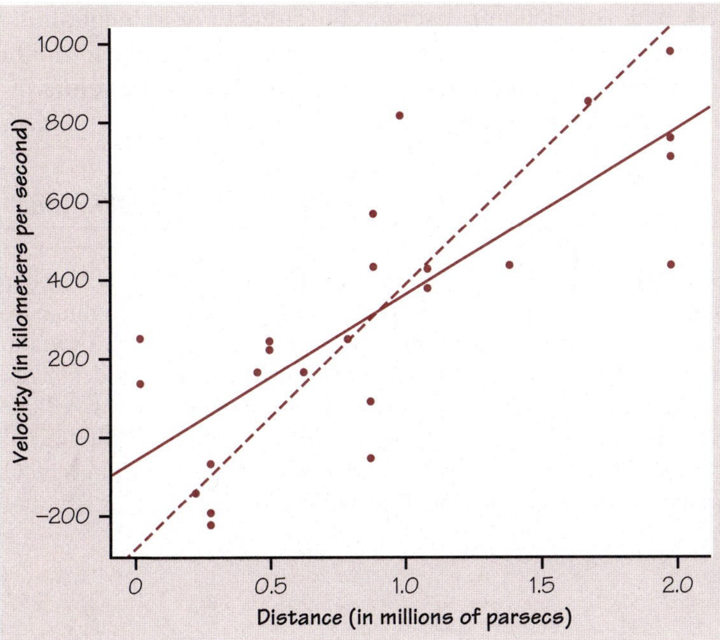

FIGURE 3.15 Scatterplot of Hubble's data on the distance from earth of 24 galaxies and the velocity at which they are moving away from us. The two lines are the two least-squares regression lines: of velocity on distance (solid) and of distance on velocity (dashed).

The two lines on the plot are the two least-squares regression lines. The regression line of velocity on distance is solid. The regression line of distance on velocity is dashed. *Regression of velocity on distance and regression of distance on velocity give different lines.* In the regression setting you must know clearly which variable is explanatory.

Fact 2. There is a close connection between correlation and the slope of the least-squares line. The slope is

$$b = r\frac{s_y}{s_x}$$

This equation says that along the regression line, **a change of one standard deviation in x corresponds to a change of r standard deviations in y.** When the variables are perfectly correlated ($r = 1$ or $r = -1$), the change in the predicted response $\hat{y}$ is the same (in standard deviation units) as the change in x. Otherwise, because $-1 \leq r \leq 1$, the change in $\hat{y}$ is less than the change in x. As the correlation grows less strong, the prediction $\hat{y}$ moves less in response to changes in x.

Fact 3. The least-squares regression line always passes through the point $(\bar{x}, \bar{y})$ on the graph of y against x. So the least-squares regression line of y on x is the line with slope rs_y/s_x that passes through the point $(\bar{x}, \bar{y})$. We can describe regression entirely in terms of the basic descriptive measures $\bar{x}$, s_x, $\bar{y}$, s_y, and r.

Fact 4. The correlation r describes the strength of a straight-line relationship. In the regression setting, this description takes a specific form: **the square of the correlation, r^2, is the fraction of the variation in the values of y that is explained by the least-squares regression of y on x.**

EXAMPLE 3.13 COMPARING r^2 VALUES

First consider the Sanchez gas consumption data in Figure 3.16(a). There is a lot of variation in the observed y's, the gas consumption data. They range from a low of about 1 to a high of 11. The scatterplot shows that most of this variation in y is accounted for by the fact that outdoor temperature (measured by degree-days x) was changing and pulled gas consumption along with it. There is only a little remaining variation in y, which appears in the scatter of points about the line.The correlation is very strong: $r = 0.9953$, and $r^2 = 0.9906$. Our interpretation is that over 99% of the variation in gas consumption is accounted for by the linear relationship with degree-days.

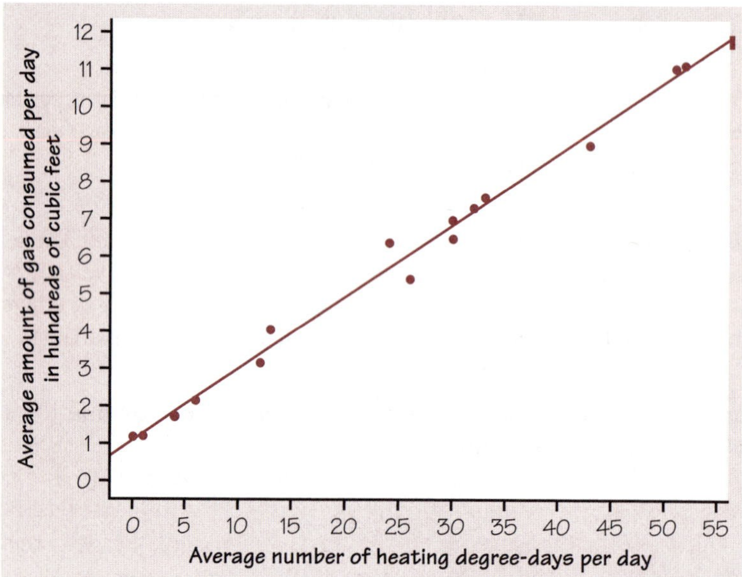

FIGURE 3.16(a) The Sanchez household gas consumption data.

The points in Figure 3.16(b), on the other hand, are more scattered. Linear dependence on distance does explain some of the observed variation in velocity. You would guess a higher value for the velocity y knowing that $x = 2$ than you would if you were told that $x = 0$. But there is still considerable variation in y even when x is held fixed—look at the four points in Figure 3.16(b) with $x = 2$. For the Hubble data, $r = 0.7842$ and $r^2 = 0.6150$. The linear relationship between distance and velocity explains 61.5% of the variation *in either variable.* There are two regression lines, but just one correlation, and r^2 helps interpret both regressions.

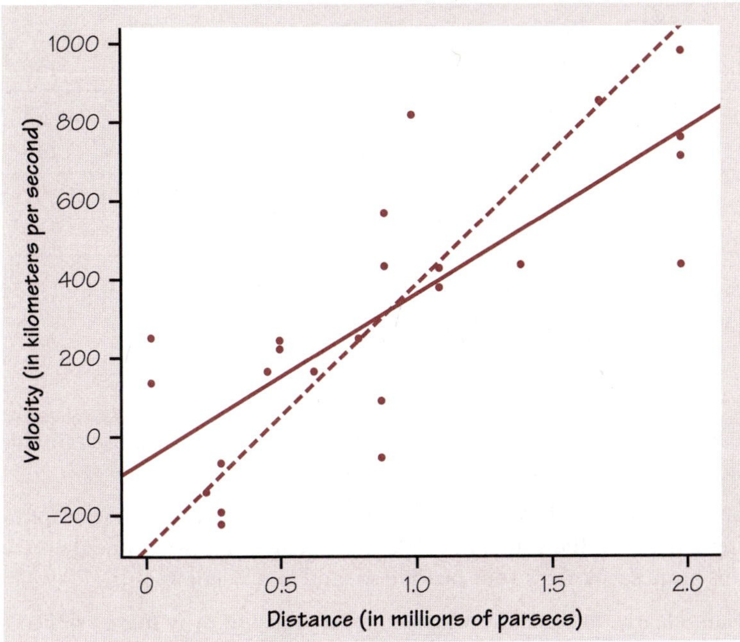

FIGURE 3.16(b) Hubble's data on the distance from earth of 24 galaxies and the velocity at which they are moving away from us.

When you report a regression, give r^2 as a measure of how successful the regression was in explaining the response. When you see a correlation, square it to get a better feel for the strength of the association. Perfect correlation ($r = -1$ or $r = 1$) means the points lie exactly on a line. Then $r^2 = 1$ and all of the variation in one variable is accounted for by the linear relationship with the other variable. If $r = -0.7$ or $r = 0.7$, $r^2 = 0.49$ and about half the variation is accounted for by the linear relationship. In the r^2 scale, correlation ± 0.7 is about halfway between 0 and ± 1.

These connections with correlation are special properties of least-squares regression. They are not true for other methods of fitting a line to data. Another reason that least-squares is the most common method for fitting a regression line to data is that it has many of these convenient special properties.

EXERCISES

3.42 CLASS ATTENDANCE AND GRADES A study of class attendance and grades among first-year students at a state university showed that in general students who attended a higher percent of their classes earned higher grades. Class attendance explained 16% of the variation in grade index among the students. What is the numerical value of the correlation between percent of classes attended and grade index?

3.43 THE PROFESSOR SWIMS Here are Professor Moore's times (in minutes) to swim 2000 yards and his pulse rate after swimming (in beats per minute) for 23 sessions in the pool:

Time:	34.12	35.72	34.72	34.05	34.13	35.72	36.17	35.57
Pulse:	152	124	140	152	146	128	136	144
Time:	35.37	35.57	35.43	36.05	34.85	34.70	34.75	33.93
Pulse:	148	144	136	124	148	144	140	156
Time:	34.60	34.00	34.35	35.62	35.68	35.28	35.97	
Pulse:	136	148	148	132	124	132	139	

(a) A scatterplot shows a moderately strong negative linear relationship. Use your calculator or software to verify that the least-squares regression line is

$$\text{pulse} = 479.9 - (9.695 \times \text{time})$$

(b) The next day's time is 34.30 minutes. Predict the professor's pulse rate. In fact, his pulse rate was 152. How accurate is your prediction?

(c) Suppose you were told only that the pulse rate was 152. You now want to predict swimming time. Find the equation of the least-squares regression line that is appropriate for this purpose. What is your prediction, and how accurate is it?

(d) Explain clearly, to someone who knows no statistics, why there are two different regression lines.

3.44 PREDICTING THE STOCK MARKET Some people think that the behavior of the stock market in January predicts its behavior for the rest of the year. Take the explanatory variable x to be the percent change in a stock market index in January and the response variable y to be the change in the index for the entire year. We expect a positive correlation between x and y because the change during January contributes to the full year's change. Calculation from data for the years 1960 to 1997 gives

$$\bar{x} = 1.75\% \qquad s_x = 5.36\% \qquad r = 0.596$$
$$\bar{y} = 9.07\% \qquad s_y = 15.35\%$$

(a) What percent of the observed variation in yearly changes in the index is explained by a straight-line relationship with the change during January?

(b) What is the equation of the least-squares line for predicting full-year change from January change?

(c) The mean change in January is $\bar{x} = 1.75\%$. Use your regression line to predict the change in the index in a year in which the index rises 1.75% in January. Why could you have given this result (up to roundoff error) without doing the calculation?

3.45 BEAVERS AND BEETLES Ecologists sometimes find rather strange relationships in our environment. One study seems to show that beavers benefit beetles. The researchers laid out 23 circular plots, each four meters in diameter, in an area where beavers were cutting down cottonwood trees. In each plot, they counted the number of stumps from trees cut by beavers and the number of clusters of beetle larvae. Here are the data:[15]

Stumps:	2	2	1	3	3	4	3	1	2	5	1	3
Beetle larvae:	10	30	12	24	36	40	43	11	27	56	18	40
Stumps:	2	1	2	2	1	1	4	1	2	1	4	
Beetle larvae:	25	8	21	14	16	6	54	9	13	14	50	

(a) Make a scatterplot that shows how the number of beaver-caused stumps influences the number of beetle larvae clusters. What does your plot show? (Ecologists think that the new sprouts from stumps are more tender than other cottonwood growth, so that beetles prefer them.)

(b) Find the least-squares regression line and draw it on your plot.

(c) What percent of the observed variation in beetle larvae counts can be explained by straight-line dependence on stump counts?

Residuals

A regression line is a mathematical model for the overall pattern of a linear relationship between an explanatory variable and a response variable. Deviations from the overall pattern are also important. In the regression setting, we see deviations by looking at the scatter of the data points about the regression line. The vertical distances from the points to the least-squares regression line are as small as possible, in the sense that they have the smallest possible sum of squares. Because they represent "left-over" variation in the response after fitting the regression line, these distances are called *residuals.*

RESIDUALS

A **residual** is the difference between an observed value of the response variable and the value predicted by the regression line. That is,

$$\begin{aligned}\text{residual} &= \text{observed } y - \text{predicted } y \\ &= y - \hat{y}\end{aligned}$$

EXAMPLE 3.14 GESELL SCORES

Does the age at which a child begins to talk predict later score on a test of mental ability? A study of the development of young children recorded the age in months at which each of the 21 children spoke their first word and Gesell Adaptive Score, the result of an aptitude test taken much later. The data appear in Table 3.4.

TABLE 3.4 Age at first word and Gesell score

Child	Age	Score	Child	Age	Score	Child	Age	Score
1	15	95	8	11	100	15	11	102
2	26	71	9	8	104	16	10	100
3	10	83	10	20	94	17	12	105
4	9	91	11	7	113	18	42	57
5	15	102	12	9	96	19	17	121
6	20	87	13	10	83	20	11	86
7	18	93	14	11	84	21	10	100

Source: These data were originally collected by L. M. Linde of UCLA but were first published by M. R. Mickey, O. J. Dunn, and V. Clark, "Note on the use of stepwise regression in detecting outliers," *Computers and Biomedical Research,* 1 (1967), pp. 105–111. The data have been used by several authors. We found them in N. R. Draper and J. A. John, "Influential observations and outliers in regression," *Technometrics,* 23 (1981), pp. 21–26.

Figure 3.17 is a scatterplot, with age at first word as the explanatory variable x and Gesell score as the response variable y. Children 3 and 13, and also Children 16 and 21, have identical values of both variables. We use a different plotting symbol to show that one point stands for two individuals. The plot shows a negative association. That is, children who begin to speak later tend to have lower test scores than early talkers. The overall pattern is moderately linear. The correlation describes both the direction and strength of the linear relationship. It is $r = -0.640$.

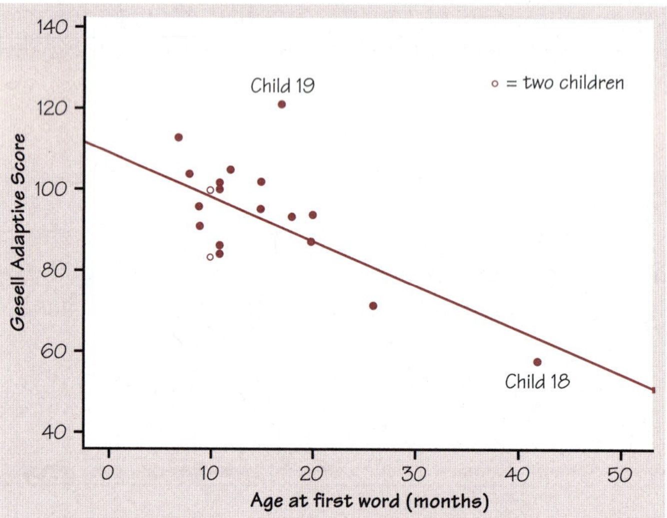

FIGURE 3.17 Scatterplot of Gesell Adaptive Scores versus the age at first word for 21 children from Table 3.4. The line is the least-squares regression line for predicting Gesell score from age at first word.

The line on the plot is the least-squares regression line of Gesell score on age at first word. Its equation is

$$\hat{y} = 109.8738 - 1.1270x$$

For Child 1, who first spoke at 15 months, we predict the score

$$\hat{y} = 109.8738 - (1.1270)(15) = 92.97$$

This child's actual score was 95. The residual is

$$\begin{aligned}\text{residual} &= \text{observed } y - \text{predicted } y \\ &= 95 - 92.97 = 2.03\end{aligned}$$

The residual is positive because the data point lies above the line.

There is a residual for each data point. Here are the 21 residuals for the Gesell data, from Example 3.14, as output by a statistical software package:

```
residuals:
 2.0310   -9.5721  -15.6040 -8.7309   9.0310    -0.3341    3.4120
 2.5230    3.1421    6.6659 11.0151  -3.7309  -15.6040  -13.4770
 4.5230    1.3960    8.6500 -5.5403  30.2850  -11.4770    1.3960
```

Because the residuals show how far the data fall from our regression line, examining the residuals helps assess how well the line describes the data. Although residuals can be calculated from any model fitted to the data, the residuals from the least-squares line have a special property: **the mean of the least-squares residuals is always zero.** You can check that the sum of the residuals above is –0.0002. The sum is not exactly 0 because the software rounded the residuals to four decimal places. This is ***roundoff error.***

roundoff error

Compare the scatterplot in Figure 3.17 with the *residual plot* for the same data in Figure 3.18. The horizontal line at zero in Figure 3.18 helps orient us. It corresponds to the regression line in Figure 3.17.

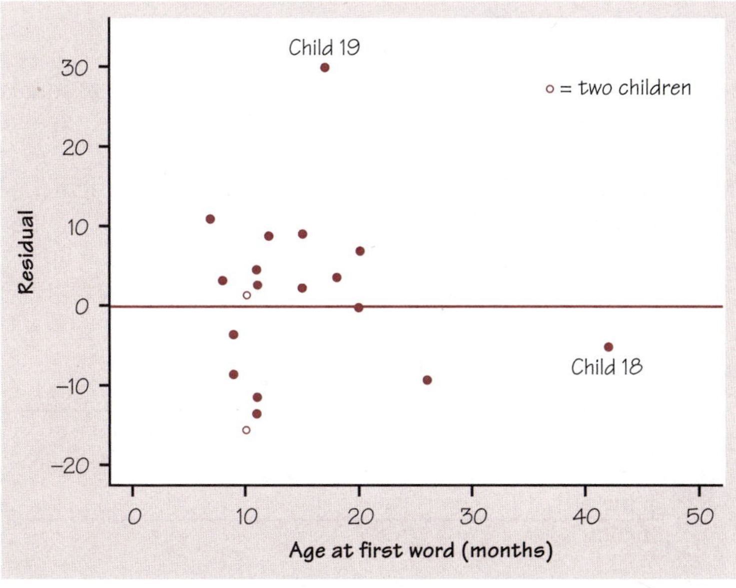

FIGURE 3.18 Residual plot for the regression of Gesell score on age at first word. Child 19 is an outlier, and Child 18 is an influential observation that does not have a large residual.

RESIDUAL PLOTS

A **residual plot** is a scatterplot of the regression residuals against the explanatory variable. Residual plots help us assess the fit of a regression line.

You should be aware that some computer utilities, such as Data Desk, prefer to plot the residuals against the fitted values $\hat{y}_i$ instead of against the values x_i of the explanatory variable. The information in the two plots is the same because $\hat{y}$ is linearly related to x.

If the regression line captures the overall relationship between x and y, the residuals should have no systematic pattern. The residual plot will look something like the simplified pattern in Figure 3.19(a). That plot shows a uniform scatter of the points about the fitted line, with no unusual individual observations.

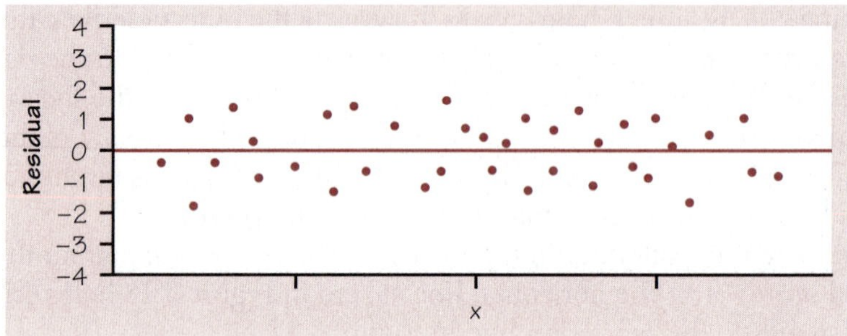

FIGURE 3.19(a) The uniform scatter of points indicates that the regression line fits the data well, so the line is a good model.

Here are some things to look for when you examine the residuals, using either a scatterplot of the data or a residual plot.

- **A curved pattern** shows that the relationship is not linear. Figure 3.19(b) is a simplified example. A straight line is not a good summary for such data.

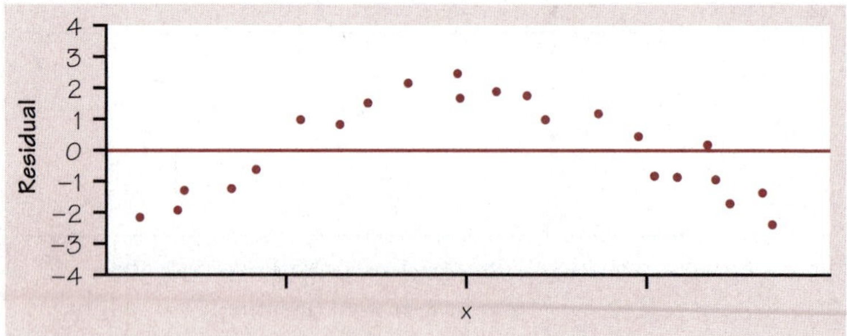

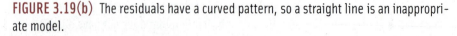

FIGURE 3.19(b) The residuals have a curved pattern, so a straight line is an inappropriate model.

• **Increasing or decreasing spread about the line** as x increases indicates that prediction of y will be less accurate for larger x. Figure 3.19(c) is a simplified example.

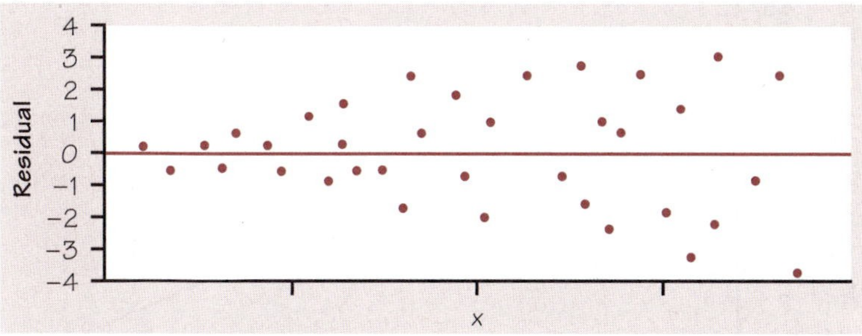

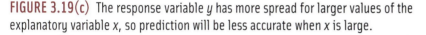

FIGURE 3.19(c) The response variable y has more spread for larger values of the explanatory variable x, so prediction will be less accurate when x is large.

• **Individual points with large residuals,** like Child 19 in Figures 3.17 and 3.18 are outliers in the vertical (y) direction because they lie far from the line that describes the overall pattern.

• **Individual points that are extreme in the x direction,** like Child 18 in Figures 3.17 and 3.18, may not have large residuals, but they can be very important. We address such points next.

Influential observations

Children 18 and 19 are both unusual in the Gesell example. They are unusual in different ways. Child 19 lies far from the regression line. This child's Gesell score is so high that we should check for a mistake in recording it. In fact, the score is correct. Child 18 is close to the line but far out in the x direction. He or she began to speak much later than any of the other children. *Because of its extreme position on the age scale, this point has a strong influence on the position of the regression line.* Figure 3.20 adds a second regression line, calculated after leaving out Child 18. You can see that this one point moves the line quite a bit. We call such points *influential.*

OUTLIERS AND INFLUENTIAL OBSERVATIONS IN REGRESSION

An **outlier** is an observation that lies outside the overall pattern of the other observations.

An observation is **influential** for a statistical calculation if removing it would markedly change the result of the calculation. Points that are outliers in the x direction of a scatterplot are often influential for the least-squares regression line.

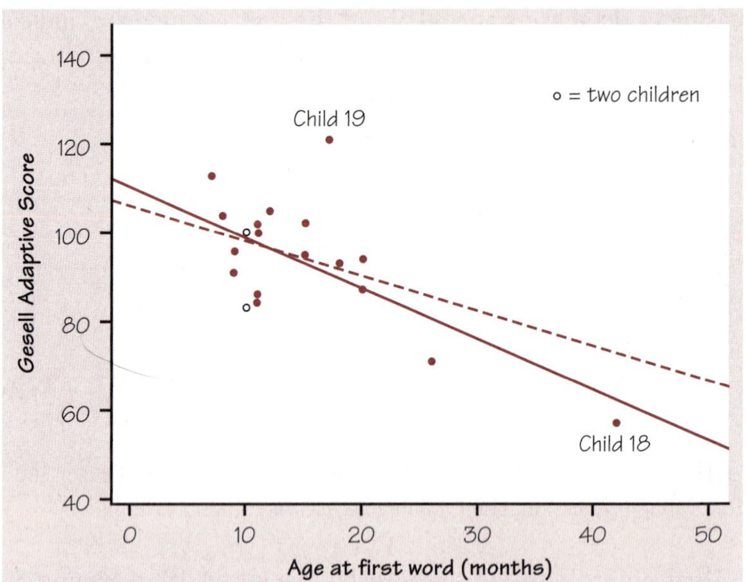

FIGURE 3.20 Two least-squares regression lines of Gesell score on age at first word. The solid line is calculated from all the data. The dashed line is calculated leaving out Child 18. Child 18 is an influential observation because leaving out this point moves the regression line quite a bit.

Children 18 and 19 are both outliers in Figure 3.20. Child 18 is an outlier in the x direction and influences the least-squares line. Child 19 is an outlier in the y direction. It has less influence on the regression line because the many other points with similar values of x anchor the line well below the outlying point. Influential points often have small residuals, because they pull the regression line toward themselves. If you just look at residuals, you will miss influential points. Influential observations can greatly change the interpretation of data.

EXAMPLE 3.15 AN INFLUENTIAL OBSERVATION

The strong influence of Child 18 makes the original regression of Gesell score on age at first word misleading. The original data have $r^2 = 0.41$. That is, the age at which a child begins to talk explains 41% of the variation on a later test of mental ability. This relationship is strong enough to be interesting to parents. If we leave out Child 18, r^2 drops to only 11%. The apparent strength of the association was largely due to a single influential observation.

What should the child development researcher do? She must decide whether Child 18 is so slow to speak that this individual should not be allowed to influence the analysis. If she excludes Child 18, much of the evidence for a connection between the age at which a child begins to talk and later ability score vanishes. If she keeps Child 18, she needs data on other children who were also slow to begin talking, so that the analysis no longer depends so heavily on just one child.

EXERCISES

3.46 DRIVING SPEED AND FUEL CONSUMPTION Exercise 3.11 (page 129) gives data on the fuel consumption y of a car at various speeds x. Fuel consumption is measured in liters of gasoline per 100 kilometers driven and speed is measured in kilometers per hour. A statistical software package gives the least-squares regression line and also the residuals. The regression line is

$$\hat{y} = 11.058 - 0.01466x$$

The residuals, in the same order as the observations, are

10.09	2.24	−0.62	−2.47	−3.33	−4.28	−3.73	−2.94
−2.17	−1.32	−0.42	0.57	1.64	2.76	3.97	

(a) Make a scatterplot of the observations and draw the regression line on your plot.

(b) Would you use the regression line to predict y from x? Explain your answer.

(c) Check that the residuals have sum zero (up to roundoff error).

(d) Make a plot of residuals against the values of x. Draw a horizontal line at height zero on your plot. Notice that the residuals show the same pattern about this line as the data points show about the regression line in the scatterplot in (a). What do you conclude about the residual plot?

3.47 HOW MANY CALORIES? Exercise 3.20 (page 138) gives data on the true calories in ten foods and the average guesses made by a large group of people. Exercise 3.31 (page 147) explored the influence of two outlying observations on the correlation.

(a) Make a scatterplot suitable for predicting guessed calories from true calories. Circle the points for spaghetti and snack cake on your plot. These points lie outside the linear pattern of the other eight points.

(b) Use your calculator to find the least-squares regression line of guessed calories on true calories. Do this twice, first for all ten data points and then leaving out spaghetti and snack cake.

(c) Plot both lines on your graph. (Make one dashed so that you can tell them apart.) Are spaghetti and snack cake, taken together, influential observations? Explain your answer.

3.48 INFLUENTIAL OR NOT? The discussion of Example 3.15 shows that Child 18 in the Gesell data in Table 3.4 is an influential observation. Now we will examine the effect of Child 19, who is also an outlier in Figure 3.20.

(a) Find the least-squares regression line of Gesell score on age at first word, leaving out Child 19. Example 3.14 gives the regression line from all the children. Plot both lines on the same graph. (You do not have to make a scatterplot of all the points—just plot the two lines.) Would you call Child 19 very influential? Why?

(b) How does removing Child 19 change the r^2 for this regression? Explain why r^2 changes in this direction when you drop Child 19.

TECHNOLOGY TOOLBOX *Residual plots by calculator*

Here is a procedure for calculating residuals on your TI-83/89 and then displaying a residual plot.

- Enter the ages and Gesell scores from Table 3.4 (page 168). Plot the scatterplot and perform the linear regression. Store the regression equation in Y_1 (y1(x) on the TI-89) and superimpose the LSRL on the scatterplot.

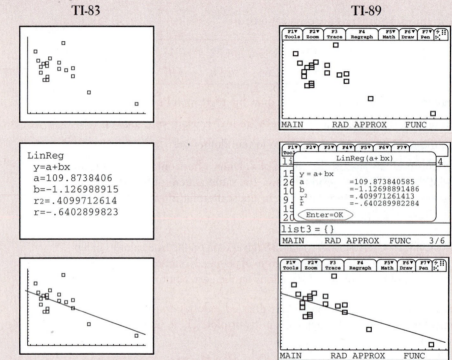

TI-83

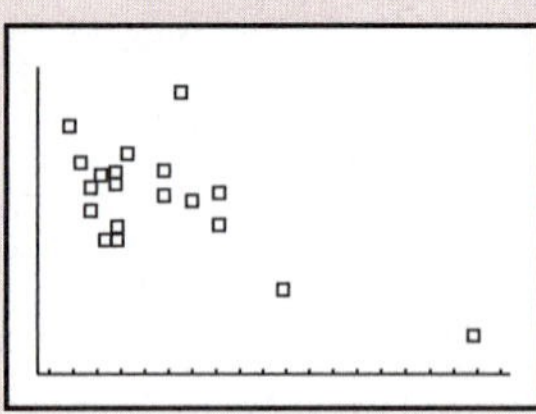

```
LinReg
 y=a+bx
 a=109.8738406
 b=-1.126988915
 r2=.4099712614
 r=-.6402899823
```

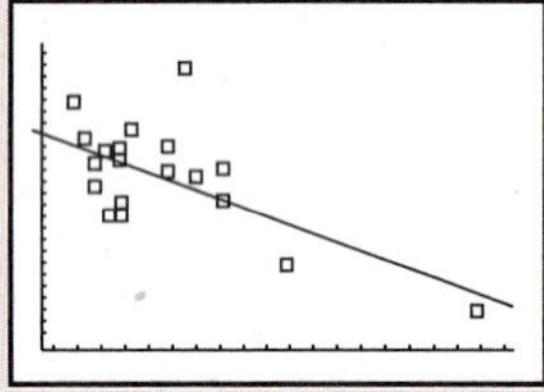

TI-89

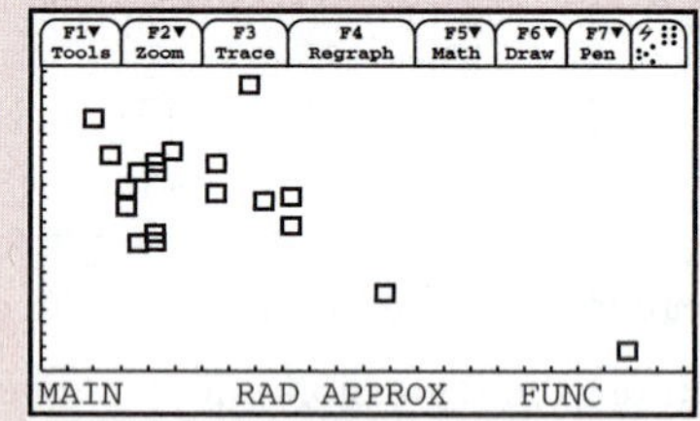

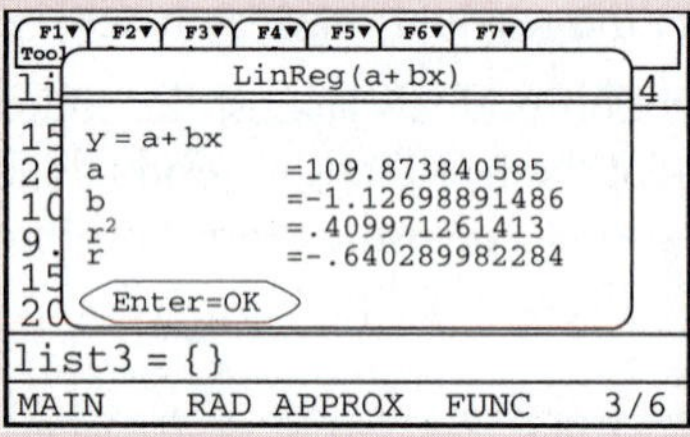

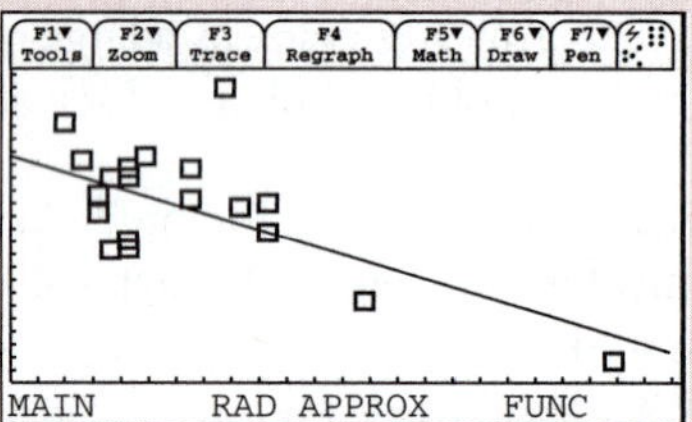

This sets the stage. To graph the residual plot:

- Restore the six default lists using the SetUpEditor command.
- Press STAT, choose `5:SetUpEditor`, and press ENTER.
- Press CATALOG, choose `SetUpEd(` , type), and press ENTER.
- Define L_3/list3 as the observed value minus the predicted value.
- With L_3 highlighted, enter the command $L_2 - Y_1(L_1)$. Press ENTER to show the residuals.
- With list3 highlighted, enter the command `list2 -y1(list1)`. Press ENTER to show the residuals.

L1	L2	L3 3
15	95	2.031
26	71	⁻9.572
10	83	⁻15.6
9	91	⁻8.731
15	102	9.031
20	87	⁻.3341
18	93	3.412

L3(1)=2.030993137...

list1	list2	list3	list4
15.	95.	2.031	----
26.	71.	⁻9.572	
10.	83.	⁻15.6	
9.	91.	⁻8.731	
15.	102.	9.031	
20.	87.	⁻.3341	

list3[1]=2.030994

MAIN RAD APPROX FUNC 3/7

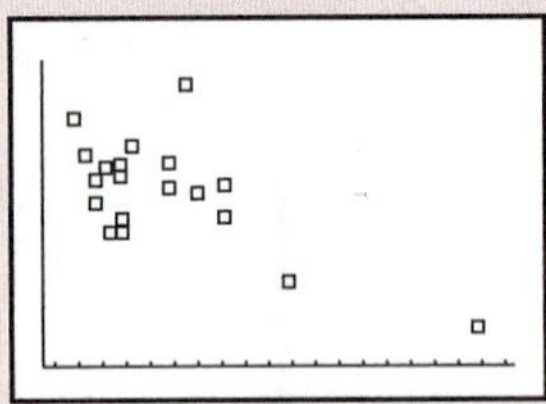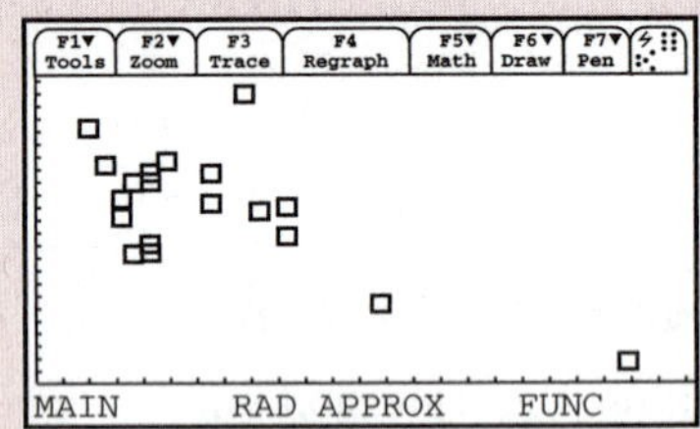

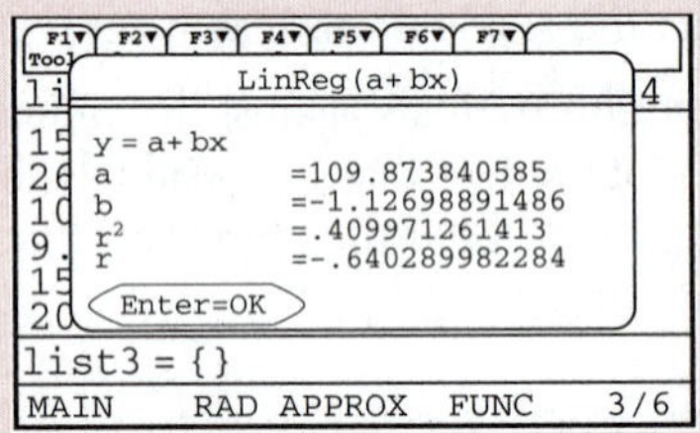

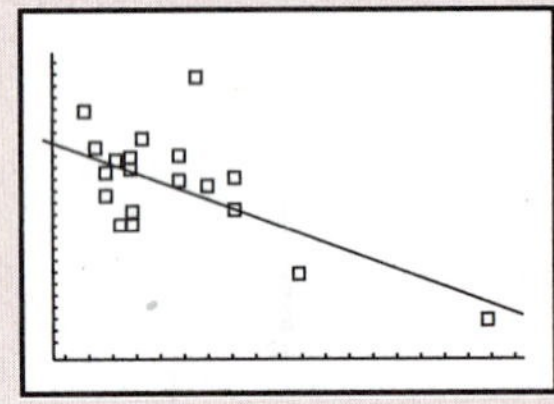

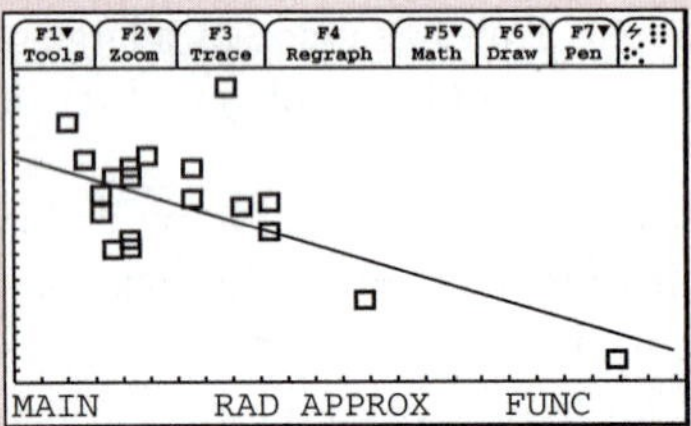

TECHNOLOGY TOOLBOX *Residual plots by calculator (continued)*

- Turn off Plot1 and deselect the regression equation. Specify Plot2 with L_1/list1 as the *x* variable and L_3/list3 as the *y* variable. Use ZoomStat (ZoomData) to see the residual plot.

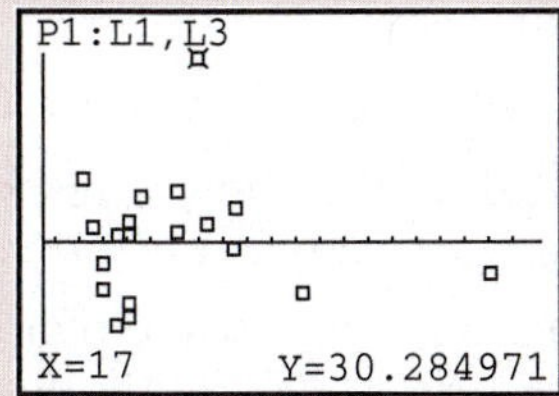

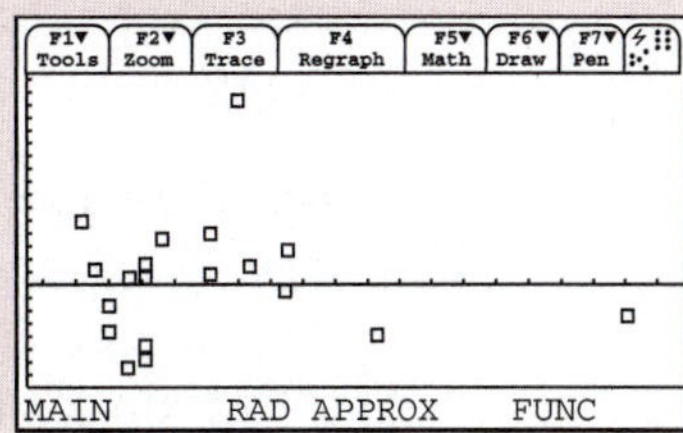

The *x* axis in the residual plot serves as a reference line, with points above this line corresponding to positive residuals and points below the line corresponding to negative residuals. We used TRACE to see the regression outlier at $x = 17$.

- Finally, we have previously noted that an important property of residuals is that their sum is zero. Calculate one-variable statistics on the residuals list to verify that ● (residuals) = 0 and that, consequently, the mean of the residuals is also 0.

```
1-Var Stats
 x̄=3.857143E-12
 Σx=8.1E-11
 Σx²=2308.58578
 Sx=10.74380235
 σx=10.4848775
↓n=21
```

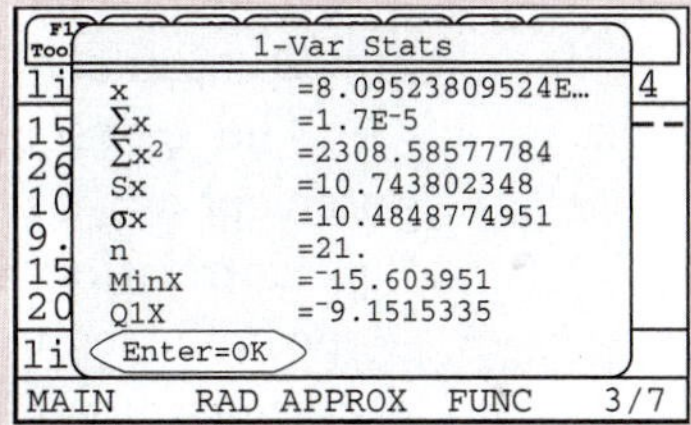

Note that the calculator is showing some roundoff error. You should recognize these peculiar looking numbers as equivalent to 0.

3.49 LEAN BODY MASS AS A PREDICTOR OF METABOLIC RATE Exercise 3.12 (page 132) provides data from a study of dieting for 12 women and 7 men subjects. We will explore the women's data further.

(a) Define two lists on your calculator, MASSF for female mass and METF for female metabolic rate. Then transfer the data to lists 1 and 2. Define Plot1 using the □ plotting symbol, and plot the scatterplot.

(b) Perform least-squares regression on your calculator and record the equation and the correlation. Lean body mass explains what percent of the variation in metabolic rate for the women?

(c) Does the least-squares line provide an adequate model for the data? Define Plot2 to be a residual plot on your calculator with residuals on the vertical axis and lean body mass (*x*-values) on the horizontal axis. Use the □ plotting symbol. Use ZoomStat/ZoomData to see the plot. Copy the plot onto your paper. Label both axes appropriately.

(d) Define list3 to be the predicted y-values: $Y_1(L_1)$ on the TI-83 or Y_1(list1) on the TI-89. Define Plot3 to be a residual plot on your calculator with residuals on the vertical axis and predicted metabolic rate on the horizontal axis. Use the + plotting symbol. Use ZoomStat/ZoomData to see the plot. Copy the plot onto your paper. Label both axes. Compare the two residual plots.

SUMMARY

A **regression line** is a straight line that describes how a response variable y changes as an explanatory variable x changes.

The most common method of fitting a line to a scatterplot is least squares. The **least-squares regression line** is the straight line $\hat{y} = a + bx$ that minimizes the sum of the squares of the vertical distances of the observed points from the line.

You can use a regression line to **predict** the value of y for any value of x by substituting this x into the equation of the line.

The **slope** b of a regression line $\hat{y} = a + bx$ is the rate at which the predicted response $\hat{y}$ changes along the line as the explanatory variable x changes. Specifically, b is the change in $\hat{y}$ when x increases by 1.

The **intercept** a of a regression line $\hat{y} = a + bx$ is the predicted response $\hat{y}$ when the explanatory variable $x = 0$. This prediction is of no statistical use unless x can actually take values near 0.

The least-squares regression line of y on x is the line with slope rs_y/s_x and intercept $a = \bar{y} - b\bar{x}$. This line always passes through the point $(\bar{x}, \bar{y})$.

Correlation and regression are closely connected. The correlation r is the slope of the least-squares regression line when we measure both x and y in standardized units. The square of the correlation r^2 is the fraction of the variance of one variable that is explained by least-squares regression on the other variable.

You can examine the fit of a regression line by studying the **residuals,** which are the differences between the observed and predicted values of y. Be on the lookout for outlying points with unusually large residuals and also for nonlinear patterns and uneven variation about the line.

Also look for **influential observations,** individual points that substantially change the regression line. Influential observations are often outliers in the x direction, but they need not have large residuals.

SECTION 3.3 EXERCISES

3.50 REVIEW OF STRAIGHT LINES Fred keeps his savings under his mattress. He began with \$500 from his mother and adds \$100 each year. His total savings y after x years are given by the equation

$$y = 500 + 100x$$

(a) Draw a graph of this equation. (Choose two values of x, such as 0 and 10. Compute the corresponding values of y from the equation. Plot these two points on graph paper and draw the straight line joining them.)

(b) After 20 years, how much will Fred have under his mattress?

(c) If Fred had added $200 instead of $100 each year to his initial $500, what is the equation that describes his savings after x years?

3.51 REVIEW OF STRAIGHT LINES During the period after birth, a male white rat gains exactly 40 grams (g) per week. (This rat is unusually regular in his growth, but 40 g per week is a realistic rate.)

(a) If the rat weighed 100 g at birth, give an equation for his weight after x weeks. What is the slope of this line?

(b) Draw a graph of this line between birth and 10 weeks of age.

(c) Would you be willing to use this line to predict the rat's weight at age 2 years? Do the prediction and think about the reasonableness of the result. (There are 454 grams in a pound. To help you assess the result, note that a large cat weighs about 10 pounds.)

3.52 IQ AND SCHOOL GPA Figure 3.5 (page 135) plots school grade point average (GPA) against IQ test score for 78 seventh-grade students. Calculation shows that the mean and standard deviation of the IQ scores are

$$\bar{x} = 108.9 \quad s_x = 13.17$$

For the grade point averages,

$$\bar{y} = 7.447 \quad s_y = 2.10$$

The correlation between IQ and GPA is $r = 0.6337$.

(a) Find the equation of the least-squares line for predicting GPA from IQ.

(b) What percent of the observed variation in these students' GPAs can be explained by the linear relationship between GPA and IQ?

(c) One student has an IQ of 103 but a very low GPA of 0.53. What is the predicted GPA for a student with IQ = 103? What is the residual for this particular student?

3.53 TAKE ME OUT TO THE BALL GAME What is the relationship between the price charged for a hot dog and the price charged for a 16-ounce soda in major league baseball stadiums? Here are some data:[16]

Team	Hot dog	Soda	Team	Hot dog	Soda	Team	Hot dog	Soda
Angels	2.50	1.75	Giants	2.75	2.17	Rangers	2.00	2.00
Astros	2.00	2.00	Indians	2.00	2.00	Red Sox	2.25	2.29
Braves	2.50	1.79	Marlins	2.25	1.80	Rockies	2.25	2.25
Brewers	2.00	2.00	Mets	2.50	2.50	Royals	1.75	1.99
Cardinals	3.50	2.00	Padres	1.75	2.25	Tigers	2.00	2.00
Dodgers	2.75	2.00	Phillies	2.75	2.20	Twins	2.50	2.22
Expos	1.75	2.00	Pirates	1.75	1.75	White Sox	2.00	2.00

(a) Make a scatterplot appropriate for predicting soda price from hot dog price. Describe the relationship that you see. Are there any outliers?

(b) Find the correlation between hot dog price and soda price. What percent of the variation in soda price does a linear relationship account for?

(c) Find the equation of the least-squares line for predicting soda price from hot dog price. Draw the line on your scatterplot. Based on your findings in (b), explain why it is not surprising that the line is nearly horizontal (slope near zero).

(d) Circle the observation that is potentially the most influential. What team is this? Find the least-squares line without this one observation and draw it on your scatterplot. Was the observation in fact influential?

3.54 KEEPING WATER CLEAN Keeping water supplies clean requires regular measurement of levels of pollutants. The measurements are indirect—a typical analysis involves forming a dye by a chemical reaction with the dissolved pollutant, then passing light through the solution and measuring its "absorbence." To calibrate such measurements, the laboratory measures known standard solutions and uses regression to relate absorbence to pollutant concentration. This is usually done every day. Here is one series of data on the absorbence for different levels of nitrates. Nitrates are measured in milligrams per liter of water.[17]

Nitrates:	50	50	100	200	400	800	1200	1600	2000	2000
Absorbence:	7.0	7.5	12.8	24.0	47.0	93.0	138.0	183.0	230.0	226.0

(a) Chemical theory says that these data should lie on a straight line. If the correlation is not at least 0.997, something went wrong and the calibration procedure is repeated. Plot the data and find the correlation. Must the calibration be done again?

(b) What is the equation of the least-squares line for predicting absorbence from concentration? If the lab analyzed a specimen with 500 milligrams of nitrates per liter, what do you expect the absorbence to be? Based on your plot and the correlation, do you expect your predicted absorbence to be very accurate?

3.55 A GROWING CHILD Sarah's parents are concerned that she seems short for her age. Their doctor has the following record of Sarah's height:

Age (months):	36	48	51	54	57	60
Height (cm):	86	90	91	93	94	95

(a) Make a scatterplot of these data. Note the strong linear pattern.

(b) Using your calculator, find the equation of the least-squares regression line of height on age.

(c) Predict Sarah's height at 40 months and at 60 months. Use your results to draw the regression line on your scatterplot.

(d) What is Sarah's rate of growth, in centimeters per month? Normally growing girls gain about 6 cm in height between ages 4 (48 months) and 5 (60 months). What rate of growth is this in centimeters per month? Is Sarah growing more slowly than normal?

3.56 INVESTING AT HOME AND OVERSEAS Investors ask about the relationship between returns on investments in the United States and on investments overseas. Table 3.5 gives the total returns on U.S. and overseas common stocks over a 26-year period. (The total return is change in price plus any dividends paid, converted into U.S. dollars. Both returns are averages over many individual stocks.)

TABLE 3.5 Annual total return on overseas and U.S. stocks

Year	Overseas % return	U.S. % return	Year	Overseas % return	U.S. % return	Year	Overseas % return	U.S. % return
1971	29.6	14.6	1980	22.6	32.3	1989	10.6	31.5
1972	36.3	18.9	1981	–2.3	–5.0	1990	–23.0	–3.1
1973	–14.9	–14.8	1982	–1.9	21.5	1991	12.8	30.4
1974	–23.2	–26.4	1983	23.7	22.4	1992	–12.1	7.6
1975	35.4	37.2	1984	7.4	6.1	1993	32.9	10.1
1976	2.5	23.6	1985	56.2	31.6	1994	6.2	1.3
1977	18.1	–7.4	1986	69.4	18.6	1995	11.2	37.6
1978	32.6	6.4	1987	24.6	5.1	1996	6.4	23.0
1979	4.8	18.2	1988	28.5	16.8	1997	2.1	33.4

Source: The U.S. returns are for the Standard & Poor's 500 Index. The overseas returns are for the Morgan Stanley Europe, Australasia, Far East (EAFE) index.

(a) Make a scatterplot suitable for predicting overseas returns from U.S. returns.

(b) Find the correlation and r^2. Describe the relationship between U.S. and overseas returns in words, using r and r^2 to make your description more precise.

(c) Find the least-squares regression line of overseas returns on U.S. returns. Draw the line on the scatterplot.

(d) In 1997, the return on U.S. stocks was 33.4%. Use the regression line to predict the return on overseas stocks. The actual overseas return was 2.1%. Are you confident that predictions using the regression line will be quite accurate? Why?

(e) Circle the point that has the largest residual (either positive or negative). What year is this? Are there any points that seem likely to be very influential?

3.57 WHAT'S MY GRADE? In Professor Friedman's economics course the correlation between the students' total scores prior to the final examination and their final examination scores is $r = 0.6$. The pre-exam totals for all students in the course have mean 280 and standard deviation 30. The final exam scores have mean 75 and standard deviation 8. Professor Friedman has lost Julie's final exam but knows that her total before the exam was 300. He decides to predict her final exam score from her pre-exam total.

(a) What is the slope of the least-squares regression line of final exam scores on pre-exam total scores in this course? What is the intercept?

(b) Use the regression line to predict Julie's final exam score.

(c) Julie doesn't think this method accurately predicts how well she did on the final exam. Calculate r^2 and use the value you get to argue that her actual score could have been much higher (or much lower) than the predicted value.

3.58 A NONSENSE PREDICTION Use the least-squares regression line for the data in Exercise 3.55 to predict Sarah's height at age 40 years (480 months). Your prediction is in centimeters. Convert it to inches using the fact that a centimeter is 0.3937 inch.

The prediction is impossibly large. It is not reasonable to use data for 36 to 60 months to predict height at 480 months.

3.59 INVESTING AT HOME AND OVERSEAS Exercise 3.56 examined the relationship between returns on U.S. and overseas stocks. Investors also want to know what typical returns are and how much year-to-year variability (called *volatility* in finance) there is. Regression and correlation do not answer these questions.

(a) Find the five-number summaries for both U.S. and overseas returns, and make side-by-side boxplots to compare the two distributions.

(b) Were returns generally higher in the United States or overseas during this period? Explain your answer.

(c) Were returns more volatile (more variable) in the United States or overseas during this period? Explain your answer.

3.60 WILL I BOMB THE FINAL? We expect that students who do well on the midterm exam in a course will usually also do well on the final exam. Gary Smith of Pomona College looked at the exam scores of all 346 students who took his statistics class over a 10-year period.[18] The least-squares line for predicting final exam score from midterm exam score was $\hat{y} = 46.6 + 0.41x$.

Octavio scores 10 points above the class mean on the midterm. How many points above the class mean do you predict that he will score on the final? (*Hint*: Use the fact that the least-squares line passes through the point $(\bar{x}, \bar{y})$ and the fact that Octavio's midterm score is $\bar{x} + 10$. This is an example of the phenomenon that gave "regression" its name: students who do well on the midterm will on the average do less well, but still above average, on the final.)

3.61 NAHYA INFANT WEIGHTS A study of nutrition in developing countries collected data from the Egyptian village of Nahya. Here are the mean weights (in kilograms) for 170 infants in Nahya who were weighed each month during their first year of life.

Age (months):	1	2	3	4	5	6	7	8	9	10	11	12
Weight (kg):	4.3	5.1	5.7	6.3	6.8	7.1	7.2	7.2	7.2	7.2	7.5	7.8

(a) Plot the weight against time.

(b) A hasty user of statistics enters the data into software and computes the least-squares line without plotting the data. The result is

```
THE REGRESSION EQUATION IS
WEIGHT = 4.88 + 0.267 AGE
```

Plot this line on your graph. Is it an acceptable summary of the overall pattern of growth? Remember that you can calculate the least-squares line for *any* set of two-variable data. It's up to you to decide if it makes sense to fit a line.

(c) Fortunately, the software also prints out the residuals from the least-squares line. In order of age along the rows, they are

−0.85	−0.31	0.02	0.35	0.58	0.62
0.45	0.18	−0.08	−0.35	−0.32	−0.28

Verify that the residuals have sum 0 (except for roundoff error). Plot the residuals against age and add a horizontal line at 0. Describe carefully the pattern that you see.

CHAPTER REVIEW

Chapters 1 and 2 dealt with data analysis for a single variable. In this chapter, we have studied analysis of data for two or more variables. The proper analysis depends on whether the variables are categorical or quantitative and on whether one is an explanatory variable and the other a response variable.

Data analysis begins with graphs and then adds numerical summaries of specific aspects of the data.

This chapter concentrates on relations between two quantitative variables. Scatterplots show the relationship, whether or not there is an explanatory-response distinction. Correlation describes the strength of a linear relationship, and least-squares regression fits a line to data that have an explanatory-response relation.

ANALYZING DATA FOR TWO VARIABLES

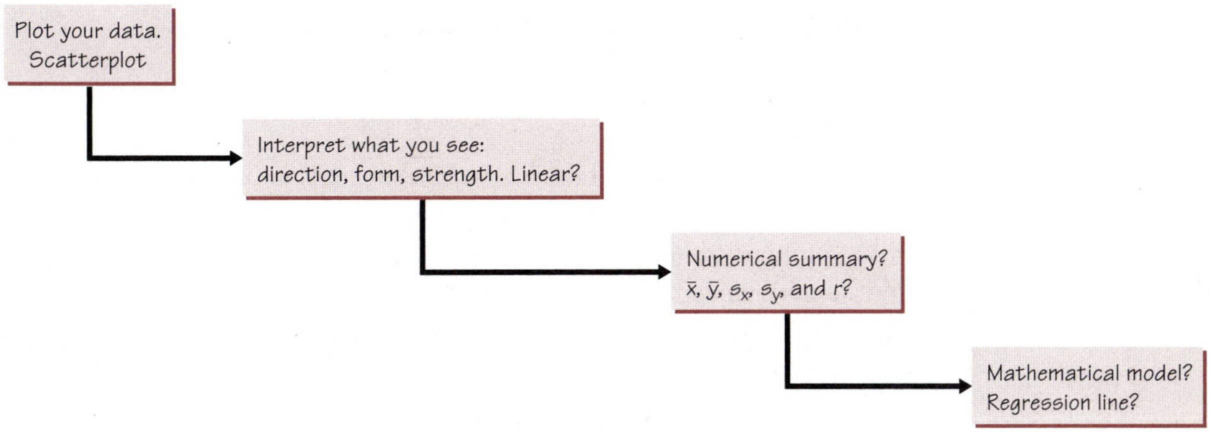

Plot your data.
Scatterplot

→ Interpret what you see:
direction, form, strength. Linear?

→ Numerical summary?
$\bar{x}$, $\bar{y}$, s_x, s_y, and r?

→ Mathematical model?
Regression line?

Here is a review list of the most important skills you should have gained from studying this chapter.

A. DATA

1. Recognize whether each variable is quantitative or categorical.

2. Identify the explanatory and response variables in situations where one variable explains or influences another.

B. SCATTERPLOTS

1. Make a scatterplot to display the relationship between two quantitative variables. Place the explanatory variable (if any) on the horizontal scale of the plot.

2. Add a categorical variable to a scatterplot by using a different plotting symbol or color.

3. Describe the form, direction, and strength of the overall pattern of a scatterplot. In particular, recognize positive or negative association and linear (straight-line) patterns. Recognize outliers in a scatterplot.

C. CORRELATION

1. Using a calculator, find the correlation r between two quantitative variables.

2. Know the basic properties of correlation: r measures the strength and direction of only linear relationships; $-1 \leq r \leq 1$ always; $r = \pm 1$ only for perfect straight-line relations; r moves away from 0 toward ± 1 as the linear relation gets stronger.

D. STRAIGHT LINES

1. Explain what the slope b and the intercept a mean in the equation $y = a + bx$ of a straight line.

2. Draw a graph of the straight line when you are given its equation.

E. REGRESSION

1. Using a calculator, find the least-squares regression line of a response variable y on an explanatory variable x from data.

2. Find the slope and intercept of the least-squares regression line from the means and standard deviations of x and y and their correlation.

3. Use the regression line to predict y for a given x. Recognize extrapolation and be aware of its dangers.

4. Use r^2 to describe how much of the variation in one variable can be accounted for by a straight-line relationship with another variable.

5. Recognize outliers and potentially influential observations from a scatterplot with the regression line drawn on it.

6. Calculate the residuals and plot them against the explanatory variable x or against other variables. Recognize unusual patterns.

CHAPTER 3 REVIEW EXERCISES

3.62 Figure 3.21 is a scatterplot that displays the heights of 53 pairs of parents. The mother's height is plotted on the vertical axis and the father's height on the horizontal axis.[20]

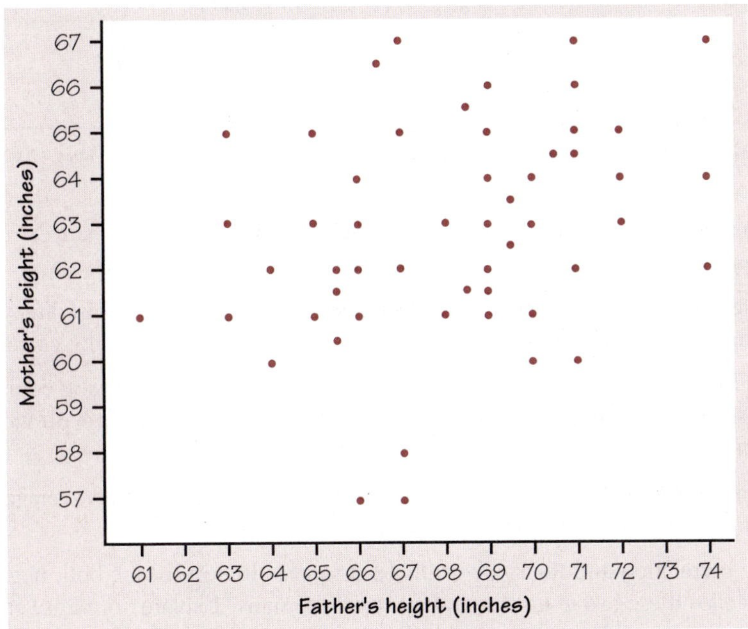

FIGURE 3.21 Scatterplot of the heights of the mother and father in 53 pairs of parents.

(a) What is the smallest height of any mother in the group? How many mothers have that height? What are the heights of the fathers in these pairs?

(b) What is the greatest height of any father in the group? How many fathers have that height? How tall are the mothers in these pairs?

(c) Are there clear explanatory and response variables, or could we freely choose which variable to plot horizontally?

(d) Say in words what a positive association between these variables means. The scatterplot shows a weak positive association. Why do we say the association is weak?

3.63 IS WINE GOOD FOR YOUR HEART? Table 3.6 below gives data on average per capita wine consumption and heart disease death rates in 19 countries.

TABLE 3.6 Wine consumption and heart disease

Country	Alcohol from wine (liters/year)	Heart disease death rate (per 100,000)	Country	Alcohol from wine (liters/year)	Heart disease death rate (per 100,000)
Australia	2.5	211	Netherlands	1.8	167
Austria	3.9	167	New Zealand	1.9	266
Belgium/Lux.	2.9	131	Norway	0.8	227
Canada	2.4	191	Spain	6.5	86
Denmark	2.9	220	Sweden	1.6	207
Finland	0.8	297	Switzerland	5.8	115
France	9.1	71	United Kingdom	1.3	285
Iceland	0.8	211	United States	1.2	199
Ireland	0.7	300	West Germany	2.7	172
Italy	7.9	107			

Source: M. H. Criqui, University of California, San Diego, reported in the *New York Times*, December 28, 1994.

(a) Construct a scatterplot for these data. Describe the relationship between the two variables.

(b) Determine the equation of the least-squares line for predicting heart disease death rate from wine consumption using the data in Table 3.6. Determine the correlation.

(c) Interpret the correlation. About what percent of the variation among countries in heart disease death rates is explained by the straight-line relationship with wine consumption?

(d) Predict the heart disease death rate in another country where adults average 4 liters of alcohol from wine each year.

(e) The correlation and the slope of the least-squares line in (b) are both negative. Is it possible for these two quantities to have opposite signs? Explain your answer.

3.64 AGE AND EDUCATION IN THE STATES Because older people as a group have less education than younger people, we might suspect a relationship between the percent of state residents aged 65 and over and the percent who are not high school graduates. Figure 3.22 is a scatterplot of these variables. The data appear in Tables 1.5 and 1.15 (pages 23 and 70).

(a) There are at least two and perhaps three outliers in the plot. Identify these states, and give plausible reasons for why they might be outliers.

(b) If we ignore the outliers, does the relationship have a clear form and direction? Explain your answer.

(c) If we calculate the correlation with and without the three outliers, we get $r = 0.067$ and $r = 0.267$. Which of these is the correlation without the outliers? Explain your answer.

3.65 ALWAYS PLOT YOUR DATA! Table 3.7 presents four sets of data prepared by the statistician Frank Anscombe to illustrate the dangers of calculating without first plotting the data.

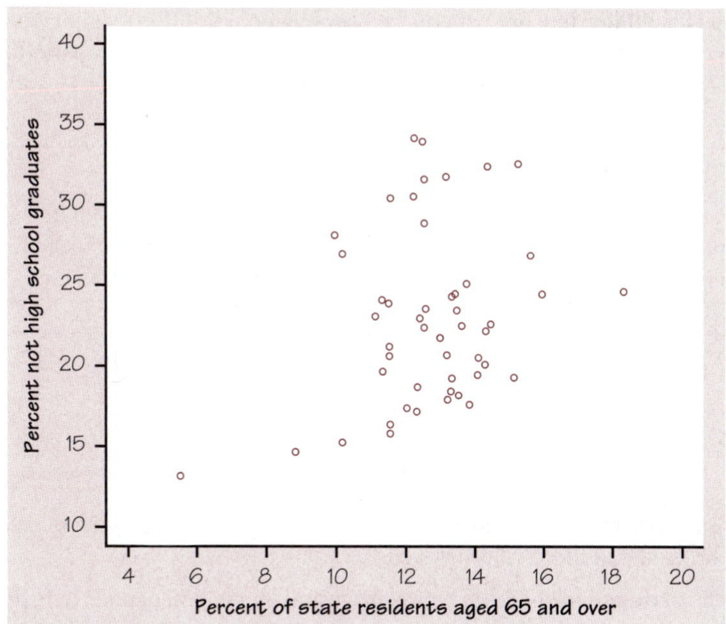

FIGURE 3.22 Scatterplot of the percent of residents who are not high school graduates against the percent of residents aged 65 and over in the 50 states, for Exercise 3.64.

TABLE 3.7 Four data sets for exploring correlation and regression

Data Set A											
x	10	8	13	9	11	14	6	4	12	7	5
y	8.04	6.95	7.58	8.81	8.33	9.96	7.24	4.26	10.84	4.82	5.68
Data Set B											
x	10	8	13	9	11	14	6	4	12	7	5
y	9.14	8.14	8.74	8.77	9.26	8.10	6.13	3.10	9.13	7.26	4.74
Data Set C											
x	10	8	13	9	11	14	6	4	12	7	5
y	7.46	6.77	12.74	7.11	7.81	8.84	6.08	5.39	8.15	6.42	5.73
Data Set D											
x	8	8	8	8	8	8	8	8	8	8	19
y	6.58	5.76	7.71	8.84	8.47	7.04	5.25	5.56	7.91	6.89	12.50

Source: Frank J. Anscombe, "Graphs in statistical analysis," *American Statistician*, 27 (1973), pp. 17–21.

(a) Without making scatterplots, find the correlation and the least-squares regression line for all four data sets. What do you notice? Use the regression line to predict y for $x = 10$.

(b) Make a scatterplot for each of the data sets and add the regression line to each plot.

(c) In which of the four cases would you be willing to use the regression line to describe the dependence of y on x? Explain your answer in each case.

3.66 FOOD POISONING Here are data on 18 people who fell ill from an incident of food poisoning.[21] The data give each person's age in years, the incubation period (the time in hours between eating the infected food and the first signs of illness), and whether the victim survived (S) or died (D).

Person:	1	2	3	4	5	6	7	8	9
Age:	29	39	44	37	42	17	38	43	51
Incubation:	13	46	43	34	20	20	18	72	19
Outcome:	D	S	S	D	D	S	D	S	D

Person:	10	11	12	13	14	15	16	17	18
Age:	30	32	59	33	31	32	32	36	50
Incubation:	36	48	44	21	32	86	48	28	16
Outcome:	D	D	S	D	D	S	D	S	D

(a) Make a scatterplot of incubation period against age, using different symbols for people who died and those who survived.

(b) Is there an overall relationship between age and incubation period? If so, describe it.

(c) More important, is there a relationship between either age or incubation period and whether the victim survived? Describe any relations that seem important here.

(d) Are there any unusual cases that may require individual investigation?

3.67 NEMATODES AND TOMATOES Nematodes are microscopic worms. Here are data from an experiment to study the effect of nematodes in the soil on plant growth. The experimenter prepared 16 planting pots and introduced different numbers of nematodes. Then he placed a tomato seedling in each pot and measured its growth (in centimeters) after 16 days.[22]

Nematodes	Seedling growth (cm)			
0	10.8	9.1	13.5	9.2
1,000	11.1	11.1	8.2	11.3
5,000	5.4	4.6	7.4	5.0
10,000	5.8	5.3	3.2	7.5

Analyze these data and give your conclusions about the effects of nematodes on plant growth.

3.68 A HOT STOCK? It is usual in finance to describe the returns from investing in a single stock by regressing the stock's returns on the returns from the stock market as a whole. This helps us see how closely the stock follows the market. We analyzed the monthly percent total return y on Philip Morris common stock and the monthly return x on the Standard & Poor's 500 Index, which represents the market, for the period between July 1990 and May 1997. Here are the results:

$$\bar{x} = 1.304 \quad s_x = 3.392 \quad r = 0.5251$$
$$\bar{y} = 1.878 \quad s_y = 7.554$$

A scatterplot shows no very influential observations.

(a) Find the equation of the least-squares line from this information. What percent of the variation in Philip Morris stock is explained by the linear relationship with the market as a whole?

(b) Explain carefully what the slope of the line tells us about how Philip Morris stock responds to changes in the market. This slope is called "beta" in investment theory.

(c) Returns on most individual stocks have a positive correlation with returns on the entire market. That is, when the market goes up, an individual stock tends to also go up. Explain why an investor should prefer stocks with beta > 1 when the market is rising and stocks with beta < 1 when the market is falling.

3.69 HUSBANDS AND WIVES The mean height of American women in their early twenties is about 64.5 inches and the standard deviation is about 2.5 inches. The mean height of men the same age is about 68.5 inches, with standard deviation about 2.7 inches. If the correlation between the heights of husbands and wives is about $r = 0.5$, what is the slope of the regression line of the husband's height on the wife's height in young couples? Draw a graph of this regression line. Predict the height of the husband of a woman who is 67 inches tall.

3.70 MEASURING ROAD STRENGTH Concrete road pavement gains strength over time as it cures. Highway builders use regression lines to predict the strength after 28 days (when curing is complete) from measurements made after 7 days. Let x be strength after 7 days (in pounds per square inch) and y the strength after 28 days. One set of data gives this least-squares regression line:

$$\hat{y} = 1389 + 0.96x$$

(a) Draw a graph of this line, with x running from 3000 to 4000 pounds per square inch.

(b) Explain what the slope $b = 0.96$ in this equation says about how concrete gains strength as it cures.

(c) A test of some new pavement after 7 days shows that its strength is 3300 pounds per square inch. Use the equation of the regression line to predict the strength of this pavement after 28 days. Also draw the "up and over" lines from $x = 3300$ on your graph (as in Figure 3.10. page 150).

3.71 COMPETITIVE RUNNERS Good runners take more steps per second as they speed up. Here are the average numbers of steps per second for a group of top female runners at different speeds. The speeds are in feet per second.[23]

Speed (ft/s):	15.86	16.88	17.50	18.62	19.97	21.06	22.11
Steps per second:	3.05	3.12	3.17	3.25	3.36	3.46	3.55

(a) You want to predict steps per second from running speed. Make a scatterplot of the data with this goal in mind.

(b) Describe the pattern of the data and find the correlation.

(c) Find the least-squares regression line of steps per second on running speed. Draw this line on your scatterplot.

(d) Does running speed explain most of the variation in the number of steps a runner takes per second? Calculate r^2 and use it to answer this question.

(e) If you wanted to predict running speed from a runner's steps per second, would you use the same line? Explain your answer. Would r^2 stay the same?

3.72 RESISTANCE REVISITED

(a) Is correlation a resistant measure? Give an example to support your answer.

(b) Is the least-squares regression line resistant? Give an example to support your answer.

3.73 BANK FAILURES The Franklin National Bank failed in 1974. Franklin was one of the 20 largest banks in the nation, and the largest ever to fail. Could Franklin's weakened condition have been detected in advance by simple data analysis? The table below gives the total assets (in billions of dollars) and net income (in millions of dollars) for the 20 largest banks in 1973, the year before Franklin failed.[24] Franklin is bank number 19.

Bank:	1	2	3	4	5	6	7	8	9	10
Assets:	49.0	42.3	36.3	16.4	14.9	14.2	13.5	13.4	13.2	11.8
Income:	218.8	265.6	170.9	85.9	88.1	63.6	96.9	60.9	144.2	53.6
Bank:	11	12	13	14	15	16	17	18	19	20
Assets:	11.6	9.5	9.4	7.5	7.2	6.7	6.0	4.6	3.8	3.4
Income:	42.9	32.4	68.3	48.6	32.2	42.7	28.9	40.7	13.8	22.2

(a) We expect banks with more assets to earn higher income. Make a scatterplot of these data that displays the relation between assets and income. Mark Franklin (Bank 19) with a separate symbol.

(b) Describe the overall pattern of your plot. Are there any banks with unusually high or low income relative to their assets? Does Franklin stand out from other banks in your plot?

(c) Find the least-squares regression line for predicting a bank's income from its assets. Draw the regression line on your scatterplot.

(d) Use the regression line to predict Franklin's income. Was the actual income higher or lower than predicted? What is the residual?

3.74 CAN YOU THINK OF A SCATTERPLOT?

(a) Draw a scatterplot that has a positive correlation such that when one point is added, the correlation becomes negative. Circle the influential point.

(b) Draw a scatterplot that has a correlation close to 0 (say less than 0.1) such that when one point is added, the correlation is close to 1 (say greater than 0.9). Circle the influential point.

3.75 WILL WOMEN SOON OUTRUN MEN? Table 3.8 shows the men's and women's world records in the 800-meter run.

TABLE 3.8 Men's and women's world records in the 800-meter run

Year	Men's record	Women's record	Year	Men's record	Women's record
1905	113.4	—	1955	105.7	125.0
1915	111.9	—	1965	104.3	118.0
1925	111.9	144.0	1975	104.1	117.5
1935	109.7	135.6	1985	101.73	113.28
1945	106.6	132.0	1995	101.73	113.28

Source: This exercise was suggested in an article by Edward Wallace in *Mathematics Teacher*, 86, no. 9 (December 1993), p. 741.

(a) For each gender separately, do the following: Enter the data into your calculator or computer package and then plot a scatterplot. (Use the box plotting symbol for the men, and use the + plotting symbol for the women.) Describe the trend, if there is one. Perform least-squares regression and calculate the correlation. Comment on the suitability of the LSRL as a model for the data and interpret the correlation. Identify any regression outliers and influential observations.

(b) Brian Whipp and Susan Ward wrote an article based on the 800-meter run data entitled "Will Women Soon Outrun Men?" which appeared in the British journal *Nature* in 1992. They suggested in the article that women have made more progress in track events over the last half-century than men, hence the title of the article. Extend your calculator viewing window so that you can see both data sets and least-squares lines, and determine the intersection of the two LSRLs. Then comment on the premise of the *Nature* article.

3.76 MORE ON MANATEES Exercises 3.6 (page 125), 3.9 (page 129) and 3.41 (page 157) investigated the association between manatees killed and the number of powerboat registrations. For this exercise, you are to use the data for the years 1977 to 1994. Here is part of the output from the regression command in the Minitab statistical software:

```
The regression equation is
Killed = -35.2 + 0.113 Boats

Unusual Observations
Obs.  Boats  Killed    Fit  Stdev.Fit   Residual  St.Resid
 17    716   35.00  45.51        1.92     -10.51    -2.08R

R denotes an obs. with a large st. resid.
```

(a) Minitab checks for large residuals and influential observations. It calls attention to one observation that has a somewhat large residual. Circle this observation on your plot. We have no reason to remove it.

(b) Residuals from least-squares regression often have a distribution that is roughly normal. So Minitab reports the *standardized* residuals—that's what `St.Resid` means. Use the 68–95–99.7 rule for normal distributions to say how surprising a residual with standardized value –2.08 is.

3.77 JET SKI FATALITIES Exercise 3.7 (page 125) examined the association between the number of jet ski accidents and the number of jet skis in use during the period 1987 to 1996. The data also included the number of fatalities during those years.

(a) Use the methods of this chapter to investigate a possible association between the number of fatalities and the number of jet skis in use. Report your findings and support them with the appropriate numerical and graphical analyses.

(b) Use a search engine on the Internet to see which states have passed laws to regulate the use of jet skis in an attempt to reduce the number of accidents and fatalities. Are there any federal regulations for the operation of jet skis?

NOTES AND DATA SOURCES

1. Data from Personal Watercraft Industry Association, U.S. Coast Guard.
2. Based on T. N. Lam, "Estimating fuel consumption from engine size," *Journal of Transportation Engineering,* 111 (1985), pp. 339–357. The data for 10 to 50 km/h are measured; those for 60 and higher are calculated from a model given in the paper and are therefore smoothed.
3. A sophisticated treatment of improvements and additions to scatterplots is W. S. Cleveland and R. McGill, "The many faces of a scatterplot," *Journal of the American Statistical Association,* 79 (1984), pp. 807–822.
4. Data provided by Darlene Gordon, Purdue University.
5. Data from *Consumer Reports,* June 1986, pp. 366–367.
6. Data for 1995, from the 1997 *Statistical Abstract of the United States.*
7. The data are from M. A. Houck et al., "Allometric scaling in the earliest fossil bird, *Archaeopteryx lithographica,*" *Science,* 247 (1990), pp. 195–198. The authors conclude from a variety of evidence that all specimens represent the same species.
8. From a survey by the Wheat Industry Council reported in *USA Today,* October 20, 1983.
9. The data are from W. L. Colville and D. P. McGill, "Effect of rate and method of planting on several plant characters and yield of irrigated corn," *Agronomy Journal,* 54 (1962), pp. 235–238.
10. Modified from M. C. Wilson and R. E. Shade, "Relative attractiveness of various luminescent colors to the cereal leaf beetle and the meadow spittlebug," *Journal of Economic Entomology,* 60 (1967), pp. 578–580.
11. A careful study of this phenomenon is W. S. Cleveland, P. Diaconis, and R. McGill, "Variables on scatterplots look more highly correlated when the scales are increased," *Science,* 216 (1982), pp. 1138–1141.
12. *T. Rowe Price Report,* winter 1997, p. 4.
13. From W. M. Lewis and M. C. Grant, "Acid precipitation in the western United States," *Science,* 207 (1980), pp. 176–177.
14. Data from E. P. Hubble, "A relation between distance and radial velocity among extra-galactic nebulae," *Proceedings of the National Academy of Sciences,* 15 (1929), pp. 168–173.
15. Based on a plot in G. D. Martinsen, E. M. Driebe, and T. G. Whitham, "Indirect interactions mediated by changing plant chemistry: beaver browsing benefits beetles," *Ecology,* 79 (1998), pp. 192–200.

16. From the *Philadelphia City Paper*, May 23–29, 1997. Because the sodas served vary in size, we have converted soda prices to the price of a 16-ounce soda at each price per ounce.
17. From a presentation by Charles Knauf, Monroe County (New York) Environmental Health Laboratory.
18. Gary Smith, "Do statistics test scores regress toward the mean?" *Chance*, 10, No. 4(1997), pp. 42–45.
19. Data provided by Peter Cook, Purdue University.
20. The data are a random sample of 53 from the 1079 pairs recorded by K. Pearson and A. Lee, "On the laws of inheritance in man," *Biometrika*, November 1903, p. 408.
21. Modified from data provided by Dana Quade, University of North Carolina.
22. Data provided by Matthew Moore.
23. Data from R.C. Nelson, C.M. Brooks, and N.L. Pike, "Biomechanical comparison of male and female distance runners," in P. Milvy (ed.), *The Marathon: Physiological, Medical, Epidemiological, and Psychological Studies*, New York Academy of Sciences, 1977, pp. 793–807.
24. Data from D.E. Booth, *Regression Methods and Problem Banks*, COMAP, Inc., 1986.

The Granger Collection, New York

CARL FRIEDRICH GAUSS

The Gaussian Distributions

By age 18, *Carl Friedrich Gauss (1777–1855)* had independently discovered the binomial theorem, the arithmetic-geometric mean, the law of quadratic reciprocity, and the prime-number theorem. By age 21, he had made one of his most important discoveries: the construction of a regular 17-sided polygon by ruler and compasses, the first advance in the field since the early Greeks.

Gauss's contributions to the field of statistics include the method of least squares and the normal distribution, frequently called a Gaussian distribution in his honor. The normal distribution arose as a result of his attempts to account for the variation in individual observations of stellar locations. In 1801, Gauss predicted the position of a newly discovered asteroid, Ceres. Although he did not disclose his methods at the time, Gauss had used his least-squares approximation method. When the French mathematician Legendre published his version of the method of least-squares in 1805, Gauss's response was that he had known the method for years but had never felt the need to publish. This was his frequent response to the discoveries of fellow scientists. Gauss was not being boastful; rather, he cared little for fame.

In 1807, Gauss was appointed director of the University of Göttingen Observatory, where he worked for the rest of his life. He made important discoveries in number theory, algebra, conic sections and elliptic orbits, hypergeometric functions, infinite series, differential equations, differential geometry, physics, and astronomy. Five years before Samuel Morse, Gauss built a primitive telegraph device that could send messages up to a mile away. It is probably fair to say that Archimedes, Newton, and Gauss are in a league of their own among the great mathematicians.

Gauss's contributions to the field of statistics include the method of least-squares and the normal distribution, frequently called a Gaussian distribution in his honor.

chapter 4

More on Two-Variable Data

ACTIVITY 4 Modeling the Spread of Cancer in the Body

Materials: a regular six-sided die for each student; transparency grid; copy of grid for each student

Cancer begins with one cell, which divides into two cells.[1] Then these two cells divide and produce four cells. All the cancer cells produced are exactly like the original cell. This process continues until there is some intervention such as radiation or chemotherapy to interrupt the spread of the disease or until the patient dies. In this activity you will simulate the spread of cancer cells in the body.

1. Select one student to represent the original bad cell. That person rolls the die repeatedly, each roll representing a year. The number 5 will signal a cell division. When a 5 is rolled, a new student from the class will receive a die and join the original student (bad cell), so that there are now two cancer cells. These two students should be physically separated from the rest of the class, perhaps in a corner of the room.

2. As the die is rolled, another student will plot points on a transparency grid on the overhead projector. "Time," from 0 to 25 years, is marked on the horizontal axis, and the "Number of cancer cells," from 0 to 50, is on the vertical axis. The points on the grid will form a scatterplot.

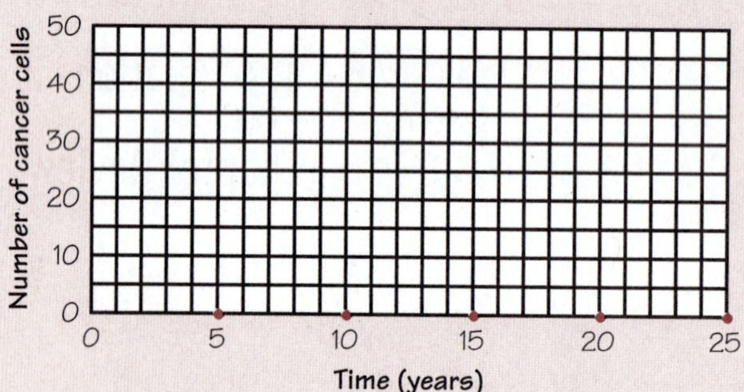

3. At a signal from the teacher, each "cancer cell" will roll his or her die. If anyone rolls the number 5, a new student from the class receives a die and joins the circle of cancer cells. The total number of cancer cells is counted, and the next point on the grid is plotted. The simulation continues until all students in the class have become cancer cells.

Questions:
Do the points show a pattern? If so, is the pattern linear? Is it a curved pattern? What mathematical function would best describe the pattern of points?

Each student should keep a copy of the transparency grid with the plotted points. We will analyze the results later in the chapter, after establishing some principles.

4.1 TRANSFORMING RELATIONSHIPS

How is the weight of an animal's brain related to the weight of its body? Figure 4.1 is a scatterplot of brain weight against body weight for 96 species of mammals.[2] This line is the least-squares regression line for predicting brain weight from body weight. The outliers are interesting. We might say that dolphins and humans are smart, hippos are dumb, and elephants are just big. That's because dolphins and humans have larger brains than their body weights suggest, hippos have smaller brains, and the elephant is much heavier than any other mammal in both body and brain.

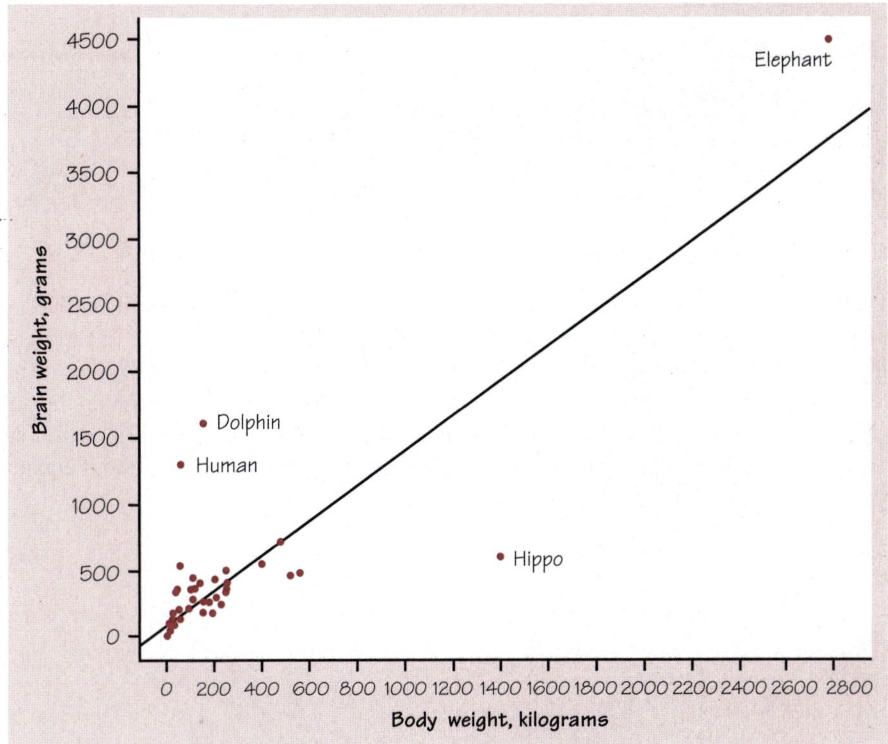

FIGURE 4.1 Scatterplot of brain weight against body weight for 96 species of mammals.

EXAMPLE 4.1 MODELING MAMMAL BRAIN WEIGHT VERSUS BODY WEIGHT

The plot in Figure 4.1 is not very satisfactory. Most mammals are so small relative to elephants and hippos that their points overlap to form a blob in the lower-left corner of the plot. The correlation between brain weight and body weight is $r = 0.86$, but this is misleading. If we remove the elephant, the correlation for the other 95 species is $r = 0.50$. Figure 4.2 is a scatterplot of the data with the four outliers removed to allow a closer look at the other 92 observations. We can now see that the relationship is not linear. It bends to the right as body weight increases.

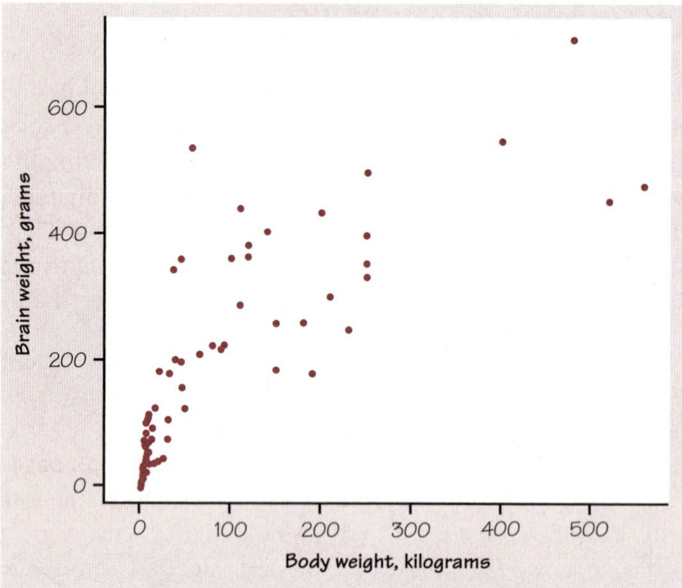

FIGURE 4.2 Brain weight against body weight for mammals, with outliers removed.

Biologists know that data on sizes often behave better if we take logarithms before doing more analysis. Figure 4.3 plots the logarithm of brain weight against the logarithm of body weight for all 96 species. The effect is almost magical. There are no longer any extreme outliers or very influential observations. The pattern is very linear, with correlation $r = 0.96$. The vertical spread about the least-squares line is similar everywhere, so that predictions of brain weight from body weight will be about equally precise for any body weight (in the log scale).

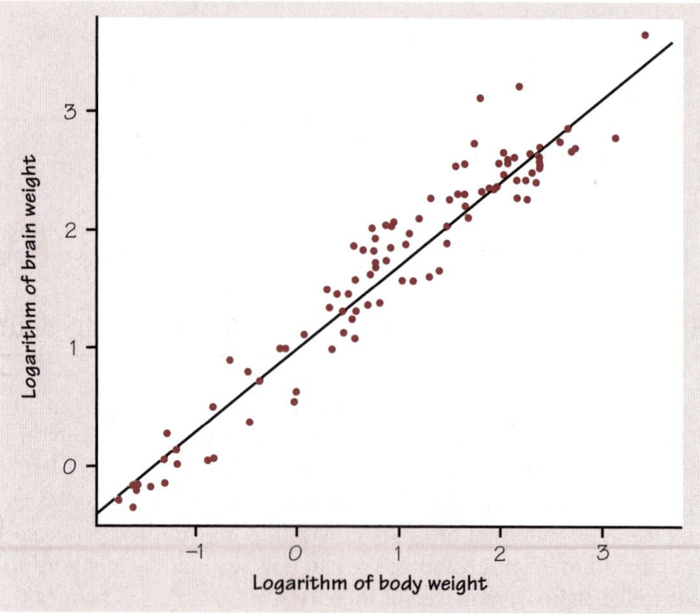

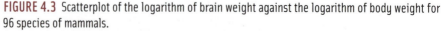

FIGURE 4.3 Scatterplot of the logarithm of brain weight against the logarithm of body weight for 96 species of mammals.

Example 4.1 shows that working with a *function* of our original measurements can greatly simplify statistical analysis. Applying a function such as the logarithm or square root to a quantitative transforming variable is called ***transforming*** or ***reexpressing*** the data. We will see in this section that understanding how simple functions work helps us choose and use transformations. Because we may want to transform either the explanatory variable x or the response variable y in a scatterplot, or both, we will call the variable t when talking about transforming in general.

transforming
reexpressing

First steps in transforming

Transforming data amounts to changing the scale of measurement that was used when the data were collected. We can choose to measure temperature in degrees Fahrenheit or in degrees Celsius, distance in miles or in kilometers. These changes of units are *linear transformations,* discussed on pages 53 to 55. **Linear transformations cannot straighten a curved relationship between two variables.** To do that, we resort to functions that are not linear. The logarithm, applied in Example 4.1, is a nonlinear function. Here are some others.

- How shall we measure the size of a sphere or of such roughly spherical objects as grains of sand or bubbles in a liquid? The size of a sphere can be expressed in terms of the diameter t, in terms of surface area (proportional to t^2), or in terms of volume (proportional to t^3). Any one of these *powers* of the diameter may be natural in a particular application.
- We commonly measure the fuel consumption of a car in miles per gallon, which is how many miles the car travels on 1 gallon of fuel. Engineers prefer to measure in gallons per mile, which is how many gallons of fuel the car needs to travel 1 mile. This is a *reciprocal* transformation. A car that gets 25 miles per gallon uses

$$\frac{1}{\text{miles per gallon}} = \frac{1}{25} = 0.04 \text{ gallons per mile}$$

The reciprocal is a *negative power* $1/t = t^{-1}$.

The transformations we have mentioned—linear, positive and negative powers, and logarithms—are those used in most statistical problems. They are all *monotonic.*

MONOTONIC FUNCTIONS

A **monotonic function** $f(t)$ moves in one direction as its argument t increases.

A **monotonic increasing function** preserves the order of data. That is, if $a > b$, then $f(a) > f(b)$.

A **monotonic decreasing function** reverses the order of data. That is, if $a > b$, then $f(a) < f(b)$.

The graph of a linear function is a straight line. The graph of a monotonic increasing function is increasing everywhere. A monotonic decreasing function has a graph that is decreasing everywhere. A function can be monotonic over some range of t without being everywhere monotonic. For example, the square function t^2 is monotonic increasing for $t \geq 0$. If the range of t includes both positive and negative values, the square is not monotonic—it decreases as t increases for negative values of t and increases as t increases for positive values.

Figure 4.4 compares three monotonic increasing functions and three monotonic decreasing functions for positive values of the argument t. Many variables take only 0 or positive values, so we are particularly interested in how functions behave for positive values of t. The increasing functions for $t > 0$ are

Linear	$a + bt$, slope $b > 0$
Square	t^2
Logarithm	$\log t$

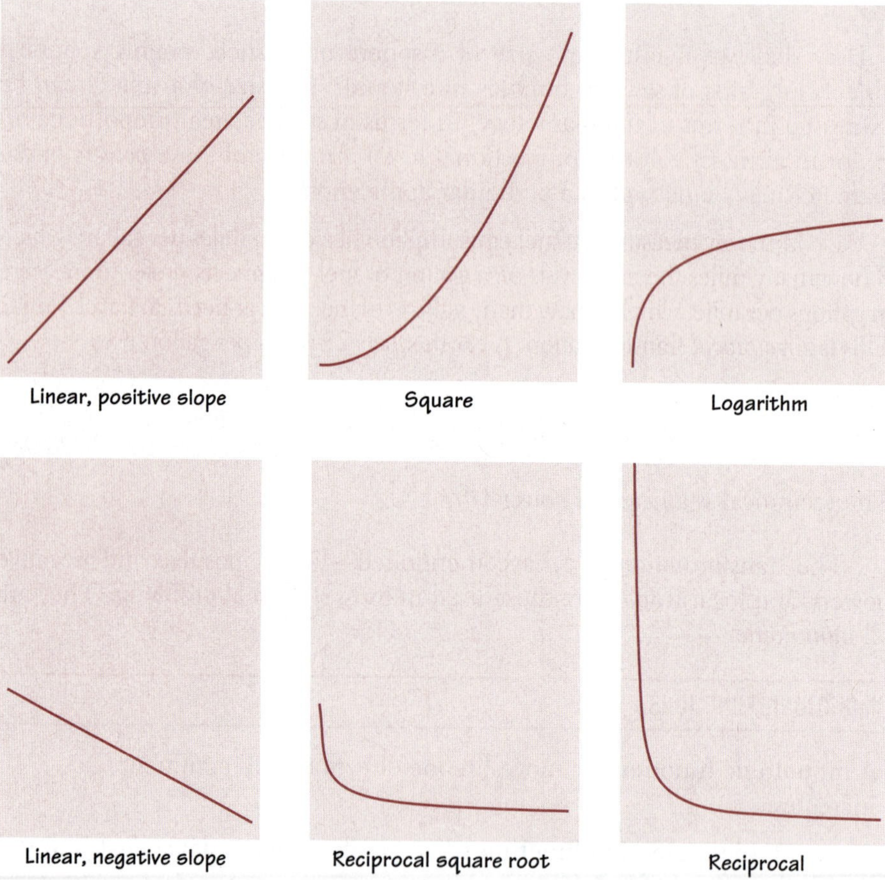

FIGURE 4.4 Three monotonic increasing functions and three monotonic decreasing functions.

The decreasing functions for $t > 0$ in the lower panel of Figure 4.4 are

Linear	$a + bt$, slope $b < 0$
Reciprocal square root	$1/\sqrt{t}$, or $t^{-1/2}$
Reciprocal	$1/t$, or t^{-1}

Nonlinear monotonic transformations change data enough to alter the shape of distributions and the form of relations between two variables, yet are simple enough to preserve order and allow recovery of the original data. We will concentrate on powers and logarithms. The even-numbered powers t^2, t^4, and so on are monotonic increasing for $t \geq 0$, but not when t can take both negative and positive values. The logarithm is not even defined unless $t > 0$. Our strategy for transforming data is therefore as follows:

1. If the variable to be transformed takes values that are 0 or negative, first apply a linear transformation to make the values all positive. Often we just add a constant to all the observations.

2. Then choose a power or logarithmic transformation that simplifies the data, for example, one that approximately straightens a scatterplot.

EXERCISES

4.1 Which of these transformations are monotonic increasing? Monotonic decreasing? Not monotonic? Give an equation for each transformation.

(a) You transform height in inches to height in centimeters.

(b) You transform typing speed in words per minute into seconds needed to type a word.

(c) You transform the diameter of a coin to its circumference.

(d) A composer insists that her new piece of music should take exactly 5 minutes to play. You time several performances, then transform the time in minutes into squared error, the square of the difference between 5 minutes and the actual time.

4.2 Suppose that t is an angle, measured in degrees between 0° and 180°. On what part of this range is the function $\sin t$ monotonic increasing? Monotonic decreasing?

The ladder of power transformations

Though simple in algebraic form and easy to compute with a calculator, the power and logarithm functions are varied in their behavior. It is natural to think of powers such as

$$\ldots, t^{-1}, t^{-1/2}, t^{1/2}, t, t^2, \ldots$$

as a hierarchy or ladder. Some facts about this ladder will help us choose transformations. In all cases, we look only at positive values of the argument t.

MONOTONICITY OF POWER FUNCTIONS

Power functions t^p for positive powers p are monotonic increasing for values $t > 0$. They preserve the order of observations. This is also true of the logarithm.

Power functions t^p for negative powers p are monotonic decreasing for values $t > 0$. They reverse the order of the observations.

It is hard to interpret graphs when the order of the original observations has been reversed. We can make a negative power such as the reciprocal $1/t$ monotonic increasing rather than monotonic decreasing by using $-1/t$ instead. Figure 4.5 takes this idea a step farther. This graph compares the ladder of power functions in the form

$$\frac{t^p - 1}{p}$$

The reciprocal (power $p = -1$), for example, is graphed as

$$\frac{1/x - 1}{-1} = 1 - \frac{1}{x}$$

This linear transformation does not change the nature of the power functions t^p, except that all are now monotonic increasing. It is chosen so that every power has the value 0 at $t = 1$ and also has slope 1 at that point. So the graphs in Figure 4.5 all touch at $t = 1$ and go through that point at the same slope.

Look at the $p = 0$ graph in Figure 4.5. The 0th power t^0 is just the constant 1, which is not very useful. The $p = 0$ entry in the figure is not constant. In fact, it is the logarithm, log t. That is, **the logarithm fits into the ladder of power transformations at $p = 0$.**[3]

Figure 4.5 displays another key fact about these functions. The graph of a linear function (power $p = 1$) is a straight line. Powers greater than 1 give graphs that bend upward. That is, the transformed variable grows ever faster as t gets larger. Powers less than 1 give graphs that bend downward. The transformed values continue to grow with t, but at a rate that decreases as t increases. What is more, the sharpness of the bend increases as we move away from $p = 1$ in either direction.

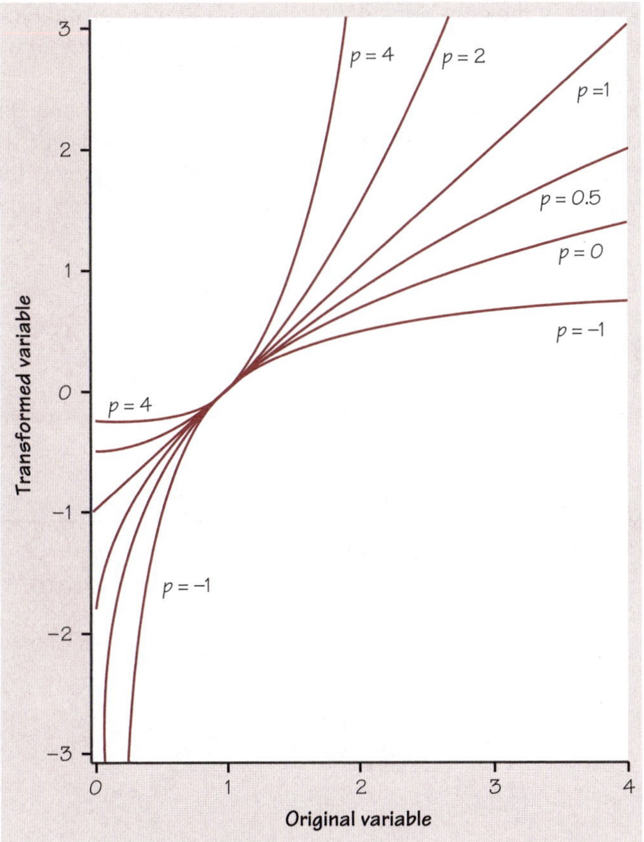

FIGURE 4.5 The ladder of power functions in the form $(t^p - 1)/p$.

> **CONCAVITY OF POWER FUNCTIONS**
>
> Power transformations t^p for powers p greater than 1 are **concave up**; that is, they have the shape ⌡. These transformations push out the right tail of a distribution and pull in the left tail. This effect gets stronger as the power p moves up away from 1.
>
> Power transformations t^p for powers p less than 1 (and the logarithm for $p = 0$) are **concave down**; that is, they have the shape ⌢. These transformations pull in the right tail of a distribution and push out the left tail. This effect gets stronger as the power p moves down away from 1.

EXAMPLE 4.2 A COUNTRY'S GDP AND LIFE EXPECTANCY

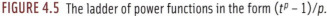

Figure 4.6(a) is a scatterplot of data from the World Bank.[4] The individuals are all the world's nations for which data are available. The explanatory variable x is a measure of

how rich a country is: the gross domestic product (GDP) per person. GDP is the total value of the goods and services produced in a country, converted into dollars. The response variable y is life expectancy at birth.

Life expectancy increases in richer nations, but only up to a point. The pattern in Figure 4.6(a) at first rises rapidly as GDP increases but then levels out. Three African nations (Botswana, Gabon, and Namibia) are outliers with much lower life expectancy than the overall pattern suggests. Can we straighten the overall pattern by transforming?

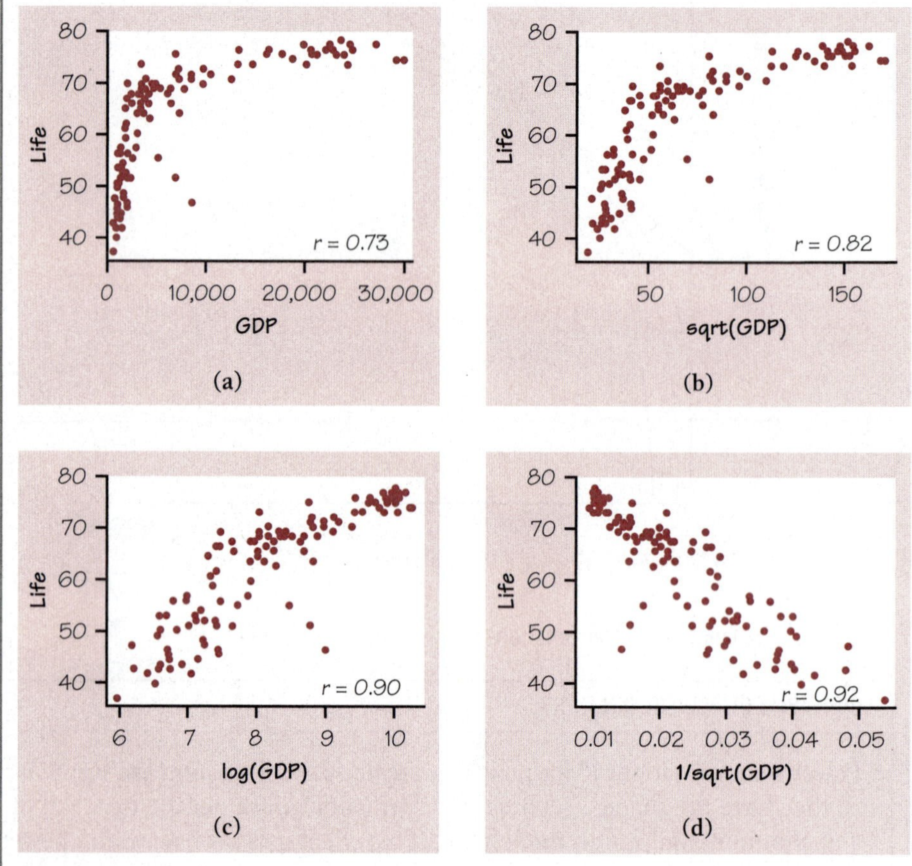

FIGURE 4.6 The ladder of transformations at work. The data are life expectancy and gross domestic product (GDP) for 115 nations. Panel (a) displays the original data. Panels (b), (c), and (d) transform GDP, moving down the ladder away from linear functions.

Life expectancy does not have a large range, but we can see that the distribution of GDP is right-skewed and very spread out. So GDP is a better candidate for transformation. We want to pull in the long right tail, so we try transformations with $p < 1$. Figures 4.6(b), (c), and (d) show the results of three transformations of GDP. The r-value in each figure is the correlation when the three outliers are omitted.

The square root $\sqrt{x}$, with $p = 1/2$, reduces the curvature of the scatterplot, but not enough. The logarithm $\log x$ ($p = 0$) straightens the pattern more, but it still bends to the right. The reciprocal square root $1/\sqrt{x}$, with $p = -1/2$, gives a pattern that is quite straight except for the outliers. To avoid reversing the order of the observations, we actually used $-1/\sqrt{x}$.

EXERCISES

4.3 MUSCLE STRENGTH AND WEIGHT, I Bigger people are generally stronger than smaller people, though there's a lot of individual variation. Let's find a theoretical model. Body weight increases as the cube of height. The strength of a muscle increases with its cross-sectional area, which we expect to go up as the square of height. Put these together: What power law should describe how muscle strength increases with weight?

4.4 MUSCLE STRENGTH AND WEIGHT, II Let's apply your result from the previous problem. Graph the power law relation between strength and body weight for weights from (say) 1 to 1000. (Constants in the power law just reflect the units of measurement used, so we can ignore them.) Use the graph to explain why a person 1 million times as heavy as an ant can't lift a million times as much as an ant can lift.

4.5 HEART RATE AND BODY RATE Physiologists say that resting heart rate of humans is related to our body weight by a power law. Specifically, average heart rate y (beats per minute) is found from body weight x (kilograms) by[5]

$$y = 241 \times x^{-1/4}$$

Let's try to make sense of this. Kleiber's law says that energy use in animals, including humans, increases as the 3/4 power of body weight. But the weight of human hearts and lungs and the volume of blood in the body are directly proportional to body weight. Given these facts, you should not be surprised that heart rate is proportional to the $-1/4$ power of body weight. Why not?

Example 4.2 shows the ladder of powers at work. As we move down the ladder from linear transformations (power $p = 1$), the scatterplot gets straighter. Moving farther down the ladder, to the reciprocal $1/x = x^{-1}$, begins to bend the plot in the other direction. But this "try it and see" approach isn't very satisfactory. That life expectancy depends linearly on $1/\sqrt{\text{GDP}}$ does not increase our understanding of the relationship between the health and wealth of nations. We don't recommend just pushing buttons on your calculator to try to straighten a scatterplot.

It is much more satisfactory to begin with a theory or mathematical model that we expect to describe a relationship. The transformation needed to make the relationship linear is then a consequence of the model. One of the most common models is *exponential growth*.

Exponential growth

A variable grows linearly over time if it *adds* a fixed increment in each equal time period. Exponential growth occurs when a variable is *multiplied* by a fixed number in each time period. To grasp the effect of multiplicative growth, consider a population of bacteria in which each bacterium splits into two each hour. Beginning with a single bacterium, we have 2 after one hour, 4 at the end of two hours, 8 after three hours, then 16, 32, 64, 128, and so on. These first few numbers are deceiving. After 1 day of doubling each hour, there are 2^{24} (16,777,216) bacteria in the population. That number then doubles the next hour! Try successive multiplications by 2 on your calculator to see for yourself the very rapid increase after a slow start. Figure 4.7 shows the growth of the bacteria population over 24

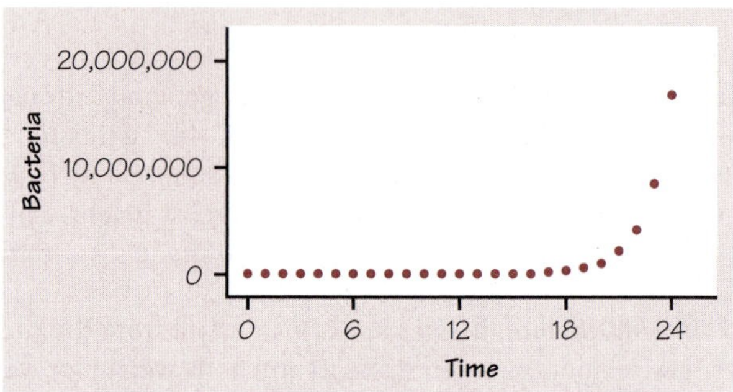

FIGURE 4.7 Growth of a bacteria population over a 24-hour period.

hours. For the first 15 hours, the population is too small to rise visibly above the zero level on the graph. It is characteristic of exponential growth that the increase appears slow for a long period, then seems to explode.

LINEAR VERSUS EXPONENTIAL GROWTH

Linear growth increases by a fixed *amount* in each equal time period. **Exponential growth** increases by a fixed *percentage* of the previous total.

Populations of living things—like bacteria and the malignant cancer cells in Activity 4—tend to grow exponentially if not restrained by outside limits such as lack of food or space. More pleasantly, money also displays exponential growth when returns to an investment are compounded. Compounding means that last period's income earns income this period.

EXAMPLE 4.3 THE GROWTH OF MONEY

A dollar invested at an annual rate of 6% turns into \$1.06 in a year. The original dollar remains and has earned \$0.06 in interest. That is, 6% annual interest means that any amount on deposit for the entire year is multiplied by 1.06. If the \$1.06 remains invested for a second year, the new amount is therefore 1.06×1.06, or 1.06^2. That is only \$1.12, but this in turn is multiplied by 1.06 during the third year, and so on. After x years, the dollar has become 1.06^x dollars.

If the Native Americans who sold Manhattan Island for \$24 in 1626 had deposited the \$24 in a savings account at 6% annual interest, they would now have almost \$80 billion. Our savings accounts don't make us billionaires, because we don't stay around long enough. A century of growth at 6% per year turns \$24 into \$8143. That's 1.06^{100} times \$24. By 1826, two centuries after the sale, the account would hold a bit over \$2.7 million. Only after a patient 302 years do we finally reach \$1 billion. That's real money, but 302 years is a long time.

exponential growth model

The count of bacteria after x hours is 2^x. The value of \$24 invested for x years at 6% interest is 24×1.06^x. Both are examples of the ***exponential growth model*** $y = a \times b^x$ for different constants a and b. In this model, the response y is multiplied by b in each time period.

EXAMPLE 4.4 GROWTH OF CELL PHONE USE

Does the exponential growth model sometimes describe real data that don't arise from any obvious process of multiplying by a fixed number over and over again? Let's look at the cell phone phenomenon in the United States. Cell phones have revolutionized the communications industry, the way we do business, and the way we stay in touch with friends and family. The industry enjoyed substantial growth in the 1990s. One way to measure cell phone growth in the 1990s is to look at the number of subscribers. Table 4.1 and Figure 4.8 show the growth of cell phone subscribers from 1990 to 1999.

TABLE 4.1 The number of cell phone subscribers in the United States, 1990–1999

Year	1990	1993	1994	1995	1996	1997	1998	1999
Subscribers (thousands)	5283	16,009	24,134	33,786	44,043	55,312	69,209	86,047

Source: Statistical Abstract of the United States, 2000 and the Cellular Telecommunications Industry Association, Washington, D.C.

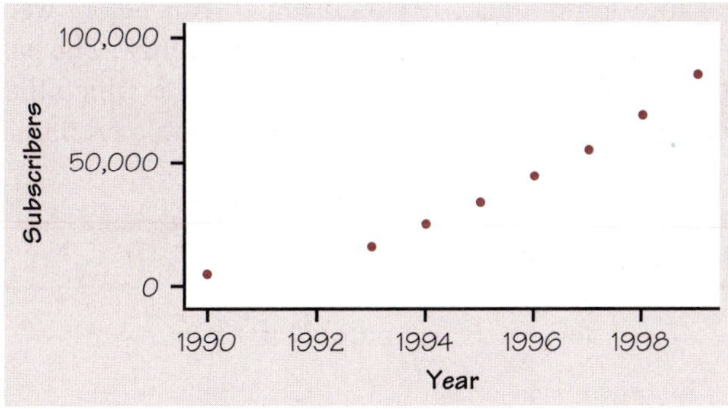

FIGURE 4.8 Scatterplot of cell phone growth versus year, 1990–1999.

There is an increasing trend, but the overall pattern is not linear. The number of cell phone subscribers has increased much faster than linear growth. The pattern of growth follows a smooth curve, and it looks a lot like an exponential curve. Is this exponential growth?

The logarithm transformation

The growth curve for the number of cell phone subscribers does look somewhat like the exponential curve in Figure 4.7, but our eyes are not very good at comparing curves of roughly similar shape. We need a better way to check whether growth is exponential. If you suspect exponential growth, you should first calculate ratios of consecutive terms. In Table 4.2, we have divided each entry in the "Subscribers" column (the y variable) by its predecessor, leaving out both the first value of y, because it doesn't have a predecessor, and the second value, because the x increment is not 1. Notice that the ratios are not *exactly* the same, but they are *approximately* the same.

TABLE 4.2 Ratios of consecutive y-values and the logarithms of the y-values for the cell phone data of Example 4.4

Year	Subscribers	Ratios	log(y)
1990	5,283	—	3.72288
1993	16,009	—	4.20436
1994	24,134	1.51	4.38263
1995	33,786	1.40	4.52874
1996	44,043	1.30	4.64388
1997	55,312	1.26	4.74282
1998	69,209	1.25	4.84016
1999	86,047	1.24	4.93474

The next step is to apply a mathematical transformation that changes exponential growth into linear growth—and patterns of growth that are not exponential into something other than linear. But before we do the transformation, we need to review the properties of logarithms. The basic idea of a logarithm is this: $\log_2 8 = 3$ because 3 is the exponent to which the base 2 must be raised to yield 8. Here is a quick summary of algebraic properties of logarithms:

ALGEBRAIC PROPERTIES OF LOGARITHMS

$$\log_b x = y \quad \text{if and only if} \quad b^y = x$$

The rules for logarithms are

1. $\log(AB) = \log A + \log B$
2. $\log(A/B) = \log A - \log B$
3. $\log X^p = p \log X$

EXAMPLE 4.5 TRANSFORMING CELL PHONE GROWTH

Returning to the cell phone growth model, we hypothesize an exponential model of the form $y = ab^x$ where a and b represent constants. The necessary transformation is carried out by taking the logarithm of both sides of this equation:

$$\begin{aligned} \log y &= \log(ab^x) \\ &= \log a + \log b^x && \text{using Rule 1} \\ &= \log a + (\log b)x && \text{using Rule 3} \end{aligned}$$

Notice that log a and log b are constants because a and b are constants. So the right side of the equation looks like the form for a straight line. That is, if our data really are growing exponentially and we plot log y versus x, we should observe a straight line for the transformed data. Table 4.2 includes the logarithms of the y-values. Figure 4.9 plots points in the form (x, log y).

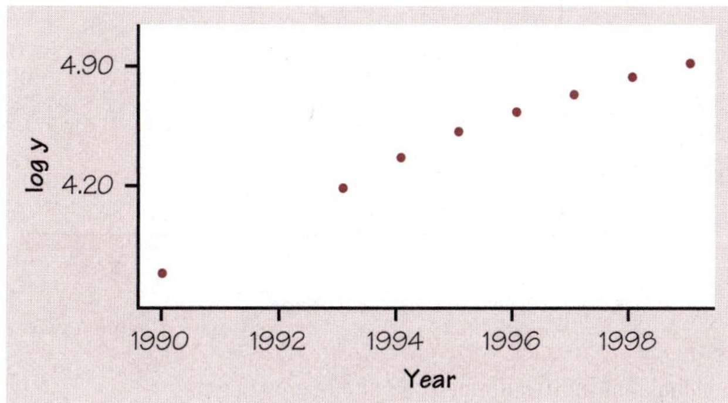

FIGURE 4.9 Scatterplot of log(subscribers) versus year.

The plot appears to be slightly concave down, but it is more linear than our original scatterplot. Applying least-squares regression to the transformed data, Minitab reports:

```
LOG(Y) = - 263 + 0.134 YEAR

Predictor        Coef        Stdev    t-ratio        p
Constant      -263.20        14.63     -17.99    0.000
YEAR         0.134170     0.007331      18.30    0.000

s = 0.05655    R-sq = 98.2%     R-sq(adj) = 97.9%
```

As is usually the case, Minitab tells us more than we want to know, but observe that the value of r^2 is 0.982. That means that 98.2% of the variation in log y is explained by least-squares regression of log y on x. That's pretty impressive. Let's continue. Figure 4.10 is a plot of the transformed data along with the fitted line.

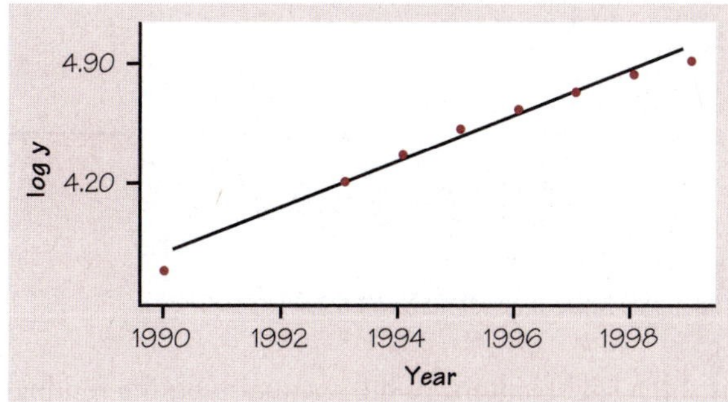

FIGURE 4.10 Plot of transformed data with least-squares line.

This appears to be a useful model for prediction purposes. Although the r^2-value is high, one should always inspect the residual plot to further assess the quality of the model. Figure 4.11 is a residual plot.

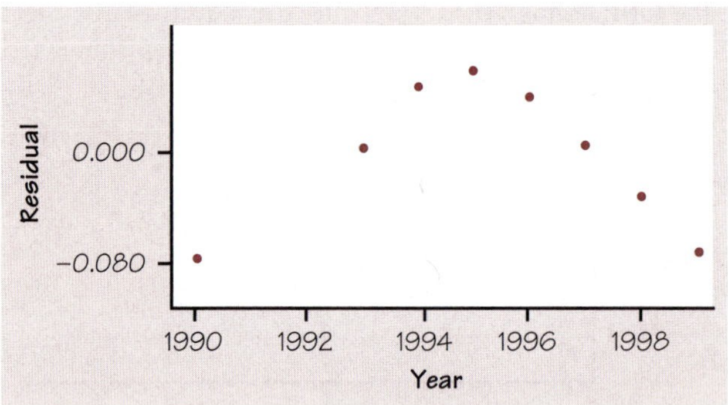

FIGURE 4.11 Residual plot for transformed cell phone growth data.

This is a surprise. But it also suggests an adjustment. The very regular pattern of the last four points really does look linear. So if the purpose is to be able to predict the number of subscribers in the year 2000, then one approach would be to discard the first four points, because they are the oldest and furthest removed from the year 2000, and retain the last four points. If you do this, the least-squares line for the four transformed points (years 1996 through 1999) is

```
log NewY = -189 + 0.0970 NewX
```

and the r^2-value improves to 1. The actual r^2-value is 0.999897 to six decimal places. The residual plot is shown in Figure 4.12.

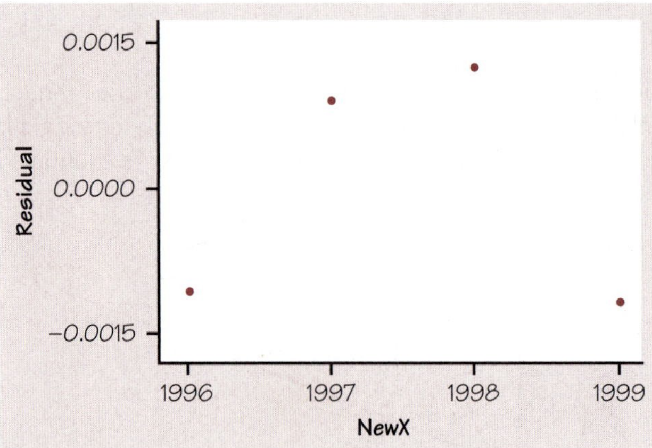

FIGURE 4.12 Residual plot for reduced transformed data set.

Although there is still a slight pattern in the residual plot, the residuals are very small in magnitude, and the r^2 value is nearly 1.

Prediction in the exponential growth model

Regression is often used for prediction. When we fit a least-squares regression line, we find the predicted response y for any value of the explanatory variable x by substituting our x-value into the equation of the line. In the case of exponential growth, the logarithms rather than the actual responses follow a linear pattern. To do prediction, we need to "undo" the logarithm transformation to return to the original units of measurement. The same idea works for any monotonic transformation. There is always exactly one original value behind any transformed value, so we can always go back to our original scale.

EXAMPLE 4.6 PREDICTING CELL PHONE GROWTH FOR 2000

Our examination of cell phone growth left us with four transformed data points and a least-squares line with equation

```
log(subscribers) = -189 + 0.0970(year)
```

To perform the back-transformation, we need to do the inverse operation. The inverse operation of the logarithmic function is raising 10 to a power. If we raise 10 to the left side of the equation, and set that equal to 10 raised to the right side of the equation, we will eliminate the log() on the left;

$$10^{\log(\text{subscribers})} = 10^{-189 + 0.0970(\text{year})}$$

Then

$$\text{subscribers} = (10^{-189})(10^{0.0970(\text{year})})$$

To then predict the number of subscribers in the year 2000, we substitute 2000 for year and solve for number of subscribers. The problem is that the first factor is too small a quantity for the calculator, and it will evaluate to 0. To get around this machine difficulty, if you have installed the equation of the least-squares line in the calculator as Y1, then define Y2 to be 10^Y1. Doing this, we find that the predicted number of subscribers for the year 2000 is Y2(2000) = 10,7864.5. Alternatively, we could have coded the years to avoid the overflow problem.

Postscript: The stock market tumbled in 2000, the economy floundered, unemployment increased, and the cell phone industry in particular had a very poor year. So predicting the number of cell phone subscribers in 2000 is risky indeed.

Make sure that you understand the big idea here. The necessary transformation is carried out by taking the logarithm of the response variable. Your calculator and most statistical software will calculate the logarithms of all the values of a variable with a single command. The essential property of the logarithm for our purposes is that it straightens an exponential growth curve. **If a variable grows exponentially, its logarithm grows linearly.**

EXAMPLE 4.7 TRANSFORMING BACTERIA COUNTS

Figure 4.13 plots the logarithms of the bacteria counts in Figure 4.7 (page 204). Sure enough, exact exponential growth turns into an exact straight line when we plot the logarithms. After 15 hours, for example, the population contains $2^{15} = 32{,}768$ bacteria. The logarithm of 32,768 is 4.515, and this point appears above the 15-hour mark in Figure 4.13.

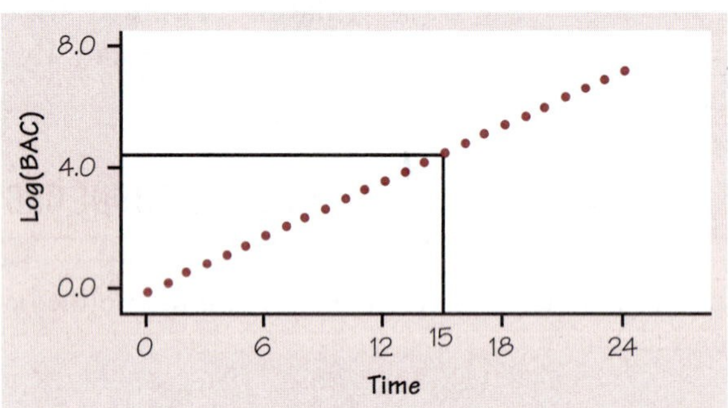

FIGURE 4.13 Logarithms of the bacteria counts.

TECHNOLOGY TOOLBOX *Modeling exponential growth with the TI-83/89*

The Census Bureau classifies residents of the United States as being either white; black; Hispanic origin; American Indian, Eskimo, Aleut; or Asian, Pacific Islander. The population totals for these last two categories, from 1950 to 1990, are[6]

Year:	1950	1960	1970	1980	1990
Population (thousands):	1131	1620	2557	5150	9534

- Code the years using 1900 as the reference year, 0. Then 1950 is coded as 50, and so forth. Enter the coded years and population, in thousands, in L_1/list1 and L_2/list2. Then plot the scatterplot.

TI-83

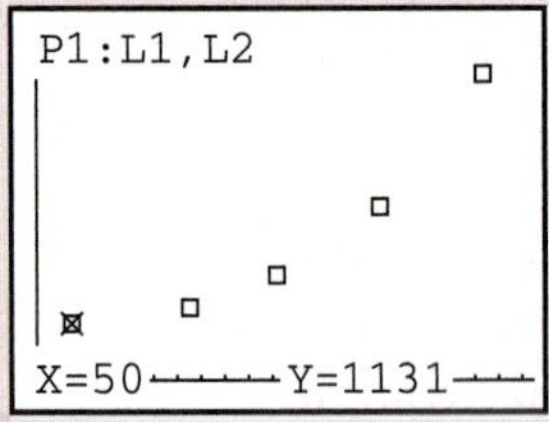

TI-89

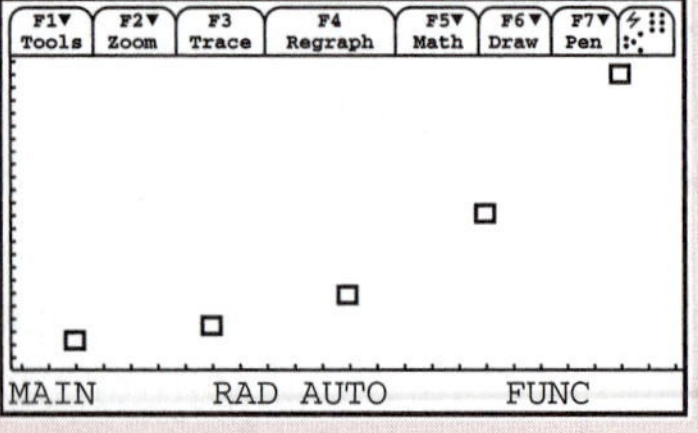

- Assuming an exponential model, here is a plot of log(POP), in L_3, versus YEAR on the TI-83. We'll plot ln(POP) versus YEAR on the TI-89 since the natural logarithm key is more accessible on the TI-89. The pattern is the same, but the regression equation numbers will be different.

TECHNOLOGY TOOLBOX *Modeling exponential growth with the TI-83/89 (continued)*

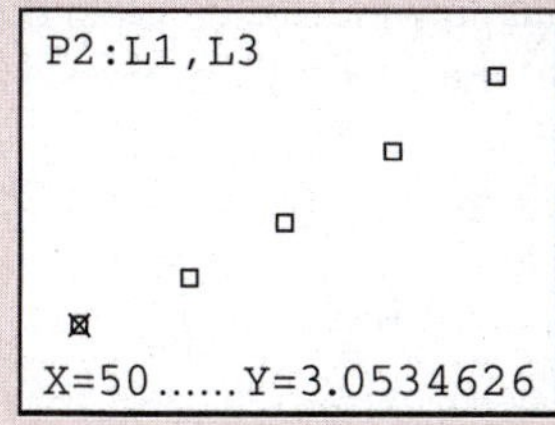

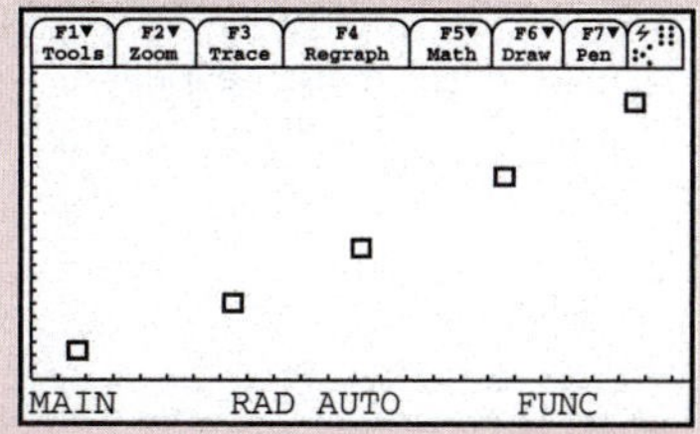

- The plot still shows a little upward concavity, and the residual plot will confirm this. Next, we perform least-squares regression on the transformed data.

```
LinReg
 y=a+bx
 a=1.824616035
 b=.023539173
 r2=.9841134289
 r=.9920249135
```

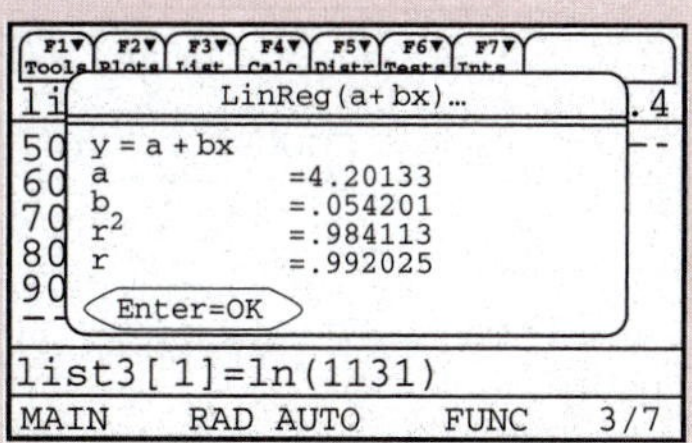

- Notice that the values of *a* and *b* in the equation of the least-squares line are different for the two calculators. That's because we use base 10 (log) on the TI-83 and we used base *e* (ln) on the TI-89. The final predicted values will be the same regardless of which route we take. Here are the scatterplots with the least-squares lines:

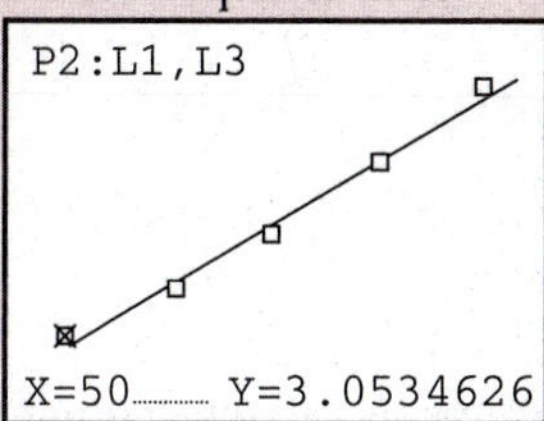

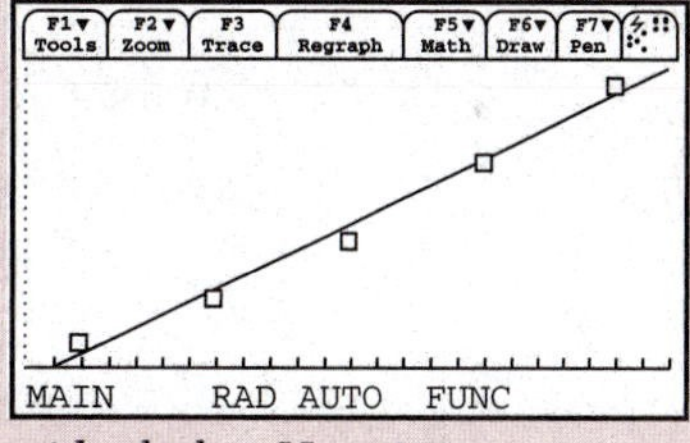

- Despite the high r^2-value, you should always inspect the residual plot. Here it is:

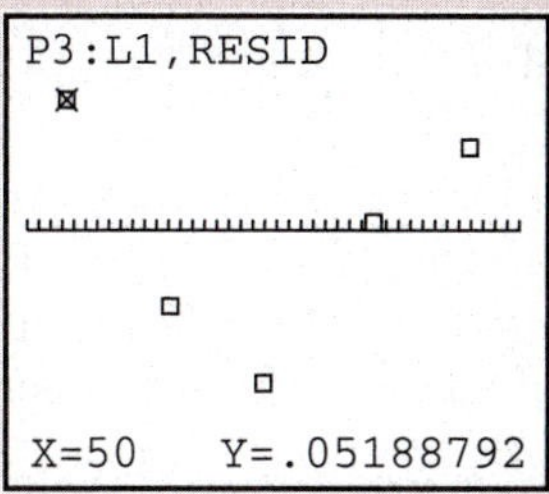

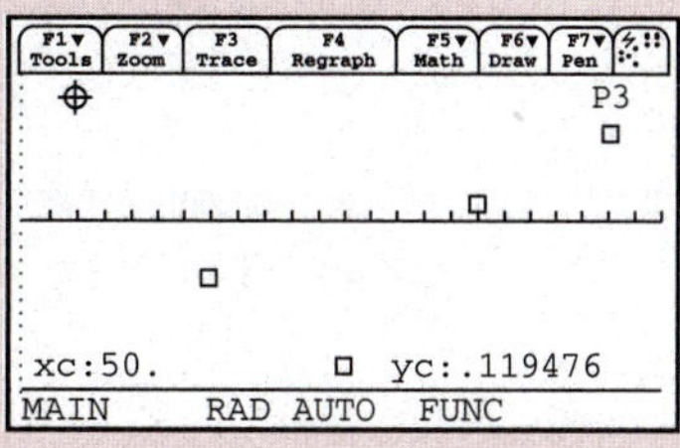

Ideally, the residual plot should show random scatter about the y = 0 reference line. The fact that the residual plot still shows a clearly curved pattern tells us that some improvement is still possible. For now, though, we will accept the exponential model on the basis of the high r^2-value ($r^2 = 0.992$).

- Now we're ready to predict the population of American Indians, Eskimos, Aleuts, Asians, and Pacific Islanders for the year 2000. With the regression equation installed as Y1, define Y2 = 10^Y1 on the TI-83, and Y2 = e^Y1 on the TI-89. The predicted population in year 2000 is then Y2(100) = 15,084.584 on the TI-83, and 15,084.7 on the TI-89. The difference is due to roundoff error. Since the table entries are in thousands, the actual predicted population is approximately 15,085,000. Looking at the plots, do you think this prediction will be too high or too low? Why?

EXERCISES

4.6 GYPSY MOTHS Biological populations can grow exponentially if not restrained by predators or lack of food. The gypsy moth outbreaks that occasionally devastate the forests of the Northeast illustrate approximate exponential growth. It is easier to count the number of acres defoliated by the moths than to count the moths themselves. Here are data on an outbreak in Massachusetts:[7]

Year	Acres
1978	63,042
1979	226,260
1980	907,075
1981	2,826,095

(a) Plot the number of acres defoliated y against the year x. The pattern of growth appears exponential.

(b) Verify that y is being multiplied by about 4 each year by calculating the ratio of acres defoliated each year to the previous year. (Start with 1979 to 1978, when the ratio is 226,260/63,042 = 3.6.)

(c) Take the logarithm of each number y and plot the logarithms against the year x. The linear pattern confirms that the growth is exponential.

(d) Verify that the least-squares line fitted to the transformed data is

$$\log \hat{y} = -1094.51 + 0.5558 \times \text{year}$$

(e) Construct and interpret a residual plot for $\log \hat{y}$ on year.

(f) Perform the inverse transformation to express $\hat{y}$ as an exponential equation. Display a scatterplot of the original data with the exponential curve model superimposed. Is your exponential function a satisfactory model for the data?

(g) Use your model to predict the number of acres defoliated in 1982.

(*Postscript:* A viral disease reduced the gypsy moth population between the readings in 1981 and 1982. The actual count of defoliated acres in 1982 was 1,383,265.)

4.7 MOORE'S LAW, I Gordon Moore, one of the founders of Intel Corporation, predicted in 1965 that the number of transistors on an integrated circuit chip would double every 18 months. This is "Moore's law," one way to measure the revolution in computing. Here are data on the dates and number of transistors for Intel microprocessors:[8]

Processor	Date	Transistors	Processor	Date	Transistors
4004	1971	2,250	486 DX	1989	1,180,000
8008	1972	2,500	Pentium	1993	3,100,000
8080	1974	5,000	Pentium II	1997	7,500,000
8086	1978	29,000	Pentium III	1999	24,000,000
286	1982	120,000	Pentium 4	2000	42,000,000
386	1985	275,000			

(a) Explain why Moore's law says that the number of transistors grows exponentially over time.

(b) Make a plot suitable to check for exponential growth. Does it appear that the number of transistors on a chip has in fact grown approximately exponentially?

4.8 MOORE'S LAW, II Return to Moore's law, described in Exercise 4.7.

(a) Find the least-squares regression line for predicting the logarithm of the number of transistors on a chip from the date. Before calculating your line, subtract 1970 from all the dates so that 1971 becomes year 1, 1972 is year 2, and so on.

(b) Suppose that Moore's law is exactly correct. That is, the number of transistors is 2250 in year 1 (1971) and doubles every 18 months (1.5 years) thereafter. Write the model for predicting transistors in year x after 1970. What is the equation of the line that, according to your model, connects the logarithm of transistors with x? Explain why a comparison of this line with your regression line from (a) shows that although transistor counts have grown exponentially, they have grown a bit more slowly than Moore's law predicts.

4.9 *E. COLI* (Exact exponential growth) The common intestinal bacterium *E. coli* is one of the fastest-growing bacteria. Under ideal conditions, the number of *E. coli* in a colony doubles about every 15 minutes until restrained by lack of resources. Starting from a single bacterium, how many *E. coli* will there be in 1 hour? In 5 hours?

4.10 GUN VIOLENCE (Exact exponential growth) A paper in a scholarly journal once claimed (I am not making this up), "Every year since 1950, the number of American children gunned down has doubled."[9] To see that this is silly, suppose that in 1950 just 1 child was "gunned down" and suppose that the paper's claim is exactly right.

(a) Make a table of the number of children killed in each of the next 10 years, 1951 to 1960.

(b) Plot the number of deaths against the year and connect the points with a smooth curve. This is an exponential curve.

(c) The paper appeared in 1995, 45 years after 1950. How many children were killed in 1995, according to the paper?

(d) Take the logarithm of each of your counts from (a). Plot these logarithms against the year. You should get a straight line.

(e) From your graph in (d) find the approximate values of the slope b and the intercept a for the line. Use the equation $y = a + bx$ to predict the logarithm of the count for the 45th year. Check your result by taking the logarithm of the count you found in (c).

4.11 U.S. POPULATION The following table gives the resident population of the United States from 1790 to 2000, in millions of persons:

Date	Pop.	Date	Pop.	Date	Pop.	Date	Pop.
1790	3.9	1850	23.2	1910	92.0	1970	203.3
1800	5.3	1860	31.4	1920	105.7	1980	226.5
1810	7.2	1870	39.8	1930	122.8	1990	248.7
1820	9.6	1880	50.2	1940	131.7	2000	281.4
1830	12.9	1890	62.9	1950	151.3		
1840	17.1	1900	76.0	1960	179.3		

(a) Plot population against time. The growth of the American population appears roughly exponential.

(b) Plot the logarithms of population against time. The pattern of growth is now clear. An expert says that "the population of the United States increased exponentially from 1790 to about 1880. After 1880 growth was still approximately exponential, but at a slower rate." Explain how this description is obtained from the graph.

(c) Use part or all the data to construct an exponential model for the purpose of predicting the population in 2010. Justify your modeling decision. Then predict the population in the year 2010. Do you think your prediction will be too low or too high? Explain.

(d) Construct a residual plot for the transformed data. What is the value of r^2 for the transformed data?

(e) Comment on the quality of your model.

Power law models

When you visit a pizza parlor, you order a pizza by its diameter, say 10 inches, 12 inches, or 14 inches. But the amount you get to eat depends on the *area* of the pizza. The area of a circle is π times the square of its radius. So the area of a round pizza with diameter x is

$$\text{area} = \pi r^2 = \pi(x/2)^2 = \pi(x^2/4) = (\pi/4)x^2$$

power law model

This is a ***power law model*** of the form

$$y = a \times x^p$$

When we are dealing with things of the same general form, whether circles or fish or people, we expect area to go up with the square of a dimension such as diameter or height. Volume should go up with the cube of a linear dimension. That is, geometry tells us to expect power laws in some settings.

Biologists have found that many characteristics of living things are described quite closely by power laws. There are more mice than elephants, and more flies than mice—the abundance of species follows a power law with body weight as the explanatory variable. So do pulse rate, length of life, the number of eggs a bird lays, and so on. Sometimes the powers can be predicted from geometry, but sometimes they are mysterious. Why, for example, does the rate at which animals use energy go up as the 3/4 power of their body weight? Biologists call this relationship *Kleiber's law*. It has been found to work all the way from bacteria to whales. The search goes on for some physical or geometrical explanation for why life follows power laws. There is as yet no general explanation, but power laws are a good place to start in simplifying relationships for living things.

Exponential growth models become linear when we apply the logarithm transformation to the response variable y. **Power law models become linear when we apply the logarithm transformation to both variables.** Here are the details:

1. The power law model is

$$y = a \times x^p$$

2. Take the logarithm of both sides of this equation. You see that

$$\log y = \log a + p \log x$$

That is, taking the logarithm of both variables straightens the scatterplot of y against x.

3. Look carefully: The *power* p in the power law becomes the *slope* of the straight line that links $\log y$ to $\log x$.

Prediction in power law models

If taking the logarithms of both variables makes a scatterplot linear, a power law is a reasonable model for the original data. We can even roughly estimate what power p the law involves by regressing $\log y$ on $\log x$ and using the slope of the regression line as an estimate of the power. Remember that the slope is only an estimate of the p in an underlying power model. The greater the scatter of the points in the scatterplot about the fitted line, the smaller our confidence that this estimate is accurate.

EXAMPLE 4.8 PREDICTING BRAIN WEIGHT

The magical success of the logarithm transformation in Example 4.1 on page 195 would not surprise a biologist. We suspect that a power law governs this relationship. Least-squares regression for the scatterplot in Figure 4.3 on page 196 gives the line

$$\log \hat{y} = 1.01 + 0.72 \times \log x$$

for predicting the logarithm of brain weight from the logarithm of body weight. To undo the logarithm transformation, remember that for common logarithms with base 10, $y = 10^{\log y}$. We see that

$$\begin{aligned}\hat{y} &= 10^{1.01 + 0.72 \log x} \\ &= 10^{1.01} \times 10^{0.72 \log x} \\ &= 10.2 \times (10^{\log x})^{0.72}\end{aligned}$$

Because $10^{\log x} = x$, the estimated power model connecting predicted brain weight $\hat{y}$ with body weight x for mammals is

$$\hat{y} = 10.2 \times x^{0.72}$$

Based on footprints and some other sketchy evidence, some people think that a large apelike animal, called Sasquatch or Bigfoot, lives in the Pacific Northwest. His weight is estimated to be about 280 pounds, or 127 kilograms. How big is Bigfoot's brain? Based on the power law estimated from data on other mammals, we predict

$$\begin{aligned}\hat{y} &= 10.2 \times 127^{0.72} \\ &= 10.2 \times 32.7 \\ &= 333.7 \text{ grams}\end{aligned}$$

For comparison, gorillas have an average body weight of about 140 kilograms and an average brain weight of about 406 grams. Of course, Bigfoot may have a larger brain than his weight predicts—after all, he has avoided being captured, shot, or videotaped for many years.

EXAMPLE 4.9 FISHING TOURNAMENT

Imagine that you have been put in charge of organizing a fishing tournament in which prizes will be given for the heaviest fish caught. You know that many of the fish caught during the tournament will be measured and released. You are also aware that trying to weigh a fish that is flipping around, in a boat that is rolling with the swells, using delicate scales will probably not yield very reliable results.

It would be much easier to measure the *length* of the fish on the boat. What you need is a way to convert the length of the fish to its weight. You reason that since length is one-dimensional and weight is three-dimensional, and since a fish 0 units long would weigh 0 pounds, the weight of a fish should be proportional to the cube of its length. Thus, a model of the form weight = $a \times \text{length}^3$ should work. You contact the nearby marine research laboratory and they provide the average length and weight catch data for the Atlantic Ocean rockfish *Sebastes mentella* (Table 4.3).[10] The lab also advises you that the model relationship between body length and weight has been found to be accurate for most fish species growing under normal feeding conditions.

TABLE 4.3 Average length and weight at different ages for Atlantic Ocean rockfish, *Sebastes mentella*

Age (yr)	Length (cm)	Weight (g)	Age (yr)	Length (cm)	Weight (g)
1	5.2	2	11	28.2	318
2	8.5	8	12	29.6	371
3	11.5	21	13	30.8	455
4	14.3	38	14	32.0	504
5	16.8	69	15	33.0	518
6	19.2	117	16	34.0	537
7	21.3	148	17	34.9	651
8	23.3	190	18	36.4	719
9	25.0	264	19	37.1	726
10	26.7	293	20	37.7	810

Figure 4.14 is a scatterplot of weight in grams versus height in centimeters. Although the growth might appear to be exponential, we know that it is frequently misleading to trust too much to the eye. Moreover, we have already decided on a model that makes sense in this context: weight $= a \times \text{length}^3$.

If we take the $\log_{10}$ of both sides, we obtain

$$\log(\text{weight}) = \log a + [3 \times \log(\text{length})]$$

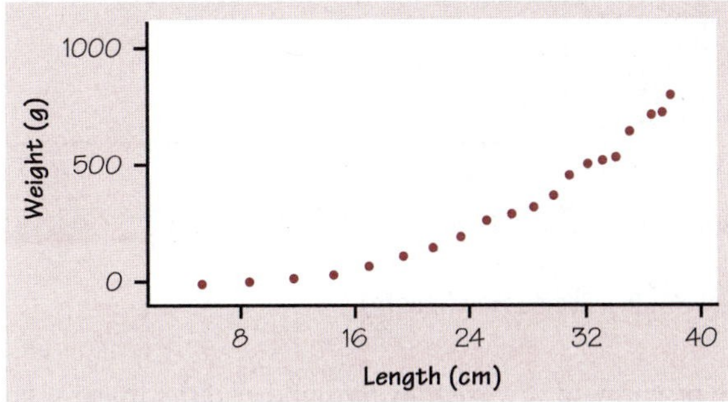

FIGURE 4.14 Scatterplots of Atlantic Ocean rockfish weight versus length.

This equation looks like a linear equation

$$Y = A + BX$$

so we plot log (weight) against log (length). See Figure 4.15.

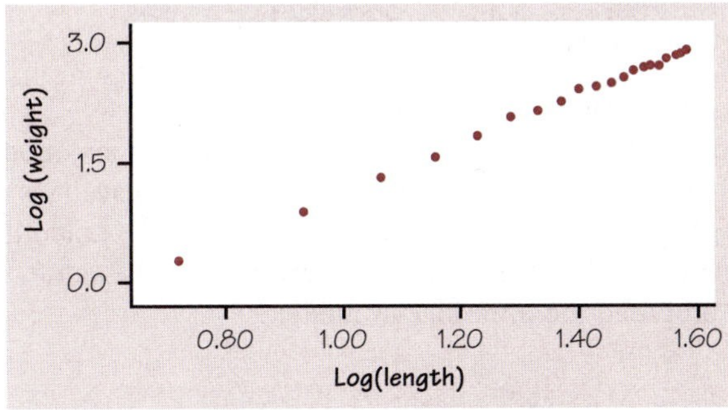

FIGURE 4.15 Scatterplot of log (weight) versus log (length).

We visually confirm that the relationship appears very linear. We perform a least-squares regression on the transformed points [log (length), log (weight)].

The least-squares regression line equation is

$$\log(\text{weight}) = -1.8994 + 3.0494 \log(\text{length})$$

$r = 0.99926$ and $r^2 = 0.9985$. We see that the correlation r of the logarithms of length and weight is virtually 1. (Remember, however, that correlation was defined only for linear fits.) Despite the very high r-value, it's still important to look at a residual plot. The random scatter of the points in Figure 4.16 tells us that the line is a good model for the logs of length and weight.

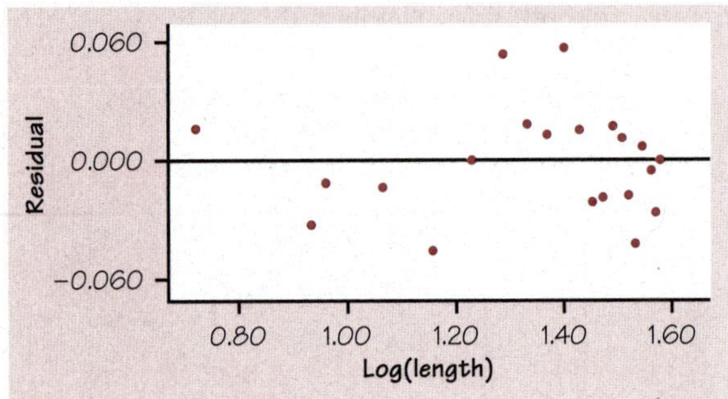

FIGURE 4.16 Plot of residuals versus log (length).

The last step is to perform an inverse transformation on the linear regression equation:

$$\begin{aligned}\log(\text{weight}) &= -1.8994 + [3.0494 \log(\text{length})] \\ &= -1.8994 + \log(\text{length})^{3.0494}\end{aligned}$$

This is the critical step: to remember to use a property of logarithms to write the multiplicative constant 3.0494 as an exponent. Let's continue. Raise 10 to the left side of the equation and set this equal to 10 raised to the right side:

$$\begin{aligned}10^{\log(\text{weight})} &= 10^{-1.8994+\log(\text{length})^{3.0494}} \\ \text{weight} &= 10^{-1.8994} \times \text{length}^{3.0494}\end{aligned}$$

This is the final power equation for the original data.

The scatterplot of the original data along with the power law model appears in Figure 4.17. The fit of this model has visual appeal. We will leave it as an exercise to calculate the sum of the squares of the deviations. It should be noted that the power of x that we obtained for the model, 3.0494, is very close to the value 3 that we conjectured when we proposed the form for our model.

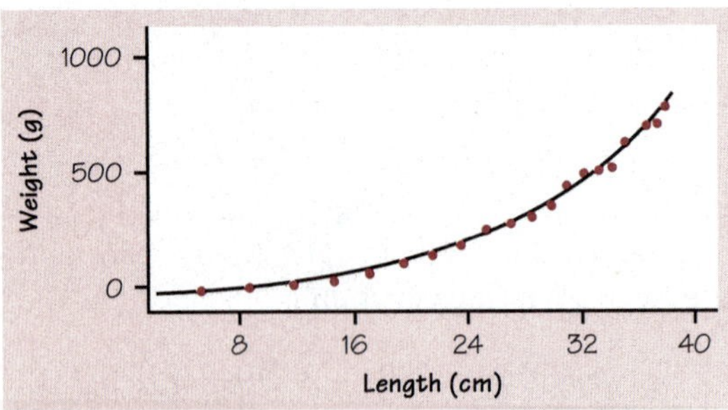

FIGURE 4.17 Atlantic Ocean rockfish data with power law model.

The original purpose for developing this model was to approximate the weight of a fish given its length. Suppose your catch measured 36 centimeters. Our model predicts a weight of Y2(36) = 702.0836281, or about 702 grams. If you entered a fishing contest, would you be comfortable with this procedure for determining the weights of the fish caught, and hence for determining the winner of the contest?

TECHNOLOGY TOOLBOX *Power law modeling*

- Enter the x data (explanatory) into L_1/list1 and the y data (response) into L_2/list2.
- Produce a scatterplot of y versus x. Confirm a nonlinear trend that could be modeled by a power function in the form $y = ax^b$.
- Define L_3/list3 to be log(L_1) or log(list1), and define L_4/list4 to be log(L_2) or log(list2).
- Plot log y versus log x. Verify that the pattern is approximately linear.
- Regress log y on log x. The command line should read LinReg a+bx,L3,L4,Y1. This stores the regression equation as Y1. Remember that Y1 is really log y. Check the r^2-value.
- Construct a residual plot, in the form of either RESID versus x or RESID versus predicted values (fits). Ideally, the points in a residual plot should be randomly scattered above and below the $y = 0$ reference line.
- Perform the back-transformation to find the power function $y = ax^b$ that models the original data. Define Y2 to be (10^a)(x^b). The calculator has stored the values of a and b for the most recent regression performed. Deselect Y1 and plot Y2 and the scatterplot for the original data together.
- To make a prediction for the value $x = k$, evaluate Y2(k) in the Home screen.

EXERCISES

4.12 FISH WEIGHTS

(a) Use the model we derived for approximating the weight of *Sebastes mentella*, $\hat{y} = 10^{-1.8994}x^{3.0494}$, to determine the sum of the squares of the deviations between the observed weights (in grams) and the predicted values. Did we minimize this quantity in the process of constructing our model? If not, what quantity was minimized?

(b) When we performed least-squares regression of log (weight) on log (length) on the calculator, residuals were calculated and stored in a list named RESID. Use this list and the 1-Var Stats command to calculate the sum of the squares of the residuals. Compare this sum of squares with the sum of squares you calculated in (a).

(c) Would you expect the answers in (a) and (b) to be the same or different? Explain.

4.13 BODY WEIGHT AND LIFETIME Table 4.4 gives the average weight and average life span in captivity for several species of mammals. Some writers on power laws in biology claim that life span depends on body weight according to a power law with power

TABLE 4.4 Body weight and lifetime for several species of mammals

Species	Weight (kg)	Life span (years)	Species	Weight (kg)	Life span (years)
Baboon	32	20	Guinea pig	1	4
Beaver	25	5	Hippopotamus	1400	41
Cat, domestic	2.5	12	Horse	480	20
Chimpanzee	45	20	Lion	180	15
Dog	8.5	12	Mouse, house	0.024	3
Elephant	2800	35	Pig, domestic	190	10
Goat, domestic	30	8	Red fox	6	7
Gorilla	140	20	Sheep, domestic	30	12
Grizzly bear	250	25			

Source: G. A. Sacher and E. F. Staffelt, "Relation of gestation time to brain weight for placental mammals: implications for the theory of vertebrate growth," *American Naturalist*, 108 (1974), pp. 593–613. We found these data in F. L. Ramsey and D. W. Schafer, *The Statistical Sleuth: A Course in Methods of Data Analysis*, Duxbury, 1997.

$p = 0.2$. Fit a power law model to these data (using logarithms). Does this small set of data appear to follow a power law with power close to 0.2? Use your fitted model to predict the average life span for humans (average weight 143 kilograms). Humans are an exception to the rule.

4.14 HEART WEIGHTS OF MAMMALS Use the methods discussed in this section to analyze the following data on the hearts of various mammals.[11] Write your findings and conclusions in a short narrative.

Mammal	Heart weight (grams)	Length of cavity of left ventricle (centimeters)
Mouse	0.13	0.55
Rat	0.64	1.0
Rabbit	5.8	2.2
Dog	102	4.0
Sheep	210	6.5
Ox	2030	12.0
Horse	3900	16.0

4.15 The U.S. Department of Health and Human Services characterizes adults as "seriously overweight" if they meet certain criterion for their height as shown in the table below (only a portion of the chart is reproduced here).

Height (ft, in)	Height (in)	Severely overweight (lb)	Height (ft, in)	Height (in)	Severely overweight (lb)
4′10″	58	138	5′8″	68	190
5′0″	60	148	6′0″	72	213
5′2″	62	158	6′2″	74	225
5′4″	64	169	6′4″	76	238
5′6″	66	179	6′6″	78	250

Weights are given in pounds, without clothes. Height is measured without shoes. There is no distinction between men and women; a note accompanying the table states, "The higher weights apply to people with more muscle and bone, such as many men." Despite any reservations you may have about the department's common standards for both genders, do the following:

(a) Without looking at the data, hypothesize a relationship between height and weight of U.S. adults. That is, write a general form of an equation that you believe will model the relationship.

(b) Which variable would you select as explanatory and which would be the response? Plot the data from the table.

(c) Perform a transformation to linearize the data. Do a least-squares regression on the transformed data and check the correlation coefficient.

(d) Construct a residual plot of the transformed data. Interpret the residual plot.

(e) Perform the inverse transformation and write the equation for your model. Use your model to predict how many pounds a 5′10″ adult would have to weigh in order to be classified by the department as "seriously overweight." Do the same for a 7-foot tall individual.

4.16 THE PRICE OF PIZZAS The new manager of a pizza restaurant wants to add variety to the pizza offerings at the restaurant. She also wants to determine if the prices for existing sizes of pizzas are consistent. Prices for plain (cheese only) pizzas are shown below:

Size	Diameter (inches)	Cost
Small	10	$4.00
Medium	12	$6.00
Large	14	$8.00
Giant	18	$10.00

(a) Construct an appropriate model for these data. Comment on your choice of model.

(b) Based on your analysis, would you advise the manager to adjust the price on any of the pizza sizes? If so, explain briefly.

(c) Use your model to suggest a price for a new "personal pizza," with a 6-inch diameter.

(d) Use your model to suggest a price for a new "soccer team" size, with a 24-inch diameter (assuming the oven is large enough to hold it).

SUMMARY

Nonlinear relationships between two quantitative variables can sometimes be changed into linear relationships by **transforming** one or both of the variables.

The most common transformations belong to the family of **power transformations** t^p. The logarithm $\log t$ fits into the power family at position $p = 0$.

When the variable being transformed takes only positive values, the power transformations are all **monotonic.** This implies that there is an inverse transformation that returns to the original data from the transformed values. The effect of the power transformations on data becomes stronger as we move away from linear transformations ($p = 1$) in either direction.

Transformation is particularly effective when there is reason to think that the data are governed by some mathematical model. The **exponential growth model** $y = ab^x$ becomes linear when we plot $\log y$ against x. The **power law model** $y = ax^p$ becomes linear when we plot $\log y$ against $\log x$.

We can fit exponential growth and power models to data by finding the least-squares regression line for the transformed data, then doing the inverse transformation.

SECTION 4.1 EXERCISES

4.17 EXACT EXPONENTIAL GROWTH, I Maria is given a savings bond at birth. The bond is initially worth \$500 and earns interest at 7.5% each year. This means that the value is multiplied by 1.075 each year.

(a) Find the value of the bond at the end of 1 year, 2 years, and so on up to 10 years.

(b) Plot the value y against years x. Connect the points with a smooth curve. This is an exponential curve.

(c) Take the logarithm of each of the values y that you found in (a). Plot the logarithm $\log y$ against years x. You should obtain a straight line.

4.18 EXACT EXPONENTIAL GROWTH, II Fred and Alice were born the same year, and each began life with \$500. Fred added \$100 each year, but earned no interest. Alice added nothing, but earned interest at 7.5% annually. After 25 years, Fred and Alice are getting married. Who has more money?

4.19 FISH IN FINLAND, I Here are data for 12 perch caught in a lake in Finland:[12]

Weight (grams)	Length (cm)	Width (cm)	Weight (grams)	Length (cm)	Width (cm)
5.9	8.8	1.4	300.0	28.7	5.1
100.0	19.2	3.3	300.0	30.1	4.6
110.0	22.5	3.6	685.0	39.0	6.9
120.0	23.5	3.5	650.0	41.4	6.0
150.0	24.0	3.6	820.0	42.5	6.6
145.0	25.5	3.8	1000.0	46.6	7.6

(a) Make a scatterplot of weight against length. Describe the pattern you see.

(b) How do you expect the weight of animals of the same species to change as their length increases? Make a transformation of weight that should straighten the plot if your expectation is correct. Plot the transformed weights against length. Is the plot now roughly linear?

4.20 FISH IN FINLAND, II Plot the widths of the 12 perch in the previous problem against their lengths. What is the pattern of the plot? Explain why we should expect this pattern.

4.21 HOW MOLD GROWS, I Do mold colonies grow exponentially? In an investigation of the growth of molds, biologists inoculated flasks containing a growth medium with equal amounts of spores of the mold *Aspergillus nidulans*. They measured the size of a colony by analyzing how much remains of a radioactive tracer substance that is consumed by the mold as it grows. Each size measurement requires destroying that colony, so that the data below refer to 30 separate colonies. To smooth the pattern, we take the mean size of the three colonies measured at each time.[13]

Hours	Colony sizes			Mean
0	1.25	1.60	0.85	1.23
3	1.18	1.05	1.32	1.18
6	0.80	1.01	1.02	0.94
9	1.28	1.46	2.37	1.70
12	2.12	2.09	2.17	2.13
15	4.18	3.94	3.85	3.99
18	9.95	7.42	9.68	9.02
21	16.36	13.66	12.78	14.27
24	25.01	36.82	39.83	33.89
36	138.34	116.84	111.60	122.26

(a) Graph the mean colony size against time. Then graph the logarithm of the mean colony size against time.

(b) On the basis of data such as these, microbiologists divide the growth of mold colonies into three phases that follow each other in time. Exponential growth occurs during only one of these phases. Briefly describe the three phases, making specific reference to the graphs to support your description.

(c) The exponential growth phase for these data lasts from about 6 hours to about 24 hours. Find the least-squares regression line of the logarithms of mean size on hours for only the data between 6 and 24 hours. Use this line to predict the size of a colony 10 hours after inoculation. (The line predicts the logarithm. You must obtain the size from its logarithm.)

4.22 DETERMINING TREE BIOMASS It is easy to measure the "diameter at breast height" of a tree. It's hard to measure the total "aboveground biomass" of a tree, because to do this you must cut and weigh the tree. The biomass is important for studies of ecology, so ecologists commonly estimate it using a power law. Combining data on 378 trees in tropical rain forests gives this relationship between biomass y measured in kilograms and diameter x measured in centimeters:[14]

$$\log_e y = -2.00 + 2.42 \log_e x$$

Note that the investigators chose to use *natural logarithms*, with base $e = 2.71828$, rather than common logarithms with base 10.

(a) Translate the line given into a power model. Use the fact that for natural logarithms,

$$y = e^{\log_e y}$$

(b) Estimate the biomass of a tropical tree 30 centimeters in diameter.

4.23 HOW MOLD GROWS, II Find the correlation between the logarithm of mean size and hours for the data between 6 and 24 hours in Exercise 4.21. Make a scatterplot of the logarithms of the individual size measurements against hours for this same period and find the correlation. Why do we expect the second r to be smaller? Is it in fact smaller?

4.24 BE LIKE GALILEO Galileo studied motion by rolling balls down ramps. Newton later showed how Galileo's data fit his general laws of motion. Imagine that you are Galileo, without Newton's laws to guide you. He rolled a ball down a ramp at different heights above the floor and measured the horizontal distance the ball traveled before it hit the floor. Here are Galileo's data when he placed a horizontal shelf at the end of the ramp so that the ball is moving horizontally when it starts to fall. (We won't try to describe the obscure seventeenth-century units Galileo used to measure distance.)[15]

Distance	Height
1500	1000
1340	828
1328	800
1172	600
800	300

Plot distance y against height x. The pattern is very regular, as befits data described by a physical law. We want to find distance as a function of height. That is, we want to transform x to straighten the graph.

(a) Think before you calculate: Will powers x^p for $p < 1$ or $p > 1$ tend to straighten the graph. Why?

(b) Move along the ladder of transformations in the direction you have chosen until the graph is nearly straight. What transformation do you suggest?

4.25 SEED PRODUCTION Table 4.5 gives data on the mean number of seeds produced in a year by several common tree species and the mean weight (in milligrams) of the seeds produced. (Some species appear twice because their seeds were counted in two locations.) We might expect that trees with heavy seeds produce fewer of them, but what is the form of the relationship?

TABLE 4.5 Count and weight of seeds produced by common tree species

Tree species	Seed count	Seed weight (mg)	Tree species	Seed count	Seed weight (mg)
Paper birch	27,239	0.6	American beech	463	247
Yellow birch	12,158	1.6	American beech	1,892	247
White spruce	7,202	2.0	Black oak	93	1,851
Engelmann spruce	3,671	3.3	Scarlet oak	525	1,930
Red spruce	5,051	3.4	Red oak	411	2,475
Tulip tree	13,509	9.1	Red oak	253	2,475
Ponderosa pine	2,667	37.7	Pignut hickory	40	3,423
White fir	5,196	40.0	White oak	184	3,669
Sugar maple	1,751	48.0	Chestnut oak	107	4,535
Sugar pine	1,159	216.0			

Source: Data from many studies compiled in D. F. Greene and E. A. Johnson, "Estimating the mean annual seed production of trees," *Ecology*, 75 (1994), pp. 642–647.

(a) Make a scatterplot showing how the weight of tree seeds helps explain how many seeds the tree produces. Describe the form, direction, and strength of the relationship.

(b) If a power law holds for this relationship, the logarithms of the original data will display a linear pattern. Use your calculator or software to obtain the logarithms of both the seed weights and the seed counts in Table 4.5. Make a new scatterplot using these new variables. Now what are the form, direction, and strength of the relationship?

4.26 ACTIVITY 4: THE SPREAD OF CANCER CELLS

(a) Using the data you and your class collected in the chapter-opening activity, use transformation methods to construct an appropriate model. Show the important numerical and graphical steps you go through to develop your model, and tie these together with explanatory narrative to support your choice of a model.

(b) A theoretical analysis might begin as follows: The probability that an individual malignant cell reproduces is 1/6 each year. Let P = population of cancer cells at time t and let P_0 = population of cancer cells at time $t = 0$. At the end of Year 1, the population is $P = P_0 + (1/6)P_0 = P_0(7/6)$. At the end of Year 2, the population is $P = P_0(7/6) + P_0(1/6)(7/6) = P_0(7/6)^2$. Continue this line of reasoning to show that the growth equation after n years is $P = P_0(7/6)^n$.

(c) Enter the growth equation into your calculator as Y3, and plot it along with your exponential model calculated in (a). Specify a thick plotting line for one of the curves. How do the two exponential curves compare?

4.2 CAUTIONS ABOUT CORRELATION AND REGRESSION

Correlation and regression are powerful tools for describing the relationship between two variables. When you use these tools, you must be aware of their limitations, beginning with the fact that **correlation and regression describe only linear relationships.** Also remember that **the correlation r and the least-squares**

regression line are not resistant. One influential observation or incorrectly entered data point can greatly change these measures. Always plot your data before interpreting regression or correlation. Here are some other cautions to keep in mind when you apply correlation and regression or read accounts of their use.

Extrapolation

Suppose that you have data on a child's growth between 3 and 8 years of age. You find a strong linear relationship between age x and height y. If you fit a regression line to these data and use it to predict height at age 25 years, you will predict that the child will be 8 feet tall. Growth slows down and stops at maturity, so extending the straight line to adult ages is foolish. Few relationships are linear for all values of x. So don't stray far from the domain of x that actually appears in your data.

EXTRAPOLATION

Extrapolation is the use of a regression line for prediction far outside the domain of values of the explanatory variable x that you used to obtain the line or curve. Such predictions are often not accurate.

Lurking variables

In our study of correlation and regression we looked at just two variables at a time. Often the relationship between two variables is strongly influenced by other variables. More advanced statistical methods allow the study of many variables together, so that we can take other variables into account. But sometimes the relationship between two variables is influenced by other variables that we did not measure or even think about. Because these variables are lurking in the background, we call them *lurking variables.*

LURKING VARIABLE

A **lurking variable** is a variable that is not among the explanatory or response variables in a study and yet may influence the interpretation of relationships among those variables.

A lurking variable can falsely suggest a strong relationship between x and y, or it can hide a relationship that is really there. Here are examples of each of these effects.

EXAMPLE 4.10 DISCRIMINATION IN MEDICAL TREATMENT?

Studies show that men who complain of chest pain are more likely to get detailed tests and aggressive treatment such as bypass surgery than are women with similar complaints. Is this association between gender and treatment due to discrimination?

Perhaps not. Men and women develop heart problems at different ages—women are on the average between 10 and 15 years older than men. Aggressive treatments are more risky for older patients, so doctors may hesitate to advise them. Lurking variables—the patient's age and condition—may explain the relationship between gender and doctors' decisions. As the author of one study of the issue said, "When men and women are otherwise the same and the only difference is gender, you find that treatments are very similar."[16]

EXAMPLE 4.11 MEASURING INADEQUATE HOUSING

A study of housing conditions in the city of Hull, England, measured a large number of variables for each of the wards in the city. Two of the variables were a measure x of overcrowding and a measure y of the lack of indoor toilets. Because x and y are both measures of inadequate housing, we expect a high correlation. In fact the correlation was only $r = 0.08$. How can this be?

Investigation found that some poor wards had a lot of public housing. These wards had high values of x but low values of y because public housing always includes indoor toilets. Other poor wards lacked public housing, and these wards had high values of both x and y. Within wards of each type, there was a strong positive association between x and y. Analyzing all wards together ignored the lurking variable—amount of public housing—and hid the nature of the relationship between x and y.[17]

Figure 4.18 shows in simplified form how groups formed by a lurking variable can make correlation and regression misleading. The groups appear as clusters of points in the scatterplot. There is a strong relationship between x and y within each of the clusters. In fact, $r = 0.85$ and $r = 0.91$ in the two clusters. However, because similar values of x correspond to quite different values of y in the two clusters, x alone is of little value for predicting y. The correlation for all the points together is only $r = 0.14$.

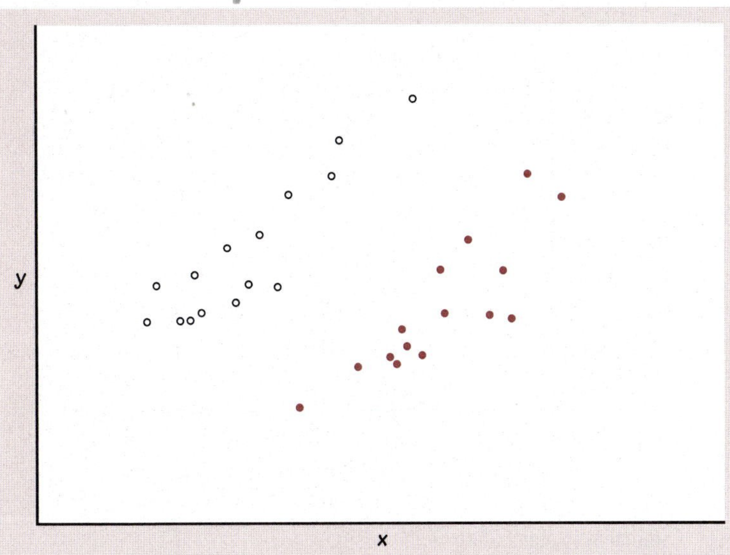

FIGURE 4.18 The variables in this scatterplot have a small correlation even though there is a strong correlation within each of the clusters.

Never forget that the relationship between two variables can be strongly influenced by other variables that are lurking in the background. Lurking variables can dramatically change the conclusions of a regression study. Because lurking variables are often unrecognized and unmeasured, detecting their effect is a challenge. Many lurking variables change systematically over time. One useful method for detecting lurking variables is therefore to *plot both the response variable and the regression residuals against the time order of the observations* whenever the time order is available. An understanding of the background of the data then allows you to guess what lurking variables might be present. Here is an example of plotting and interpreting residuals that uncovered a lurking variable.

EXAMPLE 4.12 PREDICTING ENROLLMENT

The mathematics department of a large state university must plan the number of sections and instructors required for its elementary courses. The department hopes that the number of students in these courses can be predicted from the number of first-year students, which is known before the new students actually choose courses. The table below contains data for several years.[18] The explanatory variable x is the number of first-year students. The response variable y is the number of students who enroll in elementary mathematics courses.

Year	1993	1994	1995	1996	1997	1998	1999	2000
x	4595	4827	4427	4258	3995	4330	4265	4351
y	7364	7547	7099	6894	6572	7156	7232	7450

A scatterplot (Figure 4.19) shows a reasonably linear pattern with a cluster of points near the center. We use regression software to obtain the equation of the least-squares regression line:

$$\hat{y} = 2492.69 + 1.0663x$$

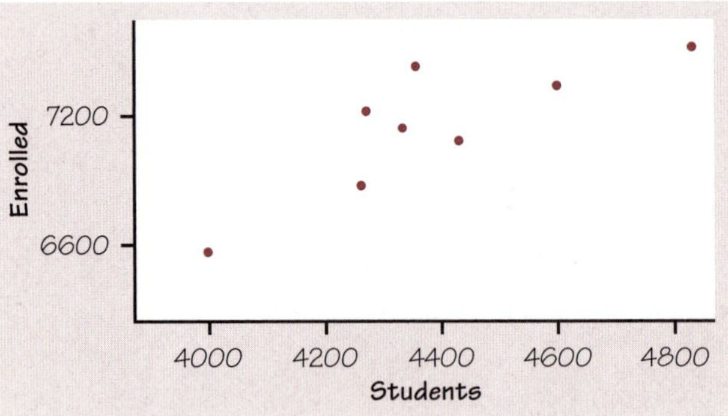

FIGURE 4.19 Enrollment in elementary math classes.

The software also tells us that $r^2 = 0.694$. That is, linear dependence on x explains about 70% of the variation in y. The line appears to fit reasonably well.

A plot of the residuals against x (Figure 4.20) magnifies the vertical deviations of the points from the line. We can see that a somewhat different line would fit the five lower points well. The three points above the line represent a different relation between the number of first-year students x and mathematics enrollments y.

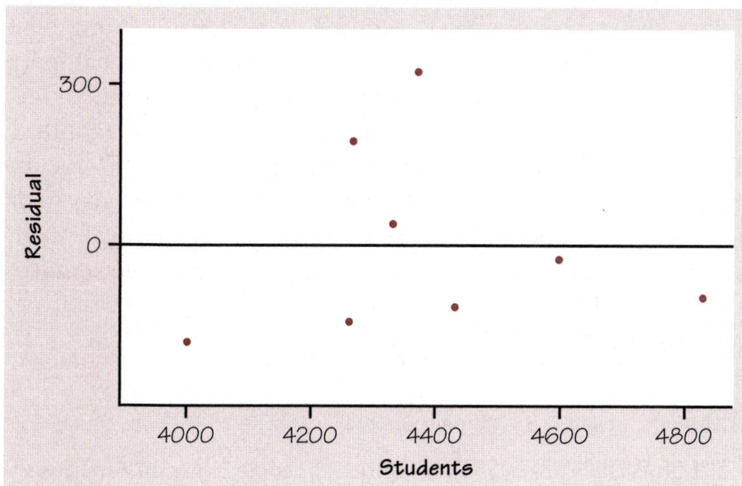

FIGURE 4.20 Residual plot.

A second plot of the residuals clarifies the situation. Figure 4.21 is a plot of the residuals against year. We now see that the five negative residuals are from the years 1993 to 1997, and the three positive residuals represent the years 1998 to 2000. This plot suggests that a change took place between 1997 and 1998 that caused a higher proportion of students to take mathematics courses beginning in 1998. In fact, one of the schools in the university changed its program to require that entering students take another mathematics course. This change is the lurking variable that explains the pattern we observed. The mathematics department should not use data from years before 1998 for predicting future enrollment.

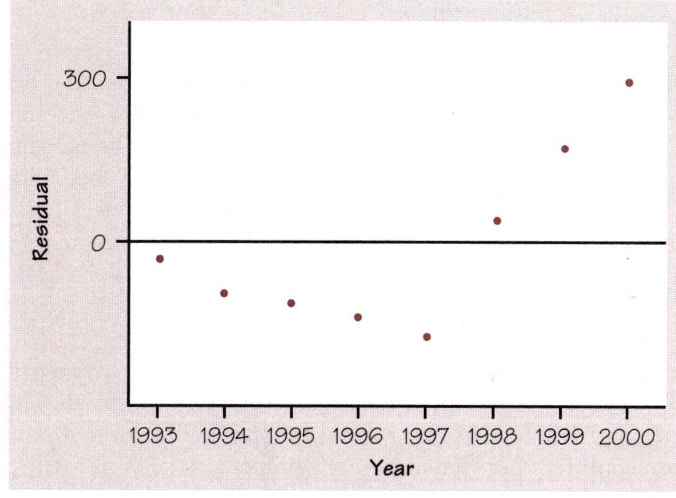

FIGURE 4.21 Plot of residuals versus year.

Using averaged data

Many regression or correlation studies work with averages or other measures that combine information from many individuals. You should note this carefully and resist the temptation to apply the results of such studies to individuals. We have seen, starting with Figure 3.2 (page 128), a strong relationship between outside temperature and the Sanchez household's natural gas consumption. Each point on the scatterplot represents a month. Both degree-days and gas consumed are averages over all the days in the month. Data for individual days would show more scatter about the regression line and lower correlation. Averaging over an entire month smooths out the day-to-day variation due to doors left open, houseguests using more gas to heat water, and so on. *Correlations based on averages are usually too high when applied to individuals.* This is another reminder that it is important to note exactly what variables were measured in a statistical study.

EXERCISES

4.27 THE SIZE OF AMERICAN FARMS The number of people living on American farms has declined steadily during this century. Here are data on the farm population (millions of persons) from 1935 to 1980.

Year:	1935	1940	1945	1950	1955	1960	1965	1970	1975	1980
Population:	32.1	30.5	24.4	23.0	19.1	15.6	12.4	9.7	8.9	7.2

(a) Make a scatterplot of these data and find the least-squares regression line of farm population on year.

(b) According to the regression line, how much did the farm population decline each year on the average during this period? What percent of the observed variation in farm population is accounted for by linear change over time?

(c) Use the regression equation to predict the number of people living on farms in 1990. Is this result reasonable? Why?

4.28 THE POWER OF HERBAL TEA A group of college students believes that herbal tea has remarkable powers. To test this belief, they make weekly visits to a local nursing home, where they visit with the residents and serve them herbal tea. The nursing home staff reports that after several months many of the residents are more cheerful and healthy. A skeptical sociologist commends the students for their good deeds but scoffs at the idea that herbal tea helped the residents. Identify the explanatory and response variables in this informal study. Then explain what lurking variables account for the observed association.

4.29 STRIDE RATE The data in Exercise 3.71 (page 187) give the average steps per second for a group of top female runners at each of several running speeds. There is a high positive correlation between steps per second and speed. Suppose that you had the full data, which record steps per second for each runner separately at each speed. If you plotted each individual observation and computed the correlation, would you expect the correlation to be lower than, about the same as, or higher than the correlation for the published data? Why?

4.30 HOW TO SHORTEN A HOSPITAL STAY A study shows that there is a positive correlation between the size of a hospital (measured by its number of beds x) and the median number of days y that patients remain in the hospital. Does this mean that you can shorten a hospital stay by choosing a small hospital?

4.31 STOCK MARKET INDEXES The Standard & Poor's 500-stock index is an average of the price of 500 stocks. There is a moderately strong correlation (roughly $r = 0.6$) between how much this index changes in January and how much it changes during the entire year. If we looked instead at data on all 500 individual stocks, we would find a quite different correlation. Would the correlation be higher or lower? Why?

4.32 GOLF SCORES Here are the golf scores of 11 members of a women's golf team in two rounds of tournament play:

Player	1	2	3	4	5	6	7	8	9	10	11
Round 1	89	90	87	95	86	81	105	83	88	91	79
Round 2	94	85	89	89	81	76	89	87	91	88	80

(a) Plot the data with the Round 1 scores on the x axis and the Round 2 scores on the y axis. There is a generally linear pattern except for one potentially influential observation. Circle this observation on your graph.

(b) Here are the equations of two least-squares lines. One of them is calculated from all 11 data points and the other omits the influential observation.

$$\hat{y} = 20.49 + 0.754x$$
$$\hat{y} = 50.01 + 0.410x$$

Draw both lines on your scatterplot. Which line omits the influential observation? How do you know this?

The question of causation

In many studies of the relationship between two variables, the goal is to establish that changes in the explanatory variable *cause* changes in the response variable. Even when a strong association is present, the conclusion that this association is due to a causal link between the variables is often elusive. What ties between two variables (and others lurking in the background) can explain an observed association? What constitutes good evidence for causation? We begin our consideration of these questions with a set of examples. In each case, there is a clear association between an explanatory variable x and a response variable y. Moreover, the association is positive whenever the direction makes sense.

EXAMPLE 4.13 ASSOCIATIONS

The following are some examples of observed associations between x and y:

1. x = mother's body mass index
   y = daughter's body mass index
2. x = amount of the artificial sweetener saccharin in a rat's diet
   y = count of tumors in the rat's bladder
3. x = a high school senior's SAT score
   y = the student's first-year college grade point average
4. x = monthly flow of money into stock mutual funds
   y = monthly rate of return for the stock market
5. x = whether a person regularly attends religious services
   y = how long the person lives
6. x = the number of years of education a worker has
   y = the worker's income

Explaining association: causation

Figure 4.22 shows in outline form how a variety of underlying links between variables can explain association. The dashed line represents an observed association between the variables x and y. Some associations are explained by a direct cause-and-effect link between these variables. The first diagram in Figure 4.22 shows "x causes y" by a solid arrow running from x to y.

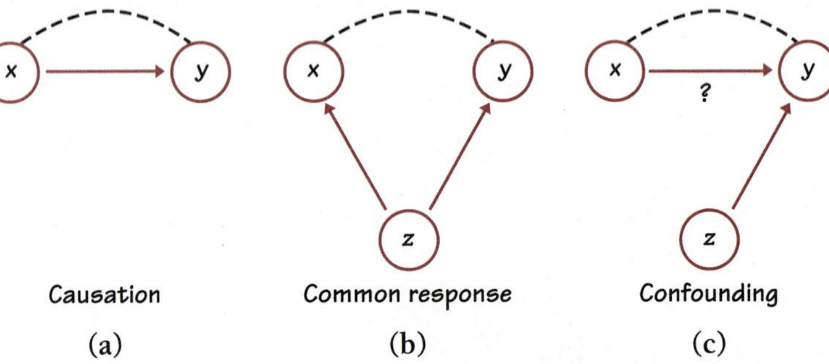

FIGURE 4.22 Variables *x* and *y* show a strong association (*dashed line*). **This association may be the result of any of several causal relationships** (*solid arrow*). (a) Causation: Changes in x cause changes in y. (b) Common response: Changes in both x and y are caused by changes in a lurking variable z. (c) Confounding: The effect (if any) of x on y is confounded with the effect of a lurking variable z.

EXAMPLE 4.14 CAUSATION?

Items 1 and 2 in Example 4.13 are examples of direct causation. Thinking about these examples, however, shows that "causation" is not a simple idea.

1. A study of Mexican American girls aged 9 to 12 years recorded body mass index (BMI), a measure of weight relative to height, for both the girls and their mothers. People with high BMI are overweight or obese. The study also measured hours of television, minutes of physical activity, and intake of several kinds of food. The strongest correlation ($r = 0.506$) was between the BMI of daughters and the BMI of their mothers.[19]

Body type is in part determined by heredity. Daughters inherit half their genes from their mothers. There is therefore a direct causal link between the BMI of mothers and daughters. Yet the mothers' BMIs explain only 25.6% (that's r^2 again) of the variation among the daughters' BMIs. Other factors, such as diet and exercise, also influence BMI. **Even when direct causation is present, it is rarely a complete explanation of an association between two variables.**

2. The best evidence for causation comes from experiments that actually change x while holding all other factors fixed. If y changes, we have good reason to think that x caused the change in y. Experiments show conclusively that large amounts of saccharin in the diet cause bladder tumors in rats. Should we avoid saccharin as a replacement for sugar in food? Rats are not people. Although we can't experiment with people, studies of people who consume different amounts of saccharin show little association between saccharin and bladder tumors.[20] **Even well-established causal relations may not generalize to other settings.**

Explaining association: common response

common response

"Beware the lurking variable" is good advice when thinking about an association between two variables. The second diagram in Figure 4.22 illustrates ***common response.*** The observed association between the variables x and y is explained by a lurking variable z. Both x and y change in response to changes in z. This common response creates an association even though there may be no direct causal link between x and y.

EXAMPLE 4.15 COMMON RESPONSE

The third and fourth items in Example 4.13 illustrate how common response can create an association.

3. Students who are smart and who have learned a lot tend to have both high SAT scores and high college grades. The positive correlation is explained by this common response to students' ability and knowledge.

4. There is a strong positive correlation between how much money individuals add to mutual funds each month and how well the stock market does the same month. Is the new money driving the market up? The correlation may be explained in part by common response to underlying investor sentiment: when optimism reigns, individuals send money to funds and large institutions also invest more. The institutions would drive up prices even if individuals did nothing. In addition, what causation there is may operate in the other direction: when the market is doing well, individuals rush to add money to their mutual funds.[21]

Explaining association: confounding

We noted in Example 4.14 that inheritance no doubt explains part of the association between the body mass indexes (BMIs) of daughters and their mothers. Can we use r or r^2 to say how much inheritance contributes to the daughters' BMIs? No. It may well be that mothers who are overweight also set an example of little exercise, poor eating habits, and lots of television. Their daughters pick up these habits to some extent, so the influence of heredity is mixed up with influences from the girls' environment. We call this mixing of influences *confounding*.

CONFOUNDING

Two variables are **confounded** when their effects on a response variable cannot be distinguished from each other. The confounded variables may be either explanatory variables or lurking variables.

When many variables interact with each other, confounding of several variables often prevents us from drawing conclusions about causation. The third diagram in Figure 4.22 illustrates confounding. Both the explanatory variable x and the lurking variable z may influence the response variable y. Because x is confounded with z, we cannot distinguish the influence of x from the influence of z. We cannot say how strong the direct effect of x on y is. In fact, it can be hard to say if x influences y at all.

EXAMPLE 4.16 CONFOUNDING

The last two associations in Example 4.13 (Items 5 and 6) are explained in part by confounding.

5. Many studies have found that people who are active in their religion live longer than nonreligious people. But people who attend church or mosque or synagogue also take better care of themselves than nonattenders. They are less likely to smoke, more likely to exercise, and less likely to be overweight. The effects of these good habits are confounded with the direct effects of attending religious services.

6. It is likely that more education is a cause of higher income—many highly paid professions require advanced education. However, confounding is also present. People who have high ability and come from prosperous homes are more likely to get many years of education than people who are less able or poorer. Of course, people who start out able and rich are more likely to have high earnings even without much education. We can't say how much of the higher income of well-educated people is actually caused by their education.

Many observed associations are at least partly explained by lurking variables. Both common response and confounding involve the influence of a lurking variable (or variables) z on the response variable y. The distinction between these two types of relationships is less important than the common element, the influence of lurking variables. The most important lesson of these examples is one we have already emphasized: **even a very strong association between two variables is not by itself good evidence that there is a cause-and-effect link between the variables.**

Establishing causation

How can a direct causal link between x and y be established? The best method—indeed, the only fully compelling method—of establishing causation is to conduct a carefully designed experiment in which the effects of possible lurking variables are controlled. Much of Chapter 5 is devoted to the art of designing convincing experiments.

Many of the sharpest disputes in which statistics plays a role involve questions of causation that cannot be settled by experiment. Does gun control reduce violent crime? Does living near power lines cause cancer? Has increased free trade helped to increase the gap between the incomes of more educated and less educated American workers? All of these questions have become public issues. All concern associations among variables. And all have this in common: they try to pinpoint cause and effect in a setting involving complex relations among many interacting variables. Common response and confounding, along with the number of potential lurking variables, make observed associations misleading. Experiments are not possible for ethical or practical reasons. We can't assign some people to live near power lines or compare the same nation with and without free-trade agreements.

EXAMPLE 4.17 DO POWER LINES INCREASE THE RISK OF LEUKEMIA?

Electric currents generate magnetic fields. So living with electricity exposes people to magnetic fields. Living near power lines increases exposure to these fields. Really strong fields can disturb living cells in laboratory studies. What about the weaker fields we experience if we live near power lines?

It isn't ethical to do experiments that expose children to magnetic fields. It's hard to compare cancer rates among children who happen to live in more and less exposed locations, because leukemia is rare and locations vary in many ways other than magnetic fields. We must rely on studies that compare children who have leukemia with children who don't.

A careful study of the effect of magnetic fields on children took five years and cost \$5 million. The researchers compared 638 children who had leukemia and 620 who did not. They went into the homes and actually measured the magnetic fields in the children's bedrooms, in other rooms, and at the front door. They recorded facts about nearby power lines for the family home and also for the mother's residence when she was pregnant. Result: no evidence of more than a chance connection between magnetic fields and childhood leukemia.[22]

"No evidence" that magnetic fields are connected with childhood leukemia doesn't prove that there is no risk. It says only that a careful study could not find any risk that stands out from the play of chance that distributes leukemia cases across the landscape. Critics continue to argue that the study failed to measure some lurking variables, or that the children studied don't fairly represent all children. Nonetheless, a carefully designed study comparing children with and without leukemia is a great advance over haphazard and sometimes emotional counting of cancer cases.

EXAMPLE 4.18 DOES SMOKING CAUSE LUNG CANCER?

Despite the difficulties, it is sometimes possible to build a strong case for causation in the absence of experiments. The evidence that smoking causes lung cancer is about as strong as nonexperimental evidence can be.

Doctors had long observed that most lung cancer patients were smokers. Comparison of smokers and similar nonsmokers showed a very strong association between smoking and death from lung cancer. Could the association be due to common response? Might there be, for example, a genetic factor that predisposes people both to nicotine addiction and to lung cancer? Smoking and lung cancer would then be positively associated even if smoking had no direct effect on the lungs. Or perhaps confounding is to blame. It might be that smokers live unhealthy lives in other ways (diet, alcohol, lack of exercise) and that some other habit confounded with smoking is a cause of lung cancer. How were these objections overcome?

Let's answer this question in general terms: What are the criteria for establishing causation when we cannot do an experiment?

- *The association is strong.* The association between smoking and lung cancer is very strong.
- *The association is consistent.* Many studies of different kinds of people in many countries link smoking to lung cancer. That reduces the chance that a lurking variable specific to one group or one study explains the association.
- *Higher doses are associated with stronger responses.* People who smoke more cigarettes per day or who smoke over a longer period get lung cancer more often. People who stop smoking reduce their risk.
- *The alleged cause precedes the effect in time.* Lung cancer develops after years of smoking. The number of men dying of lung cancer rose as smoking became more common, with a lag of about 30 years. Lung cancer kills more men than any other form of cancer. Lung cancer was rare among women until women began to smoke. Lung cancer in women rose along with smoking, again with a lag of about 30 years, and has now passed breast cancer as the leading cause of cancer death among women.
- *The alleged cause is plausible.* Experiments with animals show that tars from cigarette smoke do cause cancer.

Medical authorities do not hesitate to say that smoking causes lung cancer. The U.S. Surgeon General states that cigarette smoking is "the largest avoidable cause of death and disability in the United States."[23] The evidence for causation is overwhelming—but it is not as strong as the evidence provided by well-designed experiments.

EXERCISES

For Exercises 4.33 through 4.37, answer the question. State whether the relationship between the two variables involves causation, common response, or confounding. Identify possible lurking variable(s). Draw a diagram of the relationship in which each circle represents a variable. Write a brief description of the variable by each circle.

4.33 FIGHTING FIRES Someone says, "There is a strong positive correlation between the number of firefighters at a fire and the amount of damage the fire does. So sending lots of firefighters just causes more damage." Why is this reasoning wrong?

4.34 HOW'S YOUR SELF-ESTEEM? People who do well tend to feel good about themselves. Perhaps helping people feel good about themselves will help them do better in school and life. Raising self-esteem became for a time a goal in many schools. California even created a state commission to advance the cause. Can you think of explanations for the association between high self-esteem and good school performance other than "Self-esteem causes better work in school"?

4.35 SAT MATH AND VERBAL SCORES Table 1.15 (page 70) gives education data for the states. The correlation between the average SAT math scores and the average SAT verbal scores for the states is $r = 0.962$

(a) Find r^2 and explain in simple language what this number tells us.

(b) If you calculated the correlation between the SAT math and verbal scores of a large number of individual students, would you expect the correlation to be about 0.96 or quite different? Explain your answer.

4.36 BETTER READERS A study of elementary school children, ages 6 to 11, finds a high positive correlation between shoe size x and score y on a test of reading comprehension. What explains this correlation?

4.37 THE BENEFITS OF FOREIGN LANGUAGE STUDY Members of a high school language club believe that study of a foreign language improves a student's command of English. From school records, they obtain the scores on an English achievement test given to all seniors. The mean score of seniors who studied a foreign language for at least two years is much higher than the mean score of seniors who studied no foreign language. These data are not good evidence that language study strengthens English skills. Identify the explanatory and response variables in this study. Then explain what lurking variable prevents the conclusion that language study improves students' English scores.

SUMMARY

Correlation and regression must be **interpreted with caution.** Plot the data to be sure that the relationship is roughly linear and to detect outliers and influential observations. Remember that correlation and regression describe **only linear** relations.

Avoid **extrapolation,** which is the use of a regression line or curve for prediction for values of the explanatory variable outside the domain of the data from which the line was calculated.

Remember that **correlations based on averages** are usually too high when applied to individuals.

Lurking variables may explain the relationship between the explanatory and response variables. Correlation and regression can be misleading if you ignore important lurking variables.

The effect of lurking variables can operate through **common response** if changes in both the explanatory and response variables are caused by changes in lurking variables. **Confounding** of two variables (either explanatory or lurking variables) means that we cannot distinguish their effects on the response variable.

Most of all, be careful not to conclude that there is a cause-and-effect relationship between two variables just because they are strongly associated. The relationship could involve common response or confounding. **High correlation does not imply causation.** The best evidence that an association is due to causation comes from an **experiment** in which the explanatory variable is directly changed and other influences on the response are controlled.

In the absence of experimental evidence be cautious in accepting claims of causation. Good evidence of causation requires a strong association that appears consistently in many studies, a clear explanation for the alleged causal link, and careful examination of possible lurking variables.

SECTION 4.2 EXERCISES

For Exercises 4.38 through 4.45, carry out the instructions. Then state whether the relationship between the two variables involves causation, common response, or confounding. Then identify possible lurking variable(s). Draw a diagram of the relationship in which each circle represents a variable. By each circle, write a brief description of the variable.

4.38 DO ARTIFICIAL SWEETENERS CAUSE WEIGHT GAIN? People who use artificial sweeteners in place of sugar tend to be heavier than people who use sugar. Does this mean that artificial sweeteners cause weight gain? Give a more plausible explanation for this association.

4.39 DOES EXPOSURE TO INDUSTRIAL CHEMICALS CAUSE MISCARRIAGES? A study showed that women who work in the production of computer chips have abnormally high numbers of miscarriages. The union claimed that exposure to chemicals used in production causes the miscarriages. Another possible explanation is that these workers spend most of their time standing up.

4.40 IS MATH THE KEY TO SUCCESS IN COLLEGE? Here is the opening of a newspaper account of a College Board study of 15,941 high school graduates:

Minority students who take high school algebra and geometry succeed in college at almost the same rate as whites, a new study says.

The link between high school math and college graduation is "almost magical," says College Board President Donald Stewart, suggesting "math is the gatekeeper for success in college."

"These findings," he says, "justify serious consideration of a national policy to ensure that all students take algebra and geometry."[24]

What lurking variables might explain the association between taking several math courses in high school and success in college? Explain why requiring algebra and geometry may have little effect on who succeeds in college.

4.41 ARE GRADES AND TV WATCHING LINKED? Children who watch many hours of television get lower grades in school on the average than those who watch less TV. Explain clearly why this fact does not show that watching TV *causes* poor grades. In particular, suggest some other variables that may be confounded with heavy TV viewing and may contribute to poor grades.

4.42 MOZART FOR MINORS In 1998, the Kalamazoo (Michigan) Symphony advertised a "Mozart for Minors" program with this statement: "Question: Which students scored 51 points higher in verbal skills and 39 points higher in math? Answer: Students who had experience in music."[25] What do you think of the claim that "experience in music" causes higher test scores?

4.43 RAISING SAT SCORES A study finds that high school students who take the SAT, enroll in an SAT coaching course, and then take the SAT a second time raise their SAT mathematics scores from a mean of 521 to a mean of 561.[26] What factors other than "taking the course causes higher scores" might explain this improvement?

4.44 ECONOMISTS' EDUCATION AND INCOME There is a strong positive correlation between years of education and income for economists employed by business firms. (In particular, economists with doctorates earn more than economists with only a bachelor's degree.) There is also a strong positive correlation between years of education and income for economists employed by colleges and universities. But when all economists are considered, there is a *negative* correlation between education and income. The explanation for this is that business pays high salaries and employs mostly economists with bachelor's degrees, while colleges pay lower salaries and employ mostly economists with doctorates. Sketch a scatterplot with two groups of cases (business and academic) that illustrates how a strong positive correlation within each group and a negative overall correlation can occur together. (*Hint:* Begin by studying Figure 4.18 on page 227.)

4.45 TV AND OBESITY Over the last 20 years there has developed a positive association between sales of television sets and the number of obese adolescents in the United States. Do more TVs cause more children to put on weight, or are there other factors involved? List some of the possible lurking variables.

4.46 THE S&P 500 The Standard & Poor's 500-stock index is an average of the price of 500 stocks. There is a moderately strong correlation (roughly $r = 0.6$) between how much this index changes in January and how much it changes during the entire year.

If we looked instead at data on all 500 individual stocks, we would find a quite different correlation. Would the correlation be higher or lower? Why?

4.47 THE LINK BETWEEN HEALTH AND INCOME An article entitled "The Health and Wealth of Nations" says: 'The positive correlation between health and income per capita is one of the best-known relations in international development. This correlation is commonly thought to reflect a causal link running from income to health. . . . Recently, however, another intriguing possibility has emerged: that the health-income correlation is partly explained by a causal link running the other way—from health to income."[27]

Explain how higher income in a nation can cause better health. Then explain how better health can cause higher income. There is no simple way to determine the direction of the link.

4.48 RETURNS FOR U.S. AND OVERSEAS STOCKS Exercise 3.56 (page 179) examined the relationship between returns on U.S. and overseas stocks. Return to the scatterplot and regression line for predicting overseas returns from U.S. returns.

(a) Circle the point that has the largest residual (either positive or negative). What year is this? Redo the regression without this point and add the new regression line to your plot. Was this observation very influential?

(b) Whenever we regress two variables that both change over time, we should plot the residuals against time as a check for time-related lurking variables. Make this plot for the stock returns data. Are there any suspicious patterns in the residuals?

4.49 HEART ATTACKS AND HOSPITALS If you need medical care, should you go to a hospital that handles many cases like yours? Figure 4.23 presents some data for heart attacks.

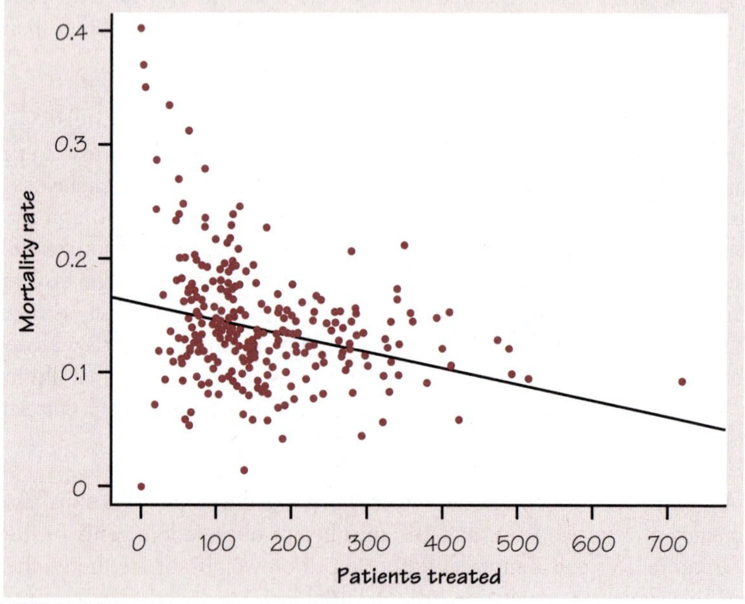

FIGURE 4.23 Mortality of heart attack patients and number of heart attack cases treated for a large group of hospitals.

The figure plots mortality rate (the proportion of patients who died) against the number of heart attack patients treated for a large number of hospitals in a recent year. The line on the plot is the least-squares regression line for predicting mortality from number of patients.

(a) Do the plot and regression generally support the thesis that mortality is lower at hospitals that treat more heart attacks? Is the relationship very strong?

(b) In what way is the pattern of the plot nonlinear? Does the nonlinearity strengthen or weaken the conclusion that heart attack patients should avoid hospitals that treat few heart attacks? Why?

4.3 RELATIONS IN CATEGORICAL DATA

To this point we have concentrated on relationships in which at least the response variable was quantitative. Now we will shift to describing relationships between two or more categorical variables. Some variables—such as sex, race, and occupation—are inherently categorical. Other categorical variables are created by grouping values of a quantitative variable into classes. Published data are often reported in grouped form to save space. To analyze categorical data, we use the *counts* or *percents* of individuals that fall into various categories.

EXAMPLE 4.19 EDUCATION AND AGE

Table 4.6 presents Census Bureau data on the years of school completed by Americans of different ages. Many people under 25 years of age have not completed their education, so they are left out of the table. Both variables, age and education, are grouped into categories. This is a ***two-way table*** because it describes two categorical variables. Education is the ***row variable*** because each row in the table describes people with one level of education. Age is the ***column variable*** because each column describes one age group. The entries in the table are the counts of persons in each age-by-education class. Although both age and education in this table are categorical variables, both have a natural order from least to most. The order of the rows and the columns in Table 4.6 reflects the order of the categories.

two-way table
row variable
column variable

TABLE 4.6 Years of school completed, by age, 2000 (thousands of persons)

	Age group			
Education	25 to 34	35 to 54	55+	Total
Did not complete high school	4,474	9,155	14,224	27,853
Completed high school	11,546	26,481	20,060	58,087
1 to 3 years of college	10,700	22,618	11,127	44,445
4 or more years of college	11,066	23,183	10,596	44,845
Total	37,786	81,435	56,008	175,230

Marginal distributions

How can we best grasp the information contained in Table 4.6 First, *look at the distribution of each variable separately.* The distribution of a categorical variable just says how often each outcome occurred. The "Total" column at the right of the table contains the totals for each of the rows. These row totals give the distribution of education level (the row variable) among all people over 25 years of age: 27,853,000 did not complete high school, 58,087,000 finished high school but did not attend college, and so on. In the same way, the "Total" row on the bottom gives the age distribution. If the row and column totals are missing, the first thing to do in studying a two-way table is to calculate them. The distributions of education alone and age alone are often called ***marginal distributions*** because they appear at the right and bottom margins of the two-way table.

marginal distributions

If you check the column totals in Table 4.6, you will notice a few discrepancies. For example, the sum of the entries in the "35 to 54" column is 81,437. The entry in the "Total" row for that column is 81,435. The explanation is ***roundoff error.*** The table entries are in the thousands of persons, and each is rounded to the nearest thousand. The Census Bureau obtained the "Total" entry by rounding the exact number of people aged 35 to 54 to the nearest thousand. The result was 81,435,000. Adding the column entries, each of which is already rounded, gives a slightly different result.

roundoff error

Percents are often more informative than counts. We can display the marginal distribution of education level in terms of percents by dividing each row total by the table total and converting to a percent.

EXAMPLE 4.20 MARGINAL DISTRIBUTION

The percent of people 25 years of age or older who have at least 4 years of college is

$$\frac{\text{total with four years of college}}{\text{table total}} = \frac{44,845}{175,230} = 0.256 = 25.6\%$$

Do three more such calculations to obtain the marginal distribution of education level in percents. Here it is.

Education:	Did not finish high school	Completed high school	1–3 years of college	≥ 4 years of college
Percent:	15.9	33.1	25.4	25.6

The total is 100% because everyone is in one of the four education categories.

Each marginal distribution from a two-way table is a distribution for a single categorical variable. As we saw in Chapter 1, we can use a bar graph or a pie chart to display such a distribution. Figure 4.24 is a bar graph of the distribu-

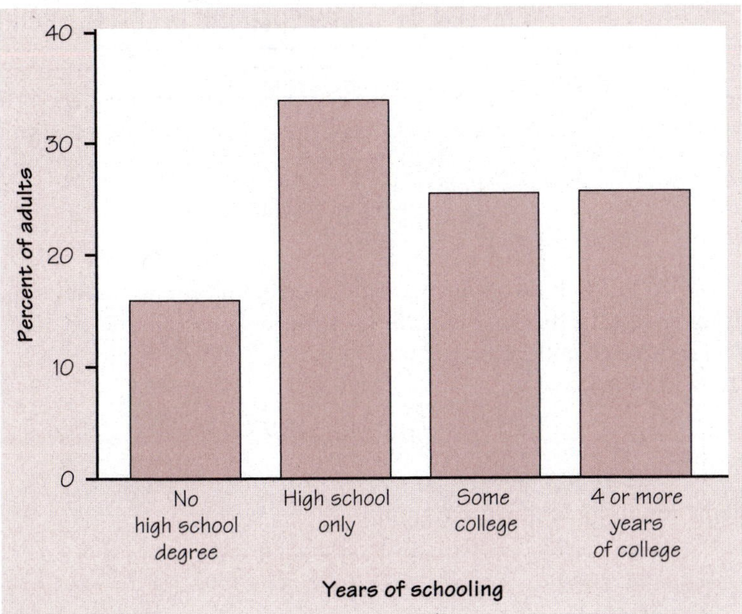

FIGURE 4.24 A bar graph of the distribution of years of schooling completed among people aged 25 years and over. This is one of the marginal distributions for Table 4.6.

tion of years of schooling. We see that people with at least some college education make up about half of the 25-or-older population.

In working with two-way tables, you must calculate lots of percents. Here's a tip to help decide what fraction gives the percent you want. Ask, "What group represents the total that I want a percent of?" The count for that group is the denominator of the fraction that leads to the percent. In Example 4.20, we wanted a percent "of people 25 or older years of age," so the count of people 25 or older (the table total) is the denominator.

Describing relationships

The marginal distributions of age and of education separately do not tell us how the two variables are related. That information is in the body of the table. How can we describe the relationship between age and years of school completed? No single graph (such as a scatterplot) portrays the form of the relationship between categorical variables, and no single numerical measure (such as the correlation) summarizes the strength of an association. *To describe relationships among categorical variables, calculate appropriate percents from the counts given.* We use percents because counts are often hard to compare. For example, 11,066,000 people age 25 to 34 have completed college, and only 10,596,000 people in the 55 and over age group have done so. But the older age group is larger, so we can't directly compare these counts.

EXAMPLE 4.21 HOW COMMON IS COLLEGE EDUCATION?

What percent of people aged 25 to 34 have completed 4 years of college? This is the count who are 25 to 34 and have 4 years of college as a percent of the age group total:

$$\frac{11,066}{37,786} = 0.293 = 29.3\%$$

"People aged 25 to 34" is the group we want a percent of, so the count for that group is the denominator. In the same way, the percent of people in the 55 and over age group who completed college is

$$\frac{10,596}{56,008} = 0.189 = 18.9\%$$

Here are the results for all three age groups:

Age group:	25 to 34	35 to 54	55+
Percent with 4 years of college:	29.3	28.5	18.9

These percents help us see how the education of Americans varies with age. Older people are less likely to have completed college.

Although graphs are not as useful for describing categorical variables as they are for quantitative variables, a graph still helps an audience to grasp the data quickly. The bar graph in Figure 4.25 presents the information in Example 4.20.

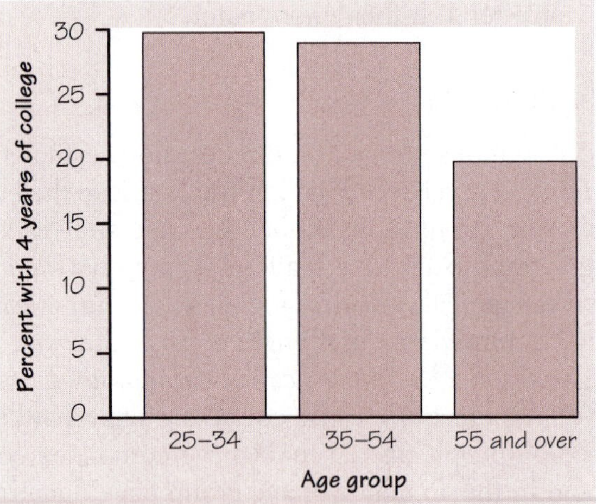

FIGURE 4.25 Bar graph comparing the percents of three age groups who have completed 4 or more years of college. The height of each bar is the percent of people in one age group who have completed at least 4 years of college.

Each bar represents one age group. The height of the bar is the percent of that age group with at least 4 years of college. Although bar graphs look a bit like histograms, their details and uses are different. A histogram shows the distribution of the values of a quantitative variable. A bar graph compares the sizes of different items. The horizontal axis of a bar graph need not have any measurement scale but may simply identify the items being compared. The items compared in Figure 4.25 are the three age groups. Because each bar in a bar graph describes a different item, we draw the bars with space between them.

EXERCISES

4.50 Sum the counts in the "55+" age column in Table 4.6 (page 241). Then explain why the sum is not the same as the entry for this column in the "Total" row.

4.51 Give the marginal distribution of age among people 25 years of age or older in percents, starting from the counts in Table 4.6 (page 241).

4.52 Using the counts in Table 4.6 (page 241), find the percent of people in each age group who did not complete high school. Draw a bar graph that compares these percents. State briefly what the data show.

4.53 SMOKING BY STUDENTS AND THEIR PARENTS Here are data from eight high schools on smoking among students and among their parents:[28]

	Neither parent smokes	One parent smokes	Both parents smoke
Student does not smoke	1168	1823	1380
Student smokes	188	416	400

(a) How many students do these data describe?

(b) What percent of these students smoke?

(c) Give the marginal distribution of parents' smoking behavior, both in counts and in percents.

4.54 PYTHON EGGS How is the hatching of water python eggs influenced by the temperature of the snake's nest? Researchers assigned newly laid eggs to one of three temperatures: hot, neutral, or cold. Hot duplicates the extra warmth provided by the mother python, and cold duplicates the absence of the mother. Here are the data on the number of eggs and the number that hatched:[29]

	Cold	Neutral	Hot
Number of eggs	27	56	104
Number hatched	16	38	75

(a) Make a two-way table of temperature by outcome (hatched or not).

(b) Calculate the percent of eggs in each group that hatched. The researchers anticipated that eggs would not hatch in cold water. Do the data support that anticipation?

4.55 IS HIGH BLOOD PRESSURE DANGEROUS? Medical researchers classified each of a group of men as "high" or "low" blood pressure, then watched them for 5 years. (Men with systolic blood pressure 140 mm Hg or higher were "high"; the others, "low.") The following two-way table gives the results of the study:[30]

	Died	Survived
Low blood pressure	21	2655
High blood pressure	55	3283

(a) How many men took part in the study? What percent of these men died during the 5 years of the study?

(b) The two categorical variables in the table are blood pressure (high or low) and outcome (died or survived). Which is the explanatory variable?

(c) Is high blood pressure associated with a higher death rate? Calculate and compare percents to answer this question.

Conditional distributions

Example 4.21 does not compare the complete distributions of years of schooling in the three age groups. It compares only the percents who finished college. Let's look at the complete picture.

EXAMPLE 4.22 CONDITIONAL DISTRIBUTION

Information about the 25 to 34 age group occupies the first column in Table 4.6. To find the complete distribution of education in this age group, look only at that column. Compute each count as a percent of the column total: 37,786. Here is the distribution:

Education:	Did not finish high school	Completed high school	1–3 years of college	≥ 4 years of college
Percent:	11.8	30.6	28.3	29.3

conditional distribution

These percents add to 100% because all 25- to 34-year-olds fall in one of the educational categories. The four percents together are the ***conditional distribution*** of education, given that a person is 25 to 34 years of age. We use the term "conditional" because the distribution refers only to people who satisfy the condition that they are 25 to 34 years old.

For comparison, here is the conditional distribution of years of school completed among people age 55 and over. To find these percents, look only at the "55+" column in Table 4.6. The column total is the denominator for each percent calculation.

Education:	Did not finish high school	Completed high school	1–3 years of college	≥ 4 years of college
Percent:	25.4	35.8	19.9	18.9

The percent who did not finish high school is much higher in the older age group, and the percents with some college and who finished college are much lower. Comparing the conditional distributions of education in different age groups describes the association between age and education. There are three different conditional distributions of education given age, one for each of the three age groups. All of these conditional distributions differ from the marginal distribution of education found in Example 4.20.

Statistical software can speed the task of finding each entry in a two-way table as a percent of its column total. Figure 4.26 displays the result. The software found the row and column totals from the table entries, so they may differ slightly from those in Table 4.6.

TABLE OF EDU BY AGE

EDU / AGE Frequency Col Pct	25-34	35-54	55 over	Total
NoHS	4474 11.84	9155 11.24	14224 25.40	27853
HSonly	11546 30.56	26481 32.52	20060 35.82	58087
SomeColl	10700 28.32	22618 27.77	11127 19.87	44445
Coll4yrs	11066 29.29	23183 28.47	10596 18.92	44845
Total	37786	81435	56008	175230

FIGURE 4.26 SAS output of the two-way table of education by age with the three conditional distributions of education, one for each age group. The percents in each column add to 100%.

Each cell in this table contains a count from Table 4.6 along with that count as a percent of the column total. The percents in each column form the conditional distribution of years of schooling for one age group.

The percents in each column add to 100% because everyone in the age group is accounted for. Comparing the conditional distributions reveals the nature of the association between age and education. The distributions of education in the two younger groups are quite similar, but higher education is less common in the 55 and over group.

Bar graphs can help make the association visible. We could make three side-by-side bar graphs, each resembling Figure 4.24 (page 243), to present the three conditional distributions. Figure 4.27 shows an alternative form of bar graph. Each set of three bars compares the percents in the three age groups who have reached a specific educational level.

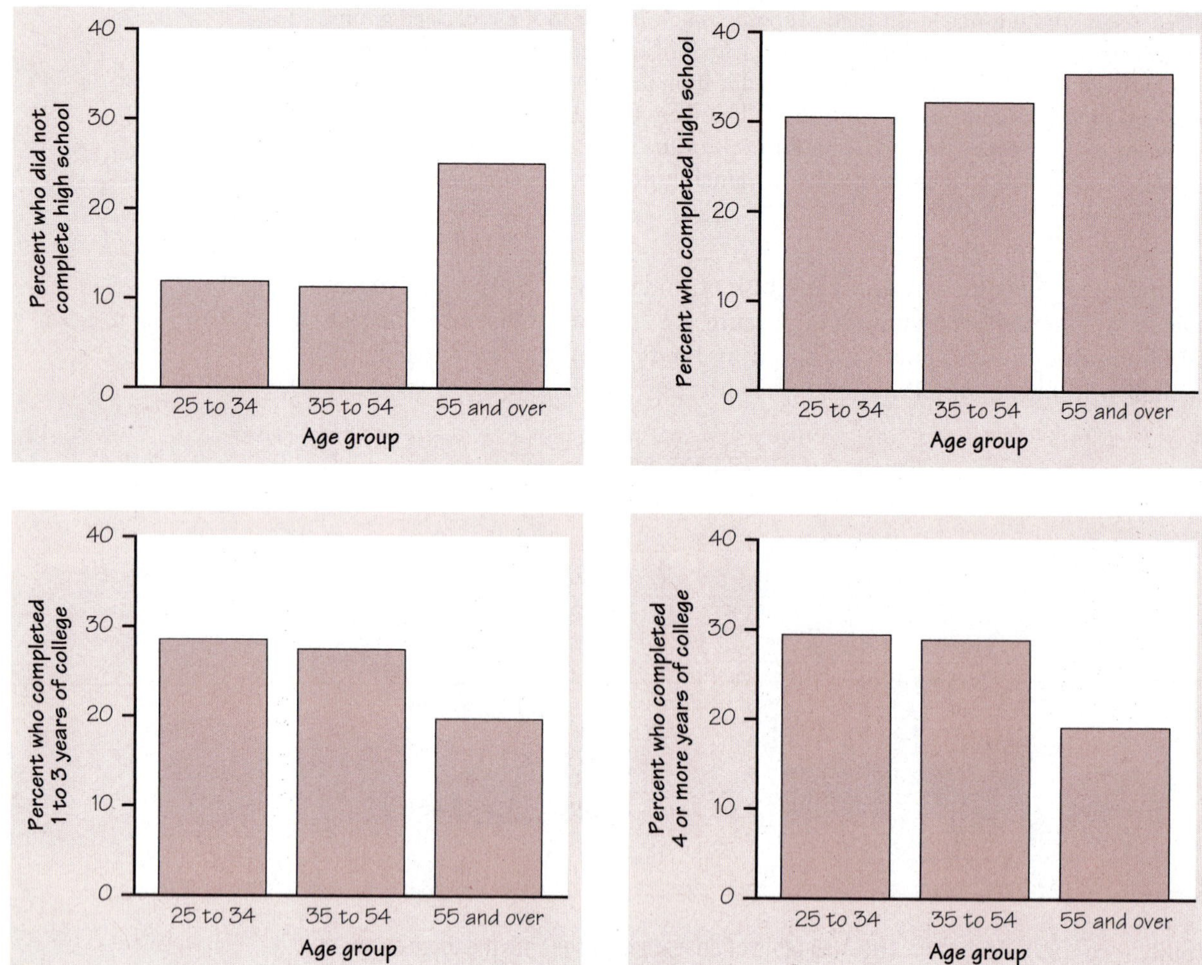

FIGURE 4.27 Bar graphs to compare the education levels of three age groups. Each graph compares the percents of three groups who fall in one of the four education levels.

We see at once that the "25 to 34" and "35 to 54" bars are similar for all four levels of education, and that the "55 and over" bars show that many more people in this group did not finish high school and that many fewer have any college.

No single graph (such as a scatterplot) portrays the form of the relationship between categorical variables. No single numerical measure (such as the correlation) summarizes the strength of the association. Bar graphs are flexible enough to be helpful, but you must think about what comparisons you want to display. For numerical measures, we rely on well-chosen percents. You must decide which percents you need. Here is a hint: compare the conditional distributions of the response variable (education) for the separate values of the explanatory variable (age). That's what we did in Figure 4.26.

In Example 4.22 we compared the education of different age groups. That is, we thought of age as the explanatory variable and education as the response variable. We might also be interested in the distribution of age among persons having a certain level of education. To do this, look only at one row in Table 4.6. Calculate each entry in that row as a percent of the row total, the total of that education group. The result is another conditional distribution, the conditional distribution of age given a certain level of education.

A two-way table contains a great deal of information in compact form. Making that information clear almost always requires finding percents. You must decide which percents you need. If you are studying trends in the training of the American workforce, comparing the distributions of education for different age groups reveals the more extensive education of younger people. If, on the other hand, you are planning a program to improve the skills of people who did not finish high school, the age distribution within this educational group is important information.

Simpson's paradox

As is the case with quantitative variables, the effects of lurking variables can change or even reverse relationships between two categorical variables. Here is a hypothetical example that demonstrates the surprises that can await the unsuspecting user of data.

EXAMPLE 4.23 PATIENT OUTCOMES IN HOSPITALS

To help consumers make informed decisions about health care, the government releases data about patient outcomes in hospitals. You want to compare Hospital A and Hospital B, which serve your community. Here is a two-way table of data on the survival of patients after surgery in these two hospitals. All patients undergoing surgery in a recent time period are included. "Survived" means that the patient lived at least 6 weeks following surgery.

	Hospital A	Hospital B
Died	63	16
Survived	2037	784
Total	2100	800

The evidence seems clear: Hospital A loses 3% (63/2100) of its surgery patients, and Hospital B loses only 2% (16/800). It seems that you should choose Hospital B if you need surgery.

Not all surgery cases are equally serious, however. Patients are classified as being in either "poor" or "good" condition before surgery. Here are the data broken down by patient condition. Check that the entries in the original two-way table are just the sums of the "poor" and "good" entries in this pair of tables.

Good Condition		
	Hospital A	Hospital B
Died	6	8
Survived	594	592
Total	600	600

Poor Condition		
	Hospital A	Hospital B
Died	57	8
Survived	1443	192
Total	1500	200

Hospital A beats Hospital B for patients in good condition: only 1% (6/600) died in Hospital A, compared with 1.3% (8/600) in Hospital B. And Hospital A wins again for patients in poor condition, losing 3.8% (57/1500) to Hospital B's 4% (8/200). So Hospital A is safer for both patients in good condition and patients in poor condition. If you are facing surgery, you should choose Hospital A.

The patient's condition is a lurking variable when we compare the death rates at the two hospitals. When we ignore the lurking variable, Hospital B seems safer, even though Hospital A does better for both classes of patients. How can A do better in each group, yet do worse overall? Look at the data. Hospital A is a medical center that attracts seriously ill patients from a wide region. It had 1500 patients in poor condition. Hospital B had only 200 such cases. Because patients in poor condition are more likely to die, Hospital A has a higher death rate despite its superior performance for each class of patients. The original two-way table, which did not take account of the condition of the patients, was misleading. Example 4.23 illustrates *Simpson's paradox.*

SIMPSON'S PARADOX

Simpson's paradox refers to the reversal of the direction of a comparison or an association when data from several groups are combined to form a single group.

The lurking variables in Simpson's paradox are categorical. That is, they break the individuals into groups, as when surgery patients are classified as "good condition" or "poor condition." Simpson's paradox is just an extreme form of the fact that observed associations can be misleading when there are lurking variables.

EXERCISES

4.56 Verify that the results for the conditional distribution of education level among people aged 55 and over given in Example 4.22 (page 246) are correct.

4.57 Example 4.22 (page 246) gives the conditional distributions of education level among 25- to 34-year-olds and among people 55 and over. Find the conditional distribution of education level among 35- to 54-year-olds in percents. Is this distribution more like the distribution for 25- to 34-year-olds or the distribution for people 55 and over?

4.58 Find the conditional distribution of age among people with at least 4 years of college using the data from Example 4.22 (page 246).

4.59 MAJORS FOR MEN AND WOMEN IN BUSINESS A study of the career plans of young women and men sent questionnaires to all 722 members of the senior class in the College of Business Administration at the University of Illinois. One question asked which major within the business program the student had chosen. Here are the data from the students who responded:[31]

	Female	Male
Accounting	68	56
Administration	91	40
Economics	5	6
Finance	61	59

(a) Find the two conditional distributions of major, one for women and one for men. Based on your calculations, describe the differences between women and men with a graph and in words.

(b) What percent of the students did not respond to the questionnaire? The nonresponse weakens conclusions drawn from these data.

4.60 COLLEGE ADMISSIONS PARADOX Upper Wabash Tech has two professional schools, business and law. Here are two-way tables of applicants to both schools, categorized by gender and admission decision. (Although these data are made up, similar situations occur in reality.)[32]

Business			Law		
	Admit	Deny		Admit	Deny
Male	480	120	Male	10	90
Female	180	20	Female	100	200

(a) Make a two-way table of gender by admission decision for the two professional schools together by summing entries in this table.

(b) From the two-way table, calculate the percent of male applicants who are admitted and the percent of female applicants who are admitted. Wabash admits a higher percent of male applicants.

(c) Now compute separately the percents of male and female applicants admitted by the business school and by the law school. Each school admits a higher percent of female applicants.

(d) This is Simpson's paradox: both schools admit a higher percent of the women who apply, but overall Wabash admits a lower percent of female applicants than of male applicants. Explain carefully, as if speaking to a skeptical reporter, how it can happen that Wabash appears to favor males when each school individually favors females.

4.61 RACE AND THE DEATH PENALTY Whether a convicted murderer gets the death penalty seems to be influenced by the race of the victim. Here are data on 326 cases in which the defendant was convicted of murder:[33]

	White defendant			Black defendant	
	White victim	Black victim		White victim	Black victim
Death	19	0	Death	11	6
Not	132	9	Not	52	97

(a) Use these data to make a two-way table of defendant's race (white or black) versus death penalty (yes or no).

(b) Show that Simpson's paradox holds: a higher percent of white defendants are sentenced to death overall, but for both black and white victims a higher percent of black defendants are sentenced to death.

(c) Use the data to explain why the paradox holds in language that a judge could understand.

SUMMARY

A **two-way table** of counts organizes data about two categorical variables. Values of the **row variable** label the rows that run across the table, and values of the **column variable** label the columns that run down the table. Two-way tables are often used to summarize large amounts of data by grouping outcomes into categories.

The **row totals** and **column totals** in a two-way table give the **marginal distributions** of the two individual variables. It is clearer to present these distributions as percents of the table total. Marginal distributions tell us nothing about the relationship between the variables.

To find the **conditional distribution** of the row variable for one specific value of the column variable, look only at that one column in the table. Find each entry in the column as a percent of the column total.

There is a conditional distribution of the row variable for each column in the table. Comparing these conditional distributions is one way to describe the association between the row and the column variables. It is particularly useful when the column variable is the explanatory variable.

Bar graphs are a flexible means of presenting categorical data. There is no single best way to describe an association between two categorical variables.

A comparison between two variables that holds for each individual value of a third variable can be changed or even reversed when the data for all values of the third variable are combined. This is **Simpson's paradox.** Simpson's paradox is an example of the effect of lurking variables on an observed association.

SECTION 4.3 EXERCISES

COLLEGE UNDERGRADUATES Exercises 4.62 to 4.66 are based on Table 4.7. This two-way table reports data on all undergraduate students enrolled in U.S. colleges and universities in the fall of 1995 whose age was known.

TABLE 4.7 Undergraduate college enrollment, fall 1995 (thousands of students)

Age	2-year full-time	2-year part-time	4-year full-time	4-year part-time
under 18	41	125	75	45
18 to 24	1378	1198	4607	588
25 to 39	428	1427	1212	1321
40 and up	119	723	225	605
Total	1966	3472	6119	2559

Source: *Digest of Education Statistics 1997*, accessed on the National Center for Education Statistics Web site, http://www.ed.gov/NCES.

4.62

(a) How many undergraduate students were enrolled in colleges and universities?

(b) What percent of all undergraduate students were 18 to 24 years old in the fall of the academic year?

(c) Find the percent of the undergraduates enrolled in each of the four types of program who were 18 to 24 years old. Make a bar graph to compare these percents.

(d) The 18 to 24 group is the traditional age group for college students. Briefly summarize what you have learned from the data about the extent to which this group predominates in different kinds of college programs.

4.63

(a) An association of two-year colleges asks: "What percent of students enrolled part-time at 2-year colleges are 25 to 39 years old?"

(b) A bank that makes education loans to adults asks: "What percent of all 25- to 39-year-old students are enrolled part-time at 2-year colleges?"

4.64

(a) Find the marginal distribution of age among all undergraduate students, first in counts and then in percents. Make a bar graph of the distribution in percents.

(b) Find the conditional distribution of age (in percents) among students enrolled part-time in 2-year colleges and make a bar graph of this distribution.

(c) Briefly describe the most important differences between the two age distributions.

(d) The sum of the entries in the "2-year part-time" column is not the same as the total given for that column. Why is this?

4.65 Call students aged 40 and up "older students." Compare the presence of older students in the four types of program with numbers, a graph, and a brief summary of your findings.

4.66 With a little thought, you can extract from Table 4.7 information other than marginal and conditional distributions. The traditional college age group is ages 18 to 24 years.

(a) What percent of all undergraduates fall in this age group?

(b) What percent of students at 2-year colleges fall in this age group?

(c) What percent of part-time students fall in this group?

4.67 FIREARM DEATHS Firearms are second to motor vehicles as a cause of nondisease deaths in the United States. Here are counts from a study of all firearm-related deaths in Milwaukee, Wisconsin, between 1990 and 1994.[34] We want to compare the types of firearms used in homicides and in suicides. We suspect that long guns (shotguns and rifles) will more often be used in suicides because many people keep them at home for hunting. Make a careful comparison of homicides and suicides, with a bar graph. What do you find about long guns versus handguns?

	Handgun	Shotgun	Rifle	Unknown	Total
Homicides	468	28	15	13	524
Suicides	124	22	24	5	175

4.68 HELPING COCAINE ADDICTS Cocaine addiction is hard to break. Addicts need cocaine to feel any pleasure, so perhaps giving them an antidepressant drug will help. A 3-year study with 72 chronic cocaine users compared an antidepressant drug called desipramine with lithium and a placebo. (Lithium is a standard drug to treat cocaine addiction. A placebo is a dummy drug, used so that the effect of being in the study but not taking any drug can be seen.) One-third of the subjects, chosen at random, received each drug. Here are the results:[35]

	Desipramine	Lithium	Placebo
Relapse	10	18	20
No relapse	14	6	4
Total	24	24	24

(a) Compare the effectiveness of the three treatments in preventing relapse. Use percents and draw a bar graph.

(b) Do you think that this study gives good evidence that desipramine actually *causes* a reduction in relapses?

4.69 SEAT BELTS AND CHILDREN Do child restraints and seat belts prevent injuries to young passengers in automobile accidents? Here are data on the 26,971 passengers under the age of 15 in accidents reported in North Carolina during two years before the law required restraints:[36]

	Restrained	Unrestrained
Injured	197	3,844
Uninjured	1,749	21,181

(a) What percent of these young passengers were restrained?

(b) Do the data provide evidence that young passengers are less likely to be injured in an accident if they wear restraints? Calculate and compare percents to answer this question.

4.70 **BASEBALL PARADOX** Most baseball hitters perform differently against right-handed and left-handed pitching. Consider two players, Joe and Moe, both of whom bat right-handed. The table below records their performance against right-handed and left-handed pitchers.

Player	Pitcher	Hits	At bats
Joe	Right	40	100
	Left	80	400
Moe	Right	120	400
	Left	10	100

(a) Make a two-way table of player (Joe or Moe) versus outcome (hit or no hit) by summing over both kinds of pitcher.

(b) Find the overall batting average (hits divided by total times at bat) for each player. Who has the higher batting average?

(c) Make a separate two-way table of player versus outcome for each kind of pitcher. From these tables, find the batting averages of Joe and Moe against right-handed pitching. Who does better? Do the same for left-handed pitching. Who does better?

(d) The manager doesn't believe that one player can hit better against both left-handers and right-handers yet have a lower overall batting average. Explain in simple language why this happens to Joe and Moe.

4.71 **OBESITY AND HEALTH** Recent studies have shown that earlier reports underestimated the health risks associated with being overweight. The error was due to overlooking lurking variables. In particular, smoking tends both to reduce weight and to lead to earlier death. Illustrate Simpson's paradox by a simplified version of this situation. That is, make up tables of overweight (yes or no) by early death (yes or no) by smoker (yes or no) such that

- Overweight smokers and overweight nonsmokers both tend to die earlier than those not overweight.
- But when smokers and nonsmokers are combined into a two-way table of overweight by early death, persons who are not overweight tend to die earlier.

CHAPTER REVIEW

In Chapter 3, we learned how to analyze two-variable data that show a linear pattern. We learned about positive and negative associations and how to measure the strength of association between two variables. We also developed a procedure for constructing a model (the least-squares regression line) that captures the trend of the data. This LSRL is useful for prediction purposes. A recurring theme is that data analysis begins with graphs and then adds numerical summaries of specific aspects of the data.

In this chapter we learned how to construct mathematical models for data that fit a curve, such as an exponential function or a power function. We also learned that although correlation and regression are powerful tools for understanding two-variable data when both variables are quantitative, both correlation and regression have their limitations. In particular, we are cautioned that a strong observed association between two variables may exist without a cause-and-effect link between them. If both variables are categorical, there is no satisfactory graph for displaying the data, although bar graphs can be helpful. We describe the relationship by comparing percents.

Here is a review list of the most important skills you should have gained from studying this chapter.

A. MODELING NONLINEAR DATA

1. Recognize that when a variable is multiplied by a fixed number greater than 1 in each equal time period, exponential growth results; when the ratio is a positive number less than 1, it's called exponential decay.

2. Recognize that when one variable is proportional to a power of a second variable, the result is a power function.

3. In the case of both exponential growth and power function, perform a logarithmic transformation and obtain points that lie in a linear pattern. Then use least-squares regression on the transformed points. An inverse transformation then produces a curve that is a model for the original points.

4. Know that deviations from the overall pattern are most easily examined by fitting a line to the transformed points and plotting the residuals from this line against the explanatory variable (or fitted values).

B. INTERPRETING CORRELATION AND REGRESSION

1. Understand that both r and the least-squares regression line can be strongly influenced by a few extreme observations.

2. Recognize possible lurking variables that may explain the observed association between two variables x and y.

3. Understand that even a strong correlation does not mean that there is a cause-and-effect relationship between x and y.

C. RELATIONS IN CATEGORICAL DATA

1. From a two-way table of counts, find the marginal distributions of both variables by obtaining the row sums and column sums.

2. Express any distribution in percents by dividing the category counts by their total.

3. Describe the relationship between two categorical variables by computing and comparing percents. Often this involves comparing the conditional distributions of one variable for the different categories of the other variable.

4. Recognize Simpson's paradox and be able to explain it.

CHAPTER 4 REVIEW EXERCISES

4.72 **LIGHT INTENSITY** In physics class, the intensity of a 100-watt light bulb was measured by a sensing device at various distances from the light source, and the following data were collected. Note that a *candela* (cd) is an international unit of luminous intensity.

Distance (meters)	Intensity (candelas)
1.0	0.2965
1.1	0.2522
1.2	0.2055
1.3	0.1746
1.4	0.1534
1.5	0.1352
1.6	0.1145
1.7	0.1024
1.8	0.0923
1.9	0.0832
2.0	0.0734

(a) Plot the data. Based on the pattern of points, propose a model form for the data. Then use a transformation followed by linear regression and then an inverse transformation to construct a model.

(b) Report the equation, and plot the original data with the model on the same axes.

(c) Describe the relationship between the intensity and the distance from the light source.

(d) Consult the physics textbooks used in your school and find the formula for the intensity of light as a function of distance from the light source. How do your experimental results compare with the theoretical formula?

4.73 **PENDULUM** An experiment was conducted with a pendulum of variable length. The *period*, or length of time to complete one complete oscillation, was recorded for several lengths. Here are the data:

Length (feet):	1	2	3	4	5	6	7
Period (seconds):	1.10	1.56	1.92	2.20	2.50	2.71	2.93

(a) Make a plot of period against length. Describe the pattern that you see.

(b) Propose a model form. Then use a transformation to construct a model for the data. Report the equation, and plot the original data with the model on the same axes.

(c) Describe the relationship between the length of a pendulum and its period.

4.74 EXACT EXPONENTIAL GROWTH, I A clever courtier, offered a reward by an ancient king of Persia, asked for a grain of rice on the first square of a chess board, 2 grains on the second square, then 4, 8, 16, and so on.

(a) Make a table of the number of grains on each of the first 10 squares of the board.

(b) Plot the number of grains on each square against the number of the square for squares 1 to 10, and connect the points with a smooth curve. This is an exponential curve.

(c) How many grains of rice should the king deliver for the 64th (and final) square?

(d) Take the logarithm of each of your numbers of grains from (a). Plot these logarithms against the number of squares from 1 to 10. You should get a straight line.

(e) From your graph in (d) find the approximate values of the slope b and the intercept a for the line. Use the equation $y = a + bx$ to predict the logarithm of the amount for the 64th square. Check your result by taking the logarithm of the amount you found in (c).

4.75 800-METER RUN Return to the 800-meter world record times for men and women of Exercise 3.75 (page 188). Suppose you are uncomfortable with the linear model for the decline in winning times that will eventually intersect the horizontal axis.

(a) Construct exponential and power regression models for the *men's* record times. Which do you consider to be a better model?

(b) Based on your answer to (a), construct a similar model for the *women's* record times.

(c) Will either of these curves eventually reach zero? Will the curves intersect each other? If so, in what year will the curves intersect?

(d) Is this a satisfactory model, or is there a better model for these data?

4.76 SOCIAL INSURANCE Federal expenditures on social insurance (chiefly social security and Medicare) increased rapidly after 1960. Here are the amounts spent, in millions of dollars:

Year:	1960	1965	1970	1975	1980	1985	1990
Spending:	14,307	21,807	45,246	99,715	191,162	310,175	422,257

(a) Plot social insurance expenditures against time. Does the pattern appear closer to linear growth or to exponential growth?

(b) Take the logarithm of the amounts spent. Plot these logarithms against time. Do you think that the exponential growth model fits well?

(c) After entering the data into the Minitab statistical system, with year as C1 and expenditures as C2, we obtain the least-squares line for the logarithms as follows:

```
MTB> LET C3 = LOGT(C2)
MTB> REGRESS C3 ON 1, C1

The regression equation is
C3 = -98.63833 + 0.05244 C1
```

That is, the least-squares line is

$$\log y = -98.63833 + (0.05244 \times \text{year})$$

Draw this line on your graph from (b).

(d) Use this line to predict the logarithm of social insurance outlays for 1988. Then compute

$$y = 10^{\log y}$$

to predict the amount y spent in 1988.

(e) The actual amount (in millions) spent in 1988 was $358,412. Take the logarithm of this amount and add the 1988 point to your graph in (b). Does it fall close to the line? When President Reagan took office in 1981, he advocated a policy of slowing growth in spending on social programs. Did the trend of exponential growth in spending for social insurance change in a major way during the Reagan years, 1981 to 1988?

4.77 KILLING BACTERIA Expose marine bacteria to X-rays for time periods from 1 to 15 minutes. Here are the number of surviving bacteria (in hundreds) on a culture plate after each exposure time:[37]

Time t	Count y	Time t	Count y
1	355	9	56
2	211	10	38
3	197	11	36
4	166	12	32
5	142	13	21
6	106	14	19
7	104	15	15
8	60		

Theory suggests an exponential growth or decay model. Do the data appear to conform to this theory?

4.78 BANK CARDS Electronic fund transfers, from bank automatic teller machines and the use of debit cards by consumers, have grown rapidly in the United States. Here are data on the number of such transfers (in millions).[38]

Year	EFT	Year	EFT	Year	EFT
1985	3,579	1991	6,642	1996	11,780
1987	4,108	1992	7,537	1997	12,580
1988	4,581	1993	8,135	1998	13,160
1989	5,274	1994	9,078	1999	13,316
1990	5,942	1995	10,464		

Write a clear account of the pattern of growth of electronic transfers over time, supporting your description with plots and calculations as needed. Has the pattern changed in the most recent years?

4.79 ICE CREAM AND FLU There is a negative correlation between the number of flu cases reported each week throughout the year and the amount of ice cream sold in that particular week. It's unlikely that ice cream prevents flu. What is a more plausible explanation for this observed correlation?

4.80 VOTING FOR PRESIDENT The following table gives the U.S. resident population of voting age and the votes cast for president, both in thousands, for presidential elections between 1960 and 2000:

Year	Population	Votes	Year	Population	Votes
1960	109,672	68,838	1984	173,995	92,653
1964	114,090	70,645	1988	181,956	91,595
1968	120,285	73,212	1992	189,524	104,425
1972	140,777	77,719	1996	196,511	96,456
1976	152,308	81,556	2000	209,128	105,363
1980	163,945	86,515			

(a) For each year compute the percent of people who voted. Make a time plot of the percent who voted. Describe the change over time in participation in presidential elections.

(b) Before proposing political explanations for this change, we should examine possible lurking variables. The minimum voting age in presidential elections dropped from 21 to 18 years in 1970. Use this fact to propose a partial explanation for the trend you saw in (a).

4.81 WOMEN AND MARITAL STATUS The following two-way table describes the age and marital status of American women in 2000. The table entries are in thousands of women.

	Marital status				
Age	Single	Married	Widowed	Divorced	Total
15–24	16,121	2,694	21	203	19,040
25–39	7,409	19,925	212	2,965	30,510
40–64	3,553	29,687	2,338	6,797	42,373
≥65	680	8,223	8,490	1,344	18,735
Total					110,660

(a) Find the sum of the entries in the 15–24 row. Why does this sum differ from the "Total" entry for that row?

(b) Give the marginal distribution of marital status for all adult women (use percents). Draw a bar graph to display this distribution.

(c) Compare the conditional distributions of marital status for women aged 15 to 24 and women aged 40 to 64. Briefly describe the most important differences between the two groups of women, and back up your description with percents.

(d) You are planning a magazine aimed at single women who have never been married. (That's what "single" means in government data.) Find the conditional distribution of ages among single women.

4.82 WOMEN SCIENTISTS A study by the National Science Foundation[39] found that the median salary of newly graduated female engineers and scientists was only 73% of the median salary for males. When the new graduates were broken down by field, however, the picture changed. Women's median salaries as a percent of the male median in the 16 fields studied were

94%	96%	98%	95%	85%	85%	84%	100%
103%	100%	107%	93%	104%	93%	106%	100%

How can women do nearly as well as men in every field yet fall far behind men when we look at all young engineers and scientists?

4.83 SMOKING AND STAYING ALIVE In the mid-1970s, a medical study contacted randomly chosen people in a district in England. Here are data on the 1314 women contacted who were either current smokers or who had never smoked. The table classifies these women by their smoking status and age at the time of the survey and whether they were still alive 20 years later.[40]

Age 18 to 44

	Smoker	Not
Dead	19	13
Alive	269	327

Age 45 to 64

	Smoker	Not
Dead	78	52
Alive	167	147

Age 65+

	Smoker	Not
Dead	42	165
Alive	7	28

(a) Make a two-way table of smoking (yes or no) by dead or alive. What percent of the smokers stayed alive for 20 years? What percent of the nonsmokers survived? It seems surprising that a higher percent of smokers stayed alive.

(b) The age of the women at the time of the study is a lurking variable. Show that within each of the three age groups in the data, a higher percent of nonsmokers remained alive 20 years later. This is another example of Simpson's paradox.

(c) The study authors give this explanation: "Few of the older women (over 65 at the original survey) were smokers, but many of them had died by the time of follow-up." Compare the percent of smokers in the three age groups to verify the explanation.

NOTES AND DATA SOURCES

1. This activity was described in Elizabeth B. Applebaum, "A simulation to model exponential growth," *Mathematics Teacher*, 93, No.7 (October 2000), pp. 614–615.

2. Data from G. A. Sacher and E. F. Staffelt, "Relation of gestation time to brain weight for placental mammals: implications for the theory of vertebrate growth," *American Naturalist*, 108 (1974), pp. 593–613. We found these data in F. L. Ramsey and D. W. Schafer, *The Statistical Sleuth: A Course in Methods of Data Analysis*, Duxbury, 1997.

3. There are several mathematical ways to show that log t fits into the power family at $p = 0$. Here's one. For powers $p \neq 0$, the indefinite integral $\int t^{p-1}\,dt$ is a multiple of t^p. When $p = 0$, $\int t^{-1}dt$ is log t.

4. Data from the World Bank's *1999 World Development Indicators*. Life expectancy is estimated for 1997, and GDP per capita (purchasing-power parity basis) is estimated for 1998.

5. The power law connecting heart rate with body weight was found online at "The Worldwide Anaesthetist," www.anaesthetist.com. Anesthesiologists are interested in power laws because they must judge how drug doses should increase in bigger patients.

6. Data from *Statistical Abstract of the United States, 2000*. Data for Alaska and Hawaii were included for the first time in 1950.

7. Gypsy moth data provided by Chuck Schwalbe, U.S. Department of Agriculture.

8. From Intel Web site, www.intel.com/research/silicon/mooreslaw.htm.

9. From Joel Best, *Damned Lies and Statistics: Untangling Numbers from the Media, Politicians, and Activists*, University of California Press, Berkeley and Los Angeles, 2001.

10. Fish data from Gordon L. Swartzman and Stephen P. Kaluzny, *Ecological Simulation Primer*, Macmillan, New York, 1987, p. 98.

11. Data originally from A. J. Clark, *Comparative Physiology of the Heart*, Macmillan, New York, 1927, p. 84. Obtained from Frank R. Giordano and Maurice D. Weir, *A First Course in Mathematical Modeling*, Brooks/Cole, Belmont, Calif., 1985, p. 56.

12. Data on a sample of 12 of 56 perch in a data set contributed to the *Journal of Statistics Education* data archive (www.amstat.org/publications/jse/) by Juha Puranen of the University of Helsinki.

13. Similar experiments are described by A. P. J. Trinci, "A kinetic study of the growth of *Aspergillus nidulans* and other fungi," *Journal of General Microbiology*, 57 (1969), pp. 11–24. These data were provided by Thomas Kuczek, Purdue University.

14. Jérôme Chave, Bernard Riéra, and Marc-A. Dubois, "Estimation of biomass in a neotropical forest of French Guiana: spatial and temporal variability," *Journal of Tropical Ecology*, 17 (2001), pp. 79–96.

15. Data from Stillman Drake, *Galileo at Work*, University of Chicago Press, 1978. We found these data in D. A. Dickey and J. T. Arnold, "Teaching statistics with data of historic significance," *Journal of Statistics Education*, 3 (1995), www.amstat.org/publications/jse/.

16. The quotation is from Dr. Daniel Mark of Duke University, in "Age, not bias, may explain differences in treatment," *New York Times*, April 26, 1994.

17. This example is drawn from M. Goldstein, "Preliminary inspection of multivariate data," *The American Statistician*, 36(1982), pp. 358–362.
18. Data provided by Peter Cook, Department of Mathematics, Purdue University.
19. Laura L. Calderon *et al.*, "Risk factors for obesity in Mexican-American girls: dietary factors, anthropometric factors, physical activity, and hours of television viewing," *Journal of the American Dietetic Association*, 96 (1996), pp. 1177–1179.
20. Saccharin appears on the National Institute of Health's list of suspected carcinogens but remains in wide use. In October 1997, an expert panel recommended by a 4 to 3 vote to keep it on the list, despite recommendations from other scientific groups that it be removed.
21. A detailed study of this correlation appears in E. M. Remolona, P. Kleinman, and D. Gruenstein, "Market returns and mutual fund flows," *Federal Reserve Bank of New York Economic Policy Review*, 3, No. 2(1997), pp. 33–52.
22. M. S. Linet *et al.*, "Residential exposure to magnetic fields and acute lymphoblastic leukemia in children," *New England Journal of Medicine*, 337 (1997), pp.1–7.
23. *The Health Consequences of Smoking: 1983*, U.S. Health Service, 1983.
24. From a Gannett News Service article appearing in the Lafayette (Indiana) *Journal and Courier*, April 23, 1994.
25. Contributed by Marigene Arnold of Kalamazoo College.
26. D. E. Powers and D. A. Rock, *Effects of Coaching on SAT I: Reasoning Test Scores*, Educational Testing Service Research Report 98-6, College Entrance Examination Board, 1998.
27. David E. Bloom and David Canning, "The health and wealth of nations," *Science*, 287 (2000), pp. 1207–1208.
28. From S. V. Zagona (ed.), *Studies and Issues in Smoking Behavior*, University of Arizona Press, Tucson, 1967, pp. 157–180.
29. R. Shine, T. R. L. Madsen, M. J. Elphick, and P. S. Harlow, "The influence of nest temperature and maternal brooding on hatchling phenotypes in water python," *Ecology*, 78 (1997), pp. 1713–1721.
30. From J. Stamler, "The mass treatment of hypertensive disease: defining the problem," *Mild Hypertension: To Treat or Not to Treat*, New York Academy of Sciences, 1978, pp. 333–358.
31. From F. D. Blau and M. A. Ferber, "Career plans and expectations of young women and men," *Journal of Human Resources*, 26 (1991), pp. 581–607.
32. See P. J. Bickel and J. W. O'Connell, "Is there a sex bias in graduate admissions?" *Science*, 187 (1975), pp. 398–404.
33. From M. Radelet, "Racial characteristics and imposition of the death penalty," *American Sociological Review*, 46 (1981), pp. 918–927.
34. S. W. Hargarten *et al.*, "Characteristics of firearms involved in fatalities," *Journal of the American Medical Association*, 275 (1996), pp. 42–45.
35. From D. M. Barnes, "Breaking the cycle of addiction," *Science*, 241 (1988), pp. 1029–1030.
36. Adapted from data of Williams and Zador in *Accident Analysis and Prevention*, 9 (1977), pp. 69–76.
37. S. Chatterjee and B. Price, *Regression Analysis by Example*, Wiley, New York, 1977.
38. From several editions of the *Statistical Abstract of the United States*.

39. National Science Board, *Science and Engineering Indicators, 1991,* U.S. Government Printing Office, Washington, D.C., 1991. The detailed data appear in Appendix Table 3-5, p. 274.
40. Condensed from D. R. Appleton, J. M. French, and M. P. J. Vanderpump, "Ignoring a covariate: an example of Simpson's paradox," *The American Statistician,* 50 (1996), pp. 340–341.

PART II

Producing Data:
Samples, Experiments, and Simulations

RONALD A. FISHER

The Father of Statistics

The ideas and methods that we study as "statistics" were invented in the nineteenth and twentieth centuries by people working on problems that required analysis of data. Astronomy, biology, social science, and even surveying can claim a role in the birth of statistics. But if anyone can claim to be "the father of statistics," that honor belongs to *Sir Ronald A. Fisher (1890–1962).*

Fisher's writings helped organize statistics as a distinct field of study whose methods apply to practical problems across many disciplines. He systematized the mathematical theory of statistics and invented many new techniques. The randomized comparative experiment is perhaps Fisher's greatest contribution.

Like other statistical pioneers, Fisher was driven by the demands of practical problems. Beginning in 1919, he worked on agricultural field experiments at Rothamsted in England. How should we arrange the planting of different crop varieties or the application of different fertilizers to get a fair comparison among them? Because fertility and other variables change as we move across a field, experiments used elaborate checkerboard planting arrangements to obtain fair comparisons. Fisher had a better idea: "arrange the plots deliberately at random."

This chapter explores statistical design for producing data to answer specific questions like "Which crop variety has the highest mean yield?" Fisher's innovation, the deliberate use of chance in producing data, is the central theme of the chapter and one of the most important ideas in statistics.

Like other statistical pioneers, Fisher was driven by the demands of practical problems.

chapter 5

Producing Data

ACTIVITY 5A A Class Survey

A class survey is a quick way to collect interesting data. Certainly there are things about the class as a group that you would like to know. Your task here is to construct a *draft* of a class survey, a questionnaire that would be used to gather data about the members of your class. Here are the steps to take:

1. As a class, discuss the questions you would like to include on the survey. In addition to *what* you want to ask, you should also consider *how many questions* you want to ask. Have one student serve as recorder and make a list on the blackboard or overhead projector of topics to include.

2. Once you have identified the topics, then work on the wording of the questions. Try to achieve as much consensus as possible. If there is a computer in the room, a student could use a word-processing program to enter the questions as they are developed.

3. Make one copy of the final draft of the survey for each student, but do not distribute the surveys at this time. The surveys are to be put aside for the time being. As you complete this chapter, you will return to take another look at the survey you have constructed, make final adjustments, and then administer the survey to all of the members of your class. This survey should provide some interesting data that can be analyzed during the remainder of the course.

As a starting point, here is a sample of a short survey:

CLASS SURVEY

Your answers to the questions below will help describe your class. DO NOT PUT YOUR NAME ON THIS PAPER. Your answers are completely private. They just help us describe the entire class.

1. Are you MALE or FEMALE? (Circle one.)

2. How many brothers and sisters do you have? ________

3. How tall are you in inches, to the nearest inch? ________

4. Estimate the number of pairs of shoes you own. ________

5. How much money in coins are you carrying right now? (Don't count any paper money, just coins.) ________

6. On a typical school night, how much time do you spend doing homework? (Answer in minutes. For example, 2 hours is 120 minutes.) ________

7. On a typical school night, how much time do you spend watching television? (Answer in minutes.) ________

INTRODUCTION

Exploratory data analysis seeks to discover and describe what data say by using graphs and numerical summaries. The conclusions we draw from data analysis apply to the specific data that we examine. Often, however, we want to answer questions about some large group of individuals. To get sound answers, we must produce data in a way that is designed to answer our questions.

Suppose our question is "What percent of American adults agree that the United Nations should continue to have its headquarters in the United States?" To answer the question, we interview American adults. We can't afford to ask all adults, so we put the question to a ***sample*** chosen to represent the entire adult population. How shall we choose a sample that truly represents the opinions of the entire population? Statistical designs for choosing samples are the topic of Section 5.1.

sample

Our goal in choosing a sample is a picture of the population, disturbed as little as possible by the act of gathering information. Sample surveys are one kind of *observational study*. In other settings, we gather data from an *experiment*. In doing an experiment, we don't just observe individuals or ask them questions. We actively impose some treatment in order to observe the response. Experiments can answer questions such as "Does aspirin reduce the chance of a heart attack?" and "Does a majority of college students prefer Pepsi to Coke when they taste both without knowing which they are drinking?" Experiments, like samples, provide useful data only when properly designed. We will discuss statistical design of experiments in Section 5.2. The distinction between experiments and observational studies is one of the most important ideas in statistics.

OBSERVATION VERSUS EXPERIMENT

An **observational study** observes individuals and measures variables of interest but does not attempt to influence the responses.

An **experiment,** on the other hand, deliberately imposes some treatment on individuals in order to observe their responses.

Observational studies are essential sources of data about topics from the opinions of voters to the behavior of animals in the wild. But an observational study, even one based on a statistical sample, is a poor way to gauge the effect of an intervention. To see the response to a change, we must actually impose the change. When our goal is to understand cause and effect, experiments are the only source of fully convincing data.

EXAMPLE 5.1 HELPING WELFARE MOTHERS FIND JOBS

Most adult recipients of welfare are mothers of young children. Observational studies of welfare mothers show that many are able to increase their earnings and leave the welfare system. Some take advantage of voluntary job-training programs to improve their skills. Should participation in job-training and job-search programs be required of all able-bodied welfare mothers? Observational studies cannot tell us what the effects of such a policy would be. Even if the mothers studied are a properly chosen sample of all welfare recipients, those who seek out training and find jobs may differ in many ways from those who do not. They are observed to have more education, for example, but they may also differ in values and motivation, things that cannot be observed.

To see if a required jobs program will help mothers escape welfare, such a program must actually be tried. Choose two similar groups of mothers when they apply for welfare. Require one group to participate in a job-training program, but do not offer the program to the other group. This is an experiment. Comparing the income and work record of the two groups after several years will show whether requiring training has the desired effect.

When we simply observe welfare mothers, the effect of job-training programs on success in finding work is *confounded* with (mixed up with) the characteristics of mothers who seek out training on their own. Recall that two variables (explanatory variables or lurking variables) are said to be **confounded** when their effects on a response variable cannot be distinguished from each other.

Observational studies of the effect of one variable on another often fail because the explanatory variable is confounded with lurking variables. We will see that well-designed experiments take steps to defeat confounding. Because experiments allow us to pin down the effects of specific variables of interest to us, they are the preferred method of gaining knowledge in science, medicine, and industry.

simulation

In some situations, it may not be possible to observe individuals directly or to perform an experiment. In other cases, it may be logistically difficult or simply inconvenient to obtain a sample or to impose a treatment. ***Simulations*** provide an alternative method for producing data in such circumstances. Section 5.3 introduces techniques for simulating experiments.

statistical inference

Statistical techniques for producing data open the door to formal ***statistical inference,*** which answers specific questions with a known degree of confidence. The later chapters of this book are devoted to inference. We will see that careful design of data production is the most important prerequisite for trustworthy inference.

5.1 DESIGNING SAMPLES

A political scientist wants to know what percent of the voting-age population consider themselves conservatives. An automaker hires a market research firm to learn what percent of adults aged 18 to 35 recall seeing television advertisements for a new sport utility vehicle. Government economists inquire about average household income. In all these cases, we want to gather information about a large group of individuals. We will not, as in an experiment, impose a treatment in order to observe the response. Time, cost, and inconvenience forbid contacting every individual. In such cases, we gather information about only part of the group in order to draw conclusions about the whole.

POPULATION AND SAMPLE

The entire group of individuals that we want information about is called the **population.**

A **sample** is a part of the population that we actually examine in order to gather information.

Notice that "population" is defined in terms of our desire for knowledge. If we wish to draw conclusions about all U.S. college students, that group is our population even if only local students are available for questioning. The sample is the part from which we draw conclusions about the whole. *Sampling* and conducting a *census* are two distinct ways of collecting data.

SAMPLING VERSUS A CENSUS

Sampling involves studying a part in order to gain information about the whole.

A **census** attempts to contact every individual in the entire population.

We want information on current unemployment and public opinion next week, not next year. Moreover, a carefully conducted sample is often more accurate than a census. Accountants, for example, sample a firm's inventory to verify the accuracy of the records. Attempting to count every last item in the warehouse would be not only expensive but inaccurate. Bored people do not count carefully.

If conclusions based on a sample are to be valid for the entire population, a sound design for selecting the sample is required. The ***design*** of a sample refers to the method used to choose the sample from the population. Poor sample designs can produce misleading conclusions, as the following examples illustrate.

sample design

EXAMPLE 5.2 CALL-IN OPINION POLLS

Television news programs like to conduct call-in polls of public opinion. The program announces a question and asks viewers to call one telephone number to respond "Yes" and another for "No." Telephone companies charge for these calls. The ABC network program *Nightline* once asked whether the United Nations should continue to have its headquarters in the United States. More than 186,000 callers responded, and 67% said "No."

People who spend the time and money to respond to call-in polls are not representative of the entire adult population. In fact, they tend to be the same people who call radio talk shows. People who feel strongly, especially those with strong negative opinions, are more likely to call. It is not surprising that a properly designed sample showed that 72% of adults want the UN to stay.[1]

Call-in opinion polls are an example of *voluntary response sampling*. A voluntary response sample can easily produce 67% "No" when the truth about the population is close to 72% "Yes."

VOLUNTARY RESPONSE SAMPLE

A **voluntary response sample** consists of people who choose themselves by responding to a general appeal. Voluntary response samples are biased because people with strong opinions, especially negative opinions, are most likely to respond.

convenience sampling

Voluntary response is one common type of bad sample design. Another is ***convenience sampling***, which chooses the individuals easiest to reach. Here is an example of convenience sampling.

EXAMPLE 5.3 INTERVIEWING AT THE MALL

Manufacturers and advertising agencies often use interviews at shopping malls to gather information about the habits of consumers and the effectiveness of ads. A sample of mall shoppers is fast and cheap. "Mall interviewing is being propelled primarily as a budget issue," one expert told the *New York Times*. But people contacted at shopping malls are not representative of the entire U.S. population. They are richer, for example, and more likely to be teenagers or retired. Moreover, mall interviewers tend to select neat, safe-looking individuals from the stream of customers. Decisions based on mall interviews may not reflect the preferences of all consumers.[2]

Both voluntary response samples and convenience samples choose a sample that is almost guaranteed not to represent the entire population. These sampling methods display *bias*, or systematic error, in favoring some parts of the population over others.

BIAS

The design of a study is **biased** if it systematically favors certain outcomes.

EXERCISES

5.1 FUNDING FOR DAY CARE A sociologist wants to know the opinions of employed adult women about government funding for day care. She obtains a list of the 520 members of a local business and professional women's club and mails a questionnaire to 100 of these women selected at random. Only 48 questionnaires are returned. What is the population in this study? What is the sample?

5.2 WHAT IS THE POPULATION? For each of the following sampling situations, identify the population as exactly as possible. That is, say what kind of individuals the population consists of and say exactly which individuals fall in the population. If the information given is not complete, complete the description of the population in a reasonable way.

(a) Each week, the Gallup Poll questions a sample of about 1500 adult U.S. residents to determine national opinion on a wide variety of issues.

(b) The 2000 census tried to gather basic information from every household in the United States. But a "long form" requesting much additional information was sent to a sample of about 17% of households.

(c) A machinery manufacturer purchases voltage regulators from a supplier. There are reports that variation in the output voltage of the regulators is affecting the performance of the finished products. To assess the quality of the supplier's production, the manufacturer sends a sample of 5 regulators from the last shipment to a laboratory for study.

5.3 TEACHING READING An educator wants to compare the effectiveness of computer software that teaches reading with that of a standard reading curriculum. He tests the reading ability of each student in a class of fourth graders, then divides them into two groups. One group uses the computer regularly, while the other studies a standard curriculum. At the end of the year, he retests all the students and compares the increase in reading ability in the two groups. Is this an experiment? Why or why not? What are the explanatory and response variables?

5.4 THE EFFECTS OF PROPAGANDA In 1940, a psychologist conducted an experiment to study the effect of propaganda on attitude toward a foreign government. He administered a test of attitude toward the German government to a group of American students. After the students read German propaganda for several months, he tested them again to see if their attitudes had changed.

Unfortunately, Germany attacked and conquered France while the experiment was in progress. Explain clearly why confounding makes it impossible to determine the effect of reading the propaganda.

5.5 ALCOHOL AND HEART ATTACKS Many studies have found that people who drink alcohol in moderation have lower risk of heart attacks than either nondrinkers or heavy drinkers. Does alcohol consumption also improve survival after a heart attack? One study followed 1913 people who were hospitalized after severe heart attacks. In the year before their heart attack, 47% of these people did not drink, 36% drank moderately, and 17% drank heavily. After four years, fewer of the moderate drinkers had died.[3] Is this an observational study or an experiment? Why? What are the explanatory and response variables?

5.6 ARE ANESTHETICS SAFE? The National Halothane Study was a major investigation of the safety of anesthetics used in surgery. Records of over 850,000 operations performed in 34 major hospitals showed the following death rates for four common anesthetics:[4]

Anesthetic:	A	B	C	D
Death rate:	1.7%	1.7%	3.4%	1.9%

There is a clear association between the anesthetic used and the death rate of patients. Anesthetic C appears to be dangerous.

(a) Explain why we call the National Halothane Study an observational study rather than an experiment, even though it compared the results of using different anesthetics in actual surgery.

(b) When the study looked at other variables that are confounded with a doctor's choice of anesthetic, it found that Anesthetic C was not causing extra deaths. Suggest several variables that are mixed up with what anesthetic a patient receives.

5.7 CALL THE SHOTS A newspaper advertisement for *USA Today: The Television Show* once said:

Should handgun control be tougher? You call the shots in a special call-in poll tonight. If yes, call 1-900-720-6181. If no, call 1-900-720-6182. Charge is 50 cents for the first minute.

Explain why this opinion poll is almost certainly biased.

5.8 EXPLAIN IT TO THE CONGRESSWOMAN You are on the staff of a member of Congress who is considering a bill that would provide government-sponsored insurance for nursing home care. You report that 1128 letters have been received on the issue, of which 871 oppose the legislation. "I'm surprised that most of my constituents oppose the bill. I thought it would be quite popular," says the congresswoman. Are you convinced that a majority of the voters oppose the bill? How would you explain the statistical issue to the congresswoman?

Simple random samples

In a voluntary response sample, people choose whether to respond. In a convenience sample, the interviewer makes the choice. In both cases, personal choice produces bias. The statistician's remedy is to allow impersonal chance to choose the sample. A sample chosen by chance allows neither favoritism by the sampler nor self-selection by respondents. Choosing a sample by chance attacks bias by giving all individuals an equal chance to be chosen. Rich and poor, young and old, black and white, all have the same chance to be in the sample.

The simplest way to use chance to select a sample is to place names in a hat (the population) and draw out a handful (the sample). This is the idea of *simple random sampling*.

SIMPLE RANDOM SAMPLE

A **simple random sample (SRS)** of size n consists of n individuals from the population chosen in such a way that every set of n individuals has an equal chance to be the sample actually selected.

An SRS not only gives each individual an equal chance to be chosen (thus avoiding bias in the choice) but also gives every possible sample an equal chance to be chosen. There are other random sampling designs that give each individual, but not each sample, an equal chance. Exercise 5.30 describes one such design, called systematic random sampling.

The idea of an SRS is to choose our sample by drawing names from a hat. In practice, computer software can choose an SRS almost instantly from a list of the individuals in the population. If you don't use software, you can randomize by using a *table of random digits*.

RANDOM DIGITS

A **table of random digits** is a long string of the digits 0, 1, 2, 3, 4, 5, 6, 7, 8, 9 with these two properties:

1. Each entry in the table is equally likely to be any of the 10 digits 0 through 9.

2. The entries are independent of each other. That is, knowledge of one part of the table gives no information about any other part.

Table B at the back of the book is a table of random digits. You can think of Table B as the result of asking an assistant (or a computer) to mix the digits 0 to 9 in a hat, draw one, then replace the digit drawn, mix again, draw a second digit, and so on. The assistant's mixing and drawing save us the work of mixing and drawing when we need to randomize. Table B begins with the digits 19223950340575628713. To make the table easier to read, the digits appear in groups of five and in numbered rows. The groups and rows have no meaning—the table is just a long list of randomly chosen digits. Because the digits in Table B are random:

- Each entry is equally likely to be any of the 10 possibilities 0, 1, . . . , 9.
- Each pair of entries is equally likely to be any of the 100 possible pairs 00, 01, . . . , 99.
- Each triple of entries is equally likely to be any of the 1000 possibilities 000, 001, . . . , 999, and so on.

These "equally likely" facts make it easy to use Table B to choose an SRS. Here is an example that shows how.

EXAMPLE 5.4 HOW TO CHOOSE AN SRS

Joan's small accounting firm serves 30 business clients. Joan wants to interview a sample of 5 clients in detail to find ways to improve client satisfaction. To avoid bias, she chooses an SRS of size 5.

Step 1: **Label.** Give each client a numerical label, using as few digits as possible. Two digits are needed to label 30 clients, so we use labels

01, 02, 03, . . . , 29, 30

It is also correct to use labels 00 to 29 or even another choice of 30 two-digit labels. Here is the list of clients, with labels attached:

01	A-1 Plumbing	16	JL Records
02	Accent Printing	17	Johnson Commodities
03	Action Sport Shop	18	Keiser Construction
04	Anderson Construction	19	Liu's Chinese Restaurant
05	Bailey Trucking	20	MagicTan
06	Balloons Inc.	21	Peerless Machine
07	Bennett Hardware	22	Photo Arts
08	Best's Camera Shop	23	River City Books
09	Blue Print Specialties	24	Riverside Tavern
10	Central Tree Service	25	Rustic Boutique
11	Classic Flowers	26	Satellite Services
12	Computer Answers	27	Scotch Wash
13	Darlene's Dolls	28	Sewer's Center
14	Fleisch Realty	29	Tire Specialties
15	Hernandez Electronics	30	Von's Video Store

Step 2: **Table.** Enter Table B anywhere and read two-digit groups. Suppose we enter at line 130, which is

69051 64817 87174 09517 84534 06489 87201 97245

The first 10 two-digit groups in this line are

69 05 16 48 17 87 17 40 95 17

Each successive two-digit group is a label. The labels 00 and 31 to 99 are not used in this example, so we ignore them. The first 5 labels between 01 and 30 that we encounter in the table choose our sample. Of the first 10 labels in line 130, we ignore 5 because they are too high (over 30). The others are 05, 16, 17, 17, and 17. The clients labeled 05, 16, and 17 go into the sample. Ignore the second and third 17s because that client is already in the sample. Now run your finger across line 130 (and continue to line 131 if needed) until 5 clients are chosen.

The sample is the clients labeled 05, 16, 17, 20, 19. These are Bailey Trucking, JL Records, Johnson Commodities, MagicTan, and Liu's Chinese Restaurant.

CHOOSING AN SRS

Choose an SRS in two steps:

Step 1: **Label.** Assign a numerical label to every individual in the population.

Step 2: **Table.** Use Table B to select labels at random.

You can assign labels in any convenient manner, such as alphabetical order for names of people. Be certain that all labels have the same number of digits. Only then will all individuals have the same chance to be chosen. Use the shortest possible labels: one digit for a population of up to 10 members, 2 digits for 11 to 100 members, three digits for 101 to 1000 members, and so on. As standard practice, we recommend that you begin with label 1 (or 01 or 001, as needed). You can read digits from Table B in any order—across a row, down a column, and so on—because the table has no order. As standard practice, we recommend reading across rows.

Other sampling designs

The general framework for designs that use chance to choose a sample is a *probability sample.*

PROBABILITY SAMPLE

A **probability sample** is a sample chosen by chance. We must know what samples are possible and what chance, or probability, each possible sample has.

Some probability sampling designs (such as an SRS) give each member of the population an *equal* chance to be selected. This may not be true in more elaborate sampling designs. In every case, however, **the use of chance to select the sample is the essential principle of statistical sampling.**

Designs for sampling from large populations spread out over a wide area are usually more complex than an SRS. For example, it is common to sample important groups within the population separately, then combine these samples. This is the idea of a *stratified sample.*

STRATIFIED RANDOM SAMPLE

To select a **stratified random sample,** first divide the population into groups of similar individuals, called **strata.** Then choose a separate SRS in each stratum and combine these SRSs to form the full sample.

Choose the strata based on facts known before the sample is taken. For example, a population of election districts might be divided into urban, suburban, and rural strata. A stratified design can produce more exact information than an SRS of the same size by taking advantage of the fact that individuals in the same stratum are similar to one another. If all individuals in each stratum are identical, for example, just one individual from each stratum is enough to completely describe the population.

EXAMPLE 5.5 WHO WROTE THAT SONG?

A radio station that broadcasts a piece of music owes a royalty to the composer. The organization of composers (called ASCAP) collects these royalties for all its members by charging stations a license fee for the right to play members' songs. ASCAP has four million songs in its catalog and collects $435 million in fees each year. How should ASCAP distribute this income among its members? By sampling: ASCAP tapes about 60,000 hours from the 53 million hours of local radio programs across the country each year.

Radio stations are stratified by type of community (metropolitan, rural), geographic location (New England, Pacific, etc.), and the size of the license fee paid to ASCAP, which reflects the size of the audience. In all, there are 432 strata. Tapes are made at random hours for randomly selected members of each stratum. The tapes are reviewed by experts who can recognize almost every piece of music ever written, and the composers are then paid according to their popularity.[5]

Another common means of restricting random selection is to choose the sample in stages. This is usual practice for national samples of households or people. For example, data on employment and unemployment are gathered by the government's Current Population Survey, which conducts interviews in about 55,000 households each month. It is not practical to maintain a list of all U.S. households from which to select an SRS. Moreover, the cost of sending interviewers to the widely scattered households in an SRS would be too high. The Current Population Survey therefore uses a ***multistage sampling design.*** The final sample consists of clusters of near-

multistage sample

by households that an interviewer can easily visit. Most opinion polls and other national samples are also multistage, though interviewing in most national samples today is done by telephone rather than in person, eliminating the economic need for clustering. The Current Population Survey sampling design is roughly as follows:[6]

Stage 1: Divide the United States into 2007 geographical areas called Primary Sampling Units, or PSUs. Select a sample of 756 PSUs. This sample includes the 428 PSUs with the largest population and a stratified sample of 328 of the others.

Stage 2: Divide each PSU selected into smaller areas called "neighborhoods." Stratify the neighborhoods using ethnic and other information and take a stratified sample of the neighborhoods in each PSU.

Stage 3: Sort the housing units in each neighborhood into clusters of four nearby units. Interview the households in a random sample of these clusters.

Analysis of data from sampling designs more complex than an SRS takes us beyond basic statistics. But the SRS is the building block of more elaborate designs, and analysis of other designs differs more in complexity of detail than in fundamental concepts.

EXERCISES

5.9 CHOOSE YOUR SAMPLE You must choose an SRS of 10 of the 440 retail outlets in New York that sell your company's products. How would you label this population? Use Table B, starting at line 105, to choose your sample.

5.10 WHO SHOULD BE INTERVIEWED? A firm wants to understand the attitudes of its minority managers toward its system for assessing management performance. Below is a list of all the firm's managers who are members of minority groups. Use Table B at line 139 to choose 6 to be interviewed in detail about the performance appraisal system.

Agarwal	Gates	Peters
Anderson	Goel	Pliego
Baxter	Gomez	Puri
Bonds	Hernandez	Richards
Bowman	Huang	Rodriguez
Castillo	Kim	Santiago
Cross	Liao	Shen
Dewald	Mourning	Vega
Fernandez	Naber	Wang
Fleming		

5.11 WHO GOES TO THE CONVENTION? A club has 30 student members and 10 faculty members. The students are

Abel	Fisher	Huber	Miranda	Reinmann
Carson	Ghosh	Jimenez	Moskowitz	Santos
Chen	Griswold	Jones	Neyman	Shaw
David	Hein	Kim	O'Brien	Thompson
Deming	Hernandez	Klotz	Pearl	Utts
Elashoff	Holland	Liu	Potter	Varga

The faculty members are

Andrews	Fernandez	Kim	Moore	West
Besicovitch	Gupta	Lightman	Phillips	Yang

The club can send 4 students and 2 faculty members to a convention. It decides to choose those who will go by random selection. Use Table B, beginning at line 106, to choose a stratified random sample of 4 students and 2 faculty members.

5.12 SAMPLING BY ACCOUNTANTS Accountants often use stratified samples during audits to verify a company's records of such things as accounts receivable. The stratification is based on the dollar amount of the item and often includes 100% sampling of the largest items. One company reports 5000 accounts receivable. Of these, 100 are in amounts over \$50,000; 500 are in amounts between \$1000 and \$50,000; and the remaining 4400 are in amounts under \$1000. Using these groups as strata, you decide to verify all of the largest accounts and to sample 5% of the midsize accounts and 1% of the small accounts. How would you label the two strata from which you will sample? Use Table B, starting at line 115, to select *only the first* 5 accounts from each of these strata.

Cautions about sample surveys

Random selection eliminates bias in the choice of a sample from a list of the population. When the population consists of human beings, however, accurate information from a sample requires much more than a good sampling design.[7] To begin, we need an accurate and complete list of the population. Because such a list is rarely available, most samples suffer from some degree of *undercoverage.* A sample survey of households, for example, will miss not only homeless people but prison inmates and students in dormitories. An opinion poll conducted by telephone will miss the 7% to 8% of American households without residential phones. The results of national sample surveys therefore have some bias if the people not covered—who most often are poor people—differ from the rest of the population.

A more serious source of bias in most sample surveys is *nonresponse,* which occurs when a selected individual cannot be contacted or refuses to cooperate. Nonresponse to sample surveys often reaches 30% or more, even with careful planning and several callbacks. Because nonresponse is higher in urban areas, most sample surveys substitute other people in the same area to avoid favoring rural areas in the final sample. If the people contacted differ from those who are rarely at home or who refuse to answer questions, some bias remains.

UNDERCOVERAGE AND NONRESPONSE

Undercoverage occurs when some groups in the population are left out of the process of choosing the sample.

Nonresponse occurs when an individual chosen for the sample can't be contacted or does not cooperate.

EXAMPLE 5.6 THE CENSUS UNDERCOUNT

Even the U.S. census, backed by the resources of the federal government, suffers from undercoverage and nonresponse. The census begins by mailing forms to every household in the country. The Census Bureau's list of addresses is incomplete, resulting in undercoverage. Despite special efforts to count homeless people (who can't be reached at any address), homelessness causes more undercoverage.

In 1990, about 35% of households that were mailed census forms did not mail them back. In New York City, 47% did not return the form. That's nonresponse. The Census Bureau sent interviewers to these households. In inner-city areas, the interviewers could not contact about one in five of the nonresponders, even after six tries.

The Census Bureau estimates that the 1990 census missed about 1.8% of the total population due to undercoverage and nonresponse. Because the undercount was greater in the poorer sections of large cities, the Census Bureau estimates that it failed to count 4.4% of blacks and 5.0% of Hispanics.[8]

For the 2000 census, the Bureau planned to replace follow-up of all nonresponders with more intense pursuit of a probability sample of nonresponding households plus a national sample of 750,000 households. The final counts would be based on comparing the national sample with the original responses. This idea was politically controversial. The Supreme Court ruled that the sampling could be used for most purposes, but not for dividing seats in Congress among the states.

response bias

In addition, the behavior of the respondent or of the interviewer can cause ***response bias*** in sample results. Respondents may lie, especially if asked about illegal or unpopular behavior. The sample then underestimates the presence of such behavior in the population. An interviewer whose attitude suggests that some answers are more desirable than others will get these answers more often. The race or sex of the interviewer can influence responses to questions about race relations or attitudes toward feminism. Answers to questions that ask respondents to recall past events are often inaccurate because of faulty memory. For example, many people "telescope" events in the past, bringing them forward in memory to more recent time periods. "Have you visited a dentist in the last 6 months?" will often draw a "Yes" from someone who last visited a dentist 8 months ago.[9] Careful training of interviewers and careful supervision to avoid variation among the interviewers can greatly reduce response bias. Good interviewing technique is another aspect of a well-done sample survey.

wording effects

The ***wording of questions*** is the most important influence on the answers given to a sample survey. Confusing or leading questions can introduce strong bias, and even minor changes in wording can change a survey's outcome. Here are two examples.

EXAMPLE 5.7 SHOULD WE BAN DISPOSABLE DIAPERS?

A survey paid for by makers of disposable diapers found that 84% of the sample opposed banning disposable diapers. Here is the actual question:

> *It is estimated that disposable diapers account for less than 2% of the trash in today's landfills. In contrast, beverage containers, third-class mail and yard wastes are estimated to account for about 21% of the trash in landfills. Given this, in your opinion, would it be fair to ban disposable diapers?*[10]

This question gives information on only one side of an issue, then asks an opinion. That's a sure way to bias the responses. A different question that described how long disposable diapers take to decay and how many tons they contribute to landfills each year would draw a quite different response.

EXAMPLE 5.8 DOUBTING THE HOLOCAUST

An opinion poll conducted in 1992 for the American Jewish Committee asked: "Does it seem possible or does it seem impossible to you that the Nazi extermination of the Jews never happened?" When 22% of the sample said "possible," the news media wondered how so many Americans could be uncertain that the Holocaust happened. Then a second poll asked the question in different words: "Does it seem possible to you that the Nazi extermination of the Jews never happened, or do you feel certain that it happened?" Now only 1% of the sample said "possible." The complicated wording of the first question confused many respondents.[11]

Never trust the results of a sample survey until you have read the exact questions posed. The sampling design, the amount of nonresponse, and the date of the survey are also important. Good statistical design is a part, but only a part, of a trustworthy survey.

Inference about the population

Despite the many practical difficulties in carrying out a sample survey, using chance to choose a sample does eliminate bias in the actual selection of the sample from the list of available individuals. But it is unlikely that results from a sample are exactly the same as for the entire population. Sample results, like the official unemployment rate obtained from the monthly Current Population Survey, are only estimates of the truth about the population. If we select two samples at random from the same population, we will draw different individuals. So the sample results will almost certainly differ somewhat. Two runs of the Current Population Survey would produce somewhat different unemployment rates. Properly designed samples avoid systematic bias, but their results are rarely exactly correct and they vary from sample to sample.

How accurate is a sample result like the monthly unemployment rate? We can't say for sure, because the result would be different if we took another sample. But the results of random sampling don't change haphazardly from sample to sample. Because we deliberately use chance, the results obey the laws of ***probability*** that govern chance behavior. We can say how large an error we are likely to make in drawing conclusions about the population from a sample. Results from a sample survey usually come with a margin of error that sets bounds on the size of the likely error. How to do this is part of the business of statistical inference. We will describe the reasoning in Chapter 10.

probability

One point is worth making now: **larger random samples give more accurate results than smaller samples.** By taking a very large sample, you can be confident that the sample result is very close to the truth about the population. The Current Population Survey's sample of 50,000 households estimates the national unemployment rate very accurately. Of course, only probability samples carry this guarantee. *Nightline's* voluntary response sample is worthless even though 186,000 people called in. Using a probability sampling design and taking care to deal with practical difficulties reduce bias in a sample. The size of the sample then determines how close to the population truth the sample result is likely to fall.

EXERCISES

5.13 SAMPLING FRAME The list of individuals from which a sample is actually selected is called the ***sampling frame.*** Ideally, the frame should list every individual in the population, but in practice this is often difficult. A frame that leaves out part of the population is a common source of undercoverage.

sampling frame

(a) Suppose that a sample of households in a community is selected at random from the telephone directory. What households are omitted from this frame? What types of people do you think are likely to live in these households? These people will probably be underrepresented in the sample.

(b) It is more common in telephone surveys to use random digit dialing equipment that selects the last four digits of a telephone number at random after being given the exchange (the first three digits). Which of the households you mentioned in your answer to (a) will be included in the sampling frame by random digit dialing?

5.14 RING-NO-ANSWER A common form of nonresponse in telephone surveys is "ring-no-answer." That is, a call is made to an active number but no one answers. The Italian National Statistical Institute looked at nonresponse to a government survey of households in Italy during the periods January 1 to Easter and July 1 to August 31. All calls were made between 7 and 10 p.m., but 21.4% gave "ring-no-answer" in one period versus 41.5% "ring-no-answer" in the other period.[12] Which period do you think had the higher rate of no answers? Why? Explain why a high rate of nonresponse makes sample results less reliable.

5.15 QUESTION WORDING During the 2000 presidential campaign, the candidates debated what to do with the large government surplus. The Pew Research Center asked two questions of random samples of adults. Both questions stated that social security would be "fixed." Here are the uses suggested for the remaining surplus:

> *Should the money be used for a tax cut, or should it be used to fund new government programs?*
>
> *Should the money be used for a tax cut, or should it be spent on programs for education, the environment, health care, crime-fighting and military defense?*

One of these questions drew 60% favoring a tax cut; the other, only 22%. Which wording pulls respondents toward a tax cut? Why?

5.16 GRADING THE PRESIDENT A newspaper article about an opinion poll says that "43% of Americans approve of the president's overall job performance." Toward the end of the article, you read: "The poll is based on telephone interviews with 1210 adults from around the United States, excluding Alaska and Hawaii." What variable did this poll measure? What population do you think the newspaper wants information about? What was the sample? Are there any sources of bias in the sampling method used?

5.17 EQUAL PAY FOR MALE AND FEMALE ATHLETES? The Excite Poll can be found online at http://lite.excite.com. The question appears on the screen, and you simply click buttons to vote "Yes," "No," or "Not sure." On January 25, 2000, the question was "Should female athletes be paid the same as men for the work they do?" In all, 13,147 (44%) said "Yes," another 15,182 (50%) said "No," and the remaining 1448 said "Not sure."

(a) What is the sample size for this poll?

(b) That's a much larger sample than standard sample surveys. In spite of this, we can't trust the result to give good information about any clearly defined population. Why?

(c) More men than women use the Web. How might this fact affect the poll results?

5.18 WORDING BIAS Comment on each of the following as a potential sample survey question. Is the question clear? Is it slanted toward a desired response?

(a) "Some cell phone users have developed brain cancer. Should all cell phones come with a warning label explaining the danger of using cell phones?"

(b) "Do you agree that a national system of health insurance should be favored because it would provide health insurance for everyone and would reduce administrative costs?"

(c) "In view of escalating environmental degradation and incipient resource depletion, would you favor economic incentives for recycling of resource-intensive consumer goods?"

SUMMARY

Data analysis is sometimes **exploratory** in nature. Exploratory analysis asks what the data tell us about the variables and their relations to each other. The conclusions of an exploratory analysis may not generalize beyond the specific data studied.

Statistical inference produces answers to specific questions, along with a statement of how confident we can be that the answer is correct. The conclusions of statistical inference are usually intended to apply beyond the individuals actually studied. Successful statistical inference usually requires **production of data** intended to answer the specific questions posed.

We can produce data intended to answer specific questions by sampling or experimentation. **Sampling** selects a part of a population of interest to represent the whole. **Experiments** are distinguished from **observational studies** such as sample surveys by the active imposition of some treatment on the subjects of the experiment.

A sample survey selects a **sample** from the **population** of all individuals about which we desire information. We base conclusions about the population on data about the sample.

The **design** of a sample refers to the method used to select the sample from the population. **Probability sampling designs** use impersonal chance to select a sample.

The basic probability sample is a **simple random sample (SRS)**. An SRS gives every possible sample of a given size the same chance to be chosen.

Choose an SRS by labeling the members of the population and using a **table of random digits** to select the sample. Software can automate this process.

To choose a **stratified random sample**, divide the population into **strata**, groups of individuals that are similar in some way that is important to the response. Then choose a separate SRS from each stratum and combine them to form the full sample.

Multistage samples select successively smaller groups within the population in stages, resulting in a sample consisting of clusters of individuals. Each stage may employ an SRS, a stratified sample, or another type of sample.

Failure to use probability sampling often results in **bias**, or systematic errors in the way the sample represents the population. **Voluntary response** samples, in which the respondents choose themselves, are particularly prone to large bias.

In human populations, even probability samples can suffer from bias due to **undercoverage** or **nonresponse**, from **response bias** due to the behavior of the interviewer or the respondent, or from misleading results due to **poorly worded questions**.

Larger samples give more accurate results than smaller samples.

SECTION 5.1 EXERCISES

5.19 DESCRIBE THE POPULATION For each of the following sampling situations, identify the population as exactly as possible. That is, say what kind of individuals the population consists of and say exactly which individuals fall in the population. If the information given is not complete, complete the description of the population in a reasonable way.

(a) An opinion poll contacts 1161 adults and then asks them "Which political party do you think has better ideas for leading the country in the twenty-first century?"

(b) A sociologist wants to know the opinions of employed adult women about government funding for day care. She obtains a list of the 520 members of a local business and professional women's club and mails a questionnaire to 100 of these women selected at random.

(c) The American Community Survey will contact 3 million households, including some in every county in the United States. This new Census Bureau survey will ask each household questions about their housing, economic, and social status.

5.20 THE REAGAN-CARTER ELECTION DEBATE Some television stations take quick polls of public opinion by announcing a question on the air and asking viewers to call one of two telephone numbers to register their opinion as "Yes" or "No." Telephone companies make available "900" numbers for this purpose. Dialing a 900 number results in a small charge to your telephone bill. The first major use of call-in polling was by the ABC television network in October 1980. At the end of the first Reagan-Carter presidential election debate, ABC asked its viewers which candidate won. The call-in poll proclaimed that Reagan had won the debate by a 2 to 1 margin. But a random survey by CBS News showed only a 44% to 36% margin for Reagan, with the rest undecided. Why are call-in polls likely to be biased? Can you suggest why this bias might have favored the Republican Reagan over the Democrat Carter?

5.21 TESTING CHEMICALS A manufacturer of chemicals chooses 3 from each lot of 25 containers of a reagent to test for purity and potency. Below are the control numbers stamped on the bottles in the current lot. Use Table B at line 111 to choose an SRS of 3 of these bottles.

A1096	A1097	A1098	A1101	A1108
A1112	A1113	A1117	A2109	A2211
A2220	B0986	B1011	B1096	B1101
B1102	B1103	B1110	B1119	B1137
B1189	B1223	B1277	B1286	B1299

5.22 INCREASING SAMPLE SIZE Just before a presidential election, a national opinion polling firm increases the size of its weekly sample from the usual 1500 people to 4000 people. Why do you think the firm does this?

5.23 CENSUS TRACT Figure 5.1 is a map of a census tract in a fictitious town. Census tracts are small, homogeneous areas averaging 4000 in population. On the map, each block is marked with a Census Bureau identification number. An SRS of blocks from

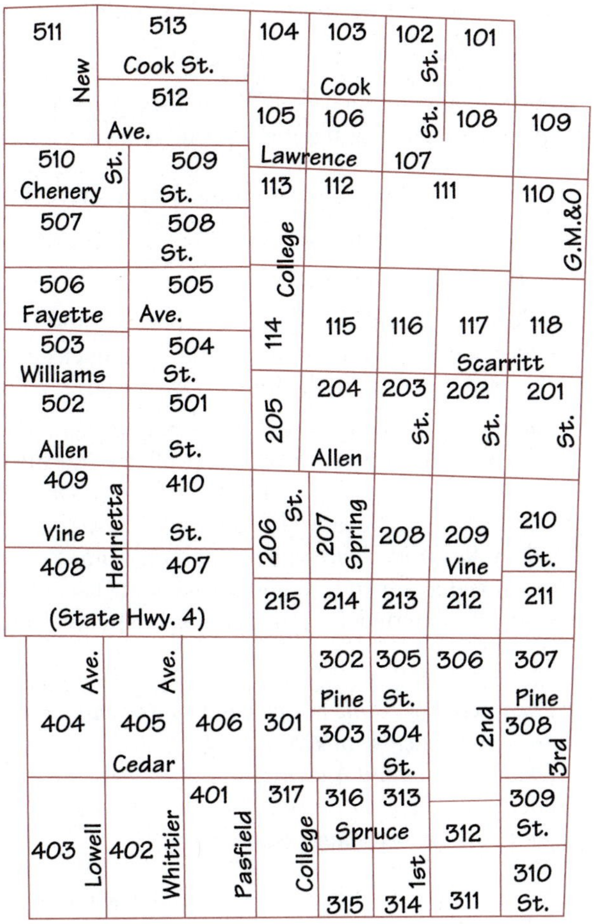

FIGURE 5.1 Map of a census tract.

a census tract is often the next-to-last stage in a multistage sample. Use Table B, beginning at line 125, to choose an SRS of 5 blocks from this census tract.

5.24 RANDOM DIGITS Which of the following statements are true of a table of random digits, and which are false? Briefly explain your answers.

(a) There are exactly four 0s in each row of 40 digits.

(b) Each pair of digits has chance 1/100 of being 00.

(c) The digits 0000 can never appear as a group, because this pattern is not random.

5.25 IS IT AN SRS? A corporation employs 2000 male and 500 female engineers. A stratified random sample of 200 male and 50 female engineers gives each engineer 1 chance in 10 to be chosen. This sample design gives every individual in the population the same chance to be chosen for the sample. Is it an SRS? Explain your answer.

5.26 CHECKING FOR BIAS Comment on each of the following as a potential sample survey question. Is the question clear? Is it slanted toward a desired response?

(a) Which of the following best represents your opinion on gun control?

1. The government should confiscate our guns.
2. We have the right to keep and bear arms.

(b) A freeze in nuclear weapons should be favored because it would begin a much-needed process to stop everyone in the world from building nuclear weapons now and reduce the possibility of nuclear war in the future. Do you agree or disagree?

(c) In view of escalating environmental degradation and incipient resource depletion, would you favor economic incentives for recycling of resource-intensive consumer goods?

5.27 SAMPLING ERROR A *New York Times* opinion poll on women's issues contacted a sample of 1025 women and 472 men by randomly selecting telephone numbers. The *Times* publishes complete descriptions of its polling methods. Here is part of the description for this poll:[13]

> *In theory, in 19 cases out of 20 the results based on the entire sample will differ by no more than three percentage points in either direction from what would have been obtained by seeking out all adult Americans.*
>
> *The potential sampling error for smaller subgroups is larger. For example, for men it is plus or minus five percentage points.*

Explain why the margin of error is larger for conclusions about men alone than for conclusions about all adults.

5.28 ATTITUDES TOWARD ALCOHOL At a party there are 30 students over age 21 and 20 students under age 21. You choose at random 3 of those over 21 and separately choose at random 2 of those under 21 to interview about attitudes toward alcohol. You have given every student at the party the same chance to be interviewed: what is the chance? Why is your sample not an SRS?

5.29 WHAT DO SCHOOLKIDS WANT? What are the most important goals of schoolchildren? Do girls and boys have different goals? Are goals different in urban, suburban, and rural areas? To find out, researchers wanted to ask children in the fourth, fifth, and sixth grades this question:

What would you most like to do at school?

A. *Make good grades.*

B. *Be good at sports.*

C. *Be popular.*

Because most children live in heavily populated urban and suburban areas, an SRS might contain few rural children. Moreover, it is too expensive to choose children at random from a large region—we must start by choosing schools rather than children. Describe a suitable sample design for this study and explain the reasoning behind your choice of design.

5.30 SYSTEMATIC RANDOM SAMPLE Sample surveys often use a ***systematic random sample*** to choose a sample of apartments in a large building or dwelling units in a block at the last stage of a multistage sample. An example will illustrate the idea of a systematic sample.

systematic random sample

Suppose that we must choose 4 addresses out of 100. Because 100/4 = 25, we can think of the list as four lists of 25 addresses. Choose 1 of the first 25 addresses at random using Table B. The sample contains this address and the addresses 25, 50, and 75 places down the list from it. If the table gives 13, for example, then the systematic random sample consists of the addresses numbered 13, 38, 63, and 88.

(a) Use Table B to choose a systematic random sample of 5 addresses from a list of 200. Enter the table at line 120.

(b) Like an SRS, a systematic random sample gives all individuals the same chance to be chosen. Explain why this is true. Then explain carefully why a systematic sample is nonetheless *not* an SRS.

Activity 5B The Class Survey Revisited

Each student should have a copy of the survey that the class constructed in Activity 5A at the beginning of the chapter. Now that you are experts on good and bad characteristics of survey questions, do the following:

1. Consider the questions in order. As you look at each item, see if the question contains bias. Does it advocate a position? Does the question contain any complicated words or phrasing that might be misinterpreted? Will any questions evoke response bias?
2. Make any changes that the group feels are needed. Remember that the survey should be *anonymous* (no names on the papers) so that students are assured that the class *as a whole* rather than themselves as individuals will be described.
3. Print the final version of the survey. Make one copy for each member of the class and an extra copy on which to tally the results.
4. Each student should complete the survey.
5. Place the completed surveys, upside down, in a pile. The last student finished should shuffle the pile of surveys to ensure anonymity.
6. Designate someone (the teacher?) to tally the responses as homework and prepare a cumulative summary. Give a copy of the results to each student in the class for later analysis.

5.2 DESIGNING EXPERIMENTS

A study is an experiment when we actually do something to people, animals, or objects in order to observe the response. Here is the basic vocabulary of experiments.

EXPERIMENTAL UNITS, SUBJECTS, TREATMENT

The individuals on which the experiment is done are the **experimental units**. When the units are human beings, they are called **subjects**. A specific experimental condition applied to the units is called a **treatment**.

Because the purpose of an experiment is to reveal the response of one variable to changes in other variables, the distinction between explanatory and response variables is important. The explanatory variables in an experiment are often called ***factors***. Many experiments study the joint effects of several factors. In such an experiment, each treatment is formed by combining a specific value (often called a ***level***) of each of the factors.

factor

level

EXAMPLE 5.9 THE PHYSICIANS' HEALTH STUDY

Does regularly taking aspirin help protect people against heart attacks? The Physicians' Health Study was a medical experiment that helped answer this question. In fact, the Physicians' Health Study looked at the effects of two drugs: aspirin and beta carotene. The body converts beta carotene into vitamin A, which may help prevent some forms of cancer. The *subjects* were 21,996 male physicians. There were two *factors*, each having two levels: aspirin (yes or no) and beta carotene (yes or no). Combinations of the levels of these factors form the four *treatments* shown in Figure 5.2. One-fourth of the subjects were assigned to each of these treatments.

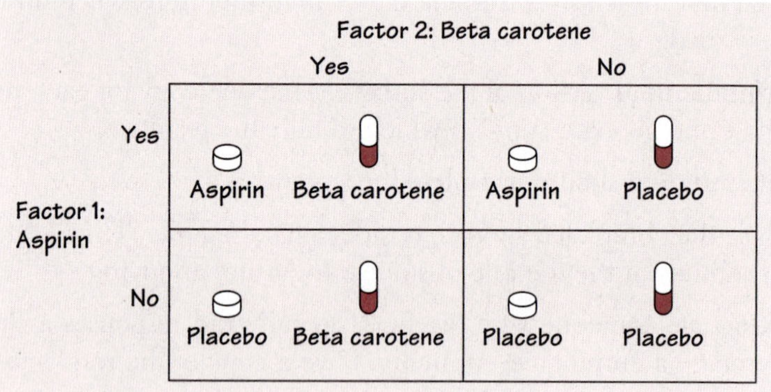

FIGURE 5.2 The treatments in the Physicians' Health Study.

On odd-numbered days, the subjects took a white tablet that contained either aspirin or a ***placebo***, a dummy pill that looked and tasted like the aspirin but had no active ingredient. On even-numbered days, they took a red capsule containing either beta carotene or a placebo. There were several *response variables*—the study looked for heart attacks, several kinds of cancer, and other medical outcomes. After several years, 239 of the placebo group but only 139 of the aspirin group had suffered heart attacks. This difference is large enough to give good evidence that taking aspirin does reduce heart attacks.[14] It did not appear, however, that beta carotene had any effect.

placebo

EXAMPLE 5.10 DOES STUDYING A FOREIGN LANGUAGE IN HIGH SCHOOL INCREASE VERBAL ABILITY IN ENGLISH?

Julie obtains lists of all seniors in her high school who did and did not study a foreign language. Then she compares their scores on a standard test of English reading and grammar given to all seniors. The average score of the students who studied a foreign language is much higher than the average score of those who did not.

This observational study gives no evidence that studying another language builds skill in English. Students decide for themselves whether or not to elect a foreign language. Those who choose to study a language are mostly students who are already better at English than most students who avoid foreign languages. The difference in average test scores just shows that students who choose to study a language differ (on the average) from those who do not. We can't say whether studying languages *causes* this difference.

Examples 5.9 and 5.10 illustrate the big advantage of experiments over observational studies. **In principle, experiments can give good evidence for causation.** All the doctors in the Physicians' Health Study took a pill every other day, and all got the same schedule of checkups and information. The only difference was the content of the pill. When one group had many fewer heart attacks, we conclude that it was the content of the pill that made the difference. Julie's observational study—a *census* of all seniors in her high school—does a good job of describing differences between seniors who have studied foreign languages and those who have not. But she can say nothing about cause and effect.

Another advantage of experiments is that they allow us to study the specific factors we are interested in, while controlling the effects of lurking variables. The subjects in the Physicians' Health Study were all middle-aged male doctors and all followed the same schedule of medical checkups. These similarities reduce variation among the subjects and make any effects of aspirin or beta carotene easier to see. Experiments also allow us to study the combined effects of several factors. The interaction of several factors can produce effects that could not be predicted from looking at the effects of each factor alone. The Physicians' Health Study tells us that aspirin helps prevent heart attacks, at least in middle-aged men, and that beta carotene taken with the aspirin neither helps nor hinders aspirin's protective powers.

Comparative experiments

Laboratory experiments in science and engineering often have a simple design with only a single treatment, which is applied to all of the experimental units. The design of such an experiment can be outlined as

Units → Treatment → Observe response

For example, we may subject a beam to a load (treatment) and measure its deflection (observation). We rely on the controlled environment of the laboratory to protect us from lurking variables. When experiments are conducted in the field or with living subjects, such simple designs often yield invalid data. That is, we cannot tell whether the response was due to the treatment or to lurking variables. Another medical example will show what can go wrong.

EXAMPLE 5.11 TREATING ULCERS

"Gastric freezing" is a clever treatment for ulcers in the upper intestine. The patient swallows a deflated balloon with tubes attached, then a refrigerated liquid is pumped through the balloon for an hour. The idea is that cooling the stomach will reduce its production of acid and so relieve ulcers. An experiment reported in the *Journal of the American Medical Association* showed that gastric freezing did reduce acid production and relieve ulcer pain. The treatment was safe and easy and was widely used for several years. The design of the experiment was

Subjects → Gastric freezing → Observe pain relief

placebo effect

The gastric freezing experiment was poorly designed. The patients' response may have been due to the ***placebo effect***. A placebo is a dummy treatment. Many patients respond favorably to any treatment, even a placebo. This may be due to trust in the doctor and expectations of a cure, or simply to the fact that medical conditions often improve without treatment. The response to a dummy treatment is the placebo effect.

A later experiment divided ulcer patients into two groups. One group was treated by gastric freezing as before. The other group received a placebo treatment in which the liquid in the balloon was at body temperature rather than freezing. The results: 34% of the 82 patients in the treatment group improved, but so did 38% of the 78 patients in the placebo group. This and other properly designed experiments showed that gastric freezing was no better than a placebo, and its use was abandoned.[15]

The first gastric freezing experiment gave misleading results because the effects of the explanatory variable were *confounded* with (mixed up with) the placebo effect. We can defeat confounding by *comparing* two groups of patients, as in the second gastric freezing experiment. The placebo effect and other lurking variables now operate on both groups. The only difference between the groups is the actual effect of gastric freezing. The group of patients who received a sham treatment is called a ***control group***, because it enables us to control the effects of outside variables on the outcome. **Control is the first basic principle of statistical design of experiments.** Comparison of several treatments in the same environment is the simplest form of control.

control group

Without control, experimental results in medicine and the behavioral sciences can be dominated by such influences as the details of the experimental arrangement, the selection of subjects, and the placebo effect. The result is often *bias*, systematic favoritism toward one outcome. An uncontrolled study of a new medical therapy, for example, is biased in favor of finding the treatment effective because of the placebo effect. It should not surprise you to learn that uncontrolled studies in medicine give new therapies a much higher success rate than proper comparative experiments. Well-designed experiments, like the Physicians' Health Study and the second gastric freezing study, usually compare several treatments.

EXERCISES

For each of the experimental situations described in Exercises 5.31 to 5.34, identify the experimental units or subjects, the factors, the treatments, and the response variables.

5.31 RESISTING DROUGHT The ability to grow in shade may help pines found in the dry forests of Arizona to resist drought. How well do these pines grow in shade? Investigators planted pine seedlings in a greenhouse in either full light or light reduced to 5% of normal by shade cloth. At the end of the study, they dried the young trees and weighed them.

5.32 PACKAGE LINERS A manufacturer of food products uses package liners that are sealed at the top by applying heated jaws after the package is filled. The customer peels the sealed pieces apart to open the package. What effect does the temperature of the jaws have on the force required to peel the liner? To answer this question, the engineers prepare 20 pairs of pieces of package liner. They seal five pairs at each of 250° F, 275° F, 300° F, and 325° F. Then they measure the strength needed to peel each seal.

5.33 IMPROVING RESPONSE RATE How can we reduce the rate of refusals in telephone surveys? Most people who answer at all listen to the interviewer's introductory remarks and then decide whether to continue. One study made telephone calls to randomly selected households to ask opinions about the next election. In some calls, the interviewer gave her name, in others she identified the university she was representing, and in still others she identified both herself and the university. For each type of call, the interviewer either did or did not offer to send a copy of the final survey results to the person interviewed. Do these differences in the introduction affect whether the interview is completed?

5.34 SICKLE-CELL DISEASE Sickle-cell disease is an inherited disorder of the red blood cells that in the United States affects mostly blacks. It can cause severe pain and many complications. Can the drug hydroxyurea reduce the severe pain caused by sickle-cell disease? A study by the National Institutes of Health gave the drug to 150 sickle-cell sufferers and a placebo (a dummy medication) to another 150. The researchers then counted the episodes of pain reported by each subject.

5.35 COMPARING LEARNING METHODS An educator wants to compare the effectiveness of computer software that teaches reading with that of a standard reading curriculum.

She tests the reading ability of each student in a class of fourth graders, then divides them into two groups. One group uses the computer regularly, while the other studies a standard curriculum. At the end of the year, she retests all the students and compares the increase in reading ability in the two groups.

(a) Is this an experiment? Why or why not?

(b) What are the explanatory and response variables?

5.36 OPTIMIZING A PRODUCTION PROCESS A chemical engineer is designing the production process for a new product. The chemical reaction that produces the product may have higher or lower yield, depending on the temperature and the stirring rate in the vessel in which the reaction takes place. The engineer decides to investigate the effects of combinations of two temperatures (50° C and 60° C) and three stirring rates (60 rpm, 90 rpm, and 120 rpm) on the yield of the process. She will process two batches of the product at each combination of temperature and stirring rate.

(a) What are the experimental units and the response variable in this experiment?

(b) How many factors are there? How many treatments? Use a diagram like that in Figure 5.2 (page 290) to lay out the treatments.

(c) How many experimental units are required for the experiment?

Randomization

The design of an experiment first describes the response variable or variables, the factors (explanatory variables), and the layout of the treatments, with *comparison* as the leading principle. Figure 5.2 illustrates this aspect of the design of the Physicians' Health Study. The second aspect of design is the rule used to assign the experimental units to the treatments. Comparison of the effects of several treatments is valid only when all treatments are applied to similar groups of experimental units. If one corn variety is planted on more fertile ground, or if one cancer drug is given to more seriously ill patients, comparisons among treatments are meaningless. Systematic differences among the groups of experimental units in a comparative experiment cause bias. How can we assign experimental units to treatments in a way that is fair to all of the treatments?

Experimenters often attempt to match groups by elaborate balancing acts. Medical researchers, for example, try to match the patients in a "new drug" experimental group and a "standard drug" control group by age, sex, physical condition, smoker or not, and so on. Matching is helpful but not adequate—there are too many lurking variables that might affect the outcome. The experimenter is unable to measure some of these variables and will not think of others until after the experiment. Some important variables, such as how advanced a cancer patient's disease is, are so subjective that an experimenter might bias the study by, for example, assigning more advanced cancer cases to a promising new treatment in the unconscious hope that it will help them.

The statistician's remedy is to rely on chance to make an assignment that does not depend on any characteristic of the experimental units and that does not rely on the judgment of the experimenter in any way. The use of chance can be combined with matching, but the simplest design creates groups by chance alone. Here is an example.

EXAMPLE 5.12 TESTING A BREAKFAST FOOD

A food company assesses the nutritional quality of a new "instant breakfast" product by feeding it to newly weaned male white rats. The response variable is a rat's weight gain over a 28-day period. A control group of rats eats a standard diet but otherwise receives exactly the same treatment as the experimental group.

This experiment has one factor (the diet) with two levels. The researchers use 30 rats for the experiment and so must divide them into two groups of 15. To do this in an unbiased fashion, put the cage numbers of the 30 rats in a hat, mix them up, and draw 15. These rats form the experimental group and the remaining 15 make up the control group. That is, *each group is an SRS of the available rats.* Figure 5.3 outlines the design of this experiment.

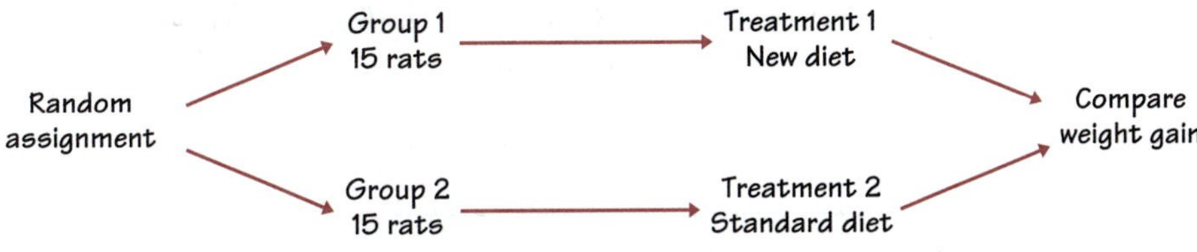

FIGURE 5.3 Outline of a randomized comparative experiment.

We can use software or the table of random digits to randomize. Label the rats 01 to 30. Enter Table B at (say) line 130. Run your finger along this line (and continue to lines 131 and 132 as needed) until 15 rats are chosen. They are the rats labeled

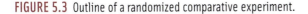

05, 16, 17, 20, 19, 04, 25, 29, 18, 07, 13, 02, 23, 27, 21

These rats form the experimental group; the remaining 15 are the control group.

Randomization, the use of chance to divide experimental units into groups, is an essential ingredient for a good experimental design. The design in Figure 5.3 combines comparison and randomization to arrive at the simplest randomized comparative design. This "flowchart" outline presents all the essentials: randomization, the sizes of the groups and which treatment they receive, and the response variable. There are, as we will see later, statistical reasons for generally using treatment groups about equal in size.

Randomized comparative experiments

The logic behind the randomized comparative design in Figure 5.3 is as follows:

- Randomization produces groups of rats that should be similar in all respects before the treatments are applied.

• Comparative design ensures that influences other than the diets operate equally on both groups.

• Therefore, differences in average weight gain must be due either to the diets or to the play of chance in the random assignment of rats to the two diets.

That "either-or" deserves more thought. We cannot say that *any* difference in the average weight gains of rats fed the two diets must be caused by a difference between the diets. There would be some difference even if both groups received the same diet, because the natural variability among rats means that some grow faster than others. Chance assigns the faster-growing rats to one group or the other, and this creates a chance difference between the groups. We would not trust an experiment with just one rat in each group, for example. The results would depend too much on which group got lucky and received the faster-growing rat. If we assign many rats to each diet, however, **the effects of chance will average out** and there will be little difference in the average weight gains in the two groups unless the diets themselves cause a difference. "**Use enough experimental units to reduce chance variation**" is the third big idea of statistical design of experiments.

PRINCIPLES OF EXPERIMENTAL DESIGN

The basic principles of statistical design of experiments are

1. **Control** the effects of lurking variables on the response, most simply by comparing two or more treatments.
2. **Randomize**—use impersonal chance to assign experimental units to treatments.
3. **Replicate** each treatment on many units to reduce chance variation in the results.

We hope to see a difference in the responses so large that it is unlikely to happen just because of chance variation. We can use the laws of probability, which give a mathematical description of chance behavior, to learn if the treatment effects are larger than we would expect to see if only chance were operating. If they are, we call them *statistically significant.*

STATISTICAL SIGNIFICANCE

An observed effect so large that it would rarely occur by chance is called **statistically significant**.

You will often see the phrase "statistically significant" in reports of investigations in many fields of study. It tells you that the investigators found good evidence for the effect they were seeking. The Physicians' Health Study, for example, reported statistically significant evidence that aspirin reduces the number of heart attacks compared with a placebo.

EXAMPLE 5.13 ENCOURAGING ENERGY CONSERVATION

Many utility companies have programs to encourage their customers to conserve energy. An electric company is considering placing electronic meters in households to show what the cost would be if the electricity use at that moment continued for a month. Will meters reduce electricity use? Would cheaper methods work almost as well? The company decides to design an experiment.

One cheaper approach is to give customers a chart and information about monitoring their electricity use. The experiment compares these two approaches (meter, chart) with each other and also with a control group of customers who receive no help in monitoring electricity use. The response variable is total electricity used in a year. The company finds 60 single-family residences in the same city willing to participate, so it assigns 20 residences at random to each of the three treatments. The outline of the design appears in Figure 5.4.

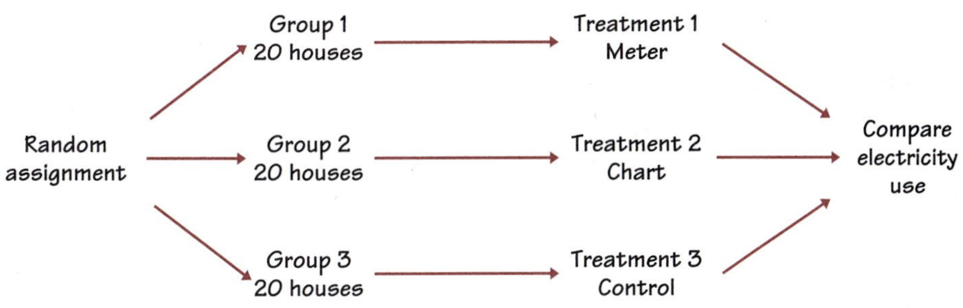

FIGURE 5.4 Outline of a completely randomized design comparing three treatments.

To carry out the random assignment, label the 60 houses 01 to 60. Then enter Table B and read two-digit groups until you have selected 20 houses to receive the meters. Continue in Table B to select 20 more to receive charts. The remaining 20 form the control group. The process is simple but tedious.

completely randomized design

When all experimental units are allocated at random among all treatments, the experimental design is ***completely randomized.*** The designs in Figures 5.3 (page 295) and 5.4 are both completely randomized. Completely randomized designs can compare any number of treatments. In Example 5.13, we compared the three levels of a single factor: the method used to encourage energy conservation. The treatments can be formed by more than one factor. The Physicians' Health Study had two factors, which combine to form the four treatments shown in Figure 5.2 (page 290). The study used a completely randomized design that assigned 5499 of the 21,996 subjects to each of the four treatments.

EXERCISES

5.37 TREATING PROSTATE DISEASE A large study used records from Canada's national health care system to compare the effectiveness of two ways to treat prostate disease. The two treatments are traditional surgery and a new method that does not require surgery. The records described many patients whose doctors had chosen each method. The study found that patients treated by the new method were significantly more likely to die within 8 years.[16]

(a) Further study of the data showed that this conclusion was wrong. The extra deaths among patients who got the new method could be explained by lurking variables. What lurking variables might be confounded with a doctor's choice of surgical or non-surgical treatment?

(b) You have 300 prostate patients who are willing to serve as subjects in an experiment to compare the two methods. Use a diagram to outline the design of a randomized comparative experiment. (When using a diagram to outline the design of an experiment, be sure to indicate the size of the treatment groups and the response variable. The diagrams in Examples 5.12 (page 295) and 5.13 (page 297) are models.)

5.38 PACKAGE LINERS

(a) Use a diagram to describe a completely randomized experimental design for the package liner experiment of Exercise 5.32. (When using a diagram to outline the design of an experiment, be sure to indicate the size of the treatment groups and the response variable. The diagrams in Examples 5.12 (page 295) and 5.13 (page 297) are models.)

(b) Use Table B, starting at line 120, to do the randomization required by your design.

5.39 RECRUITING FEMALE EMPLOYEES Will providing child care for employees make a company more attractive to women, even those who are unmarried? You are designing an experiment to answer this question. You prepare recruiting material for two fictitious companies, both in similar businesses in the same location. Company A's brochure does not mention child care. There are two versions of Company B's material, identical except that one describes the company's on-site child-care facility. Your subjects are 40 unmarried women who are college seniors seeking employment. Each subject will read recruiting material for both companies and choose the one she would prefer to work for. You will give each version of Company B's brochure to half the women. You expect that a higher percentage of those who read the description that includes child care will choose Company B.

(a) Outline an appropriate design for the experiment.

(b) The names of the subjects appear below. Use Table B, beginning at line 131, to do the randomization required by your design. List the subjects who will read the version that mentions child care.

Abrams	Danielson	Gutierrez	Lippman	Rosen
Adamson	Durr	Howard	Martinez	Sugiwara
Afifi	Edwards	Hwang	McNeill	Thompson
Brown	Fluharty	Iselin	Morse	Travers
Cansico	Garcia	Janle	Ng	Turing
Chen	Gerson	Kaplan	Quinones	Ullmann
Cortez	Green	Kim	Rivera	Williams
Curzakis	Gupta	Lattimore	Roberts	Wong

5.40 ENCOURAGING ENERGY CONSERVATION Example 5.13 (page 297) describes an experiment to learn whether providing households with electronic indicators or charts will reduce their electricity consumption. An executive of the electric company objects to including a control group. He says, "It would be simpler to just compare electricity use last year (before the indicator or chart was provided) with consumption in the same period this year. If households use less electricity this year, the indicator or chart must be working." Explain clearly why this design is inferior to that in Example 5.13.

5.41 EXERCISE AND HEART ATTACKS Does regular exercise reduce the risk of a heart attack? Here are two ways to study this question. Explain clearly why the second design will produce more trustworthy data.

1. A researcher finds 2000 men over 40 who exercise regularly and have not had heart attacks. She matches each with a similar man who does not exercise regularly, and she follows both groups for 5 years.
2. Another researcher finds 4000 men over 40 who have not had heart attacks and are willing to participate in a study. She assigns 2000 of the men to a regular program of supervised exercise. The other 2000 continue their usual habits. The researcher follows both groups for 5 years.

5.42 STOCKS DECLINE ON MONDAYS Puzzling but true: stocks tend to go down on Mondays. There is no convincing explanation for this fact. A recent study looked at this "Monday effect" in more detail, using data of the daily returns of stocks on several U.S. exchanges over a 30-year period. Here are some of the findings:

> *To summarize, our results indicate that the well-known Monday effect is caused largely by the Mondays of the last two weeks of the month. The mean Monday return of the first three weeks of the month is, in general, not significantly different from zero and is generally significantly higher than the mean Monday return of the last two weeks. Our finding seems to make it more difficult to explain the Monday effect.*[17]

A friend thinks that "significantly" in this article has its plain English meaning, roughly "I think this is important." Explain in simple language what "significantly higher" and "not significantly different from zero" actually tell us here.

Cautions about experimentation

The logic of a randomized comparative experiment depends on our ability to treat all the experimental units identically in every way except for the actual treatments being compared. Good experiments therefore require careful attention to details. For example, the subjects in both the Physicians' Health Study (Example 5.9, page 290) and the second gastric freezing experiment (Example 5.11, page 292) all got the same medical attention over the several years the studies continued. Moreover, these studies were ***double-blind***—neither the subjects themselves nor the medical personnel who worked with them knew which treatment any subject had received. The double-blind method avoids unconscious bias by, for example, a doctor who doesn't think that "just a placebo" can benefit a patient.

double-blind

DOUBLE-BLIND EXPERIMENT

In a double-blind experiment, neither the subjects nor the people who have contact with them know which treatment a subject received.

lack of realism

The most serious potential weakness of experiments is ***lack of realism***. The subjects or treatments or setting of an experiment may not realistically duplicate the conditions we really want to study. Here are some examples.

EXAMPLE 5.14 RESPONSE TO ADVERTISING

A study compares two television advertisements by showing TV programs to student subjects. The students know it's "just an experiment." We can't be sure that the results apply to everyday television viewers. Many behavioral science experiments use as subjects students who know they are subjects in an experiment. That's not a realistic setting.

EXAMPLE 5.15 CENTER BRAKE LIGHTS

Do those high center brake lights, required on all cars sold in the United States since 1986, really reduce rear-end collisions? Randomized comparative experiments with fleets of rental and business cars, done before the lights were required, showed that the third brake light reduced rear-end collisions by as much as 50%. Alas, requiring the third light in all cars led to only a 5% drop.

What happened? Most cars did not have the extra brake light when the experiments were carried out, so it caught the eye of following drivers. Now that almost all cars have the third light, they no longer capture attention.

Lack of realism can limit our ability to apply the conclusions of an experiment to the settings of greatest interest. Most experimenters want to generalize their conclusions to some setting wider than that of the actual experiment. Statistical analysis of the original experiment cannot tell us how far the results will generalize. Nonetheless, the randomized comparative experiment, because of its ability to give convincing evidence for causation, is one of the most important ideas in statistics.

Matched pairs designs

Completely randomized designs are the simplest statistical designs for experiments. They illustrate clearly the principles of control, randomization, and replication. However, completely randomized designs are often inferior to more elaborate statistical designs. In particular, matching the subjects in various ways can produce more precise results than simple randomization.

EXAMPLE 5.16 CEREAL LEAF BEETLES

Are cereal leaf beetles more strongly attracted by the color yellow or by the color green? Agriculture researchers want to know, because they detect the presence of the pests in farm fields by mounting sticky boards to trap insects that land on them. The board color should attract beetles as strongly as possible. We must design an experiment to compare yellow and green by mounting boards on poles in a large field of oats.

The experimental units are locations within the field far enough apart to represent independent observations. We erect a pole at each location to hold the boards. We might employ a completely randomized design in which we randomly select half the poles to receive a yellow board while the remaining poles receive green. The locations vary widely in the number of beetles present. For example, the alfalfa that borders the oats on one side is a natural host of the beetles, so locations near the alfalfa will have extra beetles. This variation among experimental units can hide the systematic effect of the board color.

It is more efficient to use a ***matched pairs design*** in which we mount boards of both colors on each pole. The observations (numbers of beetles trapped) are matched in pairs from the same poles. We compare the number of trapped beetles on a yellow board with the number trapped by the green board on the same pole. Because the boards are mounted one above the other, we select the color of the top board at random. Just toss a coin for each board—if the coin falls heads, the yellow board is mounted above the green board.

matched pairs design

Matched pairs designs compare just two treatments. We choose *blocks* of two units that are as closely matched as possible. In Example 5.16, two boards on the same pole form a block. We assign one of the treatments to each unit by tossing a coin or reading odd and even digits from Table B. Alternatively, each block in a matched pairs design may consist of just one subject, who gets both treatments one after the other. Each subject serves as his or her own control. The *order* of the treatments can influence the subject's response, so we randomize the order for each subject, again by a coin toss.

Block designs

The matched pairs design of Example 5.16 uses the principles of comparison of treatments, randomization, and replication on several experimental units. However, the randomization is not complete (all locations randomly assigned to treatment groups) but restricted to assigning the order of the boards at each location. The matched pairs design reduces the effect of variation among locations in the field by comparing the pair of boards at each location. Matched pairs are an example of *block designs*.

BLOCK DESIGN

A **block** is a group of experimental units or subjects that are known before the experiment to be similar in some way that is expected to affect the response to the treatments. In a **block design**, the random assignment of units to treatments is carried out separately within each block.

Block designs can have blocks of any size. A block design combines the idea of creating equivalent treatment groups by matching with the principle of forming treatment groups at random. Blocks are another form of *control*. They control the effects of some outside variables by bringing those variables into the experiment to form the blocks. Here are some typical examples of block designs.

EXAMPLE 5.17 COMPARING CANCER THERAPIES

The progress of a type of cancer differs in women and men. A clinical experiment to compare three therapies for this cancer therefore treats sex as a blocking variable. Two separate randomizations are done, one assigning the female subjects to the treatments and the other assigning the male subjects. Figure 5.5 outlines the design of this experiment. Note that there is no randomization involved in making up the blocks. They are groups of subjects who differ in some way (sex in this case) that is apparent before the experiment begins.

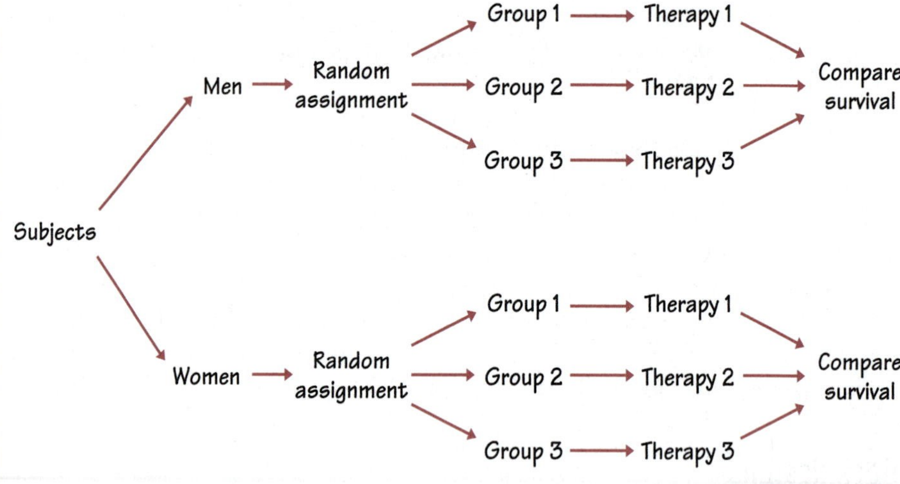

FIGURE 5.5 Outline of a block design. The blocks consist of male and female subjects. The treatments are three therapies for cancer.

EXAMPLE 5.18 SOYBEANS

The soil type and fertility of farmland differ by location. Because of this, a test of the effect of tillage type (two types) and pesticide application (three application schedules) on soybean yields uses small fields as blocks. Each block is divided into six plots, and the six treatments are randomly assigned to plots separately within each block.

EXAMPLE 5.19 STUDYING WELFARE SYSTEMS

A social policy experiment will assess the effect on family income of several proposed new welfare systems and compare them with the present welfare system. Because the income of a family under any welfare system is strongly related to its present income, the families who agree to participate are divided into blocks of similar income levels. The families in each block are then allocated at random among the welfare systems.

Blocks allow us to draw separate conclusions about each block, for example, about men and women in the cancer study in Example 5.17. Blocking also allows more precise overall conclusions, because the systematic differences between men and women can be removed when we study the overall effects of the three therapies. The idea of blocking is an important additional principle of statistical design of experiments. A wise experimenter will form blocks based on the most important unavoidable sources of variability among the experimental units. Randomization will then average out the effects of the remaining variation and allow an unbiased comparison of the treatments.

EXERCISES

5.43 **MEDITATION FOR ANXIETY** An experiment that claimed to show that meditation lowers anxiety proceeded as follows. The experimenter interviewed the subjects and rated their level of anxiety. Then the subjects were randomly assigned to two groups. The experimenter taught one group how to meditate and they meditated daily for a month. The other group was simply told to relax more. At the end of the month, the experimenter interviewed all the subjects again and rated their anxiety level. The meditation group now had less anxiety. Psychologists said that the results were suspect because the ratings were not blind. Explain what this means and how lack of blindness could bias the reported results.

5.44 **PAIN RELIEF STUDY** Fizz Laboratories, a pharmaceutical company, has developed a new pain-relief medication. Sixty patients suffering from arthritis and needing pain relief are available. Each patient will be treated and asked an hour later, "About what percentage of pain relief did you experience?"

(a) Why should Fizz not simply administer the new drug and record the patients' responses?

(b) Outline the design of an experiment to compare the drug's effectiveness with that of aspirin and of a placebo.

(c) Should patients be told which drug they are receiving? How would this knowledge probably affect their reactions?

(d) If patients are not told which treatment they are receiving, the experiment is single-blind. Should this experiment be double-blind also? Explain.

5.45 COMPARING WEIGHT-LOSS TREATMENTS Twenty overweight females have agreed to participate in a study of the effectiveness of four weight-loss treatments: A, B, C, and D. The researcher first calculates how overweight each subject is by comparing the subject's actual weight with her "ideal" weight. The subjects and their excess weights in pounds are

Birnbaum	35	Hernandez	25	Moses	25	Smith	29
Brown	34	Jackson	33	Nevesky	39	Stall	33
Brunk	30	Kendall	28	Obrach	30	Tran	35
Cruz	34	Loren	32	Rodriguez	30	Wilansky	42
Deng	24	Mann	28	Santiago	27	Williams	22

The response variable is the weight lost after 8 weeks of treatment. Because a subject's excess weight will influence the response, a block design is appropriate.

(a) Arrange the subjects in order of increasing excess weight. Form 5 blocks of 4 subjects each by grouping the 4 least overweight, then the next 4, and so on.

(b) Use Table B to randomly assign the 4 subjects in each block to the 4 weight-loss treatments. Be sure to explain exactly how you used the table.

5.46 CARBON DIOXIDE AND TREE GROWTH The concentration of carbon dioxide (CO_2) in the atmosphere is increasing rapidly due to our use of fossil fuels. Because plants use CO_2 to fuel photosynthesis, more CO_2 may cause trees and other plants to grow faster. An elaborate apparatus allows researchers to pipe extra CO_2 to a 30-meter circle of forest. We want to compare the growth in base area of trees in treated and untreated areas to see if extra CO_2 does in fact increase growth. We can afford to treat three circular areas.[18]

(a) Describe the design of a completely randomized experiment using 6 well-separated 30-meter circular areas in a pine forest. Sketch the circles and carry out the randomization your design calls for.

(b) Areas within the forest may differ in soil fertility. Describe a matched pairs design using three pairs of circles that will reduce the extra variation due to different fertility. Sketch the circles and carry out the randomization your design calls for.

5.47 DOES ROOM TEMPERATURE AFFECT MANUAL DEXTERITY? An expert on worker performance is interested in the effect of room temperature on the performance of tasks requiring manual dexterity. She chooses temperatures of 70° F and 90° F as treatments. The response variable is the number of correct insertions, during a 30-minute period, in a peg-and-hole apparatus that requires the use of both hands simultaneously. Each subject is trained on the apparatus and then asked to make as many insertions as possible in 30 minutes of continuous effort.

(a) Outline a completely randomized design to compare dexterity at 70° and 90°. Twenty subjects are available.

(b) Because individuals differ greatly in dexterity, the wide variation in individual scores may hide the systematic effect of temperature unless there are many subjects in each group. Describe in detail the design of a matched pairs experiment in which each subject serves as his or her own control.

5.48 CHARTING AS AN INVESTMENT STRATEGY Some investment advisors believe that charts of past trends in the prices of securities can help predict future prices. Most economists disagree. In an experiment to examine the effects of using charts, business students trade (hypothetically) a foreign currency at computer screens. There are 20 student subjects available, named for convenience A, B, C, . . . , T. Their goal is to make as much money as possible, and the best performances are rewarded with small prizes. The student traders have the price history of the foreign currency in dollars in their computers. They may or may not also have software that highlights trends. Describe *two* designs for this experiment, a completely randomized design and a matched pairs design in which each student serves as his or her own control. In both cases, carry out the randomization required by the design.

SUMMARY

In an experiment, one or more **treatments** are imposed on the **experimental units** or **subjects**. Each treatment is a combination of **levels** of the explanatory variables, which we call **factors**.

The **design** of an experiment refers to the choice of treatments and the manner in which the experimental units or subjects are assigned to the treatments.

The basic principles of statistical design of experiments are **control**, **randomization**, and **replication**.

The simplest form of control is **comparison**. Experiments should compare two or more treatments in order to prevent **confounding** the effect of a treatment with other influences, such as lurking variables.

Randomization uses chance to assign subjects to the treatments. Randomization creates treatment groups that are similar (except for chance variation) before the treatments are applied. Randomization and comparison together prevent **bias**, or systematic favoritism, in experiments.

You can carry out randomization by giving numerical labels to the experimental units and using a **table of random digits** to choose treatment groups.

Replication of the treatments on many units reduces the role of chance variation and makes the experiment more sensitive to differences among the treatments.

Good experiments require attention to detail as well as good statistical design. Many behavioral and medical experiments are **double-blind**. **Lack of realism** in an experiment can prevent us from generalizing its results.

In addition to comparison, a second form of control is to restrict randomization by forming **blocks** of experimental units that are similar in some way that is important to the response. Randomization is then carried out separately within each block.

Matched pairs are a common form of blocking for comparing just two treatments. In some matched pairs designs, each subject receives both treatments in a random order. In others, the subjects are matched in pairs as closely as possible, and one subject in each pair receives each treatment.

SECTION 5.2 EXERCISES

5.49 DOES SAINT-JOHN'S-WORT RELIEVE MAJOR DEPRESSION? Here are some excerpts from the report of a study of this issue.[19] The study concluded that the herb is no more effective than a placebo.

(a) "Design: Randomized, double-blind, placebo-controlled clinical trial. . . ." Explain the meaning of each of the terms in this description.

(b) "Participants . . . were randomly assigned to receive either Saint-John's-wort extract ($n = 98$) or placebo ($n = 102$). . . . The primary outcome measure was the rate of change in the Hamilton Rating Scale for Depression over the treatment period." Based on this information, use a diagram to outline the design of this clinical trial.

5.50 MARKETING TO CHILDREN, I If children are given more choices within a class of products, will they tend to prefer that product to a competing product that offers fewer choices? Marketers want to know. An experiment prepared three "choice sets" of beverages. The first contained two milk drinks and two fruit drinks. The second had the same two fruit drinks but four milk drinks. The third contained four fruit drinks but only the original two milk drinks. The researchers divided 210 children aged 4 to 12 years into 3 groups at random. They offered each group one of the choice sets. As each child chose a beverage to drink from the choice set presented, the researchers noted whether the choice was a milk drink or a fruit drink.

(a) What are the experimental units or subjects?

(b) What is the factor, and what are its levels?

(c) What is the response variable?

5.51 BODY TEMPERATURE AND SURGERY Surgery patients are often cold because the operating room is kept cool and the body's temperature regulation is disturbed by anesthetics. Will warming patients to maintain normal body temperature reduce infections after surgery? In one experiment, patients undergoing colon surgery received intravenous fluids from a warming machine and were covered with a blanket through which air circulated. For some patients, the fluid and the air were warmed; for others, they were not. The patients received identical treatment in all other respects.[20]

(a) Identify the experimental subjects, the factor and its levels, and the response variables.

(b) Draw a diagram to outline the design of a randomized comparative experiment for this study.

(c) The following subjects have given consent to participate in this study. Do the random assignment required by your design. (If you use Table B, begin at line 121.)

Abbott	Decker	Gutierrez	Lucero	Rosen
Adamson	Devlin	Howard	Masters	Sugiwara
Afifi	Engel	Hwang	McNeill	Thompson
Brown	Fluharty	Iselin	Morse	Travers
Cansico	Garcia	Janle	Ng	Turing
Chen	Gerson	Kaplan	Quinones	Ullmann
Cordoba	Green	Kim	Rivera	Williams
Curzakis	Gupta	Lattimore	Roberts	Wong

(d) To simplify the setup of the study, we might warm the fluids and air blanket for one operating team and not for another doing the same kind of surgery. Why might this design result in bias?

(e) The operating team did not know whether fluids and air blanket were heated, nor did the doctors who followed the patients after surgery. What is this practice called? Why was it used here?

5.52 MARKETING TO CHILDREN, II Use a diagram to outline a completely randomized design for the children's choice study of Exercise 5.50.

5.53 DOES CALCIUM REDUCE BLOOD PRESSURE? You are participating in the design of a medical experiment to investigate whether a calcium supplement in the diet will reduce the blood pressure of middle-aged men. Preliminary work suggests that calcium may be effective and that the effect may be greater for black men than for white men. You have available 40 men with high blood pressure who are willing to serve as subjects.

(a) Outline an appropriate design for the experiment.

(b) The names of the subjects appear below. Use Table B, beginning at line 119, to do the randomization required by your design, and list the subjects to whom you will give the drug.

Alomar	Denman	Han	Liang	Rosen
Asihiro	Durr	Howard	Maldonado	Solomon
Bennett	Edwards	Hruska	Marsden	Tompkins
Bikalis	Farouk	Imrani	Moore	Townsend
Chen	Fratianna	James	O'Brian	Tullock
Clemente	George	Kaplan	Ogle	Underwood
Cranston	Green	Krushchev	Plochman	Willis
Curtis	Guillen	Lawless	Rodriguez	Zhang

(c) Choosing the sizes of the treatment groups requires more statistical expertise. We will learn more about this aspect of design in later chapters. Explain in plain language the advantage of using larger groups of subjects.

5.54 MARKETING TO CHILDREN, III The children's choice experiment in Exercise 5.50 has 210 subjects. Explain how you would assign labels to the 210 children in the actual experiment. Then use Table B at line 125 to choose *only the first* 5 children assigned to the first treatment.

5.55 PLACEBO EFFECT A survey of physicians found that some doctors give a placebo to a patient who complains of pain for which the physician can find no cause. If the patient's pain improves, these doctors conclude that it had no physical basis. The medical school researchers who conducted the survey claimed that these doctors do not understand the placebo effect. Why?

5.56 WILL TAKING ANTIOXIDANTS HELP PREVENT COLON CANCER? People who eat lots of fruits and vegetables have lower rates of colon cancer than those who eat little of these foods. Fruits and vegetables are rich in "antioxidants" such as vitamins A, C, and E. Will taking antioxidants help prevent colon cancer? A clinical trial studied 864 people who were at risk of colon cancer. The subjects were divided into four groups: daily beta carotene, daily vitamins C and E, all three vitamins every day, and daily placebo. After four years, the researchers were surprised to find no significant difference in colon cancer among the groups.[21]

(a) What are the explanatory and response variables in this experiment?

(b) Outline the design of the experiment. Use your judgment in choosing the group sizes.

(c) Assign labels to the 864 subjects and use Table B, starting at line 118, to choose the first 5 subjects for the beta carotene group.

(d) The study was double-blind. What does this mean?

(e) What does "no significant difference" mean in describing the outcome of the study?

(f) Suggest some lurking variables that could explain why people who eat lots of fruits and vegetables have lower rates of colon cancer. The experiment suggests that these variables, rather than the antioxidants, may be responsible for the observed benefits of fruits and vegetables.

5.57 TREATING DRUNK DRIVERS Once a person has been convicted of drunk driving, one purpose of court-mandated treatment or punishment is to prevent future offenses of the same kind. Suggest three different treatments that a court might require. Then outline the design of an experiment to compare their effectiveness. Be sure to specify the response variables you will measure.

5.58 ACCULTURATION RATING There are several psychological tests that measure the extent to which Mexican Americans are oriented toward Mexican/Spanish or Anglo/English culture. Two such tests are the Bicultural Inventory (BI) and the Acculturation Rating Scale for Mexican Americans (ARSMA). To study the correlation between the scores on these two tests, researchers will give both tests to a group of 22 Mexican Americans.

(a) Briefly describe a matched pairs design for this study. In particular, how will you use randomization in your design?

(b) You have an alphabetized list of the subjects (numbered 1 to 22). Carry out the randomization required by your design and report the result.

5.3 SIMULATING EXPERIMENTS

Toss a coin 10 times. What is the likelihood of a run of 3 or more consecutive heads or tails? A couple plans to have children until they have a girl or until they have four children, whichever comes first. What are the chances that they will have a girl among their children? An airline knows from past experience that a certain percentage of customers who have purchased tickets will not show up to board the airplane. If the airline "overbooks" a particular flight (i.e., sells more tickets than they have seats), what are the chances that the airline will encounter more ticketed passengers than they have seats for? There are three methods we can use to answer questions involving chance like these:

1. Try to estimate the likelihood of a result of interest by actually carrying out the experiment many times and calculating the result's relative frequency. That's slow, sometimes costly, and often impractical or logistically difficult.

2. Develop a ***probability model*** and use it to calculate a theoretical answer. This requires that we know something about the rules of probability and therefore may not be feasible. (We will develop a probability model in the next chapter.)

probability model

3. Start with a model that, in some fashion, reflects the truth about the experiment, and then develop a procedure for imitating—or simulating—a number of repetitions of the experiment. This is quicker than repeating the real experiment, especially if we can use the TI-83/89 or a computer, and it allows us to do problems that are hard when done with formal mathematical analysis.

Here is an example of a simulation.

EXAMPLE 5.20 A GIRL IN THE FAMILY

Suppose we are interested in estimating the likelihood of a couple's having a girl among their first four children. Let a flip of a fair coin represent a birth, with heads corresponding to a girl and tails a boy. Since girls and boys are equally likely to occur on any birth, the coin flip is an accurate imitation of the situation. Flip the coin until a head appears or until the coin has been flipped 4 times, whichever comes first. The appearance of a head within the first 4 flips corresponds to the couple's having a girl among their first four children.

If this coin-flipping procedure is repeated many times, to represent the births in a large number of families, then the proportion of times that a head appears within the first 4 flips should be a good estimate of the true likelihood of the couple's having a girl.

A single die (one of a pair of dice) could also be used to simulate the birth of a son or daughter. Let an even number of spots (called pips) represent a girl, and let an odd number of spots represent a boy.

SIMULATION

The imitation of chance behavior, based on a model that accurately reflects the experiment under consideration, is called a **simulation**.

Simulation is an effective tool for finding likelihoods of complex results once we have a trustworthy model. In particular, we can use random digits from a table, graphing calculator, or computer software to simulate many repetitions quickly. The proportion of repetitions on which a result occurs will eventually be close to its true likelihood, so simulation can give good estimates of probabilities. The art of random digit simulation can be illustrated by a series of examples.

EXAMPLE 5.21 SIMULATION STEPS

Step 1: **State the problem or describe the experiment.** Toss a coin 10 times. What is the likelihood of a run of at least 3 consecutive heads or 3 consecutive tails?

Step 2: **State the assumptions.** There are two:

- A head or a tail is equally likely to occur on each toss.
- Tosses are independent of each other (i.e., what happens on one toss will not influence the next toss).

Step 3: **Assign digits to represent outcomes.** In a random number table, such as Table B in the back of the book, the digits 0, 1, 2, 3, 4, 5, 6, 7, 8, and 9 occur with the same long-term relative frequency (1/10). We also know that the successive digits in the table are independent. It follows that even digits and odd digits occur with the same long-term relative frequency, 50%. Here is one assignment of digits for coin tossing:

- One digit simulates one toss of the coin.
- Odd digits represent heads; even digits represent tails.

Successive digits in the table simulate independent tosses.

Step 4: **Simulate many repetitions.** Looking at 10 consecutive digits in Table B simulates one repetition. Read many groups of 10 digits from the table to simulate many repetitions. Be sure to keep track of whether or not the event we want (a run of 3 heads or 3 tails) occurs on each repetition.

Here are the first three repetitions, starting at line 101 in Table B. Runs of 3 or more heads or tails have been underlined.

Digits	1 9 2 2 3 9 5 0 3 4	0 5 7 5 6 2 8 7 1 3	9 6 4 0 9 1 2 5 3 1
Heads/tails	H H T T H H H T H T	T H H H T T T H H H	H T T T H H T H H H
Run of 3	YES	YES	YES

Twenty-two additional repetitions were done for a total of 25 repetitions; 23 of them did have a run of 3 or more heads or tails.

Step 5: **State your conclusions.** We estimate the probability of a run by the proportion

$$\text{estimated probability} = \frac{23}{25} = 0.92$$

Of course, 25 repetitions are not enough to be confident that our estimate is accurate. Now that we understand how to do the simulation, we can tell a computer to do many thousands of repetitions. A long simulation (or mathematical analysis) finds that the true probability is about 0.826.

Once you have gained some experience in simulation, establishing a correspondence between random numbers and outcomes in the experiment is usually the hardest part, and must be done carefully. Although coin tossing may not fascinate you, the model in Example 5.21 is typical of many probability problems because it consists of independent trials (the tosses) all having the same possible outcomes and probabilities. The coin tosses are said to be ***independent*** because the result of one toss has no effect or influence over the next coin toss. Shooting 10 free throws and observing the sexes of 10 children have similar models and are simulated in much the same way.

independent

The idea is to state the basic structure of the random phenomenon and then use simulation to move from this model to the probabilities of more complicated events. The model is based on opinion and past experience. If it does not correctly describe the random phenomenon, the probabilities derived from it by simulation will also be incorrect.

Step 3 (assigning digits) can usually be done in several different ways, but some assignments are more efficient than others. Here are some examples of this step.

EXAMPLE 5.22 ASSIGNING DIGITS

(a) Choose a person at random from a group of which 70% are employed. One digit simulates one person:

0, 1, 2, 3, 4, 5, 6 = employed
7, 8, 9 = not employed

The following correspondence is also satisfactory:

00, 01, . . . , 69 = employed
70, 71, . . . , 99 = not employed

This assignment is less efficient, however, because it requires twice as many digits and ten times as many numbers.

(b) Choose one person at random from a group of which 73% are employed. Now *two* digits simulate one person:

00, 01, 02, . . . , 72 = employed
73, 74, 75, . . . , 99 = not employed

We assigned 73 of the 100 two-digit pairs to "employed" to get probability 0.73. Representing "employed" by 01, 02, . . . , 73 would also be correct.

(c) Choose one person at random from a group of which 50% are employed, 20% are unemployed, and 30% are not in the labor force. There are now three possible outcomes, but the principle is the same. One digit simulates one person:

0, 1, 2, 3, 4 = employed
5, 6 = unemployed
7, 8, 9 = not in the labor force

Another valid assignment of digits might be

0, 1 = unemployed
2, 3, 4 = not in the labor force
5, 6, 7, 8, 9 = employed

What is important is the number of digits assigned to each outcome, not the order of the digits.

As the last example shows, simulation methods work just as easily when outcomes are not equally likely. Consider the following slightly more complicated example.

EXAMPLE 5.23 FROZEN YOGURT SALES

Orders of frozen yogurt flavors (based on sales) have the following relative frequencies: 38% chocolate, 42% vanilla, and 20% strawberry. The experiment consists of customers entering the store and ordering yogurt. The task is to simulate 10 frozen yogurt sales based on this recent history. Instead of considering the random number table to be made up of single digits, we now consider it to be made up of pairs of digits. This is because the relative frequencies of interest have a maximum of *two* significant digits. The range of the pairs of digits is 00 to 99, and since all the pairs are equally likely to occur, the pairs 00, 01, 02, . . . , 99 all have relative frequency 0.01.

Thus we may assign the numbers in the random number table as follows:

- 00 to 37 to correspond to the outcome chocolate (C)
- 38 to 79 to correspond to the outcome vanilla (V)
- 80 to 99 to correspond to the outcome strawberry (S)

The sequence of random numbers (starting at the 21st column of row 112 in Table B) is as follows:

19352 73089 84898 45785

This yields the following two-digit numbers:

19 35 27 30 89 84 89 84 57 85

which correspond to the outcomes

C C C C S S S S V S

EXAMPLE 5.24 A GIRL OR FOUR

A couple plans to have children until they have a girl or until they have four children, whichever comes first. We will show how to use random digits to estimate the likelihood that they will have a girl.

The model is the same as for coin tossing. We will assume that each child has probability 0.5 of being a girl and 0.5 of being a boy, and the sexes of successive children are independent.

Assigning digits is also easy. One digit simulates the sex of one child:

0, 1, 2, 3, 4 = girl
5, 6, 7, 8, 9 = boy

To simulate one repetition of this child-bearing strategy, read digits from Table B until the couple has either a girl or four children. Notice that the number of digits needed to simulate one repetition depends on how quickly the couple gets a girl. Here is the simulation, using line 130 of Table B. To interpret the digits, G for girl and B for boy are written under them, space separates repetitions, and under each repetition "+" indicates if a girl was born and "–" indicates one was not.

690	51	64	81	7871	74	0
BBG	BG	BG	BG	BBBG	BG	G
+	+	+	+	+	+	+
951	784	53	4	0	64	8987
BBG	BBG	BG	G	G	BG	BBBB
+	+	+	+	+	+	–

In these 14 repetitions, a girl was born 13 times. Our estimate of the probability that this strategy will produce a girl is therefore

$$\text{estimated probability} = \frac{13}{14} = 0.93$$

Some mathematics shows that if our probability model is correct, the true likelihood of having a girl is 0.938. Our simulated answer came quite close. Unless the couple is unlucky, they will succeed in having a girl.

EXERCISES

5.59 ESTABLISHING A CORRESPONDENCE State how you would use the following aids to establish a correspondence in a simulation that involves a 75% chance:

(a) a coin

(b) a six-sided die

(c) a random digit table (Table B)

(d) a standard deck of playing cards

5.60 THE CLEVER COINS Suppose you left your statistics textbook and calculator in your locker, and you need to simulate a random phenomenon that has a 25% chance of a desired outcome. You discover two nickels in your pocket that are left over from your lunch money. Describe how you could use the two coins to set up your simulation.

5.61 ABOLISH EVENING EXAMS? Suppose that 84% of a university's students favor abolishing evening exams. You ask 10 students chosen at random. What is the likelihood that all 10 favor abolishing evening exams?

(a) Describe how you would pose this question to 10 students independently of each other. How would you model the procedure?

(b) Assign digits to represent the answers "Yes" and "No."

(c) Simulate 5 repetitions, starting at line 129 of Table B. Then combine your results with those of the rest of your class. What is your estimate of the likelihood of the desired result?

5.62 SHOOTING FREE THROWS A basketball player makes 70% of her free throws in a long season. In a tournament game she shoots 5 free throws late in the game and misses 3 of them. The fans think she was nervous, but the misses may simply be chance. You will shed some light by estimating a probability.

(a) Describe how to simulate a single shot if the probability of making each shot is 0.7. Then describe how to simulate 5 independent shots.

(b) Simulate 50 repetitions of the 5 shots and record the number missed on each repetition. Use Table B starting at line 125. What is the approximate likelihood that the player will miss 3 or more of the 5 shots?

5.63 A POLITICAL POLL, I An opinion poll selects adult Americans at random and asks them, "Which political party, Democratic or Republican, do you think is better able to manage the economy?" Explain carefully how you would assign digits from Table B to simulate the response of one person in each of the following situations.

(a) Of all adult Americans, 50% would choose the Democrats and 50% the Republicans.

(b) Of all adult Americans, 60% would choose the Democrats and 40% the Republicans.

(c) Of all adult Americans, 40% would choose the Democrats, 40% would choose the Republicans, and 20% would be undecided.

(d) Of all adult Americans, 53% would choose the Democrats and 47% the Republicans.

5.64 A POLITICAL POLL, II Use Table B to simulate the responses of 10 independently chosen adults in each of the four situations of Exercise 5.63.

(a) For situation (a), use line 110.

(b) For situation (b), use line 111.

(c) For situation (c), use line 112.

(d) For situation (d), use line 113.

Simulations with the calculator or computer

The calculator and computer can be extremely useful in conducting simulations because they can be easily programmed to quickly perform a large number of repetitions. Study the reasoning and the steps involved in the following example so that you may become adept at using the capabilities of the TI-83/89 to design and carry out simulations.

EXAMPLE 5.25 RANDOMIZING WITH THE CALCULATOR

The command `randInt` (found under MATH/PRB/5:randInt on the TI-83, and under CATALOG F3 (Flash Apps) on the TI-89) can be used to generate random digits between any two specified values. Here are three applications.

The command `randInt(0,9,5)` generates 5 random integers between 0 and 9. This could serve as a block of 5 random digits in the random number table. The command `randInt(1,6,7)` could be used to simulate rolling a die 7 times. Generating 10 two-digit numbers between 00 and 99 from Example 5.23 could be done with the command `randInt(0,99,10)`.

```
randInt(0,9,5)
       {5 6 5 7 1}
randInt(1,6,7)
     {5 6 5 5 3 4 1}
randInt(0,99,10)

{81 23 86 2 40...
```

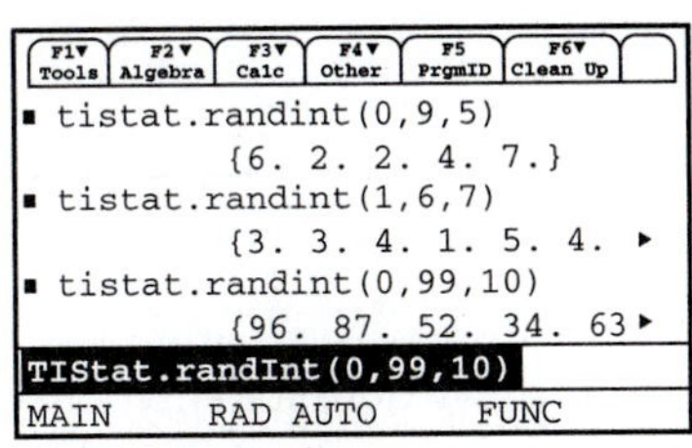

Using the statistical software package Minitab, the following set of commands will generate a set of 10 random numbers in the range 00 to 99 and store these numbers in column C1.

```
MTB > random 10 c1;
SUBC> integer 0 99.
MTB > Print C1

C1
   38   93   14   30   50   92   16   18   84   20
```

When you combine the power and simplicity of simulations with the power of technology, you have formidable tools for answering questions involving chance behavior.

EXERCISES

5.65 A GIRL OR FOUR Use your calculator to simulate a couple's having children until they have a girl or until they have four children, whichever comes first. (See Example 5.24.) Use the simulation to estimate the probability that they will have a girl among their children. Compare your calculator results with those of Example 5.24.

5.66 WORLD SERIES Suppose that in a particular year the American League baseball team is considered to have a 60% chance of beating the National League team in any given World Series game. (This assumption ignores any possible home-field advantage, which is probably not very realistic.) To win the World Series, a team must win 4 out of 7 games in the series. Further assume that the outcome of each game is not influenced by the outcome of any other game (that is, who wins one game is independent of who wins any other game).

(a) Use simulation methods to approximate the number of games that would have to be played in order to determine the world champion.

(b) The so-called home-field advantage is one factor that might be an explanatory variable in determining the winner of a game. What are some other possible factors?

5.67 TENNIS RACQUETS Professional tennis players bring multiple racquets to each match. They know that high string tension, the force with which they hit the ball, and occasional "racquet abuse" are all reasons why racquets break during a match. Brian Lob's coach tells him that he has a 15% chance of breaking a racquet in any given match. How many matches, on average, can Brian expect to play until he breaks a racquet and needs to use a backup? Use simulation methods to answer this question.

SUMMARY

There are times when actually carrying out an experiment is too costly, too slow, or simply impractical. In situations like these, a carefully designed **simulation** can provide approximate answers to our questions.

A simulation is an imitation of chance behavior, most often carried out with random numbers. The **steps of a simulation** are:

1. State the problem or describe the experiment.
2. State the assumptions.
3. Assign digits to represent outcomes.
4. Simulate many repetitions.
5. State your conclusions.

Programmable calculators, like the TI-83/89, and computers are particularly useful for conducting simulations because they can perform many repetitions quickly.

SECTION 5.3 EXERCISES

5.68 GAME OF CHANCE, I Amarillo Slim is a cardsharp who likes to play the following game. Draw 2 cards from the deck of 52 cards. If at least one of the cards is a heart, then you win $1. If neither card is a heart, then you lose $1.

(a) Describe a correspondence between random numbers and possible outcomes in this game.

(b) Simulate playing the game for 25 rounds. Shuffle the cards after each round. See if you can beat Amarillo Slim at his own game. Remember to write down the results of each game. When you finish, combine your results with those of 3 other students to obtain a total of 100 trials. Report your cumulative proportion of wins. Do you think this is a "fair" game? That is, do both you and Slim have an equal chance of winning?

5.69 GAME OF CHANCE, II A certain game of chance is based on randomly selecting three numbers from 00 to 99, inclusive (allowing repetitions), and adding the numbers. A person wins the game if the resulting sum is a multiple of 5.

(a) Describe your scheme for assigning random numbers to outcomes in this game.

(b) Use simulation to estimate the proportion of times a person wins the game.

5.70 THE BIRTHDAY PROBLEM Use your calculator and the simulation method to show that in a class of 23 unrelated students, the chances of at least 2 students with the same birthday are about 50%. Show that in a room of 41 people, the chances of at least 2 people having the same birthday are about 90%. What assumptions are you using in your simulations?

5.71 BATTER UP! Suppose a major league baseball player has a current batting average of .320. Note that the batting average = (number of hits)/(number of at-bats).

(a) Describe an assignment of random numbers to possible results in order to simulate the player's next 20 at-bats.

(b) Carry out the simulation for 20 repetitions, and report your results. What is the relative frequency of at-bats in which the player gets a hit?

(c) Compare your simulated experimental results with the player's actual batting average of .320.

5.72 NUCLEAR SAFETY A nuclear reactor is equipped with two independent automatic shutdown systems to shut down the reactor when the core temperature reaches the danger level. Neither system is perfect. System A shuts down the reactor 90% of the time when the danger level is reached. System B does so 80% of the time. The reactor is shut down if *either* system works.

(a) Explain how to simulate the response of System A to a dangerous temperature level.

(b) Explain how to simulate the response of System B to a dangerous temperature level.

(c) Both systems are in operation simultaneously. Combine your answers to (a) and (b) to simulate the response of both systems to a dangerous temperature level. Explain why you cannot use the same entry in Table B to simulate both responses.

(d) Now simulate 100 trials of the reactor's response to an emergency of this kind. Estimate the probability that it will shut down. This probability is higher than the probability that either system working alone will shut down the reactor.

5.73 SPREADING A RUMOR On a small island there are 25 inhabitants. One of these inhabitants, named Jack, starts a rumor which spreads around the isle. Any person who hears the rumor continues spreading it until he or she meets someone who has heard the story before. At that point, the person stops spreading it, since nobody likes to spread stale news.

(a) Do you think that all 25 inhabitants will eventually hear the rumor or will the rumor die out before that happens? Estimate the proportion of inhabitants who will hear the rumor.

(b) In the first time increment, Jack randomly selects one of the other inhabitants, named Jill, to tell the rumor to. In the second time increment, both Jack and Jill each randomly select one of the remaining 24 inhabitants to tell the rumor to. (*Note:* They could conceivably pick each other again.) In the next time increment, there are 4 rumor spreaders, and so on. If a randomly selected person has already heard the rumor, that rumor teller stops spreading the rumor. Design a record-keeping chart, and simulate this procedure. Use your TI-83/89 to help with the random selection. Continue until all 25 inhabitants hear the rumor or the rumor dies out. How many inhabitants out of 25 eventually heard the rumor?

(c) Combine your results with those of other students in the class. What is the mean number of inhabitants who hear the rumor?

CHAPTER REVIEW

Designs for producing data are essential parts of statistics in practice. Random sampling and randomized comparative experiments are perhaps the most important statistical inventions in this century. Both were slow to gain acceptance, and you will still see many voluntary response samples and uncontrolled experiments. This chapter has explained good techniques for producing data and has also explained why bad techniques often produce worthless data.The deliberate use of chance in producing data is a central idea in statistics. It allows use of the laws of probability to analyze data, as we will see in the following chapters. Here are the major skills you should have now that you have studied this chapter.

A. SAMPLING

1. Identify the population in a sampling situation.
2. Recognize bias due to voluntary response samples and other inferior sampling methods.
3. Use Table B of random digits to select a simple random sample (SRS) from a population.
4. Recognize the presence of undercoverage and nonresponse as sources of error in a sample survey. Recognize the effect of the wording of questions on the response.
5. Use random digits to select a stratified random sample from a population when the strata are identified.

B. EXPERIMENTS

1. Recognize whether a study is an observational study or an experiment.
2. Recognize bias due to confounding of explanatory variables with lurking variables in either an observational study or an experiment.
3. Identify the factors (explanatory variables), treatments, response variables, and experimental units or subjects in an experiment.
4. Outline the design of a completely randomized experiment using a diagram like those in Examples 5.12 and 5.13. The diagram in a specific case should show the sizes of the groups, the specific treatments, and the response variable.
5. Use Table B of random digits to carry out the random assignment of subjects to groups in a completely randomized experiment.
6. Recognize the placebo effect. Recognize when the double-blind technique should be used.
7. Recognize a block design when it would be appropriate. Know when a matched pairs design would be appropriate and how to design a matched pairs experiment.
8. Explain why a randomized comparative experiment can give good evidence for cause-and-effect relationships.

C. SIMULATIONS

1. Recognize that many random phenomena can be investigated by means of a carefully designed simulation.
2. Use the following steps to construct and run a simulation:
   a. State the problem or describe the experiment.
   b. State the assumptions.
   c. Assign digits to represent outcomes.
   d. Simulate many repetitions.
   e. Calculate relative frequencies and state your conclusions.
3. Use a random number table, the TI-83/89, or a computer utility such as Minitab, Data Desk, or a spreadsheet to conduct simulations.

CHAPTER 5 REVIEW EXERCISES

5.74 **ONTARIO HEALTH SURVEY** The Ministry of Health in the Province of Ontario, Canada, wants to know whether the national health care system is achieving its goals in the province. Much information about health care comes from patient records, but that source doesn't allow us to compare people who use health services with those who don't. So the Ministry of Health conducted the Ontario Health Survey, which interviewed a random sample of 61,239 people who live in the Province of Ontario.[22]

(a) What is the population for this sample survey? What is the sample?

(b) The survey found that 76% of males and 86% of females in the sample had visited a general practitioner at least once in the past year. Do you think these estimates are close to the truth about the entire population? Why?

5.75 TREATING BREAST CANCER What is the preferred treatment for breast cancer that is detected in its early stages? The most common treatment was once removal of the breast. It is now usual to remove only the tumor and nearby lymph nodes, followed by radiation. To study whether these treatments differ in their effectiveness, a medical team examines the records of 25 large hospitals and compares the survival times after surgery of all women who have had either treatment.

(a) What are the explanatory and response variables?

(b) Explain carefully why this study is not an experiment.

(c) Explain why confounding will prevent this study from discovering which treatment is more effective. (The current treatment was in fact recommended after a large randomized comparative experiment.)

5.76 WHICH DESIGN? What is the best way to answer each of the questions below: an experiment, a sample survey, or an observational study that is not a sample survey? Explain your choices.

(a) Are people generally satisfied with how things are going in the country right now?

(b) Do college students learn basic accounting better in the classroom or using an online course?

(c) How long do your teachers wait on the average after they ask their class a question?

5.77 COACH, I NEED OXYGEN! We often see players on the sidelines of a football game inhaling oxygen. Their coaches think it will speed their recovery. We might measure recovery from intense exercise as follows: Have a football player run 100 yards three times in quick succession. Then allow three minutes to rest before running 100 yards again. Time the final run. Because players vary greatly in speed, you plan a matched pairs experiment using 25 football players as subjects. Discuss the design of such an experiment to investigate the effect of inhaling oxygen during the rest period.

5.78 POLLING THE FACULTY A labor organization wants to study the attitudes of college faculty members toward collective bargaining. These attitudes appear to be different depending on the type of college. The American Association of University Professors classifies colleges as follows:

Class I. Offer doctorate degrees and award at least 15 per year.

Class IIA. Award degrees above the bachelor's but are not in Class I.

Class IIB. Award no degrees beyond the bachelor's.

Class III. Two-year colleges.

Discuss the design of a sample of faculty from colleges in your state, with total sample size about 200.

5.79 FOOD FOR CHICKS New varieties of corn with altered amino acid content may have higher nutritional value than standard corn, which is low in the amino acid lysine. An experiment compares two new varieties, called opaque-2 and floury-2, with normal corn. The researchers mix corn-soybean meal diets using each type of corn at each of three protein levels, 12% protein, 16% protein, and 20% protein. They feed each diet to 10 one-day-old male chicks and record their weight gains after 21 days. The weight gain of the chicks is a measure of the nutritional value of their diet.

(a) What are the experimental units and the response variable in this experiment?

(b) How many factors are there? How many treatments? Use a diagram like Figure 5.2 to describe the treatments. How many experimental units does the experiment require?

(c) Use a diagram to describe a completely randomized design for this experiment. (You do not need to actually do the randomization.)

5.80 VITAMIN C FOR MARATHON RUNNERS An ultramarathon, as you might guess, is a footrace longer than the 26.2 miles of a marathon. Runners commonly develop respiratory infections after an ultramarathon. Will taking 600 milligrams of vitamin C daily reduce those infections? Researchers randomly assigned ultramarathon runners to receive either vitamin C or a placebo. Separately, they also randomly assigned these treatments to a group of nonrunners the same age as the runners. All subjects were watched for 14 days after the big race to see if infections developed.[23]

(a) What is the name for this experimental design?

(b) Use a diagram to outline the design.

(c) The report of the study said:

> *Sixty-eight percent of the runners in the placebo group reported the development of symptoms of upper respiratory tract infection after the race; this was significantly more than that reported by the vitamin C–supplemented group (33%).*

Explain to someone who knows no statistics why "significantly more" means there is good reason to think that vitamin C works.

5.81 DELIVERING THE MAIL Is the number of days a letter takes to reach another city affected by the time of day it is mailed and whether or not the zip code is used? Describe briefly the design of a two-factor experiment to investigate this question. Be sure to specify the treatments exactly and to tell how you will handle lurking variables such as the day of the week on which the letter is mailed.

5.82 McDONALD'S VERSUS WENDY'S Do consumers prefer the taste of a cheeseburger from McDonald's or from Wendy's in a blind test in which neither burger is identified? Describe briefly the design of a matched pairs experiment to investigate this question.

5.83 REPAIRING KNEES IN COMFORT Knee injuries are routinely repaired by arthroscopic surgery that does not require opening up the knee. Can we reduce patient discomfort by giving them a nonsteroidal anti-inflammatory drug (NSAID)? Eighty-three patients were placed in three groups. Group A received the NSAID both before and after the surgery. Group B was given a placebo before and the NSAID after. Group C received a placebo both before and after surgery. The patients recorded a pain score by answering questions one day after the surgery.[24]

(a) Outline the design of this experiment. You do not need to do the randomization that your design requires.

(b) You read that "the patients, physicians and physical therapists were blinded" during the study. What does this mean?

(c) You also read that "the pain scores for Group A were significantly lower than Group C but not significantly lower than Group B." What does this mean? What does this finding lead you to conclude about the use of NSAIDs?

5.84 A SPINNER GAME OF CHANCE A game of chance is based on spinning a 1–10 spinner like the one shown in the illustration two times in succession. The player wins if the larger of the two numbers is greater than 5.

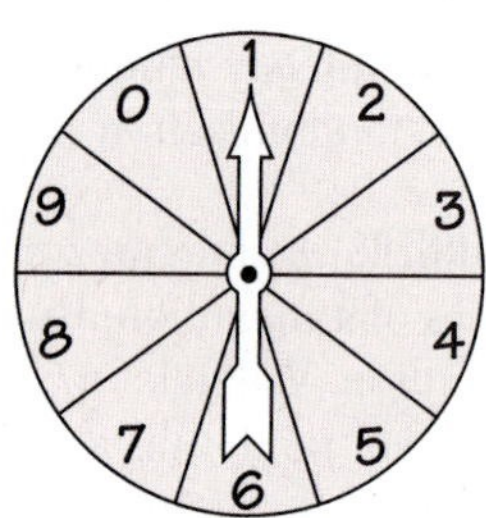

(a) What constitutes a single run of this experiment? What are the possible outcomes resulting in win or lose?

(b) Describe a correspondence between random digits from a random number table and outcomes in the game.

(c) Describe a technique using the `randInt` command on the TI-83/89 to simulate the result of a single run of the experiment.

(d) Use either the random number table or your calculator to simulate 20 trials. Report the proportion of times you win the game. Then combine your results with those of other students to obtain results for a large number of trials.

5.85 GAUGING THE DEMAND FOR CHEESECAKE The owner of a bakery knows that the daily demand for a highly perishable cheesecake is as follows:

Number/day:	0	1	2	3	4	5
Relative frequency:	0.05	0.15	0.25	0.25	0.20	0.10

(a) Use simulation to find the demand for the cheesecake on 30 consecutive business days.

(b) Suppose that it cost the baker \$5 to produce a cheesecake, and that the unused cheesecakes must be discarded at the end of the business day. Suppose also that the selling price of a cheesecake is \$13. Use simulation to estimate the number of cheesecakes that he should produce each day in order to maximize his profit.

5.86 HOT STREAKS IN FOUL SHOOTING Joey is interested in investigating so-called hot streaks in foul shooting among basketball players. He's a fan of Carla, who has been making approximately 80% of her free throws. Specifically, Joey wants to use simulation methods to determine Carla's longest *run* of baskets on average, for 20 consecutive free throws.

(a) Describe a correspondence between random numbers and outcomes.

(b) What will constitute one repetition in this simulation? Carry out 20 repetitions and record the longest run for each repetition. Combine your results with those of 4 other students to obtain at least 100 replications.

(c) What is the mean run length? Are you surprised? Determine the five-number summary for the data.

(d) Construct a histogram of the results.

5.87 SELF-PACED LEARNING, I Elaine is enrolled in a self-paced course that allows three attempts to pass an examination on the material. She does not study and has 2 out of 10 chances of passing on any one attempt by luck. What is Elaine's likelihood of passing on at least one of the three attempts? (Assume the attempts are independent because she takes a different examination on each attempt.)

(a) Explain how you would use random digits to simulate one attempt at the exam. Elaine will of course stop taking the exam as soon as she passes.

(b) Simulate 50 repetitions. What is your estimate of Elaine's likelihood of passing the course?

(c) Do you think the assumption that Elaine's likelihood of passing the exam is the same on each trial is realistic? Why?

5.88 SELF-PACED LEARNING, II A more realistic model for Elaine's attempts to pass an exam in the previous exercise is as follows: On the first try she has probability 0.2 of passing. If she fails on the first try, her probability on the second try increases to 0.3 because she learned something from her first attempt. If she fails on two attempts, the probability of passing on a third attempt is 0.4. She will stop as soon as she passes. The course rules force her to stop after three attempts in any case.

(a) Explain how to simulate one repetition of Elaine's tries at the exam. Notice that she has different probabilities of passing on each successive try.

(b) Simulate 50 repetitions and estimate the probability that Elaine eventually passes the exam.

NOTES AND DATA SOURCES

1. Reported by D. Horvitz in his contribution to "Pseudo–opinion polls: SLOP or useful data?" *Chance*, 8, No. 2 (1995), pp. 16–25.
2. Based in part on Randall Rothenberger, "The trouble with mall interviewing," *New York Times*, August 16, 1989.
3. K. J. Mukamal et al., "Prior alcohol consumption and mortality following acute myocardial infarction," *Journal of the American Medical Association*, 285 (2001), pp. 1965–1970.
4. L. E. Moses and F. Mosteller, "Safety of anesthetics," in J. M. Tanur et al. (eds.), *Statistics: A Guide to the Unknown*, 3rd ed., Wadsworth, 1989, pp. 15–24.
5. The information in this example is taken from *The ASCAP Survey and Your Royalties*, ASCAP, New York, undated.
6. The most recent account of the design of the CPS is Bureau of Labor Statistics, *Design and Methodology*, Current Population Survey Technical Paper 63, March 2000 (available in print or online at www.bls.census.gov/cps/tp/tp63.htm). The account here omits many complications, such as the need to separately sample "group quarters" like college dormitories.

7. For more detail on the material of this section and complete references, see P. E. Converse and M. W. Traugott, "Assessing the accuracy of polls and surveys," *Science*, 234 (1986), pp. 1094–1098.
8. The estimates of the census undercount come from Howard Hogan, "The 1990 post-enumeration survey: operations and results," *Journal of the American Statistical Association*, 88 (1993), pp. 1047–1060. The information about nonresponse appears in Eugene P. Eriksen and Teresa K. DeFonso, "Beyond the net undercount: how to measure census error, *Chance*, 6, No. 4 (1993), pp. 38–43 and 14.
9. For more detail on the limits of memory in surveys, see N. M. Bradburn, L. J. Rips, and S. K. Shevell, "Answering autobiographical questions: the impact of memory and inference on surveys," *Science*, 236 (1987), pp. 157–161.
10. Cynthia Crossen, "Margin of error: studies galore support products and positions, but are they reliable?" *Wall Street Journal*, November 14, 1991.
11. M. R. Kagay, "Poll on doubt of Holocaust is corrected," *New York Times*, July 8, 1994.
12. Giuliana Coccia, "An overview of non-response in Italian telephone surveys," *Proceedings of the 99th Session of the International Statistics Institute*, 1993, Book 3, pp. 271–272.
13. From the *New York Times* of August 21, 1989.
14. Steering Committee of the Physicians' Health Study Research Group, "Final report on the aspirin component of the ongoing Physicians' Health Study," *New England Journal of Medicine*, 321 (1989), pp. 129–135.
15. L. L. Miao, "Gastric freezing: an example of the evaluation of medical therapy by randomized clinical trials," in J. P. Bunker, B. A. Barnes, and F. Mosteller (eds.), *Costs, Risks, and Benefits of Surgery*, Oxford University Press, New York, 1977, pp. 198–211.
16. Based on Christopher Anderson, "Measuring what works in health care," *Science*, 263 (1994), pp. 1080–1082.
17. K. Wang, Y. Li, and J. Erickson, "A new look at the Monday effect," *Journal of Finance*, 52 (1997), pp. 2171–2186.
18. Based on Evan H. DeLucia et al., "Net primary production of a forest ecosystem with experimental CO_2 enhancement," *Science*, 284 (1999), pp. 1177–1179. The investigators used the block design.
19. R. C. Shelton et al., "Effectiveness of St.-John's-wort in major depression," *Journal of the American Medical Association*, 285 (2001), pp. 1978–1986.
20. Based on the Electronic Encyclopedia of Statistical Examples and Exercises (EESEE) story "Surgery in a Blanket," found on the TPS Web site www.whfreeman.com/tps.
21. The study is described in G. Kolata, "New study finds vitamins are not cancer preventers," *New York Times*, July 21, 1994. Look in the *Journal of the American Medical Association* of the same date for the details.
22. Information from Warren McIsaac and Vivek Goel, "Is access to physician services in Ontario equitable?" Institute for Clinical Evaluative Sciences in Ontario, October 18, 1993.
23. E. M. Peters et al., "Vitamin C supplementation reduces the incidence of post-race symptoms of upper-respiratory tract infection in ultramarathon runners," *American Journal of Clinical Nutrition*, 57 (1993), pp. 170–174.
24. This exercise is based on the EESEE story "Blinded Knee Doctors." The study was reported in W. E. Nelson, R. C. Henderson, L. C. Almekinders, R. A. DeMasi, and T. N. Taft, "An evaluation of pre- and postoperative nonsteroidal antiinflammatory drugs in patients undergoing knee arthroscopy," *Journal of Sports Medicine*, 21 (1994), pp. 510–516.

PART III

Probability:
Foundations for Inference

The Granger Collection, New York

A. N. KOLMOGOROV

General Laws of Probability

There are national styles in science as well as in cuisine. Statistics, the science of data, was created mainly by British and Americans. Probability, the mathematics of chance, was long led by French and Russians. *Andrei Nikolaevich Kolmogorov (1903–1987)* was the greatest of the Russian probabilists and one of the most influential mathematicians of the twentieth century. His more than 500 mathematical publications shaped several areas of modern mathematics and applied mathematical ideas to areas as far afield as the rhythms and meters of poetry.

Kolmogorov entered Moscow State University as a student in 1920 and remained there until his death. He was named a Hero of Socialist Labor in 1963, a rare honor for someone whose career was devoted entirely to scholarship.

Kolmogorov's first work in probability concerned the behavior of strings of random observations. The law of large numbers is the starting point for these studies, and Kolmogorov discovered many extensions of that law. Kolmogorov effectively established probability as a field of mathematics in 1933, when he placed it on a firm mathematical foundation by starting with a few general laws from which all else follows. The general laws of probability in this chapter are in the spirit of Kolmogorov.

Statistics, the science of data, was created mainly by British and Americans. Probability, the mathematics of chance, was long led by French and Russians.

chapter 6

Probability: The Study of Randomness

ACTIVITY 6 The Spinning Wheel

Materials: Margarine tub spinner or graphing calculator or table of random numbers

Imagine a spinner with three sectors, all the same size, marked 1, 2, and 3 as shown.

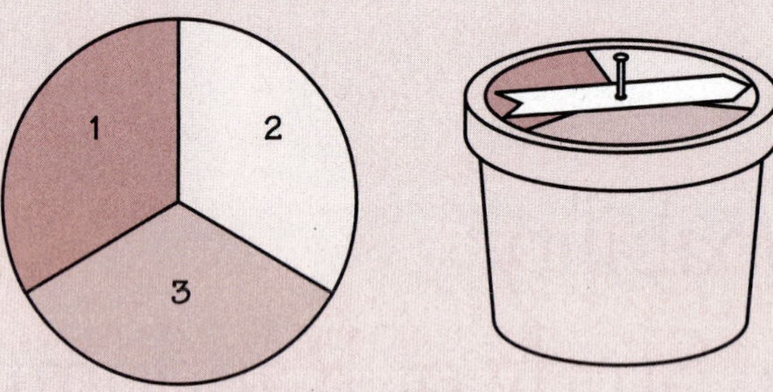

The experiment consists of spinning the spinner three times and recording the numbers as they occur (e.g., 123). We want to determine the proportion of times that *at least one digit occurs in its correct position*. For example, in the number 123, all of the digits are in their proper positions, but in the number 331, none are. For this activity, use a spinner like the one in the illustration, a table of random digits, or your calculator.

1. Guess the proportion of times at least one digit will occur in its proper place.

2. To use your calculator to randomly generate the three-digit number, enter the command `randInt(1,3,3)`. Continue to press ENTER to generate more three-digit numbers. Use a tally mark to record the results in a table like the one below. Do 20 trials and then calculate the relative frequency for the event "at least one digit in the correct position."

At least one digit in the correct position	
Not	

To use a random number table, select a row, and discarding digits 4 to 9 and 0, record digits in the 1 to 3 range in groups of three.

ACTIVITY 6 The Spinning Wheel (*continued*)

3. Combine your results with those of your classmates to obtain as many trials as possible (at least 100 randomly generated three-digit numbers; 200 would be better).

4. Count the number of times at least one digit occurred in its correct position, and calculate the proportion.

5. The program SPIN123 implements the experiment for the TI-83/89. The key step uses the calculator's Boolean logic to count the number of "hits." Enter the program or link it from a classmate or your teacher.

TI-83

```
PROGRAM:SPIN123
:ClrHome
:ClrList L1,L2
:Disp "HOW MANY TRIALS"
:Prompt N
:1→C
:While C≤N
:randInt(1,3,3)→L1
:(L1(1)=1  or  L1(2)=2  or
 L1(3)=3)→L2(C)
:1+C→C
:End
:Disp "REL FREQ="
:Disp sum(L2=1)/N
```

TI-89

```
spin123()
Prgm
ClrHome
tistat.clrlist(list1,
list2)
Disp "how many trials"
Prompt n
1→c
While c≤n
tistat.randint(1,3,3)→
 list1
list1[1]=1 or list1[2]=2
 or list1[3]=3→list2[c]
1+c→c
EndWhile
Disp "rel freq="
0→s
For i,1,n
If list2[i]=true
s+1→s
EndFor
Disp s/n
```

Execute the program for 25, 50, and 100 repetitions. Compare the calculator results with the results you obtained in steps 2 to 4.

Later in the chapter we will calculate the theoretical probability of this event happening, so keep your data at hand so that you can compare the theoretical probability with your experimental results.

INTRODUCTION

Chance is all around us. Sometimes chance results from human design, as in the casino's games of chance and the statistician's random samples. Sometimes nature uses chance, as in choosing the sex of a child. Sometimes the reasons for chance behavior are mysterious, as when the number of deaths each year in a large population is as regular as the number of heads in many tosses of a coin. Probability is the branch of mathematics that describes the pattern of chance outcomes.

The reasoning of statistical inference rests on asking, "How often would this method give a correct answer if I used it very many times?" When we produce data by random sampling or randomized comparative experiments, the laws of probability answer the question "What would happen if we did this many times?" This chapter presents the fundamental concepts of probability. Probability calculations are the basis for inference. The tools you acquire in this chapter will help you describe the behavior of statistics from random samples and randomized comparative experiments in later chapters. Even our brief acquaintance with probability will enable us to answer questions like these:

- If we know the blood types of a man and a woman, what can we say about the blood types of their future children?
- Give a test for the AIDS virus to the employees of a small company. What is the chance of at least one positive test if all the people tested are free of the virus?
- An opinion poll asks a sample of 1500 adults what they consider the most serious problem facing our schools. How often will the poll percent who answer "drugs" come within two percentage points of the truth about the entire population?

6.1 THE IDEA OF PROBABILITY

The mathematics of probability begins with the observed fact that some phenomena are random—that is, the relative frequencies of their outcomes seem to settle down to fixed values in the long run. Consider tossing a single coin. The relative frequency of heads is quite erratic in 2 or 5 or 10 tosses. But after several thousand tosses it remains stable, changing very little over further thousands of tosses. The big idea is this: **chance behavior is unpredictable in the short run but has a regular and predictable pattern in the long run.**

Toss a coin, or choose an SRS. The result can't be predicted in advance, because the result will vary when you toss the coin or choose the sample repeatedly. But there is still a regular pattern in the results, a pattern that emerges clearly only after many repetitions. This remarkable fact is the basis for the idea of probability.

EXAMPLE 6.1 COIN TOSSING

When you toss a coin, there are only two possible outcomes, heads or tails. Figure 6.1 shows the results of tossing a coin 1000 times. For each number of tosses from 1 to 1000, we have plotted the proportion of those tosses that gave a head. The first toss was a head, so the proportion of heads starts at 1. The second toss was a tail, reducing the proportion of heads to 0.5 after two tosses. The next three tosses gave a tail followed by two heads, so the proportion of heads after five tosses is 3/5, or 0.6.

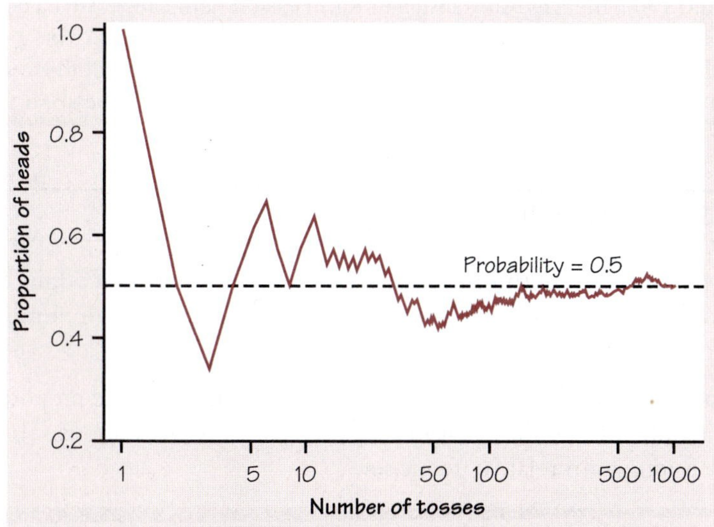

FIGURE 6.1 The behavior of the proportion of coin tosses that give a head, from 1 to 1000 tosses of a coin. In the long run, the proportion of heads approaches 0.5, the probability of a head.

The proportion of tosses that produce heads is quite variable at first, but it settles down as we make more and more tosses. Eventually this proportion gets close to 0.5 and stays there. We say that 0.5 is the *probability* of a head. The probability 0.5 appears as a horizontal line on the graph.

"Random" in statistics is not a synonym for "haphazard" but a description of a kind of order that emerges only in the long run. We often encounter the unpredictable side of randomness in our everyday experience, but we rarely see enough repetitions of the same random phenomenon to observe the long-term regularity that probability describes. You can see that regularity emerging in Figure 6.1. In the very long run, the proportion of tosses that give a head is 0.5. This is the intuitive idea of probability. Probability 0.5 means "occurs half the time in a very large number of trials."

We might suspect that a coin has probability 0.5 of coming up heads just because the coin has two sides. As Exercise 6.1 illustrates, such suspicions are not always correct. The idea of probability is empirical. That is, it is based on observation rather than theorizing. Probability describes what happens in very many trials, and we must actually observe many trials to pin down a probability. In the case of tossing a coin, some diligent people have in fact made thousands of tosses.

EXAMPLE 6.2 SOME COIN TOSSERS

The French naturalist Count Buffon (1707–1788) tossed a coin 4040 times. Result: 2048 heads, or proportion 2048/4040 = 0.5069 for heads.

Around 1900, the English statistician Karl Pearson heroically tossed a coin 24,000 times. Result: 12,012 heads, a proportion of 0.5005.

While imprisoned by the Germans during World War II, the South African mathematician John Kerrich tossed a coin 10,000 times. Result: 5067 heads, a proportion of 0.5067.

RANDOMNESS AND PROBABILITY

We call a phenomenon **random** if individual outcomes are uncertain but there is nonetheless a regular distribution of outcomes in a large number of repetitions.

The **probability** of any outcome of a random phenomenon is the proportion of times the outcome would occur in a very long series of repetitions. That is, probability is long-term relative frequency.

Thinking about randomness

That some things are random is an observed fact about the world. The outcome of a coin toss, the time between emissions of particles by a radioactive source, and the sexes of the next litter of lab rats are all random. So is the outcome of a random sample or a randomized experiment. Probability theory is the branch of mathematics that describes random behavior. Of course, we can never observe a probability exactly. We could always continue tossing the coin, for example. Mathematical probability is an idealization based on imagining what would happen in an indefinitely long series of trials.

The best way to understand randomness is to observe random behavior—not only the long-run regularity but the unpredictable results of short runs. You can do this with physical devices, as in Exercises 6.1, 6.2, 6.6, and 6.7, but computer simulations (imitations) of random behavior allow faster exploration. Exercises 6.3 and 6.10 suggest some simulations of random behavior. As you explore randomness, remember:

independence

- You must have a long series of ***independent*** trials. That is, the outcome of one trial must not influence the outcome of any other. Imagine a crooked gambling house where the operator of a roulette wheel can stop it where she chooses—she can prevent the proportion of "red" from settling down to a fixed number. These trials are not independent.

• The idea of probability is empirical. Computer simulations start with given probabilities and imitate random behavior, but we can estimate a real-world probability only by actually observing many trials.

• Nonetheless, computer simulations are very useful because we need long runs of trials. In situations such as coin tossing, the proportion of an outcome often requires several hundred trials to settle down to the probability of that outcome. The kinds of physical random devices suggested in the exercises are too slow for this. Short runs give only rough estimates of a probability.

The uses of probability

Probability theory originated in the study of games of chance. Tossing dice, dealing shuffled cards, and spinning a roulette wheel are examples of deliberate randomization that are similar to random sampling. Although games of chance are ancient, they were not studied by mathematicians until the sixteenth and seventeenth centuries. It is only a mild simplification to say that probability as a branch of mathematics arose when seventeenth-century French gamblers asked the mathematicians Blaise Pascal and Pierre de Fermat for help. Gambling is still with us, in casinos and state lotteries. We will make use of games of chance as simple examples that illustrate the principles of probability.

Careful measurements in astronomy and surveying led to further advances in probability in the eighteenth and nineteenth centuries because the results of repeated measurements are random and can be described by distributions much like those arising from random sampling. Similar distributions appear in data on human life span (mortality tables) and in data on lengths or weights in a population of skulls, leaves, or cockroaches.[1] In the twentieth century, we employ the mathematics of probability to describe the flow of traffic through a highway system, a telephone interchange, or a computer processor; the genetic makeup of individuals or populations; the energy states of subatomic particles; the spread of epidemics or rumors; and the rate of return on risky investments. Although we are interested in probability because of its usefulness in statistics, the mathematics of chance is important in many fields of study.

SECTION 6.1 EXERCISES

6.1 PENNIES SPINNING Hold a penny upright on its edge under your forefinger on a hard surface, then snap it with your other forefinger so that it spins for some time before falling. Based on 50 spins, estimate the probability of heads.

6.2 A GAME OF CHANCE In the game of Heads or Tails, Betty and Bob toss a coin four times. Betty wins a dollar from Bob for each head and pays Bob a dollar for each tail—that is, she wins or loses the difference between the number of heads and the number of tails. For example, if there are one head and three tails, Betty loses \$2. You can check that Betty's possible outcomes are

$$\{-4, -2, 0, 2, 4\}$$

Assign probabilities to these outcomes by playing the game 20 times and using the proportions of the outcomes as estimates of the probabilities. If possible, combine your trials with those of other students to obtain long-run proportions that are closer to the probabilities.

6.3 SHAQ The basketball player Shaquille O'Neal makes about half of his free throws over an entire season. We will use the calculator to simulate 100 free throws shot independently by a player who has probability 0.5 of making each shot. We let the number 1 represent the outcome "Hit" and 0 represent a "Miss."

(a) Enter the command `randInt(0,1,100)→SHAQ`. (`randInt` is found in the CATALOG under Flash Apps on the TI-89.) This tells the calculator to randomly select a hit (1) or a miss (0), do this 100 times in succession, and store the results in the list named SHAQ.

(b) What percent of the 100 shots are hits?

(c) Examine the sequence of hits and misses. How long was the longest run of shots made? Of shots missed? (Sequences of random outcomes often show runs longer than our intuition thinks likely.)

6.4 MATCHING PROBABILITIES Probability is a measure of how likely an event is to occur. Match one of the probabilities that follow with each statement about an event. (The probability is usually a much more exact measure of likelihood than is the verbal statement.)

$$0, 0.01, 0.3, 0.6, 0.99, 1$$

(a) This event is impossible. It can never occur.

(b) This event is certain. It will occur on every trial of the random phenomenon.

(c) This event is very unlikely, but it will occur once in a while in a long sequence of trials.

(d) This event will occur more often than not.

6.5 RANDOM DIGITS The table of random digits (Table B) was produced by a random mechanism that gives each digit probability 0.1 of being a 0. What proportion of the first 200 digits in the table are 0s? This proportion is an estimate, based on 200 repetitions, of the true probability, which in this case is known to be 0.1.

6.6 HOW MANY TOSSES TO GET A HEAD? When we toss a penny, experience shows that the probability (long-term proportion) of a head is close to 1/2. Suppose now that we toss the penny repeatedly until we get a head. What is the probability that the first head comes up in an odd number of tosses (1, 3, 5, and so on)? To find out, repeat this experiment 50 times, and keep a record of the number of tosses needed to get a head on each of your 50 trials.

(a) From your experiment, estimate the probability of a head on the first toss. What value should we expect this probability to have?

(b) Use your results to estimate the probability that the first head appears on an odd-numbered toss.

6.7 TOSSING A THUMBTACK Toss a thumbtack on a hard surface 100 times. How many times did it land with the point up? What is the approximate probability of landing point up?

6.8 THREE OF A KIND You read in a book on poker that the probability of being dealt three of a kind in a five-card poker hand is 1/50. Explain in simple language what this means.

6.9 WINNING A BASEBALL GAME A study of the home-field advantage in baseball found that over the period from 1969 to 1989 the league champions won 63% of their home games.[2] The two league champions meet in the baseball World Series. Would you use the study results to assign probability 0.63 to the event that the home team wins in a World Series game? Explain your answer.

6.10 SIMULATING AN OPINION POLL A recent opinion poll showed that about 73% of married women agree that their husbands do at least their fair share of household chores. Suppose that this is exactly true. Choosing a married woman at random then has probability 0.73 of getting one who agrees that her husband does his share. Use software or your calculator to simulate choosing many women independently. (In most software, the key phrase to look for is "Bernoulli trials." This is the technical term for independent trials with Yes/No outcomes. Our outcomes here are "Agree" or not.)

(a) Simulate drawing 20 women, then 80 women, then 320 women. What proportion agree in each case? We expect (but because of chance variation we can't be sure) that the proportion will be closer to 0.73 in longer runs of trials.

(b) Simulate drawing 20 women 10 times and record the percents in each trial who agree. Then simulate drawing 320 women 10 times and again record the 10 percents. Which set of 10 results is less variable? We expect the results of 320 trials to be more predictable (less variable) than the results of 20 trials. That is "long-run regularity" showing itself.

6.2 PROBABILITY MODELS

Earlier chapters gave mathematical models for linear relationships (in the form of the equation of a line) and for some distributions of data (in the form of normal density curves). Now we must give a mathematical description or model for randomness. To see how to proceed, think first about a very simple random phenomenon, tossing a coin once. When we toss a coin, we cannot know the outcome in advance. What do we know? We are willing to say that the outcome will be either heads or tails. We believe that each of these outcomes has probability 1/2. This description of coin tossing has two parts:

- A list of possible outcomes.
- A probability for each outcome.

Such a description is the basis for all probability models. Here is the basic vocabulary we use.

PROBABILITY MODELS

The **sample space S** of a random phenomenon is the set of all possible outcomes.

An **event** is any outcome or a set of outcomes of a random phenomenon. That is, an event is a subset of the sample space.

A **probability model** is a mathematical description of a random phenomenon consisting of two parts: a sample space S and a way of assigning probabilities to events.

The sample space S can be very simple or very complex. When we toss a coin once, there are only two outcomes, heads and tails. The sample space is S = {H, T}. If we draw a random sample of 50,000 U.S. households, as the Current Population Survey does, the sample space contains all possible choices of 50,000 of the 103 million households in the country. This S is extremely large. Each member of S is a possible sample, which explains the term *sample space*.

EXAMPLE 6.3 ROLLING DICE

Rolling two dice is a common way to lose money in casinos. There are 36 possible outcomes when we roll two dice and record the up-faces in order (first die, second die). Figure 6.2 displays these outcomes. They make up the sample space S.

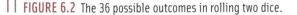

FIGURE 6.2 The 36 possible outcomes in rolling two dice.

"Roll a 5" is an event, call it A, that contains four of these 36 outcomes:

A = { ⚀⚃ ⚁⚂ ⚂⚁ ⚃⚀ }

Gamblers care only about the number of pips on the up-faces of the dice. The sample space for rolling two dice and counting the pips is

S = {2, 3, 4, 5, 6, 7, 8, 9, 10, 11, 12}

Comparing this S with Figure 6.2 reminds us that we can change S by changing the detailed description of the random phenomenon we are describing.

The name "sample space" is natural in random sampling, where each possible outcome is a sample and the sample space contains all possible samples.

To specify S, we must state what constitutes an individual outcome and then state which outcomes can occur. We often have some freedom in defining the sample space, so the choice of S is a matter of convenience as well as correctness. The idea of a sample space, and the freedom we may have in specifying it, are best illustrated by examples.

EXAMPLE 6.4 RANDOM DIGIT

Let your pencil point fall blindly into Table B of random digits; record the value of the digit it lands on. The possible outcomes are

S = {0, 1, 2, 3, 4, 5, 6, 7, 8, 9}

EXAMPLE 6.5 FLIP A COIN AND ROLL A DIE

An experiment consists of flipping a coin and rolling a die. Possible outcomes are a head (H) followed by any of the digits 1 to 6, or a tail (T) followed by any of the digits 1 to 6. The sample space contains 12 outcomes:

S = {H1, H2, H3, H4, H5, H6, T1, T2, T3, T4, T5, T6}

tree diagram

Being able to properly enumerate the outcomes in a sample space will be critical to determining probabilities. Two techniques are very helpful in making sure you don't accidentally overlook any outcomes. The first is called a ***tree diagram*** because it resembles the branches of a tree. The first action in Example 6.5 is to toss a coin. To construct the tree diagram, begin with a point and draw a line from the point to H and a second line from the point to T. The second action is to roll a die; there are six possible faces that can come up on the die. So draw a line from each of H and T to these six outcomes. See Figure 6.3.

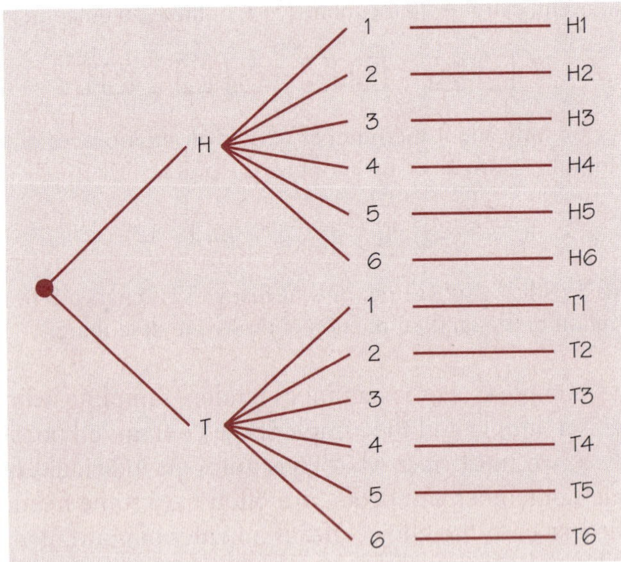

FIGURE 6.3 Tree diagram.

The second technique is to make use of the following rule.

MULTIPLICATION PRINCIPLE

If you can do one task in a number of ways and a second task in b number of ways, then both tasks can be done in $a \times b$ number of ways.

To determine the number of outcomes in the sample space for Example 6.5, there are 2 ways the coin can come up, and there are 6 ways the die can come up, so there are 2×6 possible outcomes in the sample space. To see why this is true, just sketch a tree diagram.

EXAMPLE 6.6 FLIP FOUR COINS

An experiment consists of flipping four coins. You can think of either tossing four coins onto the table all at once or flipping a coin four times in succession and recording the four outcomes. One possible outcome is HHTH. Because there are two ways each coin can come up, the multiplication principle says that the total number of outcomes is $2 \times 2 \times 2 \times 2 = 16$. This is the easy part. Listing all 16 outcomes requires a scheme or systematic method so that you don't leave out any possibilities. One way is to list all the ways you can obtain 0 heads, then list all the ways you can get 1 head, 2 heads, 3 heads, and finally all 4 heads. Here is an enumeration:

0 heads	1 head	2 heads	3 heads	4 heads
TTTT	HTTT	HHTT	HHHT	HHHH
	THTT	HTHT	HHTH	
	TTHT	HTTH	HTHH	
	TTTH	THHT	THHH	
		THTH		
		TTHH		

Suppose that our only interest is the number of heads in four tosses. Now we can be exact in a simpler fashion. The random phenomenon is to toss a coin four times and count the number of heads. The sample space contains only five outcomes:

$$S = \{0, 1, 2, 3, 4\}$$

This example also illustrates the importance of carefully specifying what constitutes an individual outcome.

Although these examples seem remote from the practice of statistics, the connection is surprisingly close. Suppose that in the course of conducting an opinion poll you select four people at random from a large population and ask each if he or she favors reducing federal spending on low-interest student loans. The possible outcomes—the sample space—are the answers "Yes" or "No." Similarly, the possible outcomes of an SRS of 1500 people are the same in principle as the possible outcomes of tossing a coin 1500 times. One of the great advantages of mathematics is that the essential features of quite different phenomena can be described by the same mathematical model.

Of course, some sample spaces are simply too large to allow all of the possible outcomes to be listed, as the next example shows.

EXAMPLE 6.7 GENERATE A RANDOM DECIMAL NUMBER

Many computing systems have a function that will generate a random number between 0 and 1. The sample space is

$$S = \{\text{all numbers between 0 and 1}\}$$

This S is a mathematical idealization. Any specific random number generator produces numbers with some limited number of decimal places so that, strictly speaking, not all numbers between 0 and 1 are possible outcomes. The entire interval from 0 to 1 is easier to think about. It also has the advantage of being a suitable sample space for different computers that produce random numbers with different numbers of significant digits.

If you are selecting objects from a collection of distinct choices, such as drawing playing cards from a standard deck of 52 cards, then much depends on whether each choice is exactly like the previous choice. If you are selecting random digits by drawing numbered slips of paper from a hat, and you want all ten digits to be equally likely to be selected each draw, then after you draw a digit and record it, you must put it back into the hat. Then the second draw will be exactly like the first. This is referred to as sampling ***with replacement.*** If you do not replace the slips you draw, however, there are only nine choices for the second slip picked, and eight for the third. This is called sampling ***without replacement***. So if the question is "How many three-digit numbers can you make?" the answer is, by the multiplication principle, $10 \times 10 \times 10 = 1000$, providing all ten numbers are eligible for each of the three positions in the number. On the other had, there are $10 \times 9 \times 8 = 720$ different ways to construct a three-digit number *without replacement*. You should be able to determine from the context of the problem whether the selection is with or without replacement, and this will help you properly identify the sample space.

replacement

EXERCISES

6.11 DESCRIBE THE SAMPLE SPACE In each of the following situations, describe a sample space S for the random phenomenon. In some cases, you have some freedom in your choice of S.

(a) A seed is planted in the ground. It either germinates or fails to grow.

(b) A patient with a usually fatal form of cancer is given a new treatment. The response variable is the length of time that the patient lives after treatment.

(c) A student enrolls in a statistics course and at the end of the semester receives a letter grade.

(d) A basketball player shoots four free throws. You record the sequence of hits and misses.

(e) A basketball player shoots four free throws. You record the number of baskets she makes.

6.12 DESCRIBE THE SAMPLE SPACE In each of the following situations, describe a sample space S for the random phenomenon. In some cases you have some freedom in specifying S, especially in setting the largest and the smallest value in S.

(a) Choose a student in your class at random. Ask how much time that student spent studying during the past 24 hours.

(b) The Physicians' Health Study asked 11,000 physicians to take an aspirin every other day and observed how many of them had a heart attack in a five-year period.

(c) In a test of a new package design, you drop a carton of a dozen eggs from a height of 1 foot and count the number of broken eggs.

(d) Choose a student in your class at random. Ask how much cash that student is carrying.

(e) A nutrition researcher feeds a new diet to a young male white rat. The response variable is the weight (in grams) that the rat gains in 8 weeks.

6.13 CALORIES IN HOT DOGS Give a reasonable sample space for the number of calories in a hot dog. (Table 1.10 on page 59 contains some typical values to guide you.)

6.14 LISTING OUTCOMES, I For each of the following, use a tree diagram or the multiplication principle to determine the number of outcomes in the sample space. Then write the sample space using set notation.

(a) Toss 2 coins.

(b) Toss 3 coins.

(c) Toss 4 coins.

6.15 LISTING OUTCOMES, II For each of the following, use a tree diagram or the multiplication principle to determine the number of outcomes in the sample space.

(a) Suppose a county license tag has a four-digit number for identification. If any digit can occupy any of the four positions, how many county license tags can you have?

(b) If the county license tags described in (a) do not allow duplicate digits, how many county license tags can you have?

(c) Suppose the county license tags described in (a) can have *up* to four digits. How many county license tags will this scheme allow?

6.16 SPIN 123 Refer to the experiment described in Activity 6.

(a) Determine the number of outcomes in the sample space.

(b) List the outcomes in the sample space.

6.17 ROLLING TWO DICE Example 6.3 (page 336) showed the 36 outcomes when we roll two dice. Another way to summarize these results is to make a table like this:

Number of ways	Sum	Outcomes
1	2	1,1
2	3	1,2 2,1
...	...	...

(a) Complete the table.

(b) In how many ways can you get an even sum?

(c) In how many ways can you get a sum of 5? A sum of 8?

(d) Describe any patterns that you see in the table.

6.18 PICK A CARD Suppose you select a card from a standard deck of 52 playing cards. In how many ways can the selected card be

(a) a red card?

(b) a heart?

(c) a queen and a heart?

(d) a queen or a heart?

(e) a queen that is not a heart?

Probability rules

The true probability of any outcome—say, "roll a 5 when we toss two dice"—can be found only by actually tossing two dice many times, and then only approximately. How then can we describe probability mathematically? Rather than try to give "correct" probabilities, we start by laying down facts that must be true for any assignment of probabilities. These facts follow from the idea of probability as "the long-run proportion of repetitions on which an event occurs."

1. **Any probability is a number between 0 and 1.** Any proportion is a number between 0 and 1, so any probability is also a number between 0 and 1. An event with probability 0 never occurs, and an event with probability 1 occurs on every trial. An event with probability 0.5 occurs in half the trials in the long run.

2. **All possible outcomes together must have probability 1.** Because some outcome must occur on every trial, the sum of the probabilities for all possible outcomes must be exactly 1.

3. **The probability that an event does not occur is 1 minus the probability that the event does occur.** If an event occurs in (say) 70% of all trials, it fails to occur in the other 30%. The probability that an event occurs and the probability that it does not occur always add to 100%, or 1.

4. **If two events have no outcomes in common, the probability that one or the other occurs is the sum of their individual probabilities.** If one event occurs in 40% of all trials, a different event occurs in 25% of all trials, and the two can never occur together, then one or the other occurs on 65% of all trials because 40% + 25% = 65%.

We can use mathematical notation to state Facts 1 to 4 more concisely. Capital letters near the beginning of the alphabet denote events. If A is any event, we write its probability as $P(A)$. Here are our probability facts in formal language. As you apply these rules, remember that they are just another form of intuitively true facts about long-run proportions.

PROBABILITY RULES

Rule 1. The probability $P(A)$ of any event A satisfies $0 \leq P(A) \leq 1$.

Rule 2. If S is the sample space in a probability model, then $P(S) = 1$.

Rule 3. The **complement** of any event A is the event that A does not occur, written as A^c. The **complement rule** states that

$$P(A^c) = 1 - P(A)$$

Rule 4. Two events A and B are **disjoint** (also called mutually exclusive) if they have no outcomes in common and so can never occur simultaneously. If A and B are disjoint,

$$P(A \text{ or } B) = P(A) + P(B)$$

This is the **addition rule** for disjoint events.

Sometime we use set notation to describe events. The event $\{A \cup B\}$, read "A ***union*** B," is the set of all outcomes that are either in A or in B. So $\{A \cup B\}$ is just another way to indicate the event {A or B}. We will use these two notations interchangeably. The symbol $\varnothing$ is used for the ***empty event***, that is, the event that has no outcomes in it. If two events A and B are disjoint (mutually exclusive), we can write $A \cap B = \varnothing$, read "A ***intersect*** B is empty." Sometimes we emphasize that we are describing a compound event by enclosing it within braces.

union

empty event

intersect

You may find it helpful to draw a picture to remind yourself of the meaning of complements and disjoint events. A picture like Figure 6.4 that shows the sample space S as a rectangular area and events as areas within S is called a ***Venn diagram***. The events A and B in Figure 6.4 are disjoint because they do not overlap; that is, they have no outcomes in common. Their intersection is the empty event, $\varnothing$. Their union consists of the two shaded regions.

Venn diagram

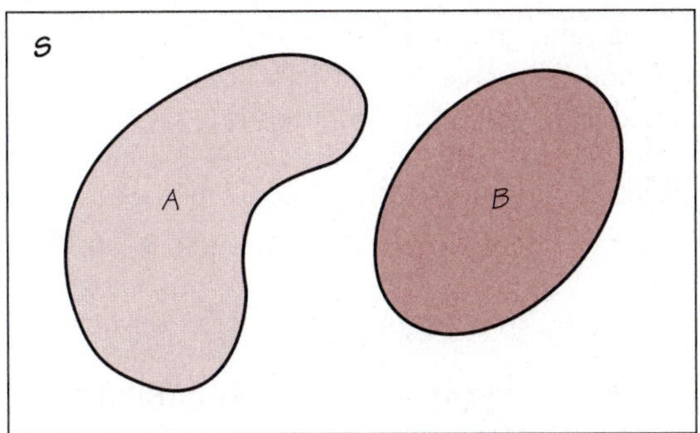

FIGURE 6.4 Venn diagram showing disjoint (mutually exclusive) events *A* and *B*.

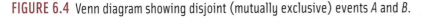

The complement A^c in Figure 6.5 contains exactly the outcomes that are not in A. Note that we could write $A \cup A^c = S$ and $A \cap A^c = \varnothing$.

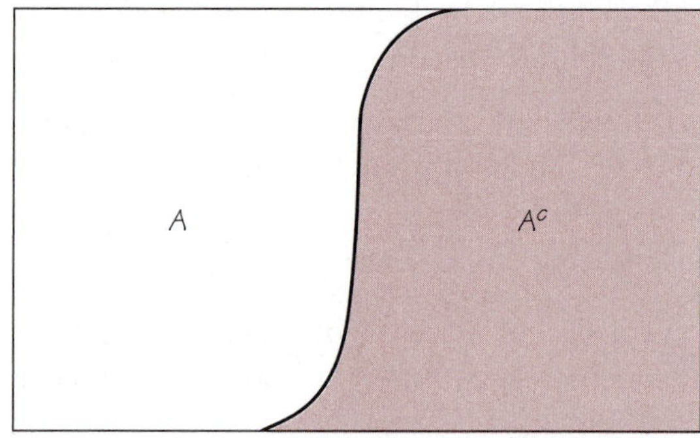

FIGURE 6.5 Venn diagram showing the complement A^c of an event A.

EXAMPLE 6.8 MARITAL STATUS OF YOUNG WOMEN

Draw a woman aged 25 to 34 years old at random and record her marital status. "At random" means that we give every such woman the same chance to be the one we choose. That is, we choose an SRS of size 1. The probability of any marital status is just the proportion of all women aged 25 to 34 who have that status—if we drew many women, this is the proportion we would get. Here is the probability model:

Marital status:	Never married	Married	Widowed	Divorced
Probability:	0.298	0.622	0.005	0.075

Each probability is between 0 and 1. The probabilities add to 1 because these outcomes together make up the sample space S.

The probability that the woman we draw is not married is, by the complement rule,

$$\begin{aligned} P(\text{not married}) &= 1 - P(\text{married}) \\ &= 1 - 0.622 = 0.378 \end{aligned}$$

That is, if 62.2% are married, then the remaining 37.8% are not married.

"Never married" and "divorced" are disjoint events, because no woman can be both never married and divorced. So the addition rule says that

$$\begin{aligned} P(\text{never married or divorced}) &= P(\text{never married}) + P(\text{divorced}) \\ &= 0.298 + 0.075 = 0.373 \end{aligned}$$

That is, 37.3% of women in this age group are either never married or divorced.

EXAMPLE 6.9 PROBABILITIES FOR ROLLING DICE

Figure 6.2 (page 336) displays the 36 possible outcomes of rolling two dice. What probabilities should we assign to these outcomes?

Casino dice are carefully made. Their spots are not hollowed out, which would give the faces different weights, but are filled with white plastic of the same density as the colored plastic of the body. For casino dice it is reasonable to assign the same probability to each of the 36 outcomes in Figure 6.2. Because all 36 outcomes together must have probability 1 (Rule 2), each outcome must have probability 1/36.

Gamblers are often interested in the sum of the pips on the up-faces. What is the probability of rolling a 5? Because the event "roll a 5" contains the four outcomes displayed in Example 6.3, the addition rule (Rule 4) says that its probability is

$$\begin{aligned} P(\text{roll a 5}) &= P(\text{⚀ ⚃}) + P(\text{⚂ ⚁}) + P(\text{⚁ ⚂}) + P(\text{⚃ ⚀}) \\ &= \frac{1}{36} + \frac{1}{36} + \frac{1}{36} + \frac{1}{36} \\ &= \frac{4}{36} = 0.111 \end{aligned}$$

What about the probability of rolling a 7? In Figure 6.2 you will find six outcomes for which the sum of the pips is 7. The probability is 6/36, or about 0.167.

Assigning probabilities: finite number of outcomes

Examples 6.8 and 6.9 illustrate one way to assign probabilities to events: assign a probability to every individual outcome, then add these probabilities to find the probability of any event. If such an assignment is to satisfy the rules of probability, the probabilities of all the individual outcomes must sum to exactly 1.

PROBABILITIES IN A FINITE SAMPLE SPACE

Assign a probability to each individual outcome. These probabilities must be numbers between 0 and 1 and must have sum 1.

The probability of any event is the sum of the probabilities of the outcomes making up the event.

EXAMPLE 6.10 BENFORD'S LAW

Faked numbers in tax returns, payment records, invoices, expense account claims, and many other settings often display patterns that aren't present in legitimate records. Some patterns, like too many round numbers, are obvious and easily avoided by a clever crook. Others are more subtle. It is a striking fact that the first digits of numbers in legitimate records often follow a distribution known as *Benford's Law*. Here it is (note that a first digit can't be 0):[3]

First digit:	1	2	3	4	5	6	7	8	9
Probability:	0.301	0.176	0.125	0.097	0.079	0.067	0.058	0.051	0.046

Benford's Law usually applies to the first digits of the sizes of similar quantities, such as invoices, expense account claims, and county populations. Investigators can detect fraud by comparing these probabilities with the first digits in records such as invoices paid by a business.

Consider the events

$$A = \{\text{first digit is 1}\}$$
$$B = \{\text{first digit is 6 or greater}\}$$

From the table of probabilities,

$$\begin{aligned} P(A) &= P(1) = 0.301 \\ P(B) &= P(6) + P(7) + P(8) + P(9) \\ &= 0.067 + 0.058 + 0.051 + 0.046 = 0.222 \end{aligned}$$

Note that $P(B)$ is not the same as the probability that a random digit is greater than 6. The probability $P(6)$ that a first digit is 6 is included in "6 or greater" but not in "greater than 6."

The probability that a first digit is anything other than a 1 is, by the complement rule,

$$\begin{aligned} P(A^c) &= 1 - P(A) \\ &= 1 - 0.301 = 0.699 \end{aligned}$$

The events A and B are disjoint, so the probability that a first digit either is 1 or is 6 or greater is, by the addition rule,

$$\begin{aligned} P(A \text{ or } B) &= P(A) + P(B) \\ &= 0.301 + 0.222 = 0.523 \end{aligned}$$

Be careful to apply the addition rule only to disjoint events. Check that the probability of the event C that a first digit is odd is

$$P(C) = P(1) + P(3) + P(5) + P(7) + P(9) = 0.609$$

The probability

$$P(B \text{ or } C) = P(1) + P(3) + P(5) + P(6) + P(7) + P(8) + P(9) = 0.727$$

is *not* the sum of $P(B)$ and $P(C)$, because events B and C are not disjoint. Outcomes 7 and 9 are common to both events.

Assigning probabilities: equally likely outcomes

Assigning correct probabilities to individual outcomes often requires long observation of the random phenomenon. In some special circumstances, however, we are willing to assume that individual outcomes are equally likely because of some balance in the phenomenon. Ordinary coins have a physical balance that should make heads and tails equally likely, for example, and the table of random digits comes from a deliberate randomization.

EXAMPLE 6.11 RANDOM DIGITS

You might think that first digits are distributed "at random" among the digits 1 to 9. The 9 possible outcomes would then be equally likely. The sample space for a single digit is

$$S = \{1, 2, 3, 4, 5, 6, 7, 8, 9\}$$

Because the total probability must be 1, the probability of each of the 9 outcomes must be 1/9. That is, the assignment of probabilities to outcomes is

First digit:	1	2	3	4	5	6	7	8	9
Probability:	1/9	1/9	1/9	1/9	1/9	1/9	1/9	1/9	1/9

The probability of the event *B* that a randomly chosen first digit is 6 or greater is

$$\begin{aligned} P(B) &= P(6) + P(7) + P(8) + P(9) \\ &= \frac{1}{9} + \frac{1}{9} + \frac{1}{9} + \frac{1}{9} = \frac{4}{9} = 0.444 \end{aligned}$$

Compare this with the Benford's Law probability in Example 6.10. A crook who fakes data by using "random" digits will end up with too many first digits 6 or greater and too few 1s and 2s.

In Example 6.11 all outcomes have the same probability. Because there are 9 equally likely outcomes, each must have probability 1/9. Because exactly 4 of the 9 equally likely outcomes are 6 or greater, the probability of this event is 4/9. In the special situation where all outcomes are equally likely, we have a simple rule for assigning probabilities to events.

EQUALLY LIKELY OUTCOMES

If a random phenomenon has k possible outcomes, all equally likely, then each individual outcome has probability $1/k$. The probability of any event A is

$$\begin{aligned} P(A) &= \frac{\text{count of outcomes in A}}{\text{count of outcomes in S}} \\ &= \frac{\text{count of outcomes in A}}{k} \end{aligned}$$

Most random phenomena do not have equally likely outcomes, so the general rule for finite sample spaces is more important than the special rule for equally likely outcomes.

EXERCISES

6.19 BLOOD TYPES All human blood can be typed as one of O, A, B, or AB, but the distribution of the types varies a bit with race. Here is the distribution of the blood type of a randomly chosen black American:

Blood type:	O	A	B	AB
Probability:	0.49	0.27	0.20	?

(a) What is the probability of type AB blood? Why?

(b) Maria has type B blood. She can safely receive blood transfusions from people with blood types O and B. What is the probability that a randomly chosen black American can donate blood to Maria?

6.20 DISTRIBUTION OF M&M COLORS If you draw an M&M candy at random from a bag of the candies, the candy you draw will have one of six colors. The probability of drawing each color depends on the proportion of each color among all candies made.

(a) The table below gives the probability of each color for a randomly chosen plain M&M:

Color:	Brown	Red	Yellow	Green	Orange	Blue
Probability:	0.3	0.2	0.2	0.1	0.1	?

What must be the probability of drawing a blue candy?

(b) The probabilities for peanut M&Ms are a bit different. Here they are:

Color:	Brown	Red	Yellow	Green	Orange	Blue
Probability:	0.2	0.1	0.2	0.1	0.1	?

What is the probability that a peanut M&M chosen at random is blue?

(c) What is the probability that a plain M&M is any of red, yellow, or orange? What is the probability that a peanut M&M has one of these colors?

6.21 HEART DISEASE AND CANCER Government data assign a single cause for each death that occurs in the United States. The data show that the probability is 0.45 that a randomly chosen death was due to cardiovascular (mainly heart) disease, and 0.22 that it was due to cancer. What is the probability that a death was due either to cardiovascular disease or to cancer? What is the probability that the death was due to some other cause?

6.22 DO HUSBANDS DO THEIR SHARE? The *New York Times* (August 21, 1989) reported a poll that interviewed a random sample of 1025 women. The married women in the sample were asked whether their husbands did their fair share of household chores. Here are the results:

Outcome	Probability
Does more than his fair share	0.12
Does his fair share	0.61
Does less than his fair share	?

These proportions are probabilities for the random phenomenon of choosing a married woman at random and asking her opinion.

(a) What must be the probability that the woman chosen says that her husband does less than his fair share? Why?

(b) The event "I think my husband does at least his fair share" contains the first two outcomes. What is its probability?

6.23 ACADEMIC RANK Select a first-year college student at random and ask what his or her academic rank was in high school. Here are the probabilities, based on proportions from a large sample survey of first-year students:

Rank:	Top 20%	Second 20%	Third 20%	Fourth 20%	Lowest 20%
Probability:	0.41	0.23	0.29	0.06	0.01

(a) What is the sum of these probabilities? Why do you expect the sum to have this value?

(b) What is the probability that a randomly chosen first-year college student was not in the top 20% of his or her high school class?

(c) What is the probability that a first-year student was in the top 40% in high school?

6.24 SPIN 123 Refer to the experiment described in Activity 6 and Exercise 6.16 (page 341).

(a) Determine the theoretical probability that at least one digit will occur in its correct place.

(b) Compare the theoretical probability with your experimental (empirical) results.

6.25 TETRAHEDRAL DICE Psychologists sometimes use tetrahedral dice to study our intuition about chance behavior. A tetrahedron is a pyramid (think of Egypt) with four identical faces, each a triangle with all sides equal in length. Label the four faces of a tetrahedral die with 1, 2, 3, and 4 spots.

(a) Give a probability model for rolling such a die and recording the number of spots on the down-face. Explain why you think your model is at least close to correct.

(b) Give a probability model for rolling two such dice. That is, write down all possible outcomes and give a probability to each. What is the probability that the sum of the down-faces is 5?

6.26 BENFORD'S LAW Example 6.10 (page 345) states that the first digits of numbers in legitimate records often follow a distribution known as Benford's Law. Here is the distribution:

First digit:	1	2	3	4	5	6	7	8	9
Probability:	0.301	0.176	0.125	0.097	0.079	0.067	0.058	0.051	0.046

It was shown in Example 6.10 that

$$P(A) = P(\text{first digit is } 1) = 0.301$$
$$P(B) = P(\text{first digit is 6 or greater}) = 0.222$$
$$P(C) = P(\text{first digit is odd}) = 0.609$$

We will define event D to be {first digit is less than 4}. Using the union and intersection notation, find the following probabilities.

(a) $P(D)$

(b) $P(B \cup D)$

(c) $P(D^c)$

(d) $P(C \cap D)$

(e) $P(B \cap C)$

Independence and the multiplication rule

Rule 4, the addition rule for disjoint events, describes the probability that *one or the other* of two events A and B will occur in the special situation when A and B cannot occur together because they are disjoint. Now we will describe the probability that *both* events A and B occur, again only in a special situation. More general rules appear in Section 6.3.

Suppose that you toss a balanced coin twice. You are counting heads, so two events of interest are

$$A = \text{first toss is a head}$$
$$B = \text{second toss is a head}$$

The events A and B are not disjoint. They occur together whenever both tosses give heads. We want to compute the probability of the event {A and B} that *both* tosses are heads. The Venn diagram in Figure 6.6 illustrates the event {A and B} as the overlapping area that is common to both A and B.

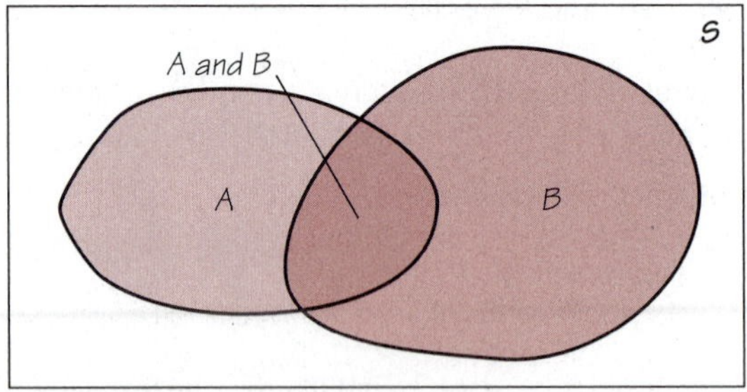

FIGURE 6.6 Venn diagram showing the event {*A* and *B*}.

The coin tossing of Buffon, Pearson, and Kerrich described at the beginning of this chapter makes us willing to assign probability 1/2 to a head when we toss a coin. So

$$P(A) = 0.5$$
$$P(B) = 0.5$$

What is $P(A \text{ and } B)$? Our common sense says that it is 1/4. The first toss will give a head half the time and then the second will give a head on half of those trials, so both tosses will give heads on $1/2 \times 1/2 = 1/4$ of all trials in the long run. This reasoning assumes that the second toss still has probability 1/2 of a head after the first has given a head. This is true—we can verify it by performing many trials of two tosses and observing the proportion of heads on the second toss after the first toss has produced a head. We say that the events "head on the first toss" and "head on the second toss" are **independent.** Here is our next probability rule.

independent

THE MULTIPLICATION RULE FOR INDEPENDENT EVENTS

Rule 5. Two events A and B are **independent** if knowing that one occurs does not change the probability that the other occurs. If A and B are independent,

$$P(A \text{ and } B) = P(A)P(B)$$

This is the **multiplication rule** for independent events.

Our definition of independence is rather informal. A more precise definition appears in Section 6.3. In practice, though, we rarely need a precise definition of independence, because independence is usually *assumed* as part of a probability model when we want to describe random phenomena that seem to be physically unrelated to each other.

EXAMPLE 6.12 INDEPENDENT OR NOT INDEPENDENT?

Because a coin has no memory and most coin tossers cannot influence the fall of the coin, it is safe to assume that successive coin tosses are independent. For a balanced coin this means that after we see the outcome of the first toss, we still assign probability 1/2 to heads on the second toss.

On the other hand, the colors of successive cards dealt from the same deck are not independent. A standard 52-card deck contains 26 red and 26 black cards. For the first card dealt from a shuffled deck, the probability of a red card is $26/52 = 0.50$ because the 52 possible cards are equally likely. Once we see that the first card is red, we know that there are only 25 reds among the remaining 51 cards. The probability that the second card is red is therefore only $25/51 = 0.49$. Knowing the outcome of the first deal changes the probability for the second.

If a doctor measures your blood pressure twice, it is reasonable to assume that the two results are independent because the first result does not influence the instrument that makes the second reading. But if you take an IQ test or other mental test twice in succession, the two test scores are not independent. The learning that occurs on the first attempt influences your second attempt.

When independence is part of a probability model, the multiplication rule applies. Here is an example.

EXAMPLE 6.13 MENDEL'S PEAS

Gregor Mendel used garden peas in some of the experiments that revealed that inheritance operates randomly. The seed color of Mendel's peas can be either green or yellow. Two parent plants are "crossed" (one pollinates the other) to produce seeds. Each parent plant carries two genes for seed color, and each of these genes has probability 1/2 of being passed to a seed. The two genes that the seed receives, one from each parent, determine its color. The parents contribute their genes independently of each other.

Suppose that both parents carry the G and the Y genes. The seed will be green if both parents contribute a G gene; otherwise it will be yellow. If M is the event that the male contributes a G gene and F is the event that the female contributes a G gene, then the probability of a green seed is

$$\begin{aligned} P(M \text{ and } F) &= P(M)P(F) \\ &= (0.5)(0.5) = 0.25 \end{aligned}$$

In the long run, 1/4 of all seeds produced by crossing these plants will be green.

The multiplication rule $P(A \text{ and } B) = P(A)P(B)$ holds if A and B are independent but not otherwise. The addition rule $P(A \text{ or } B) = P(A) + P(B)$ holds if A and B are disjoint but not otherwise. Resist the temptation to use these simple formulas when the circumstances that justify them are not present. You must also be certain not to confuse disjointness and independence. If A and B are disjoint, then the fact that A occurs tells us that B cannot occur—look again at Figure 6.4. So **disjoint events are not independent.** Unlike disjointness or complements, independence cannot be pictured by a Venn diagram, because it involves the probabilities of the events rather than just the outcomes that make up the events.

Applying the probability rules

If two events A and B are independent, then their complements A^c and B^c are also independent and A^c is independent of B. Suppose, for example, that 75% of all registered voters in a suburban district are Republicans. If an opinion poll interviews two voters chosen independently, the probability that the first is a Republican and the second is not a Republican is (0.75)(0.25) = 0.1875. The multiplication rule also extends to collections of more than two events, provided that all are independent. Independence of events A, B, and C means that no information about any one or any two can change the probability of the remaining events. The formal definition is a bit messy. Fortunately, independence is usually assumed in setting up a probability model. We can then use the multiplication rule freely, as in this example.

EXAMPLE 6.14 ATLANTIC TELEPHONE CABLE

The first successful transatlantic telegraph cable was laid in 1866. The first telephone cable across the Atlantic did not appear until 1956—the barrier was designing "repeaters," amplifiers needed to boost the signal, that could operate for years on the sea bottom. This first cable had 52 repeaters. The copper cable, laid in 1983 and retired in 1994, had 662 repeaters. The first fiber optic cable was laid in 1988 and has 109 repeaters. There are now more than 400,000 miles of undersea cable, with more being laid every year to handle the flood of Internet traffic.

Repeaters in undersea cables must be very reliable. To see why, suppose that each repeater has probability 0.999 of functioning without failure for 25 years. Repeaters fail independently of each other. (This assumption means that there are no "common causes" such as earthquakes that would affect several repeaters at once.) Denote by A_i the event that the ith repeater operates successfully for 25 years.

The probability that two repeaters both last 25 years is

$$\begin{aligned} P(A_1 \text{ and } A_2) &= P(A_1)P(A_2) \\ &= 0.999 \times 0.999 = 0.998 \end{aligned}$$

For a cable with 10 repeaters the probability of no failures in 25 years is

$$\begin{aligned} P(A_1 \text{ and } A_2 \text{ and} \ldots \text{and } A_{10}) &= P(A_1)P(A_2) \cdots P(A_{10}) \\ &= 0.999 \times 0.999 \times \cdots \times 0.999 \\ &= 0.999^{10} = 0.990 \end{aligned}$$

Cables with 2 or 10 repeaters would be quite reliable. Unfortunately, the last copper transatlantic cable had 662 repeaters. The probability that all 662 work for 25 years is

$$P(A_1 \text{ and } A_2 \text{ and} \ldots \text{and } A_{662}) = 0.999^{662} = 0.516$$

This cable will fail to reach its 25-year design life about half the time if each repeater is 99.9% reliable over that period. The multiplication rule for probabilities shows that repeaters must be much more than 99.9% reliable.

By combining the rules we have learned, we can compute probabilities for rather complex events. Here is an example.

EXAMPLE 6.15 AIDS TESTING

Screening large numbers of blood samples for HIV, the virus that causes AIDS, uses an enzyme immunoassay (EIA) test that detects antibodies to the virus. Samples that test positive are retested using a more accurate "western blots" test. Applied to people who have no HIV antibodies, EIA has probability about 0.006 of producing a false positive. If the 140 employees of a medical clinic are tested and all 140 are free of HIV antibodies, what is the probability that at least one false positive will occur?

It is reasonable to assume as part of the probability model that the test results for different individuals are independent. The probability that the test is positive for a single person is 0.006, so the probability of a negative result is 1 – 0.006 = 0.994 by the complement rule. The probability of at least one false positive among the 140 people tested is therefore

$$\begin{aligned} P(\text{at least one positive}) &= 1 - P(\text{no positives}) \\ &= 1 - P\,(140 \text{ negatives}) \\ &= 1 - 0.994^{140} \\ &= 1 - 0.431 = 0.569 \end{aligned}$$

The probability is greater than 1/2 that at least one of the 140 people will test positive for HIV even though no one has the virus.

EXERCISES

6.27 A BATTLE PLAN A general can plan a campaign to fight one major battle or three small battles. He believes that he has probability 0.6 of winning the large battle and probability 0.8 of winning each of the small battles. Victories or defeats in the small battles are independent. The general must win either the large battle or all three small battles to win the campaign. Which strategy should he choose?

6.28 DEFECTIVE CHIPS An automobile manufacturer buys computer chips from a supplier. The supplier sends a shipment containing 5% defective chips. Each chip chosen from this shipment has probability 0.05 of being defective, and each automobile uses 12 chips selected independently. What is the probability that all 12 chips in a car will work properly?

6.29 COLLEGE-EDUCATED LABORERS? Government data show that 26% of the civilian labor force have at least 4 years of college and that 16% of the labor force work as laborers or operators of machines or vehicles. Can you conclude that because (0.26)(0.16) = 0.0416, about 4% of the labor force are college-educated laborers or operators? Explain your answer.

6.30 Choose at random a U.S. resident at least 25 years of age. We are interested in the events

$$A = \{\text{The person chosen completed 4 years of college}\}$$
$$B = \{\text{The person chosen is 55 years old or older}\}$$

Government data recorded in Table 4.6 on page 241 allow us to assign probabilities to these events.

(a) Explain why $P(A) = 0.230$.

(b) Find $P(B)$.

(c) Find the probability that the person chosen is at least 55 years old *and* has 4 years of college education, $P(A \text{ and } B)$. Are the events A and B independent?

6.31 BRIGHT LIGHTS? A string of Christmas lights contains 20 lights. The lights are wired in series, so that if any light fails the whole string will go dark. Each light has probability 0.02 of failing during a 3-year period. The lights fail independently of each other. What is the probability that the string of lights will remain bright for 3 years?

6.32 DETECTING STEROIDS An athlete suspected of having used steroids is given two tests that operate independently of each other. Test A has probability 0.9 of being positive if steroids have been used. Test B has probability 0.8 of being positive if steroids have been used. What is the probability that *neither* test is positive if steroids have been used?

6.33 TELEPHONE SUCCESS Most sample surveys use random digit dialing equipment to call residential telephone numbers at random. The telephone polling firm Zogby International reports that the probability that a call reaches a live person is 0.2.[4] Calls are independent.

(a) A polling firm places 5 calls. What is the probability that none of them reaches a person?

(b) When calls are made to New York City, the probability of reaching a person is only 0.08. What is the probability that none of 5 calls made to New York City reaches a person?

SUMMARY

A **random phenomenon** has outcomes that we cannot predict but that nonetheless have a regular distribution in very many repetitions.

The **probability** of an event is the proportion of times the event occurs in many repeated trials of a random phenomenon.

A **probability model** for a random phenomenon consists of a sample space S and an assignment of probabilities P.

The **sample space S** is the set of all possible outcomes of the random phenomenon. Sets of outcomes are called **events.** P assigns a number $P(A)$ to an event A as its probability.

The **complement** A^c of an event A consists of exactly the outcomes that are not in A. Events A and B are **disjoint** (mutually exclusive) if they have no outcomes in common. Events A and B are **independent** if knowing that one event occurs does not change the probability we would assign to the other event.

Any assignment of probability must obey the rules that state the basic properties of probability:

1. $0 \le P(A) \le 1$ for any event A.

2. $P(S) = 1$.

3. **Complement rule:** For any event A, $P(A^c) = 1 - P(A)$.

4. **Addition rule:** If events A and B are **disjoint,** then $P(A \text{ or } B) = P(A \cup B) = P(A) + P(B)$.

5. **Multiplication rule:** If events A and B are **independent,** then $P(A \text{ and } B) = P(A \cap B) = P(A)P(B)$.

SECTION 6.2 EXERCISES

6.34 LEGITIMATE PROBABILITY MODEL? Figure 6.7 displays several assignments of probabilities to the six faces of a die. We can learn which assignment is actually *accurate* for a particular die only by rolling the die many times. However, some of the assignments are not *legitimate* assignments of probability. That is, they do not obey the rules. Which are legitimate and which are not? In the case of the illegitimate models, explain what is wrong.

Outcome	Model 1	Model 2	Model 3	Model 4
⚀	$\frac{1}{3}$	$\frac{1}{6}$	$\frac{1}{7}$	$\frac{1}{3}$
⚁	0	$\frac{1}{6}$	$\frac{1}{7}$	$\frac{1}{3}$
⚂	$\frac{1}{6}$	$\frac{1}{6}$	$\frac{1}{7}$	$-\frac{1}{6}$
⚃	0	$\frac{1}{6}$	$\frac{1}{7}$	$-\frac{1}{6}$
⚄	$\frac{1}{6}$	$\frac{1}{6}$	$\frac{1}{7}$	$\frac{1}{3}$
⚅	$\frac{1}{3}$	$\frac{1}{6}$	$\frac{1}{7}$	$\frac{1}{3}$

FIGURE 6.7 Four assignments of probabilities to the six faces of a die.

6.35 LEGITIMATE ASSIGNMENT OF PROBABILITIES? In each of the following situations, state whether or not the given assignment of probabilities to individual outcomes is legitimate, that is, satisfies the rules of probability. If not, give specific reasons for your answer.

(a) When a coin is spun, $P(\text{H}) = 0.55$ and $P(\text{T}) = 0.45$.

(b) When two coins are tossed, $P(\text{HH}) = 0.4$, $P(\text{HT}) = 0.4$, $P(\text{TH}) = 0.4$, and $P(\text{TT}) = 0.4$.

(c) When a die is rolled, the number of spots on the up-face has $P(1) = 1/2$, $P(4) = 1/6$, $P(5) = 1/6$, and $P(6) = 1/6$.

6.36 CAR COLORS Choose a new car or light truck at random and note its color. Here are the probabilities of the most popular colors for vehicles made in North America in 2000:[5]

Color:	Silver	White	Black	Dark green	Dark blue	Medium red
Probability:	0.176	0.172	0.113	0.089	0.088	0.067

(a) What is the probability that the vehicle you choose has any color other than the six listed?

(b) What is the probability that a randomly chosen vehicle is either silver or white?

(c) Choose two vehicles at random. What is the probability that both are silver or white?

6.37 NEW CENSUS CATEGORIES The 2000 census allowed each person to choose one or more from a long list of races. That is, in the eyes of the Census Bureau, you belong to whatever race or races you say you belong to. "Hispanic/Latino" is a separate category; Hispanics may be of any race. If we choose a resident of the United States at random, the 2000 census gives these probabilities:

	Hispanic	Not Hispanic
Asian	0.000	0.036
Black	0.003	0.121
White	0.060	0.691
Other	0.062	0.027

Let A be the event that a randomly chosen American is Hispanic, and let B be the event that the person chosen is white.

(a) Verify that the table gives a legitimate assignment of probabilities.

(b) What is $P(A)$?

(c) Describe B^c in words and find $P(B^c)$ by the complement rule.

(d) Express "the person chosen is a non-Hispanic white" in terms of events A and B. What is the probability of this event?

6.38 BEING HISPANIC Exercise 6.37 assigns probabilities for the ethnic background of a randomly chosen resident of the United States. Let A be the event that the person chosen is Hispanic, and let B be the event that he or she is white. Are events A and B independent? How do you know?

6.39 PREPARING FOR THE GMAT A company that offers courses to prepare would-be M.B.A. students for the GMAT examination finds that 40% of its customers are currently undergraduate students and 60% are college graduates. After completing the course, 50% of the undergraduates and 70% of the graduates achieve scores of at least 600 on the GMAT.

(a) What percent of customers are undergraduates *and* score at least 600? What percent of customers are graduates *and* score at least 600?

(b) What percent of all customers score at least 600 on the GMAT?

6.40 THE RISE AND FALL OF PORTFOLIO VALUES The "random walk" theory of securities prices holds that price movements in disjoint time periods are independent of each other. Suppose that we record only whether the price is up or down each year, and that the probability that our portfolio rises in price in any one year is 0.65. (This probability is approximately correct for a portfolio containing equal dollar amounts of all common stocks listed on the New York Stock Exchange.)

(a) What is the probability that our portfolio goes up for 3 consecutive years?

(b) If you know that the portfolio has risen in price 2 years in a row, what probability do you assign to the event that it will go down next year?

(c) What is the probability that the portfolio's value moves in the same direction in both of the next 2 years?

6.41 USING A TABLE TO FIND PROBABILITIES The type of medical care a patient receives may vary with the age of the patient. A large study of women who had a breast lump investigated whether or not each woman received a mammogram and a biopsy when the lump was discovered. Here are some probabilities estimated by the study. The entries in the table are the probabilities that *both* of two events occur; for example, 0.321 is the probability that a patient is under 65 years of age *and* the tests were done. The four probabilities in the table have sum 1 because the table lists all possible outcomes.

	Tests done?	
	Yes	No
Age under 65:	0.321	0.124
Age 65 or over:	0.365	0.190

(a) What is the probability that a patient in this study is under 65? That a patient is 65 or over?

(b) What is the probability that the tests were done for a patient? That they were not done?

(c) Are the events A = {patient was 65 or older} and B = {the tests were done} independent? Were the tests omitted on older patients more or less frequently than would be the case if testing were independent of age?

6.42 ROULETTE A roulette wheel has 38 slots, numbered 0, 00, and 1 to 36. The slots 0 and 00 are colored green, 18 of the others are red, and 18 are black. The dealer spins the wheel and at the same time rolls a small ball along the wheel in the opposite direction. The wheel is carefully balanced so that the ball is equally likely to land in any slot when the wheel slows. Gamblers can bet on various combinations of numbers and colors.

(a) What is the probability that the ball will land in any one slot?

(b) If you bet on "red," you win if the ball lands in a red slot. What is the probability of winning?

(c) The slot numbers are laid out on a board on which gamblers place their bets. One column of numbers on the board contains all multiples of 3, that is, 3, 6, 9, . . . , 36. You place a "column bet" that wins if any of these numbers comes up. What is your probability of winning?

6.43 WHICH IS MOST LIKELY? A six-sided die has four green and two red faces and is balanced so that each face is equally likely to come up. The die will be rolled several times. You must choose one of the following three sequences of colors; you will win \$25 if the first rolls of the die give the sequence that you have chosen.

RGRRR
RGRRRG
GRRRRR

Which sequence do you choose? Explain your choice.[6]

6.44 ALBINISM IN GENETICS The gene for albinism in humans is recessive. That is, carriers of this gene have probability 1/2 of passing it to a child, and the child is albino only if both parents pass the albinism gene. Parents pass their genes independently of each other. If both parents carry the albinism gene, what is the probability that their first child is albino? If they have two children (who inherit independently of each other), what is the probability that both are albino? That neither is albino?

6.45 DISJOINT VERSUS INDEPENDENT EVENTS This exercise explores the relationship between mutually exclusive and independent events.

(a) Assume that events A and B are non-empty, independent events. Show that A and B must intersect (i.e., that $A \cap B \neq \varnothing$).

(b) Use the results of (a) to argue that if A and B are disjoint, then they cannot be independent.

(c) Find an example of two events that are neither disjoint nor independent.

6.3 GENERAL PROBABILITY RULES

In this section we will consider some additional laws that govern any assignment of probabilities. The purpose of learning more laws of probability is to be able to give probability models for more complex random phenomena. We have already met and used five rules.

RULES OF PROBABILITY

Rule 1. $0 \le P(A) \le 1$ for any event A.

Rule 2. $P(S) = 1$.

Rule 3. **Complement rule:** For any event A,

$$P(A^c) = 1 - P(A)$$

Rule 4. **Addition rule:** If A and B are **disjoint** events, then

$$P(A \text{ or } B) = P(A) + P(B)$$

Rule 5. **Multiplication rule:** If A and B are **independent** events, then

$$P(A \text{ and } B) = P(A)P(B)$$

General addition rules

Probability has the property that if A and B are disjoint events, then $P(A \text{ or } B) = P(A) + P(B)$. What if there are more than two events, or if the events are not disjoint? These circumstances are covered by more general addition rules for probability.

UNION

The **union** of any collection of events is the event that at least one of the collection occurs.

For two events A and B, the union is the event {A or B} that A or B or both occur. From the addition rule for two disjoint events, we can obtain rules for more general unions. Suppose first that we have several events—say A, B, and C—that are disjoint in pairs. That is, no two can occur simultaneously. The Venn diagram in Figure 6.8 illustrates three disjoint events.

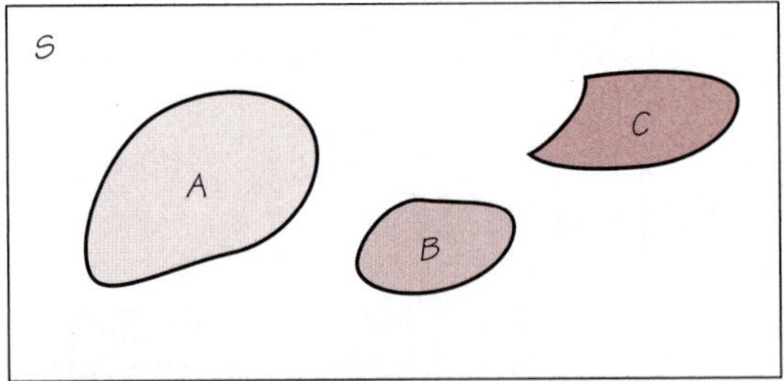

FIGURE 6.8 The addition rule for disjoint events: $P(A \text{ or } B \text{ or } C) = P(A) + P(B) + P(C)$ when events A, B, and C are disjoint.

The addition rule for two disjoint events extends to the following law.

ADDITION RULE FOR DISJOINT EVENTS

If events A, B, and C are disjoint in the sense that no two have any outcomes in common, then

$$P(\text{one or more of } A, B, C) = P(A) + P(B) + P(C)$$

This rule extends to any number of disjoint events.

EXAMPLE 6.16 UNIFORM DISTRIBUTION

Generate a random number X between 0 and 1. What is the probability that the first digit will be odd? We will learn in Chapter 7 that the variable X has the density curve of a uniform distribution (see Exercise 2.2, page 83.). This density curve has constant height 1 between 0 and 1 and is 0 elsewhere. The event that the first digit of X is odd is the union of five disjoint events. These events are

$$0.10 \le X < 0.20$$
$$0.30 \le X < 0.40$$
$$0.50 \le X < 0.60$$
$$0.70 \le X < 0.80$$
$$0.90 \le X < 1.00$$

Figure 6.9 illustrates the probabilities of these events as areas under the density curve. Each has probability 0.1 equal to its length. The union of the five therefore has probability equal to the sum, or 0.5. As we should expect, a random number is equally likely to begin with an odd or an even digit.

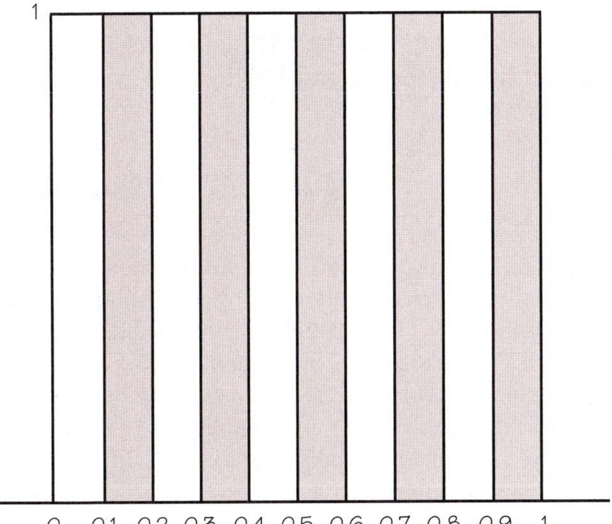

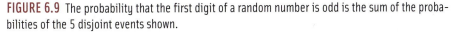

FIGURE 6.9 The probability that the first digit of a random number is odd is the sum of the probabilities of the 5 disjoint events shown.

If events A and B are *not* disjoint, they can occur simultaneously. The probability of their union is then *less* than the sum of their probabilities. As Figure 6.10 suggests, the outcomes common to both are counted twice when we add probabilities, so we must subtract this probability once.

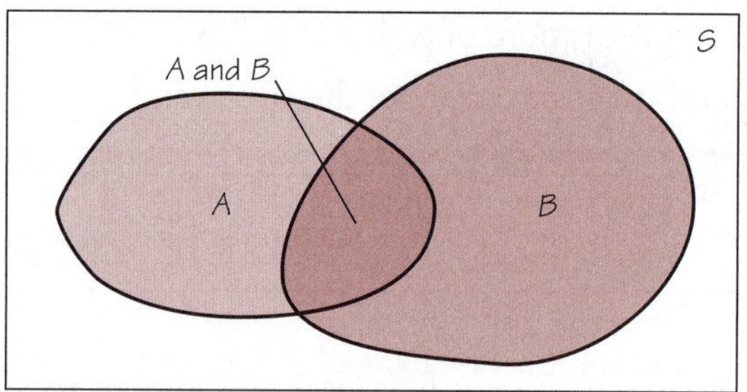

FIGURE 6.10 The general addition rule for the union of two events: $P(A \text{ or } B) = P(A) + P(B) - P(A \text{ and } B)$ for any events A and B.

Here is the addition rule for the union of any two events, disjoint or not.

GENERAL ADDITION RULE FOR UNIONS OF TWO EVENTS

For any two events A and B,

$$P(\text{A or B}) = P(\text{A}) + P(\text{B}) - P(\text{A and B})$$

Equivalently,

$$P(\text{A} \cup \text{B}) = P(\text{A}) + P(\text{B}) - P(\text{A} \cap \text{B})$$

If A and B are disjoint, the event {A and B} that both occur has no outcomes in it. This *empty event* ∅ is the complement of the sample space S and must have probability 0. So the general addition rule includes Rule 4, the addition rule for disjoint events.

EXAMPLE 6.17 PROBABILITY OF PROMOTION

Deborah and Matthew are anxiously awaiting word on whether they have been made partners of their law firm. Deborah guesses that her probability of making partner is 0.7 and that Matthew's is 0.5. (These are personal probabilities reflecting Deborah's assessment of chance.) This assignment of probabilities does not give us enough information to compute the probability that at least one of the two is promoted. In particular, adding the individual probabilities of promotion gives the impossible result 1.2. If Deborah also guesses that the probability that *both* she and Matthew are made partners is 0.3, then by the addition rule for unions

$$P(\text{at least one is promoted}) = 0.7 + 0.5 - 0.3 = 0.9$$

The probability that *neither* is promoted is then 0.1 by the complement rule.

Venn diagrams are a great help in finding probabilities for unions, because you can just think of adding and subtracting areas. Figure 6.11 shows some events and their probabilities for Example 6.17. What is the probability that Deborah is promoted and Matthew is not? The Venn diagram shows that this is the probability that Deborah is promoted minus the probability that both are promoted, 0.7 – 0.3 = 0.4. Similarly, the probability that Matthew is promoted and Deborah is not is 0.5 – 0.3 = 0.2. The four probabilities that appear in the figure add to 1 because they refer to four disjoint events whose union is the entire sample space.

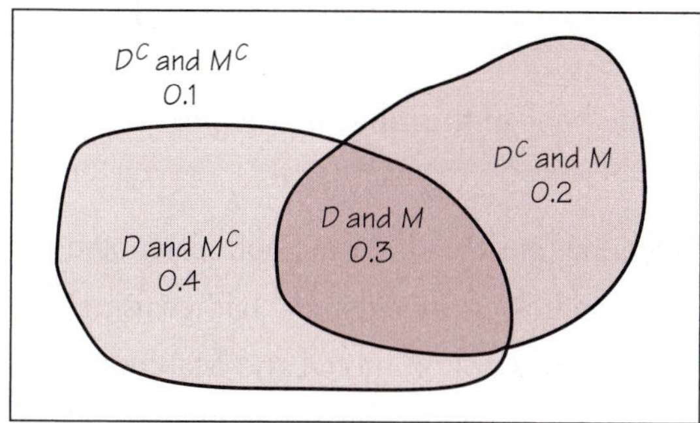

FIGURE 6.11 Venn diagram and probabilities.

The simultaneous occurrence of two events, such as A = Deborah is promoted *and* B = Matthew is promoted, is called a ***joint event.*** The probability of a joint event, such as *P*(Deborah is promoted *and* Matthew is promoted) = *P*(A and *B*), is called a ***joint probability.*** Determining joint probabilities when you have equally likely outcomes can be as easy as counting outcomes. For most situations, however, we will need more powerful methods, which will be developed later in this section.

Here's another way to work with joint events. We have two variables. One variable is employee, which has two values: Deborah and Matthew. The other variable is promotion, which also has two values: promoted and not promoted.

$$D = \{\text{Deborah promoted}\}$$
$$D^c = \{\text{Deborah not promoted}\}$$
$$M = \{\text{Matthew promoted}\}$$
$$M^c = \{\text{Matthew not promoted}\}$$

We can construct a table and write in the probabilities that Deborah assumes:

		Matthew		
		Promoted	Not promoted	Total
Deborah	Promoted	0.3		0.7
	Not promoted			
	Total	0.5		1

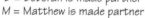

The rows and columns have to add to the totals shown, so we can fill in the rest of the table to produce the completed table:

		Matthew		
		Promoted	Not promoted	Total
Deborah	Promoted	0.3	0.4	0.7
	Not promoted	0.2	0.1	0.3
Total		0.5	0.5	1

The four entries in the body of the table are the probabilities of the joint events of interest:

$P(D \text{ and } M) = P(\text{Deborah and Matthew are both promoted}) = 0.3$

$P(D \text{ and } M^c) = P(\text{Deborah is promoted and Matthew is not promoted}) = 0.4$

$P(D^c \text{ and } M) = P(\text{Deborah is not promoted and Matthew is promoted}) = 0.2$

$P(D^c \text{ and } M^c) = P(\text{Deborah is not promoted and Matthew is not promoted}) = 0.1$

Note that these joint probabilities add to 1.

We will continue our discussion of tables and joint events in the next section.

EXERCISES

6.46 PROSPERITY AND EDUCATION Call a household prosperous if its income exceeds \$100,000. Call the household educated if the householder completed college. Select an American household at random, and let A be the event that the selected household is prosperous and B the event that it is educated. According to the Census Bureau, $P(A) = 0.134$, $P(B) = 0.254$, and the joint probability that a household is both prosperous and educated is $P(A \text{ and } B) = 0.080$. What is the probability $P(A \text{ or } B)$ that the household selected is either prosperous or educated?

6.47 Draw a Venn diagram that shows the relation between events A and B in Exercise 6.46. Indicate each of the following events on your diagram and use the information in Exercise 6.46 to calculate the probability of each event. Finally, describe in words what each event is.

(a) $\{A \text{ and } B\}$

(b) $\{A \text{ and } B^c\}$

(c) $\{A^c \text{ and } B\}$

(d) $\{A^c \text{ and } B^c\}$

6.48 WINNING CONTRACTS Consolidated Builders has bid on two large construction projects. The company president believes that the probability of winning the first contract (event A) is 0.6, that the probability of winning the second (event B) is 0.5, and that the joint probability of winning both jobs (event $\{A \text{ and } B\}$) is 0.3. What is the probability of the event $\{A \text{ or } B\}$ that Consolidated will win at least one of the jobs?

6.49 In the setting of the previous exercise, are events A and B independent? Do a calculation that proves your answer.

6.50 Draw a Venn diagram that illustrates the relation between events A and B in Exercise 6.48. Write each of the following events in terms of A, B, A^c, and B^c. Indicate the events on your diagram and use the information in Exercise 6.48 to calculate the probability of each.

(a) Consolidated wins both jobs.

(b) Consolidated wins the first job but not the second.

(c) Consolidated does not win the first job but does win the second.

(d) Consolidated does not win either job.

6.51 CAFFEINE IN THE DIET Common sources of caffeine are coffee, tea, and cola drinks. Suppose that

55% of adults drink coffee

25% of adults drink tea

45% of adults drink cola

and also that

15% drink both coffee and tea

5% drink all three beverages

25% drink both coffee and cola

5% drink only tea

Draw a Venn diagram marked with this information. Use it along with the addition rules to answer the following questions.

(a) What percent of adults drink only cola?

(b) What percent drink none of these beverages?

6.52 TASTES IN MUSIC Musical styles other than rock and pop are becoming more popular. A survey of college students finds that 40% like country music, 30% like gospel music, and 10% like both.

(a) Make a Venn diagram with these results.

(b) What percent of college students like country but not gospel?

(c) What percent like neither country nor gospel?

6.53 GETTING INTO COLLEGE Ramon has applied to both Princeton and Stanford. He thinks the probability that Princeton will admit him is 0.4, the probability that Stanford will admit him is 0.5, and the probability that both will admit him is 0.2.

(a) Make a Venn diagram with the probabilities given marked.

(b) What is the probability that neither university admits Ramon?

(c) What is the probability that he gets into Stanford but not Princeton?

Conditional probability

The probability we assign to an event can change if we know that some other event has occurred. This idea is the key to many applications of probability.

EXAMPLE 6.18 AMARILLO SLIM WANTS AN ACE

Slim is a professional poker player. He stares at the dealer, who prepares to deal. What is the probability that the card dealt to Slim is an ace? There are 52 cards in the deck. Because the deck was carefully shuffled, the next card dealt is equally likely to be any of the cards. Four of the 52 cards are aces. So

$$P(\text{ace}) = \frac{4}{52} = \frac{1}{13}$$

This calculation assumes that Slim knows nothing about any cards already dealt. Suppose now that he is looking at 4 cards already in his hand, and that 1 of them is an ace. He knows nothing about the other 48 cards except that exactly 3 aces are among them. Slim's probability of being dealt an ace, *given what he knows*, is now

$$P(\text{ace} \mid 1 \text{ ace in 4 visible cards}) = \frac{3}{48} = \frac{1}{16}$$

Knowing that there is one ace among the four cards Slim can see changes the probability that the next card dealt is an ace.

conditional probability

The new notation $P(A \mid B)$ is a ***conditional probability***. That is, it gives the probability of one event (the next card dealt is an ace) under the condition that we know another event (exactly one of the four visible cards is an ace). You can read the bar | as "given the information that."

In Example 6.18 we could find probabilities because we were willing to use an equally likely probability model for a shuffled deck of cards. Here is an example based on data.

EXAMPLE 6.19 MARITAL STATUS OF WOMEN

Table 6.1 shows the marital status of adult women broken down by age group.

TABLE 6.1 Age and marital status of women (thousands of women)

	Age			
	18–29	30–64	65 and over	Total
Married	7,842	43,808	8,270	59,920
Never married	13,930	7,184	751	21,865
Widowed	36	2,523	8,385	10,944
Divorced	704	9,174	1,263	11,141
Total	22,512	62,689	18,669	103,870

Source: Data for 1999 from the 2000 *Statistical Abstract of the United States.*

We are interested in the probability that a randomly chosen woman is married. It is common sense that knowing her age group will change the probability: many young women have not married, most middle-aged women are married, and older women are more likely to be widows. To help us think carefully, let's define two events:

$$A = \text{the woman chosen is young, ages 18 to 29}$$
$$B = \text{the woman chosen is married}$$

There are (in thousands) 103,870 adult women in the United States. Of these women, 22,512 are aged 18 to 29. Choosing at random gives each woman an equal chance, so the probability of choosing a young woman is

$$P(A) = \frac{22{,}512}{103{,}870} = 0.217$$

The table shows that there are 7842 thousand young married women. So the probability that we choose a woman who is both young and married is

$$P(A \text{ and } B) = \frac{7842}{103{,}870} = 0.075$$

To find the *conditional* probability that a woman is married *given the information* that she is young, look only at the "18–29" column. The young women are all in this column, so the information given says that only this column is relevant. The conditional probability is

$$P(B \mid A) = \frac{7842}{22{,}512} = 0.348$$

As we expected, the conditional probability that a woman is married when we know she is under age 30 is much higher than the probability for a randomly chosen woman.

It is easy to confuse the three probabilities in Example 6.19. Look carefully at Table 6.1 and be sure you understand the example. There is a relationship among these three probabilities. The probability that a woman is both young *and* married is the product of the probabilities that she is young and that she is married *given* that she is young. That is,

$$\begin{aligned} P(A \text{ and } B) &= P(A) \times P(B \mid A) \\ &= \frac{22{,}512}{103{,}870} \times \frac{7842}{22{,}512} \\ &= \frac{7842}{103{,}870} = 0.075 \qquad \text{(as before)} \end{aligned}$$

Try to think your way through this in words: First, the woman is young; then, given that she is young, she is married. We have just discovered the fundamental multiplication rule of probability.

GENERAL MULTIPLICATION RULE FOR ANY TWO EVENTS

The probability that both of two events A and B happen together can be found by

$$P(\text{A and B}) = P(\text{A})P(\text{B} \mid \text{A})$$

Here $P(\text{B} \mid \text{A})$ is the conditional probability that B occurs given the information that A occurs.

In words, this rule says that for both of two events to occur, first one must occur and then, given that the first event has occurred, the second must occur. In our example, the joint probability that a randomly chosen woman is both age 18 to 29 (event A) and married (event B) is

$$\begin{aligned} P(\text{A and B}) &= P(\text{A})P(\text{B} \mid \text{A}) \\ &= (0.217)(0.348) = 0.076 \end{aligned}$$

EXAMPLE 6.20 SLIM WANTS DIAMONDS

Slim is still at the poker table. At the moment, he wants very much to draw 2 diamonds in a row. As he looks at his hand and at the upturned cards on the table, Slim sees 11 cards. Of these, 4 are diamonds. The full deck contains 13 diamonds among its 52 cards, so 9 of the 41 unseen cards are diamonds. To find Slim's probability of drawing two diamonds, first calculate

$$\begin{aligned} P(\text{first card diamond}) &= \frac{9}{41} \\ P(\text{second card diamond} \mid \text{first card diamond}) &= \frac{8}{40} \end{aligned}$$

Slim finds both probabilities by counting cards. The probability that the first card drawn is a diamond is 9/41 because 9 of the 41 unseen cards are diamonds. If the first card is a diamond, that leaves 8 diamonds among the 40 remaining cards. So the *conditional* probability of another diamond is 8/40. The multiplication rule now says that

$$P(\text{both cards diamonds}) = \frac{9}{41} \times \frac{8}{40} = 0.044$$

Slim will need luck to draw his diamonds.

If we know $P(\text{A})$ and $P(\text{A and B})$, we can rearrange the general multiplication rule to produce a *definition* of the conditional probability $P(\text{B} \mid \text{A})$ in terms of unconditional probabilities.

DEFINITION OF CONDITIONAL PROBABILITY

When $P(A) > 0$, the conditional probability of B given A is

$$P(B \mid A) = \frac{P(A \text{ and } B)}{P(A)}$$

Be sure to keep in mind the distinct roles in $P(B \mid A)$ of the event B whose probability we are computing and the event A that represents the information we are given. The conditional probability $P(B \mid A)$ makes no sense if the event A can never occur, so we require that $P(A) > 0$ whenever we talk about $P(B \mid A)$.

EXAMPLE 6.21 FINDING CONDITIONAL PROBABILITIES

What is the conditional probability that a woman is a widow, given that she is at least 65 years old? We see from Table 6.1 that

$$P(\text{at least } 65) = \frac{18{,}669}{103{,}870} = 0.180$$

$$P(\text{widowed } \textit{and} \text{ at least } 65) = \frac{8385}{103{,}870} = 0.081$$

The conditional probability is therefore

$$P(\text{widowed} \mid \text{at least } 65) = \frac{P(\text{widowed } \textit{and} \text{ at least } 65)}{P(\text{at least } 65)}$$

$$= \frac{0.081}{0.180} = 0.450$$

Check that this agrees (up to roundoff error) with the result obtained from the "65 and over" column of Table 6.1:

$$P(\text{widowed} \mid \text{at least } 65) = \frac{8385}{18{,}669} = 0.449$$

EXERCISES

6.54 **AMERICAN WOMEN, I** Choose an adult American woman at random. Table 6.1 describes the population from which we draw. Use the information in that table to answer the following questions.

(a) What is the probability that the woman chosen is 65 years old or older?

(b) What is the conditional probability that the woman chosen is married, given that she is 65 or over?

(c) How many women are *both* married and in the over-65 age group? What is the probability that the woman we choose is a married woman at least 65 years old?

(d) Verify that the three probabilities you found in (a), (b), and (c) satisfy the multiplication rule.

6.55 AMERICAN WOMEN, II Choose an adult American woman at random. Table 6.1 describes the population from which we draw.

(a) What is the conditional probability that the woman chosen is 18 to 29 years old, given that she is married?

(b) In Example 6.19 we found that $P(\text{married} \mid \text{age 18 to 29}) = 0.348$. Complete this sentence: 0.348 is the proportion of women who are _____ among those women who are _____.

(c) In (a), you found $P(\text{age 18 to 29} \mid \text{married})$. Write a sentence of the form given in (b) that describes the meaning of this result. The two conditional probabilities give us very different information.

6.56 WOMAN MANAGERS Choose an employed person at random. Let A be the event that the person chosen is a woman, and B the event that the person holds a managerial or professional job. Government data tell us that $P(A) = 0.46$ and the probability of managerial and professional jobs among women is $P(B \mid A) = 0.32$. Find the probability that a randomly chosen employed person is a woman holding a managerial or professional position.

6.57 BUYING FROM JAPAN Functional Robotics Corporation buys electrical controllers from a Japanese supplier. The company's treasurer thinks that there is probability 0.4 that the dollar will fall in value against the Japanese yen in the next month. The treasurer also believes that *if* the dollar falls there is probability 0.8 that the supplier will demand renegotiation of the contract. What probability has the treasurer assigned to the event that the dollar falls and the supplier demands renegotiation?

6.58 THE PROBABILITY OF A FLUSH A poker player holds a flush when all 5 cards in the hand belong to the same suit. We will find the probability of a flush when 5 cards are dealt. Remember that a deck contains 52 cards, 13 of each suit, and that when the deck is well shuffled, each card dealt is equally likely to be any of those that remain in the deck.

(a) We will concentrate on spades. What is the probability that the first card dealt is a spade? What is the conditional probability that the second card is a spade, given that the first is a spade?

(b) Continue to count the remaining cards to find the conditional probabilities of a spade on the third, the fourth, and the fifth card, given in each case that all previous cards are spades.

(c) The probability of being dealt 5 spades is the product of the five probabilities you have found. Why? What is this probability?

(d) The probability of being dealt 5 hearts or 5 diamonds or 5 clubs is the same as the probability of being dealt 5 spades. What is the probability of being dealt a flush?

6.59 THE PROBABILITY OF A ROYAL FLUSH A royal flush is the highest hand possible in poker. It consists of the ace, king, queen, jack, and ten of the same suit. Modify the outline given in Exercise 6.58 to find the probability of being dealt a royal flush in a five-card deal.

6.60 INCOME TAX RETURNS Here is the distribution of the adjusted gross income (in thousands of dollars) reported on individual federal income tax returns in 1994:

Income:	<10	10–29	30–49	50–99	≥100
Probability:	0.12	0.39	0.24	0.20	0.05

(a) What is the probability that a randomly chosen return shows an adjusted gross income of \$50,000 or more?

(b) Given that a return shows an income of at least \$50,000, what is the conditional probability that the income is at least \$100,000?

6.61 TASTES IN MUSIC Musical styles other than rock and pop are becoming more popular. A survey of college students finds that 40% like country music, 30% like gospel music, and 10% like both.

(a) What is the conditional probability that a student likes gospel music if we know that he or she likes country music?

(b) What is the conditional probability that a student who does not like country music likes gospel music? (A Venn diagram may help you.)

Extended multiplication rules

The definition of conditional probability reminds us that in principle all probabilities, including conditional probabilities, can be found from the assignment of probabilities to events that describe a random phenomenon. More often, however, conditional probabilities are part of the information given to us in a probability model, and the multiplication rule is used to compute $P(A \text{ and } B)$.

The union of a collection of events is the event that *any* of them occur. Here is the corresponding term for the event that *all* of them occur.

INTERSECTION

The **intersection** of any collection of events is the event that *all* of the events occur.

To extend the multiplication rule to the probability that all of several events occur, the key is to condition each event on the occurrence of *all* of the preceding events. For example, the intersection of three events A, B, and C has probability

$$P(A \text{ and } B \text{ and } C) = P(A)P(B \mid A)P(C \mid A \text{ and } B)$$

EXAMPLE 6.22 THE FUTURE OF HIGH SCHOOL ATHLETES

Only 5% of male high school basketball, baseball, and football players go on to play at the college level. Of these, only 1.7% enter major league professional sports. About 40% of the athletes who compete in college and then reach the pros have a career of more than 3 years.[7] Define these events:

$$A = \{\text{competes in college}\}$$
$$B = \{\text{competes professionally}\}$$
$$C = \{\text{pro career longer than 3 years}\}$$

What is the probability that a high school athlete competes in college and then goes on to have a pro career of more than 3 years? We know that

$$P(A) = 0.05$$
$$P(B \mid A) = 0.017$$
$$P(C \mid A \text{ and } B) = 0.4$$

The probability we want is therefore

$$\begin{aligned} P(A \text{ and } B \text{ and } C) &= P(A)P(B \mid A)P(C \mid A \text{ and } B) \\ &= 0.05 \times 0.017 \times 0.40 = 0.00034 \end{aligned}$$

Only about 3 of every 10,000 high school athletes can expect to compete in college and have a professional career of more than 3 years. High school students would be wise to concentrate on studies rather than on unrealistic hopes of fortune from pro sports.

Tree diagrams revisited

Probability problems often require us to combine several of the basic rules into a more elaborate calculation. Here is an example that illustrates how to solve problems that have several stages.

EXAMPLE 6.23 A FUTURE IN PROFESSIONAL SPORTS?

What is the probability that a male high school athlete will go on to professional sports? In the notation of Example 6.22, this is $P(B)$. To find $P(B)$ from the information in Example 6.22, use the tree diagram in Figure 6.12 to organize your thinking.

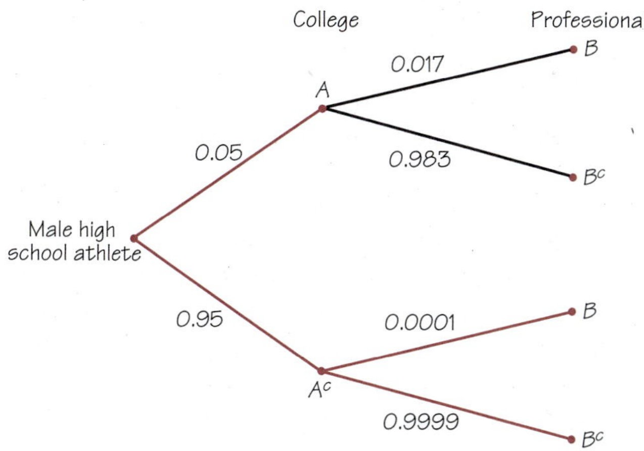

FIGURE 6.12 Tree diagram. The probability $P(B)$ is the sum of the probabilities of the two branches ending at B.

Each segment in the tree is one stage of the problem. Each complete branch shows a path that an athlete can take. The probability written on each segment is the conditional probability that an athlete follows that segment given that he has reached the point from which it branches. Starting at the left, high school athletes either do or do not compete in college. We know that the probability of competing in college is $P(A) = 0.05$, so the probability of not competing is $P(A^c) = 0.95$. These probabilities mark the leftmost branches in the tree.

Conditional on competing in college, the probability of playing professionally is $P(B \mid A) = 0.017$. So the conditional probability of *not* playing professionally is

$$P(B^c \mid A) = 1 - P(B \mid A) = 1 - 0.017 = 0.983$$

These conditional probabilities mark the paths branching out from A in Figure 6.12.

The lower half of the tree diagram describes athletes who do not compete in college (A^c). It is unusual for these athletes to play professionally, but a few go straight from high school to professional leagues. Suppose that the conditional probability that a high school athlete reaches professional play given that he does not compete in college is $P(B \mid A^c) = 0.0001$. We can now mark the two paths branching from A^c in Figure 6.12.

There are two disjoint paths to B (professional play). By the addition rule, $P(B)$ is the sum of their probabilities. The probability of reaching B through college (top half of the tree) is

$$\begin{aligned} P(B \text{ and } A) &= P(A)P(B \mid A) \\ &= 0.05 \times 0.017 = 0.00085 \end{aligned}$$

The probability of reaching B without college is

$$\begin{aligned} P(B \text{ and } A^c) &= P(A^c)P(B \mid A^c) \\ &= 0.95 \times 0.0001 = 0.000095 \end{aligned}$$

The final result is

$$P(B) = 0.00085 + 0.000095 = 0.000945$$

About 9 high school athletes out of 10,000 will play professional sports.

Tree diagrams combine the addition and multiplication rules. The multiplication rule says that **the probability of reaching the end of any complete branch is the product of the probabilities written on its segments.** The probability of any outcome, such as the event B that an athlete reaches professional sports, is then found by adding the probabilities of all branches that are part of that event.

Bayes's rule

There is another kind of probability question that we might ask in the context of studies of athletes. Our earlier calculations look forward toward professional sports as the final stage of an athlete's career. Now let's concentrate on professional athletes and look back at their earlier careers.

EXAMPLE 6.24 LOOKING BACK

What proportion of professional athletes competed in college? In the notation of Examples 6.22 and 6.23 this is the conditional probability $P(A \mid B)$. We start from the definition of conditional probability and then apply the results of Example 6.23:

$$\begin{aligned} P(A \mid B) &= \frac{P(A \text{ and } B)}{P(B)} \\ &= \frac{0.00085}{0.000945} = 0.8995 \end{aligned}$$

Almost 90% of professional athletes competed in college.

We know the probabilities $P(A)$ and $P(A^c)$ that a high school athlete does and does not compete in college. We also know the conditional probabilities $P(B \mid A)$ and $P(B \mid A^c)$ that an athlete from each group reaches professional sports. Example 6.23 shows how to use this information to calculate $P(B)$. The method can be summarized in a single expression that adds the probabilities of the two paths to B in the tree diagram:

$$P(B) = P(A)P(B \mid A) + P(A^c)P(B \mid A^c)$$

In Example 6.24 we calculated the "reverse" conditional probability $P(A \mid B)$. The denominator 0.000945 in that example came from the expression just above. Put in this general notation, we have another probability law.

BAYES'S RULE

If A and B are any events whose probabilities are not 0 or 1,

$$P(A \mid B) = \frac{P(B \mid A)P(A)}{P(B \mid A)P(A) + P(B \mid A^c)P(A^c)}$$

Bayes's rule is named after Thomas Bayes, who wrestled with arguing from outcomes like B back to antecedents like A in a book published in 1763. It is far better to think your way through problems like Examples 6.23 and 6.24 rather than memorize these formal expressions.

Independence again

The conditional probability $P(B \mid A)$ is generally not equal to the unconditional probability $P(B)$. That is because the occurrence of event A generally gives us some additional information about whether or not event B occurs. If knowing that A occurs gives no additional information about B, then A and B are independent events. The formal definition of independence is expressed in terms of conditional probability.

INDEPENDENT EVENTS

Two events A and B that both have positive probability are **independent** if

$$P(B \mid A) = P(B)$$

This definition makes precise the informal description of independence given in Section 6.2. We now see that the multiplication rule for independent events, $P(A \text{ and } B) = P(A)P(B)$, is a special case of the general multiplication rule, $P(A \text{ and } B) = P(A)P(B \mid A)$, just as the addition rule for disjoint events is a special case of the general addition rule.

Decision analysis

One kind of decision making in the presence of uncertainty seeks to make the probability of a favorable outcome as large as possible. Here is an example that illustrates how the multiplication and addition rules, organized with the help of a tree diagram, apply to a decision problem.

EXAMPLE 6.25 TRANSPLANT OR DIALYSIS?

Lynn has end-stage kidney disease: her kidneys have failed so that she cannot survive unaided. Only about 52% of patients survive for 3 years with kidney dialysis. Fortunately, a kidney is available for transplant. Lynn's doctor gives her the following information for patients in her condition.

Transplant operations usually succeed. After 1 month, 96% of the transplanted kidneys are functioning. Three percent fail to function, and the patient must return to dialysis. The remaining 1% of the patients die within a month. Patients who return to dialysis have the same chance (52%) of surviving 3 years as if they had not attempted a transplant.

Of the successful transplants, however, only 82% continue to function for 3 years. Another 8% of these patients must return to dialysis, and 70% of these survive to the 3-year mark. The remaining 10% of "successful" patients die without returning to dialysis.[8]

There is too much information here to sort through without a tree diagram. The key is to realize that most of the percentages that Lynn's doctor gives her are conditional probabilities given that a patient has some specific prior history. Figure 6.13 is a tree diagram that organizes the information.

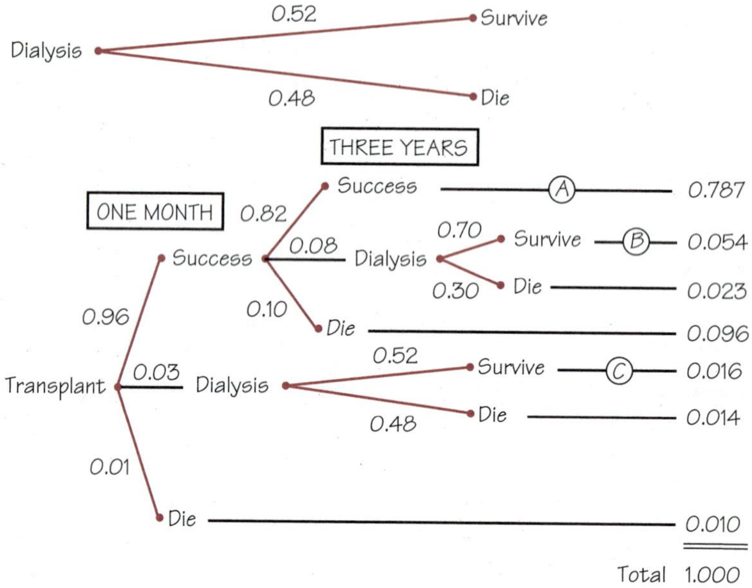

FIGURE 6.13 Tree diagram for the kidney failure decision problem.

Each path through the tree represents a possible outcome of Lynn's case. The probability written beside each branch after the first stage is the conditional probability of the next step given that Lynn has reached this point. For example, 0.82 is the conditional probability that a patient whose transplant succeeded survives 3 years with the transplant still functioning. The conditional probabilities of the other 3-year outcomes for a successful transplant are 0.08 and 0.10. They appear on the other branches from the "Success" node.

These three conditional probabilities add to 1 because these are all the possible outcomes following a successful transplant. Study the tree to convince yourself that it organizes all the information available.

The multiplication rule says that the probability of reaching the end of any path is the product of all the probabilities along that path. For example, look at the path marked A. The probability that a transplant succeeds and endures for 3 years is

$$\begin{aligned} P(\text{succeeds and lasts 3 years}) &= P(\text{succeeds})P(\text{lasts 3 years} \mid \text{succeeds}) \\ &= (0.96)(0.82) = 0.787 \end{aligned}$$

Similarly, the path marked B is the event that a patient's transplant succeeds at the 1-month stage, fails before 3 years, and the patient nonetheless survives to 3 years after returning to dialysis. The probability of this is

$$P(B) = (0.96)(0.08)(0.70) = 0.054$$

The probabilities at the end of all the paths in Figure 6.13 add to 1 because these are all the possible 3-year outcomes.

What is the probability that Lynn will survive for 3 years if she has a transplant? This is the union of the three disjoint events marked A, B, and C in Figure 6.13. By the addition rule,

$$\begin{aligned} P(\text{survive}) &= P(A) + P(B) + P(C) \\ &= 0.787 + 0.054 + 0.016 = 0.857 \end{aligned}$$

Lynn's decision is easy: 0.857 is much higher than the probability 0.52 of surviving 3 years on dialysis. She will elect the transplant.

Where do the conditional probabilities in Example 6.25 come from? They are based in part on data—that is, on studies of many patients with kidney disease. But an individual's chances of survival depend on her age, general health, and other factors. Lynn's doctor considered her individual situation before giving her these particular probabilities. It is characteristic of most decision analysis problems that *personal probabilities* are used to describe the uncertainty of an informed decision maker.

EXERCISES

6.62 IRS RETURNS In 1999, the Internal Revenue Service received 127,075,145 individual tax returns. Of these, 9,534,653 reported an adjusted gross income of at least \$100,000 and 205,124 reported at least \$1 million.

(a) What is the probability that a randomly chosen individual tax return reports an income of at least \$100,000? At least \$1 million?

(b) If you know that the return chosen shows an income of \$100,000 or more, what is the conditional probability that the income is at least \$1 million?

6.63 SURGERY RISKS You have torn a tendon and are facing surgery to repair it. The orthopedic surgeon explains the risks to you. Infection occurs in 3% of such operations, the repair fails in 14%, and both infection and failure occur together in 1%. What percent of these operations succeed and are free from infection?

6.64 HIV TESTING Enzyme immunoassay (EIA) tests are used to screen blood specimens for the presence of antibodies to HIV, the virus that causes AIDS. Antibodies indicate the presence of the virus. The test is quite accurate but is not always correct. Here are approximate probabilities of positive and negative EIA outcomes when the blood tested does and does not actually contain antibodies to HIV.[9]

	Test result	
	+	−
Antibodies present:	0.9985	0.0015
Antibodies absent:	0.006	0.994

Suppose that 1% of a large population carries antibodies to HIV in their blood.

(a) Draw a tree diagram for selecting a person from this population (outcomes: antibodies present or absent) and for testing his or her blood (outcomes: EIA positive or negative).

(b) What is the probability that the EIA is positive for a randomly chosen person from this population?

(c) What is the probability that a person has the antibody given that the EIA test is positive?

(This exercise illustrates a fact that is important when considering proposals for widespread testing for HIV, illegal drugs, or agents of biological warfare: if the condition being tested is uncommon in the population, many positives will be false positives.)

6.65 The previous exercise gives data on the results of EIA tests for the presence of antibodies to HIV. Repeat part (c) of that exercise for two different populations:

(a) Blood donors are prescreened for HIV risk factors, so perhaps only 0.1% (0.001) of this population carries HIV antibodies.

(b) Clients of a drug rehab clinic are a high-risk group, so perhaps 10% of this population carries HIV antibodies.

(c) What general lesson do your calculations illustrate?

SUMMARY

The **complement** A^c of an event A contains all outcomes that are not in A. The **union** {A **or** B} of events A and B contains all outcomes in A, in B, or in both A and B. The **intersection** {A and B} contains all outcomes that are in both A and B, but not outcomes in A alone or B alone.

The essential general rules of elementary probability are

Legitimate values: $0 \leq P(A) \leq 1$ for any event A

Total probability 1: $P(S) = 1$

Complement rule: $P(A^c) = 1 - P(A)$

Addition rule: $P(A \text{ or } B) = P(A) + P(B) - P(A \text{ and } B)$

Multiplication rule: $P(A \text{ and } B) = P(A)P(B \mid A)$

The **conditional probability** $P(B \mid A)$ of an event B given an event A is defined by

$$P(B \mid A) = \frac{P(A \text{ and } B)}{P(A)}$$

when $P(A) > 0$ but in practice is most often found from directly available information.

If A and B are **disjoint** (mutually exclusive), then $P(A \text{ and } B) = 0$. The general addition rule for unions then becomes the special addition rule, $P(A \text{ or } B) = P(A) + P(B)$.

A and B are **independent** when $P(B \mid A) = P(B)$. The multiplication rule for intersections then becomes $P(A \text{ and } B) = P(A)P(B)$.

A Venn diagram, together with the general addition rule, can be helpful in finding probabilities of the union of two events $P(A \text{ or } B)$ or the joint probability $P(A \text{ and } B)$. The joint probability $P(A \text{ and } B)$ can also be found using the general multiplication rule: $P(A \text{ and } B) = P(A)P(B \mid A) = P(B)P(A \mid B)$.

Constructing a table is a good approach for determining a conditional probability.

In problems with several stages, draw a **tree diagram** to organize use of the multiplication and addition rules.

SECTION 6.3 EXERCISES

6.66 NOBEL PRIZE WINNERS The numbers of Nobel Prize laureates in selected sciences, 1901 to 1998, are shown in the following table by location of award-winning research:[10]

Country	Physics	Chemistry	Physiology/medicine
United States	70	46	82
United Kingdom	21	26	24
Germany	61	17	29
France	25	11	7
Soviet Union	10	7	1
Japan	4	3	1

If a laureate is selected at random, what is the probability that

(a) his or her award was in chemistry?

(b) the award was won by someone from the United States?

(c) the awardee was from the United States, given that the award was for physiology/medicine?

(d) the award was for physiology/medicine, given that the awardee was from the United States?

(e) Interpret each of your results in parts (a) through (d) in terms of percents.

6.67 ACADEMIC DEGREES Here are the counts (in thousands) of earned degrees in the United States in a recent year, classified by level and by the sex of the degree recipient:

	Bachelor's	Master's	Professional	Doctorate	Total
Female	616	194	30	16	856
Male	529	171	44	26	770
Total	1145	365	74	42	1626

(a) If you choose a degree recipient at random, what is the probability that the person you choose is a woman?

(b) What is the conditional probability that you choose a woman, given that the person chosen received a professional degree?

(c) Are the events "choose a woman" and "choose a professional degree recipient" independent? How do you know?

6.68 PICK A CARD The suit of 13 hearts (A, 2 to 10, J, Q, K) from a standard deck of cards is placed in a hat. The cards are thoroughly mixed and a student reaches into the hat and selects two cards without replacement.

(a) What is the probability that the first card selected is the jack?

(b) Given that the first card selected is the jack, what is the probability that the second card is the 5?

(c) What is the probability of selecting the jack on the first draw and then the 5?

(d) What is the probability that both cards selected are greater than 5 (when the ace is considered "low")?

6.69 ACADEMIC DEGREES, II Exercise 6.67 gives the counts (in thousands) of earned degrees in the United States in a recent year. Use these data to answer the following questions.

(a) What is the probability that a randomly chosen degree recipient is a man?

(b) What is the conditional probability that the person chosen received a bachelor's degree, given that he is a man?

(c) Use the multiplication rule to find the joint probability of choosing a male bachelor's degree recipient. Check your result by finding this probability directly from the table of counts.

6.70 TEENAGE DRIVERS An insurance company has the following information about drivers aged 16 to 18 years: 20% are involved in accidents each year; 10% in this age group are A students; among those involved in an accident, 5% are A students.

(a) Let A be the event that a young driver is an A student and C the event that a young driver is involved in an accident this year. State the information given in terms of probabilities and conditional probabilities for the events A and C.

(b) What is the probability that a randomly chosen young driver is an A student and is involved in an accident?

6.71 MORE ON TEENAGE DRIVERS Use your work from Exercise 6.70 to find the percent of A students who are involved in accidents. (Start by expressing this as a conditional probability.)

6.72 Suppose that in Exercise 6.57 (page 370) the treasurer also feels that if the dollar does not fall, there is probability 0.2 that the Japanese supplier will demand that the contract be renegotiated. What is the probability that the supplier will demand renegotiation?

6.73 MULTIPLE-CHOICE EXAM STRATEGIES An examination consists of multiple-choice questions, each having five possible answers. Linda estimates that she has probability 0.75 of knowing the answer to any question that may be asked. If she does not know the answer, she will guess, with conditional probability 1/5 of being correct. What is the probability that Linda gives the correct answer to a question? (Draw a tree diagram to guide the calculation.)

6.74 ELECTION MATH The voters in a large city are 40% white, 40% black, and 20% Hispanic. (Hispanics may be of any race in official statistics, but in this case we are speaking of political blocks.) A black mayoral candidate anticipates attracting 30% of the white vote, 90% of the black vote, and 50% of the Hispanic vote. Draw a tree diagram with probabilities for the race (white, black, or Hispanic) and vote (for or against the candidate) of a randomly chosen voter. What percent of the overall vote does the candidate expect to get?

6.75 In the setting of Exercise 6.73, find the conditional probability that Linda knows the answer, given that she supplies the correct answer. (*Hint:* Use the result of Exercise 6.73 and the definition of conditional probability.)

6.76 GEOMETRIC PROBABILITY Choose a point at random in the square □ with sides $0 \leq x \leq 1$ and $0 \leq y \leq 1$. This means that the probability that the point falls in any region within the square is the area of that region. Let X be the x coordinate and Y the y coordinate of the point chosen. Find the conditional probability $P(Y < 1/2 \mid Y > X)$. (*Hint:* Draw a diagram of the square and the events $Y < 1/2$ and $Y > X$.)

6.77 INSPECTING SWITCHES A shipment contains 10,000 switches. Of these, 1000 are bad. An inspector draws switches at random, so that each switch has the same chance to be drawn.

(a) Draw one switch. What is the probability that the switch you draw is bad? What is the probability that it is not bad?

(b) Suppose the first switch drawn is bad. How many switches remain? How many of them are bad? Draw a second switch at random. What is the conditional probability that this switch is bad?

(c) Answer the questions in (b) again, but now suppose that the first switch drawn is not bad.

Comment: Knowing the result of the first trial changes the conditional probability for the second trial, so the trials are not independent. But because the shipment is large, the probabilities change very little. The trials are almost independent.

CHAPTER REVIEW

Probability describes the pattern of chance outcomes. Probability calculations provide the basis for inference. When data are produced by random sampling or randomized comparative experiments, the laws of probability answer the question, "What would happen if we did this very many times?" Probability is used to describe the long-term regularity that results from many repetitions of the same random phenomenon. The reasoning of statistical inference rests on asking "How often would this method give a correct answer if I used it very many times?" This chapter developed a probability model, including rules and tools that will help you describe the behavior of statistics from random samples in later chapters. Here are the most important things you should be able to do after studying this chapter.

PROBABILITY RULES

1. Describe the sample space of a random phenomenon. For a finite number of outcomes, use the multiplication principle to determine the number of outcomes, and use counting techniques, Venn diagrams, and tree diagrams to determine simple probabilities. For the continuous case, use geometric areas to find probabilities (areas under simple density curves) of events (intervals on the horizontal axis).

2. Know the probability rules and be able to apply them to determine probabilities of defined events. In particular, determine if a given assignment of probabilities is valid.

3. Determine if two events are disjoint, complementary, or independent. Find unions and intersections of two or more events.

4. Use Venn diagrams to picture relationships among several events.

5. Use the general addition rule to find probabilities that involve overlapping events.

6. Understand the idea of independence. Judge when it is reasonable to assume independence as part of a probability model.

7. Use the multiplication rule for independent events to find the probability that all of several independent events occur.

8. Use the multiplication rule for independent events in combination with other probability rules to find the probabilities of complex events.

9. Understand the idea of conditional probability. Find conditional probabilities for individuals chosen at random from a table of counts of possible outcomes.

10. Use the general multiplication rule to find the joint probability $P(A \text{ and } B)$ from $P(A)$ and the conditional probability $P(B \mid A)$.

11. Construct tree diagrams to organize the use of the multiplication and addition rules to solve problems with several stages.

CHAPTER 6 REVIEW EXERCISES

6.78 WHO GETS TO GO? Abby, Deborah, Julie, Sam, and Roberto work in a firm's public relations office. Their employer must choose two of them to attend a conference in Paris. To avoid unfairness, the choice will be made by drawing two names from a hat. (This is an SRS of size 2.)

(a) Write down all possible choices of two of the five names. This is the sample space.

(b) The random drawing makes all choices equally likely. What is the probability of each choice?

(c) What is the probability that Julie is chosen?

(d) What is the probability that neither of the two men (Sam and Roberto) is chosen?

6.79 ARE YOU MY (BLOOD) TYPE? All human blood can be "ABO-typed" as one of O, A, B, or AB, but the distribution of the types varies a bit among groups of people. Here is the distribution of blood types for a randomly chosen person in the United States:

Blood type:	O	A	B	AB
U.S. probability:	0.45	0.40	0.11	?

(a) What is the probability of type AB blood in the United States?

(b) An individual with type B blood can safely receive transfusions only from persons with type B or type O blood. What is the probability that the husband of a woman with type B blood is an acceptable blood donor for her?

(c) What is the probability that in a randomly chosen couple the wife has type B blood and the husband has type A?

(d) What is the probability that one of a randomly chosen couple has type A blood and the other has type B?

(e) What is the probability that at least one of a randomly chosen couple has type O blood?

6.80 The distribution of blood types in China differs from the U.S. distribution given in the previous exercise:

Blood type:	O	A	B	AB
China probability:	0.35	0.27	0.26	0.12

Choose an American and a Chinese at random, independently of each other.

(a) What is the probability that both have type O blood?

(b) What is the probability that both have the same blood type?

6.81 INCOME AND SAVINGS A sample survey chooses a sample of households and measures their annual income and their savings. Some events of interest are

A = the household chosen has income at least $100,000

C = the household chosen has at least $50,000 in savings

Based on this sample survey, we estimate that $P(A) = 0.07$ and $P(C) = 0.2$.

(a) We want to find the probability that a household either has income at least $100,000 *or* savings at least $50,000. Explain why we do not have enough information to find this probability. What additional information is needed?

(b) We want to find the probability that a household has income at least $100,000 *and* savings at least $50,000. Explain why we do not have enough information to find this probability. What additional information is needed?

6.82 SCREENING JOB APPLICANTS A company retains a psychologist to assess whether job applicants are suited for assembly-line work. The psychologist classifies applicants as A (well suited), B (marginal), or C (not suited). The company is concerned about event *D*: an employee leaves the company within a year of being hired. Data on all people hired in the past 5 years give these probabilities:

$P(A) = 0.4$	$P(B) = 0.3$	$P(C) = 0.3$
$P(A \text{ and } D) = 0.1$	$P(B \text{ and } D) = 0.1$	$P(C \text{ and } D) = 0.2$

Sketch a Venn diagram of the events A, B, C, and D and mark on your diagram the probabilities of all combinations of psychological assessment and leaving (or not) within a year. What is *P*(*D*), the probability that an employee leaves within a year?

6.83 SUICIDES Here is a two-way table of suicides committed in a recent year, classified by the gender of the victim and whether or not a firearm was used:

	Male	Female	Total
Firearm	16,381	2,559	18,940
Other	9,034	3,536	12,570
Total	25,415	6,095	31,510

Choose a suicide at random. Find the following probabilities.

(a) *P*(a firearm was used)

(b) *P*(firearm | female)

(c) P(female and firearm)

(d) P(firearm | male)

(e) P(male | firearm)

6.84 AT THE GYM Many conditional probability calculations are just common sense made automatic. For example, 10% of adults belong to health clubs, and 40% of these health club members go to the club at least twice a week. What percent of all adults go to a health club at least twice a week? Write the information in terms of probabilities and use the general multiplication rule.

6.85 TOSS TWO COINS Independence of events is not always obvious. Toss two balanced coins independently. The four possible combinations of heads and tails in order each have probability 0.25. The events

A = head on the first toss

B = both tosses have the same outcome

may seem intuitively related. Show that $P(B \mid A) = P(B)$, so that A and B are in fact independent.

6.86 BYPASS SURGERY John has coronary artery disease. He and his doctor must decide between medical management of the disease and coronary bypass surgery. Because John has been quite active, he is concerned about his quality of life as well as length of life. He wants to make the decision that will maximize the probability of the event A that he survives for 5 years and is able to carry on moderate activity during that time. The doctor makes the following probability estimates for patients of John's age and condition:

- Under medical management, $P(A) = 0.7$.
- There is probability 0.05 that John will not survive bypass surgery, probability 0.10 that he will survive with serious complications, and probability 0.85 that he will survive the surgery without complications.
- If he survives with complications, the conditional probability of the desired outcome A is 0.73. If there are no serious complications, the conditional probability of A is 0.76.

Draw a tree diagram that summarizes this information. Then calculate $P(A)$ assuming that John chooses the surgery. Does surgery or medical management offer him a better chance of achieving his goal?

6.87 POLL ON SENSITIVE ISSUES It is difficult to conduct sample surveys on sensitive issues because many people will not answer questions if the answers might embarrass them. "Randomized response" is an effective way to guarantee anonymity while collecting information on topics such as student cheating or sexual behavior. Here is the idea. To ask a sample of students whether they have plagiarized a term paper while in college, have each student toss a coin in private. If the coin lands "heads" *and* they have not plagiarized, they are to answer "No." Otherwise they are to give "Yes" as their answer. Only the student knows whether the answer reflects the truth or just the coin toss, but the researchers can use a proper random sample with follow-up for nonresponse and other good sampling practices.

Suppose that in fact the probability is 0.3 that a randomly chosen student has plagiarized a paper. Draw a tree diagram in which the first stage is tossing the coin and the second is the truth about plagiarism. The outcome at the end of each branch is the answer given to the randomized-response question. What is the probability of a "No" answer in the randomized-response poll? If the probability of plagiarism were 0.2, what would be the probability of a "No" response on the poll? Now suppose that you get 39% "No" answers in a randomized-response poll of a large sample of students at your college. What do you estimate to be the percent of the population who have plagiarized a paper?

NOTES AND DATA SOURCES

1. An informative and entertaining account of the origins of probability theory is Florence N. David, *Games, Gods and Gambling*, Charles Griffin, London, 1962.
2. From the EESEE story "Home-Field Advantage." The study is W. Hurley, "What sort of tournament should the World Series be?" *Chance*, 6, No. 2 (1993), pp. 31–33.
3. You can find a mathematical explanation of Benford's Law in Ted Hill, "The first-digit phenomenon," *American Scientist*, 86 (1996), pp. 358–363, and Ted Hill, "The difficulty of faking data," *Chance*, 12, No. 3 (1999), pp. 27–31. Applications to fraud detection are discussed in the second paper by Hill and in Mark A. Nigrini, "I've got your number," *Journal of Accountancy*, May 1999, available online at www.aicpa.org/pubs/jofa/joaiss.htm.
4. Corey Kilgannon, "When New York is on the end of the line," *New York Times*, November 7, 1999.
5. From the Dupont Automotive North America Color Popularity Survey, reported at www.dupont.com/automotive/.
6. This and similar psychology experiments are reported by A. Tversky and D. Kahneman, "Extensional versus intuitive reasoning: the conjunction fallacy in probability judgement," *Psychological Review*, 90 (1983), pp. 293–315.
7. These probabilities come from studies by the sociologist Harry Edwards, reported in the *New York Times*, February 25, 1986.
8. This example is modeled on Benjamin A. Barnes, "An overview of the treatment of end-stage renal disease and a consideration of some of the consequences," in J. P. Bunker, B. A. Barnes, and F. W. Mosteller (eds.), *Costs, Risks and Benefits of Surgery*, Oxford University Press, New York, 1977, pp. 325–341. The probabilities are recent estimates based on data from the United Network for Organ Sharing (www.unos.org) and Rebecca D. Williams, "Living day-to-day with kidney dialysis," Food and Drug Administration, www.fda.gov.
9. Probabilities from trials with 2897 people known to be free of HIV antibodies and 673 people known to be infected are reported in J. Richard George, "Alternative

specimen sources: methods for confirming positives," 1998 Conference on the Laboratory Science of HIV, found online at the Centers for Disease Control and Prevention, www.cdc.gov.

10. Data from the National Science Foundation, as reported in the *Statistical Abstract of the United States, 2000.*

The Granger Collection, New York

JAKOB BERNOULLI

The Law of Large Numbers

In three generations, the remarkable Bernoulli family of Basel, Switzerland, produced eight mathematicians, several of them outstanding. Five of them, including *Jakob (1654–1705)* and his brother Johann, made significant contributions to the early study of probability. By 1689 Jakob had published his *law of large numbers* in probability theory. The law of large numbers, which we will meet in this chapter, says that if an experiment is repeated many times, then the relative frequency with which an event occurs equals the probability of the event. Although he may be best known for his work in probability theory, Jakob made contributions in other areas as well. He published important work on the connections between logic and algebra, on geometry, and on infinite series.

Jakob Bernoulli's most important work was *Ars Conjectandi* (The Art of Conjecture), published in Basel in 1713, eight years after his death. The book was incomplete at the time of his death but it is still a significant accomplishment in the theory of probability. In the book Bernoulli reviewed the work of others on probability and gave many examples on how much one could expect to win playing various games of chance. He also offered the first proof of the binomial theorem for arbitrary positive integral powers.

Jakob Bernoulli held the chair of mathematics at the University of Basel from 1687 until his death in 1705, when he was succeeded by his brother Johann. Jakob had always been fascinated by the logarithmic spiral, several of whose properties he discovered, and he directed that it be carved on his tombstone with the Latin inscription *Eadem Mutata Resurgo*, "Though changed I shall arise the same."

If an experiment is repeated many times, then the relative frequency with which an event occurs equals the probability of the event.

chapter 7

Random Variables

ACTIVITY 7 The Game of Craps

Materials: Pair of dice for each pair of students

The game of craps is one of the most famous (or notorious) of all gambling games played with dice. In this game, the player rolls a pair of six-sided dice, and the *sum* of the numbers that turn up on the two faces is noted. If the sum is 7 or 11, then the player wins immediately. If the sum is 2, 3, or 12, then the player loses immediately. If any other sum is obtained, then the player continues to throw the dice until he either wins by repeating the first sum he obtained or loses by rolling a 7. Your mission in this activity is to estimate the probability of a player winning at craps. But first, let's get a feel for the game. For this activity, your class will be divided into groups of two. Your instructor will provide a pair of dice for each group of two students.

1. In your group of two students, play a total of 20 games of craps. One person will roll the dice; the other will keep track of the sums and record the end result (win or lose). If you like, you can switch jobs after 10 games have been completed. How many times out of 20 does the player win? What is the relative frequency (i.e., percentage, written as a decimal) of wins?

2. Combine your results with those of the other two-student groups in the class. What is the relative frequency of wins for the entire class?

3. Use simulation techniques to represent 25 games of craps, using either the table of random numbers or the random number generating feature of your TI-83/89. What is the relative frequency of wins based on the 25 simulations? How does this number compare to the relative frequency you found in step 2?

4. One of the ways you can win at craps is to roll a sum of 7 or 11 on your first roll. Using your results and those of your fellow students, determine the number of times a player won by rolling a sum of 7 of the first roll. What is the relative frequency of rolling a sum of 7? Repeat these calculations for a sum of 11. Which of these sums appears more likely to occur than the other, based on the class results?

5. One of the ways you can lose at craps is to roll a sum of 2, 3, or 12 on your first roll. Using your results and those of your fellow students, determine the number of times a player lost by rolling a sum of 2 on the first roll. What is the relative frequency of rolling a sum of 2? Repeat these calculations for a sum of 3 and a sum of 12. Which of these sums appears more likely to occur than the others, based on the class results?

6. Clearly, the key quantity of interest in craps is the *sum* of the numbers on the two dice. Let's try to get a better idea of how this sum behaves in general by conducting a simulation. First, determine how you would simulate

ACTIVITY 7 The Game of Craps (*continued*)

the roll of a single fair die. (*Hint*: Just use digits 1 to 6 and ignore the others.) Then determine how you would simulate a roll of two fair dice. Using this model, simulate 36 rolls of a pair of dice and determine the relative frequency of each of the possible sums.

7. Construct a relative frequency histogram of the relative frequency results in step 6. What is the approximate shape of the distribution? What sum appears most likely to occur? Which appears least likely to occur?

8. From the relative frequency data in step 6, compute the relative frequency of winning and the relative frequency of losing on your first roll in craps. How do these simulated results compare with what the class obtained?

INTRODUCTION

Sample spaces need not consist of numbers. When we toss four coins, we can record the outcome as a string of heads and tails, such as HTTH. In statistics, however, we are most often interested in numerical outcomes such as the count of heads in the four tosses. It is convenient to use a shorthand notation: Let X be the number of heads. If our outcome is HTTH, then $X = 2$. If the next outcome is TTTH, the value of X changes to $X = 1$. The possible values of X are 0, 1, 2, 3, and 4. Tossing a coin four times will give X one of these possible values. Tossing four more times will give X another and probably different value. We call X a *random variable* because its values vary when the coin tossing is repeated. We usually denote random variables by capital letters near the end of the alphabet, such as X or Y. Of course, the random variables of greatest interest to us are

RANDOM VARIABLE

A **random variable** is a variable whose value is a numerical outcome of a random phenomenon.

outcomes such as the mean $\bar{x}$ of a random sample, for which we will keep the familiar notation.[1] As we progress from general rules of probability toward statistical inference, we will concentrate on random variables. When a random variable X describes a random phenomenon, the sample space S just lists the possible values of the random variable. We usually do not mention S separately. There remains the second part of any probability model, the assignment of probabilities to events. In this section, we will learn two ways of assigning probabilities to the values of a random variable. The two types of probability models that result will dominate our application of probability to statistical inference.

7.1 DISCRETE AND CONTINUOUS RANDOM VARIABLES

Discrete random variables

We have learned several rules of probability but only one method of assigning probabilities: state the probabilities of the individual outcomes and assign probabilities to events by summing over the outcomes. The outcome probabilities must be between 0 and 1 and have sum 1. When the outcomes are numerical, they are values of a random variable. We will now attach a name to random variables having probability assigned in this way.[2]

DISCRETE RANDOM VARIABLE

A **discrete random variable** X has a countable number of possible values. The **probability distribution** of X lists the values and their probabilities:

Value of X:	x_1	x_2	x_3	$\cdots$	x_k
Probability:	p_1	p_2	p_3	$\cdots$	p_k

The probabilities p_i must satisfy two requirements:

1. Every probability p_i is a number between 0 and 1.
2. $p_1 + p_2 + \cdots + p_k = 1$.

Find the probability of any event by adding the probabilities p_i of the particular values x_i that make up the event.

EXAMPLE 7.1 GETTING GOOD GRADES

The instructor of a large class gives 15% each of A's and D's, 30% each of B's and C's, and 10% F's. Choose a student at random from this class. To "choose at random" means to give every student the same chance to be chosen. The student's grade on a four-point scale (A = 4) is a random variable X.

The value of X changes when we repeatedly choose students at random, but it is always one of 0, 1, 2, 3, or 4. Here is the distribution of X:

Grade:	0	1	2	3	4
Probability:	0.10	0.15	0.30	0.30	0.15

The probability that the student got a B or better is the sum of the probabilities of an A and a B:

$$\begin{aligned} P(\text{grade is 3 or 4}) &= P(X = 3) + P(X = 4) \\ &= 0.30 + 0.15 = 0.45 \end{aligned}$$

We can use histograms to display probability distributions as well as distributions of data. Figure 7.1 displays **probability histograms** that compare the probability model for random digits (Example 6.11 page 347) with the model given by Benford's law (Example 6.10 page 345)

probability histogram

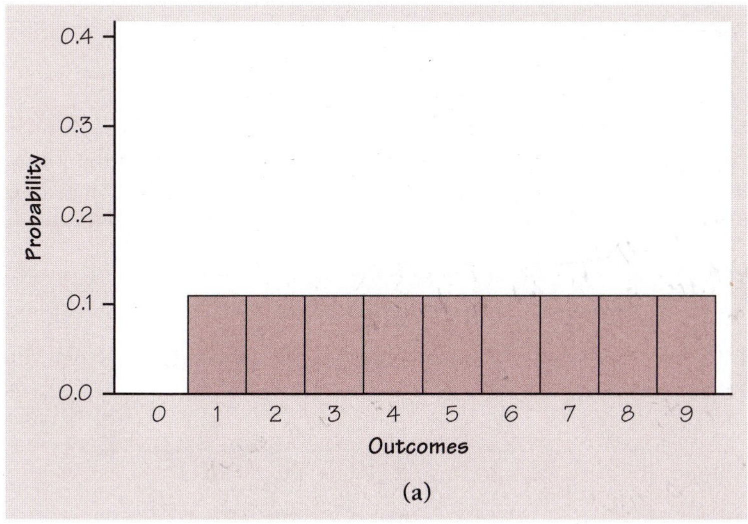

(a)

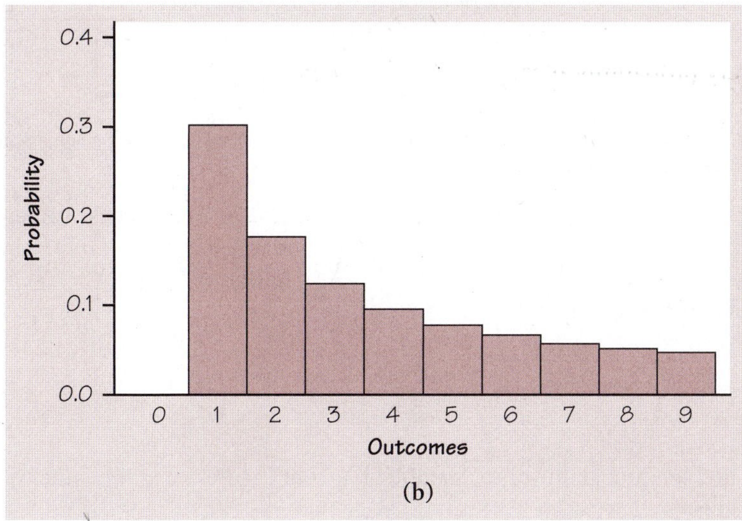

(b)

FIGURE 7.1 Probability histograms for (a) random digits 1 to 9 and (b) Benford's law. The height of each bar shows the probability assigned to a single outcome.

The height of each bar shows the probability of the outcome at its base. Because the heights are probabilities, they add to 1. As usual, all the bars in a histogram have the same width. So the areas of the bars also display the assignment of probability to outcomes. Think of these histograms as idealized pictures of the results of very many trials. The histograms make it easy to quickly compare the two distributions.

EXAMPLE 7.2 TOSSING COINS

What is the probability distribution of the discrete random variable X that counts the number of heads in four tosses of a coin? We can derive this distribution if we make two reasonable assumptions:

1. The coin is balanced, so each toss is equally likely to give H or T.
2. The coin has no memory, so tosses are independent.

The outcome of four tosses is a sequence of heads and tails such as HTTH. There are 16 possible outcomes in all. Figure 7.2 lists these outcomes along with the value of X for each outcome. The multiplication rule for independent events tells us that, for example,

$$P(\text{HTTH}) = \frac{1}{2} \times \frac{1}{2} \times \frac{1}{2} \times \frac{1}{2} = \frac{1}{16}$$

Each of the 16 possible outcomes similarly has probability 1/16. That is, these outcomes are equally likely.

		HTTH		
		HTHT		
	HTTT	THTH	HHHT	
	THTT	HHTT	HHTH	
	TTHT	THHT	HTHH	
TTTT	TTTH	TTHH	THHH	HHHH
X = 0	X = 1	X = 2	X = 3	X = 4

FIGURE 7.2 Possible outcomes in four tosses of a coin. The random variable X is the number of heads.

The number of heads X has possible values 0, 1, 2, 3, and 4. These values are *not* equally likely. As Figure 7.2 shows, there is only one way that X = 0 can occur: namely when the outcome is TTTT. So $P(X = 0) = 1/16$. But the event {X = 2} can occur in six different ways, so that

$$P(X = 2) = \frac{\text{count of ways X = 2 can occur}}{16} = \frac{6}{16}$$

We can find the probability of each value of X from Figure 7.2 in the same way. Here is the result:

$$P(X = 0) = \frac{1}{16} = 0.0625 \qquad P(X = 1) = \frac{4}{16} = 0.25 \qquad P(X = 2) = \frac{6}{16} = 0.37$$

$$P(X = 3) = \frac{4}{16} = 0.25 \qquad P(X = 4) = \frac{1}{16} = 0.0625$$

These probabilities have sum 1, so this is a legitimate probability distribution. In table form the distribution is

Number of heads:	0	1	2	3	4
Probability:	0.0625	0.25	0.375	0.25	0.0625

Figure 7.3 is a probability histogram for this distribution. The probability distribution is exactly symmetric. It is an idealization of the relative frequency distribution of the number of heads after many tosses of four coins, which would be nearly symmetric but is unlikely to be exactly symmetric.

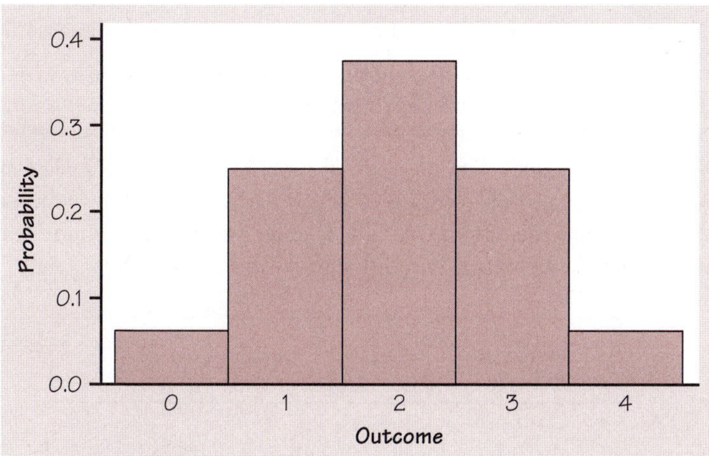

FIGURE 7.3 Probability histogram for the number of heads in four tosses of a coin.

Any event involving the number of heads observed can be expressed in terms of X, and its probability can be found from the distribution of X. For example, the probability of tossing at least two heads is

$$P(X \geq 2) = 0.375 + 0.25 + 0.0625 = 0.6875$$

The probability of at least one head is most simply found by use of the complement rule:

$$\begin{aligned} P(X \geq 1) &= 1 - P(X = 0) \\ &= 1 - 0.0625 = 0.9375 \end{aligned}$$

Recall that tossing a coin n times is similar to choosing an SRS of size n from a large population and asking a yes-or-no question. We will extend the results of Example 7.2 when we return to sampling distributions in the next two chapters.

EXERCISES

7.1 ROLL OF THE DIE If a carefully made die is rolled once, it is reasonable to assign probability 1/6 to each of the six faces.

(a) What is the probability of rolling a number less than 3?

(b) Use your TI-83/89 to simulate rolling a die 100 times, and assign the values to L_1/list1. Sort the list in ascending order, and then count the outcomes that are either 1s or 2s. Record the relative frequency.

(c) Repeat part (b) four more times, and then average the five relative frequencies. Is this number close to your result in (a)?

7.2 THREE CHILDREN A couple plans to have three children. There are 8 possible arrangements of girls and boys. For example, GGB means the first two children are girls and the third child is a boy. All 8 arrangements are (approximately) equally likely.

(a) Write down all 8 arrangements of the sexes of three children. What is the probability of any one of these arrangements?

(b) Let X be the number of girls the couple has. What is the probability that $X = 2$?

(c) Starting from your work in (a), find the distribution of X. That is, what values can X take, and what are the probabilities for each value?

7.3 SOCIAL CLASS IN ENGLAND A study of social mobility in England looked at the social class reached by the sons of lower-class fathers. Social classes are numbered from 1 (low) to 5 (high). Take the random variable X to be the class of a randomly chosen son of a father in Class 1. The study found that the distribution of X is

Son's class:	1	2	3	4	5
Probability:	0.48	0.38	0.08	0.05	0.01

(a) What percent of the sons of lower-class fathers reach the highest class, Class 5?

(b) Check that this distribution satisfies the requirements for a discrete probability distribution.

(c) What is $P(X \leq 3)$?

(d) What is $P(X < 3)$?

(e) Write the event "a son of a lower-class father reaches one of the two highest classes" in terms of values of X. What is the probability of this event?

(f) Briefly describe how you would use simulation to answer the question in (c).

7.4 HOUSING IN SAN JOSE, I How do rented housing units differ from units occupied by their owners? Here are the distributions of the number of rooms for owner-occupied units and renter-occupied units in San Jose, California:[3]

Rooms:	1	2	3	4	5	6	7	8	9	10
Owned:	0.003	0.002	0.023	0.104	0.210	0.224	0.197	0.149	0.053	0.035
Rented:	0.008	0.027	0.287	0.363	0.164	0.093	0.039	0.013	0.003	0.003

Make probability histograms of these two distributions, using the same scales. What are the most important differences between the distributions of owner-occupied and rented housing units?

7.5 HOUSING IN SAN JOSE, II Let the random variable X be the number of rooms in a randomly chosen owner-occupied housing unit in San Jose, California. Exercise 7.4 gives the distribution of X.

(a) Express "the unit has five or more rooms" in terms of X. What is the probability of this event?

(b) Express the event $\{X > 5\}$ in words. What is its probability?

(c) What important fact about discrete random variables does comparing your answers to (a) and (b) illustrate?

Continuous random variables

When we use the table of random digits to select a digit between 0 and 9, the result is a discrete random variable. The probability model assigns probability 1/10 to each of the 10 possible outcomes, as Figure 7.1(a) shows. Suppose that we want to choose a number at random between 0 and 1, allowing *any* number between 0 and 1 as the outcome. Software random number generators will do this. You can visualize such a random number by thinking of a spinner (Figure 7.4) that turns freely on its axis and slowly comes to a stop. The pointer can come to rest anywhere on a circle that is marked from 0 to 1. The sample space is now an entire interval of numbers:

$$S = \{\text{all numbers } x \text{ such that } 0 \leq x \leq 1\}$$

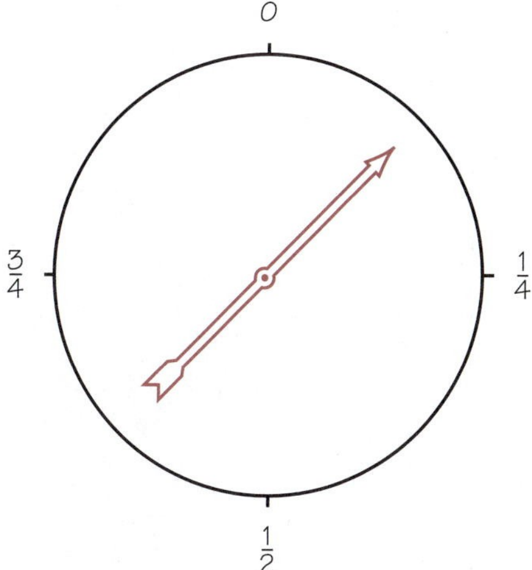

FIGURE 7.4 A spinner that generates a random number between 0 and 1.

How can we assign probabilities to such events as $0.3 \leq x \leq 0.7$? As in the case of selecting a random digit, we would like all possible outcomes to be equally likely. But we cannot assign probabilities to each individual value of x and then sum, because there are infinitely many possible values. Instead we use a new way of assigning probabilities directly to events—as *areas under a density curve*. Any density curve has area exactly 1 underneath it, corresponding to total probability 1.

EXAMPLE 7.3 RANDOM NUMBERS AND THE UNIFORM DISTRIBUTION

uniform distribution

The random number generator will spread its output uniformly across the entire interval from 0 to 1 as we allow it to generate a long sequence of numbers. The results of many trials are represented by the density curve of a ***uniform distribution*** (Figure 7.5). This density curve has height 1 over the interval from 0 to 1. The area under the density curve is 1, and the probability of any event is the area under the density curve and above the event in question.

As Figure 7.5(a) illustrates, the probability that the random number generator produces a number X between 0.3 and 0.7 is

$$P(0.3 \leq X \leq 0.7) = 0.4$$

because the area under the density curve and above the interval from 0.3 to 0.7 is 0.4. The height of the density curve is 1 and the area of a rectangle is the product of height and length, so the probability of any interval of outcomes is just the length of the interval. Similarly,

$$P(X \leq 0.5) = 0.5$$
$$P(X > 0.8) = 0.2$$
$$P(X \leq 0.5 \text{ or } X > 0.8) = 0.7$$

Notice that the last event consists of two nonoverlapping intervals, so the total area above the event is found by adding two areas, as illustrated by Figure 7.5(b). This assignment of probabilities obeys all of our rules for probability.

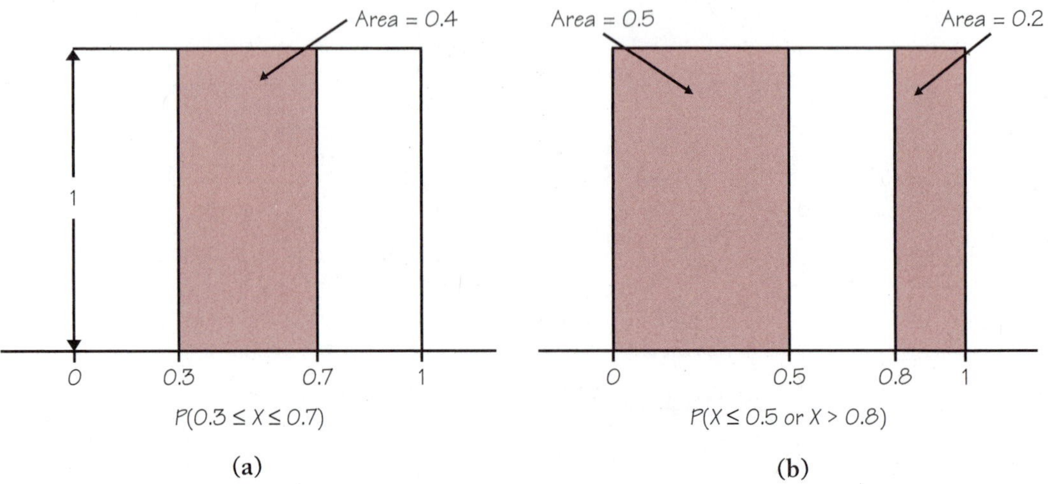

FIGURE 7.5 Assigning probability for generating a random number between 0 and 1. The probability of any interval of numbers is the area above the interval and under the curve.

Probability as area under a density curve is a second important way of assigning probabilities to events. Figure 7.6 illustrates this idea in general form.

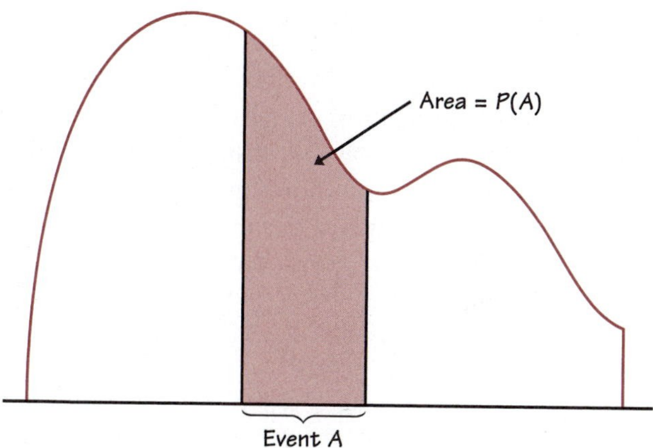

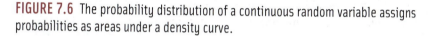

FIGURE 7.6 The probability distribution of a continuous random variable assigns probabilities as areas under a density curve.

We call X in Example 7.3 a *continuous random variable* because its values are not isolated numbers but an entire interval of numbers.

CONTINUOUS RANDOM VARIABLE

A **continuous random variable** X takes all values in an interval of numbers. The **probability distribution** of X is described by a density curve. The probability of any event is the area under the density curve and above the values of X that make up the event.

The probability model for a continuous random variable assigns probabilities to intervals of outcomes rather than to individual outcomes. In fact, **all continuous probability distributions assign probability 0 to every individual outcome.** Only intervals of values have positive probability. To see that this is true, consider a specific outcome such as $P(X = 0.8)$ in Example 7.3. The probability of any interval is the same as its length. The point 0.8 has no length, so its probability is 0. Although this fact may seem odd at first glance, it does make intuitive as well as mathematical sense. The random number generator produces a number between 0.79 and 0.81 with probability 0.02. An outcome between 0.799 and 0.801 has probability 0.002, and a result between 0.7999 and 0.8001 has probability 0.0002. Continuing to home in on 0.8, we can see why an outcome *exactly* equal to 0.8 should have probability 0. Because there is no probability exactly at $X = 0.8$, the two events $\{X > 0.8\}$ and $\{X \geq 0.8\}$ have the same probability. We can ignore the distinction between $>$ and $\geq$ when finding probabilities for continuous (but not discrete) random variables.

Normal distributions as probability distributions

The density curves that are most familiar to us are the normal curves. (We discussed normal curves in Section 2.1.) Because any density curve describes an assignment of probabilities, *normal distributions are probability distributions.* Recall that $N(\mu, \sigma)$ is our shorthand notation for the normal distribution having mean μ and standard deviation σ. In the language of random variables, if X has the $N(\mu, \sigma)$ distribution, then the standardized variable

$$Z = \frac{X - \mu}{\sigma}$$

is a standard normal random variable having the distribution $N(0, 1)$.

EXAMPLE 7.4 DRUGS IN SCHOOLS

An opinion poll asks an SRS of 1500 American adults what they consider to be the most serious problem facing our schools. Suppose that if we could ask all adults this question, 30% would say "drugs." We will learn in Chapter 9 that the proportion $p = 0.3$ is a population parameter and that the proportion $\hat{p}$ of the sample who answer "drugs" is a statistic used to estimate p. We will see in Chapter 9 that $\hat{p}$ is a random variable that has approximately the $N(0.3, 0.0118)$ distribution. The mean 0.3 of this distribution is the same as the population parameter because $\hat{p}$ is an unbiased estimate of p. The standard deviation is controlled mainly by the sample size, which is 1500 in this case.

What is the probability that the poll result differs from the truth about the population by more than two percentage points? Figure 7.7 shows this probability as an area under a normal density curve.

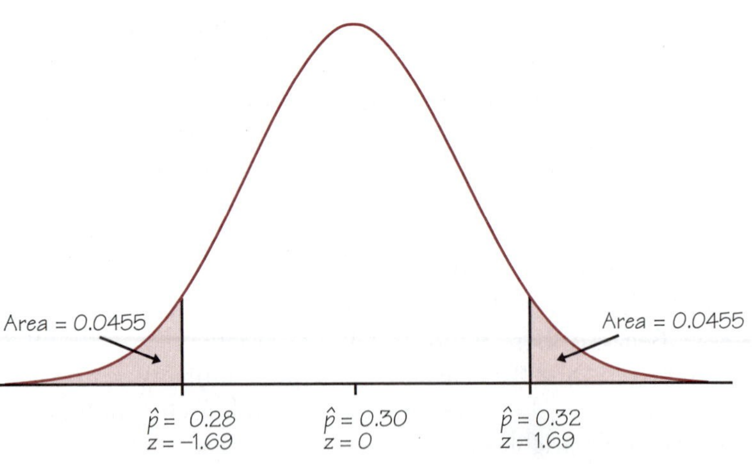

FIGURE 7.7 Probability in Example 7.4 as area under a normal density curve.

By the addition rule for disjoint events, the desired probability is

$$P(\hat{p} < 0.28 \text{ or } \hat{p} > 0.32) = P(\hat{p} < 0.28) + P(\hat{p} > 0.32)$$

You can find the two individual probabilities from software or by standardizing and using Table A.

$$\begin{aligned} P(\hat{p} < 0.28) &= P\left(Z < \frac{0.28 - 0.3}{0.0118}\right) \\ &= P(Z < -1.69) = 0.0455 \\ P(\hat{p} > 0.32) &= P\left(Z > \frac{0.32 - 0.3}{0.0118}\right) \\ &= P(Z > 1.69) = 0.0455 \end{aligned}$$

Therefore,

$$P(\hat{p} < 0.28 \text{ or } \hat{p} > 0.32) = 0.0455 + 0.0455 = 0.0910$$

The probability that the sample result will miss the truth by more than two percentage points is 0.091. The arrangement of this calculation is familiar from our earlier work with normal distributions. Only the language of probability is new.

We could also do the calculation by first finding the probability of the complement:

$$\begin{aligned} P(0.28 \le \hat{p} \le 0.32) &= P\left(\frac{0.28 - 0.3}{0.0118} \le Z \le \frac{0.32 - 0.3}{0.0118}\right) \\ &= P(-1.69 \le Z \le 1.69) \\ &= 0.9545 - 0.0455 = 0.9090 \end{aligned}$$

Then by the complement rule,

$$\begin{aligned} P(\hat{p} < 0.28 \text{ or } \hat{p} > 0.32) &= 1 - P(0.28 \le \hat{p} \le 0.32) \\ &= 1 - 0.9090 = 0.0910 \end{aligned}$$

There is often more than one correct way to use the rules of probability to answer a question.

EXERCISES

7.6 CONTINUOUS RANDOM VARIABLE, I Let X be a random number between 0 and 1 produced by the idealized uniform random number generator described in Example 7.3 and Figure 7.5. Find the following probabilities:

(a) $P(0 \le X \le 0.4)$

(b) $P(0.4 \le X \le 1)$

(c) $P(0.3 \le X \le 0.5)$

(d) $P(0.3 < X < 0.5)$

(e) $P(0.226 \le X \le 0.713)$

(f) What important fact about continuous random variables does comparing your answers to (c) and (d) illustrate?

7.7 CONTINUOUS RANDOM VARIABLE, II Let the random variable X be a random number with the uniform density curve in Figure 7.5, as in the previous exercise. Find the following probabilities:

(a) $P(X \leq 0.49)$

(b) $P(X \geq 0.27)$

(c) $P(0.27 < X < 1.27)$

(d) $P(0.1 \leq X \leq 0.2 \text{ or } 0.8 \leq X \leq 0.9)$

(e) The probability that X is not in the interval 0.3 to 0.8.

(f) $P(X = 0.5)$

7.8 VIOLENCE IN SCHOOLS, I An SRS of 400 American adults is asked, "What do you think is the most serious problem facing our schools?" Suppose that in fact 40% of all adults would answer "violence" if asked this question. The proportion $\hat{p}$ of the sample who answer "violence" will vary in repeated sampling. In fact, we can assign probabilities to values of $\hat{p}$ using the normal density curve with mean 0.4 and standard deviation 0.023. Use this density curve to find the probabilities of the following events:

(a) At least 45% of the sample believes that violence is the schools' most serious problem.

(b) Less than 35% of the sample believes that violence is the most serious problem.

(c) The sample proportion is between 0.35 and 0.45.

7.9 VIOLENCE IN SCHOOLS, II How could you design a simulation to answer part (b) of Exercise 7.8? What we need to do is simulate 400 observations from the N(0.4, 0.023) distribution. This is easily done on the calculator. Here's one way: Clear L_1/list1 and enter the following commands (randNorm is found under the MATH/PRB menu on the TI-83, and in the CATALOG under FlashApps on the TI-89):

TI-83	TI-89
• `randNorm(0.4,.023,400)→L₁`	• `tistat.randNorm(0.4,.023,400)→list1`

This will select 400 random observations from the N(0.4, 0.023) distribution.

• `SortA(L₁)`	• `SortA list1`

This will sort the 400 observations in L_1/list1 in ascending order.

Then scroll through L_1/list1. How many entries (observations) are less than 0.25? What is the relative frequency of this event? Compare the results of your simulation with your answer to Exercise 7.8(b).

SUMMARY

The previous chapter included a general discussion of the idea of probability and the properties of probability models. Two very useful specific types of probability models are distributions of discrete and continuous random variables. In our study of statistics we will employ only these two types of probability models.

A **random variable** is a variable taking numerical values determined by the outcome of a random phenomenon. The **probability distribution** of a random variable X tells us what the possible values of X are and how probabilities are assigned to those values.

A random variable X and its distribution can be discrete or continuous.

A **discrete random variable** has a countable number of possible values. The probability distribution assigns each of these values a probability between 0 and 1 such that the sum of all the probabilities is exactly 1. The probability of any event is the sum of the probabilities of all the values that make up the event.

A **continuous random variable** takes all values in some interval of numbers. A **density curve** describes the probability distribution of a continuous random variable. The probability of any event is the area under the curve above the values that make up the event.

Normal distributions are one type of continuous probability distribution.

You can picture a probability distribution by drawing a **probability histogram** in the discrete case or by graphing the density curve in the continuous case.

When you work problems, get in the habit of first identifying the random variable of interest. X = number of ______ for discrete random variables, and X = amount of ______ for continuous random variables.

SECTION 7.1 EXERCISES

7.10 SIZE OF AMERICAN HOUSEHOLDS, I In government data, a household consists of all occupants of a dwelling unit, while a family consists of two or more persons who live together and are related by blood or marriage. So all families form households, but some households are not families. Here are the distributions of household size and family size in the United States:

Number of persons:	1	2	3	4	5	6	7
Household probability:	0.25	0.32	0.17	0.15	0.07	0.03	0.01
Family probability:	0	0.42	0.23	0.21	0.09	0.03	0.02

(a) Verify that each is a legitimate discrete probability distribution function.

(b) Make probability histograms for these two discrete distributions, using the same scales. What are the most important differences between the sizes of households and families?

7.11 SIZE OF AMERICAN HOUSEHOLDS, II Choose an American household at random and let the random variable Y be the number of persons living in the household. Exercise 7.10 gives the distribution of Y.

(a) Express "more than one person lives in this household" in terms of Y. What is the probability of this event?

(b) What is $P(2 < Y \leq 4)$?

(c) What is $P(Y \neq 2)$?

7.12 CAR OWNERSHIP Choose an American household at random and let the random variable X be the number of cars (including SUVs and light trucks) they own. Here is the probability model if we ignore the few households that own more than 5 cars:

Number of cars X:	0	1	2	3	4	5
Probability:	0.09	0.36	0.35	0.13	0.05	0.02

(a) Verify that this is a legitimate discrete distribution. Display the the distribution in a probability histogram.

(b) Say in words what the event $\{X \geq 1\}$ is. Find $P(X \geq 1)$.

(c) A housing company builds houses with two-car garages. What percent of households have more cars than the garage can hold?

7.13 ROLLING TWO DICE Some games of chance rely on tossing two dice. Each die has six faces, marked with 1, 2, . . ., 6 spots called pips. The dice used in casinos are carefully balanced so that each face is equally likely to come up. When two dice are tossed, each of the 36 possible pairs of faces is equally likely to come up. The outcome of interest to a gambler is the sum of the pips on the two up-faces. Call this random variable X.

(a) Write down all 36 possible pairs of faces.

(b) If all pairs have the same probability, what must be the probability of each pair?

(c) Define the random variable X. Then write the value of X next to each pair of faces and use this information with the result of (b) to give the probability distribution of X. Draw a probability histogram to display the distribution.

(d) One bet available in craps wins if a 7 or 11 comes up on the next roll of two dice. What is the probability of rolling a 7 or 11 on the next roll? Compare your answer with your experimental results (relative frequency) in Activity 7, step 4.

(e) After the dice are rolled the first time, several bets lose if a 7 is then rolled. If any outcome other than a 7 occurs, these bets either win or continue to the next roll. What is the probability that anything other than a 7 is rolled?

7.14 WEIRD DICE Nonstandard dice can produce interesting distributions of outcomes. You have two balanced, six-sided dice. One is a standard die, with faces having 1, 2, 3, 4, 5, and 6 spots. The other die has three faces with 0 spots and three faces with 6 spots. Find the probability distribution for the total number of spots Y on the up-faces when you roll these two dice.

7.15 EDUCATION LEVELS A study of education followed a large group of fifth-grade children to see how many years of school they eventually completed. Let X be the highest year of school that a randomly chosen fifth grader completes. (Students who go on to college are included in the outcome X = 12.) The study found this probability distribution for X:

Years:	4	5	6	7	8	9	10	11	12
Probability:	0.010	0.007	0.007	0.013	0.032	0.068	0.070	0.041	0.752

(a) What percent of fifth graders eventually finished twelfth grade?

(b) Check that this is a legitimate discrete probability distribution.

(c) Find $P(X \geq 6)$.

(d) Find $P(X > 6)$.

(e) What values of X make up the event "the student completed at least one year of high school"? (High school begins with the ninth grade.) What is the probability of this event?

7.16 HOW STUDENT FEES ARE USED Weary of the low turnout in student elections, a college administration decides to choose an SRS of three students to form an advisory board that represents student opinion. Suppose that 40% of all students oppose the use of student fees to fund student interest groups and that the opinions of the three students on the board are independent. Then the probability is 0.4 that each opposes the funding of interest groups.

(a) Call the three students A, B, and C. What is the probability that A and B support funding and C opposes it?

(b) List all possible combinations of opinions that can be held by students A, B, and C. (*Hint:* There are eight possibilities.) Then give the probability of each of these outcomes. Note that they are not equally likely.

(c) Let the random variable X be the number of student representatives who oppose the funding of interest groups. Give the probability distribution of X.

(d) Express the event "a majority of the advisory board opposes funding" in terms of X and find its probability.

7.17 A UNIFORM DISTRIBUTION Many random number generators allow users to specify the range of the random numbers to be produced. Suppose that you specify that the range is to be $0 \leq Y \leq 2$. Then the density curve of the outcomes has constant height between 0 and 2, and height 0 elsewhere.

(a) What is the height of the density curve between 0 and 2? Draw a graph of the density curve.

(b) Use your graph from (a) and the fact that probability is area under the curve to find $P(Y \leq 1)$.

(c) Find $P(0.5 < Y < 1.3)$.

(d) Find $P(Y \geq 0.8)$.

7.18 THE SUM OF TWO RANDOM DECIMALS Generate *two* random numbers between 0 and 1 and take Y to be their sum. Then Y is a continuous random variable that can take any value between 0 and 2. The density curve of Y is the triangle shown in Figure 7.8.

(a) Verify that the area under this curve is 1.

(b) What is the probability that Y is less than 1? (Sketch the density curve, shade the area that represents the probability, then find that area. Do this for (c) also.)

(c) What is the probability that Y is less than 0.5?

(d) Use simulation methods to answer the questions in (b) and (c). Here's one way using the TI-83/89. Clear L_1/list1, L_2/list2, and L_3/list3 and enter these commands:

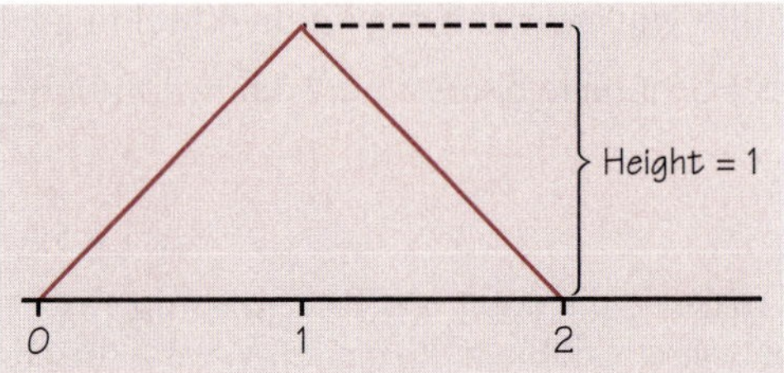

FIGURE 7.8 The density curve for the sum of two random numbers. This continuous random variable takes values between 0 and 2.

TI-83	TI-89	
`rand(200)→L1`	`tistat.rand83(200) →list1`	Generates 200 random numbers and stores them in L_1/list1
`rand(200)→L2`	`tistat.rand83(200) →list2`	Generates 200 random numbers and stores them in L_2/list2
`L1+L2→L3`	`list1+list2→list3`	Adds the first number in L_1/list1 and the first number in L_2/list2 and stores the sum in L_3/list3, and so forth, from $i = 1$ to $i = 200$
`SortA(L3)`	`SortA list3`	Sorts the sums in L_3/list3 in ascending order

Now simply scroll through L_3/list3 and count the number of sums that satisfy the conditions stated in (b) and (c), and determine the relative frequency.

7.19 **PICTURING A DISTRIBUTION** This is a continuation of the previous exercise. If you carried out the simulation in 7.18(d), you can picture the distribution as follows: Deselect any active functions in the Y = screen, and turn off all STAT PLOTs. Define Plot1 to be a histogram using list L_3/list3. On the TI-89, set the Hist. Bucket Width at 0.1. Set WINDOW dimensions as follows: $X[0, 2]_{0.1}$ and $Y[-6, 25]_5$. Then press GRAPH. Does the resulting histogram resemble the triangle in Figure 7.8? Can you imagine the triangle superimposed on top of the histogram? Of course, some bars will be too short and others will be too long, but this is due to chance variation. To overlay the triangle, define Y_1 to be:

- TI-83: $Y_1 = (25X)(X \geq 0 \text{ and } X \leq 1) + (-25X + 50)(X \geq 1 \text{ and } X \leq 2)$
- TI-89: when ($x \geq 0$ and $x \leq 2$, when ($x \leq 1$, $25x$, $-25x + 50$), 0)

Then press GRAPH again. How well does this "curve" fit your histogram?

7.20 **JOGGERS, I** An opinion poll asks an SRS of 1500 adults, "Do you happen to jog?" Suppose that the population proportion who jog is $p = 0.15$. To estimate p, we use the proportion $\hat{p}$ in the sample who answer "Yes." The statistic $\hat{p}$ is a random variable that is approximately normally distributed with mean $\mu = 0.15$ and standard deviation $\sigma = 0.0092$. Find the following probabilities:

(a) $P(\hat{p} \geq 0.16)$

(b) $P(0.14 \leq \hat{p} \leq 0.16)$

7.21 JOGGERS, II Describe the details of a simulation you could carry out to approximate an answer to Exercise 7.20(a). Then carry out the simulation. About how many repetitions do you need to get a result close to your answer to Exercise 7.20(a)?

7.2 MEANS AND VARIANCES OF RANDOM VARIABLES

Probability is the mathematical language that describes the long-run regular behavior of random phenomena. The probability distribution of a random variable is an idealized relative frequency distribution. The probability histograms and density curves that picture probability distributions resemble our earlier pictures of distributions of data. In describing data, we moved from graphs to numerical measures such as means and standard deviations. Now we will make the same move to expand our descriptions of the distributions of random variables. We can speak of the mean winnings in a game of chance or the standard deviation of the randomly varying number of calls a travel agency receives in an hour. In this section we will learn more about how to compute these descriptive measures and about the laws they obey.

The mean of a random variable

The mean $\bar{x}$ of a set of observations is their ordinary average. The mean of a random variable X is also an average of the possible values of X, but with an essential change to take into account the fact that not all outcomes need be equally likely. An example will show what we must do.

EXAMPLE 7.5 THE TRI-STATE PICK 3

Most states and Canadian provinces have government-sponsored lotteries. Here is a simple lottery wager, from the Tri-State Pick 3 game that New Hampshire shares with Maine and Vermont. You choose a three-digit number; the state chooses a three-digit winning number at random and pays you \$500 if your number is chosen. Because there are 1000 three-digit numbers, you have probability 1/1000 of winning. Taking X to be the amount your ticket pays you, the probability distribution of X is

Payoff X:	\$0	\$500
Probability:	0.999	0.001

What is your average payoff from many tickets? The ordinary average of the two possible outcomes \$0 and \$500 is \$250, but that makes no sense as the average because \$500 is much less likely than \$0. In the long run you receive \$500 once in every 1000 tickets and \$0 on the remaining 999 of 1000 tickets. The long-run average payoff is

$$\$500\frac{1}{1000} + \$0\frac{999}{1000} = \$0.50$$

or fifty cents. That number is the mean of the random variable X. (Tickets cost \$1, so in the long run the state keeps half the money you wager.)

If you play Tri-State Pick 3 several times, we would as usual call the mean of the actual amounts you win $\bar{x}$. The mean in Example 7.5 is a different quantity—it is the long-run average winnings you expect if you play a very large number of times. Just as probabilities are an idealized description of long-run proportions, the mean of a probability distribution describes the long-run average outcome. We can't call this mean $\bar{x}$, so we need a different symbol. The common symbol for the mean of a probability distribution is μ, the Greek letter mu. We used μ in Chapter 2 for the mean of a normal distribution, so this is not a new notation. We will often be interested in several random variables, each having a different probability distribution with a different mean. To remind ourselves that we are talking about the mean of X we often write μ_X rather than simply μ. In Example 7.5, $\mu_X = \$0.50$. Notice that, as often happens, the mean is not a possible value of X. You will often find the mean of a random variable X called the ***expected value*** of X. This term can be misleading, for we don't necessarily expect one observation on X to be close to its expected value.

mean μ

expected value

The mean of any discrete random variable is found just as in Example 7.5. It is an average of the possible outcomes, but a weighted average in which each outcome is weighted by its probability. Because the probabilities add to 1, we have total weight 1 to distribute among the outcomes. An outcome that occurs half the time has probability one-half and so gets one-half the weight in calculating the mean. Here is the general definition.

MEAN OF A DISCRETE RANDOM VARIABLE

Suppose that X is a discrete random variable whose distribution is

Value of X:	x_1	x_2	x_3	...	x_k
Probability:	p_1	p_2	p_3	...	p_k

To find the **mean** of X, multiply each possible value by its probability, then add all the products:

$$\mu_X = x_1p_1 + x_2p_2 + \cdots + x_kp_k$$
$$= \Sigma x_i pm_i$$

EXAMPLE 7.6 BENFORD'S LAW

If first digits in a set of data appear "at random," the nine possible digits 1 to 9 all have the same probability. The probability distribution of the first digit X is then

First digit X:	1	2	3	4	5	6	7	8	9
Probability:	1/9	1/9	1/9	1/9	1/9	1/9	1/9	1/9	1/9

The mean of this distribution is

$$\mu_x = 1\times\frac{1}{9}+2\times\frac{1}{9}+3\times\frac{1}{9}+4\times\frac{1}{9}+5\times\frac{1}{9}+6\times\frac{1}{9}+7\times\frac{1}{9}+8\times\frac{1}{9}+9\times\frac{1}{9}$$
$$= 45\times\frac{1}{9} = 5$$

If, on the other hand, the data obey Benford's law, the distribution of the first digit V is

First digit V:	1	2	3	4	5	6	7	8	9
Probability:	0.301	0.176	0.125	0.097	0.079	0.067	0.058	0.051	0.046

The mean of V is

$$\mu_V = (1)(0.301) + (2)(0.176) + (3)(0.125) + (4)(0.097) + (5)(0.079) + (6)(0.067)$$
$$+ (7)(0.058) + (8)(0.051) + (9)(0.046)$$
$$= 3.441$$

The means reflect the greater probability of smaller first digits under Benford's law.

Figure 7.9 locates the means of X and V on the two probability histograms. Because the discrete uniform distribution of Figure 7.9(a) is symmetric, the mean lies at the center of symmetry. We can't locate the mean of the right-skewed distribution of Figure 7.9(b) by eye—calculation is needed.

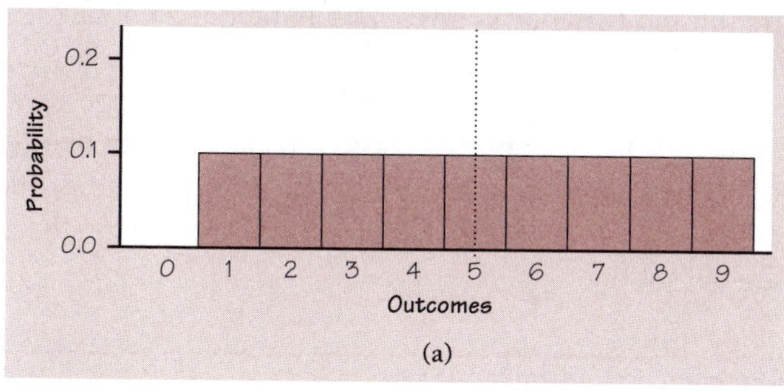

(a)

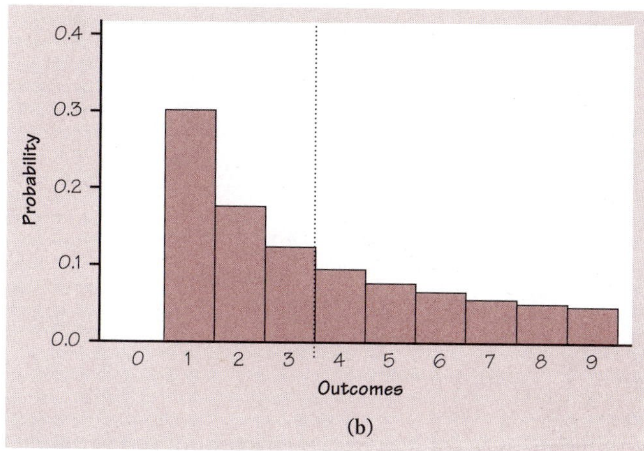

(b)

FIGURE 7.9 Locating the mean of a discrete random variable on the probability histogram for (a) digits between 1 and 9 chosen at random; (b) digits between 1 and 9 chosen from records that obey Benford's law.

What about continuous random variables? The probability distribution of a continuous random variable X is described by a density curve. Chapter 2 showed how to find the mean of the distribution: it is the point at which the area under the density curve would balance if it were made out of solid material. The mean lies at the center of symmetric density curves such as the normal curves. Exact calculation of the mean of a distribution with a skewed density curve requires advanced mathematics.[4]

The idea that the mean is the balance point of the distribution applies to discrete random variables as well, but in the discrete case we have a formula that gives us this point.

The variance of a random variable

The mean is a measure of the center of a distribution. Even the most basic numerical description requires in addition a measure of the spread or variability of the distribution. The variance and the standard deviation are the measures of spread that accompany the choice of the mean to measure center. Just as for the mean, we need a distinct symbol to distinguish the variance of a random variable from the variance s^2 of a data set. We write the variance of a random variable X as σ_X^2. Once again the subscript reminds us which variable we have in mind. The definition of the variance σ_X^2 of a random variable is similar to the definition of the sample variance s^2 given in Chapter 1. That is, the variance is an average of the squared deviation $(X - \mu_X)^2$ of the variable X from its mean μ_X. As for the mean, the average we use is a weighted average in which each outcome is weighted by its probability in order to take account of outcomes that are not equally likely. Calculating this weighted average is straightforward for discrete random variables but requires advanced mathematics in the continuous case. Here is the definition.

VARIANCE OF A DISCRETE RANDOM VARIABLE

Suppose that X is a discrete random variable whose distribution is

Value of X:	x_1	x_2	x_3	...	x_k
Probability:	p_1	p_2	p_3	...	p_k

and that μ is the mean of X. The **variance** of X is

$$\begin{aligned}\sigma_X^2 &= (x_1 - \mu_X)^2 p_1 + (x_2 - \mu_X)^2 p_2 + \cdots + (x_k - \mu_X)^2 p_k \\ &= \Sigma (x_i - \mu_X)^2 p_i\end{aligned}$$

The **standard deviation** σ_X of X is the square root of the variance.

EXAMPLE 7.7 SELLING AIRCRAFT PARTS

Gain Communications sells aircraft communications units to both the military and the civilian markets. Next year's sales depend on market conditions that cannot be predicted exactly. Gain follows the modern practice of using probability estimates of sales. The military division estimates its sales as follows:

Units sold:	1000	3000	5000	10,000
Probability:	0.1	0.3	0.4	0.2

These are personal probabilities that express the informed opinion of Gain's executives. Take X to be the number of military units sold. From the probability distribution we compute that

$$\begin{aligned}\mu_X &= (1000)(0.1) + (3000)(0.3) + (5000)(0.4) + (10{,}000)(0.2)\\ &= 100 + 900 + 2000 + 2000\\ &= 5000 \text{ units}\end{aligned}$$

The variance of X is calculated as

$$\begin{aligned}\sigma_X^2 = \Sigma(x_i - \mu_X)^2 p_i &= (1000 - 5000)^2(0.1) + (3000 - 5000)^2(0.3) + (5000 - 5000)^2(0.4)\\ &\quad + (10{,}000 - 5000)^2(0.2)\\ &= 1{,}600{,}000 + 1{,}200{,}000 + 0 + 5{,}000{,}000\\ &= 7{,}800{,}000\end{aligned}$$

The calculations can be arranged in the form of a table. Both μ_X and σ_X^2 are sums of columns in this table.

x_i	p_i	$x_i p_i$	$(x_i - \mu_X)^2 p_i$
1,000	0.1	100	$(1{,}000 - 5{,}000)^2\ (0.1) = 1{,}600{,}000$
3,000	0.3	900	$(3{,}000 - 5{,}000)^2\ (0.3) = 1{,}200{,}000$
5,000	0.4	2,000	$(5{,}000 - 5{,}000)^2\ (0.4) = 0$
10,000	0.2	2,000	$(10{,}000 - 5{,}000)^2\ (0.2) = 5{,}000{,}000$
		$\mu_X = 5{,}000$	$\sigma_X^2 = 7{,}800{,}000$

We see that $\sigma_X^2 = 7{,}800{,}000$. The standard deviation of X is $\sigma_X = \sqrt{7{,}800{,}000} = 2792.8$. The standard deviation is a measure of how variable the number of units sold is. As in the case of distributions for data, the standard deviation of a probability distribution is easiest to understand for normal distributions.

EXERCISES

7.22 A GRADE DISTRIBUTION Example 7.1 gives the distribution of grades (A = 4, B = 3, and so on) in a large class as

Grade:	0	1	2	3	4
Probability:	0.10	0.15	0.30	0.30	0.15

Find the average (that is, the mean) grade in this course.

7.23 OWNED AND RENTED HOUSING, I How do rented housing units differ from units occupied by their owners? Exercise 7.4 (page 396) gives the distributions of the number of rooms for owner-occupied units and renter-occupied units in San Jose, California. Find the mean number of rooms for both types of housing unit. How do the means reflect the differences between the distributions that you found in Exercise 7.4?

7.24 PICK 3 The Tri-State Pick 3 lottery game offers a choice of several bets. You choose a three-digit number. The lottery commission announces the winning three-digit number, chosen at random, at the end of each day. The "box" pays $83.33 if the number you choose has the same digits as the winning number, in any order. Find the expected payoff for a $1 bet on the box. (Assume that you chose a number having three different digits.)

7.25 KENO Keno is a favorite game in casinos, and similar games are popular with the states that operate lotteries. Balls numbered 1 to 80 are tumbled in a machine as the bets are placed, then 20 of the balls are chosen at random. Players select numbers by marking a card. The simplest of the many wagers available is "Mark 1 Number." Your payoff is $3 on a $1 bet if the number you select is one of those chosen. Because 20 of 80 numbers are chosen, your probability of winning is 20/80, or 0.25.

(a) What is the probability distribution (the outcomes and their probabilities) of the payoff X on a single play?

(b) What is the mean payoff μ_X?

(c) In the long run, how much does the casino keep from each dollar bet?

7.26 GRADE DISTRIBUTION, II Find the standard deviation σ_X of the distribution of grades in Exercise 7.22.

7.27 HOUSEHOLDS AND FAMILIES Exercise 7.10 (page 403) gives the distributions of the number of people in households and in families in the United States. An important difference is that many households consist of one person living alone, whereas a family must have at least two members. Some households may contain families along with other people, and so will be larger than the family. These differences make it hard to compare the distributions without calculations. Find the mean and standard deviation of both household size and family size. Combine these with your descriptions from Exercise 7.10 to give a comparison of the two distributions.

7.28 OWNED AND RENTED HOUSING, II Which of the two distributions for room counts in Exercises 7.4 (page 396) and 7.23 appears more spread out in the probability histograms? Why? Find the standard deviation for both distributions. The standard deviation provides a numerical measure of spread.

7.29 KIDS AND TOYS In an experiment on the behavior of young children, each subject is placed in an area with five toys. The response of interest is the number of toys that the child plays with. Past experiments with many subjects have shown that the probability distribution of the number X of toys played with is as follows:

Number of toys x_i:	0	1	2	3	4	5
Probability p_i:	0.03	0.16	0.30	0.23	0.17	0.11

(a) Calculate the mean μ_X and the standard deviation σ_X.

(b) Describe the details of a simulation you could carry out to approximate the mean number of toys μ_X and the standard deviation σ_X. Then carry out your simulation. Are the mean and standard deviation produced from your simulation close to the values you calculated in (a)?

Statistical estimation and the law of large numbers

We would like to estimate the mean height μ of the population of all American women between the ages of 18 and 24 years. This μ is the mean μ_X of the random variable X obtained by choosing a young woman at random and measuring her height. To estimate μ, we choose an SRS of young women and use the sample mean $\bar{x}$ to estimate the unknown population mean μ. Statistics obtained from probability samples are random variables because their values would vary in repeated sampling. The sampling distributions of statistics are just the probability distributions of these random variables. We will study sampling distributions in Chapter 9.

It seems reasonable to use $\bar{x}$ to estimate μ. An SRS should fairly represent the population, so the mean $\bar{x}$ of the sample should be somewhere near the mean μ of the population. Of course, we don't expect $\bar{x}$ to be exactly equal to μ, and we realize that if we choose another SRS, the luck of the draw will probably produce a different $\bar{x}$.

If $\bar{x}$ is rarely exactly right and varies from sample to sample, why is it nonetheless a reasonable estimate of the population mean μ? If we keep on adding observations to our random sample, the statistic $\bar{x}$ is *guaranteed* to get as close as we wish to the parameter μ and then stay that close. We have the comfort of knowing that if we can afford to keep on measuring more young women, eventually we will estimate the mean height of all young women very accurately. This remarkable fact is called the *law of large numbers*. It is remarkable because it holds for *any* population, not just for some special class such as normal distributions.

LAW OF LARGE NUMBERS

Draw independent observations at random from any population with finite mean μ. Decide how accurately you would like to estimate μ. As the number of observations drawn increases, the mean $\bar{x}$ of the observed values eventually approaches the mean μ of the population as closely as you specified and then stays that close.

The behavior of $\overline{x}$ is similar to the idea of probability. In the long run, the proportion of outcomes taking any value gets close to the probability of that value, and the average outcome gets close to the distribution mean. Figure 6.1 (page 331) shows how proportions approach probability in one example. Here is an example of how sample means approach the distribution mean.

EXAMPLE 7.8 HEIGHTS OF YOUNG WOMEN

The distribution of the heights of all young women is close to the normal distribution with mean 64.5 inches and standard deviation 2.5 inches. Suppose that $\mu = 64.5$ were exactly true. Figure 7.10 shows the behavior of the mean height $\overline{x}$ of n women chosen at random from a population whose heights follow the N(64.5, 2.5) distribution. The graph plots the values of $\overline{x}$ as we add women to our sample. The first woman drawn had height 64.21 inches, so the line starts there. The second had height 64.35 inches, so for $n = 2$ the mean is

$$\overline{x} = \frac{64.21 + 64.35}{2} = 64.28$$

This is the second point on the line in the graph.

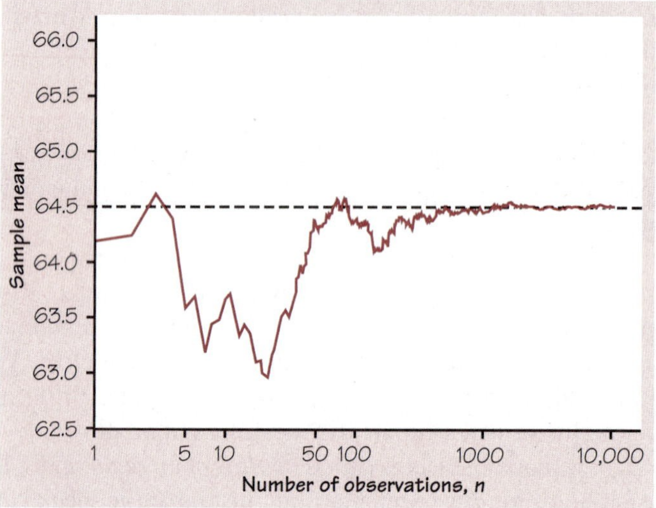

FIGURE 7.10 The law of large numbers in action. As we increase the size of our sample, the sample mean $\overline{x}$ always approaches the mean μ of the population.

At first, the graph shows that the mean of the sample changes as we take more observations. Eventually, however, the mean of the observations gets close to the population mean $\mu = 64.5$ and settles down at that value. The law of large numbers says that this *always* happens.

The mean μ of a random variable is the average value of the variable in two senses. By its definition, μ is the average of the possible values, weighted by their probability of occurring. The law of large numbers says that μ is also the long-run average of many independent observations on the variable. The law of large numbers can be proved mathematically starting from the basic laws of probability.

Thinking about the law of large numbers

The law of large numbers says broadly that the average results of many independent observations are stable and predictable. Casinos are not the only businesses that base forecasts on this fact. An insurance company deciding how much to charge for life insurance and a fast-food restaurant deciding how many beef patties to prepare rely on the fact that averaging over many individuals produces a stable result. It is worth the effort to think a bit more closely about so important a fact.

The "law of small numbers"

Both the rules of probability and the law of large numbers describe the regular behavior of chance phenomena *in the long run*. Psychologists have discovered that the popular understanding of randomness is quite different from the true laws of chance.[5] Most people believe in an incorrect "law of small numbers." That is, we expect even short sequences of random events to show the kind of average behavior that in fact appears only in the long run.

Try this experiment: Write down a sequence of heads and tails that you think imitates 10 tosses of a balanced coin. How long was the longest string (called a *run*) of consecutive heads or consecutive tails in your tosses? Most people will write a sequence with no runs of more than two consecutive heads or tails. Longer runs don't seem "random" to us. In fact, the probability of a run of three or more consecutive heads or tails in 10 tosses is greater than 0.8, and the probability of *both* a run of three or more heads and a run of three or more tails is almost 0.2.[6] This and other probability calculations suggest that a short sequence of coin tosses will often not appear random to us. The runs of consecutive heads or consecutive tails that appear in real coin tossing (and that are predicted by the mathematics of probability) seem surprising to us. Because we don't expect to see long runs, we may conclude that the coin tosses are not independent or that some influence is disturbing the random behavior of the coin.

EXAMPLE 7.9 THE "HOT HAND" IN BASKETBALL

Belief in the law of small numbers influences behavior. If a basketball player makes several consecutive shots, both the fans and her teammates believe that she has the "hot hand" and is more likely to make the next shot. This is doubtful. Careful study suggests that runs of baskets made or missed are no more frequent in basketball than it would be expected if each shot were independent of the player's previous shots. Baskets made or missed are just like heads and tails in tossing a coin. (Of course, some players make 30% of their shots in the long run and others make 50%, so a coin-toss model for basketball must allow coins with different probabilities of a head.) Our perception of hot or cold streaks simply shows that we don't perceive random behavior very well.[7]

Gamblers often follow the hot-hand theory, betting that a run will continue. At other times, however, they draw the opposite conclusion when confronted with a run of outcomes. If a coin gives 10 straight heads, some gamblers feel that it must now produce some extra tails to get back to the average of half heads and half tails. Not so. If the next 10,000 tosses give about 50% tails, those 10 straight heads will be swamped by the later thousands of heads and tails. No compensation is needed to get back to the average in the long run. Remember that it is *only* in the long run that the regularity described by probability and the law of large numbers takes over.

Our inability to accurately distinguish random behavior from systematic influences points out once more the need for statistical inference to supplement exploratory analysis of data. Probability calculations can help verify that what we see in the data is more than a random pattern.

How large is a large number?

The law of large numbers says that the actual mean outcome of many trials gets close to the distribution mean μ as more trials are made. It doesn't say how many trials are needed to guarantee a mean outcome close to μ. That depends on the *variability* of the random outcomes. The more variable the outcomes, the more trials are needed to ensure that the mean outcome $\overline{x}$ is close to the distribution mean μ.

The law of large numbers is the foundation of such business enterprises as gambling casinos and insurance companies. Games of chance must be quite variable if they are to hold the interest of gamblers. Even a long evening in a casino has an unpredictable outcome. Gambles with extremely variable outcomes, like state lottos with their very large but very improbable jackpots, require impossibly large numbers of trials to ensure that the average outcome is close to the expected value. Though most forms of gambling are less variable than lotto, the layman's answer to the applicability of the law of large numbers is usually that the house plays often enough to rely on it, but you don't. Much of the psychological allure of gambling is its unpredictability for the player. The business of gambling rests on the fact that the result is not unpredictable for the house. The average winnings of the house on tens of thousands of bets will be very close to the mean of the distribution of winnings. Needless to say, this mean guarantees the house a profit.

EXERCISES

7.30 LAW OF LARGE NUMBERS SIMULATION This exercise is based on Example 7.8 and uses the TI-83/89 to simulate the law of large numbers and the sampling process. Begin by clearing L_1/list1, L_2/list2, L_3/list3, and L_4/list4. Then enter the commands from the table on the following page.

Specify Plot1 as follows: xyLine (2nd Type icon on the TI-83); Xlist: L_1/list1; Ylist: L_4/list4; Mark: . Set the viewing WINDOW as follows: $X[1,10]_{10}$. To set the Y dimensions, scan the values in L_4/list4. Or start with $Y[60,69]_1$ and adjust as necessary. Press

TI-83	TI-89	
seq(X,X,1,200)→L_1	seq(X,X,1,200) →list1	Enters the positive integers 1 to 200 into L_1/list1 (for seq, look under 2nd / LIST/ OPS on the TI-83 and under CATALOG on the TI-89).
randNorm(64.5,2.5, 200)→L_2	tistat.randNorm (64.5,2.5,200) →list2	Generates 200 random heights (in inches) from the N(64.5, 2.5) distribution and stores these values in L_2/list2 (for randNorm, look under MATH / PRB on the TI-83 and under CATALOG on the TI-89) .
cumSum(L_2)→L_3	cumSum(list2) →list3	Provides a cumulative sum of the observations and stores these values in L_3/list3 (for cumSum, look under 2nd/ LIST/OPS on the TI-83 and under CATALOG on the TI-89).
L_3/L_1→L_4	list3/list1 →list4	Calculates the average heights of the women and stores these values in L_4/list4

GRAPH. In the WINDOW screen, change Xmax to 100, and press GRAPH again. In your own words, write a short description of the principle that this exercise demonstrates.

7.31 A GAME OF CHANCE One consequence of the law of large numbers is that once we have a probability distribution for a random variable, we can find its mean by simulating many outcomes and averaging them. The law of large numbers says that if we take enough outcomes, their average value is sure to approach the mean of the distribution.

I have a little bet to offer you. Toss a coin ten times. If there is no run of three or more straight heads or tails in the ten outcomes, I'll pay you \$2. If there is a run of three or more, you pay me just \$1. Surely you will want to take advantage of me and play this game?

Simulate enough plays of this game (the outcomes are +\$2 if you win and –\$1 if you lose) to estimate the mean outcome. Is it to your advantage to play?

7.32

(a) A gambler knows that red and black are equally likely to occur on each spin of a roulette wheel. He observes five consecutive reds and bets heavily on red at the next spin. Asked why, he says that "red is hot" and that the run of reds is likely to continue. Explain to the gambler what is wrong with this reasoning.

(b) After hearing you explain why red and black remain equally probable after five reds on the roulette wheel, the gambler moves to a poker game. He is dealt five straight red cards. He remembers what you said and assumes that the next card dealt in the same hand is equally likely to be red or black. Is the gambler right or wrong? Why?

7.33 OVERDUE FOR A HIT Retired baseball player Tony Gwynn got a hit about 35% of the time over an entire season. After he failed to hit safely in six straight at-bats, a TV commentator said, "Tony is due for a hit by the law of averages." Is that right? Why?

Rules for means

You are studying flaws in the painted finish of refrigerators made by your firm. Dimples and paint sags are two kinds of surface flaw. Not all refrigerators have the same number of dimples: many have none, some have one, some two, and so on. You ask for the average number of imperfections on a refrigerator. How many total imperfections of both kinds (on the average) are there on a refrigerator? That's easy: If the average number of dimples is 0.7 and the average number of sags is 1.4, then counting both gives an average of $0.7 + 1.4 = 2.1$ flaws.

In more formal language, the number of dimples on a refrigerator is a random variable X that takes values 0, 1, 2, and so on. X varies as we inspect one refrigerator after another. Only the mean number of dimples $\mu_X = 0.7$ was reported to you. The number of paint sags is a second random variable Y having mean $\mu_Y = 1.4$. (You see how the subscripts keep straight which variable we are talking about.) The total number of both dimples and sags is the sum $X + Y$. That sum is another random variable that varies from refrigerator to refrigerator. Its mean μ_{X+Y} is the average number of dimples and sags together and is just the sum of the individual means μ_X and μ_Y. That is an important rule for how means of random variables behave.

Here's another rule. The crickets living in a field have mean length 1.2 inches. What is the mean in centimeters? There are 2.54 centimeters in an inch, so the length of a cricket in centimeters is 2.54 times its length in inches. If we multiply every observation by 2.54, we also multiply their average by 2.54. The mean in centimeters must be 2.54×1.2, or about 3.05 centimeters. More formally, the length in inches of a cricket chosen at random from the field is a random variable X with mean μ_X. The length in centimeters is 2.54X, and this new random variable has mean $2.54\mu_X$.

The point of these examples is that means behave like averages. Here are the rules we need.

RULES FOR MEANS

Rule 1. If X is a random variable and a and b are fixed numbers, then

$$\mu_{a+bX} = a + b\mu_X$$

Rule 2. If X and Y are random variables, then

$$\mu_{X+Y} = \mu_X + \mu_Y$$

Here is an example that applies these rules.

EXAMPLE 7.10 GAIN COMMUNICATIONS

In Example 7.7 (page 411) we saw that the number X of communications units sold by the Gain Communications *military* division has distribution

X = units sold:	1000	3000	5000	10,000
Probability:	0.1	0.3	0.4	0.2

The corresponding sales estimates for the *civilian* division are

Y = units sold:	300	500	750
Probability:	0.4	0.5	0.1

In Example 7.7, we calculated $\mu_X = 5000$. In similar fashion, we calculate

$$\begin{aligned}\mu_Y &= (300)(0.4) + (500)(0.5) + (750)(0.1)\\ &= 445 \text{ units}\end{aligned}$$

Gain makes a profit of \$2000 on each military unit sold and \$3500 on each civilian unit. Next year's profit from military sales will be 2000X, \$2000 times the number X of units sold. By Rule 1, the mean military profit is

$$\mu_{2000X} = 2000\mu_X = (2000)(5000) = \$10{,}000{,}000$$

Similarly, the civilian profit is 3500Y and the mean profit from civilian sales is

$$\mu_{3500Y} = 3500\mu_Y = (3500)(445) = \$1{,}557{,}500$$

The total profit is the sum of the military and civilian profit:

$$Z = 2000X + 3500Y$$

Rule 2 says that the mean of this sum of two variables is the sum of the two individual means:

$$\begin{aligned}\mu_Z &= \mu_{2000X} + \mu_{3500Y}\\ &= 10{,}000{,}000 + 1{,}557{,}500\\ &= \$11{,}557{,}500\end{aligned}$$

This mean is the company's best estimate of next year's profit, combining the probability estimates of the two divisions. We can do this calculation more quickly by combining Rules 1 and 2:

$$\begin{aligned}\mu_Z &= \mu_{2000X+3500Y}\\ &= 2000\mu_X + 3500\mu_Y\\ &= (2000)(5000) + (3500)(445) = \$11{,}557{,}500\end{aligned}$$

Rules for variances

What are the facts for variances that parallel Rules 1 and 2 for means? The mean of a sum of random variables is always the sum of their means, but this addition rule is not always true for variances. To understand why, take X to be the percent of a family's after-tax income that is spent and Y the percent that is saved. When X increases, Y decreases by the same amount. Though X and Y may vary widely from year to year, their sum X + Y is always 100% and does not vary at all. It is the association between the variables X and Y that prevents their variances from adding. If random variables are independent, this kind of association between their values is ruled out and their variances do add. Two random variables X and Y are ***independent*** if knowing that any event involving X alone did or did not occur tells us nothing about the occurrence of any event involving Y alone. Probability models often assume independence when the random variables describe outcomes that appear unrelated to each other. You should ask in each instance whether the assumption of independence seems reasonable.

independent

When random variables are not independent, the variance of their sum depends on the ***correlation*** between them as well as on their individual variances. In Chapter 3, we met the correlation r between two observed variables measured on the same individuals. We defined (page 140) the correlation r as an average of the products of the standardized x and y observations. The correlation between two random variables is defined in the same way, once again using a weighted average with probabilities as weights. We won't give the details—it is enough to know that the correlation between two random variables has the same basic properties as the correlation r calculated from data. We use ρ, the Greek letter rho, for the correlation between two random variables. The correlation ρ is a number between –1 and 1 that measures the direction and strength of the linear relationship between two variables. **The correlation between two independent random variables is zero.**

correlation

Returning to family finances, if X is the percent of a family's after-tax income that is spent and Y the percent that is saved, then Y = 100 – X. This is a perfect linear relationship with a negative slope, so the correlation between X and Y is $\rho = -1$. With the correlation at hand, we can state the rules for manipulating variances.

RULES FOR VARIANCES

Rule 1. If X is a random variable and a and b are fixed numbers, then

$$\sigma^2_{a+bX} = b^2\sigma^2_X$$

Rule 2. If X and Y are independent random variables, then

$$\sigma^2_{X+Y} = \sigma^2_X + \sigma^2_Y$$

$$\sigma^2_{X-Y} = \sigma^2_X + \sigma^2_Y$$

RULES FOR VARIANCES (*continued*)

This is the **addition rule for variances of independent random variables.**

Rule 3. If X and Y have correlation ρ, then

$$\sigma^2_{X+Y} = \sigma^2_X + \sigma^2_Y + 2\rho\sigma_X\sigma_Y$$
$$\sigma^2_{X-Y} = \sigma^2_X + \sigma^2_Y + 2\rho\sigma_X\sigma_Y$$

This is the **general addition rule for variances of random variables.**

Notice that because a variance is the average of *squared* deviations from the mean, multiplying X by a constant b multiplies σ^2_X by the *square* of the constant. Adding a constant a to a random variable changes its mean but does not change its variability. The variance of $X + a$ is therefore the same as the variance of X. Because the square of –1 is 1, the addition rule says that the variance of a difference is the *sum* of the variances. For independent random variables, the difference $X - Y$ is more variable than either X or Y alone because variations in both X and Y contribute to variation in their difference.

As with data, we prefer the standard deviation to the variance as a measure of variability. The addition rule for variances implies that standard deviations do *not* generally add. Standard deviations are most easily combined by using the rules for variances rather than by giving separate rules for standard deviations. For example, the standard deviations of 2X and –2X are both equal to $2\sigma_X$ because this is the square root of the variance $4\sigma^2_X$.

EXAMPLE 7.11 WINNING THE LOTTERY

The payoff X of a $1 ticket in the Tri-State Pick 3 game is $500 with probability 1/1000 and $0 the rest of the time. Here is the combined calculation of mean and variance:

x_i	p_i	$x_i p_i$	$(x_i - \mu_X)^2 p_i$		
0	0.999	0	$(0 - 0.5)^2\ (0.999)$	=	0.24975
500	0.001	0.5	$(500 - 0.5)^2\ (0.001)$	=	249.50025
		$\mu_X = 0.5$	σ^2_X	=	249.75

The standard deviation is $\sigma_X = \sqrt{249.75} = \$15.80$. It is usual for games of chance to have large standard deviations, because large variability makes gambling exciting.

If you buy a Pick 3 ticket, your winnings are $W = X - 1$ because the dollar you paid for the ticket must be subtracted from the payoff. By the rules for means, the mean amount you win is

$$\mu_W = \mu_X - 1 = -\$0.50$$

That is, you lose an average of 50 cents on a ticket. The rules for variances remind us that the variance and standard deviation of the winnings $W = X - 1$ are the same as those of X. Subtracting a fixed number changes the mean but not the variance.

Suppose now that you buy a $1 ticket on each of two different days. The payoffs X and Y on the two tickets are independent because separate drawings are held each day. Your total payoff X + Y has mean

$$\mu_{X+Y} = \mu_X + \mu_Y = \$0.50 + \$0.50 = \$1.00$$

Because X and Y are independent, the variance of X + Y is

$$\sigma^2_{X+Y} = \sigma^2_X + \sigma^2_Y = 249.75 + 249.75 = 499.5$$

The standard deviation of the total payoff is

$$\sigma_{X+Y} = \sqrt{499.5} = \$22.35$$

This is not the same as the sum of the individual standard deviations, which is $15.80 + $15.80 = $31.60. Variances of independent random variables add; standard deviations do not.

If you buy a ticket every day (365 tickets in a year), your mean payoff is the sum of 365 daily payoffs. That's 365 times 50 cents, or $182.50. Of course, it costs $365 to play, so the state's mean take from a daily Pick 3 player is $182.50. Results for individual players will vary, but the law of large numbers assures the state its profit.

EXAMPLE 7.12 SAT SCORES

A college uses SAT scores as one criterion for admission. Experience has shown that the distribution of SAT scores among its entire population of applicants is such that

SAT Math score X	$\mu_X = 625$	$\sigma_X = 90$
SAT Verbal score Y	$\mu_Y = 590$	$\sigma_Y = 100$

What are the mean and standard deviation of the total score X + Y among students applying to this college?

The mean overall SAT score is

$$\mu_{X+Y} = \mu_X + \mu_Y = 625 + 590 = 1215$$

The variance and standard deviation of the total *cannot be computed* from the information given. SAT verbal and math scores are not independent, because students who score high on one exam tend to score high on the other also. Therefore, Rule 2 does not apply and we need to know ρ, the correlation between X and Y, to apply Rule 3.

Nationally, the correlation between SAT Math and Verbal scores is about $\rho = 0.7$. If this is true for these students,

$$\begin{aligned}\sigma^2_{X+Y} &= \sigma^2_X + \sigma^2_Y + 2\rho\sigma_X\sigma_Y \\ &= (90)^2 + (100)^2 + (2)(0.7)(90)(100) \\ &= 30{,}700\end{aligned}$$

The variance of the sum X + Y is greater than the sum of the variances $\sigma_X^2 + \sigma_Y^2$ because of the positive correlation between SAT Math scores and SAT Verbal scores. That is, X and Y tend to move up together and down together, which increases the variability of their sum. We find the standard deviation from the variance,

$$\sigma_{X+Y} = \sqrt{30{,}700} = 175$$

EXAMPLE 7.13 INVESTING IN STOCKS AND T-BILLS

Zadie has invested 20% of her funds in Treasury bills and 80% in an "index fund" that represents all U.S. common stocks. The rate of return of an investment over a time period is the percent change in the price during the time period, plus any income received. If X is the annual return on T-bills and Y the annual return on stocks, the portfolio rate of return is

$$R = 0.2X + 0.8Y$$

The returns X and Y are random variables because they vary from year to year. Based on annual returns between 1950 and 2000, we have[8]

X = annual return on T-bills	$\mu_X = 5.2\%$	$\sigma_X = 2.9\%$
Y = annual return on stocks	$\mu_Y = 13.3\%$	$\sigma_Y = 17.0\%$
Correlation between X and Y	$\rho = -0.1$	

Stocks had higher returns than T-bills on the average, but the standard deviations show that returns on stocks varied much more from year to year. That is, the risk of investing in stocks is greater than the risk for T-bills because their returns are less predictable.

For the return *R* on Zadie's portfolio of 20% T-bills and 80% stocks,

$$\begin{aligned} R &= 0.2X + 0.8Y \\ \mu_R &= 0.2\mu_X + 0.8\mu_Y \\ &= (0.2 \times 5.2) + (0.8 \times 13.3) = 11.68\% \end{aligned}$$

To find the variance of the portfolio return, combine Rules 1 and 3:

$$\begin{aligned} \sigma_R^2 &= \sigma_{0.2X}^2 + \sigma_{0.8Y}^2 + 2\rho\sigma_{0.2X}\sigma_{0.8Y} \\ &= (0.2)^2\sigma_X^2 + 0.8^2\sigma_Y^2 + 2\rho(0.2\sigma_X)(0.8\sigma_Y) \\ &= (0.2)^2(2.9)^2 + (0.8)^2(17.0)^2 + (2)(-0.1)(0.2 \times 2.9)(0.8 \times 17.0) \\ &= 183.719 \\ \sigma_R &= \sqrt{183.719} = 13.55\% \end{aligned}$$

The portfolio has a smaller mean return than an all-stock portfolio, but it is also less risky. As a proportion of the all-stock values, the reduction in standard deviation is greater than the reduction in mean return. That's why Zadie put some funds into Treasury bills.

Combining normal random variables

So far, we have concentrated on finding rules for means and variances of random variables. If a random variable is normally distributed, we can use its mean and variance to compute probabilities. Example 7.4 (page 400) shows the method. What if we combine two normal random variables?

Any linear combination of independent normal random variables is also normally distributed. That is, if X and Y are independent normal random variables and a and b are any fixed numbers, $aX + bY$ is also normally distributed. In particular, the sum or difference of independent normal random variables has a normal distribution. The mean and standard deviation of $aX + bY$ are found as usual from the addition rules for means and variances. These facts are often used in statistical calculations.

EXAMPLE 7.14 A ROUND OF GOLF

Tom and George are playing in the club golf tournament. Their scores vary as they play the course repeatedly. Tom's score X has the N(110,10) distribution, and George's score Y varies from round to round according to the N(100,8) distribution. If they play independently, what is the probability that Tom will score lower than George and thus do better in the tournament? The difference X – Y between their scores is normally distributed, with mean and variance

$$\mu_{X-Y} = \mu_X - \mu_Y = 110 - 100 = 10$$
$$\sigma^2_{X-Y} = \sigma^2_X + \sigma^2_Y = 10^2 + 8^2 = 164$$

Because $\sqrt{164} = 12.8$, X – Y has the N(10,12.8) distribution. Figure 7.11 illustrates the probability computation:

$$\begin{aligned} P(X < Y) &= P(X - Y < 0) \\ &= P\left(\frac{(X-Y)-10}{12.8} < \frac{0-10}{12.8}\right) \\ &= P(Z < -0.78) = 0.2177 \end{aligned}$$

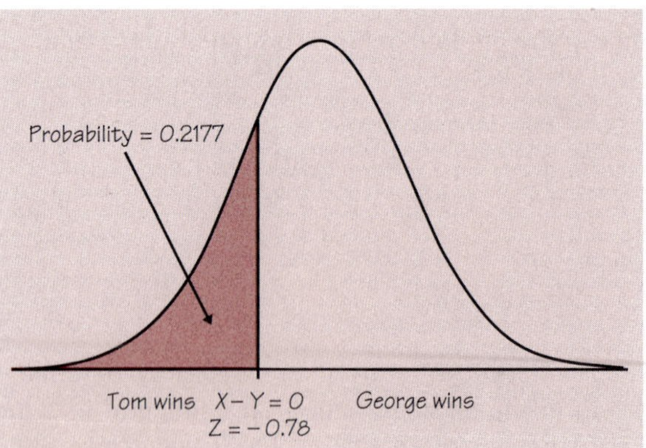

FIGURE 7.11 The normal probability calculation for Example 7.14.

Although George's score is 10 strokes lower on the average, Tom will have the lower score in about one of every five matches.

EXERCISES

7.34 CHECKING INDEPENDENCE, I For each of the following situations, would you expect the random variables X and Y to be independent? Explain your answers.

(a) X is the rainfall (in inches) on November 6 of this year, and Y is the rainfall at the same location on November 6 of next year.

(b) X is the amount of rainfall today, and Y is the rainfall at the same location tomorrow.

(c) X is today's rainfall at the airport in Orlando, Florida, and Y is today's rainfall at Disney World just outside Orlando.

7.35 CHECKING INDEPENDENCE, II In which of the following games of chance would you be willing to assume independence of X and Y in making a probability model? Explain your answer in each case.

(a) In blackjack, you are dealt two cards and examine the total points X on the cards (face cards count 10 points). You can choose to be dealt another card and compete based on the total points Y on all three cards.

(b) In craps, the betting is based on successive rolls of two dice. X is the sum of the faces on the first roll, and Y is the sum of the faces on the next roll.

7.36 CHEMICAL REACTIONS, I Laboratory data show that the time required to complete two chemical reactions in a production process varies. The first reaction has a mean time of 40 minutes and a standard deviation of 2 minutes; the second has a mean time of 25 minutes and a standard deviation of 1 minute. The two reactions are run in sequence during production. There is a fixed period of 5 minutes between them as the product of the first reaction is pumped into the vessel where the second reaction will take place. What is the mean time required for the entire process?

7.37 TIME AND MOTION, I A time and motion study measures the time required for an assembly-line worker to perform a repetitive task. The data show that the time required to bring a part from a bin to its position on an automobile chassis varies from car to car with mean 11 seconds and standard deviation 2 seconds. The time required to attach the part to the chassis varies with mean 20 seconds and standard deviation 4 seconds.

(a) What is the mean time required for the entire operation of positioning and attaching the part?

(b) If the variation in the worker's performance is reduced by better training, the standard deviations will decrease. Will this decrease change the mean you found in (a) if the mean times for the two steps remain as before?

(c) The study finds that the times required for the two steps are independent. A part that takes a long time to position, for example, does not take more or less time to attach than other parts. How would your answers to (a) and (b) change if the two variables were dependent with correlation 0.8? With correlation 0.3?

7.38 TIME AND MOTION, II Find the standard deviation of the time required for the two-step assembly operation studied in Exercise 7.37, assuming that the study shows the two times to be independent. Redo the calculation assuming that the two times are dependent, with correlation 0.3. Can you explain in nontechnical language why positive correlation increases the variability of the total time?

7.39 CHEMICAL REACTIONS, II The times for the two reactions in the chemical production process described in Exercise 7.36 are independent. Find the standard deviation of the time required to complete the process.

7.40 Examples 7.7 (page 411) and 7.10 (page 419) concern a probabilistic projection of sales and profits by an electronics firm, Gain Communications.

(a) Find the variance and standard deviation of the estimated sales X of Gain's civilian unit, using the distribution and mean from Example 7.10.

(b) Because the military budget and the civilian economy are not closely linked, Gain is willing to assume that its military and civilian sales vary independently. Combine your result from (a) with the results for the military unit from Example 7.10 to obtain the standard deviation of the total sales X + Y.

(c) Find the standard deviation of the estimated profit, Z = 2000X + 3500Y.

7.41 Leona and Fred are friendly competitors in high school. Both are about to take the ACT college entrance examination. They agree that if one of them scores 5 or more points better than the other, the loser will buy the winner a pizza. Suppose that in fact Fred and Leona have equal ability, so that each score varies normally with mean 24 and standard deviation 2. (The variation is due to luck in guessing and the accident of the specific questions being familiar to the student.) The two scores are independent. What is the probability that the scores differ by 5 or more points in either direction?

SUMMARY

The probability distribution of a random variable X, like a distribution of data, has a **mean** μ_X and a **standard deviation** σ_X.

The **mean** μ is the balance point of the probability histogram or density curve. If X is discrete with possible values x_i having probabilities p_i, the mean is the average of the values of X, each weighted by its probability:

$$\mu_X = x_1p_1 + x_2p_2 + \cdots + x_kp_k$$

The **variance** σ_X^2 is the average squared deviation of the values of the variable from their mean. For a discrete random variable,

$$\sigma_X^2 = (x_1 - \mu)^2p_1 + (x_2 - \mu)^2p_2 + \cdots + (x_k - \mu)^2p_k$$

The **standard deviation** σ_X is the square root of the variance. The standard deviation measures the variability of the distribution about the mean. It is easiest to interpret for normal distributions.

The mean and variance of a continuous random variable can be computed from the density curve, but to do so requires more advanced mathematics.

The **law of large numbers** says that the average of the values of X observed in many trials must approach μ.

The means and variances of random variables obey the following rules. If a and b are fixed numbers, then

$$\mu_{a+bX} = a + b\mu_X$$
$$\sigma^2_{a+bX} = b^2\sigma^2_X$$

If X and Y are any two random variables, then

$$\mu_{X+Y} = \mu_X + \mu_Y$$

and if X and Y are independent, then

$$\sigma^2_{X+Y} = \sigma^2_X + \sigma^2_Y$$
$$\sigma^2_{X-Y} = \sigma^2_X + \sigma^2_Y$$

Any linear combination of independent random variables is also normally distributed.

SECTION 7.2 EXERCISES

7.42 BUYING STOCK You purchase a hot stock for \$1000. The stock either gains 30% or loses 25% each day, each with probability 0.5. Its returns on consecutive days are independent of each other. You plan to sell the stock after two days.

(a) What are the possible values of the stock after two days, and what is the probability for each value? What is the probability that the stock is worth more after two days than the \$1000 you paid for it?

(b) What is the mean value of the stock after two days? You see that these two criteria give different answers to the question, "Should I invest?"

7.43 APPLYING BENFORD'S LAW It is easier to use Benford's law (Example 7.6, page 408) to spot suspicious patterns when you have very many items (for example, many invoices from the same vendor) than when you have only a few. Explain why this is true.

7.44 WEIRD DICE You have two balanced, six-sided dice. The first has 1, 3, 4, 5, 6, and 8 spots on its six faces. The second die has 1, 2, 2, 3, 3, and 4 spots on its faces.

(a) What is the mean number of spots on the up-face when you roll each of these dice?

(b) Write the probability model for the outcomes when you roll both dice independently. From this, find the probability distribution of the sum of the spots on the up-faces of the two dice.

(c) Find the mean number of spots on the two up-faces in two ways: from the distribution you found in (b) and by applying the addition rule to your results in (a). You should of course get the same answer.

7.45 SSHA The academic motivation and study habits of female students as a group are better than those of males. The Survey of Study Habits and Attitudes (SSHA) is a psychological test that measures these factors. The distribution of SSHA scores among the women at a college has mean 120 and standard deviation 28, and the distribution of scores among men students has mean 105 and standard deviation 35. You select a single male student and a single female student at random and give them the SSHA test.

(a) Explain why it is reasonable to assume that the scores of the two students are independent.

(b) What are the mean and standard deviation of the difference (female minus male) between their scores?

(c) From the information given, can you find the probability that the woman chosen scores higher than the man? If so, find this probability. If not, explain why you cannot.

7.46 A GLASS ACT, I In a process for manufacturing glassware, glass stems are sealed by heating them in a flame. The temperature of the flame varies a bit. Here is the distribution of the temperature X measured in degrees Celsius:

Temperature:	540°	545°	550°	555°	560°
Probability:	0.1	0.25	0.3	0.25	0.1

(a) Find the mean temperature μ_X and the standard deviation σ_X.

(b) The target temperature is 550° C. What are the mean and standard deviation of the number of degrees off target X – 550?

(c) A manager asks for results in degrees Fahrenheit. The conversion of X into degrees Fahrenheit is given by

$$Y = \frac{9}{5}X + 32$$

What are the mean μ_Y and the standard deviation σ_Y of the temperature of the flame in the Fahrenheit scale?

7.47 A GLASS ACT, II In continuation of the previous exercise, describe the details of a simulation you could carry out to approximate the mean temperature and the standard deviation in degrees Celsius. Then carry out your simulation. Are the mean and standard deviation produced from your simulation close to the values you calculated in 7.46 (a)?

7.48 A machine fastens plastic screw-on caps onto containers of motor oil. If the machine applies more torque than the cap can withstand, the cap will break. Both the torque applied and the strength of the caps vary. The capping-machine torque has the normal distribution with mean 7 inch-pounds and standard deviation 0.9 inch-pounds. The cap strength (the torque that would break the cap) has the normal distribution with mean 10 inch-pounds and standard deviation 1.2 inch-pounds.

(a) Explain why it is reasonable to assume that the cap strength and the torque applied by the machine are independent.

(b) What is the probability that a cap will break while being fastened by the capping machine?

7.49 A study of working couples measures the income X of the husband and the income Y of the wife in a large number of couples in which both partners are employed. Suppose that you knew the means μ_X and μ_Y and the variances σ_X^2 and σ_Y^2 of both variables in the population.

(a) Is it reasonable to take the mean of the total income $X + Y$ to be $\mu_X + \mu_Y$? Explain your answer.

(b) Is it reasonable to take the variance of the total income to be $\sigma_X^2 + \sigma_Y^2$? Explain your answer.

7.50 The design of an electronic circuit calls for a 100-ohm resistor and a 250-ohm resistor connected in series so that their resistances add. The components used are not perfectly uniform, so that the actual resistances vary independently according to normal distributions. The resistance of 100-ohm resistors has mean 100 ohms and standard deviation 2.5 ohms, while that of 250-ohm resistors has mean 250 ohms and standard deviation 2.8 ohms.

(a) What is the distribution of the total resistance of the two components in series?

(b) What is the probability that the total resistance lies between 345 and 355 ohms?

Portfolio analysis. Here are the means, standard deviations, and correlations for the monthly returns from three Fidelity mutual funds for the 36 months ending in December 2000.[9] Because there are three random variables, there are three correlations. We use subscripts to show which pair of random variables a correlation refers to.

W = monthly return on Magellan Fund	$\mu_W = 1.14\%$	$\sigma_W = 4.64\%$
X = monthly return on Real Estate Fund	$\mu_X = 0.16\%$	$\sigma_X = 3.61\%$
Y = monthly return on Japan Fund	$\mu_Y = 1.59\%$	$\sigma_Y = 6.75\%$

Correlations

$\rho_{WX} = 0.19 \qquad \rho_{WY} = 0.54 \qquad \rho_{XY} = -0.17$

Exercises 7.51 to 7.53 make use of these historical data.

7.51 Many advisors recommend using roughly 20% foreign stocks to diversify portfolios of U.S. stocks. Michael owns Fidelity Magellan Fund, which concentrates on stocks of large American companies. He decides to move to a portfolio of 80% Magellan and 20% Fidelity Japan Fund. Show that (based on historical data) this portfolio has both a *higher* mean return and *less* volatility (variability) than Magellan alone. This illustrates the beneficial effects of diversifying among investments.

7.52 Diversification works better when the investments in a portfolio have small correlations. To demonstrate this, suppose that returns on Magellan Fund and Japan

Fund had the means and standard deviations we have given but were uncorrelated ($\rho_{WY} = 0$). Show that the standard deviation of a portfolio that combines 80% Magellan with 20% Japan is smaller than your result from the previous exercise. What happens to the mean return if the correlation is 0?

7.53 Portfolios often contain more than two investments. The rules for means and variances continue to apply, though the arithmetic gets messier. A portfolio containing proportions a of Magellan Fund, b of Real Estate Fund, and c of Japan Fund has return $R = aW + bX + cY$. Because a, b, and c are the proportions invested in the three funds, $a + b + c = 1$. The mean and variance of the portfolio return are

$$\mu_R = a\mu_W + b\mu_X + c\mu_Y$$
$$\sigma_R^2 = a^2\sigma_W^2 + b^2\sigma_X^2 + c^2\sigma_Y^2 + 2ab\rho_{WX}\sigma_W\sigma_X + 2ac\rho_{WY}\sigma_W\sigma_Y + 2bc\rho_{XY}\sigma_X\sigma_Y$$

Having seen the advantages of diversification, Michael decides to invest his funds 60% in Magellan, 20% in Real Estate, and 20% in Japan. What are the (historical) mean and standard deviation of the monthly returns for this portfolio?

CHAPTER REVIEW

A random variable defines what is counted or measured in a statistics application. If the random variable X is a count, such as the number of heads in four tosses of a coin, then X is discrete, and its distribution can be pictured as a histogram. If X is measured, as in the number of inches of rainfall in Richmond in April, then X is continuous, and its distribution is pictured as a density curve. Among the continuous random variables, the normal random variable is the most important. First introduced in Chapter 2, the normal distribution is revisited, with emphasis this time on it as a probability distribution. The mean and variance of a random variable are calculated, and rules for the sum or difference of two random variables are developed. Here is a checklist of the major skills you should have acquired by studying this chapter.

A. RANDOM VARIABLES

1. Recognize and define a discrete random variable, and construct a probability distribution table and a probability histogram for the random variable.

2. Recognize and define a continuous random variable, and determine probabilities of events as areas under density curves.

3. Given a normal random variable, use the standard normal table or a graphing calculator to find probabilities of events as areas under the standard normal distribution curve.

B. MEANS AND VARIANCES OF RANDOM VARIABLES

1. Calculate the mean and variance of a discrete random variable. Find the expected payout in a raffle or similar game of chance.

2. Use simulation methods and the law of large numbers to approximate the mean of a distribution.

3. Use rules for means and rules for variances to solve problems involving sums, differences, and linear combinations of random variables.

CHAPTER 7 REVIEW EXERCISES

7.54 TWO-FINGER MORRA Ann and Bob are playing the game Two-Finger Morra. Each player shows either one or two fingers and at the same time calls out a guess for the number of fingers the other player will show. If a player guesses correctly and the other player does not, the player wins a number of dollars equal to the total number of fingers shown by both players. If both or neither guesses correctly, no money changes hands. On each play both Ann and Bob choose one of the following options:

Choice	Show	Guess
A	1	1
B	1	2
C	2	1
D	2	2

(a) Give the sample space S by writing all possible choices for both players on a single play of this game.

(b) Let X be Ann's winnings on a play. (If Ann loses \$2, then $X = -2$; when no money changes hands, $X = 0$.) Write the value of the random variable X next to each of the outcomes you listed in (a). This is another choice of sample space.

(c) Now assume that Ann and Bob choose independently of each other. Moreover, they both play so that all four choices listed above are equally likely. Find the probability distribution of X.

(d) If the game is fair, X should have mean zero. Does it? What is the standard deviation of X?

Insurance. The business of selling insurance is based on probability and the law of large numbers. Consumers (including businesses) buy insurance because we all face risks that are unlikely but carry high cost. Think of a fire destroying your home. So we form a group to share the risk: we all pay a small amount, and the insurance policy pays a large amount to those few of us whose homes burn down. The insurance company sells many policies, so it can rely on the law of large numbers. Exercises 7.55 to 7.58 explore aspects of insurance.

7.55 LIFE INSURANCE, I A life insurance company sells a term insurance policy to a 21-year-old male that pays \$100,000 if the insured dies within the next 5 years. The probability that a randomly chosen male will die each year can be found in mortality tables. The company collects a premium of \$250 each year as payment for the insurance. The amount X that the company earns on this policy is \$250 per year, less the \$100,000 that it must pay if the insured dies. Here is the distribution of X. Fill in the missing probability in the table and calculate the mean profit μ_X.

Age at death:	21	22	23	24	25	≥ 26
Profit:	–\$99,750	–\$99,500	–\$99,250	–\$99,000	–\$98,750	\$1250
Probability:	0.00183	0.00186	0.00189	0.00191	0.00193	

7.56 LIFE INSURANCE, II It would be quite risky for you to insure the life of a 21-year-old friend under the terms of the previous exercise. There is a high probability that your friend would live and you would gain \$1250 in premiums. But if he were to die, you would lose almost \$100,000. Explain carefully why selling insurance is not risky for an insurance company that insures many thousands of 21-year-old men.

7.57 LIFE INSURANCE, III The risk of an investment is often measured by the standard deviation of the return on the investment. The more variable the return is (the larger σ is), the riskier the investment. We can measure the great risk of insuring a single person's life in Exercise 7.55 by computing the standard deviation of the income X that the insurer will receive. Find σ_X, using the distribution and mean found in Exercise 7.55.

7.58 LIFE INSURANCE, IV The risk of insuring one person's life is reduced if we insure many people. Use the result of the previous exercise and rules for means and variances to answer the following questions.

(a) Suppose that we insure two 21-year-old males, and that their ages at death are independent. If X and Y are the insurer's income from the two insurance policies, the insurer's average income on the two policies is

$$Z = \frac{X+Y}{2} = 0.5X + 0.5Y$$

Find the mean and standard deviation of Z. You see that the mean income is the same as for a single policy but the standard deviation is less.

(b) If four 21-year-old men are insured, the insurer's average income is

$$Z = \frac{1}{4}(X_1 + X_2 + X_3 + X_4)$$

where X_i is the income from insuring one man. The X_i are independent and each has the same distribution as before. Find the mean and standard deviation of Z. Compare your results with the results of (a). We see that averaging over many insured individuals reduces risk.

7.59 AUTO EMISSIONS The amount of nitrogen oxides (NOX) present in the exhaust of a particular type of car varies from car to car according to the normal distribution with mean 1.4 grams per mile (g/mi) and standard deviation 0.3 g/mi. Two cars of this type are tested. One has 1.1 g/mi of NOX, the other 1.9. The test station attendant finds this much variation between two similar cars surprising. If X and Y are independent NOX levels for cars of this type, find the probability

$$P(X - Y \geq 0.8 \text{ or } X - Y \leq -0.8)$$

that the difference is at least as large as the value the attendant observed.

7.60 MAKING A PROFIT Rotter Partners is planning a major investment. The amount of profit X is uncertain but a probabilistic estimate gives the following distribution (in millions of dollars):

Profit:	1	1.5	2	4	10
Probability:	0.1	0.2	0.4	0.2	0.1

(a) Find the mean profit μ_X and the standard deviation of the profit.

(b) Rotter Partners owes its source of capital a fee of $200,000 plus 10% of the profits X. So the firm actually retains

$$Y = 0.9X - 0.2$$

from the investment. Find the mean and standard deviation of Y.

7.61 A BALANCED SCALE You have two scales for measuring weights in a chemistry lab. Both scales give answers that vary a bit in repeated weighings of the same item. If the true weight of a compound is 2.00 grams (g), the first scale produces readings X that have a mean 2.000 g and standard deviation 0.002 g. The second scale's readings Y have a mean 2.001 g and standard deviation 0.001 g.

(a) What are the mean and standard deviation of the difference Y – X between the readings? (The readings X and Y are independent.)

(b) You measure once with each scale and average the readings. Your result is $Z = (X + Y)/2$. What are μ_Z and σ_Z? Is the average Z more or less variable than the reading Y of the less variable scale?

7.62 IT'S A GIRL! A couple plans to have children until they have a girl or until they have four children, whichever comes first. Example 5.24 (page 313) estimated the probability that they will have a girl among their children. Now we ask a different question: How many children, on the average, will couples who follow this plan have?

(a) To answer this question, construct a simulation similar to that in Example 5.24 but this time keep track of the number of children in each repetition. Carry out 25 repetitions and then average the results to estimate the expected value.

(b) Construct the probability distribution table for the random variable X = number of children.

(c) Use the table from (b) to calculate the expected value of X. Compare this number with the result from your simulation in (a).

7.63 SLIM AGAIN Amarillo Slim is back and he's got another deal for you. We have a fair coin (heads and tails each have probability 1/2). Toss it twice. If two heads come up, you win. If you get any other result, you get another chance: toss the coin twice more, and if you get two heads, you win. If you fail to get two heads on the second try, you lose. You pay a dollar to play. If you win, you get your dollar back plus another dollar.

(a) Explain how to simulate one play of this game using Table B. How could you simulate one play using your calculator? Simulate two tosses of a fair coin.

(b) Simulate 50 plays, using Table B or your calculator. Use your simulation to estimate the expected value of the game.

(c) There are two outcomes in this game: win or lose. Let the random variable X be the (monetary) outcome. What are the two values X can take? Calculate the actual probabilities of each value of X. Then calculate μ_X. How does this compare with your estimate from the simulation in (b)?

7.64 BE CREATIVE Here is a simple way to create a random variable X that has mean μ and standard deviation σ: X takes only the two values $\mu - \sigma$ and $\mu + \sigma$, each with probability 0.5. Use the definition of the mean and variance for discrete random variables to show that X does have mean μ and standard deviation σ.

7.65 WHEN STANDARD DEVIATIONS ADD We know that variances add if the random variables involved are uncorrelated ($\rho = 0$), but not otherwise. The opposite extreme is perfect positive correlation ($\rho = 1$). Show by using the general addition rule for variances that in this case the standard deviations add. That is, $\sigma_{X+Y} = \sigma_X + \sigma_Y$ if $\rho_{XY} = 1$.

7.66 A MECHANICAL ASSEMBLY A mechanical assembly (Figure 7.12) consists of a shaft with a bearing at each end. The total length of the assembly is the sum X + Y + Z of the shaft length X and the lengths Y and Z of the bearings. These lengths vary from part to part in production, independently of each other and with normal distributions. The shaft length X has mean 11.2 inches and standard deviation 0.002 inch, while each bearing length Y and Z has mean 0.4 inch and standard deviation 0.001 inch.

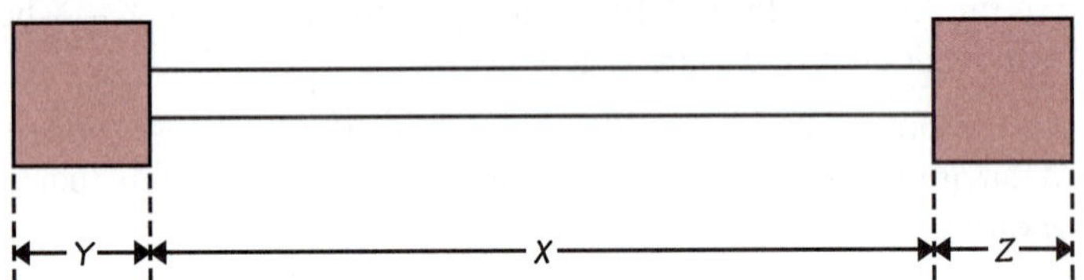

FIGURE 7.12 The dimensions of a mechanical assembly.

(a) According to the 68–95–99.7 rule, about 95% of all shafts have lengths in the range $11.2 \pm d_1$ inches. What is the value of d_1? Similarly, about 95% of the bearing lengths fall in the range of $0.4 \pm d_2$. What is the value of d_2?

(b) It is common practice in industry to state the "natural tolerance" of parts in the form used in (a). An engineer who knows no statistics thinks that tolerances add, so that the natural tolerance for the total length of the assembly (shaft and two bearings) is $12 \pm d$ inches, where $d = d_1 + 2d_2$. Find the standard deviation of the total length X + Y + Z. Then find the value d such that about 95% of all assemblies have lengths in the range $12 \pm d$. Was the engineer correct?

7.67 SWEDISH BRAINS A study of the weights of the brains of Swedish men found that the weight X was a random variable with mean 1400 grams and standard deviation 20 grams. Find positive numbers a and b such that $Y = a + bX$ has mean 0 and standard deviation 1.

7.68 ROLLING THE DICE You are playing a board game in which the severity of a penalty is determined by rolling three dice and adding the spots on the up-faces. The dice are all balanced so that each face is equally likely, and the three dice fall independently.

(a) Give a sample space for the sum X of the spots.

(b) Find $P(X = 5)$.

(c) If X_1, X_2, and X_3 are the number of spots on the up-faces of the three dice, then $X = X_1 + X_2 + X_3$. Use this fact to find the mean μ_X and the standard deviation σ_X without finding the distribution of X. (Start with the distribution of each of the X_i.)

NOTES AND DATA SOURCES

1. We use $\bar{x}$ both for the random variable, which takes different values in repeated sampling, and for the numerical value of the random variable in a particular sample. Similarly, s and $\hat{p}$ stand both for random variables and for specific values. This notation is mathematically imprecise but statistically convenient.

2. In most applications X takes a finite number of possible values. The same ideas, implemented with more advanced mathematics, apply to random variables with an infinite but still countable collection of values. An example is a geometric random variable, considered in Section 8.2.

3. From the Census Bureau's 1998 American Housing Survey.

4. The mean of a continuous random variable X with density function $f(x)$ can be found by integration:

$$\mu_X = \int x f(x) dx$$

This integral is a kind of weighted average, analogous to the discrete-case mean

$$\mu_X = \Sigma x P(X = x)$$

The variance of a continuous random variable X is the average squared deviation of the values of X from their mean, found by the integral

$$\sigma_X^2 = \int (x - \mu)^2 f(x) dx$$

5. See A. Tversky and D. Kahneman, "Belief in the law of small numbers," *Psychological Bulletin*, 76 (1971), pp. 105–110, and other writings of these authors for a full account of our misperception of randomness.

6. Probabilities involving runs can be quite difficult to compute. That the probability of a run of three or more heads in 10 independent tosses of a fair coin is (1/2) + (1/128) = 0.508 can be found by clever counting, as can the other results given in the text. A general treatment using advanced methods appears in Section XIII.7 of William Feller, *An Introduction to Probability Theory and Its Applications*, vol. 1, 3rd ed., Wiley, New York, 1968.

7. R. Vallone and A. Tversky, "The hot hand in basketball: on the misperception of random sequences," *Cognitive Psychology*, 17 (1985), pp. 295–314. A later series of articles that debate the independence question is A. Tversky and T. Gilovich, "The cold facts about the 'hot hand' in basketball," *Chance*, 2, no. 1 (1989), pp. 16–21; P. D. Larkey, R. A. Smith, and J. B. Kadane, "It's OK to believe in the 'hot hand,'" *Chance*, 2, no. 4 (1989), pp. 22–30; and A. Tversky and T. Gilovich, "The 'hot hand': statistical reality or cognitive illusion?" *Chance*, 2, no. 4 (1989), pp. 31–34.

8. The data on returns are from several sources, especially the *Fidelity Insight* newsletter, fidelity.kobren.com.

9. See Note 8

The Granger Collection, New York

PIERRE-SIMON LAPLACE

The Best Mathematician in France

Pierre-Simon Laplace (1749–1827) may be best remembered for his work on mathematical astronomy and the theory of probability. Before Laplace, probability theory was solely concerned with the mathematical analysis of games of chance. Laplace applied probabilistic ideas to many scientific and practical problems.

In 1812 he published the first of a series of four books on probability theory and its applications. In the first book, he studied generating functions and approximations to various expressions occurring in probability theory. The second book included Laplace's definition of probability, Bayes's rule, and remarks on moral and mathematical expectation, on methods of finding probabilities of compound events, on the method of least squares, on Buffon's needle problem, and on inverse probability. He also included work on probability in legal matters and applications to mortality, life expectancy, and the length of marriages. Later editions applied probability to errors in observations, to determining the masses of several planets, to triangulation methods in surveying, and to problems in geodesy.

Laplace survived the French Revolution by changing his views with the changing political events of the time. His colleague Lavoisier was a casualty. Despite Laplace's important contributions to science, he was not well liked by his colleagues. He was not modest about his abilities and achievements, and he let it be known widely that he considered himself the best mathematician in France. And he was! Laplace is now widely regarded as one of the greatest and most influential scientists of all time.

Laplace applied probabilistic ideas to many scientific and practical problems.

chapter 8

The Binomial and Geometric Distributions

ACTIVITY 8 A Gaggle of Girls

The Ferrells have 3 children: Jennifer, Jessica, and Jaclyn. If we assume that a couple is equally likely to have a girl or a boy, then how unusual is it for a family like the Ferrells to have 3 children who are all girls? We have encountered problems like this in an earlier chapter. But this time we're going to use the method of simulation. If success = girl, and failure = boy, then $p(\text{success}) = 0.5$. We will define the random variable X as the number of girls. Then we want to simulate families with 3 children. Our goal is to determine the long-term relative frequency of a family with 3 girls, that is, $P(X = 3)$.

1. Using a random number table, let even digits represent "girl" and odd digits represent "boy." Select a row, and beginning at that row, read off numbers 3 digits at a time. Each 3 digits will constitute one trial. Use tally marks in a table like this one to record the results:

3 girls	
Not 3 girls	

Do at least 40 trials. Then combine your results with those of other students in the class to obtain at least 200 trials. Calculate the relative frequency of the event {3 girls}.

2. For variety, do the same thing as before, but this time using the calculator. Using the codes 1 = girl and 0 = boy, enter the command `randInt(0,1,3)`. This command instructs the calculator to randomly pick a whole number from the set {0, 1} and to do this 3 times. The outcome {0, 0, 1}, using our codes, means {*boy, boy, girl*}, in that order. Continue to press ENTER and count until you have 40 trials. Use a tally mark to record each time you observe a {1, 1, 1} result. Calculate the relative frequency for the event {3 girls}.

3. *Extra for programming experts:* Write a calculator program to carry out the process described above. Allow the user to specify the number of trials, and have the calculator report the relative frequency of {3 girls} as a decimal number.

4. Determine the total number of outcomes for this experiment. List the outcomes in the sample space. Then complete the probability distribution table for the random variable X = number of girls.

X	0 1 2 3
P(X)	

Do the results of your simulations come close to the theoretical value for $P(X = 3)$?

INTRODUCTION

In practice, we frequently encounter experimental situations where there are two outcomes of interest. Some examples are:

- We use a coin toss to see which of the two football teams gets the choice of kicking off or receiving to begin the game.
- A basketball player shoots a free throw; the outcomes of interest are {she makes the shot; she misses}.
- A young couple prepares for their first child; the possible outcomes are {boy; girl}.
- A quality control inspector selects a widget coming off the assembly line; he is interested in whether or not the widget meets production requirements.

In this chapter we will explore two important classes of distributions—the binomial distributions and the geometric distributions—and learn some of their properties. We will use what we have learned about probability and random variables from previous chapters, with the view toward completing the necessary foundation to study inference.

8.1 THE BINOMIAL DISTRIBUTIONS

In Activity 8, we simulated families with 3 children to discover how often the children would be all girls. Flipping a fair coin 3 times and letting heads represent having a girl and tails represent having a boy would produce exactly the same results. The characterizing features of this experiment are as follows: A *trial* consists of flipping the coin once. There are two outcomes: heads = girl (success), and tails = boy (failure). We will flip the coin 3 times. The coin flips are independent in the sense that the outcome of one coin flip has no influence on the outcome of the next flip. And last, the probability of success (girl) is the same for each coin flip (trial). A situation where these four conditions are satisfied is said to be a ***binomial setting.***

binomial setting

THE BINOMIAL SETTING

1. Each observation falls into one of just two categories, which for convenience we call "success" or "failure."
2. There is a fixed number n of observations.
3. The n observations are all **independent**. That is, knowing the result of one observation tells you nothing about the other observations.
4. The probability of success, call it p, is the same for each observation.

If you are presented with an experimental setting, it is important to be able to recognize it as a binomial setting or a geometric setting (covered in the next section) or neither. If you can verify that each of these four conditions is satisfied, you will be able to make use of known properties of binomial situations to gain more insights.

binomial random variable

If data are produced in a binomial setting, then the random variable X = number of successes is called a ***binomial random variable***, and the probability distribution of X is called a *binomial distribution*.

BINOMIAL DISTRIBUTION

The distribution of the count X of successes in the binomial setting is the **binomial distribution** with parameters n and p. The parameter n is the number of observations, and p is the probability of a success on any one observation. The possible values of X are the whole numbers from 0 to n. As an abbreviation, we say that X is $B(n, p)$.

The binomial distributions are an important class of discrete probability distributions. Pay attention to the binomial setting because not all counts have binomial distributions.

EXAMPLE 8.1 BLOOD TYPES

Blood type is inherited. If both parents carry genes for the O and A blood types, each child has probability 0.25 of getting two O genes and so of having blood type O. Different children inherit independently of each other. The number of O blood types among 5 children of these parents is the count X of successes in 5 independent observations with probability 0.25 of a success on each observation. So X has the binomial distribution with $n = 5$ and $p = 0.25$. We say that X is $B(5, 0.25)$.

EXAMPLE 8.2 DEALING CARDS

Deal 10 cards from a shuffled deck and count the number X of red cards. There are 10 observations, and each gives either a red or a black card. A "success" is a red card. But the observations are *not* independent. If the first card is black, the second is more likely to be red because there are more red cards than black cards left in the deck. The count X does *not* have a binomial distribution.

EXAMPLE 8.3 INSPECTING SWITCHES

An engineer chooses an SRS of 10 switches from a shipment of 10,000 switches. Suppose that (unknown to the engineer) 10% of the switches in the shipment are bad. The engineer counts the number X of bad switches in the sample.

This is not quite a binomial setting. Just as removing one card in Example 8.2 changed the makeup of the deck, removing one switch changes the proportion of bad switches remaining in the shipment. So the state of the second switch chosen is not independent of the first. But removing one switch from a shipment of 10,000 changes the makeup of the remaining 9999 switches very little. In practice, the distribution of X is very close to the binomial distribution with $n = 10$ and $p = 0.1$.

Example 8.3 shows how we can use the binomial distributions in the statistical setting of selecting an SRS. When the population is much larger than the sample, a count of successes in an SRS of size n has approximately the binomial distribution with n equal to the sample size and p equal to the proportion of successes in the population.

EXAMPLE 8.4 AIRCRAFT ENGINE RELIABILITY

Engineers define reliability as the probability that an item will perform its function under specific conditions for a specific period of time. If an aircraft engine turbine has probability 0.999 of performing properly for an hour of flight, the number of turbines in a fleet of 350 engines that fly for an hour without failure has the $B(350, 0.999)$ distribution. This binomial distribution is obtained by assuming, as seems reasonable, that the turbines fail independently of each other. A common cause of failure, such as sabotage, would destroy the independence and make the binomial model inappropriate.

EXERCISES

8.1 BINOMIAL SETTING? In each situation below, is it reasonable to use a binomial distribution for the random variable X? Give reasons for your answer in each case.

(a) An auto manufacturer chooses one car from each hour's production for a detailed quality inspection. One variable recorded is the count X of finish defects (dimples, ripples, etc.) in the car's paint.

(b) The pool of potential jurors for a murder case contains 100 persons chosen at random from the adult residents of a large city. Each person in the pool is asked whether he or she opposes the death penalty; X is the number who say "Yes."

(c) Joe buys a ticket in his state's "Pick 3" lottery game every week; X is the number of times in a year that he wins a prize.

8.2 BINOMIAL SETTING? In each of the following cases, decide whether or not a binomial distribution is an appropriate model, and give your reasons.

(a) Fifty students are taught about binomial distributions by a television program. After completing their study, all students take the same examination. The number of students who pass is counted.

(b) A student studies binomial distributions using computer-assisted instruction. After the initial instruction is completed, the computer presents 10 problems. The student solves each problem and enters the answer; the computer gives additional instruction between problems if the student's answer is wrong. The number of problems that the student solves correctly is counted.

(c) A chemist repeats a solubility test 10 times on the same substance. Each test is conducted at a temperature 10° higher than the previous test. She counts the number of times that the substance dissolves completely.

Finding binomial probabilities

We will give a formula later for the probability that a binomial random variable takes any of its values. In practice, you will rarely have to use this formula for calculations. The TI-83/89 and most statistical software packages calculate binomial probabilities.

EXAMPLE 8.5 INSPECTING SWITCHES

A quality engineer selects an SRS of 10 switches from a large shipment for detailed inspection. Unknown to the engineer, 10% of the switches in the shipment fail to meet the specifications. What is the probability that no more than 1 of the 10 switches in the sample fail inspection?

The count X of bad switches in the sample has approximately the $B(10, 0.1)$ distribution. Figure 8.1 is a probability histogram for this distribution.

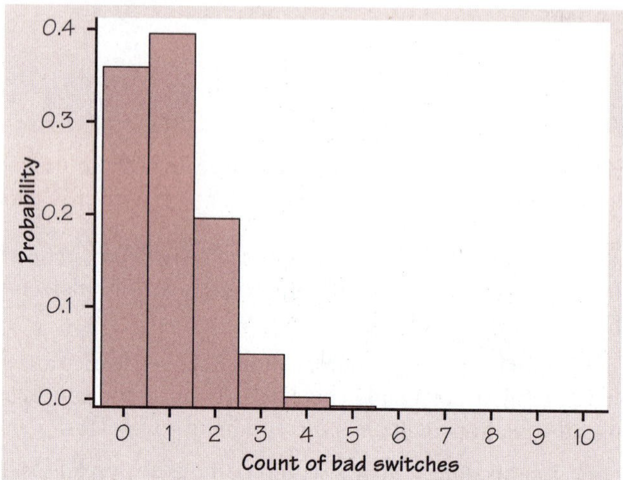

FIGURE 8.1 Probability histogram for the binomial distribution with $n = 10$ and $p = 0.1$.

The distribution is strongly skewed. Although X can take any whole-number value from 0 to 10, the probabilities of values larger than 5 are so small that they do not appear in the histogram. We want to calculate

$$P(X \leq 1) = P(X = 0) + P(X = 1)$$

when X is $B(10, 0.1)$. The TI-83 command `binompdf(n,p,X)` and the TI-89 command `tistat.binomPdf(n,p,X)` calculate the binomial probability of the value X. The suffix `pdf` stands for "probability distribution function." We met the probability distribution in Chapter 7.

pdf

Given a discrete random variable X, the **probability distribution function** assigns a probability to each value of X. The probabilities must satisfy the rules for probabilities given in Chapter 6.

The `binompdf` command is found under 2nd (DISTR) / `0:binompdf` on the TI-83. On the TI-89, it's in the CATALOG under Flash Apps. The TI-83 command `binompdf(10,.1,0)` and the TI-89 command `tistat.binompdf(10,.1,0)` calculate the binomial probability that X = 0 to be 0.3486784401. The command `binompdf(10,.1,1)` returns probability 0.387420489. Thus,

$$\begin{aligned} P(X \leq 1) &= P(X = 0) + P(X = 1) \\ &= 0.3487 + 0.3874 = 0.7361 \end{aligned}$$

About 74% of all samples will contain more than 1 bad switch. A sample of size 10 cannot be trusted to alert the engineer to the presence of unacceptable items in the shipment.

EXAMPLE 8.6 CORINNE'S FREE THROWS

Corinne is a basketball player who makes 75% of her free throws over the course of a season. In a key game, Corinne shoots 12 free throws and makes only 7 of them. The fans think that she failed because she was nervous. Is it unusual for Corinne to perform this poorly? To answer this question, assume that free throws are independent with probability 0.75 of a success on each shot. (Studies of long sequences of free throws have found no evidence that they are dependent, so this is a reasonable assumption.) The number X of baskets (successes) in 12 attempts has the $B(12, 0.75)$ distribution.

We want the probability of making a basket on at most 7 free throws. This is

$$\begin{aligned} P(X \leq 7) &= P(X = 0) + P(X = 1) + P(X = 2) + \cdots + P(X = 7) \\ &= 0.0000 + 0.0000 + 0.0000 + 0.0004 + 0.0024 + 0.0115 + 0.0401 \\ &\quad + 0.1032 \\ &= 0.1576 \end{aligned}$$

Corinne will make at most 7 of her 12 free throws about 16% of the time, or roughly in one of every six games. While below her average level, this performance is well within the range of the usual chance variation in her shooting.

EXAMPLE 8.7 THREE GIRLS

In Activity 8 we wanted to determine the probability that all 3 children in a family are girls. In this case, the random variable of interest, X = the number of girls, has the $B(3, 0.5)$ distribution. We want to find the probability that the number of girls is 3, that is, $P(X = 3)$. The TI-83 command `binompdf(3,.5,3)` and the TI-89 command `tistat.binomPdf(3,.5,3)` return the probability 0.125.

In applications we frequently want to find the probability that a random variable takes a range of values. The *cumulative* binomial probability is useful in these cases.

cdf

Given a random variable X, the **cumulative distribution function** (cdf) of X calculates the sum of the probabilities for 0, 1, 2, ..., up to the value X. That is, it calculates the probability of obtaining at most X successes in n trials.

For the count X of defective switches in Example 8.5, the command `binomcdf(10,.1,1)` and the TI-89 command `tistat.binomCdf(10,.1,1)` output 0.736098903 for the cumulative probability $P(X \leq 1)$.

EXAMPLE 8.8 IS CORINNE IN A SLUMP?

In Example 8.6. Corinne shoots $n = 12$ free throws and makes only 7 of them. Since she is a 75% free-throw shooter ($p = 0.75$), we wanted to know if it was unusual for Corinne to perform this poorly. If X = number of baskets made on free throws, then X has the $B(12, 0.75)$ distribution, and we need to find the probability that she makes at most 7 of her free throws, that is, $P(X \leq 7)$. The TI-83 command `binomcdf(12,.75,7)` and the TI-89 command `tistat.binomCdf(12,.75,7)` calculates the cumulative probability $P(X \leq 7)$ to be 0.1576436761. We round the answer to four decimal places and report that the probability that Corinne makes *at most* 7 of her 12 free throws is 0.1576.

The pdf table for Corinne's shots looks like this:

X	0	1	2	3	4	5	6
P(X)	0.000	0.000	0.000	0.000	0.002	0.011	0.040

X	7	8	9	10	11	12
P(X)	0.103	0.194	0.258	0.232	0.127	0.032

If we denote the cumulative distribution function by $F(X)$, we can record the cumulative sum of the probabilities in a third row of the table:

X	0	1	2	3	4	5	6
P(X)	0.000	0.000	0.000	0.000	0.002	0.011	0.040
F(X)	0.000	0.000	0.000	0.000	0.003	0.014	0.054

X	7	8	9	10	11	12
P(X)	0.103	0.194	0.258	0.232	0.127	0.032
F(X)	0.158	0.351	0.609	0.842	0.968	1

Notice that terms sometimes don't appear to add up as they should. The cumulative function $F(4)$, for example, should equal $P(0) + P(1) + P(2) + P(3) + P(4)$. Of course, the culprit is roundoff error. With your calculator, enter the integers 0 to 12 into L_1/list1, the corresponding binomial probabilities into L_2/list2, and use the command `binomcdf(12,.75,L1)→L3` (`tistat.binomCdf(12,.75,list1)→list3` on the TI-89)to enter the cumulative probabilities into L_3/list3.

In addition to being helpful in answering questions involving wording such as "find the probability that it takes at most 6 trials," the cdf is also particularly useful for calculating the probability that it takes *more* than a certain number of trials to see the first success. This calculation uses the complement rule:

$$P(X > n) = 1 - P(X \leq n) \quad n = 2, 3, 4, \ldots$$

EXERCISES

Use your calculator's binomial pdf or cdf commands to find the following probabilities:

8.3 INHERITING BLOOD TYPE Each child born to a particular set of parents has probability 0.25 of having blood type O. Suppose these parents have 5 children. Let X = number of children who have type O blood. Then X is $B(5, 0.25)$.

(a) What is the probability that exactly 2 of the children have type O blood?

(b) Make a table for the pdf of the random variable X. Then use the calculator to find the probabilities of all possible values of X, and complete the table.

(c) Verify that the sum of the probabilities is 1.

(d) Construct a histogram of the pdf.

(e) Use the calculator to find the cumulative probabilities, and add these values to your pdf table. Then construct a cumulative distribution histogram. How is this histogram different from the histogram for Corinne's free throws?

8.4 GUESSING ON A TRUE-FALSE QUIZ Suppose that James guesses on each question of a 50-item true-false quiz. Find the probability that James passes if

(a) a score of 25 or more correct is needed to pass.

(b) a score of 30 or more correct is needed to pass.

(c) a score of 32 or more correct is needed to pass.

8.5 GUESSING ON A MULTIPLE-CHOICE QUIZ Suppose that Erin guesses on each question of a multiple-choice quiz.

(a) If each question has four different choices, find the probability that Erin gets one or more correct answers on a 10-item quiz.

(b) If the quiz consists of three questions, question 1 has 3 possible answers, question two has 4 possible answers, and question 3 has 5 possible answers, find the probability that Erin gets one or more correct answers.

8.6 DAD'S IN THE POKEY According to a 2000 study by the Bureau of Justice Statistics, approximately 2% of the nation's 72 million children had a parent behind bars—nearly 1.5 million minors. Let X be the number of children who had an incarcerated parent. Suppose that 100 children are randomly selected.

(a) Does X satisfy the requirements for a binomial setting? Explain. If $X = B(n, p)$, what are n and p?

(b) Describe $P(X = 0)$ in words. Then find $P(X = 0)$ and $P(X = 1)$.

(c) What is the probability that 2 or more of the 100 children have a parent behind bars?

8.7 DO OUR ATHLETES GRADUATE? A university claims that 80% of its basketball players get degrees. An investigation examines the fate of all 20 players who entered the program over a period of several years that ended six years ago. Of these players, 11 graduated and the remaining 9 are no longer in school. If the university's claim is true, the number of players among the 20 who graduate should have the binomial distribution with $n = 20$ and $p = 0.8$. What is the probability that exactly 11 out of 20 players graduate?

8.8 MARITAL STATUS Among employed women, 25% have never been married. Select 10 employed women at random.

(a) The number in your sample who have never been married has a binomial distribution. What are n and p?

(b) What is the probability that exactly 2 of the 10 women in your sample have never been married?

(c) What is the probability that 2 or fewer have never been married?

Binomial formulas

We can find a formula for the probability that a binomial random variable takes any value by adding probabilities for the different ways of getting exactly that many successes in n observations. Here is the example we will use to show the idea.

EXAMPLE 8.9 INHERITING BLOOD TYPE

Each child born to a particular set of parents has probability 0.25 of having blood type O. If these parents have 5 children, what is the probability that exactly 2 of them have type O blood?

The count of children with type O blood is a binomial random variable X with $n = 5$ tries and probability $p = 0.25$ of a success on each try. We want $P(X = 2)$.

Because the method doesn't depend on the specific example, let's use "S" for success and "F" for failure for short. Do the work in two steps.

Step 1. Find the probability that a specific 2 of the 5 tries give successes, say the first and the third. This is the outcome SFSFF. Here's how to find the probability of this outcome:

- The probability that the first try is a success is 0.25. That is, in many repetitions, we succeed on the first try 25% of the time.
- Out of all the repetitions with a success on the first try, 75% have a failure on the second try. So the proportion of repetitions on which the first two tries are SF is (0.25)(0.75). We can multiply here because the tries are *independent*. That is, the first try has no influence on the second.
- Keep going: Of these repetitions, the proportion 0.25 have S on the third try. So the probability of SFS is (0.25)(0.75)(0.25). After two more tries, the probability of SFSFF is the product of the try-by-try probabilities:

$$(0.25)(0.75)(0.25)(0.75)(0.75) = (0.25)^2(0.75)^3$$

Step 2. Observe that the probability of *any one* arrangement of 2 S's and 3 F's has this same probability. That's true because we multiply together 0.25 twice and 0.75 three times whenever we have 2 S's and 3 F's. The probability that X = 2 is the probability of getting 2 S's and 3 F's in any arrangement whatsoever. Here are all the possible arrangements:

SSFFF SFSFF SFFSF SFFFS FSSFF
FSFSF FSFFS FFSSF FFSFS FFFSS

There are 10 of them, all with the same probability. The overall probability of 2 successes is therefore

$$P(X = 2) = 10(0.25)^2(0.75)^3 = 0.2637$$

The pattern of this calculation works for any binomial probability. To use it, we need to be able to count the number of arrangements of k successes in n observations without actually listing them. We use the following fact to do the counting:

BINOMIAL COEFFICIENT

The number of ways of arranging k successes among n observations is given by the **binomial coefficient**

$$\binom{n}{k} = \frac{n!}{k!(n-k)!}$$

for $k = 0, 1, 2, \ldots, n$.

factorial

The formula for binomial coefficients uses the ***factorial*** notation. For any positive whole number n, its factorial $n!$ is

$$n! = n \times (n-1) \times (n-2) \times \cdots \times 3 \times 2 \times 1$$

Also, $0! = 1$.

Notice that the larger of the two factorials in the denominator of a binomial coefficient will cancel much of the $n!$ in the numerator. For example, the binomial coefficient we need for Example 8.9 is

$$\begin{aligned}\binom{5}{2} &= \frac{5!}{2!\,3!} \\ &= \frac{(5)(4)(3)(2)(1)}{(2)(1) \times (3)(2)(1)} \\ &= \frac{(5)(4)}{(2)(1)} = \frac{20}{2} = 10\end{aligned}$$

The notation $\binom{n}{k}$ is *not* related to the fraction $\frac{n}{k}$. A helpful way to remember its meaning is to read it as "binomial coefficient n choose k." Binomial coefficients have many uses in mathematics, but we are interested in them only as an aid to finding binomial probabilities. The binomial coefficient $\binom{n}{k}$ counts the number of ways in which k successes can be distributed among n observations. The binomial probability $P(X = k)$ is this count multiplied by the probability of any specific arrangement of the k successes. Here is the formula we seek:

BINOMIAL PROBABILITY

If X has the binomial distribution with n observations and probability p of success on each observation, the possible values of X are $0, 1, 2, \ldots, n$. If k is any one of these values,

$$P(X = k) = \binom{n}{k} p^k (1-p)^{n-k}$$

EXAMPLE 8.10 DEFECTIVE SWITCHES

The number X of switches that fail inspection in Example 8.3 has approximately the binomial distribution with $n = 10$ and $p = 0.1$. The probability that no more than 1 switch fails is

$$\begin{aligned}
P(X \le 1) &= P(X = 1) + P(X = 0) \\
&= \binom{10}{1}(0.1)^1(0.9)^9 + \binom{10}{0}(0.1)^0(0.9)^{10} \\
&= \frac{10!}{1!9!}(0.1)(0.3874) + \frac{10!}{0!10!}(1)(0.3487) \\
&= (10)(0.1)(0.3874) + (1)(1)(0.3487) \\
&= 0.3874 + 0.3487 \\
&= 0.7361
\end{aligned}$$

Notice that the calculation uses the facts that $0! = 1$ and that $a^0 = 1$ for any number a other than 0.

EXERCISES

In each of the following exercises, you are to use the binomial probability formula to answer the question. You may not use the binomial pdf command on your calculator. Begin with the formula, and show substitution into the formula.

8.9 BLOOD TYPES The count X of children with type O blood among 5 children whose parents carry genes for both the O and the A blood types is $B(5, 0.25)$. See Example 8.1 on page 440. Use the binomial probability formula to find $P(X = 3)$.

8.10 BROCCOLI PLANTS Suppose you purchase a bundle of 10 bare-root broccoli plants. The sales clerk tells you that on average you can expect 5% of the plants to die before producing any broccoli. Assume that the bundle is a random sample of plants. Use the binomial formula to find the probability that you will lose at most one of the broccoli plants.

8.11 MORE ON BLOOD TYPES Use the binomial probability formula to find the probability that at least one of the children in Exercise 8.9 has blood type O. (*Hint:* Do not calculate more than one binomial formula.)

8.12 GRADUATION RATE FOR ATHLETES See Exercise 8.7 on page 446. The number of athletes who graduate is $B(20, 0.8)$. Use the binomial probability formula to find the probability that all 20 graduate. What's the probability that not all of the 20 graduate?

8.13 HISPANIC REPRESENTATION A factory employs several thousand workers, of whom 30% are Hispanic. If the 15 members of the union executive committee were chosen from the workers at random, the number of Hispanics on the committee would have the binomial distribution with $n = 15$ and $p = 0.3$.

(a) What is the probability that exactly 3 members of the committee are Hispanic?

(b) What is the probability that none of the committee members are Hispanic?

8.14 CORINNE'S FREE THROWS. Use the binomial probability formula to show that the probability that Corinne makes exactly 7 of her 12 free throws is 0.1032 (see Example 8.6 on page 443).

Binomial mean and standard deviation

If a count X has the binomial distribution based on n observations with probability p of success, what is its mean μ? We can guess the answer. If a basketball player makes 75% of her free throws, the mean number made in 12 tries should be 75% of 12, or 9. In general, the mean of a binomial distribution should be $\mu = np$. To derive the expressions for the mean and standard deviation in the general case, let X represent the number of successes in a single trial. Then X takes two values, 1 (for success) and 0 (for failure). We'll let p be the probability of success on a single trial, and introduce $q = 1 - p$ as the probability of failure. This is common notation. Then the probability distribution is simply

X	0	1
P(X)	q	p

The expected value for this one trial is

$$E(X) = \mu_X = 0\ (q) + 1\ (p) = p$$

The variance is

$$\sigma_X^2 = (0 - p)^2 q + (1 - p)^2 p = p^2 q + pq^2 = pq(p + q) = pq$$

Now define a new random variable Y to be the number of successes in n independent trials. Then $Y = X_1 + X_2 + \cdots + X_n$. Using the rules for means and variances of linear combinations of *independent* random variables, we can say that

$$\begin{aligned}\mu_Y &= \mu_{(X_1+X_2+\cdots+X_n)} = \mu_{X_1} + \mu_{X_2} + \cdots + \mu_{X_n}\\ &= p + p + \cdots + p\\ &= np\end{aligned}$$

and

$$\begin{aligned}\sigma_Y^2 &= \sigma_{X_1+X_2+\cdots+X_n}^2 = \sigma_{X_1}^2 + \sigma_{X_2}^2 + \cdots + \sigma_{X_n}^2\\ &= pq + pq + \cdots + pq\\ &= npq = np(1-p)\end{aligned}$$

and the standard deviation of Y is $\sqrt{np(1-p)}$. Here is what we have shown:

MEAN AND STANDARD DEVIATION OF A BINOMIAL RANDOM VARIABLE

If a count X has the binomial distribution with number of observations n and probability of success p, the mean and standard deviation of X are

$$\mu = np$$

$$\sigma = \sqrt{np(1-p)}$$

Important note: These short formulas are good only for binomial distributions. They can't be used for other discrete random variables.

EXAMPLE 8.11 BAD SWITCHES

Continuing Example 8.10, the count X of bad switches is binomial with $n = 10$ and $p = 0.1$. This is the sampling distribution the engineer would see if she drew all possible SRSs of 10 switches from the shipment and recorded the value of X for each sample.

The mean and standard deviation of the binomial distribution are

$$\begin{aligned}\mu &= np \\ &= (10)(0.1) = 1 \\ \sigma &= \sqrt{np(1-p)} \\ &= \sqrt{(10)(0.1)(0.9)} = \sqrt{0.9} = 0.9487\end{aligned}$$

The mean is marked on the probability histogram in Figure 8.2.

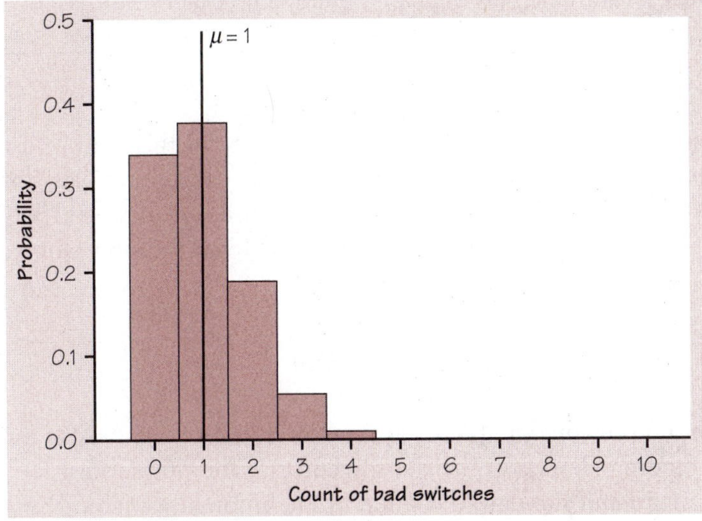

FIGURE 8.2 Probability histogram for the binomial distribution with $n = 10$ and $p = 0.1$.

The normal approximation to binomial distributions

The formula for binomial probabilities becomes awkward as the number of trials n increases. You can use software or a statistical calculator to handle some problems for which the formula is not practical. Here is another alternative: *as the number of trials n gets larger, the binomial distribution gets close to a normal distribution.* When n is large, we can use normal probability calculations to approximate hard-to-calculate binomial probabilities.

EXAMPLE 8.12 ATTITUDES TOWARD SHOPPING

Are attitudes toward shopping changing? Sample surveys show that fewer people enjoy shopping than in the past. A recent survey asked a nationwide random sample of 2500 adults if they agreed or disagreed that "I like buying new clothes, but shopping is often frustrating and time-consuming."[1] The population that the poll wants to draw conclusions about is all U.S. residents aged 18 and over. Suppose that in fact 60% of all adult U.S. residents would say "Agree" if asked the same question. What is the probability that 1520 or more of the sample agree?

Because there are more than 195 million adults, we can take the responses of 2500 randomly chosen adults to be independent. So the number in our sample who agree that shopping is frustrating is a random variable X having the binomial distribution with $n = 2500$ and $p = 0.6$. To find the probability that at least 1520 of the people in the sample find shopping frustrating, we must add the binomial probabilities of all outcomes from $X = 1520$ to $X = 2500$. This isn't practical. Here are three ways to do this problem.

1. Statistical software can do the calculation. The exact result is

$$P(X \geq 1520) = 0.2131$$

2. We can simulate a large number of repetitions of the sample. Figure 8.3 displays a histogram of the counts X from 1000 samples of size 2500 when the truth about the population is $p = 0.6$. Because 221 of these 1000 samples have X at least 1520, the probability estimated from the simulation is

$$P(X \geq 1520) = \frac{221}{1000} = 0.221$$

3. Both of the previous methods require software. Instead, look at the normal curve in Figure 8.3. This is the density curve of the normal distribution with the same mean and standard deviation as the binomial variable X:

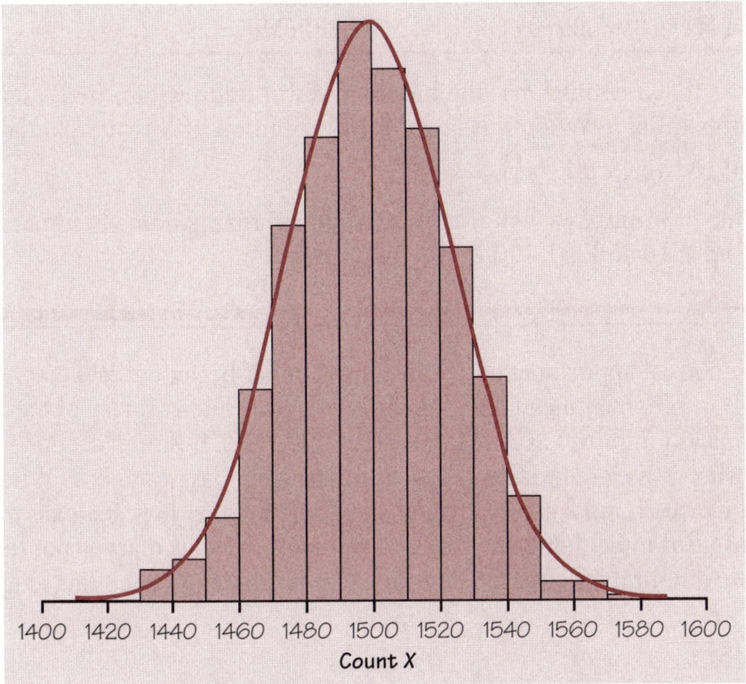

FIGURE 8.3 Histogram of 1000 binomial counts (n = 2500, p = 0.6) and the normal density curve that approximates this binomial distribution.

$$\mu = np = (2500)(0.6) = 1500$$
$$\sigma = \sqrt{np(1-p)} = \sqrt{(2500)(0.6)(0.4)} = 24.49$$

As the figure shows, this normal distribution approximates the binomial distribution quite well. So we can do a normal calculation.

EXAMPLE 8.13 NORMAL CALCULATION OF A BINOMIAL PROBABILITY

If we act as though the count X has the N(1500, 24.49) distribution, here is the probability we want, using Table A:

$$\begin{aligned} P(X \geq 1520) &= P\left(\frac{X-1500}{24.49} \geq \frac{1520-1500}{24.49}\right) \\ &= P(Z \geq 0.82) \\ &= 1 - 0.7939 = 0.2061 \end{aligned}$$

The normal approximation 0.2061 differs from the software result 0.2131 by only 0.007.

NORMAL APPROXIMATION FOR BINOMIAL DISTRIBUTIONS

Suppose that a count X has the binomial distribution with n trials and success probability p. When n is large, the distribution of X is approximately normal, $N\left(np, \sqrt{np(1-p)}\right)$.

As a rule of thumb, we will use the normal approximation when n and p satisfy $np \geq 10$ and $n(1-p) \geq 10$.

The normal approximation is easy to remember because it says that X is normal with its binomial mean and standard deviation. The accuracy of the normal approximation improves as the sample size n increases. It is most accurate for any fixed n when p is close to 1/2 and least accurate when p is near 0 or 1. Whether or not you use the normal approximation should depend on how accurate your calculations need to be. For most statistical purposes great accuracy is not required. Our "rule of thumb" for use of the normal approximation reflects this judgment.

EXERCISES

8.15 ATTITUDES ON SHOPPING Refer to Example 8.12 on attitudes toward shopping.

(a) Verify that the rule of thumb conditions are satisfied for using the normal approximation to the binomial distribution.

(b) Use your calculator and the cumulative binomial function to verify the exact answer for the probability that at least 1520 people in the sample find shopping frustrating is 0.2131. What is the probability correct to 6 decimal places?

(c) What is the probability that at most 1468 people in the sample would agree with the statement that shopping is frustrating?

8.16 HISPANIC COMMITTEE MEMBERS

(a) What is the mean number of Hispanics on randomly chosen committees of 15 workers in Exercise 8.13 (page 449)?

(b) What is the standard deviation σ of the count X of Hispanic members?

(c) Suppose that 10% of the factory workers were Hispanic. Then $p = 0.1$. What is σ in this case? What is σ if $p = 0.01$? What does your work show about the behavior of the standard deviation of a binomial distribution as the probability of a success gets closer to 0?

8.17 DO OUR ATHLETES GRADUATE?

(a) Find the mean number of graduates out of 20 players in the setting of Exercise 8.12 (page 449).

(b) Find the standard deviation σ of the count X.

(c) Suppose that the 20 players came from a population of which $p = 0.9$ graduated. What is the standard deviation σ of the count of graduates? If $p = 0.99$, what is σ? What does your work show about the behavior of the standard deviation of a binomial distribution as the probability p of success gets closer to 1?

8.18 MARITAL STATUS OF EMPLOYED WOMEN You choose 10 employed women at random, as in Exercise 8.8 (page 446). What is the mean number of women in such a sample who have never been married? What is the standard deviation?

8.19 POLLING Many local polls of public opinion use samples of size 400 to 800. Consider a poll of 400 adults in Richmond that asks the question "Do you approve of President George W. Bush's response to the World Trade Center terrorists attacks in September 2001?" Suppose we know that President Bush's approval rating on this issue nationally is 92% a week after the incident.

(a) What is the random variable X? Is X binomial? Explain.

(b) Calculate the binomial probability that at most 358 of the 400 adults in the Richmond poll answer "Yes" to this question.

(c) Find the expected number of people in the sample who indicate approval. Find the standard deviation of X.

(d) Perform a normal approximation to answer the question in (b), and compare the results of the binomial calculation and the normal approximation. Is the normal approximation satisfactory?

8.20 A MARKET RESEARCH SURVEY You operate a restaurant. You read that a sample survey by the National Restaurant Association shows that 40% of adults are committed to eating nutritious food when eating away from home. To help plan your menu, you decide to conduct a sample survey in your own area. You will use random digit dialing to contact an SRS of 200 households by telephone.

(a) If the national result holds in your area, it is reasonable to use the binomial distribution with $n = 200$ and $p = 0.4$ to describe the count X of respondents who seek nutritious food when eating out. Explain why.

(b) What is the mean number of nutrition-conscious people in your sample if $p = 0.4$ is true? What is the standard deviation?

(c) What is the probability that X lies between 75 and 85? Make sure that the rule of thumb conditions are satisfied, and then use a normal approximation to answer the question.

Binomial distribution with the calculator

The following Technology Toolbox summarizes some important calculator techniques when working in a binomial setting:

TECHNOLOGY TOOLBOX *Exploring binomial distributions*

For illustration purposes, we will use the sample of $n = 10$ switches with probability $p = 0.10$ of a defective switch from Example 8.3 (page 440). The random variable X is the number of defective switches (success) and $X = B(10, 0.1)$. To have the calculator make the probability distribution table and plot a histogram for the distribution of defective switches in a sample of 10 switches, proceed as follows:

TECHNOLOGY TOOLBOX *Exploring binomial distributions (continued)*

TI-83

1. Enter the values of X into list L_1, either through the STAT/EDIT mode or by entering the command `seq(X,X,0,10,1)→L1`. (The `seq` command is under 2nd/LIST/OPS/5:seq. The syntax is: the first X is the function, the second X is the counting variable, the next two numbers define the starting and ending values, and the last number is the increment.)

2. Enter the binomial probabilities into list L_2. Highlight L_2 and press 2nd VARS (DISTR)/ `0:binompdf(`. Then complete the command: `binompdf(10,0.1,L1)`. Note that the largest probability listed is about 0.3874. This will help us define our viewing window.

L1	L2	L3 2
0	.34868	------
1	.38742	
2	.19371	
3	.0574	
4	.01116	
5	.00149	
6	1.4E-4	

L2(1)=.3486784401...

3. Deselect or delete any active defined functions in the Y = window.

4. Define Plot1 to be a histogram with Xlist:L_1 and Freq:L_2.

5. Set the viewing window to be $X[0,11]_1$ and $Y[-.15,.5]_1$. Press TRACE and use the left and right cursor keys to inspect heights of various bars in the histogram.

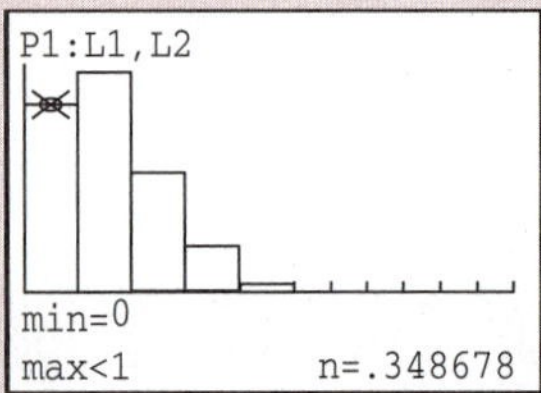

TI-89

1. Enter the values of X into list1, either through the Statistics/List Editor or by entering the command `seq(X,X,0,10,1)→list1`. The `seq` command is in the CATALOG. The syntax is: the first X is the function, the second X is the counting variable, the next two numbers define the starting and ending values, and the last number is the increment.)

2. Highlight list2 and press CATALOG, type F3 (Flash Apps), choose `binompdf(`. Press ENTER and complete the command: `tistat.binomPdf(10,0.1,list1)` and then ENTER.

F1▼ Tools | F2▼ Plots | F3▼ List | F4▼ Calc | F5▼ Distr | F6▼ Tests | F7▼ Ints

list1	list2	list3	list4
0	.34868	----	----
1	.38742		
2	.19371		
3	.0574		
4	.01116		
5	.00149		

list2="tistat.binompdf<10...

MAIN RAD AUTO FUNC 2/7

3. Deselect or delete any active defined functions in the Y = window.

4. Define Plot1 to be a histogram using list1 for *x* and list2 for frequency.

5. Define the viewing window: ◆ F2 (WINDOW). Specify $X[0,11]_1$ and $Y[-.15,.5]_1$. Press ◆ F3. Here is the pdf.

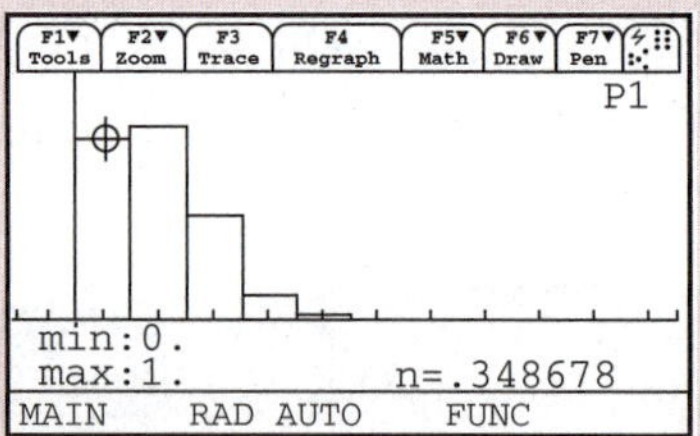

Outcomes larger than 6 do not have probability exactly 0, but their probabilities are so small that the rounded values are 0.0000. Verify that the sum of the probabilities is 1.

TECHNOLOGY TOOLBOX *Exploring binomial distributions (continued)*

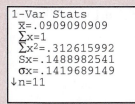

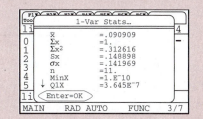

```
1-Var Stats
 x̄=.0909090909
 Σx=1
 Σx²=.312615992
 Sx=.1488982541
 σx=.1419689149
↓n=11
```

6. To calculate the cumulative probabilities, highlight list L_3. Press 2nd DISTR, and select `A:binomcdf(`. Complete the command: `binomcdf(10,.1,L1)`. Press ENTER. The cumulative probabilities are in L_3.

L1	L2	L3 3
0	.34868	.34868
1	.38742	.7361
2	.19371	.92981
3	.0574	.9872
4	.01116	.99837
5	.00149	.99985
6	1.4E-4	.99999

L3(1)=.3486784400...

7. Turn off Plot1 and turn on Plot2. Define Plot2 to be a histogram with Xlist: L_1 and Ylist: L_3. In the viewing window, set Ymin = -.3 and Ymax = 1.2. Here is the histogram for the cdf:

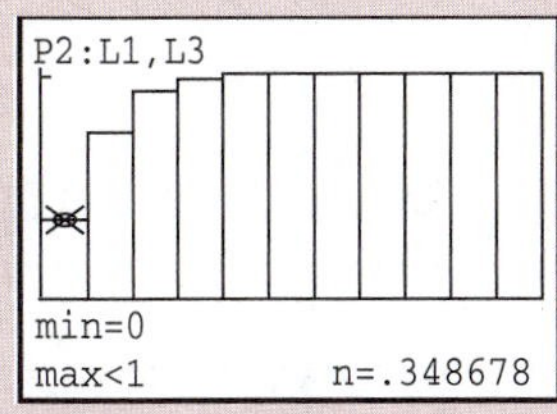


```
1-Var Stats...
 x̄      =.090909
 Σx     =1.
 Σx²    =.312616
 Sx     =.148898
 σx     =.141969
 n      =11.
 MinX   =1.E-10
↓ Q1X   =3.645E-7
Enter=OK
MAIN   RAD AUTO   FUNC   3/7
```

6. To calculate the cumulative distribution values, highlight list3 and press CATALOG F3 (Flash Apps), and choose `binomcdf(`. Press ENTER and complete the command: `tistat.binomCdf(10,.1,list1)` and then ENTER.

list1	list2	list3	list4
0	.34868	.34868	----
1	.38742	.7361	
2	.19371	.92981	
3	.0574	.9872	
4	.01116	.99837	
5	.00149	.99985	

list3="tistat.binomcdf<10...

MAIN RAD AUTO FUNC 3/7

7. Turn off Plot1 (F2 (Plot Setup)), highlight Plot1, then F4 (✓) to deselect Plot1. Define Plot2 to be a histogram, except this time specify list3 for the frequency. In the viewing window, set Ymin = -.3 and Ymax = 1.2. Here is the histogram for the cdf:

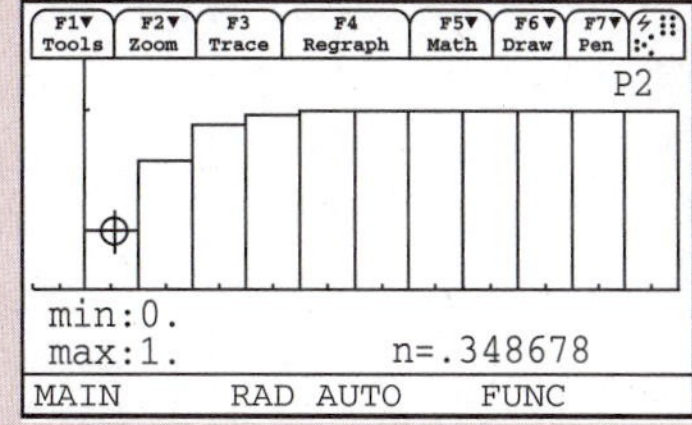

Simulating binomial experiments

In order to simulate a binomial experiment, you need to know how the random variable X and "success" are defined, the probability of success, and the number of trials. But if you know these things, you can apply the rules learned in this section to calculate the probabilities of events exactly. So perhaps simulation methods are not as important in a binomial setting as they are in other settings. On the other hand, being able to simulate a binomial experiment can give credence to results obtained by applying formulas and rules when the results may be less than convincing to someone who knows no statistics.

EXAMPLE 8.14 CORINNE'S FREE THROWS

Recall that Corinne's free throw percentage was 75% (see Example 8.6). In a particular game, she had 12 attempts and she made only 7. The question was, "How unusual was it for Corinne to make at most 7 shots out of 12 attempts?" In Example 8.6, we calculated this binomial probability to be $P(X \leq 7) = 0.1576$. Now we will use the calculator to simulate 12 attempted shots and we will count the number of hits (baskets). Let X = number of hits in 12 free throw attempts. Note that the probability of "success" is 0.75. To set up the simulation, we will assign the digit 0 to a miss and a 1 to a hit. The command `randBin(1,.75,12)` simulates 12 free throw attempts. In the long run, this random function will select the number 1 75% of the time and the number 0 25% of the time.

Here are the results of one simulated game on the TI-83: the first three shots were hits, the fourth was a miss, the fifth was a hit, and the next two were misses, and so forth. If we repeated this many times and counted the proportion of times Corinne had 7 or fewer hits, that would give an estimate of the probability $P(X \leq 7)$ that Corinne made at most 7 of her 12 attempts. One way to automate this more is to assign these results to list L_1/list1 and then sum the entries in the list. Enter the TI-83 command `randBin(1,.75,12)→L₁:sum(L₁)` (or for the TI-89, press CATALOG F3 (Flash Apps) and select `randbin(` and then complete the command: `tistat.randbin(1,.75,12)`).

TI-83 **TI-89**

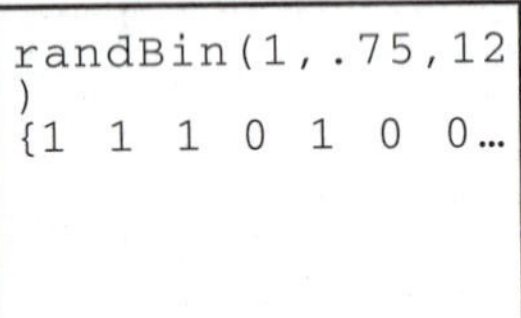

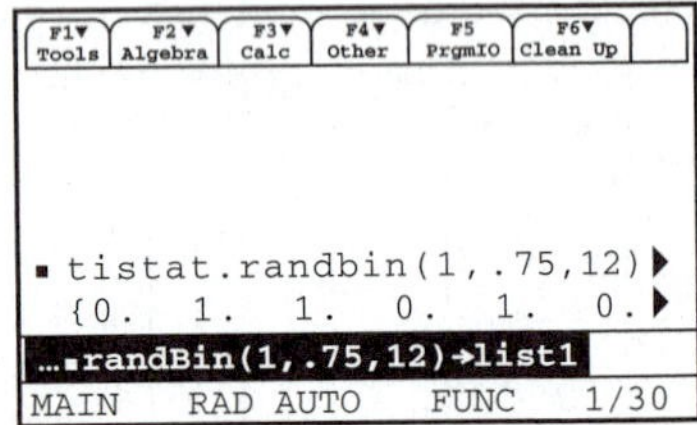


```
randBin(1,.75,12
)→L1:sum(L1)
                10
```

```
F1▾ Tools | F2▾ Algebra | F3▾ Calc | F4▾ Other | F5 PrgmIO | F6▾ Clean Up

▪ tistat.randbin(1,.75,12)▸
                          10.
...▪75,12)→list1:sum(list1)
MAIN   RAD AUTO   FUNC   1/30
```

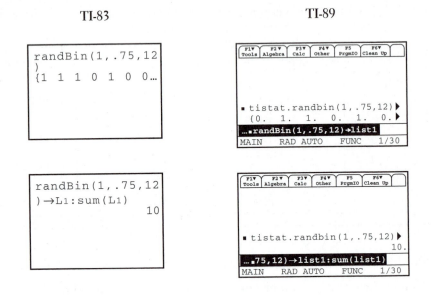

Continue pressing the ENTER key until you have 10 numbers. Record these numbers.

This makes 10 repetitions (i.e., simulates 10 games) for both calculators. (The TI-89 results were 10, 10, 10, 9, 10, 12, 9, 7, 11, 9.) So far Corinne has made 7 or fewer shots in 1 out of 10 games, for a relative frequency of 0.10. Compare this with the binomial probability of 0.1576 for this event. Continue to press ENTER to simulate 10 more

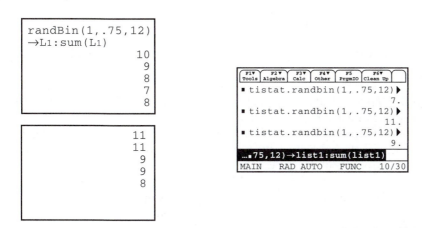

```
randBin(1,.75,12)
→L1:sum(L1)
                 10
                  9
                  8
                  7
                  8
```

```
                 11
                 11
                  9
                  9
                  8
```

```
F1▾ Tools | F2▾ Algebra | F3▾ Calc | F4▾ Other | F5 PrgmIO | F6▾ Clean Up
■ tistat.randbin(1,.75,12)▶
                           7.
■ tistat.randbin(1,.75,12)▶
                          11.
■ tistat.randbin(1,.75,12)▶
                           9.
...■75,12)→list1:sum(list1)
MAIN    RAD AUTO    FUNC    10/30
```

games. Calculate the relative frequency for 20 games and so on. According to the law of large numbers, these relative frequencies should get closer to 0.1576 as the number of simulated games increases. Continue in this fashion until you have simulated 50 games. Are your cumulative results close to 0.1576?

EXERCISES

8.21 CORINNE'S FREE THROWS Use lists L_1/list1 and L_2/list2 on your calculator to construct a pdf for Corinne's free throw probabilities. (Refer to Examples 8.6 and 8.8 on page 443 and 444.) Use the random variable X = number of baskets made on free throws. Then on both calculators, execute the command `cumSum(L₂)→L₃`. (`cumSum` is found under 2nd / LIST / OPS / `6:cumSum` on the TI-83; `cumSum` is found in the CATALOG on the TI-89). What do you think this command does? Then use the `binomcdf` command to enter the cumulative probabilities into list L_4/list4. Compare L_3/list3 and L_4/list4. Are they the same?

8.22 SIMULATING DEFECTIVE SWITCHES

(a) Use the calculator's `randBin` function to simulate the random selection of 10 switches from the $B(10, 0.1)$ distribution, and assign these 10 results to L_1/list1. Then use the 1-Var Stats function to find the mean number of defective switches among the 10. Compare this result with the known mean $\mu = 1$. Repeat these steps to find the mean of 25 randomly selected switches and then the mean of 50 randomly selected switches. What effect, if any, does the number of switches sampled have on the mean number of defective switches?

(b) Do the same as in (a) for the distribution $B(12, 0.75)$ of Corinne's free throws made in 12 attempts (see Examples 8.6 and 8.8). How do your results for samples of size 10, 25, and 50 compare with the true mean number of successes?

8.23 SIMULATING COMMITTEE SELECTION Refer to Exercise 8.13 (page 449). Construct a simulation to estimate the probability that in a committee of 15 members, 3 or fewer members are Hispanic. Describe the design of your experiment, including the correspondence between digits and outcomes in the experiment, and report the relative frequency for 30 repetitions.

8.24 STUDENT INDEBTEDNESS According to the General Accounting Office and the student loan agency Nellie Mae, the average college student credit-card debt in 2000 was $2,748, and a third of students have four or more credit cards. Assume that a randomly selected student has probability 0.33 of having four or more credit cards. Use simulation methods to determine the probability that more than 12 students in a sample of 30 have four or more credit cards.

8.25 SIMULATING MARRIAGE Refer to Exercise 8.8 (page 446). Construct a simulation to estimate the probability that 2 or fewer of a random sample of 10 employed women have never been married. Describe the design of your experiment, including the correspondence between digits and outcomes in the experiment and the number of repetitions you carried out. Report your results.

8.26 DRAWING POKER CHIPS There are 50 poker chips in a container, 25 of which are red, 15 white, and 10 blue. You draw a chip without looking 25 times, each time returning the chip to the container.

(a) What is the expected number of white chips you will draw in 25 draws?

(b) What is the standard deviation of the number of blue chips that you will draw?

(c) Simulate 25 draws by hand or by calculator. Repeat the process as many times as you think necessary.

(d) Based on your answers to parts (a) to (c), is it likely or unlikely that you will draw 9 or fewer blue chips?

(e) Is it likely or unlikely that you will draw 15 or fewer blue chips?

SUMMARY

A count X of successes has a binomial distribution in the **binomial setting**: there are n observations; the observations are **independent** of each other; each observation results in a success or a failure; and each observation has the same probability p of a success.

If X has the binomial distribution with parameters n and p, the possible values of X are the whole numbers 0, 1, 2, ..., n. The **binomial probability** that X takes any value is

$$P(X = k) = \binom{n}{k} p^k (1-p)^{n-k}$$

The **binomial coefficient**

$$\binom{n}{k} = \frac{n!}{k!(n-k)!}$$

counts the number of ways k successes can be arranged among n observations. Here the **factorial** $n!$ is

$$n! = n \times (n-1) \times (n-2) \times \cdots \times 3 \times 2 \times 1$$

for positive whole numbers n, and $0! = 1$.

Given a random variable X, the **probability distribution function** (pdf) assigns a probability to each value of X. For each value of X, the **cumulative distribution function** (cdf) assigns the sum of the probabilities for values less than or equal to X.

The **mean** and **standard deviation** of a binomial count X are

$$\mu = np$$

$$\sigma = \sqrt{np(1-p)}$$

The **normal approximation** to the binomial distribution says that if X is a count having the binomial distribution with parameters n and p, then when n is large, X is approximately $N\left(np, \sqrt{np(1-p)}\right)$. We will use this approximation when $np \geq 10$ and $n(1 - p) \geq 10$.

SECTION 8.1 EXERCISES

8.27 RANDOM DIGITS Each entry in a table of random digits like Table B has probability 0.1 of being a 0, and digits are independent of each other.

(a) What is the probability that a group of five digits from the table will contain at least one 0?

(b) What is the mean number of 0s in lines 40 digits long?

8.28 TESTING ESP In a test for ESP (extrasensory perception), a subject is told that cards the experimenter can see but the subject cannot contain either a star, a circle, a wave, or a square. As the experimenter looks at each of 20 cards in turn, the subject names the shape on the card. A subject who is just guessing has probability 0.25 of guessing correctly on each card.

(a) The count of correct guesses in 20 cards has a binomial distribution. What are n and p?

(b) What is the mean number of correct guesses in many repetitions?

(c) What is the probability of exactly 5 correct guesses?

8.29 RANDOM STOCK PRICES A believer in the "random walk" theory of stock markets thinks that an index of stock prices has probability 0.65 of increasing in any year. Moreover, the change in the index in any given year is not influenced by whether it rose or fell in earlier years. Let X be the number of years among the next 5 years in which the index rises.

(a) X has a binomial distribution. What are n and p?

(b) What are the possible values that X can take?

(c) Find the probability of each value X. Draw a probability histogram for the distribution of X.

(d) What are the mean and standard deviation of this distribution? Mark the location of the mean on the histogram.

8.30 LIE DETECTORS A federal report finds that lie detector tests given to truthful persons have probability about 0.2 of suggesting that the person is deceptive.[2]

(a) A company asks 12 job applicants about thefts from previous employers, using a lie detector to assess their truthfulness. Suppose that all 12 answer truthfully. What is the probability that the lie detector says all 12 are truthful? What is the probability that the lie detector says at least 1 is deceptive?

(b) What is the mean number among 12 truthful persons who will be classified as deceptive? What is the standard deviation of this number?

(c) What is the probability that the number classified as deceptive is less than the mean?

8.31 A MARKET RESEARCH SURVEY Return to the restaurant sample described in Exercise 8.20 (page 455). You find 100 of your 200 respondents concerned about nutrition. Is this reason to believe that the percent in your area is higher than the national 40%? To answer this question, find the probability that X is 100 or larger if $p = 0.4$ is true. If this probability is very small, that is reason to think that p is actually greater than 0.4.

8.32 PLANNING A SURVEY You are planning a sample survey of small businesses in your area. You will choose an SRS of businesses listed in the telephone book's Yellow Pages. Experience shows that only about half the businesses you contact will respond.

(a) If you contact 150 businesses, it is reasonable to use the binomial distribution with $n = 150$ and $p = 0.5$ for the number X who respond. Explain why.

(b) What is the expected number (the mean) who will respond?

(c) What is the probability that 70 or fewer will respond? (Use the normal approximation.)

(d) How large a sample must you take to increase the mean number of respondents to 100?

8.33 ARE WE SHIPPING ON TIME? Your mail-order company advertises that it ships 90% of its orders within three working days. You select an SRS of 100 of the 5000 orders received in the past week for an audit. The audit reveals that 86 of these orders were shipped on time.

(a) If the company really ships 90% of its orders on time, what is the probability that 86 or fewer in an SRS of 100 orders are shipped on time?

(b) A critic says, "Aha! You claim 90%, but in your sample the on-time percentage is only 86%. So the 90% claim is wrong." Explain in simple language why your probability calculation in (a) shows that the result of the sample does not refute the 90% claim.

8.34 AIDS TEST A test for the presence of antibodies to the AIDS virus in blood has probability 0.99 of detecting the antibodies when they are present. Suppose that during a year 20 units of blood with AIDS antibodies pass through a blood bank.

(a) Take X to be the number of these 20 units that the test detects. What is the distribution of X?

(b) What is the probability that the test detects all 20 contaminated units? What is the probability that at least 1 unit is not detected?

(c) What is the mean number of units among the 20 that will be detected? What is the standard deviation of the number detected?

8.35 SIMULATING GRADUATION Refer to Exercise 8.7 (page 446). Construct a simulation to estimate the probability that at most 11 of the 20 basketball players graduated. Describe the design of your experiment, including the correspondence between digits and outcomes in the experiment and the number of repetitions you carried out. Report your results.

8.36 ALLERGY RELIEF Clinical trials of the popular allergy medicine Allegra-D (fexofenadine HCl 60 mg/pseudoephedrine HCl 120 mg Extended Release Tablets) found that 13% of the 215 subjects reported headache as an adverse reaction to the drug.[3] Assume that in fact 13% of all users of this medicine experience headaches after taking this medicine. Suppose that 8 allergy sufferers are selected at random. This exercise uses the statistical software Minitab to answer several questions. To calculate binomial probabilities with Minitab, begin by entering the integers 0 to 8 in column 1 and naming this column VALUES. Then select **Calc > Probability Distributions > Binomial.** Then select **Probability** to indicate that you want individual probabilities. Specify the **Number of trials** and the **Probability of success.** Select **input column** and specify VALUES in column C1 to tell Minitab to calculate binomial probabilities for each of the values in that column. Then click **OK.** The following results are produced:

```
MTB >  PDF 'VALUES';
SUBC>    Binomial 8 .13.
           K            P( X = K)
        0.00               0.3282
        1.00               0.3923
        2.00               0.2052
        3.00               0.0613
        4.00               0.0115
        5.00               0.0014
        6.00               0.0001
        7.00               0.0000
        8.00               0.0000
MTB>
```

To calculate the cumulative distribution, make the same menu choices, except this time select **Cumulative probability** instead of **Probability.** The following output is produced:

```
MTB >  CDF 'VALUES';
SUBC>    Binomial 8 .13.
           K            P( X LESS OR = K)
        0.00               0.3282
        1.00               0.7206
        2.00               0.9257
        3.00               0.9871
        4.00               0.9985
        5.00               0.9999
        6.00               1.0000
        7.00               1.0000
        8.00               1.0000
```

From the printouts, calculate the probability that out of the 8 randomly selected subjects, the number experiencing headaches is

(a) exactly 3
(b) at most 2
(c) less than 2
(d) at least 3 but no more than 5
(e) either less than 2 or more than 5

8.2 THE GEOMETRIC DISTRIBUTIONS

In the case of a binomial random variable, the number of trials is fixed beforehand, and the binomial variable X counts the number of successes in that fixed number of trials. If there are n trials then the possible values of X are 0, 1, 2, ..., n. By way of comparison, there are situations in which the goal is to obtain a fixed number of successes. In particular, if the goal is to obtain one success, a random variable X can be defined that counts the number of trials needed to obtain that first success. A random variable that satisfies the above description is called *geometric*, and the distribution produced by this random variable is called a ***geometric distribution***. The possible values of a geometric random variable are 1, 2, 3, . . . , that is, an infinite set, because it is theoretically possible to proceed indefinitely without ever obtaining a success. Consider the following situations:

geometric distribution

- Flip a coin until you get a *head*.
- Roll a die until you get a 3.
- In basketball, attempt a three-point shot until you make a *basket*.

Notice that all of these situations involve counting the number of trials until an event of interest happens. We are now ready to characterize the geometric setting.

A random variable X is geometric provided that the following conditions are met:

THE GEOMETRIC SETTING

1. Each observation falls into one of just two categories, which for convenience we call "success" or "failure."

2. The probability of a success, call it p, is the same for each observation.

3. The observations are all **independent**.

4. The variable of interest is the number of trials required to obtain the first success.

EXAMPLE 8.15 ROLL A DIE

An experiment consists of rolling a single die. The event of interest is rolling a 3; this event is called a success. The random variable is defined as X = the number of trials until a 3 occurs. To verify that this is a geometric setting, note that rolling a 3 will represent a success, and rolling any other number will represent a failure. The probability of rolling a 3 on each roll is the same: 1/6. The observations are independent. A trial consists of rolling the die once. We roll the die until a 3 appears. Since all of the requirements are satisfied, this experiment describes a geometric setting.

EXAMPLE 8.16 DRAW AN ACE

Suppose you repeatedly draw cards without replacement from a deck of 52 cards until you draw an ace. There are two categories of interest: ace = success; not ace = failure. But is the probability of success the same for each trial? No. The probability of an ace on the first card is 4/52. If you don't draw an ace on the first card, then the probability of an ace on the second card is 4/51. Since the result of the first draw affects probabilities on the second draw (and on all successive draws required), the trials are not independent. So this is not a geometric setting.

Using the setting of Example 8.15, let's calculate some probabilities.

$$
\begin{aligned}
X = 1: \quad P(X = 1) &= P(\text{success on first roll}) = 1/6 \\
X = 2: \quad P(X = 2) &= P(\text{success on second roll}) \\
&= P(\text{failure on first roll and success on second roll}) \\
&= P(\text{failure on first roll}) \times P(\text{success on second roll}) \\
&= (5/6) \times (1/6)
\end{aligned}
$$

(since trials are independent).

$$
\begin{aligned}
X = 3: \quad P(X = 3) &= P(\text{failure on first roll}) \times P(\text{failure on second roll}) \\
&\quad \times \text{P(success on third roll)} \\
&= (5/6) \times (5/6) \times (1/6)
\end{aligned}
$$

Continue the process. The pattern suggests that a general formula for the variable X is

$$P(X = n) = (5/6)^{n-1}(1/6)$$

Now we can state the following principle:

RULE FOR CALCULATING GEOMETRIC PROBABILITIES

If X has a geometric distribution with probability p of success and $(1 - p)$ of failure on each observation, the possible values of X are 1, 2, 3, ... If n is any

RULE FOR CALCULATING GEOMETRIC PROBABILITIES (*continued*)

one of these values, the probability that the first success occurs on the nth trial is

$$P(X = n) = (1 - p)^{n-1}p$$

Although the setting for the geometric distribution is very similar to the binomial setting, there are some striking differences. In rolling a die, for example, it is possible that you will have to roll the die many times before you roll a 3. In fact, it is theoretically possible to roll the die forever without rolling a 3 (although the probability gets closer and closer to 0 the longer you roll the die without getting a 3). The probability of observing the first 3 on the fiftieth roll of the die is $P(X = 50) = 0.0000$.

A probability distribution table for the geometric random variable is strange indeed because it never ends; that is, the number of table entries is infinite. The rule for calculating geometric probabilities shown above can be used to construct the table:

X	1	2	3	4	5	6	7...
P(X)	p	$(1-p)p$	$(1-p)^2p$	$(1-p)^3p$	$(1-p)^4p$	$(1-p)^5p$	$(1-p)^6p$...

The probabilities (i.e., the entries in the second row) are the terms of a *geometric sequence* (hence the name for this random variable). You may recall from your study of algebra that the general form for a geometric sequence is

$$a, ar, ar^2, ar^3, \ldots, ar^{n-1}, \ldots$$

where a is the first term, r is the ratio of one term in the sequence to the next, and the nth term is ar^{n-1}. You may also recall that even though the sequence continues forever, and even though you could never finish adding the terms, the sequence does have a sum (one of the implausible truths of the infinite!). This sum is

$$\frac{a}{1-r}$$

In order for the geometric random variable to have a valid pdf, the probabilities in the second row of the table must add to 1. Using the formula for the sum of a geometric sequence, we have

$$\sum_{i=1}^{\infty} P(x_i) = p + (1-p)p + (1-p)^2p + \cdots$$

$$= \frac{p}{1-(1-p)} = \frac{p}{p} = 1$$

EXAMPLE 8.17 ROLL A DIE

The rule for calculating geometric probabilities can be used to construct a probability distribution table for X = number of rolls of a die until a 3 occurs:

X	1	2	3	4	5	6	7	...
P(X)	0.1667	0.1389	0.1157	0.0965	0.0804	0.0670	0.0558	...

Here's one way to find these probabilities with your calculator:

1. Enter the probability of success, 1/6. Press ENTER.
2. Enter $*(5/6)$ and press ENTER.
3. Continue to press ENTER repeatedly.

```
1/6
         .1666666667
Ans*(5/6)
         .1388888889
         .1157407407
         .0964506173
         .0803755144
```

Verify that the entries in the second row are as shown:

X	1	2	3	4
P(X)	1/6	5/36	25/216	125/1296

Figure 8.4 is a graph of the distribution of X. As you might expect, the probability distribution histogram is strongly skewed to the right with a peak at the leftmost value, 1. It is easy to see why this must be so, since the height of each bar after the first is the height of the previous bar times the probability of failure $1 - p$. Since you're multiplying the height of each bar by a number less than 1, each new bar will be shorter than the previous bar, and hence the histogram will be right-skewed. Always.

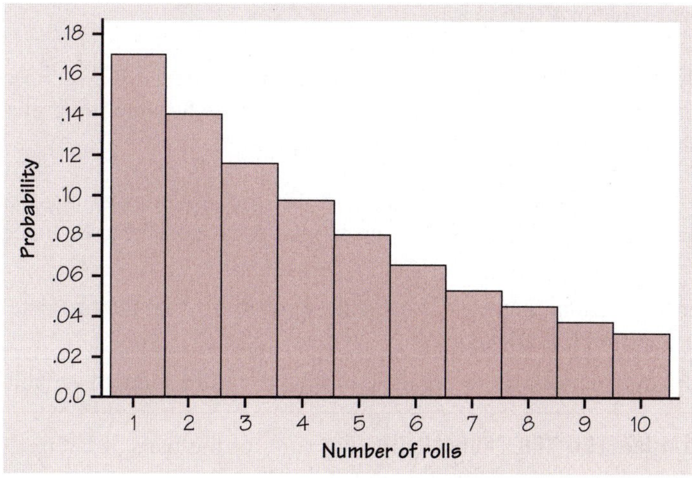

FIGURE 8.4 Probability histogram for the geometric distribution.

EXERCISES

8.37 GEOMETRIC SETTING? For each of the following, determine if the experiment describes a geometric distribution. If it does, describe the two events of interest (success and failure), what constitutes a trial, and the probability of success on one trial. If the random variable is not geometric, identify a condition of the geometric setting that is not satisfied.

(a) Flip a coin until you observe a tail.

(b) Record the number of times a player makes both shots in a one-and-one foul-shooting situation. (In this situation, you get to attempt a second shot only if you make your first shot.)

(c) Draw a card from a deck, observe the card, and replace the card within the deck. Count the number of times you draw a card in this manner until you observe a jack.

(d) Buy a "Match 6" lottery ticket every day until you win the lottery. (In a "Match 6" lottery, a player chooses 6 different numbers from the set $\{1, 2, 3, ..., 44\}$. A lottery representative draws 6 different numbers from this set. To win, the player must match all 6 numbers, in any order.)

(e) There are 10 red marbles and 5 blue marbles in a jar. You reach in and, without looking, select a marble. You want to know how many marbles you will have to draw (without replacement), on average, in order to be sure that you have 3 red marbles.

8.38 ROLL A PRIME An experiment consists of rolling a die until a prime number (2, 3, or 5) is observed. Let X = number of rolls required to get the first prime number.

(a) Verify that X has a geometric distribution.

(b) Construct a probability distribution table to include at least 5 entries for the probabilities of X. Record probabilities to four decimal places.

(c) Construct a graph of the pdf of X.

(d) Compute the cdf of X and plot its histogram.

(e) Use the formula for the sum of a geometric sequence to show that the probabilities in the pdf table of X add to 1.

8.39 TESTING HARD DRIVES Suppose we have data that suggest that 3% of a company's hard disk drives are defective. You have been asked to determine the probability that the first defective hard drive is the fifth unit tested.

(a) Verify that this is a geometric setting. Identify the random variable; that is, write X = number of ________ and fill in the blank. What constitutes a success in this situation?

(b) Answer the original question: What is the probability that the first defective hard drive is the fifth unit tested?

(c) Find the first four entries in the table of the pdf for the random variable X.

8.40 CALCULATING GEOMETRIC PROBABILITIES For each of the parts of Exercise 8.37 that describes a geometric setting, find the probability that X = 4.

The expected value and other properties of the geometric random variable

If you're flipping a fair coin, how many times would you expect to have to flip the coin in order to observe the first head? If you're rolling a die, how many times would you expect to have to roll the die in order to observe the first 3? If you said 2 coin tosses and 6 rolls of the die, then your intuition is serving you well. To derive an expression for the mean (expected value) of a geometric random variable, we begin with the probability distribution table. The notation will be simplified if we let p = probability of success and let q = probability of failure. Then $q = 1 - p$ and the probability distribution table looks like this:

X	1	2	3	4	...
P(X)	p	pq	pq^2	pq^3	...

The mean (expected value) of X is calculated as follows:

$$\mu_X = 1(p) + 2(pq) + 3(pq^2) + 4(pq^3) + \cdots$$
$$= p(1 + 2q + 3q^2 + 4q^3 + \cdots)$$

Multiplying both sides by q, we have

$$q\mu_X = p(q + 2q^2 + 3q^3 + 4q^4 + \cdots)$$

Now subtract this equation from the previous equation, and group like terms on the right.

$$\mu_X - q\mu_X = p(1 + q + q^2 + q^3 + \cdots)$$
$$\mu_X(1-q) = p\left(\frac{1}{1-q}\right)$$
$$\mu_X = \frac{p}{(1-q)^2} = \frac{p}{p^2} = \frac{1}{p}$$

Deriving the variance and standard deviation of the geometric random variable X is considerably more work and would take us too far afield.

Here are the facts:

THE MEAN AND STANDARD DEVIATION OF A GEOMETRIC RANDOM VARIABLE

If X is a geometric random variable with probability of success p on each trial, then the **mean,** or **expected value,** of the random variable, that is, the expected number of trials required to get the first success, is $\mu = 1/p$. The variance of X is $(1 - p)/p^2$.

EXAMPLE 8.18 ARCADE GAME

Glenn likes the game at the state fair where you toss a coin into a saucer. You win if the coin comes to rest in the saucer without sliding off. Glenn has played this game many times and has determined that on average he wins 1 out of every 12 times he plays. He believes that his chances of winning are the same for each toss. He has no reason to think that his tosses are not independent. Let X be the number of tosses until a win. Glenn believes that this describes a geometric setting.

Since $E(X) = 12 = 1/p$, the probability of success on any given trial is

$$p = 1/12 = 0.0833$$

The variance of X is

$$\sigma_X^2 = \frac{1-p}{p^2} = \frac{11/12}{1/144} = 132$$

And the standard deviation is $\sigma_X \simeq 11.5$

There is another interesting result that relates to the probability that it takes *more* than a certain number of trials to achieve success. Here are the steps:

$$\begin{aligned}
P(X > n) &= 1 - P(X \leq n) \\
&= 1 - (p + qp + q^2p + \cdots + q^{n-1}p) \\
&= 1 - p(1 + q + q^2 + \cdots + q^{n-1}) \\
&= 1 - p\left(\frac{1-q^n}{1-q}\right) \\
&= 1 - p\left(\frac{1-q^n}{p}\right) \\
&= 1 - (1 - q^n) \\
&= q^n = (1-p)^n
\end{aligned}$$

We summarize as follows:

$P(X > n)$

The probability that it takes *more* than n trials to see the first success is

$$P(X > n) = (1-p)^n$$

EXAMPLE 8.19 APPLYING THE FORMULA

Roll a die until a 3 is observed. The probability that it takes more than 6 rolls to observe a 3 is

$$P(X > 6) = (1 - p)^n = (5/6)^6 \cong 0.335$$

Let Y be the number of Glenn's coin tosses until a coin stays in the saucer (see Example 8.18). The expected number is 12. The probability that it takes more than 12 tosses to win a stuffed animal is

$$P(X > 12) = (11/12)^{12} \cong 0.352$$

The probability that it takes more than 24 tosses to win a stuffed animal is

$$P(X > 24) = (11/12)^{24} \cong 0.124$$

The following Technology Toolbox summarizes some calculator techniques when working in a geometric setting:

TECHNOLOGY TOOLBOX *Exploring geometric distributions*

For illustration purposes, we will use the roll of a die with $n = 6$ equally likely outcomes and probability $p = 1/6$ of rolling a 3, from Example 8.15 (page 465). The random variable X is the number of rolls until a 3 is observed.

To have the calculator calculate the probability distribution table and plot a histogram for the distribution, proceed as follows:

TI-83

1. Enter the numbers 1 to 10 in list L_1. Next, enter the probabilities into L_2 by first highlighting L_2. Then press 2nd VARS (DISTR). Scroll down and select `D:geometpdf(`. Complete the command: `geometpdf(1/6,L1)`, and press ENTER. Here are the results:

L1	L2	L3 2
1	.16667	------
2	.13889	
3	.11574	
4	.09645	
5	.08038	
6	.06698	
7	.05582	

L2(1)=.1666666666...

2. Specify the dimensions of an appropriate viewing window. Scanning the list of values gives you insight into reasonable dimensions for the window. Specify $X[0,11]_1$ and $Y[-.05,.2]_{.1}$.

TI-89

1. Enter the numbers 1 to 10 in list1. Next, enter the probabilities into list2 by first highlighting list2. Then press CATALOG F3 (Flash Apps) and scroll down to select `geomPdf(`. Complete the command: `TIStat.geomPdf(1/6,list1)`. Here are the results:

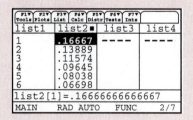

F1 Tools | F2 Plots | F3 List | F4 Calc | F5 Distr | F6 Tests | F7 Ints

list1	list2	list3	list4
1	.16667	----	----
2	.13889		
3	.11574		
4	.09645		
5	.08038		
6	.06698		

list2[1]=.16666666666667

MAIN RAD AUTO FUNC 2/7

2. Specify these dimensions for the viewing window: $X[0,11]_1$ and $Y[-.05,.2]_{.1}$.

TECHNOLOGY TOOLBOX *Exploring geometric distributions (continued)*

3. When you define a histogram for Plot1, specify Xlist: L_1 and Freq: L_2. The resulting plot shows that the distribution is strongly right-skewed.

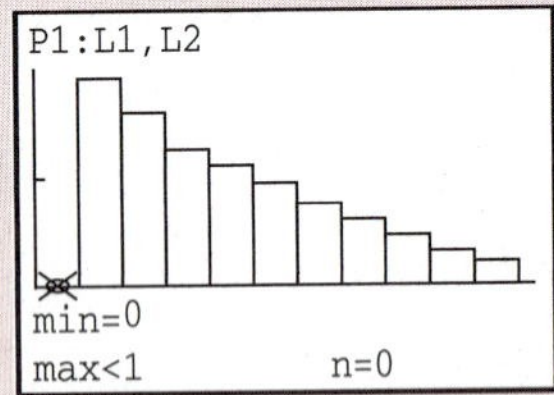

4. Next install the cdf as list L_3. In the STAT/ Edit window, place the cursor on list L_3. Enter the formula `geometcdf(1/6,L`$_1$`)` and press ENTER.

L1	L2	L3 3
1	.16667	.16667
2	.13889	.30556
3	.11574	.4213
4	.09645	.51775
5	.08038	.59812
6	.06698	.6651
7	.05582	.72092

L3(1)=.1666666666...

5. To plot the cumulative distribution histogram, first specify the viewing window: $X[0,11]_1$ and $Y[-.3,1]_1$. The deselect Plot1 and define Plot2 to be a histogram with Xlist: L_1 and Freq: L_3. Press GRAPH. Here is the cdf histogram:

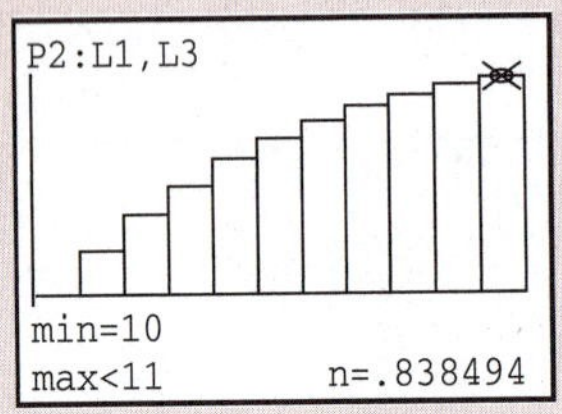

3. From the Statistics/List Editor, press F2 (Plots). Select `1:Plot Setup`. Define Plot 1 to be a histogram using list1 for the X-values and list2 for the frequency. To plot the histogram, press ◆ GRAPH. Here is the pdf histogram:

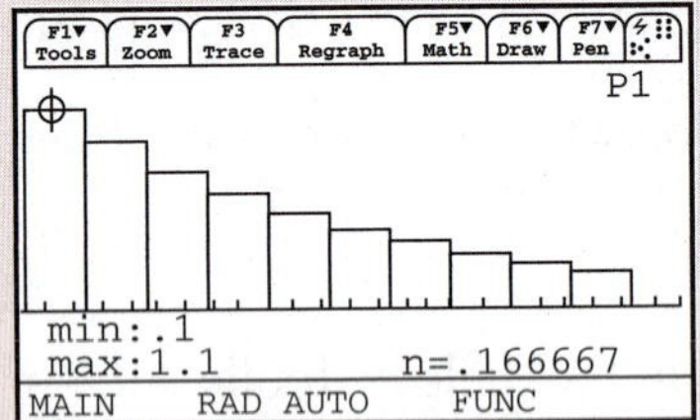

4. To calculate the cumulative distribution values, highlight list3 and press CATALOG F3 (Flash Apps), then scroll down and select `geomCdf(`. Complete the command `tistat.binomCdf(1/6,.1,list1)`. Here is the geometric cdf:

list1	list2	list3	list4
1	.16667	.16667	----
2	.13889	.30556	
3	.11574	.4213	
4	.09645	.51775	
5	.08038	.59812	
6	.06698	.6651	

list3[1]=.16666666666667

MAIN RAD AUTO FUNC 3/7

5. Deselect Plot 1. Then define Plot 2 to be a histogram using list3 for the frequency. Here is the histogram for the cdf:

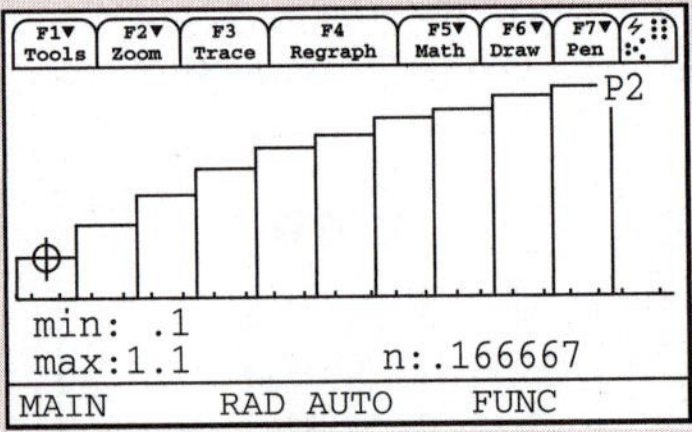

Simulating geometric experiments

Geometric simulations are frequently called "waiting time" simulations because you continue to conduct trials and wait until a "success" is observed.

Conducting a geometric simulation by hand is generally pretty easy but tedious. Conducting a geometric simulation by calculator or computer usually takes more effort initially, but the payoff is that you can quickly run a rather large number of repetitions and frequently get results that are quite respectable. Here is an example:

EXAMPLE 8.20 SHOW ME THE MONEY!

In 1986–1987, Cheerios™ cereal boxes displayed a dollar bill on the front of the box and a cartoon character who said, "Free $1 bill in every 20th box." Here is a simulation to determine the number of boxes of Cheerios you would expect to buy in order to get one of the "free" dollar bills.

Let a two-digit number, 00 to 99, represent a box of Cheerios, and let the digits 01 to 05 represent a box of Cheerios with a $1 bill in it. The digits 00 and 06 to 99 represent boxes without the $1 bill. Starting at the third block of digits in line 127 of Table B, we select digits in pairs:

23	33	06	43	59	40	08	61	69	25
85	11	73	60	71	15	68	91	42	27
06	56	51	43	74	13	35	24	93	67
81	98	28	72	09	36	75	95	89	84
68	28	82	29	13	18	63	84	43	[03]

So in our first run of this simulation, we had to buy 50 boxes of Cheerios until we found one with a $1 bill in it! If you don't usually buy Cheerios, would this promotion induce you to buy a box in hopes of getting one with a dollar in it?

Why did it take so many boxes (50) to achieve success? Since the probability of success on a single trial is $p = 1/20 = 0.05$, we know that the mean (expected value) is $E(X) = 1/p = 20$, so a value of 50 in our simulation seems high. But the variance is

$$\sigma^2(X) = \frac{1-p}{p^2} = \frac{0.95}{0.0025} = 380$$

and

$$\sigma(X) = \sqrt{380} \cong 19.49$$

Our simulated result of 50 is about 1.5 standard deviations to the right of the mean, 20. So perhaps we should not be too surprised with our 50. (Keep in mind that the standard deviation is not an appropriate measure of spread for strongly skewed distributions, and this geometric distribution *is* strongly skewed right.)

EXERCISES

8.41 FLIP A COIN Consider the following experiment: flip a coin until a head appears.

(a) Identify the random variable X.

(b) Construct the pdf table for X. Then plot the probability histogram.

(c) Compute the cdf and plot its histogram.

8.42 ARCADE GAME Refer to Example 8.19 on page 471.

(a) Use the formula for calculating $P(X > n)$ on page 470 to find the probability that it takes more than 10 tosses until Glenn wins a stuffed animal.

(b) Find the answer to (a) by calculating the probability of the complementary event: $1 - P(X \leq 10)$. Your results should agree, of course.

(*Note*: The formula for $P(X > n)$ is not practically important since there are other ways to answer the question. But it's a nice little result, and it's quite easy to derive.)

8.43 ROLL A DIE

(a) Plot the cumulative distribution histogram for the die-rolling experiment described in Example 8.15 with the pdf table in Example 8.17.

(b) Find the probability that it takes more than 10 rolls to observe a 3.

(c) Find the smallest positive integer k for which $P(X \leq k) > 0.99$.

8.44 LANGUAGE SKILLS The State Department is trying to identify an individual who speaks Farsi to fill a foreign embassy position. They have determined that 4% of the applicant pool are fluent in Farsi.

(a) If applicants are contacted randomly, how many individuals can they expect to interview in order to find one who is fluent in Farsi?

(b) What is the probability that they will have to interview more than 25 until they find one who speaks Farsi? More than 40?

8.45 SHOOTING FREE THROWS A basketball player makes 80% of her free throws. We put her on the free-throw line and ask her to shoot free throws until she misses one. Let X = the number of free throws the player takes until she misses.

(a) What assumption do you need to make in order for the geometric model to apply? With this assumption, verify that X has a geometric distribution. What action constitutes "success" in this context?

(b) What is the probability that the player will make 5 shots before she misses?

(c) What is the probability that she will make at most 5 shots before she misses?

8.46 GAME OF CHANCE Three friends each toss a coin. The odd man wins; that is, if one coin comes up different from the other two, that person wins that round. If the coins all match, then no one wins and they toss again. We're interested in the number of times the players will have to toss the coins until someone wins.

(a) What is the probability that no one will win on a given coin toss?

(b) Define a success as "someone wins on a given coin toss." What is the probability of a success?

(c) Define the random variable of interest: X = number of ________. Is X binomial? Geometric? Justify your answer.

(d) Construct a probability distribution table for X. Then extend your table by the addition of cumulative probabilities in a third row.

(e) What is the probability that it takes no more than 2 rounds for someone to win?

(f) What is the probability that it takes more than 4 rounds for someone to win?

(g) What is the expected number of tosses needed for someone to win?

(h) Use the `randInt` function on your calculator to simulate 25 rounds of play. Then calculate the relative frequencies for X = 1, 2, 3, ... Compare the results of your simulation with the theoretical probabilities you calculated in (d).

SUMMARY

A count X of successes has a **geometric distribution** in the geometric setting if the following are satisfied: each observation results in a success or a failure; each observation has the same probability p of success; observations are independent; and X counts the number of trials required to obtain the first success. A geometric random variable differs from a binomial variable because in the geometric setting the number of trials varies and the desired number of defined successes (1) is fixed in advance.

If X has the geometric distribution with probability of success p, the possible values of X are the positive integers 1, 2, 3, ... The **geometric probability** that X takes any value is

$$P(X = n) = (1 - p)^{n-1}p$$

The **mean** (expected value) of a geometric count X is $1/p$.

The standard deviation is

$$\sqrt{\frac{(1-p)}{p^2}}$$

The probability that it takes *more* than n trials to see the first success is

$$P(X > n) = (1 - p)^n$$

SECTION 8.2 EXERCISES

8.47 DRAWING MARBLES There are 20 red marbles, 10 blue marbles, and 5 white marbles in a jar. An experiment consists of selecting a marble without looking, noting the color, and then replacing the marble in the jar. We're interested in the number of marbles you would have to draw in order to be sure you have a red marble.

(a) Is this a binomial or a geometric setting? Explain your choice, and write a description of the random variable X.

(b) Calculate the probability of drawing a red marble on the second draw. Calculate the probability of drawing a red marble by the second draw. Calculate the probability that it would take more than 2 draws to get a red marble.

(c) What single calculator command will install the first 20 values of X into L_1/list1? What single command will install the corresponding probabilities into L_2/list2? What single command will install the cumulative probabilities into L_3/list3? Enter these commands in the Home screen. Copy this information from your calculator onto your paper to make an expanded probability distribution table (with the cdf as the third row).

(d) Construct a probability distribution histogram as STAT PLOT1, and then construct a cumulative distribution histogram as STAT PLOT2.

8.48 DRAWING MARBLES II This is a continuation of Exercise 8.47. Given the jar containing red, white, and blue marbles, Joey thinks a more interesting problem would be to find the number of marbles you would have to draw, without replacing them in the jar, to be sure that you have 2 red marbles.

(a) Does this experiment describe a geometric setting? Why or why not?

(b) Would your answer to (a) change if the marble was replaced after each draw? Explain.

(c) Design and carry out a simulation to determine the number of marbles you would have to draw, with replacement, until you get 2 red marbles. Compare the results from your simulation with the results from the previous exercise.

8.49 MULTIPLE-CHOICE Carla makes random guesses on a multiple-choice test that has five choices for each question. We want to know how many questions Carla answers until she gets one correct.

(a) Define a success in this context, and define the random variable X of interest. What is the probability of success?

(b) What is the probability that Carla's first correct answer occurs on problem 5?

(c) What is the probability that it takes more than 4 questions before Carla answers one correctly?

(d) Construct a probability distribution table for X.

(e) If Carla took a test like this test many times and randomly guessed at each question, what would be the average number of questions she would have to answer before she answered one correctly?

8.50 IT'S A BOY! In some cultures, it is considered very important to have a son to carry on the family name. Suppose that a couple in one of these cultures plans to have children until they have exactly one son.

(a) Find the average number of children per family in such a culture.

(b) What is the expected number of girls in this family?

(c) Describe a simulation that could be used to find approximate answers to the questions in (a) and (b).

8.51 FAMILY PLANNING, I Example 5.24 (page 313) used simulation techniques to explore the following situation: A couple plans to have children until they have a girl or until they have four children, whichever comes first.

(a) List the outcomes in the sample space for this "experiment." What event represents a success?

(b) Let X = the number of boys in this family. What values can X take? Use an appropriate probability rule to calculate the probability for each value of X, and make a probability distribution table for X. Then show that the sum of the probabilities is 1.

(c) Let Y = the number of children produced in this family until a girl is produced. Show that Y starts out as a geometric distribution but then is stopped abruptly. Make a probability distribution table for Y.

(d) What is the expected number of children for this couple?

(e) What is the probability that this couple will have more than the expected number of children?

(f) At the end of Example 5.24, it states that the probability of having a girl in this situation is 0.938. How can you prove this?

8.52 FAMILY PLANNING, II This is a continuation of Exercise 8.51. A couple plans to have children until they have a girl or until they have four children, whichever comes first. Use the random number table (Table B), beginning on line 130, to simulate 25 repetitions of this childbearing strategy. As in Example 5.24, since a girl and boy are equally likely, let the digits 0 to 4 represent a girl, and let digits 5 to 9 represent a boy. Write the digits in a string until you observe a girl, write B or G under each digit, and write the number of children noted at the bottom. The first two repetitions would be recorded as

```
6 9 0     5 1
B B G     B G
  3        2
```

Then find the mean of the 25 repetitions. How do your results compare with the theoretical expected value of 1.8 children?

8.53 FAMILY PLANNING, III This is a continuation of Exercises 8.51 and 8.52. Devise a simulation procedure for the calculator to approximate the expected number of children. List the steps and commands you use as well as the number of repetitions and the results. Alternatively, incorporate these steps into a calculator program similar to the programs SPIN123 (page 329) or FLIP50 (page 92).

8.54 MAKING THE CONNECTION This exercise provides visual reinforcement of the relationship between the probability of success and the mean (expected value) of a geometric random variable.

(a) Begin by completing the table below, where X = probability of success and Y = expected value.

X	0.10	0.20	0.30	0.40	0.50	0.60	0.70	0.80	0.90
Y									

(b) Make a scatterplot of the points (X, Y).

(c) Enter the data into your calculator, and transform the data assuming a power function model.

(d) Remember that the purpose of transforming data is to make the data points linear so that the method of least squares can be employed. Sketch the plot of the transformed data.

(e) What is the correlation, r, for the transformed data?

(f) Write the equation of the power function. Draw the power function curve on your scatterplot.

(g) Briefly explain the connection between this curve and what you have learned about the expected value of a geometric random variable.

CHAPTER REVIEW

The previous chapter introduced discrete and continuous random variables and described methods for finding means and variances, as well as rules for means and variances. This chapter focused on two important classes of discrete random variables, each of which involves two outcomes or events of interest. Both require independent trials and the same probability of success on each trial. The **binomial** random variable requires a fixed number of trials; the **geometric** random variable has the property that the number of trials varies. Both the binomial and the geometric settings occur sufficiently often in applications that they deserve special attention. Here is a checklist of the major skills you should have acquired by studying this chapter:

A. BINOMIAL

1. Identify a random variable as binomial by verifying four conditions: two outcomes (success and failure); fixed number of trials; independent trials; and the same probability of success for each trial.

2. Use a TI-83/89 or the formula to determine binomial probabilities and to construct probability distribution tables and histograms.

3. Calculate cumulative distribution functions for binomial random variables, and construct cumulative distribution tables and histograms.

4. Calculate means (expected values) and standard deviations of binomial random variables.

5. Use a normal approximation to the binomial distribution to compute probabilities.

B. GEOMETRIC

1. Identify a random variable as geometric by verifying four conditions: two outcomes (success and failure); the same probability of success for each trial; independent trials; and the count of interest is the number of trials required to get the first success.

2. Use formulas or a TI-83/89 to determine geometric probabilities and to construct probability distribution tables and histograms.

3. Calculate cumulative distribution functions for geometric random variables, and construct cumulative distribution tables and histograms.

4. Calculate expected values and standard deviations of geometric random variables.

CHAPTER 8 REVIEW EXERCISES

8.55 BINOMIAL SETTING? In each of the following cases, decide whether or not a binomial distribution is an appropriate model, and give your reasons.

(a) You want to know what percent of married people believe that mothers of young children should not be employed outside the home. You plan to interview 50 people, and for the sake of convenience you decide to interview both the husband and wife in 25 married couples. The random variable X is the number among the 50 persons interviewed who think mothers should not be employed.

(b) You are interested in attitudes toward drinking among the 75 members of a fraternity. You choose 25 members at random to interview. One question is "Have you had five or more drinks at one time during the last week?" Suppose that in fact 20% of the 75 members would say "Yes." Explain why you cannot safely use the $B(25, 0.2)$ distribution for the count X in your sample who say "Yes."

8.56 VIRGINIA ROAD FATALITIES In 2001 there were 930 road fatalities in Virginia, according to the Virginia Department of Motor Vehicles. Of these, 355 were alcohol-related. A DMV analyst wants to randomly select several groups of 25 road fatalities for further study. Find the mean and standard deviation for the number of alcohol-related road fatalities in such groups of 25. What is the probability that such a group will have no more than 5 alcohol-related road fatalities?

8.57 SEVEN BROTHERS! This exercise is an extension of Activity 8. There's a movie classic entitled *Seven Brides for Seven Brothers.* Even if these brothers had a few sisters, this many brothers is unusual. We will assume that there are no sisters.

(a) Let X = number of boys in a family of 7 children. Assume that sons and daughters are equally likely outcomes. Do you think the distribution of X will be skewed left, symmetric, or skewed right? The answer to this question depends on what fact?

(b) Use the `binompdf` command to construct a pdf table for X. Then construct a probability distribution histogram and a cumulative distribution histogram for X. Keep a written record of your numerical results as they are produced by your calculator, as well as sketches of the histograms.

(c) What is the probability that all of the 7 children are boys?

8.58 GET A HEAD Suppose we toss a penny repeatedly until we get a head. We want to determine the probability that the first head comes up in an *odd* number of tosses (1, 3, 5, and so on).

(a) Toss a penny until the first head occurs, and repeat the procedure 50 times. Keep a record of the results of the first toss and of the number of tosses needed to get a head on each of your 50 repetitions.

(b) Based on the result of your first toss in the 50 repetitions, estimate the probability of getting a head on the first toss.

(c) Use your 50 repetitions to estimate the probability that the first head appears on an odd-numbered toss.

8.59 ARMED AND DANGEROUS According to a 1997 Centers for Disease Control study of risky behavior, roughly one in five teenagers carries a weapon. Hispanics were most likely to arm themselves, with 23% carrying a gun, knife, or club, compared with 22% of blacks and 17% of whites. Suppose that 4 teenagers are selected at random and subjected to a search. Suppose that success is defined as "the teen is carrying a gun, knife, or club."

(a) What is the probability, p, of success?

(b) Make a list of all of the possible results of the search of the 4 teenagers. Use S to represent success, and F for failure. For each of these responses, write a product of four factors for that combination of successes and failures. For example, SSFS is one such response, and the probability of that outcome is $(0.2)(0.2)(0.8)(0.2) = 0.0064$. Display the probabilities to four decimal places.

(c) Draw a tree diagram to show the possible outcomes.

(d) List the outcomes in which exactly 2 of the 4 students are found to carry a gun, knife, or club.

(e) What are the probabilities of the outcomes in part (d)? Briefly explain why all of these probabilities are the same.

8.60 TOOTH DECAY AND GUM DISEASE Dentists are increasingly concerned about the growing trend of local school districts to grant soft drink companies exclusive rights to install soda pop machines in schools in return for money—usually millions—that goes directly into school coffers. According to a recent study by the National Soft Drink Association, 62% of schools nationally already have such contracts. This comes at a time when dentists are seeing an alarming increase in horribly decayed teeth and eroded enamel in the mouths of teenagers and young adults. With ready access to soft drinks, children tend to drink them all day. That, combined with no opportunity to brush, leads to disaster, dentists say. Suppose that 20 schools around the country are randomly selected and asked if they have a soft drink contract. Find the probability that the number of "Yes" answers is

(a) exactly 8

(b) at most 8

(c) at least 4

(d) between 4 and 12, inclusive

(e) Identify the random variable of interest, X. Then write the probability distribution table for X.

(f) Draw a probability histogram for X.

8.61 FAITH AND HEALING A higher percentage of southerners believe in God and prayer, according to a 1998 study by the University of North Carolina's Institute for Research in Social Science. The survey was conducted by means of telephone interviews with 844 adults in 12 southern states and 413 adults in other states. One of the findings was that 46% of southerners believe they have been healed by prayer, compared with 28% of others. Assume that the results of the UNC survey are true for the region. Suppose that 20 southerners are selected at random and asked if they believe they have been healed by prayer. Find the probability that the number who answer "Yes" to this question is

(a) exactly 10

(b) between 10 and 15

(c) over 75% of the 20

(d) less than 8

8.62 CONTAMINATED SUPERMARKET MEAT In an October 2001 study by the University of Maryland and the Food and Drug Administration, 200 samples of ground beef, ground chicken, ground turkey, and ground pork were collected from supermarkets around Washington, D.C. Forty-one samples, or about 20%, were contaminated by salmonella. Salmonella is a microorganism that can produce flu-like symptoms of fever, diarrhea, and vomiting within 12 to 36 hours after eating improperly cooked food contaminated by it. Assume that 20% of all supermarket ground meat and poultry is contaminated by salmonella. Suppose 7 ground meat and poultry samples are selected from supermarkets at random. Let X denote the number of those chosen that are contaminated by salmonella. The Minitab printout below provides the probability distribution for the random variable X.

```
MTB > PDF 'Values';
SUBC > Binomial 7 .2.
      K           P(X = K)
    0.00            0.2097
    1.00            0.3670
    2.00            0.2753
    3.00            0.1147
    4.00            0.0287
    5.00            0.0043
    6.00            0.0004
    7.00            0.0000
MTB >
```

From the printout, determine the probability that of the 7 meat and poultry samples chosen, the number contaminated by salmonella is

(a) exactly 2

(b) at least 2

(c) less than 2

(d) between 2 and 5, inclusive

8.63 FIRST HIT OF THE SEASON Suppose that Roberto, a well-known major league baseball player, finished last season with a .325 batting average. He wants to calculate the probability that he will get his first hit of this new season in his first at-bat. You define a success as getting a hit and define the random variable X = number of at-bats until Roberto gets his first hit.

(a) What is the probability that Roberto will get a hit on his first at-bat (i.e., that X = 1)?

(b) What is the probability that it will take him at most 3 at-bats to get his first hit?

(c) What is the probability that it will take him more than 4 at-bats to get his first hit?

(d) Roberto wants to know the expected number of at-bats until he gets a hit. What would you tell him?

(e) Enter the first 10 values of X into L_1/list1, the corresponding geometric probabilities into L_2/list2, and the cumulative probabilities into L_3/list3.

(f) Construct a probability distribution histogram as STAT PLOT1, and then construct a cumulative distribution histogram as STAT PLOT2.

You show this analysis to Roberto, and he is so impressed he gives you two free tickets to his first game.

8.64 QUALITY CONTROL Many manufacturing companies use statistical techniques to ensure that the products they make meet standards. One common way to do this is to take a random sample of products at regular intervals throughout the production shift. Assuming that the process is working properly, the mean measurements from these random samples will vary normally around the target mean μ, with a standard deviation of σ.

(a) If the process is working properly, what is the probability that 4 out of 5 consecutive sample means fall within the interval $(\mu - \sigma, \mu + \sigma)$?

(b) If the process is working properly, what is the probability that the first sample mean that is greater than $\mu + 2\sigma$ is the one from the fourth sample taken?

8.65 A MINI-THEOREM Suppose that $X = B(n, p)$. Show that $P(X \geq 1) = 1 - (1 - p)^n$.

NOTES AND DATA SOURCES

1. The survey question is reported in Trish Hall, "Shop? Many say 'Only if I must,'" *New York Times*, November 28, 1990. In fact, 66% (1650 of 2500) in the sample said "Agree."

2. Office of Technology Assessment, *Scientific Validity of Polygraph Testing: A Research Review and Evaluation*, Government Printing Office, Washington, D.C., 1983.

3. Prescribing information, including results of clinical trials, as of November 2000, appear frequently in newspaper and magazine advertisements and can be found at the Web site www.allegra.com.

Courtesy Dr. David Blackwell, UCA, Berkeley

DAVID BLACKWELL

Mathematics in the Service of Statistics

Statistical practice rests in part on statistical theory. Statistics has been advanced not only by people concerned with practical problems, from Florence Nightingale to R. A. Fisher and John Tukey, but also by people whose first love is mathematics for its own sake. *David Blackwell (1919–)* is one of the major contemporary contributors to the mathematical study of statistics.

Blackwell grew up in Illinois, earned a doctorate in mathematics at the age of 22, and in 1944 joined the faculty of Howard University in Washington, D.C. "It was the ambition of every black scholar in those days to get a job at Howard University," he says. "That was the best job you could hope for." Society changed, and in 1954 Blackwell became professor of statistics at the University of California at Berkeley.

Washington, D.C., had an active statistical community, and the young mathematician Blackwell soon began to work on mathematical aspects of statistics. He explored the behavior of statistical procedures that, rather than working with a fixed sample, keep taking observations until there is enough information to reach a firm conclusion. He found insights into statistical inference by thinking of inference as a game in which nature plays against the statistician. Blackwell's work uses probability theory, the mathematics that describes chance behavior. This chapter presents the probabilistic ideas needed to understand the reasoning of statistical inference.

Statistics has been advanced not only by people concerned with practical problems but also by people whose first love is mathematics for its own sake.

chapter 9

Sampling Distributions

ACTIVITY 9A Young Women's Heights

Materials: Several 3" × 3" or 3" × 5" Post-it Notes

The height of young women varies approximately according to the N(64.5, 2.5) distribution. That is to say, the population of young women is normally distributed with mean μ = 64.5 inches and standard deviation σ = 2.5 inches. The random variable measured is X = the height of a randomly selected young woman. In this activity you will use the TI-83/TI-89 to sample from this distribution and then use Post-it Notes to construct a distribution of averages.

1. If we choose one woman at random, the heights we get in repeated choices follow the N(64.5, 2.5) distribution. On your calculator, go into the Statistics/List Editor and clear L_1/list1. Simulate the heights of 100 randomly selected young women and store these heights in L_1/list1 as follows:

- Place your cursor at the top of L_1/list1 (on the list name, not below it).
- TI-83: Press MATH, choose PRB, choose 6:randNorm(.
  TI-89: Press F4, choose 4:Probability, choose 6:.randNorm(.
- Complete the command: `randNorm(64.5,2.5,100)` and press ENTER.

2. Plot a histogram of the 100 heights as follows. Deselect active functions in the Y = window, and turn off all STAT PLOTS. Set WINDOW dimensions to: $X[57,72]_{2.5}$ and $Y[-10,45]_5$ to extend three standard deviations to either side of the mean, 64.5. Define PLOT 1 to be a histogram using the heights in L_1/list1. (You must set the Hist. Bucket Width to 2.5 in the TI-89 Plot Setup.) Press GRAPH (on the TI-89 press ♦ F3) to plot the histogram. Describe the approximate shape of your histogram. Is it fairly symmetric or clearly skewed?

3. Approximately how many heights should there be within 3σ of the mean (i.e., between 57 and 72)? Use TRACE to count the number of heights within 3σ. How many heights should there be within 1σ of the mean? Within 2σ of the mean? Again use TRACE to find these counts, and compare them with the numbers you would expect.

4. Use 1-Var Stats to find the mean, median, and standard deviation for your data. Compare $\bar{x}$ with the population mean μ = 64.5. Compare the sample standard deviation *s* with σ = 2.5. How do the mean and median for your 100 heights compare? Recall that the closer the mean and the median are, the more symmetric the distribution.

5. Define PLOT 2 to be a boxplot using L_1/list1 and then GRAPH again. The boxplot will be plotted above the histogram. Does the boxplot appear symmetric? How close is the median in the boxplot to the mean of the his-

ACTIVITY 9A Young Women's Heights (*continued*)

togram? Based on the appearance of the histogram and the boxplot, and a comparison of the mean and median, would you say that the distribution is nonsymmetric, moderately symmetric, or very symmetric?

6. Repeat steps 1 to 5 two or three more times. Each time, record the mean $\bar{x}$, median, and standard deviation s. (*Note:* While this is going on, your teacher will draw a baseline at the bottom of a clean blackboard and mark a scale from 63 to 66 with tick marks at 0.25 intervals. The tick marks should be spaced about an inch wider apart than the width of the Post-it Notes. Each tick mark will represent the center of a bar in a histogram.)

7. Write (big and neat) the mean $\bar{x}$ for each sample on a different Post-it Note. Next, you will build a "Post-it Note histogram" of the distribution of the sample means $\bar{x}$. When instructed, go to the blackboard and stick each of your notes directly above the tick mark that is closest to the mean written on the note. When the Post-it Note histogram is complete, answer the following questions:

(a) What is the approximate shape of the distribution of $\bar{x}$?

(b) Where is the center of the distribution of $\bar{x}$? How does this center compare with the mean of heights of the population of *all* young women?

(c) Roughly, how does the spread of the distribution of $\bar{x}$ compare with the spread of the original distribution ($\sigma = 2.5$)?

8. While someone calls out the values of $\bar{x}$ from the Post-it Notes, enter these values into L_2/list2 in your calculator. Turn off PLOT 1 and define PLOT 3 to be a boxplot of the $\bar{x}$ data. How do these distributions of X and $\bar{x}$ compare visually? Use 1-Var Stats to calculate the standard deviation $s_{\bar{x}}$ for the distribution of $\bar{x}$. Compare this value with $\sigma / \sqrt{100}$.

9. Fill in the blanks in the following statement with a function of μ or σ: "The distribution of $\bar{x}$ is approximately normal with mean $\mu(\bar{x})$ = ________ and standard deviation $\sigma(\bar{x})$ = ________."

INTRODUCTION

The reasoning of statistical inference rests on asking, "How often would this method give a correct answer if I used it very many times?" If it doesn't make sense to imagine repeatedly producing your data in the same circumstances, statistical inference is not possible.[1] Exploratory data analysis makes sense for any data, but formal inference does not. Even experts can disagree about how widely statistical inference should be used. But all agree that inference is most secure when we produce data by random sampling or randomized comparative experiments. The reason is that when we use chance to choose respondents or assign subjects, the laws of probability answer the question "What would happen if we did this many times?" The purpose of this chapter is to prepare for the study of statistical inference by looking at the probability distributions of some very common statistics: sample proportions and sample means.

9.1 SAMPLING DISTRIBUTIONS

What is the mean income of households in the United States? The government's Current Population Survey contacted a sample of 50,000 households in 2000. Their mean income was $\bar{x} = \$57{,}045$.[2] That \$57,045 describes the sample, but we use it to estimate the mean income of all households. We must now take care to keep straight whether a number describes a sample or a population. Here is the vocabulary we use.

PARAMETER, STATISTIC

A **parameter** is a number that describes the population. A parameter is a fixed number, but in practice we do not know its value because we cannot examine the entire population.

A **statistic** is a number that describes a sample. The value of a statistic is known when we have taken a sample, but it can change from sample to sample. We often use a statistic to estimate an unknown parameter.

EXAMPLE 9.1 MAKING MONEY

The mean income of the sample of households contacted by the Current Population Survey was $\bar{x} = \$57{,}045$. The number \$57,045 is a *statistic* because it describes this one Current Population Survey sample. The population that the poll wants to draw conclusions about is all 106 million U.S. households. The *parameter* of interest is the mean income of all of these households. We don't know the value of this parameter.

Remember: statistics come from samples, and parameters come from populations. As long as we were just doing data analysis, the distinction between population and sample was not important. Now, however, it is essential. The notation we use must reflect this distinction. We write μ (the Greek letter mu) for the **mean of a population**. This is a fixed parameter that is unknown when we use a sample for inference. The **mean of the sample** is the familiar $\bar{x}$, the average of the observations in the sample. This is a statistic that would almost certainly take a different value if we chose another sample from the same population. The sample mean $\bar{x}$ from a sample or an experiment is an estimate of the mean μ of the underlying population.

sampling variability

How can $\bar{x}$, based on a sample of only a few of the 100 million American households, be an accurate estimate of μ? After all, a second random sample taken at the same time would choose different households and no doubt produce a different value of $\bar{x}$. This basic fact is called ***sampling variability***: the value of a statistic varies in repeated random sampling.

EXAMPLE 9.2 DO YOU BELIEVE IN GHOSTS?

The Gallup Poll asked a random sample of 515 U.S. adults whether they believe in ghosts. Of the respondents, 160 said "Yes."[3] So the proportion of the sample who say they believe in ghosts is

$$\hat{p} = \frac{160}{515} = 0.31$$

The number 0.31 is a *statistic*. We can use it to estimate the proportion of all U.S. adults who believe in ghosts. This is our *parameter* of interest.

We use p to represent a population proportion. The sample proportion $\hat{p}$ estimates the unknown parameter p. Based on the sample survey of Example 9.2, we might conclude that the proportion of all U.S. adults who believe in ghosts *is* 0.31. That would be a mistake. After all, a second random sample of 515 adults would probably yield a different value of $\hat{p}$. Sampling variability strikes again!

EXERCISES

For each boldface number in Exercises 9.1 to 9.4, (a) state whether it is a parameter or a statistic and (b) use appropriate notation to describe each number; for example, $p = 0.65$.

9.1 MAKING BALL BEARINGS A carload lot of ball bearings has mean diameter **2.5003** centimeters (cm). This is within the specifications for acceptance of the lot by the purchaser. By chance, an inspector chooses 100 bearings from the lot that have mean diameter **2.5009** cm. Because this is outside the specified limits, the lot is mistakenly rejected.

9.2 UNEMPLOYMENT The Bureau of Labor Statistics last month interviewed 60,000 members of the U.S. labor force, of whom **7.2%** were unemployed.

9.3 TELEMARKETING A telemarketing firm in Los Angeles uses a device that dials residential telephone numbers in that city at random. Of the first 100 numbers dialed, **48%** are unlisted. This is not surprising because **52%** of all Los Angeles residential phones are unlisted.

9.4 WELL-FED RATS A researcher carries out a randomized comparative experiment with young rats to investigate the effects of a toxic compound in food. She feeds the control group a normal diet. The experimental group receives a diet with 2500 parts per million of the toxic material. After 8 weeks, the mean weight gain is **335** grams for the control group and **289** grams for the experimental group.

Sampling variability

To understand why sampling variability is not fatal, we ask, "What would happen if we took many samples?" Here's how to answer that question:

- Take a large number of samples from the same population.
- Calculate the sample mean $\overline{x}$ or sample proportion $\hat{p}$ for each sample.
- Make a histogram of the values of $\overline{x}$ or $\hat{p}$.
- Examine the distribution displayed in the histogram for shape, center, and spread, as well as outliers or other deviations.

In practice it is too expensive to take many samples from a population like all adult U.S. residents. But we can imitate many samples by using simulation.

EXAMPLE 9.3 BAGGAGE CHECK!

Thousands of travelers pass through Guadalajara airport each day. Before leaving the airport, each passenger must pass through the Customs inspection area. Customs officials want to be sure that passengers do not bring illegal items into the country. But they do not have time to search every traveler's luggage. Instead, they require each person to press a button that activates a modified "stoplight." When the button is pressed, either a red or a green bulb lights up. If the red light shows, the passenger will be searched by Customs agents. A green light means "go ahead." Customs officers claim that the probability that the light turns green on any press of the button is 0.70.

We will simulate drawing simple random samples (SRSs) of size 100 from the population of travelers passing through Guadalajara airport. The parameter of interest is the proportion of travelers who get a green light at the Customs station. Assuming the Customs officials are telling the truth, we know that $p = 0.70$.

We can imitate the population by a huge table of random digits, such as Table B at the back of the book, with each entry standing for a traveler. Seven of the ten digits (say 0 to 6) stand for passengers who get a green light at Customs. The remaining three digits, 7 to 9, stand for those who get a red light and are searched. Because all digits in a random number table are equally likely, this assignment produces a population proportion of passengers who get the green light equal to $p = 0.7$. We then imitate an SRS of 100 travelers from the population by taking 100 consecutive digits from Table B. The statistic $\hat{p}$ is the proportion of 0s to 6s in the sample.

For example, if we begin at line 101 in Table B:

GRGGG	RGGGG	GGRGG	GRRGG	...
19223	95034	05756	28713	...

71 of the first 100 entries are between 0 and 6, so $\hat{p} = 71/100 = 0.71$. A second SRS based on the second 100 entries in Table B gives a different result, $\hat{p} = 0.62$. The two sample results are different, and neither is equal to the true population value $p = 0.7$. That's sampling variability.

Simulation is a powerful tool for studying chance. It is much faster to use Table B than to actually draw repeated SRSs, and much faster yet to use a computer programmed to produce random digits. Figure 9.1 is the histogram of values of $\hat{p}$ from 1000 separate SRSs of size 100 drawn from a population with $p = 0.7$. This histogram shows what would happen if we drew many samples. It approximates the *sampling distribution of* $\hat{p}$.

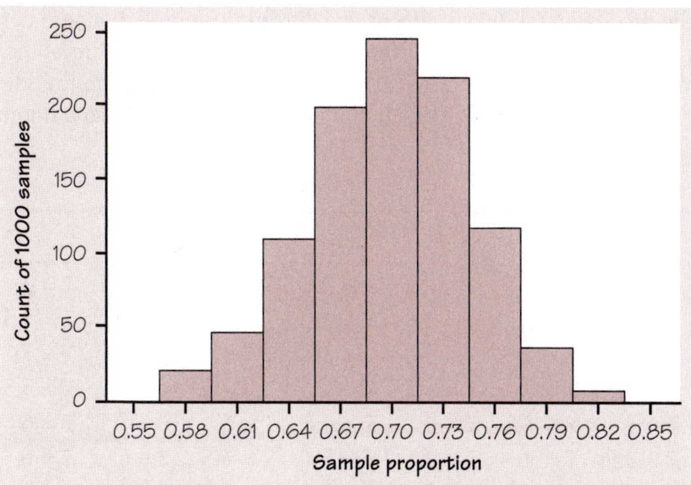

FIGURE 9.1 The distribution of the sample proportion $\hat{p}$ from SRSs of size 100 drawn from a population with population proportion $p = 0.7$. The histogram shows the results of drawing 1000 SRSs.

SAMPLING DISTRIBUTION

The **sampling distribution** of a statistic is the distribution of values taken by the statistic in all possible samples of the same size from the same population.

Strictly speaking, the sampling distribution is the ideal pattern that would emerge if we looked at all possible samples of size 100 from our population. A distribution obtained from a fixed number of trials, like the 1000 trials in Figure 9.1, is only an approximation to the sampling distribution. One of the uses of probability theory in statistics is to obtain exact sampling distributions without simulation. The interpretation of a sampling distribution is the same, however, whether we obtain it by simulation or by the mathematics of probability.

EXAMPLE 9.4 RANDOM DIGITS

The population used to construct the random digits table (Table B) can be described by the probability distribution shown in Figure 9.2.

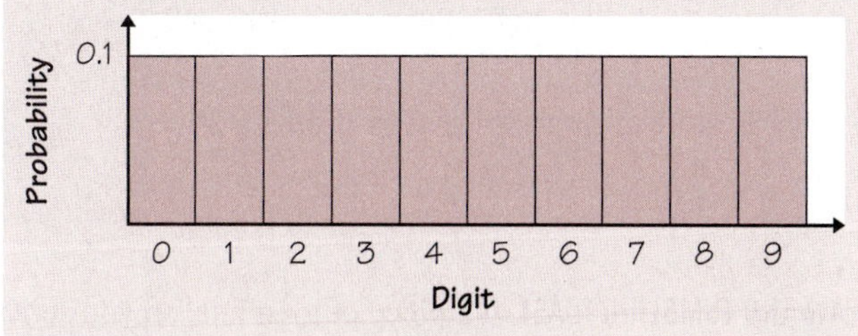

FIGURE 9.2 Probability distribution used to construct Table B.

Consider the process of taking an SRS of size 2 from this population and computing $\bar{x}$ for the sample. We could perform a simulation to get a rough picture of the sampling distribution of $\bar{x}$. But in this case, we can construct the actual sampling distribution. Figure 9.3 displays the values of $\bar{x}$ for all 100 possible samples of two random digits.

First digit \ Second digit	0	1	2	3	4	5	6	7	8	9
0	$\bar{x}=0$	$\bar{x}=0.5$	$\bar{x}=1$	$\bar{x}=1.5$	$\bar{x}=2$	$\bar{x}=2.5$	$\bar{x}=3$	$\bar{x}=3.5$	$\bar{x}=4$	$\bar{x}=4.5$
1	$\bar{x}=0.5$	$\bar{x}=1$	$\bar{x}=1.5$	$\bar{x}=2$	$\bar{x}=2.5$	$\bar{x}=3$	$\bar{x}=3.5$	$\bar{x}=4$	$\bar{x}=4.5$	$\bar{x}=5$
2	$\bar{x}=1$	$\bar{x}=1.5$	$\bar{x}=2$	$\bar{x}=2.5$	$\bar{x}=3$	$\bar{x}=3.5$	$\bar{x}=4$	$\bar{x}=4.5$	$\bar{x}=5$	$\bar{x}=5.5$
3	$\bar{x}=1.5$	$\bar{x}=2$	$\bar{x}=2.5$	$\bar{x}=3$	$\bar{x}=3.5$	$\bar{x}=4$	$\bar{x}=4.5$	$\bar{x}=5$	$\bar{x}=5.5$	$\bar{x}=6$
4	$\bar{x}=2$	$\bar{x}=2.5$	$\bar{x}=3$	$\bar{x}=3.5$	$\bar{x}=4$	$\bar{x}=4.5$	$\bar{x}=5$	$\bar{x}=5.5$	$\bar{x}=6$	$\bar{x}=6.5$
5	$\bar{x}=2.5$	$\bar{x}=3$	$\bar{x}=3.5$	$\bar{x}=4$	$\bar{x}=4.5$	$\bar{x}=5$	$\bar{x}=5.5$	$\bar{x}=6$	$\bar{x}=6.5$	$\bar{x}=7$
6	$\bar{x}=3$	$\bar{x}=3.5$	$\bar{x}=4$	$\bar{x}=4.5$	$\bar{x}=5$	$\bar{x}=5.5$	$\bar{x}=6$	$\bar{x}=6.5$	$\bar{x}=7$	$\bar{x}=7.5$
7	$\bar{x}=3.5$	$\bar{x}=4$	$\bar{x}=4.5$	$\bar{x}=5$	$\bar{x}=5.5$	$\bar{x}=6$	$\bar{x}=6.5$	$\bar{x}=7$	$\bar{x}=7.5$	$\bar{x}=8$
8	$\bar{x}=4$	$\bar{x}=4.5$	$\bar{x}=5$	$\bar{x}=5.5$	$\bar{x}=6$	$\bar{x}=6.5$	$\bar{x}=7$	$\bar{x}=7.5$	$\bar{x}=8$	$\bar{x}=8.5$
9	$\bar{x}=4.5$	$\bar{x}=5$	$\bar{x}=5.5$	$\bar{x}=6$	$\bar{x}=6.5$	$\bar{x}=7$	$\bar{x}=7.5$	$\bar{x}=8$	$\bar{x}=8.5$	$\bar{x}=9$

FIGURE 9.3 Values of $\bar{x}$ in all possible samples of two random digits.

The distribution of $\bar{x}$ can be summarized by the histogram shown in Figure 9.4. Since this graph displays all possible values of $\bar{x}$ from SRSs of size $n = 2$ from the population, it is the *sampling distribution of* $\bar{x}$.

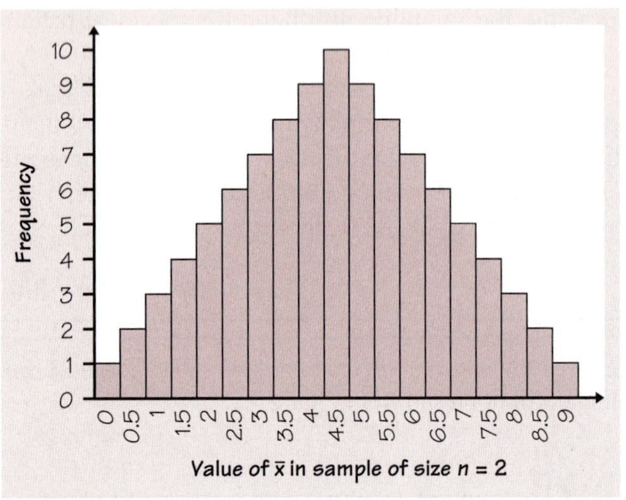

FIGURE 9.4 The sampling distribution of $\bar{x}$ for samples of size $n = 2$.

EXERCISES

9.5 MURPHY'S LAW AND TUMBLING TOAST If a piece of toast falls off your breakfast plate, is it more likely to land with the buttered side down? According to Murphy's Law—the assumption that if anything can go wrong, it will—the answer is "Yes." Most scientists would argue that by the laws of probability, the toast is equally likely to land butter-side up or butter-side down. Robert Matthews, science correspondent of the *Sunday Telegraph*, disagrees. He claims that when toast falls off a plate that is being carried at a "typical height," the toast has just enough time to rotate once (landing butter-side down) before it lands. To test his claim, Mr. Matthews has arranged for 150,000 students in Great Britain to carry out an experiment with tumbling toast.[4]

Assuming scientists are correct, the proportion of times that the toast will land butter-side down is $p = 0.5$. We can use a coin toss to simulate the experiment. Let heads represent the toast landing butter-side down.

(a) Toss a coin 20 times and record the proportion of heads obtained, $\hat{p}$ = (number of heads)/20. Explain how your result relates to the tumbling-toast experiment.

(b) Repeat this sampling process 10 times. Make a histogram of the 10 values of $\hat{p}$. Is the center of this distribution close to 0.5?

(c) Ten repetitions give a very crude approximation to the sampling distribution. Pool your work with that of other students to obtain several hundred repetitions. Make a histogram of all the values of $\hat{p}$. Is the center close to 0.5? Is the shape approximately normal?

(d) How much sampling variability is present? That is, how much do your values of $\hat{p}$ based on samples of size 20 differ from the actual population proportion, $p = 0.5$?

(e) Why do you think Mr. Matthews is asking so many students to participate in his experiment?

9.6 MORE TUMBLING TOAST Use your calculator to replicate Exercise 9.5 as follows. The command `randBin(20,.5)` simulates tossing a coin 20 times. The output is the number of heads in 20 tosses. The command `randBin(20,.5,10)/20` simulates 10 repetitions of tossing a coin 20 times and finding the proportions of heads. Go into your Statistics/List Editor and place your cursor on the top of L_1/list1. Execute the command `randBin(20,.5,10)/20` as follows:

- TI-83: Press MATH, choose PRB, choose 7:randBin(. Complete the command and press ENTER.
- TI-89: Press F4, choose 4:Probability, choose 7:randBin(. Complete the command and press ENTER.

(a) Plot a histogram of the 10 values of $\hat{p}$. Set WINDOW parameters to $X[-.05,1.05]_{.1}$ and $Y[-2,6]_1$ and then TRACE. Is the center of the histogram close to 0.5? Do this several times to see if you get similar results each time.

(b) Increase the number of repetitions to 100. The command should read `randBin(20,.5,100)/20`. Execute the command (be patient!) and then plot a histogram using these 100 values. Don't change the XMIN and XMAX values, but do adjust the Y-values to $Y[-20, 50]_{10}$ to accommodate the taller bars. Is the center close to 0.5? Describe the shape of the distribution.

(c) Define PLOT 2 to be a boxplot using L_1/list1, and TRACE again. How close is the median (in the boxplot) to the mean (balance point) of the histogram?

(d) Note that we didn't increase the sample size, only the number of repetitions. Did the spread of the distribution change? What would you change to decrease the spread of the distribution?

9.7 SAMPLING TEST SCORES Let us illustrate the idea of a sampling distribution of $\bar{x}$ in the case of a very small sample from a very small population. The population is the scores of 10 students on an exam:

Student:	0	1	2	3	4	5	6	7	8	9
Score:	82	62	80	58	72	73	65	66	74	62

The parameter of interest is the mean score in this population, which is 69.4. The sample is an SRS drawn from the population. Because the students are labeled 0 to 9, a single random digit from Table B chooses one student for the sample.

(a) Use Table B to draw an SRS of size $n = 4$ from this population. Write the four scores in your sample and calculate the mean $\bar{x}$ of the sample scores. This statistic is an estimate of the population parameter.

(b) Repeat this process 10 times. Make a histogram of the 10 values of $\bar{x}$. You are constructing the sampling distribution of $\bar{x}$. Is the center of your histogram close to 69.4?

(c) Ten repetitions give a very crude approximation to the sampling distribution. Pool your work with that of other students—using different parts of Table B—to obtain several hundred repetitions. Make a histogram of all the values of $\bar{x}$. Is the center close to 69.4? Describe the shape of the distribution. This histogram is a better approximation to the sampling distribution.

(d) It is possible to construct the actual sampling distribution of $\bar{x}$ for samples of size $n = 2$ taken from this population. (Refer to Example 9.4.) Draw this sampling distribution.

(e) Compare the sampling distributions of $\bar{x}$ for samples of size 2 and size 4. Are the shapes, centers, and spreads similar or different?

Describing sampling distributions

We can use the tools of data analysis to describe any distribution. Let's apply these tools in the world of television.

EXAMPLE 9.5 ARE YOU A *SURVIVOR* FAN?

Television executives and companies who advertise on TV are interested in how many viewers watch particular television shows. According to 2001 Nielsen ratings, *Survivor II* was one of the most-watched television shows in the United States during every week that it aired. Suppose that the true proportion of U.S. adults who watched *Survivor II* is $p = 0.37$. Figure 9.5 shows the results of drawing 1000 SRSs of size $n = 100$ from a population with $p = 0.37$.

From the figure, we can see that:

- The overall *shape* of the distribution is symmetric and approximately normal.
- The *center* of the distribution is very close to the true value $p = 0.37$ for the population from which the samples were drawn. In fact, the mean of the 1000 $\hat{p}$'s is 0.372 and their median is exactly 0.370.

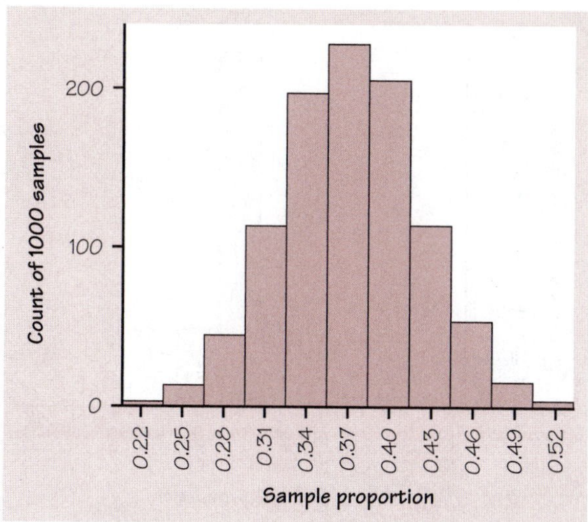

FIGURE 9.5 Proportion of sample who watched *Survivor II* in samples of size $n = 100$.

- The values of $\hat{p}$ have a large *spread*. They range from 0.22 to 0.54. Because the distribution is close to normal, we can use the standard deviation to describe its spread. The standard deviation is about 0.05.
- There are no *outliers* or other important deviations from the overall pattern.

Figure 9.5 shows that a sample of 100 people often gave a $\hat{p}$ quite far from the population parameter $p = 0.37$. That is, a sample of 100 people does not produce a trustworthy estimate of the population proportion. That is why a March 11, 2001, Gallup Poll asked, not 100, but 1000 people whether they had watched *Survivor II*.[5] Let's repeat our simulation, this time taking 1000 SRSs of size 1000 from a population with proportion $p = 0.37$ who have watched *Survivor II*.

Figure 9.6 displays the distribution of the 1000 values of $\hat{p}$ from these new samples. Figure 9.6 uses the same horizontal scale as Figure 9.5 to make comparison easy. Here's what we see:

- The *center* of the distribution is again close to 0.37. In fact, the mean is 0.3697 and the median is exactly 0.37.
- The *spread* of Figure 9.6 is much less than that of Figure 9.5. The range of the values of $\hat{p}$ from 1000 samples is only 0.321 to 0.421. The standard deviation is about 0.016. Almost all samples of 1000 people give a $\hat{p}$ that is close to the population parameter $p = 0.37$.
- Because the values of $\hat{p}$ cluster so tightly about 0.37, it is hard to see the *shape* of the distribution in Figure 9.6. Figure 9.7 displays the same 1000 values of $\hat{p}$ on an expanded scale that makes the shape clearer. The distribution is again approximately normal in shape.

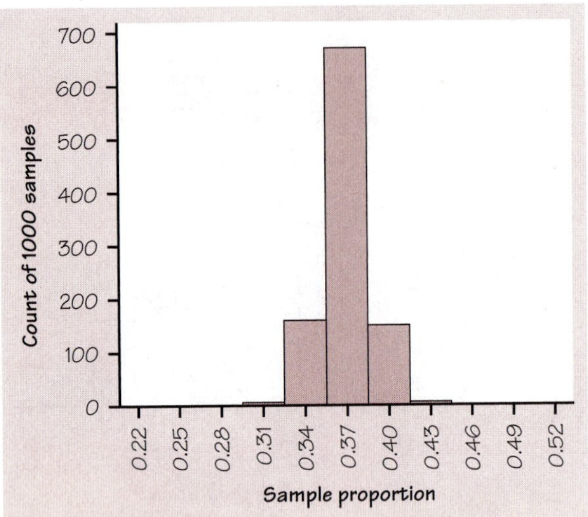

FIGURE 9.6 The approximate sampling distribution of the sample proportion $\hat{p}$ from SRSs of size 1000 drawn from a population with population proportion $p = 0.37$. The histogram shows the results of 1000 SRSs. The scale is the same as in Figure 9.5.

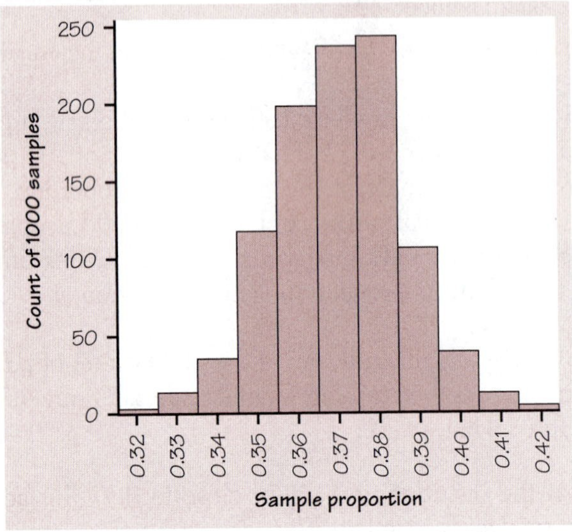

FIGURE 9.7 The approximate sampling distribution from Figure 9.6, for samples of size 1000, redrawn on an expanded scale to better display the shape.

The appearance of the approximate sampling distributions in Figures 9.5 to 9.7 is a consequence of random sampling. Haphazard sampling does not give such regular and predictable results. When randomization is used in a design for producing data, statistics computed from the data have a definite pattern of behavior over many repetitions, even though the result of a single repetition is uncertain.

The bias of a statistic

The fact that statistics from random samples have definite sampling distributions allows a more careful answer to the question of how trustworthy a statistic is as an estimate of a parameter. Figure 9.8 shows the two sampling distributions of $\hat{p}$ for samples of 100 people and samples of 1000 people, side by side and drawn to the same scale. Both distributions are approximately normal, so we have also drawn normal curves for both. How trustworthy is the sample proportion $\hat{p}$ as an estimator of the population proportion p in each case?

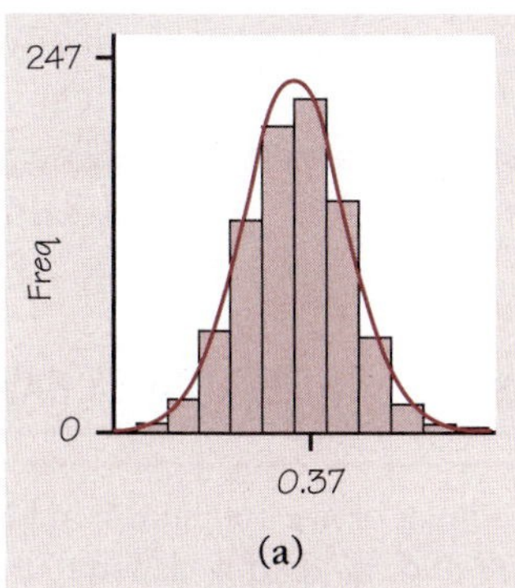

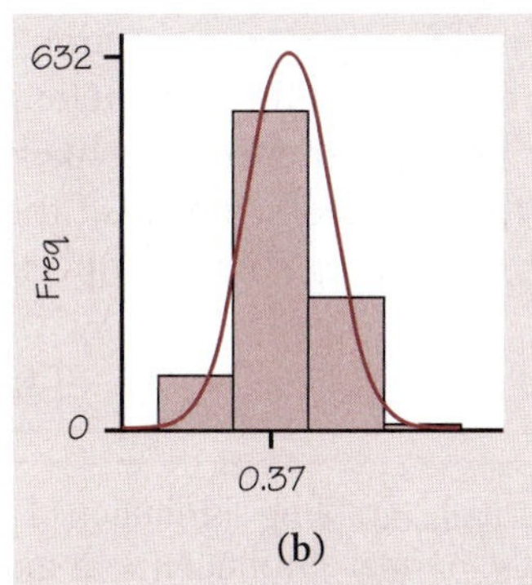

FIGURE 9.8 The approximate sampling distributions for sample proportions $\hat{p}$ for SRSs of two sizes drawn from a population with $p = 0.37$. (a) Sample size 100. (b) Sample size 1000. Both statistics are *unbiased* because the means of their distributions equal the true population value $p = 0.37$. The statistic from the larger sample is less variable.

bias

Sampling distributions allow us to describe ***bias*** more precisely by speaking of the bias of a statistic rather than bias in a sampling method. Bias concerns the center of the sampling distribution. The centers of the sampling distributions in Figure 9.8 are very close to the true value of the population parameter. Those distributions show the results of 1000 samples. In fact, the mean of the sampling distribution (think of taking all possible samples, not just 1000 samples) is *exactly* equal to 0.37, the parameter in the population.

UNBIASED STATISTIC

A statistic used to estimate a parameter is **unbiased** if the mean of its sampling distribution is equal to the true value of the parameter being estimated.

An unbiased statistic will sometimes fall above the true value of the parameter and sometimes below if we take many samples. Because its sampling distribution is centered at the true value, however, there is no systematic tendency to overestimate or underestimate the parameter. This makes the idea of lack of bias in the sense of "no favoritism" more precise. The sample proportion $\hat{p}$ from an SRS is an unbiased estimator of the population proportion p. If we draw an SRS from a population in which 37% have watched *Survivor II*, the mean of the sampling distribution of $\hat{p}$ is 0.37. If we draw an SRS from a population in which 50% have seen *Survivor II*, the mean of the sampling distribution of $\hat{p}$ is then 0.5.

The variability of a statistic

The statistics whose sampling distributions appear in Figure 9.8 are both unbiased. That is, both distributions are centered at 0.37, the true population proportion. The sample proportion $\hat{p}$ from a random sample of any size is an unbiased estimate of the parameter p. Larger samples have a clear advantage, however. They are much more likely to produce an estimate close to the true value of the parameter because there is much less variability among large samples than among small samples.

EXAMPLE 9.6 THE STATISTICS HAVE SPOKEN

The approximate sampling distribution of $\hat{p}$ for samples of size 100, shown in Figure 9.8(a), is close to the normal distribution with mean 0.37 and standard deviation 0.05. Recall the 68–95–99.7 rule for normal distributions. It says that 95% of values of $\hat{p}$ will fall within two standard deviations of the mean of the distribution, $p = 0.37$. So 95% of all samples give an estimate $\hat{p}$ between

$$\text{mean} \pm (2 \times \text{standard deviation}) = 0.37 \pm (2 \times 0.05) = 0.37 \pm 0.1$$

If in fact 37% of U.S. adults have seen *Survivor II*, the estimates from repeated SRSs of size 100 will usually fall between 27% and 47%. That's not very satisfactory.

For samples of size 1000, Figure 9.8(b) shows that the standard deviation is only about 0.01. So 95% of these samples will give an estimate within about 0.02 of the true parameter, that is, between 0.35 and 0.39. An SRS of size 1000 can be trusted to give sample estimates that are very close to the truth about the entire population.

In Section 9.2 we will give the standard deviation of $\hat{p}$ for any size sample. We will then see Example 9.6 as part of a general rule that shows exactly how the variability of sample results decreases for larger samples. One important and surprising fact is that the spread of the sampling distribution does *not* depend very much on the size of the *population*.

Why does the size of the population have little influence on the behavior of statistics from random samples? To see that this is plausible, imagine sampling harvested corn by thrusting a scoop into a lot of corn kernels. The scoop doesn't know whether it is surrounded by a bag of corn or by an entire truckload. As long as the corn is well mixed (so that the scoop selects a random sample), the variability of the result depends only on the size of the scoop.

The fact that the variability of sample results is controlled by the size of the sample has important consequences for sampling design. A statistic from an SRS of size 2500 from the more than 280,000,000 residents of the United

VARIABILITY OF A STATISTIC

The **variability of a statistic** is described by the spread of its sampling distribution. This spread is determined by the sampling design and the size of the sample. Larger samples give smaller spread.

As long as the population is much larger than the sample (say, at least 10 times as large), the spread of the sampling distribution is approximately the same for any population size.

States is just as precise as an SRS of size 2500 from the 775,000 inhabitants of San Francisco. This is good news for designers of national samples but bad news for those who want accurate information about the citizens of San Francisco. If both use an SRS, both must use the same size sample to obtain equally trustworthy results.

Bias and variability

We can think of the true value of the population parameter as the bull's-eye on a target and of the sample statistic as an arrow fired at the target. Both bias and variability describe what happens when we take many shots at the target. *Bias* means that our aim is off and we consistently miss the bull's-eye in the same direction. Our sample values do not center on the population value. *High variability* means that repeated shots are widely scattered on the target. Repeated samples do not give very similar results. Figure 9.9 (page 500) shows this target illustration of the two types of error.

Notice that low variability (shots are close together) can accompany high bias (shots are consistently away from the bull's-eye in one direction). And low bias (shots center on the bull's-eye) can accompany high variability (shots are widely scattered). Properly chosen statistics computed from random samples of sufficient size will have low bias and low variability.

EXERCISES

9.8 BEARING DOWN The table below contains the results of simulating on a computer 100 repetitions of the drawing of an SRS of size 200 from a large lot of ball bearings. Ten percent of the bearings in the lot do not conform to the specifications. That is, $p = 0.10$ for this population. The numbers in the table are the counts of nonconforming bearings in each sample of 200.

17	23	18	27	15	17	18	13	16	18	20	15	18	16	21	17	18	19	16	23
20	18	18	17	19	13	27	22	23	26	17	13	16	14	24	22	16	21	24	21
30	24	17	14	16	16	17	24	21	16	17	23	18	23	22	24	23	23	20	19
20	18	20	25	16	24	24	24	15	22	22	16	28	15	22	9	19	16	19	19
25	24	20	15	21	25	24	19	19	20	28	18	17	17	25	17	17	18	19	18

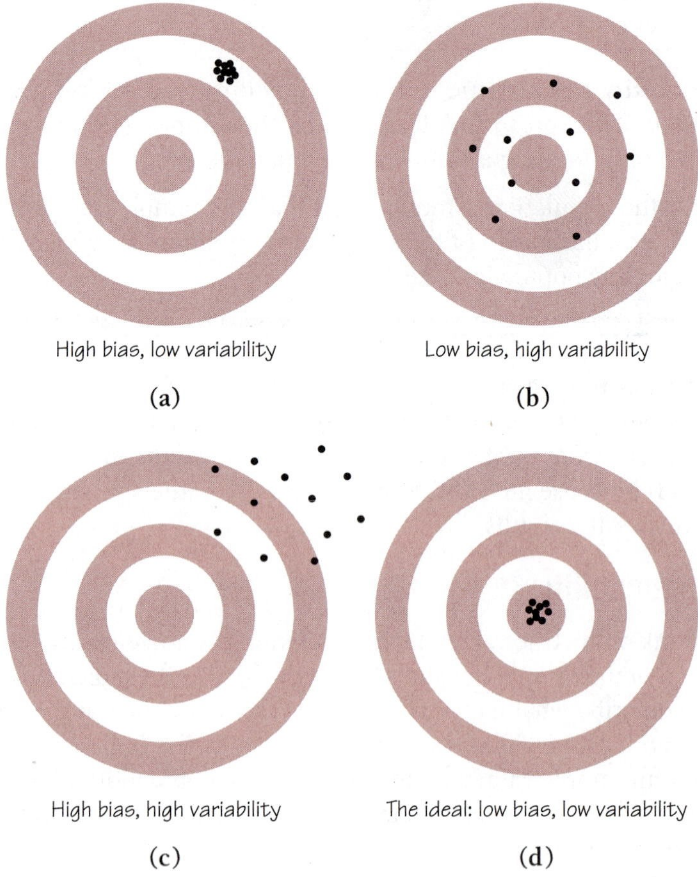

FIGURE 9.9 Bias and variability. (a) High bias, low variability. (b) Low bias, high variability. (c) High bias, high variability. (d) The ideal: low bias, low variability.

(a) Make a table that shows how often each count occurs. For each count in your table, give the corresponding value of the sample proportion $\hat{p}$ = count/200. Then draw a histogram for the values of the statistic $\hat{p}$.

(b) Describe the shape of the distribution.

(c) Find the mean of the 100 observations of $\hat{p}$. Mark the mean on your histogram to show its center. Does the statistic $\hat{p}$ appear to have large or small bias as an estimate of the population proportion p?

(d) The sampling distribution of $\hat{p}$ is the distribution of the values of $\hat{p}$ from all possible samples of size 200 from this population. What is the mean of this distribution?

(e) If we repeatedly selected SRSs of size 1000 instead of 200 from this same population, what would be the mean of the sampling distribution of the sample proportion $\hat{p}$? Would the spread be larger, smaller, or about the same when compared with the spread of your histogram in (a)?

9.9 GUINEA PIGS Table 9.1 gives the survival times of 72 guinea pigs in a medical experiment. Consider these 72 animals to be the population of interest.

TABLE 9.1 Survival time (days) of guinea pigs in a medical experiment

43	45	53	56	56	57	58	66	67	73	74	79
80	80	81	81	81	82	83	83	84	88	89	91
91	92	92	97	99	99	100	100	101	102	102	102
103	104	107	108	109	113	114	118	121	123	126	128
137	138	139	144	145	147	156	162	174	178	179	184
191	198	211	214	243	249	329	380	403	511	522	598

Source: T. Bjerkedal, "Acquisition of resistance in guinea pigs infected with different doses of virulent tubercle bacilli," *American Journal of Hygiene*, 72 (1960), pp. 130–148.

(a) Make a histogram of the 72 survival times. This is the population distribution. It is strongly skewed to the right.

(b) Find the mean of the 72 survival times. This is the population mean μ. Mark μ on the x axis of your histogram.

(c) Label the members of the population 01 to 72 and use Table B to choose an SRS of size $n = 12$. What is the mean survival time $\bar{x}$ for your sample? Mark the value of $\bar{x}$ with a point on the axis of your histogram from (a).

(d) Choose four more SRSs of size 12, using different parts of Table B. Find $\bar{x}$ for each sample and mark the values on the axis of your histogram from (a). Would you be surprised if all five $\bar{x}$'s fell on the same side of μ? Why?

(e) If you chose all possible SRSs of size 12 from this population and made a histogram of the $\bar{x}$-values, where would you expect the center of this sampling distribution to lie?

(f) Pool your results with those of your classmates to construct a histogram of the $\bar{x}$-values you obtained. Describe the shape, center, and spread of this distribution. Is the histogram approximately normal?

9.10 BIAS AND VARIABILITY Figure 9.10 shows histograms of four sampling distributions of statistics intended to estimate the same parameter. Label each distribution relative to the others as having large or small bias and as having large or small variability.

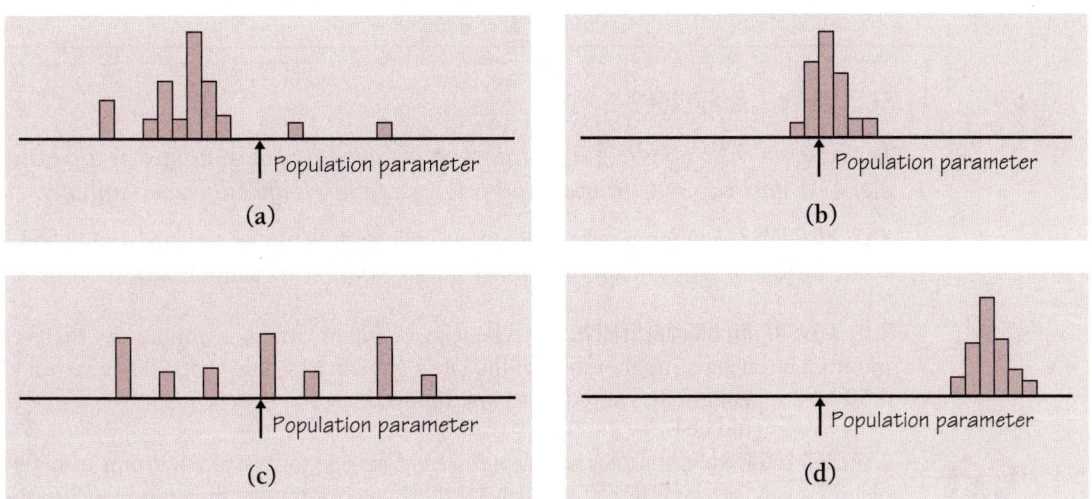

FIGURE 9.10 Which of these sampling distributions displays large or small bias and large or small variability?

9.11 IRS AUDITS The Internal Revenue Service plans to examine an SRS of individual federal income tax returns from each state. One variable of interest is the proportion of returns claiming itemized deductions. The total number of tax returns in each state varies from almost 14 million in California to fewer than 210,000 in Wyoming.

(a) Will the sampling variability of the sample proportion change from state to state if an SRS of 2000 tax returns is selected in each state? Explain your answer.

(b) Will the sampling variability of the sample proportion change from state to state if an SRS of 1% of all tax returns is selected in each state? Explain your answer.

SUMMARY

A number that describes a population is called a **parameter**. A number that can be computed from the sample data is called a **statistic**. The purpose of sampling or experimentation is usually to use statistics to make statements about unknown parameters.

A statistic from a probability sample or randomized experiment has a **sampling distribution** that describes how the statistic varies in repeated data production. The sampling distribution answers the question, "What would happen if we repeated the sample or experiment many times?" Formal statistical inference is based on the sampling distributions of statistics.

A statistic as an estimator of a parameter may suffer from **bias** or from high **variability**. Bias means that the center of the sampling distribution is not equal to the true value of the parameter. The variability of the statistic is described by the spread of its sampling distribution.

Properly chosen statistics from randomized data production designs have no bias resulting from the way the sample is selected or the way the experimental units are assigned to treatments. The variability of the statistic is determined by the size of the sample or by the size of the experimental groups. Statistics from larger samples have less variability.

SECTION 9.1 EXERCISES

In Exercises 9.12 and 9.13, (a) state whether each boldface number is a parameter or a statistic, and (b) use appropriate notation to describe each number.

9.12 HOW TALL? A random sample of female college students has a mean height of **64.5** inches, which is greater than the **63**-inch mean height of all adult American women.

9.13 MEASURING UNEMPLOYMENT The Bureau of Labor Statistics announces that last month it interviewed all members of the labor force in a sample of 50,000 households; **4.5%** of the people interviewed were unemployed.

9.14 BAD EGGS An entomologist samples a field for egg masses of a harmful insect by placing a yard-square frame at random locations and examining the ground within the frame carefully. He wants to estimate the proportion of square yards in which egg masses are present. Suppose that in a large field egg masses are present in 20% of all possible yard-square areas. That is, $p = 0.2$ in this population.

(a) Use Table B to simulate the presence or absence of egg masses in each square yard of an SRS of 10 square yards from the field. Be sure to explain clearly which digits you used to represent the presence and the absence of egg masses. What proportion of your 10 sample areas had egg masses? This is the statistic $\hat{p}$.

(b) Repeat (a) with different lines from Table B, until you have simulated the results of 20 SRSs of size 10. What proportion of the square yards in each of your 20 samples had egg masses? Make a stemplot from these 20 values to display the distribution of your 20 observations on $\hat{p}$. What is the mean of this distribution? What is its shape?

(c) If you looked at all possible SRSs of size 10, rather than just 20 SRSs, what would be the mean of the values of $\hat{p}$? This is the mean of the sampling distribution of $\hat{p}$.

(d) In another field, 40% of all square-yard areas contain egg masses. What is the mean of the sampling distribution of $\hat{p}$ in samples from this field?

9.15 ROLLING THE DICE, I Consider the population of all rolls of a fair, six-sided die.

(a) Draw a histogram that shows the population distribution. Find the mean μ and standard deviation σ of this population.

(b) If you took an SRS of size $n = 2$ from this population, what would you actually be doing?

(c) List all possible SRSs of size 2 from this population, and compute $\bar{x}$ for each sample.

(d) Draw the sampling distribution of $\bar{x}$ for samples of size $n = 2$. Describe its shape, center, and spread. How do these characteristics compare with those of the population distribution?

9.16 ROLLING THE DICE, II In Exercise 9.15, you constructed the sampling distribution of $\bar{x}$ in samples of size $n = 2$ from the population of rolls of a fair, six-sided die. What would happen if we increased the sample size to $n = 3$? For starters, it would take you a long time to list all possible SRSs for $n = 3$. Instead, you can use your calculator to simulate rolling the die three times.

(a) Generate L_1/list1 using the command `randInt(1,6,100)+randInt(1,6,100)+randInt(1,6,100)`. This will run 100 simulations of rolling the die three times and calculating the sum of the three rolls.

(b) Define L_2/list2 as $L_1/3$ (list1/3). Now L_2/list2 contains the values of $\bar{x}$ for the 100 simulations.

(c) Plot a histogram of the $\bar{x}$-values.

9.17 SCHOOL VOUCHERS A national opinion poll recently estimated that 44% ($\hat{p} = 0.44$) of all adults agree that parents of school-age children should be given vouchers good for education at any public or private school of their choice. The polling organization used a probability sampling method for which the sample proportion $\hat{p}$ has a normal distribution with standard deviation about 0.015. If a sample were drawn by the same method from the state of New Jersey (population 7.8 million) instead of from the entire United States (population 280 million), would this standard deviation be larger, about the same, or smaller? Explain your answer.

9.18 SIMULATING *SURVIVOR* Suppose the true proportion of U.S. adults who have watched *Survivor II* is 0.41. Here is a short program that simulates sampling from this population.

TI-83	TI-89

```PROGRAM: SURVIVOR```
```ClrHome```
```ClrList L1```
```Disp "HOW MANY TRIALS?"```
```Prompt N```
```randInt(1,100,N)→L1```
```0→M```
```For(X,1,N,1)```
```If (L1(X)≥1 and L1(X)≤41)```
```M+1→M```
```End```
```Disp "SAMP PROPORTION="```
```Disp M/N```
	```survivor()```
	```Prgm```
	```ClrHome```
	```tistat.clrlist(list1)```
	```Disp "How many trials"```
	```Prompt n```
	```tistat.randint(1,100,n)→list1```
	```0→m```
	```For x,1,n,1```
	```If list1[x] ≥1 and```
	```list1[x]≤41```
	```m+1→m```
	```EndFor```
	```Disp "samp proportion="```
	```Disp approx(m/n)```
	```EndPrgm```

Enter this program or link it from your teacher or a classmate.

(a) In the program, what digits are assigned to U.S. adults? What digits are assigned to U.S. adults who say they have watched *Survivor II*? Does the program output a count of adults who answer "Yes," a percent, or a proportion?

(b) Execute the program and specify 5 trials (sample size = 5). Do this 10 times, and record the 10 numbers.

(c) Execute the program 10 more times, specifying a sample size of 25. Record the 10 results for sample size = 25.

(d) Execute the program 10 more times, specifying a sample size of 100. Record the 10 results for sample size = 100.

(e) Enter the 10 outputs for sample size = 5 in L_1/list1, the 10 results for sample size = 25 in L_2/list2, and the 10 results for sample size = 100 in L_3/list3. Then do 1-Var Stats for L_1/list 1, L_2/list2, and L_3/list3, and record the means and sample standard deviations s_x for each sample size. Complete the sentence "As the sample size increases, the variability __________."

9.2 SAMPLE PROPORTIONS

What proportion of U.S. teens know that 1492 was the year in which Columbus "discovered" America? A Gallup Poll found that 210 out of a random sample of 501 American teens aged 13 to 17 knew this historically important date.[6] The sample proportion

$$\hat{p} = \frac{210}{501} = 0.42$$

is the statistic that we use to gain information about the unknown population parameter p. We may say that "42% of U.S. teenagers know that Columbus discovered America in 1492." Statistical recipes work with proportions expressed as decimals, so 42% becomes 0.42.

The sampling distribution of $\hat{p}$

How good is the statistic $\hat{p}$ as an estimate of the parameter p? To find out, we ask, "What would happen if we took many samples?" The sampling distribution of $\hat{p}$ answers this question. How do we determine the center, shape, and spread of the sampling distribution of $\hat{p}$? By making an important connection between proportions and counts. We want to estimate the proportion of "successes" in the population. We take an SRS from the population of interest. Our estimator is the sample proportion of successes:

$$\hat{p} = \frac{\text{count of "successes" in sample}}{\text{size of sample}} = \frac{X}{n}$$

Since values of X and $\hat{p}$ will vary in repeated samples, both X and $\hat{p}$ are random variables. Provided that the population is much larger than the sample (say at least 10 times), the count X will follow a binomial distribution. The proportion $\hat{p}$ does not have a binomial distribution.

From Chapter 8, we know that

$$\mu_X = np \text{ and } \sigma_X = \sqrt{np(1-p)}$$

give the mean and standard deviation of the random variable X. Since $\hat{p} = X/n = (1/n)X$, we can use the rules from Chapter 7 to find the mean and standard deviation of the random variable $\hat{p}$. Recall that if $Y = a + bX$, then $\mu_Y = a + b\mu_X$ and $\sigma_Y = b\sigma_X$. In this case, $\hat{p} = 0 + (1/n)X$, so

$$\mu_{\hat{p}} = 0 + \frac{1}{n}np = p$$

$$\sigma_{\hat{p}} = \frac{1}{n}\sqrt{np(1-p)} = \sqrt{\frac{np(1-p)}{n^2}} = \sqrt{\frac{p(1-p)}{n}}$$

SAMPLING DISTRIBUTION OF A SAMPLE PROPORTION

Choose an SRS of size n from a large population with population proportion p having some characteristic of interest. Let $\hat{p}$ be the proportion of the sample having that characteristic. Then:

- The **mean** of the sampling distribution is exactly p.

SAMPLING DISTRIBUTION OF A SAMPLE PROPORTION (*continued*)

- The **standard deviation** of the sampling distribution is

$$\sqrt{\frac{p(1-p)}{n}}$$

Because the mean of the sampling distribution of $\hat{p}$ is always equal to the parameter p, the sample proportion $\hat{p}$ is an unbiased estimator of p. The standard deviation of $\hat{p}$ gets smaller as the sample size n increases because n appears in the denominator of the formula for the standard deviation. That is, $\hat{p}$ is less variable in larger samples. What is more, the formula shows just how quickly the standard deviation decreases as n increases. The sample size n is under the square root sign, so to cut the standard deviation in half, we must take a sample four times as large, not just twice as large.

The formula for the standard deviation of $\hat{p}$ doesn't apply when the sample is a large part of the population. You can't use this recipe if you choose an SRS of 50 of the 100 people in a class, for example. In practice, we usually take a sample only when the population is large. Otherwise, we could examine the entire population. Here is a practical guide.[7]

RULE OF THUMB 1

Use the recipe for the standard deviation of $\hat{p}$ only when the population is at least 10 times as large as the sample.

Using the normal approximation for $\hat{p}$

What about the shape of the sampling distribution of $\hat{p}$? In the simulation examples in Section 9.1, we found that the sampling distribution of $\hat{p}$ is **approximately normal** and is closer to a normal distribution when the sample size n is large. For example, if we sample 100 individuals, the only possible values of $\hat{p}$ are 0, 1/100, 2/100, and so on. The statistic has only 101 possible values, so its distribution cannot be exactly normal. The accuracy of the normal approximation improves as the sample size n increases. For a fixed sample size n, the normal approximation is most accurate when p is close to 1/2, and least accurate when p is near 0 or 1. If $p = 1$, for example, then $\hat{p} = 1$ in every sample because every individual in the population has the characteristic we are counting. The normal approximation is no good at all when $p = 1$ or $p = 0$. Here is a rule of thumb that ensures that normal calculations are accurate enough for most statistical purposes. Unlike the first rule of thumb, this one rules out some settings of practical interest.

RULE OF THUMB 2

We will use the normal approximation to the sampling distribution of $\hat{p}$ for values of n and p that satisfy $np \geq 10$ and $n(1-p) \geq 10$.

Using what we have learned about the sampling distribution of $\hat{p}$, we can determine the likelihood of obtaining an SRS in which $\hat{p}$ is close to p. This is especially useful to college admissions officers, as the following example shows.

EXAMPLE 9.7 APPLYING TO COLLEGE

A polling organization asks an SRS of 1500 first-year college students whether they applied for admission to any other college. In fact, 35% of all first-year students applied to colleges besides the one they are attending. What is the probability that the random sample of 1500 students will give a result within 2 percentage points of this true value?

We have an SRS of size $n = 1500$ drawn from a population in which the proportion $\hat{p} = 0.35$ applied to other colleges. The sampling distribution of $\hat{p}$ has mean $\mu_{\hat{p}} = 0.35$. What about its standard deviation? By the first "rule of thumb," the population must contain at least 10(1500) = 15,000 people for us to use the standard deviation formula we derived. There are over 1.7 million first-year college students, so

$$\sigma_{\hat{p}} = \sqrt{\frac{p(1-p)}{n}} = \sqrt{\frac{(0.35)(0.65)}{1500}} = 0.0123$$

Can we use a normal distribution to approximate the sampling distribution of $\hat{p}$? Checking the second "rule of thumb": $np = 1500(0.35) = 525$ and $n(1-p) = 1500(0.65) = 975$. Both are much larger than 10, so the normal approximation will be quite accurate.

We want to find the probability that $\hat{p}$ falls between 0.33 and 0.37 (within 2 percentage points, or 0.02, of 0.35). This is a normal distribution calculation. Figure 9.11 shows the normal distribution that approximates the sampling distribution of $\hat{p}$. The area of the shaded region corresponds to the probability that $0.33 \leq \hat{p} \leq 0.37$.

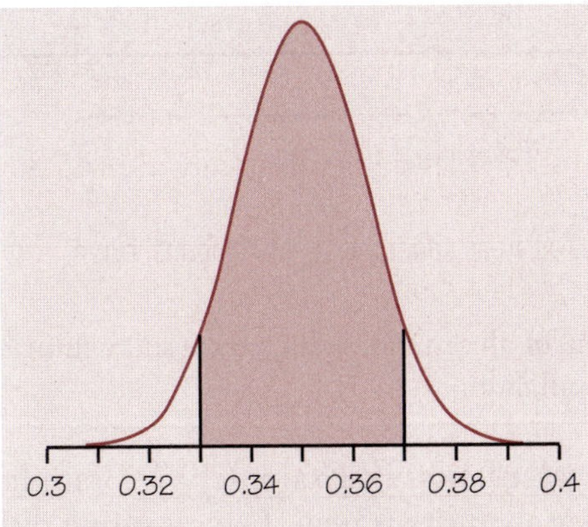

FIGURE 9.11 The normal approximation to the sampling distribution of $\hat{p}$.

Step 1: Standardize $\hat{p}$ by subtracting its mean 0.35 and dividing by its standard deviation 0.0123. That produces a new statistic that has the standard normal distribution. It is usual to call such a statistic z:

$$z = \frac{\hat{p}-0.3}{0.012}$$

Step 2: Find the standardized values (z-scores) of $\hat{p} = 0.33$ and $\hat{p} = 0.37$. For $\hat{p} = 0.33$:

$$z = \frac{0.33-0.35}{0.0123} = -1.63$$

For $\hat{p} = 0.37$:

$$z = \frac{0.37-0.35}{0.0123} = 1.63$$

Step 3: Draw a picture of the area under the standard normal curve corresponding to these standardized values (Figure 9.12). Then use Table A to find the shaded area. Here is the calculation:

$$P(0.33 \le \hat{p} \le 0.37) = P(-1.63 \le z \le 1.63) = 0.9484 - 0.0516 = 0.8968$$

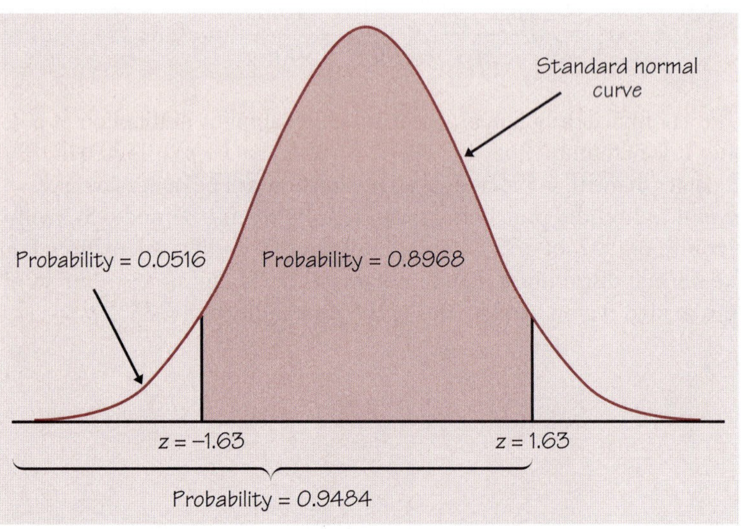

FIGURE 9.12 Probabilities as areas under the standard normal curve.

We see that almost 90% of all samples will give a result within 2 percentage points of the truth about the population.

The outline of the calculation in Example 9.7 is familiar from Chapter 2, but the language of probability is new. The sampling distribution of $\hat{p}$ gives probabilities for its values, so the entries in Table A are now probabilities.

We used a brief notation that is common in statistics. The capital P in $P(0.33 \le \hat{p} \le 0.37)$ stands for "probability." The expression inside the parentheses tells us what event we are finding the probability of. This entire expression is a short way of writing "the probability that $\hat{p}$ lies between 0.33 and 0.37."

EXAMPLE 9.8 SURVEY UNDERCOVERAGE?

One way of checking the effect of undercoverage, nonresponse, and other sources of error in a sample survey is to compare the sample with known facts about the population. About 11% of American adults are black. The proportion $\hat{p}$ of blacks in an SRS of 1500 adults should therefore be close to 0.11. It is unlikely to be exactly 0.11 because of sampling variability. If a national sample contains only 9.2% blacks, should we suspect that the sampling procedure is somehow underrepresenting blacks? We will find the probability that a sample contains no more than 9.2% blacks when the population is 11% black.

The mean of the sampling distribution of $\hat{p}$ is $p = 0.11$. Since the population of all black American adults is larger than 10(1500) = 15,000, the standard deviation of $\hat{p}$ is

$$\sqrt{\frac{p(1-p)}{n}} = \sqrt{\frac{(0.11)(0.89)}{1500}} = 0.00808$$

(by rule of thumb 1). Next, we check to see that $np = (1500)(0.11) = 165$ and $n(1 - p) = (1500)(0.89) = 1335$. So rule of thumb 2 tells us that we can use the normal approximation to the sampling distribution of $\hat{p}$. Figure 9.13(a) shows the normal distribution with the area corresponding to $\hat{p} \le 0.092$ shaded.

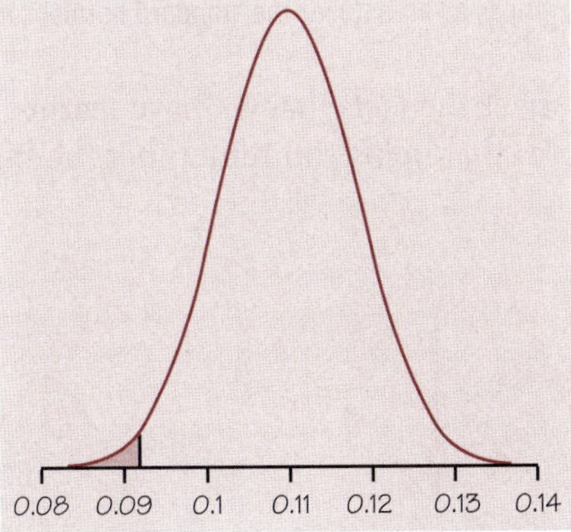

FIGURE 9.13(a) The normal approximation to the sampling distribution of $\hat{p}$.

Step 1: Standardize $\hat{p}$.

$$z = \frac{\hat{p} - 0.11}{0.00808}$$

has the standard normal distribution.

Step 2: Find the standardized value (z-score) of $\hat{p} = 0.092$.

$$z = \frac{0.092 - 0.11}{0.00808} = -2.23$$

Step 3: Draw a picture of the area under the standard normal curve corresponding to the standardized value (Figure 9.13(b)). Then use Table A to find the shaded area.

$$P(\hat{p} \le 0.092) = P(z \le -2.23) = 0.0129$$

Only 1.29% of all samples would have so few blacks. Because it is unlikely that a sample would include so few blacks, we have good reason to suspect that the sampling procedure underrepresents blacks.

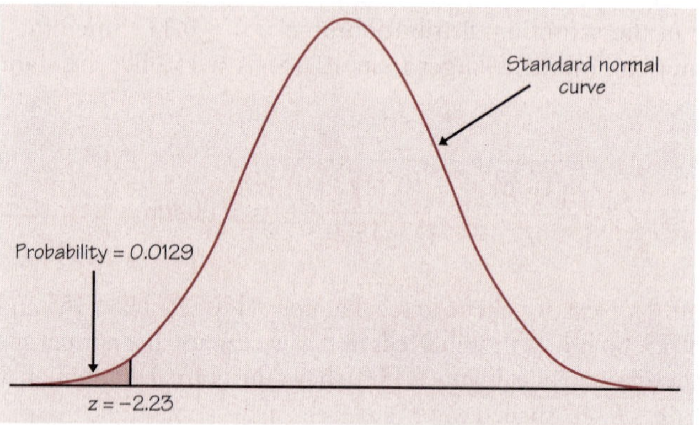

FIGURE 9.13(b) The probability as an area under the standard normal curve.

Figure 9.14 summarizes the facts that we have learned about the sampling distribution of $\hat{p}$ in a form that helps you remember the big idea of a sampling distribution.

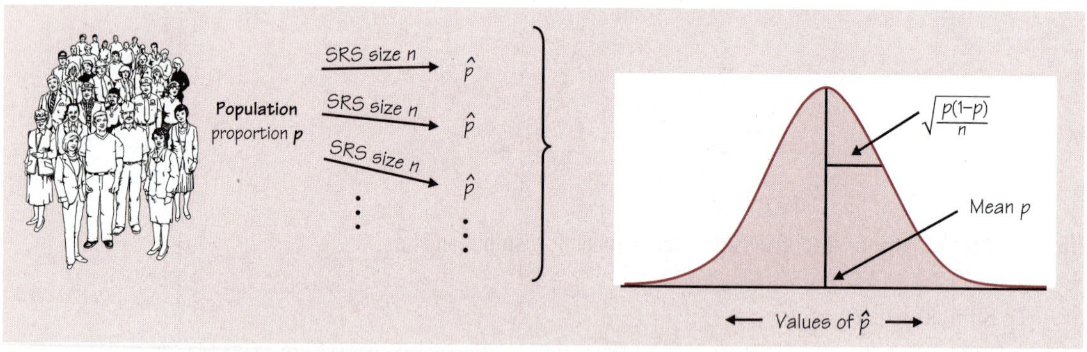

FIGURE 9.14 Select a large SRS from a population of which the proportion p are successes. The sampling distribution of the proportion $\hat{p}$ of successes in the sample is approximately normal. The mean is p and the standard deviation is $\sqrt{p(1-p)/n}$.

EXERCISES

9.19 DO YOU DRINK THE CEREAL MILK? A *USA Today* poll asked a random sample of 1012 U.S. adults what they do with the milk in the bowl after they have eaten the cereal. Of the respondents, 67% said that they drink it. Suppose that 70% of U.S. adults actually drink the cereal milk.

(a) Find the mean and standard deviation of the proportion $\hat{p}$ of the sample that say they drink the cereal milk.

(b) Explain why you can use the formula for the standard deviation of $\hat{p}$ in this setting (rule of thumb 1).

(c) Check that you can use the normal approximation for the distribution of $\hat{p}$ (rule of thumb 2).

(d) Find the probability of obtaining a sample of 1012 adults in which 67% or fewer say they drink the cereal milk. Do you have any doubts about the result of this poll?

(e) What sample size would be required to reduce the standard deviation of the sample proportion to one-half the value you found in (a)?

(f) If the pollsters had surveyed 1012 teenagers instead of 1012 adults, do you think the sample proportion $\hat{p}$ would have been greater than, equal to, or less than 0.67? Explain.

9.20 DO YOU GO TO CHURCH? The Gallup Poll asked a probability sample of 1785 adults whether they attended church or synagogue during the past week. Suppose that 40% of the adult population did attend. We would like to know the probability that an SRS of size 1785 would come within plus or minus 3 percentage points of this true value.

(a) If $\hat{p}$ is the proportion of the sample who did attend church or synagogue, what is the mean of the sampling distribution of $\hat{p}$? What is its standard deviation?

(b) Explain why you can use the formula for the standard deviation of $\hat{p}$ in this setting (rule of thumb 1).

(c) Check that you can use the normal approximation for the distribution of $\hat{p}$ (rule of thumb 2).

(d) Find the probability that $\hat{p}$ takes a value between 0.37 and 0.43. Will an SRS of size 1785 usually give a result $\hat{p}$ within plus or minus 3 percentage points of the true population proportion? Explain.

9.21 DO YOU GO TO CHURCH? Suppose that 40% of the adult population attended church or synagogue last week. Exercise 9.20 asks the probability that $\hat{p}$ from an SRS estimates $p = 0.4$ within 3 percentage points. Find this probability for SRSs of sizes 300, 1200, and 4800. What general fact do your results illustrate?

9.22 HARLEY MOTORCYCLES Harley-Davidson motorcycles make up 14% of all the motorcycles registered in the United States. You plan to interview an SRS of 500 motorcycle owners.

(a) What is the approximate distribution of your sample who own Harleys?

(b) How likely is your sample to contain 20% or more who own Harleys? Do a normal probability calculation to answer this question.

(c) How likely is your sample to contain at least 15% who own Harleys? Do a normal probability calculation to answer this question.

9.23 ON-TIME SHIPPING Your mail-order company advertises that it ships 90% of its orders within three working days. You select an SRS of 100 of the 5000 orders received in the past week for an audit. The audit reveals that 86 of these orders were shipped on time.

(a) What is the sample proportion of orders shipped on time?

(b) If the company really ships 90% of its orders on time, what is the probability that the proportion in an SRS of 100 orders is as small as the proportion in your sample or smaller?

(c) Compare your answer to (b) with your results in Exercise 8.33 (page 462) where you used a normal approximation to the binomial to solve this problem.

9.24 Exercise 9.22 asks for probability calculations about Harley-Davidson motorcycle ownership. Exercise 9.23 asks for a similar calculation about a random sample of mail orders. For which calculation does the normal approximation to the sampling distribution of $\hat{p}$ give a more accurate answer? Why? (You need not actually do either calculation.)

SUMMARY

When we want information about the **population proportion** p of individuals with some special characteristic, we often take an SRS and use the **sample proportion** $\hat{p}$ to estimate the unknown parameter p.

The **sampling distribution** of $\hat{p}$ describes how the statistic varies in all possible samples from the population.

The **mean** of the sampling distribution is equal to the population proportion p. That is, $\hat{p}$ is an unbiased estimator of p.

The **standard deviation** of the sampling distribution is $\sqrt{p(1-p)/n}$ for an SRS of size n. This recipe can be used if the population is at least 10 times as large as the sample.

The standard deviation of $\hat{p}$ gets smaller as the sample size n gets larger. Because of the square root, a sample four times larger is needed to cut the standard deviation in half.

When the sample size n is large, the sampling distribution of $\hat{p}$ is close to a normal distribution with mean p and standard deviation $\sqrt{p(1-p)/n}$. In practice, use this **normal approximation** when both $np \geq 10$ and $n(1-p) \geq 10$.

SECTION 9.2 EXERCISES

9.25 DO YOU JOG? The Gallup Poll once asked a random sample of 1540 adults, "Do you happen to jog?" Suppose that in fact 15% of all adults jog.

(a) Find the mean and standard deviation of the proportion $\hat{p}$ of the sample who jog. (Assume the sample is an SRS.)

(b) Explain why you can use the formula for the standard deviation of $\hat{p}$ in this setting.

(c) Check that you can use the normal approximation for the distribution of $\hat{p}$.

(d) Find the probability that between 13% and 17% of the sample jog.

(e) What sample size would be required to reduce the standard deviation of the sample proportion to one-third the value you found in (a)?

9.26 MORE JOGGING! Suppose that 15% of all adults jog. Exercise 9.25 asks the probability that the sample proportion $\hat{p}$ from an SRS estimates $p = 0.15$ within 2 percentage points. Find this probability for SRSs of sizes 200, 800, and 3200. What general conclusion can you draw from your calculations?

9.27 LET'S GO SHOPPING Are attitudes toward shopping changing? Sample surveys show that fewer people enjoy shopping than in the past. A recent survey asked a nationwide random sample of 2500 adults if they agreed or disagreed that "I like buying new clothes, but shopping is often frustrating and time-consuming."[8] The population that the poll wants to draw conclusions about is all U.S. residents aged 18 and over. Suppose that in fact 60% of all adult U.S. residents would say "Agree" if asked the same question. What is the probability that 1520 or more of the sample agree?

9.28 UNLISTED NUMBERS According to a market research firm, 52% of all residential telephone numbers in Los Angeles are unlisted. A telephone sales firm uses random digit dialing equipment that dials residential numbers at random, whether or not they are listed in the telephone directory. The firm calls 500 numbers in Los Angeles.

(a) What are the mean and standard deviation of the proportion of unlisted numbers in the sample?

(b) What is the probability that at least half the numbers dialed are unlisted? (Remember to check that you can use the normal approximation.)

9.29 MULTIPLE-CHOICE TESTS Here is a simple probability model for multiple-choice tests. Suppose that a student has probability p of correctly answering a question chosen at random from a universe of possible questions. (A good student has a higher p than a poor student.) The correctness of an answer to any specific question doesn't depend on other questions. A test contains n questions. Then the proportion of correct answers that a student gives is a sample proportion $\hat{p}$ from an SRS of size n drawn from a population with population proportion p.

(a) Julie is a good student for whom $p = 0.75$. Find the probability that Julie scores 70% or lower on a 100-question test.

(b) If the test contains 250 questions, what is the probability that Julie will score 70% or lower?

(c) How many questions must the test contain in order to reduce the standard deviation of Julie's proportion of correct answers to one-fourth its value for a 100-item test?

(d) Laura is a weaker student for whom $p = 0.6$. Does the answer you gave in (c) for the standard deviation of Julie's score apply to Laura's standard deviation also? Explain.

9.30 RULES OF THUMB Explain why you cannot use the methods of this section to find the following probabilities.

(a) A factory employs 3000 unionized workers, of whom 30% are Hispanic. The 15-member union executive committee contains 3 Hispanics. What would be the probability of 3 or fewer Hispanics if the executive committee were chosen at random from all the workers?

(b) A university is concerned about the academic standing of its intercollegiate athletes. A study committee chooses an SRS of 50 of the 316 athletes to interview in detail. Suppose that in fact 40% of the athletes have been told by coaches to neglect their studies on at least one occasion. What is the probability that at least 15 in the sample are among this group?

(c) Use what you learned in Chapter 8 to find the probability described in part (a).

9.3 SAMPLE MEANS

Sample proportions arise most often when we are interested in categorical variables. We then ask questions like "What proportion of U.S. adults have watched *Survivor II?*" or "What percent of the adult population attended church last week?" When we record quantitative variables—the income of a household, the lifetime of a car brake pad, the blood pressure of a patient—we are interested in other statistics, such as the median or mean or standard deviation of the variable. Because sample means are just averages of observations, they are among the most common statistics. This section describes the sampling distribution of the mean of the responses in an SRS.

EXAMPLE 9.9 BULL MARKET OR BEAR MARKET?

A basic principle of investment is that diversification reduces risk. That is, buying several securities rather than just one reduces the variability of the return on an investment. Figure 9.15 illustrates this principle in the case of common stocks listed on the New York Stock Exchange. Figure 9.15(a) shows the distribution of returns for all 1815 stocks listed on the Exchange for the entire year 1987.[9] This was a year of extreme swings in stock prices, including a record loss of over 20% in a single day. The mean return for all 1815 stocks was $\mu = -3.5\%$ and the distribution shows a very wide spread.

Figure 9.15(b) shows the distribution of returns for all possible portfolios that invested equal amounts in each of 5 stocks. A portfolio is just a sample of 5 stocks and its return is the average return for the 5 stocks chosen. The mean return for all portfolios is still –3.5%, but the variation among portfolios is much less than the variation among individual stocks. For example, 11% of all individual stocks had a loss of more than 40%, but only 1% of the portfolios had a loss that large.

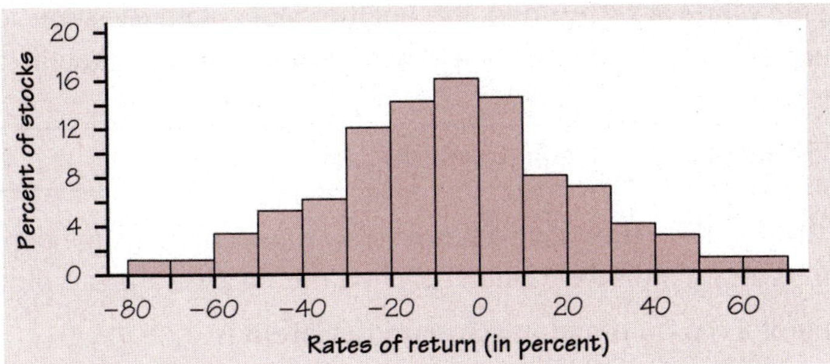

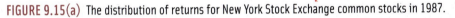

FIGURE 9.15(a) The distribution of returns for New York Stock Exchange common stocks in 1987.

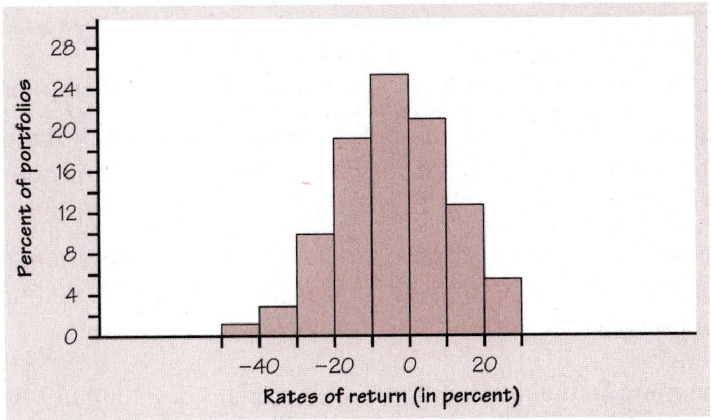

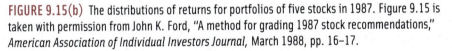

FIGURE 9.15(b) The distributions of returns for portfolios of five stocks in 1987. Figure 9.15 is taken with permission from John K. Ford, "A method for grading 1987 stock recommendations," *American Association of Individual Investors Journal*, March 1988, pp. 16–17.

The histograms in Figure 9.15 emphasize a principle that we will make precise in this section:

- Averages are less variable than individual observations.

More detailed examination of the distributions would point to a second principle:

- Averages are more normal than individual observations.

These two facts contribute to the popularity of sample means in statistical inference.

The mean and the standard deviation of $\bar{x}$

The sampling distribution of $\bar{x}$ is the distribution of the values of $\bar{x}$ in all possible samples of the same size from the population. Figure 9.15(a) shows the distribution of a population, with mean $\mu = -3.5\%$. Figure 9.15(b) is the sampling distribution of the sample mean $\bar{x}$ from all samples of size $n = 5$ from this population. The mean of all the values of $\bar{x}$ is again –3.5%, but the values of $\bar{x}$ are less spread out than the individual values in the population. This is an example of a general fact.

MEAN AND STANDARD DEVIATION OF A SAMPLE MEAN

Suppose that $\bar{x}$ is the mean of an SRS of size n drawn from a large population with mean μ and standard deviation σ. Then the **mean** of the sampling distribution of $\bar{x}$ is $\mu_{\bar{x}} = \mu$ and its **standard deviation** is $\sigma_{\bar{x}} = \sigma / \sqrt{n}$.

The behavior of $\bar{x}$ in repeated samples is much like that of the sample proportion $\hat{p}$:

- The sample mean $\bar{x}$ is an unbiased estimator of the population mean μ.
- The values of $\bar{x}$ are less spread out for larger samples. Their standard deviation decreases at the rate $\sqrt{n}$, so you must take a sample four times as large to cut the standard deviation of $\bar{x}$ in half.
- You should only use the recipe $\sigma / \sqrt{n}$ for the standard deviation of $\bar{x}$ when the population is at least 10 times as large as the sample. This is almost always the case in practice.

Notice that these facts about the mean and standard deviation of $\bar{x}$ are true no matter what the shape of the population distribution is.

EXAMPLE 9.10 YOUNG WOMEN'S HEIGHTS

The height of young women varies approximately according to the N(64.5, 2.5) distribution. This is a population distribution with $\mu = 64.5$ and $\sigma = 2.5$. If we choose one young woman at random, the heights we get in repeated choices follow this distribution. That is, the distribution of the population is also the distribution of one observation chosen at random. So we can think of the population distribution as a distribution of probabilities, just like a sampling distribution.

Now measure the height of an SRS of 10 young women. The sampling distribution of their sample mean height $\bar{x}$ will have mean $\mu_{\bar{x}} = \mu = 64.5$ inches and standard deviation

$$\sigma_{\bar{x}} = \frac{\sigma}{\sqrt{n}} = \frac{2.5}{\sqrt{10}} = 0.79 \text{ inch}$$

The heights of individual women vary widely about the population mean, but the average height of a sample of 10 women is less variable.

In Activity 9A, you plotted the distribution of $\bar{x}$ for samples of size $n = 100$, so the standard deviation of x is $\sigma / \sqrt{100} = 2.5/10 = 0.25$. How close did your class come to this number?

We have described the mean and standard deviation of the sampling distribution of a sample mean $\overline{x}$, but not its shape. The shape of the distribution of $\overline{x}$ depends on the shape of the population distribution. In particular, if the population distribution is normal, then so is the distribution of the sample mean.

SAMPLING DISTRIBUTION OF A SAMPLE MEAN FROM A NORMAL POPULATION

Draw an SRS of size n from a population that has the normal distribution with mean μ and standard deviation σ. Then the sample mean $\overline{x}$ has the normal distribution $N(\mu, \sigma/\sqrt{n})$ with mean μ and standard deviation $\sigma/\sqrt{n}$.

We already knew the mean and standard deviation of the sampling distribution. All that we have added now is the normal shape. In Activity 9A, we began with a normal distribution, N(64.5, 2.5). The center (mean) of the approximate sampling distribution of $\overline{x}$ should have been very close to the mean of the population: 64.5 inches. Was it? The spread of the distribution of $\overline{x}$ should have been very close to $\sigma/\sqrt{n}$. Was it? The reason that you don't observe exact agreement is sampling variability.

EXAMPLE 9.11 MORE ON YOUNG WOMEN'S HEIGHTS

What is the probability that a randomly selected young woman is taller than 66.5 inches? What is the probability that the mean height of an SRS of 10 young women is greater than 66.5 inches? We can answer both of these questions using normal calculations.

If we let X = the height of a randomly selected young woman, then the random variable X follows a normal distribution with μ = 64.5 inches and σ = 2.5 inches. To find $P(X > 66.5)$, we first standardize the values of X by setting

$$z = \frac{X - \mu}{\sigma}$$

The random variable z follows the standard normal distribution. When X = 66.5,

$$z = \frac{66.5 - 64.5}{2.5} = 0.80$$

From Table A,

$$P(X > 66.5) = P(z > 0.80) = 1 - 0.7881 = 0.2119$$

The probability of choosing a young woman at random whose height exceeds 66.5 inches is about 0.21.

Now let's take an SRS of 10 young women from this population and compute $\overline{x}$ for the sample. In Example 9.10, we saw that in repeated samples of size n = 10, the values of $\bar{x}$ will follow a $N(64.5, 0.79)$ distribution. To find the probability that $\bar{x} > 66.5$ inches, we start by standardizing:

$$z = \frac{\bar{x} - \mu_{\bar{x}}}{\sigma_{\bar{x}}}$$

A sample mean of 66.5 inches yields a z-score of

$$z = \frac{66.5 - 64.5}{0.79} = 2.53$$

Finally,

$$P(\bar{x} > 66.5) = P(z > 2.53) = 1 - 0.9943 = 0.0057$$

It is very unlikely (less than a 1% chance) that we would draw an SRS of 10 young women whose average height exceeds 66.5 inches.

Figure 9.16 compares the population distribution and the sampling distribution of $\bar{x}$. It also shows the areas corresponding to the probabilities that we just computed.

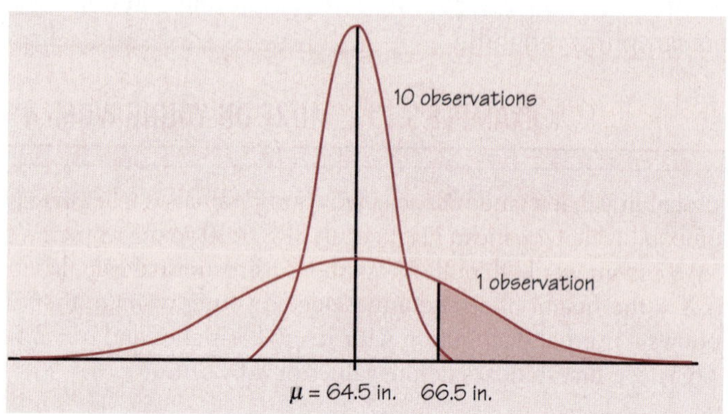

FIGURE 9.16 The sampling distribution of the mean height $\bar{x}$ for samples of 10 young women compared with the distribution of the height of a single woman chosen at random.

The fact that averages of several observations are less variable than individual observations is important in many settings. For example, it is common practice to repeat a careful measurement several times and report the average of the results. Think of the results of n repeated measurements as an SRS from the population of outcomes we would get if we repeated the measurement forever. The average of the n results (the sample mean $\bar{x}$) is less variable than a single measurement.

EXERCISES

9.31 BULL MARKET OR BEAR MARKET? Investors remember 1987 as the year stocks lost 20% of their value in a single day. For 1987 as a whole, the mean return of all common stocks on the New York Stock Exchange was $\mu = -3.5\%$. (That is, these stocks lost an average of 3.5% of their value in 1987.) The standard deviation of the returns was about $\sigma = 26\%$. Figure 9.15(a) on page 515 shows the distribution of returns. Figure 9.15(b) is the sampling distribution of the mean returns $\bar{x}$ for all possible samples of 5 stocks.

(a) What are the mean and the standard deviation of the distribution in Figure 9.15(b)?

(b) Assuming that the population distribution of returns on individual common stocks is normal, what is the probability that a randomly chosen stock showed a return of at least 5% in 1987?

(c) Assuming that the population distribution of returns on individual common stocks is normal, what is the probability that a randomly chosen portfolio of 5 stocks showed a return of at least 5% in 1987?

(d) What percentage of 5-stock portfolios lost money in 1987?

9.32 ACT SCORES The scores of individual students on the American College Testing (ACT) composite college entrance examination have a normal distribution with mean 18.6 and standard deviation 5.9.

(a) What is the probability that a single student randomly chosen from all those taking the test scores 21 or higher?

(b) Now take an SRS of 50 students who took the test. What are the mean and standard deviation of the average (sample mean) score for the 50 students? Do your results depend on the fact that individual scores have a normal distribution?

(c) What is the probability that the mean score $\bar{x}$ of these students is 21 or higher?

9.33 MEASUREMENTS IN THE LAB Juan makes a measurement in a chemistry laboratory and records the result in his lab report. The standard deviation of students' lab measurements is $\sigma = 10$ milligrams. Juan repeats the measurement 3 times and records the mean $\bar{x}$ of his 3 measurements.

(a) What is the standard deviation of Juan's mean result? (That is, if Juan kept on making 3 measurements and averaging them, what would be the standard deviation of all his $\bar{x}$'s?)

(b) How many times must Juan repeat the measurement to reduce the standard deviation of $\bar{x}$ to 3 milligrams? Explain to someone who knows no statistics the advantage of reporting the average of several measurements rather than the result of a single measurement.

9.34 MEASURING BLOOD CHOLESTEROL A study of the health of teenagers plans to measure the blood cholesterol level of an SRS of youth of ages 13 to 16 years. The researchers will report the mean $\bar{x}$ from their sample as an estimate of the mean cholesterol level μ in this population.

(a) Explain to someone who knows no statistics what it means to say that $\bar{x}$ is an "unbiased" estimator of μ.

(b) The sample result $\bar{x}$ is an unbiased estimator of the population parameter μ no matter what size SRS the study chooses. Explain to someone who knows no statistics why a large sample gives more trustworthy results than a small sample.

ACTIVITY 9B Sampling Pennies

Materials: For a week or so prior to this experiment, you should collect 25 pennies from current circulation. You should bring to class your 25 pennies, as well as 2 nickels, 2 dimes, 1 quarter, and a small container such as a Styrofoam coffee cup or a margarine tub.

1. This activity[10] begins by plotting the distribution of ages (in years) of the pennies you have brought to class. Sketch a density curve that you think will capture the shape of the distribution of ages of the pennies.

2. Make a table of years, beginning with the current year and counting backward. Make the second column the age of the penny. For the age, subtract the date on the penny from the current year. Make the third column the frequency, and use tally marks to record the number of pennies of each age. For example, if it is 2002:

Year	Age	Frequency
2002	0	|||
2001	1	|
2000	2	||||
...		

3. Put your 25 pennies in a cup, and randomly select 5 pennies. Find the average age of the 5 pennies in your sample, and record the mean age as $\bar{x}(5)$. If you are in a small class (fewer than about 15), you should repeat this step. Replace the pennies in the cup, stir so they are randomly distributed, and then repeat the process.

4. Repeat step 3, except this time randomly select 10 pennies. Calculate the average age of the sample of 10 pennies, and record this as $\bar{x}(10)$. If your class is small, do this twice to obtain two means.

5. Repeat step 3 but take all 25 pennies. Record the mean age as $\bar{x}(25)$.

6. Select a flat surface (or clear a space on the floor), and use masking tape to make a number line (axis) with ages marked from 0 to about 30 on the axis. Each interval should be a little more than the width of a penny. You should place your 25 pennies on the axis according to age. When everyone has done this, look at the shape of the histogram. Are you surprised? How would you explain the shape?

7. Make a second axis on the floor, and label it 0, 0.5, 1, 1.5, etc. This time, use nickels to plot the means for samples of size 5. What is the shape of this histogram for the distribution of $\bar{x}(5)$?

ACTIVITY 9B Sampling Pennies (*continued*)

8. Make a third histogram for the means of samples of size 10. Use dimes to make this histogram.

9. Finally, use the quarters to make a histogram of the distribution of means for samples of size 25. Describe the shape of this histogram. Are you surprised?

10. Write a short description of what you have discovered by doing this activity.

The central limit theorem

Although many populations have roughly normal distributions, very few indeed are exactly normal. What happens to $\bar{x}$ when the population distribution is not normal? In Activity 9B, the distribution of ages of pennies should have been right-skewed, but as the sample size increased from 1 to 5 to 10 and then to 25, the distribution should have gotten closer and closer to a normal distribution. This is true no matter what shape the population distribution has, as long as the population has a finite standard deviation σ. This famous fact of probability is called the *central limit theorem.* It is much more useful than the fact that the distribution of $\bar{x}$ is exactly normal if the population is exactly normal.

CENTRAL LIMIT THEOREM

Draw an SRS of size n from any population whatsoever with mean μ and finite standard deviation σ. When n is large, the sampling distribution of the sample mean $\bar{x}$ is close to the normal distribution $N(\mu, \sigma/\sqrt{n})$ with mean μ and standard deviation $\sigma/\sqrt{n}$.

How large a sample size n is needed for $\bar{x}$ to be close to normal depends on the population distribution. More observations are required if the shape of the population distribution is far from normal.

EXAMPLE 9.12 EXPONENTIAL DISTRIBUTION

Figure 9.17 shows the central limit theorem in action for a very nonnormal population. Figure 9.17(a) displays the density curve for the distribution of the population. The distribution is strongly right-skewed, and the most probable outcomes are near 0 at one end of the range of possible values. The mean μ of this distribution is 1 and its standard deviation σ is also 1. This particular distribution is called an ***exponential distribution*** from the shape of its density curve. Exponential distributions are used to describe the lifetime in service of electronic components and the time required to serve a customer or repair a machine.

exponential distribution

Figures 9.17(b), (c), and (d) are the density curves of the sample means of 2, 10, and 25 observations from this population. As n increases, the shape becomes more normal. The mean remains at $\mu = 1$ and the standard deviation decreases, taking the value $1/\sqrt{n}$. The density curve for 10 observations is still somewhat skewed to the right but already resembles a normal curve with $\mu = 1$ and $\sigma = 1/\sqrt{10} = 0.32$. The density curve for $n = 25$ is yet more normal. The contrast between the shape of the population distribution and the shape of the distribution of the mean of 10 or 25 observations is striking.

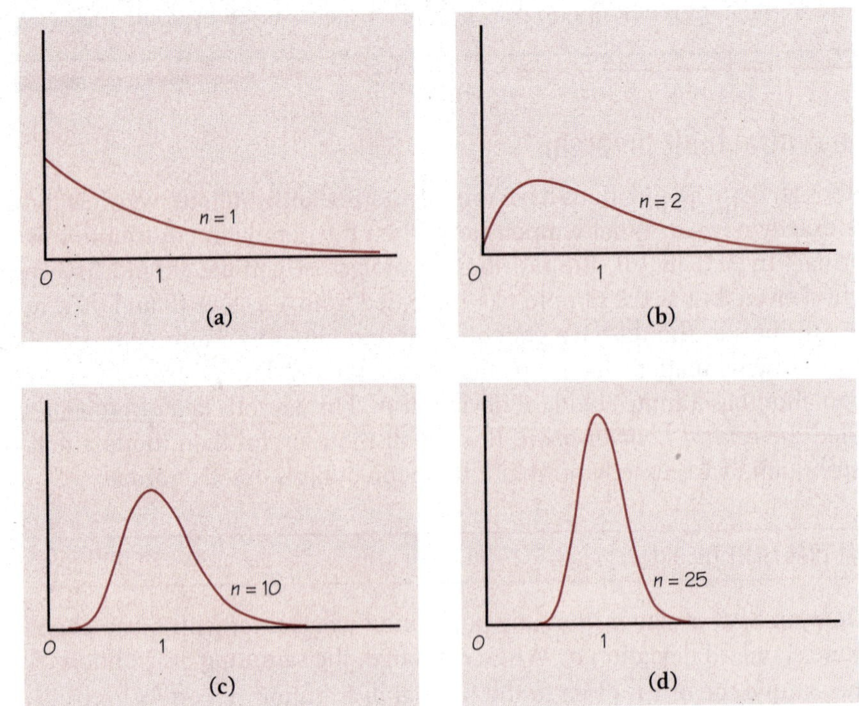

FIGURE 9.17 The central limit theorem in action: the distribution of sample means $\bar{x}$ from a strongly nonnormal population becomes more normal as the sample size increases. (a) The distribution of 1 observation. (b) The distribution of $\bar{x}$ for 2 observations. (c) The distribution of $\bar{x}$ for 10 observations. (d) The distribution of $\bar{x}$ for 25 observations.

The central limit theorem allows us to use normal probability calculations to answer questions about sample means from many observations even when the population distribution is not normal.

EXAMPLE 9.13 SERVICING AIR CONDITIONERS

The time that a technician requires to perform preventive maintenance on an air-conditioning unit is governed by the exponential distribution whose density curve appears in Figure 9.17(a). The mean time is $\mu = 1$ hour and the standard deviation is $\sigma = 1$ hour. Your company operates 70 of these units. What is the probability that their average maintenance time exceeds 50 minutes?

The central limit theorem says that the sample mean time $\bar{x}$ (in hours) spent working on 70 units has approximately the normal distribution with mean equal to the population mean $\mu = 1$ hour and standard deviation

$$\frac{\sigma}{\sqrt{70}} = \frac{1}{\sqrt{70}} = 0.120 \text{ hour}$$

The distribution of $\bar{x}$ is therefore approximately N(1, 0.120). Figure 9.18 shows this normal curve (solid) and also the actual density curve of $\bar{x}$ (dashed).

Because 50 minutes is 50/60 of an hour, or 0.833 hour, the probability we want is $P(\bar{x} > 0.83)$. Since

$$z = \frac{\bar{x} - \mu_{\bar{x}}}{\sigma_{\bar{x}}} = \frac{0.83-1}{0.120} = -1.42$$

$$P(\bar{x} > 0.83) = P(z > -1.42) = 1-0.0778 = 0.9222$$

This is the area to the right of 0.83 under the solid normal curve in Figure 9.18. The exactly correct probability is the area under the dashed density curve in the figure. It is 0.9294. The central limit theorem normal approximation is off by only about 0.007.

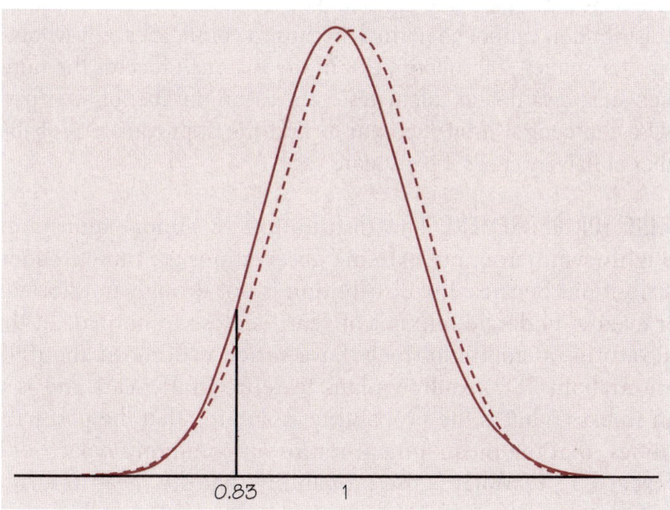

FIGURE 9.18 The exact distribution (*dashed*) and the normal approximation from the central limit theorem (*solid*) for the average time needed to maintain an air conditioner.

Figure 9.19 summarizes the facts about the sampling distribution of $\bar{x}$. It reminds us of the big idea of a sampling distribution. Keep taking random samples of size n from a population with mean μ. Find the sample mean $\bar{x}$ for each sample. Collect all the $\bar{x}$'s and display their distribution. That's the sampling distribution of $\bar{x}$. Sampling distributions are the key to understanding statistical inference. Keep this figure in mind as you go forward.

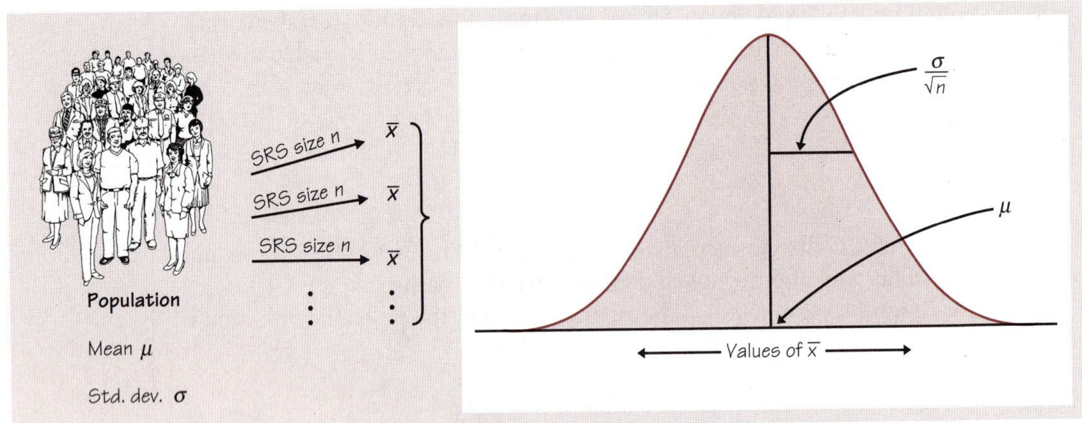

FIGURE 9.19 The sampling distribution of a sample mean $\bar{x}$ has mean μ and standard deviation $\sigma / \sqrt{n}$. The distribution is normal if the population distribution is normal; it is approximately normal for large samples in any case.

EXERCISES

9.35 BAD CARPET The number of flaws per square yard in a type of carpet material varies with mean 1.6 flaws per square yard and standard deviation 1.2 flaws per square yard. The population distribution cannot be normal, because a count takes only whole-number values. An inspector studies 200 square yards of the material, records the number of flaws found in each square yard, and calculates $\bar{x}$, the mean number of flaws per square yard inspected. Use the central limit theorem to find the approximate probability that the mean number of flaws exceeds 2 per square yard.

9.36 INVESTING FOR RETIREMENT The distribution of annual returns on common stocks is roughly symmetric, but extreme observations are more frequent than in a normal distribution. Because the distribution is not strongly nonnormal, the mean return over even a moderate number of years is close to normal. In the long run, annual real returns on common stocks have varied with mean about 9% and standard deviation about 28%. Andrew plans to retire in 45 years and is considering investing in stocks. What is the probability (assuming that the past pattern of variation continues) that the mean annual return on common stocks over the next 45 years will exceed 15%? What is the probability that the mean return will be less than 5%?

9.37 COAL MINER'S DUST A laboratory weighs filters from a coal mine to measure the amount of dust in the mine atmosphere. Repeated measurements of the weight of dust on the same filter vary normally with standard deviation $\sigma = 0.08$ milligrams (mg) because the weighing is not perfectly precise. The dust on a particular filter actually weighs 123 mg. Repeated weighings will then have the normal distribution with mean 123 mg and standard deviation 0.08 mg.

(a) The laboratory reports the mean of 3 weighings. What is the distribution of this mean?

(b) What is the probability that the laboratory reports a weight of 124 mg or higher for this filter?

9.38 MAKING AUTO PARTS An automatic grinding machine in an auto parts plant prepares axles with a target diameter $\mu = 40.125$ millimeters (mm). The machine has some variability, so the standard deviation of the diameters is $\sigma = 0.002$ mm. The machine operator inspects a sample of 4 axles each hour for quality control purposes and records the sample mean diameter.

(a) What will be the mean and standard deviation of the numbers recorded? Do your results depend on whether or not the axle diameters have a normal distribution?

(b) Can you find the probability that an SRS of 4 axles has a mean diameter greater than 40.127 mm? If so, do it. If not, explain why not.

SUMMARY

When we want information about the **population mean** $\boldsymbol{\mu}$ for some variable, we often take an SRS and use the **sample mean** $\bar{x}$ to estimate the unknown parameter μ.

The **sampling distribution of** $\bar{x}$ describes how the statistic $\bar{x}$ varies in all possible samples from the population.

The **mean** of the sampling distribution is μ, so that $\bar{x}$ is an unbiased estimator of μ.

The **standard deviation** of the sampling distribution of $\bar{x}$ is $\sigma / \sqrt{n}$ for an SRS of size n if the population has standard deviation σ. This recipe can be used if the population is at least 10 times as large as the sample.

If the population has a normal distribution, so does $\bar{x}$.

The **central limit theorem** states that for large n the sampling distribution of $\bar{x}$ is approximately normal for any population with finite standard deviation σ. The mean and standard deviation of the normal distribution are the mean μ and standard deviation $\sigma / \sqrt{n}$ of $\bar{x}$ itself.

SECTION 9.3 EXERCISES

9.39 BOTTLING COLA A bottling company uses a filling machine to fill plastic bottles with cola. The bottles are supposed to contain 300 milliliters (ml). In fact, the contents vary according to a normal distribution with mean $\mu = 298$ ml and standard deviation $\sigma = 3$ ml.

(a) What is the probability that an individual bottle contains less than 295 ml?

(b) What is the probability that the mean contents of the bottles in a six-pack is less than 295 ml?

9.40 STOP THE CAR! A company that owns and services a fleet of cars for its sales force has found that the service lifetime of disc brake pads varies from car to car according to a normal distribution with mean $\mu = 55{,}000$ miles and standard deviation $\sigma = 4500$ miles. The company installs a new brand of brake pads on 8 cars.

(a) If the new brand has the same lifetime distribution as the previous type, what is the distribution of the sample mean lifetime for the 8 cars?

(b) The average life of the pads on these 8 cars turns out to be $\overline{x} = 51{,}800$ miles. What is the probability that the sample mean lifetime is 51,800 miles or less if the lifetime distribution is unchanged? (The company takes this probability as evidence that the average lifetime of the new brand of pads is less than 55,000 miles.)

9.41 WHAT A WRECK! The number of traffic accidents per week at an intersection varies with mean 2.2 and standard deviation 1.4. The number of accidents in a week must be a whole number, so the population distribution is not normal.

(a) Let $\overline{x}$ be the mean number of accidents per week at the intersection during a year (52 weeks). What is the approximate distribution of $\overline{x}$ according to the central limit theorem?

(b) What is the approximate probability that $\overline{x}$ is less than 2?

(c) What is the approximate probability that there are fewer than 100 accidents at the intersection in a year? (*Hint:* Restate this event in terms of $\overline{x}$.)

9.42 TESTING KINDERGARTEN CHILDREN Children in kindergarten are sometimes given the Ravin Progressive Matrices Test (RPMT) to assess their readiness for learning. Experience at Southwark Elementary School suggests that the RPMT scores for its kindergarten pupils have mean 13.6 and standard deviation 3.1. The distribution is close to normal. Mr. Lavin has 22 children in his kindergarten class this year. He suspects that their RPMT scores will be unusually low because the test was interrupted by a fire drill. To check this suspicion, he wants to find the level L such that there is probability only 0.05 that the mean score of 22 children falls below L when the usual Southwark distribution remains true. What is the value of L? (*Hint:* This requires a backward normal calculation. See Chapter 2 if you need to review.)

CHAPTER REVIEW

This chapter lays the foundations for the study of statistical inference. Statistical inference uses data to draw conclusions about the population or process from which the data come. What is special about inference is that the conclusions include a statement, in the language of probability, about how reliable they are. The statement gives a probability that answers the question "What would happen if I used this method very many times?"

This chapter introduced sampling distributions of statistics. A sampling distribution describes the values a statistic would take in very many repetitions of a sample or an experiment under the same conditions. Understanding that idea is the key to understanding statistical inference. The chapter gave details about the sampling distributions of two important statistics: a sample proportion $\hat{p}$ and a sample mean $\overline{x}$. These statistics behave much the same. In particular, their sampling distributions are approximately normal if the sample is large. This is a main reason why normal distributions are so important in statistics. We can use everything we know about normal distributions to study the sampling distributions of proportions and means.

Here is a review list of the most important things you should be able to do after studying this chapter.

A. SAMPLING DISTRIBUTIONS

1. Identify parameters and statistics in a sample or experiment.

2. Recognize the fact of sampling variability: a statistic will take different values when you repeat a sample or experiment.

3. Interpret a sampling distribution as describing the values taken by a statistic in all possible repetitions of a sample or experiment under the same conditions.

4. Describe the bias and variability of a statistic in terms of the mean and spread of its sampling distribution.

5. Understand that the variability of a statistic is controlled by the size of the sample. Statistics from larger samples are less variable.

B. SAMPLE PROPORTIONS

1. Recognize when a problem involves a sample proportion $\hat{p}$.

2. Find the mean and standard deviation of the sampling distribution of a sample proportion $\hat{p}$ for an SRS of size n from a population having population proportion p.

3. Know that the standard deviation (spread) of the sampling distribution of $\hat{p}$ gets smaller at the rate $\sqrt{n}$ as the sample size n gets larger.

4. Recognize when you can use the normal approximation to the sampling distribution of $\hat{p}$. Use the normal approximation to calculate probabilities that concern $\hat{p}$.

C. SAMPLE MEANS

1. Recognize when a problem involves the mean $\bar{x}$ of a sample.

2. Find the mean and standard deviation of the sampling distribution of a sample mean $\bar{x}$ from an SRS of size n when the mean μ and standard deviation σ of the population are known.

3. Know that the standard deviation (spread) of the sampling distribution of $\bar{x}$ gets smaller at the rate $\sqrt{n}$ as the sample size n gets larger.

4. Understand that $\bar{x}$ has approximately a normal distribution when the sample is large (central limit theorem). Use this normal distribution to calculate probabilities that concern $\bar{x}$.

CHAPTER 9 REVIEW EXERCISES

9.43 REPUBLICAN VOTERS Voter registration records show that 68% of all voters in Indianapolis are registered as Republicans. To test whether the numbers dialed by a random digit dialing device really are random, you use the device to call 150 randomly chosen residential telephones in Indianapolis. Of the registered voters contacted, **73%** are registered Republicans.

(a) Is each of the boldface numbers a parameter or a statistic? Give the appropriate notation for each.

(b) What are the mean and the standard deviation of the sample proportion of registered Republicans in samples of size 150 from Indianapolis?

(c) Find the probability of obtaining an SRS of size 150 from the population of Indianapolis voters in which 73% or more are registered Republicans. How well is your random digit device working?

9.44 BAGGAGE CHECK! In Example 9.3, we performed a simulation to determine what proportion of a sample of 100 travelers would get the "green light" in Customs at Guadalajara airport. Suppose the Customs agents say that the probability that the light shows green is 0.7 on each push of the button. You observe 100 passengers at the Customs "stoplight." Only 65 get a green light. Does this give you reason to doubt the Customs officials?

(a) Use your calculator to simulate 50 groups of 100 passengers activating the Customs stoplight. Generate L_1/list1 with the command `randBin(100,0.7,50)/100`. L_1/list1 will contain 50 values of $\hat{p}$, the proportion of the 100 passengers who got a green light.

(b) Sort L_1/list1 in descending order. In how many of the 50 simulations did you obtain a value of $\hat{p}$ that is less than or equal to 0.65? Do you believe the Customs agents?

(c) Describe the shape, center, and spread of the sampling distribution of $\hat{p}$ for samples of $n = 100$ passengers.

(d) Use the sampling distribution from part (c) to find the probability of getting a sample proportion of 0.65 or less if $p = 0.7$ is actually true. How does this compare with the results of your simulation in part (b)?

(e) Repeat parts (c) and (d) for samples of size $n = 1000$ passengers.

9.45 THIS WINE STINKS! Sulfur compounds such as dimethyl sulfide (DMS) are sometimes present in wine. DMS causes "off-odors" in wine, so winemakers want to know the odor threshold, the lowest concentration of DMS that the human nose can detect. Different people have different thresholds, so we start by asking about the DMS threshold in the population of all adults. Extensive studies have found that the DMS odor threshold of adults follows roughly a normal distribution with mean $\mu = 25$ micrograms per liter and standard deviation $\sigma = 7$ micrograms per liter.

In an experiment, we present tasters with both natural wine and the same wine spiked with DMS at different concentrations to find the lowest concentration at which they identify the spiked wine. Here are the odor thresholds (measured in micrograms of DMS per liter of wine) for 10 randomly chosen subjects:

28	40	28	33	20	31	29	27	17	21

The mean threshold for these subjects is $\bar{x} = 27.4$. Find the probability of getting a sample mean even farther away from $\mu = 25$ than $\bar{x} = 27.4$.

9.46 POLLING WOMEN Suppose that 47% of all adult women think they do not get enough time for themselves. An opinion poll interviews 1025 randomly chosen women and records the sample proportion who feel they don't get enough time for themselves.

(a) Describe the sampling distribution of $\hat{p}$.

(b) The truth about the population is $p = 0.47$. In what range will the middle 95% of all sample results fall?

(c) What is the probability that the poll gets a sample in which fewer than 45% say they do not get enough time for themselves?

9.47 INSURANCE The idea of insurance is that we all face risks that are unlikely but carry high cost. Think of a fire destroying your home. So we form a group to share the risk: we all pay a small amount, and the insurance policy pays a large amount to those few of us whose homes burn down. An insurance company looks at the records for millions of homeowners and sees that the mean loss from fire in a year is $\mu = \$250$ per person. (Most of us have no loss, but a few lose their homes. The \$250 is the average loss.) The company plans to sell fire insurance for \$250 plus enough to cover its costs and profit. Explain clearly why it would be a poor practice to sell only 12 policies. Then explain why selling thousands of such policies is a safe business practice.

9.48 MORE ON INSURANCE The insurance company sees that in the entire population of homeowners, the mean loss from fire is $\mu = \$250$ and the standard deviation of the loss is $\sigma = \$300$. The distribution of losses is strongly right-skewed: many policies have \$0 loss, but a few have large losses. If the company sells 10,000 policies, what is the approximate probability that the average loss will be greater than \$260?

9.49 IQ TESTS The Wechsler Adult Intelligence Scale (WAIS) is a common "IQ test" for adults. The distribution of WAIS scores for persons over 16 years of age is approximately normal with mean 100 and standard deviation 15.

(a) What is the probability that a randomly chosen individual has a WAIS score of 105 or higher?

(b) What are the mean and standard deviation of the sampling distribution of the average WAIS score $\bar{x}$ for an SRS of 60 people?

(c) What is the probability that the average WAIS score of an SRS of 60 people is 105 or higher?

(d) Would your answers to any of (a), (b), or (c) be affected if the distribution of WAIS scores in the adult population were distinctly nonnormal?

9.50 AUTO ACCIDENTS A study of rush-hour traffic in San Francisco counts the number of people in each car entering a freeway at a suburban interchange. Suppose that this count has mean 1.5 and standard deviation 0.75 in the population of all cars that enter at this interchange during rush hours.

(a) Could the exact distribution of the count be normal? Why or why not?

(b) Traffic engineers estimate that the capacity of the interchange is 700 cars per hour. According to the central limit theorem, what is the approximate distribution of the mean number of persons $\bar{x}$ in 700 randomly selected cars at this interchange?

(c) What is the probability that 700 cars will carry more than 1075 people? (*Hint:* Restate this event in terms of the mean number of people $\bar{x}$ per car.)

9.51 POLLUTANTS IN AUTO EXHAUST The level of nitrogen oxide (NOX) in the exhaust of a particular car model varies with mean 1.4 grams per mile (g/mi) and standard deviation

0.3 g/mi. A company has 125 cars of this model in its fleet. If $\bar{x}$ is the mean NOX emission level for these cars, what is the level L such that the probability that $\bar{x}$ is greater than L is only 0.01? (*Hint*: This requires a backward normal calculation. See Chapter 2 if you need to review.)

9.52 HIGH SCHOOL DROPOUTS High school dropouts make up 14.1% of all Americans aged 18 to 24. A vocational school that wants to attract dropouts mails an advertising flyer to 25,000 persons between the ages of 18 and 24.

(a) If the mailing list can be considered a random sample of the population, what is the mean number of high school dropouts who will receive the flyer?

(b) What is the probability that at least 3500 dropouts will receive the flyer?

9.53 WEIGHT OF EGGS The weight of the eggs produced by a certain breed of hen is normally distributed with mean 65 grams (g) and standard deviation 5 g. Think of cartons of such eggs as SRSs of size 12 from the population of all eggs. What is the probability that the weight of a carton falls between 750 g and 825 g?

NOTES AND DATA SOURCES

1. In this book we discuss only the most widely used kind of statistical inference. This is sometimes called *frequentist* because it is based on answering the question "What would happen in many repetitions?" Another approach to inference, called *Bayesian*, can be used even for one-time situations. Bayesian inference is important but is conceptually complex and much less widely used in practice.
2. From the Current Population Survey Web site: www.bls.census.gov/cps.
3. Data from October 25–28, 2000, Gallup Poll Surveys from www.gallup.com.
4. "Pupils to judge Murphy's Law with toast test," *Sunday Telegraph* (UK), March 4, 2001.
5. From the Gallup Poll Web site: www.gallup.com.
6. Results from a poll taken January–April 2000 and reported at www.gallup.com.
7. Strictly speaking, the recipes we give for the standard deviations of $\bar{x}$ and $\hat{p}$ assume that an SRS of size n is drawn from an *infinite* population. If the population has finite size N, the standard deviations in the recipes are multiplied by $\sqrt{1-(n/N)}$. This "finite population correction" approaches 1 as N increases. When $n/N \leq 0.1$, it is ≥ 0.948.
8. The survey question and result are reported in Trish Hall, "Shop? Many say 'Only if I must,'" *New York Times*, November 28, 1990.
9. From John K. Ford, "A method for grading 1987 stock recommendations," *American Association of Individual Investors Journal*, March 1988, pp. 16–17.
10. This activity is suggested in Richard L. Schaeffer, Ann Watkins, Mrudulla Gnanadeskian, and Jeffrey A. Witmer, *Activity-Based Statistics*, Springer, New York, 1996.

Inference:
Conclusions with Confidence

Photo by G. Paul Bishop. Courtesy of Paul Bishop, Jr., Berkeley, CA

JERZY NEYMAN

Statistical Confidence

The most-used methods of statistical inference are confidence intervals and tests of significance. Both are products of the twentieth century. From complex and sometimes confusing origins, statistical tests took their current form in the writings of R. A. Fisher, whom we met at the beginning of Chapter 5. Confidence intervals appeared in 1934, the brainchild of *Jerzy Neyman (1894–1981)*.

Neyman was trained in Poland and, like Fisher, worked at an agricultural research institute. He moved to London in 1934 and in 1938 joined the University of California at Berkeley. He founded Berkeley's Statistical Laboratory and remained its head even after his official retirement as a professor in 1961. Age did not slow Neyman's work—he remained active until the end of his long life and almost doubled his list of publications after "retiring." Statistical problems arising from astronomy, biology, and attempts to modify the weather attracted his attention.

Neyman ranks with Fisher as a founder of modern statistical practice. In addition to introducing confidence intervals, he helped systematize the theory of sample surveys and reworked significance tests from a new point of view. Fisher, who was very argumentative, disliked Neyman's approach to tests and said so. Neyman, who wasn't shy, replied vigorously.

Age did not slow down Neyman's work—he remained active until the end of his long life and almost doubled his list of publications after "retiring."

Tests and confidence intervals are our topic in this chapter. Like most users of statistics, we will stay close to Fisher's approach to tests. You can find some of Neyman's ideas in the final section of this chapter.

chapter 10

Introduction to Inference

ACTIVITY 10 A Little Tacky!

Materials: Small box of thumbtacks

When you flip a fair coin, it is equally likely to land "heads" or "tails." Do thumbtacks behave in the same way? In this activity, you will toss a thumbtack several times and observe whether it comes to rest with the point up (U) or point down (D). The question you are trying to answer is: what proportion of the time does a tossed thumbtack settle with its point up (U)?

1. Before you begin the activity, make a guess about what will happen. If you could toss your thumbtack over and over and over, what proportion of all tosses do you think would settle with the point up (U)?

2. Toss your thumbtack 50 times. Record the result of each toss (U or D) in a table like the one shown. In the third column, calculate the proportion of point up (U) tosses you have obtained so far.

Toss	Outcome	Cumulative proportion of U's
1	U	1/1 = 1.00
2	D	1/2 = 0.50
3	D	1/3 = 0.33
. . .	. . .	. . .

3. Make a scatterplot with the number of tosses on the horizontal axis and the cumulative proportion of U's on the vertical axis. Connect consecutive points with a line segment. Does the overall proportion of U's seem to be approaching a single value?

4. Your set of 50 tosses can be thought of as a simple random sample from the population of all possible tosses of your thumbtack. The parameter p is the (unknown) population proportion of tosses that would land point up (U). What is your best estimate for p? It's $\hat{p}$, the proportion of U's in your 50 thumbtack tosses. Record your value of $\hat{p}$. How does it compare with the conjecture you made in step 1?

5. If you tossed your thumbtack 50 more times (don't do it!), would you expect to get the same value of $\hat{p}$? In chapter 9, we learned that the values of $\hat{p}$ in repeated samples could be described by a sampling distribution. The mean of the sampling distribution $\mu_{\hat{p}}$ is equal to the population proportion p. How far will your sample proportion $\hat{p}$ be from the true value p? If the sampling distribution is approximately normal, then the 68–95–99.7 rule tells us that about 95% of all $\hat{p}$-values will be within two standard deviations of p.

ACTIVITY 10 A Little Tacky! (*continued*)

6. The sampling distribution of $\hat{p}$ will be approximately normal if $n\hat{p} \geq 10$ and $n(1 - \hat{p}) \geq 10$. Verify that these conditions are satisfied for your sample.

7. Estimate the standard deviation of the sampling distribution by computing $\sqrt{\frac{\hat{p}(1-\hat{p})}{n}}$ using your value of $\hat{p}$. This is the formula we developed in Chapter 9 with p replaced by $\hat{p}$.

8. Construct the interval $\hat{p} \pm 2\sqrt{\frac{\hat{p}(1-\hat{p})}{n}}$ based on your sample of 50 tosses.

This is called a *confidence interval* for $\hat{p}$.

9. Your teacher will draw a number line with a scale marked off from 0 to 1 that has tick marks every 0.05 units. Draw your *confidence interval* above the number line. Your classmates will do the same. Do most of the intervals overlap? If so, what values are contained in all of the overlapping intervals?

10. About 95% of the time, the sample proportion $\hat{p}$ of point up (U) tosses will be within two standard deviations of the actual population proportion of point up(U) tosses of a thumbtack. But if $\hat{p}$ is within two standard deviations of p, then p is within two standard deviations of $\hat{p}$. So about 95% of the time, the interval $\hat{p} \pm 2\sqrt{\frac{\hat{p}(1-\hat{p})}{n}}$ will contain the true proportion p.

11. There is no way to know whether the *confidence interval* you constructed in step 8 actually "catches" the true proportion p of times that your thumbtack will land point up. What we can say is that the method you used in step 8 will succeed in capturing the unknown population parameter about 95% of the time. Likewise, we would expect about 95% of all the confidence intervals drawn by the members of your class in step 9 to capture p.

This activity shows you how sample statistics can be used to estimate unknown population parameters. This is one of the two types of statistical inference that you will meet in this chapter.

INTRODUCTION

When we select a sample, we know the responses of the individuals in the sample. Often we are not content with information about the sample. We want to *infer* from the sample data some conclusion about a wider population that the sample represents.

STATISTICAL INFERENCE

Statistical inference provides methods for drawing conclusions about a population from sample data.

We have, of course, been drawing conclusions from data all along. What is new in formal inference is that we use probability to express the strength of our conclusions. Probability allows us to take chance variation into account and so to correct our judgment by calculation. Here are two examples of how probability can correct our judgment.

EXAMPLE 10.1 DRAFT LOTTERIES AND DRUG STUDIES

In the Vietnam War years, a lottery determined the order in which men were drafted for army service. The lottery assigned draft numbers by choosing birth dates in random order. We expect a correlation near zero between birth dates and draft numbers if the draft numbers come from random choice. The actual correlation between birth date and draft number in the first draft lottery was $r = -0.226$. That is, men born later in the year tended to get lower draft numbers. Is this small correlation evidence that the lottery was biased? Our unaided judgment can't tell because any two variables will have some association in practice, just by chance. So we calculate that a correlation this far from zero has probability less than 0.001 in a truly random lottery. Because a correlation as strong as that observed would almost never occur in a random lottery, there is strong evidence that the lottery was unfair.

Probability calculations can also protect us from jumping to a conclusion when only chance variation is at work. Give a new drug and a placebo to 20 patients each; 12 of those taking the drug show improvement, but only 8 of the placebo patients improve. Is the drug more effective than the placebo? Perhaps, but a difference this large or larger between the results in the two groups would occur about one time in five simply because of chance variation. An effect that could so easily be just chance is not convincing.

In this chapter, we will meet the two most common types of formal statistical inference. Section 10.1 concerns *confidence intervals* for estimating the value of a population parameter. Section 10.2 presents *tests of significance*, which assess the evidence for a claim about a population. Both types of inference are based on the sampling distributions of statistics. That is, both report probabilities that state *what would happen if we used the inference method many times.* This kind of probability statement is characteristic of statistical inference. Users of statistics must understand the meaning of the probability statements that appear, for example, on computer output for statistical procedures.

The methods of formal inference require the long-run regular behavior that probability describes. Inference is most reliable when the data are produced by a properly randomized design. *When you use statistical inference you are acting as if the data are a random sample or come from a randomized experiment.* If this is not true, your conclusions may be open to challenge. Do not be overly impressed by the complex details of formal inference. This elaborate machinery cannot remedy basic flaws in producing the data, such as voluntary response samples and uncontrolled experiments. Use the common sense developed in your study of the first nine chapters of this book, and proceed to formal inference only when you are satisfied that the data deserve such analysis.

The purpose of this chapter is to describe the reasoning used in statistical inference. We will illustrate the reasoning by a few specific inference techniques, but these are oversimplified so that they are not very useful in practice. Later chapters will first show how to modify these techniques to make them practically useful and will then introduce inference methods for use in most of the settings we met in learning to explore data. There are libraries–both of books and of computer software–full of more elaborate statistical techniques. Informed use of any of these methods requires an understanding of the underlying reasoning. *A computer will do the arithmetic, but you must still exercise judgment based on understanding.*

10.1 ESTIMATING WITH CONFIDENCE

What decides whether you will gain admission to the college or university of your choice? By taking challenging courses (AP courses, for example) and earning high grades in those courses, you certainly improve your chances. Many schools also look closely at your performance on standardized tests, such as the SAT I: Reasoning Test (more commonly referred to as the SAT). This test has two parts, one for verbal reasoning ability and one for mathematical reasoning ability. In 2000, 1,260,278 college-bound seniors took the SAT. Their mean SAT Math score was 514 with a standard deviation of 113. For the SAT Verbal, the mean was 505 with a standard deviation of 111.

In early 2000, University of California President Richard Atkinson stirred considerable controversy when he suggested that the University of California system drop the SAT I: Reasoning Test as a factor in college admissions decisions. He suggested replacing this test with tests that reflect course content better.

EXAMPLE 10.2 SAT MATH SCORES IN CALIFORNIA

Suppose you want to estimate the mean SAT Math score for the more than 350,000 high school seniors in California. Only about 49% of California students take the SAT. These self-selected seniors are planning to attend college and so are not representative of all California seniors. You know better than to make inferences about the population based on any sample data. At considerable effort and expense, you give the test to a simple random sample (SRS) of 500 California high school seniors. The mean for your sample is $\overline{x} = 461$. What can you say about the mean score μ in the population of all 350,000 seniors?

The law of large numbers tells us that the sample mean $\overline{x}$ from a large SRS will be close to the unknown population mean μ. Because $\overline{x} = 461$, we guess that μ is "somewhere around 461." To make "somewhere around 461" more precise, we ask: *How would the sample mean $\overline{x}$ vary if we took many samples of 500 seniors from this same population?*

Recall the essential facts about the sampling distribution of $\overline{x}$:

- The central limit theorem tells us that the mean $\bar{x}$ of 500 scores has a distribution that is close to normal.
- The mean of this normal sampling distribution is the same as the unknown mean μ of the entire population.
- The standard deviation of $\bar{x}$ for an SRS of 500 students is $\sigma/\sqrt{500}$, where σ is the standard deviation of individual SAT Math scores among all California high school seniors.

Let us suppose that we know that the standard deviation of SAT Math scores in the population of all California seniors is $\sigma = 100$. The standard deviation of $\bar{x}$ is then

$$\frac{\sigma}{\sqrt{n}} = \frac{100}{\sqrt{500}} = 4.5$$

(It is usually not realistic to assume we know σ. We will see in the next chapter how to proceed when σ is not known. For now, we are more interested in statistical reasoning than in details of realistic methods.)

If we choose many samples of size 500 and find the mean SAT Math score for each sample, we might get mean $\bar{x} = 461$ from the first sample, $\bar{x} = 455$ from the second, $\bar{x} = 463$ from the third sample, and so on. If we collect all these sample means and display their distribution, we get the normal distribution with mean equal to the unknown μ and standard deviation 4.5. Inference about the unknown μ starts from this sampling distribution. Figure 10.1 displays the distribution. The different values of $\bar{x}$ appear along the axis in the figure, and the normal curve shows how probable these values are.

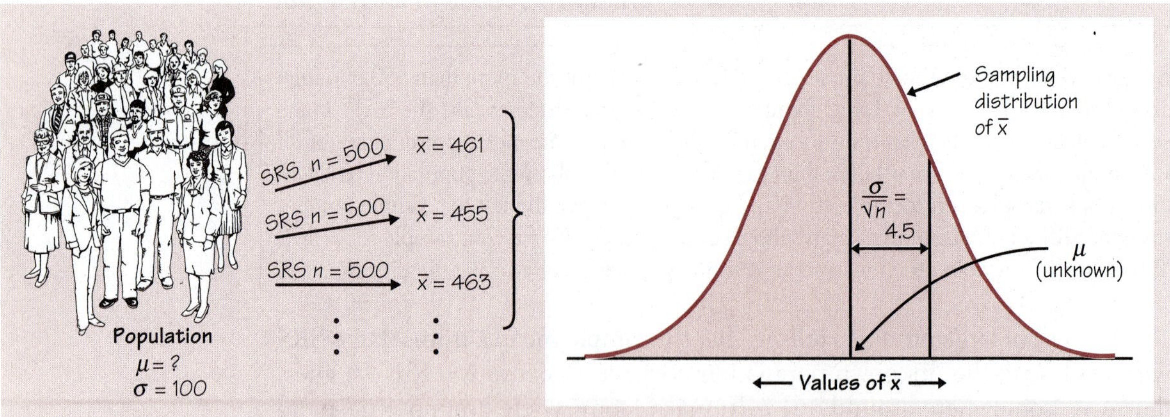

FIGURE 10.1 The sampling distribution of the mean score $\bar{x}$ of an SRS of 500 California seniors on the SAT Math quantitative test.

Statistical confidence

Figure 10.2 is another picture of the same sampling distribution. It illustrates the following line of thought:

- The 68–95–99.7 rule says that in 95% of all samples, the mean score $\bar{x}$ for the sample will be within two standard deviations of the population mean score μ. So the mean $\bar{x}$ of 500 SAT Math scores will be within 9 points of μ in 95% of all samples.
- Whenever $\bar{x}$ is within 9 points of the unknown μ, μ is within 9 points of the observed $\bar{x}$. This happens in 95% of all samples.
- So in 95% of all samples, the unknown μ lies between $\bar{x} - 9$ and $\bar{x} + 9$. Figure 10.3 displays this fact in picture form.

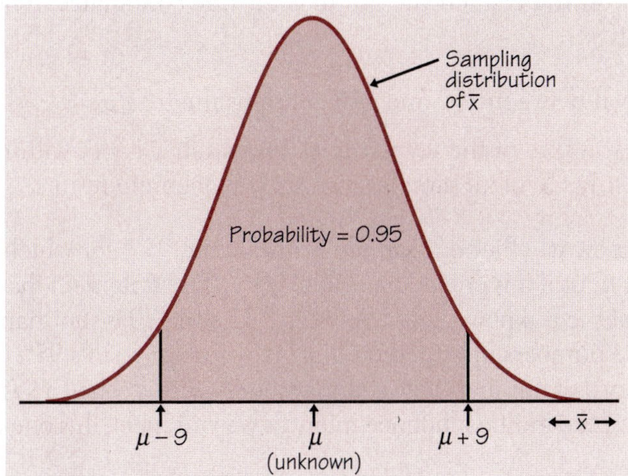

FIGURE 10.2 In 95% of all samples, $\bar{x}$ lies within ±9 of the unknown population mean μ. So μ also lies within ±9 of $\bar{x}$ in those samples.

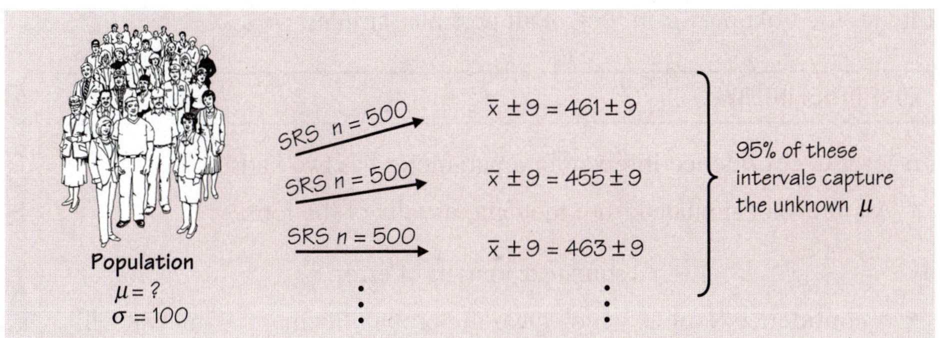

FIGURE 10.3 To say that $\bar{x} \pm 9$ is a 95% confidence interval for the population mean μ is to say that in repeated samples, 95% of these intervals capture μ.

This conclusion just restates a fact about the sampling distribution of $\bar{x}$. The language of statistical inference uses this fact about what would happen in the long run to express our confidence in the results of any one sample.

EXAMPLE 10.3 95% CONFIDENCE

Our sample of 500 California seniors gave $\bar{x} = 461$. We say that we are 95% *confident* that the unknown mean SAT Math score for all California high school seniors lies between

$$\bar{x} - 9 = 461 - 9 = 452$$

and

$$\bar{x} + 9 = 461 + 9 = 470$$

Be sure you understand the grounds for our confidence. There are only two possibilities:

1. The interval between 452 and 470 contains the true μ.
2. Our SRS was one of the few samples for which $\bar{x}$ is not within 9 points of the true μ. Only 5% of all samples give such inaccurate results.

We cannot know whether our sample is one of the 95% for which the interval $\bar{x} \pm 9$ catches μ, or if it is one of the unlucky 5%. The statement that we are 95% confident that the unknown μ lies between 452 and 470 is shorthand for saying, "We got these numbers by a method that gives correct results 95% of the time."

The interval of numbers between the values $\bar{x} \pm 9$ is called a 95% *confidence interval* for μ. Like most confidence intervals we will meet, this one has the form

$$\text{estimate} \pm \text{margin of error}$$

margin of error

The estimate ($\bar{x}$ in this case) is our guess for the value of the unknown parameter. The ***margin of error*** ± 9 shows how accurate we believe our guess is, based on the variability of the estimate. This is a 95% confidence interval because it catches the unknown μ in 95% of all possible samples.

CONFIDENCE INTERVAL

A **level C confidence interval** for a parameter has two parts:

- An interval calculated from the data, usually of the form

$$\text{estimate} \pm \text{margin of error}$$

- A **confidence level** C, which gives the probability that the interval will capture the true parameter value in repeated samples.

Users can choose the confidence level, most often 90% or higher because we most often want to be quite sure of our conclusions. We will use C to stand for the confidence level in decimal form. For example, a 95% confidence level corresponds to $C = 0.95$.

Figure 10.3 is one way to picture the idea of a 95% confidence interval. Figure 10.4 illustrates the idea in a different form. Study these figures carefully. If you understand what they say, you have mastered one of the big ideas of statistics. Figure 10.4 shows the result of drawing many SRSs from the same population and calculating a 95% confidence interval from each sample. The center of each interval is at $\bar{x}$ and therefore varies from sample to sample. The sampling distribution of $\bar{x}$ appears at the top of the figure to show the long-term pattern of this variation. The 95% confidence intervals from 25 SRSs appear below. The center $\bar{x}$ of each interval is marked by a dot. The arrows on either side of the dot span the confidence interval. All except one of these 25 intervals cover the true value of μ. In a very large number of samples, 95% of the confidence intervals would contain μ.

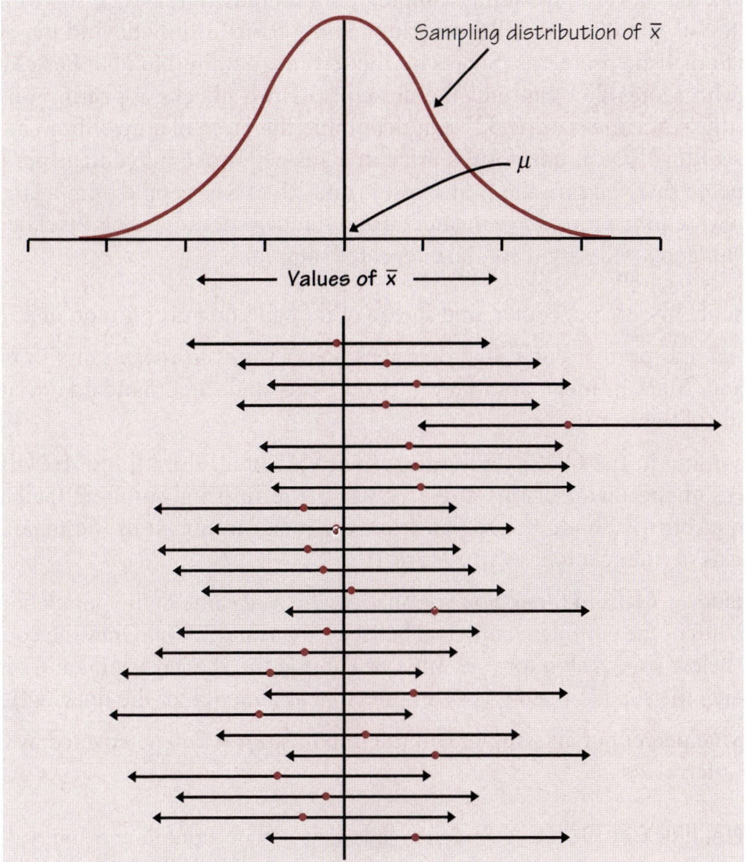

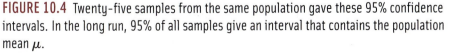

FIGURE 10.4 Twenty-five samples from the same population gave these 95% confidence intervals. In the long run, 95% of all samples give an interval that contains the population mean μ.

EXERCISES

10.1 POLLING WOMEN A *New York Times* poll on women's issues interviewed 1025 women randomly selected from the United States, excluding Alaska and Hawaii. The poll found that 47% of the women said they do not get enough time for themselves.

(a) The poll announced a margin of error of ±3 percentage points for 95% confidence in its conclusions. What is the 95% confidence interval for the percent of all adult women who think they do not get enough time for themselves?

(b) Explain to someone who knows no statistics why we can't just say that 47% of all adult women do not get enough time for themselves.

(c) Then explain clearly what "95% confidence" means.

10.2 NAEP SCORES Young people have a better chance of full-time employment and good wages if they are good with numbers. How strong are the quantitative skills of young Americans of working age? One source of data is the National Assessment of Educational Progress (NAEP) Young Adult Literacy Assessment Survey, which is based on a nationwide probability sample of households. The NAEP survey includes a short test of quantitative skills, covering mainly basic arithmetic and the ability to apply it to realistic problems. Scores on the test range from 0 to 500. For example, a person who scores 233 can add the amounts of two checks appearing on a bank deposit slip; someone scoring 325 can determine the price of a meal from a menu; a person scoring 375 can transform a price in cents per ounce into dollars per pound.[1]

Suppose that you give the NAEP test to an SRS of 840 people from a large population in which the scores have mean 280 and standard deviation $\sigma = 60$. The mean $\bar{x}$ of the 840 scores will vary if you take repeated samples.

(a) Describe the shape, center, and spread of the sampling distribution of $\bar{x}$.

(b) Sketch the normal curve that describes how $\bar{x}$ varies in many samples from this population. Mark its mean and the values one, two, and three standard deviations on either side of the mean.

(c) According to the 68–95–99.7 rule, about 95% of all the values of $\bar{x}$ fall within ________ of the mean of this curve. What is the missing number? Call it m for "margin of error." Shade the region from the mean minus m to the mean plus m on the axis of your sketch, as in Figure 10.2.

(d) Whenever $\bar{x}$ falls in the region you shaded, the true value of the population mean, $\mu = 280$, lies in the confidence interval between $\bar{x} - m$ and $\bar{x} + m$. Draw the confidence interval below your sketch for one value of $\bar{x}$ inside the shaded region and one value of $\bar{x}$ outside the shaded region. (Use Figure 10.4 as a model for the drawing.)

(e) In what percent of all samples will the true mean $\mu = 280$ be covered by the confidence interval $\bar{x} \pm m$?

10.3 EXPLAINING CONFIDENCE A student reads that a 95% confidence interval for the mean NAEP quantitative score for men of ages 21 to 25 is 267.8 to 276.2. Asked to explain the meaning of this interval, the student says, "95% of all young men have scores between 267.8 and 276.2." Is the student right? Justify your answer.

10.4 AUTO EMISSIONS Oxides of nitrogen (called NOX for short) emitted by cars and trucks are important contributors to air pollution. The amount of NOX emitted by a particular model varies from vehicle to vehicle. For one light truck model, NOX emissions vary with mean μ that is unknown and standard deviation $\sigma = 0.4$ grams per mile. You test an SRS of 50 of these trucks. The sample mean NOX level $\bar{x}$ estimates the unknown μ. You will get different values of $\bar{x}$ if you repeat your sampling.

(a) Describe the shape, center, and spread of the sampling distribution of $\bar{x}$.

(b) Sketch the normal curve for the sampling distribution of $\bar{x}$. Mark its mean and the values one, two, and three standard deviations on either side of the mean.

(c) According to the 68–95–99.7 rule, about 95% of all values of $\bar{x}$ lie within a distance m of the mean of the sampling distribution. What is m? Shade the region on the axis of your sketch that is within m of the mean, as in Figure 10.2.

(d) Whenever $\bar{x}$ falls in the region you shaded, the unknown population mean μ lies in the confidence interval $\bar{x} \pm m$. For what percent of all possible samples does this happen?

(e) Following the style of Figure 10.4, draw the confidence intervals below your sketch for two values of $\bar{x}$, one that falls within the shaded region and one that falls outside it.

Confidence interval for a population mean

We can now give the recipe for a level C confidence interval for the mean μ of a population when the data come from an SRS of size n. The construction of the interval depends on the fact that the sampling distribution of the sample mean $\bar{x}$ is at least approximately normal. This distribution is exactly normal if the population distribution is normal. When the population distribution is not normal, the central limit theorem tells us that the sampling distribution of $\bar{x}$ will be approximately normal if n is sufficiently large. Be sure to check that these conditions are satisfied before you construct a confidence interval.

CONDITIONS FOR CONSTRUCTING A CONFIDENCE INTERVAL FOR μ

The construction of a confidence interval for a population mean μ is appropriate when

- the data come from an SRS from the population of interest, and
- the sampling distribution of $\bar{x}$ is approximately normal.

Our construction of a 95% confidence interval for the mean SAT Math score began by noting that any normal distribution has probability about 0.95 within 2 standard deviations of its mean. To construct a level C confidence interval, we want to catch the central probability C under a normal curve. To do that, we must go out z^* standard deviations on either side of the mean. Since any normal distribution can be standardized, we can get the value z^* from the standard normal table. Here is an example of how to find z^*.

EXAMPLE 10.4 FINDING z^*

To find an 80% confidence interval, we must catch the central 80% of the normal sampling distribution of $\bar{x}$. In catching the central 80% we leave out 20%, or 10% in each tail. So z^* is the point with area 0.1 to its right (and 0.9 to its left) under the standard normal curve. Search the body of Table A to find the point with area 0.9 to its left. The closest entry is $z^* = 1.28$. There is area 0.8 under the standard normal curve between –1.28 and 1.28. Figure 10.5 shows how z^* is related to areas under the curve.

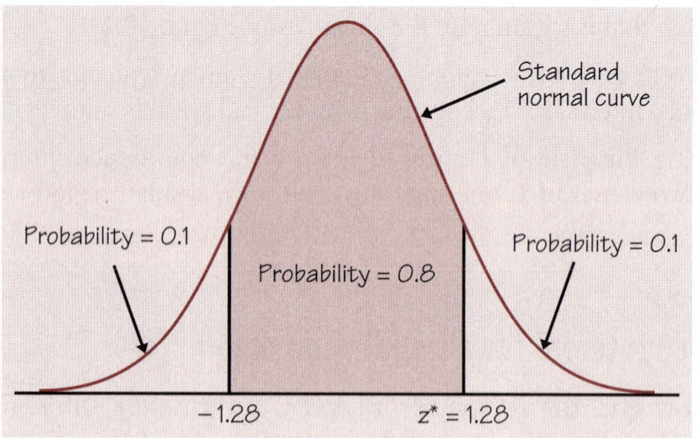

FIGURE 10.5 The central probability 0.8 under a standard normal curve lies between –1.28 and 1.28. That is, there is area 0.1 to the right of 1.28 under the curve.

Figure 10.6 shows the general situation for any confidence level C. If we catch the central area C, the leftover tail area is 1 – C, or (1 – C)/2 on each side. You can find z^* for any C by searching Table A. Here are the results for the most common confidence levels:

Confidence level	Tail Area	z^*
90%	0.05	1.645
95%	0.025	1.960
99%	0.005	2.576

Notice that for 95% confidence we use $z^* = 1.960$. This is more exact than the approximate value $z^* = 2$ given by the 68–95–99.7 rule. The bottom row in Table C gives the values of z^* for many confidence levels C. This row is labeled z^*. (You can find Table C inside the rear cover. We will use the other rows of the table in the next chapter.) Values z^* that mark off a specified area under the standard normal curve are often called *critical values* of the distribution.

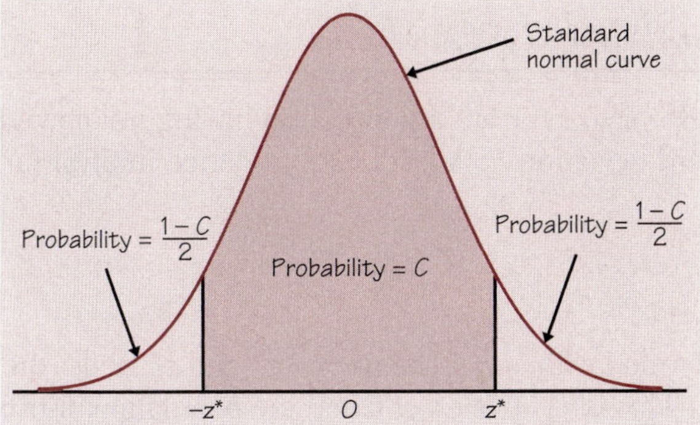

FIGURE 10.6 In general, the central probability C under a standard normal curve lies between $-z^*$ and z^*. Because z^* has area $(1-C)/2$ to its right under the curve, we call it the upper $(1-C)/2$ critical value.

CRITICAL VALUES

The number z^* with probability p lying to its right under the standard normal curve is called the **upper p critical value** of the standard normal distribution.

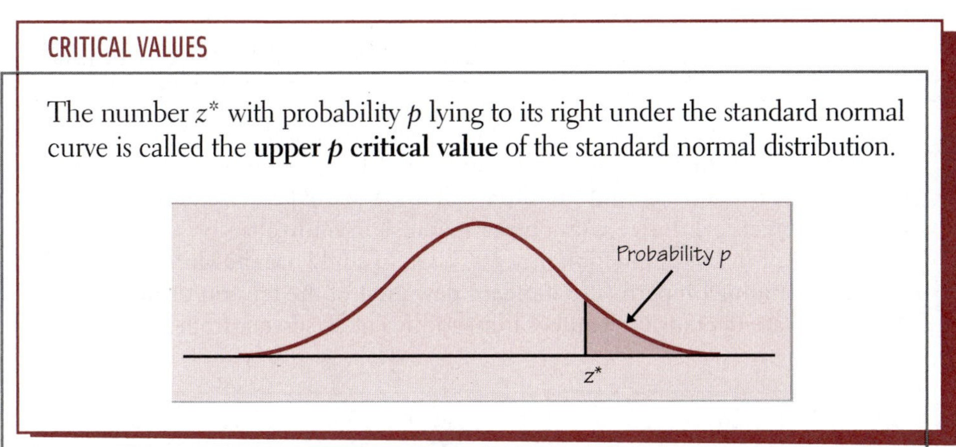

Here's the thinking that leads to the level C confidence interval:

- Any normal curve has probability C between the point z^* standard deviations below its mean and the point z^* standard deviations above its mean.
- The standard deviation of the sampling distribution of $\bar{x}$ is $\sigma/\sqrt{n}$, and its mean is the population mean μ. So there is probability C that the observed sample mean $\bar{x}$ takes a value between

$$\mu - z^* \frac{\sigma}{\sqrt{n}} \quad \text{and} \quad \mu + z^* \frac{\sigma}{\sqrt{n}}$$

- Whenever this happens, the population mean μ is contained between

$$\bar{x} - z^* \frac{\sigma}{\sqrt{n}} \quad \text{and} \quad \bar{x} + z^* \frac{\sigma}{\sqrt{n}}$$

That is our confidence interval. The estimate of the unknown μ is $\bar{x}$, and the margin of error is $z^*\sigma/\sqrt{n}$.

CONFIDENCE INTERVAL FOR A POPULATION MEAN

Choose an SRS of size n from a population having unknown mean μ and known standard deviation σ. A level C confidence interval for μ is

$$\bar{x} \pm z^* \frac{\sigma}{\sqrt{n}}$$

Here z^* is the value with area C between $-z^*$ and z^* under the standard normal curve. This interval is exact when the population distribution is normal and is approximately correct for large n in other cases.

EXAMPLE 10.5 VIDEO SCREEN TENSION

A manufacturer of high-resolution video terminals must control the tension on the mesh of fine wires that lies behind the surface of the viewing screen. Too much tension will tear the mesh and too little will allow wrinkles. The tension is measured by an electrical device with output readings in millivolts (mV). Some variation is inherent in the production process. Careful study has shown that when the process is operating properly, the standard deviation of the tension readings is σ = 43 mV. Here are the tension readings from an SRS of 20 screens from a single day's production.

269.5	297.0	269.6	283.3	304.8	280.4	233.5	257.4	317.5	327.4
264.7	307.7	310.0	343.3	328.1	342.6	338.8	340.1	374.6	336.1

Construct a 90% confidence interval for the mean tension μ of all the screens produced on this day.

Step 1: Identify the population of interest and the parameter you want to draw conclusions about. The population of interest is all of the video terminals produced on the day in question. We want to estimate μ, the mean tension for all of these screens.

Step 2: Choose the appropriate inference procedure. Verify the conditions for using the selected procedure. Since we know σ, we should use the confidence interval for a population mean that was just introduced to estimate μ. Now we must check that the two required conditions—(1) SRS from the population of interest and (2) sampling distribution of $\bar{x}$ approximately normal—are met.

- The data come from an SRS of 20 screens from the population of all screens produced that day.

- Is the sampling distribution of $\bar{x}$ approximately normal? Past experience suggests that the tension readings of screens produced on a single day follow a normal distribution quite closely. If the population distribution is normal, the sampling distribution of $\bar{x}$ will be normal. Let's examine the sample data.

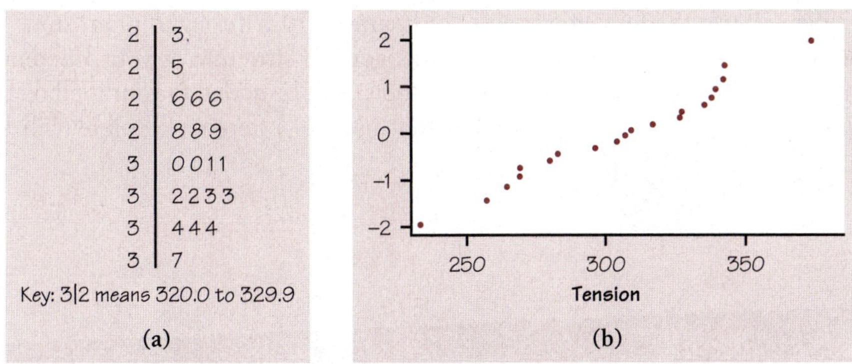

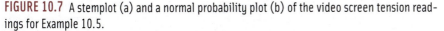

FIGURE 10.7 A stemplot (a) and a normal probability plot (b) of the video screen tension readings for Example 10.5.

A stemplot of the tension readings (Figure 10.7(a)) shows no outliers or strong skewness. The norman probability plot in Figure 10.7(b) tells us that the sample data are approximately normally distributed. These data give us no reason to doubt the normality of the population from which they came.

Step 3: If the conditions are met, carry out the inference procedure. You can check that the mean tension reading for the 20 screens in our sample is $\bar{x} = 306.3$ mV. The confidence interval formula is $\bar{x} \pm z^* \sigma / \sqrt{n}$. For a 90% confidence level, the critical value is $z^* = 1.645$. So the 90% confidence interval for μ is

$$\bar{x} \pm z^* \frac{\sigma}{\sqrt{n}} = 306.3 \pm 1.645 \frac{43}{\sqrt{20}} = 306.3 \pm 15.8 = (290.5,\ 322.1)$$

Step 4: Interpret your results in the context of the problem. We are 90% confident that the true mean tension in the entire batch of video terminals produced that day is between 290.5 and 322.1 mV.

Suppose that a single computer screen had a tension reading of 306.3 mV, the same value as the sample mean in Example 10.5. Repeating the calculation with $n = 1$ shows that the 90% confidence interval based on a single measurement is

$$\bar{x} \pm z^* \frac{\sigma}{\sqrt{n}} = 306.3 \pm 1.645 \frac{43}{\sqrt{1}} = 306.3 \pm 70.7 = (235.6,\ 377.0)$$

The mean of twenty measurements gives a smaller margin of error and therefore a shorter interval than a single measurement. Figure 10.8 illustrates the gain from using 20 observations.

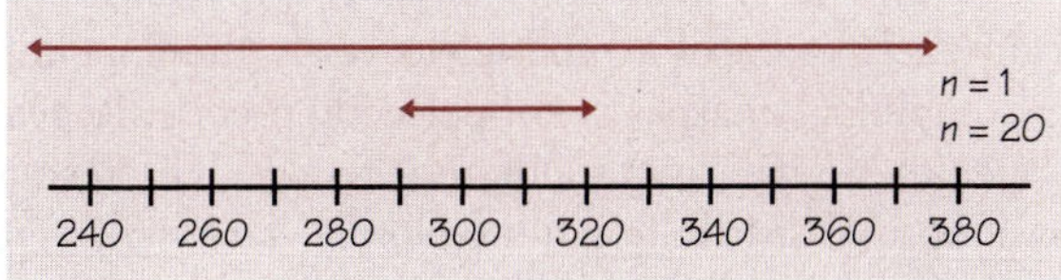

FIGURE 10.8 Confidence intervals for $n = 20$ and $n = 1$ for Example 10.5. Larger samples give shorter intervals.

We will use the four-step process of Example 10.5 throughout our study of statistical inference. You can think of this general structure as your Inference Toolbox. Specific inference procedures (tools) will be added to your toolbox for use in a variety of settings. Examples that use the Inference Toolbox will be marked with a [toolbox icon].

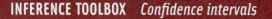

INFERENCE TOOLBOX ***Confidence intervals***

To construct a confidence interval:

Step 1: Identify the population of interest and the parameter you want to draw conclusions about.

Step 2: Choose the appropriate inference procedure. Verify the conditions for using the selected procedure.

Step 3: If the conditions are met, carry out the inference procedure.

$$\text{CI} = \text{estimate} \pm \text{margin of error}$$

Step 4: Interpret your results in the context of the problem.

The form of confidence intervals for the population mean μ rests on the fact that the statistic $\bar{x}$ used to estimate μ has a normal distribution. Because many sample statistics have normal distributions (at least approximately), it is useful to notice that the confidence interval has the form

$$\text{estimate} \pm z^*\sigma_{\text{estimate}}$$

The estimate based on the sample is the center of the confidence interval. The margin of error is $z^*\sigma_{\text{estimate}}$. The desired confidence level determines z^* from Table A. The standard deviation of the estimate, σ_{estimate}, depends on the particular estimate we use. When the estimate is $\bar{x}$ from an SRS, the standard deviation of the estimate is $\sigma / \sqrt{n}$.

EXERCISES

10.5 ANALYZING PHARMACEUTICALS A manufacturer of pharmaceutical products analyzes a specimen from each batch of a product to verify the concentration of the active ingredient. The chemical analysis is not perfectly precise. Repeated measurements on the same specimen give slightly different results. The results of repeated measurements follow a normal distribution quite closely. The analysis procedure has no bias, so the mean μ of the population of all measurements is the true concentration in the specimen. The standard deviation of this distribution is known to be $\sigma = 0.0068$ grams per liter. The laboratory analyzes each specimen three times and reports the mean result.

Three analyses of one specimen give concentrations

0.8403 0.8363 0.8447

Construct a 99% confidence interval for the true concentration μ. Use the Inference Toolbox as a guide.

10.6 SURVEYING HOTEL MANAGERS A study of the career paths of hotel general managers sent questionnaires to an SRS of 160 hotels belonging to major U.S. hotel chains. There were 114 responses. The average time these 114 general managers had spent with their current company was 11.78 years. Give a 99% confidence interval for the mean number of years general managers of major-chain hotels have spent with their current company. (Take it as known that the standard deviation of time with the company for all general managers is 3.2 years.) Use the Inference Toolbox as a guide.

10.7 ENGINE CRANKSHAFTS Here are measurements (in millimeters) of a critical dimension on a sample of auto engine crankshafts:

224.120	224.001	224.017	223.982	223.989	223.961	223.960	224.089	223.987
223.976	223.902	223.980	224.098	224.057	223.913	223.999		

The data come from a production process that is known to have standard deviation σ = 0.060 mm. The process mean is supposed to be μ = 224 mm but can drift away from this target during production.

(a) We expect the distribution of the dimension to be close to normal. Make a plot of these data and describe the shape of the distribution.

(b) Give a 95% confidence interval for the process mean at the time these crankshafts were produced.

How confidence intervals behave

The confidence interval $\bar{x} \pm z^* \sigma / \sqrt{n}$ for the mean of a normal population illustrates several important properties that are shared by all confidence intervals in common use. The user chooses the confidence level, and the margin of error follows from this choice. We would like high confidence and also a small margin of error. High confidence says that our method almost always gives correct answers. A small margin of error says that we have pinned down the parameter quite precisely. The margin of error is

$$\text{margin of error} = z^* \frac{\sigma}{\sqrt{n}}$$

This expression has z^* and σ in the numerator and $\sqrt{n}$ in the denominator. So the margin of error gets smaller when

- z^* gets smaller. Smaller z^* is the same as smaller confidence level C (look at Figure 10.6 again). There is a trade-off between the confidence level and the margin of error. To obtain a smaller margin of error from the same data, you must be willing to accept lower confidence.

- σ gets smaller. The standard deviation σ measures the variation in the population. You can think of the variation among individuals in the population as noise that obscures the average value μ. It is easier to pin down μ when σ is small.
- n gets larger. Increasing the sample size n reduces the margin of error for any fixed confidence level. Because n appears under a square root sign, we must take four times as many observations in order to cut the margin of error in half.

EXAMPLE 10.6 CHANGING THE CONFIDENCE LEVEL

Suppose that the manufacturer in Example 10.5 wants 99% confidence rather than 90%. Table C gives the critical value for 99% confidence as $z^* = 2.575$. The 99% confidence interval for μ based on an SRS of 20 video monitors with mean $\bar{x} = 306.3$ is

$$\bar{x} \pm z^* \frac{\sigma}{\sqrt{n}} = 306.3 \pm 2.575 \frac{43}{\sqrt{20}} = 306.3 \pm 24.8 = (281.5,\ 331.1$$

Demanding 99% confidence instead of 90% confidence has increased the margin of error from 15.8 to 24.8. Figure 10.9 compares these two measurements.

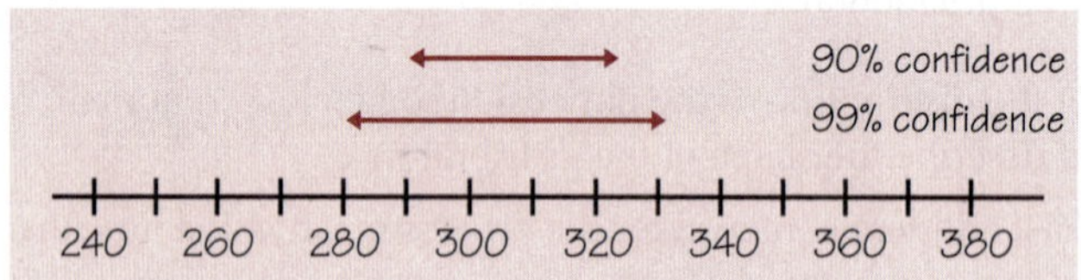

FIGURE 10.9 90% and 99% confidence intervals for Example 10.6. Higher confidence requires a longer interval.

EXERCISES

10.8 CORN YIELD Crop researchers plant 15 plots with a new variety of corn. The yields in bushels per acre are

138.0	139.1	113.0	132.5	140.7	109.7	118.9	134.8	109.6
127.3	115.6	130.4	130.2	111.7	105.5			

Assume that $\sigma = 10$ bushels per acre.

(a) Find the 90% confidence interval for the mean yield μ for this variety of corn. Use your Inference Toolbox.

(b) Find the 95% confidence interval.

(c) Find the 99% confidence interval.

(d) How do the margins of error in (a), (b), and (c) change as the confidence level increases?

10.9 MORE CORN Suppose that the crop researchers in Exercise 10.8 obtained the same value of $\bar{x}$ from a sample of 60 plots rather than 15.

(a) Compute the 95% confidence interval for the mean yield μ.

(b) Is the margin of error larger or smaller than the margin of error found for the sample of 15 plots in Exercise 10.8? Explain in plain language why the change occurs.

(c) Will the 90% and 99% intervals for a sample of size 60 be wider or narrower than those for $n = 15$? (You need not actually calculate these intervals.)

10.10 CONFIDENCE LEVEL AND INTERVAL LENGTH Examples 10.5 and 10.6 give confidence intervals for the screen tension μ based on 20 measurements with $\bar{x} = 306.3$ and $\sigma = 43$. The 99% confidence interval is 281.6 to 331.1 and the 90% confidence interval is 290.5 to 322.1.

(a) Find the 80% confidence interval for μ.

(b) Find the 99.9% confidence interval for μ.

(c) Make a sketch like Figure 10.9 to compare all four intervals. How does increasing the confidence level affect the length of the confidence interval?

10.11 SAMPLE SIZE AND MARGIN OF ERROR Find the margin of error for 90% confidence in Example 10.5 if the manufacturer measures the tension of 80 video monitors. Check that your result is half as large as the margin of error based on 20 measurements in Example 10.5.

Choosing the sample size

A wise user of statistics never plans data collection without planning the inference at the same time. You can arrange to have both high confidence and a small margin of error by taking enough observations. The margin of error of the confidence interval for the mean of a normally distributed population is $m = z^* \sigma / \sqrt{n}$. To obtain a desired margin of error m, substitute the value of z^* for your desired confidence level, set the expression for m less than or equal to the specified margin of error, and solve the inequality for n. The procedure is best illustrated with an example.

EXAMPLE 10.7 DETERMINING SAMPLE SIZE

Company management wants a report of the mean screen tension for the day's production accurate to within ±5 mV with 95% confidence. How large a sample of video monitors must be measured to comply with this request?

For 95% confidence, Table C gives $z^* = 1.96$. We know that $\sigma = 43$. Set the margin of error to be at most 5:

$$m \le 5$$

$$z^* \frac{\sigma}{\sqrt{n}} \le 5$$

$$1.96\frac{43}{\sqrt{n}} \le 5$$
$$\sqrt{n} \ge \frac{(1.96)(43)}{5}$$
$$\sqrt{n} \ge 16.856$$
$$n \ge 284.125 \text{ so take } n = 285$$

Because n is a whole number, the company must measure the tension of 285 video screens to meet management's demand. On learning the cost of this many measurements, management may reconsider this request!

Here is the principle:

SAMPLE SIZE FOR DESIRED MARGIN OF ERROR

To determine the sample size n that will yield a confidence interval for a population mean with a specified margin of error m, set the expression for the margin of error to be less than or equal to m and solve for n:

$$z^* \frac{\sigma}{\sqrt{n}} \le m$$

In practice, taking observations costs time and money. The required sample size may be impossibly expensive. Do notice once again that it is the size of the *sample* that determines the margin of error. The size of the *population* (as long as the population is much larger than the sample) does not influence the sample size we need.

EXERCISES

10.12 A BALANCED SCALE? To assess the accuracy of a laboratory scale, a standard weight known to weigh 10 grams is weighed repeatedly. The scale readings are normally distributed with unknown mean (this mean is 10 grams if the scale has no bias). The standard deviation of the scale readings is known to be 0.0002 gram.

(a) The weight is weighed five times. The mean result is 10.0023 grams. Give a 98% confidence interval for the mean of repeated measurements of the weight.

(b) How many measurements must be averaged to get a margin of error of ±0.0001 with 98% confidence?

10.13 SURVEYING HOTEL MANAGERS, II How large a sample of the hotel managers in Exercise 10.6 (page 549) would be needed to estimate the mean μ within ±1 year with 99% confidence?

10.14 ENGINE CRANKSHAFTS, II How large a sample of the crankshafts in Exercise 10.7 (page 549) would be needed to estimate the mean μ within ±0.020 mm with 95% confidence?

Some cautions

Any formula for inference is correct only in specific circumstances. If statistical procedures carried warning labels like those on drugs, most inference methods would have long labels indeed. Our handy formula $\bar{x} \pm z^* \sigma / \sqrt{n}$ for estimating a normal mean comes with the following list of warnings for the user:

- The data must be an SRS from the population. We are completely safe if we actually carried out the random selection of an SRS. We are not in great danger if the data can plausibly be thought of as observations taken at random from a population. That is the case in Exercise 10.5 (page 548), where we have in mind the population resulting from a very large number of repeated analyses of the same specimen.
- The formula is not correct for probability sampling designs more complex than an SRS. Correct methods for other designs are available. We will not discuss confidence intervals based on multistage or stratified samples. If you plan such samples, be sure that you (or your statistical consultant) know how to carry out the inference you desire.
- There is no correct method for inference from data haphazardly collected with bias of unknown size. Fancy formulas cannot rescue badly produced data.
- Because $\bar{x}$ is strongly influenced by a few extreme observations, outliers can have a large effect on the confidence interval. You should search for outliers and try to correct them or justify their removal before computing the interval. If the outliers cannot be removed, ask your statistical consultant about procedures that are not sensitive to outliers.
- If the sample size is small and the population is not normal, the true confidence level will be different from the value C used in computing the interval. Examine your data carefully for skewness and other signs of nonnormality. The interval relies only on the distribution of $\bar{x}$, which even for quite small sample sizes is much closer to normal than the individual observations. When $n \geq 15$, the confidence level is not greatly disturbed by nonnormal populations unless extreme outliers or quite strong skewness are present. We will discuss this issue in more detail in the next chapter.
- You must know the standard deviation σ of the population. This unrealistic requirement renders the interval $\bar{x} \pm z^* \sigma / \sqrt{n}$ of little use in statistical practice. We will learn in the next chapter what to do when σ is unknown. However, if the sample is large, the sample standard deviation s will be close to the unknown σ. Then $\bar{x} \pm z^* s / \sqrt{n}$ is an approximate confidence interval for μ.

The most important caution concerning confidence intervals is a consequence of the first of these warnings. *The margin of error in a confidence interval covers only random sampling errors.* The margin of error is obtained from the sampling distribution and indicates how much error can be expected because of chance variation in randomized data production. Practical difficulties, such as undercoverage and nonresponse in a sample survey, can cause additional errors that may be larger than the random sampling error. Remember this unpleasant fact when reading the results of an opinion poll or other sample survey. The practical conduct of the survey influences the trustworthiness of its results in ways that are not included in the announced margin of error.

Every inference procedure that we will meet has its own list of warnings. Because many of the warnings are similar to those above, we will not print the full warning label each time. It is easy to state (from the mathematics of probability) conditions under which a method of inference is exactly correct. These conditions are *never* fully met in practice. For example, no population is exactly normal. Deciding when a statistical procedure should be used in practice often requires judgment assisted by exploratory analysis of the data.

Finally, you should understand what statistical confidence does not say. We are 95% confident that the mean SAT Math score for all California high school seniors lies between 452 and 470. That is, these numbers were calculated by a method that gives correct results in 95% of all possible samples. We cannot say that the probability is 95% that the true mean falls between 452 and 470. No randomness remains after we draw one particular sample and get from it one particular interval. The true mean either is or is not between 452 and 470. The probability calculations of standard statistical inference describe how often the *method* gives correct answers.

EXERCISES

10.15 A TALK SHOW OPINION POLL A radio talk show invites listeners to enter a dispute about a proposed pay increase for city council members. "What yearly pay do you think council members should get? Call us with your number." In all, 958 people call. The mean pay they suggest is $\overline{x} = \$8740$ per year, and the standard deviation of the responses is $s = \$1125$. For a large sample such as this, s is very close to the unknown population σ. The station calculates the 95% confidence interval for the mean pay μ that all citizens would propose for council members to be \$8669 to \$8811.

(a) Is the station's calculation correct?

(b) Does their conclusion describe the population of all the city's citizens? Explain your answer.

10.16 THE 2000 PRESIDENTIAL ELECTION A closely contested presidential election pitted George W. Bush against Al Gore in 2000. A poll taken immediately before the 2000 election showed that 51% of the sample intended to vote for Gore. The polling organization announced that they were 95% confident that the sample result was within ±2 points of the true percent of all voters who favored Gore.

(a) Explain in plain language to someone who knows no statistics what "95% confident" means in this announcement.

(b) The poll showed Gore leading. Yet the polling organization said the election was too close to call. Explain why.

(c) On hearing of the poll, a nervous politician asked, "What is the probability that over half the voters prefer Gore?" A statistician replied that this question can't be answered from the poll results, and that it doesn't even make sense to talk about such a probability. Explain why.

10.17 PRAYER IN THE SCHOOLS The *New York Times*/CBS News Poll asked the question, "Do you favor an amendment to the Constitution that would permit organized prayer in public schools?" Sixty-six percent of the sample answered "Yes." The article describing the poll says that it "is based on telephone interviews conducted from Sept. 13 to Sept. 18 with 1,664 adults around the United States, excluding Alaska and Hawaii. . . . the telephone numbers were formed by random digits, thus permitting access to both listed and unlisted residential numbers."

(a) The article gives the margin of error as 3 percentage points. Make a confidence statement about the percent of all adults who favor a school prayer amendment.

(b) The news article goes on to say: "The theoretical errors do not take into account a margin of additional error resulting from the various practical difficulties in taking any survey of public opinion." List some of the "practical difficulties" that may cause errors in addition to the ±3% margin of error. Pay particular attention to the news article's description of the sampling method.

10.18 95% CONFIDENCE A student reads that a 95% confidence interval for the mean SAT Math score of California high school seniors is 452 to 470. Asked to explain the meaning of this interval, the student says, "95% of California high school seniors have SAT Math scores between 452 and 470." Is the student right? Justify your answer.

SUMMARY

A **confidence interval** uses sample data to estimate an unknown population parameter with an indication of how accurate the estimate is and of how confident we are that the result is correct.

Any confidence interval has two parts: an interval computed from the data and a confidence level. The **interval** often has the form

$$\text{estimate} \pm \text{margin of error}$$

The **confidence level** states the probability that the method will give a correct answer. That is, if you use 95% confidence intervals often, in the long run 95% of your intervals will contain the true parameter value. You do not know whether a 95% confidence interval calculated from a particular set of data contains the true parameter value.

A level C **confidence interval for the mean** μ of a normal population with known standard deviation σ, based on an SRS of size n, is given by

$$\bar{x} \pm z^* \frac{\sigma}{\sqrt{n}}$$

The **critical value** z^* is chosen so that the standard normal curve has area C between $-z^*$ and z^*. Because of the central limit theorem, this interval is approximately correct for large samples when the population is not normal.

Other things being equal, the **margin of error** of a confidence interval gets smaller as

- the confidence level C decreases
- the population standard deviation σ decreases
- the sample size n increases

The sample size required to obtain a confidence interval with specified margin of error m for a normal mean is found by setting

$$z^* \frac{\sigma}{\sqrt{n}} \leq m$$

and solving for n, where z^* is the critical value for the desired level of confidence. Always round n up when you use this formula.

A specific confidence interval recipe is correct only under specific conditions. The most important conditions concern the method used to produce the data. Other factors such as the form of the population distribution may also be important.

Use the Inference Toolbox (page 548) as a guide when you construct a confidence interval.

SECTION 10.1 EXERCISES

10.19 WHO SHOULD GET WELFARE? A news article on a Gallup Poll noted that "28 percent of the 1548 adults questioned felt that those who were able to work should be taken off welfare." The article also said, "The margin of error for a sample size of 1548 is plus or minus three percentage points." Opinion polls usually announce margins of error for 95% confidence. Using this fact, explain to someone who knows no statistics what "margin of error plus or minus three percentage points" means.

10.20 HOTEL COMPUTER SYSTEMS How satisfied are hotel managers with the computer systems their hotels use? A survey was sent to 560 managers in hotels of size 200 to 500 rooms in Chicago and Detroit.[2] In all, 135 managers returned the survey. Two questions concerned their degree of satisfaction with the ease of use of their computer systems and with the level of computer training they had received. The managers responded using a seven-point scale, with 1 meaning "not satisfied," 4 meaning "moderately satisfied," and 7 meaning "very satisfied."

(a) What do you think is the population for this study? There are some major shortcomings in the data production. What are they? These shortcomings reduce the value of the formal inference you are about to do.

(b) The measurements of satisfaction are certainly not normally distributed, because they take only whole-number values from 1 to 7. Nonetheless, the use of confidence intervals based on the normal distribution is justified for this study. Why?

(c) The mean response for satisfaction with ease of use was $\bar{x} = 5.396$. Give a 95% confidence interval for the mean in the entire population. (Assume that the population standard deviation is $\sigma = 1.75$.) Use your Inference Toolbox.

(d) For satisfaction with training, the mean response was $\bar{x} = 4.398$. Taking $\sigma = 1.75$, give a 99% confidence interval for the population mean.

10.21 A NEWSPAPER POLL A *New York Times* poll on women's issues interviewed 1025 women and 472 men randomly selected from the United States, excluding Alaska and Hawaii. The poll announced a margin of error of ±3 percentage points for 95% confidence in conclusions about women. The margin of error for results concerning men was ±4 percentage points. Why is this larger than the margin of error for women?

10.22 HEALING OF SKIN WOUNDS Biologists studying the healing of skin wounds measured the rate at which new cells closed a razor cut made in the skin of an anesthetized newt. Here are data from 18 newts, measured in micrometers (millionths of a meter) per hour:[3]

29	27	34	40	22	28	14	35	26
35	12	30	23	18	11	22	23	33

(a) Make a stemplot of the healing rates (split the stems). It is difficult to assess normality from 18 observations, but look for outliers or extreme skewness. Now make a normal probability plot. What do you find?

(b) Scientists usually assume that animal subjects are SRSs from their species or genetic type. Treat these newts as an SRS and suppose you know that the standard deviation of healing rates for this species of newt is 8 micrometers per hour. Give a 90% confidence interval for the mean healing rate for the species.

(c) A friend who knows almost no statistics follows the formula $\bar{x} \pm z^* \sigma / \sqrt{n}$ in a biology lab manual to get a 95% confidence interval for the mean. Is her interval wider or narrower than yours? Explain to her why it makes sense that higher confidence changes the length of the interval.

10.23 MORE NEWTS! How large a sample would enable you to estimate the mean healing rate of skin wounds in newts (see Exercise 10.22) within a margin of error of 1 micrometer per hour with 90% confidence?

10.24 DETERMINING SAMPLE SIZE Researchers planning a study of the reading ability of third-grade children want to obtain a 95% confidence interval for the population mean score on a reading test, with margin of error no greater than 3 points. They carry out a small pilot study to estimate the variability of test scores. The sample standard deviation is $s = 12$ points in the pilot study, so in preliminary calculations the researchers take the population standard deviation to be $\sigma = 12$.

(a) The study budget will allow as many as 100 students. Calculate the margin of error of the 95% confidence interval for the population mean based on $n = 100$.

(b) There are many other demands on the research budget. If all of these demands were met, there would be funds to measure only 10 children. What is the margin of error of the confidence interval based on $n = 10$ measurements?

(c) Find the smallest value of n that would satisfy the goal of a 95% confidence interval with margin of error 3 or less. Is this sample size within the limits of the budget?

10.25 HOW THE POLL WAS CONDUCTED The *New York Times* includes a box entitled "How the poll was conducted" in news articles about its own opinion polls. Here are quotations from one such box (March 26, 1995). The box also announced a margin of error of ±3%, with 95% confidence.

The latest New York Times/CBS News poll is based on telephone interviews conducted March 9 through 12 with 1,156 adults around the United States, excluding Alaska and Hawaii. [The box then describes random digit dialing, the method used to select the sample.]

In addition to sampling error, the practical difficulties of conducting any survey of public opinion may introduce other sources of error into the poll. Variations in question wording or in the order of questions, for instance, can lead to somewhat different results.

(a) This account mentions several sources of possible errors in the poll's results. List these sources.

(b) Which of the sources of error you listed in (a) are covered by the announced margin of error?

10.26 CALCULATING CONFIDENCE INTERVALS

(a) Use your TI-83/89 to find the confidence interval for the mean rate of healing for newts in Exercise 10.22. Follow the Technology Toolbox below.

(b) If you have summary statistics but not raw data, you would select "Stats" as your input method and then provide the sample mean $\bar{x}$, the population standard deviation σ, and the number n of observations. Use your calculator in this way to find the confidence interval for the mean number of years hotel managers have spent with their current company in Exercise 10.6 (page 549).

Caution: Calculating the confidence interval is only one part of the inference process. Follow the steps in your Inference Toolbox.

TECHNOLOGY TOOLBOX *Calculator confidence intervals*

You can use your TI-83/89 to construct confidence intervals, using either data stored in a list or summary statistics. In Exercise 10.5, for example, we would begin by entering the three specimen concentrations into L_1/list1.

TI-83	TI-89
• Press STAT, choose TESTS, and then choose 7:Zinterval... • Adjust your settings as shown. • Then choose "Calculate."	• Press 2nd F2 (F7), choose 1:Zinterval... • Choose Input Method = Data. • Adjust your settings as shown. Press ENTER.

TECHNOLOGY TOOLBOX *Calculator confidence intervals (continued)*

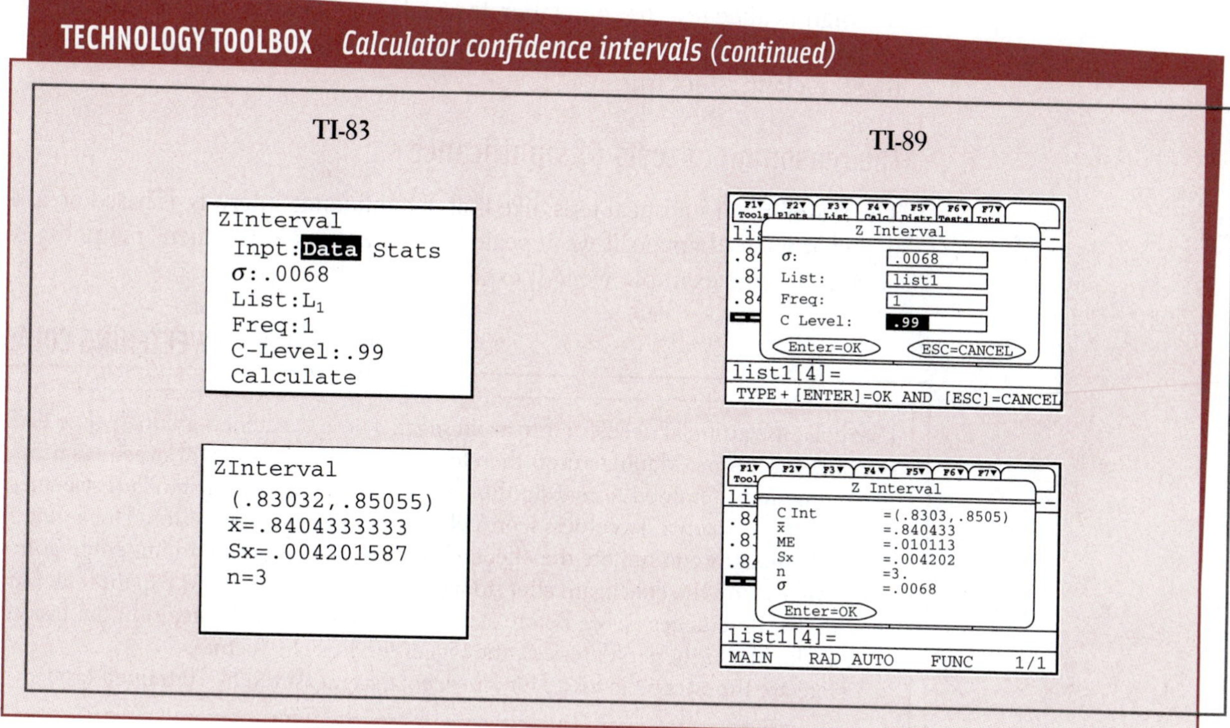

10.2 TESTS OF SIGNIFICANCE

Confidence intervals are one of the two most common types of statistical inference. Use a confidence interval when your goal is to estimate a population parameter. The second common type of inference, called *tests of significance*, has a different goal: to assess the evidence provided by data about some claim concerning a population. Here is the reasoning of statistical tests in a nutshell.

EXAMPLE 10.8 I'M A GREAT FREE-THROW SHOOTER

I claim that I make 80% of my basketball free throws. To test my claim, you ask me to shoot 20 free throws. I make only 8 of the 20. "Aha!" you say. "Someone who makes 80% of his free throws would almost never make only 8 out of 20. So I don't believe your claim."

Your reasoning is based on asking what would happen if my claim were true and we repeated the sample of 20 free throws many times—I would almost never make as few as 8. This outcome is so unlikely that it gives strong evidence that my claim is not true.

You can say how strong the evidence against my claim is by giving the probability that I would make as few as 8 out of 20 free throws if I really make 80% in the long run. This probability is 0.0001. I would make as few as 8 of 20 only once in 10,000 tries in the long run if my claim to make 80% is true. The small probability convinces you that my claim is false.

Significance tests use an elaborate vocabulary, but the basic idea is simple: an outcome that would rarely happen if a claim were true is good evidence that the claim is not true.

The reasoning of tests of significance

The reasoning of statistical tests, like that of confidence intervals, is based on asking what would happen if we repeated the sample or experiment many times. Here is the first example we will explore.

EXAMPLE 10.9 SWEETENING COLAS

Diet colas use artificial sweeteners to avoid sugar. These sweeteners gradually lose their sweetness over time. Manufacturers therefore test new colas for loss of sweetness before marketing them. Trained tasters sip the cola along with drinks of standard sweetness and score the cola on a "sweetness score" of 1 to 10. The cola is then stored for a month at high temperature to imitate the effect of four months' storage at room temperature. Each taster scores the cola again after storage. This is a matched pairs experiment. Our data are the differences (score before storage minus score after storage) in the tasters' scores. The bigger these differences, the bigger the loss of sweetness.

Here are the sweetness losses for a new cola, as measured by 10 trained tasters:

2.0	0.4	0.7	2.0	−0.4	2.2	−1.3	1.2	1.1	2.3

Most are positive. That is, most tasters found a loss of sweetness. But the losses are small, and two tasters (the negative scores) thought the cola gained sweetness. *Are these data good evidence that the cola lost sweetness in storage?*

The average sweetness loss for our cola is given by the sample mean,

$$\bar{x} = \frac{2.0 + 0.4 + \cdots + 2.3}{10} = 1.02$$

That's not a large loss. Ten different tasters would almost surely give a different result. Maybe it's just chance that produced this result. A ***test of significance*** asks

significance test

Does the sample result $\bar{x} = 1.02$ reflect a real loss of sweetness?

OR

Could we easily get the outcome $\bar{x} = 1.02$ just by chance?

The significance test starts with a careful statement of these alternatives. First, we always draw conclusions about some parameter of the population, so we must identify this parameter. In this case, it's the population mean μ. The mean μ is the average loss in sweetness that a very large number of tasters would detect in the cola. Our 10 tasters are a sample from this population.

null hypothesis

Next, state the ***null hypothesis.*** The null hypothesis says that there is *no effect* or *no change* in the population. If the null hypothesis is true, the sample result is just chance at work. Here, the null hypothesis says that the cola does not lose sweetness (no change). We can write that in terms of the mean sweetness loss μ in the population as

$$H_0: \mu = 0$$

We write H_0, read as "H-nought," to indicate the null hypothesis.

The effect we suspect is true, the alternative to "no effect" or "no change," is described by the ***alternative hypothesis.*** We suspect that the cola does lose sweetness. In terms of the mean sweetness loss μ, the alternative hypothesis is

alternative hypothesis

$$H_a: \mu > 0$$

The reasoning of a significance test goes like this:

- Suppose for the sake of argument that the null hypothesis is true, and that on the average there is no loss of sweetness.
- *Is the sample outcome* $\bar{x} = 1.02$ *surprisingly large under that supposition?* If it is, that's evidence against H_0 and in favor of H_a.

To answer the question, we use our knowledge of how the sample mean $\bar{x}$ would vary in repeated samples if H_0 really were true. That's the sampling distribution of $\bar{x}$ once again.

From long experience we know that individual tasters' scores vary according to a normal distribution. The mean of this distribution is the parameter μ. We're asking what would happen if there is really no change in sweetness on the average, so μ is 0. That's just what the null hypothesis says. From long experience we also know that the standard deviation for all individual tasters is $\sigma = 1$. (It is not realistic to suppose that we know the population standard deviation σ. We will eliminate this assumption in the next chapter.) The sampling distribution of $\bar{x}$ from 10 tasters is then normal with mean $\mu = 0$ and standard deviation

$$\frac{\sigma}{\sqrt{n}} = \frac{1}{\sqrt{10}} = 0.316$$

We can judge whether any observed $\bar{x}$ is surprising by locating it on this distribution. Figure 10.10 shows the sampling distribution with the observed values of $\bar{x}$ for two types of cola.

- One cola had $\bar{x} = 0.3$ for a sample of 10 tasters. It is clear from Figure 10.10 that an $\bar{x}$ this large could easily occur just by chance when the population mean is $\mu = 0$. That 10 tasters find $\bar{x} = 0.3$ is not evidence of a sweetness loss.
- The taste test for our cola produced $\bar{x} = 1.02$. That's way out on the normal curve in Figure 10.10, so far out that an observed value this large would almost never occur just by chance if the true μ were 0. This

observed value is good evidence that in fact the true μ is greater than 0, that is, that the cola lost sweetness. The manufacturer must reformulate the cola and try again.

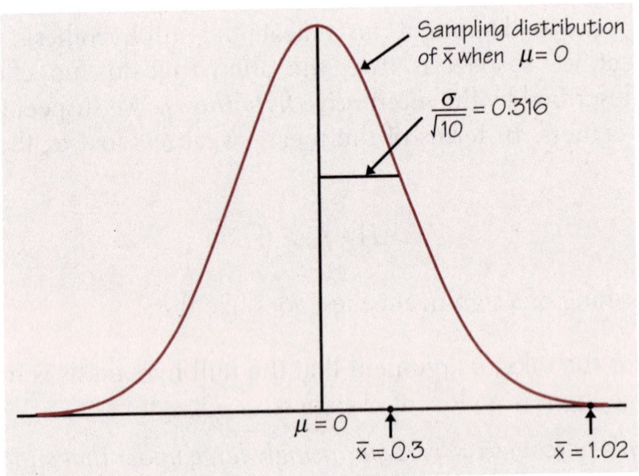

FIGURE 10.10 If a cola does not lose sweetness in storage, the mean score $\bar{x}$ for 10 tasters will have this sampling distribution. The actual result for one cola was $\bar{x} = 0.3$. That could easily happen just by chance. Another cola had $\bar{x} = 1.02$. That's so far out on the normal curve that it is good evidence that this cola did lose sweetness.

A significance test works by asking how unlikely the observed outcome would be if the null hypothesis were really true. The final step in our test is to assign a number to measure how unlikely our observed $\bar{x}$ is if H_0 is true. The less likely this outcome is, the stronger is the evidence against H_0.

Look again at Figure 10.10. If the alternative hypothesis is true, there is a sweetness loss and we expect the mean loss $\bar{x}$ found by the tasters to be positive. The farther out $\bar{x}$ is in the positive direction, the more convinced we are that the population mean μ is not zero but positive. We measure the strength of the evidence against H_0 by the probability under the normal curve in Figure 10.10 to the right of the observed $\bar{x}$. This probability is called the ***P*-value.** It is the probability of a result at least as far out as the result we actually got. The lower this probability, the more surprising our result, and the stronger the evidence against the null hypothesis.

P-value

- For one new cola, our 10 tasters gave $\bar{x} = 0.3$. Figure 10.11 shows the *P*-value for this outcome. It is the probability to the right of 0.3. This probability is about 0.17. That is, 17% of all samples would give a mean score as large or larger than 0.3 just by chance when the true population mean is 0. An outcome this likely to occur just by chance is not good evidence against the null hypothesis.

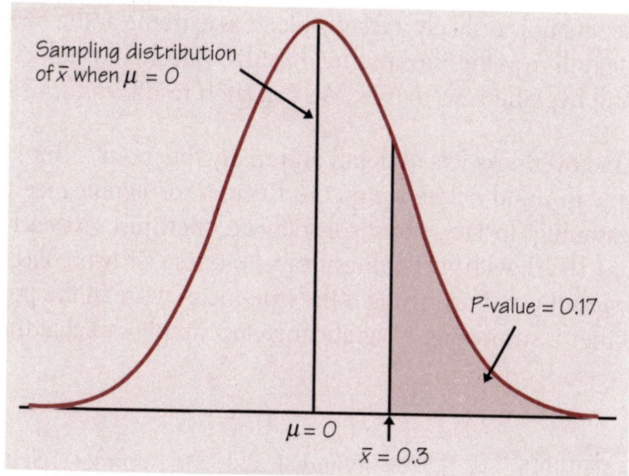

FIGURE 10.11 The *P*-value for the result $\bar{x} = 0.3$ in the cola taste test. The *P*-value is the probability (when H_0 is true) that $\bar{x}$ takes a value as large or larger than the actually observed value.

- Our cola showed a larger sweetness loss, $\bar{x} = 1.02$. The probability of a result this large or larger is only 0.0006. This probability is the *P*-value. Ten tasters would have an average score as large as 1.02 only 6 times in 10,000 tries if the true mean sweetness change were 0. An outcome this unlikely convinces us that the true mean is really greater than 0.

Small *P*-values are evidence against H_0 because they say that the observed result is unlikely to occur just by chance. Large *P*-values fail to give evidence against H_0. How small must a *P*-value be in order to persuade us? There's no fixed rule. But the level 0.05 (a result that would occur no more than once in 20 tries just by chance) is a common rule of thumb. A result with a small *P*-value, say less than 0.05, is called ***statistically significant.*** That's just a way of saying that chance alone would rarely produce so extreme a result.

statistically significant

Outline of a test

Here is the reasoning of a significance test in outline form:

- Describe the effect you are searching for in terms of a population parameter like the mean μ. (Never state a hypothesis in terms of a sample statistic like $\bar{x}$.) The null hypothesis is the statement that this effect is *not* present in the population, whereas the alternative hypothesis states that it is.
- From the data, calculate a statistic like $\bar{x}$ that estimates the parameter. Is the value of this statistic far from the parameter value stated by the null hypothesis? If so, the data give evidence that the null hypothesis is false and that the effect you are looking for is really there.

- The *P*-value says how unlikely a result at least as extreme as the one we observed would be if the null hypothesis were true. Results with small *P*-values would rarely occur if the null hypothesis were true. We call such results *statistically significant.*

This outline overlooks lots of detail and many fine points. But it is important to have it firmly in mind before we go on. Recipes for significance tests hide the underlying reasoning. In fact, statistical software often just gives a *P*-value. Look again at Figure 10.10, with its $\bar{x}$-values from taste tests of two colas. It should be clear that one result is not surprising if the true mean score in the population is 0, and that the other is surprising. A significance test simply says that more precisely.

EXERCISES

10.27 STUDENT ATTITUDES The Survey of Study Habits and Attitudes (SSHA) is a psychological test that measures the attitude toward school and study habits of students. Scores range from 0 to 200. The mean score for U.S. college students is about 115, and the standard deviation is about 30. A teacher suspects that older students have better attitudes toward school. She gives the SSHA to 25 students who are at least 30 years of age. Assume that scores in the population of older students are normally distributed with standard deviation $\sigma = 30$. The teacher wants to test the hypotheses

$$H_0: \mu = 115$$
$$H_a: \mu > 115$$

(a) What is the sampling distribution of the mean score $\bar{x}$ of a sample of 25 older students if the null hypothesis is true? Sketch the density curve of this distribution. (*Hint:* Sketch a normal curve first, then mark the axis using what you know about locating μ and σ on a normal curve.)

(b) Suppose that the sample data give $\bar{x} = 118.6$. Mark this point on the axis of your sketch. In fact, the result was $\bar{x} = 125.7$. Mark this point on your sketch. Using your sketch, explain in simple language why one result is good evidence that the mean score of all older students is greater than 115 and why the other outcome is not.

(c) Shade the area under the curve that is the *P*-value for the sample result $\bar{x} = 118.6$.

10.28 SPENDING ON HOUSING The Census Bureau reports that households spend an average of 31% of their total spending on housing. A homebuilders association in Cleveland believes that this average is lower in their area. They interview a sample of 40 households in the Cleveland metropolitan area to learn what percent of their spending goes toward housing. Take μ to be the mean percent of spending devoted to housing among all Cleveland households. We want to test the hypotheses

$$H_0: \mu = 31\%$$
$$H_a: \mu < 31\%$$

The population standard deviation is $\sigma = 9.6\%$.

(a) What is the sampling distribution of the mean percent $\bar{x}$ that the sample spends on housing if the null hypothesis is true? Sketch the density curve of the sampling distribution. (*Hint:* Sketch a normal curve first, then mark the axis using what you know about locating μ and σ on a normal curve.)

(b) Suppose that the study finds $\bar{x} = 30.2\%$ for the 40 households in the sample. Mark this point on the axis in your sketch. Then suppose that the study result is $\bar{x} = 27.6\%$. Mark this point on your sketch. Referring to your sketch, explain in simple language why one result is good evidence that average Cleveland spending on housing is less than 31%, whereas the other result is not.

(c) Shade the area under the curve that gives the *P*-value for the result $\bar{x} = 30.2\%$. (Note that we are looking for evidence that spending is *less* than the null hypothesis states.)

More detail: stating hypotheses

We will now look in more detail at some aspects of significance tests. The first step in a test of significance is to state a claim that we will try to find evidence *against.* This claim is our null hypothesis.

NULL HYPOTHESIS H_0

The statement being tested in a test of significance is called the **null hypothesis.** The test of significance is designed to assess the strength of the evidence against the null hypothesis. Usually the null hypothesis is a statement of "no effect" or "no difference."

The alternative hypothesis H_a is the claim about the population that we are trying to find evidence *for.* In Example 10.9, we were seeking evidence of a loss in sweetness. The null hypothesis says "no loss" on the average in a large population of tasters. The alternative hypothesis says "there is a loss." So the hypotheses are

$$H_0: \mu = 0$$
$$H_a: \mu > 0$$

This alternative hypothesis is ***one-sided*** because we are interested only in deviations from the null hypothesis in one direction. Here is another example. *one-sided alternative*

EXAMPLE 10.10 STUDYING JOB SATISFACTION

Does the job satisfaction of assembly workers differ when their work is machine-paced rather than self-paced? One study chose 28 subjects at random from a group of women who worked at assembling electronic devices. Half of the subjects were assigned at random to each of two groups. Both groups did similar assembly work, but one work setup allowed workers to pace themselves and the other featured an assembly line that moved at fixed time intervals so that the workers were paced by machine. After two weeks, all subjects took the Job Diagnosis Survey (JDS), a test of job satisfaction. Then they switched work setups, and took the JDS again after two more weeks. This is another matched pairs design. The response variable is the difference in JDS scores, self-paced minus machine-paced.[4]

The parameter of interest is the mean μ of the differences in JDS scores in the population of all female assembly workers. The null hypothesis says that there is no difference between self-paced and machine-paced work, that is,

$$H_0: \mu = 0$$

The authors of the study wanted to know if the two work conditions have different levels of job satisfaction. They did not specify the direction of the difference. The alternative hypothesis is therefore ***two-sided***,

two-sided alternative

$$H_a: \mu \neq 0$$

Hypotheses always refer to some population, not to a particular outcome. For this reason, always state H_0 and H_a in terms of population parameters. Because H_a expresses the effect that we hope to find evidence *for*, it is often easier to begin by stating H_a and then set up H_0 as the statement that the hoped-for effect is not present.

It is not always easy to decide whether H_a should be one-sided or two-sided. In Example 10.10, the alternative H_a: $\mu \neq 0$ is two-sided. That is, it simply says there is a difference in job satisfaction without specifying the direction of the difference. The alternative H_a: $\mu > 0$ in the taste test example is one-sided. Because colas can only lose sweetness in storage, we are interested only in detecting an upward shift in μ. The alternative hypothesis should express the hopes or suspicions we bring to the data. It is cheating to first look at the data and then frame H_a to fit what the data show. Thus the fact that the workers in the study of Example 10.10 were more satisfied with self-paced work should not influence our choice of H_a. If you do not have a specific direction firmly in mind in advance, use a two-sided alternative.

The choice of the hypotheses in Example 10.9 as

$$H_0: \mu = 0$$
$$H_a: \mu > 0$$

deserves a final comment. The cola maker is not concerned with the possibility that the tasters may detect a gain in sweetness, indicated by a negative mean loss μ. However, we can allow for the possibility that μ is less than zero by including this case in the null hypothesis. Then we would write

$$H_0: \mu \leq 0$$
$$H_a: \mu > 0$$

This statement is logically satisfying because the hypotheses account for all possible values of μ. However, only the parameter value in H_0 that is closest to H_a influences the form of the test in all common significance testing situations. We will therefore take H_0 to be the simpler statement that the parameter has a specific value, in this case H_0: $\mu = 0$.

EXERCISES

Each of the following situations calls for a significance test for a population mean μ. State the null hypothesis H_0 and the alternative hypothesis H_a in each case.

10.29 MOTORS The diameter of a spindle in a small motor is supposed to be 5 mm. If the spindle is either too small or too large, the motor will not work properly. The manufacturer measures the diameter in a sample of motors to determine whether the mean diameter has moved away from the target.

10.30 HOUSEHOLD INCOME Census Bureau data show that the mean household income in the area served by a shopping mall is $42,500 per year. A market research firm questions shoppers at the mall. The researchers suspect the mean household income of mall shoppers is higher than that of the general population.

10.31 BAD TEACHING? The examinations in a large accounting class are scaled after grading so that the mean score is 50. The professor thinks that one teaching assistant is a poor teacher and suspects that his students have a lower mean score than the class as a whole. The TA's students this semester can be considered a sample from the population of all students in the course, so the professor compares their mean score with 50.

10.32 SERVICE TECHNICIANS Last year, your company's service technicians took an average of 2.6 hours to respond to trouble calls from business customers who had purchased service contracts. Do this year's data show a different average response time?

More detail: *P*-values and statistical significance

A test of significance assesses the evidence against the null hypothesis by giving a probability, the *P*-value. If the sample statistic falls far from the value of the population parameter suggested by the null hypothesis in the direction specified by the alternative hypothesis, it is good evidence against H_0 and in favor of H_a. The *P*-value describes how strong the evidence is because it is the probability of getting an outcome *as extreme or more extreme than the actually observed outcome.* "Extreme" means "far from what we would expect if H_0 were true." The direction or directions that count as "far from what we would expect" are determined by the alternative hypothesis H_a.

Computer software that carries out tests of significance usually calculates the *P*-value for us. In some cases we can find *P*-values from our knowledge of sampling distributions.

> ***P*-VALUE**
>
> The probability, computed assuming that H_0 is true, that the observed outcome would take a value as extreme or more extreme than that actually observed is called the ***P*-value** of the test. The smaller the *P*-value is, the stronger is the evidence against H_0 provided by the data.

EXAMPLE 10.11 CALCULATING A ONE-SIDED *P*-VALUE

In Example 10.9 the observations are an SRS of size $n = 10$ from a normal population with $\sigma = 1$. The observed mean sweetness loss for one cola was $\bar{x} = 0.3$. The *P*-value for testing

$$H_0: \mu = 0$$
$$H_a: \mu > 0$$

is therefore

$$P(\bar{x} \geq 0.3)$$

calculated assuming that H_0 is true. When H_0 is true, $\bar{x}$ has the normal distribution with mean 0 and standard deviation

$$\frac{\sigma}{\sqrt{n}} = \frac{1}{\sqrt{10}} = 0.316$$

Find the *P*-value by a normal probability calculation. Start by drawing a picture that shows the *P*-value as an area under a normal curve. Figure 10.12 is the picture for this example. Then standardize $\bar{x}$ to get a standard normal Z and use Table A,

$$\begin{aligned} P(\bar{x} \geq 0.3) &= P\left(\frac{\bar{x} - 0}{0.316} \geq \frac{0.3 - 0}{0.316}\right) \\ &= P(Z \geq 0.95) \\ &= 1 - 0.8289 = 0.1711 \end{aligned}$$

So if H_0 is true, and the mean sweetness loss for this cola is 0, there is about a 17% chance that we will obtain a sample of 10 sweetness loss values whose mean is 0.3 or greater. Such a sample could occur quite easily by chance alone. The evidence against H_0 is not that strong.

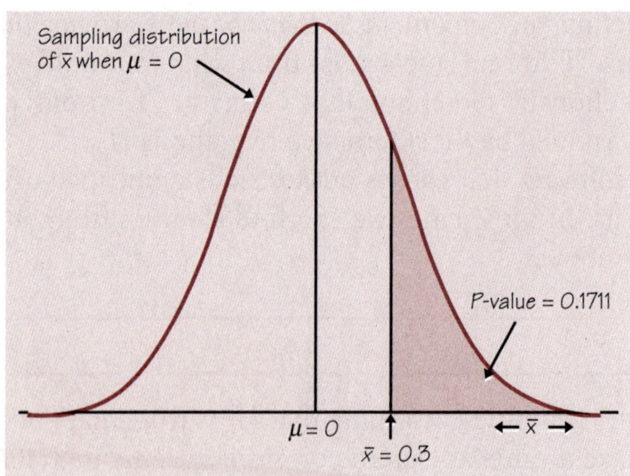

FIGURE 10.12 The *P*-value for the one-sided test in Example 10.11.

We sometimes take one final step to assess the evidence against H_0. We can compare the P-value with a fixed value that we regard as decisive. This amounts to announcing in advance how much evidence against H_0 we will insist on. The decisive value of P is called the ***significance level.*** We write it as α, the Greek letter alpha. If we choose $\alpha = 0.05$, we are requiring that the data give evidence against H_0 so strong that it would happen no more than 5% of the time (1 time in 20 samples in the long run) when H_0 is true. If we choose $\alpha = 0.01$, we are insisting on stronger evidence against H_0, evidence so strong that it would appear only 1% of the time (1 time in 100 samples) if H_0 is in fact true.

significance level

STATISTICAL SIGNIFICANCE

If the P-value is as small or smaller than alpha, we say that the data are **statistically significant at level α.**

"Significant" in the statistical sense does not mean "important." It means simply "not likely to happen just by chance." The significance level α makes "not likely" more exact. Significance at level 0.01 is often expressed by the statement "The results were significant ($P < 0.01$)." Here P stands for the P-value. The P-value is more informative than a statement of significance because it allows us to assess significance at any level we choose. For example, a result with $P = 0.03$ is significant at the $\alpha = 0.05$ level but is not significant at the $\alpha = 0.01$ level.

EXERCISES

10.33 Return to Exercise 10.27 (page 564).

(a) Starting from the picture you drew there, calculate the P-values for both $\bar{x} = 118.6$ and $\bar{x} = 125.7$. The two P-values express in numbers the comparison you made informally in Exercise 10.27.

(b) Which of the two observed values of $\bar{x}$ are statistically significant at the $\alpha = 0.05$ level? At the $\alpha = 0.01$ level?

10.34 Return to Exercise 10.28 (page 564).

(a) Starting from the picture you drew there, calculate the P-values for both $\bar{x} = 30.2\%$ and $\bar{x} = 27.6\%$. The two P-values express in numbers the comparison you made informally in Exercise 10.28.

(b) Is the result $\bar{x} = 27.6$ statistically significant at the $\alpha = 0.05$ level? Is it significant at the $\alpha = 0.01$ level?

10.35 COFFEE SALES Weekly sales of regular ground coffee at a supermarket have in the recent past varied according to a normal distribution with mean $\mu = 354$ units per week and standard deviation $\sigma = 33$ units. The store reduces the price by 5%. Sales in the next three weeks are 405, 378, and 411 units. Is this good evidence that average sales are now higher? The hypotheses are

$$H_0: \mu = 354$$
$$H_a: \mu > 354$$

Assume that the standard deviation of the population of weekly sales remains $\sigma = 33$.

(a) Find the value of the test statistic $\bar{x}$.

(b) Sketch the normal curve for the sampling distribution of $\bar{x}$ when H_0 is true. Why is the sampling distribution normal?

(c) Shade the area that represents the *P*-value for the observed outcome. Calculate the *P*-value.

(d) Is the result statistically significant at the $\alpha = 0.05$ level? Is it significant at the $\alpha = 0.01$ level? Do you think there is convincing evidence that mean sales are higher?

10.36 CEO PAY A study of the pay of corporate chief executive officers (CEOs) examined the increase in cash compensation of the CEOs of 104 companies, adjusted for inflation, in a recent year. The mean increase in real compensation was $\bar{x} = 6.9\%$ and the standard deviation of the increases was $s = 55\%$. Is this good evidence that the mean real compensation μ of all CEOs increased that year? The hypotheses are

$$H_0: \mu = 0 \quad \text{(no increase)}$$
$$H_a: \mu > 0 \quad \text{(an increase)}$$

Because the sample size is large, the sample s is close to the population σ, so take $\sigma = 55\%$.

(a) Sketch the normal curve for the sampling distribution of $\bar{x}$ when H_0 is true. Why is the sampling distribution approximately normal?

(b) Shade the area that represents the *P*-value for the observed outcome $\bar{x} = 6.9\%$. Calculate the *P*-value.

(c) Is the result significant at the $\alpha = 0.05$ level? Do you think the study gives strong evidence that the mean compensation of all CEOs went up?

10.37 STUDENT EARNINGS The financial aid office of a university asks a sample of students about their employment and earnings. The report says that "for academic year earnings, a significant difference ($P = 0.038$) was found between the sexes, with men earning more on the average. No difference ($P = 0.476$) was found between the earnings of black and white students." Explain both of these conclusions for the effects of sex and of race on mean earnings in language understandable to someone who knows no statistics.[5]

Tests for a population mean

Although the reasoning of significance testing isn't simple, carrying out a test is. The process is very similar to the one we followed when constructing a confidence interval. With a few minor changes, the four-step Inference Toolbox will once again guide us through the inference procedure.

Once you have completed the first two steps, your calculator or computer can do step 3 by following a recipe. We now develop the recipe for the test we have used in our examples.

INFERENCE TOOLBOX ***Significance tests***

To test a claim about an unknown population parameter:

Step 1: Identify the population of interest and the parameter you want to draw conclusions about. State null and alternative hypotheses in words and symbols.

Step 2: Choose the appropriate inference procedure. Verify the conditions for using the selected procedure.

Step 3: If the conditions are met, carry out the inference procedure.

- Calculate the test statistic.
- Find the *P*-value.

Step 4: Interpret your results in the context of the problem.

We have an SRS of size n drawn from a normal population with unknown mean μ. We want to test the hypothesis that μ has a specified value. Call the specified value μ_0. The null hypothesis is

$$H_0: \mu = \mu_0$$

The test is based on the sample mean $\overline{x}$. Because normal calculations require standardized variables, we will use as our test statistic the *standardized* sample mean

$$z = \frac{\overline{x} - \mu_0}{\sigma / \sqrt{n}}$$

This ***one-sample z statistic*** has the standard normal distribution when H_0 is true. If the alternative is one-sided on the high side

one-sample z statistic

$$H_a: \mu > \mu_0$$

then the *P*-value is the probability that a standard normal variable Z takes a value at least as large as the observed z. That is,

$$P = P(Z \geq z)$$

Example 10.11 calculates this *P*-value for the cola taste test. There, $\mu_0 = 0$, the standardized sample mean was $z = 0.95$, and the *P*-value was $P(Z \geq 0.95) = 0.1711$. Similar reasoning applies when the alternative hypothesis states that the true μ lies below the hypothesized μ_0 (one-sided).

When H_a states that μ is simply unequal to μ_0 (two-sided), values of z away from zero in either direction count against the null hypothesis. The P-value is the probability that a standard normal Z is at least as far from zero *in either direction* as the observed z.

EXAMPLE 10.12 CALCULATING A TWO-SIDED *P*-VALUE

Suppose that the z test statistic for a two-sided test is $z = 1.7$. The two-sided P-value is the probability that $Z \le -1.7$ or $Z \ge 1.7$. Figure 10.13 shows this probability as areas under the standard normal curve. Because the standard normal distribution is symmetric, we can calculate this probability by finding $P(Z \ge 1.7)$ and *doubling* it.

$$P(Z \le -1.7 \text{ or } Z \ge 1.7) = 2P(Z \ge 1.7) = 2(1 - 0.9554) = 0.0892$$

We would make exactly the same calculation if we observed $z = -1.7$. It is the absolute value $|z|$ that matters, not whether z is positive or negative.

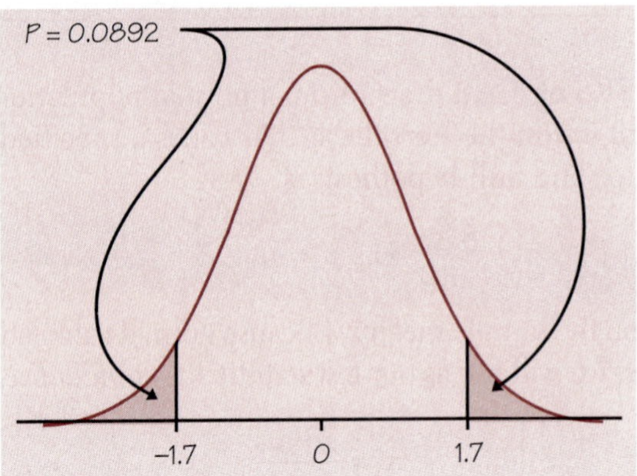

FIGURE 10.13 The P-value for the two-sided test in Example 10.12 is the sum of the area above 1.7 and below −1.7.

z TEST FOR A POPULATION MEAN

To test the hypothesis H_0: $\mu = \mu_0$ based on an SRS of size n from a population with unknown mean μ and known standard deviation σ, compute the **one-sample z statistic**

$$z = \frac{\bar{x} - \mu_0}{\sigma / \sqrt{n}}$$

In terms of a variable Z having the standard normal distribution, the P-value for a test of H_0 against

z TEST FOR A POPULATION MEAN (*continued*)

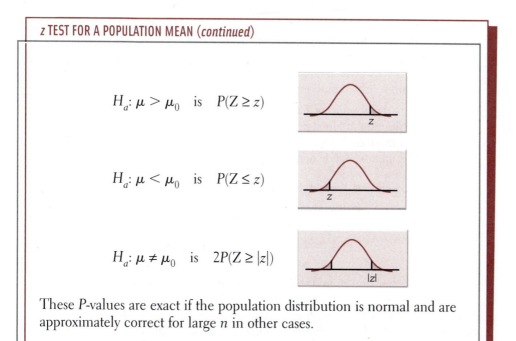

$H_a: \mu > \mu_0$ is $P(Z \geq z)$

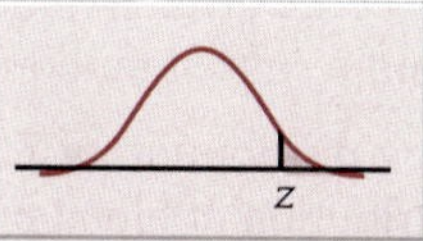

$H_a: \mu < \mu_0$ is $P(Z \leq z)$

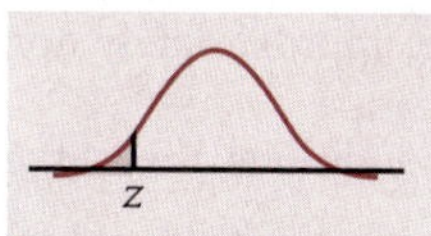

$H_a: \mu \neq \mu_0$ is $2P(Z \geq |z|)$

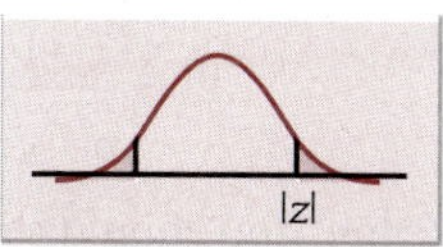

These *P*-values are exact if the population distribution is normal and are approximately correct for large *n* in other cases.

EXAMPLE 10.13 EXECUTIVES' BLOOD PRESSURES

The National Center for Health Statistics reports that the mean systolic blood pressure for males 35 to 44 years of age is 128 and the standard deviation in this population is 15. The medical director of a large company looks at the medical records of 72 executives in this age group and finds that the mean systolic blood pressure in this sample is $\bar{x} = 126.07$. Is this evidence that the company's executives have a different mean blood pressure from the general population? As usual in this chapter, we make the unrealistic assumption that we know the population standard deviation. Assume that executives have the same $\sigma = 15$ as the general population of middle-aged males.

Step 1: *Identify the population of interest and the parameter you want to draw conclusions about.* The population of interest is all middle-aged male executives in this company. We want to test a claim about the mean blood pressure μ for these executives. The null hypothesis is "no difference" from the national mean $\mu_0 = 128$. The alternative is two-sided because the medical director did not have a particular direction in mind before examining the data. So the hypotheses about the unknown mean μ of the executive population are

$H_0: \mu = 128$ Company executives' mean blood pressure is 128.

$H_a: \mu \neq 128$ Company executives' mean blood pressure differs from the national mean of 128.

Step 2: *Choose the appropriate inference procedure. Verify the conditions for using the selected procedure.* Since σ is known, we will use a one-sample *z* test for a population mean. Now we check conditions.

- The data come from an SRS from the population of interest. The *z* test assumes that the 72 executives in the sample are an SRS from the population of all middle-aged male executives in the company. We should check this assumption by asking how the data were produced. If medical records are available only for executives with recent medical problems, for example, the data are of little value for our purpose. It turns out that all executives are given a free annual medical exam and that the medical director selected 72 exam results at random.

- The sampling distribution of $\bar{x}$ is approximately normal. We do not know that the population distribution of blood pressures among the company executives is normally distributed. But the large sample size ($n = 72$) guarantees that the sampling distribution of $\bar{x}$ will be approximately normal (central limit theorem).

Step 3: If the conditions are met, carry out the inference procedure:

- Calculate the test statistic. The one-sample z statistic is

$$z = \frac{\bar{x} - \mu_0}{\sigma / \sqrt{n}} = \frac{126.07 - 128}{15 / \sqrt{72}} = -1.09$$

- Find the *P*-value. You should still draw a picture to help find the *P*-value, but now you can sketch the standard normal curve with the observed value of z. Figure 10.14 shows that the *P*-value is the probability that a standard variable Z takes a value at least 1.09 away from zero. From Table A we find that this probability is

$$P\text{-value} = 2P(Z \geq 1.09) = 2(1 - 0.862) = 0.2758$$

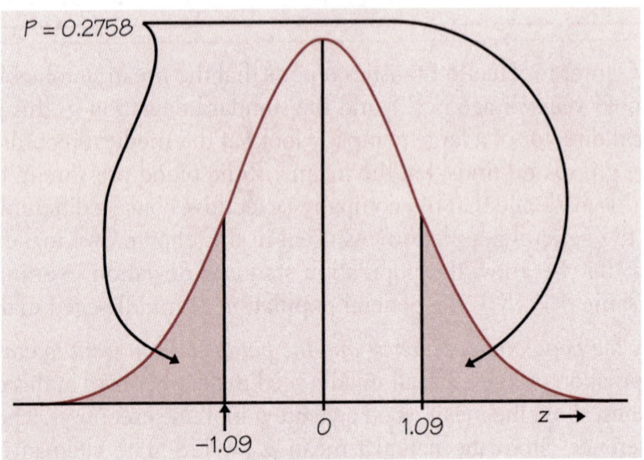

FIGURE 10.14 The *P*-value for the two-sided test.

Step 4: Interpret your results in the context of the problem. More than 27% of the time, an SRS of size 72 from the general male population would have a mean blood pressure at least as far from 128 as that of the executive sample. The observed $\bar{x} = 126.07$ is therefore not good evidence that executives differ from other men. We fail to reject our null hypothesis (H_0).

The data in Example 10.13 do *not* establish that the mean blood pressure μ for this company's executives is 128. We sought evidence that μ differed from 128 and failed to find convincing evidence. That is all we can say. No doubt the mean blood pressure of the entire executive population is not exactly equal to 128. A large enough sample would give evidence of the difference, even if it is very small. Tests of significance assess the evidence against H_0. If the evidence is strong, we can confidently reject H_0 in favor of the alternative. Failing to find evidence against H_0 means only that the data are consistent with H_0, not that we have clear evidence that H_0 is true.

When you interpret your results in context (step 4 of the Inference Toolbox), be sure to link your comments directly to your *P*-value or significance level. Do not simply say "reject H_0." Provide a basis for any decision that you make about the claim expressed in your hypotheses.

EXAMPLE 10.14 CAN YOU BALANCE YOUR CHECKBOOK?

In a discussion of the education level of the American workforce, someone says, "The average young person can't even balance a checkbook." The NAEP survey says that a score of 275 or higher on its quantitative test (see Exercise 10.2 on page 542) reflects the skill needed to balance a checkbook. The NAEP random sample of 840 young Americans had a mean score of $\bar{x} = 272$, a bit below the checkbook-balancing level. Is this sample result good evidence that the mean for *all* young men is less than 275? As in Exercise 10.2, assume that $\sigma = 60$.

Step 1: Identify the population of interest and the parameter you want to draw conclusions about. We want to test a claim about the mean NAEP score μ of all young Americans. The hypotheses are

H_0: $\mu = 275$ The mean NAEP score for all young Americans is at the checkbook-balancing level.

H_a: $\mu < 275$ The mean NAEP score for all young Americans is below the checkbook-balancing level.

The alternative hypothesis is one-sided because we believe that the mean NAEP score in the population might fall below 275.

Step 2: Choose the appropriate inference procedure. Verify the conditions for using the selected procedure. Since σ is known, we will use a one-sample z test for a population mean. Checking conditions:

- The data come from an SRS from the population of interest. We were told this in Exercise 10.2.

- The sampling distribution of $\bar{x}$ is approximately normal. We do not know that the population distribution of NAEP scores for young Americans is normally distributed. But since $n = 840$, the central limit theorem tells us that the sampling distribution of $\bar{x}$ will be approximately normal.

Step 3: If the conditions are met, carry out the inference procedure:

- Calculate the test statistic. The one-sample z statistic is

$$z = \frac{\bar{x} - \mu_0}{\sigma / \sqrt{n}} = \frac{272 - 275}{60 / \sqrt{840}} = -1.45$$

- Find the *P*-value. Because H_a is one-sided on the low side, small values of z count against H_0. Figure 10.15 illustrates the *P*-value. Using Table A, we find that

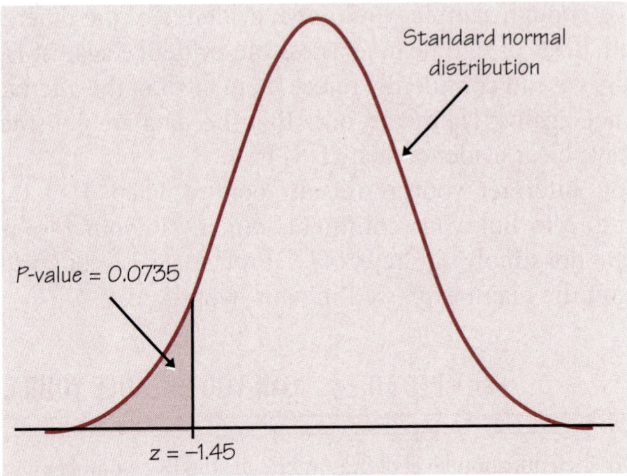

FIGURE 10.15 The *P*-value for the one-sided test.

$$P\text{-value} = P(Z \leq -1.45) = 0.0735$$

Step 4: Interpret your results in the context of the problem. A mean score as low as 272 would occur about 7 times in 100 samples if the population mean were 275. This is modest evidence that the mean NAEP score for all young Americans is less than 275, but it is not significant at the $\alpha = 0.05$ level. We fail to reject the null hypothesis.

EXERCISES

10.38 PRESSING PILLS A pharmaceutical manufacturer forms tablets by compressing a granular material that contains the active ingredient and various fillers. The hardness of a sample from each lot of tablets produced is measured in order to control the compression process. The target values for the hardness are $\mu = 11.5$ and $\sigma = 0.2$. The hardness data for a sample of 20 tablets are

11.627	11.613	11.493	11.602	11.360	11.374	11.592	11.458	11.552	11.463
11.383	11.715	11.485	11.509	11.429	11.477	11.570	11.623	11.472	11.531

Is there significant evidence at the 5% level that the mean hardness of the tablets is different from the target value? Use the Inference Toolbox.

10.39 FILLING COLA BOTTLES Bottles of a popular cola are supposed to contain 300 milliliters (ml) of cola. There is some variation from bottle to bottle because the filling machinery is not perfectly precise. The distribution of the contents is normal with standard deviation $\sigma = 3$ ml. An inspector who suspects that the bottler is underfilling measures the contents of six bottles. The results are

299.4	297.7	301.0
298.9	300.2	297.0

Is this convincing evidence that the mean contents of cola bottles is less than the advertised 300 ml?

Tests with fixed significance level

Sometimes we demand a specific degree of evidence in order to reject the null hypothesis. A level of significance α says how much evidence we require. In terms of the *P*-value, the outcome of a test is significant at level α if $P \leq \alpha$. Significance at any level is easy to assess once you have the *P*-value. The following example illustrates how to assess significance at a fixed level α by using a table of critical values, the same table used to obtain confidence intervals.

EXAMPLE 10.15 DETERMINING SIGNIFICANCE

In Example 10.14, we examined whether the mean NAEP quantitative score of young Americans is less than 275. The hypotheses are

$$H_0: \mu = 275$$
$$H_a: \mu < 275$$

The z statistic takes the value $z = -1.45$. Is the evidence against H_0 statistically significant at the 5% level?

To determine significance, we need only compare the observed $z = -1.45$ with the 5% critical value $z^* = 1.645$ from Table C. Because $z = -1.45$ is *not* farther from 0 than -1.645, it is *not* significant at level $\alpha = 0.05$.

Here is why. The *P*-value is the area to the left of -1.45 under the standard normal curve, shown in Figure 10.16. The result $z = -1.45$ is significant at the 5% level exactly when this area is no more than 5%. The area to the left of the critical value -1.645 is exactly 5%. So -1.645 separates values of z that are significant from those that are not. Figure 10.16 illustrates the procedure.

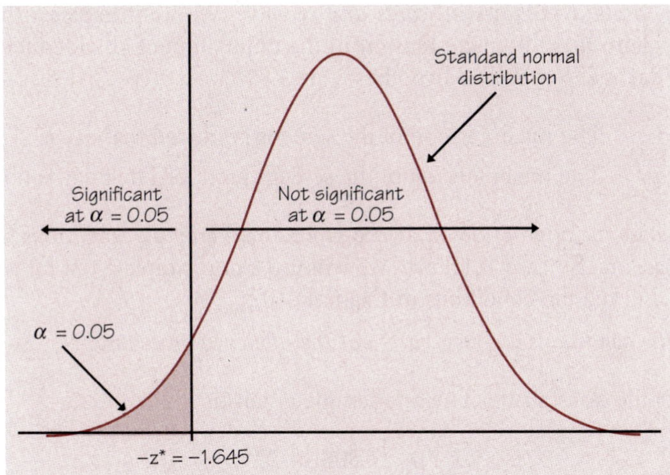

FIGURE 10.16 Deciding whether a z statistic is significant at the $\alpha = 0.05$ level in the one-sided test of Example 10.15.

FIXED SIGNIFICANCE LEVEL z TESTS FOR A POPULATION MEAN

To test the hypothesis H_0: $\mu = \mu_0$ based on an SRS of size n from a population with unknown mean μ and known standard deviation σ, compute the z test statistic

$$z = \frac{\bar{x} - \mu_0}{\sigma / \sqrt{n}}$$

Reject H_0 at significance level α against a one-sided alternative

$$H_a\text{: } \mu > \mu_0 \text{ if } z \geq z^*$$
$$H_a\text{: } \mu < \mu_0 \text{ if } z \leq -z^*$$

where z^* is the upper α critical value from Table C. Reject H_0 at significance level α against a two-sided alternative

$$H_a\text{: } \mu \neq \mu_0 \text{ if } |z| \geq z^*$$

where z^* is the upper $\alpha/2$ critical value from Table C.

EXAMPLE 10.16 IS THE SCREEN TENSION OK?

The manufacturer in Example 10.5 (page 546) knows from careful study that the proper tension of the mesh in a video terminal is 275 mV. Is there significant evidence at the 1% level that $\mu \neq 275$?

Step 1: *Identify the population of interest and the parameter you want to draw conclusions about. State hypotheses in words and symbols.* We want to assess the evidence against the claim that the mean tension in the population of all video terminals produced that day is 275 mV. The hypotheses are

H_0: $\mu = 275$	The mean tension of the screens produced that day is 275 mV.
H_a: $\mu \neq 275$	The mean tension of the screens produced that day is not 275 mV.

Step 2: *Choose the appropriate inference procedure. Verify the conditions for using the selected procedure.* Since σ is known, we will use a one-sample z test for a population mean. We checked the conditions in Example 10.5.

Step 3: *If the conditions are met, carry out the inference procedure:*

- Calculate the test statistic. The one-sample z statistic is

$$z = \frac{\bar{x} - \mu_0}{\sigma / \sqrt{n}} = \frac{306.3 - 275}{43 / \sqrt{20}} = 3.26$$

- Determine significance. Because the alternative is two-sided, we compare $|z| = 3.26$ with the $\alpha/2 = 0.005$ critical value from Table C. This critical value is $z^* = 2.576$. Figure 10.17

shows how this critical value separates values of z that are statistically significant from those that are not significant. Because $|z| > 2.576$, our result is statistically significant.

Step 4: Interpret your results in the context of the problem. We reject the null hypothesis at the $\alpha = 0.01$ significance level and conclude that the screen tension for the day's production is not at the desired 275 mV level.

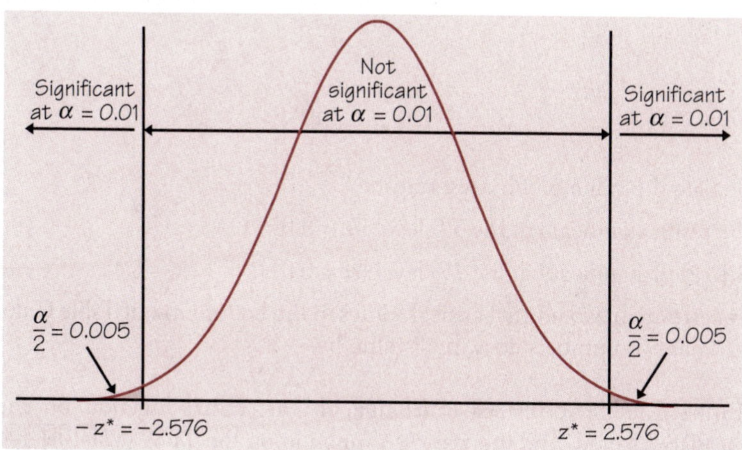

FIGURE 10.17 Deciding whether a z statistic is significant at the $\alpha = 0.01$ level in the two-sided test of Example 10.16.

The observed result in Example 10.16 was $z = 3.26$. The conclusion that this result is significant at the 1% level does not tell the whole story. The observed z is far beyond the 1% critical value, and the evidence against H_0 is much stronger that 1% significance indicates. The P-value

$$P\text{-value} = 2P(Z \geq 3.26) = 0.00111$$

gives a better sense of how strong the evidence is. The P-value is the smallest level α at which the data are significant. Knowing the P-value allows us to assess significance at any level.

In addition to assessing significance at fixed levels, a table of critical values allows us to estimate P-values without a calculation. In Example 10.16, compare the observed $z = 3.26$ with the normal critical values in the bottom row of Table C. This value falls between 3.091 and 3.291, so the corresponding upper tail probability is between 0.0005 and 0.001. So we know that for the two-sided test, $0.001 < P < 0.002$. In Example 10.15, $z = -1.45$ lies between the 0.05 and 0.10 entries in the table. So the P-value for the one-sided test lies between 0.05 and 0.10. This approximation is accurate enough for most purposes.

Because the practice of statistics almost always employs software that calculates P-values automatically, tables of critical values are becoming outdated. Tables of critical values, such as Table C, appear in this book for learning purposes and to rescue students without good computing facilities.

EXERCISES

10.40 TESTING A RANDOM NUMBER GENERATOR A random number generator is supposed to produce random numbers that are uniformly distributed on the interval from 0 to 1. If this is true, the numbers generated come from a population with $\mu = 0.5$ and $\sigma = 0.2887$. A command to generate 100 random numbers gives outcomes with mean $\bar{x} = 0.4365$. Assume that the population σ remains fixed. We want to test

$$H_0: \mu = 0.5$$
$$H_a: \mu \neq 0.5$$

(a) Calculate the value of the z test statistic.

(b) Is the result significant at the 5% level ($\alpha = 0.05$)?

(c) Is the result significant at the 1% level ($\alpha = 0.01$)?

(d) Between which two normal critical values in the bottom row of Table C does z lie? Between what two numbers does the P-value lie?

10.41 TESTING A RANDOM NUMBER GENERATOR, II The `rand` function on the TI-83 (MATH / PRB / 1:rand) and the `rand83` function on the TI-89 (CATALOG F3) generate a pseudorandom real number in the interval [0, 1)—that is, in the interval $0 \leq X < 1$. The command `rand(100)` (`tistat.rand83(100)` on the TI-89) generates 100 random real numbers in the interval [0, 1). Describe how you would use your calculator to carry out the experiment described in the previous exercise, using the sample mean calculated from your 100 calculator values. (*Hint:* Store the 100 values in a list.) As in Exercise 10.40, take $\sigma = 0.2887$. Then carry out your plan and answer the questions in Exercise 10.40.

10.42 NICOTINE IN CIGARETTES To determine whether the mean nicotine content of a brand of cigarettes is greater than the advertised value of 1.4 milligrams, a health advocacy group tests

$$H_0: \mu = 1.4$$
$$H_a: \mu > 1.4$$

The calculated value of the test statistic is $z = 2.42$.

(a) Is the result significant at the 5% level?

(b) Is the result significant at the 1% level?

(c) By comparing z with the critical values in the bottom row of Table C, give two numbers that catch the P-value between them.

Tests from confidence intervals

The calculation in Example 10.16 for a 1% significance test is very similar to that in Example 10.6 (page 550) for a 99% confidence interval. In fact, a two-sided test at significance level α can be carried out directly from a confidence interval with confidence level $C = 1 - \alpha$.

CONFIDENCE INTERVALS AND TWO-SIDED TESTS

A level α two-sided significance test rejects a hypothesis H_0: $\mu = \mu_0$ exactly when the value μ_0 falls outside a level $1 - \alpha$ confidence interval for μ.

EXAMPLE 10.17 TESTS FROM A CONFIDENCE INTERVAL

The 99% confidence interval for the mean screen tension μ in Example 10.16 is

$$\bar{x} \pm z^* \frac{\sigma}{\sqrt{n}} = 306.3 \pm 2.576 \frac{43}{\sqrt{20}} = 306.3 \pm 24.8$$
$$= (281.5,\ 331.1)$$

Our reasoning goes like this. We are 99% confident that this interval captures the true population mean, μ. But our hypothesized population mean, μ_0 =275, is not the interval. So we conclude that our null hypothesis that μ = 275 is implausible. Thus, we conclude that μ is different from 275. Note that this is consistent with our conclusion in Example 10.16

If our null hypothesis had been μ_0 = 290, then that value would have been consistent with our 99% confidence interval, and we would not have been able to reject H_0. Figure 10.18 illustrates both situations.

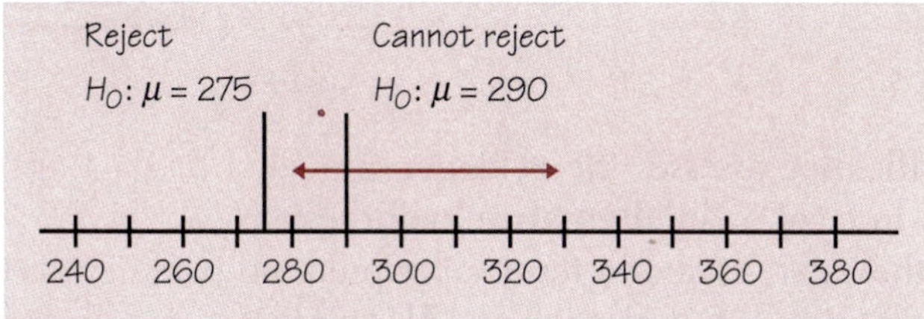

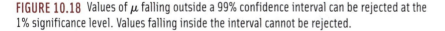

FIGURE 10.18 Values of μ falling outside a 99% confidence interval can be rejected at the 1% significance level. Values falling inside the interval cannot be rejected.

EXERCISES

10.43 RADON DETECTORS Radon is a colorless, odorless gas that is naturally released by rocks and soils and may concentrate in tightly closed houses. Because radon is slightly radioactive, there is some concern that it may be a health hazard. Radon detectors are sold to homeowners worried about this risk, but the detectors may be inaccurate. University researchers placed 12 detectors in a chamber where they were exposed to 105 picocuries per liter of radon over 3 days. Here are the readings given by the detectors:[6]

91.9	97.8	111.4	122.3	105.4	95.0
103.8	99.6	96.6	119.3	104.8	101.7

Assume (unrealistically) that you know that the standard deviation of readings for all detectors of this type is $\sigma = 9$.

(a) Give a 90% confidence interval for the mean reading μ for this type of detector. Use the Inference Toolbox.

(b) Is there significant evidence at the 10% level that the mean reading differs from the true value 105? State hypotheses and base a test on your confidence interval from (a).

10.44 IQ TEST SCORES Here are the IQ test scores of 31 seventh-grade girls in a Midwest school district:[7]

114	100	104	89	102	91	114	114	103	105	108	130	120	132	111	128
118	119	86	72	111	103	74	112	107	103	98	96	112	112	93	

Treat the 31 girls as an SRS of all seventh-grade girls in the school district. Suppose that the standard deviation of IQ scores in this population is known to be $\sigma = 15$.

(a) Give a 95% confidence interval for the mean IQ score μ in the population. Use the Inference Toolbox.

(b) Is there significant evidence at the 5% level that the mean IQ score in the population differs from 100? Give appropriate statistical evidence to support your conclusion.

(c) In fact, the scores are those of all seventh-grade girls in one of the several schools in the district. Explain carefully why your results from (a) and (b) cannot be trusted.

SUMMARY

A **test of significance** assesses the evidence provided by data against a **null hypothesis** H_0 in favor of an **alternative hypothesis** H_a.

The hypotheses are stated in terms of population parameters. Usually H_0 is a statement that no effect is present, and H_a says that a parameter differs from its null value in a specific direction (**one-sided alternative**) or in either direction (**two-sided alternative**).

The essential reasoning of a significance test is as follows. Suppose for the sake of argument that the null hypothesis is true. If we repeated our data production many times, would we often get data as inconsistent with H_0 as the data we actually have? If the data are unlikely when H_0 is true, they provide evidence against H_0.

A test is based on a **test statistic.** The **P-value** is the probability, computed supposing H_0 to be true, that the test statistic will take a value at least as extreme as that actually observed. Small P-values indicate strong evidence against H_0. Calculating P-values requires knowledge of the sampling distribution of the test statistic when H_0 is true.

If the P-value is as small or smaller than a specified value α, the data are **statistically significant** at significance level α.

Significance tests for the hypothesis H_0: $\mu = \mu_0$ concerning the unknown mean μ of a population are based on the **one-sample z statistic**

$$z = \frac{\bar{x} - \mu_0}{\sigma / \sqrt{n}}$$

The z test assumes an SRS of size n, known population standard deviation σ, and either a normal population or a large sample. P-values are computed from the normal distribution (Table A). Fixed α tests use the table of standard normal **critical values** (bottom row of Table C).

Use the Inference Toolbox as a guide when you perform a significance test.

SECTION 10.2 EXERCISES

10.45 STUDYING JOB SATISFACTION The job satisfaction study of Example 10.10 (page 565) measured the JDS job satisfaction score of 28 female assemblers doing both self-paced and machine-paced work. The parameter μ is the mean amount by which the self-paced score exceeds the machine-paced score in the population of all such workers. Scores are normally distributed. The population standard deviation is $\sigma = 0.60$. The hypotheses are

$$H_0: \mu = 0$$
$$H_a: \mu \neq 0$$

(a) What is the sampling distribution of the mean JDS score $\bar{x}$ for 28 workers if the null hypothesis is true? Sketch the density curve of this distribution. (*Hint:* Sketch a normal curve first, then mark the axis using what you know about locating μ and σ on a normal curve.)

(b) Suppose that the study had found $\bar{x} = 0.09$. Mark this point on the axis in your sketch. In fact, the study found $\bar{x} = 0.27$ for these 28 workers. Mark this point on your sketch. Referring to your sketch, explain in simple language why one result is good evidence that H_0 is not true, and why the other is not.

(c) Make another copy of your sketch. Shade the area under the curve that gives the P-value for the result $\bar{x} = 0.09$. Then calculate this P-value. (Note that H_a is two-sided.)

(d) Calculate the P-value for the result $\bar{x} = 0.27$ also. The two P-values express your explanation in (b) in numbers.

10.46 SMALL APARTMENTS? The mean area of the several thousand apartments in a new development is advertised to be 1250 square feet. A tenant group thinks that the apartments are smaller than advertised. They hire an engineer to measure a sample of apartments to test their suspicion. What are the null hypothesis H_0 and the alternative hypothesis H_a?

10.47 MICE IN A MAZE Experiments on learning in animals sometimes measure how long it takes mice to find their way through a maze. The mean time is 18 seconds for one particular maze. A researcher thinks that a loud noise will cause the mice to complete the maze faster. She measures how long each of 10 mice takes with a noise as stimulus. What are the null hypothesis H_0 and the alternative hypothesis H_a?

10.48 IS THIS MILK WATERED DOWN? Cobra Cheese Company buys milk from several suppliers. Cobra suspects that some producers are adding water to their milk to increase their profits. Excess water can be detected by measuring the freezing point of the milk. The freezing temperature of natural milk varies normally, with mean $\mu = -0.545°$ Celsius (C) and standard deviation $\sigma = 0.008°$ C. Added water raises the freezing temperature toward 0° C, the freezing point of water. Cobra's laboratory manager measures the freezing temperature of five consecutive lots of milk from one producer. The mean measurement is $\bar{x} = -0.538°$ C. Is this good evidence that the producer is adding water to the milk?

10.49 INTERPRETING z STATISTICS There are other z statistics that we have not yet studied. You can use Table C to assess the significance of any z statistic. A study compares American-Japanese joint ventures in which the U.S. company is larger than its Japanese partner with joint ventures in which the U.S. company is smaller. One variable measured is the excess returns earned by shareholders in the American company. The null hypothesis is "no difference" between the means for the two populations. The alternative hypothesis is two-sided. The value of the test statistic is $z = -1.37$.

(a) Is this result significant at the 5% level?

(b) Is the result significant at the 10% level?

10.50 BENEFITS OF PATENT PROTECTION Market pioneers, companies that are among the first to develop a new product or service, tend to have higher market shares than latecomers to the market. What accounts for this advantage? Here is an excerpt from the conclusions of a study of a sample of 1209 manufacturers of industrial goods:

Can patent protection explain pioneer share advantages? Only 21% of the pioneers claim a significant benefit from either a product patent or a trade secret. Though their average share is two points higher than that of pioneers without this benefit, the increase is not statistically significant (z = 1.13). Thus, at least in mature industrial markets, product patents and trade secrets have little connection to pioneer share advantages.[8]

Find the P-value for the given z. Then explain to someone who knows no statistics what "not statistically significant" in the study's conclusion means. Why does the author conclude that patents and trade secrets don't help, even though they contributed 2 percentage points to average market share?

10.51 STATISTICAL SIGNIFICANCE, I Explain in plain language why a significance test that is significant at the 1% level must always be significant at the 5% level. If a test is significant at the 5% level, what can you say about its significance at the 1% level?

10.52 STATISTICAL SIGNIFICANCE, II Asked to explain the meaning of "statistically significant at the $\alpha = 0.05$ level," a student says: "This means that the probability that the null hypothesis is true is less than 0.05." Is this explanation correct? Why or why not?

10.53 SIGNIFICANCE TESTS You will perform a significance test of

$$H_0: \mu = 0$$

versus

$$H_a: \mu > 0$$

(a) What values of z would lead you to reject H_0 at the 5% level?

(b) If the alternative hypothesis was

$$H_a: \mu \neq 0$$

what values of z would lead you to reject H_0 at the 5% level?

(c) Explain why your answers to parts (a) and (b) are different.

10.54 COCKROACHES An understanding of cockroach biology may lead to an effective control strategy for these annoying insects. Researchers studying the absorption of sugar by insects feed cockroaches a diet containing measured amounts of a particular sugar. After 10 hours, the cockroaches are killed and the concentration of the sugar in various body parts is determined by a chemical analysis. The paper that reports the research states that a 95% confidence interval for the mean amount (in milligrams) of the sugar in the hindguts of the cockroaches is 4.2 ± 2.3.[9]

(a) Does this paper give evidence that the mean amount of sugar in the hindguts under these conditions is not equal to 7 mg? State H_0 and H_a and base a test on the confidence interval.

(b) Would the hypothesis that $\mu = 5$ mg be rejected at the 5% level in favor of a two-sided alternative?

10.55 CALIFORNIA SAT SCORES In a discussion of SAT scores, someone comments: "Because only a minority of high school students take the test, the scores overestimate the ability of typical high school seniors. The mean SAT Mathematics score is about 475, but I think that if all seniors took the test, the mean score would be no more than 450." You gave the test to an SRS of 500 seniors from California (Example 10.2). These students had a mean score of $\bar{x} = 461$. Is this good evidence against the claim that the mean for all California seniors is no more than 450?

TECHNOLOGY TOOLBOX *Performing significance tests on the TI-83/89*

The TI-83 and TI-89 can be used to conduct z tests of inference, using either data stored in a list or summary statistics.

TI-83

- Press STAT and choose TESTS and 1:Z-Test to access the Z-Test screen, as shown.

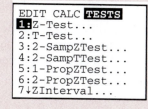

```
EDIT CALC TESTS
1:Z-Test...
2:T-Test...
3:2-SampZTest...
4:2-SampTTest...
5:1-PropZTest...
6:2-PropZTest...
7↓ZInterval...
```

TI-89

- Press 2nd F1 ([F6]) and choose 1:Z-Test.
- Choose "Stats" as the input method.

```
F1▾    F2▾   F3▾  F4▾  F5▾   F6    F7▾
Tools  Plots List Calc Distr Tests Ints
list1   --   1:Z-Test...
-----        2:T-Test...
             3:2-SampZTest...
             4:2-SampTTest...
             5:1-PropZTest...
             6:2-PropZTest...
             7:Chi2 GOF...
             8↓Chi2 2-way...
list1={}
TYPE 0* USE ←→↑↓ + [ENTER] OR [ESC]
```

In the NAEP testing of Example 10.14 (page 575), for example, you would enter 275 for the null hypothesized mean μ_0. Next enter 60 for σ, 272 for $\bar{x}$, and 840 for n. Select $< \mu_0$ for the alternative hypothesis, and choose "Calculate."

TECHNOLOGY TOOLBOX *Performing significance tests on the TI-83/89 (continued)*

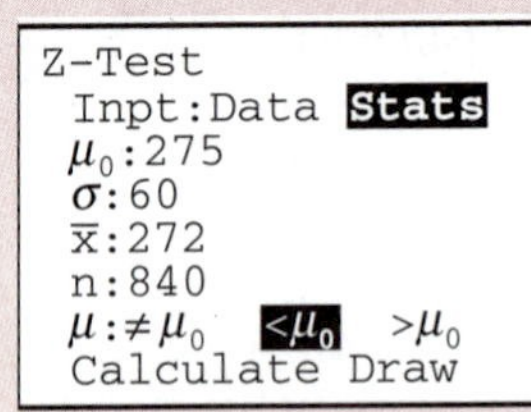

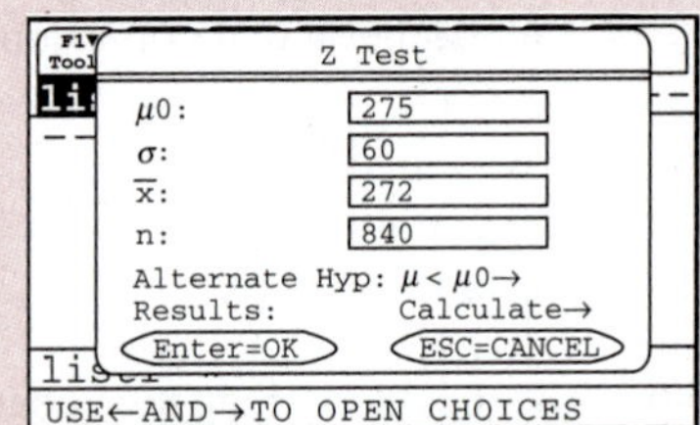

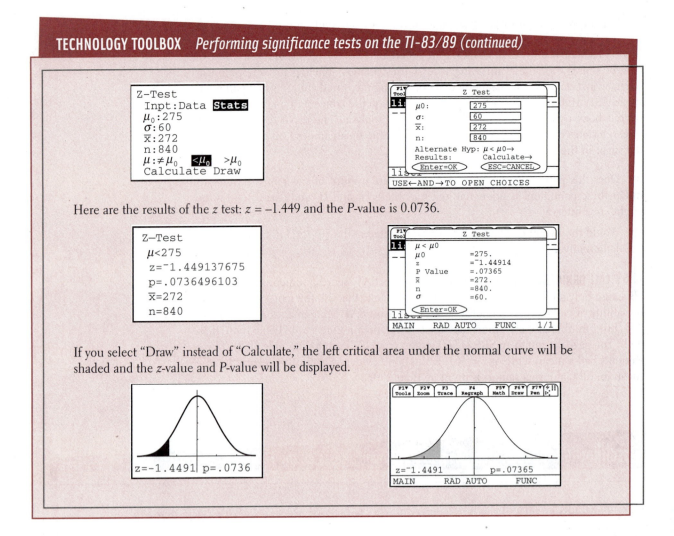

Here are the results of the z test: $z = -1.449$ and the P-value is 0.0736.

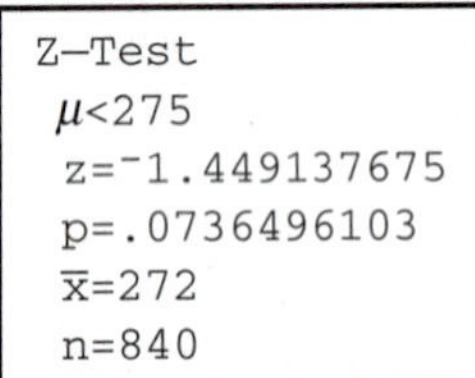

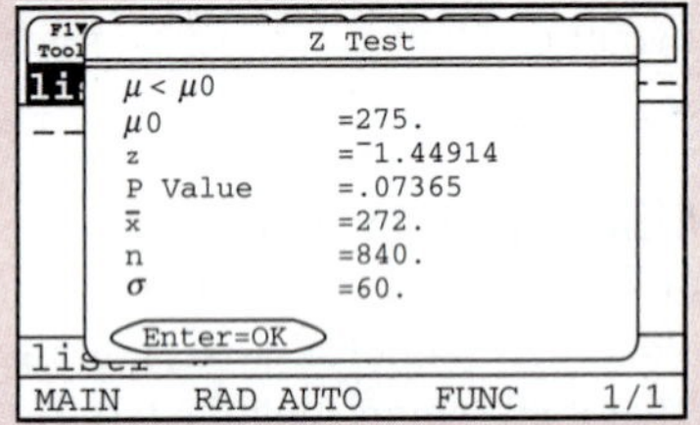

If you select "Draw" instead of "Calculate," the left critical area under the normal curve will be shaded and the z-value and P-value will be displayed.

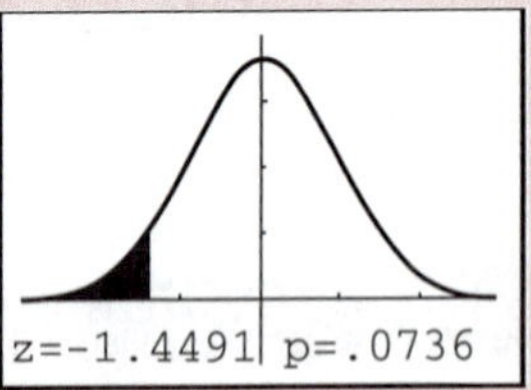

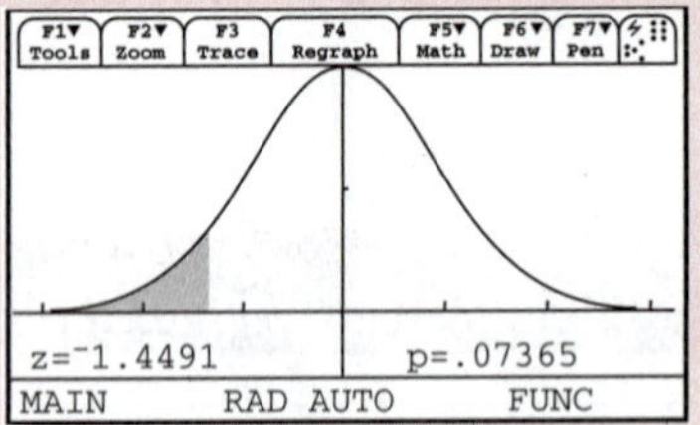

10.56 CALCULATOR SIGNIFICANCE TESTS

(a) Use your calculator in this way to find the z-value and P-value for the true value of the mean μ in Example 10.14 (575). Check your calculator findings with the numbers given in Example 10.14.

(b) If you have data, enter the data into any list and then specify a z test. In Example 10.5 (page 546), for example, you would enter the 20 screen tensions into L_1/list1. Then from the Z-test screen, specify "Data" and then enter the appropriate values.

(c) Use your TI-83/89 to perform the test of significance described in Exercise 10.39 (page 576).

10.3 MAKING SENSE OF STATISTICAL SIGNIFICANCE

Significance tests are widely used in reporting the results of research in many fields of applied science and in industry. New pharmaceutical products require significant evidence of effectiveness and safety. Courts inquire about statistical significance in hearing class action discrimination cases. Marketers want to know whether a new ad campaign significantly outperforms the old one, and medical researchers want to know whether a new therapy performs significantly better. In all these uses, statistical significance is valued because it points to an effect that is unlikely to occur simply by chance.

Carrying out a test of significance is often quite simple, especially if you get a *P*-value effortlessly from a calculator or computer. Using tests wisely is not so simple. Here are some points to keep in mind when using or interpreting significance tests.

Choosing a level of significance

The purpose of a test of significance is to give a clear statement of the degree of evidence provided by the sample against the null hypothesis. The *P*-value does this. But sometimes you will make some decision or take some action if your evidence reaches a certain standard. A level of significance α sets such a standard. Perhaps you will publish a research finding if the effect is significant at the $\alpha = 0.01$ level. Or perhaps your company will lose a lawsuit alleging racial discrimination if the percent of blacks hired is significantly below the percent of blacks in the pool of potential employees at the $\alpha = 0.05$ level. Courts have in fact tended to accept this standard in discrimination cases.[10]

Making a decision is different in spirit from testing significance, though the two are often mixed in practice. Choosing a level α in advance makes sense if you must make a decision, but not if you wish only to describe the strength of your evidence. Using tests with fixed α for decision making is discussed at greater length in Section 10.4.

If you do use a fixed α significance test to make a decision, choose α by asking how much evidence is required to reject H_0. This depends mainly on two circumstances:

- *How plausible is H_0?* If H_0 represents an assumption that the people you must convince have believed for years, strong evidence (small α) will be needed to persuade them.
- *What are the consequences of rejecting H_0?* If rejecting H_0 in favor of H_a means making an expensive changeover from one type of product packaging to another, you need strong evidence that the new packaging will boost sales.

Both the plausibility of H_0 and H_a and the consequences of any action that rejection may lead to are somewhat subjective. Different people may want to use different levels of significance. It is better to report the *P*-value, which allows each of us to decide individually if the evidence is sufficiently strong.

Users of statistics have often emphasized certain standard levels of significance, such as 10%, 5%, and 1%. This emphasis reflects the time when tables of critical values rather than computer programs dominated statistical practice. The 5% level ($\alpha = 0.05$) is particularly common. *There is no sharp border between "significant" and "insignificant," only increasingly strong evidence as the P-value decreases.* There is no practical distinction between the P-values 0.049 and 0.051. It makes no sense to treat $\alpha = 0.05$ as a universal rule for what is significant.

EXERCISE

10.57 IS IT SIGNIFICANT? Suppose that in the absence of special preparation SAT Math scores vary normally with mean $\mu = 475$ and $\sigma = 100$. One hundred students go through a rigorous training program designed to raise their SAT Math scores by improving their mathematics skills. Carry out a test of

$$H_0: \mu = 475$$
$$H_a: \mu > 475$$

in each of the following situations:

(a) The students' average score is $\bar{x} = 491.4$. Is this result significant at the 5% level?

(b) The average score is $\bar{x} = 491.5$. Is this result significant at the 5% level?

The difference between the two outcomes in (a) and (b) is of no importance. Beware attempts to treat $\alpha = 0.05$ as sacred.

Statistical significance and practical significance

When a null hypothesis ("no effect" or "no difference") can be rejected at the usual levels, $\alpha = 0.05$ or $\alpha = 0.01$, there is good evidence that an effect is present. But that effect may be very small. When large samples are available, even tiny deviations from the null hypothesis will be significant.

EXAMPLE 10.18 WOUND HEALING TIME

Suppose we're testing a new antibacterial cream, "Formulation NS," on a small cut made on the inner forearm. We know from previous research that with no medication, the mean healing time (defined as the time for the scab to fall off) is 7.6 days, with a standard deviation of 1.4 days. The claim we want to test here is that Formulation NS speeds healing. We will use a 5% significance level.

Procedure: We cut 25 volunteer college students and apply Formulation NS to the wound. The mean healing time for these subjects is $\bar{x} = 7.1$ days. We will assume that $\sigma = 1.4$ days.

Step 1: Identify the population of interest and the parameter you want to draw conclusions about. We want to test a claim about the mean healing time μ in the population of people who treat cuts with Formulation NS. Our hypotheses are

$H_0: \mu = 7.6$	The mean healing time for cuts treated with Formulation NS is 7.6 days.
$H_a: \mu < 7.6$	The mean healing time for cuts treated with Formulation NS is less than 7.6 days. Formulation NS reduces healing time.

Step 2: Choose the appropriate inference procedure. Verify the conditions for using the selected procedure. Since we are assuming $\sigma = 1.4$ days, we should use a one-sample z test. We are told that the 25 subjects participating in this experiment are volunteers, so they are certainly not a true SRS from the population of interest. As a result, we may not be able to generalize our conclusions about Formulation NS to the larger population. Likewise, we do not know that the population of wound healing times with Formulation NS would be normally distributed. Unless the population distribution is severely nonnormal, the sampling distribution of $\bar{x}$ based on samples of size $n = 25$ should be approximately normal. We proceed with caution.

Step 3: If the conditions are met, carry out the inference procedure. If we assume H_0 is true, the probability of obtaining a sample of 25 subjects with a mean healing time at least as unusual as our sample is $P(\bar{x} < 7.1)$. Standardizing, we find that

$$P(\bar{x} < 7.1) = P\left(\frac{\bar{x} - 7.6}{1.4/\sqrt{25}} < \frac{7.1 - 7.6}{1.4/\sqrt{25}}\right) = P(Z < -1.79) = 0.0367$$

Step 4: Interpret your results in the context of the problem. Since $0.0367 < \alpha = 0.05$, we reject H_0 and conclude that Formulation NS's healing effect is *statistically significant.* However, this result is not *practically important.* Having your scab fall off half a day sooner is no big deal.

Remember the wise saying: *Statistical significance is not the same thing as practical significance.* Exercise 10.58 demonstrates in detail the effect on *P* of increasing the sample size.

The remedy for attaching too much importance to statistical significance is to pay attention to the actual data as well as to the *P*-value. Plot your data and examine them carefully. Are there outliers or other deviations from a consistent pattern? A few outlying observations can produce highly significant results if you blindly apply common tests of significance. Outliers can also destroy the significance of otherwise convincing data. The foolish user of statistics who feeds the data to a computer without exploratory analysis will often be embarrassed. Is the effect you are seeking visible in your plots? If not, ask yourself if the effect is large enough to be practically important. It is usually wise to give a confidence interval for the parameter in which you are interested. A confidence interval actually estimates the size of an effect, rather than simply asking if it is too large to reasonably occur by chance alone. Confidence intervals are not used as often as they should be, whereas tests of significance are perhaps overused.

EXERCISES

10.58 COACHING AND THE SAT Let us suppose that SAT Math scores in the absence of coaching vary normally with mean $\mu = 475$ and $\sigma = 100$. Suppose also that coaching may change μ but does not change σ. An increase in the SAT Math score from 475 to 478 is of no importance in seeking admission to college, but this unimportant change can be statistically very significant. To see this, calculate the *P*-value for the test of

$$H_0: \mu = 475$$
$$H_a: \mu > 475$$

in each of the following situations:

(a) A coaching service coaches 100 students. Their SAT Math scores average $\bar{x} = 478$.

(b) By the next year, the service has coached 1000 students. Their SAT Math scores average $\bar{x} = 478$.

(c) An advertising campaign brings the number of students coached to 10,000. Their average score is still $\bar{x} = 478$.

10.59 COACHING AND THE SAT, II Give a 99% confidence interval for the mean SAT Math score μ after coaching in each part of the previous exercise. For large samples, the confidence interval tells us, "Yes, the mean score is higher than 475 after coaching, but only by a small amount."

Statistical inference is not valid for all sets of data

We emphasize again that badly designed surveys or experiments often produce invalid results. Formal statistical inference cannot correct basic flaws in the design. Each test is valid only in certain circumstances, with properly produced data being particularly important. The z test, for example, should bear the same warning label that we attached on page 553 to the z confidence interval. Similar warnings accompany the other tests that we will learn.

EXAMPLE 10.19 DOES MUSIC INCREASE WORKER PRODUCTIVITY?

Hawthorne effect

You wonder whether background music would improve the productivity of the staff who process mail orders in your business. After discussing the idea with the workers, you add music and find a significant increase. You should not be impressed. In fact, almost any change in the work environment together with knowledge that a study is under way will produce a short-term productivity increase. This is the ***Hawthorne effect***, named after the Western Electric manufacturing plant where it was first noted.

The significance test correctly informs you that an increase has occurred that is larger than would often arise by chance alone. It does not tell you *what* other than chance caused the increase. The most plausible explanation is that workers change their behavior when they know they are being studied. Your experiment was uncontrolled, so the significant result cannot be interpreted. A randomized comparative experiment would isolate the actual effect of background music and so make significance meaningful.

Tests of significance and confidence intervals are based on the laws of probability. Randomization in sampling or experimentation ensures that these laws apply. Yet we must often analyze data that do not arise from randomized samples or experiments. To apply statistical inference to such data, we must have confidence in the use of probability to describe the data. The diameters of successive holes bored in auto engine blocks during production, for example, may behave like a random sample from a normal distribution. We can check this probability model by examining the data. If the model appears correct, we can apply the recipes of this chapter to do inference about the process mean diameter μ. Always ask how the data were produced, and don't be too impressed by *P*-values on a printout until you are confident that the data deserve a formal analysis.

EXERCISE

10.60 A CALL-IN OPINION POLL A local television station announces a question for a call-in opinion poll on the six o'clock news, and then gives the response on the eleven o'clock news. Today's question concerns a proposed gun-control ordinance. Of the 2372 calls received, 1921 oppose the new law. The station, following standard statistical practice, makes a confidence statement: "81% of the Channel 13 Pulse Poll sample oppose gun control. We can be 95% confident that the proportion of all viewers who oppose the law is within 1.6% of the sample result." Is the station's conclusion justified? Explain your answer.

Beware of multiple analyses

Statistical significance ought to mean that you have found an effect that you were looking for. The reasoning behind statistical significance works well if you decide what effect you are seeking, design a study to search for it, and use a test of significance to weigh the evidence you get. In other settings, significance may have little meaning.

EXAMPLE 10.20 PREDICTING SUCCESSFUL TRAINEES

You want to learn what distinguishes managerial trainees who eventually become executives from those who, after expensive training, don't succeed, and who leave the company. You have abundant data on past trainees—data on their personalities and goals, their college preparation and performance, even their family backgrounds and their hobbies. Statistical software makes it easy to perform dozens of significance tests on these dozens of variables to see which ones best predict later success. Aha! You find that future executives are significantly more likely than washouts to have an urban or suburban upbringing and an undergraduate degree in a technical field.

Before basing future recruiting on these findings, pause for a moment of reflection. When you make dozens of tests at the 5% level, you expect a few of them to be significant by chance alone. After all, results significant at the 5% level do occur 5 times in 100 in the long run even when H_0 is true. Running one test and reaching the $\alpha = 0.05$ level is reasonably good evidence that you have found something. Running several dozen tests and reaching that level once or twice is not.

There are methods for testing many hypotheses simultaneously while controlling the risk of false findings of significance. But if you carry out many individual tests without these special methods, finding a few small *P*-values is only suggestive, not conclusive. The same is true of less formal analyses. Searching the trainee data for the variable with the biggest difference between future washouts and future executives, then testing whether that difference is significant, is bad statistics. The P-value assumes you had that specific difference in mind before you looked at the data. It is very misleading when applied to the largest of many differences.

Searching data for suggestive patterns is certainly legitimate. Exploratory data analysis is an important aspect of statistics. But the reasoning of formal inference does not apply when your search for a striking effect in the data is successful. The remedy is clear. Once you have a hypothesis, design a study to search specifically for the effect you now think is there. If the result of this study is statistically significant, you have real evidence.

EXERCISE

10.61 DO YOU HAVE ESP? A researcher looking for evidence of extrasensory perception (ESP) tests 500 subjects. Four of these subjects do significantly better ($P < 0.01$) than random guessing.

(a) Is it proper to conclude that these four people have ESP? Explain your answer.

(b) What should the researcher now do to test whether any of these four subjects have ESP?

SUMMARY

P-values are more informative than the reject-or-not result of a fixed level α test. Beware of placing too much weight on traditional values of α, such as $\alpha = 0.05$.

Very small effects can be highly significant (small P), especially when a test is based on a large sample. A statistically significant effect need not be practically important. Plot the data to display the effect you are seeking, and use confidence intervals to estimate the actual value of parameters.

On the other hand, lack of significance does not imply that H_0 is true, especially when the test is based on just a few observations.

Significance tests are not always valid. Faulty data collection, outliers in the data, and testing a hypothesis on the same data that suggested the hypothesis can invalidate a test.

Many tests run at once will probably produce some significant results by chance alone, even if all the null hypotheses are true.

SECTION 10.3 EXERCISES

10.62 Which of the following questions does a test of significance answer?

(a) Is the sample or experiment properly designed?

(b) Is the observed effect due to chance?

(c) Is the observed effect important?

10.63 PACKAGE DESIGNS A company compares two package designs for a laundry detergent by placing bottles with both designs on the shelves of several markets. Checkout scanner data on more than 5000 bottles bought show that more shoppers bought Design A than Design B. The difference is statistically significant ($P = 0.02$). Can we conclude that consumers strongly prefer Design A? Explain your answer.

10.64 SCHIZOPHRENICS A group of psychologists once measured 77 variables on a sample of schizophrenic people and a sample of people who were not schizophrenic. They compared the two samples using 77 separate significance tests. Two of these tests were significant at the 5% level. Suppose that there is in fact no difference on any of the 77 variables between people who are and people who are not schizophrenic in the adult population. Then all 77 null hypotheses are true.

(a) What is the probability that one specific test shows a difference significant at the 5% level?

(b) Why is it not surprising that 2 of the 77 tests were significant at the 5% level?

10.65 RADAR DETECTORS AND SPEEDING Researchers observed the speed of cars on a rural highway (speed limit 55 miles per hour) before and after police radar was directed at them. They compared the speed of cars that had radar detectors with the speed of cars without detectors. Here are the mean speeds (miles per hour) observed for 22 cars with radar detectors and 46 cars without:[11]

	Radar detector?	
	Yes	No
Before radar	70	68
At radar	59	67

The study report says: "Those vehicles with radar detectors were significantly faster ($P < 0.01$) than those without them before radar exposure. . . and significantly slower immediately after ($P < 0.0001$)."

(a) Explain in simple language why these P-values are good evidence that drivers with radar detectors do behave differently.

(b) Despite the fact that $P < 0.01$, there is little difference in the mean speeds of drivers with and without radar detectors before the radar is turned on. Explain in simple language how so small a difference can be statistically significant.

10.4 INFERENCE AS DECISION

Tests of significance assess the strength of evidence against the null hypothesis. We measure evidence by the P-value, which is a probability computed under the assumption that H_0 is true. The alternative hypothesis (the statement we seek evidence for) enters the test only to help us see what outcomes count against the null hypothesis.

Using significance tests with fixed level α, however, suggests another way of thinking. A level of significance α chosen in advance points to the outcome of the test as a *decision*. If our result is significant at level α, we reject H_0 in favor of H_a. Otherwise, we fail to reject H_0. The transition from measuring the strength of evidence to making a decision is not a small step. Many statisticians feel that making decisions should be left to the user rather than built into the statistical test. A test result is only one among many factors that influence a decision.

acceptance sampling

Yet there are circumstances that call for a decision or action as the end result of inference. ***Acceptance sampling*** is one such circumstance. A potato chip manufacturer agrees that each batch of potato chips produced must meet certain quality standards. When a batch of chips is produced, a company employee inspects a sample of the chips. On the basis of the sample outcome, the company either accepts or rejects the entire batch of chips. We will use acceptance sampling to show how a different concept—inference as decision—changes the reasoning used in in tests of significance.

Type I and Type II errors

Tests of significance concentrate on H_0, the null hypothesis. If a decision is called for, however, there is no reason to single out H_0. There are simply two hypotheses, and we must accept one and reject the other. It is convenient to continue to call the two hypotheses H_0 and H_a, but H_0 no longer has the special status (the statement we try to find evidence against) that it had in tests of significance. In the acceptance sampling problem, we must decide between

H_0: the batch of potato chips meets standards

H_a: the potato chips do not meet standards

on the basis of a sample of potato chips.

We hope that our decision will be correct, but sometimes it will be wrong. There are two types of incorrect decisions. We can accept a bad batch of chips, or we can reject a good batch. Accepting a bad batch may upset consumers, whereas rejecting a good batch hurts the company. To distinguish these two types of error, we give them specific names.

TYPE I AND TYPE II ERRORS

If we reject H_0 (accept H_a) when in fact H_0 is true, this is a **Type I error.**

If we accept (reject H_a) H_0 when in fact H_a is true, this is a **Type II error.**

The possibilities are summed up in Figure 10.19. If H_0 is true, our decision is either correct (if we accept H_0) or is a Type I error. If H_a is true, our decision is either correct or is a Type II error. Only one error is possible at one time.

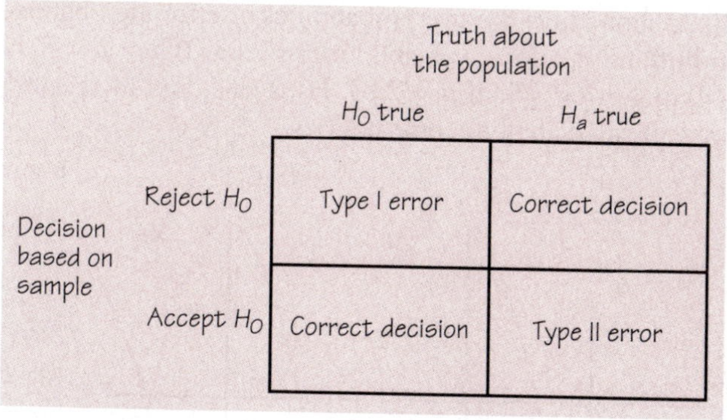

		Truth about the population	
		H_0 true	H_a true
Decision based on sample	Reject H_0	Type I error	Correct decision
	Accept H_0	Correct decision	Type II error

FIGURE 10.19 The two types of error in testing hypotheses.

Error probabilities

We assess any rule for making decisions by looking at the probabilities of the two types of error. This is in keeping with the idea that statistical inference is based on asking, "What would happen if I used this procedure many times?"

Significance tests with fixed level α give a rule for making decisions, because the test either rejects H_0 or fails to reject it. If we adopt the decision-making way of thought, failing to reject H_0 means deciding that H_0 is true. We can then describe the performance of a test by the probabilities of Type I and Type II errors.

EXAMPLE 10.21 ARE THESE POTATO SHIPS TOO SALTY?

The mean salt content of a certain type of potato chip is supposed to be 2.0 milligrams (mg). The salt content of these chips varies normally with standard deviation $\sigma = 0.1$ mg. From each batch produced, an inspector takes a sample of 50 chips and measures the salt content of each chip. The inspector rejects the entire batch if the sample mean salt content is significantly different from 2 mg at the 5% significance level.

This is a test of the hypotheses

$$H_0: \mu = 2$$
$$H_a: \mu \neq 2$$

To carry out the test, the company statistician computes the z statistic $z = \dfrac{\bar{x} - 2}{0.1/\sqrt{50}}$ and rejects H_0 if $z < -1.96$ or $z > 1.96$. A Type I error is to reject H_0 when in fact $\mu = 2$.

What about Type II errors? Because there are many values of μ in H_a, we will concentrate on one value. The potato chip company decides that any batch with a mean salt content as far away from 2 as 2.05 should be rejected. So a particular Type II error is to accept H_0 when in fact $\mu = 2.05$.

Figure 10.20 shows how the two probabilities of error are obtained from the *two* sampling distributions of $\bar{x}$, for $\mu = 2$ and for $\mu = 2.05$. When $\mu = 2$, H_0 is true and to reject H_0 is a Type I error. When $\mu = 2.05$, H_a is true, and to accept H_0 is a Type II error. Next, we will calculate these probabilities.

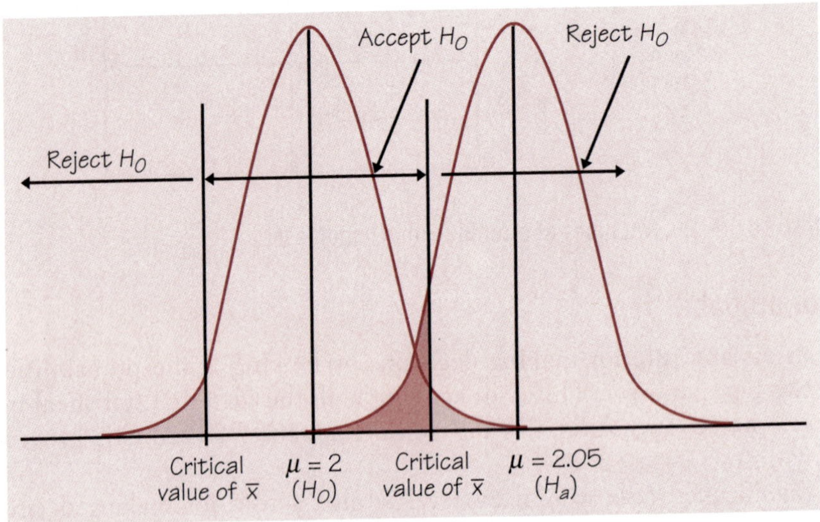

FIGURE 10.20 The two error probabilities for Example 10.21. The probability of a Type I error (light shaded area) is the probability of rejecting H_0: $\mu = 2$ when in fact $\mu = 2$. The probability of a Type II error (dark shaded area) is the probability of accepting H_0 when in fact $\mu = 2.05$.

The probability of a Type I error is the probability of rejecting H_0 when it is really true. This is the probability that $|z| \geq 1.96$ when $\mu = 2$. But this is exactly the significance level of the test. The critical value 1.96 was chosen to make this probability 0.05, so we do not have to compute it again. The definition of "significance level 0.05" is that values of z this extreme will occur with probability 0.05 when H_0 is true.

SIGNIFICANCE AND TYPE I ERROR

The significance level α of any fixed level test is the probability of a Type I error. That is, α is the probability that the test will reject the null hypothesis H_0 when H_0 is in fact true.

The probability of a Type II error for the particular alternative $\mu = 2.05$ in Example 10.21 is the probability that the test will accept H_0 when μ has this alternative value. This is the probability that the test statistic z falls between -1.96 and 1.96, calculated assuming that $\mu = 2.05$. This probability is not $1 - 0.05$ because the probability 0.05 was found assuming that $\mu = 2$. Here is the calculation for Type II error.

EXAMPLE 10.22 CALCULATING TYPE II ERROR

To calculate the probability of a Type II error:

Step 1: Write the rule for accepting H_0 in terms of $\bar{x}$. The test accepts H_0 when

$$-1.96 \le \frac{\bar{x}-2}{0.1\sqrt{50}} \le 1.96$$

This is the same as

$$2 - 1.96\left(\frac{0.1}{\sqrt{50}}\right) \le \bar{x} \le 2 + 1.96\left(\frac{0.1}{\sqrt{50}}\right)$$

or, doing the arithmetic,

$$1.9723 \le \bar{x} \le 2.0277$$

This step does not involve the particular alternative $\mu = 2.05$.

Step 2: Find the probability of accepting H_0 assuming that the alternative is true. Take $\mu = 2.05$ and standardize to find the probability.

$$\begin{aligned} P(\text{Type II error}) &= P(1.9723 \le \bar{x} \le 2.0277) \\ &= P\left(\frac{1.9723-2.05}{0.1/\sqrt{50}} \le \frac{\bar{x}-2.05}{0.1/\sqrt{50}} \le \frac{2.0277-2.05}{0.1/\sqrt{50}}\right) \\ &= P(-5.49 \le Z \le -1.58) \\ &= 0.0571 \end{aligned}$$

Figure 10.21 illustrates this error probability in terms of the sampling distribution of $\bar{x}$ when $\mu = 2.05$. The test will wrongly accept the hypothesis that $\mu = 2$ in about 6% of all samples when in fact $\mu = 2.05$.

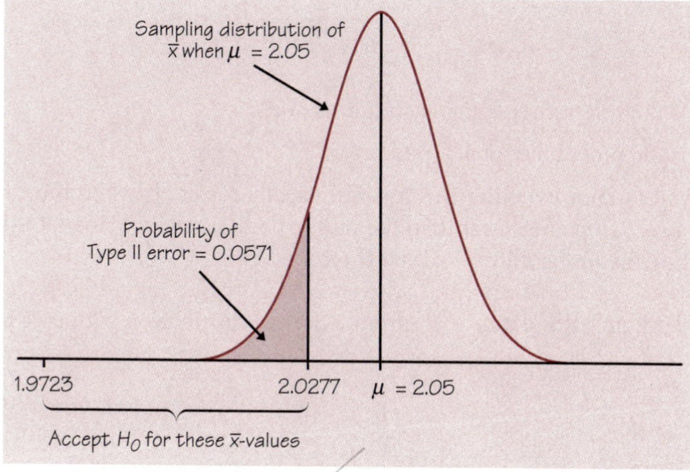

FIGURE 10.21 The probability of a Type II error for Example 10.22. This is the probability that the test accepts H_0 when the alternative hypothesis is true.

How do we interpret the result of the calculations performed in Example 10.22? The probability of 0.0571 tells us that this test will lead us to fail to reject H_0: $\mu = 2$ for about 6% of all batches of chips with $\mu = 2.05$. In other words, we will accept 6% of batches of potato chips so bad that their mean salt content is 2.05 mg. Since we used $\alpha = 0.05$ in our calculations, the probability of a Type I error is 0.05. This means that our test will reject 5% of all good batches of potato chips for which $\mu = 2$. Calculations of error probabilities help the potato chip manufacturer decide whether the test is satisfactory.

EXERCISES

10.66 MEDICAL SCREENING Your company markets a computerized medical diagnostic program. The program scans the results of routine medical tests (pulse rate, blood tests, etc.) and either clears the patient or refers the case to a doctor. The program is used to screen thousands of people who do not have specific medical complaints. The program makes a decision about each person.

(a) What are the two hypotheses and the two types of error that the program can make? Describe the two types of error in terms of "false positive" and "false negative" test results.

(b) The program can be adjusted to decrease one error probability at the cost of an increase in the other error probability. Which error probability would you choose to make smaller, and why? (This is a matter of judgment. There is no single correct answer.)

10.67 NAEP SCORES You have the NAEP quantitative scores for an SRS of 840 young Americans. You plan to test hypotheses about the population mean score,

$$H_0: \mu = 275$$
$$H_a: \mu < 275$$

at the 1% level of significance. The population standard deviation is known to be $\sigma = 60$. The z test statistic is

$$z = \frac{\bar{x} - 275}{60 / \sqrt{840}}$$

(a) What is the rule for rejecting H_0 in terms of z?

(b) What is the probability of a Type I error?

(c) You want to know whether this test will usually reject H_0 when the true population mean is 270, 5 points lower than the null hypothesis claims. Answer this question by calculating the probability of a Type II error when $\mu = 270$.

10.68 You have an SRS of size $n = 9$ from a normal distribution with $\sigma = 1$. You wish to test

$$H_0: \mu = 0$$
$$H_a: \mu > 0$$

You decide to reject H_0 if $\bar{x} > 0$ and to accept H_0 otherwise.

(a) Find the probability of a Type I error. That is, find the probability that the test rejects H_0 when in fact $\mu = 0$.

(b) Find the probability of a Type II error when $\mu = 0.3$. This is the probability that the test accepts H_0 when in fact $\mu = 0.3$.

(c) Find the probability of a Type II error when $\mu = 1$.

10.69 OPENING A RESTAURANT You are thinking about opening a restaurant and are searching for a good location. From research you have done, you know that the mean income of those living near the restaurant must be over \$45,000 to support the type of upscale restaurant you wish to open. You decide to take a simple random sample of 50 people living near one potential location. Based on the mean income of this sample, you will decide whether to open a restaurant there. A number of similar studies have shown that σ = \$5,000.[12]

(a) Describe the two types of errors that you might make. Identify which is a Type I error and which is a Type II error.

(b) Which of the two types of error is most serious? Explain!

(c) State the null and alternative hypotheses.

(d) If you had to choose one of the "standard" significance levels for your significance test, would you choose $\alpha = 0.01$, 0.05, or 0.10? Justify your choice.

(e) Based on your choice in part (d), how high will the sample mean need to be before you decide to open a restaurant in that area?

(f) If the mean income in a certain area is \$47,000, how likely are you *not* to open a restaurant in that area? What is the probability of a Type II error?

Power

A test makes a Type II error when it fails to reject a null hypothesis that really is false. A high probability of a Type II error for a particular alternative means that the test is not sensitive enough to usually detect that alternative. Calculations of the probability of Type II errors are therefore useful even if you don't think that a statistical test should be viewed as making a decision. The language used is a bit different in the significance test setting. It is usual to report the probability that a test *does* reject H_0 when an alternative is true. The higher this probability is, the more sensitive the test is.

POWER

The probability that a fixed level α significance test will reject H_0 when a particular alternative value of the parameter is true is called the **power** of the test against that alternative.

The power of a test against any alternative is 1 minus the probability of a Type II error for that alternative.

Calculations of power are essentially the same as calculations of the probability of Type II error. In Example 10.22, the power is the probability of *rejecting* H_0 in step 2 of the calculation. It is 1 – 0.0571, or 0.9429.

Calculations of *P*-values and calculations of power both say what would happen if we repeated the test many times. A *P*-value describes what would happen supposing that the null hypothesis is true. Power describes what would happen supposing that a particular alternative is true.

In planning an investigation that will include a test of significance, a careful user of statistics decides what alternatives the test should detect and checks that the power is adequate. The power depends on which particular parameter value in H_a we are interested in. Values of the mean μ that are in H_a but lie close to the hypothesized value μ_0 are harder to detect (lower power) than values of μ that are far from μ_0. If the power is too low, a larger sample size will increase the power for the same significance level α. In order to calculate power, we must fix an α so that there is a fixed rule for rejecting H_0. We prefer to report *P*-values rather than to use a fixed significance level. The usual practice is to calculate the power at a common significance level, such as $\alpha = 0.05$, even though you intend to report a *P*-value.

EXAMPLE 10.23 EXERCISE IS GOOD

Can a 6-month exercise program increase the total body bone mineral content (TBBMC) of young women? A team of researchers is planning a study to examine this question. Based on the results of a previous study, they are willing to assume that $\sigma = 2$ for the percent change in TBBMC over the 6-month period. A change in TBBMC of 1% would be considered important, and the researchers would like to have a reasonable chance of detecting a change this large or larger. Is 25 subjects a large enough sample for this project?

We will answer this question by calculating the power of the significance test that will be used to evaluate the data to be collected. The calculation consists of three steps:

1. State H_0, H_a, the particular alternative we want to detect, and the significance level α.

2. Find the values of $\bar{x}$ that will lead us to reject H_0.

3. Calculate the probability of observing these values of $\bar{x}$ when the alternative is true.

Step 1: The null hypothesis is that the exercise program has no effect on TBBMC. In other words, the mean percent change is zero. The alternative is that exercise is beneficial; that is, the mean change is positive. Formally, we have

$$H_0: \mu = 0$$
$$H_a: \mu > 0$$

The alternative of interest is $\mu = 1\%$. A 5% test of significance will be used.

Step 2: The z test rejects H_0 at the $\alpha = 0.05$ level whenever

$$z = \frac{\bar{x} - \mu_0}{\sigma / \sqrt{n}} = \frac{\bar{x} - 0}{2 / \sqrt{25}} \geq 1.645$$

Be sure you understand why we use 1.645. Rewrite this in terms of $\bar{x}$:

$$x \geq 1.645 \frac{2}{\sqrt{25}}$$

$$\bar{x} \geq 0.658$$

Because the significance level is $\alpha = 0.05$, this event has probability 0.05 of occurring *when the population mean* μ is 0.

Step 3: The power against the alternative $\mu = 1\%$ increase in TBBMC is the probability that H_0 will be rejected *when in fact* $\mu = 1$. We calculate this probability by standardizing $\bar{x}$, using the value $\mu = 1$, the population standard deviation $\sigma = 2$, and the sample size $n = 25$. The power is

$$P(\bar{x} \geq 0.658 \text{ when } \mu = 1) = P\left(\frac{\bar{x} - \mu}{\sigma / \sqrt{n}} \geq \frac{0.658 - 1}{2 / \sqrt{25}}\right)$$
$$= P(Z \geq -0.855) = 0.80$$

Figure 10.22 illustrates the power with the sampling distribution of $\bar{x}$ when $\mu = 1$. This significance test rejects the null hypothesis that exercise has no effect on TBBMC 80% of the time if the true effect of exercise is a 1% increase in TBBMC. If the true effect of exercise is a greater percent increase, the test will have greater power; it will reject with a higher probability.

High power is desirable. Along with 95% confidence intervals and 5% significance tests, 80% power is becoming a standard. Many U.S. government agencies that provide research funds require that the sample size for the funded studies be sufficient to detect important results 80% of the time using a 5% test of significance.

Increasing the power

Suppose you have performed a power calculation and found that the power is too small. What can you do to increase it? Here are four ways:

- Increase α. A 5% test of significance will have a greater chance of rejecting the alternative than a 1% test because the strength of evidence required for rejection is less.

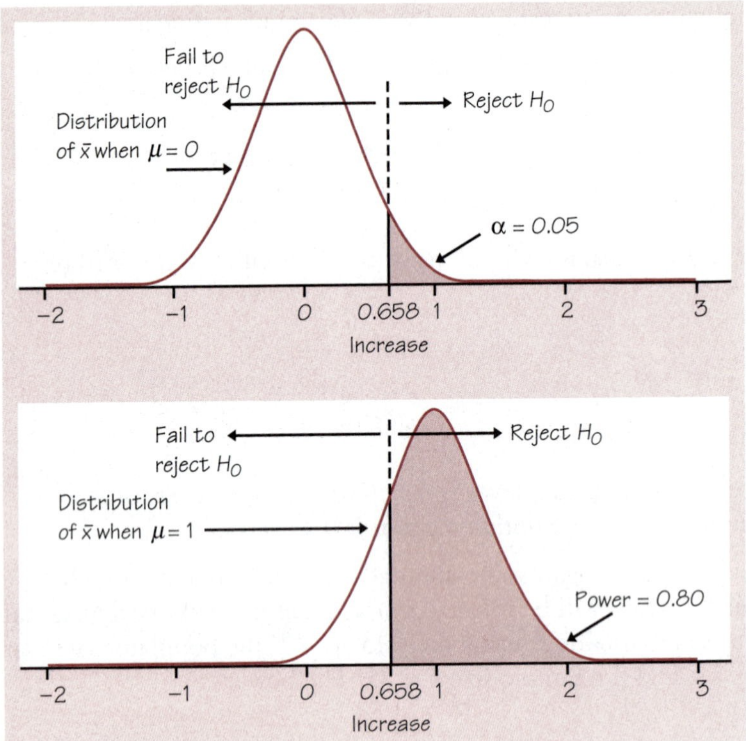

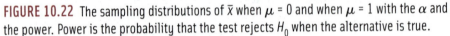

FIGURE 10.22 The sampling distributions of $\bar{x}$ when $\mu = 0$ and when $\mu = 1$ with the α and the power. Power is the probability that the test rejects H_0 when the alternative is true.

- Consider a particular alternative that is farther away from μ_0. Values of μ that are in H_a but lie close to the hypothesized value μ_0 are harder to detect (lower power) than values of μ that are far from μ_0.
- Increase the sample size. More data will provide more information about $\bar{x}$ so we have a better chance of distinguishing values of μ.
- Decrease σ. This has the same effect as increasing the sample size: more information about μ. Improving the measurement process and restricting attention to a subpopulation are two common ways to decrease σ.

Power calculations are important in planning studies. Using a significance test with low power makes it unlikely that you will find a significant effect even if the truth is far from the null hypothesis. A null hypothesis that is in fact false can become widely believed if repeated attempts to find evidence against it fail because of low power.

EXERCISES

10.70 COLA SWEETNESS: POWER The cola maker of Example 10.9 (page 560) determines that a sweetness loss is too large to accept if the mean response for all tasters is $\mu = 1.1$. Will a 5% significance test of the hypotheses

$$H_0\colon \mu = 0$$
$$H_a\colon \mu > 0$$

based on a sample of 10 tasters usually detect a change this great?

We want the power of the test against the alternative $\mu = 1.1$. This is the probability that the test rejects H_0 when $\mu = 1.1$ is true. The calculation method is similar to that for Type II error.

(a) *Step 1: Write the rule for rejecting H_0 in terms of $\bar{x}$.* We know that $\sigma = 1$, so the test rejects H_0 at the $\alpha = 0.05$ level when

$$z = \frac{\bar{x} - 0}{1/\sqrt{10}} \geq 1.645$$

Restate this in terms of $\bar{x}$.

(b) *Step 2: The power is the probability of this event supposing that the alternative is true.* Standardize using $\mu = 1.1$ to find the probability that $\bar{x}$ takes a value that leads to rejection of H_0.

10.71 SAT MATH SCORES: POWER Exercise 10.55 (page 585) gives a test of a hypothesis about the SAT Math scores of California high school students based on an SRS of 500 students. The hypotheses are

$$H_0\colon \mu = 450$$
$$H_a\colon \mu > 450$$

Assume that the population standard deviation is $\sigma = 100$. The test rejects H_0 at the 1% level of significance when $z \geq 2.326$, where

$$z = \frac{\bar{x} - 450}{100/\sqrt{500}}$$

Is this test sufficiently sensitive to usually detect an increase of 10 points in the population mean SAT Math score? Answer this question by calculating the power of the test against the alternative $\mu = 460$.

10.72 FILLING COLA BOTTLES: POWER Exercise 10.39 (page 576) concerns a test about the mean contents of cola bottles. The hypotheses are

$$H_0\colon \mu = 300$$
$$H_a\colon \mu < 300$$

The sample size is $n = 6$, and the population is assumed to have a normal distribution with $\sigma = 3$. A 5% significance test rejects H_0 if $z \leq -1.645$, where the test statistic z is

$$z = \frac{\bar{x} - 300}{3/\sqrt{6}}$$

Power calculations help us see how large a shortfall in the bottle contents the test can be expected to detect.

(a) Find the power of this test against the alternative $\mu = 299$.

(b) Find the power against the alternative $\mu = 295$.

(c) Is the power against $\mu = 290$ higher or lower than the value you found in (b)? (Don't actually calculate that power.) Explain your answer.

10.73 FILLING COLA BOTTLES: POWER, II Increasing the sample size increases the power of a test when the level α is unchanged. Suppose that in the previous exercise a sample of n bottles had been measured. In that exercise, $n = 6$. The 5% significance test still rejects H_0 when $z \leq -1.645$, but the z statistic is now

$$z = \frac{\bar{x} - 300}{3/\sqrt{n}}$$

(a) Find the power of this test against the alternative $\mu = 299$ when $n = 25$.

(b) Find the power against $\mu = 299$ when $n = 100$.

Different views of statistical tests

The distinction between tests of significance and tests as rules for deciding between two hypotheses does not lie in the calculations but in the reasoning that motivates the calculations. In a test of significance we focus on a single hypothesis (H_0) and a single probability (the P-value). The goal is to measure the strength of the sample evidence against H_0. Calculations of power are done to check the sensitivity of the test. If we cannot reject H_0, we conclude only that there is not sufficient evidence against H_0, not that H_0 is actually true. If the same inference problem is thought of as a decision problem, we focus on two hypotheses and give a rule for deciding between them based on the sample evidence. We therefore must focus equally on two probabilities, the probabilities of the two types of error. We must choose one or the other hypothesis and cannot abstain on grounds of insufficient evidence.

testing hypotheses

There are clear distinctions between the two ways of thinking about statistical tests. But sometimes the two approaches merge. Jerzy Neyman advocated an approach called ***testing hypotheses*** that mixes the reasoning of significance tests and decision rules as follows:

1. State H_0 and H_a just as in a test of significance. In particular, we are seeking evidence against H_0.

2. Think of the problem as a decision problem, so that the probabilities of Type I and Type II errors are relevant.

3. Because of Step 1, Type I errors are more serious. So choose an α (significance level) and consider only tests with probability of Type I error no greater than α.

4. Among these tests, select one that makes the probability of a Type II error as small as possible (that is, power as large as possible). If this probability is too large, you will have to take a larger sample to reduce the chance of an error.

Hypothesis testing is often emphasized in mathematical presentations of statistics because Neyman developed an impressive mathematical theory. In simple settings, this theory shows how to find the test that has the smallest possible probability of a Type II error among all tests with a given probability (like 0.05) of a Type I error. In part because such pleasing results aren't available for many practical settings, the significance test way of thinking prevails in statistical practice.

SUMMARY

An alternative to significance testing regards H_0 and H_a as two statements of equal status that we must decide between. This **decision analysis** point of view regards statistical inference in general as giving rules for making decisions in the presence of uncertainty.

In the case of testing H_0 versus H_a, decision analysis chooses a decision rule on the basis of the probabilities of two types of error. A **Type I error** occurs if we reject H_0 when it is in fact true. A **Type II error** occurs if we accept H_0 when in fact H_a is true.

The **power** of a significance test measures its ability to detect an alternative hypothesis. The power against a specific alternative is the probability that the test will reject H_0 when the alternative is true.

In a fixed level α significance test, the significance level α is the probability of a Type I error, and the power against a specific alternative is 1 minus the probability of a Type II error for that alternative.

Increasing the size of the sample increases the power (reduces the probability of a Type II error) when the significance level remains fixed.

SECTION 10.4 EXERCISES

10.74 VIDEO SCREEN TENSION: POWER Power calculations for two-sided tests follow the same outline as for one-sided tests. Example 10.16 (page 578) presents a test of

$$H_0: \mu = 275$$
$$H_a: \mu \neq 275$$

at the 1% level of significance. The sample size is $n = 20$ and $\sigma = 43$. We will find the power of this test against the alternative $\mu = 240$.

(a) The test in Example 10.16 rejects H_0 when $|z| \geq 2.576$. The test statistic z is

$$z = \frac{\bar{x} - 275}{43/\sqrt{20}}$$

Write the rule for rejecting H_0 in terms of the values of $\bar{x}$. (Because the test is two-sided, it rejects when $\bar{x}$ is either too large or too small.)

(b) Now find the probability that $\bar{x}$ takes values that lead to rejecting H_0 if the true mean is $\mu = 240$. This probability is the power.

(c) What is the probability that this test makes a Type II error when $\mu = 240$?

10.75 BLOOD PRESSURE: POWER In Example 10.13 (page 573), a company medical director failed to find significant evidence that the mean blood pressure of a population of executives differed from the national mean $\mu = 128$. The medical director now wonders if the test used would detect an important difference if one were present. For the SRS of size 72 from a population with standard deviation $\sigma = 15$, the z statistic is

$$z = \frac{\bar{x} - 128}{15/\sqrt{72}}$$

The two-sided test rejects H_0: $\mu = 128$ at the 5% level of significance when $|z| \geq 1.96$.

(a) Find the power of the test against the alternative $\mu = 134$.

(b) Find the power of the test against $\mu = 122$. Can the test be relied on to detect a mean that differs from 128 by 6?

(c) If the alternative were farther from H_0, say $\mu = 136$, would the power be higher or lower than the values calculated in (a) and (b)?

10.76 IS THE STOCK MARKET EFFICIENT? You are reading an article in a business journal that discusses the "efficient market hypothesis" for the behavior of securities prices. The author admits that most tests of this hypothesis have failed to find significant evidence against it. But he says this failure is a result of the fact that the tests used have low power. "The widespread impression that there is strong evidence for market efficiency may be due just to a lack of appreciation of the low power of many statistical tests."[13]

Explain in simple language why tests having low power often fail to give evidence against a hypothesis even when the hypothesis is really false.

10.77 ANALYZING PHARMACEUTICALS: POWER Refer to Exercise 10.5 (page 548).

(a) Perform a test of

$$H_0: \mu = 0.86$$
$$H_a: \mu \neq 0.86$$

at the 1% significance level.

(b) What is the power of this test against the specific alternative $\mu = 0.845$?

CHAPTER REVIEW

Statistical inference draws conclusions about a population on the basis of sample data and uses probability to indicate how reliable the conclusions are. A confidence interval estimates an unknown parameter. A significance test shows how strong the evidence is for some claim about a parameter.

The probabilities in both confidence intervals and tests tell us what would happen if we used the recipe for the interval or test very many times. A confidence level is the probability that the recipe for a confidence interval actually produces an interval that contains the unknown parameter. A 95% confidence interval gives a correct result 95% of the time when we use it repeatedly. A *P*-value is the probability that the test would produce a result at least as extreme as the observed result if the null hypothesis really were true. That is, a *P*-value tells us how surprising the observed outcome is. Very surprising outcomes (small *P*-values) are good evidence that the null hypothesis is not true.

The figures on the next page present the ideas of confidence intervals and tests in picture form. These ideas are the foundation for the rest of this book. We will have much to say about many statistical methods and their use in practice. In every case, the basic reasoning of confidence intervals and significance tests remains the same. Here are the most important things you should be able to do after studying this chapter.

A. CONFIDENCE INTERVALS

1. State in nontechnical language what is meant by "95% confidence" or other statements of confidence in statistical reports.

2. Calculate a confidence interval for the mean μ of a normal population with known standard deviation σ, using the recipe $\bar{x} \pm z^* \sigma / \sqrt{n}$.

3. Recognize when you can safely use this confidence interval recipe and when the data collection design or a small sample from a skewed population makes it inaccurate. Understand that the margin of error does not include the effects of undercoverage, nonresponse, or other practical difficulties.

4. Understand how the margin of error of a confidence interval changes with the sample size and the level of confidence C.

5. Find the sample size required to obtain a confidence interval of specified margin of error m when the confidence level and other information are given.

B. SIGNIFICANCE TESTS

1. State the null and alternative hypotheses in a testing situation when the parameter in question is a population mean μ.

2. Explain in nontechnical language the meaning of the *P*-value when you are given the numerical value of *P* for a test.

3. Calculate the one-sample z statistic and the *P*-value for both one-sided and two-sided tests about the mean μ of a normal population.

4. Assess statistical significance at standard levels α, either by comparing P to α or by comparing z to standard normal critical values.

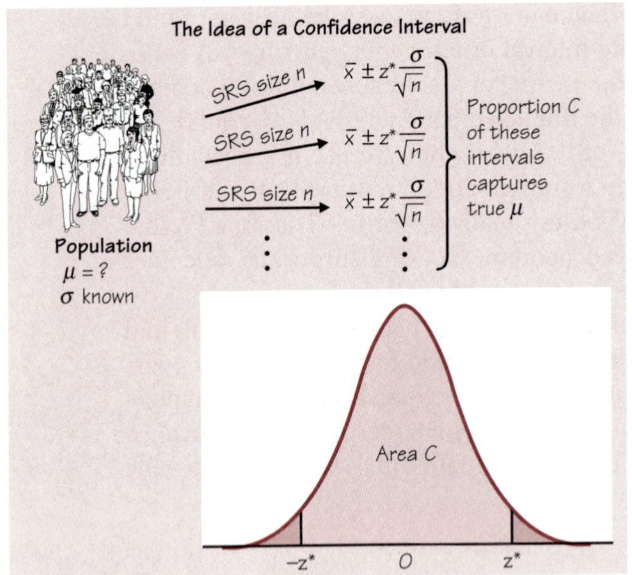

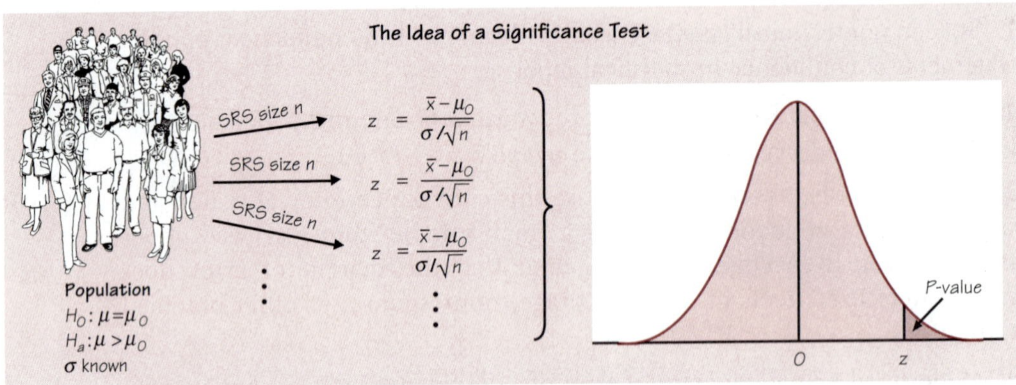

5. Recognize that significance testing does not measure the size or importance of an effect.

6. Recognize when you can use the z test and when the data collection design or a small sample from a skewed population makes it inappropriate.

7. Explain Type I error, Type II error, and power in a significance testing problem.

CHAPTER 10 REVIEW EXERCISES

10.78 STATING HYPOTHESES Each of the following situations requires a significance test about a population mean μ. State the appropriate null hypothesis H_0 and alternative hypothesis H_a in each case.

(a) Larry's car averages 32 miles per gallon on the highway. He now switches to a new motor oil that is advertised as increasing gas mileage. After driving 3000 highway miles with the new oil, he wants to determine if his gas mileage actually has increased.

(b) A university gives credit in french language courses to students who pass a placement test. The language department wants to see if students who get credit in this way differ in their understanding of spoken French from students who actually take the French courses. Some faculty think the students who test out of the courses are better, but others argue that they are weaker in oral comprehension. Experience has shown that the mean score of students in the courses on a standard listening test is 24. The language department gives the same listening test to a sample of 40 students who passed the credit examination to see if their performance is different.

10.79 THIS WINE STINKS Sulfur compounds cause "off-odors" in wine, so winemakers want to know the odor threshold, the lowest concentration of a compound that the human nose can detect. The odor threshold for dimethyl sulfide (DMS) in trained wine tasters is about 25 micrograms per liter of wine (μg/l). The untrained noses of consumers may be less sensitive, however. Here are the DMS odor thresholds for 10 untrained students:

31	31	43	36	23	34	32	30	20	24

Assume that the standard deviation of the odor threshold for untrained noses is known to be $\sigma = 7$ μg/l.

(a) Make a stemplot to verify that the distribution is roughly symmetric with no outliers. (More data confirm that there are no systematic departures from normality.)

(b) Give a 95% confidence interval for the mean DMS odor threshold among all students.

(c) Are you convinced that the mean odor threshold for students is higher than the published threshold, 25 μg/l? Carry out a significance test to justify your answer.

10.80 IRON DEFICIENCY IN INFANTS Researchers studying iron deficiency in infants examined infants who were following different feeding patterns. One group of 26 infants was being breast-fed. At 6 months of age, these children had mean hemoglobin level $\bar{x} =$ 12.9 grams per 100 milliliters of blood. Assume that the population standard deviation is $\sigma = 1.6$. Give a 90% confidence interval for the mean hemoglobin level of breast-fed infants. What conditions (other than the unrealistic assumption that we know σ) does the method you used to get the confidence interval require?

10.81 READING TEST SCORES Here are the Degree of Reading Power (DRP) scores for an SRS of 44 third-grade students from a suburban school district:[14]

40	26	39	14	42	18	25	43	46	27	19
47	19	26	35	34	15	44	40	38	31	46
52	25	35	35	33	29	34	41	49	28	52
47	35	48	22	33	41	51	27	14	54	45

DRP scores are approximately normal. Suppose that the standard deviation of scores in this school district is known to be $\sigma = 11$. The researcher believes that the mean score μ of all third graders in this district is higher than the national mean, which is 32.

(a) State H_0 and H_a to test this suspicion.

(b) Carry out the test.

10.82 WHY ARE LARGE SAMPLES BETTER? Statisticians prefer large samples. Describe briefly the effect of increasing the size of a sample (or the number of subjects in an experiment) on each of the following:

(a) The margin of error of a 95% confidence interval.

(b) The *P*-value of a test, when H_0 is false and all facts about the population remain unchanged as n increases.

(c) The power of a fixed level α test, when α, the alternative hypothesis, and all facts about the population remain unchanged.

10.83 EXPLAINING *P*-VALUES When asked to explain the meaning of "the *P*-value was $P = 0.03$," a student says, "This means there is only probability 0.03 that the null hypothesis is true." Is this an essentially correct explanation? Explain your answer.

10.84 EXPLAINING STATISTICAL SIGNIFICANCE Another student, when asked why statistical significance appears so often in research reports, says, "Because saying that results are significant tells us that they cannot easily be explained by chance variation alone." Do you think that this statement is essentially correct? Explain your answer.

10.85 WELFARE PROGRAMS A study compares two groups of mothers with young children who were on welfare two years ago. One group attended a voluntary training program offered free of charge at a local vocational school and advertised in the local news media. The other group did not choose to attend the training program. The study finds a significant difference ($P < 0.01$) between the proportions of the mothers in the two groups who are still on welfare. The difference is not only significant but quite large. The report says that with 95% confidence the percent of the nonattending group still on welfare is 21% ±4% higher than that of the group who attended the program. You are on the staff of a member of Congress who is interested in the plight of welfare mothers, and who asks you about the report.

(a) Explain in simple language what "a significant difference ($P < 0.01$)" means.

(b) Explain clearly and briefly what "95% confidence" means.

(c) Is this study good evidence that requiring job training of all welfare mothers would greatly reduce the percent who remain on welfare for several years?

10.86 STUDENT ATTITUDES The Survey of Study Habits and Attitudes (SSHA) is a psychological test that measures the motivation, attitude toward school, and study habits of students. Scores range from 0 to 200. The mean score for U.S. college students is about 115, and the standard deviation is about 30. A teacher who suspects that older students have better attitudes toward school gives the SSHA to 20 students who are at least 30 years of age. Their mean score is $\bar{x} = 135.2$.

(a) Assuming that $\sigma = 30$ for the population of older students, carry out a test of

$$H_0: \mu = 115$$
$$H_a: \mu > 115$$

Report the *P*-value of your test, and state your conclusion clearly.

(b) Your test in (a) required two important conditions in addition to the assumption that the value of σ is known. What are they? Which of these conditions is most important to the validity of your conclusion in (a)?

10.87 CORN YIELD The mean yield of corn in the United States is about 120 bushels per acre. A survey of 40 farmers this year gives a sample mean yield of $\bar{x} = 123.8$ bushels per acre. We want to know whether this is good evidence that the national mean this year is not 120 bushels per acre. Assume that the farmers surveyed are an SRS from the population of all commercial corn growers and that the standard deviation of the yield in this population is $\sigma = 10$ bushels per acre. Give the *P*-value for the test of

$$H_0: \mu = 120$$
$$H_a: \mu \neq 120$$

Are you convinced that the population mean is not 120 bushels per acre? Is your conclusion correct if the distribution of corn yields is somewhat nonnormal? Why?

10.88 CELL PHONES AND BRAIN CANCER Could the radiation from cell phones be harmful to users? Many studies have found little or no connection between using cell phones and various illnesses. Here is part of a news account of one study:[15]

A hospital study that compared brain cancer patients and a similar group without brain cancer found no statistically significant association between cell phone use and a group of brain cancers known as gliomas. But when 20 types of glioma were considered separately an association was found between phone use and one rare form. Puzzlingly, however, this risk appeared to decrease rather than increase with greater mobile phone use.

This is a pretty weak study. Let's dissect it.

(a) Explain why this study is not an experiment.

(b) What does "no statistically significant association" mean in plain language?

(c) Why are you not surprised that in 20 separate tests for 20 types of cancer, one was significant at the 5% level?

10.89 STRONG CHAIRS? A company that manufacturers classroom chairs for high school students claims that the mean braking strength of the chairs they make is 300 pounds. From years of production, they have seen that $\sigma = 15$ pounds. One of the chairs collapsed beneath a 220-pound student last week. You wonder whether the manufacturer is exaggerating about the breaking strength of their chairs.

(a) State null and alternative hypotheses in words and symbols.

(b) Describe a Type I error and a Type II error in this situation. Which is more serious?

(c) There are 30 chairs in your classroom. You decide to determine the breaking strength of each chair, and then to find the mean of those values. What values of $\overline{x}$ would cause you to reject H_0 at the 5% significance level?

(d) If the truth is that $\mu = 270$ pounds, find the probability that you will make a Type II error.

NOTES AND DATA SOURCES

1. Information from Francisco L. Rivera-Batiz, "Quantitative literacy and the likelihood of employment among young adults," *Journal of Human Resources*, 27 (1992), pp. 313–328.
2. Data provided by John Rousselle and Huei-Ru Shieh, Department of Restaurant, Hotel, and Institutional Management, Purdue University.
3. Data provided by Drina Iglesia, Department of Biological Sciences, Purdue University. The data are part of a larger study reported in D.D.S. Iglesia, E.J. Cragoe, Jr., and J.W. Vanable, "Electric field strength and epithelization in the newt (*Notophthalmus viridescens*)," *Journal of Experimental Zoology*, 274 (1996), pp. 56–62.
4. Based on G. Salvendy, G. P. McCabe, S. G. Sanders, J. L. Knight, and E. J. McCormick, "Impact of personality and intelligence on job satisfaction of assembly line and bench work–an industrial study," *Applied Ergonomics*, 13 (1982), pp. 293–299.
5. Data provided by Mugdha Gore and Joseph Thomas, Purdue University School of Pharmacy.
6. Data provided by Diana Schellenberg, Purdue University School of Health Sciences.
7. Data provided by Darlene Gordon, Purdue University.
8. William T. Robinson, "Sources of market pioneer advantages: the case of industrial goods industries," *Journal of Marketing Research*, 25 (February 1988), pp. 87–94.
9. D.L. Shankland et al., "The effect of 5-thio-D-glucose on insect development and its absorption by insects," *Journal of Insect Physiology*, 14 (1968), pp. 63–72.
10. For a discussion of statistical significance in the legal setting, see D. H. Kaye, "Is proof of statistical significance relevant?" *Washington Law Review*, 61 (1986), pp. 1333–1365. Kaye argues: "Presenting the *P*-value without characterizing the evidence by a significance test is a step in the right direction. Interval estimation, in turn, is an improvement over *P*-values."
11. These are some of the data from the EESEE story "Radar Detectors and Speeding." The study is reported in N. Teed, K. L. Adrian, and R. Knoblouch, "The duration of speed reductions attributable to radar detectors," *Accident Analysis and Prevention*, 25 (1991), pp. 131–137.
12. The idea for this problem was provided by Michael Legacy and Susan McGann.
13. Robert J. Schiller, "The volatility of stock market prices," *Science*, 235 (1987), pp. 33–36.

14. Data provided by Maribeth Cassidy Schmitt, from her Ph.D. dissertation, "The Effects of an Elaborated Directed Reading Activity on the Metacomprehension Skills of Third Graders," Purdue University, 1987.
15. Warren E. Leary, "Cell phones: questions but no answers," *New York Times*, October 26, 1999.

University College Library, London

WILLIAM S. GOSSET

Brewing Better Beer, Brewing New Statistics What would cause the head brewer of the famous Guinness brewery in Dublin, Ireland, not only to use statistics but to invent new statistical methods? The search for better beer, of course.

William S. Gosset (1876–1937), fresh from Oxford University, joined Guinness as a brewer in 1899. He soon became involved in experiments and in statistics to understand the data from those experiments. What are the best varieties of barley and hops for brewing? How should they be grown, dried, and stored? The results of the field experiments, as you can guess, varied. Statistical inference can uncover the pattern behind the variation. The statistical methods available at the turn of the century ended with a version of the z test for means—even confidence intervals were not yet available.

Gosset faced the problem we noted in using the z test to introduce the reasoning of statistical tests: he didn't know the population standard deviation σ. What is more, field experiments give only small numbers of observations. Just replacing σ by s in the z statistic and calling the result roughly normal wasn't accurate enough. So Gosset asked the key question: What is the exact sampling distribution of the statistic $(\bar{x} - \mu)/s$?

By 1907 Gosset was brewer-in-charge of Guinness's experimental brewery. He also had the answer to his question and had calculated a table of critical values for his new distribution. We call it the t distribution. The new t test identified the best barley variety, and Guinness promptly bought up all the available seed. Guinness allowed Gosset to publish his discoveries, but not under his own name. He used the name "Student," and the t test is sometimes called "Student's t" in his honor. Gosset's statistical work helped him become head brewer, a more interesting title than professor of statistics.

The new t test identified the best barley variety, and Guinness promptly bought up all the available seed.

chapter 11

Inference for Distributions

ACTIVITY 11 Paper Airplane Experiment

Materials: Two paper airplane pattern sheets (in the Teacher Resource Binder), scissors, masking tape, 25-meter steel tape measure, graphing calculator

The Experiment The purpose of this activity is to see which of two prototype paper airplane models has superior flight characteristics. Specifically, the object is to determine the average distance flown for each prototype plane and to compare these average distances flown. The null hypothesis will be that there is no difference between the average distance flown by Prototype A and the average distance flown by Prototype B. So H_0: $\mu_A = \mu_B$. Equivalently, we could write H_0: $\mu_A - \mu_B = 0$. What form should the alternative hypothesis take? (Remember that the alternative hypothesis should be stated *before* you conduct the experiment.)

The Task Your task, as a class, is to design an experiment to determine which of the two prototype paper airplanes has the superior flight characteristics (i.e., flies the farthest). Then you will carry out your plan and gather the necessary data. You may want to explore the data both numerically and graphically prior to conducting formal inference. Unfortunately, we don't know the population standard deviation σ for either prototype, so we can't apply the methods of Chapter 10 to conduct significance tests. We will develop methods in this chapter that will enable us to calculate a test statistic so that we can answer the question about which prototype paper airplane flies the farthest. We will also calculate confidence intervals for the true population difference in flight distances. Keep your data at hand so that you can perform this analysis later. *Note:* These data will also be used in Activity 15 (in Chapter 15).

INTRODUCTION

With the principles in hand, we proceed to practice. This chapter describes confidence intervals and significance tests for the mean of a single population and for comparing the means of two populations. Later chapters present procedures for inference about population proportions, for comparing the means of more than two populations, and for studying relationships among variables.

11.1 INFERENCE FOR THE MEAN OF A POPULATION

Confidence intervals and tests of significance for the mean μ of a normal population are based on the sample mean $\bar{x}$. The sampling distribution of $\bar{x}$ has μ as its mean. That is, $\bar{x}$ is an unbiased estimator of the unknown μ. The spread of $\bar{x}$ depends on the sample size and also on the population standard deviation σ. In the previous chapter we made the unrealistic assumption that we knew the value of σ. In practice, σ is unknown. We must estimate σ from the data even though we are primarily interested in μ. The need to estimate σ changes some details of tests and confidence intervals for μ, but not their interpretation.

Here are the conditions we need to verify before we do inference about a population mean.

CONDITIONS FOR INFERENCE ABOUT A MEAN

- Our data are a **simple random sample** (SRS) of size n from the population of interest. This condition is very important.
- Observations from the population have a **normal distribution** with mean μ and standard deviation σ. In practice, it is enough that the distribution be symmetric and single-peaked unless the sample is very small. Both μ and σ are unknown parameters.

In this setting, the sample mean $\bar{x}$ has the normal distribution with mean μ and standard deviation $\sigma/\sqrt{n}$ Because we don't know σ, we estimate it by the sample standard deviation s. We then estimate the standard deviation of $\bar{x}$ by $s/\sqrt{n}$. This quantity is called the *standard error* of the sample mean $\bar{x}$.

STANDARD ERROR

When the standard deviation of a statistic is estimated from the data, the result is called the **standard error** of the statistic. The standard error of the sample mean $\bar{x}$ is $s/\sqrt{n}$.

The *t* distributions

When we know the value of σ, we base confidence intervals and tests for μ on the one-sample z statistic

$$z = \frac{\bar{x} - \mu}{\sigma/\sqrt{n}}$$

This z statistic has the standard normal distribution $N(0,1)$. When we do not know σ, we substitute the standard error $s/\sqrt{n}$ of $\bar{x}$ for its standard deviation $\sigma/\sqrt{n}$. The statistic that results does not have a normal distribution. It has a distribution that is new to us, called a *t distribution*.

THE ONE-SAMPLE t STATISTIC AND THE t DISTRIBUTIONS

Draw an SRS of size n from a population that has the normal distribution with mean μ and standard deviation σ. The **one-sample t statistic**

$$t = \frac{\bar{x} - \mu}{s / \sqrt{n}}$$

has the **t distribution** with $n - 1$ degrees of freedom.

The t statistic has the same interpretation as any standardized statistic: it says how far $\bar{x}$ is from its mean μ in standard deviation units. There is a different t distribution for each sample size. We specify a particular t distribution by giving its ***degrees of freedom.*** The degrees of freedom for the one-sample t statistic come from the sample standard deviation s in the denominator of t. We saw in Chapter 1 that s has $n - 1$ degrees of freedom. There are other t statistics with different degrees of freedom, some of which we will meet later. We will write the t distribution with k degrees of freedom as $t(k)$ for short.

degrees of freedom

Figure 11.1 compares the density curves of the standard normal distribution and the t distributions with 2 and 9 degrees of freedom. The figure illustrates these facts about the t distributions:

- The density curves of the t distributions are similar in shape to the standard normal curve. They are symmetric about zero, single-peaked, and bell-shaped.

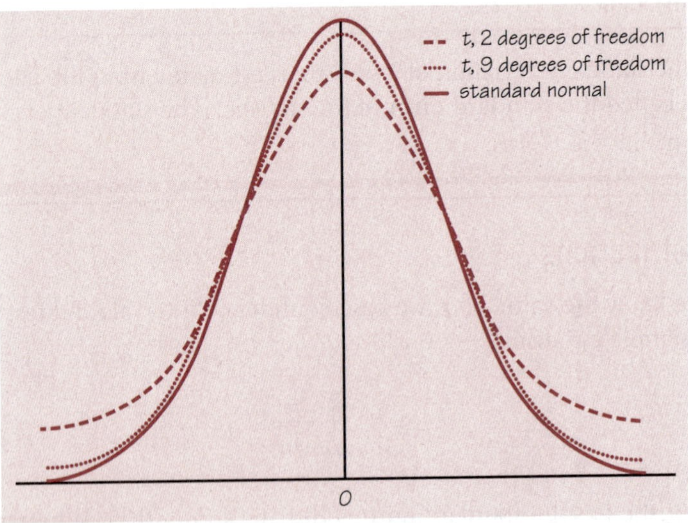

FIGURE 11.1 Density curves for the t distributions with 2 and 9 degrees of freedom and the standard normal distribution. All are symmetric with center 0. The t distributions have more probability in the tails than does the standard normal.

- The spread of the t distributions is a bit greater than that of the standard normal distribution. The t distributions in Figure 11.1 have more probability in the tails and less in the center than does the standard normal. This is true because substituting the estimate s for the fixed parameter σ introduces more variation into the statistic.

- As the degrees of freedom k increase, the $t(k)$ density curve approaches the $N(0,1)$ curve ever more closely. This happens because s estimates σ more accurately as the sample size increases. So using s in place of σ causes little extra variation when the sample is large.

Table C in the back of the book gives critical values for the t distributions. Each row in the table contains critical values for one of the t distributions; the degrees of freedom appear at the left of the row. For convenience, we label the table entries both by p, the upper tail probability needed for significance tests, and by the confidence level C (in percent) required for confidence intervals. You have already used the standard normal critical values z^* in the bottom row of Table C. By looking down any column, you can check that the t critical values approach the normal values as the degrees of freedom increase. As in the case of the normal table, statistical software often makes Table C unnecessary.

EXAMPLE 11.1 USING THE "*t* TABLE"

What critical value t^* from Table C (often referred to as the "t table") would you use for a t distribution with 18 degrees of freedom having probability 0.90 to the left of t^*?

Table C allows you to find a critical value t^* with known probability to its right. We want t^* with probability 0.90 to its left. That same t^* critical value has probability 0.10 to its right. So we look on the df = 18 row for the entry under the column corresponding to an upper tail probability of 0.10. The desired critical value is $t^* = 1.330$.

Now suppose you want to construct a 95% confidence interval for the mean μ of a population based on an SRS of size $n = 12$. What critical value t^* should you use?

In Table C, we consult the row corresponding to df $= n - 1 = 11$. We move across that row to the entry that is directly above 95% confidence level on the bottom of the chart. The desired critical value is $t^* = 2.201$. Notice that the corresponding z critical value is $z^* = 1.96$.

EXERCISES

11.1 Writers in some fields often summarize data by giving $\bar{x}$ and its standard error rather than $\bar{x}$ and s. The standard error of the mean $\bar{x}$ is often abbreviated as SEM.

(a) A medical study finds that $\bar{x} = 114.9$ and $s = 9.3$ for the seated systolic blood pressure of the 27 members of one treatment group. What is the standard error of the mean?

(b) Biologists studying the levels of several compounds in shrimp embryos reported their results in a table, with the note, "Values are means ± SEM for three independent samples." The table entry for the compound ATP was 0.84 ± 0.01. The researchers made three measurements of ATP, which had $\bar{x} = 0.84$. What was the sample standard deviation s for these measurements?

11.2 What critical value t^* from Table C satisfies each of the following conditions?

(a) The t distribution with 5 degrees of freedom has probability 0.05 to the right of t^*.

(b) The t distribution with 21 degrees of freedom has probability 0.99 to the left of t^*.

11.3 What critical value t^* from Table C satisfies each of the following conditions?

(a) The one-sample t statistic from a sample of 15 observations has probability 0.025 to the right of t^*.

(b) The one-sample t statistic from an SRS of 20 observations has probability 0.75 to the left of t^*.

11.4 What critical value t^* from Table C should be used for a confidence interval for the mean of the population in each of the following situations?

(a) A 90% confidence interval based on $n = 12$ observations.

(b) A 95% confidence interval from an SRS of 30 observations.

(c) An 80% confidence interval from a sample of size 18.

11.5 COMPARING THE *z* AND *t* DISTRIBUTIONS This exercise uses the TI-83/89 to compare the standard normal distribution and the two t distribution curves shown in Figure 11.1. Begin by clearing any functions in Y_1, Y_2, and Y_3. Turn off all STAT PLOTS and clear the graphics screen (ClrDraw).

(a) Define Y_1 = `normalpdf(X)`. Change the graph style to a thick line. (On the TI-83, move to the left of Y_1 and press ENTER. On the TI-89, press 2nd F1 (F6) and choose 4: Thick.)

(b) Next define Y_2 = `tpdf(X, 2)`. Note that tpdf is found under the DISTR menu on the TI-83 and in the CATALOG on the TI-89 under Flash Apps. The second parameter, 2, specifies the degrees of freedom.

(c) Set your WINDOW to $X[-3,3]_1$ and $Y[-0.1,0.4]_{0.1}$ and then GRAPH. Sketch the graphs of the two curves on your paper and write a brief description of the similarities and differences you see.

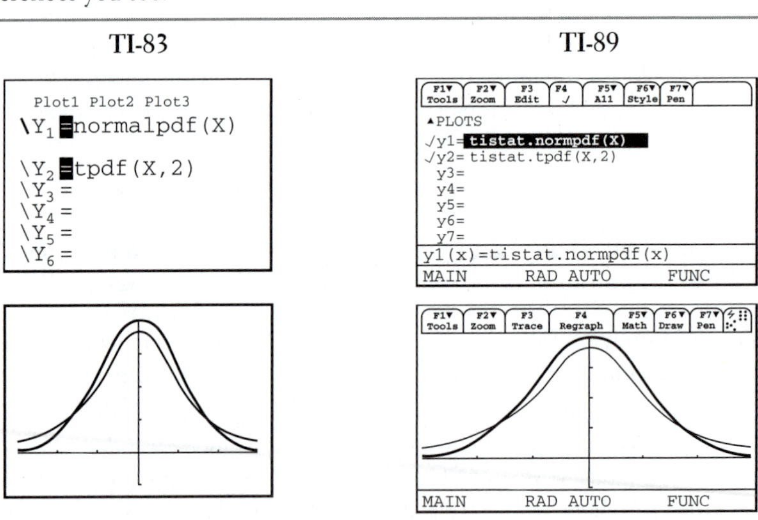

(d) Change the graph style for Y_1 to a dotted line. Deselect Y_2 and define Y_3 = `tpdf(X,9)`. GRAPH these two functions. How do these two curves compare?

(e) Deselect Y_3 and define Y_4 = `tpdf(X,30)`. GRAPH these two functions. What appears to be happening to the shape of the t distribution curve as the number of degrees of freedom increases?

11.6 UPPER TAIL PROBABILITIES IN THE *z* AND *t* DISTRIBUTIONS This exercise uses the TI-83/89 to compare upper tail probabilities in the standard normal and several representative t distributions. Begin by clearing the graphics screen (ClrDraw). Set your WINDOW to $X[-3,3]_1$ and $Y[-0.1,0.4]_{0.1}$.

(a) Enter `ShadeNorm(2,100)` to shade the area under the standard normal curve to the right of $Z = 2$. Record this area (probability), rounded to four decimal places.

(b) Make a table like the one following.

d	$P(t > 2)$	Absolute difference
2		
10		
30		
50		
100		

(c) Clear the graphics screen (ClrDraw). Enter `Shade_t(2,100,2)`. The syntax is Shade_t(leftendpoint, rightendpoint, df). Round the probability value to four decimal places and enter it in the second column of the table. Calculate the absolute value of the difference between the upper tail probabilities for the normal curve and the $t(2)$ curve, and enter this value in column 3.

(d) Calculate the areas to the right of $t = 2$ for the $t(10)$, $t(30)$, $t(50)$, and $t(100)$ curves, and record your answers in the table.

(e) Describe what happens to the area to the right of $t = 2$ under the $t(k)$ distribution as the degrees of freedom increase.

The *t* confidence intervals and tests

To analyze samples from normal populations with unknown σ, just replace the standard deviation $\sigma/\sqrt{n}$ of $\bar{x}$ by its standard error $s/\sqrt{n}$ in the z procedures of Chapter 10. The z procedures then become *one-sample t procedures.* Use P-values or critical values from the t distribution with $n - 1$ degrees of freedom in place of the normal values. The one-sample t procedures are similar in both reasoning and computational detail to the z procedures of Chapter 10. So we will now pay more attention to questions about using these methods in practice.

THE ONE-SAMPLE *t* PROCEDURES

Draw an SRS of size n from a population having unknown mean μ. A level C confidence interval for μ is

$$\bar{x} \pm t^* \frac{s}{\sqrt{n}}$$

where t^* is the upper $(1 - C)/2$ critical value for the $t(n - 1)$ distribution. This interval is exact when the population distribution is normal and is approximately correct for large n in other cases.

To test the hypothesis H_0: $\mu = \mu_0$ based on an SRS of size n, compute the one-sample t statistic

$$t = \frac{\bar{x} - \mu_0}{s / \sqrt{n}}$$

In terms of a variable T having the $t(n - 1)$ distribution, the P-value for a test of H_0 against

H_a: $\mu > \mu_0$ is $P(T \geq t)$

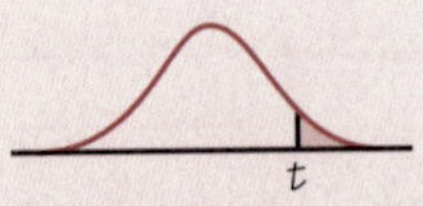

H_a: $\mu < \mu_0$ is $P(T \leq t)$

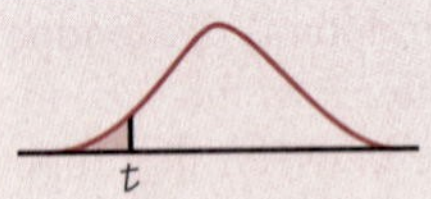

H_a: $\mu \uparrow \mu_0$ is $2P(T \geq |t|)$

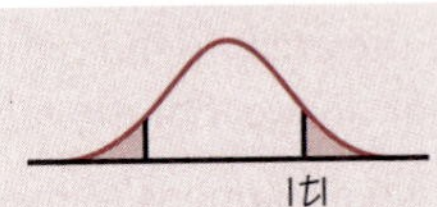

These P-values are exact if the population distribution is normal and are approximately correct for large n in other cases.

The following example shows you how to construct a confidence interval for a population mean using t procedures. You should recognize the four-step process as the Inference Toolbox developed in Chapter 10.

EXAMPLE 11.2 AUTO POLLUTION

Environmentalists, government officials, and vehicle manufacturers are all interested in studying the auto exhaust emissions produced by motor vehicles. The major pollutants in auto exhaust are hydrocarbons, monoxide, and nitrogen oxides (NOX). Table 11.1 gives the NOX levels (in grams/mile) for a sample of light-duty engines of the same type.

TABLE 11.1 Amount of nitrogen oxides (NOX) emitted by light-duty engines (grams/mile)

1.28	1.17	1.16	1.08	0.60	1.32	1.24	0.71	0.49	1.38	1.20	0.78
0.95	2.20	1.78	1.83	1.26	1.73	1.31	1.80	1.15	0.97	1.12	0.72
1.31	1.45	1.22	1.32	1.47	1.44	0.51	1.49	1.33	0.86	0.57	1.79
2.27	1.87	2.94	1.16	1.45	1.51	1.47	1.06	2.01	1.39		

Source: T. J. Lorenzen, "Determining statistical characteristics of a vehicle emissions audit procedure," *Technometrics*, 22 (1980), pp. 483–493.

Construct a 95% confidence interval for the mean amount of NOX emitted by light-duty engines of this type.

Step 1: Identify the population of interest and the parameter you want to draw conclusions about. The population of interest is all light-duty engines of this type. We want to estimate μ, the mean amount of the pollutant NOX emitted, for all of these engines.

Step 2: Choose the appropriate inference procedure. Verify the conditions for using the selected procedure. Since we do not know σ, we should use the one-sample t procedures to construct a confidence interval for the mean NOX level μ. Now we proceed to check the required conditions.

- The data come from a sample of 46 engines from the population of all light-duty engines of this type. We are not told that the sample is an SRS, however. If the data do not come from a random sample of engines, then we should hesitate to draw conclusions about the population mean.
- Is the population distribution of NOX emissions normal? We do not know from the problem statement. Let's examine the sample data.

Figure 11.2 is a Minitab stemplot of the data. The distribution of NOX values in the sample is fairly symmetric if we ignore the one extremely high value. Figure 11.3 shows a normal probability plot from a TI-83 calculator. The plot is somewhat linear, although the one engine with the extremely high NOX reading is obvious once again.

```
Stem-and-leaf of NOX          N = 46
Leaf Unit = 0.10

    3   0   455
    7   0   6777
   10   0   899
   17   1   0011111
  (12)  1   222223333333
   17   1   4444445
   10   1   777
    7   1   888
    4   2   0
    3   2   22
    1   2
    1   2
    1   2   9
```

FIGURE 11.2 Minitab stemplot of NOX emissions in a sample of 46 light-duty engines. Note the roughly symmetric shape and the one high outlier.

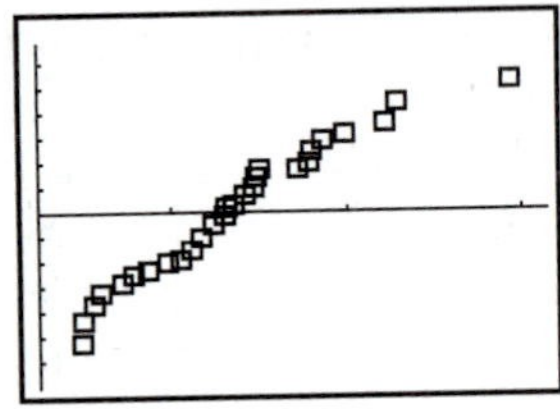

FIGURE 11.3 Calculator normal probability plot of the NOX sample data. If the data are normally distributed, the plot will be roughly linear.

Since the sample size is large (n = 46), the central limit theorem tells us that the distribution of sample means will be approximately normal. Use of the t procedures is justified in this case.

Step 3: If the conditions are met, carry out the inference procedure. Check that the mean NOX emission reading for the 46 light-duty engines in our sample is $\bar{x} = 1.329$ grams per mile. The confidence interval formula is $\bar{x} \pm t^* s/\sqrt{n}$. We use the t distribution with df = 46 − 1 = 45. Unfortunately, there is no row corresponding to 45 degrees of freedom in Table C. Instead, we use the df = 40 row, which will yield a higher critical value t^*, and thus a wider confidence interval. At a 95% confidence level, the critical value is $t^* = 2.021$. So the 95% confidence interval for μ is

$$\bar{x} \pm t^* \frac{s}{\sqrt{n}} = 1.329 \pm 2.021 \frac{0.484}{\sqrt{46}} = 1.329 \pm 0.144 = (1.185, 1.473)$$

Step 4: Interpret your results in the context of the problem. We are 95% confident that the true mean level of nitrogen oxides emitted by this type of light-duty engine is between 1.185 grams/mile and 1.473 grams/mile.

The one-sample t confidence interval has the form

$$\text{estimate} \pm t^* \text{SE}_{\text{estimate}}$$

where "SE" stands for "standard error." We will meet a number of confidence intervals that have this common form. Like the confidence interval, t tests are close in form to the z tests we met earlier. Here is an example. In Chapter 10 we used the z test on these data. That required the unrealistic assumption that we knew the population standard deviation σ. Now we can do a realistic analysis. Once again, we follow the steps in our Inference Toolbox.

EXAMPLE 11.3 SIGNIFICANCE TEST FOR μ WHEN σ IS UNKNOWN

Cola makers test new recipes for loss of sweetness during storage. Trained tasters rate the sweetness before and after storage. Here are the sweetness losses (sweetness before storage minus sweetness after storage) found by 10 tasters for one new cola recipe:

2.0	0.4	0.7	2.0	−0.4	2.2	−1.3	1.2	1.1	2.3

Are these data good evidence that the cola lost sweetness?

Step 1: Identify the population of interest and the parameter you want to draw conclusions about. State null and alternative hypotheses in words and symbols. Tasters vary in their perception of sweetness loss. So we ask the question in terms of the mean loss $\mu_{BEFORE-AFTER} = \mu_{DIFF}$ for a large population of tasters. The null hypothesis is "no loss" and the alternative hypothesis says " there is a loss."

$H_0: \mu_{DIFF} = 0$	The mean sweetness loss for the population of tasters is 0.
$H_a: \mu_{DIFF} > 0$	The mean sweetness loss for the population of tasters is positive. The cola seems to be losing sweetness in storage.

Step 2: Choose the appropriate inference procedure. Verify the conditions for using the selected procedure. Since we do not know the standard deviation of sweetness loss in the population of tasters, we must use a one-sample t test. Now we must check the two required conditions.

- We must be willing to treat our 10 tasters as an SRS from the population of tasters if we want to draw conclusions about tasters in general. The tasters all have the same training. So even though we don't actually have an SRS from the population of interest, we are willing to act as if we did. This is a matter of judgment.
- The assumption that the population distribution is normal cannot be effectively checked with only 10 observations. In part, the researchers rely on experience with similar variables. They also look at the data. We can construct a stemplot of the sweetness loss data:

−1	3
−0	4
0	47
1	12
2	0023

The distribution is somewhat left-skewed, but there are no gaps or outliers or other signs of nonnormal behavior. So we proceed with caution.

Step 3: If the conditions are met, carry out the inference procedure:

- Calculate the test statistic. The basic statistics are

$$\bar{x}_{DIFF} = 1.02 \text{ and } s_{DIFF} = 1.196$$

The one-sample t test statistic is

$$\frac{\bar{x}_{DIFF} - \mu_0}{s_{DIFF}/\sqrt{n}} = \frac{1.02 - 0}{1.196/\sqrt{10}} = 2.70$$

- Find the P-value. The P-value for $t = 2.70$ is the area to the right of 2.70 under the t distribution curve with degrees of freedom $n - 1 = 9$. Figure 11.4 shows this area. We can't find the exact value of P without a calculator or computer. But we

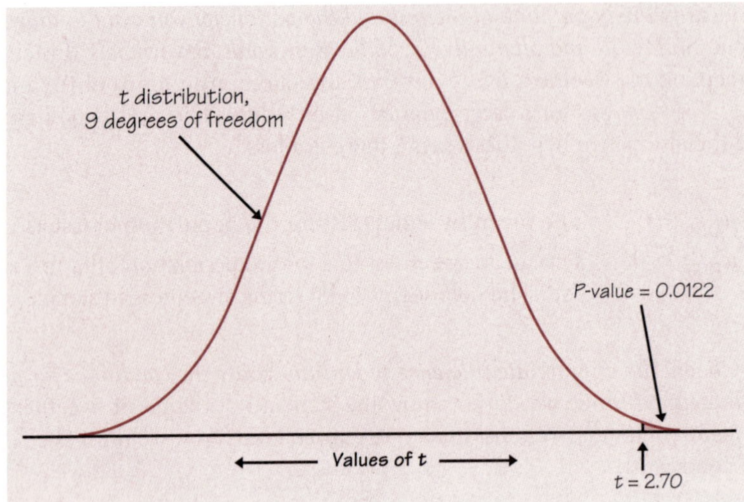

FIGURE 11.4 The P-value for the one-sided t test.

df = 9

p	.02	.01
t^*	2.398	2.821

can pin P between two values by using Table C. Search the df = 9 row of Table C for entries that bracket t = 2.70. Because the observed t lies between 2.398 and 2.821, the P-value lies between 0.01 and 0.02. Computer software gives the more exact result P = 0.012.

Step 4: Interpret your results in the context of the problem. A P-value this low gives quite strong evidence against the null hypothesis. We reject H_0 and conclude that the cola has lost sweetness during storage.

Notice the linkage between the P-value computed in step 3 and the conclusion drawn in step 4. If you keep in mind the three C's—conclusion, connection, context—then you will include all the important elements in your interpretation of results.

The procedures used in Examples 11.2 and 11.3 depend on conditions that are often reasonable but not always easy to check: random sampling and a normal population distribution. Fortunately, we will see that confidence levels and P-values from the t procedures are not very sensitive to lack of normality. Violation of the random sampling condition is more serious.

Because the t procedures are so common, all statistical software systems will do the calculations for you. Figure 11.5 shows the output from three statistical packages: Data Desk, Minitab, and S-PLUS. In each case, we entered the 10 sweetness losses as values of a variable called "cola" and asked for the one-sample t test of H_0: $\mu = 0$ against H_a: $\mu > 0$. The three outputs report slightly different information, but all include the basic facts: $\overline{x}$ = 1.02, t = 2.70, P = 0.012. These are the results we found in Example 11.3.

```
                    Data Desk
cola:
Test Ho: mu (cola) = 0 vs Ha: mu (cola) > 0
Sample Mean = 1.02000 t-Statistic = 2.697 w/9 df
Reject Ho at Alpha = 0.0500
p = 0.0123
                    Minitab

TEST OF MU = 0.000 VS  MU  G. T.  0.000
         N   MEAN   STDEV   SE  MEA    T    P  VALUE
cola    10   1.02   1.196        0.3   2.7     0.012

                    S-PLUS
data:    cola
t = 2.6967,  df = 9,  p-value = 0.0123
alternative hypothesis: true mean is greater than 0
sample estimates:
   mean of x
        1.02
```

FIGURE 11.5 Output for the one-sample *t* test of Example 11.3 from three statistical software packages. You can easily locate the basic results in output from any statistical software.

EXERCISES

11.7 The one-sample *t* statistic for testing

$$H_0\colon \mu = 0$$
$$H_a\colon \mu > 0$$

from a sample of $n = 15$ observations has the value $t = 1.82$.

(a) What are the degrees of freedom for this statistic?

(b) Give the two critical values t^* from Table C that bracket t. What are the right-tail probabilities p for these two entries?

(c) Between what two values does the *P*-value of the test fall?

(d) Is the value $t = 1.82$ significant at the 5% level? Is it significant at the 1% level?

11.8 The one-sample *t* statistic from a sample of $n = 25$ observations for the two-sided test of

$$H_0\colon \mu = 64$$
$$H_a\colon \mu \uparrow 64$$

has the value $t = 1.12$.

(a) What are the degrees of freedom for t?

(b) Locate the two critical values t^* from Table C that bracket t. What are the right-tail probabilities p for these two values?

(c) Between what two values does the *P*-value of the test fall? (Note that H_a is two-sided.)

(d) Is the value $t = 1.12$ statistically significant at the 10% level? At the 5% level?

11.9 VITAMIN C CONTENT In fiscal year 1996, the U.S. Agency for International Development provided 238,300 metric tons of corn soy blend (CSB) for development programs and emergency relief in countries throughout the world. CSB is a highly nutritious, low-cost fortified food that is partially precooked and can be incorporated into different food preparations by the recipients. As part of a study to evaluate appropriate vitamin C levels in this commodity, measurements were taken on samples of CSB produced in a factory.[1]

The following data are the amounts of vitamin C, measured in milligrams per 100 grams (mg/100 g) of blend (dry basis), for a random sample of size 8 from a production run:

26	31	23	22	11	22	14	31

(a) What conditions must be satisfied in order to make inferences about μ, the mean vitamin C content of the CSB produced during this run? Check whether each of the conditions is met in this case.

(b) Construct a 95% confidence interval for μ. Use the Inference Toolbox.

(c) The specifications for the CSB state that the mixture should produce a mean (μ) vitamin C content in the final product of 40 mg/100 g. Does the CSB produced in this production run conform to these specifications? Perform a significance test to answer this question.

11.10 HEALTHY BONES, I Here are estimates of the daily intakes of calcium (in milligrams) for 38 women between the ages of 18 and 24 years who participated in a study of women's bone health:

808	882	1062	970	909	802	374	416	784	997
651	716	438	1420	1425	948	1050	976	572	403
626	774	1253	549	1325	446	465	1269	671	696
1156	684	1933	748	1203	2433	1255	1100		

(a) Display the data using a stemplot and make a normal probability plot. Describe the distribution of calcium intakes for these women.

(b) Calculate the mean, the standard deviation, and the standard error.

(c) Find a 95% confidence interval for the mean. Use the Inference Toolbox.

(d) Eliminate the two largest values and recompute the 95% confidence interval. What do you notice?

11.11 HEALTHY BONES, II Refer to Exercise 11.10. Suppose that the recommended daily allowance (RDA) of calcium for women in this age range is 1200 milligrams. Doctors involved in the study suspected that participating subjects had lower calcium intakes than the RDA. Test this claim at the $\alpha = 0.05$ significance level.

Matched pairs *t* procedures

The taste test in Example 11.3 was a matched pairs study in which the same 10 tasters rated before-and-after sweetness. Comparative studies are more convincing than single-sample investigations. For that reason, one-sample inference is less common than comparative inference. One common design to compare two treatments makes use of one-sample procedures. In a ***matched pairs design,*** subjects are matched in pairs and each treatment is given to one subject in each pair. The experimenter can toss a coin to assign two treatments to the two subjects in each pair.

matched pairs design

MATCHED PAIRS *t* PROCEDURES

To compare the responses to the two treatments in a matched pairs design, apply the one-sample *t* procedures to the observed differences.

The parameter μ in a matched pairs *t* procedure is the mean difference in the responses to the two treatments within matched pairs of subjects in the entire population.

EXAMPLE 11.4 FLORAL SCENTS AND LEARNING

We hear that listening to Mozart improves students' performance on tests. Perhaps pleasant odors have a similar effect. To test this idea, 21 subjects worked a paper-and-pencil maze while wearing a mask. The mask was either unscented or carried a floral scent. The response variable is their average time on three trials. Each subject worked the maze with both masks, in a random order. The randomization is important because subjects tend to improve their times as they work a maze repeatedly. Table 11.2 gives the subjects' average times with both masks.

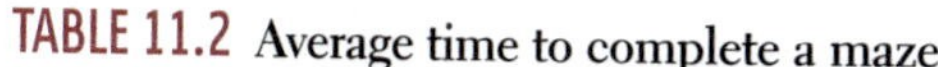

TABLE 11.2 Average time to complete a maze

Subject	Unscented (seconds)	Scented (seconds)	Difference	Subject	Unscented (seconds)	Scented (seconds)	Difference
1	30.60	37.97	−7.37	12	58.93	83.50	−24.57
2	48.43	51.57	−3.14	13	54.47	38.30	16.17
3	60.77	56.67	4.10	14	43.53	51.37	−7.84
4	36.07	40.47	−4.40	15	37.93	29.33	8.60
5	68.47	49.00	19.47	16	43.50	54.27	−10.77
6	32.43	43.23	−10.80	17	87.70	62.73	24.97
7	43.70	44.57	−0.87	18	53.53	58.00	−4.47
8	37.10	28.40	8.70	19	64.30	52.40	11.90
9	31.17	28.23	2.94	20	47.37	53.63	−6.26
10	51.23	68.47	−17.24	21	53.67	47.00	6.67
11	65.40	51.10	14.30				

Source: A. R. Hirsch and L. H. Johnston, "Odors and learning," *Journal of Neurological and Orthopedic Medicine and Surgery*, 17 (1996), pp. 119–126.

To analyze these data, subtract the scented time from the unscented time for each subject. The 21 differences form a single sample. They appear in the "Difference" column in Table 11.2. The first subject, for example, was 7.37 seconds slower wearing the scented mask, so the difference is negative. Because shorter times represent better performance, positive differences show that the subject did better when wearing the scented mask.

Step 1: *Identify the population of interest and the parameter you want to draw conclusions about. State null and alternative hypotheses in words and symbols.* To assess whether the floral scent significantly improved performance, we test

$$H_0: \mu = 0$$
$$H_a: \mu > 0$$

Here μ is the mean difference in the population from which the subjects were drawn. The null hypothesis says that no improvement occurs, and H_a says that unscented times are longer than scented times on the average.

Step 2: *Choose the appropriate inference procedure, and verify the conditions for using the selected procedure.* We do not know the standard deviation of the population of differences, so we should perform a one-sample *t* test. Next, we check the required conditions.

- The data come from a randomized, matched pairs experiment. But we can generalize the results of this study to the population of interest only if we view our 21 subjects as an SRS from the population.
- A stemplot of the differences shows that their distribution is symmetric and appears reasonably normal in shape. We have no reason to question the normality of the population of differences.

Stem	Leaves
−2	5
−1	711
−0	8764431
0	34799
1	2469
2	5

FIGURE 11.6 Stemplot of the differences in time to complete a maze for 21 subjects. The data are rounded to the nearest whole second. Notice that the stem 0 must appear twice, to display differences between −9 and 0 and between 0 and 9.

df = 20

p	.25	.20
t^*	0.687	0.860

Step 3: *If the conditions are met, carry out the inference procedure:*

- Calculate the test statistic. The 21 differences have

$$\bar{x} = 0.9567 \quad \text{and} \quad s = 12.5479$$

The one-sample *t* statistic is therefore

$$t = \frac{\bar{x} - 0}{s/\sqrt{n}} = \frac{0.9567 - 0}{12.5479/\sqrt{21}} = 0.349$$

- Find the *P*-value from the $t(20)$ distribution. (Remember that the degrees of freedom are 1 less than the sample size.) Table C shows that 0.349 is less than the 0.25 critical value of the $t(20)$ distribution. The *P*-value is therefore greater than 0.25. Statistical software gives the value $P = 0.3652$.

Step 4: *Interpret your results in the context of the problem.* The data do not support the claim that floral scents improve performance. The average improvement is small, just 0.96 seconds over the 50 seconds that the average subject took when wearing the unscented mask. This small improvement is not statistically significant at even the 25% level.

Example 11.4 illustrates how to turn matched pairs data into single-sample data by taking differences within each pair. We are making inferences about a single population, the population of all differences within matched pairs. It is incorrect to ignore the pairs and analyze the data as if we had two samples, one from subjects who wore unscented masks and a second from subjects who wore scented masks. Inference procedures for comparing two samples assume that the samples are selected independently of each other. This assumption does not hold when the same subjects are measured twice. The proper analysis depends on the design used to produce the data.

Because the t procedures are so common, all statistical software will do the calculations for you. If you are familiar with the t procedures, you can understand the output from any software. Figure 11.7 shows the output for Example 11.4 from Minitab, Data Desk, and Excel. In each case, we entered the data and asked for the one-sided matched pairs t test. The three outputs report slightly different information, but all include the basic facts: $t = 0.349$, $P = 0.365$.

Paired T-Test and Confidence Level

```
Paired T for Unscented - Scented

               N          Mean        StDev       SE Mean
Unscented     21         50.01        14.36          3.13
Scented       21         49.06        13.39          2.92
Difference    21          0.96        12.55          2.74

95% CI for mean difference:   (-4.76, 6.67)
T-Test of mean difference = 0 (vs > 0): T-Value = 0.35  P-Value = 0.365
```

(a) Minitab

```
Unscented - Scented:
Test Ho: μ(Unscented-Scented) = 0 vs Ha: μ(Unscented-Scented) > 0
Mean of Paired Differences = 0.956667 t-Statistic = 0.349 w/20 df
Fail to reject Ho at Alpha = 0.0500
p = 0.3652
```

(b) Data Desk

t-Test: Paired Two Sample for Means

	Variable 1	*Variable 2*
Mean	50.01429	49.05762
Variance	206.3097	179.1748
Observations	21	21
Pearson Correlation	0.593026	
Hypothesized Mean Difference	0	
df	20	
t Stat	0.349381	
P(T<=t) one-tail	0.365227	
t Critical one-tail	1.724718	
P(T<=t) two-tail	0.730455	
t Critical two-tail	2.085962	

(c) Excel

FIGURE 11.7 Output for the matched pairs t test of Example 11.4 from three statistical software packages. You can easily locate the basic results in output from any statistical software.

EXAMPLE 11.5 IS CAFFEINE DEPENDENCE REAL?

Our subjects are 11 people diagnosed as being dependent on caffeine. Each subject was barred from coffee, colas, and other substances containing caffeine. Instead, they took capsules containing their normal caffeine intake. During a different time period, they took placebo capsules. The order in which subjects took caffeine and the placebo was randomized. Table 11.3 contains data on two of several tests given to the subjects. "Depression" is the score on the Beck Depression Inventory. Higher scores show more symptoms of depression. "Beats" is the beats per minute the subject achieved when asked to press a button 200 times as quickly as possible. We are interested in whether being deprived of caffeine affects these outcomes.

TABLE 11.3 Results of a caffeine-deprivation study

Subject	Depression (caffeine)	Depression (placebo)	Beats (caffeine)	Beats (placebo)
1	5	16	281	201
2	5	23	284	262
3	4	5	300	283
4	3	7	421	290
5	8	14	240	259
6	5	24	294	291
7	0	6	377	354
8	0	3	345	346
9	2	15	303	283
10	11	12	340	391
11	1	0	408	411

Source: E. C. Strain et al., "Caffeine dependence syndrome: evidence from case histories and experimental evaluation," *Journal of the American Medical Association*, 272 (1994), pp. 1604–1607.

Let's construct a 90% confidence interval for the mean change in depression score.

Step 1: Identify the population of interest and the parameter you want to draw conclusions about. The population of interest is all people who are dependent on caffeine. We want to estimate the mean difference $\mu_{DIFF} = \mu_{PLACEBO-CAFFEINE}$ in depression score that would be reported if all individuals in the population took both the caffeine capsule and the placebo.

Step 2: Choose the appropriate inference procedure, and verify the conditions for using the selected procedure. We will use one-sample *t* procedures to construct a confidence interval for μ_{DIFF} since the population standard deviation of the differences in depression scores is unknown. Next, we check the required conditions.

- The data come from an SRS from the population of interest. This is probably not the case. We may have trouble generalizing the results of this study to the population of caffeine-dependent people.
- The population distribution is normal. Figure 11.8 is a stemplot of the differences (PLACEBO – CAFFEINE) in depression scores for our 11 subjects. There are no

obvious outliers or other departures from normality in the sample data. We are given no reason to doubt the normality of the population of differences.

−0	1
0	1134
0	66
1	13
1	89

FIGURE 11.8 Stemplot of the differences in Beck Depression Inventory scores for the subjects in the matched pairs experiment.

Step 3: If the conditions are met, carry out the inference procedure. For our 11 subjects, $\bar{x}_{PLACEBO-CAFFEINE} = \bar{x}_{DIFF} = 7.364$ and $s_{DIFF} = 6.918$. The t critical value for a 90% confidence interval with $11 - 1 = 10$ degrees of freedom is $t^* = 1.812$. So the desired confidence interval is

$$\bar{x}_{DIFF} \pm t^* \frac{s_{DIFF}}{\sqrt{n}} = 7.364 \pm 1.812 \frac{6.928}{\sqrt{11}} = 7.364 \pm 3.785 = (3.579, 11.149)$$

Step 4: Interpret your results in the context of the problem. We are 90% confident that the actual mean difference in depression score for the population is between 3.579 and 11.149 points. That is, we estimate that caffeine-dependent individuals would score, on average, between 3.6 and 11.1 points higher on the Beck Depression Inventory when they are given placebo instead of caffeine. This study provides evidence that withholding caffeine from caffeine-dependent individuals may lead to depression.

EXERCISES

11.12 **GROWING TOMATOES** An agricultural field trial compares the yield of two varieties of tomatoes for commercial use. The researchers divide in half each of 10 small plots of land in different locations and plant each tomato variety on one half of each plot. After harvest, they compare the yields in pounds per plant at each location. The 10 differences (Variety A – Variety B) give $\bar{x} = 0.34$ and $s = 0.83$. Is there convincing evidence that Variety A has the higher mean yield?

(a) Describe in words what the parameter μ is in this setting.

(b) Perform a significance test to answer the question. Follow the Inference Toolbox.

11.13 **SPANISH TEACHERS WORKSHOP** The National Endowment for the Humanities sponsors summer institutes to improve the skills of high school language teachers. One institute hosted 20 Spanish teachers for four weeks. At the beginning of the period, the teachers took the Modern Language Association's listening test of understanding of spoken Spanish. After four weeks of immersion in Spanish in and out of class, they took the listening test again. (The actual spoken Spanish in the two tests was different, so that simply taking the first test should not improve the score on the second test.) Table 11.4 gives the pretest and posttest scores. The maximum possible score on the test is 36.

TABLE 11.4 MLA listening scores for 20 Spanish teachers

Subject	Pretest	Posttest	Subject	Pretest	Posttest
1	30	29	11	30	32
2	28	30	12	29	28
3	31	32	13	31	34
4	26	30	14	29	32
5	20	16	15	34	32
6	30	25	16	20	27
7	34	31	17	26	28
8	15	18	18	25	29
9	28	33	19	31	32
10	20	25	20	29	32

Source: Data provided by Joseph A. Wipf, Department of Foreign Languages and Literatures, Purdue University.

(a) We hope to show that attending the institute improves listening skills. State an appropriate H_0 and H_a. Be sure to identify the parameters appearing in the hypotheses.

(b) Make a graphical check for outliers or strong skewness in the data that you will use in your statistical test, and report your conclusions on the validity of the test.

(c) Carry out a test. Can you reject H_0 at the 5% significance level? At the 1% significance level?

(d) Give a 90% confidence interval for the mean increase in listening score due to attending the summer institute.

11.14 CAFFEINE DEPENDENCE

(a) The study in Example 11.5 was double-blind. What does this mean?

(b) Examine the differences in beats per minute with and without caffeine. What conclusions can you draw?

11.15 DOES PLAYING THE PIANO MAKE YOU SMARTER? Do piano lessons improve the spatial-temporal reasoning of preschool children? Neurobiological arguments suggest that this may be true. A study designed to test this hypothesis measured the spatial-temporal reasoning of 34 preschool children before and after six months of piano lessons.[2] (The study also included children who took computer lessons and a control group; but we are not concerned with those here.) The changes in the reasoning scores are

2	5	7	−2	2	7	4	1	0	7	3	4	3	4	9	4	5
2	9	6	0	3	6	−1	3	4	6	7	−2	7	−3	3	4	4

(a) Display the data and summarize the distribution.

(b) Find the mean, the standard deviation, and the standard error of the mean.

(c) Give a 95% confidence interval for the mean improvement in reasoning scores.

11.16 PIANO PLAYING, II Refer to the previous exercise. Test the null hypothesis that there is no improvement versus the alternative suggested by the neurobiological arguments. What do you conclude?

Robustness of *t* procedures

The *t* confidence interval and test are exactly correct when the distribution of the population is exactly normal. No real data are exactly normal. The usefulness of the *t* procedures in practice therefore depends on how strongly they are affected by lack of normality.

ROBUST PROCEDURES

A confidence interval or significance test is called **robust** if the confidence level or *P*-value does not change very much when the assumptions of the procedure are violated.

Because the tails of normal curves drop off quickly, samples from normal distributions will have very few outliers. Outliers suggest that your data are not a sample from a normal population. **Like $\bar{x}$ and s, the *t* procedures are strongly influenced by outliers.**

EXAMPLE 11.6 THE EFFECTS OF OUTLIERS

In Example 11.2, we constructed a confidence interval for the mean level of NOX emitted by a specific type of light-duty car engine. One of the 46 engines in our sample emitted an unusually high amount (2.94 grams/mile) of NOX. If we remove that single data point, $\bar{x} = 1.293$ and $s = 0.424$ for the remaining 45 sample values. The confidence interval based on this sample of 45 engines would be

$$\bar{x} \pm t^* \frac{s}{\sqrt{n}} = 1.293 \pm 2.021 \frac{0.424}{\sqrt{46}} = 1.293 \pm 0.126 = (1.167, 1.419)$$

Our new confidence interval is narrower and is centered at a lower value than our original interval of 1.185 to 1.473.

Fortunately, the *t* procedures are quite robust against nonnormality of the population when there are no outliers, especially when the distribution is roughly symmetric. Larger samples improve the accuracy of *P*-values and critical values from the t distributions when the population is not normal. The main reason for this is the central limit theorem. The t statistic is based on the sample mean $\bar{x}$, which becomes more nearly normal as the sample size gets larger even when the population does not have a normal distribution.

Always make a plot to check for skewness and outliers before you use the t procedures for small samples. For most purposes, you can safely use the one-sample t procedures when $n \geq 15$ unless an outlier or quite strong skewness is present. Here are practical guidelines for inference on a single mean.[3]

USING THE *t* PROCEDURES

- Except in the case of small samples, the assumption that the data are an SRS from the population of interest is more important than the assumption that the population distribution is normal.
- *Sample size less than 15.* Use t procedures if the data are close to normal. If the data are clearly nonnormal or if outliers are present, do not use t.
- *Sample size at least 15.* The t procedures can be used except in the presence of outliers or strong skewness.
- *Large samples.* The t procedures can be used even for clearly skewed distributions when the sample is large, roughly $n \geq 40$.

EXAMPLE 11.7 CAN WE USE *t*?

Consider several of the data sets we graphed in Chapter 1. Figure 11.9 shows the histograms.

- Figure 11.9(a) is a histogram of the percent of each state's residents who are at least 65 years of age. *We have data on the entire population of 50 states, so formal inference makes no sense.* We can calculate the exact mean for the population. There is no uncertainty due to having only a sample from the population, and no need for a confidence interval or test.
- Figure 11.9(b) shows the time of the first lightning strike each day in a mountain region in Colorado. The data contain more than 70 observations that have a symmetric distribution. You can use the t procedures to draw conclusions about the mean time of a day's first lightning strike with complete confidence.
- Figure 11.9(c) shows that the distribution of word lengths in Shakespeare's plays is skewed to the right. We aren't told how large the sample is. You can use the t procedures for a distribution like this if the sample size is roughly 40 or larger.

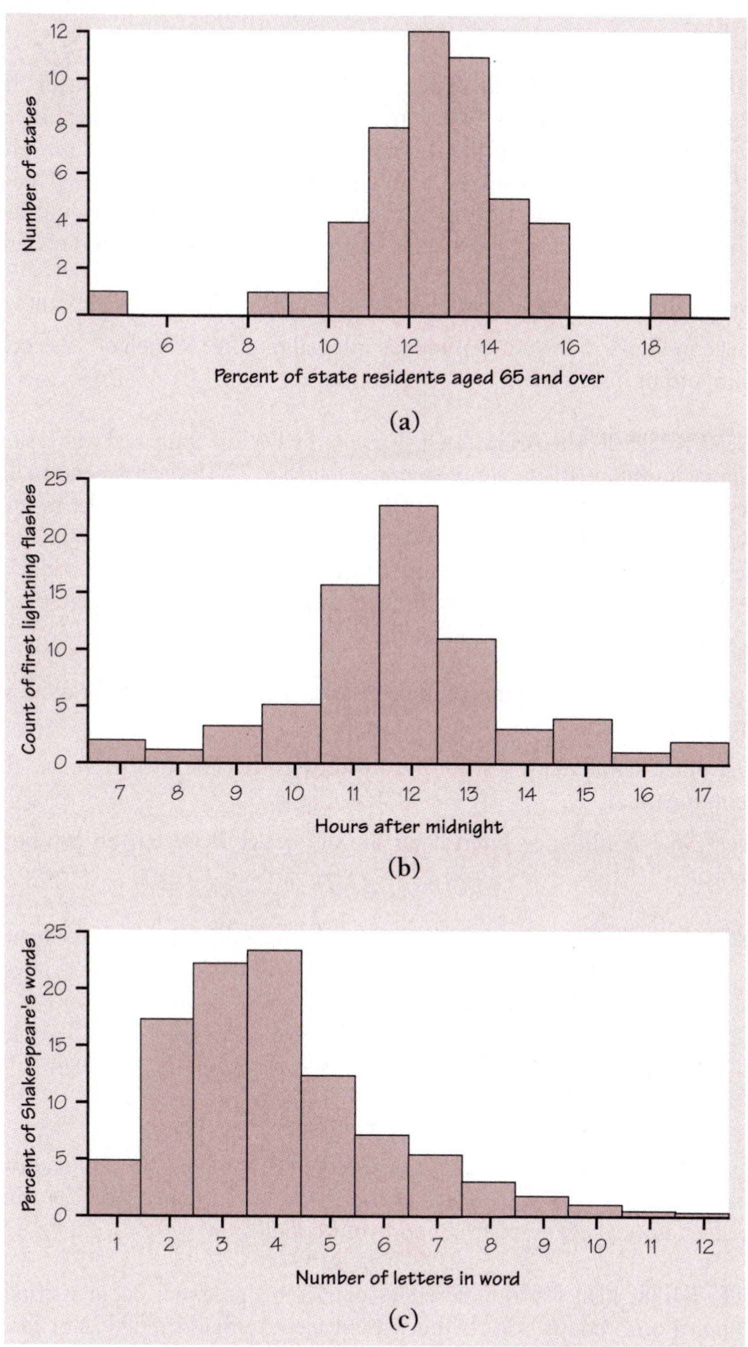

FIGURE 11.9 Can we use *t* procedures for these data? (a) Percent of residents aged 65 and over in the states. *No:* this is an entire population, not a sample. (b) Times of first lightning strikes each day at a site in Colorado. *Yes:* there are over 70 observations with a symmetric distribution. (c) Word lengths in Shakespeare's plays. *Yes, if the sample is large enough* to overcome the right skewness.

EXERCISES

11.17 MEASURING ACCULTURATION The Acculturation Rating Scale for Mexican Americans (ARSMA) measures the extent to which Mexican Americans have adopted Anglo/English culture. During the development of ARSMA, the test was given to a group of 17 Mexicans. Their scores, from a possible range of 1.00 to 5.00, had a symmetric distribution with $\bar{x} = 1.67$ and $s = 0.25$. Because low scores should indicate a Mexican cultural orientation, these results helped to establish the validity of the test.[4]

(a) Give a 95% confidence interval for the mean ARSMA score of Mexicans.

(b) What conditions does your confidence interval require? Which of these conditions is most important in this case?

11.18 ARSMA VERSUS BI The ARSMA test (Exercise 11.17) was compared with a similar test, the Bicultural Inventory (BI), by administering both tests to 22 Mexican Americans. Both tests have the same range of scores (1.00 to 5.00) and are scaled to have similar means for the groups used to develop them. There was a high correlation between the two scores, giving evidence that both are measuring the same characteristics. The researchers wanted to know whether the population mean scores for the two tests are the same. The differences in scores (ARSMA – BI) for the 22 subjects had $\bar{x} = 0.2519$ and $s = 0.2767$.

(a) Describe briefly how to arrange the administration of the two tests to the subjects, including randomization.

(b) Carry out a significance test for the hypothesis that the two tests have the same population mean.

(c) Give a 95% confidence interval for the difference between the two population mean scores.

11.19 AUTO CRANKSHAFTS Here are measurements (in millimeters) of a critical dimension for 16 auto engine crankshafts:

224.120	224.001	224.017	223.982	223.989	223.961
223.960	224.089	223.987	223.976	223.902	223.980
224.098	224.057	223.913	223.999		

The dimension is supposed to be 224 mm and the variability of the manufacturing process is unknown. Is there evidence that the mean dimension is not 224 mm? Give appropriate statistical evidence to support your conclusion.

11.20 DOES NATURE HEAL BEST? Differences of electric potential occur naturally from point to point on a body's skin. Is the natural electric field strength best for helping wounds to heal? If so, changing the field will slow healing. The research subjects are anesthetized newts.

Make a razor cut in both hind limbs. Let one heal naturally (the control). Use an electrode to change the electric field in the other to half its normal value. After two hours, measure the healing rate. Table 11.5 gives the healing rates (in micrometers per hour) for 14 newts.

TABLE 11.5 Healing rates (micrometers per hour) for newts

Newt	Experimental limb	Control limb	Difference in healing	Newt	Experimental limb	Control limb	Difference in healing
13	24	25	−1	20	33	36	−3
14	23	13	10	21	28	35	−7
15	47	44	3	22	28	38	−10
16	42	45	−3	23	21	43	−22
17	26	57	−31	24	27	31	−4
18	46	42	4	25	25	26	−1
19	38	50	−12	26	45	48	−3

Source: D. D. S. Iglesia, E. J. Cragoe, Jr., and J. W. Vanable, "Electric field strength and epithelization in the newt (*Notophthalmus viridescens*)," *Journal of Experimental Zoology*, 274 (1996), pp. 56–62.

(a) As is usual, the paper did not report these raw data. Readers are expected to be able to interpret the summaries that the paper did report. The paper summarized the differences in the table above as "−5.71 ± 2.82" and said, "All values are expressed as means ± standard error of the mean." Show carefully where the numbers −5.71 and 2.82 come from.

(b) The researchers want to know if changing the electric field reduces the mean healing rate for all newts. Carry out a test and give your conclusion. Is the result statistically significant at the 5% level? At the 1% level? (The researchers compared several field strengths and concluded that the natural strength is about right for fastest healing.)

(c) Give a 90% confidence interval for the amount by which changing the field changes the rate of healing. Then explain in a sentence what it means to say that you are "90% confident" of your result.

The power of the *t* test

The power of a statistical test measures its ability to detect deviations from the null hypothesis. In practice we carry out the test in the hope of showing that the null hypothesis is false, so higher power is important. The power of the one-sample *t* test against a specific alternative value of the population mean μ is the probability that the test will reject the null hypothesis when the mean has this alternative value. To calculate the power, we assume a fixed level of significance, usually $\alpha = 0.05$.

Calculation of the exact power of the *t* test takes into account the estimation of σ by s and is a bit complex. But an approximate calculation that acts as if σ were known is usually adequate for planning a study. This calculation is very much like that for the power of the *z* test, presented on pages 599–602.

EXAMPLE 11.8 POWER OF SPANISH TEACHERS

It is the winter before the summer language institute of Exercise 11.13. The director, thinking ahead to the report he must write, hopes that enrolling 20 teachers will enable him to be quite certain of detecting an average improvement of 2 points in the mean listening score. Is this realistic?

We wish to compute the power of the t test for

$$H_0: \mu = 0$$
$$H_a: \mu > 0$$

against the alternative $\mu = 2$ when $n = 20$. We must have a rough guess of the size of σ in order to compute the power. People planning a large study often run a small pilot study for this and other purposes. In this case, listening-score improvements in past summer language institutes have had sample standard deviations of about 3. We therefore take both $\sigma = 3$ and $s = 3$ in our approximate calculation.

Step 1: Write the rule for rejecting H_0 in terms of $\bar{x}$. The t test with 20 observations rejects H_0 at the 5% significance level if the t statistic

$$t = \frac{\bar{x} - 0}{s/\sqrt{20}}$$

exceeds the upper 5% point of $t(19)$, which is 1.729. Taking $s = 3$, the test rejects H_0 when

$$t = \frac{\bar{x}}{3/\sqrt{20}} \geq 1.729$$
$$\bar{x} \geq 1.729\frac{3}{\sqrt{20}}$$
$$\bar{x} \geq 1.160$$

Step 2: The power is the probability of rejecting H_0 assuming that the alternative is true. We want the probability that $\bar{x} \geq 1.160$ when $\mu = 2$. Taking $\sigma = 3$, standardize $\bar{x}$ to find this probability:

$$P(\bar{x} \geq 1.160) = P\left(\frac{\bar{x} - 2}{3/\sqrt{20}} \geq \frac{1.160 - 2}{3/\sqrt{20}}\right)$$
$$= P(Z \geq -1.252)$$
$$= 1 - 0.1056 = 0.8944$$

A true difference of 2 points in the population mean scores will produce significance at the 5% level in 89% of all possible samples. The director can be reasonably confident of detecting a difference this large.

EXERCISES

11.21 NO-FEE CREDIT CARD OFFER A bank wonders whether omitting the annual credit card fee for customers who charge at least \$2400 in a year would increase the amount charged on its credit cards. The bank makes this offer to an SRS of 200 of its credit card customers. It then compares how much these customers charge this year with the amount that they charged last year. The mean increase is \$332, and the standard deviation is \$108.

(a) Is there significant evidence at the 1% level that the mean amount charged increases under the no-fee offer? Give appropriate statistical evidence to support your conclusion.

(b) Give a 99% confidence interval for the mean amount of the increase.

(c) The distribution of the amount charged is skewed to the right, but outliers are prevented by the credit limit that the bank enforces on each card. Use of the t procedures is justified in this case even though the population distribution is not normal. Explain why.

(d) A critic points out that the customers would probably have charged more this year than last even without the new offer, because the economy is more prosperous and interest rates are lower. Briefly describe the design of an experiment to study the effect of the no-fee offer that would avoid this criticism.

11.22 NO-FEE CREDIT CARD OFFER, II The bank in Exercise 11.21 tested a new idea on a sample of 200 customers. The bank wants to be quite certain of detecting a mean increase of $\mu = \$100$ in the amount charged, at the $\alpha = 0.01$ significance level. Perhaps a sample of only $n = 50$ customers would accomplish this. Find the approximate power of the test with $n = 50$ against the alternative $\mu = \$100$ as follows:

(a) What is the critical value t^* for the one-sided test with $\alpha = 0.01$ and $n = 50$?

(b) Write the rule for rejecting H_0: $\mu = 0$ in terms of the t statistic. Then take $s = 108$ (an estimate based on the data in Exercise 11.21) and state the rejection rule in terms of $\bar{x}$.

(c) Assume that $\mu = 100$ (the given alternative) and that $\sigma = 108$ (an estimate from the data in Exercise 11.21). The approximate power is the probability of the event you found in (b), calculated under these assumptions. Find the power. Would you recommend that the bank do a test on 50 customers, or should more customers be included?

(d) Describe a Type I error and a Type II error in this setting. Which is more serious?

11.23 THE POWER OF TOMATOES The tomato experts who carried out the field trial described in Exercise 11.12 (page 633) suspect that the large P-value there is due to low power. They would like to be able to detect a mean difference in yields of 0.5 pound per plant at the 0.05 significance level. Based on the previous study, use 0.83 as an estimate of both the population σ and the value of s in future samples.

(a) What is the power of the test from Exercise 11.12 with $n = 10$ against the alternative $\mu = 0.5$?

(b) If the sample size is increased to $n = 25$ plots of land, what will be the power against the same alternative?

(c) Describe a Type I and a Type II error in this setting. Which is more serious?

SUMMARY

Tests and confidence intervals for the mean μ of a normal population are based on the sample mean $\bar{x}$ of an SRS. Because of the central limit theorem, the resulting procedures are approximately correct for other population distributions when the sample is large.

The standardized sample mean is the **one-sample z statistic,**

$$z = \frac{\bar{x} - \mu}{\sigma / \sqrt{n}}$$

When we know σ, we use the z statistic and the standard normal distribution.

In practice, we do not know σ. Replace the standard deviation $\sigma / \sqrt{n}$ of $\bar{x}$ by the **standard error** $s / \sqrt{n}$ to get the **one-sample t statistic**

$$t = \frac{\bar{x} - \mu}{s / \sqrt{n}}$$

The t statistic has the **t distribution** with $n - 1$ degrees of freedom.

There is a t distribution for every positive **degrees of freedom** k. All are symmetric distributions similar in shape to the standard normal distribution. The $t(k)$ distribution approaches the N(0,1) distribution as k increases.

An exact level C **confidence interval** for the mean μ of a normal population is

$$\bar{x} \pm t^* \frac{s}{\sqrt{n}}$$

where t^* is the upper (1 – C)/2 critical value of the $t(n - 1)$ distribution.

Significance tests for H_0: $\mu = \mu_0$ are based on the t statistic. Use P-values or fixed significance levels from the $t(n - 1)$ distribution.

Use these one-sample procedures to analyze **matched pairs** data by first taking the difference within each matched pair to produce a single sample.

The t procedures are relatively **robust** when the population is nonnormal, especially for larger sample sizes. The t procedures are useful for nonnormal data when $n \geq 15$ unless the data show outliers or strong skewness.

SECTION 11.1 EXERCISES

11.24 The one-sample t statistic for a test of

$$H_0: \mu = 10$$
$$H_a: \mu < 10$$

based on $n = 10$ observations has the value $t = -2.25$.

(a) What are the degrees of freedom for this statistic?

(b) Between what two probabilities p from Table C does the P-value of the test fall?

11.25 SIGNIFICANCE You are testing H_0: $\mu = 0$ against H_a: $\mu \uparrow 0$ based on an SRS of 20 observations from a normal population. What values of the t statistic are statistically significant at the $\alpha = 0.005$ level?

11.26 WHAT CRITICAL VALUE? You have an SRS of 15 observations from a normally distributed population. What critical value would you use to obtain a 98% confidence interval for the mean μ of the population?

11.27 A BIG TOE PROBLEM, I Hallux abducto valgus (call it HAV) is a deformation of the big toe that is uncommon in youth and often requires surgery. Doctors used X-rays to measure the angle (in degrees) of deformity in 38 consecutive patients under the age of 21 who came to a medical center for surgery to correct HAV. The angle is a measure of the seriousness of the deformity. Here are the data:[5]

28	32	25	34	38	26	25	18	30	26	28	13	20
21	17	16	21	23	14	32	25	21	22	20	18	26
16	30	30	20	50	25	26	28	31	38	32	21	

We are willing to consider these patients as a random sample of young patients who require HAV surgery. Give a 95% confidence interval for the mean HAV angle in the population of all such patients. Follow the Inference Toolbox.

11.28 A BIG TOE PROBLEM, II The data in the previous problem follow a normal distribution quite closely except for one patient with HAV angle 50 degrees, a high outlier.

(a) Find the 95% confidence interval for the population mean based on the 37 patients who remain after you drop the outlier.

(b) Compare your interval in (a) with your interval from the previous problem. What is the most important effect of removing the outlier?

11.29 VITAMIN C CONTENT The researchers studying vitamin C in CSB in Exercise 11.9 (page 628) were also interested in a similar commodity called wheat soy blend (WSB). A major concern was the possibility that some of the vitamin C content would be destroyed as a result of storage and shipment of the commodity to its final destination. The researchers specially marked a collection of bags at the factory and took a sample from each of these to determine the vitamin C content. Five months later in Haiti they found the specially marked bags and took samples. The data consist of two vitamin C measures for each bag, one at the time of production in the factory and the other five months later in Haiti. The units are mg/100 g as in Exercise 11.9. Here are the data:

Factory	Haiti	Factory	Haiti	Factory	Haiti
44	40	45	38	39	43
50	37	32	40	52	38
48	39	47	35	45	38
44	35	40	38	37	38
42	35	38	34	38	41
47	41	41	35	44	40
49	37	43	37	43	35
50	37	40	34	39	38
39	34	37	40	44	36

(a) Examine the question of interest to these researchers. Provide appropriate statistical evidence to justify your conclusion.

(b) Estimate the loss in vitamin C content over the five-month period. Use a 95% confidence level.

(c) Do these data provide evidence that the mean vitamin C content of all of the bags of WSB shipped to Haiti differs from the target value of 40 mg/100 g?

11.30 CALCIUM AND BLOOD PRESSURE In a randomized comparative experiment on the effect of calcium in the diet on blood pressure, researchers divided 54 healthy white males at random into two groups. One group received calcium; the other, a placebo. At the beginning of the study, the researchers measured many variables on the subjects. The paper reporting the study gives $\bar{x} = 114.9$ and $s = 9.3$ for the seated systolic blood pressure of the 27 members of the placebo group.

(a) Give a 99% confidence interval for the mean blood pressure in the population from which the subjects were recruited.

(b) What conditions about the population and the study design are required by the procedure you used in (a)? Which of these conditions are important for the validity of the procedure in this case?

11.31 RIGHT VERSUS LEFT The design of controls and instruments affects how easily people can use them. A student project investigated this effect by asking 25 right-handed students to turn a knob (with their right hands) that moved an indicator by screw action. There were two identical instruments, one with a right-hand thread (the knob turns clockwise) and the other with a left-hand thread (the knob must be turned counterclockwise). The following table gives the times in seconds each subject took to move the indicator a fixed distance:[6]

Subject	Right thread	Left thread	Subject	Right thread	Left thread
1	113	137	14	107	87
2	105	105	15	118	166
3	130	133	16	103	146
4	101	108	17	111	123
5	138	115	18	104	135
6	118	170	19	111	112
7	87	103	20	89	93
8	116	145	21	78	76
9	75	78	22	100	116
10	96	107	23	89	78
11	122	84	24	85	101
12	103	148	25	88	123
13	116	147			

(a) Each of the 25 students used both instruments. Discuss briefly how you would use randomization in arranging the experiment.

(b) The project hoped to show that right-handed people find right-hand threads easier to use. What is the parameter μ for a matched pairs t test? State H_0 and H_a in terms of μ.

(c) Carry out a test of your hypotheses. Give the P-value and report your conclusions.

11.32 RIGHT VERSUS LEFT, II Give a 90% confidence interval for the mean time advantage of right-hand over left-hand threads in the setting of Exercise 11.31. Do you think that the time saved would be of practical importance if the task were performed many times—for example, by an assembly line worker? To help answer this question, find the mean time for right-hand threads as a percent of the mean time for left-hand threads.

11.33 RADON DETECTORS Many homeowners buy detectors to check for the invisible gas radon in their homes. How accurate are these detectors? To answer this question, university researchers placed 12 radon detectors in a chamber that exposed them to 105 picocuries per liter of radon. The detector readings were as follows.[7]

91.9	97.8	111.4	122.3	105.4	95.0	103.8	99.6	96.6	119.3	104.8	101.7

(a) Make a stemplot of the data. The distribution is somewhat skewed to the right, but not strongly enough to forbid use of the t procedures.

(b) Is there convincing evidence that the mean reading of all detectors of this type differs from the true value 105? Carry out a test in detail, then write a brief conclusion.

11.34 Table 1.4 (page 19) gives the ages of U.S. presidents when they took office. It does not make sense to use the t procedures (or any other statistical procedures) to give a 95% confidence interval for the mean age of the presidents. Explain why not.

11.35 THE POWER OF A *t* TEST Exercise 11.18 (page 638) reports a small study comparing ARSMA and BI, two tests of the acculturation of Mexican Americans. Would this study usually detect a difference in mean scores of 0.2? To answer this question, calculate the approximate power of the test (with $n = 22$ subjects and $\alpha = 0.05$) of

$$H_0: \mu = 0$$
$$H_a: \mu \uparrow 0$$

against the alternative $\mu = 0.2$. We do this by acting as if σ were known.

(a) From Table C, what is the critical value for $\alpha = 0.05$?

(b) Write the rule for rejecting H_0 at the $\alpha = 0.05$ level. Then take $s = 0.3$, the approximate value observed in Exercise 11.18, and restate the rejection criterion in terms of $\bar{x}$. Note that this is a two-sided test.

(c) Find the probability of this event when $\mu = 0.2$ (the alternative given) and $\sigma = 0.3$ (estimated from the data in Exercise 11.18) by a normal probability calculation. This is the approximate power.

11.36 AP FREE-RESPONSE SCORES About 42,000 high school students took the AP Statistics exam in 2001. The free-response section of the exam consisted of five open-ended problems and an investigative task. Each free-response question is scored on a

0 to 4 scale (with 4 being the best). A random sample of 25 student papers yielded the following scores on one of the free-response questions:

1	0	1	0	0	0	3	1	1	1	0	2	0	0	2	1	1	0	2	4	1	0	2	0	3

(a) Is a sample of 25 papers large enough to provide a good estimate of the mean score of all 42,000 students on this exam problem? Justify your answer.

(b) Do you think the population of scores on this question is normally distributed? Explain why or why not.

(c) Construct a 95% confidence interval for the mean score on this exam question.

TECHNOLOGY TOOLBOX *t procedures on the TI-83/89*

Confidence intervals and *t* tests of significance can be performed on the TI-83/89, thus avoiding table lookups. Here is a brief summary of the techniques using the healing rates data from Exercise 11.20 (page 638). For reference, the difference in healing rates for the 14 newts (in micrometers per hour) are

–1	10	3	–3	–31	4	–12	–3	–7	–10	–22	–4	–1	–3

Enter these data in L_1/list1.

On the TI-83, all inference routines are found under STAT/TESTS. On the TI-89, all inference routines can be accessed from inside the Stats/List Editor APP. Choose F6 (Tests) for significance tests and F7 (Ints) for confidence intervals.

To determine a confidence interval for these data:

TI-83

- Choose 8:TInterval.

```
EDIT CALC TESTS
2↑T-Test...
3:2-SampZTest...
4:2-SampTTest...
5:1-PropZTest...
6:2-PropZTest...
7:ZInterval...
8↓TInterval...
```

TI-89

- Choose 2:TInterval.

```
F1▼    F2▼   F3▼  F4▼  F5▼   F6▼   F7▼
Tools Plots List Calc Distr Tests Ints
list1  -----  1:ZInterval...
-10           2:TInterval...
-22           3:2-SampZInt...
-4            4:2-SampTInt...
-1            5:1-PropZInt...
-3            6:2-PropZInt...
-----         7:LinRegTInt...
              8:MultRegInt...
list1[15]=
MAIN     RAD AUTO     FUNC    1/1
```

- Choose “Data” (not “Stats”) and adjust the TInterval screen as shown.

```
TInterval
 Inpt:Data Stats
 List:L1
 Freq:1
 C-Level:.95
 Calculate
```

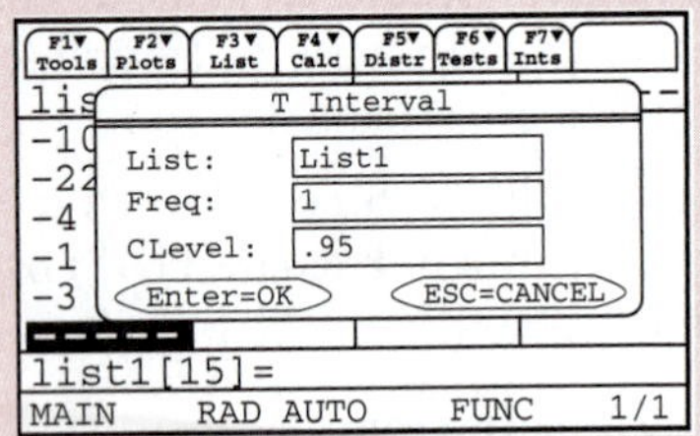

TECHNOLOGY TOOLBOX *t procedures on the TI-83/89 (continued)*

- Select "Calculate" and press ENTER.

```
TInterval
 (-11.81,.38536)
 x̄=-5.714285714
 Sx=10.56429816
 n=14
```

```
T Interval
CInt   ={-11.8,.3854}
x̄      =-5.71429
ME     =6.09964
df     =13.
Sx     =10.5643
n      =14.
Enter=OK
list1[15]=
MAIN    RAD AUTO    FUNC    1/1
```

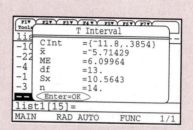

The results tell us that the 95% confidence interval for the true mean population difference in healing rate is between –11.81 and 0.385 micrometers per hour. If the researchers wanted to keep the 95% confidence level but wanted a shorter, more precise confidence interval, they would need to use more newts in the experiment (that is, increase the sample size n).

To perform tests of significance, recall that the null hypothesis in Exercise 11.20 (b) was "no change in healing rate" for the newts, while the alternative hypothesis said "changing the electric field lowers the healing rate."

In symbols,

$$H_0: \mu_{EXP-CONTROL} = \mu_{DIFF} = 0$$
$$H_a: \mu_{EXP-CONTROL} = \mu_{DIFF} < 0$$

With the data stored in L_1/list1, go to STAT/TESTS (Tests menu on the TI-89). Choose 2: T-Test. Adjust your settings as shown.

```
T-Test
 Inpt:Data Stats
 μ0:0
 List:L1
 Freq:1
 μ:≠μ0 <μ0 >μ0
 Calculate Draw
```

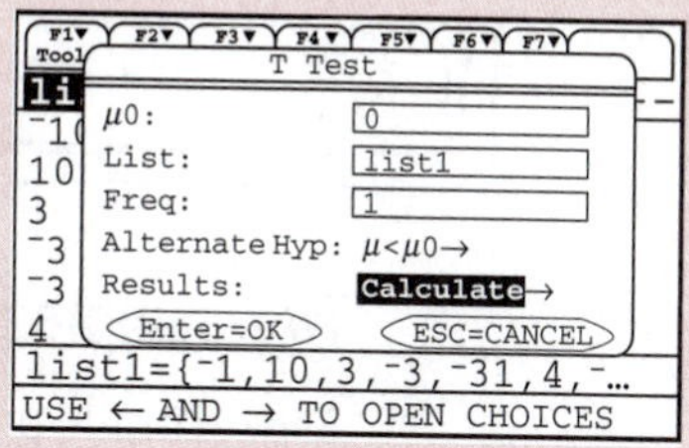

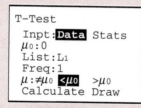

If you select "Calculate," the following screen appears:

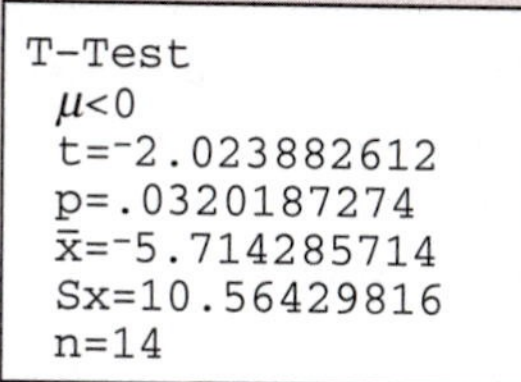

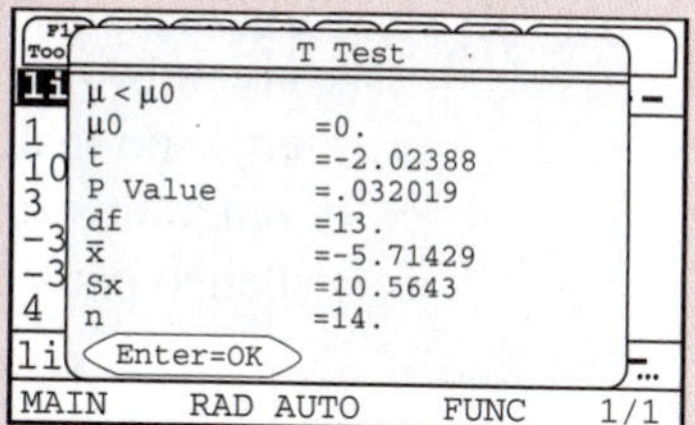

The test statistic is $t = -2.02$ and the P-value is 0.032.

If you specify "Draw," you see a $t(13)$ distribution curve with the lower critical area shaded. In either case, the P-value is 0.032.

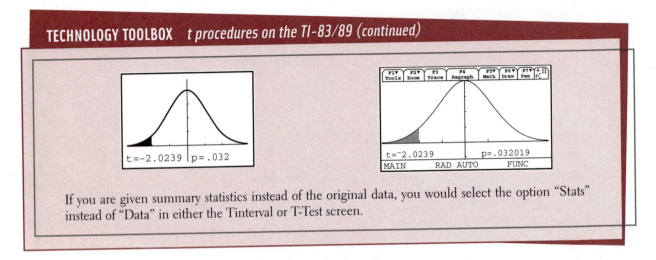

TECHNOLOGY TOOLBOX *t procedures on the TI-83/89 (continued)*

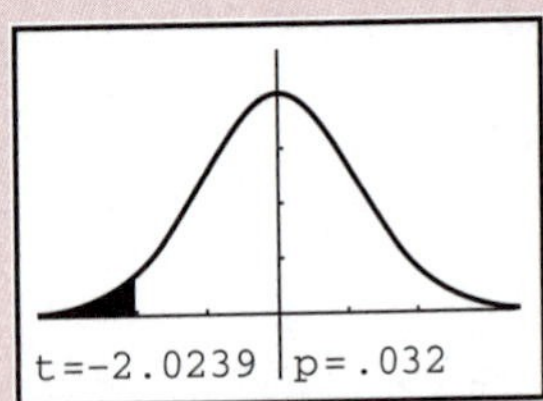

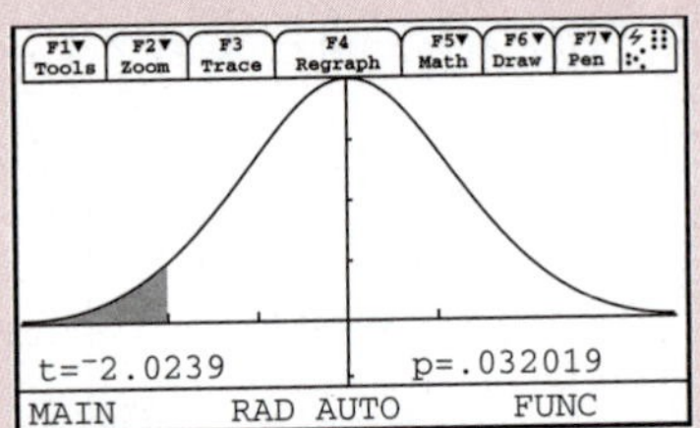

If you are given summary statistics instead of the original data, you would select the option "Stats" instead of "Data" in either the Tinterval or T-Test screen.

11.2 COMPARING TWO MEANS

Comparing two populations or two treatments is one of the most common situations encountered in statistical practice. We call such situations *two-sample problems.*

TWO-SAMPLE PROBLEMS

- The goal of inference is to compare the responses to two treatments or to compare the characteristics of two populations.
- We have a separate sample from each treatment or each population.

Two-sample problems

A two-sample problem can arise from a randomized comparative experiment that randomly divides subjects into two groups and exposes each group to a different treatment. Comparing random samples separately selected from two populations is also a two-sample problem. Unlike the matched pairs designs studied earlier, there is no matching of the units in the two samples and the two samples can be of different sizes. Inference procedures for two-sample data differ from those for matched pairs. Here are some typical two-sample problems.

EXAMPLE 11.9 TWO-SAMPLE PROBLEMS

(a) A medical researcher is interested in the effect on blood pressure of added calcium in our diet. She conducts a randomized comparative experiment in which one group of subjects receives a calcium supplement and a control group gets a placebo.

(b) A psychologist develops a test that measures social insight. He compares the social insight of male college students with that of female college students by giving the test to a sample of students of each gender.

(c) A bank wants to know which of two incentive plans will most increase the use of its credit cards. It offers each incentive to a random sample of credit card customers and compares the amount charged during the following six months.

EXERCISES

11.37 WHICH DATA DESIGN? The following situations require inference about a mean or means. Identify each as (1) single sample, (2) matched pairs, or (3) two samples. The procedures of Section 11.1 apply to cases (1) and (2). We are about to learn procedures for (3).

(a) An education researcher wants to learn whether it is more effective to put questions before or after introducing a new concept in an elementary school mathematics text. He prepares two text segments that teach the concept, one with motivating questions before and the other with review questions after. He uses each text segment to teach a separate group of children. The researcher compares the scores of the groups on a test over the material.

(b) Another researcher approaches the same issue differently. She prepares text segments on two unrelated topics. Each segment comes in two versions, one with questions before and the other with questions after. The subjects are a single group of children. Each child studies both topics, one (chosen at random) with questions before and the other with questions after. The researcher compares test scores for each child on the two topics to see which topic he or she learned better.

11.38 WHICH DATA DESIGN? The following situations require inference about a mean or means. Identify each as (1) single sample, (2) matched pairs, or (3) two samples. The procedures of Section 11.1 apply to cases (1) and (2). We are about to learn procedures for (3).

(a) To check a new analytical method, a chemist obtains a reference specimen of known concentration from the National Institute of Standards and Technology. She then makes 20 measurements of the concentration of this specimen with the new method and checks for bias by comparing the mean result with the known concentration.

(b) Another chemist is checking the same new method. He has no reference specimen, but a familiar analytic method is available. He wants to know if the new and old methods agree. He takes a specimen of unknown concentration and measures the concentration 10 times with the new method and 10 times with the old method.

Comparing two population means

We can examine two-sample data graphically by comparing stemplots (for small samples) or histograms or boxplots (for larger samples). Now we will apply the ideas of formal inference in this setting. When both population distributions are symmetric, and especially when they are at least approximately normal, a comparison of the mean responses in the two populations is the most common goal of inference. Here are the conditions that must be satisfied.

CONDITIONS FOR COMPARING TWO MEANS

- We have **two SRSs,** from two distinct populations. The samples are **independent.** That is, one sample has no influence on the other. Matching violates independence, for example. We measure the same variable for both samples.
- Both populations are **normally distributed.** The means and standard deviations of the populations are unknown.

Call the variable we measure x_1 in the first population and x_2 in the second because the variable may have different distributions in the two populations. Here is the notation we will use to describe the two populations:

Population	Variable	Mean	Standard deviation
1	x_1	μ_1	σ_1
2	x_2	μ_2	σ_2

There are four unknown parameters, the two means and the two standard deviations. The subscripts remind us which population a parameter describes. We want to compare the two population means, either by giving a confidence interval for their difference $\mu_1 - \mu_2$ or by testing the hypothesis of no difference, H_0: $\mu_1 = \mu_2$.

We use the sample means and standard deviations to estimate the unknown parameters. Again, subscripts remind us which sample a statistic comes from. Here is the notation that describes the samples:

Population	Sample size	Sample mean	Sample standard deviation
1	n_1	$\bar{x}_1$	s_1
2	n_2	$\bar{x}_2$	s_2

To do inference about the difference $\mu_1 - \mu_2$ between the means of the two populations, we start from the difference $\bar{x}_1 - \bar{x}_2$ between the means of the two samples.

EXAMPLE 11.10 CALCIUM AND BLOOD PRESSURE

Does increasing the amount of calcium in our diet reduce blood pressure? Examination of a large sample of people revealed a relationship between calcium intake and blood pressure. The relationship was strongest for black men. Such observational studies do not establish causation. Researchers therefore designed a randomized comparative experiment.

The subjects in part of the experiment were 21 healthy black men. A randomly chosen group of 10 of the men received a calcium supplement for 12 weeks. The control group of 11 men received a placebo pill that looked identical. The experiment was double-blind. The response variable is the decrease in systolic (heart contracted) blood pressure for a subject after 12 weeks, in millimeters of mercury. An increase appears as a negative response.[8]

Take Group 1 to be the calcium group and Group 2 the placebo group. Here are the data for the 10 men in Group 1 (calcium),

7	–4	18	17	–3	–5	1	10	11	–2

and for the 11 men in Group 2 (placebo),

–1	12	–1	–3	3	–5	5	2	–11	–1	–3

From the data, calculate the summary statistics:

Group	Treatment	n	$\bar{x}$	s
1	Calcium	10	5.000	8.743
2	Placebo	11	–0.273	5.901

The calcium group shows a drop in blood pressure, $\bar{x}_1 = 5.000$, while the placebo group had almost no change, $\bar{x}_2 = -0.273$. Is this outcome good evidence that calcium decreases blood pressure in the entire population of healthy black men more than a placebo does?

This example fits the two-sample setting. Since we want to test a claim, we will perform a significance test. Our Inference Toolbox provides the procedure.

Step 1: Identify the population(s) of interest and the parameter(s) you want to draw conclusions about. State hypotheses in words and symbols. We write hypotheses in terms of the mean decreases we would see in the entire population, μ_1 for men taking calcium for 12 weeks and μ_2 for men taking a placebo. The hypotheses are

$H_0: \mu_1 = \mu_2$ The mean decrease in blood pressure for those taking calcium is the same as the mean decrease in blood pressure for those taking a placebo.

$H_a: \mu_1 > \mu_2$ The mean decrease in blood pressure for those taking calcium is greater than the mean decrease in blood pressure for those taking a placebo.

Step 2: Choose the appropriate inference procedure, and verify the conditions for using the selected procedure. We do not yet know what procedure to use. Next, we check the required conditions.

- Because of the randomization, we are willing to regard the calcium and placebo groups as two independent SRSs.

- Although the samples are small, we check for serious nonnormality by examining the data. Here is a back-to-back stemplot of the responses. (We have split the stems. Notice that negative responses require –0 and 0 to be separate stems, and that the ordering of leaves out from the stems recognizes that –3 is smaller than –1.)

Calcium		Placebo
	−1	1
5	−0	5
234	−0	33111
1	0	23
7	0	5
10	1	2
87	1	

The placebo responses appear roughly normal. The calcium group has an irregular distribution, which is not unusual when we have only a few observations. There are no outliers, and no departures from normality that prevent use of t procedures.

We will continue with Steps 3 and 4 in Example 11.11.

The natural estimator of the difference $\mu_1 - \mu_2$ is the difference between the sample means:

$$\bar{x}_1 - \bar{x}_2 = 5.000 - (-0.273) = 5.273$$

This statistic measures the average advantage of calcium over a placebo. In order to use it for inference, we must know its sampling distribution.

The sampling distribution of $\bar{x}_1 - \bar{x}_2$

Here are the facts about the sampling distribution of the difference $\bar{x}_1 - \bar{x}_2$ between the sample means of two independent SRSs. These facts can be derived using the mathematics of probability or made plausible by simulation.

- The mean of $\bar{x}_1 - \bar{x}_2$ is $\mu_1 - \mu_2$. That is, the difference of sample means is an unbiased estimator of the difference of population means.
- The variance of the difference is the *sum* of the variances of $\bar{x}_1 - \bar{x}_2$, which is

$$\frac{\sigma_1^2}{n_1} + \frac{\sigma_2^2}{n_2}$$

Note that the *variances* add. The standard deviations do not.

- If the two population distributions are both normal, then the distribution of $\bar{x}_1 - \bar{x}_2$ is also normal.

two-sample z statistic

Because the statistic $\bar{x}_1 - \bar{x}_2$ has a normal distribution, we can standardize it to obtain a standard normal z statistic. Subtract its mean, then divide by its standard deviation to get the ***two-sample z statistic:***

$$z = \frac{(\bar{x}_1 - \bar{x}_2) - (\mu_1 - \mu_2)}{\sqrt{\dfrac{\sigma_1^2}{n_1} + \dfrac{\sigma_2^2}{n_2}}}$$

To assess the significance of the observed difference between the means of our two samples, we follow a familiar path. Whether an observed difference between two samples is surprising depends on the spread of the observations as well as on the two means. Widely different means can arise just by chance if the individual observations vary a great deal. To take variation into account, we would like to standardize the observed difference $\bar{x}_1 - \bar{x}_2$ by dividing by its standard deviation. This standard deviation is

$$\sqrt{\frac{\sigma_1^2}{n_1} + \frac{\sigma_2^2}{n_2}}$$

This standard deviation gets larger as either population gets more variable, that is, as σ_1 or σ_2 increases. It gets smaller as the sample sizes n_1 and n_2 increase.

Because we don't know the population standard deviations, we estimate them by the sample standard deviations from our two samples. The result is the ***standard error***, or estimated standard deviation, of the difference in sample means:

standard error

$$\text{SE} = \sqrt{\frac{s_1^2}{n_1} + \frac{s_2^2}{n_2}}$$

When we standardize the estimate by dividing it by its standard error, the result is the ***two-sample t statistic***:

two-sample t statistic

$$t = \frac{(\bar{x}_1 - \bar{x}_2) - (\mu_1 - \mu_2)}{\sqrt{\dfrac{s_1^2}{n_1} + \dfrac{s_2^2}{n_2}}}$$

The statistic t has the same interpretation as any z or t statistic: it says how far $\bar{x}_1 - \bar{x}_2$ is from its mean in standard deviation units. Unfortunately, the two-sample t statistic does *not* have a t distribution. A t distribution replaces a N(0,1) distribution when we replace just one standard deviation in a z statistic by a standard error. In this case, we replaced two standard deviations by the corresponding standard errors. This does not produce a statistic having a t distribution.

Nonetheless, the two-sample t statistic is used with t critical values in inference for two-sample problems. There are two ways to do this.

Option 1: Use procedures based on the statistic t with critical values from a t distribution with degrees of freedom computed from the data. The degrees of freedom are generally not a whole number. This is a very accurate approximation to the distribution of t.

Option 2: Use procedures based on the statistic t with critical values from the t distribution with degrees of freedom equal to the smaller of $n_1 - 1$ and $n_2 - 1$. These procedures are always conservative for any two normal populations.

Most statistical software systems and the TI-83/89 use the two-sample t statistic with Option 1 for two-sample problems unless the user requests another method. Using this option without software is a bit complicated. We will therefore present the second, simpler, option first. We recommend that you use Option 2 when doing calculations without a calculator or computer. If you use a computer package, it should automatically do the calculations for Option 1. Here is a statement of the Option 2 procedures that includes a statement of just how they are "conservative."

THE TWO-SAMPLE t PROCEDURES

Draw an SRS of size n_1 from a normal population with unknown mean μ_1, and draw an independent SRS of size n_2 from another normal population with unknown mean μ_2. The confidence interval for $\mu_1 - \mu_2$ given by

$$(\bar{x}_1 - \bar{x}_2) \pm t^* \sqrt{\frac{s_1^2}{n_1} + \frac{s_2^2}{n_2}}$$

has confidence level *at least* C no matter what the population standard deviations may be. Here t^* is the upper $(1 - C)/2$ critical value for the $t(k)$ distribution with k the smaller of $n_1 - 1$ and $n_2 - 1$.

To test the hypothesis H_0: $\mu_1 = \mu_2$, compute the two-sample t statistic

$$t = \frac{\bar{x}_1 - \bar{x}_2}{\sqrt{\frac{s_1^2}{n_1} + \frac{s_2^2}{n_2}}}$$

and use P-values or critical values for the $t(k)$ distribution. The true P-value or fixed significance level will always be *equal to or less than* the value calculated from $t(k)$ no matter what values the unknown population standard deviations have.

These two-sample t procedures always err on the safe side, reporting *higher* P-values and *lower* confidence than are actually true. The gap between what is reported and the truth is quite small unless the sample sizes are both small and unequal. As the sample sizes increase, probability values based on t with degrees of freedom equal to the smaller of $n_1 - 1$ and $n_2 - 1$ become more accurate.[9] The following examples illustrate the two-sample t procedures.

EXAMPLE 11.11 CALCIUM AND BLOOD PRESSURE, CONTINUED

The medical researchers in Example 11.10 can use the two-sample t procedures to compare calcium with a placebo.

Step 3: Compute the test statistic and the P-value:

$$t = \frac{\bar{x}_1 - \bar{x}_2}{\sqrt{\frac{s_1^2}{n_1} + \frac{s_2^2}{n_2}}}$$

$$= \frac{5.000 - (-0.273)}{\sqrt{\frac{8.743^2}{10} + \frac{5.901^2}{11}}}$$

$$= \frac{5.273}{3.2878} = 1.604$$

There are 9 degrees of freedom, the smaller of $n_1 - 1 = 9$ and $n_2 - 1 = 10$. Because H_a is one-sided on the high side, the P-value is the area to the right of $t = 1.604$ under the $t(9)$ curve. Figure 11.10 illustrates this P-value. Table C shows that it lies between 0.05 and 0.10.

df = 9

p	.10	.05
t^*	1.383	1.833

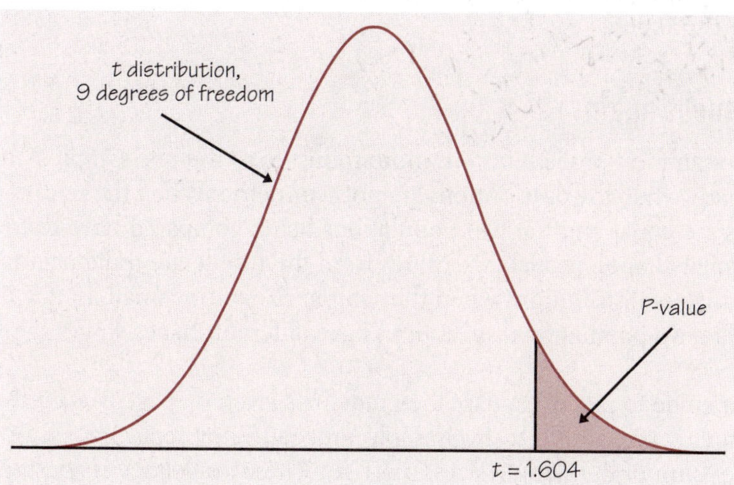

FIGURE 11.10 The P-value. This example uses the conservative method, which leads to the t distribution with 9 degrees of freedom.

Step 4: Interpret your results in the context of the problem. The experiment found evidence that calcium reduces blood pressure, but the evidence falls a bit short of the traditional 5% and 1% levels. We would fail to reject H_0 at either of these significance levels.

We can estimate the difference in the mean decreases in blood pressure for the calcium and placebo populations using a two-sample t interval. The next example shows how.

EXAMPLE 11.12 TWO-SAMPLE t CONFIDENCE INTERVAL

For a 90% confidence interval, Table C shows that the $t(9)$ critical value is $t^* = 1.833$. We are 90% confident that the mean advantage of calcium over a placebo, $\mu_1 - \mu_2$, lies in the interval

$$\begin{aligned}(\bar{x}_1 - \bar{x}_2) \pm t^* \sqrt{\frac{s_1^2}{n_1} + \frac{s_2^2}{n_2}} &= [5.000 - (-0.273)] \pm 1.833\sqrt{\frac{8.743^2}{10} + \frac{5.901^2}{11}} \\ &= 5.273 \pm 6.026 \\ &= (-0.753, 11.299)\end{aligned}$$

That the 90% confidence interval covers 0 tells us that we cannot reject H_0: $\mu_1 = \mu_2$ against the two-sided alternative at the $\alpha = 0.10$ level of significance.

Sample size strongly influences the P-value of a test. An effect that fails to be significant at a specified level α in a small sample will be significant in a larger sample. In the light of the rather small samples in Example 11.11, we suspect that more data might show that calcium has a significant effect. The published account of the study combined these results for blacks with results for whites and adjusted for pretest differences among the subjects. Using this more detailed analysis, the researchers were able to report the P-value $P = 0.008$.

Robustness again

The two-sample t procedures are more robust than the one-sample t methods, particularly when the distributions are not symmetric. When the sizes of the two samples are equal and the two populations being compared have distributions with similar shapes, probability values from the t table are quite accurate for a broad range of distributions when the sample sizes are as small as $n_1 = n_2 = 5$.[10] When the two population distributions have different shapes, larger samples are needed.

As a guide to practice, adapt the guidelines given on page 636 for the use of one-sample t procedures to two-sample procedures by replacing "sample size" with the "sum of the sample sizes," $n_1 + n_2$. These guidelines err on the side of safety, especially when the two samples are of equal size. In planning a two-sample study, you should usually choose equal sample sizes. The two-sample t procedures are most robust against nonnormality in this case, and the conservative probability values are most accurate.

EXERCISES

11.39 SOCIAL INSIGHT AMONG MEN AND WOMEN The Chapin Social Insight Test is a psychological test designed to measure how accurately a person appraises other people. The possible scores on the test range from 0 to 41. During the development of the Chapin test, it was given to several different groups of people. Here are the results for male and female college students majoring in the liberal arts:[11]

Group	Sex	n	$\bar{x}$	s
1	Male	133	25.34	5.05
2	Female	162	24.94	5.44

Do these data support the contention that female and male students differ in average social insight? Perform a significance test to help you answer this question.

11.40 THE EFFECT OF LOGGING How badly does logging damage tropical rainforests? One study compared forest plots in Borneo that had never been logged with similar plots nearby that had been logged 8 years earlier. The study found that the effects of logging were somewhat less severe than expected. Here are the data on the number of tree species in 12 unlogged plots and 9 logged plots:[12]

Unlogged:	22	18	22	20	15	21	13	13	19	13	19	15
Logged:	17	4	18	14	18	15	15	10	12			

(a) The study report says, "Loggers were unaware that the effects of logging would be assessed." Why is this important? The study report also explains why the plots can be considered to be randomly assigned.

(b) Does logging significantly reduce the mean number of species in a plot after 8 years? Give appropriate statistical evidence to support your conclusion.

(c) Give a 90% confidence interval for the difference in mean number of species between unlogged and logged plots.

11.41 SURGERY IN A BLANKET When patients undergo surgery, the operating room is kept cool so that the physicians in heavy gowns will not be overheated. The patient may pay the price for the surgeon's comfort. The exposure to cold, in addition to impairment of temperature regulation caused by anesthesia and altered distribution of body heat, may result in mild hypothermia (approximately 2° C below the normal core body temperature). As a result of the hypothermia, patients may have an increased susceptibility to wound infections or even heart attacks. In 1996, researchers in Austria investigated whether maintaining a patient's body temperature close to normal by heating the patient during surgery decreases wound infection rates. Patients were assigned at random to two groups: the normothermic group (patients' core temperatures were maintained at near normal 36.5° C with heating blankets) and the hypothermic group (patients' core temperatures were allowed to decrease to about 34.5° C). If keeping patients warm during surgery reduces the chance of infection, then patients in the normothermic group should have shorter hospital stays than those in the hypothermic group.

Here are summary statistics on length of hospital stay for the two treatment groups.[13]

Group	n	$\bar{x}$	s
Normothermic	104	12.1	4.4
Hypothermic	96	14.7	6.5

(a) Do these data provide evidence that the use of warming blankets reduces the length of a patient's hospital stay?

(b) Construct a 95% confidence interval for the difference between the means for length of stay in the hospital for the normothermic and hypothermic groups. What does this interval tell you about the effect of the treatment?

11.42 PAYING FOR COLLEGE College financial aid offices expect students to use summer earnings to help pay for college. But how large are these earnings? One college studied this question by asking a sample of students how much they earned. Omitting students who were not employed, there were 1296 responses. Here are the data in summary form:[14]

Group	n	$\bar{x}$	s
Males	675	\$1884.52	\$1368.37
Females	621	\$1360.39	\$1037.46

(a) The distribution of earnings is strongly skewed to the right. Nevertheless, use of t procedures is justified. Why?

(b) Give a 90% confidence interval for the difference between the mean summer earnings of male and female students.

(c) Once the sample size was decided, the sample was chosen by taking every 20th name from an alphabetical list of all undergraduates. Is it reasonable to consider the samples as SRSs chosen from the male and female undergraduate populations?

(d) What other information about the study would you request before accepting the results as describing all undergraduates?

11.43 BEETLES IN OATS In a study of cereal leaf beetle damage on oats, researchers measured the number of beetle larvae per stem in small plots of oats after randomly applying one of two treatments: no pesticide, or malathion at the rate of 0.25 pound per acre. The data appear roughly normal. Here are the summary statistics.[15]

Group	Treatment	n	$\bar{x}$	s
1	Control	13	3.47	1.21
2	Malathion	14	1.36	0.52

Is there significant evidence at the 1% level that malathion reduces the mean number of larvae per stem? Be sure to state H_0 and H_a.

More accurate levels in the *t* procedures

The two-sample t statistic does not have a t distribution. Moreover, the exact distribution changes as the unknown population standard deviations σ_1 and σ_2 change. However, an excellent approximation is available.

APPROXIMATE DISTRIBUTION OF THE TWO-SAMPLE t STATISTIC

The distribution of the two-sample t statistic is close to the t distribution with degrees of freedom df given by

$$\text{df} = \frac{\left(\frac{s_1^2}{n_1} + \frac{s_2^2}{n_2}\right)^2}{\frac{1}{n_1 - 1}\left(\frac{s_1^2}{n_1}\right)^2 + \frac{1}{n_2 - 1}\left(\frac{s_2^2}{n_2}\right)^2}$$

This approximation is quite accurate when both sample sizes n_1 and n_2 are 5 or larger.

The t procedures remain exactly as before except that we use the t distribution with df degrees of freedom to give critical values and P-values.

EXAMPLE 11.13 CALCIUM AND BLOOD PRESSURE, CONTINUED

In the calcium experiment of Examples 11.10 to 11.12 the data gave

Group	Treatment	n	$\bar{x}$	s
1	Calcium	10	5.000	8.743
2	Placebo	11	−0.273	5.901

For improved accuracy, we can use critical points from the t distribution with degrees of freedom df given by

$$\text{df} = \frac{\left(\frac{8.743^2}{10} + \frac{5.901^2}{11}\right)^2}{\frac{1}{9}\left(\frac{8.743^2}{10}\right)^2 + \frac{1}{10}\left(\frac{5.901^2}{11}\right)^2}$$

$$= \frac{116.848}{7.494} = 15.59$$

Notice that the degrees of freedom df is not a whole number.

The conservative 90% confidence interval for $\mu_1 - \mu_2$ in Example 11.12 used the critical value $t^* = 1.833$ based on 9 degrees of freedom. A more exact confidence interval replaces this critical value with the critical value for df = 15.59 degrees of freedom. We cannot find this critical value exactly without using a computer package. For a close approximation, use the next smaller entry (15 degrees of freedom) in Table C. The critical value is $t^* = 1.753$. The 90% confidence interval is now

$$\left(\bar{x}_1 - \bar{x}_2\right) \pm t^* \sqrt{\frac{s_1^2}{n_1} + \frac{s_2^2}{n_2}}$$

$$= \left[5.000 - (-0.273)\right] \pm 1.753\sqrt{\frac{8.743^2}{10} + \frac{5.901^2}{11}}$$

$$= 5.273 \pm 5.764$$

$$= (-0.491,\ 11.037)$$

This confidence interval is a bit shorter (margin of error 5.764 rather than 6.026) than the conservative interval in Example 11.12.

As Example 11.13 illustrates, the two-sample t procedures are exactly as before, except that we use a t distribution with more degrees of freedom. The number df from the box on page 659 is always at least as large as the smaller of $n_1 - 1$ and $n_2 - 1$. On the other hand, df is never larger than the sum $n_1 + n_2 - 2$ of the two individual degrees of freedom. The number of degrees of freedom df is generally not a whole number. There is a t distribution for any positive degrees of freedom, even though Table C contains entries only for whole-number degrees of freedom. Some software packages find df and then use the t distribution with the next smaller whole-number degrees of freedom. Others take care to use t(df) even when df is not a whole number. We do not recommend regular use of this method unless a computer is doing the arithmetic. With a TI-83/89 or computer, the more accurate procedures are painless, as the following Technology Toolbox illustrates.

TECHNOLOGY TOOLBOX *Two-sample inference with the TI-83/89*

Constructing confidence intervals and t tests of significance for two-sample models on the TI-83/89 is very similar to the one-sample case. To illustrate, we will use the data on calcium supplements to lower blood pressure from Examples 11.10 to 11.12. The data represent a decrease in systolic blood pressure after 12 weeks, in millimeters of mercury. The data for the 10 men in Group 1 (calcium) were

7	−4	18	17	−3	−5	1	10	11	−2

and for the 11 men in Group 2 (placebo),

−1	12	−1	−3	3	−5	5	2	−11	−1	−3

Tests of significance

- Enter the Group 1 (calcium) data into L_1/list1 and the Group 2 (placebo) data into L_2/list2.
- To perform the significance test, go to STAT/TESTS (Tests menu in the Statistics/List Editor APP on the TI-89) and choose 4:2-SampTTest.
- In the 2-SampTTest screen, specify "Data" and adjust your other settings as shown.

TECHNOLOGY TOOLBOX *Two-sample inference with the TI-83/89 (continued)*

TI-83 | **TI-89**

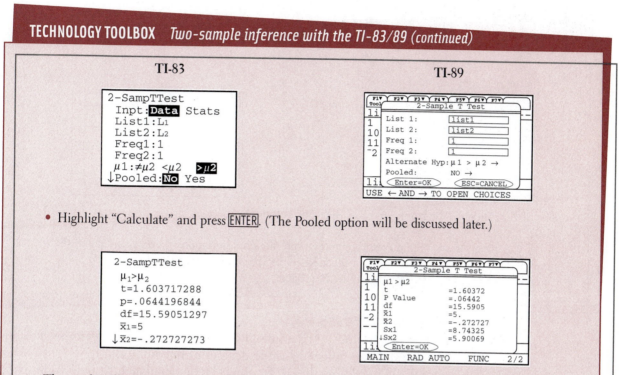

```
2-SampTTest
 Inpt:Data Stats
 List1:L1
 List2:L2
 Freq1:1
 Freq2:1
 μ1:≠μ2 <μ2 >μ2
↓Pooled:No Yes
```

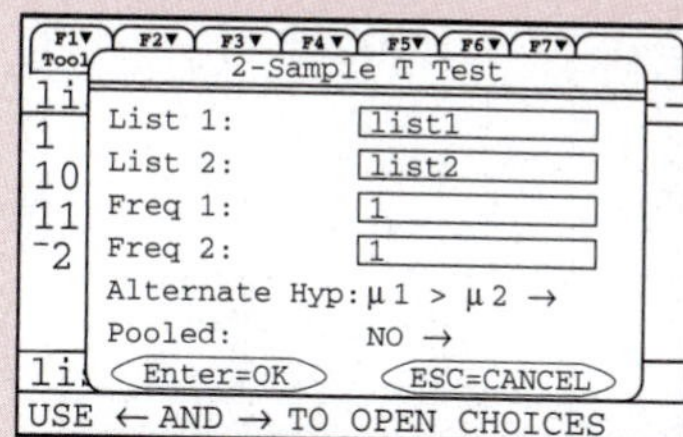

• Highlight "Calculate" and press ENTER. (The Pooled option will be discussed later.)

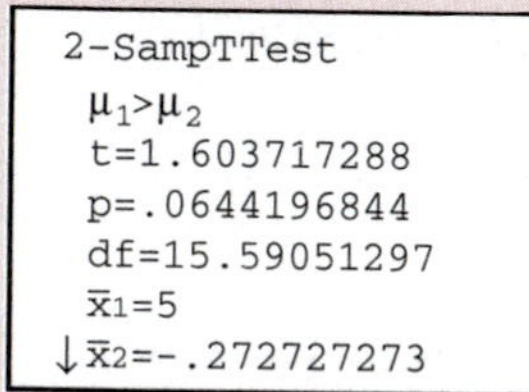

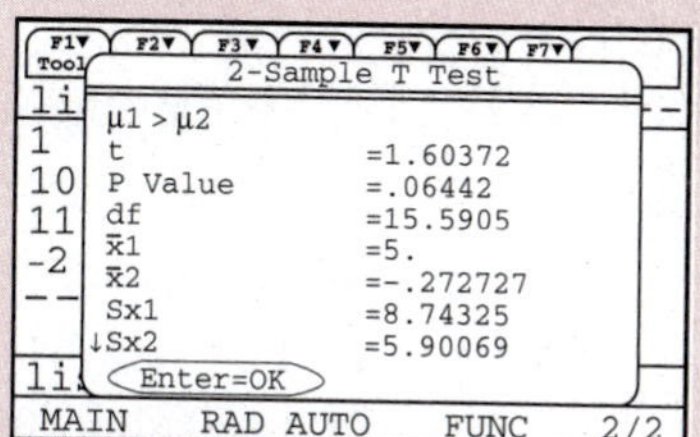

The results tell us that the t test statistic is $t = 1.6037$, and the P-value is $P = 0.0644$. This represents only modest evidence against H_0.

If you select "Draw" in the 2-SampTTest screen instead of "Calculate," the $t(k)$ distribution will be displayed, showing the t test statistic $t = 1.6037$ and the upper 0.0644 critical area shaded.

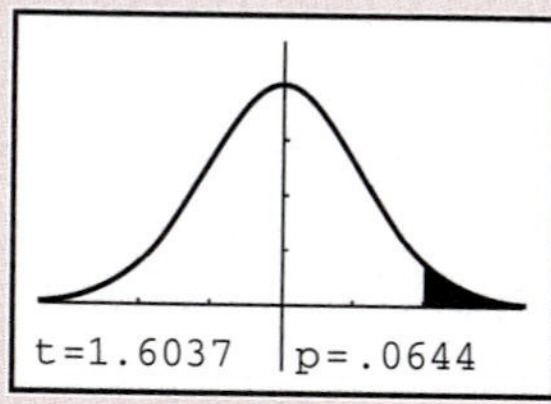

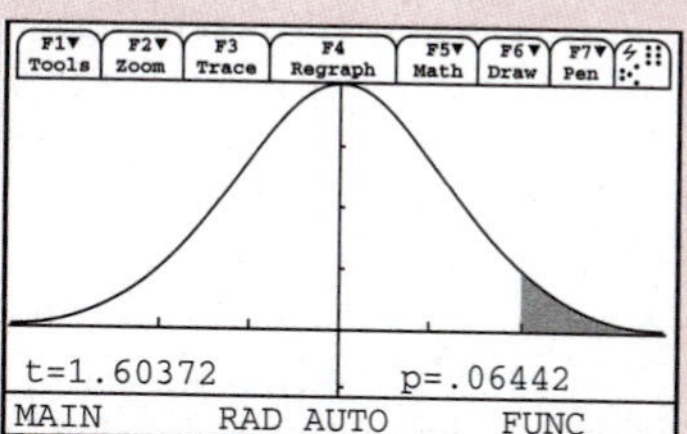

Confidence intervals

• With the data still stored in L_1/list1 and L_2/list2, select STAT/TESTS (Ints menu in the Stats/List Editor APP on the TI-89). Choose 2-SampTInt.

• In the 2-SampTInt screen, choose "Data" and adjust your settings as shown.

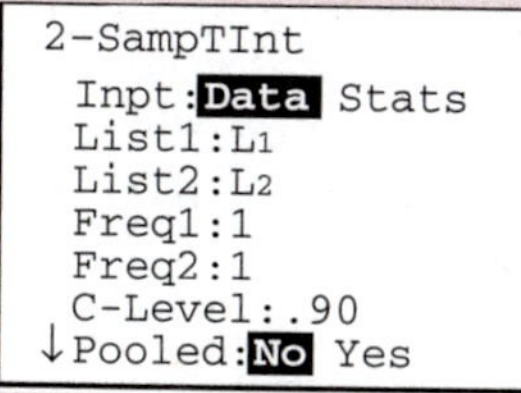

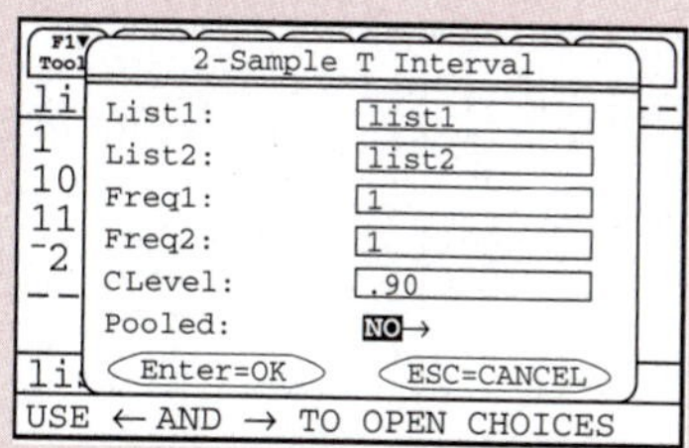

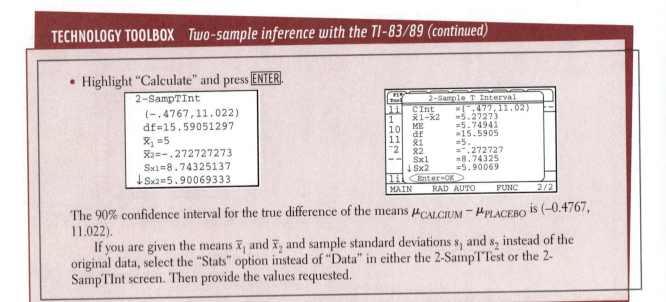

TECHNOLOGY TOOLBOX *Two-sample inference with the TI-83/89 (continued)*

- Highlight "Calculate" and press ENTER.

```
2-SampTInt
 (-.4767,11.022)
 df=15.59051297
 x̄1=5
 x̄2=-.272727273
 Sx1=8.74325137
↓Sx2=5.90069333
```

```
2-Sample T Interval
 CInt    ={⁻.477,11.02}
 x̄1-x̄2   =5.27273
 ME      =5.74941
 df      =15.5905
 x̄1      =5.
 x̄2      =⁻.272727
 Sx1     =8.74325
↓Sx2     =5.90069
 Enter=OK
MAIN    RAD AUTO    FUNC    2/2
```

The 90% confidence interval for the true difference of the means $\mu_{CALCIUM} - \mu_{PLACEBO}$ is (–0.4767, 11.022).

If you are given the means $\bar{x}_1$ and $\bar{x}_2$ and sample standard deviations s_1 and s_2 instead of the original data, select the "Stats" option instead of "Data" in either the 2-SampTTest or the 2-SampTInt screen. Then provide the values requested.

In the calcium study of Example 11.12, the pencil-and-paper-with-tables solution selected the following as the degrees of freedom: $\min(n_1 - 1, n_2 - 1) = \min(9,10) = 9$. Using df = 9, Example 11.12 calculated the 90% confidence interval for $\mu_1 - \mu_2$ to be (–0.753, 11.299). In Example 11.13, the more accurate fractional degrees of freedom were calculated by the formula to be 15.59, and using the smaller whole-number value df = 15 and Table C, the critical value was determined to be $t^* = 1.753$. With this t^*-value, the 90% confidence interval was calculated to be (–0.491, 11.037). By comparison, the TI-83/89 calculates this same 90% confidence interval to be (–0.4767, 11.022). The confidence interval that the calculator produces is shorter and more precise. The reason, of course, is that the calculator has been programmed to use the formula on page 659 to give the more accurate fractional degrees of freedom (15.59) and to calculate the t test statistic using this fractional df value.

EXAMPLE 11.14 DDT POISONING

Poisoning by the pesticide DDT causes convulsions in humans and other mammals. Researchers seek to understand how the convulsions are caused. In a randomized comparative experiment, they compared 6 white rats poisoned with DDT with a control group of 6 unpoisoned rats. Electrical measurements of nerve activity are the main clue to the nature of DDT poisoning. When a nerve is stimulated, its electrical response shows a sharp spike followed by a much smaller second spike. The experiment found that the second spike is larger in rats fed DDT than in normal rats. This finding helped biologists understand how DDT poisoning works.[16]

The researchers measured the height of the second spike as a percent of the first spike when a nerve in the rat's leg was stimulated. For the poisoned rats the results were

12.207	16.869	25.050	22.429	8.456	20.589

The control group data were

11.074	9.686	12.064	9.351	8.182	6.642

Here is the output from the SAS statistical software system for these data:[17]

```
                      TTEST PROCEDURE

Variable: SPIKE

GROUP         N          Mean             Std Dev        Std Error
-------------------------------------------------------------------
DDT           6      17.60000000       6.34014839       2.58835474
CONTROL       6       9.49983333       1.95005932       0.79610839

Variances       T        DF      Prob>|T|
------------------------------------------
Unequal      2.9912      5.9      0.0247
Equal        2.9912     10.0      0.0135
```

The difference in means for the two groups is quite large, but in such small samples the sample mean is highly variable. A significance test can help confirm that we are seeing a real effect.

Step 1: Identify the populations of interest and the parameters you want to draw conclusions about. State hypotheses in words and symbols. We want to compare the mean height μ_{DDT} of the second-spike electrical response in the population of rats fed DDT to $\mu_{CONTROL}$, the population mean second-spike height for normal rats. Because the researchers did not conjecture in advance that the size of the second spike would be higher in rats fed DDT, we use the two-sided alternative:

$$H_0: \mu_1 = \mu_2 \qquad \text{or, equivalently,} \qquad H_0: \mu_1 - \mu_2 = 0$$

$$H_a: \mu_1 \uparrow \mu_2 \qquad\qquad H_a: \mu_1 - \mu_2 \uparrow 0$$

Step 2: Choose the appropriate inference procedure, and verify the conditions for using the selected procedure. Since both population standard deviations are unknown, we should use a two-sample t test.

The DDT data are much more spread out than the control data.

- The researchers are willing to treat both samples as SRSs from their respective populations.
- Normal probability plots (Figure 11.11) show no evidence of outliers or strong skewness. Both populations are plausibly normal, as far as can be judged from 6 observations.

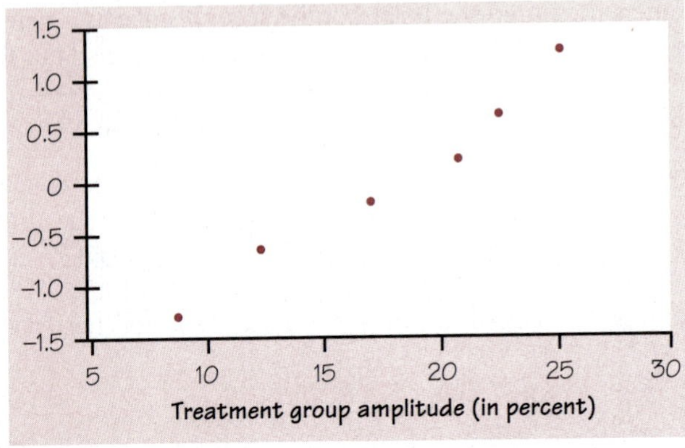

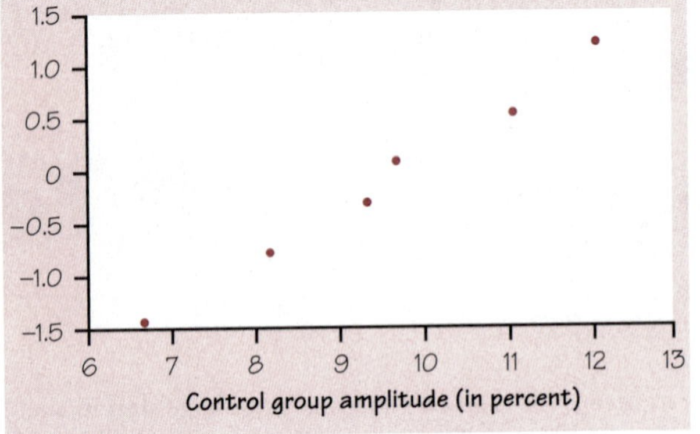

FIGURE 11.11 Normal probability plots for the amplitude of the second spike as a percent of the first spike.

Step 3: Carry out the procedure. The SAS printout reports the results of two t procedures: the general two-sample procedure ("unequal" variances) and a special procedure that assumes the two population variances are equal. We are interested in the first of these procedures. The two-sample t statistic has the value $t = 2.9912$, the degrees of freedom are df = 5.9, and the P-value from the $t(5.9)$ distribution is 0.0247.

Step 4: Interpret your results in the context of the problem. The low P-value provides strong evidence against H_0. We reject H_0 and conclude that the mean size of the secondary spike is larger in rats fed DDT.

Would the conservative test based on 5 degrees of freedom (both $n_1 - 1$ and $n_2 - 1$ are 5) have given a different result in Example 11.14? The statistic is exactly the same: $t = 2.9912$. The conservative P-value is $2P(T \geq 2.9912)$, where T has the $t(5)$ distribution. Table C shows that 2.9912 lies between the 0.02 and 0.01 upper critical values of the $t(5)$ distribution, so P for the two-sided test lies between 0.02 and 0.04. For practical purposes this is the same result as that given by the software. As this example and Example 11.13 suggest, the difference between the t procedures using the conservative and the approximately correct distributions is rarely of practical importance. That is why we recommend the simpler conservative procedure for inference without a computer.

EXERCISES

11.44 Example 11.13 demonstrates that if all other statistics stay the same, a higher number of degrees of freedom will produce a narrower (and hence more precise) confidence interval. Briefly explain why this is so.

11.45 Use your TI-83/89 and the two-sample procedures to replicate the results of the DDT–nerve stimulus experiment in Example 11.14. Verify that you get the same *t* test statistic and *P*-value.

11.46 Example 11.14 reports the analysis of data on the effects of DDT poisoning. The software uses the two-sample *t* test with degrees of freedom given in the box on page 659. Starting from the computer's results for $\bar{x}_i$ and s_i, verify the computer's values for the test statistic $t = 2.99$ and the degrees of freedom df = 5.9.

11.47 COMPETITIVE ROWERS What aspects of rowing technique distinguish between novice and skilled competitive rowers? Researchers compared two groups of female competitive rowers: a group of skilled rowers and a group of novices. The researchers measured many mechanical aspects of rowing style as the subjects rowed on a Stanford Rowing Ergometer. One important variable is the angular velocity of the knee, which describes the rate at which the knee joint opens as the legs push the body back on the sliding seat. The data show no outliers or strong skewness. Here is the SAS computer output:[18]

```
                    TTEST PROCEDURE

Variable: KNEE

GROUP       N        Mean          Std Dev        Std Error
-------------------------------------------------------------
SKILLED    10     4.18283335     0.47905935     0.15149187
NOVICE      8     3.01000000     0.95894830     0.33903942

Variances      T        DF     Prob>|T|
----------------------------------------
Unequal     3.1583     9.8     0.0104
Equal       3.3918    16.0     0.0037
```

(a) The researchers believed that the knee velocity would be higher for skilled rowers. State H_0 and H_a.

(b) What is the value of the two-sample *t* statistic and its *P*-value? (Note that SAS provides two-sided *P*-values. If you need a one-sided *P*-value, divide the two-sided value by 2.) What do you conclude?

(c) Give a 90% confidence interval for the mean difference between the knee velocities of skilled and novice female rowers.

11.48 COMPETITIVE ROWERS, II The research in the previous exercise also wondered whether skilled and novice rowers differ in weight or other physical characteristics. Here is the SAS computer output for weight in kilograms:

```
                    TTEST PROCEDURE

Variable: WEIGHT

GROUP            N          Mean          Std Dev       Std Error
------------------------------------------------------------------
SKILLED         10     70.3700000      6.10034898      1.92909973
NOVICE           8     68.4500000      9.03999930      3.19612240

Variances        T          DF     Prob>|T|
-------------------------------------------
Unequal     0.5143        11.8      0.6165
Equal       0.5376        16.0      0.5982
```

Is there significant evidence of a difference in the mean weights of skilled and novice rowers? State H_0 and H_a, report the two-sample t statistic and its P-value, and state your conclusion. (Note that SAS provides two-sided P-values. If you need a one-sided P-value, divide the two-sided value by 2.)

11.49 NICOTINE AND GUINEA PIGS Many studies have shown that smoking during pregnancy adversely affects the baby's health. Researchers investigated the behavior of guinea pig offspring whose mothers had been randomly assigned to receive either a normal saline or nicotine saline injection throughout pregnancy. Each group consisted of 15 randomly chosen male and female guinea pigs. At 85 days of age, 10 subjects from each group were randomly chosen to run a maze and choose a black door rather than a white door at the end of the maze. The number of trials it took each guinea pig to complete the task successfully with no more than one mistake in two consecutive days was recorded.

Here are the summary statistics on number of trials to successful completion for the nicotine group and the control (normal saline) group:[19]

Group	n	$\bar{x}$	s
Nicotine	10	111.9	50.28
Control	10	75.6	27.512

(a) Is there a significant difference in the mean number of trials recorded between the treatment group and the control group?

(b) Calculate *two* 95% confidence intervals for the difference in the mean number of trials required to complete the task for the treatment group and the control groups. Use the conservative number of degrees of freedom for the first interval. For the second interval, use the more precise df given by the formula on page 659. Comment on what you notice.

The pooled two-sample *t* procedures

In Example 11.14 the software offered a choice between two t tests. One is labeled for "unequal" variances, the other for "equal" variances. The "unequal" variance procedure is our two-sample t. This test is valid whether or not the population variances are equal. The other choice is a special version of the two-sample t statistic that assumes that the two populations have the same variance. This procedure averages (the statistical term is "pools") the two sample variances to estimate the common population variance. The resulting statistic is called the pooled two-sample t statistic. It is equal to our t statistic if the two sample sizes are the same, but not otherwise. We could choose to use the pooled t for both tests and confidence intervals.

The pooled t statistic has the advantage that it has exactly the t distribution with $n_1 + n_2 - 2$ degrees of freedom *if* the two population variances really are equal. Of course, the population variances are often not equal. Moreover, the assumption of equal variances is hard to check from the data. The pooled t was in common use before software made it easy to use the accurate approximation to the distribution of our two-sample t statistic. Now it is useful only in special situations. We cannot use the pooled t in Example 11.14 for example, because it is clear that the variance is much larger among rats fed DDT.

SUMMARY

The data in a **two-sample problem** are two independent SRSs, each drawn from a separate normally distributed population.

Tests and confidence intervals for the difference between the means μ_1 and μ_2 of the two populations start from the difference $\bar{x}_1 - \bar{x}_2$ of the two sample means. Because of the central limit theorem, the resulting procedures are approximately correct for other population distributions when the sample sizes are large.

Draw independent SRSs of sizes n_1 and n_2 from two normal populations with parameters μ_1, σ_1 and μ_2, σ_2. The **two-sample t statistic** is

$$t = \frac{(\bar{x}_1 - \bar{x}_2) - (\mu_1 - \mu_2)}{\sqrt{\dfrac{s_1^2}{n_1} + \dfrac{s_2^2}{n_2}}}$$

The statistic t does *not* have exactly a t distribution.

For conservative inference procedures to compare μ_1 and μ_2, use the two-sample t statistic with the $t(k)$ distribution. The degrees of freedom k is the smaller of $n_1 - 1$ and $n_2 - 1$. For more accurate probability values, use the $t(k)$ distribution with degrees of freedom k estimated from the data. This is the usual procedure in statistical software.

The **confidence interval** for $\mu_1 - \mu_2$ given by

$$(\bar{x}_1 - \bar{x}_2) \pm t^* \sqrt{\frac{s_1^2}{n_1} + \frac{s_2^2}{n_2}}$$

has confidence level at least C if t^* is the upper $(1 - C)/2$ critical value for $t(k)$ with k the smaller of $n_1 - 1$ and $n_2 - 1$.

Significance tests for H_0: $\mu_1 = \mu_2$ based on

$$t = \frac{\bar{x}_1 - \bar{x}_2}{\sqrt{\frac{s_1^2}{n_1} + \frac{s_2^2}{n_2}}}$$

have a true P-value no higher than that calculated from $t(k)$.

The guidelines for practical use of two-sample t procedures are similar to those for one-sample t procedures. Equal sample sizes are recommended.

SECTION 11.2 EXERCISES

In exercises that call for two-sample t procedures, you may use as the degrees of freedom either the smaller of $n_1 - 1$ and $n_2 - 1$ or the more exact value df given in the box on page 659. We recommend the first choice unless you are using a computer. Many of these exercises ask you to think about issues of statistical practice as well as to carry out t procedures.

11.50 INDEPENDENT SAMPLES VERSUS PAIRED SAMPLES Deciding whether to perform a matched pairs t test or a two-sample t test can be tricky.[20] Your decision should be based on the design that produced the data. Which procedure would you choose in each of the following situations?

(a) To test the wear characteristics of two tire brands, A and B, Brand A is mounted on 50 cars and Brand B on 50 other cars.

(b) To test the wear characteristics of two tire brands, A and B, one Brand A tire is mounted on one side of each car in the rear, while a Brand B tire is mounted on the other side. Which side gets which brand is determined by flipping a coin. The same procedure is used on the front.

(c) To test the effect of background music on productivity, factory workers are observed. For 1 month they had no background music. For another month they had background music.

(d) A random sample of 10 workers in Plant A are to be compared to a random sample of 10 workers in Plant B in terms of productivity.

(e) A new weight-reducing diet was tried on 10 women. The weight of each woman was measured before the diet and again after 10 weeks on the diet.

11.51 TREATING SCRAPIE IN HAMSTERS Scrapie is a degenerative disease of the nervous system. A study of the substance IDX as a treatment for scrapie used as subjects 20 infected hamsters. Ten, chosen at random, were injected with IDX. The other 10 were untreated. The researchers recorded how long each hamster lived. They reported, "Thus, although all infected control hamsters had died by 94 days after infection (mean ± SEM = 88.5 ± 1.9 days), IDX-treated hamsters lived up to 128 days (mean ± SEM = 116 ± 5.6 days)."[21]

(a) Fill in the values in this summary table:

Group	Treatment	n	$\bar{x}$	s
1	IDX	?	?	?
2	Untreated	?	?	?

(b) What degrees of freedom would you use in the conservative two-sample t procedures to compare the two treatments?

11.52 TREATING SCRAPIE, II Exercise 11.51 contains the results of a study to determine whether IDX is an effective treatment of scrapie.

(a) Is there good evidence that hamsters treated with IDX live longer on the average?

(b) Give a 95% confidence interval for the mean amount by which IDX prolongs life.

11.53 TEACHING READING An educator believes that new reading activities in the classroom will help elementary school pupils improve their reading ability. She arranges for a third-grade class of 21 students to follow these activities for an 8-week period. A control classroom of 23 third graders follows the same curriculum without the activities. At the end of the 8 weeks, all students are given the Degree of Reading Power (DRP) test, which measures the aspects of reading ability that the treatment is designed to improve. Here are the data:[22]

Treatment					Control				
24	43	58	71	43	42	43	55	26	62
49	61	44	67	49	37	33	41	19	54
53	56	59	52	62	20	85	46	10	17
54	57	33	46	43	60	53	42	37	42
57					55	28	48		

(a) Examine the data with a graph. Are there strong outliers or skewness that could prevent use of the t procedures?

(b) Is there good evidence that the new activities improve the mean DRP score? Carry out a test and report your conclusions.

(c) Although this study is an experiment, its design is not ideal because it had to be done in a school without disrupting classes. What aspect of good experimental design is missing?

11.54 WEIGHT LOSS PROGRAM In a study of the effectiveness of a weight loss program, 47 subjects who were at least 20% overweight took part in the program for 10 weeks. Private weighings determined each subject's weight at the beginning of the program and 6 months after the program's end. The matched pairs t test was used to assess the significance of the average weight loss. The paper reporting the study said, "The subjects lost a significant amount of weight over time, $t(46) = 4.68$, $p < .01$." It is common to report the results of statistical tests in this abbreviated style.[23]

(a) Why was the matched pairs t test appropriate?

(b) Explain to someone who knows no statistics but is interested in weight-loss programs what the practical conclusion is.

(c) The paper follows the tradition of reporting significance only at fixed levels such as $\alpha = 0.01$. In fact, the results are more significant than "$p < .01$" suggests. Use Table C to say more about the P-value of the t test.

11.55 COMPARING TWO DRUGS Makers of generic drugs must show that they do not differ significantly from the "reference" drug that they imitate. One aspect in which drugs might differ is their extent of absorption in the blood. Table 11.6 gives data taken from 20 healthy nonsmoking male subjects for one pair of drugs. This is a matched pairs design. Subjects 1 to 10 received the generic drug first, and Subjects 11 to 20 received the reference drug first. In all cases, a washout period separated the two drugs so that the first had disappeared from the blood before the subject took the second. The subject numbers in the table were assigned at random to decide the order of the drugs for each subject.

(a) Do a data analysis of the differences between the absorption measures for the generic and reference drugs. Is there any reason not to apply t procedures?

(b) Give a 90% confidence interval for the mean difference in gains between treatment and control.

TABLE 11.6 Absorption extent for two versions of a drug

Subject	Reference drug	Generic drug	Subject	Reference drug	Generic drug
15	4108	1755	4	2344	2738
3	2526	1138	16	1864	2302
9	2779	1613	6	1022	1284
13	3852	2254	10	2256	3052
12	1833	1310	5	938	1287
8	2463	2120	7	1339	1930
18	2059	1851	14	1262	1964
20	1709	1878	11	1438	2549
17	1829	1682	1	1735	3340
2	2594	2613	19	1020	3050

Source: Data from Lianng Yuh, "A biopharmaceutical example for undergraduate students," unpublished manuscript.

11.56 COACHING AND SAT SCORES Coaching companies claim that their courses can raise the SAT scores of high school students. Of course, students who retake the SAT without paying for coaching generally raise their scores. A random sample of students who took the SAT twice found 427 who were coached and 2,733 who were uncoached.[24] Starting with their Verbal scores on the first and second tries, we have these summary statistics:

	Try 1		Try 2		Gain	
	Mean	Std. dev.	Mean	Std. dev.	Mean	Std.dev.
Coached	500	92	529	97	29	59
Uncoached	506	101	527	101	21	52

Let's first ask if students who are coached increased their scores significantly.

(a) You could use the information given to carry out either a two-sample t test comparing Try 1 with Try 2 for coached students or a matched pairs t test using Gain. Which is the correct test? Why?

(b) Carry out the proper test. What do you conclude?

(c) Give a 99% confidence interval for the mean gain of all students who are coached.

11.57 COACHING AND SAT SCORES, II What we really want to know is whether coached students improve more than uncoached students, and whether any advantage is large enough to be worth paying for. Use the information in the previous problem to answer these questions.

(a) Is there good evidence that coached students gained more on the average than uncoached students?

(b) How much more do coached students gain on the average? Give a 99% confidence interval.

(c) Based on your work, what is your opinion: do you think coaching courses are worth paying for?

11.58 COACHING AND SAT SCORES: CRITIQUE The data you used in the previous two exercises came from a random sample of students who took the SAT twice. The response rate was 63%, which is pretty good for nongovernment surveys, so let's accept that the respondents do represent all students who took the exam twice. Nonetheless, we can't be sure that coaching actually *caused* the coached students to gain more than the uncoached students. Explain briefly but clearly why this is so.

11.59 STUDENTS' SELF-CONCEPT Here is SAS output for a study of the self-concept of seventh-grade students. The variable SC is the score on the Piers-Harris Self Concept Scale. The analysis was done to see if male and female students differ in mean self-concept score.[25]

```
                     TTEST PROCEDURE

Variable: SC

SEX     N          Mean            Std Dev          Std Error
-------------------------------------------------------------
F      31     55.51612903     12.69611743        2.28029001
M      47     57.91489362     12.26488410        1.78901722

Variances      T          DF    Prob>|T|
----------------------------------------
Unequal     -0.8276     62.8    0.4110
Equal       -0.8336     76.0    0.4071
```

Write a sentence or two summarizing the comparison of females and males, as if you were preparing a report for publication.

The remaining exercises concern the power of the two-sample t test, an optional topic. If you have read Section 10.4 and the discussion of the power of the one-sample t test on pages 639–640, Exercise 11.64 guides you in finding the power of the two-sample t.

11.60 In Example 11.10 (page 650), a small study of black men suggested that a calcium supplement can reduce blood pressure. Now we are planning a larger clinical trial of this effect. We plan to use 100 subjects in each of the two groups. Are these sample sizes large enough to make it very likely that the study will give strong evidence ($\alpha = 0.01$) of the effect of calcium if in fact calcium lowers blood pressure by 5 millimeters more than a placebo? To answer this question, we will compute the power of the two-sample t test of

$$H_0: \mu_1 = \mu_2$$
$$H_a: \mu_1 > \mu_2$$

against the specific alternative $\mu_1 - \mu_2 = 5$. Based on the pilot study reported in Example 11.10, we take 8, the larger of the two observed s-values, as a rough estimate of both the population σ's and future sample s's.

(a) What is the approximate value of the $\alpha = 0.01$ critical value t^* for the two-sample t statistic when $n_1 = n_2 = 100$?

(b) **Step 1:** *Write the rule for rejecting H_0 in terms of $\bar{x}_1 - \bar{x}_2$.* The test rejects H_0 when

$$\frac{\bar{x}_1 - \bar{x}_2}{\sqrt{\dfrac{s_1^2}{n_1} + \dfrac{s_2^2}{n_2}}} \geq t^*$$

Take both s_1 and s_2 to be 8, and n_1 and n_2 to be 100. Find the number c such that the test rejects H_0 when $\bar{x}_1 - \bar{x}_2 \geq c$.

(c) **Step 2:** *The power is the probability of rejecting H_0 when the alternative is true.* Suppose that $\mu_1 - \mu_2 = 5$ and that both σ_1 and σ_2 are 8. The power we seek is the probability that $\bar{x}_1 - \bar{x}_2 \geq c$ under these assumptions. Calculate the power.

(d) Describe a Type I and a Type II error in this experiment. Which is more serious?

11.61 A bank asks you to compare two ways to increase the use of its credit cards. Plan A would offer customers a cash-back rebate based on their total amount charged. Plan B would reduce the interest rate charged on card balances. The response variable is the total amount a customer charges during the test period. You decide to offer each of Plan A and Plan B to a separate SRS of the bank's credit card customers. In the past, the mean amount charged in a six-month period has been about \$1100, with a standard deviation of \$400. Will a two-sample t test based on SRSs of 350 customers in each group detect a difference of \$100 in the mean amounts charged under the two plans?

(a) State H_0 and H_a, and write the formula for the test statistic.

(b) Give the $\alpha = 0.05$ critical value for the test when $n_1 = n_2 = 350$.

(c) Calculate the power of the test with $\alpha = 0.05$, using \$400 as a rough estimate of all standard deviations.

(d) Describe a Type I error and a Type II error in this setting. Which is of more concern to the bank?

CHAPTER REVIEW

This chapter presents *t* tests and confidence intervals for inference about the mean of a single population and for comparing the means of two populations. The one-sample *t* procedures do inference about one mean and the two-sample *t* procedures compare two means. Matched pairs studies use one-sample procedures because you first create a single sample by taking the differences in the responses within each pair. These *t* procedures are among the most common methods of statistical inference. The figure below helps you decide when to use them. Before you use any inference method, think about the design of the study and examine the data for outliers and other problems.

The *t* Procedures for Means

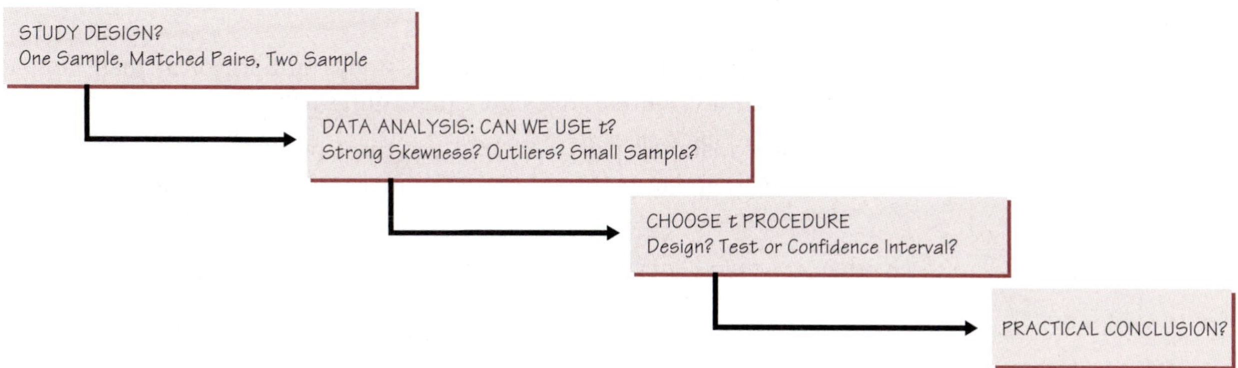

The *t* procedures require that the data be random samples and that the distribution of the population or populations be normal. One reason for the wide use of *t* procedures is that they are not very strongly affected by lack of normality. If you can't regard your data as a random sample, however, the results of inference may be of little value.

Chapter 10 concentrated on the reasoning of confidence intervals and tests. Understanding the reasoning is essential for wise use of the *t* and other inference methods. The discussion in this chapter paid more attention to practical aspects of using the methods. We saw that there are several versions of the two-sample *t*, for example. Which one you use depends largely on whether or not you use statistical software. Before you use any inference method, think about the design of the study and examine the data for outliers and other problems.

The chapter exercises are important in this and later chapters. You must now recognize problem settings and decide which of the methods presented in the chapter fits. In this chapter, you must recognize one-sample studies, matched pairs studies, and two-sample studies. Here are the most important skills you should have after reading this chapter.

A. RECOGNITION

1. Recognize when a problem requires inference about a mean or comparing two means.

2. Recognize from the design of a study whether one-sample, matched pairs, or two-sample procedures are needed.

B. ONE-SAMPLE *t* PROCEDURES

1. Use the *t* procedure to obtain a confidence interval at a stated level of confidence for the mean μ of a population.

2. Carry out a *t* test for the hypothesis that a population mean μ has a specified value against either a one-sided or a two-sided alternative. Use Table C of *t* critical values to approximate the *P*-value or carry out a fixed α test.

3. Recognize when the *t* procedures are appropriate in practice, in particular that they are quite robust against lack of normality but are influenced by outliers.

4. Also recognize when the design of the study, outliers, or a small sample from a skewed distribution make the *t* procedures risky.

5. Recognize matched pairs data and use the *t* procedures to obtain confidence intervals and to perform tests of significance for such data.

C. TWO-SAMPLE *t* PROCEDURES

1. Give a confidence interval for the difference between two means. Use the two-sample *t* statistic with conservative degrees of freedom if you do not have statistical software. Use the TI-83/89 or software if you have it.

2. Test the hypothesis that two populations have equal means against either a one-sided or a two-sided alternative. Use the two-sample *t* test with conservative degrees of freedom if you do not have statistical software. Use the TI-83/89 or software if you have it.

3. Recognize when the two-sample *t* procedures are appropriate in practice.

CHAPTER 11 REVIEW EXERCISES

11.62 EXPENSIVE ADS Consumers who think a product's advertising is expensive often also think the product must be of high quality. Can other information undermine this effect? To find out, marketing researchers did an experiment. The subjects were 90 women from the clerical and administrative staff of a large organization. All subjects read an ad that described a fictional line of food products called "Five Chefs." The ad also described the major TV commercials that would soon be shown, an unusual expense for this type of product. The 45 women in the control group read nothing else. The 45 in the "undermine group" also read a news story headlined "No Link between Advertising Spending and New Product Quality."

All the subjects then rated the quality of Five Chefs products on a seven-point scale. The study report said, "The mean quality ratings were significantly lower in the undermine treatment ($\bar{X}_A = 4.56$) than in the control treatment ($\bar{X}_C = 5.05$; $t = 2.64$, $p < .01$)."[26]

(a) Is the matched pairs t test or the two-sample t test the right test in this setting? Why?

(b) What degrees of freedom would you use for the t statistic you chose in (a)?

(c) The distribution of individual responses is not normal, because there is only a seven-point scale. Why is it nonetheless proper to use a t test?

11.63 SHARKS Great white sharks are big and hungry. Here are the lengths in feet of 44 great whites:[27]

18.7	12.3	18.6	16.4	15.7	18.3	14.6	15.8	14.9	17.6	12.1
16.4	16.7	17.8	16.2	12.6	17.8	13.8	12.2	15.2	14.7	12.4
13.2	15.8	14.3	16.6	9.4	18.2	13.2	13.6	15.3	16.1	13.5
19.1	16.2	22.8	16.8	13.6	13.2	15.7	19.7	18.7	13.2	16.8

(a) Examine these data for shape, center, spread, and outliers. The distribution is reasonably normal except for one outlier in each direction. Because these are not extreme and preserve the symmetry of the distribution, use of the t procedures is safe with 44 observations.

(b) Give a 95% confidence interval for the mean length of great white sharks. Based on this interval, is there significant evidence at the 5% level to reject the claim "Great white sharks average 20 feet in length"?

(c) It isn't clear exactly what parameter μ you estimated in (b). What information do you need to say what μ is?

11.64 INDEPENDENT SAMPLES VERSUS PAIRED SAMPLES Deciding whether to perform a matched pairs t test or a two-sample t test can be tricky.[28] Your decision should be based on the design that produced the data. Which procedure would you choose in each of the following situations?

(a) To compare the average weight gain of pigs fed two different rations, nine pairs of pigs were used. The pigs in each pair were littermates.

(b) To test the effects of a new fertilizer, 100 plots are treated with the new fertilizer, and 100 plots are treated with another fertilizer.

(c) A sample of college teachers is taken. We wish to compare the average salaries of male and female teachers.

(d) A new fertilizer is tested on 100 plots. Each plot is divided in half. Fertilizer A is applied to one half and B to the other.

(e) Consumers Union wants to compare two types of calculators. They get 100 volunteers and ask them to carry out a series of 50 routine calculations (such as figuring discounts, sales tax, totaling a bill, etc.). Each calculation is done on each type of calculator, and the time required for each calculation is recorded.

11.65 KICKING A HELIUM-FILLED FOOTBALL On a calm, clear Saturday in 1993, the Auburn Tigers were faced with fourth down deep in their own territory. Their opposition, the Mississippi State Bulldogs, looked for good field position following a punt. The football was snapped, kicked, and eyed in disbelief as it sailed an estimated 71 yards through the air. Shocked, the Mississippi State coaches cried foul and the football was immediately seized by the officials. The football was later tested to see if it had been filled with helium, as many thought that this might explain its unusually long flight. No helium was found in that football, but the possible benefits of filling a football with gas lighter than air would be kicked around both science and sports communities in the weeks to come. Many devised their own experiments to see if helium-filled balls traveled farther than footballs filled with air.

The *Columbus Dispatch* conducted one such study. Two identical footballs, one air-filled and one helium-filled, were used outdoors on a windless day at Ohio State University's athletic complex. The kicker was a novice punter and was not informed which football contained the helium. Each football was kicked 39 times and the two footballs were alternated with each kick. Table 11.7 provides the data from this experiment.

TABLE 11.7 Distance traveled (in yards) by two kicked footballs, one filled with helium and one filled with air

Trial	Air	Helium	Trial	Air	Helium	Trial	Air	Helium	Trial	Air	Helium
1	25	25	11	25	12	21	31	31	31	27	26
2	23	16	12	19	28	22	27	34	32	26	32
3	18	25	13	27	28	23	22	39	33	28	30
4	16	14	14	25	31	24	29	32	34	32	29
5	35	23	15	34	22	25	28	14	35	28	30
6	15	29	16	26	29	26	29	28	36	25	29
7	26	25	17	20	23	27	22	30	37	31	29
8	24	26	18	22	26	28	31	27	38	28	30
9	24	22	19	33	35	29	25	33	39	28	26
10	28	26	20	29	24	30	20	11			

Source: Data from the EESEE story "Kicking a Helium-Filled Football."

Based on the summary statistics, the researcher concluded that there is "not much difference" in the results for the two footballs.

(a) Perform an appropriate statistical test of this statement.

(b) Conduct your test from (a) with any outliers in the data set removed. Compare the two results.

(c) The researcher also stated: "The kicker changed footballs on each kick, guaranteeing that his leg would play no favorites if he tired. However, it appears he improved with practice." Perform an appropriate statistical analysis to address this claim.

11.66 LEARNING TO SOLVE A MAZE Table 11.2 (page 629) contains the times required to complete a maze for 21 subjects wearing scented and unscented masks. Example 11.4 used the matched pairs t test to show that the scent makes no significant difference in the time. Now we ask whether there is a learning effect, so that subjects complete the maze faster on their second trial. All of the odd-numbered subjects in Table 11.2 first worked the maze wearing the unscented mask. Even-numbered subjects wore the scented mask first. The numbers were assigned at random.

(a) We will compare the unscented times for "unscented first" subjects with the unscented times for the "scented first" subjects. Explain why this comparison requires two-sample procedures.

(b) We suspect that on the average subjects are slower when the unscented time is their first trial. Make a back-to-back stemplot of unscented times for "scented first" and "unscented first" subjects. Find the mean unscented times for these two groups. Do the data appear to support our suspicion? Do the data have features that prevent use of the *t* procedures?

(c) Do the data give statistically significant support to our suspicion? State hypotheses, carry out a test, and report your conclusion.

11.67 COMPARING WELFARE PROGRAMS A major study of alternative welfare programs randomly assigned women on welfare to one of two programs, called "WIN" and "Options." WIN was the existing program. The new Options program gave more incentives to work. An important question was how much more (on the average) women in Options earned than those in WIN. Here is Minitab output for earnings in dollars over a 3-year period:[29]

```
TWOSAMPLE T FOR 'OPT' VS 'WIN'

           N     MEAN      STDEV      SE MEAN
OPT     1362     7638        289       7.8309
WIN     1395     6595        247       6.6132

95 PCT CI FOR MU OPT - MU WIN: (1022.90, 1063.10)
```

(a) Give a 99% confidence interval for the amount by which the mean earnings of Options participants exceeded the mean earnings of WIN subjects. (Minitab will give a 99% confidence interval if you instruct it to do so. Here we have only the basic output, which includes the 95% confidence interval.)

(b) The distribution of incomes is strongly skewed to the right but includes no extreme outliers because all the subjects were on welfare. What fact about these data allows us to use *t* procedures despite the strong skewness?

11.68 EACH DAY I AM GETTING BETTER IN MATH A "subliminal" message is below our threshold of awareness but may nonetheless influence us. Can subliminal messages help students learn math? A group of students who had failed the mathematics part of the City University of New York Skills Assessment Test agreed to participate in a study to find out.

All received a daily subliminal message, flashed on a screen too rapidly to be consciously read. The treatment group of 10 students (chosen at random) was exposed to "Each day I am getting better in math." The control group of 8 students was exposed to a neutral message, "People are walking on the street." All students participated in a summer program designed to raise their math skills, and all took the assessment test again at the end of the program. Table 11.8 gives data on the subjects' scores before and after the program.

(a) Is there good evidence that the treatment brought about a greater improvement in math scores than the neutral message? State hypotheses, carry out a test, and state your conclusion. Is your result significant at the 5% level? At the 10% level?

TABLE 11.8 Mathematics skills scores before and after a subliminal message

Treatment Group		Control Group	
Pre-test	Post-test	Pre-test	Post-test
18	24	18	29
18	25	24	29
21	33	20	24
18	29	18	26
18	33	24	38
20	36	22	27
23	34	15	22
23	36	19	31
21	34		
17	27		

Source: Data provided by Warren Page, New York City Technical College, from a study done by John Hudesman.

(b) Give a 90% confidence interval for the mean difference in gains between treatment and control.

11.69 STRESS AMONG PETS AND FRIENDS Stress is a fact of everyday life. Researchers explored how the presence of others can affect certain stress indicators when a person performs a stressful task. In this study, the researchers asked 45 women to perform mental arithmetic in the presence of their pet dog (P), a good female friend (F), or alone (C, for control). To record the participants' stress levels during the task and rest periods, the experimenters measured maximum heart rate (beats/minute). The researchers were interested in exploring whether the fact that a human friend could evaluate the subject's performance at arithmetic while a dog could not would affect the participant's stress level.[30]

Condition	Max. heart rate	Condition	Max. heart rate	Condition	Max. heart rate
C	115	F	128	P	72
C	110	F	122	P	72
C	113	F	108	P	74
C	103	F	128	P	68
C	114	F	131	P	61
C	112	F	118	P	82
C	115	F	83	P	72
C	96	F	127	P	78
C	107	F	132	P	92
C	103	F	103	P	127
C	95	F	126	P	87
C	115	F	116	P	73
C	120	F	110	P	74
C	96	F	113	P	76
C	84	F	120	P	70

Use these data to examine the researchers' question of interest. If you find statistically significant difference between two of the groups, estimate the size of that difference.

11.70 You look up a census report that gives the populations of all 92 counties in the state of Indiana. Is it proper to apply the one-sample t method to these data to give a 95% confidence interval for the mean population of an Indiana county? Explain your answer.

11.71 Exercise 1.28 (page 35) gives 29 measurements of the density of the earth, made in 1798 by Henry Cavendish. Display the data graphically to check for skewness and outliers. Then give an estimate for the density of the earth from Cavendish's data and a margin of error for your estimate.

11.72 CHOLESTEROL IN DOGS High levels of cholesterol in the blood are not healthy in either humans or dogs. Because a diet rich in saturated fats raises the cholesterol level, it is plausible that dogs owned as pets have higher cholesterol levels than dogs owned by a veterinary research clinic. "Normal" levels of cholesterol based on the clinic's dogs would then be misleading. A clinic compared healthy dogs it owned with healthy pets brought to the clinic to be neutered. The summary statistics for blood cholesterol levels (milligrams per deciliter of blood) appear below.[31]

Group	n	$\bar{x}$	s
Pets	26	193	68
Clinic	23	174	44

(a) Is there strong evidence that pets have higher mean cholesterol level than clinic dogs? State the H_0 and H_a and carry out an appropriate test. Give the P-value and state your conclusion.

(b) Give a 95% confidence interval for the difference in mean cholesterol levels between pets and clinic dogs.

(c) Give a 95% confidence interval for the mean cholesterol level in pets.

(d) What conditions must be satisfied to justify the procedures you used in (a), (b), and (c)? Assuming that the cholesterol measurements have no outliers and are not strongly skewed, what is the chief threat to the validity of the results of this study?

11.73 ACTIVE VERSUS PASSIVE LEARNING A study of computer-assisted learning examined the learning of "Blissymbols" by children. Blissymbols are pictographs (think of Egyptian hieroglyphs) that are sometimes used to help learning-impaired children communicate. The researcher designed two computer lessons that taught the same content using the same examples. One lesson required the children to interact with the material, while in the other the children controlled only the pace of the lesson. Call these two styles "Active" and "Passive." After the lesson, the computer presented a quiz that asked the children to identify 56 Blissymbols. Here are the numbers of correct identifications by the 24 children in the Active group:[32]

29	28	24	31	15	24	27	23	20	22	23	21
24	35	21	24	44	28	17	21	21	20	28	16

The 24 children in the Passive group had these counts of correct identifications:

16	14	17	15	26	17	12	25	21	20	18	21
20	16	18	15	26	15	13	17	21	19	15	12

(a) Is there good evidence that active learning is superior to passive learning? Give appropriate statistical justification for your answer.

(b) Give a 90% confidence interval for the mean number of Blissymbols identified correctly in a large population of children after the Active computer lesson.

NOTES AND DATA SOURCES

1. These data are from "Results report on the vitamin C pilot program," prepared by SUSTAIN (Sharing United States Technology to Aid in the Improvement of Nutrition) for the U.S. Agency for International Development. The report was used by the Committee on International Nutrition of the National Academy of Sciences/Institute of Medicine (NAS/IOM) to make recommendations on whether or not the vitamin C content of food commodities used in U.S. food aid programs should be increased. The program was directed by Peter Ranum and Françoise Chomé.
2. F. H. Rauscher et al., "Music training causes long-term enhancement of preschool children's spatial-temporal reasoning," *Neurological Research*, 19 (1997), pp. 2–8.
3. These recommendations are based on extensive computer work. See, for example, Harry O. Posten, "The robustness of the one-sample *t*-test over the Pearson system," *Journal of Statistical Computation and Simulation*, 9 (1979), pp. 133–149, and E. S. Pearson and N. W. Please, "Relation between the shape of population distribution and the robustness of four simple test statistics," *Biometrika*, 62 (1975), pp. 223–241.
4. Based on I. Cuellar, L. C. Harris, and R. Jasso, "An acculturation scale for Mexican American normal and clinical populations," *Hispanic Journal of Behavioral Sciences*, 2 (1980), pp. 199–217.
5. Alan S. Banks et al., "Juvenile hallux abducto valgus association with metatarsus adductus," *Journal of the American Podiatric Medical Association*, 84 (1994), pp. 219–224.
6. Data provided by Timothy Sturm.
7. Data provided by Diana Schellenberg, Purdue University School of Health.
8. This study is reported in Roseann M. Lyle et al., "Blood pressure and metabolic effects of calcium supplementation in normotensive white and black men," *Journal of the American Medical Association*, 257 (1987), pp. 1772–1776. The data were provided by Dr. Lyle.
9. Detailed information about the conservative *t* procedures can be found in Paul Leaverton and John W. Birch, "Small sample power curves for the two sample location problem," *Technometrics*, 11 (1969), pp. 299–307; in Henry Scheffé, "Practical solutions of the Beherns-Fisher problem," *Journal of the American Statistical Association*, 65 (1970), pp. 1501–1508; and in D. J. Best and J. C. W. Rayner, "Welch's approximate solution for the Beherns-Fisher problem," *Technometrics*, 29 (1987), pp. 205–210.
10. See the extensive simulation studies in Harry O. Posten, "The robustness of the two-sample *t*-test over the Pearson system," *Journal of Statistical Computation and Simulation*, 6 (1978), pp. 295–311, and in Harry O. Posten, H. Yeh, and Donald B. Owen, "Robustness of the two-sample *t*-test under violations of the homogeneity assumption," *Communications in Statistics*, 11 (1982), pp. 109–126.

11. From H. G. Gough, *The Chapin Social Insight Test,* Consulting Psychologists Press, Palo Alto, Calif., 1968.
12. Data provided by Charles Cannon, Duke University. The study report is C. H. Cannon, D. R. Peart, and M. Leighton, "Tree species diversity in commercially logged Bornean rainforest," *Science*, 281 (1998), pp. 1366–1367.
13. From the EESEE story "Surgery in a blanket."
14. Data for 1982, provided by Marvin Schlatter, Division of Financial Aid, Purdue University.
15. Based on M. C. Wilson et al., "Impact of cereal leaf beetle larvae on yields of oats," *Journal of Economic Entomology*, 62 (1969), pp. 699–702.
16. This example is loosely based on D. L. Shankland, "Involvement of spinal cord and peripheral nerves in DDT-poisoning syndrome in albino rats," *Toxicology and Applied Pharmacology*, 6 (1964), pp. 197–213.
17. We did not use Minitab or Data Desk in Example 11.16 because these packages shortcut the two-sample t procedure. They calculate the degrees of freedom df using the formula in the box on page 659 but then truncate to the next lower whole-number degrees of freedom to obtain the P-value. The result is slightly less accurate than the P-value from the t(df) distribution.
18. Based on W. N. Nelson and C. J. Widule, "Kinematic analysis and efficiency estimate of intercollegiate female rowers," unpublished manuscript, 1983.
19. From the EESEE story "Nicotine and Guinea Pigs."
20. The idea for this exercise was provided by R. W. W. Taylor.
21. F. Tagliavini et al., "Effectiveness of anthracycline against experimental prion disease in Syrian hamsters," *Science*, 276 (1997), pp. 1119–1121.
22. Adapted from Maribeth Cassidy Schmitt, "The Effects of an Elaborated Directed Reading Activity on the Metacomprehension Skills of Third Graders," Ph.D. dissertation, Purdue University, 1987.
23. Loosely based on D. R. Black et al., "Minimal interventions for weight control: a cost-effective alternative," *Addictive Behaviors*, 9 (1984), pp. 279–285.
24. Wayne J. Camera and Donald Powers, "Coaching and the SAT I," TIP (online journal: www.siop.org/tip), July 1999.
25. Data provided by Darlene Gordon, School of Education, Purdue University.
26. Based on Amna Kirmani and Peter Wright, "Money talks: perceived advertising expense and expected product quality," *Journal of Consumer Research*, 16 (1989), pp. 344–353.
27. Data provided by Chris Olsen, who found the information in scuba diving magazines.
28. The idea for this exercise was provided by R. W. W. Taylor.
29. Based on D. Friedlander, *Supplemental Report on the Baltimore Options Program,* Manpower Demonstration Research Corporation, 1987.
30. Data from the EESEE story "Stress among Pets and Friends."
31. From V. D. Bass, W. E. Hoffmann, and J. L. Dorner, "Normal canine lipid profiles and effects of experimentally induced pancreatitis and hepatic necrosis on lipids." *American Journal of Veterinary Research*, 37 (1976), pp. 1355–1357.
32. Data from Orit E. Hetzroni, "The effects of active versus passive computer-assisted instruction on the acquisition, retention, and generalization of Blissymbols while using elements for teaching compounds," Ph.D. thesis, Purdue University, 1995.
33. Data provided by Matthew Moore.

Lara Harris, AP, Wide World Photos

JANET NORWOOD

The Government's Statistician

Modern governments run on statistics. They need data on economic and social trends that are accurate, timely, and free of political influence. Unlike most nations, the United States does not have a single statistical agency such as Statistics Canada. The Bureau of Labor Statistics is one of the government's major statistical offices, and its head, the commissioner of labor statistics, is one of the nations most influential statisticians.

The data collected by the Bureau of Labor Statistics are often politically sensitive, as when a report released just before an election shows rising unemployment. For this reason, the bureau must remain objective and independent of political influence. To safeguard the bureau's independence, the commissioner is appointed by the President and confirmed by the Senate for a fixed term of four years. The commissioner must have statistical skill, administrative ability, and a facility for working with both Congress and the President.

Janet Norwood served three terms as commissioner, from 1979 to 1991, under three presidents. When she retired, the *New York Times* said (December 31, 1991) that she left with "a near-legendary reputation for nonpartisanship and plaudits that include one senator's designation of her as a 'national treasure.'" Norwood says, "There have been times in the past when commissioners have been in open disagreement with the Secretary of Labor or, in some cases, with the President. We have guarded our professionalism with great care."

The data collected by the Bureau of Labor Statistics are often politically sensitive, as when a report released just before an election shows rising unemployment.

Some of the most important statistics produced by the Bureau of Labor Statistics are proportions. The monthly unemployment rate, for example, is the proportion of the labor force that is unemployed this month. Methods for inference about proportions are the topic of this chapter.

chapter 12

Inference for Proportions

ACTIVITY 12 Is One Side of a Coin Heavier?

Materials: 20 pennies for each student

Using a coin to randomly determine an outcome, most people would flip the coin. Is it equivalent to hold the coin vertically on a tabletop and spin the coin with a quick flick of your finger? In this activity, we will try a third variation. We will stand pennies on edge and then bang the table to make the pennies fall. We are interested in the proportion of times the pennies fall heads up. If the pennies are equally heavy on both sides of the coin, then it would be reasonable to expect the long-term proportion of heads to be about 0.5. We state the following hypotheses:

$$H_0: p = 0.5$$
$$H_a: p \uparrow 0.5$$

Procedure

1. Stand 20 pennies on edge on a horizontal tabletop. Take your time—this may take a steady hand and some patience.
2. Bang the table just hard enough to make all of the pennies fall.
3. Count the number of pennies that fall heads up.
4. Combine your results with those of other students in the class.

Questions

- Are the results about what you expected? Or are you surprised by the results?
- Do you think it is likely, by chance alone, to obtain results like the results you actually observed if H_0 is true?

Keep these results handy. As soon as we develop the necessary theory, you will test to see if your results are significant, and you will construct a confidence interval for the true proportion of heads obtained in this manner.

INTRODUCTION

Our discussion of statistical inference to this point has concerned making inferences about population *means*. But we often want to answer questions about the proportion of some outcome in a population, or to compare proportions across several populations. Here are some examples that call for inference about population proportions.

EXAMPLE 12.1 RISKY BEHAVIOR IN THE AGE OF AIDS

How common is behavior that puts people at risk of AIDS? The National AIDS Behavioral Surveys interviewed a random sample of 2673 adult heterosexuals. Of these, 170 had more than one sexual partner in the past year. That's 6.36% of the sample.[1] Based on these data, what can we say about the percent of all adult heterosexuals who have multiple partners? We want to *estimate a single population proportion.*

EXAMPLE 12.2 DOES PRESCHOOL MAKE A DIFFERENCE?

Do preschool programs for poor children make a difference in later life? A study looked at 62 children who were enrolled in a Michigan preschool in the late 1960s and at a control group of 61 similar children who were not enrolled. At 27 years of age, 61% of the preschool group and 80% of the control group had required the help of a social service agency (mainly welfare) in the previous ten years.[2] Is this significant evidence that preschool for poor children reduces later use of social services? We want to *compare two population proportions.*

EXAMPLE 12.3 EXTRACURRICULARS AND GRADES

What is the relationship between time spent in extracurricular activities and success in a tough course in college? North Carolina State University looked at the 123 students in an introductory chemical engineering course. Students needed a grade of C or better to advance to the next course. The passing rates were 55% for students who spent less than 2 hours per week in extracurricular activities, 75% for those who spent between 2 and 12 hours per week, and 38% for those who spent more than 12 hours per week.[3] Are the differences in passing rates statistically significant? We must *compare more than two population proportions.*

Our study of inference for proportions will follow the same pattern as these examples. Section 12.1 discusses inference for one population proportion, and Section 12.2 presents methods for comparing two proportions. Comparing more than two proportions raises new issues and requires more elaborate methods that also apply to some other inference problems. These methods are the topic of Chapter 13.

12.1 INFERENCE FOR A POPULATION PROPORTION

We are interested in the unknown proportion p of a population that has some outcome. For convenience, call the outcome we are looking for a "success." In Example 12.1, the population is adult heterosexuals, and the parameter p is the proportion who have had more than one sexual partner in the past year. To estimate p, the National AIDS Behavioral Surveys used random dialing of telephone numbers to contact a sample of 2673 people. Of these, 170 said they

sample proportion

had multiple sexual partners. The statistic that estimates the parameter p is the ***sample proportion***

$$\hat{p} = \frac{\text{count of successes in the sample}}{\text{count of observations in the sample}}$$
$$= \frac{170}{2673} = 0.0636$$

Read the sample proportion $\hat{p}$ as "p-hat."

EXERCISES

In each of the following settings: (a) Describe the population and explain in words what the parameter p is. (b) Give the numerical value of the statistic $\hat{p}$ that estimates p.

12.1 Tonya wants to estimate what proportion of the students in her dormitory like the dorm food. She interviews an SRS of 50 of the 175 students living in the dormitory. She finds that 14 think the dorm food is good.

12.2 Glenn wonders what proportion of the students at his school think that tuition is too high. He interviews an SRS of 50 of the 2400 students at his college. Thirty-eight of those interviewed think tuition is too high.

12.3 A college president says, "99% of the alumni support my firing of Coach Boggs." You contact an SRS of 200 of the college's 15,000 living alumni and find that 76 of them support firing the coach.

Conditions for inference

As always, inference is based on the sampling distribution of a statistic. We described the sampling distribution of a sample proportion $\hat{p}$ in Section 2 of Chapter 9. The mean is p. That is, the sample proportion $\hat{p}$ is an unbiased estimator of the population proportion p. The standard deviation of $\hat{p}$ is $\sqrt{p(1-p)/n}$, provided that the population is at least 10 times as large as the sample. If the sample size is large enough that both np and $n(1 - p)$ are at least 10, the distribution of $\hat{p}$ is approximately normal. Figure 12.1 displays this sampling distribution.

Standardize $\hat{p}$ by subtracting its mean and dividing by its standard deviation. The result is a z statistic:

$$z = \frac{\hat{p} - p}{\sqrt{\frac{p(1-p)}{n}}}$$

The statistic z has approximately the standard normal distribution $N(0,1)$ if the sample is not too small and the sample is not a large part of the population. Inference about p uses this z statistic and standard normal critical values.

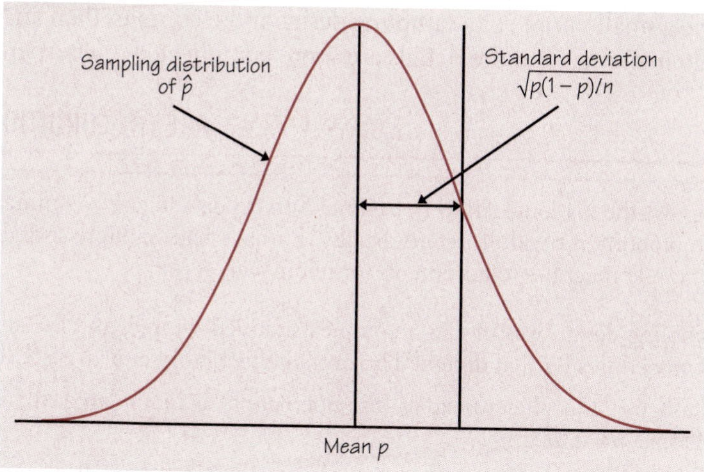

FIGURE 12.1 Select a large SRS from a population that contains proportion p of successes. The sampling distribution of the proportion $\hat{p}$ of successes in the sample is approximately normal. The mean is p and the standard deviation is $\sqrt{p(1-p)/n}$.

In practice, of course, we don't know the value of p. So we cannot calculate z or check whether np and $n(1-p)$ are 10 or greater. Here's what we do:

- To test the null hypothesis H_0: $p = p_0$ that the unknown p has a specific value p_0, just replace p by p_0 in the z statistic and in checking the values of np and $n(1-p)$.
- In a confidence interval for p, we have no specific value to substitute. In large samples, $\hat{p}$ will be close to p. So we replace p by $\hat{p}$ in determining the values of np and $n(1-p)$. We also replace the standard deviation by the ***standard error of*** $\hat{p}$

standard error of $\hat{p}$

$$\text{SE} = \sqrt{\frac{\hat{p}(1-\hat{p})}{n}}$$

to get a confidence interval of the form

$$\text{estimate} \pm z^* \text{SE}_{\text{estimate}}$$

The requirements for using the z procedures for inference about a proportion are stated in terms of p_0 or $\hat{p}$.

CONDITIONS FOR INFERENCE ABOUT A PROPORTION

- The data are an SRS from the population of interest.
- The population is at least 10 times as large as the sample.
- For a test of H_0: $p = p_0$, the sample size n is so large that both np_0 and $n(1-p_0)$ are 10 or more. For a confidence interval, n is so large that both the count of successes $n\hat{p}$ and the count of failures $n(1-\hat{p})$ are 10 or more.

If you have a small sample or a sampling design more complex than an SRS, you can still do inference, but the details are more complicated. Get expert advice.

EXAMPLE 12.4 ARE THE CONDITIONS MET?

We want to use the National AIDS Behavioral Surveys data to give a confidence interval for the proportion of adult heterosexuals who have had multiple sexual partners. Does the sample meet the requirements for inference?

- The sampling design was in fact a complex stratified sample, and the survey used inference procedures for that design. The overall effect is close to an SRS, however.
- The number of adult heterosexuals (the population) is much larger than 10 times the sample size, $n = 2673$.
- The counts of "Yes" and "No" responses are much greater than 10:

$$n\hat{p} = (2673)(0.0636) = 170$$
$$n(1-\hat{p}) = (2673)(0.9364) = 2503$$

The second and third requirements are easily met. The first requirement, that the sample be an SRS, is only approximately met.

As usual, the practical problems of a large sample survey pose a greater threat to the AIDS survey's conclusions. Only people in households with telephones could be reached. This is acceptable for surveys of the general population, because about 94% of American households have telephones. However, some groups at high risk for AIDS, like intravenous drug users, often don't live in settled households and are underrepresented in the sample. About 30% of the people reached refused to cooperate. A nonresponse rate of 30% is not unusual in large sample surveys, but it may cause some bias if those who refuse differ systematically from those who cooperate. The survey used statistical methods that adjust for unequal response rates in different groups. Finally, some respondents may not have told the truth when asked about their sexual behavior. The survey team tried hard to make respondents feel comfortable. For example, Hispanic women were interviewed only by Hispanic women, and Spanish speakers were interviewed by Spanish speakers with the same regional accent (Cuban, Mexican, or Puerto Rican). Nonetheless, the survey report says that some bias is probably present:

> *It is more likely that the present figures are underestimates; some respondents may underreport their numbers of sexual partners and intravenous drug use because of embarrassment and fear of reprisal, or they may forget or not know details of their own or of their partner's HIV risk and their antibody testing history.*[4]

Reading the report of a large study like the National AIDS Behavioral Surveys reminds us that statistics in practice involves much more than recipes for inference.

EXERCISES

12.4 In which of the following situations can you safely use the methods of this section to get a confidence interval for the population proportion p? Explain your answers.

(a) Tonya wants to estimate what proportion of the students in her dormitory like the dorm food. She interviews an SRS of 50 of the 175 students living in the dormitory. She finds that 14 think the dorm food is good.

(b) Glenn wonders what proportion of the students at his school think that tuition is too high. He interviews an SRS of 50 of the 2400 students at his college. Thirty-eight of those interviewed think tuition is too high.

(c) In the National AIDS Behavioral Surveys sample of 2673 adult heterosexuals, 0.2% (that's 0.002 as a decimal fraction) had both received a blood transfusion and had a sexual partner from a group at high risk of AIDS. (We want to estimate the proportion p in the population who share these two risk factors.)

12.5 In which of the following situations can you safely use the methods of this section for a significance test? Explain your answers.

(a) You toss a coin 10 times in order to test the hypothesis H_0: $p = 0.5$ that the coin is balanced.

(b) A college president says, "99% of the alumni support my firing of Coach Boggs." You contact an SRS of 200 of the college's 15,000 living alumni to test the hypothesis H_0: $p = 0.99$.

(c) Do a majority of the 250 students in a statistics course agree that knowing statistics will help them in their future careers? You interview an SRS of 20 students to test H_0: $p = 0.5$.

The *z* procedures

Here are the z procedures for inference about p when our conditions are satisfied.

INFERENCE FOR A POPULATION PROPORTION

Draw an SRS of size n from a large population with unknown proportion p of successes. An approximate level C confidence interval for p is

$$\hat{p} \pm z^* \sqrt{\frac{\hat{p}(1-\hat{p})}{n}}$$

where z^* is the upper $(1 - C)/2$ standard normal critical value.

To test the hypothesis H_0: $p = p_0$, compute the z statistic

$$z = \frac{\hat{p} - p_0}{\sqrt{\frac{p_0(1-p_0)}{n}}}$$

INFERENCE FOR A POPULATION PROPORTION (*continued*)

In terms of a variable Z having the standard normal distribution, the approximate P-value for a test of H_0 against

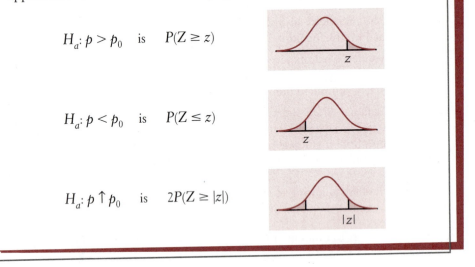

$H_a: p > p_0$ is $P(Z \geq z)$

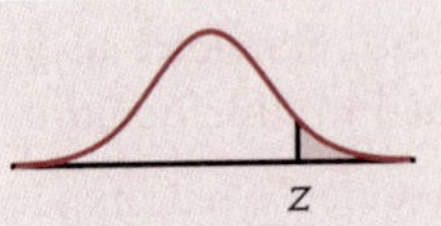

$H_a: p < p_0$ is $P(Z \leq z)$

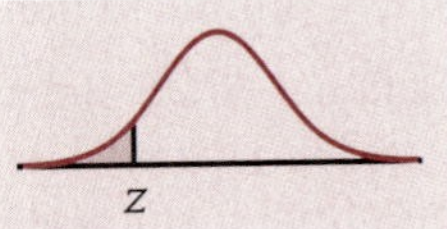

$H_a: p \uparrow p_0$ is $2P(Z \geq |z|)$

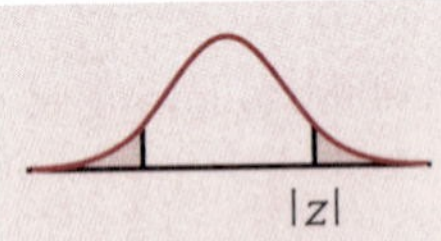

EXAMPLE 12.5 ESTIMATING RISKY BEHAVIOR

The National AIDS Behavioral Surveys found that 170 of a sample of 2673 adult heterosexuals had multiple partners. That is, $\hat{p} = 0.0636$. We will act as if the sample were an SRS.

A 99% confidence interval for the proportion p of all adult heterosexuals with multiple partners uses the standard normal critical value $z^* = 2.576$. (Look in the bottom row of Table C for standard normal critical values.) The confidence interval is

$$
\begin{aligned}
p \pm z^* \sqrt{\frac{\hat{p}(1-\hat{p})}{n}} &= 0.0636 \pm 2.576\sqrt{\frac{(0.0636)(0.9364)}{2673}} \\
&= 0.0636 \pm 0.0122 \\
&= (0.0514, 0.0758)
\end{aligned}
$$

We are 99% confident that the percent of adult heterosexuals who had more than one sexual partner in the past year lies between about 5.1% and 7.6%.

Taken together, Examples 12.4 and 12.5 show you how to construct a confidence interval for an unknown population proportion p. You can organize the inference process using the four-step Inference Toolbox, as the next two examples illustrate.

EXAMPLE 12.6 BINGE DRINKING IN COLLEGE

The 1995 Harvard School of Public Health College Alcohol Study examined alcohol use among college students, including the practice called "binge drinking." Binge drinking for men was defined as consuming five or more drinks on at least one occasion during the two weeks prior to the survey (four drinks for women). Binge drinkers experience a higher percentage of alcohol-related problems such as disciplinary problems, violence, irresponsible sexual activity, personal injury, and poor academic performance. In a representative sample of 140 colleges and 17,592 students, 7741 students identified themselves as binge drinkers. Considering this an SRS of 17,592 from the population of all U.S. college students, does this constitute strong evidence that more than 40% of all college students engaged in binge drinking?

Step 1: *Identify the population of interest and the parameter you want to draw conclusions about. State hypotheses in words and symbols.* We want to test a claim about the proportion p of all U.S. college students who have engaged in binge drinking. Our hypotheses are

H_0: $p = 0.40$	40% of U.S. college students are binge drinkers.
H_a: $p > 0.40$	More than 40% of all U.S. college students have engaged in binge drinking.

The sample proportion of binge drinkers is $\hat{p} = \dfrac{7741}{17{,}592} = 0.44$

Step 2: *Choose the appropriate inference procedure. Verify the conditions for using the selected procedure.* For testing the claim $p > 0.40$, we will use a one-proportion z test. Now we check the conditions.

- We are told that the survey design allows us to consider the sample of 17,592 students as an SRS from the population of U.S. college students.
- There are certainly more than 10(17,592) = 175,920 college students in the United States.
- $np_0 = 17{,}592(0.40) = 7036.8 \geq 10$ and $n(1 - p_0) = 17{,}592(0.60) = 10{,}555.2 \geq 10$.

So we are safe using normal approximation.

Step 3: *If conditions are met, carry out the selected procedure:*

- The z test statistic is

$$z = \frac{\hat{p} - p_0}{\sqrt{\dfrac{p_0(1-p_0)}{n}}} = \frac{0.44 - 0.40}{\sqrt{\dfrac{0.40(0.60)}{17{,}592}}} = 10.83$$

- With a z-score this large, the P-value is approximately 0.

Step 4: *Interpret your results in the context of the problem.* The P-value tells us that there is virtually no chance of obtaining a sample proportion as far away from 0.40 as $\hat{p} = 0.44$. We reject H_0 and conclude that more than 40% of U.S. college students have engaged in binge drinking.

EXAMPLE 12.7 IS THAT COIN FAIR?

A coin that is balanced should come up heads half the time in the long run. The French naturalist Count Buffon (1707–1788) tossed a coin 4040 times. He got 2048 heads. The sample proportion of heads is

$$\hat{p} = \frac{2048}{4040} = 0.5069$$

That's a bit more than one-half. Is this evidence that Buffon's coin was not balanced? This is a job for a significance test.

Step 1: Identify the population of interest and the parameter you want to draw conclusions about. State hypotheses in words and symbols.

The population for coin tossing contains the results of tossing the coin forever. The parameter p is the probability of a head, which is the proportion of all tosses that give a head. The null hypothesis says that the coin is balanced ($p = 0.5$). The alternative hypothesis is two-sided, because we did not suspect before seeing the data that the coin favored either heads or tails. We therefore test the hypotheses

$$H_0: p = 0.5$$
$$H_a: p \uparrow 0.5$$

The null hypothesis gives p the value $p_0 = 0.5$.

Step 2: Choose the appropriate inference procedure. Verify conditions. We will use a one-proportion z test to assess the evidence against H_0: $p = 0.5$. We first check the conditions:

- The tosses we make can be considered an SRS from the population of all tosses.
- The population of tosses is infinite.

$$np_0 = 4040(0.5) = 2020 \geq 10$$
$$n(1 - p_0) = 4040(0.5) = 2020 \geq 10$$

Step 3: Carry out the selected procedure:

- The z test statistic is

$$z = \frac{\hat{p} - p_0}{\sqrt{\frac{p_0(1 - p_0)}{n}}} = \frac{0.5069 - 0.5}{\sqrt{\frac{(0.5)(0.5)}{4040}}} = 0.88$$

- Because the test is two-sided, the P-value is the area under the standard normal curve more than 0.88 away from 0 in either direction. Figure 12.2 shows this area. From Table A we find that the area below –0.88 is 0.1894. The P-value is twice this area:

$$P = 2(0.1894) = 0.3788$$

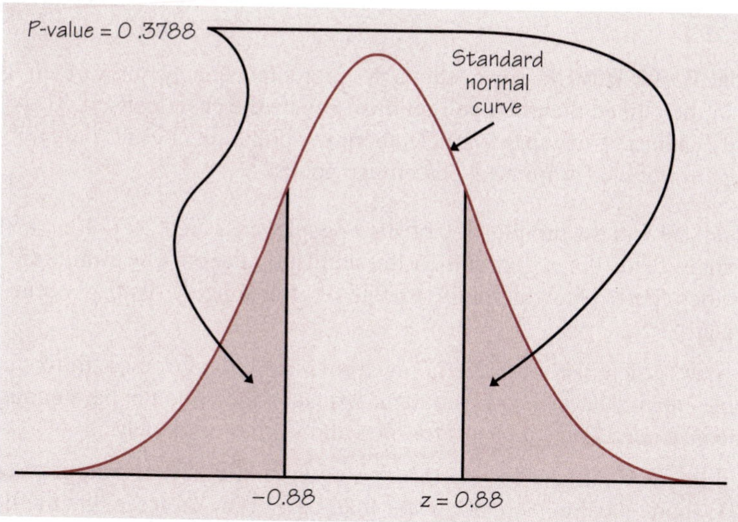

FIGURE 12.2 The *P*-value for the two-sided test.

Step 4: Interpret your results in the context of the problem. A proportion of heads as far from one-half as Buffon's would happen 38% of the time when a balanced coin is tossed 4040 times. This provides little evidence against H_0. Buffon's result doesn't show that his coin is unbalanced.

In Example 12.7, we failed to find good evidence against H_0: $p = 0.5$. We *cannot* conclude that H_0 is true, that is, that the coin is perfectly balanced. No doubt p is not exactly 0.5. The test of significance only shows that the results of Buffon's 4040 tosses can't distinguish this coin from one that is perfectly balanced. To see what values of p are consistent with the sample results, use a confidence interval.

EXAMPLE 12.8 CONFIDENCE INTERVAL FOR p

The 95% confidence interval for the probability p that Buffon's coin gives a head is

$$\begin{aligned}\hat{p} \pm z^* \sqrt{\frac{\hat{p}(1-\hat{p})}{n}} &= 0.5069 \pm 1.960\sqrt{\frac{(0.5069)(0.4931)}{4040}} \\ &= 0.5069 \pm 0.0154 \\ &= (0.4915, 0.5223)\end{aligned}$$

We are 95% confident that the probability of a head is between 0.4915 and 0.5223.

The confidence interval is more informative than the test in Example 12.7. It tells us that any null hypothesis H_0: $p = p_0$ for a p_0 between 0.4915 and 0.5223 would not be rejected at the $\alpha = 0.05$ level of significance. We would not be surprised if the true probability of a head for Buffon's coin were something like 0.51.

EXERCISES

12.6 EQUALITY FOR WOMEN? Have efforts to promote equality for women gone far enough in the United States? A poll on this issue by the cable network MSNBC contacted 1019 adults. A newspaper article about the poll said, "Results have a margin of sampling error of plus or minus 3 percentage points."[5]

(a) Overall, 54% of the sample (550 of 1019 people) answered "Yes." Find a 95% confidence interval for the proportion in the adult population who would say "Yes" if asked. Is the report's claim about the margin of error roughly right? (Assume that the sample is an SRS.)

(b) The news article said that 65% of men, but only 43% of women, think that efforts to promote equality have gone far enough. Explain why we do not have enough information to give confidence intervals for men and women separately.

(c) Would a 95% confidence interval for women alone have a margin of error less than 0.03, about equal to 0.03, or greater than 0.03? Why? You see that the news article's statement about the margin of error for poll results is a bit misleading.

12.7 TEENS AND THEIR TV SETS *The New York Times* and CBS News conducted a nationwide poll of 1048 randomly selected 13- to 17-year-olds. Of these teenagers, 692 had a television in their room and 189 named Fox as their favorite television network.[6] We will act as if the sample were an SRS.

(a) Give 95% confidence intervals for the proportion of all people in this age group who have a TV in their room and the proportion who would choose Fox as their favorite network. Check that we can use our methods.

(b) The news article says, "In theory, in 19 cases out of 20, the poll results will differ by no more than three percentage points in either direction from what would have been obtained by seeking out all American teenagers." Explain how your results agree with this statement.

(c) Is there good evidence that more than half of all teenagers have a TV in their room? Follow the Inference Toolbox.

12.8 WE WANT TO BE RICH In a recent year, 73% of first-year college students responding to a national survey identified "being very well-off financially" as an important personal goal. A state university finds that 132 of an SRS of 200 of its first-year students say that this goal is important.

(a) Give a 95% confidence interval for the proportion of all first-year students at the university who would identify being well-off as an important personal goal. Follow the Inference Toolbox.

(b) Is there good evidence that the proportion of all first-year students at this university who think being very well-off is important differs from the national value, 73%?

12.9 DO JOB APPLICANTS LIE? When trying to hire managers and executives, companies sometimes verify the academic credentials described by the applicants. One company that performs these checks summarized their findings for a six-month period. Of the 84 applicants whose credentials were checked, 15 lied about having a degree.[7]

(a) Find the proportion of applicants who lied about having a degree, and find the standard error.

(b) Consider these data to be a random sample of credentials from a large collection of similar applicants. Give a 90% confidence interval for the true proportion of applicants who lie about having a degree.

Choosing the sample size

In planning a study, we may want to choose a sample size that will allow us to estimate the parameter within a given margin of error. We saw earlier how to do this for a population mean. The method is similar for estimating a population proportion.

The margin of error in the approximate confidence interval for p is

$$m = z^* \sqrt{\frac{\hat{p}(1-\hat{p})}{n}}$$

Here z^* is the standard normal critical value for the level of confidence we want. Because the margin of error involves the sample proportion of successes $\hat{p}$, we need to guess this value when choosing n. Call our guess p^*. Here are two ways to get p^*:

1. Use a guess p^* based on a pilot study or on past experience with similar studies. You should do several calculations that cover the range of $\hat{p}$-values you might get.

2. Use $p^* = 0.5$ as the guess. The margin of error m is largest when $\hat{p} = 0.5$, so this guess is conservative in the sense that if we get any other $\hat{p}$ when we do our study, we will get a margin of error smaller than planned.

Once you have a guess p^*, the recipe for the margin of error can be solved to give the sample size n needed. Here is the result.

SAMPLE SIZE FOR DESIRED MARGIN OF ERROR

To determine the sample size n that will yield a level C confidence interval for a population proportion p with a specified margin of error m, set the following expression for the margin of error to be less than or equal to m, and solve for n:

$$z^* \sqrt{\frac{p^*(1-p^*)}{n}} \leq m$$

where p^* is a guessed value for the sample proportion. The margin of error will be less than or equal to m if you take the guess p^* to be 0.5.

Which method for finding the guess p^* should you use? The n you get doesn't change much when you change p^* as long as p^* is not too far from 0.5.

So use the conservative guess $p^* = 0.5$ if you expect the true $\hat{p}$ to be roughly between 0.3 and 0.7. If the true $\hat{p}$ is close to 0 or 1, using $p^* = 0.5$ as your guess will give a sample much larger than you need. So try to use a better guess from a pilot study when you suspect that $\hat{p}$ will be less than 0.3 or greater than 0.7.

EXAMPLE 12.9 DETERMINING SAMPLE SIZE FOR ELECTION POLLING

Gloria Chavez and Ronald Flynn are the candidates for mayor in a large city. You are planning a sample survey to determine what percent of the voters plan to vote for Chavez. This is a population proportion p. You will contact an SRS of registered voters in the city. You want to estimate p with 95% confidence and a margin of error no greater than 3%, or 0.03. How large a sample do you need?

The winner's share in all but the most lopsided elections is between 30% and 70% of the vote. So use the guess $p^* = 0.5$. Then you want

$$z^* \sqrt{\frac{p^*(1-p^*)}{n}} \leq 0.03$$

$$\frac{1.960\sqrt{0.5(0.5)}}{\sqrt{n}} \leq 0.03$$

$$\frac{1.960(0.5)}{0.03} \leq \sqrt{n}$$

$$\sqrt{n} \geq 32.6\overline{6}$$

$$n \geq (32.66)^2 = 1067.1$$

Since the number of people in the sample must be a whole number, n must be 1068 to satisfy the inequality. If you want a 2.5% margin of error, you can show in similar fashion that $n = 1537$ is the required sample size. For a 2% margin of error, the sample size you need is 2401. (Work these out for practice!) As usual, smaller margins of error call for larger samples.

EXERCISES

12.10 STARTING A NIGHT CLUB A college student organization wants to start a nightclub for students under the age of 21. To assess support for this proposal, they will select an SRS of students and ask each respondent if he or she would patronize this type of establishment. They expect that about 70% of the student body would respond favorably. What sample size is required to obtain a 90% confidence interval with an approximate margin of error of 0.04? Suppose that 50% of the sample responds favorably. Calculate the margin of error of the 90% confidence interval.

12.11 SCHOOL VOUCHERS A national opinion poll found that 44% of all American adults agree that parents should be given vouchers good for education at any public or private school of their choice. The result was based on a small sample. How large an SRS is required to obtain a margin of error of 0.03 (that is, ±3%) in a 95% confidence interval?

(a) Answer this question using the previous poll's result as the guessed value p^*.

(b) Do the problem again using the conservative guess $p^* = 0.5$. By how much do the two sample sizes differ?

12.12 CAN YOU TASTE PTC? PTC is a substance that has a strong bitter taste for some people and is tasteless for others. The ability to taste PTC is inherited. About 75% of Italians can taste PTC, for example. You want to estimate the proportion of Americans with at least one Italian grandparent who can taste PTC. Starting with the 75% estimate for Italians, how large a sample must you test in order to estimate the proportion of PTC tasters within ±0.04 with 95% confidence?

SUMMARY

Tests and confidence intervals for a population proportion p when the data are an SRS of size n are based on the **sample proportion** $\hat{p}$.

When n is large, $\hat{p}$ has approximately the normal distribution with mean p and standard deviation $\sqrt{p(1-p)/n}$.

The level C **confidence interval** for p is

$$\hat{p} \pm z^* \sqrt{\frac{\hat{p}(1-\hat{p})}{n}}$$

where z^* is the upper $(1 - C)/2$ standard normal critical value.

Tests of $H_0: p = p_0$ are based on the **z statistic**

$$z = \frac{\hat{p} - p_0}{\sqrt{\frac{p_0(1-p_0)}{n}}}$$

with *P*-values calculated from the standard normal distribution.

These inference procedures are approximately correct when the population is at least 10 times as large as the sample and the sample is large enough to satisfy $n\hat{p} \geq 10$ and $n(1-\hat{p}) \geq 10$ for a confidence interval or $np_0 \geq 10$ and $n(1-p_0) \geq 10$ for a test of $H_0: p = p_0$.

The **sample size** needed to obtain a confidence interval with approximate margin of error m for a population proportion involves solving

$$z^* \sqrt{\frac{p^*(1-p^*)}{n}} \leq m$$

for n, where p^* is a guessed value for the sample proportion $\hat{p}$, and z^* is the standard normal critical point for the level of confidence you want. If you use $p^* = 0.5$ in this formula, the margin of error of the interval will be less than or equal to m no matter what the value of $\hat{p}$ is.

SECTION 12.1 EXERCISES

12.13 DRUNKEN CYCLISTS? In the United States approximately 900 people die in bicycle accidents each year. One study examined the records of 1711 bicyclists aged 15 or older who were fatally injured in bicycle accidents between 1987 and 1991 and were tested for alcohol. Of these, 542 tested positive for alcohol (blood alcohol concentration of 0.01% or higher).[8]

(a) Find a 95% confidence interval for p. Follow the Inference Toolbox.

(b) Can you conclude from your statistical analysis of this study that alcohol causes fatal bicycle accidents? Explain.

12.14 SIDE EFFECTS An experiment on the side effects of pain relievers assigned arthritis patients to one of several over-the-counter pain medications. Of the 440 patients who took one brand of pain reliever, 23 suffered some "adverse symptom." Does the experiment provide strong evidence that fewer than 10% of patients who take this medication have adverse symptoms?

12.15 DO YOU GO TO CHURCH? The Gallup Poll asked a sample of 1785 adults, "Did you, yourself, happen to attend church or synagogue in the last 7 days?" Of the respondents, 750 said "Yes." Suppose (it is not, in fact, true) that Gallup's sample was an SRS of all American adults.

(a) Give a 99% confidence interval for the proportion of all adults who attended church or synagogue during the week preceding the poll.

(b) Do the results provide good evidence that less than half of the population attended church or synagogue?

(c) How large a sample would be required to obtain a margin of error of 0.01 in a 99% confidence interval for the proportion who attend church or synagogue? (Use the conservative guess $p^* = 0.5$, and explain why this method is reasonable in this situation.)

12.16 STOLEN HARLEYS Harley-Davidson motorcycles make up 14% of all motorcycles registered in the United States. In 1995, 9224 motorcycles were reported stolen; 2490 of these were Harleys. We can think of motorcycles stolen in 1995 as an SRS of motorcycles stolen in recent years.

(a) If Harleys made up 14% of motorcycles stolen, what would be the sampling distribution of the proportion of Harleys in a sample of 9224 stolen motorcycles?

(b) Is the proportion of Harleys among stolen bikes significantly higher than their share of all motorcycles?

12.17 COFFEE PREFERENCES One-sample procedures for proportions, like those for means, are used to analyze data from matched pairs designs. Here is an example.

Each of 50 subjects tastes two unmarked cups of coffee and says which he or she prefers. One cup in each pair contains instant coffee; the other, fresh-brewed coffee. Thirty-one of the subjects prefer the fresh-brewed coffee. Take p to be the proportion of the population who would prefer fresh-brewed coffee in a blind tasting.

(a) Test the claim that a majority of people prefer the taste of fresh-brewed coffee. Is your result significant at the 5% level? What is your practical conclusion?

(b) Find a 90% confidence interval for p.

(c) When you do an experiment like this, in what order should you present the two cups of coffee to the subjects?

12.18 CUSTOMER SATISFACTION An automobile manufacturer would like to know what proportion of its customers are not satisfied with the service provided by their local dealer. The customer relations department will survey a random sample of customers and compute a 99% confidence interval for the proportion who are not satisfied.

(a) From past studies, they believe that this proportion will be about 0.2. Find the sample size needed if the margin of error of the confidence interval is to be about 0.015.

(b) When the sample is actually contacted, 10% of the sample say they are not satisfied. What is the margin of error of the 99% confidence interval?

12.19 HACK-A-SHAQ Any Lost Angeles Lakers fan or archrival knows the team's very large "SHAQilles heel"—the free-throw shooting of the NBA's most valuable player during the 2000 season, Shaquille O'Neal. Over his NBA career, Shaq has made 53.3% of his free throws.

Shaquille O'Neal worked in the off-season with Assistant Coach Tex Winter on his free-throw technique. During the first two games of the next season, Shaq made 26 out of 39 free throws.

(a) Do these results provide evidence that Shaq has improved his free-throw shooting? Follow the Inference Toolbox.

(b) Describe a Type I error and a Type II error in this situation.

(c) Suppose that Shaq has actually improved his free-throw shooting percentage to 60%. What is the probability that you will correctly reject the claim that $p = 0.533$? Use a 5% significance level.

(d) Find the probability of a Type I error and a Type II error.

12.20 ACTIVITY 12 ANALYSIS

(a) Calculate the proportion of heads you obtained in Activity 12. Then use your calculator to test the null hypothesis H_0: $p = 0.5$ against the alternative hypothesis H_a: $p \uparrow 0.5$. Report the *P*-value and state your conclusion.

(b) Find the proportion of heads for your entire class. What is *n*? Using the same null and alternative hypotheses as in (a), find the new *P*-value, and compare this with the value you obtained in (a).

(c) Using the data from your experiment, find the 95% confidence interval for the true proportion of heads obtained by the method in Activity 12.

(d) This part should be done as a class activity. Draw a horizontal line at the top of the blackboard, and mark a scale wide enough to accommodate each student's confidence interval. Then below this scaled line, each student can draw his or her confidence interval. These intervals should vary somewhat. Looking at all of the confidence intervals, make a conjecture about the 95% confidence interval for the whole class.

(e) Use the cumulative data collected by the whole class to calculate the 95% confidence interval, and compare this interval with the interval conjectured in (d). Each student should also compare his or her confidence interval with the confidence interval for the whole class. Which confidence interval do you prefer, and why? What accounts for the difference in the width?

TECHNOLOGY TOOLBOX *Inference for a population proportion on the TI-83/89*

The TI-83/89 can be used to test a claim about a population proportion and to construct confidence intervals. Let's revisit Example 12.7, Buffon's coin-tossing activity.

In $n = 4040$ coin tosses, Count Buffon observed X = 2048 heads. Recall that our hypotheses were

$$H_0: p = 0.5$$
$$H_a: p \uparrow 0.5$$

To perform a significance test:

TI-83

- Press STAT, then choose TESTS and 5:1-PropZTest.

TI-89

- In the Statistics/List Editor, press 2nd F1 ([F6]) and choose 5:1-PropZTest.

- On the 1-PropZTest screen, enter the values shown: $p_0 = 0.5$, $x = 2048$, and $n = 4040$. Specify the alternative hypotheses as "prop $\uparrow p_0$."

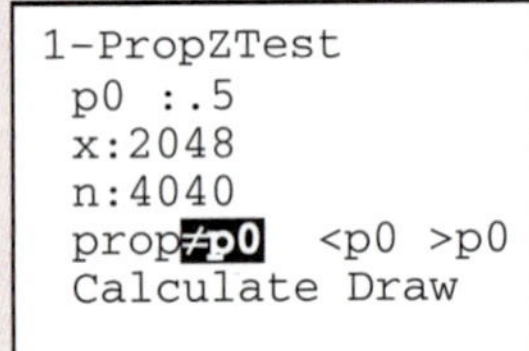

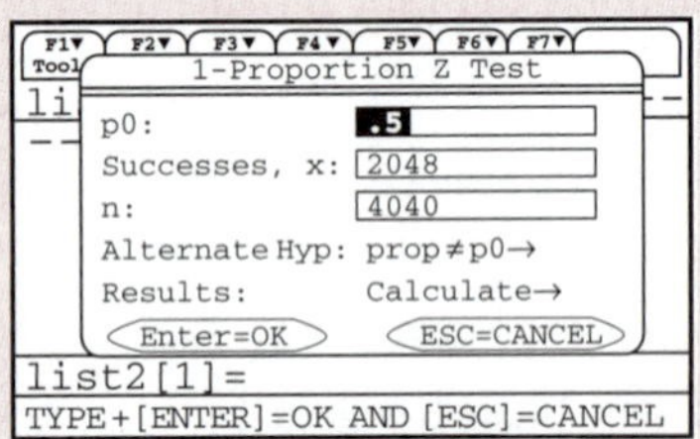

- If you select the "Calculate" choice and press ENTER, you will see that the z statistic is 0.88 and the P-value is 0.3783.

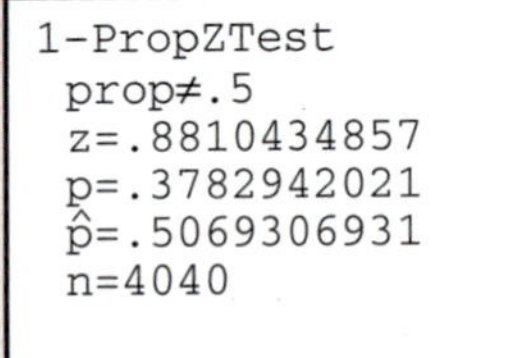

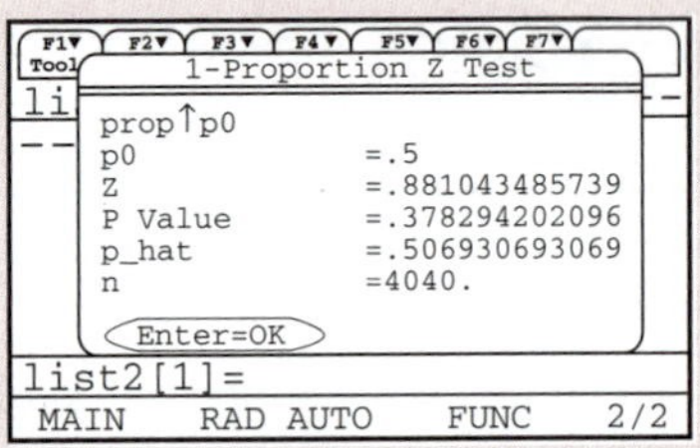

- If you select the "Draw" option, you will see the screen show here. Compare these results with those in Example 12.7.

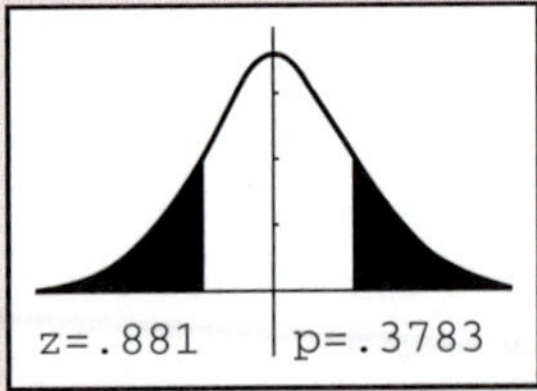

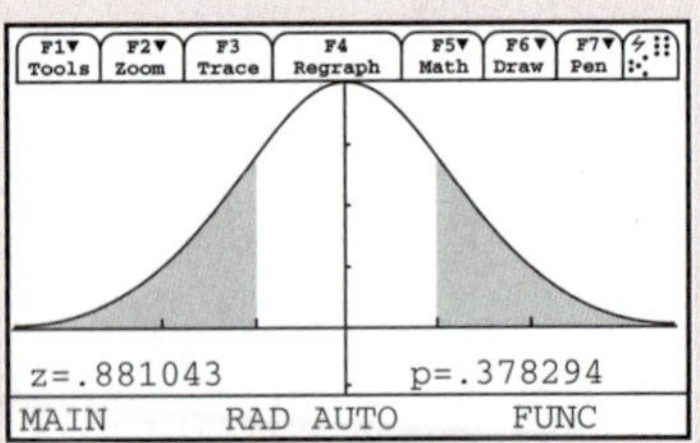

TECHNOLOGY TOOLBOX *Inference for a population proportion on the TI-83/89 (continued)*

To construct a confidence interval:

TI-83

- Press STAT, then choose TESTS and A:1-PropZInt.

TI-89

- In the Statistics/List Editor, press 2nd F2 ([F7]) and choose 5:1-PropZInt.

- When the 1-PropZInt screen appears, enter $x = 2048$, $n = 4040$, and confidence level 0.95.

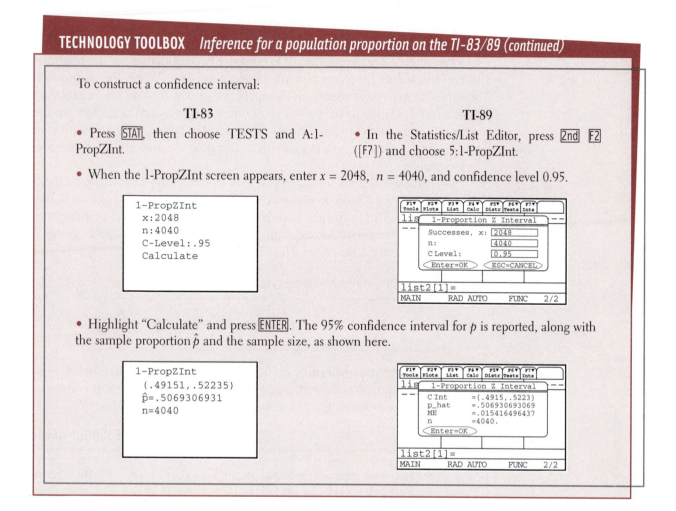

```
1-PropZInt
 x:2048
 n:4040
 C-Level:.95
 Calculate
```

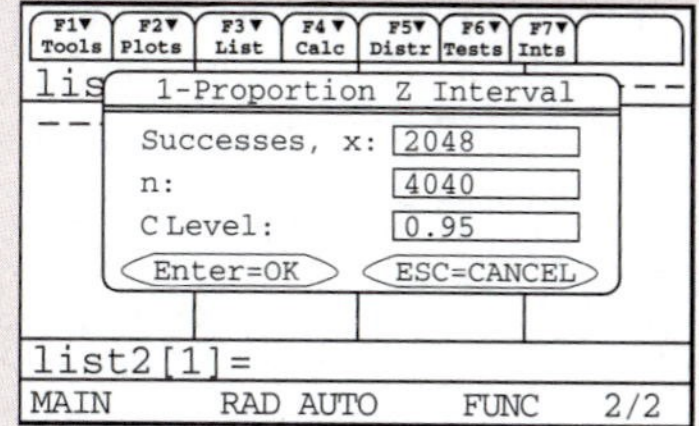

- Highlight "Calculate" and press ENTER. The 95% confidence interval for p is reported, along with the sample proportion $\hat{p}$ and the sample size, as shown here.

```
1-PropZInt
 (.49151,.52235)
 p̂=.5069306931
 n=4040
```

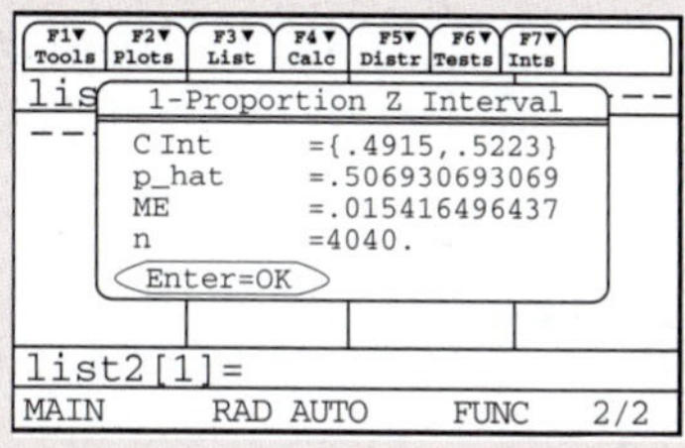

12.21 COLLEGE FOOD Tonya, Frank, and Sarah are investigating student attitudes toward college food for an assignment in their introductory statistics class. Based on comments overheard from other students, they believe that fewer than 1 in 3 students like college food. To test this hypothesis, each selects an SRS of students who regularly eat in the cafeteria, and asks them if they like college food. Fourteen in Tonya's SRS of 50 replied, "Yes," while 98 in Frank's sample of 350, and 140 in Sarah's sample of 500 said they like college food. Use your calculator to perform a test of significance on all three results and fill in a table like this:

X	n	$\hat{p}$	z	P-value
14	50			
98	350			
140	500			

Describe your findings in a short narrative.

12.2 COMPARING TWO PROPORTIONS

two-sample problem

In a ***two-sample problem,*** we want to compare two populations or the responses to two treatments based on two independent samples. When the comparison involves the mean of a quantitative variable, we use the two-sample t methods of Section 11.2. [To compare the standard deviations of a variable in two groups, we use (under restrictive conditions) the F statistic, which will be described in Section 15.1.] Now we turn to methods to compare the proportions of successes in two groups.

We will use notation similar to that used in our study of two-sample t statistics. The groups we want to compare are Population 1 and Population 2. We have a separate SRS from each population or responses from two treatments in a randomized comparative experiment. A subscript shows which group a parameter or statistic describes. Here is our notation:

Population	Population proportion	Sample size	Sample proportion
1	p_1	n_1	$\hat{p}_1$
2	p_2	n_2	$\hat{p}_2$

We compare the populations by doing inference about the difference $p_1 - p_2$ between the population proportions. The statistic that estimates this difference is the difference between the two sample proportions, $\hat{p}_1 - \hat{p}_2$.

EXAMPLE 12.10 DOES PRESCHOOL HELP?

To study the long-term effects of preschool programs for poor children, the High/Scope Educational Research Foundation has followed two groups of Michigan children since early childhood. One group of 62 attended preschool as 3- and 4-year-olds. This is a sample from Population 2, poor children who attend preschool. A control group of 61 children from the same area and similar backgrounds represents Population 1, poor children with no preschool. Thus the sample sizes are $n_1 = 61$ and $n_2 = 62$.

One response variable of interest is the need for social services as adults. In the past ten years, 38 of the preschool sample and 49 of the control sample have needed social services (mainly welfare). The sample proportions are

$$\hat{p}_1 = \frac{49}{61} = 0.803$$

$$\hat{p}_2 = \frac{38}{62} = 0.613$$

That is, about 80% of the control group uses social services, as opposed to about 61% of the preschool group.

To see if the study provides significant evidence that preschool reduces the later need for social services, we test the hypotheses

$$H_0: p_1 - p_2 = 0 \quad \text{or} \quad H_0: p_1 = p_2$$
$$H_a: p_1 - p_2 > 0 \quad \text{or} \quad H_a: p_1 > p_2$$

The sampling distribution of $\hat{p}_1 - \hat{p}_2$

Both $\hat{p}_1$ and $\hat{p}_2$ are random variables. Their values would vary if we took repeated samples of the same size. The statistic $\hat{p}_1 - \hat{p}_2$ is the difference between these two random variables. In Chapter 7, we saw that if X and Y are independent random variables,

$$\mu_{X-Y} = \mu_X - \mu_Y$$
$$\sigma^2_{X-Y} = \sigma^2_X + \sigma^2_Y$$

These results lead us to important facts about the sampling distribution of $\hat{p}_1 - \hat{p}_2$:

- The mean of $\hat{p}_1 - \hat{p}_2$ is

$$\mu_{\hat{p}_1 - \hat{p}_2} = \mu_{\hat{p}_1} - \mu_{\hat{p}_2} = p_1 - p_2$$

That is, the difference of sample proportions is an unbiased estimator of the difference of population proportions.

- The variance of $\hat{p}_1 - \hat{p}_2$ is

$$\sigma^2_{\hat{p}_1 - \hat{p}_2} = \sigma^2_{\hat{p}_1} + \sigma^2_{\hat{p}_2} = \frac{p_1(1-p_1)}{n_1} + \frac{p_2(1-p_2)}{n_2}$$

provided that the sample proportions are independent. Note that the *variances* add. The standard deviations do not.

- When the samples are large, the distribution of $\hat{p}_1 - \hat{p}_2$ is approximately normal.

Figure 12.3 displays the distribution of $\hat{p}_1 - \hat{p}_2$. The standard deviation of $\hat{p}_1 - \hat{p}_2$ involves the unknown parameters p_1 and p_2. Just as in the previous section, we must replace these by estimates in order to do inference. And just as in the previous section, we do this a bit differently for confidence intervals and for tests.

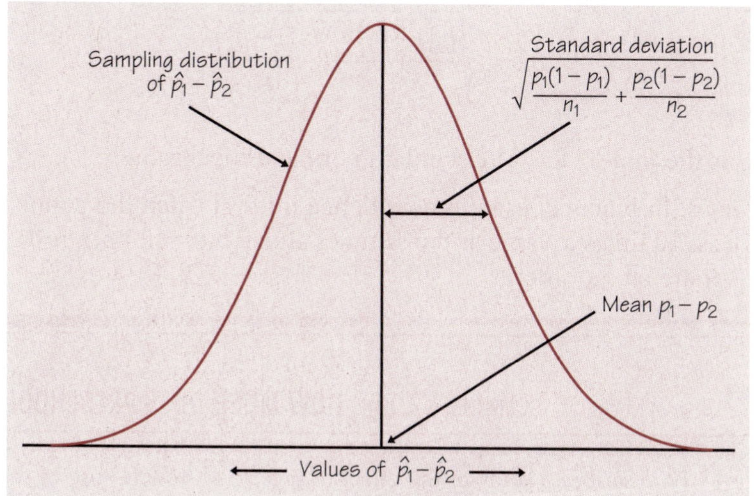

FIGURE 12.3 Select independent SRSs from two populations having proportions of successes p_1 and p_2. The proportions of successes in the two samples are $\hat{p}_1$ and $\hat{p}_2$. When the samples are large, the sampling distribution of the difference $\hat{p}_1 - \hat{p}_2$ is approximately normal.

Confidence intervals for $p_1 - p_2$

The standard deviation of $\hat{p}_1 - \hat{p}_2$ is the square root of the variance:

$$\sqrt{\frac{p_1(1-p_1)}{n_1} + \frac{p_2(1-p_2)}{n_2}}$$

standard error

To obtain a confidence interval, replace the population proportions p_1 and p_2 in this expression by the sample proportions. The result is the ***standard error*** of the statistic $\hat{p}_1 - \hat{p}_2$:

$$\text{SE} = \sqrt{\frac{\hat{p}_1(1-\hat{p}_1)}{n_1} + \frac{\hat{p}_2(1-\hat{p}_2)}{n_2}}$$

The confidence interval again has the form

$$\text{estimate} \pm z^* \text{SE}_{\text{estimate}}$$

CONFIDENCE INTERVALS FOR COMPARING TWO PROPORTIONS

Draw an SRS of size n_1 from a population having proportion p_1 of successes and draw an independent SRS of size n_2 from another population having proportion p_2 of successes. When n_1 and n_2 are large, an approximate level C confidence interval for $p_1 - p_2$ is

$$(\hat{p}_1 - \hat{p}_2) \pm z^* \text{SE}$$

In this formula the standard error SE of $\hat{p}_1 - \hat{p}_2$ is

$$\text{SE} = \sqrt{\frac{\hat{p}_1(1-\hat{p}_1)}{n_1} + \frac{\hat{p}_2(1-\hat{p}_2)}{n_2}}$$

and z^* is the upper $(1 - C)/2$ standard normal critical value.

Conditions: In practice, use this confidence interval when the populations are at least 10 times as large as the samples and $n_1\hat{p}_1$, $n_1(1 - \hat{p}_1)$, $n_2\hat{p}_2$, and $n_2(1 - \hat{p}_2)$ are all 5 or more.

EXAMPLE 12.11 HOW MUCH DOES PRESCHOOL HELP?

Example 12.10 describes a study of the effect of preschool on later use of social services. The facts are:

Population	Population description	Sample size	Sample proportion
1	Control	$n_1 = 61$	$\hat{p}_1 = 0.803$
2	Preschool	$n_2 = 62$	$\hat{p}_2 = 0.613$

We completed step 1 of the Inference Toolbox in Example 12.10.

Step 2: Choose the appropriate inference procedure. Verify conditions. To check that our approximate confidence interval is safe, look at the counts of successes and failures in the two samples. The smallest of these four quantities is

$$n_1(1 - \hat{p}_2) = (61)(1 - 0.803) = 12$$

This is larger than 5, so the interval will be accurate.

We can be fairly confident that there are at least 610 poor children who did not attend preschool and at least 620 poor children who did in our populations of interest. Since we are not told that the two samples are in fact SRSs, we must be cautious in drawing conclusions about the corresponding populations.

Step 3: Carry out the procedure. The difference $p_1 - p_2$ measures the effect of preschool in reducing the proportion of people who later need social services. To compute a 95% confidence interval for $p_1 - p_2$, first find the standard error

$$\begin{aligned} SE &= \sqrt{\frac{\hat{p}_1(1-\hat{p}_1)}{n_1} + \frac{\hat{p}_2(1-\hat{p}_2)}{n_2}} \\ &= \sqrt{\frac{(0.803)(0.197)}{61} + \frac{(0.613)(0.387)}{62}} \\ &= \sqrt{0.00642} = 0.0801 \end{aligned}$$

The 95% confidence interval is

$$\begin{aligned} (\hat{p}_1 - \hat{p}_2) \pm z^* SE &= (0.803 - 0.613) \pm (1.960)(0.0801) \\ &= 0.190 \pm 0.157 \\ &= (0.033, 0.347) \end{aligned}$$

Figure 12.4 displays Minitab output for this example. As usual, you can understand the output even without knowledge of the program that produced it.

Step 4: Interpret your results in context. We are 95% confident that the percent needing social services is somewhere between 3.3% and 34.7% lower among people who attended preschool. The confidence interval is wide because the samples are quite small.

```
Test and Confidence Interval for Two Proportions

Sample      X    N    Sample p
1          49   61    0.803279
2          38   62    0.612903

Estimate for p(1) - p(2):  0.190375
95% CI for p(1) - p(2):  (0.0333680, 0.347383)
Test for p(1) - p(2)=0(vs not=0): Z=2.32 P-Value=0.020
```

FIGURE 12.4 Minitab output.

The researchers in the study of Example 12.11 selected two separate samples from the two populations they wanted to compare. Many comparative studies start with just one sample, then divide it into two groups based on data gathered from the subjects. Exercises 12.22 and 12.24 are examples of this approach. The two-sample z procedures for comparing proportions are valid in such situations. This is an important fact about these methods.

EXERCISES

12.22 IN-LINE SKATERS A study of injuries to in-line skaters used data from the National Electronic Injury Surveillance System, which collects data from a random sample of hospital emergency rooms. In the six-month study period, 206 people came to the sample hospitals with injuries from in-line skating. We can think of these people as an SRS of all people injured while skating. Researchers were able to interview 161 of these people. Wrist injuries (mostly fractures) were the most common.[9]

(a) The interviews found that 53 people were wearing wrist guards and 6 of these had wrist injuries. Of the 108 who did not wear wrist guards, 45 had wrist injuries. What are the two sample proportions of wrist injuries?

(b) Give a 95% confidence interval for the difference between the two population proportions of wrist injuries. State carefully what populations your inference compares. We would like to draw conclusions about all in-line skaters, but we have data only for injured skaters.

(c) What was the percent of nonresponse among the original sample of 206 injured skaters? Explain why nonresponse may bias your conclusions.

12.23 LYME DISEASE Lyme disease is spread in the northeastern United States by infected ticks. The ticks are infected mainly by feeding on mice, so more mice result in more infected ticks. The mouse population in turn rises and falls with the abundance of acorns, their favored food. Experimenters studied two similar forest areas in a year when the acorn crop failed. They added hundreds of thousands of acorns to one area to imitate an abundant acorn crop, while leaving the other area untouched. The next spring, 54 of the 72 mice trapped in the first area were in breeding condition, versus 10 of the 17 mice trapped in the second area.[10] Give a 90% confidence interval for the difference between the proportion of mice ready to breed in good acorn years and bad acorn years. Follow the Inference Toolbox.

12.24 FREE SPEECH? The 1958 Detroit Area Study was an important investigation of the influence of religion on everyday life. The sample "was basically a simple random sample of the population of the metropolitan area" of Detroit, Michigan. Of the 656 respondents, 267 were white Protestants and 230 were white Catholics.

The study took place at the height of the cold war. One question asked if the right of free speech included the right to make speeches in favor of communism. Of the 267 white Protestants, 104 said "Yes," while 75 of the 230 white Catholics said "Yes."[11]

Give a 95% confidence interval for the difference between the proportion of Protestants who agreed that communist speeches are protected and the proportion of Catholics who held this opinion. Follow the Inference Toolbox.

Significance tests for $p_1 - p_2$

An observed difference between two sample proportions can reflect a difference in the populations, or it may just be due to chance variation in random sampling. Significance tests help us decide if the effect we see in the samples is really there in the populations. The null hypothesis says that there is no difference between the two populations:

$$H_0: p_1 = p_2$$

The alternative hypothesis says what kind of difference we expect.

EXAMPLE 12.12 CHOLESTEROL AND HEART ATTACKS

High levels of cholesterol in the blood are associated with higher risk of heart attacks. Will using a drug to lower blood cholesterol reduce heart attacks? The Helsinki Heart Study looked at this question. Middle-aged men were assigned at random to one of two treatments: 2051 men took the drug gemfibrozil to reduce their cholesterol levels, and a control group of 2030 men took a placebo. During the next five years, 56 men in the gemfibrozil group and 84 men in the placebo group had heart attacks.

The sample proportions who had heart attacks are

$$\hat{p}_1 = \frac{56}{2051} = 0.0273 \qquad \text{(gemfibrozil group)}$$

$$\hat{p}_2 = \frac{84}{2030} = 0.0414 \qquad \text{(placebo group)}$$

That is, about 4.1% of the men in the placebo group had heart attacks, against only about 2.7% of the men who took the drug. Is the apparent benefit of gemfibrozil statistically significant?

Step 1: Identify the populations of interest and the parameters you want to draw conclusions about. State hypotheses in words and symbols.

We want to use this comparative randomized experiment to draw conclusions about p_1 = the proportion of middle-aged men taking gemfibrozil who suffer heart attacks and p_2 = the proportion of middle-aged men taking only a placebo who suffer heart attacks. We hope to show that gemfibrozil reduces heart attacks, so we have a one-sided alternative:

$$H_0: p_1 = p_2$$
$$H_a: p_1 < p_2$$

To do a test, standardize $\hat{p}_1 - \hat{p}_2$ to get a z statistic. If H_0 is true, all the observations in both samples really come from a single population of men of whom a single unknown proportion p will have a heart attack in a five-year period. So instead of estimating p_1 and p_2 separately, we pool the two samples and use the overall sample proportion to estimate the single population parameter p. Call this the ***pooled sample proportion.*** It is

pooled sample proportion

$$\hat{p} = \frac{\text{count of successes in both samples combined}}{\text{count of observations in both samples combined}} = \frac{X_1 + X_2}{n_1 + n_2}$$

Use $\hat{p}$ in place of both $\hat{p}_1$ and $\hat{p}_2$ in the expression for the standard error SE of $\hat{p}_1 - \hat{p}_2$:

$$SE_{\hat{p}_1 - \hat{p}_2} = \sqrt{\frac{\hat{p}_1(1-\hat{p}_1)}{n_1} + \frac{\hat{p}_2(1-\hat{p}_2)}{n_2}} = \sqrt{\frac{\hat{p}(1-\hat{p})}{n_1} + \frac{\hat{p}(1-\hat{p})}{n_2}} = \sqrt{\hat{p}(1-\hat{p})\left(\frac{1}{n_1} + \frac{1}{n_2}\right)}$$

This will yield a z statistic that has the standard normal distribution when H_0 is true. Here is the test.

SIGNIFICANCE TEST FOR COMPARING TWO PROPORTIONS

To test the hypothesis

$$H_0: p_1 = p_2$$

first find the pooled proportion $\hat{p}$ of successes in both samples combined. Then compute the z statistic

$$z = \frac{\hat{p}_1 - \hat{p}_2}{\sqrt{\hat{p}(1-\hat{p})\left(\frac{1}{n_1} + \frac{1}{n_2}\right)}}$$

In terms of a variable Z having the standard normal distribution, the P-value for a test of H_0 against

$H_a: p_1 > p_2$ is $P(Z \geq z)$

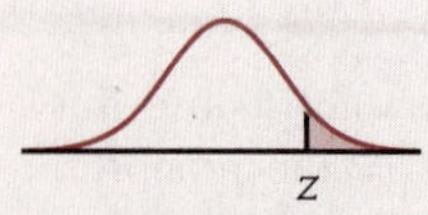

SIGNIFICANCE TEST FOR COMPARING TWO PROPORTIONS (*continued*)

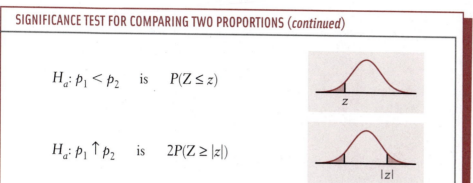

$H_a: p_1 < p_2$ is $P(Z \leq z)$

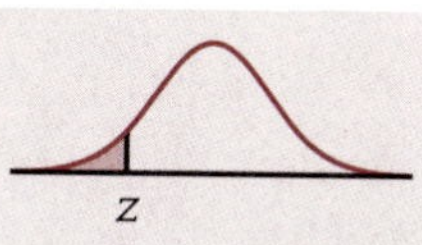

$H_a: p_1 \uparrow p_2$ is $2P(Z \geq |z|)$

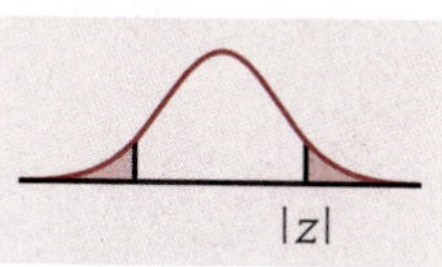

Conditions: Use these tests in practice when the populations are at least 10 times as large as the samples and $n_1\hat{p}$, $n_1(1-\hat{p})$, $n_2\hat{p}$, and $n_2(1-\hat{p})$ are all 5 or more.

EXAMPLE 12.13 CHOLESTEROL AND HEART ATTACKS, CONTINUED

Step 2: Choose the appropriate inference procedure. Verify conditions. The pooled proportion of heart attacks for the two groups in the Helsinki Heart Study is

$$\hat{p} = \frac{\text{count of heart attacks in both samples combined}}{\text{count of subjects in both samples combined}}$$
$$= \frac{56+84}{2051+2030}$$
$$= \frac{140}{4081} = 0.0343$$

Using this value, we find that

$n_1\hat{p} = 2051(0.0343) = 70.3$ $\quad$ $n_2\hat{p} = 2030(0.0343) = 69.6$

$n_1(1-\hat{p}) = 2051(0.9657) = 1980.7$ $\quad$ $n_2(1-\hat{p}) = 2030(0.9657) = 1960.4$

which are all 5 or larger. We are safe using the two-sample z procedure.

Step 3: Carry out the procedure.

- The z test statistic is

$$z = \frac{\hat{p}_1 - \hat{p}_2}{\sqrt{\hat{p}(1-\hat{p})\left(\frac{1}{n_1} + \frac{1}{n_2}\right)}}$$
$$= \frac{0.0273 - 0.0414}{\sqrt{(0.0343)(0.9657)\left(\frac{1}{2051} + \frac{1}{2030}\right)}}$$
$$= \frac{-0.0141}{0.005698} = -2.47$$

• The one-sided P-value is the area under the standard normal curve to the left of -2.47. Figure 12.5 shows this area. Table A gives $P = 0.0068$.

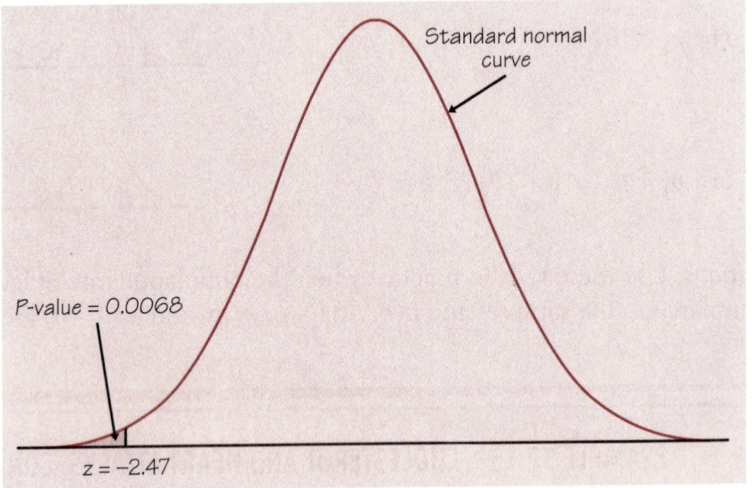

FIGURE 12.5 The P-value for the one-sided test.

Step 4: Interpret your results in context. Since $P < 0.01$, the results are statistically significant at the $\alpha = 0.01$ level. There is strong evidence that gemfibrozil reduced the rate of heart attacks. The large samples in the Helsinki Heart Study helped the study get highly significant results.

EXAMPLE 12.14 DON'T DRINK THE WATER!

The movie A *Civil Action* tells the story of a major legal battle that took place in the small town of Woburn, Massachusetts. A town well that supplied water to East Woburn residents was contaminated by industrial chemicals. During the period that residents drank water from this well, a random sample of 414 births showed 16 birth defects. On the west side of Woburn, a random sample of 228 babies born during the same time period revealed 3 with birth defects. The plaintiffs suing the companies responsible for the contamination claimed that these data show that the rate of birth defects was significantly higher in East Woburn, where the contaminated well water was in use. How strong is the evidence supporting this claim? What should the judge for this case conclude?

The proportion of babies with birth defects in the East Woburn sample is

$$\hat{p}_1 = \frac{16}{414} = 0.0386$$

For the West Woburn sample, the corresponding proportion is

$$\hat{p}_2 = \frac{3}{228} = 0.0132$$

Is the difference, $\hat{p}_1 - \hat{p}_2 = 0.0386 - 0.0132 = 0.0254$, statistically significant?

Step 1: Identify the populations of interest and the parameters you want to draw conclusions about. State hypotheses in words and symbols. Let p_1 = the proportion of East Woburn babies born with birth defects and p_2 = the proportion of West Woburn babies born with birth defects. Our hypotheses are

$$H_0: p_1 = p_2 \qquad \text{or, equivalently,} \qquad H_0: p_1 - p_2 = 0$$
$$H_a: p_1 > p_2 \qquad\qquad\qquad\qquad H_a: p_1 - p_2 > 0$$

Step 2: Choose the appropriate inference procedure. Verify conditions for using it. We will use a significant test to compare the proportions of babies born with birth defects in East and West Woburn. Since we begin by assuming that H_0: $p_1 = p_2$ is true, we use

$$\hat{p} = \frac{X_1 + X_2}{n_1 + n_2} = \frac{16+3}{414+228} = 0.0296$$

as our pooled estimate for the proportion of babies born with birth defects in all of Woburn.

We check the conditions:

- We are willing to treat the two samples of babies as SRSs from their respective populations of interest.
- Both populations are at least 10 times as large as the samples of babies.
- $n_1\hat{p} = 414(0.0296) = 12.25$ $\qquad n_2\hat{p} = 228(0.0296) = 6.75$
  $n_1(1-\hat{p}) = 414(0.9704) = 401.75$ $\qquad n_2(1-\hat{p}) = 228(0.9704) = 221.25$

Since all four of these values are larger than 5, we are safe to use a normal approximation.

Step 3: Carry out the selected procedure.

- The z statistic is

$$z = \frac{(\hat{p}_1 - \hat{p}_2)}{\sqrt{\hat{p}(1-\hat{p})\left(\frac{1}{n_1}+\frac{1}{n_2}\right)}} = \frac{0.0254}{\sqrt{0.0296(0.9704)\left(\frac{1}{414}+\frac{1}{228}\right)}}$$
$$= \frac{0.0254}{0.0140} = 1.82$$

- Figure 12.6 shows the standard normal curve with the area to the right of $z = 1.82$ shaded. From Table A, we find that the P-value = 1 – 0.9656 = 0.0344.

Step 4: Interpret your results in the context of the problem. The P-value, 0.0344, tells us that it is unlikely that we would obtain a difference in sample proportions as large as we did if the null hypothesis is true. Judges have generally adopted a 5% significance level as their standard for convincing evidence. More than likely, the judge in this case would conclude that the companies who contaminated the well water were responsible for causing a higher proportion of birth defects in East Woburn.

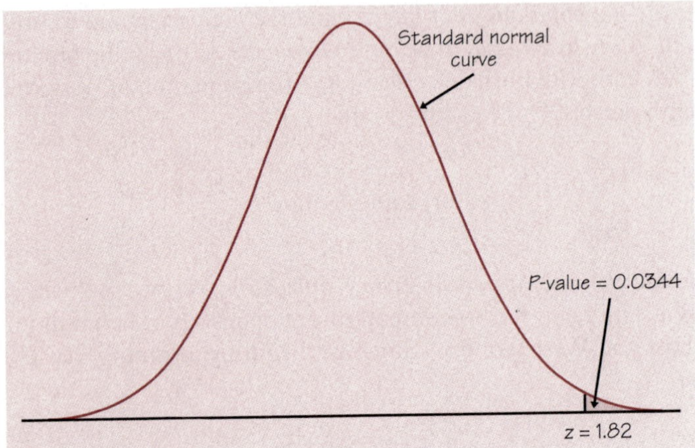

FIGURE 12.6 The *P*-value for the one-sided test.

EXERCISES

12.25 THE GOLD COAST A historian examining British colonial records for the Gold Coast in Africa suspects that the death rate was higher among African miners than among European miners. In the year 1936, there were 223 deaths among 33,809 African miners and 7 deaths among 1541 European miners on the Gold Coast.[12]

Consider this year as a sample from the pre-war era in Africa. Is there good evidence that the proportion of African miners who died was higher than the proportion of European miners who died? Follow the Inference Toolbox.

12.26 PREVENTING STROKES Aspirin prevents blood from clotting and so helps prevent strokes. The Second European Stroke Prevention Study asked whether adding another anticlotting drug named dipyridamole would be more effective for patients who had already had a stroke. Here are the data on strokes and deaths during the two years of the study:[13]

	Number of patients	Number of strokes	Number of deaths
Aspirin alone	1649	206	182
Aspirin + dipyridamole	1650	157	185

(a) The study was a randomized comparative experiment. Outline the design of the study.

(b) Is there a significant difference in the proportion of strokes in the two groups?

(c) Is there a significant difference in death rates for the two groups?

12.27 ACCESS TO COMPUTERS A sample survey by Nielsen Media Research looked at computer access and use of the Internet. Whites were significantly more likely than blacks to own a home computer, but the black-white difference in computer access at work was not significant. The study team then looked separately at the households with at least $40,000 income. The sample contained 1916 white and 131 black households in this class. Here are the sample counts for these households:[14]

	Blacks	Whites
Own home computer	86	1173
Computer access at work	100	1132

Do higher-income blacks and whites differ significantly at the 5% level in the proportion who own home computers? Do they differ significantly in the proportion who have computer access at work?

SUMMARY

We want to **compare the proportions p_1 and p_2 of successes in two populations.** The comparison is based on the difference $\hat{p}_1 - \hat{p}_2$ between the sample proportions of successes. When the sample sizes n_1 and n_2 are large enough, we can use z procedures because the sampling distribution of $\hat{p}_1 - \hat{p}_2$ is close to normal.

An approximate level C **confidence interval** for $p_1 - p_2$ is

$$(\hat{p}_1 - \hat{p}_2) \pm z^*\text{SE}$$

where the **standard error** of $\hat{p}_1 - \hat{p}_2$ is

$$\text{SE} = \sqrt{\frac{\hat{p}_1(1-\hat{p}_1)}{n_1} + \frac{\hat{p}_2(1-\hat{p}_2)}{n_2}}$$

and z^* is a standard normal critical value.

Significance tests of H_0: $p_1 = p_2$ use the **pooled sample proportion**

$$\hat{p} = \frac{\text{count of successes in both samples combined}}{\text{count of observations in both samples combined}}$$

and the **z statistic**

$$z = \frac{\hat{p}_1 - \hat{p}_2}{\sqrt{\hat{p}(1-\hat{p})\left(\frac{1}{n_1} + \frac{1}{n_2}\right)}}$$

P-values come from the standard normal table.

SECTION 12.2 EXERCISES

12.28 REDUCING NONRESPONSE? Telephone surveys often have high rates of nonresponse. When the call is handled by an answering machine, perhaps leaving a message on the machine will encourage people to respond when they are called again. Here are data from a study in which (at random) a message was or was not left when an answering machine picked up the first call from a survey:[15]

	Total households	Eventual contact	Completed survey
No message	100	58	33
Message	291	200	134

(a) Is there good evidence that leaving a message increases the proportion of households that are eventually contacted?

(b) Is there good evidence that leaving a message increases the proportion who complete the survey?

(c) If you find significant effects, look at their size. Do you think these effects are large enough to be important to survey takers?

12.29 TREATING AIDS The drug AZT was the first drug that seemed effective in delaying the onset of AIDS. Evidence for AZT's effectiveness came from a large randomized comparative experiment. The subjects were 1300 volunteers who were infected with HIV, the virus that causes AIDS, but did not yet have AIDS. The study assigned 435 of the subjects at random to take 500 milligrams of AZT each day, and another 435 to take a placebo. (The others were assigned to a third treatment, a higher dose of AZT. We will compare only two groups.) At the end of the study, 38 of the placebo subjects and 17 of the AZT subjects had developed AIDS. We want to test the claim that taking AZT lowers the proportion of infected people who will develop AIDS in a given period of time.

(a) State hypotheses, and check that you can safely use the z procedures.

(b) How significant is the evidence that AZT is effective?

(c) The experiment was double-blind. Explain what this means.

Comment: Medical experiments on treatments for AIDS and other fatal diseases raise hard ethical questions. Some people argue that because AIDS is always fatal, infected people should get any drug that has any hope of helping them. The counterargument is that we will then never find out which drugs really work. The placebo patients in this study were given AZT as soon as the results were in.

12.30 ARE URBAN STUDENTS MORE SUCCESSFUL? North Carolina State University looked at the factors that affect the success of students in a required chemical engineering course. Students must get a C or better in the course in order to continue as chemical engineering majors. There were 65 students from urban or suburban backgrounds, and 52 of these students succeeded. Another 55 students were from rural or small-town backgrounds; 30 of these students succeeded in the course.[16]

(a) Is there good evidence that the proportion of students who succeed is different for urban/suburban versus rural/small-town backgrounds?

(b) Give a 90% confidence interval for the size of the difference.

12.31 ARE GIRLS OR BOYS MORE SUCCESSFUL? The North Carolina State University study (see Exercise 12.30) also looked at possible differences in the proportions of female and male students who succeeded in the course. They found that 23 of the 34 women and 60 of the 89 men succeeded. Is there evidence of a difference between the proportions of women and men who succeed?

12.32 WHO GETS STOCK OPTIONS? Different kinds of companies compensate their key employees in different ways. Established companies may pay higher salaries, while new companies may offer stock options that will be valuable if the company succeeds. Do high-tech companies tend to offer stock options more often than other companies? One study looked at a random sample of 200 companies. Of these, 91 were listed in the *Directory of Public High Technology Corporations* and 109 were not listed. Treat these two groups as SRSs of high-tech and non-high-tech companies. Seventy-three of the high-tech companies and 75 of the non-high-tech companies offered incentive stock options to key employees.[17]

(a) Is there evidence that a higher proportion of high-tech companies offer stock options?

(b) Give a 95% confidence interval for the difference in the proportions of the two types of companies that offer stock options.

12.33 ASPIRIN AND HEART ATTACKS The Physicians' Health Study examined the effects of taking an aspirin every other day. Earlier studies suggested that aspirin might reduce the risk of heart attacks. The subjects were 22,071 healthy male physicians at least 40 years old. The study assigned 11,037 of the subjects at random to take aspirin. The others took a placebo pill. The study was double-blind. Here are the counts for some of the outcomes of interest to the researchers:

	Aspirin group	Placebo group
Fatal heart attacks	10	26
Nonfatal heart attacks	129	213
Strokes	119	98

For which outcomes is the difference between the aspirin and placebo groups significant? (Use two-sided alternatives. Check that you can apply the z test. Write a brief summary of your conclusions.)

12.34 CHILD-CARE WORKERS The Current Population Survey (CPS) is the monthly government sample survey of 60,000 households that provides data on employment in the United States. A study of child-care workers drew a sample from the CPS data tapes. We can consider this sample to be an SRS from the population of child-care workers.[18]

(a) Out of 2455 child-care workers in private households, 7% were black. Of 1191 nonhousehold child-care workers, 14% were black. Give a 99% confidence interval for the difference in the percents of these groups of workers who are black. Is the difference statistically significant at the $\alpha = 0.01$ level?

(b) The study also examined how many years of school child-care workers had. For household workers, the mean and standard deviation were $\bar{x}_1 = 11.6$ years and $s_1 = 2.2$ years. For nonhousehold workers, $\bar{x}_2 = 12.2$ years and $s_2 = 2.1$ years. Give a 99% confidence interval for the difference in mean years of education for the two groups. Is the difference significant at the $\alpha = 0.01$ level?

TECHNOLOGY TOOLBOX *Comparing proportions with the TI-83/89*

The TI-83/89 can be used to compare proportions using significance tests and confidence intervals. Here, we use the information from the Helsinki Heart Study of Examples 12.12 and 12.13.

Significance tests

In the treatment (gemfibrozil) group of 2051 middle-aged men, 56 had heart attacks. In the control (placebo) group, 84 of the 2030 men had heart attacks. The hypotheses were

$$H_0: p_1 = p_2$$
$$H_a: p_1 < p_2$$

The alternative hypothesis says that gemfibrozil reduces heart attacks. To perform a test of the null hypothesis:

- Press STAT, choose TESTS, then choose 6:2-PropZTest.
- In the Statistics/List Editor, press 2nd F1 ([F6]) and choose 6:2-PropZTest.

- When the 2-PropZTest screen appears, enter the values $x_1 = 56$, $n_1 = 2051$, $x_2 = 84$, $n_2 = 2030$. Specify the alternative hypothesis $p_1 < p_2$.

TI-83

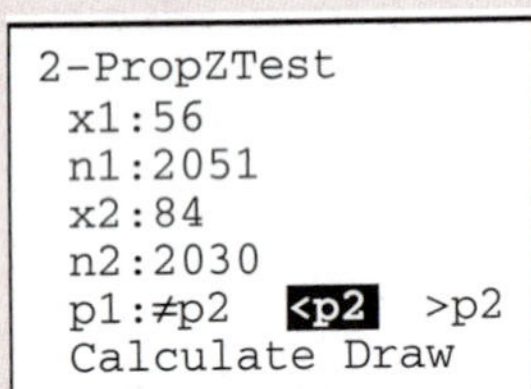

TI-89

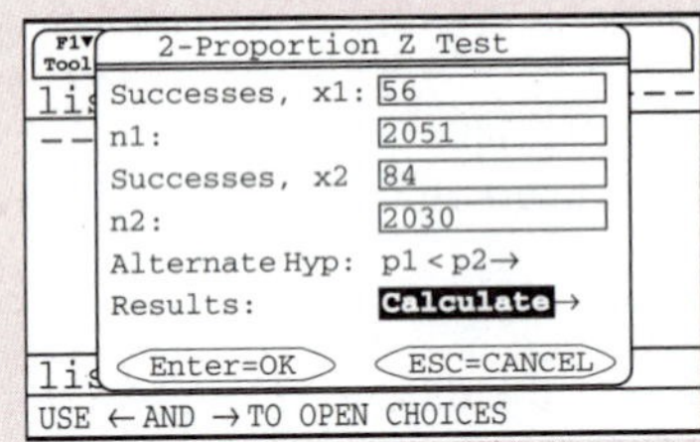

- If you select "Calculate" and press ENTER, you are told that the z statistic is $z = -2.47$ and the P-value is 0.0068, as shown here. Do you see the pooled proportion of heart attacks? Does this agree with the value calculated in Example 12.13?

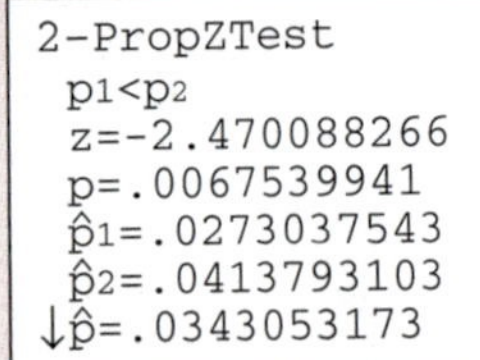

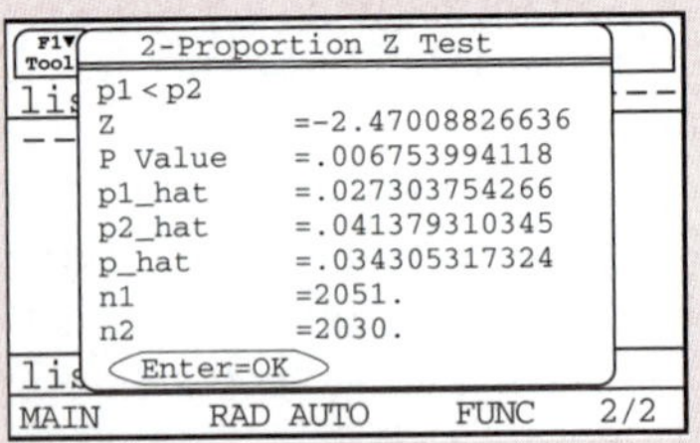

TECHNOLOGY TOOLBOX *Comparing proportions with the TI-83/89 (continued)*

- If you select the "Draw" option, you will see the screen shown here. Compare these results with those in Example 12.13.

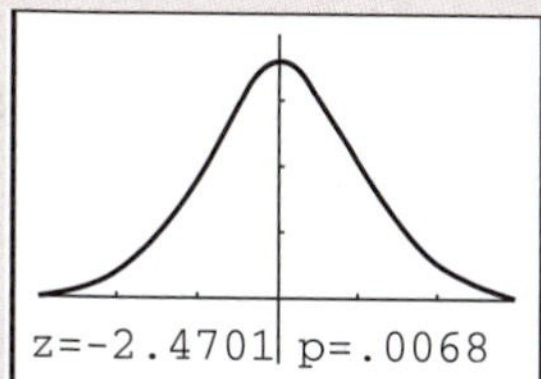

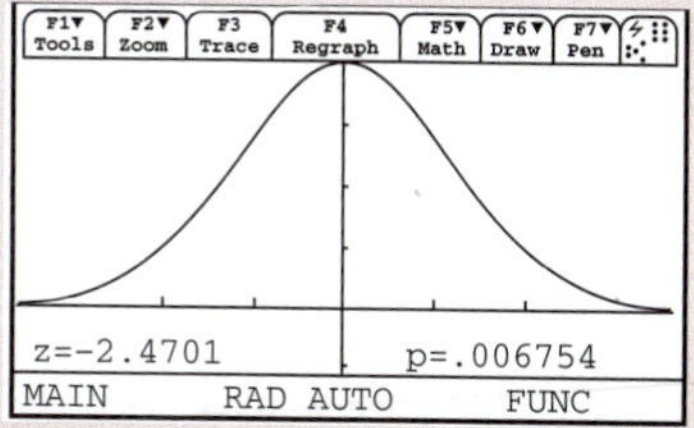

Confidence intervals

To use the TI-83/89 to construct a 95% confidence interval for the difference $p_1 - p_2$:

- Press STAT, then choose TESTS and B:2-PropZInt.
- In the Statistics/List Editor, press 2nd F2 ([F7]) and choose 6:2-PropZInt.
- When the 2-PropZInt screen appears, verify the values $x_1 = 56$, $n_1 = 2051$, $x_2 = 84$, $n_2 = 2030$, and specify the confidence level, 0.95.

```
2-PropZInt
 x1:56
 n1:2051
 x2:84
 n2:2030
 C-Level:.95
 Calculate
```

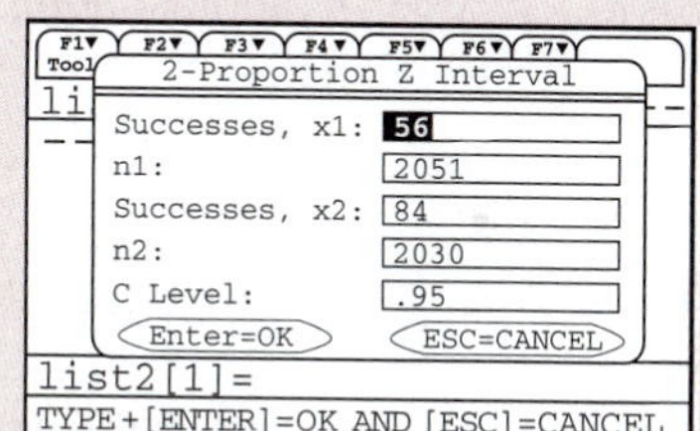

- Highlight "Calculate" and press ENTER. The 95% confidence interval for $p_1 - p_2$ is reported along with the two sample proportions and the two sample sizes, as shown here.

```
2-PropZInt
 (-.0252,-.0029)
 p̂1=.0273037543
 p̂2=.0413793103
 n1=2051
 n2=2030
```

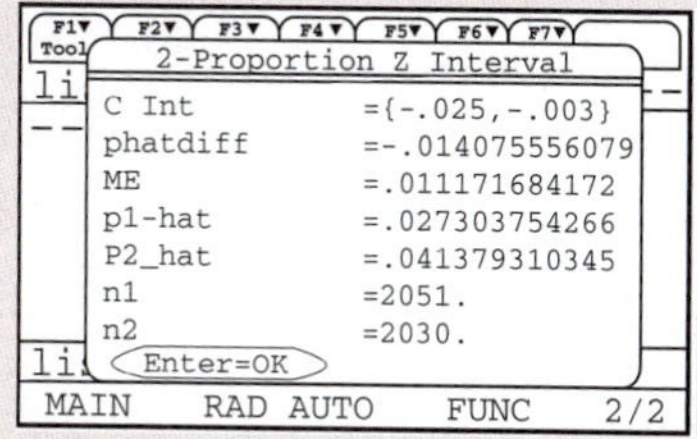

CHAPTER REVIEW

Inference about population proportions is based on sample proportions. We rely on the fact that a sample proportion has a distribution that is close to normal unless the sample is small. All the z procedures in this chapter work well when the samples are large enough. You must check this before using them. Here are the things you should now be able to do.

A. RECOGNITION

1. Recognize from the design of a study whether one-sample, matched pairs, or two-sample procedures are needed.

2. Recognize what parameter or parameters an inference problem concerns. In particular, distinguish among settings that require inference about a mean, comparing two means, inference about a proportion, or comparing two proportions.

3. Calculate from sample counts the sample proportion or proportions that estimate the parameters of interest.

B. INFERENCE ABOUT ONE PROPORTION

1. Use the z procedure to give a confidence interval for a population proportion p.

2. Use the z statistic to carry out a test of significance for the hypothesis H_0: $p = p_0$ about a population proportion p against either a one-sided or a two-sided alternative.

3. Check that you can safely use these z procedures in a particular setting.

C. COMPARING TWO PROPORTIONS

1. Use the two-sample z procedure to give a confidence interval for the difference $p_1 - p_2$ between proportions in two populations based on independent samples from the populations.

2. Use a z statistic to test the hypothesis H_0: $p_1 = p_2$ that proportions in two distinct populations are equal.

3. Check that you can safely use these z procedures in a particular setting.

Statistical inference always draws conclusions about one or more parameters of a population. When you think about doing inference, ask first what the population is and what parameter you are interested in. The t procedures of Chapter 11 allow us to give confidence intervals and carry out tests about population means. We use the z procedures of this chapter for inference about population proportions. The figure on the next page outlines the decisions you must make in choosing among the procedures we have met. First ask, "What type of population parameter does the inference concern?" Then ask, "What type of design produced the data?"

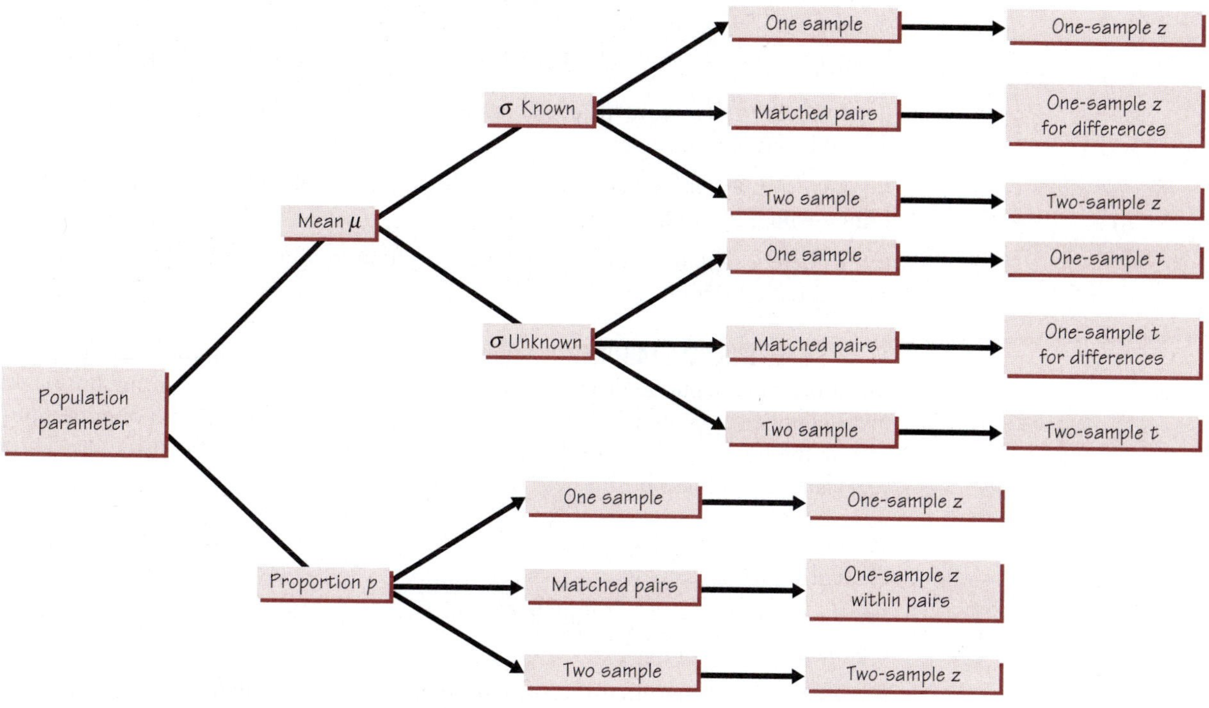

CHAPTER 12 REVIEW EXERCISES

12.35 POLICE RADAR AND SPEEDING Do drivers reduce excessive speed when they encounter police radar? Researchers studied the behavior of drivers on a rural interstate highway in Maryland where the speed limit was 55 miles per hour. They measured speed with an electronic device hidden in the pavement and, to eliminate large trucks, considered only vehicles less than 20 feet long. During some time periods, police radar was set up at the measurement location. Here are some of the data:[19]

	Number of vehicles	Number over 65 mph
No radar	12,931	5,690
Radar	3,285	1,051

(a) Give a 95% confidence interval for the proportion of vehicles going faster than 65 miles per hour when no radar is present.

(b) Give a 95% confidence interval for the effect of radar, as measured by the difference in proportions of vehicles going faster than 65 miles per hour with and without radar.

(c) The researchers chose a rural highway so that cars would be separated rather than in clusters where some cars might slow because they see other cars slowing. Explain why such clusters might make inference invalid.

12.36 STEROIDS IN HIGH SCHOOL A study by the National Athletic Trainers Association surveyed 1679 high school freshmen and 1366 high school seniors in Illinois. Results showed that 34 of the freshmen and 24 of the seniors had used anabolic steroids. Steroids, which are dangerous, are sometimes used to improve athletic performance.[20]

(a) In order to draw conclusions about all Illinois freshmen and seniors, how should the study samples be chosen?

(b) Give a 95% confidence interval for the proportion of all high school freshmen in Illinois who have used steroids.

(c) Is there a significant difference between the proportions of freshman and seniors who have used steroids?

12.37 SMALL-BUSINESS FAILURES, I A study of the survival of small businesses chose an SRS from the telephone directory's Yellow Pages listings of food-and-drink businesses in 12 counties in central Indiana. For various reasons, the study got no response from 45% of the businesses chosen. Interviews were completed with 148 businesses. Three years later, 22 of these businesses had failed.[21]

(a) Give a 95% confidence interval for the percent of all small businesses in this class that fail within three years.

(b) Based on the results of this study, how large a sample would you need to reduce the margin of error to 0.04?

(c) The authors hope that their findings describe the population of all small businesses. What about the study makes this unlikely? What population do you think the study findings describe?

12.38 SMALL-BUSINESS FAILURES, II The study of small-business failures described in the previous exercise looked at 148 food-and-drink businesses in central Indiana. Of these, 106 were headed by men and 42 were headed by women. During a three-year period, 15 of the men's businesses and 7 of the women's businesses failed. Is there a significant difference between the rate at which businesses headed by men and women fail?

12.39 SIGNIFICANT DOES NOT MEAN IMPORTANT Never forget that even small effects can be statistically significant if the samples are large. To illustrate this fact, return to the study of 148 small businesses in Exercise 12.38.

(a) Find the proportions of failures for businesses headed by women and businesses headed by men. These sample proportions are quite close to each other. Test the hypothesis that the same proportion of women's and men's businesses fail. (Use the two-sided alternative.) The test is very far from being significant.

(b) Now suppose that the same sample proportions came from a sample 30 times as large. That is, 210 out of 1260 businesses headed by women and 450 out of 3180 businesses headed by men fail. Verify that the proportions of failures are exactly the same as in (a). Repeat the z test for the new data, and show that it is now significant at the $\alpha = 0.05$ level.

(c) It is wise to use a confidence interval to estimate the size of an effect, rather than just giving a P-value. Give 95% confidence intervals for the difference between the proportions of women's and men's businesses that fail for the settings of both (a) and (b). What is the effect of larger samples on the confidence interval?

12.40 RISKY BEHAVIOR The National AIDS Behavioral Surveys (Example 12.1) also interviewed a sample of adults in the cities where AIDS is most common. This sample included 803 heterosexuals who reported having more than one sexual partner in the past year. We can consider this an SRS of size 803 from the population of all heterosexuals in high-risk cities who have multiple partners. These people risk infection with the AIDS virus. Yet 304 of the respondents said they never use condoms. Is this strong evidence that more than one-third of this population never use condoms?

12.41 MEN VERSUS WOMEN The National Assessment of Educational Progress (NAEP) Young Adult Literacy Assessment Survey interviewed a random sample of 1917 people 21 to 25 years old. The sample contained 840 men, of whom 775 were fully employed. There were 1077 women, and 680 of them were fully employed.[22]

(a) Use a 99% confidence interval to describe the difference between the proportions of young men and young women who are fully employed. Is the difference statistically significant at the 1% significance level?

(b) The mean and standard deviation of scores on the NAEP's test of quantitative skills were $\bar{x}_1 = 272.40$ and $s_1 = 59.2$ for the men in the sample. For the women, the results were $\bar{x}_2 = 274.73$ and $s_2 = 57.5$. Is the difference between the mean scores for men and women significant at the 1% level?

12.42 A TELEVISION POLL A television news program conducts a call-in poll about a proposed city ban on handgun ownership. Of the 2372 calls, 1921 oppose the ban. The station, following recommended practice, makes a confidence statement: "81% of the Channel 13 Pulse Poll sample opposed the ban. We can be 95% confident that the true proportion of citizens opposing a handgun ban is within 1.6% of the sample result." In this conclusion justified?

12.43 HOW COMMON IS SAT COACHING? A random sample of students who took the SAT college entrance examination twice found that 427 of the respondents had paid for coaching courses and that the remaining 2733 had not. Give a 99% confidence interval for the proportion of coaching among students who retake the SAT.

12.44 HOW TO QUIT SMOKING Nicotine patches are often used to help smokers quit. Does giving medicine to fight depression also help? A randomized double-blind experiment assigned 244 smokers to receive nicotine patches and another 245 to receive both a patch and the antidepressant drug bupropion. Results: After a year, 40 subjects in the nicotine patch group had abstained from smoking, as had 87 in the patch-plus-drug group.[23] Is this good evidence that adding bupropion increases the success rate?

12.45 STATISTICS GO TO COURT *Castaneda v. Partida* is an important court case in which statistical methods were used as part of a legal argument. When reviewing this case, the Supreme Court used the phrase "two or three standard deviations" as a criterion for statistical significance. This Supreme Court review has served as the basis for many subsequent applications of statistical methods in legal settings. (The two or three standard deviations referred to by the Court are values of the z statistic and correspond to P-values of approximately 0.05 and 0.0026.) In *Castaneda* the plaintiffs alleged that the method for selecting juries in a county in Texas was biased against Mexican

Americans. For the period of time at issue, there were 181,535 persons eligible for jury duty, of whom 143,611 were Mexican Americans. Of the 870 people selected for jury duty, 339 were Mexican Americans.

(a) What proportion of eligible voters were Mexican Americans? Let this value be p_0.

(b) Let p be the probability that a randomly selected juror is a Mexican American. The null hypothesis to be tested is H_0: $p = p_0$. Find the value of $\hat{p}$ for this problem, compute the z statistic, and find the P-value. What do you conclude? (A finding of statistical significance in this circumstance does not constitute a proof of discrimination. It can be used, however, to establish a prima facie case. The burden of proof then shifts to the defense.)

(c) We can reformulate this exercise as a two-sample problem. Here we wish to compare the proportion of Mexican Americans among those selected as jurors with the proportion of Mexican Americans among those not selected as jurors. Let p_1 be the probability that a randomly selected juror is a Mexican American, and let p_2 be the probability that a randomly selected nonjuror is a Mexican American. Find the z statistic and its P-value. How do your answers compare with your results in (a)?[24]

NOTES AND DATA SOURCES

1. Data from Joseph H. Catania et al., "Prevalence of AIDS-related risk factors and condom use in the United States," *Science*, 258 (1992), pp. 1101–1106.
2. The study is reported in William Celis III, "Study suggests Head Start helps beyond school," *New York Times*, April 20, 1993.
3. Data from Richard M. Felder et al., "Who gets it and who doesn't: a study of student performance in an introductory chemical engineering course," *1992 ASEE Annual Conference Proceedings*, American Society for Engineering Education, Washington, D.C., 1992, pp. 1516–1519.
4. This quotation is from page 1104 of the article cited in Note 1.
5. "Poll: men, women at odds on sexual equality," Associated Press dispatch appearing in the *Lafayette (Indiana) Journal and Courier*, October 20, 1997.
6. Laurie Goodstein and Marjorie Connelly, "Teen-age poll finds support for tradition," *New York Times*, April 30, 1998.
7. Data provided by Jude M. Werra & Associates, Brookfield, Wis.
8. Data from Guohua Li and Susan P. Baker, "Alcohol in fatally injured bicyclists," *Accident Analysis and Prevention*, 26 (1994), pp. 543–548.
9. Modified from Richard A. Schieber et al., "Risk factors for injuries from in-line skating and the effectiveness of safety gear," *New England Journal of Medicine*, 335 (1996), Internet summary.
10. Clive G. Jones, Richard S. Ostfeld, Michele P. Richard, Eric M. Schauber, and Jerry O. Wolf, "Chain reactions linking acorns to gypsy moth outbreaks and Lyme disease risk," *Science*, 279 (1998), pp. 1023–1026.
11. The Detroit Area Study is described in Gerhard Lenski, *The Religious Factor*, Doubleday, New York, 1961.
12. Data courtesy of Raymond Dumett, Department of History, Purdue University.

13. Martin Enserink, "Fraud and ethics charges hit stroke drug trial," *Science*, 274 (1996), pp. 2004–2005.
14. Donna L. Hoffman and Thomas P. Novak, "Bridging the racial divide on the Internet," *Science*, 280 (1998), pp. 390–391.
15. These are some of the data from the EESEE story "Leave Survey after the Beep." The study is reported in M. Xu, B. J. Bates, and J. C. Schweitzer, "The impact of messages on survey participation in answering machine households," *Public Opinion Quarterly*, 57 (1993), pp. 232–237.
16. Data from Richard M. Felder et al., "Who gets it and who doesn't: a study of student performance in an introductory chemical engineering course," *1992 ASEE Annual Conference Proceedings*, American Society for Engineering Education, Washington, D.C., 1992, pp. 1516–1519.
17. Based on Greg Clinch, "Employee compensation and firms' research and development activity," *Journal of Accounting Research*, 29 (1991), pp. 59–78.
18. David M. Blau, "The child care labor market," *Journal of Human Resources*, 27 (1992), pp. 9–39.
19. These are part of the data form the EESEE story "Radar Detectors and Speeding." The study is reported in N. Teed, K. L. Adrian, and R. Knoblouch, "The duration of speed reductions attributable to radar detectors," *Accident Analysis and Prevention*, 25 (1991), pp. 131–137.
20. National Athletic Trainers Association, press release dated September 30, 1994.
21. Arne L. Kalleberg and Kevin T. Leicht, "Gender and organizational performance: determinants of small business survival and success," *Academy of Management Journal*, 34 (1991), pp. 136–161.
22. Francisco L. Rivera-Batiz, "Quantitative literacy and the likelihood of employment among young adults," *Journal of Human Resources*, 27 (1992), pp. 313–328.
23. Douglass E. Jorenby et al., "A controlled trial of sustained-release bupropion, a nicotine patch, or both for smoking cessation," *New England Journal of Medicine*, 340 (1999), pp. 685–691.
24. For a further discussion of this case, see D. H. Kaye and M. Aickin (eds.), *Statistical Methods in Discrimination Litigation*, Marcel Dekker, New York, 1986.

University College, London

KARL PEARSON

The First Inference Procedure

Karl Pearson (1857–1936), a professor at University College in London, had already published nine books before he turned his abundant energy to statistics in 1893. Of course, Pearson didn't really take up statistics, which was not yet a separate field of study. He took up problems of heredity and evolution, which led him into statistics.

Pearson developed a family of curves—we would call them density curves—for describing biological data that don't follow a normal distribution. He then asked how he could test whether one of these curves actually fit a set of data well. In 1900 he invented a method, the chi-square test. Pearson's chi-square test has the honor of being the oldest inference procedure still in use. It is now most often used for problems somewhat different from the one that motivated Pearson, as we will see in this chapter.

After Pearson, statistics was a field of study. Fisher and Neyman in the 1920s and 1930s would provide much of its present form, but here is what the leading historian of statistics says about the origins: "Before 1900 we see many scientists of different fields developing and using techniques we now recognize as belonging to modern statistics. After 1900 we begin to see identifiable statisticians developing such techniques into a unified logic of empirical science that goes far beyond its component parts. There was no sharp moment of birth; but with Pearson and Yule and the growing numbers of students in Pearson's laboratory, the infant discipline may be said to have arrived."[1]

After Pearson, statistics was a field of study.

chapter 13

Inference for Tables: Chi-Square Procedures

ACTIVITY 13 "I Didn't Get Enough Blues!"

Materials needed: One 1.69-ounce bag of plain M&M's per student.

The M&M/Mars Company, headquartered in Hackettstown, New Jersey, makes plain and peanut chocolate candies. In 1995, they decided to replace the tan-colored M&M's with a new color. After conducting an extensive national preference survey, they decided to replace the tan M&M's with blue M&M's. The company's Consumer Affairs Department announced:

> *On average, the new mix of colors of M&M's Plain Chocolate Candies will contain 30 percent browns, 20 percent each of yellows and reds and 10 percent each of oranges, greens, and blues.*

They explained:

> *While we mix the colors as thoroughly as possible, the above ratios may vary somewhat, especially in the smaller bags. This is because we combine the various colors in large quantities for the last production stage (printing). The bags are then filled on high-speed packaging machines by weight, not by count.*

The purpose of this activity is to compare the color distribution of M&M's in your individual bag with the advertised distribution. We will want to see if there is sufficient evidence to dispute the company's claim for their distribution. In order to use as random a sample as possible, it is best if the bags of M&M's are purchased at different stores and not obtained from one or a few sources of supply.

1. Open your bag and carefully count the number of M&M's of each color—brown, yellow, red, orange, green, and blue—as well as the total number of M&M's in the bag.

2. Fill in the counts, by color, and the total number of M&M's for your bag in the "Observed" row in a table like this:

Color	Brown	Yellow	Red	Orange	Green	Blue	Total
Observed							
Expected							
$(O - E)^2/E$							χ^2

3. To obtain the expected counts, multiply the total number of M&M's in your bag by the company's stated percentages (expressed in decimal form) for each of the colors.

4. For each color, perform this calculation:

$$(\text{observed} - \text{expected})^2 / \text{expected}$$

ACTIVITY 13 "I Didn't Get Enough Blues!" *(continued)*

and enter the result in the last row of the table. Then add up all of these calculated values, and name the sum X^2. Keep this number handy—you will use it later in the chapter.

5. If your sample reflects the distribution advertised by the M&M/Mars Company, then there should be very little difference between the observed counts and the expected counts. Hence the calculated values making up the sum X^2 should be very small. Are the entries in the last row all about the same, or do any of the quantities stand out because they are "significantly" larger? Did you get more of a particular color than you expected? Did you get fewer of a particular color than you expected?

6. Combine the counts obtained by all the students in your class to obtain a total count of M&M's of each color. Record the results in a table like this:

Color	Brown	Yellow	Red	Orange	Green	Blue	Total
Observed							

You will need these data in the exercises.

7. Record the total number of M&M's in each student's bag in your class. How did your bag compare with those of your classmates?

INTRODUCTION

In the previous chapter, we discussed inference procedures for comparing two population proportions. Sometimes we want to examine the distribution of proportions in a single population. The *chi-square test for goodness of fit* allows us to determine whether a specified population distribution seems valid. We can compare two or more population proportions using a *chi-square test for homogeneity of populations.* In doing so, we will organize our data in a two-way table. It is also possible to use the information provided in a two-way table to determine whether the distribution of one variable has been influenced by another variable. The *chi-square test of association/independence* helps us decide this issue.

The methods of this chapter help us answer questions such as these:

- When geneticists predict the results of mating two red-eyed fruit flies, how do we use the actual offspring to evaluate the accuracy of their predicted model?
- Can an antidepressant be used to treat cocaine addiction? Researchers randomly assigned 24 cocaine addicts to each of three treatment groups in a

controlled experiment. One group was given a standard drug for treating cocaine addiction, another group took an antidepressant, and the third group received a placebo. How do the proportions of subjects who relapsed into cocaine use compare across the three treatment groups?

• Does smoking behavior vary according to the socioeconomic status (SES) of adults? To find out, researchers classify several hundred men according to their SES (high, medium, low) and also according to their smoking behavior (current smoker, former smoker, never smoked). Is there a significant relationship between SES and smoking, and if so, how can we describe it?

13.1 TEST FOR GOODNESS OF FIT

Suppose you open a 1.69-ounce bag of plain M&M chocolate candies and discover that out of 56 M&M's, there is only a single *blue* M&M. Knowing that 10% of all plain chocolate M&M's made by the M&M/Mars Company are blue, and that in your sample of size 56, the proportion of blue M&M's is only 1/56 = 0.018, you might feel that you didn't get your fair share of blues. You could use the z test described in the last chapter to test the hypotheses

$$H_0: p = 0.10$$
$$H_a: p < 0.10$$

where p is the proportion of blue M&M's. You could then perform additional tests of significance for each of the remaining colors. But this would be inefficient. More important, it wouldn't tell us how likely it is that six sample proportions differ from the values stated by M&M/Mars as much as our sample does. There is a single test that can be applied to see if the observed sample distribution is significantly different from the hypothesized population distribution. It is called the ***chi-square (χ^2) test for goodness of fit.***

chi-square (χ^2) test for goodness of fit

EXAMPLE 13.1 THE GRAYING OF AMERICA

In recent years, the expression "the graying of America" has been used to refer to the belief that with better medicine and healthier lifestyles, people are living longer, and consequently a larger percentage of the population is of retirement age. We want to investigate whether this perception is accurate. The distribution of the U.S. population in 1980 is shown in Table 13.1. We want to determine if the distribution of age groups in the United States in 1996 has changed significantly from the 1980 distribution. We will test the following hypothesis:

H_0: the age group distribution in 1996 is the *same as* the 1980 distribution

H_a: the age group distribution in 1996 is *different from* the 1980 distribution

TABLE 13.1 **U.S. population by age group, 1980**

Age group	Population (in thousands)	Percent
0 to 24	93,777	41.39
25 to 44	62,716	27.68
45 to 64	44,503	19.64
65 and older	25,550	11.28
Total	226,546	100.00

Source: *Statistical Abstract of the United States*, 1997, U.S. Department of Commerce, Bureau of the Census.

We can also state the hypotheses in terms of the proportion of the U.S. population that falls in each age group.

H_0: $p_{0-24} = 0.4139$, $p_{25-44} = 0.2768$, $p_{45-64} = 0.1964$, $p_{65+} = 0.1128$
H_a: at least one of the proportions differs from the stated values

The idea of the test is this: We compare the observed counts for a sample from the 1996 population with the counts that would be expected if the 1996 distribution were the same as the 1980 distribution, that is, if H_0 were in fact true. The 1980 distribution is the *population*. The more the observed counts differ from the expected counts, the more evidence we have to reject H_0 and to conclude that the population distribution in 1996 is significantly different from that of 1980.

A random *sample* of 500 U.S. residents in 1996 is selected and the age of each subject is recorded. The counts and percents in each age-group category are shown in Table 13.2.

TABLE 13.2 **Sample results for 500 randomly selected individuals in 1996**

Age group	Count	Percent
0 to 24	177	35.4
25 to 44	158	31.6
45 to 64	101	20.2
65 and older	64	12.8
	500	100

Before proceeding with a significance test, it's always a good idea to plot the data. In this case, a segmented bar graph allows you to compare segments of the population from 1980 to 1996. The finished graph is shown in Figure 13.1.

The next step in the test is to calculate the expected counts for each age category. If the age group distribution seen in 1980 has not changed, then in a random sample of 500 U.S. residents in 1996, we would expect 41.39% of the 500 to be in the 0 to 24 age category, 27.68% to be in the 25 to 44 age category, and so forth. For each of the categories, we would expect the appropriate percentage (from Table 13.1) of the 500 to be in the corresponding age category. The expected counts are displayed in Table 13.3.

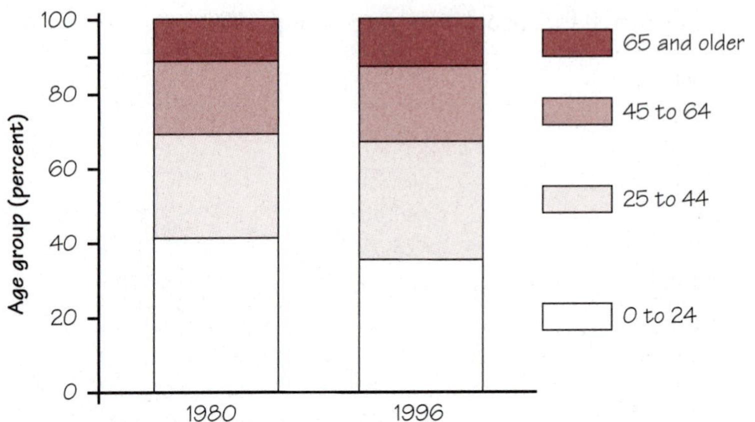

FIGURE 13.1 Segmented bar graph for the data in Tables 13.1 and 13.2.

TABLE 13.3 Expected counts

Age group	1980 population percents	Expected counts (1996)
0 to 24	41.39	500(0.4139) = 207.0
25 to 44	27.68	500(0.2768) = 138.4
45 to 64	19.64	500(0.1964) = 98.2
65 and older	11.28	500(0.1128) = 56.4
	100	500

In order to determine whether the distribution has changed since 1980, we need a way to measure how well the observed counts (O) from 1996 fit the expected counts (E) under H_0. The procedure is to calculate the quantity

$$\frac{(O-E)^2}{E}$$

chi-square statistic

for each age category and then add up these terms. The sum is labeled X^2 and is called the ***chi-square statistic.*** A summary of the calculations is shown in Table 13.4. The sum of the terms in the last column is $X^2 = 8.2275$. The larger the differences between the observed and expected values, the larger X^2 will be, and the more evidence there will be against H_0.

TABLE 13.4 Calculating the goodness of fit

Age group	Observed O	Expected E	$(O-E)^2/E$
0 to 24	177	207.0	4.3478
25 to 44	158	138.4	2.7757
45 to 64	101	98.2	0.0798
65 and older	64	56.4	1.0241
			$X^2 = 8.2275$

The χ^2 family of distribution curves is used to assess the evidence against H_0 represented in the value of X^2. The member of the family that is used is determined by the ***degrees of freedom.*** Since we are working with percentages, three of the four percentages are free to vary, but the fourth is not, since all four have to add to 100. In this case, we say that there are $4 - 1 = 3$ degrees of freedom. In the back of the book, Table E, Chi-Square Distribution Critical Values, shows a typical chi-square curve with the right-tail area shaded. The chi-square test statistic is a point on the horizontal axis, and the area to the right is the *P*-value of the test. This *P*-value is the probability of observing a value X^2 at least as extreme as the one actually observed. The larger the value of the chi-square statistic, the smaller the *P*-value, and the more evidence you have against the null hypothesis, H_0.

degrees of freedom

In Table E, for a *P*-value of 0.05 and degrees of freedom = 3, we find that the critical value is 7.81. Since our $X^2 = 8.2275$ is more extreme (larger) than the critical value, we say that the probability of observing a result as extreme as the one we actually observed, by chance alone, is less than 5%. There is sufficient evidence to reject H_0 and conclude that the population distribution in 1996 is significantly different from the 1980 distribution, at the 5% significance level.

Properties of the chi-square distributions

Example 13.1 illustrated the mechanics of the chi-square goodness of fit test. Now we turn our attention to the chi-square distributions.

THE CHI-SQUARE DISTRIBUTIONS

The **chi-square distributions** are a family of distributions that take only positive values and are skewed to the right. A specific chi-square distribution is specified by one parameter, called the **degrees of freedom.**

Figure 13.2 shows the density curves for three members of the chi-square family of distributions. As the degrees of freedom increase, the density curves become less skewed and larger values become more probable. Table E in the back of the book gives critical values for chi-square distributions. You can use Table E if software does not give you *P*-values for a chi-square test.

The chi-square density curves have the following properties:

1. The total area under a chi-square curve is equal to 1.

2. Each chi-square curve (except when df = 1) begins at 0 on the horizontal axis, increases to a peak, and then approaches the horizontal axis asymptotically from above.

3. Each chi-square curve is skewed to the right. As the number of degrees of freedom increase, the curve becomes more and more symmetrical and looks more like a normal curve.

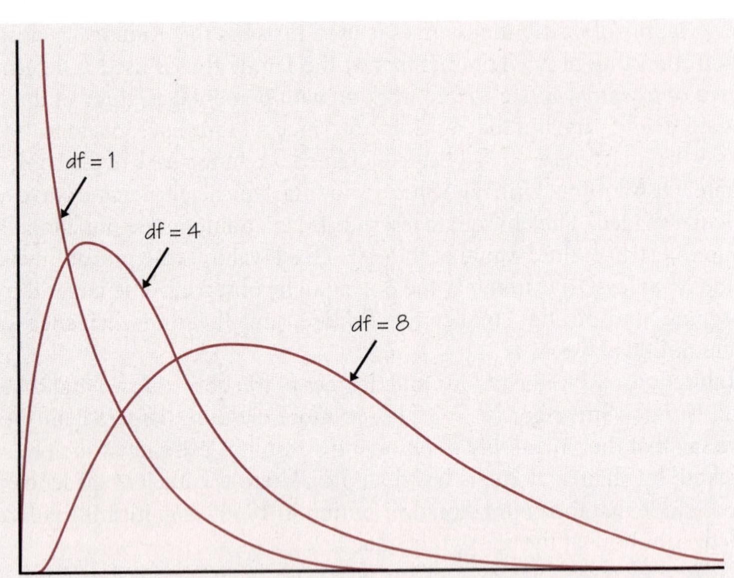

FIGURE 13.2 Density curves for the chi-square distributions with 1, 4, and 8 degrees of freedom. Chi-square distributions take only positive values.

We use the chi-square density curve with $n - 1$ degrees of freedom to calculate the *P*-value in a goodness of fit test. The following box summarizes the details.

GOODNESS OF FIT TEST

A **goodness of fit test** is used to help determine whether a population has a certain hypothesized distribution, expressed as proportions of population members falling into various outcome categories. Suppose that the hypothesized distribution has n outcome categories. To test the hypothesis

H_0: the actual population proportions are *equal to* the hypothesized proportions

first calculate the chi-square test statistic

$$X^2 = \Sigma(O - E)^2 / E$$

Then X^2 has approximately a χ^2 distribution with $(n - 1)$ degrees of freedom.

For a test of H_0 against the alternative hypothesis

H_a: the actual population proportions are *different from* the hypothesized proportions

the *P*-value is $P(\chi^2 \geq X^2)$.

GOODNESS OF FIT TEST *(continued)*

Conditions: You may use this test with critical values from the chi-square distribution when all individual expected counts are at least 1 and no more than 20% of the expected counts are less than 5.

One of the most common applications of the chi-square goodness of fit test is in the field of genetics. Scientists want to investigate the genetic characteristics of offspring that result from mating (also called "crossing") parents with known genetic makeups. They use rules about dominant and recessive genes to predict the ratio of offspring that will fall in each possible genetic category. Then, the scientists mate the parents and classify the resulting offspring. The chi-square goodness of fit test helps the scientists assess the validity of their hypothesized ratios.

EXAMPLE 13.2 RED-EYED FRUIT FLIES

Biologists wish to mate two fruit flies having genetic makeup RrCc, indicating that it has one dominant gene (R) and one recessive gene (r) for eye color, along with one dominant (C) and one recessive (c) gene for wing type. Each offspring will receive one gene for each of the two traits from both parents. The following table, often called a Punnett square, shows the possible combinations of genes received by the offspring.

		Parent 2 passes on			
		RC	Rc	rC	rc
Parent 1 passes on	RC	RRCC (x)	RRCc (x)	RrCC (x)	RrCc (x)
	Rc	RRCc (x)	RRcc (y)	RrCc (x)	Rrcc (y)
	rC	RrCC (x)	RrCc (x)	rrCC (z)	rrCc (z)
	rc	RrCc (x)	Rrcc (y)	rrCc (z)	rrcc (w)

Any offspring receiving an R gene will have red eyes, and any offspring receiving a C gene will have straight wings. So based on this Punnett square, the biologists predict a ratio of 9 red-eyed, straight-wing (x):3 red-eyed, curly wing (y):3 white-eyed, straight (z):1 white-eyed, curly (w) offspring. In order to test their hypothesis about the distribution of offspring, the biologists mate the fruit flies. Of 200 offspring, 101 had red eyes and straight wings, 42 had red eyes and curly wings, 49 had white eyes and straight wings, and 10 had white eyes and curly wings. Do these data differ significantly from what the biologists have predicted?

We return to the familiar structure of the Inference Toolbox to carry out the significance test.

Step 1: *Identify the population of interest and the parameter(s) that you want to draw conclusions about. State hypotheses in words and symbols.* The biologists are interested in the proportion of offspring that fall into each genetic category for the population of all fruit flies that would result from crossing two parents with genetic makeup RrCc. Their hypotheses are

H_0: $p_{red,straight} = 0.5625$, $p_{red,curly} = 0.1875$, $p_{white,straight} = 0.1875$, $p_{white,curly} = 0.0625$
H_a: at least one of these proportions is incorrect

Step 2: Choose the appropriate inference procedure and verify the conditions for using it. We can use a chi-square goodness of fit test to measure the strength of the evidence against the hypothesized distribution, provided that the expected cell counts are large enough. Here are the expected counts:

Red-eyed, straight-wing:	200(0.5625) = 112.5
Red-eyed, curly-wing:	200(0.1875) = 37.5
White-eyed, straight-wing:	200(0.1875) = 37.5
White-eyed, curly-wing:	200(0.0625) = 12.5

Since all the expected cell counts are greater than 5, we can proceed with the test.

Step 3: Carry out the inference procedure:

- The test statistic is

$$X^2 = \sum \frac{(O-E)^2}{E} = \frac{(101-112.5)^2}{112.5} + \frac{(42-37.5)^2}{37.5} + \frac{(49-37.5)^2}{37.5} + \frac{(10-12.5)^2}{12.5}$$
$$= 1.1756 + 0.54 + 3.5267 + 0.5 = 5.742$$

- For df = 4 – 1 = 3, Table E shows that our test statistic falls between the critical values for a 0.15 and a 0.10 significance level. Technology produces the actual *P*-value of 0.1248.

Step 4: Interpret your results in the context of the problem. The *P*-value of 0.1248 indicates that the probability of obtaining a sample of 200 fruit fly offspring in which the proportions differ from the hypothesized values by at least as much as the ones in our sample is over 12%, assuming that the null hypothesis is true. This is not sufficient evidence to reject the biologists' predicted distribution.

You can simplify the computations of Example 13.2 and other goodness of fit problems by using your calculator's list operations. The calculator also allows you to compute and visualize the *P*-value for a chi-square test.

TECHNOLOGY TOOLBOX *Goodness of fit tests on a calculator*

- Clear lists L_1/list1, L_2/list2, and L_3/list3.
- Enter the observed counts in L_1/list1. Calculate the expected counts separately and enter them in L_2/list2.

TECHNOLOGY TOOLBOX *Goodness of fit tests on a calculator (continued)*

TI-83

- Define L_3 as $(L_1 - L_2)^2 / L_2$.

L1	L2	L3 3
101	112.5	-----
42	37.5	
49	37.5	
10	12.5	
-----	-----	

L3 = (L1−L2)2/L2

- Use the command `sum(L₃)` to calculate the test statistic X^2. (sum is located in MATH/LIST.)

TI-89

- Define list3 as $[(\text{list1} - \text{list2})^2 / \text{list2}]$.

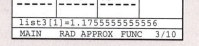

F1 Tools | F2 Plots | F3 List | F4 Calc | F5 Distr | F6 Tests | F7 Ints

list1	list2	list3	list4
101.	112.5	1.1756	-----
42.	37.5	.54	
49.	37.5	3.5267	
10.	12.5	.5	
-----	-----	-----	

list3[1]=1.1755555555556

MAIN RAD APPROX FUNC 3/10

- Use the command `sum(list3)` to calculate X^2. (sum is located in the CATALOG.)

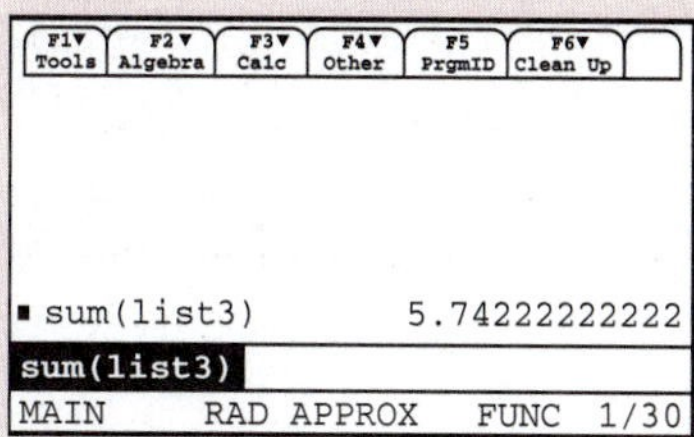

- Find the *P*-value using the χ^2 `cdf` command (in the distributions (DISTR) menu on the TI-83 and in the CATALOG under Flash Apps on the TI-89).

- We ask for the area between $X^2 = 5.742$ and a very large number (1E99) and specify the degrees of freedom, as shown.

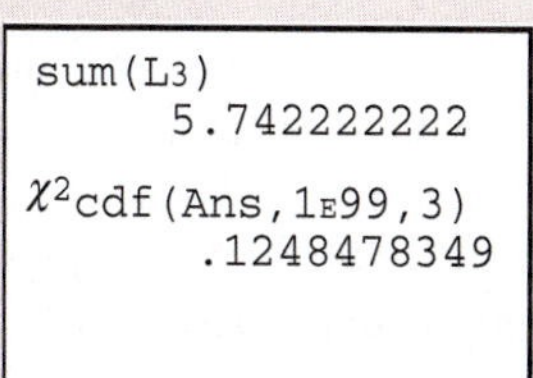

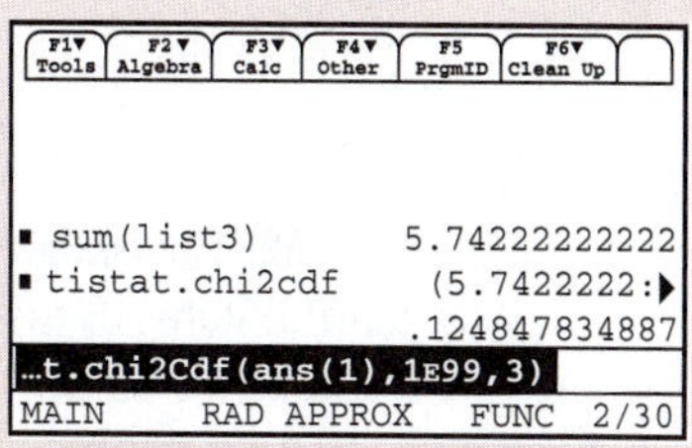

The *P*-value, 0.1248, indicates that $X^2 = 5.742$ is a possible result if H_0 is true, which does not provide strong enough evidence to reject H_0.

- To visualize this *P*-value, first set your viewing window as shown. Then enter the command `Shade` χ^2 `(5.742,1E99,3)`. (This command is located in the DISTR/DRAW menu on the TI-83 and in the CATALOG under Flash Apps on the TI-89.) Note that the `Shade` χ^2 command requires a left endpoint and a right endpoint, so take a sufficiently large right endpoint to achieve four-decimal-place accuracy.

TECHNOLOGY TOOLBOX *Goodness of fit tests on a calculator (continued)*

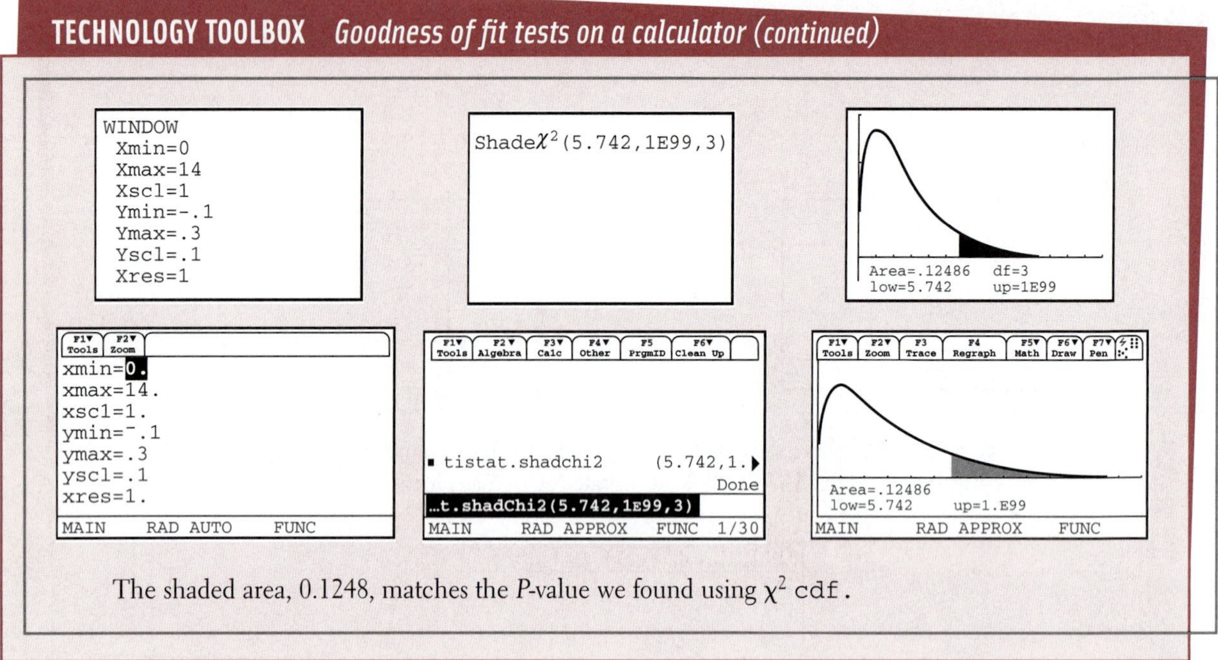

The shaded area, 0.1248, matches the *P*-value we found using χ^2 `cdf`.

EXERCISES

13.1 FINDING *P*-VALUES

(a) Find the *P*-value corresponding to $X^2 = 1.41$ for a chi-square distribution with 1 degree of freedom: (i) using Table E and (ii) with your graphing calculator.

(b) Find the area to the right of $X^2 = 19.62$ under the chi-square curve with 9 degrees of freedom: (i) using Table E and (ii) with your graphing calculator.

(c) Find the *P*-value corresponding to $X^2 = 7.04$ for a chi-square distribution with 6 degrees of freedom: (i) using Table E and (ii) with your graphing calculator.

13.2 ARE YOU MARRIED? According to the March 2000 Current Population Survey, the marital status distribution of the U.S. adult population is as follows:

Marital status:	Never married	Married	Widowed	Divorced
Percent:	28.1	56.3	6.4	9.2

A random sample of 500 U.S. males, aged 25 to 29 years old, yielded the following frequency distribution:

Marital status:	Never married	Married	Widowed	Divorced
Frequency:	260	220	0	20

Perform a goodness of fit test to determine if the marital status distribution of U.S. males 25 to 29 years old differs from that of the U.S. adult population. Use the Inference Toolbox.

13.3 GENETICS: CROSSING TOBACCO PLANTS Researchers want to cross two yellow-green tobacco plants with genetic makeup (Gg). Here is a Punnett square for this genetic experiment:

	G	g
G	GG	Gg
g	Gg	gg

This shows that the expected ratio of green (GG) to yellow-green (Gg) to albino (gg) tobacco plants is 1:2:1. When the researchers perform the experiment, the resulting offspring are 22 green, 50 yellow-green, and 12 albino seedlings. Use a chi-square goodness of fit test to assess the validity of the researchers' genetic model.

13.4 In recent years, a national effort has been made to enable more members of minority groups to have increased educational opportunities. You want to know if the policy of "affirmative action" and similar initiatives have had any effect in this regard. You obtain information on the ethnicity distribution of holders of the highest academic degree, the doctor of philosophy degree, for the year 1981:

Race/ethnicity	Percent
White, non-Hispanic	78.9
Black, non-Hispanic	3.9
Hispanic	1.4
Asian or Pacific Islander	2.7
American Indian/Alaskan Native	0.4
Nonresident alien	12.8

A random sample of 300 doctoral degree recipients in 1994 showed the following frequency distribution:

Race/ethnicity	Count
White, non-Hispanic	189
Black, non-Hispanic	10
Hispanic	6
Asian or Pacific Islander	14
American Indian/Alaskan Native	1
Nonresident alien	80

(a) Perform a goodness of fit test to determine if the distribution of doctoral degrees in 1994 is significantly different from the distribution in 1981. Don't forget to state your hypotheses, and don't forget to state your conclusion.

(b) In which categories have the greatest changes occurred, and in what direction?

13.5 M&M'S ACTIVITY Use the class M&M's data that you recorded in steps 6 and 7 of Activity 13 to answer the following questions. Consider the entire count of M&M's in the class as one large sample from the production process.

(a) Do these data give you reason to doubt the color distribution of M&M's candies advertised by the M&M/Mars Company? Give appropriate statistical evidence to support your conclusion.

(b) For which color of M&M does your sample proportion differ the most from the proportion claimed by the company? Use an appropriate statistical procedure to determine whether this difference is statistically significant.

(c) Can you use the data you collected in step 7 to construct a confidence interval for the mean number of M&M's in the population of all 1.69-ounce bags produced by M&M/Mars? If so, do it. If not, explain why not.

TECHNOLOGY TOOLBOX *Graphing a chi-square distribution*

You can use your calculator to graph chi-square distributions, like those pictured in Figure 13.2 (page 732). Here's how.

- Adjust your WINDOW settings as shown.

```
WINDOW
 Xmin=0
 Xmax=14
 Xscl=1
 Ymin=-.05
 Ymax=.3
 Yscl=.1
 Xres=1
```

- Enter Functions Y_1, Y_2, and Y_3 as illustrated below. χ^2pdf can be found in the DISTR menu (2nd VARS) on the TI-83, and in the CATALOG under Flash Apps on the TI-89.
- Change the graph style on Y_2 to a thick line.

TI-83

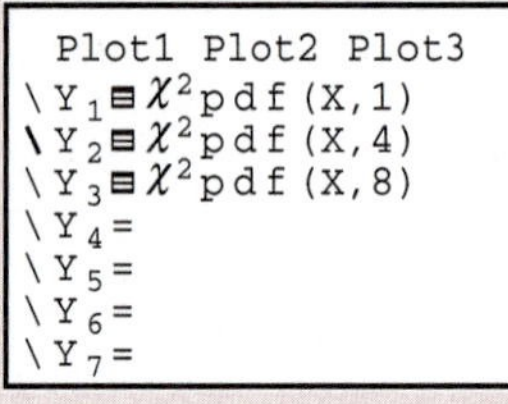

TI-89

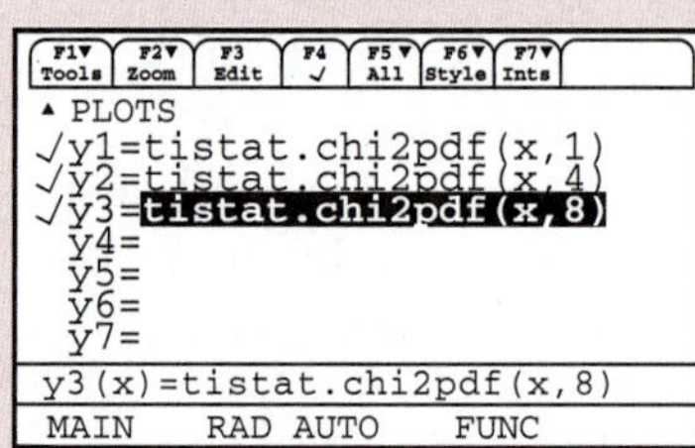

- Graph the chi-square density curves for df = 1, 4, and 8 on the same axes.

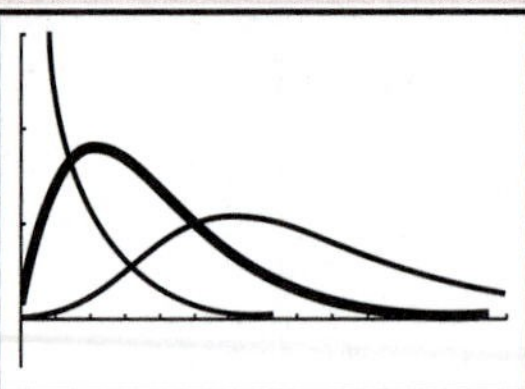

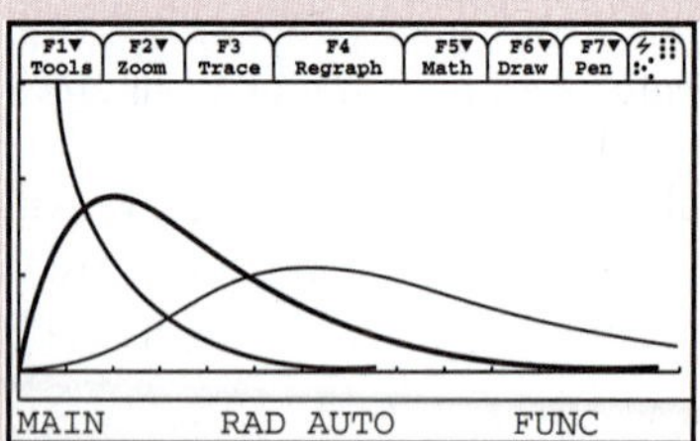

Conducting inference by simulation

Let's return to the "graying of America" problem. Suppose that we didn't have the resources to select a representative sample from the current population of the United States or that there is some other reason why we can't gather the actual data we need. We still may be able to obtain an approximate solution by means of *simulation*.

TABLE 13.5 U.S. population distribution for 1996

Age group	Population percent
0 to 24	35.4
25 to 44	31.6
45 to 64	20.2
65 and older	12.8
	100

EXAMPLE 13.3 THE GRAYING OF AMERICA, CONTINUED

In the population study of Example 13.1, we can use recent census figures to obtain percents for the age categories for the year 1996. The relative frequency distribution in Table 13.5 is calculated from data in the *Statistical Abstract of the United States, 1997*.

Step 1: Establish a correspondence between random numbers and ages. To simplify matters we will round off the percents from Table 13.5 to whole numbers: 35%, 32%, 20%, and 13%, respectively. One possible scheme is as follows. Let a number from 1 to 100 represent a randomly selected person from the U.S. population.

Let the numbers

1 to 35 (35% of randomly generated numbers) represent persons 0 to 24 years of age.

36 to 67 (32% of randomly generated numbers) represent persons 25 to 44 years of age.

68 to 87 (20% of randomly generated numbers) represent persons 45 to 64 years of age.

88 to 100 (13% of randomly generated numbers) represent persons 65 years or older.

Step 2: Determine a sample size n (typically a number between 100 and 500). If the simulation is done on a calculator, your sample size will be limited by the available memory in the calculator.

Step 3: Randomly generate n numbers in the range 1 to 100, and count the numbers that fall into each of the four age categories. The calculator program POP carries out this plan.

TI-83	TI-89

TI-83

```
PROGRAM:POP
ClrHome
ClrList L1,L2
Disp "HOW MANY TRIALS"
Prompt N
randInt(1,100,N)→L1
sum(L1≥1 and L1≤35)→L2(1)
sum(L1≥36 and L1≤67)→L2(2)
sum(L1≥68 and L1≤87)→L2(3)
sum(L1≥88 and L1≤100)→L2(4)
Disp " "
Disp "OBSERVED COUNTS"
Disp "ARE IN L2"
```

TI-89

```
pop()
()
Prgm
ClrHome
tistat.clrlist(list1,list2)
Disp "how many trials"
Prompt n
tistat.randint(1,100,n)→list1
0→b:0→c:0→d:0→f
For x,1,n,1
If list1[x]≥1 and
list1[x]≤35:b+1→b
If list1[x]≥36 and
list1[x]≤67:c+1→c
If list1[x]≥68 and
list1[x]≤87:d+1→d
If list1[x]≥88 and
list1[x]≤100:f+1→f
EndFor
b→list2[1]:c→list2[2]:d→list2[
3]:f→list2[4]
Disp " "
Disp "observed counts"
Disp "are in list3"
EndPrgm
```

The calculator randomly generates a specified number of ages and stores them in L_1/list1. It then tabulates the number of ages that fall in each of the age categories. When the simulation is finished, the four numbers in L_2/list2 are the simulated counts of people in each age category.

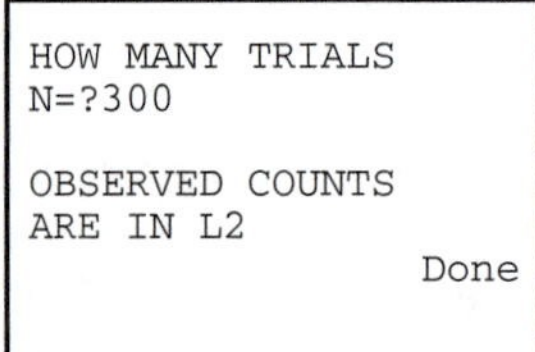

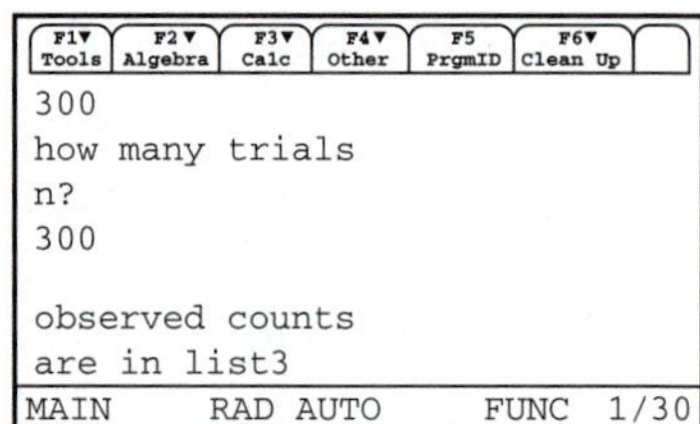

Here are the results for a sample run of the program with a specified sample size of $n = 300$.

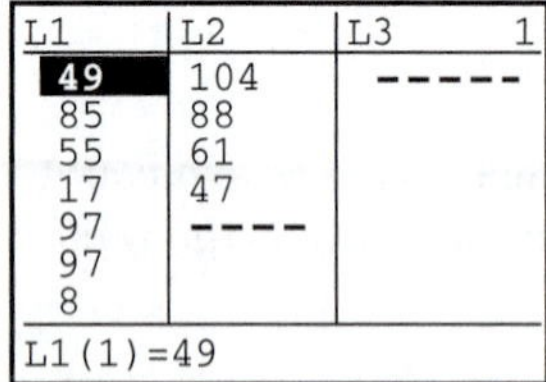

L1
L2
L3
1
49 85 55 17 97 97 8
104 88 61 47
L1(1)=49

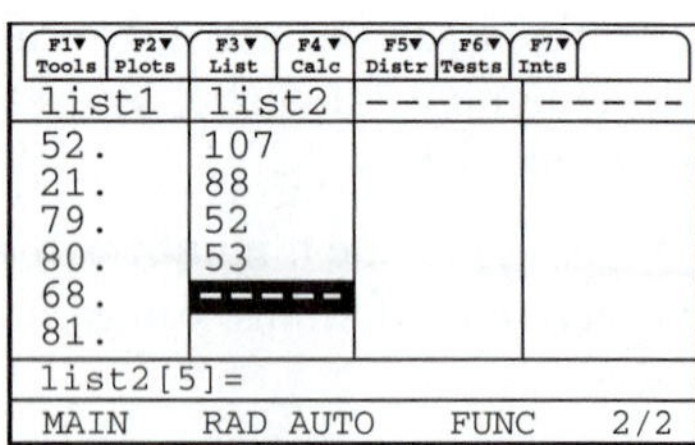

Step 4: Using the results of the simulated counts in each age category, perform a chi-square goodness of fit test. Enter the expected counts (300 times the 1980 decimals) in L_3/list3. Define L_4/list4 as $(L_2 - L_3)^2 / L_3$ ((list2 – list 3)^2 / list3).

L2	L3	L4 4
104	124.17	3.2764
88	83.04	.29626
61	58.92	.07343
47	33.84	5.1178
---	----	-----

L4(1)=3.276386405...

F1 Tools | F2 Plots | F3 List | F4 Calc | F5 Distr | F6 Tests | F7 Ints

list1	list2	list3	list4
52.	107	124.17	2.7552
21.	88	83.04	.27956
79.	52	58.92	.92089
80.	53	33.84	6.9265
68.	-----	-----	-----
81.			

list4="(list2-list3)^2/1:1i...

MAIN RAD AUTO FUNC 4/4

To obtain the value of X_2, sum the four terms (observed – expected)^2 / expected in L_4 (list4). Then clear the graphics screen (ClrDraw) and ask for a sketch of the χ^2 curve with the area to the right of X^2 shaded.

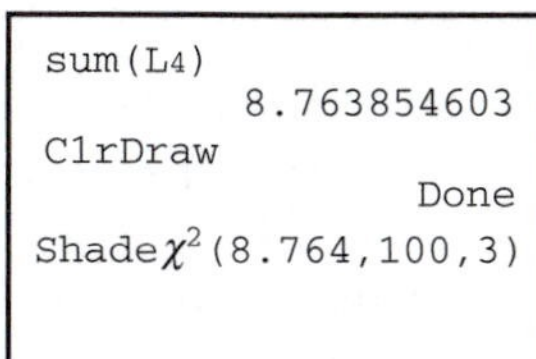

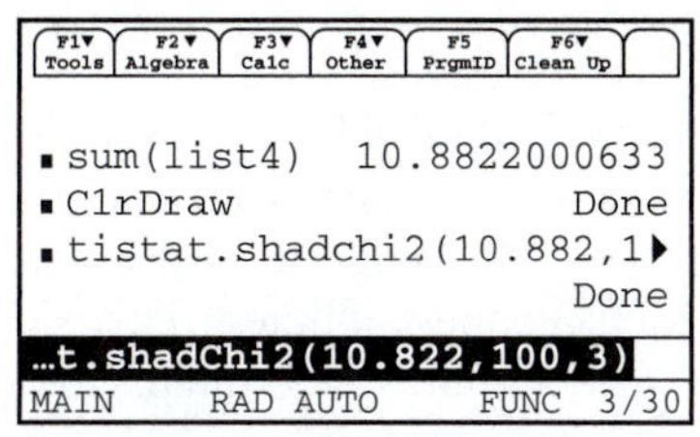

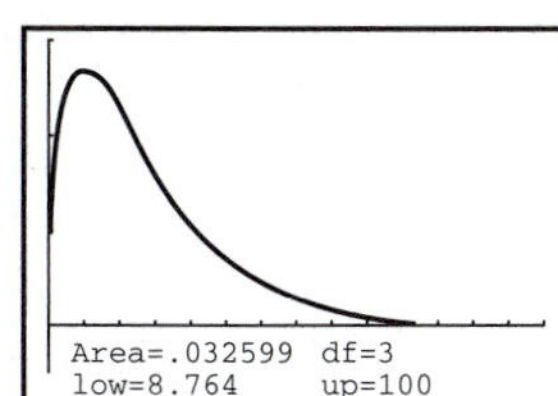

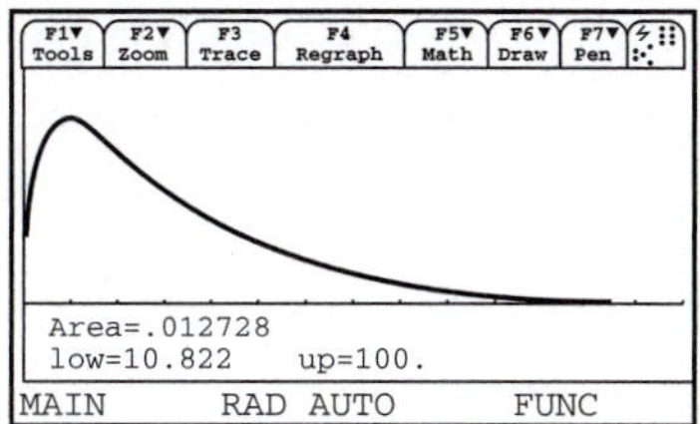

The results show that for this particular simulation, with sample size 300, the TI-83 yielded $X^2 = 8.764$ with a corresponding *P*-value of 0.0326. For the TI-89 simulation, $X^2 = 10.822$ and the *P*-value is an even lower 0.0127. Note that if you replicate this simulation on your calculator, especially with a smaller sample size (say $n = 100$), you may get very different results. Remember that the X^2 statistic will show much greater variability with smaller sample sizes. For this reason, if you simulate a sampling problem like this with a calculator, you should use as large a sample size as you can without getting a memory overflow error message.

Follow-up analysis

Do our results show the "graying of America" phenomenon? Only somewhat; the big story appears elsewhere. The sample in this example was calculated to accurately reflect the 1996 age group distribution, so a meaningful comparison with the 1980 distribution is possible. If you inspect the four terms, $(O - E)^2/E$, that are added together to give the test statistic X^2 in Example 13.1, the largest contribution to X^2 was from the first age category (see Table 13.4). The observed count for the 65 and over category was only slightly more than the expected count. So the greatest change in the distribution was in the 0 to 24 age category, where the observed (1996) population was *smaller* than the expected population size. The birthrate from 1972 until about 1987 was fairly stable at about 15.8 per 1000 residents. It increased from 1987 to 1990 and then began a steady decline. Here are the birthrates from 1987 through 1996:[2]

Year:	1987	1988	1989	1990	1991	1992	1993	1994	1995	1996
Birthrate:	15.7	16.0	16.4	16.6	16.3	15.9	15.5	15.2	14.8	14.5

Even though there is evidence that the distribution of ages has changed significantly from 1980 to 1996, one must look at the individual components of X^2 to see where the largest changes have occurred.

EXERCISES

13.6 POPULATION SIMULATION This exercise is a continuation of Example 13.3. Download the POP program to your calculator, and then execute the program. Specify a sample of size 100. While the calculator is working, specify null and alternative hypotheses for a goodness of fit test. Then complete your analysis of the results of your simulation, and determine the *P*-value. How do the results of your simulation compare with the results of this section?

13.7 IS YOUR RANDOM NUMBER GENERATOR WORKING? In this exercise you will use your calculator to simulate sampling from the following uniform distribution:

X:	0	1	2	3	4	5	6	7	8	9
P(X):	0.1	0.1	0.1	0.1	0.1	0.1	0.1	0.1	0.1	0.1

You will then perform a goodness of fit test to see if a randomly generated sample distribution comes from a population that is different from this distribution.

(a) State your null and alternative hypotheses for this test.

(b) Use the `randInt` function to randomly generate 200 digits from 0 to 9, and store these values in L_4/list 4.

(c) Plot the data as a histogram with Window dimensions set as follows: $X[-.5, 9.5]_1$ and $Y[-5,30]_5$. (You may have to increase the vertical scale.) Then TRACE to see the frequencies of each digit. Record these frequencies (observed values) in L_1/list 1.

(d) Determine the expected counts for a sample size of 200, and store them in L_2/list 2.

(e) Complete a goodness of fit test. Report your chi-square statistic, the *P*-value, and your conclusion with regard to the null and alternative hypotheses.

13.8 ROLL THE DICE Simulate rolling a fair, six-sided die 300 times on your calculator. Plot a histogram of the results, and then perform a goodness of fit test of the hypothesis that the die is fair.

13.9 IS THIS COIN FAIR? A statistics student suspected that his 1982 penny was not a fair coin, so he held it upright on a table top with a finger of one hand and spun the penny repeatedly by flicking it with the index finger of other hand. In 200 spins of the coin, it landed with tails side up 122 times.

(a) Perform a goodness of fit test to see if there is sufficient evidence to conclude that spinning the coin does not produce an equal proportion of heads and tails.

(b) Use a one-proportion inference procedure to determine whether spinning the coin is equally likely to result in heads or tails.

(c) Compare your results for parts (a) and (b).

SUMMARY

The **chi-square test for goodness of fit** tests the null hypothesis that a population distribution is the same as a reference distribution.

The **expected count** for any variable category is obtained by multiplying the proportion of the distribution for each category times the sample size.

The **chi-square statistic** is $X^2 = \sum(O - E)^2/E$, where the sum is over n variable categories.

The chi-square test compares the value of the statistic X^2 with critical values from the **chi-square distribution** with $n - 1$ **degrees of freedom.**

For a test of H_0: the population proportions equal the hypothesized values against

H_a: the population proportions *differ from* the hypothesized values

the P-value is the area under the density curve to the right of X^2. Large values of X^2 are evidence against H_0.

The chi-square distribution is an approximation to the distribution of the statistic X^2. You can safely use this approximation when all expected counts are at least 1 and no more than 20% are less than 5. If the chi-square test finds a statistically significant P-value, do a follow-up analysis that compares the observed counts with the expected counts and that looks for the largest **components of chi-square.**

SECTION 13.1 EXERCISES

13.10 A FAIR DIE? A die is tossed 200 times with the faces 1, 2, 3, 4, 5, and 6 turning up with frequencies 26, 36, 39, 30, 38, and 31, respectively. Is there reason to believe that the die is "loaded" (i.e., unfair)?

13.11 TRIX ARE FOR KIDS Trix cereal comes in five fruit flavors, and each flavor has a different shape. A curious student methodically sorted an entire box of the cereal and found the following distribution of flavors for the pieces of cereal in the box:

Flavor:	Grape	Lemon	Lime	Orange	Strawberry
Frequency:	530	470	420	610	585

Test the null hypothesis that the flavors are uniformly distributed versus the alternative that they are not.

13.12 MORE CANDY The M&M/Mars Company reports the following distribution for other M&M varieties: for Peanut Chocolate Candies, the ratio is 20% each of browns, yellows, reds, and blues, and 10% each of greens and oranges. For Peanut Butter and Almond M&M's, the distribution is 20% each of browns, yellows, reds, greens, and blues. Buy a bag of one of these varieties of M&M's, perform a goodness of fit test of the company's reported distribution, and report your results. Better still, obtain a larger sample by using multiple bags and do this problem as another class activity.

13.13 CARNIVAL GAMES A "wheel of fortune" at a carnival is divided into four equal parts:

Part I: Win a doll

Part II: Win a candy bar

Part III: Win a free ride

Part IV: Win nothing

You suspect that the wheel is unbalanced (i.e., not all parts of the wheel are equally likely to be landed upon when the wheel is spun). The results of 500 spins of the wheel are as follows:

Part:	I	II	III	IV
Frequency:	95	105	135	165

Perform a goodness of fit test. Is there evidence that the wheel is not in balance?

13.2 INFERENCE FOR TWO-WAY TABLES

The two-sample z procedures of Chapter 12 allow us to compare the proportions of success in two groups, either two populations or two treatment groups in an experiment. What if we want to compare more than two groups? We need a new statistical test. The new test starts by presenting the data in a new way, as a two-way table. Two-way tables have more general uses than comparing the proportions of successes in several groups. As we saw in Section 3 of Chapter 4, they describe relationships between any two categorical variables. The same test that compares several proportions tests whether the row and column variables are related in any two-way table. We will start with the problem of comparing several proportions.

EXAMPLE 13.4 TREATING COCAINE ADDICTION

Chronic users of cocaine need the drug to feel pleasure. Perhaps giving them a medication that fights depression will help them stay off cocaine. A three-year study compared an antidepressant called desipramine with lithium (a standard treatment for cocaine addiction) and a placebo. The subjects were 72 chronic users of cocaine who wanted to break their drug habit. Twenty-four of the subjects were randomly assigned to each treatment. Here are the counts and proportions of the subjects who avoided relapse into cocaine use during the study:[3]

Group	Treatment	Subjects	No relapse	Proportion
1	Desipramine	24	14	0.583
2	Lithium	24	6	0.250
3	Placebo	24	4	0.167

The sample proportions of subjects who stayed off cocaine are quite different. The bar graph in Figure 13.3 compares the results visually. Are these data good evidence that the proportions of successes for the three treatments differ in the population of all cocaine users?

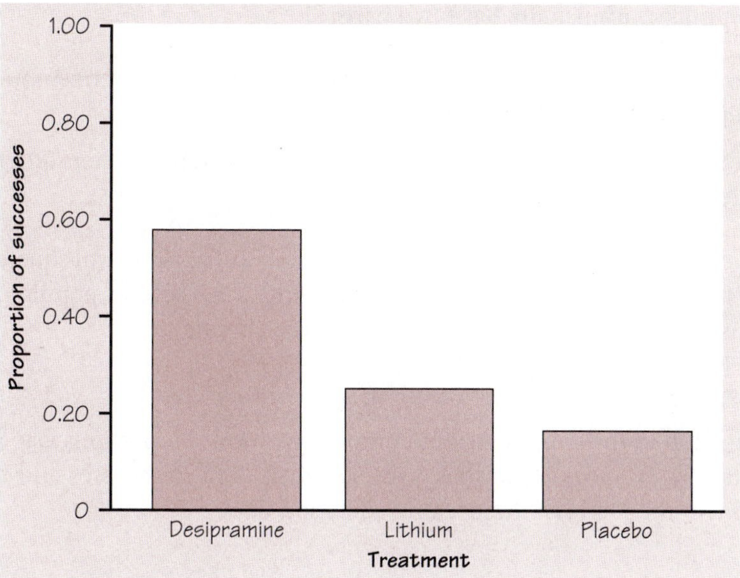

FIGURE 13.3 Bar graph comparing the success rates of three treatments for cocaine addiction.

The problem of multiple comparisons

Call the population proportions of successes in the three groups p_1, p_2, and p_3. We again use a subscript to remind us which group a parameter or statistic describes. To compare these three population proportions, we might use the two-sample z procedures several times:

- Test H_0: $p_1 = p_2$ to see if the success rate of desipramine differs from that of lithium.
- Test H_0: $p_1 = p_3$ to see if desipramine differs from a placebo.
- Test H_0: $p_2 = p_3$ to see if lithium differs from a placebo.

The weakness of doing three tests is that we get three P-values, one for each test alone. That doesn't tell us how likely it is that *three* sample proportions are spread apart as far as these are. It may be that $\hat{p}_1 = 0.583$ and $\hat{p}_3 = 0.167$ are significantly different if we look at just two groups, but not significantly different if we know that they are the smallest and largest proportions in three groups. As we look at more groups, we expect the gap between the smallest and largest sample proportion to get larger. (Think of comparing the tallest and shortest person in larger and larger groups of people.) We can't safely compare many parameters by doing tests or confidence intervals for two parameters at a time.

The problem of how to do many comparisons at once with some overall measure of confidence in all our conclusions is common in statistics. This is the problem of multiple comparisons. Statistical methods for dealing with many comparisons usually have two parts:

1. An *overall test* to see if there is good evidence of *any* differences among the parameters that we want to compare.
2. A detailed *follow-up analysis* to decide which of the parameters differ and to estimate how large the differences are.

The overall test is one with which we are familiar—the chi-square test—but in this new setting it will be used for comparing several population proportions. The follow-up analysis can be quite elaborate.

Two-way tables

two-way table

The first step in the overall test for comparing several proportions is to arrange the data in a ***two-way table*** that gives counts for both successes and failures. Here is two-way table of the cocaine addiction data:

	Relapse	
	No	Yes
Desipramine	14	10
Lithium	6	18
Placebo	4	20

r × c table

cell

We call this a 3×2 table because it has 3 rows and 2 columns. A table with r rows and c columns is an ***r × c table***. The table shows the relationship between two categorical variables. The explanatory variable is the treatment (one of three drugs). The response variable is success (no relapse) or failure (relapse). The two-way table gives the counts for all 6 combinations of values of these variables. Each of the 6 counts occupies a ***cell*** of the table.

Expected counts

We want to test the null hypothesis that there are *no differences* among the proportions of successes for addicts given the three treatments:

$$H_0: p_1 = p_2 = p_3$$

The alternative hypothesis is that there *is* some difference, that not all three proportions are equal:

$$H_a\text{: not all of } p_1, p_2, \text{ and } p_3 \text{ are equal}$$

The alternative hypothesis is no longer one-sided or two-sided. It is "many-sided," because it allows any relationship other than "all three equal." For example, H_a includes the situation in which $p_2 = p_3$ but p_1 has a different value.

To test H_0, we compare the observed counts in a two-way table with the *expected counts*, the counts we would expect—except for random variation—if H_0 were true. If the observed counts are far from the expected counts, that is evidence against H_0. It is easy to find the expected counts.

EXPECTED COUNTS

The **expected count** in any cell of a two-way table when H_0 is true is

$$\text{expected count} = \frac{\text{row total} \times \text{column total}}{\text{table total}}$$

To understand why this recipe works, think first about just one proportion.

EXAMPLE 13.5 FREE THROWS

Linda is a basketball player who makes 70% of her free throws. If she shoots 10 free throws in a game, we expect her to make 70% of them, or 7 of the 10. Of course, she won't make exactly 7 every time she shoots 10 free throws in a game. There is chance variation from game to game. But in the long run, 7 of 10 is what we expect. It is, in fact, the *mean* number of shots Linda makes when she shoots 10 times.

In more formal language, if we have n independent trials and the probability of a success on each trial is p, we expect np successes. If we draw an SRS of n individuals from a population in which the proportion of successes is p, we expect np successes in the sample. That's the fact behind the formula for expected counts in a two-way table.

Let's apply this fact to the cocaine study. The two-way table with row and column totals is

	Relapse		
	No	Yes	Total
Desipramine	14	10	24
Lithium	6	18	24
Placebo	4	20	24
Total	24	48	72

We will find the expected count for the cell in row 1 (desipramine) and column 2 (relapse). The proportion of relapses among all 72 subjects is

$$\frac{\text{count of relapses}}{\text{table total}} = \frac{\text{column 2 total}}{\text{table total}} = \frac{48}{72} = \frac{2}{3}$$

Think of this as p, the overall proportion of relapses. If H_0 is true, we expect (except for random variation) this same proportion of relapses in all three groups. So the expected count of relapses among the 24 subjects who took desipramine is

$$np = (24)\left(\frac{2}{3}\right) = 16.00$$

This expected count has the form announced in the box:

$$\frac{\text{row 1 total} \times \text{column 2 total}}{\text{table total}} = \frac{(24)(48)}{72}$$

We calculate the expected counts for the remaining cells in the same way. The results are summarized in the following example.

EXAMPLE 13.6 COMPARING OBSERVED AND EXPECTED COUNTS

Here are the observed and expected counts side by side:

	Observed		Expected	
	No	Yes	No	Yes
Desipramine	14	10	8	16
Lithium	6	18	8	16
Placebo	4	20	8	16

Because 2/3 of all subjects relapsed, we expect 2/3 of the 24 subjects in each group to relapse if there are no differences among the treatments. In fact, desipramine has fewer relapses (10) and more successes (14) than expected. The placebo has fewer successes (4) and more relapses (20). That's another way of saying what the sample proportions in Example 13.4 say more directly: desipramine does much better than the placebo, with lithium in between.

EXERCISES

13.14 HOW TO QUIT SMOKING It's hard for smokers to quit. Perhaps prescribing a drug to fight depression will work as well as the usual nicotine patch. Perhaps combining the patch and the drug will work better than either alone. Here are data from a randomized, double-blind trial that compared four treatments.[4] A "success" means that the subject did not smoke for a year following the beginning of the study.

Treatment	Subjects	Successes
Nicotine patch	244	40
Drug	244	74
Patch plus drug	245	87
Placebo	160	25

(a) Summarize these data in a two-way table.

(b) Calculate the proportion of subjects who refrain from smoking in each of the four treatment groups.

(c) Make a graph to display the association. Describe what you see.

(d) Explain in words what the null hypothesis H_0: $p_1 = p_2 = p_3 = p_4$ says about subjects' smoking habits.

(e) Find the expected counts if H_0 is true, and display them in a two-way table similar to the table of observed counts.

(f) Compare the tables of observed and expected counts. Explain how the comparison expresses the same association you see in (b) and (c).

13.15 WHY MEN AND WOMEN PLAY SPORTS Do men and women participate in sports for the same reasons? One goal for sports participants is social comparison—the desire to win or to do better than other people. Another is mastery—the desire to improve one's skills or to try one's best. A study on why students participate in sports collected data from two independent random samples of 67 male and 67 female undergraduates at a large university.[5] Each student was classified into one of four categories based on his or her responses to a questionnaire about sports goals. The four categories were high social comparison-high mastery (HSC-HM), high social comparison-low mastery (HSC-LM), low social comparison-high mastery (LSC-HM), and low social comparison-low mastery (LSC-LM). One purpose of the study was to compare the goals of male and female students. Here are the data displayed in a two-way table:

Observed counts for sports goals

	Sex	
Goal	Female	Male
HSC-HM	14	31
HSC-LM	7	18
LSC-HM	21	5
LSC-LM	25	13

(a) This is an $r \times c$ table. What numbers do r and c stand for?

(b) Calculate the proportions of females having each of the four categories of sports goals. Then do the same for males.

(c) Make a bar graph to compare the distribution of sports goals for males and females.

(d) The null hypothesis says that the proportions of females falling into the four sports goal categories are the same as the proportions of males in those categories. The overall proportion of *students* in the HSC-HM category is 45/134 = 0.336. Assuming H_0 is true, we expect that 33.6% of the females surveyed, (0.336)(67) = 22.5, will land in the HSC-HM category. Our expected count for the males in the HSC-HM category is (0.336)(67) = 22.5. Find the rest of the expected counts and display them in a two-way table.

(e) Compare the observed counts with the expected counts. Are there large deviations between them? Explain how the comparison expresses the same association you saw in (b) and (c).

The chi-square test for homogeneity of populations

Comparing the sample proportions of successes describes the differences among the three treatments for cocaine addiction. But the statistical test that tells us whether those differences are statistically significant doesn't use the sample proportions. It compares the observed and expected counts. The test statistic that makes the comparison is the *chi-square statistic.*

CHI-SQUARE STATISTIC

The **chi-square statistic** is a measure of how far the observed counts in a two-way table are from the expected counts. The formula for the statistic is

$$X^2 = \sum \frac{(\text{observed count} - \text{expected count})^2}{\text{expected count}}$$

The sum is over all $r \times c$ cells in the table.

The chi-square statistic is a sum of terms, one for each cell in the table. In the cocaine example, 14 of the desipramine group succeeded in avoiding a relapse. The expected count for this cell is 8. So the component of the chi-square statistic from this cell is

$$\frac{(\text{observed count} - \text{expected count})^2}{\text{expected count}} = \frac{(14-8)^2}{8} = \frac{36}{8} = 4.5$$

We compute the components of chi-square for the five remaining cells in Example 13.7.

As in the test for goodness of fit, you should think of the chi-square statistic X^2 as a measure of the distance of the observed counts from the expected counts. Like any distance, it is always zero or positive, and it is zero only when the observed counts are exactly equal to the expected counts. Large values of X^2 are evidence against H_0 because they say that the observed counts are far from what we would expect if H_0 were true. Although the alternative hypothesis H_a is many-sided, the chi-square test is one-sided because any violation of

H_0 tends to produce a large value of X^2. Small values of X^2 are not evidence against H_0.

In the cocaine example, we are comparing the proportion of relapses in three populations: addicts who take desipramine, addicts who take lithium, and addicts who take a placebo. The same chi-square procedure allows us to compare the distribution of proportions in several populations, provided that we take *separate and independent random samples* from each population.

CHI-SQUARE TEST FOR HOMOGENEITY OF POPULATIONS

Select independent SRSs from each of c populations. Classify each individual in a sample according to a categorical response variable with r possible values. There are c different sets of proportions to be compared, one for each population.

The null hypothesis is that the distribution of the response variable is the same in all c populations. The alternative hypothesis says that these c distributions are not all the same.

If H_0 is true, the chi-square statistic X^2 has approximately a χ^2 distribution with $(r - 1)(c - 1)$ degrees of freedom (df).

The *P*-value for the chi-square test is the area to the right of X^2 under the chi-square density curve with df degrees of freedom.

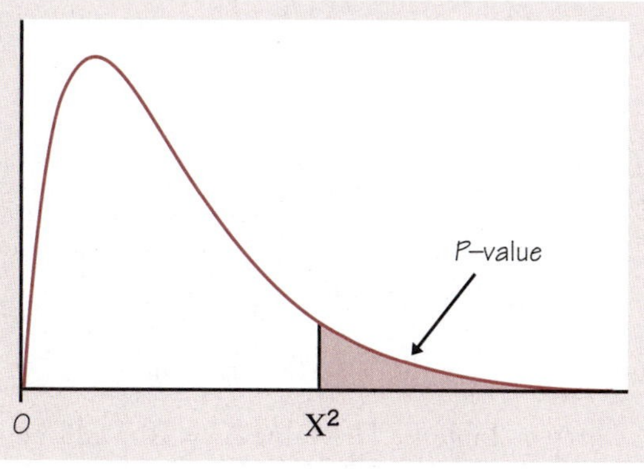

The chi-square test, like the z procedures for comparing two proportions, is an approximate method that becomes more accurate as the counts in the cells of the table get larger. Fortunately, the approximation is accurate for quite modest counts. Here is a practical guideline.[6]

CELL COUNTS REQUIRED FOR THE CHI-SQUARE TEST

You can safely use the chi-square test with critical values from the chi-square distribution when no more than 20% of the expected counts are less than 5 and all individual expected counts are 1 or greater. In particular, all four expected counts in a 2 × 2 table should be 5 or greater.

We have examined the data for the three cocaine treatment groups informally. Now we proceed to formal inference.

EXAMPLE 13.7 IS DESIPRAMINE EFFECTIVE IN TREATING COCAINE ADDICTION?

Step 1: Identify populations of interest. State hypotheses in words and symbols. We want to compare the proportions of cocaine addicts who do not relapse in the populations of patients treated with desipramine (p_1), lithium (p_2), and placebo (p_3). Our hypotheses are

$H_0: p_1 = p_2 = p_3$ The proportions of cocaine addicts who avoid relapse are the same.
H_a: Not all three of the proportions are equal.

Step 2: Choose the appropriate inference procedure and verify conditions for its use. To use the chi-square test for homogeneity of populations:

- The data must come from independent SRSs from the populations of interest. We are willing to treat the randomly allocated subjects in the three treatment groups as SRSs from their respective populations.
- All expected cell counts are greater than 1, and no more than 20% are less than 5. In Example 13.6, we saw that the smallest expected count was 8.

Step 3: Carry out the procedure.

- The test statistic is

$$X^2 = \sum \frac{(O-E)^2}{E} = \frac{(14-8)^2}{8} + \frac{(10-16)^2}{16} + \frac{(6-8)^2}{8} + \frac{(18-16)^2}{16} + \frac{(4-8)^2}{8} + \frac{(20-16)^2}{16} = 10.5$$

$$\text{df} = (r-1)(c-1) = (3-1)(2-1) = 2$$

df = 2

p	.01	.005
x^*	9.21	10.60

- Look in the df = 2 row of Table E. The value $X^2 = 10.5$ falls between the 0.01 and 0.005 critical values of the chi-square distribution with 2 degrees of freedom. Remember that the chi-square test is always one-sided. So the *P*-value of $X^2 = 10.5$ is between 0.01 and 0.005.

Step 4: Interpret your results in context. Since the *P*-value is less than 0.01, the differences among the three proportions are statistically significant at the $\alpha = 0.01$ level. We would reject the null hypothesis.

Calculating chi-square with technology

Calculating the expected counts and then the chi-square statistic by hand is a bit time-consuming. As usual, computer software saves time and always gets the arithmetic right. The TI-83 and TI-89, on a smaller scale, have also been programmed to conduct inference for two-way tables.

TECHNOLOGY TOOLBOX *Chi-square tests with Minitab*

We enter the two-way table (the 6 counts) for the cocaine study into the Minitab software package and request the chi-square test. The output appears in Figure 13.4. Most statistical software packages produce chi-square output similar to this.

Minitab repeats the two-way table of observed counts and puts the expected count for each cell below the observed count. It numbers the rows (1, 2, and 3) and the columns (C1 and C2) and also puts in the row and column totals. Then the software calculates the chi-square statistic X^2. For these data, $X^2 =$ 10.500. The statistic is a sum of 6 terms, one for each cell in the table. The `"ChiSq"` display in the output shows the individual terms, as well as their sum. The first term is 4.500, just as we calculated.

The *P*-value is the probability that X^2 would take a value as large as 10.500 if H_0 were really true. Many software systems give the *P*-value. Minitab requires us to ask for the probability of a value of 10.500 or smaller. This probability is 0.9948 (at the bottom of the output), so the *P*-value is 1 – 0.9948 = 0.0052. The small *P*-value gives us good reason to conclude that there *are* differences among the effects of the three treatments.

```
Expected counts are printed below observed counts
            C1          C2        Total
   1        14          10           24
          8.00       16.00

   2         6          18           24
          8.00       16.00

   3         4          20           24
          8.00       16.00

Total       24          48           72

ChiSq =   4.500 + 2.250 +
          0.500 + 0.250 +
          2.000 + 1.000 = 10.500
df = 2

   Chisquare 2.
   10.5000      0.9948
```

FIGURE 13.4 Minitab output for the two-way table in the cocaine study. The output gives the observed counts, the expected counts, and the value 10.500 for the chi-square statistic. The last line gives 0.9948 as the probability of a value *less than* 10.500 if the null hypothesis is true. The *P*-value is therefore 1 – 0.9948.

TECHNOLOGY TOOLBOX *Chi-square tests on the TI-83/89*

To perform a chi-square test of the cocaine study on the TI-83/89, use a matrix, say matrix [A], to store the observed counts. Here are the keystrokes, along with several calculator screens for you to check your progress.

- Enter the observed counts in the matrix [A].

TI-83

- Press 2nd x^{-1}(MATRIX), arrow to EDIT, choose 1:[A].

```
NAMES  MATH  EDIT
1:[A]  3x2
2:[B]  2x4
3:[C]  2x4
4:[D]
5:[E]
6:[F]
7↓[G]
```

TI-89

- Press APPS, select 6:Data/Matrix Editor, and then 3:New. . . .
- Adjust your settings to match those shown.

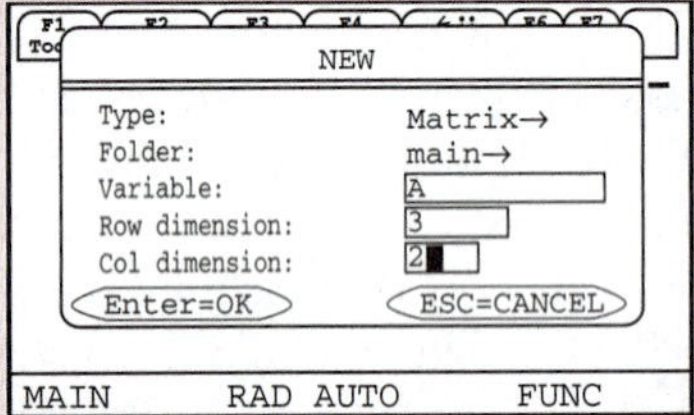

Enter the observed counts from the two-way table in the matrix in the same locations.

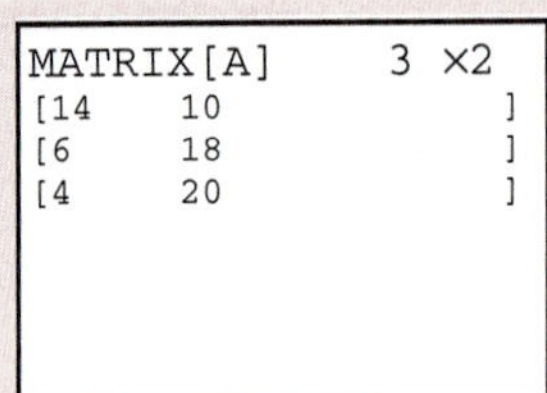

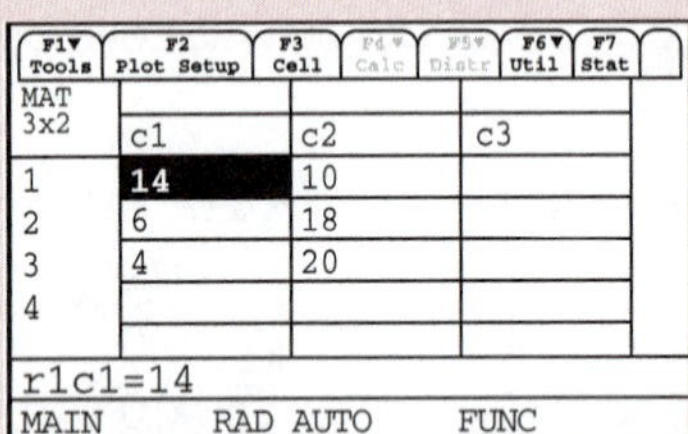

- Specify the chi-square test, the matrix where the observed counts are found, and the matrix where the expected counts will be stored.

TI-83

- Press STAT, arrow to TESTS, and choose F1 χ^2 Test. . . .

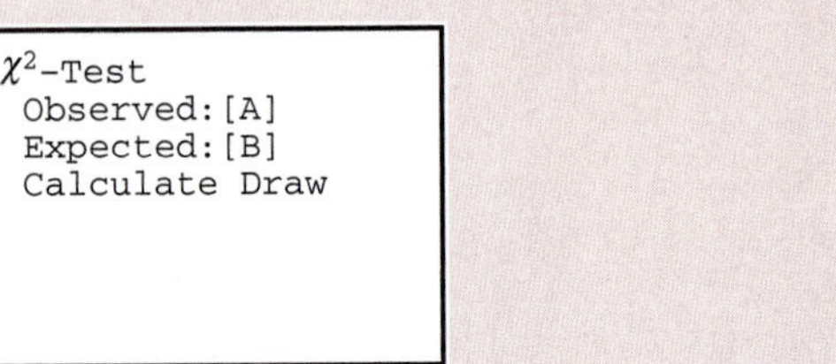

TI-89

- In the Statistics/List Editor, press 2nd F1([F6]) C: and choose 8:Chi2 2-way. . . .
- Adjust your settings as shown.

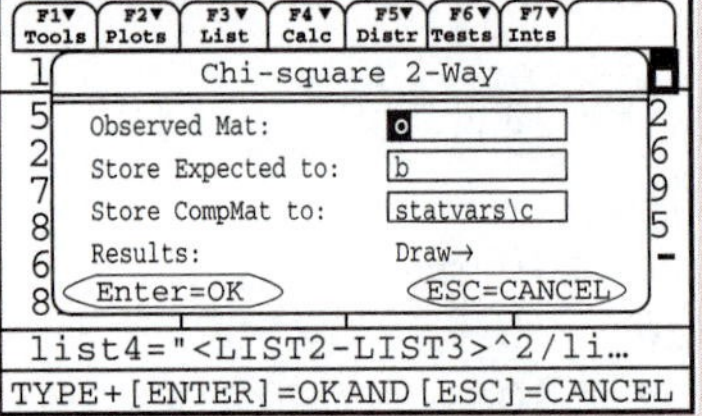

- Choose "Calculate" or "Draw" to carry out the test. If you choose "Calculate," you should get these results:

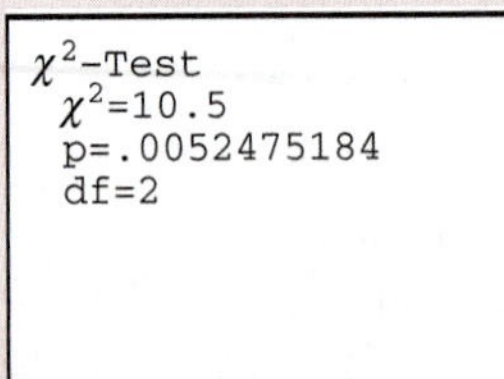

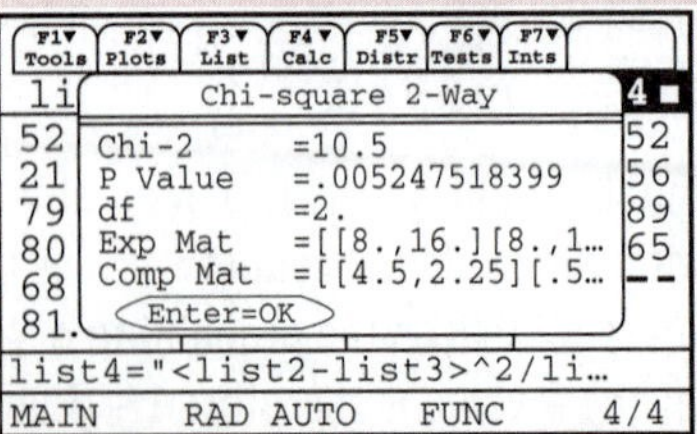

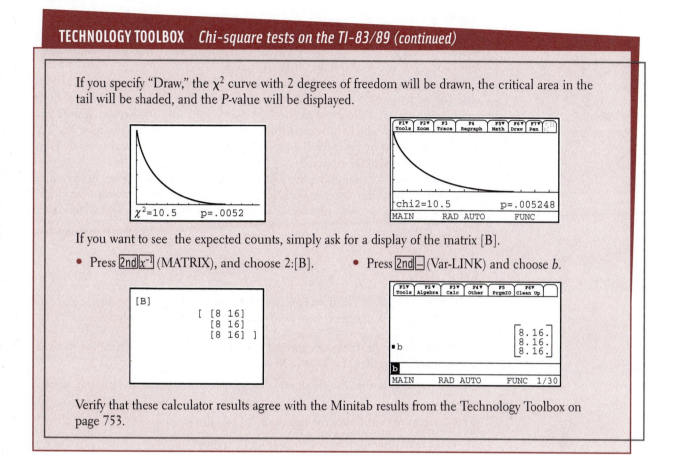

TECHNOLOGY TOOLBOX *Chi-square tests on the TI-83/89 (continued)*

If you specify "Draw," the χ^2 curve with 2 degrees of freedom will be drawn, the critical area in the tail will be shaded, and the *P*-value will be displayed.

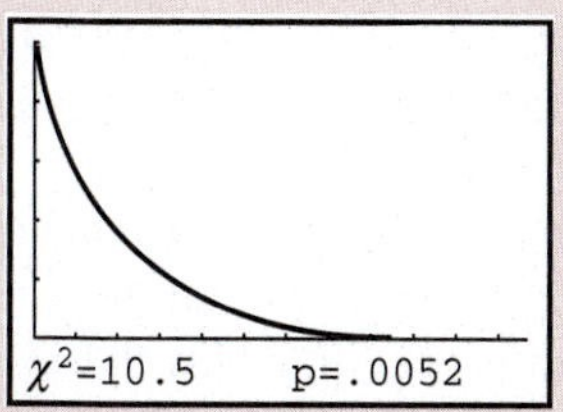

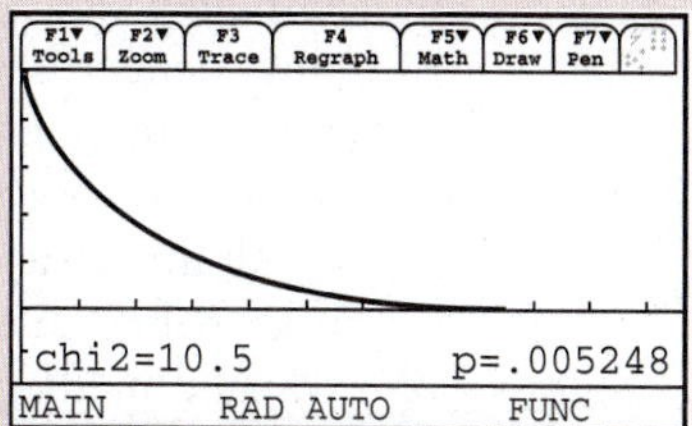

If you want to see the expected counts, simply ask for a display of the matrix [B].

- Press 2nd x^{-1} (MATRIX), and choose 2:[B].
- Press 2nd − (Var-LINK) and choose *b*.

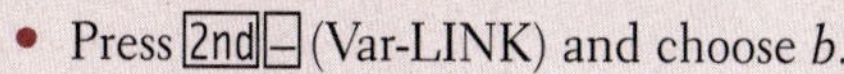

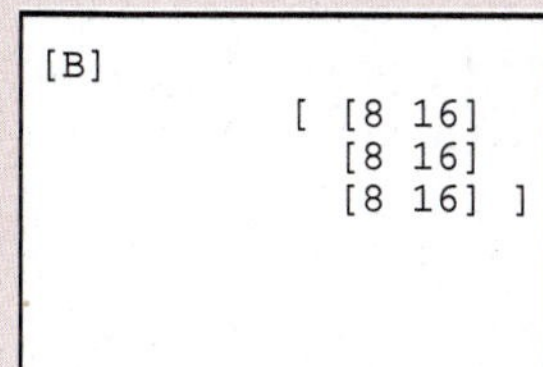

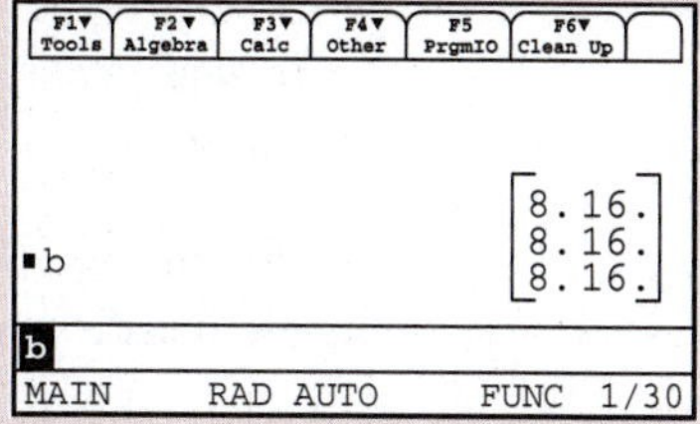

Verify that these calculator results agree with the Minitab results from the Technology Toolbox on page 753.

Follow-up analysis

The chi-square test is the overall test for comparing any number of population proportions. If the test allows us to reject the null hypothesis that all the proportions are equal, we then want to do a follow-up analysis that examines the differences in detail. We won't describe how to do a formal follow-up analysis, but you should look at the data to see what specific effects they suggest.

EXAMPLE 13.8 A FINAL LOOK AT THE COCAINE STUDY

The cocaine study found significant differences among the proportions of successes for three treatments for cocaine addiction. We can see the specific differences in three ways.

Look first at the *sample proportions:*

$$\hat{p}_1 = 0.583 \quad \hat{p}_2 = 0.250 \quad \hat{p}_3 = 0.167$$

These suggest that the major difference between the proportions is that desipramine has a much higher success rate than either lithium or a placebo. That is the effect that the study hoped to find.

Next, *compare the observed and expected counts* in Figure 13.4. Treatment 1 (desipramine) has more successes and fewer failures than we would expect if all three treatments had the same success rate in the population. The other two treatments had fewer successes and more failures than expected.

components of chi-square

Finally, Minitab prints under the table the 6 individual "distances" between the observed and expected counts that are added to get X^2. The arrangement of these ***components of X^2*** is the same as the 3×2 arrangement of the table. The largest components show which cells contribute the most to the overall distance X^2. The largest component by far is for the top left cell in the table: desipramine has more successes than would be expected.

All three ways of examining the data point to the same conclusion: desipramine works better than the other treatments. This is an informal conclusion. More advanced methods provide tests and confidence intervals that make this follow-up analysis formal.

EXERCISES

13.16 HOW TO QUIT SMOKING, II In Exercise 13.14 (page 748), you began to analyze data on the effectiveness of several treatments designed to help smokers quit.

(a) Starting from the table of expected counts, find the 8 components of the chi-square statistic and then the statistic X^2 itself.

(b) Use Table E to find the *P*-value for the test. Explain in simple language what it tells you.

(c) Which term contributes the most to X^2? Does this surprise you?

(d) What conclusion would you draw from this study?

(e) Perform the chi-square test on your calculator. Are your results the same?

13.17 WHY MEN AND WOMEN PLAY SPORTS, II In Exercise 13.15 (page 749), you began to analyze data on the reasons that men and women play sports.

(a) Starting from the table of expected counts, find the 8 components of the chi-square statistic and then the statistic X^2 itself.

(b) Use Table E to find the *P*-value for the test. What decision would you make concerning H_0? Explain what this means in plain language.

(c) Which term(s) contribute most to the X^2 statistic? What specific relation between gender and sports goals do the term(s) point to?

(d) Figure 13.5 gives Minitab output for this test. Compare your work in parts (a) through (c) with the computer output.

13.18 HOW ARE SCHOOLS DOING? The nonprofit group Public Agenda conducted telephone interviews with 3 randomly selected groups of parents of high school children. There were 202 black parents, 202 hispanic parents, and 201 white parents. One question asked was, "Are the high schools in your state doing an excellent, good, fair, or poor job, or don't you know enough to say?" Here are the survey results:[7]

```
Expected counts are printed below observed counts

          Female      Male      Total
HSC-HM      14         31         45
          22.50      22.50

HSC-LM       7         18         25
          12.50      12.50

LSC-HM      21          5         26
          13.00      13.00

LSC-LM      25         13         38
          19.00      19.00

 Total        67         67        134

 ChiSq   = 3.211 + 3.211 + 2.420 + 2.420 +
           4.923 + 4.923 + 1.895 + 1.895 = 24.898
 DF      = 3, P-Value = 0.000
```

FIGURE 13.5 Minitab output for the study of gender and sports goals, for Exercise 13.17.

	Black parents	Hispanic parents	White parents
Excellent	12	34	22
Good	69	55	81
Fair	75	61	60
Poor	24	24	24
Don't know	22	28	14
Total	202	202	201

Write a brief analysis of these results. Include a graph or graphs, a test of significance, and your own discussion of the most important findings.

The chi-square test of association/independence

Two-way tables can arise in several ways. The cocaine study is an experiment that assigned 24 addicts to each of three groups. Each group is a sample from a separate population corresponding to a separate treatment. The study design fixes the size of each sample in advance, and the data record which of two outcomes occurred for each subject. The null hypothesis of "no difference" among the treatments takes the form of "equal proportions of successes" in the three populations. The next example illustrates a different setting for a two-way table.

EXAMPLE 13.9 SMOKING AND SES

In a study of heart disease in male federal employees, researchers classified 356 volunteer subjects according to their socioeconomic status (SES) and their smoking habits.[8] There were three categories of SES: high, middle, and low. Individuals were asked whether they were current smokers, former smokers, or had never smoked, producing three categories for smoking habits as well. Here is the two-way table that summarizes the data:

Observed counts for smoking and SES

	SES			
Smoking	High	Middle	Low	Total
Current	51	22	43	116
Former	92	21	28	141
Never	68	9	22	99
Total	211	52	93	356

This is a 3×3 table, to which we have added the marginal totals obtained by summing across the rows and down the columns.

The two-way table in Example 13.9 does not compare several populations. Instead, it arises by classifying observations from a single population in two ways: by smoking habits and SES. Both of these variables have three levels, so a careful statement of the null hypothesis

$$H_0\text{: there is no association between SES and smoking habits}$$

in terms of population parameters is complicated.

The setting of Example 13.9 is very different from a comparison of several proportions. Nevertheless, we can apply a chi-square test. One of the most useful properties of chi-square is that it tests the hypothesis "the row and column variables are not related to each other" whenever this hypothesis makes sense for a two-way table.

THE CHI-SQUARE TEST OF ASSOCIATION/INDEPENDENCE

Use the chi-square test of association/independence to test the null hypothesis

$$H_0\text{: there is no relationship between two categorical variables}$$

when you have a two-way table from a single SRS, with each individual classified according to both of two categorical variables.

Computing conditional distributions

We start our analysis by computing descriptive statistics that summarize the observed relation between SES and smoking. As in Section 3 of Chapter 4, we describe a relationship between categorical variables by comparing percents. The researchers suspected that SES helps explain smoking, so in this situation SES is the explanatory variable and smoking is the response variable. We should therefore compare the column percents that give the conditional distribution of smoking within each SES category.

EXAMPLE 13.10 SMOKING HABITS IN EACH SES CATEGORY

We must calculate the column percents. For the high-SES group, there are 51 current smokers out of a total of 211 people. The column proportion for this cell is

$$\frac{51}{211} = 0.242$$

That is, 24.2% of the high-SES group are current smokers. Similarly, 92 of the 211 people in this group are former smokers. The column proportion is

$$\frac{92}{211} = 0.436$$

or 43.6%. In all, we must calculate nine percents. Here are the results:

Column percents for smoking and SES

	SES		
Smoking	High	Middle	Low
Current	24.2	42.3	46.2
Former	43.6	40.4	30.1
Never	32.2	17.3	23.7
Total	100.0	100.0	100.0

Each column of the table gives the conditional distribution of smoking habits among male federal employees having a specific SES. The sum of the percents in each column should be 100, since it accounts for all the employees in that SES category.

The bar graphs in Figure 13.6 help us compare the distributions of smoking behavior in the three SES groups. The percent of current smokers decreases as SES increases from low to middle to high; in particular, relatively few high-SES subjects smoke. The percent of former smokers increases as SES increases, suggesting that higher-SES smokers were more likely to quit. The percent of people who never smoked is highest in the high-SES group, but the middle-SES group has a somewhat lower percentage than the low-SES group. Overall, the column percents suggest that there is a negative association between smoking and SES: higher-SES people tend to smoke less.

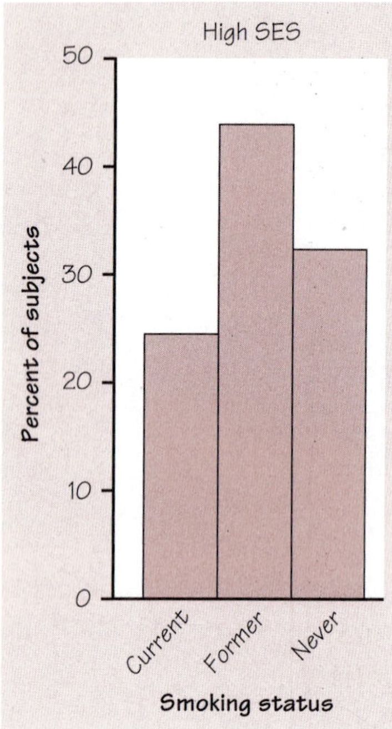

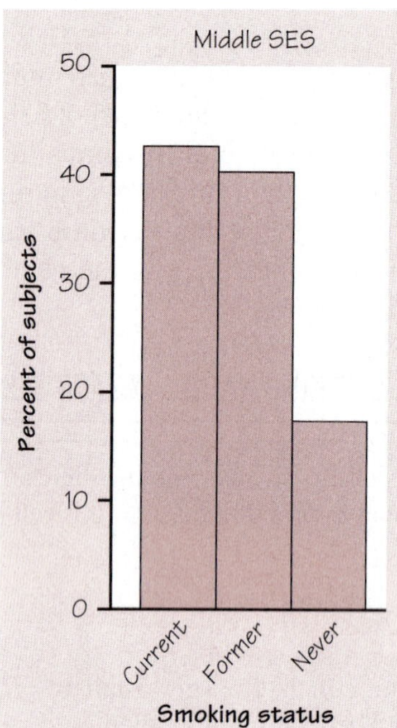

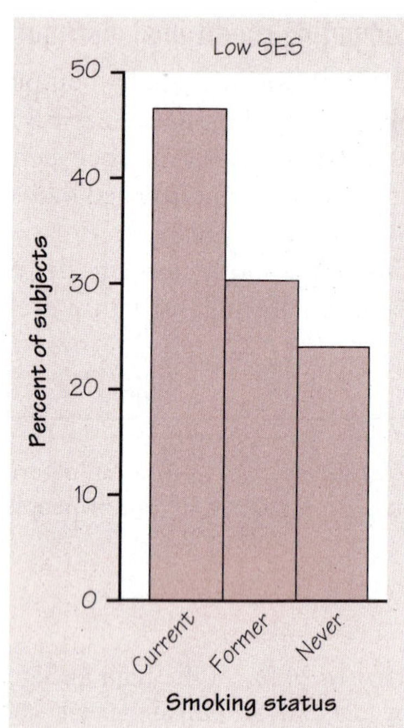

FIGURE 13.6 Comparison of smoking-behavior distributions for high, middle, and low SES, for Example 13.10.

The *chi-square test of association/independence* assesses whether this observed association is statistically significant. That is, is the SES-smoking relationship in the sample sufficiently strong for us to conclude that it is due to a relationship between these two variables in the underlying population and not merely to chance? Note that the test only asks whether there is evidence of some relationship. To explore the direction or nature of the relationship we must examine the column or row percents. Note also that in using the chi-square test we are acting as if the subjects were a simple random sample from the population of interest. If the volunteers are a biased sample—for example, if smokers are reluctant to volunteer for a study of employee health—then conclusions about the entire population of employees are not justified.

Computing expected cell counts

The null hypothesis is that there is no relationship between SES and smoking in the population. The alternative is that these two variables are related. If we assume that the null hypothesis is true, then smoking and SES are independent. We can find the expected cell counts using the multiplication rule for independent events (Chapter 6).

EXAMPLE 13.11 EXPECTED CELL COUNTS

What is the expected count for the cell corresponding to the high-SES current smokers? Under the null hypothesis that smoking and SES are independent,

$$P(\text{high SES and current smoker}) = P(\text{high SES}) \times P(\text{current smoker})$$
$$= \frac{211}{356} \times \frac{116}{356}$$

The expected count of high-SES current smokers can be found by multiplying this probability by the total number of employees in the sample:

$$356\left(\frac{211}{356} \times \frac{116}{356}\right) = 68.75$$

Simple arithmetic shows that this is the same as calculating $(211 \times 116)/356$. In other words,

$$\text{expected count} = \frac{\text{row total} \times \text{column total}}{n}$$

where n is the sample size.

Here is the completed table of expected counts:

Expected counts for smoking and SES

	SES			
Smoking	High	Middle	Low	Total
Current	68.75	16.94	30.30	115.99
Former	83.57	20.60	36.83	141.00
Never	58.68	14.46	25.86	99.00
Total	211.00	52.00	92.99	355.99

EXERCISES

13.19 EXTRACURRICULAR ACTIVITIES AND GRADES North Carolina State University studied student performance in a course required by its chemical engineering major. One question of interest is the relationship between time spent in extracurricular activities and whether a student earned a C or better in the course. Here are the data for the 119 students who answered a question about extracurricular activities:[9]

	Extracurricular activities (hours per week)		
	<2	2 to 12	>12
C or better	11	68	3
D or F	9	23	5

(a) This is an $r \times c$ table. What are the numbers r and c?

(b) Find the proportion of successful students (C or better) in each of the three extracurricular activity groups. What kind of relationship between extracurricular activities and succeeding in the course do these proportions seem to show?

(c) Make a bar graph to compare the three proportions of successes.

(d) What null hypothesis will a chi-square procedure test in this setting?

(e) Find the expected counts if this hypothesis is true, and display them in a two-way table.

(f) Compare the observed counts with the expected counts. Are there large deviations between them? These deviations are another way of describing the relationship you described in (b).

13.20 SMOKING BY STUDENTS AND THEIR PARENTS How are the smoking habits of students related to their parents' smoking? Here are data from a survey of students in eight Arizona high schools:[10]

	Student smokes	Student does not smoke
Both parents smoke	400	1380
One parent smokes	416	1823
Neither parent smokes	188	1168

(a) This is an $r \times c$ table. What are the numbers r and c?

(b) Calculate the proportion of students who smoke in each of the three parent groups. Then describe in words the association between parent smoking and student smoking.

(c) Make a graph to display the association.

(d) Explain in words what the null hypothesis H_0 says about smoking.

(e) Find the expected counts if H_0 is true, and display them in a two-way table similar to the table of observed counts.

(f) Compare the tables of observed and expected counts. Explain how the comparison expresses the same association you see in (b) and (c).

Performing the χ^2 test

We are now ready to cary out the chi-square procedure for the smoking and SES data.

EXAMPLE 13.12 CHI-SQUARE TEST FOR ASSOCIATION/INDEPENDENCE

Step 1: State hypotheses.

H_0: smoking and SES are independent
H_a: smoking and SES are dependent

or, equivalently

H_0: There is no association between smoking and SES.
H_a: There is an association between smoking and SES.

Step 2: Choose an inference procedure and verify conditions. To use the chi-square test of association/independence, we must check that all expected cell counts are at least

1, and that no more than 20% are less than 5. From Example 13.13, we can see that these conditions are easily met.

Step 3: Carry out the selected procedure.

- The test statistic is

$$
\begin{aligned}
X^2 &= \sum \frac{(\text{observed} - \text{expected})^2}{\text{expected}} \\
&= \frac{(51-68.75)^2}{68.75} + \frac{(22-16.94)^2}{16.94} + \frac{(43-30.30)^2}{30.30} \\
&\quad + \frac{(92-83.57)^2}{83.57} + \frac{(21-20.60)^2}{20.60} + \frac{(28-36.83)^2}{36.83} \\
&\quad + \frac{(68-58.68)^2}{58.68} + \frac{(9-14.46)^2}{14.46} + \frac{(22-25.86)^2}{25.86} \\
&= 4.583 + 1.511 + 5.323 + 0.850 + 0.008 + 2.117 + 1.480 + 2.062 + 0.576 \\
&= 18.51
\end{aligned}
$$

- Because there are $r = 3$ smoking categories and $c = 3$ SES groups, the degrees of freedom for this statistic are

$$(r-1)(c-1) = (3-1)(3-1) = 4$$

Under the null hypothesis that smoking and SES are independent, the test statistic X^2 has a $\chi^2(4)$ distribution. To obtain the *P*-value, refer to the row in Table E corresponding to 4 df. The calculated value $X^2 = 18.51$ lies between upper critical points corresponding to probabilities 0.001 and 0.0005. The *P*-value is therefore between 0.001 and 0.0005.

df = 4		
p	.001	.0005
χ^2	18.47	20.00

Step 4: Interpret your results in context. There is strong evidence ($X^2 = 18.51$, df = 4, $P < 0.001$) of an association between smoking and SES in the population of male federal employees. The size and nature of this association are described by the table of percents examined in Example 13.12 and the display of these percents in Figure 13.6. Of course, this association does *not* show that SES *causes* smoking behavior.

Concluding remarks

You can distinguish between the two types of chi-square tests for two-way tables by examining the design of the study. In the test of association/independence, there is a single sample from a single population. The individuals in the sample are classified according to two categorical variables. For the test of homogeneity of populations, there is a sample from each of two or more populations. Each individual is classified based on a single categorical variable. The precise statement of the hypothesis differs, depending on the sampling design.

EXERCISES

13.21 EXTRACURRICULAR ACTIVITIES AND GRADES In Exercise 13.19 (page 761), you began to analyze data on the relationship between time spent on extracurricular activities and success in a tough course. Figure 13.7 gives Minitab output (with some values deliberately omitted) for the two-way table in Exercise 13.19.

```
Chi-Square Test
Expected counts are printed below observed counts
               <2          2 to 12           >12        Total
A, B, C,       11               68             3           82
            13.78
D or F          9               23             5           37
                             28.29          2.49
Total          20               91             8          119
Chi-Sq   = 0.561 +                   +     1.145 +
            1.244 +                  +     2.538 =      6.926
DF =      , P-Value =
1 cells with expected counts less than 5.0
```

FIGURE 13.7 Minitab output for the study of extracurricular activity and success in a tough course.

(a) Starting from the table of expected counts, find the 6 components of the chi-square statistic and then the statistic X^2 itself. Copy the computer output on your paper and fill in the five related missing values.

(b) Use Table E to find the *P*-value for this test. Then use your calculator to help you fill in the df and *P*-value entries on the computer output.

(c) Which term contributes the most to X^2? What specific relation between extracurricular activities and academic success does this term point to?

(d) Does the North Carolina State study convince you that spending more or less time on extracurricular activities *causes* changes in academic success? Explain your answer.

13.22 SMOKING BY STUDENTS AND THEIR PARENTS In Exercise 13.20 (page 762), you began to analyze data on the relationship between smoking by parents and smoking by high school students.

(a) Use the Inference Toolbox to carry out the appropriate significance test.

(b) Which term(s) contribute the most to X^2? What specific relation between parent smoking and student smoking do these terms point to?

(c) Does the study convince you that parent smoking *causes* student smoking? Explain your answer.

13.23 EARLY TO BED? Is it true that "Early to bed and early to rise makes a man healthy, wealthy, and wise?" A study of older people in England suggests that Benjamin Franklin's saying no longer applies. The subjects were 1229 randomly selected adults who were at least 65 years old in 1973. The subjects were followed for 23 years to look at such things as mortality and cause of death.

The investigators call a subject an "owl" if he or she regularly goes to bed after 11 p.m. and rises at or after 8 a.m. A subject is a "lark" if he or she retires before 11 p.m. and rises before 8 a.m. The overall conclusion was that owls actually do a bit better than larks in many respects. Here is a two way table for one response variable, access to a car at the start of the study in 1973.[11]

	Access to car?	
	Yes	No
Larks	122	234
Owls	138	180
Other sleeping patterns	213	342

Write a brief report of the relationship between sleeping pattern and access to a car. Include a graph and a test of significance.

The chi-square test and the z test

We can use the chi-square test to compare any number of proportions. If we are comparing r proportions and make the columns of the table "success" and "failure," the counts form an $r \times 2$ table. P-values come from the chi-square distribution with $r - 1$ degrees of freedom. If $r = 2$, we are comparing just two proportions. We have two ways to do this: the z test from Section 12.2 and the chi-square test with 1 degree of freedom for a 2×2 table. *These two tests always agree.* In fact, the chi-square statistic X^2 is just the square of the z statistic, and the P-value for X^2 is exactly the same as the two-sided P-value for z. We recommend using the z test to compare two proportions, because it gives you the choice of a one-sided test and is related to a confidence interval for $p_1 - p_2$.

EXERCISE

13.24 TREATING ULCERS Gastric freezing was once a recommended treatment for ulcers in the upper intestine. Use of gastric freezing stopped after experiments showed it had no effect. One randomized comparative experiment found that 28 of the 82 gastric-freezing patients improved, while 30 of the 78 patients in the placebo group improved.[12] We can test the hypothesis of "no difference" between the two groups in two ways: using the two-sample z statistic or using the chi-square statistic.

(a) State the null hypothesis with a two-sided alternative and carry out the z test. What is the P-value from Table A?

(b) Present the data in a 2×2 table. Use the chi-square test to test the hypothesis from (a). Verify that the X^2 statistic is the square of the z statistic. Use your calculator to verify that the chi-square P-value agrees with the z result.

(c) What do you conclude about the effectiveness of gastric freezing as a treatment for ulcers?

SUMMARY

For two-way tables, we first compute percents or proportions that describe the relationship of interest. Then we turn to formal inference. Two different methods of generating data for two-way tables lead to the **chi-square test for homogeneity of populations** and the **chi-square test of association/independence.**

In the first design, independent SRSs are drawn from each of several populations, and each observation is classified according to a categorical variable of interest. The null hypothesis is that the distribution of this categorical variable is the same for all of the populations. We use the chi-square test for homogeneity of population to test this hypothesis.

One common use of the chi-square test for homogeneity of populations is to compare several population proportions. The null hypothesis states that all of the population proportions are equal. The alternative hypothesis states that they are not all equal but allows any other relationship among the population proportions.

In the second design, a single SRS is drawn from a population, and observations are classified according to two categorical variables. The chi-square test of association/independence tests the null hypothesis that there is no relationship between the row variable and the column variable.

The **expected count** in any cell of a two-way table when H_0 is true is

$$\text{expected count} = \frac{\text{row total} \times \text{column total}}{\text{table total}}$$

The **chi-square statistic** is

$$X^2 = \sum \frac{(\text{observed count} - \text{expected count})^2}{\text{expected count}}$$

The chi-square test compares the value of the statistic X^2 with critical values from the **chi-square distribution** with $(r-1)(c-1)$ **degrees of freedom.** Large values of X^2 are evidence against H_0, so the P-value is the area under the chi-square density curve to the right of X^2.

The chi-square distribution is an approximation to the distribution of the statistic X^2. You can safely use this approximation when all expected cell counts are at least 1 and no more than 20% are less than 5.

SECTION 13.2 EXERCISES

13.25 STRESS AND HEART ATTACKS You read a newspaper article that describes a study of whether stress management can help reduce heart attacks. The 107 subjects all had reduced blood flow to the heart and so were at risk of a heart attack. They were assigned at random to three groups. The article goes on to say:

> One group took a four-month stress management program, another underwent a four-month exercise program, and the third received usual heart care from their personal physicians.
>
> In the next three years, only three of the 33 people in the stress management group suffered "cardiac events," defined as a fatal or non-fatal heart attack or a surgical procedure such as a bypass or angioplasty. In the same period, seven of the 34 people in the exercise group and 12 out of the 40 patients in usual care suffered such events.[13]

(a) Use the information in the news article to make a two-way table that describes the study results.

(b) What are the success rates of the three treatments in avoiding cardiac events?

(c) Find the expected cell counts under the null hypothesis that there is no difference among the treatments. Verify that the expected counts meet our guideline for use of the chi-square test.

(d) Is there a significant difference among the success rates for the three treatments? Give appropriate statistical evidence to support your answer.

13.26 REGULATING GUNS The National Gun Policy Survey asked respondents, "Do you think there should be a law that would ban possession of handguns except for the police and other authorized persons?" Here are the responses, broken down by the respondent's level of education:[14]

	Yes	No
Less than high school	58	58
High school graduate	84	129
Some college	169	294
College graduate	98	135
Postgraduate degree	77	99

(a) How does the proportion of the sample who favor banning possession of handguns differ among people with different levels of education? Make a bar graph that compares the proportions, and briefly describe the relationship between education and opinion about a handgun ban.

(b) Does the sample provide good evidence that the proportion of the adult population who favor a ban on handguns changes with level of education?

13.27 DO YOU USE COCAINE? Sample surveys on sensitive issues can give different results depending on how the question is asked. A University of Wisconsin study divided 2400 respondents into 3 groups at random. All were asked if they had ever used cocaine. One group of 800 was interviewed by phone; 21% said they had used cocaine. Another 800 people were asked the question in a one-on-one personal interview; 25% said "Yes." The remaining 800 were allowed to make an anonymous written response; 28% said "Yes."[15] Are there statistically significant differences among these proportions? Give appropriate statistical evidence to support your conclusion.

13.28 CHILD-CARE WORKERS A large study of child care used samples from the data tapes of the Current Population Survey over a period of several years. The result is close to an SRS of child-care workers. The Current Population Survey has three classes of child-care workers: private household, nonhousehold, and preschool teacher. Here are data on the number of blacks among women workers in these three classes:[16]

	Total	Black
Household	2455	172
Nonhousehold	1191	167
Teachers	659	86

(a) What percent of each class of child-care workers is black?

(b) Make a two-way table of class of worker by race (black or other).

(c) Can we safely use the chi-square test? What null and alterative hypotheses does X^2 test?

(d) Calculate the chi-square statistic for this table. What are its degrees of freedom? Use Table E to approximate the *P*-value.

(e) What do you conclude from these data?

13.29 SECONDHAND STORES, I Shopping at secondhand stores is becoming more popular and has even attracted the attention of business schools. A study of customers' attitudes toward secondhand stores interviewed samples of shoppers at two secondhand stores of the same chain in two cities. The breakdown of the respondents by sex is as follows:[17]

	City 1	City 2
Men	38	68
Women	203	150
Total	241	218

Is there a significant difference between the proportions of women customers in the two cities?

(a) State the null hypothesis, find the sample proportions of women in both cities, do a two-sided z test, and give a *P*-value using Table A.

(b) Calculate the chi-square statistic X^2 and show that it is the square of the z statistic. Show that the *P*-value from Table E agrees (up to the accuracy of the table) with your result from (a).

(c) Give a 95% confidence interval for the difference between the proportions of women customers in the two cities.

13.30 SECONDHAND STORES, II The study of shoppers in secondhand stores cited in the previous exercise also compared the income distributions of shoppers in the two stores. Here is a two-way table of counts:

Income	City 1	City 2
Under \$10,000	70	62
\$10,000 to \$19,999	52	63
\$20,000 to \$24,999	69	50
\$25,000 to \$34,999	22	19
\$35,000 or more	28	24

A statistical calculator gives the chi-square statistic for this table as $X^2 = 3.955$. Is there good evidence that customers at the two stores have different income distributions?

CHAPTER REVIEW

This chapter develops several settings where a variation of the chi-square test of significance is useful. In a **goodness of fit test,** the object is to determine if a population distribution has changed. The null hypothesis states that there is no difference between two distributions, while the alternative hypothesis states that there is a difference. The chi-square test tells whether there is sufficient reason to reject the null hypothesis, but further analysis is needed to determine how and where the changes have occurred.

A goodness of fit test begins by finding the expected counts for each category, if the assumed distribution has not changed. The chi-square statistic is a measure of how much the sample distribution diverges from the hypothesized distribution. For a given number of degrees of freedom, large chi-square statistic values provide evidence to reject the null hypothesis of no difference.

The **chi-square test for homogeneity of populations** is an overall test that tells us whether the data give good reason to reject the hypothesis that the distribution of a categorical variable is the same in several populations. It can be used when the data come from independent SRSs from the populations of interest. This procedure is also useful in testing the equality of proportions of successes in any number of populations. The alternative to this hypothesis is "many-sided," because it allows any relationship other than "all equal."

Two-way tables also arise when an SRS is taken from a single population, and each individual is classified according to two categorical variables. In this setting, use a **chi-square test of association/independence.** This procedure tests the null hypothesis that there is "no relationship" between the row variable and the column variable in a two-way table.

The chi-square test is actually an approximate test that becomes more accurate as the cell counts in the two-way table increase. Fortunately, chi-square *P*-values are quite accurate even for small counts. You should always accompany a chi-square test by data analysis to see what kind of relationship is present.

After studying this chapter, you should be able to do the following.

A. CHOOSE THE APPROPRIATE CHI-SQUARE PROCEDURE

1. For goodness-of-fit tests, use percents and bar graphs to compare hypothesized and actual distributions.

2. Distinguish between tests of homogeneity of populations and tests of association/independence.

3. Organize categorical data in a two-way table. Then use percents and bar graphs to describe the relationship between the categorical variables.

B. PERFORM CHI-SQUARE TESTS

1. Explain what null hypothesis is being tested.

2. Calculate expected counts.

3. Calculate the component of the chi-square statistic for any cell, as well as the overall statistic.

4. Give the degrees of freedom of a chi-square statistic.

5. Use the chi-square critical values in Table E to approximate the *P*-values of a chi-square test.

C. INTERPRET CHI-SQUARE TESTS

1. Locate expected cell counts, the chi-square statistic, and its *P*-value in output from computer software or a calculator.

2. If the test is significant, use percents, comparison of expected and observed counts, and the components of the chi-square statistic to see what deviations from the null hypothesis are most important.

CHAPTER 13 REVIEW EXERCISES

13.31 AP EXAM SCORES The Advanced Placement (AP) Statistics examination was first administered in May 1997. Students' papers are graded on a scale of 1 to 5, with 5 being the highest score. Over 7600 students took the exam in the first year, and the distribution of scores was as follows (not including exams that were scored late):

Score:	5	4	3	2	1
Percent:	15.3	22.0	24.8	19.8	18.1

A sample of students who took the exam had the following distribution of grades:

Score:	5	4	3	2	1
Frequency:	167	158	101	79	30

Calculate marginal percents and make a segmented bar graph of the population scores and the sample scores, so that the two distributions can be compared visually. Then perform an appropriate test to determine if the distribution of scores for this particular sample is significantly different from the distribution of scores for all students who took the inaugural exam.

13.32 EFFECTS OF ALCOHOL AND NICOTINE ON CHILDREN Alcohol and nicotine consumption during pregnancy may harm children. Because drinking and smoking behaviors may be related, it is important to understand the nature of this relationship when assessing the possible effects on children. One study classified 452 mothers according to their alcohol intake prior to pregnancy recognition and their nicotine intake during pregnancy. The data are summarized in the following table:[18]

	Nicotine (milligrams/day)		
Alcohol (ounces/day)	None	1–15	16 or more
None	105	7	11
0.01–0.10	58	5	13
0.11–0.99	84	37	42
1.00 or more	57	16	17

Carry out a complete analysis of the association between alcohol and nicotine consumption. That is, describe the nature and strength of this association and assess its statistical significance. Include charts or figures to display the association.

13.33 CANCER PATIENTS' ATTITUDES It seems that the attitude of cancer patients can influence the progress of their disease. We can't experiment with humans, but here is a rat experiment on this theme. Inject 60 rats with tumor cells then divide them at random into two groups of 30. All the rats receive electric shocks, but rats in Group 1 can end the shock by pressing a lever. (Rats learn this sort of thing quickly.) The rats in Group 2 cannot control the shocks, which presumably make them feel helpless and unhappy. We suspect that the rats in Group 1 will develop fewer tumors. The results: 11 of the Group 1 rats and 22 of the Group 2 rats developed tumors.[19]

(a) State the null and alternative hypotheses for this investigation. Explain why the z test rather than the chi-square test for a 2×2 table is the proper test.

(b) Carry out the test and report your conclusion.

13.34 ALCOHOLISM IN TWINS A study of possible genetic influences on alcoholism studied pairs of adult female twins. The subjects were identified from the Virginia Twin Registry, which lists all twins born in Virginia. Each pair of twins was classified as identical or fraternal. Only identical twins share exactly the same genes. Based on an interview, each woman was classified as a problem drinker or not. Here are the data for the 1030 pairs of twins for which information was available:[20]

Problem drinker	Identical	Fraternal
Neither	443	301
One	102	113
Both	45	26
Total	590	440

(a) Is there a significant relationship between type of twin and the presence of problem drinking in the twin pair? Which cells contribute heavily to the chi-square value?

(b) Your result in (a) suggests a clearer analysis. Make a 2×2 table of "same or different" problem-drinking behavior within a twin pair by type of twin. To do this, combine the "Neither" and "Both" categories to form the "Same behavior" category. If heredity influences behavior, we would expect a higher proportion of identical twins to show the same behavior. Is there a significant effect of this kind?

13.35 PYTHON EGGS How is the hatching of water python eggs influenced by the temperature of the snake's nest? Researchers assigned newly laid eggs to one of three temperatures: hot, neutral, or cold. Hot duplicates the extra warmth provided by the mother python, and cold duplicates the absence of the mother. Here are the data on the number of eggs and the number that hatched:[21]

	Eggs	Hatched
Cold	27	16
Neutral	56	38
Hot	104	75

(a) Make a two-way table of temperature by outcome (hatched or not).

(b) Calculate the percent of eggs in each group that hatched. The researchers anticipated that eggs would not hatch in cold water. Do the data support that anticipation?

(c) Are there significant differences among the proportions of eggs that hatched in the three groups?

13.36 PEA PLANTS Much of Gregor Mendel's early genetic research was performed on pea plants. He examined the offspring resulting from "crossing" several parent plants. One discovery he made was that having green (G) seeds was a dominant trait for a pea plant, whereas having yellow seeds (g) was a recessive trait. Using the Punnett square below, Mendel could have predicted the offspring if two plants that each had one dominant gene (G) and one recessive gene (g) were crossed.

		Plant 1	
		G	g
Plant 2	G	GG	Gg
	g	Gg	gg

Based on this genetic model, we would expect a ratio of 3 green-seeded plants to 1 yellow-seeded plant.

An experiment like the one described is performed. The resulting offspring are 639 green-seeded pea plants and 241 yellow-seeded pea plants. Do these data give you any reason to doubt the hypothesized model? Give appropriate statistical evidence to support your conclusion.

13.37 DO PETS INCREASE SURVIVAL? Psychological and social factors can influence the survival of patients with serious diseases. One study examined the relationship between survival of patients with coronary heart disease (CHD) and pet ownership. Each of 92 patients was classified as having a pet or not and by whether they survived for one year. Here are the data:[22]

	Pet ownership	
Patient status	No	Yes
Alive	28	50
Dead	11	3

(a) Was this study an experiment? Why or why not?

(b) The researchers thought that having a pet might improve survival, so pet ownership is the explanatory variable. Compute appropriate percentages to describe the data and state your preliminary findings.

(c) Carry out an appropriate inference procedure to test the researchers' claim.

(d) What do you conclude? Do the data give convincing evidence that owning a pet is an effective treatment for increasing the survival of CHD patients?

(e) Did you use a χ^2 test or a z test in part (c)? Carry out the other test and compare the results.

13.38 PREVENTING STROKES Exercise 12.26 (page 712) compared aspirin plus another drug with aspirin alone as treatments for patients who had suffered a stroke. The study actually assigned stroke patients at random to four treatments. Here are the data:[23]

Treatment	Number of patients	Number of strokes	Number of deaths
Placebo	1649	250	202
Aspirin	1649	206	182
Dipyridamole	1654	211	188
Both	1650	157	185

(a) Make a two-way table of treatment by whether or not a patient had a stroke during the two-year study period. Compare the rates of strokes for the four treatments. Which treatment appears most effective in preventing strokes? Is there a significant difference among the four rates of strokes? Which components of chi-square account for most of the total?

(b) The data report two response variables: whether the patient had a stroke and whether the patient died. Repeat your analysis for patient deaths.

(c) Write a careful summary of your overall findings.

13.39 A TITANIC DISASTER In 1912 the luxury liner *Titanic*, on its first voyage across the Atlantic, struck an iceberg and sank. Some passengers got off the ship in lifeboats, but many died. Think of the Titanic disaster as an experiment in how the people of that time behaved when faced with death in a situation where only some can escape. The passengers are a sample from the population of their peers. Here is information about who lived and who died, by sex and economic status. (The data leave out a few passengers whose economic status is unknown.)[24]

Men			Women		
Status	Died	Survived	Status	Died	Survived
Highest	111	61	Highest	6	126
Middle	150	22	Middle	13	90
Lowest	419	85	Lowest	107	101
Total	680	168	Total	126	317

(a) Compare the percents of men and of women who died. Is there strong evidence that a higher proportion of men die in such situations? Why do you think this happened?

(b) Look only at the women. Describe how the three economic classes differ in the percent of women who died. Are these differences statistically significant?

(c) Now look only at the men and answer the same questions.

13.40 PROB SIM The Prob Sim APP for the TI-83/89 allows you to simulate tossing coins, rolling dice, spinning a spinner, drawing cards, and playing the lottery. If you have a TI-83/89, download this APP from your teacher.

• To run the APP, press the APPS key. On the TI-83 Plus, choose Prob Sim. On the TI-89, choose 1:FlashApps, then Prob Sim. Press ENTER. You should see the introductory screen shown at the left below. Press ENTER again to see the main menu (shown at the right below).

Probability
Simulation
Version 1.0
2000 Corey Taylor
Rusty Wagner
PRESS ANY KEY

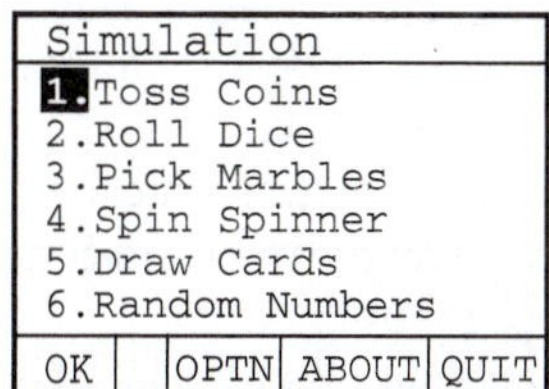

• Choose 4. Spin Spinner. Spin the spinner a total of 200 times with 4 sets of 50 spins. Record the number of times that the spinner lands in each of the four numbered sections.

• Perform a significance test of the hypothesis that this program yields an equal proportion of 1s, 2s, 3s, and 4s. If you do not have the Prob Sim APP, use the following sample data: 51 1s, 39 2s, 53 3s, 58 4s.

NOTES AND DATA SOURCES

1. Stephen M. Stigler, *The History of Statistics: The Measurement of Uncertainty before 1900*, Belknap Press, Cambridge, Mass., 1986, page 361.
2. *Statistical Abstract of the United States*, 1997, U.S. Department of Commerce, Bureau of the Census.
3. D. M. Barnes, "Breaking the cycle of addiction," *Science*, 241 (1988), pp. 1029–1030.
4. Douglas E. Jorenby et al., "A controlled trial of sustained-release bupropion, a nicotine patch, or both for smoking cessation," *New England Journal of Medicine*, 340(1990), pp. 685–691.
5. This study is reported in Joan L. Duda, "The relationship between goal perspectives, persistence, and behavioral intensity among male and female recreational sports participants," *Leisure Sciences*, 10(1988), pp. 95–106.
6. There are many computer studies of the accuracy of chi-square critical values for X^2. For a brief discussion and some references, see Section 3.2.5 of David S. Moore "Tests of chi-squared type," in Ralph B. D'Agostino and Michael A. Stephens (eds.), *Goodness-of-Fit Techniques*, Marcel Dekker, New York, 1986, pp. 63–95. If the expected cell counts are roughly equal, the chi-square approximation is adequate when the average expected counts are as small as 1 or 2. The guideline given in the text protects against unequal expected counts. For a survey of inference for smaller samples, see Alan Agresti, "A

survey of exact inference for contingency tables," *Statistical Science*, 7(1992), pp. 131–177.

7. Data compiled from a table of percents in "Americans view higher education as key to the American dream," press release by the National Center for Public Policy and Higher Education, www.highereducation.org, May 3, 2000.

8. Ray H. Rosenman et al., "A 4-year prospective study of the relationship of different habitual vocational physical activity to risk and incidence of ischemic heart disease in volunteer male federal employees," in P. Milvey (ed.), *The Marathon: Physiological, Medical, Epidemiological, and Psychological Studies*, New York Academy of Sciences, 301 (1977), pp. 627–641.

9. Richard M. Felder et al., "Who gets it and who doesn't: a study of student performance in an introductory chemical engineering course," 1992 ASEE *Annual Conference Proceedings*, American Society for Engineering Education, Washington, D.C., 1992, pp. 1516–1519.

10. S. V. Zagona (ed.), *Studies and Issues in Smoking Behavior*, University of Arizona Press, Tucson, 1967, pp. 157–180.

11. Catharine Gale and Christopher Martyn, "Larks and owls and health, wealth, and wisdom," *British Medical Journal*, 317(1998), pp. 1675–1677.

12. Lillian Lin Miao, "Gastric freezing: an example of the evaluation of medical therapy by randomized clinical trials," in John P. Bunker, Benjamin A. Barnes, and Frederick Mosteller (eds.), *Costs, Risks, and Benefits of Surgery*, Oxford University Press, New York, 1977, pp. 198–211.

13. Brenda C. Coleman, "Study: heart attack risk cut 74% by stress management," Associated Press dispatch appearing in the *Lafayette (Indiana) Journal and Courier*, October 20, 1997.

14. Based closely on Susan B. Sorenson, "Regulating firearms as a consumer product," *Science*, 286(1999), pp. 1481–1482. Because the results in the paper were "weighted to the U.S. population," I have changed some counts slightly for consistency.

15. Modified from Felicity Barringer, "Measuring sexuality through polls can be shaky," *New York Times*, April 25, 1993.

16. David M. Blau, "The child care labor market," *Journal of Human Resources*, 27 (1992), pp. 9–39.

17. William D. Darley, "Store-choice behavior for pre-owned merchandise," *Journal of Business Research*, 27 (1993), pp. 17–31.

18. Ann P. Streissguth et al., "Intrauterine alcohol and nicotine exposure: attention and reaction time in 4-year-old children," *Developmental Psychology*, 20 (1984), pp. 533–541.

19. Adapted from M. A. Visintainer, J. R. Volpicelli, and M. E. P. Seligman, "Tumor rejection in rats after inescapable or escapable shock," *Science*, 216 (1982), pp. 437–439.

20. These are part of the data from the EESEE story "Alcoholism in Twins." The study results appear in K. S. Kendler et al., "A population-based twin study of alcoholism in women," *Journal of the American Medical Association*, 268 (1992), pp. 1877–1882.

21. R. Shine, T. R. L. Madsen, M. J. Elphick, and P. S. Harlow, "The influence of nest temperatures and maternal brooding on hatchling phenotypes in water pythons," *Ecology*, 78 (1997), pp. 1713–1721.

22. Erika Friedmann et al., "Animal companions and one-year survival of patients after discharge from a coronary care unit," *Public Health Reports*, 96 (1980), pp. 307–312.

23. Martin Enserink, "Fraud and ethics charges hit stroke drug trial," *Science*, 274 (1996), pp. 2004–2005.

24. Data provided by Don Bentley, Pomona College.

The Granger Collection, New York

ADRIEN-MARIE LEGENDRE

Predicting the Paths of Comets

In 1805 the Frenchman *Adrien-Marie Legendre (1752–1833)*, made a significant contribution to statistics with his publication of the first statement of the method of least squares. In an appendix of a book on determining the orbits of comets, he described his method, which involved three observations taken at equal intervals. Assuming that a comet follows a parabolic path, he applied his methods to the data known for two comets, minimizing the sum of the squares of the residuals to fit a curve to the available data.

Later, in 1809, the younger Carl Friedrich Gauss published his version of the method of least squares and claimed it as his own even though he did acknowledge Legendre's 1805 book. This naturally caused friction between the two men, and at one point Legendre wrote about Gauss: "This excessive impudence is unbelievable in a man who has sufficient personal merit not to have need of appropriating the discoveries of others." Although Gauss may not have been first with the idea, his development of the method did enhance its usefulness. Sir Frances Galton is also recognized for popularizing the method of least squares.

Legendre was known principally for his very popular book *Elements de Geometrie*, published in 1794, in which he tried to improve on Euclid's *Elements* by extensively rearranging and simplifying many of the propositions. It is said that two years was the usual time for even the better students to master Legendre's *Geometrie*. The textbook revealed the entire structure of elementary geometry in crystal clarity and replaced Euclid's *Elements* as a textbook in most of Europe. It was translated into English in 1819 and again in 1824 and, through 33 subsequent American editions, became the prototype for later geometry textbooks in America.

In 1805, Legendre published the first statement of the method of least squares.

chapter 14

Inference for Regression

ACTIVITY 14

Materials: Fabric tape measure; calculator

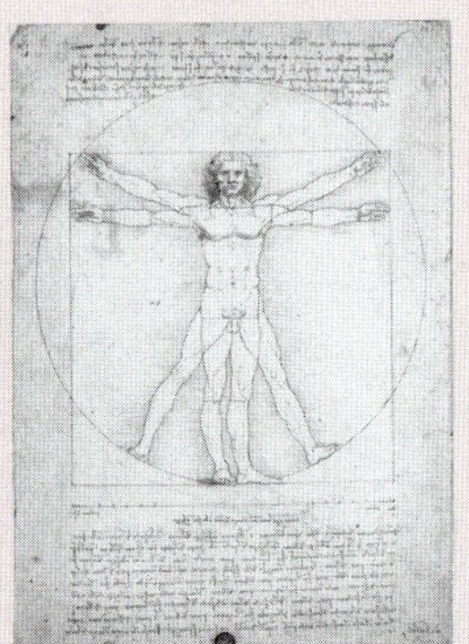
Scala/Art Resource

The architect Vitruvius said that "if you open your legs so much as to decrease your height by 1/14 and spread and raise your arms til your middle fingers touch the level of the top of your head, you must know that the center of the outspread limbs will be in the navel and the space between the legs will be an equilateral triangle. . . . The length of a man's outspread arms is equal to his height."

Leonardo da Vinci, the renowned painter, drew the illustration above for a book on the works of Vitruvius. Da Vinci believed that the human body conformed to a set of geometric proportions as shown by the lines and circles in this drawing

In this activity, we want to determine if arm span can predict height. You will need a fabric measuring tape, and you should work in teams of three: the person to be measured and two people to hold the ends of the tape. You should collect at least 18 to 20 pairs of measurements. If your class has fewer students, recruit some volunteers from other classes. Remember: the more, the better.

1. Take turns taking these two measurements and recording them. First measure your arm span: the distance between the tips of the fingers when you stretch your arms out to the sides (the x-values). Then measure your height (the y-values). Unlike Vitruvius's man, who made an equilateral triangle with his legs, you will keep your heels together and stand tall. Combine your results with those of the other groups.

2. Make a scatterplot of the data. Clearly, the association should be positive. Is it? Would you describe the association as strong, moderate, or weak?

3. Use your calculator to perform least-squares regression and find the values of r and r^2. Plot the least-squares line on your scatterplot. Write a statement that interprets the meaning, in context, of the least-squares line and value of r^2 that you found.

4. Construct a residual plot to assess whether a line is an appropriate model for these data. Write a sentence that interprets your residual plot.

Keep your data; we will use them later in the chapter.

14.1 INFERENCE ABOUT THE MODEL

When a scatterplot shows a linear relationship between a quantitative explanatory variable x and a quantitative response variable y, we can use the least-squares line fitted to the data to predict y for a given value of x. Now we want to do tests and confidence intervals in this setting.

EXAMPLE 14.1 CRYING AND IQ

Infants who cry easily may be more easily stimulated than others and this may be a sign of higher IQ. Child development researchers explored the relationship between the crying of infants four to ten days old and their later IQ test scores. A snap of a rubber band on the sole of the foot caused the infants to cry. The researchers recorded the crying and measured its intensity by the number of peaks in the most active 20 seconds. They later measured the children's IQ at age three years using the Stanford-Binet IQ test. Table 14.1 contains data on 38 infants.

TABLE 14.1 Infants' crying and IQ scores

Crying	IQ	Crying	IQ	Crying	IQ	Crying	IQ
10	87	20	90	17	94	12	94
12	97	16	100	19	103	12	103
9	103	23	103	13	104	14	106
16	106	27	108	18	109	10	109
18	109	15	112	18	112	23	113
15	114	21	114	16	118	9	119
12	119	12	120	19	120	16	124
20	132	15	133	22	135	31	135
16	136	17	141	30	155	22	157
33	159	13	162				

Source: Samuel Karelitz et al., "Relation of crying activity in early infancy to speech and intellectual development at age three years," *Child Development*, 35 (1964), pp. 769–777.

Plot and interpret. As always, we first examine the data. Figure 14.1 is a *scatterplot* of the crying data. Plot the explanatory variable (count of crying peaks) horizontally and the response variable (IQ) vertically. Look for the form, direction, and strength of the relationship as well as for outliers or other deviations. There is a moderate positive linear relationship, with no extreme outliers or potentially influential observations.

Numerical summary. Because the scatterplot shows a roughly linear (straight-line) pattern, the *correlation* describes the direction and strength of the relationship. The correlation between crying and IQ is $r = 0.455$.

Mathematical model. We are interested in predicting the response from information about the explanatory variable. So we find the *least-squares regression line* for predicting IQ from crying. This line lies as close as possible to the points (in the sense of least squares) in the vertical (y) direction. The equation of the least-squares regression line is

$$\begin{aligned}\hat{y} &= a + bx \\ &= 91.27 + 1.493x\end{aligned}$$

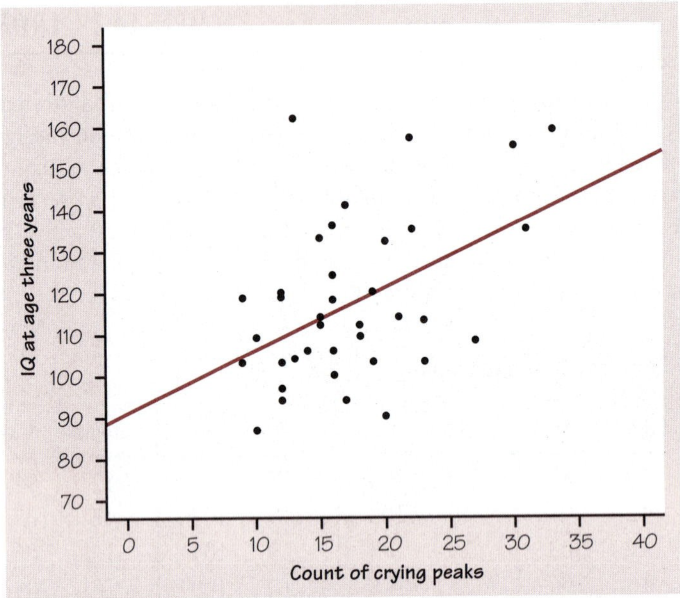

FIGURE 14.1 Scatterplot of the IQ score of infants at age three years against the intensity of their crying soon after birth, for Example 14.1.

Here are the relevant TI-83 screens. The TI-89 results are similar.

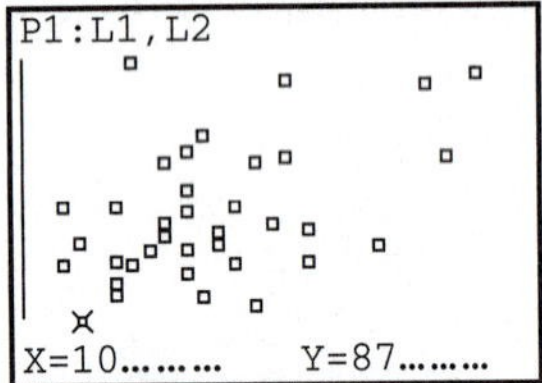

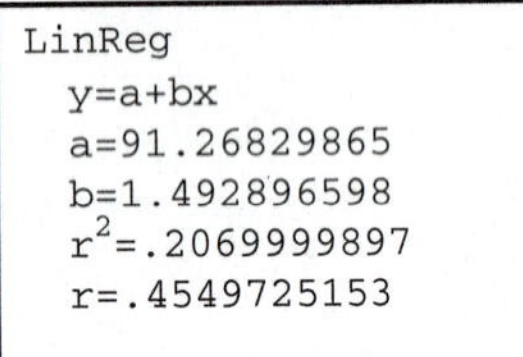

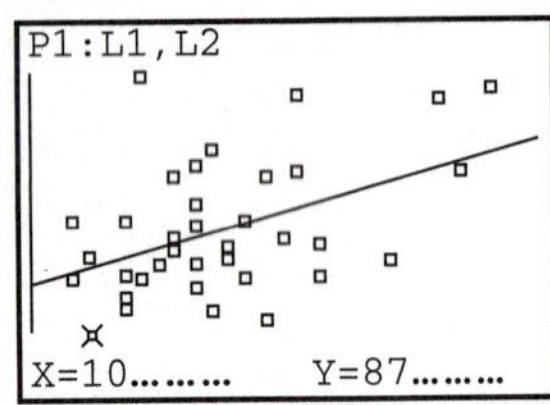

We use the notation $\hat{y}$ to remind ourselves that the regression line gives *predictions* of IQ. The predictions usually won't agree exactly with the actual values of the IQ measured several years later. Drawing the least-squares line on the scatterplot helps us see the overall pattern. Because $r^2 = 0.207$, only about 21% of the variation in IQ scores is explained by crying intensity. Prediction of IQ will not be very accurate. It is nonetheless impressive that behavior soon after birth can even partly predict IQ several years later.

The regression model

The slope b and intercept a of the least-squares line are *statistics*. That is, we calculated them from the sample data. These statistics would take somewhat different values if we repeated the study with different infants. To do formal inference, we think of a and b as estimates of unknown *parameters*. The parameters appear in a mathematical model of the process that produces our data. Here are the required conditions for performing inference about the regression model.

CONDITIONS FOR REGRESSION INFERENCE

We have n observations on an explanatory variable x and a response variable y. Our goal is to study or predict the behavior of y for given values of x.

- For any fixed value of x, the response y varies according to a normal distribution. Repeated responses y are independent of each other.
- The mean response μ_y has a straight-line relationship with x:

$$\mu_y = \alpha + \beta x$$

The slope β and intercept α are unknown parameters.

- The standard deviation of y (call it σ) is the same for all values of x. The value of σ is unknown.

The heart of this model is that there is an "on the average" straight-line relationship between y and x. The ***true regression line*** $\mu_y = \alpha + \beta x$ says that the *mean* response μ_y moves along a straight line as the explanatory variable x changes. We can't observe the true regression line. The values of y that we do observe vary about their means according to a normal distribution. If we hold x fixed and take many observations on y, the normal pattern will eventually appear in a stemplot or histogram. In practice, we observe y for many different values of x, so that we see an overall linear pattern formed by points scattered about the true line. The standard deviation σ determines whether the points fall close to the true regression line (small σ) or are widely scattered (large σ).

true regression line

Figure 14.2 shows the regression model in picture form. The line in the figure is the true regression line. The mean of the response y moves along this line as the explanatory variable x takes different values. The normal curves show how y will vary when x is held fixed at different values. All of the curves have the same σ, so the variability of y is the same for all values of x. You should check the conditions for inference when you do inference about regression. We will see later how to do that.

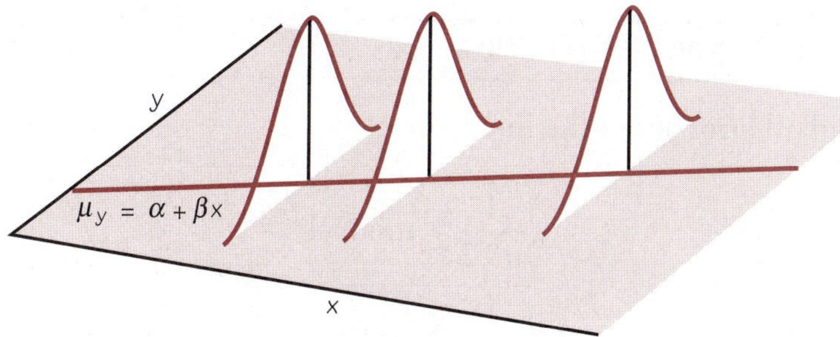

FIGURE 14.2 The regression model. The line is the true regression line, which shows how the mean response μ_y changes as the explanatory variable x changes. For any fixed value of x, the observed response y varies according to a normal distribution having mean μ_y.

Inference

The first step in inference is to estimate the unknown parameters α, β, and σ. When the regression model describes our data and we calculate the least-squares line $\hat{y} = a + bx$, **the slope b of the least-squares line is an unbiased estimator of the true slope β, and the intercept a of the least-squares line is an unbiased estimator of the true intercept α.**

EXAMPLE 14.2 SLOPE AND INTERCEPT

The data in Figure 14.1 fit the regression model of scatter about an invisible true regression line reasonably well. The least-squares line is $\hat{y} = 91.27 + 1.493x$. The slope is particularly important. A *slope is a rate of change.* The true slope β says how much higher average IQ is for children with one more peak in their crying measurement. Because $b = 1.493$ estimates the unknown β, we estimate that on the average IQ is about 1.5 points higher for each added crying peak.

We need the intercept $a = 91.27$ to draw the line, but it has no statistical meaning in this example. No child had fewer than 9 crying peaks, so we have no data near $x = 0$. We suspect that all normal children would cry when snapped with a rubber band, so that we will never observe $x = 0$.

The remaining parameter of the model is the standard deviation σ, which describes the variability of the response y about the true regression line. The least-squares line estimates the true regression line. So the *residuals* estimate how much y varies about the true line. Recall that the residuals are the vertical deviations of the data points from the least-squares line:

$$\begin{aligned}\text{residual} &= \text{observed } y - \text{predicted } y \\ &= y - \hat{y}\end{aligned}$$

There are n residuals, one for each data point. Because σ is the standard deviation of responses about the true regression line, we estimate it by a sample standard deviation of the residuals. We call this sample standard deviation a *standard error* to emphasize that it is estimated from data. The residuals from a least-squares line always have mean zero. That simplifies their standard error.

STANDARD ERROR ABOUT THE LEAST-SQUARES LINE

The **standard error about the line** is

$$\begin{aligned}s &= \sqrt{\frac{1}{n-2}\sum \text{residual}^2} \\ &= \sqrt{\frac{1}{n-2}\sum (y-\hat{y})^2}\end{aligned}$$

Use s to estimate the unknown σ in the regression model.

Because we use the standard error about the line so often in regression inference, we just call it s. Notice that s^2 is an average of the squared deviations of the data points from the line, so it qualifies as a variance. We average the squared deviations by dividing by $n - 2$, the number of data points less 2. It turns out that if we know $n - 2$ of the n residuals, the other two are determined. That is, $n - 2$ is the ***degrees of freedom*** of s. We first met the idea of degrees of freedom in the case of the ordinary sample standard deviation of n observations, which has $n - 1$ degrees of freedom. Now we observe two variables rather than one, and the proper degrees of freedom is $n - 2$ rather than $n - 1$.

degrees of freedom

Calculating s begins with finding the predicted response for each x in your data set, then the residuals, and then s. In practice you will use technology that does this arithmetic instantly. The next example shows how to use the calculator to help calculate s.

EXAMPLE 14.3 RESIDUALS AND STANDARD ERROR

Table 14.1 shows that the first infant studied had 10 crying peaks and a later IQ of 87. The predicted IQ for $x = 10$ is

$$\begin{aligned}\hat{y} &= 91.27 + 1.493x \\ &= 91.27 + 1.493(10) = 106.2\end{aligned}$$

The residual for this observation is

$$\begin{aligned}\text{residual} &= y - \hat{y} \\ &= 87 - 106.2 = -19.2\end{aligned}$$

That is, the observed IQ for this infant lies 19.2 points below the least-squares line on the scatterplot.

Repeat this calculation 37 more times, once for each subject. The 38 residuals are

−19.20	−31.13	−22.65	−15.18	−12.18	−15.15	−16.63	−6.18
−1.70	−22.60	−6.68	−6.17	−9.15	−23.58	−9.14	2.80
−9.14	−1.66	−6.14	−12.60	0.34	−8.62	2.85	14.30
9.82	10.82	0.37	8.85	10.87	19.34	10.89	−2.55
20.85	24.35	18.94	32.89	18.47	51.32		

If you haven't entered the crying and IQ data into your calculator, do that now as L_1/list1 and L_2/list2. Then on the TI-83, define list L_3 to be the observed minus the predicted values of y: $L_2 - Y1(L_1)$. On the TI-89, define list3 to be list2 − Y1(list1). Verify that the 38 residuals are as shown and that the sum of the residuals is zero:

L1	L2	L3 3
10	87	----
12	97	
9	103	
16	106	
18	109	
15	114	
12	119	

L3=L2−Y1(L1)

L1	L2	L3 3
10	87	⁻19.2
12	97	⁻12.18
9	103	⁻1.704
16	106	⁻9.155
18	109	⁻9.14
15	114	.33825
12	119	9.8169

L3(1)=⁻19.1972646...

1-Var Stats
$\bar{x}$=3.157895E⁻12
Σx=1.2E⁻10
Σx²=11023.3888
Sx=17.26063229
σx=17.03200455
↓n=38

Notice that the sum of the residuals is shown in the calculator screen as 1.2E-10, which is zero, up to roundoff error. Another reason to use technology in doing regression is that roundoff errors in hand calculation can accumulate and make the results inaccurate.

The variance about the line is

$$s^2 = \frac{1}{n-2}\sum \text{residual}^2$$
$$= \frac{1}{38-2}\left[(-19.20)^2 + (-31.13)^2 + \cdots + 51.32^2\right]$$
$$= \frac{1}{36}(11{,}023.3) = 306.20$$

Finally, the standard error about the line is

$$s = \sqrt{306.20} = 17.50$$

Software gives 17.4987 to four decimal places, so the error resulting from rounding in this hand calculation is small.

Technology tip: Here's a quick way to calculate s. With x-values in L_1/list1 and the y-values in L_2/list2, perform least-squares regression. The calculator creates or updates a list named RESID. Specify `1-Var Stats` $_L$RESID and look at the value Σx^2. That's the sum of the squares of the residuals. Divide this number by $(n - 2)$ to get s^2. Take the square root to obtain s.

We will study several kinds of inference in the regression setting. The standard error s about the line is the key measure of the variability of the responses in regression. It is part of the standard error of all the statistics we will use for inference.

EXERCISES

14.1 AN EXTINCT BEAST, I *Archaeopteryx* is an extinct beast having feathers like a bird but teeth and a long bony tail like a reptile. Here are the lengths in centimeters of the femur (a leg bone) and the humerus (a bone in the upper arm) for the five fossil specimens that preserve both bones:

Femur:	38	56	59	64	74
Humerus:	41	63	70	72	84

The strong linear relationship between the lengths of the two bones helped persuade scientists that all five specimens belong to the same species.

(a) Examine the data. Make a scatterplot with femur length as the explanatory variable. Use your calculator to obtain the correlation r and the equation of the least-squares regression line. Do you think that femur length will allow good prediction of humerus length?

(b) Explain in words what the slope β of the true regression line says about *Archaeopteryx*. What is the estimate of β from the data? What is your estimate of the intercept α of the true regression line?

(c) Calculate the residuals for the five data points. Check that their sum is 0 (up to roundoff error). Use the residuals to estimate the standard deviation σ in the regression model. You have now estimated all three parameters in the model.

14.2 SARAH'S GROWTH Sarah's growth from age 3 years to 5 years was measured as follows:

Age (months):	36	48	51	54	57	60
Height (cm):	86	90	91	93	94	95

These data were entered into a statistics package and least-squares regression of height on age was requested. Here are the results:

```
Predictor        Coef       Stdev     t-ratio       p
Constant       71.950       1.053       68.33   0.000
Age           0.38333     0.02041       18.78   0.000

s = 0.3873          R-sq = 98.9%      R-sq(adj) = 98.6%
```

(a) What is the equation of the least-squares line? (Hint: Look for the column "Coef." What is the intercept? What is the slope?)

(b) The model for regression inference has three parameters, which we call α, β, and σ. Can you determine the estimates for α and β from the computer printout? What are they?

(c) The computer output reports that $s = 0.3873$. This is an estimate of the parameter σ. Use the formula for s to verify the computer's value of s.

14.3 IDEAL PROPORTIONS, I Mr. Starnes's students measured their arm spans and heights (see Activity 14), entered their results into a Minitab worksheet, requested least-squares regression of height on arm span (both in inches), and obtained the following output:

```
Predictor        Coef       Stdev     t-ratio       p
Constant       11.547       5.600        2.06   0.056
armspan       0.84042     0.08091       10.39   0.000

s = 1.613         R-sq = 87.1%      R-sq(adj) = 86.3%
```

A residual plot for the data is shown in Figure 14.3.

(a) Determine the equation of the least-squares regression line from the "Coef" column in the printout.

(b) In your opinion, is the least-squares line an appropriate model for the data? Would you be willing to predict a student's height, if you knew that his arm span is 76 inches? Explain.

(c) Estimate the parameters of α and β.

(d) Use an appropriate formula to verify that the estimate for σ is 1.613.

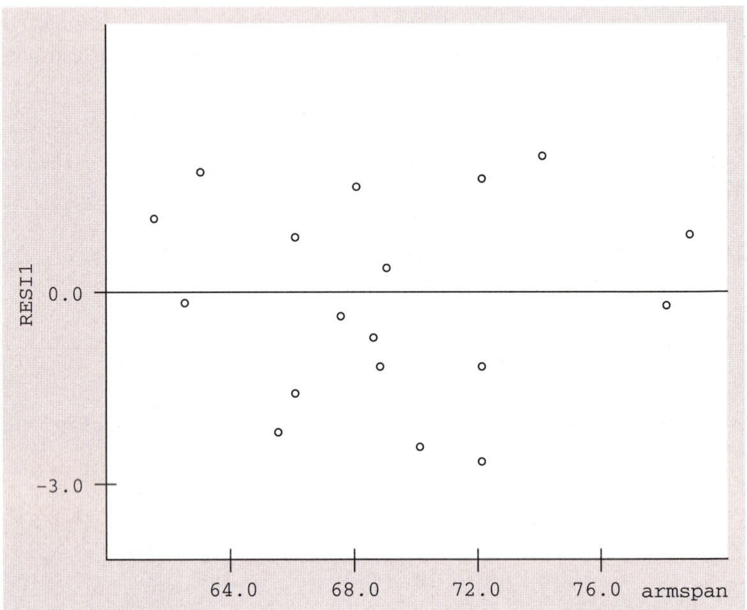

FIGURE 14.3 Residual plot for Exercise 14.3.

14.4 COMPETITIVE RUNNERS Exercise 3.71 on page 187 provided data on the speed of competitive runners and the number of steps they took per second. Good runners take more steps per second as they speed up. Here are the data again:

Speed(ft/s):	15.86	16.88	17.50	18.62	19.97	21.06	22.11
Steps per second:	3.05	3.12	3.17	3.25	3.36	3.46	3.55

(a) Enter the data into your calculator, perform least-squares regression, and plot the scatterplot with the least-squares line. What is the strength of the association between speed and steps per second?

(b) Find the residuals for all 7 data points. Check that their sum is 0 (up to roundoff error).

(c) The model for regression inference has three parameters, α, β, and σ. Estimate these parameters from the data.

14.5 IDEAL PROPORTIONS, II Estimate the parameters α, β, and σ from the arm span and height data you collected from Activity 14. For which class does the least-squares line provide a better model, your class or the class described in Exercise 14.3? Explain.

Confidence intervals for the regression slope

The slope β of the true regression line is usually the most important parameter in a regression problem. The slope is the rate of change of the mean response as the explanatory variable increases. We often want to estimate β. The slope b of the least-squares line is an unbiased estimator of β. A confidence interval is more useful because it shows how accurate the estimate b is likely to be. The confidence interval for β has the familiar form

$$\text{estimate} \pm t^* \text{SE}_{\text{estimate}}$$

Because b is our estimate, the confidence interval becomes

$$b \pm t^* \text{SE}_b$$

Here are the details.

CONFIDENCE INTERVAL FOR REGRESSION SLOPE

A level C confidence interval for the slope β of the true regression line is

$$b \pm t^* \text{SE}_b$$

In this recipe, the standard error of the least-squares slope b is

$$\text{SE}_b = \frac{s}{\sqrt{\Sigma(x-\bar{x})^2}}$$

and t^* is the upper $(1 - C)/2$ critical value from the t distribution with $n - 2$ degrees of freedom.

As advertised, the standard error of b is a multiple of s. Although we give the recipe for this standard error, you should rarely have to calculate it by hand. Regression software gives the standard error SE_b along with b itself.

EXAMPLE 14.4 REGRESSION OUTPUT: CRYING AND IQ

Figure 14.4 shows the basic output for the crying study from the regression command in the Minitab software package. Most statistical software provides similar output. (Minitab, like other software, produces more than this basic output. When you use software, just ignore the parts you don't need.)

The first line gives the equation of the least-squares regression line. The slope and intercept are rounded off there, so look in the "Coef" column of the table that follows for more accurate values. The intercept $a = 91.268$ appears in the "Constant" row. The slope $b = 1.4929$ appears in the "Crycount" row because we named the x variable "Crycount" when we entered the data.

The next column of output, headed "StDev," gives standard errors. In particular, $\text{SE}_b = 0.4870$. The standard error about the line, $s = 17.50$, appears below the table.

There are 38 data points, so the degrees of freedom are $n - 2 = 36$. For a 95% confidence interval for the true slope β, we will use the critical value $t^* = 2.042$ from the

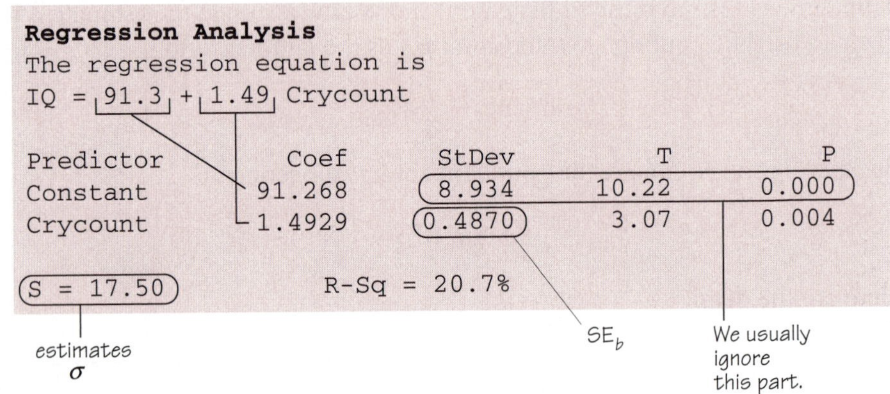

FIGURE 14.4 Minitab regression output for the crying and IQ data.

df = 30 row of Table C. This is the table degrees of freedom next smaller than 36. The interval is

$$\begin{aligned} b \pm t^* \text{SE}_b &= 1.4929 \pm (2.042)(0.4870) \\ &= 1.4929 \pm 0.9944 \\ &= 0.4985 \text{ to } 2.4873 \end{aligned}$$

We are 95% confident that mean IQ increases by between about 0.5 and 2.5 points for each additional peak in crying.

You can find a confidence interval for the intercept α of the true regression line in the same way, using a and SE_a from the "Constant" line of the printout. We rarely need to estimate α.

Testing the hypothesis of no linear relationship

We can also test hypotheses about the slope β. The most common hypothesis is

$$H_0: \beta = 0$$

A regression line with slope 0 is horizontal. That is, the mean of y does not change at all when x changes. So this H_0 says that there is *no true linear relationship* between x and y. Put another way, H_0 says that *straight-line dependence on x is of no value for predicting y*. Put yet another way, H_0 says that there is *no correlation* between x and y in the population from which we drew our data. You can use the test for zero slope to test the hypothesis of zero correlation between any two quantitative variables. That's a useful trick. Do notice that testing correlation makes sense only if the observations are a random sample. That is often not the case in regression settings, where researchers may fix the values of x they want to study.

The test statistic is just the standardized version of the least-squares slope b. It is another t statistic. Here are the details.

SIGNIFICANCE TESTS FOR REGRESSION SLOPE

To test the hypothesis H_0: $\beta = 0$, compute the t statistic

$$t = \frac{b}{\mathrm{SE}_b}$$

In terms of a random variable T having the $t(n - 2)$ distribution, the P-value for a test of H_0 against

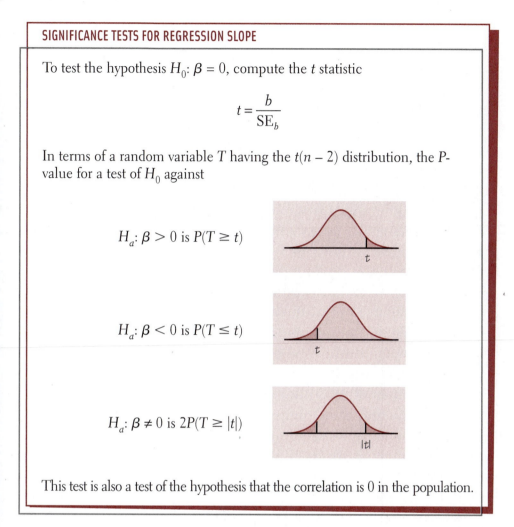

H_a: $\beta > 0$ is $P(T \geq t)$

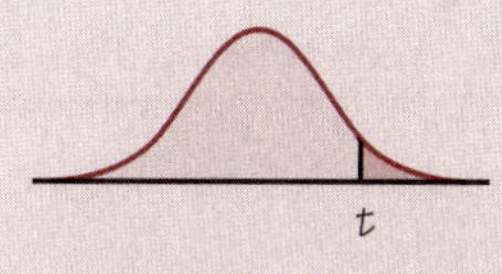

H_a: $\beta < 0$ is $P(T \leq t)$

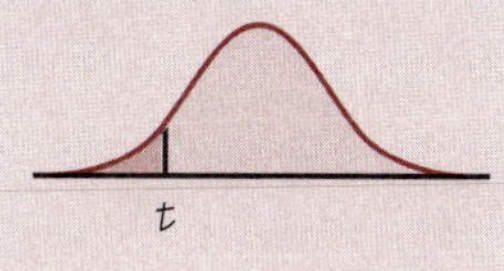

H_a: $\beta \neq 0$ is $2P(T \geq |t|)$

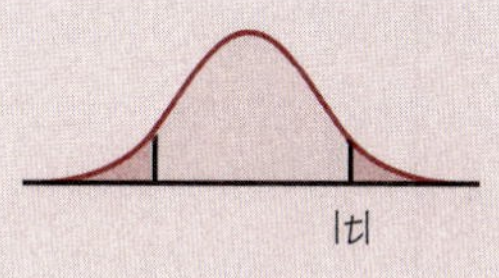

This test is also a test of the hypothesis that the correlation is 0 in the population.

Regression output from statistical software usually gives t and its *two-sided* P-value. For a one-sided test, divide the P-value in the output by 2.

EXAMPLE 14.5 TESTING REGRESSION SLOPE

The hypothesis H_0: $\beta = 0$ says that crying has no straight-line relationship with IQ. Figure 14.1 shows that there is a relationship, so it is not surprising that the computer output in Figure 14.4 gives $t = 3.07$ with two-sided P-value 0.004. There is very strong evidence that IQ is correlated with crying.

EXAMPLE 14.6 BEER AND BLOOD ALCOHOL

How well does the number of beers a student drinks predict his or her blood alcohol content? Sixteen student volunteers at Ohio State University drank a randomly assigned number of cans of beer. Thirty minutes later, a police officer measured their blood alcohol content (BAC). Here are the data:[1]

Student:	1	2	3	4	5	6	7	8
Beers:	5	2	9	8	3	7	3	5
BAC:	0.10	0.03	0.19	0.12	0.04	0.095	0.07	0.06
Student:	9	10	11	12	13	14	15	16
Beers:	3	5	4	6	5	7	1	4
BAC:	0.02	0.05	0.07	0.10	0.085	0.09	0.01	0.05

The students were equally divided between men and women and differed in weight and usual drinking habits. Because of this variation, many students don't believe that number of drinks predicts blood alcohol well. What do the data say?

The scatterplot in Figure 14.5 shows a clear linear relationship. Figure 14.6 gives part of the Minitab regression output. The solid line on the scatterplot is the least-squares line

$$\hat{y} = -0.0127 + 0.0180x$$

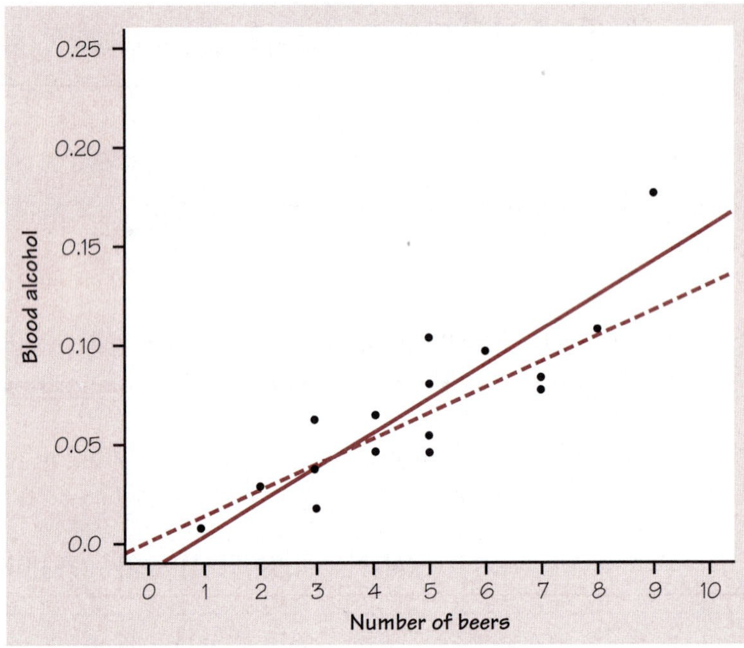

FIGURE 14.5 Scatterplot of students' blood alcohol content against the numbers of cans of beer consumed.

Because $r^2 = 0.800$, number of drinks accounts for 80% of the observed variation in BAC. That is, the data say that student opinion is wrong: the number of beers you drink predicts blood alcohol level quite well. Five beers produce an average BAC of

$$\hat{y} = -0.0127 + 0.0180(5) = 0.077$$

perilously close to the legal driving limit of 0.08 in many states.

```
The regression equation is
BAC = - 0.0127 + 0.0180 Beers


Predictor       Coef        StDev          T        P
Constant    -0.01270      0.01264      -1.00    0.332
Beers       0.017964     0.002402       7.48    0.000

S = 0.02044       R-Sq = 80.0%
```

FIGURE 14.6 Minitab output for the blood alcohol content data.

We can test the hypothesis that the number of beers has *no* effect on blood alcohol versus the one-sided alternative that more beers increases BAC. The hypotheses are

$$H_0: \beta = 0$$
$$H_a: \beta > 0$$

It is no surprise that the t statistic is $t = 7.48$ with two-sided P-value $P = 0.000$ to three decimal places. The one-sided P-value is half this value, so it is also close to 0. Check that t is the slope $b = 0.01796$ divided by its standard error, $\text{SE}_b = 0.0024$.

The scatterplot shows one unusual point: student number 3, who drank 9 beers. You can see from Figure 14.5 that this observation lies farthest from the fitted line in the y direction. That is, this point has the largest residual. Student number 3 may also be influential, though the point is not extreme in the x direction. To verify that our results are not too dependent on this one observation, do the regression again omitting student 3. The new regression line is the dashed line in Figure 14.5. Omitting student 3 decreases r^2 from 80% to 77%, and it changes the predicted BAC after 5 beers from 0.077 to 0.073. These small changes show that this observation is not very influential.

SUMMARY

Least-squares regression fits a straight line to data in order to predict a response variable y from the explanatory variable x. Inference about regression requires more assumptions.

The **regression model** says that there is a **true regression line** $\mu_y = \alpha + \beta x$ that describes how the mean response varies as x changes. The observed response y for any x has a normal distribution with mean given by the true regression line and with the same standard deviation σ for any value of x. The parameters of the regression model are the intercept α, the slope β, and the standard deviation σ.

The slope a and intercept b of the least-squares line estimate the slope α and intercept β of the true regression line. To estimate σ, use the **standard error about the line s.**

The standard error s has $n - 2$ **degrees of freedom.** All t procedures in regression inference have $n - 2$ degrees of freedom.

Confidence intervals for the slope of the true regression line have the form $b \pm t^*\text{SE}_b$. In practice, use software to find the slope b of the least-squares line and its standard error SE_b.

To test the hypothesis that the true slope is zero, use the ***t* statistic** $t = b/\text{SE}_b$, also given by software. This null hypothesis says that straight-line dependence on x has no value for predicting y. It also says that the population correlation between x and y is zero.

SECTION 14.1 EXERCISES

14.6 AN EXTINCT BEAST, II Exercise 14.1 presents data on the lengths of two bones in five fossil specimens of the extinct beast *Archaeopteryx*. Here is part of the output from the S-PLUS statistical software when we regress the length y of the humerus on the length x of the femur.

```
Coefficients:
                   Value  Std. Error     t value   Pr(>|t|)
(Intercept)      -3.6596      4.4590     -0.8207     0.4719
      Femur       1.1969      0.0751
```

(a) What is the equation of the least-squares regression line?

(b) We left out the t statistic for testing H_0: $\beta = 0$ and its P-value. Use the output to find t.

(c) How many degrees of freedom does t have? Use Table C to approximate the P-value of t against the one-sided alternative H_a: $\beta > 0$.

(d) Write a sentence to describe your conclusions about the slope of the true regression line.

(e) Determine a 99% confidence interval for the true slope of the regression line.

14.7 JET SKIS, I Data for the number of jet skis in use and number of fatalities for the years 1987 to 2000 are given in Exercise 3.7 (page 125).

(a) Formulate null and alternative hypotheses about the slope of the true regression line. State a one-sided alternative hypothesis.

(b) What conditions or assumptions are necessary in order to perform a linear regression test of significance? Are these reasonable assumptions in this situation?

(c) Perform a linear regression t test. Report the t statistic, the degrees of freedom, and the P-value. Write your conclusion in plain language.

(d) Determine a 98% confidence interval for the true slope of the regression line.

14.8 IS WINE GOOD FOR YOUR HEART? There is some evidence that drinking moderate amounts of wine helps prevent heart attacks. Exercise 3.63 (page 183) gives data on yearly wine consumption (liters of alcohol from drinking wine, per person) and yearly deaths from heart disease (deaths per 100,000 people) in 19 developed nations.

(a) Is there statistically significant evidence of a negative association between wine consumption and heart disease deaths? Carry out the appropriate test of significance and write a summary statement about your conclusions.

(b) Find a 95% confidence interval for the true slope.

14.9 DOES FAST DRIVING WASTE FUEL? Exercise 3.11 (page 129) gives data on the fuel consumption of a small car at various speeds from 10 to 150 kilometers per hour. Is there evidence of straight-line dependence between speed and fuel use? Make a scatterplot and use it to explain the result of your test.

14.10 Exercise 14.4 (page 788) presents data on the relationship between the speed of runners (x, in feet per second) and the number of steps y that they take in a second. Here is part of the Data Desk regression output for these data:

```
R squared = 99.8%
s = 0.0091 with 7 - 2 = 5 degrees of freedom

Variable   Coefficient    s.e. of Coeff  t-ratio    prob
Constant   1.76608        0.0307         57.6       <0.0001
Speed      0.080284       0.0016         49.7       <0.0001
```

(a) How can you tell from this output, even without the scatterplot, that there is a very strong straight-line relationship between running speed and steps per second?

(b) What parameter in the regression model gives the rate at which steps per second increase as running speed increases? Give a 99% confidence interval for this rate.

14.11 THE LEANING TOWER OF PISA The Leaning Tower of Pisa leans more as time passes. Here are measurements of the lean of the tower for the years 1975 to 1987.[2] The lean is the distance between where a point on the tower would be if the tower were straight and where it actually is. The distances are tenths of a millimeter in excess of 2.9 meters. For example, the 1975 lean, which was 2.9642 meters, appears in the table as 642. We use only the last two digits of the year as our time variable.

Year:	75	76	77	78	79	80	81	82	83	84	85	86	87
Lean:	642	644	656	667	673	688	696	698	713	717	725	742	757

Here is part of the output from the Data Desk regression procedure with year as the explanatory variable and lean as the response variable:

```
Variable   Coefficient    s.e. of Coeff   t-ratio   prob
Constant   -61.1209       25.13           -2.43     0.0333
year          9.31868      0.3099         30.1     <0.0001
```

(a) Plot the data. Briefly describe the shape, strength, and direction of the relationship. The tower is tilting at a steady rate.

(b) The main purpose of the study is to estimate how fast the tower is tilting. What parameter in the regression model gives the rate at which the tilt is increasing, in tenths of a millimeter per year?

(c) We want a 95% confidence interval for this rate. How many degrees of freedom does t have? Find the critical value t^* and the confidence interval.

14.2 PREDICTIONS AND CONDITIONS

One of the most common reasons to fit a line to data is to predict the response to a particular value of the explanatory variable. The method is simple: just substitute the value of x into the equation of the line. We saw in Example 14.6 that drinking 5 beers produces an average BAC of

$$\hat{y} = -0.0127 + 0.0180(5) = 0.077$$

We would like to give a confidence interval that describes how accurate this prediction is. To do that, you must answer these questions: Do you want to predict the *mean* blood alcohol level for *all students* who drink 5 beers? Or do you want to predict the BAC of *one individual student* who drinks 5 beers? Both of these predictions may be interesting, but they are two different problems. The actual prediction is the same, $\hat{y} = 0.077$. But the margin of error is different for the two kinds of prediction. Individual students who drink 5 beers don't all have the same BAC. So we need a larger margin of error to pin down one student's result than to estimate the mean BAC for all students who have 5 beers.

Write the given value of the explanatory variable x as x^*. In the example, $x^* = 5$. The distinction between predicting a single outcome and predicting the mean of all outcomes when $x = x^*$ determines what margin of error is correct. To emphasize the distinction, we use different terms for the two intervals.

- To estimate the *mean* response, we use a *confidence interval*. It is an ordinary confidence interval for the parameter

$$\mu_y = \alpha + \beta x^*$$

The regression model says that μ_y is the mean of responses y when x has the value x^*. It is a fixed number whose value we don't know.

prediction interval

- To estimate an *individual* response y, we use a ***prediction interval***. A prediction interval estimates a single random response y rather than a parameter like μ_y. The response y is not a fixed number. If we took more observations with $x = x^*$, we would get different responses.

Fortunately, the meaning of a prediction interval is very much like the meaning of a confidence interval. A 95% prediction interval, like a 95% confidence interval, is right 95% of the time in repeated use. "Repeated use" now means that we take an observation on y for each of the n values of x in the original data, and then take one more observation y with $x = x^*$. Form the prediction interval from the n observations, then see if it covers the one more y. It will in 95% of all repetitions.

The interpretation of prediction intervals is a minor point. The main point is that it is harder to predict one response than to predict a mean response. Both intervals have the usual form

$$\hat{y} \pm t^{*}\text{SE}$$

but the prediction interval is wider than the confidence interval. Here are the details.

CONFIDENCE AND PREDICTION INTERVALS FOR REGRESSION RESPONSE

A level C **confidence interval for the mean response** μ_y when x takes the value x^* is

$$\hat{y} \pm t^{*}\text{SE}_{\hat{\mu}}$$

The standard error $\text{SE}_{\hat{\mu}}$ is

$$\text{SE}_{\hat{\mu}} = s\sqrt{\frac{1}{n} + \frac{(x^{*} - \bar{x})^2}{\sum(x - \bar{x})^2}}$$

The sum runs over all the observations on the explanatory variable x.

A level C **prediction interval for a single observation** on y when x takes the value x^* is

$$\hat{y} \pm t^{*}\text{SE}_{\hat{y}}$$

The standard error for prediction $\text{SE}_{\hat{y}}$ is[3]

$$\text{SE}_{\hat{y}} = s\sqrt{1 + \frac{1}{n} + \frac{(x^{*} - \bar{x})^2}{\sum(x - \bar{x})^2}}$$

In both recipes, t^* is the upper $(1 - C)/2$ critical value of the t distribution with $n - 2$ degrees of freedom.

There are two standard errors: $\text{SE}_{\hat{\mu}}$ for estimating the mean response μ_y and $\text{SE}_{\hat{y}}$ for predicting an individual response y. The only difference between the two standard errors is the extra 1 under the square root sign in the standard error for prediction. The extra 1 makes the prediction interval wider. Both standard errors are multiples of s. The degrees of freedom are again $n - 2$, the degrees of freedom of s. Calculating these standard errors by hand is a nuisance, which technology spares us.

EXAMPLE 14.7 PREDICTING BLOOD ALCOHOL

Steve thinks he can drive legally 30 minutes after he finishes drinking 5 beers. We want to predict Steve's blood alcohol content, using no information except that he drinks 5 beers. Here is the output from the prediction option in the Minitab regression command for $x^* = 5$ when we ask for 95% intervals:

```
Predicted Values

     Fit    StDev Fit        95.0% CI              95.0% PI
0.07712      0.00513 (0.06612, 0.08812) (0.03192, 0.12232)
```

The "Fit" entry gives the predicted BAC, 0.07712. This agrees with our result in Example 14.6. Minitab gives both 95% intervals. You must choose which one you want. We are predicting a single response, so the prediction interval "95.0% PI" is the right choice. We are 95% confident that Steve's blood alcohol content will fall between about 0.032 and 0.122. The upper part of that range will get him arrested if he drives. The 95% confidence interval for the mean BAC of all students after 5 beers, given as "95.0% CI," is much narrower.

Checking the regression conditions

You can fit a least-squares line to any set of explanatory-response data when both variables are quantitative. If the scatterplot doesn't show a roughly linear pattern, the fitted line may be almost useless. But it is still the line that fits the data best in the least-squares sense. To use regression inference, however, the data must satisfy the regression model conditions. Before we do inference, we must check these conditions one by one.

The observations are independent. In particular, repeated observations on the same individual are not allowed. So we can't use ordinary regression to make inferences about the growth of a single child over time, for example.

The true relationship is linear. We can't observe the true regression line, so we will almost never see a perfect straight-line relationship in our data. Look at the scatterplot to check that the overall pattern is roughly linear. A plot of the residuals against x magnifies any unusual pattern. Draw a horizontal line at zero on the residual plot to orient your eye. Because the sum of the residuals is always zero, zero is also the mean of the residuals.

The standard deviation of the response about the true line is the same everywhere. Look at the scatterplot again. The scatter of the data points about the line should be roughly the same over the entire range of the data. A plot of the residuals against x, with a horizontal line at zero, makes this easier to check. It is quite common to find that as the response y gets larger, so does the scatter of the points about the fitted line. Rather than remaining fixed, the standard deviation σ about the line is changing with x as the mean response changes with x. You cannot safely use our inference recipes when this happens. There is no fixed σ for s to estimate.

The response varies normally about the true regression line. We can't observe the true regression line. We can observe the least-squares line and the residuals, which show the variation of the response about the fitted line. The residuals estimate the deviations of the response from the true regression line, so they should follow a normal distribution. Make a histogram or stemplot of the residuals and check for clear skewness or other major departures from normality. Like other t procedures, inference for regression is (with one exception) not very sensitive to minor lack of normality, especially when we have many observations. Do beware of influential observations, which move the regression line and can greatly affect the results of inference.

The exception is the prediction interval for a single response y. This interval relies on normality of individual observations, not just on the approximate normality of statistics like the slope a and intercept b of the least-squares line. The statistics a and b become more normal as we take more observations. This contributes to the robustness of regression inference, but it isn't enough for the prediction interval. We will not study methods that carefully check normality of the residuals, so you should regard prediction intervals as rough approximations.

The conditions for regression inference are a bit elaborate. Fortunately, it is not hard to check for gross violations. There are ways to deal with violations of any of the regression model conditions. If your data don't fit the regression model, get expert advice. Checking conditions uses the residuals. Most regression software will calculate and save the residuals for you.

EXAMPLE 14.8 BLOOD ALCOHOL RESIDUALS

Example 14.6 shows the regression of the blood alcohol content of 16 students on the number of beers they drink. The statistical software that did the regression calculations also calculates the 16 residuals. Here they are:

0.0229	0.0068	0.0410	−0.0110	−0.0012	−0.0180	0.0288	−0.0171
−0.0212	−0.0271	0.0108	0.0049	0.0079	−0.0230	0.0047	−0.0092

A residual plot appears in Figure 14.7. The values of x are on the horizontal axis. The residuals are on the vertical axis, with a horizontal line at zero.

Examine the residual plot to check that the relationship is roughly linear and that the scatter about the line is about the same from end to end. Overall, there is no clear deviation from the even scatter about the line that should occur (except for chance variation) when the regression assumptions hold.

Now examine the distribution of the residuals for signs of strong nonnormality. Here is a stemplot of the residuals after rounding to three decimal places:

−2	7 3 1
−1	8 7 1
−0	9 1
0	5 5 7 8
1	1
2	3 9
3	
4	1

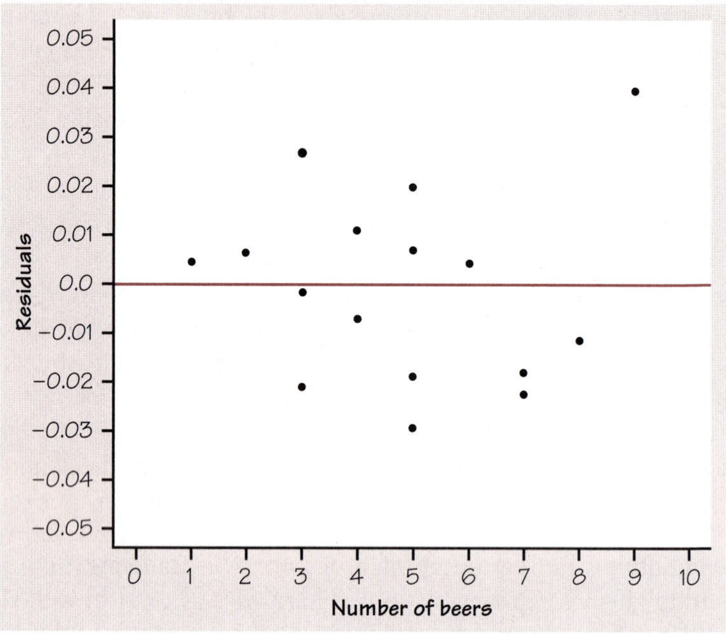

FIGURE 14.7 Plot of the regression residuals for the blood alcohol data against the explanatory variable, number of beers consumed. The mean of the residuals is always 0.

Student number 3 is a mild outlier. We saw in Example 14.6 that omitting this observation has little effect on r^2 or the fitted line. It also has little effect on inference. For example, $t = 7.58$ for the slope becomes $t = 6.57$, a change of no practical importance.

EXAMPLE 14.9 USING RESIDUAL PLOTS

The residual plots in Figure 14.8 illustrate violations of the regression assumptions that require corrective action before using regression. Both plots come from a study of the salaries of major-league baseball players.[4] Salary is the response variable.

multiple regression

There are several explanatory variables that measure the players' past performance. Regression with more than one explanatory variable is called ***multiple regression.*** Although interpreting the fitted model is more complex in multiple regression, we check conditions by examining residuals as usual.

Figure 14.8(a) is a plot of the residuals against the predicted salary $\hat{y}$, produced by the SAS statistical software. When points on the plot overlap, SAS uses letters to show how many observations each point represents. A is one observation, B stands for two observations, and so on. The plot shows a clear violation of the condition that the spread of responses about the model is everywhere the same. There is more variation among players with high salaries than among players with lower salaries.

Although we don't show a histogram, the distribution of salaries is strongly skewed to the right. Using the *logarithm* of the salary as the response variable gives a more normal distribution and also fixes the unequal-spread problem. It is common to work with some transformation of data in order to satisfy the regression conditions. But all is not yet well. Figure 14.8(b) plots the new residuals against years in the major leagues. There is a clear curved pattern. The relationship between logarithm of salary and years in the majors is not linear but curved. The statistician must take more corrective action.

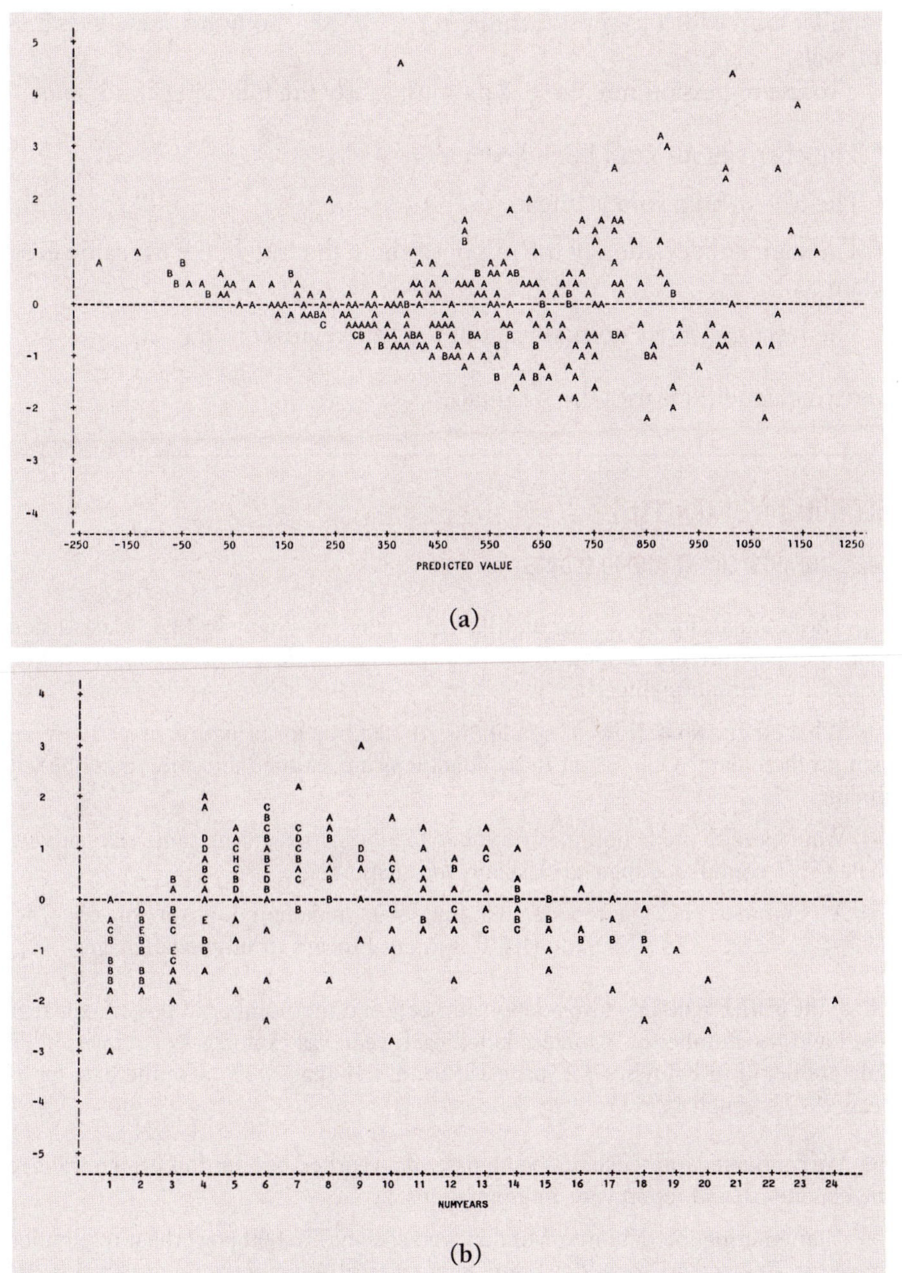

FIGURE 14.8 Two residual plots that illustrate violations of the regression conditions. (a) The variation of the residuals is not constant. (b) There is a curved relationship between the response variable and the explanatory variable.

SUMMARY

Confidence intervals for the mean response when x has value x^* have the form $\hat{y} \pm t^*\text{SE}_{\hat{\mu}}$. **Prediction intervals** for an individual future response y have a similar form with a larger standard error, $\hat{y} \pm t^* \text{SE}_{\hat{y}}$. Software often gives these intervals.

To use regression inference, data must satisfy the following conditions:

- The observations must be independent.
- The true relationship is linear.
- The standard deviation of the response about the true line is the same everywhere.
- The response varies normally about the true regression line.

Verifying conditions uses the residuals.

SECTION 14.2 EXERCISES

14.12 INFANTS' CRYING AND IQ SCORES

(a) The residuals for the crying and IQ data appear in Example 14.3. Make a stemplot to display the distribution of the residuals. Are there outliers or signs of strong departures from normality?

(b) What other assumptions or conditions are required for using inference for regression on these data? Check that these conditions are satisfied and then describe your findings.

(c) Would a 95% prediction interval for $x = 25$ be narrower, the same size, or wider than a 95% confidence interval? Explain your reasoning.

(d) A computer package reports that the 95% prediction interval for $x = 25$ is `(91.85, 165.33)`. Explain what this interval means in simple language.

14.13 THE GENTLE MANATEE The relationship between the number of powerboats registered and the number of manatees killed each year was explored in Chapter 3. The data are found in Exercises 3.6 (page 125) and 3.41 (page 157). Use the data for the years 1977 through 1994.

(a) We conducted inference on the manatee data earlier, but was this prudent? Check the conditions, and report your interpretations.

(b) After entering the data into Minitab, you specify 716,000 powerboat registrations (coded as 716) and ask for a 95% confidence interval and a prediction interval for this value. Minitab reports that the two intervals are `(41.43, 49.59)` and `(33.35, 57.66)`. Which is which, and how do you know?

14.14 PISA, PISA! In Exercise 14.11 (page 795) we regressed the lean of the Leaning Tower of Pisa on year to estimate the rate at which the tower is tilting. Here are the residuals from that regression, in order by years across the rows:

4.220	−3.099	−0.418	1.264	−2.055	3.626	2.308
−5.011	0.670	−4.648	−5.967	1.714	7.396	

Use the residuals to check the regression conditions, and describe your findings. Is the regression in Exercise 14.11 trustworthy?

14.15 DO HEAVIER PEOPLE BURN MORE ENERGY? Metabolic rate, the rate at which the body consumes energy, is important in studies of weight gain, dieting, and exercise. Lean body mass is an important influence on metabolic rate. Table 3.2 (page 132) gives data for 19 people. Because men and women showed a similar pattern, we will now ignore gender. Here are the data on mass (in kilograms) and metabolic rate (in calories):

Mass:	62.0	62.9	36.1	54.6	48.5	42.0	47.4	50.6	42.0	48.7
Rate:	1792	1666	995	1425	1396	1418	1362	1502	1256	1614
Mass:	40.3	33.1	51.9	42.4	34.5	51.1	41.2	51.9	46.9	
Rate:	1189	913	1460	1124	1052	1347	1204	1867	1439	

Use your calculator or software to analyze these data. Make a scatterplot and find the least-squares line. Give a 90% confidence interval for the slope β and explain clearly what your interval says about the relationship between lean body mass and metabolic rate. Find the residuals and examine them. Are the conditions for regression inference met?

14.16 MANATEES The 95% prediction interval in part (b) of Exercise 14.13 is quite wide. Changing to 90% confidence will give a smaller margin of error. Use the computer output in the previous exercise, along with Table C, to give a 90% interval for the mean number of manatees killed when there are 700,000 powerboats registered.

14.17 IDEAL PROPORTIONS, III Eighteen students in Mr. Starnes's class measured their arm spans and heights. Here are their measurements, in inches:

Arm span	Height	Arm span	Height	Arm span	Height
74	76	61.5	64.5	68.75	68.25
72	74	72	69.5	70	68
72	71	68	70.5	63	66.5
66	68	68.5	68.5	69	70
78	77	62.5	64	66	65.5
67.5	68	65.5	64.5	78.75	75.75

We want to predict height from arm span.

(a) Test the conditions for using inference for regression for these data. Describe your findings.

(b) When the value $x = 75$ is specified, Minitab reports that the `95% P.I.` is `(70.921, 78.238)`. Interpret this interval.

(c) Explain what a 95% confidence interval for $x = 75$ inches tells us. How is this different from the 95% prediction interval for $x = 75$ inches? Which interval is more precise (that is, shorter) and why?

TECHNOLOGY TOOLBOX *Linear regression t test*

The TI-83/89 can perform a linear regression *t* test. We will use the crying versus IQ data from Example 14.1 (page 781) to illustrate the process.

Enter the *x*-values (crying) into L_1/list1 and the *y*-values (IQ) into L_2/list2.

TI-83

- Press STAT/◄/E:LinRegTest.

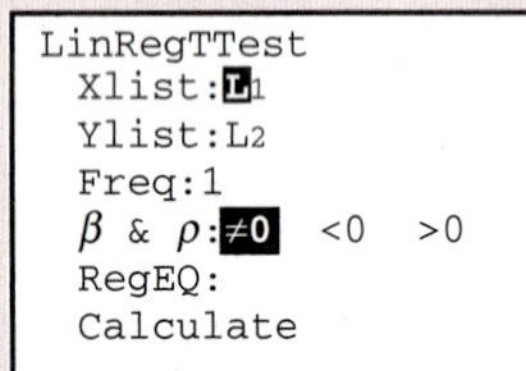

- In the LinRegTTest screen, specify L_1 for Xlist, L_2 for Ylist, and ↑ 0 for the hypothesized slope β. You can leave the RegEQ space blank. Highlight the command "Calculate" and press ENTER.

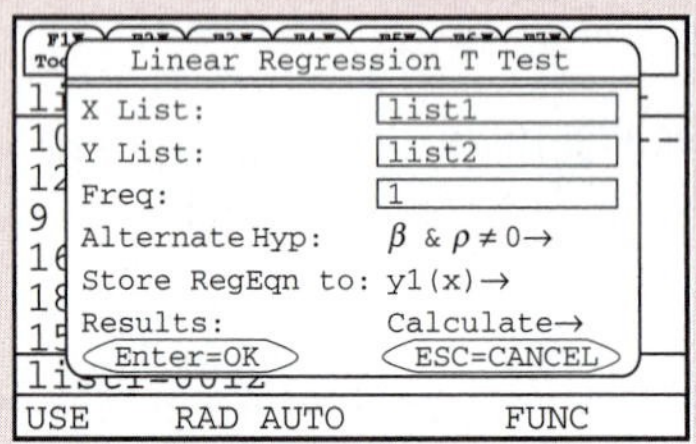

TI-89

- Press 2nd F1 (F6:Tests). Then select A:LinRegTTest.

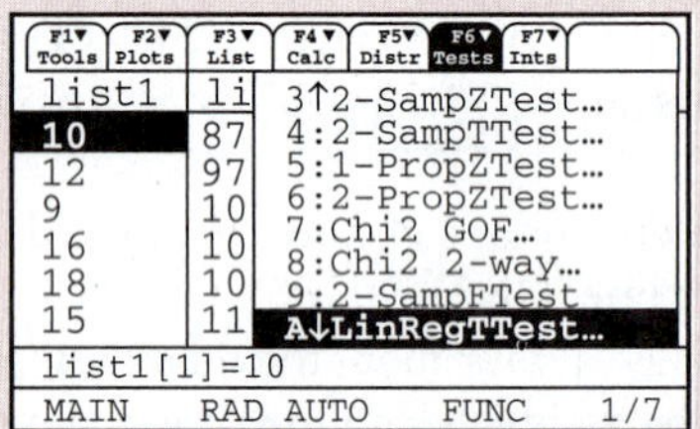

- Specify list1 for X List, list2 for Y List, and 1 for Frequency. You have three choices for Alternate Hyp: we will choose β ↑ 0. For results, choose "Calculate." (The other choice is "Draw.") Press ENTER.

The linear regression *t* test results take two screens to present.

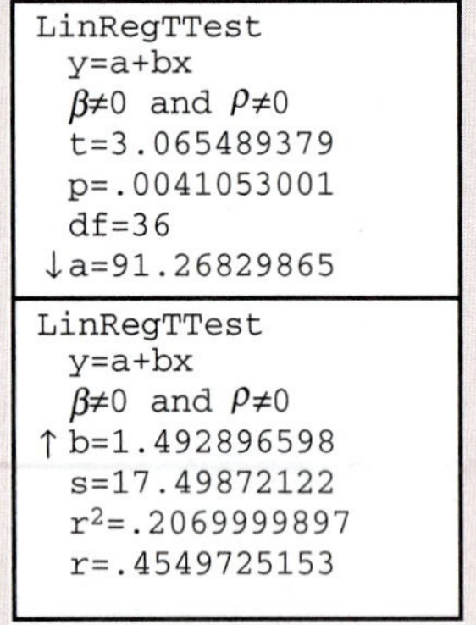

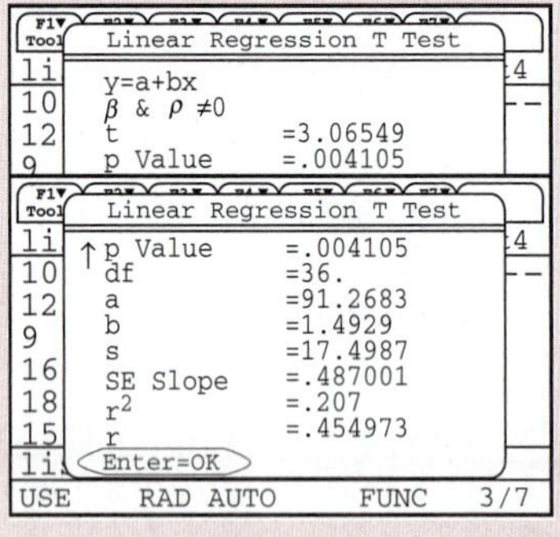

TECHNOLOGY TOOLBOX *Linear regression t test (continued)*

The first screen reports that the t statistic is 3.07 with df = 36. The P-value is 0.004. Scrolling down, you find the intercept $a = 91.2683$ and slope $b = 1.4929$ of the least-squares line, as well as the correlation $r = 0.455$, the coefficient of determination $r^2 = 0.207$, and the standard error about the line $s = 17.50$.

Note that the TI-83 does not have a provision for calculating a confidence interval for the regression slope, but the list feature can be helpful in calculating the standard error of the slope, SE_b. Usually, determining confidence intervals is provided by software.

CHAPTER REVIEW

When a scatterplot shows a straight-line relationship between an explanatory variable x and a response variable y, we often fit a least-squares regression line to describe the relationship. Use this line to predict y from x. Statistical inference in the regression setting, however, requires more than just an overall linear pattern on a scatterplot.

The regression model says that there is a true straight-line relationship between x and the mean response μ_y. We can't observe this true regression line. The responses y that we do observe vary if we take several observations at the same x. The regression model says that for any fixed x, the responses have a normal distribution. Moreover, the standard deviation σ of this distribution is the same for all values of x.

The standard deviation σ describes how much variation there is in responses y when x is held fixed. Estimating σ is the key to inference about regression. Use the standard error s (roughly, the sample standard deviation of the residuals) to estimate σ. We can then do these types of inference:

- Give confidence intervals for the slope of the true regression line.
- Test the null hypothesis that this slope is zero. This hypothesis says that a straight-line relation between x and y is of no value for predicting y. It is the same as saying that the correlation between x and y in the entire population is zero.
- Give confidence intervals for the mean response for any fixed value of x.
- Give prediction intervals for an individual response y for a fixed value of x.

A. PRELIMINARIES

1. Make a scatterplot to show the relationship between an explanatory and a response variable.

2. Use a calculator or software to find the correlation and the least-squares regression line.

B. RECOGNITION

1. Recognize the regression setting: a straight-line relationship between an explanatory variable x and a response variable y.

2. Recognize which type of inference you need in a particular regression setting.

3. Inspect the data to recognize situations in which inference isn't safe: a nonlinear relationship, influential observations, strongly skewed residuals in a small sample, or nonconstant variation of the data points about the regression line.

C. DOING INFERENCE USING SOFTWARE AND CALCULATOR OUTPUT

1. Explain in any specific regression setting the meaning of the slope β of the true regression line.

2. Understand computer output for regression. Find in the output the slope and intercept of the least-squares line, their standard errors, and the standard error about the line.

3. Use that information to carry out tests and calculate confidence intervals for β.

4. Explain the distinction between a confidence interval for the mean response and a prediction interval for an individual response.

5. If software gives output for prediction, use that output to give either confidence or prediction intervals.

CHAPTER 14 REVIEW EXERCISES

14.18 TIME AT THE TABLE Does how long young children remain at the lunch table help predict how much they eat? Here are data on 20 toddlers observed over several months at a nursery school.[5] "Time" is the average number of minutes a child spent at the table when lunch was served. "Calories" is the average number of calories the child consumed during lunch, calculated from careful observation of what the child ate each day.

Time:	21.4	30.8	37.7	33.5	32.8	39.5	22.8	34.1	33.9	43.8
Calories:	472	498	465	456	423	437	508	431	479	454
Time:	42.4	43.1	29.2	31.3	28.6	32.9	30.6	35.1	33.0	43.7
Calories:	450	410	504	437	489	436	480	439	444	408

Make a scatterplot of the data and find the equation of the least-squares line for predicting calories consumed from time at the table. Describe briefly what the data show about the behavior of children. Then give a 95% confidence interval for the slope of the true regression line.

14.19 BEAVERS AND BEETLES Ecologists sometimes find rather strange relationships in our environment. One study seems to show that beavers benefit beetles. The researchers laid out 23 circular plots, each four meters in diameter, in an area where beavers were cutting down cottonwood trees. In each plot, they measured the number of stumps from trees cut by beavers and the number of clusters of beetle larvae. Here are the data:[6]

Stumps:	2	2	1	3	3	4	3	1	2	5	1	3
Beetle larvae:	10	30	12	24	36	40	43	11	27	56	18	40
Stumps:	2	1	2	2	1	1	4	1	2	1	4	
Beetle larvae:	25	8	21	14	16	6	54	9	13	14	50	

(a) Make a scatterplot that shows how the number of beaver-caused stumps influences the number of beetle larvae clusters. What does your plot show?

(b) Here is part of the Minitab regression output for these data:

```
Predictor      Coef       StDev         T          P
Constant     -1.286       2.853     -0.45      0.657
Stumps       11.894       1.136     10.47      0.000

S = 6.419        R-Sq = 83.9%
```

Find the least-squares regression line and draw it on your plot. What percent of the observed variation in beetle larvae counts can be explained by straight-line dependence on beaver stump counts?

(c) Is there strong evidence that beaver stumps help explain beetle larvae counts? Give appropriate statistical evidence to support you conclusion.

14.20 BEAVER AND BEETLE RESIDUALS Software often calculates ***standardized residuals*** as well as the actual residuals from regression. Because the standardized residuals have the standard z-score scale, it is easier to judge whether any are extreme. Here are the standardized residuals from the previous exercise, rounded to 2 decimal places:

standardized residuals

−1.99	1.20	0.23	−1.67	0.26	−1.06	1.38	0.06	0.72	−0.40	1.21	0.90
0.40	−0.43	−0.24	−1.36	0.88	−0.75	1.30	−0.26	−1.51	0.55	0.62	

(a) Find the mean and standard deviation of the standardized residuals. Why do you expect values close to those you obtain?

(b) Make a stemplot of the standardized residuals. Are there any striking deviations from normality? The most extreme residual is $z = -1.99$. Would this be surprisingly large if the 23 observations had a normal distribution? Explain your answer.

(c) Plot the standardized residuals against the explanatory variable. Are there any suspicious patterns?

14.21 INVESTING AT HOME AND OVERSEAS Investors ask about the relationship between returns on investments in the United States and investments overseas. Exercise 3.56

(page 179) gives the percent returns on U.S. and overseas common stocks over a 27-year period.

(a) Make a scatterplot suitable for predicting overseas returns from U.S. returns.

(b) Here is part of the output from the Minitab regression command:

```
Predictor        Coef      StDev        T        P
Constant        5.683      5.144     1.10    0.280
USreturn       0.6181     0.2369        *        *

S = 19.90          R-Sq = 21.4%
```

We have omitted the t statistic for β and its P-value. What is the value of t? What are its degrees of freedom? From Table C, how strong is the evidence for a linear relationship between U.S. and overseas returns?

(c) Here is the output for prediction of overseas returns when U.S. stocks return 15%:

```
  Fit   StDev Fit         90.0% CI               90.0% PI
14.95        3.83     ( 8.41, 21.50)       ( -19.65, 49.56)
```

Verify the "Fit" by using the least-squares line from the output in (b). You think U.S. stocks will return 15% next year. Give a 90% interval for the return on foreign stocks next year if you are right about U.S. stocks.

(d) Is the regression prediction useful in practice? Use the r^2-value for this regression to help explain your finding.

14.22 STOCK RETURN RESIDUALS Exercise 14.21 presents a regression of overseas stock returns on U.S. stock returns based on 27 years' data. The residuals for this regression (in order by years across the rows) are

14.89	18.93	−11.44	−12.57	6.72	−17.77	16.99	22.96	−12.13
−3.05	−4.89	−20.87	4.17	−2.05	30.98	52.22	15.76	12.36
−14.78	−27.17	−12.04	−22.18	20.97	−0.29	−17.72	−13.80	−24.23

(a) Plot the residuals against x, the U.S. return. The plot suggests a mild violation of one of the regression conditions. Which one?

(b) Display the distribution of the residuals in a graph. In what way is the shape somewhat nonnormal? There is one possible outlier. Circle that point on the residual plot in (a). What year is this? This point is not very influential: redoing the regression without it does not greatly change the results. With 27 observations, we are willing to do regression inference for these data.

14.23 WEEDS AMONG THE CORN Lamb's-quarter is a common weed that interferes with the growth of corn. An agriculture researcher planted corn at the same rate in 16 small plots of ground, then weeded the plots by hand to allow a fixed number of lamb's-quarter plants to grow in each meter of corn row. No other weeds were allowed to grow. Here are the yields of corn (bushels per acre) in each of the plots:[7]

Weeds per meter	Corn yield	Weeds per meter	Corn yield	Weeds per meter	Corn yield	Weeds per meter	Corn yield
0	166.7	1	166.2	3	158.6	9	162.8
0	172.2	1	157.3	3	176.4	9	142.4
0	165.0	1	166.7	3	153.1	9	162.8
0	176.9	1	161.1	3	156.0	9	162.4

Use your calculator or software to analyze these data.

(a) Make a scatterplot and find the least-squares line. What percent of the observed variation in corn yield can be explained by a linear relationship between yield and weeds per meter?

(b) Is there good evidence that more weeds reduce corn yield?

(c) Explain from your findings in (a) and (b) why you expect predictions based on this regression to be quite imprecise. Predict the mean corn yield under these experimental conditions when there are 6 weeds per meter of row. Give a 95% confidence interval for this mean.

14.24 THE PROFESSOR SWIMS, I Here are data on the time (in minutes) Professor Moore takes to swim 2000 yards and his pulse rate (beats per minute) after swimming:

Time:	34.12	35.72	34.72	34.05	34.13	35.72	36.17	35.57
Pulse:	152	124	140	152	146	128	136	144
Time:	35.37	35.57	35.43	36.05	34.85	34.70	34.75	33.93
Pulse:	148	144	136	124	148	144	140	156
Time:	34.60	34.00	34.35	35.62	35.68	35.28	35.97	
Pulse:	136	148	148	132	124	132	139	

A scatterplot shows a negative linear relationship: a faster time (fewer minutes) is associated with a higher heart rate. Here is part of the output from the regression function in the Excel spreadsheet:

```
            Coefficients    Standard Error  t Stat          P-value
Intercept   479.9341457     66.22779275     7.246718119     3.87075E-07
X Variable  -9.694903394    1.888664503     -5.1332057      4.37908E-05
```

Give a 90% confidence interval for the slope of the true regression line. Explain what your result tells us about the relationship between the professor's swimming time and heart rate.

14.25 THE PROFESSOR SWIMS, II Exercise 14.24 gives data on a swimmer's time and heart rate. One day the swimmer completes his laps in 34.3 minutes but forgets to take his pulse. Minitab gives this prediction for heart rate when $x^* = 34.3$:

```
  Fit   StDev Fit         90.0% CI              90.0% PI
147.40        1.97    ( 144.02, 150.78)    ( 135.79, 159.01)
```

(a) Verify that "Fit" is the predicted heart rate from the least-squares line found in Exercise 14.24. Then choose one of the intervals from the output to estimate the swimmer's heart rate that day and explain why you chose this interval.

(b) Minitab gives only one of the two standard errors used in prediction. It is $SE_{\hat{\mu}}$, the standard error for estimating the mean response. Use this fact and a critical value from Table C to verify Minitab's 90% confidence interval for the mean heart rate on days when the swimming time is 34.3 minutes.

14.26 **FISH SIZES** Table 14.2 contains data on the size of perch caught in a lake in Finland. Statistical software will help you analyze these data.

TABLE 14.2 Measurements on 56 perch

Obs. number	Weight (grams)	Length (cm)	Width (cm)	Obs. number	Weight (grams)	Length (cm)	Width (cm)
104	5.9	8.8	1.4	132	197.0	27.0	4.2
105	32.0	14.7	2.0	133	218.0	28.0	4.1
106	40.0	16.0	2.4	134	300.0	28.7	5.1
107	51.5	17.2	2.6	135	260.0	28.9	4.3
108	70.0	18.5	2.9	136	265.0	28.9	4.3
109	100.0	19.2	3.3	137	250.0	28.9	4.6
110	78.0	19.4	3.1	138	250.0	29.4	4.2
111	80.0	20.2	3.1	139	300.0	30.1	4.6
112	85.0	20.8	3.0	140	320.0	31.6	4.8
113	85.0	21.0	2.8	141	514.0	34.0	6.0
114	110.0	22.5	3.6	142	556.0	36.5	6.4
115	115.0	22.5	3.3	143	840.0	37.3	7.8
116	125.0	22.5	3.7	144	685.0	39.0	6.9
117	130.0	22.8	3.5	145	700.0	38.3	6.7
118	120.0	23.5	3.4	146	700.0	39.4	6.3
119	120.0	23.5	3.5	147	690.0	39.3	6.4
120	130.0	23.5	3.5	148	900.0	41.4	7.5
121	135.0	23.5	3.5	149	650.0	41.4	6.0
122	110.0	23.5	4.0	150	820.0	41.3	7.4
123	130.0	24.0	3.6	151	850.0	42.3	7.1
124	150.0	24.0	3.6	152	900.0	42.5	7.2
125	145.0	24.2	3.6	153	1015.0	42.4	7.5
126	150.0	24.5	3.6	154	820.0	42.5	6.6
127	170.0	25.0	3.7	155	1100.0	44.6	6.9
128	225.0	25.5	3.7	156	1000.0	45.2	7.3
129	145.0	25.5	3.8	157	1100.0	45.5	7.4
130	188.0	26.2	4.2	158	1000.0	46.0	8.1
131	180.0	26.5	3.7	159	1000.0	46.6	7.6

Source: The data in Table 14.2 are part of a larger data set in the *Journal of Statistics Education* archive, accessible via the Internet. The original source is Pekka Brofeldt, "Bidrag till kaennedom on fiskbestondet i vaara sjoear. Laengelmaevesi," in T. H. Jaervi, *Finlands Fiskeriet,* Band 4, *Meddelanden utgivna av fiskeri-foereningen i Finland,* Helsinki, 1917. The data were contributed to the archive (with information in English) by Juha Puranen of the University of Helsinki.

(a) We want to know how well we can predict the width of a perch from its length. Make a scatterplot of width against length. There is a strong linear pattern, as expected. Perch number 143 had six newly eaten fish in its stomach. Find this fish on your scatterplot and circle the point. Is this fish an outlier in your plot of width against length?

(b) Find the least-squares regression line to predict width from length.

(c) The length of a typical perch is about $x^* = 27$ centimeters. Predict the mean width of such fish and give a 95% confidence interval.

(d) Examine the residuals. Is there any reason to mistrust inference? Does fish number 143 have an unusually large residual?

14.27 **FISH WEIGHTS** We can also use the data from Table 14.2 to study the prediction of the weight of a perch from its length.

(a) Make a scatterplot of weight versus length, with length as the explanatory variable. Describe the pattern of the data and any clear outliers.

(b) It is more reasonable to expect the one-third power of the weight to have a straight-line relationship with the length than to expect weight itself to have a straight-line relationship with length. Explain why this is true. (*Hint*: What happens to weight if length, width, and height all double?)

(c) Use your calculator or software to create a new variable that is the one-third power of weight. Make a scatterplot of this new response variable against length. Describe the pattern and any clear outliers.

(d) Is the straight-line pattern in (c) stronger or weaker than that in (a)? Compare the plots and also the values of r^2.

(e) Find the least-squares regression line to predict the new weight variable from length. Predict the mean of the new variable for perch 27 centimeters long, and give a 95% confidence interval.

(f) Examine the residuals from your regressions. Does it appear that any of the regression conditions are not met?

NOTES AND DATA SOURCES

1. These are part of the data from the EESEE story "Blood Alcohol Content."
2. Data from G. Geri and B. Palla, "Considerazioni sulle più recenti osservazioni ottiche alla Torre Pendente di Pisa," *Estratto dal Bollettino della Società Italiana di Topografia e Fotogrammetria*, 2 (1988), pp. 121–135. Professor Julia Mortera of the University of Rome provided a translation.
3. Strictly speaking, this quantity is the estimated standard deviation of $\hat{y} - y$, where y is the additional observation taken at $x = x^*$.
4. The data are for 1987 salaries and measures of past performance. They were collected and distributed by the Statistical Graphics Section of the American Statistical Association for an annual data analysis contest. The analysis here was done by Crystal Richard of Purdue University.
5. Based on Marion E. Dunshee, "A study of factors affecting the amount and kind of food eaten by nursery school children," *Child Development*, 2 (1931), pp. 163–183.

This article gives the means, standard deviations, and correlation for 37 children but does not give the actual data.

6. Based on a plot in G. D. Martinsen, E. M. Driebe, and T. G. Whitham, "Indirect interactions mediated by changing plant chemistry: beaver browsing benefits beetles," *Ecology*, 79 (1998), pp. 192–200.

7. Data provided by Samuel Phillips, Purdue University.

Post-Exam Topic

15 Analysis of Variance

Kip Brundage/Woodfin Camp & Associates

W. EDWARDS DEMING

Statistics in the Service of Quality

From one point of view, statistics is about understanding variation. Reducing variation in products and processes is the central theme of statistical quality control. So it is not surprising that a statistician should become the leading guru of quality management. In the final decades of his long life, *W. Edwards Deming (1900–1993)* was one of the world's most influential consultants to management.

Deming grew up in Wyoming and earned a doctorate in physics at Yale. Working for the U.S. Department of Agriculture in the 1930s, he became acquainted with Neyman's work on sampling design and with the new methods of statistical quality control, invented by Walter Shewhart of AT&T. In 1939 he moved to the Census Bureau as an expert on sampling. In 1943, he coined the term "*P*-value" for reporting the result of a significance test.

The work that made Deming famous began after he left the government in 1946. He visited Japan to advise on a census but returned to lecture on quality control. He earned a large following in Japan, which named its premier prize for industrial quality after him. As Japan's reputation for manufacturing excellence rose, Deming's fame rose with it. Blunt-spoken and even abrasive, he told corporate leaders that most quality problems are system problems for which management is responsible. He urged breaking down barriers to worker involvement and advocated the constant search for causes of variation. Finding the sources of variation is the theme of analysis of variance, the statistical method we introduce in this chapter.

It is not surprising that a statistician should become the leading guru of quality management.

chapter 15

Analysis of Variance

Activity 15 The Return of Paper Airplanes

Materials: Paper airplane pattern sheet (in Teacher Resource Binder), scissors, masking tape, 25-meter steel tape measure, graphing calculator

In Activity 11, you determined which of two paper airplane models flew farther. In this activity, you will test a new (third) paper airplane model and record distances flown. Later in the chapter, after sufficient theory has been presented, you will conduct a one-way analysis of variance procedure using the data from all three paper airplane models.

Task 1. Extend your experiment from Activity 11 for producing data on the flight distances of the two original paper airplane models to include the third model. Then carry out your plan and gather the necessary data. (These data will be added to the data collected in Activity 11.)

Task 2. In preparation for performing inference on the mean distances of the three models, calculate the descriptive statistics for each of the three sets of distances, and assess the normality of the data collected. Compare the three mean distances graphically. Does it appear that all of the means are about the same, or is at least one mean different from the other means?

Note: Keep these data at hand for analysis later in the chapter.

INTRODUCTION

In this chapter you will be introduced to a very useful procedure that extends the idea of comparing two means. Curiously, it is called *analysis of variance*, for reasons to be explained in Section 15.2, and it is widely known by its acronym, ANOVA. In the first part of the chapter we shall briefly consider inference procedures for comparing two standard deviations, partly to make the point that inference can be performed on population parameters other than means and proportions. Although the objectives of Section 15.1 (Inference for Population Spread) and Section 15.2 (One-Way Analysis of Variance) are quite different, and the assumptions and characteristics of the procedures are also quite different, they both involve the same distribution, the *F distribution.*

15.1 INFERENCE FOR POPULATION SPREAD

The two most basic descriptive features of a distribution are its center and spread. In a normal population, we measure center and spread by the mean and the standard deviation. We use the *t* procedures for inference about population means for normal populations, and we know that t procedures are often useful for nonnormal populations as well. It is natural to turn next to inference about the standard deviations of normal populations. Our advice here is short and clear: Don't do it without expert advice.

Avoid inference about standard deviations

There are methods for inference about the standard deviations of normal populations. We will describe the most common such method, the F test for comparing the spread of two normal populations. Unlike the t procedures for means, the F test and other procedures for standard deviations are extremely sensitive to nonnormal distributions. This lack of robustness does not improve in large samples. It is difficult in practice to tell whether a significant F-value is evidence of unequal population spreads or simply a sign that the populations are not normal.

The deeper difficulty underlying the very poor robustness of normal population procedures for inference about spread already appeared in our work on describing data. The standard deviation is a natural measure of spread for normal distributions but not for distributions in general. In fact, because skewed distributions have unequally spread tails, no single numerical measure does a good job of describing the spread of a skewed distribution. In summary, the standard deviation is not always a useful parameter, and even when it is (for symmetric distributions), the results of inference are not trustworthy. Consequently, we do not recommend trying to do inference about population standard deviations in basic statistical practice.[1]

It was once common to test equality of standard deviations as a preliminary to performing the pooled two-sample t test for equality of two population means. It is better practice to check the distributions graphically, with special attention to skewness and outliers, and to use the version of the two-sample t featured in Section 11.2. This test does not require equal standard deviations.

The F test for comparing two standard deviations

Because of the limited usefulness of procedures for inference about the standard deviations of normal distributions, we will present only one such procedure. Suppose that we have independent SRSs from two normal populations, a sample of size n_1 from $N(\mu_1, \sigma_1)$ and a sample of size n_2 from $N(\mu_2, \sigma_2)$. The population means and standard deviations are all unknown. The two-sample t test examines whether the means are equal in this setting. To test the hypothesis of equal spread,

$$H_0: \sigma_1 = \sigma_2$$
$$H_a: \sigma_1 \neq \sigma_2$$

we use the ratio of sample variances. This is the F *statistic*.

THE *F* STATISTIC AND *F* DISTRIBUTIONS

When s_1^2 and s_2^2 are sample variances from independent SRSs of sizes n_1 and n_2 drawn from normal populations, the **F statistic**

$$F = \frac{s_1^2}{s_2^2}$$

has the **F distribution** with $n_1 - 1$ and $n_2 - 1$ degrees of freedom when H_0: $\sigma_1 = \sigma_2$ is true.

The F distributions are a family of distributions with two parameters. The parameters are the degrees of freedom of the sample variances in the numerator and denominator of the F statistic. The numerator degrees of freedom are always mentioned first. Interchanging the degrees of freedom changes the distribution, so the order is important. Our brief notation will be $F(j, k)$ for the F distribution with j degrees of freedom in the numerator and k in the denominator. The F distributions are not symmetric but are right-skewed. The density curve in Figure 15.1 illustrates the shape. Because sample variances cannot be negative, the F statistic takes only positive values, and the F distribution has no probability below 0. The peak of the F density curve is near 1. When the two populations have the same standard deviation, we expect the two sample variances to be close in size, so that F takes a value near 1. Values of F far from 1 in either direction provide evidence against the hypothesis of equal standard deviations.

Tables of F critical values are awkward, because we need a separate table for every pair of degrees of freedom j and k. Table D in the back of the book gives upper p critical values of the F distributions for p = 0.10, 0.05, 0.025, 0.01, and 0.001. For example, these critical values for the $F(9, 10)$ distribution shown in Figure 15.1 are

p	0.10	0.05	0.025	0.01	0.001
F^*	2.35	3.02	3.78	4.94	8.96

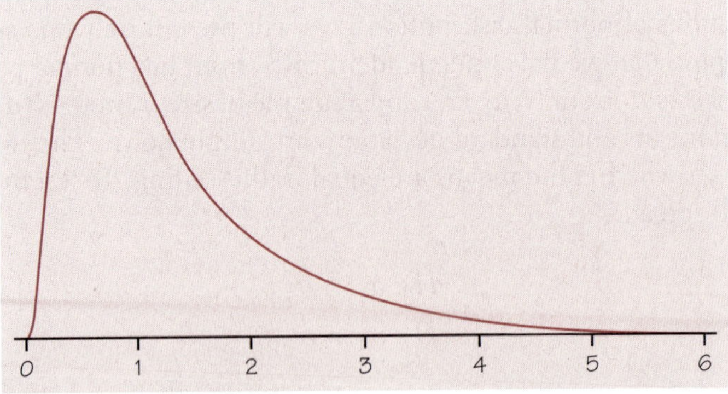

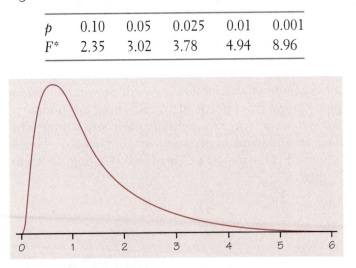

FIGURE 15.1 The density curve for the $F(9,10)$ distribution. The F distributions are skewed to the right.

The skewness of the F distributions causes additional complications. In the symmetric normal and t distributions, the point with probability 0.05 below it is just the negative of the point with probability 0.05 above it. This is not true for F distributions. We therefore need either tables of both the upper and lower tails or some way to eliminate the need for lower tail critical values. The TI-83/89 graphing calculator and statistical software both do away with the need for tables. If you do not use your calculator or statistical software, arrange the two-sided F test as follows:

CARRYING OUT THE *F* TEST

Step 1: Take the test statistic to be

$$F = \frac{\text{larger } s^2}{\text{smaller } s^2}$$

This amounts to naming the populations so that Population 1 has the larger of the observed sample variances. The resulting F is always 1 or greater.

Step 2: Compare the value of F with critical values from Table D. Then *double* the significance levels from the table to obtain the significance level for the two-sided F test.

The idea is that we calculate the probability in the upper tail and double it to obtain the probability of all ratios on either side of 1 that are at least as improbable as that observed. Remember that the order of the degrees of freedom is important in using Table D.

EXAMPLE 15.1 CALCIUM AND BLOOD PRESSURE

Example 11.10 (p. 650) describes a medical experiment to compare the mean effects of calcium and a placebo on the blood pressure of black men. We might also compare the standard deviations to see whether calcium changes the spread of blood pressures among black men. As in previous chapters, we organize our work using the Inference Toolbox.

Step 1: Identify parameters of interest. State hypotheses. We want to test

$$H_0: \sigma_1 = \sigma_2$$
$$H_a: \sigma_1 \neq \sigma_2$$

where σ_1 = the standard deviation of blood pressure readings in the population of healthy black men taking a calcium supplement, and σ_2 = the standard deviation in the (hypothetical) placebo population.

Step 2: Choose an inference procedure and verify the conditions. We should use an F test to compare the population standard deviations. This test requires that both populations are approximately normal. The back-to-back stemplot in Example 11.10 raises questions about the normality of the calcium population. We proceed with caution.

Step 3: Carry out the procedure.

• The larger of the two sample standard deviations is $s = 8.743$ from 10 observations. The smaller is 5.901 from 11 observations. The F test statistic is therefore

$$F = \frac{\text{larger } s^2}{\text{smaller } s^2} = \frac{(8.743)^2}{(5.901)^2} = 2.195$$

• Compare the calculated value $F = 2.20$ with critical points for the $F(9, 10)$ distribution. Table D shows that 2.20 is less than the 0.10 critical value of the $F(9, 10)$ distribution, which is $F^* = 2.35$. Doubling 0.10, we know that the P-value for the two-sided test is greater than 0.20. The results are not significant at the 20% level (or any lower level). Statistical software shows that the exact upper tail probability is 0.118, and hence $P = 0.24$.

Step 4: Interpret your results in context. If the populations were normal, the observed standard deviations would give little reason to suspect unequal population standard deviations. Because one of the populations shows some nonnormality, we cannot be fully confident of this conclusion.

Although the computations for a two-sample F test and the table lookup procedure are not complicated, performing an F test on the TI-83/89 has the advantage of not requiring a table of areas under the F distribution. The following Technology Toolbox shows the steps.

TECHNOLOGY TOOLBOX *F test on the TI-83/89*

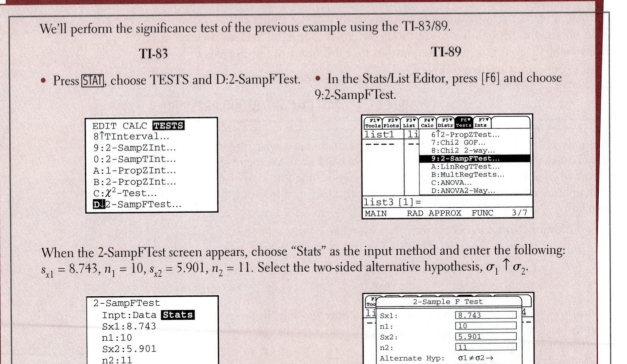

We'll perform the significance test of the previous example using the TI-83/89.

TI-83

• Press STAT, choose TESTS and D:2-SampFTest.

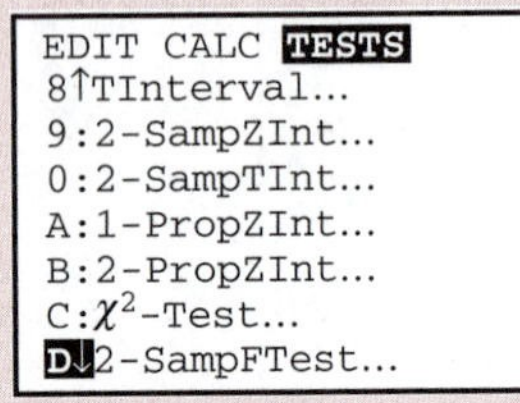

TI-89

• In the Stats/List Editor, press [F6] and choose 9:2-SampFTest.

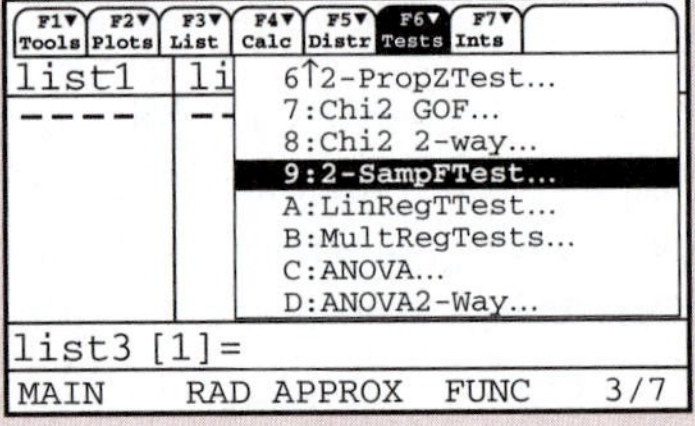

When the 2-SampFTest screen appears, choose "Stats" as the input method and enter the following: $s_{x1} = 8.743$, $n_1 = 10$, $s_{x2} = 5.901$, $n_2 = 11$. Select the two-sided alternative hypothesis, $\sigma_1 \uparrow \sigma_2$.

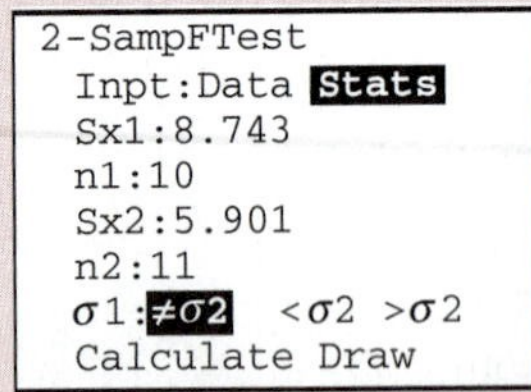

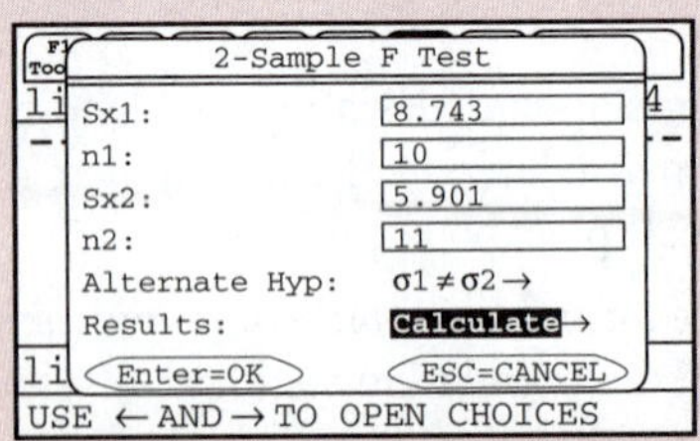

TECHNOLOGY TOOLBOX *F test on the TI-83/89 (continued)*

If you select "Calculate," you will see the following results. Compare these results with those from Example 15.1.

```
2-SampFTest
 σ1≠σ2
 F=2.195177929
 p=.2365479014
 Sx1=8.743
 Sx2=5.901
↓n1=10
```

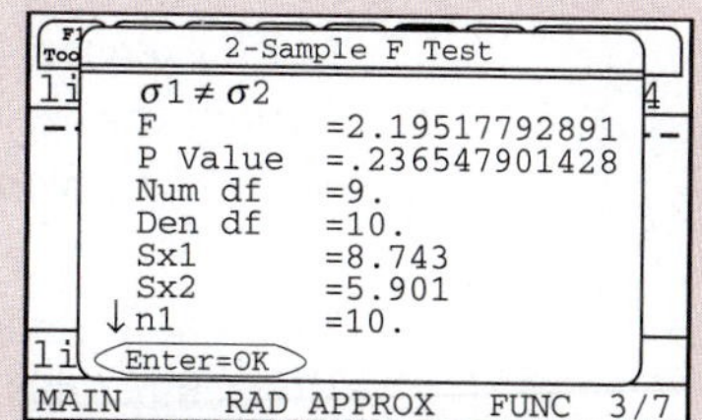

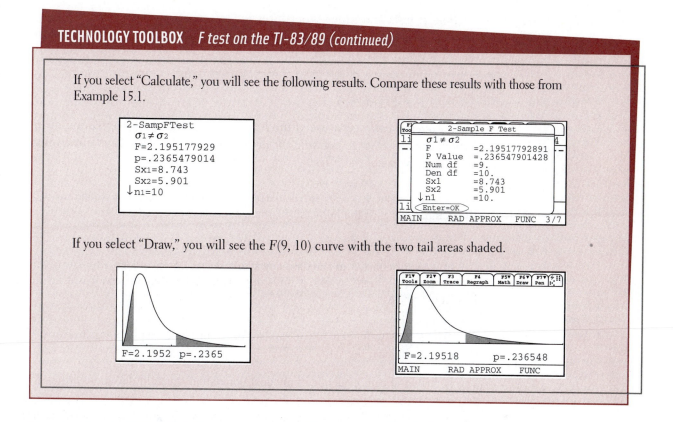

If you select "Draw," you will see the $F(9, 10)$ curve with the two tail areas shaded.

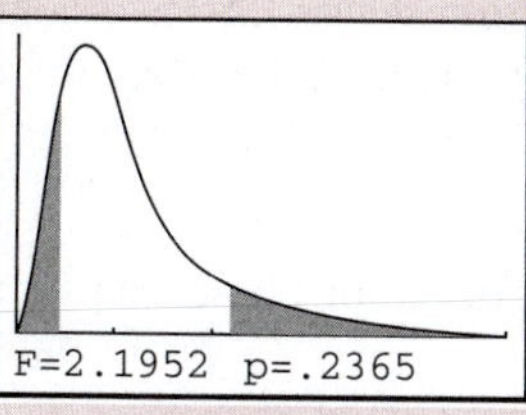

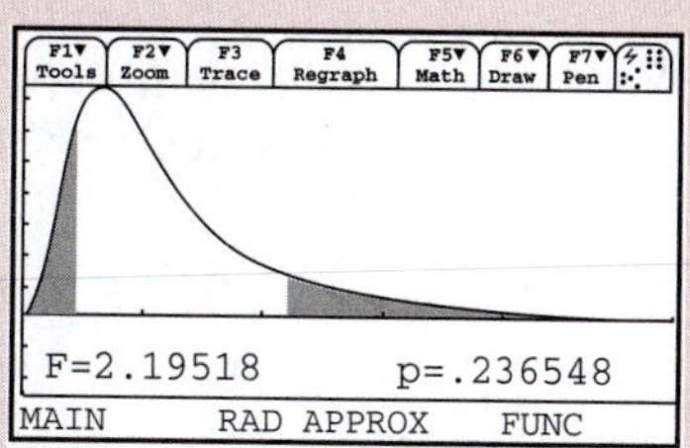

SUMMARY

Inference procedures for comparing the standard deviations of two normal populations are based on the **F statistic,** which is the ratio of sample variances

$$F = \frac{s_1^2}{s_2^2}$$

If an SRS of size n_1 is drawn from Population 1 and an independent SRS of size n_2 is drawn from Population 2, the F statistic has the **F distribution** $F(n_1 - 1, n_2 - 1)$ if the two population standard deviations σ_1 and σ_2 are in fact equal.

The **F distributions** are skewed to the right and take only values greater than 0. A specific F distribution $F(j, k)$ is fixed by the two **degrees of freedom** j and k.

The two-sided **F test** of H_0: $\sigma_1 = \sigma_2$ uses the statistic

$$F = \frac{\text{larger } s^2}{\text{smaller } s^2}$$

and doubles the upper tail probability to obtain the P-value.

The F tests and other procedures for inference on the spread of one or more normal distributions are so strongly affected by lack of normality that we do not recommend them for regular use.

SECTION 15.1 EXERCISES

In all exercises calling for use of the F test, assume that both population distributions are very close to normal. The actual data are not always sufficiently normal to justify use of the F test.

15.1 The F statistic $F = s_1^2/s_2^2$ is calculated from samples of size $n_1 = 10$ and $n_2 = 8$. (Remember that n_1 is the numerator sample size.)

(a) What is the upper 5% critical value for this F?

(b) In a test of equality of standard deviations against the two-sided alternative, this statistic has the value $F = 3.45$. Is this value significant at the 10% level? Is it significant at the 5% level?

15.2 The F statistic for equality of standard deviations based on samples of sizes $n_1 = 21$ and $n_2 = 16$ takes the value $F = 2.78$.

(a) Is this significant evidence of unequal population standard deviations at the 5% level? At the 1% level?

(b) Between which two values obtained from Table D does the P-value of the test fall?

15.3 DDT POISONING The sample variance for the treatment group in the DDT experiment of Example 11.14 (page 662) is more than 10 times as large as the sample variance for the control group. Can you reject the hypothesis of equal population standard deviations at the 5% significance level? At the 1% level? Give appropriate statistical evidence to support your answer.

15.4 COMPETITIVE ROWING Exercise 11.47 (page 665) records the results of comparing a measure of rowing style for skilled and novice female competitive rowers. Is there significant evidence of inequality between the standard deviations of the two populations? Follow the Inference Toolbox.

15.5 TREATING SCRAPIE The report in Exercise 11.51 (page 668) suggests that hamsters in the longer-lived group also had more variation in their length of life. It is common to see s increase with $\bar{x}$ when we compare groups. Do the data give significant evidence of unequal standard deviations?

15.6 SOCIAL INSIGHT AMONG MEN AND WOMEN The data in Exercise 11.39 (page 657) show that scores among men and women on the Chapin Social Insight Test have similar sample standard deviations. The samples are large, however, so we might nonetheless find evidence of a significant difference between the population standard deviations. Is this the case?

15.7 STUDENTS' ATTITUDES Return to the SSHA data in Exercise 1.36 (page 47). We want to know if the spread of SSHA scores is different among women and among men at this college. Use the *F* test to obtain a conclusion.

15.8 EXPLORING THE *F* DISTRIBUTIONS This exercise investigates the effect of increasing both sample sizes on the shape of the *F* distribution curve.

(a) Using the TI-83/89, define Y_1 = Fpdf(X,2,25). Note that Fpdf is found in the DISTR menu on the TI-83 and in the CATALOG under Flash Apps on the TI-89. Set the Window as follows: $X[0,5]_1$ and $Y[-.2,1.2]_1$. Briefly describe the shape of the curve.

(b) Define Y_2 = Fpdf(X,5,50). How does this curve differ from Y_1?

(c) Define Y_3 = Fpdf(X,12,100). How does this curve differ from Y_1 and Y_2?

(d) What values for df1 and df2 will yield a plot of Y = Fpdf(X,df1,df2) that is approximately the shape of a normal distribution?

(e) Describe what happens to the shape of the *F* distribution curve as the sample sizes increase.

15.2 ONE-WAY ANALYSIS OF VARIANCE

Which of four advertising offers mailed to sample households produces the highest sales in dollars? Which of ten brands of automobile tires wears longest? How long do cancer patients live under each of three therapies for their cancer? In each of these settings we wish to compare several treatments. In each case the data are subject to sampling variability—if we mailed the advertising offers to another set of households, we would get different data. We therefore pose the question for inference in terms of the *mean* response. We compare, for example, the mean tread lifetime of different brands of tires. In Chapter 11 we used the two-sample *t* procedures to compare the means of two populations or the mean responses to two treatments in an experiment. We are now ready to extend those methods to problems involving more than two populations. The statistical technique for comparing several means is called ***analysis of variance***, or simply **ANOVA.** ANOVA

EXAMPLE 15.2 CARS, PICKUP TRUCKS, AND SUVs

Pickup trucks and four-wheel-drive sport utility vehicles are replacing cars in American driveways. Do trucks and SUVs have lower gas mileage than cars? Table 15.1 contains data on the highway gas mileage (in miles per gallon) for 28 midsize cars, 8 standard size pickup trucks, and 26 SUVs.

TABLE 15.1 Highway gas mileage for 1998 model vehicles

Midsize cars		Pickup trucks		Sport utility vehicles	
Model	MPG	Model	MPG	Model	MPG
Acura 3.5RL	25	Chevrolet C1500	20	Acura SLX	19
Audi A6 Quattro	26	Dodge Dakota	25	Chevrolet Blazer	20
BMW 740i	24	Dodge Ram	20	Chevrolet Tahoe	19
Buick Century	29	Ford F150	21	Chrysler Town & Country	23
Cadillac Catera	24	Ford Ranger	27	Dodge Durango	17
Cadillac Seville	26	Mazda B2000	25	Ford Expedition	18
Chevrolet Lumina	29	Nissan Frontier	24	Ford Explorer	19
Chevrolet Malibu	32	Toyota T100	23	Geo Tracker	26
Chrysler Cirrus	30			GMC Jimmy	21
Ford Taurus	28			Infiniti QX4	19
Honda Accord	29			Isuzu Rodeo	20
Hyundai Sonata	27			Isuzu Trooper	19
Infiniti I30	28			Jeep Grand Cherokee	21
Infiniti Q45	23			Jeep Wrangler	19
Jaguar XJ8L	24			Kia Sportage	23
Lexus GS300	25			Land Rover Discovery	17
Lexus LS400	25			Lincoln Navigator	16
Lincoln Mark VIII	26			Mazda MPV	19
Mazda 626	29			Mercedes ML320	21
Mercedes-Benz E320	29			Mitsubishi Montero	20
Mitsubishi Diamante	24			Nissan Pathfinder	19
Nissan Maxima	28			Range Rover	17
Oldsmobile Aurora	26			Suburu Forester	27
Oldsmobile Intrigue	30			Suzuki Sidekick	24
Plymouth Breeze	33			Toyota RAV4	26
Saab 900S	25			Toyota 4Runner	22
Toyota Camry	30				
Volvo S70	25				

Source: Environmental Protection Agency, *Model Year 1998 Fuel Economy Guide*, available on-line at www.epa.gov. The table gives data for the basic engine/transmission combination for each model. Models that are essentially identical (such as the Ford Taurus and Mercury Sable) appear only once.

Figure 15.2 shows side-by-side stemplots of the gas mileages for the three types of vehicles. We used the same stems in all three for easier comparison. It does appear that gas mileage decreases as we move from cars to pickups to SUVs.

Midsize		Pickup		SUV	
16		16		16	0
17		17		17	000
18		18		18	0
19		19		19	00000000
20		20	00	20	000
21		21	0	21	000
22		22		22	0
23	0	23	0	23	00
24	0000	24	0	24	0
25	00000	25	00	25	
26	0000	26		26	00
27	0	27	0	27	0
28	000	28		28	
29	00000	29		29	
30	000	30		30	
31		31		31	
32	0	32		32	
33	0	33		33	

FIGURE 15.2 Side-by-side stemplots comparing the highway gas mileages of midsize cars, standard pickup trucks, and SUVs from Table 15.1.

Here are the means, standard deviations, and five-number summaries for the three car types, from statistical software:

	N	MEAN	MEDIAN	STDEV	MIN	MAX	Q1	Q3
Midsize	28	27.107	26.500	2.629	23.000	33.000	25.000	29.000
Pickup	8	23.125	23.500	2.588	20.000	27.000	20.250	25.000
SUV	26	20.423	19.500	2.914	16.000	27.000	19.000	22.250

We will use the mean to describe the center of the gas mileage distributions. As we expect, mean gas mileage goes down as we move from midsize cars to pickup trucks to SUVs. The differences among the means are not large. Are they statistically significant?

The problem of multiple comparisons

Call the mean highway gas mileages for the three populations of vehicles μ_1 for midsize cars, μ_2 for pickups, and μ_3 for SUVs. The subscript reminds us which group a parameter or statistic describes. To compare these three population means, we might use the two-sample t test several times:

- Test H_0: $\mu_1 = \mu_2$ to see if the mean miles per gallon for midsize cars differs from the mean for pickup trucks.
- Test H_0: $\mu_1 = \mu_3$ to see if midsize cars differ from SUVs.
- Test H_0: $\mu_2 = \mu_3$ to see if pickups differ from SUVs.

The weakness of doing three tests is that we get three *P*-values, one for each test alone. That doesn't tell us how likely it is that *three* sample means are spread apart as far as these are. It may be that $\bar{x}_1 = 27.107$ and $\bar{x}_3 = 20.423$ are significantly different if we look at just two groups but not significantly different if we know that they are the smallest and largest means in three groups. As we look at more groups, we expect the gap between the smallest and largest sample mean to get larger. (Think of comparing the tallest and shortest person in larger and larger groups of people.) We can't safely compare many parameters by doing tests or confidence intervals for two parameters at a time.

The problem of how to do many comparisons at once with some overall measure of confidence in all our conclusions is common in statistics. This is the problem of ***multiple comparisons.*** Statistical methods for dealing with multiple comparisons usually have two steps:

multiple comparisons

1. An *overall test* to see if there is good evidence of *any* differences among the parameters that we want to compare.
2. A detailed *follow-up analysis* to decide which of the parameters differ and to estimate how large the differences are.

The overall test, though more complex than the tests we met earlier, is often reasonably straightforward. The follow-up analysis can be quite elaborate. In our basic introduction to statistical practice, we will look only at some overall tests. Chapter 13 describes an overall test to compare several population proportions. In this chapter we present a test for comparing several population means.

The analysis of variance *F* test

We want to test the null hypothesis that there are *no differences* among the mean highway gas mileages for the three vehicle types:

$$H_0: \mu_1 = \mu_2 = \mu_3$$

The alternative hypothesis is that there is some difference, that not all three population means are equal:

$$H_a: \text{not all of } \mu_1, \mu_2, \text{ and } \mu_3 \text{ are equal}$$

The alternative hypothesis is no longer one-sided or two-sided. It is "many-sided," because it allows any relationship other than "all three equal." For example, H_a includes the case in which $\mu_2 = \mu_3$ but μ_1 has a different value. The test of H_0 against H_a is called the ***analysis of variance F test***. (Don't confuse the ANOVA F, which compares several means, with the F statistic of Section 15.1, which compares two standard deviations and is *not* robust against nonnormality.)

analysis of variance F test

EXAMPLE 15.3 INTERPRETING ANOVA FROM SOFTWARE

We entered the gas mileage data from Table 15.1 into the Minitab software package and requested analysis of variance. The output appears in Figure 15.3. Most statistical software packages produce an ANOVA output similar to this one.

First, check that the sample sizes, means, and standard deviations agree with those in Example 15.2. Then find the F test statistic, $F = 40.12$, and its P-value. The P-value is given as 0.000. This means that P is zero to three decimal places, or $P < 0.0005$. There is very strong evidence that the three types of vehicles do not all have the same mean gas mileage.

The F test does not say *which* of the three means are significantly different. It appears from our preliminary analysis of the data that all three may differ. The computer output includes confidence intervals for all three means that suggest the same conclusion. The SUV and pickup intervals overlap only slightly, and the midsize interval lies well above them on the mileage scale. These are 95% confidence intervals for each mean separately. We are not 95% confident that *all three* intervals cover the three means. There are follow-up procedures that provide 95% confidence that we have caught all three means at once, but we won't study them.

Our conclusion: There is strong evidence ($P < 0.0005$) that the means are not all equal. The most important difference among the means is that midsize cars have better gas mileage than pickups and SUVs.

```
Analysis of Variance for Mileage
Source      DF       SS        MS         F        P
Vehicle      2    606.37    303.19     40.12    0.000
Error       59    445.90      7.56
Total       61   1052.27

                                    Individual 95% CIs For Mean
                                    Based on Pooled StDev
Level        N      Mean     StDev --+---------+---------+---------+---
Midsize     28    27.107     2.629                         (---*----)
Pickup       8    23.125     2.588             (-------*------)
SUV         26    20.423     2.914   (----*---)
                                   --+---------+---------+---------+---
Pooled StDev  =    2.749            20.0      22.5      25.0      27.5
```

FIGURE 15.3 Minitab output for analysis of variance of the gas mileage data.

Example 15.3 illustrates our approach to comparing means. The ANOVA F test (often done by software) assesses the evidence for *some* difference among the population means. In most cases, we expect the F test to be significant. We would not undertake a study if we did not expect to find some effect. The formal test is nonetheless important to guard against being misled by chance variation. We will not do the formal follow-up analysis that is often the most useful part of an ANOVA study. Follow-up analysis would allow us to say which means differ and by how much, with (say) 95% confidence that all our conclusions are correct. We rely instead on examination of the data to show what differences are present and whether they are large enough to be interesting. The gap of almost 7 miles per gallon between midsize cars and SUVs is large enough to be of practical interest.

EXERCISES

15.9 DOGS, FRIENDS, AND STRESS If you are a dog lover, perhaps having your dog along reduces the effect of stress. To examine the effect of pets in stressful situations, researchers recruited 45 women who said they were dog lovers. Fifteen of the subjects were randomly assigned to each of three groups to do a stressful task alone, with a good friend present, or with their dog present. (The stressful task was to count backward by 13s or 17s.) The subject's mean heart rate during the task is one measure of the effect of stress. Table 15.2 contains the data.

(a) Make stemplots of the heart rates for the three groups (round to the nearest whole number of beats). Do any of the groups show outliers or extreme skewness?

TABLE 15.2 Mean heart rates during stress with a pet (P), with a friend (F), and for the control group (C)

Group	Rate	Group	Rate	Group	Rate	Group	Rate
P	69.169	P	68.862	C	84.738	C	75.477
F	99.692	C	87.231	C	84.877	C	62.646
P	70.169	P	64.169	P	58.692	P	70.077
C	80.369	C	91.754	P	79.662	F	88.015
C	87.446	C	87.785	P	69.231	F	81.600
P	75.985	F	91.354	C	73.277	F	86.985
F	83.400	F	100.877	C	84.523	F	92.492
F	102.154	C	77.800	C	70.877	P	72.262
P	86.446	P	97.538	F	89.815	P	65.446
F	80.277	P	85.000	F	98.200		
C	90.015	F	101.062	F	76.908		
C	99.046	F	97.046	P	69.538		

Source: These are some of the data from the EESEE story "Stress among Pets and Friends." The study results appear in K. Allen, J. Blascovich, J. Tomaka, and R. M. Kelsey, "Presence of human friends and pet dogs as moderators of autonomic responses to stress in women," *Journal of Personality and Social Psychology*, 83 (1988), pp. 582–589.

(b) Figure 15.4 gives the Minitab ANOVA output for these data. Do the mean heart rates for the groups appear to show that the presence of a pet or a friend reduces heart rate during a stressful task?

```
Analysis of Variance for Beats
Source       DF        SS          MS          F         P
Group         2    2387.7      1193.8      14.08     0.000
Error        42    3561.3        84.8
Total        44    5949.0

                                          Individual 95% CIs For Mean
                                          Based on Pooled StDev

Level         N      Mean       StDev   ----+---------+---------+---------+--
Control      15    82.524       9.242             (-----*-----)
Friend       15    91.325       8.341                       (-----*-----)
Pet          15    73.483       9.970   (-----*-----)
                                        ----+---------+---------+---------+--
Pooled StDev =      9.208                  72.0      80.0      88.0      96.0
```

FIGURE 15.4 Minitab output for the data in Table 15.2 on heart rates (beats per minute) during stress. The control group worked alone, the "Friend" group had a friend present, and the "Pet" group had a pet dog present.

(c) What are the values of the ANOVA F statistic and its P-value? What hypotheses does F test? Briefly describe the conclusions you draw from these data. Did you find anything surprising?

15.10 HOW MUCH CORN SHOULD I PLANT? How much corn per acre should a farmer plant to obtain the highest yield? Too few plants will give a low yield. On the other hand, if there are too many plants, they will compete with each other for moisture and nutrients, and yields will fall. To find out, plant at different rates on several plots of ground and measure the harvest. (Be sure to treat all the plots the same except for the planting rate.) Here are data from such an experiment:[2]

Plants per acre	Yield (bushels per acre)			
12,000	150.1	113.0	118.4	142.6
16,000	166.9	120.7	135.2	149.8
20,000	165.3	130.1	139.6	149.9
24,000	134.7	138.4	156.1	
28,000	119.0	150.5		

(a) Make side-by-side stemplots of yield for each number of plants per acre. What do the data appear to show about the influence of plants per acre on yield?

(b) ANOVA will assess the statistical significance of the observed differences in yield. What are H_0 and H_a for the ANOVA F test in this situation?

(c) The Minitab ANOVA output for these data appears in Figure 15.5. What is the sample mean yield for each planting rate? What does the ANOVA F test say about the significance of the effects you observe?

```
Analysis of Variance for Yield
Source      DF        SS         MS        F        P
Group        4       600        150      0.50    0.736
Error       12      3597        300
Total       16      4197

                                       Individual 95% CIs For Mean
                                       Based on Pooled StDev
Level        N      Mean      StDev  ---+---------+---------+---------+---
12,000       4    131.03      18.09     (-----------*-----------)
16,000       4    143.15      19.79              (-----------*-----------)
20,000       4    146.22      15.07                (----------*----------)
24,000       3    143.07      11.44            (-------------*-------------)
28,000       2    134.75      22.27  (----------------*----------------)
                                     ---+---------+---------+---------+---
Pooled StDev =    17.31                112       128       144       160
```

FIGURE 15.5 Minitab output for yields of corn at five planting rates.

(d) The observed differences among the mean yields in the sample are quite large. Why are they not statistically significant?

The idea of analysis of variance

Here is the main idea for comparing means: what matters is not how far apart the sample means are but how far apart they are *relative to the variability of individual observations.* Look at the two sets of boxplots in Figure 15.6. For simplicity, these distributions are all symmetric, so that the mean and median are the same. The centerline in each boxplot is therefore the sample mean. Both figures compare three samples with the same three means. Like the three vehicle types in Example 15.3, the means are different but not very different. Could differences this large easily arise just due to chance, or are they statistically significant?

- The boxplots in Figure 15.6(a) have tall boxes, which show lots of variation among the individuals in each group. With this much variation among individuals, we would not be surprised if another set of samples gave quite different sample means. The observed differences among the sample means could easily happen just by chance.

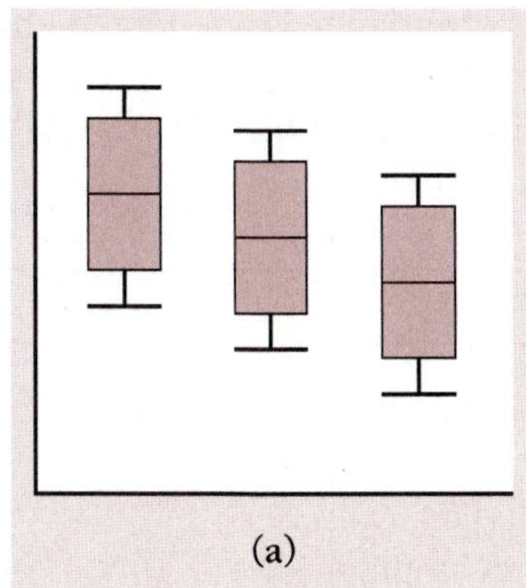

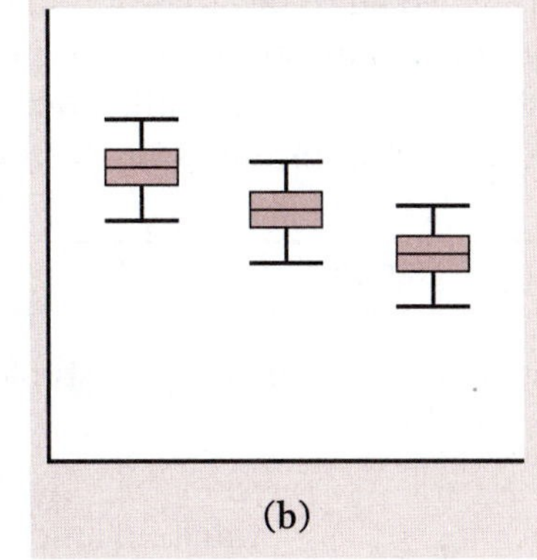

FIGURE 15.6 Boxplots for two sets of three samples each. The sample means are the same in (a) and (b). Analysis of variance will find a more significant difference among the means in (b) because there is less variation among the individuals within those samples.

• The boxplots in Figure 15.6(b) have the same centers as those in Figure 15.6(a), but the boxes are much shorter. That is, there is much less variation among the individuals in each group. It is unlikely that any sample from the first group would have a mean as small as the mean of the second group. Because means as far apart as those observed would rarely arise just by chance in repeated sampling, they are good evidence of real differences among the means of the three populations we are sampling from.

This comparison of the two parts of Figure 15.6 is too simple in one way. It ignores the effect of the sample sizes, an effect that boxplots do not show. Small differences among sample means can be significant if the samples are very large. Large differences among sample means may fail to be significant if the samples are very small. All we can be sure of is that for the same sample size, Figure 15.6(b) will give a much smaller *P*-value than Figure 15.6(a). Despite this qualification, the big idea remains: if sample means are far apart relative to the variation among individuals in the same group, that's evidence that something other than chance is at work.

THE ANALYSIS OF VARIANCE IDEA

Analysis of variance compares the variation due to specific sources with the variation among individuals who should be similar. In particular, ANOVA tests whether several populations have the same mean by comparing how far apart the sample means are with how much variation there is within the samples.

It is one of the oddities of statistical language that methods for comparing means are named after the variance. The reason is that the test works by comparing two kinds of variation. Analysis of variance is a general method for studying sources of variation in responses. Comparing several means is the simplest form of ANOVA, called ***one-way ANOVA.*** One-way ANOVA is the only form of ANOVA that we will study.

one-way ANOVA

THE ANOVA *F* STATISTIC

The **analysis of variance F statistic** for testing the equality of several means has this form:

$$F = \frac{\text{variation among the sample means}}{\text{variation among individuals in the same sample}}$$

We give more detail later. Because ANOVA is in practice done by software, the idea is more important than the detail. The F statistic can only take values that are zero or positive. It is zero only when all the sample means are identical and gets larger as they move farther apart. Large values of F are evidence against the null hypothesis H_0 that all population means are the same. Although the alternative hypothesis H_a is many-sided, the ANOVA F test is one-sided because any violation of H_0 tends to produce a large value of F.

How large must F be to provide significant evidence against H_0? To answer questions of statistical significance, compare the F statistic with critical values from an ***F distribution.*** Recall from Section 15.1 that a specific F distribution is specified by two parameters: a numerator degrees of freedom and a denominator degrees of freedom. Table D in the back of the book contains critical values for F distributions with various degrees of freedom.

F distribution

EXAMPLE 15.4 USING THE *F* TABLE

Look again at the computer output for the gas mileage data in Figure 15.3. The degrees of freedom for the F test appear in the first two rows of the "DF" column. There are 2 degrees of freedom in the numerator and 59 in the denominator.

In Table D, find the numerator degrees of freedom 2 at the top of the table. Then look for the denominator degrees of freedom 59 at the left of the table. There is no entry for 59, so we use the next smaller entry, 50 degrees of freedom. The upper critical values for 2 and 50 degrees of freedom are

p	Critical value
.100	2.41
.050	3.18
.025	3.97
.010	5.06
.001	7.96

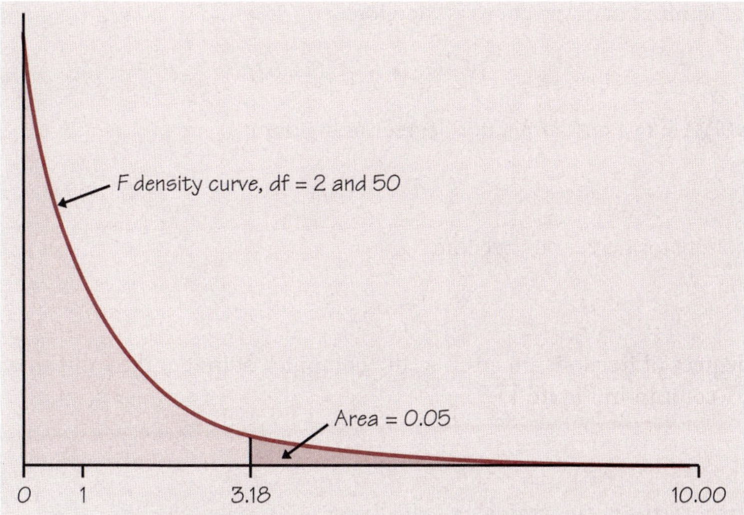

FIGURE 15.7 The density curve of the *F* distribution with 2 degrees of freedom in the numerator and 50 degrees of freedom in the denominator. The value $F^* = 3.18$ is the 0.05 critical value.

Figure 15.7 shows the F density curve with 2 and 50 degrees of freedom and the upper 5% critical value 3.18. The observed $F = 40.12$ lies far to the right on this curve. We see from the table that $F = 40.12$ is larger than the 0.001 critical value, so $P < 0.001$.

The degrees of freedom of the F statistic depend on the number of means we are comparing and the number of observations in each sample. That is, the F test does take into account the number of observations. Here are the details.

DEGREES OF FREEDOM FOR THE *F* TEST

We want to compare the means of I populations. We have an SRS of size n_i from the ith population, so that the total number of observations in all samples combined is

$$N = n_1 + n_2 + \cdots + n_I$$

If the null hypothesis that all population means are equal is true, the ANOVA F statistic has the F distribution with $I - 1$ degrees of freedom in the numerator and $N - I$ degrees of freedom in the denominator.

EXAMPLE 15.5 DEGREES OF FREEDOM FOR *F*

In Examples 15.2 and 15.3, we compared the mean highway gas mileage for three types of vehicles, so $I = 3$. The three sample sizes are

$$n_1 = 28 \qquad n_2 = 8 \qquad n_3 = 26$$

The total number of observations is therefore

$$N = 28 + 8 + 26 = 62$$

The ANOVA F test has numerator degrees of freedom

$$I - 1 = 3 - 1 = 2$$

and denominator degrees of freedom

$$N - I = 62 - 3 = 59$$

These degrees of freedom are given in the computer output as the first two entries in the "DF" column in Figure 15.3.

EXERCISES

15.11 DOGS, FRIENDS, AND STRESS Exercise 15.9 (page 828) compares the mean heart rates for women performing a stressful task under three conditions.

(a) What are I, the n_i, and N for these data? State in words what each of these numbers means.

(b) Find both degrees of freedom for the ANOVA F statistic. Check your results against the output in Figure 15.4.

(c) The output shows that $F = 14.08$. What does Table D tell you about the significance of this result?

15.12 HOW MUCH CORN SHOULD I PLANT? Exercise 15.10 (page 829) compares the yields for several planting rates for corn.

(a) What are I, the n_i, and N for these data? Identify these quantities in words and give their numerical values.

(b) Find the degrees of freedom for the ANOVA F statistic. Check your work against the computer output in Figure 15.5.

(c) For these data, $F = 0.50$. What does Table D tell you about the P-value of this statistic?

15.13 In each of the following situations, we want to compare the mean response in several populations. For each setting, identify the populations and the response variable. Then give I, the n_i, and N. Finally, give the degrees of freedom of the ANOVA F statistic.

(a) Do four tomato varieties differ in mean yield? Grow ten plants of each variety and record the yield of each plant in pounds of tomatoes.

(b) A maker of detergents wants to compare the attractiveness to consumers of six package designs. Each package is shown to 120 different consumers who rate the attractiveness of the design on a 1 to 10 scale.

(c) An experiment to compare the effectiveness of three weight-loss programs has 32 subjects who want to lose weight. Ten subjects are assigned at random to each of two programs, and the remaining 12 subjects follow the third program. After six months, each subject's change in weight is recorded.

Conditions for performing ANOVA

Like all inference procedures, ANOVA is valid only in some circumstances. Here are the requirements for using ANOVA to compare population means.

ANOVA CONDITIONS

- We have ***I*** **independent SRSs,** one from each of I populations.
- Each population has a **normal distribution** with unknown mean μ_i. The means may be different in the different populations. The ANOVA F statistic tests the null hypothesis that all of the populations have the same mean:

$$H_0\colon \mu_1 = \mu_2 = \cdots = \mu_I$$
$$H_a\colon \text{not all of the } \mu_i \text{ are equal}$$

- All of the populations have the **same standard deviation** σ, whose value is unknown.

The first two requirements are familiar from our study of the two-sample t procedures for comparing two means. As usual, the design of the data production is the most important foundation for inference. Biased sampling or confounding can make any inference meaningless. If we do not actually draw separate SRSs from each population or carry out a randomized comparative experiment, it is often unclear to what population the conclusions of inference apply. This is the case in Example 15.2, for example. ANOVA, like other inference procedures, is often used when random samples are not available. You must judge each use on its merits, a judgment that usually requires some knowledge of the subject of the study in addition to some knowledge of statistics. We might consider the 1998 vehicles in Example 15.2 to be samples from vehicles of their types produced in recent years.

Because no real population has exactly a normal distribution, the usefulness of inference procedures that assume normality depends on how sensitive they are to departures from normality. Fortunately, procedures for comparing means are not very sensitive to lack of normality. The ANOVA F test, like the t procedures, is ***robust.*** What matters is normality of the sample means, so ANOVA becomes safer as the sample sizes get larger, because of the central limit theorem effect. Remember to check for outliers that change the value of sample means and for extreme skewness. When there are no outliers and the

robustness

distributions are roughly symmetric, you can safely use ANOVA for sample sizes as small as 4 or 5.

The third condition is annoying: ANOVA assumes that the variability of observations, measured by the standard deviation, is the same in all populations. You may recall from Chapter 11 (page 666) that there is a special version of the two-sample t test that assumes equal standard deviations in both populations. The ANOVA F for comparing two means is exactly the square of this special t statistic. We prefer the t test that does not assume equal standard deviations, but for comparing more than two means there is no general alternative to the ANOVA F. It is not easy to check the assumption that the populations have equal standard deviations. Statistical tests for equality of standard deviations are very sensitive to lack of normality, so much so that they are of little practical value. You must either seek expert advice or rely on the robustness of ANOVA.

How serious are unequal standard deviations? ANOVA is not too sensitive to violations of the condition, especially when all samples have the same or similar sizes and no sample is very small. When designing a study, try to take samples of the same size from all the groups you want to compare. The sample standard deviations estimate the population standard deviations, so check before doing ANOVA that the sample standard deviations are similar to each other. We expect some variation among them due to chance. Here is a rule of thumb that is safe in almost all situations.

CHECKING STANDARD DEVIATIONS IN ANOVA

The results of the ANOVA F test are approximately correct when the largest sample standard deviation is no more than twice as large as the smallest sample standard deviation.

EXAMPLE 15.6 DO THE STANDARD DEVIATIONS ALLOW ANOVA?

In the gas mileage study, the sample standard deviations for midsize cars, pickups, and SUVs are

$$s_1 = 2.629 \qquad s_2 = 2.588 \qquad s_3 = 2.914$$

These standard deviations easily satisfy our rule of thumb. We can safely use ANOVA to compare the mean gas mileage for the three vehicle types.

The report from which Table 15.1 was taken also contained data on 25 subcompact cars. Can we use ANOVA to compare the means for all four types of vehicles? The standard deviation for subcompact cars is $s_4 = 5.240$ miles per gallon. This is slightly more than twice the smallest standard deviation:

$$\frac{\text{largest } s}{\text{smallest } s} = \frac{5.240}{2.588} = 2.02$$

A large standard deviation is often due to skewness or an outlier. Here is a stemplot of the subcompact car gas mileages:

```
2 | 44
2 | 5778899999
3 | 00011334
3 | 6667
4 |
4 | 9
```

The Geo Metro gets 49 miles per gallon. The standard deviation drops from 5.24 to 3.73 if we omit this one car. The Metro is a subcompact car based on its interior volume but has a much smaller engine than any other car in the class. If we decide to drop the Metro, ANOVA is safe.

EXAMPLE 15.7 WHICH COLOR ATTRACTS BEETLES BEST?

To detect the presence of harmful insects in farm fields, we can put up boards covered with a sticky material and examine the insects trapped on the boards. Which colors attract insects best? Experimenters placed six boards of each of four colors at random locations in a field of oats and measured the number of cereal leaf beetles trapped. Here are the data:[3]

Board color	Insects trapped					
Blue	16	11	20	21	14	7
Green	37	32	20	29	37	32
White	21	12	14	17	13	20
Yellow	45	59	48	46	38	47

We would like to use ANOVA to compare the mean numbers of beetles that would be trapped by all boards of each color. Because the samples are small, we plot the data in side-by-side stemplots in Figure 15.8. Computer output for ANOVA appears in Figure 15.9. The yellow boards attract by far the most insects ($\bar{x}_4 = 47.167$), with green next ($\bar{x}_2 = 31.167$) and blue and white far behind.

Check that we can safely use ANOVA to test equality of the four means. The largest of the four sample standard deviations is 6.795 and the smallest is 3.764. The ratio

$$\frac{\text{largest } s}{\text{smallest } s} = \frac{6.795}{3.764} = 1.8$$

is less than 2, so these data satisfy our rule of thumb. The shapes of the four distributions are irregular, as we expect with only 6 observations in each group, but there are no outliers. The ANOVA results will be approximately correct.

Blue		Green		White		Yellow	
0	7	0		0		0	
1	146	1		1	2347	1	
2	01	2	09	2	01	2	
3		3	2277	3		3	8
4		4		4		4	5678
5		5		5		5	9

FIGURE 15.8 Side-by-side stemplots comparing the counts of insects attracted by six boards for each of four board colors.

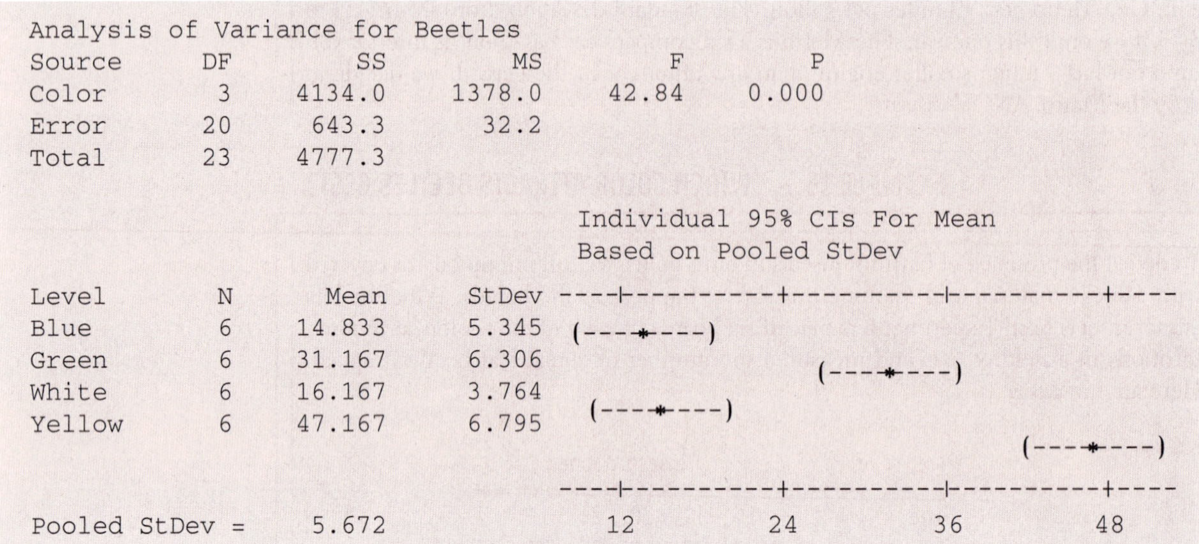

```
Analysis of Variance for Beetles
Source      DF        SS        MS        F        P
Color        3    4134.0    1378.0    42.84    0.000
Error       20     643.3      32.2
Total       23    4777.3

                                  Individual 95% CIs For Mean
                                  Based on Pooled StDev
Level       N      Mean     StDev ---+---------+---------+---------+----
Blue        6    14.833     5.345 (---*---)
Green       6    31.167     6.306                 (---*---)
White       6    16.167     3.764  (---*---)
Yellow      6    47.167     6.795                           (---*---)
                                  ---+---------+---------+---------+----
Pooled StDev =    5.672             12        24        36        48
```

FIGURE 15.9 Minitab ANOVA output for comparing the four board colors.

There are $I = 4$ groups and $N = 24$ observations overall, so the degrees of freedom for F are

$$\text{numerator:} \quad I - 1 = 4 - 1 = 3$$
$$\text{denominator:} \quad N - I = 24 - 4 = 20$$

This agrees with the computer results. The F statistic is $F = 42.84$, a large F with P-value $P < 0.001$. Despite the small samples, the experiment gives very strong evidence of differences among the colors. Yellow boards appear best at attracting leaf beetles.

A one-way ANOVA test can be performed with the TI-83/89, as long as you are comparing no more than 6 means. The following Technology Toolbox shows you how.

TECHNOLOGY TOOLBOX *One-way ANOVA on the TI-83/89*

We will repeat the ANOVA procedure of Example 15.7 on the TI-83/89. Begin by entering the data into lists: Blue ↔ L_1/list1, Green ↔ L_2/list2, White ↔ L_3/list3, and Yellow ↔ L_4/list4.

TI-83

- Press STAT, choose TESTS, and F:ANOVA(. Then enter the lists that contain the data. Here, the command is ANOVA(L_1,L_2,L_3,L_4).

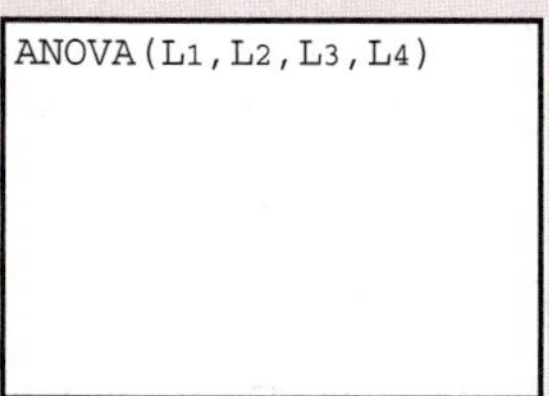

TI-89

- In the Statistics/List Editor, choose [F6] (Tests), then C:ANOVA. Set Data Input Method to "Data" and Number of Groups to "4."

The results of the one-way ANOVA are shown on the following screens. The calculator reports that the *F* statistic is 42.84 and the *P*-value is 6.8×10^{-9}. The numerator degrees of freedom are $I - 1 = 3$, and by scrolling down, you see that the denominator degrees of freedom are $N - I = 24 - 4 = 20$.

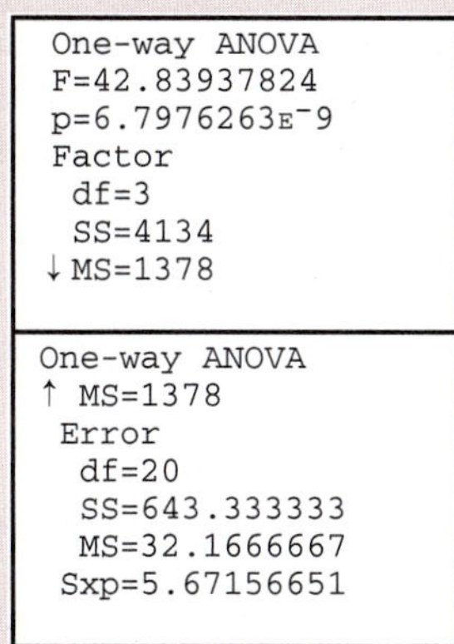

One-way ANOVA
F=42.83937824
p=6.7976263E-9
Factor
df=3
SS=4134
↓MS=1378
One-way ANOVA
↑ MS=1378
Error
df=20
SS=643.333333
MS=32.1666667
Sxp=5.67156651

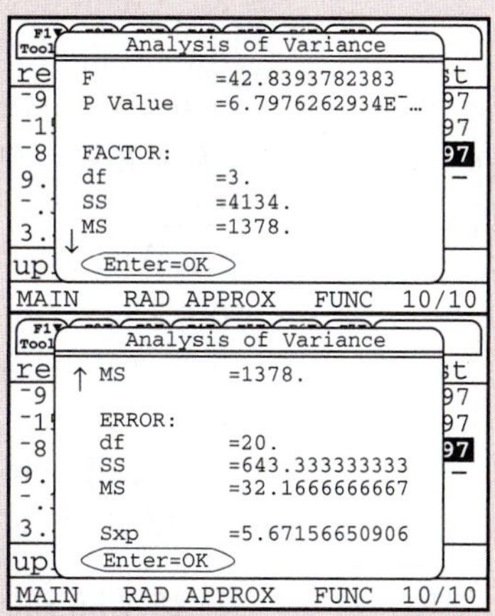

If you know the *F* statistic and the numerator and denominator degrees of freedom, you can find the *P*-value with the command `Fcdf`, under the DISTR menu on the TI-83, and in the CATALOG on the TI-89. The syntax is `Fcdf`(leftendpoint, rightendpoint, df numerator, df denominator). See the following screen shots.

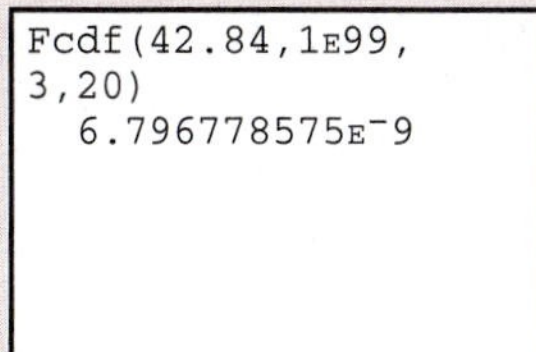

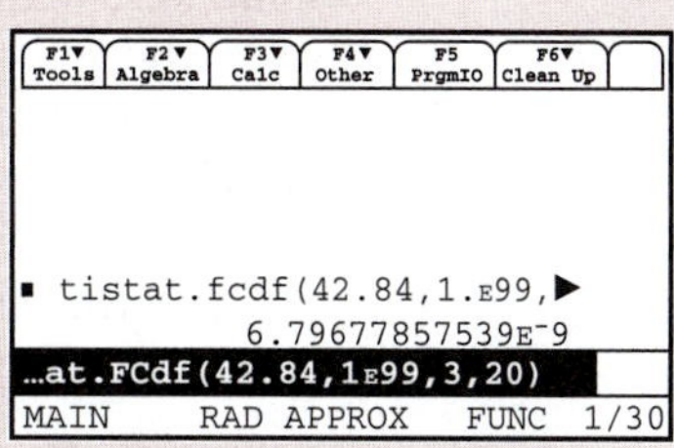

EXERCISES

15.14 Verify that the sample standard deviations for these sets of data do allow use of ANOVA to compare the population means.

(a) The heart rates of Exercise 15.9 (page 828) and Figure 15.4.

(b) The corn yields of Exercise 15.10 (page 829) and Figure 15.5.

15.15 MARITAL STATUS AND SALARY Married men tend to earn more than single men. An investigation of the relationship between marital status and income collected data on all 8235 men employed as managers or professionals by a large manufacturing firm in 1976. Suppose (this is risky) we regard these men as a random sample from the population of all men employed in managerial or professional positions in large companies. Here are descriptive statistics for the salaries of these men:[4]

	Single	Married	Divorced	Widowed
n_i	337	7,730	126	42
$\bar{x}_i$	\$21,384	\$26,873	\$25,594	\$26,936
s_i	\$5,731	\$7,159	\$6,347	\$8,119

(a) Briefly describe the relationship between marital status and salary.

(b) Do the sample standard deviations allow use of the ANOVA F test? (The distributions are skewed to the right. We expect right skewness in income distributions. The investigators actually applied ANOVA to the logarithms of the salaries, which are more symmetric.)

(c) What are the degrees of freedom of the ANOVA F test?

(d) The F test is a formality for these data, because we are sure that the P-value will be very small. Why are we sure?

(e) Single men earn less on the average than men who are or have been married. Do the highly significant differences in mean salary show that getting married raises men's mean income? Explain your answer.

15.16 WHO SUCCEEDS IN COLLEGE? What factors influence the success of college students? Look at all 256 students who entered a university planning to study computer science (CS) in a specific year. We are willing to regard these students as a random sample of the students the university CS program will attract in subsequent years. After three semesters of study, some of these students were CS majors, some were majors in another field of science or engineering, and some had left science and engineering or left the university. The table below gives the sample means and standard deviations and the ANOVA F statistics for three variables that describe the students' high school performance. These are three separate ANOVA F tests.[5]

The first variable is a student's rank in the high school class, given as a percentile (so rank 50 is the middle of the class and rank 100 is the top). The next variable is the number of semester courses in mathematics the student took in high school. The third variable is the student's average grade in high school mathematics. The mean and standard deviation appear in a form common in published reports, with the standard deviation in parentheses following the mean.

		Mean (standard deviation)		
Group	n	High school class rank	Semesters of HS math	Average grade in HS math
CS majors	103	88.0 (10.5)	8.74 (1.28)	3.61 (0.46)
Sci./eng. majors	31	89.2 (10.8)	8.65 (1.31)	3.62 (0.40)
Other	122	85.8 (10.8)	8.25 (1.17)	3.35 (0.55)
F statistic		1.95	4.56	9.38

(a) What null and alternative hypotheses does F test for rank in the high school class? Express the hypotheses both in symbols and in words. The hypotheses are similar for the other two variables.

(b) What are the degrees of freedom for each F?

(c) Check that the standard deviations allow use of all three F tests. The shapes of the distributions also allow use of F. How significant is F for each of these variables?

(d) Write a brief summary of the differences among the three groups of students, taking into account both the significance of the F tests and the values of the means.

15.17 ACTIVITY 15 ANOVA Use your TI-83/89 to analyze the paper airplane flight distance data from Activity 15.

(a) First, enter the data into your calculator and explore the data graphically. Does it appear that at least one of the means is different from the other means?

(b) Check the ANOVA conditions. Can you state with some assurance that these conditions are satisfied?

(c) Apply the ANOVA standard deviation test. Is the quotient (largest standard deviation)/(smallest standard deviation) ≤ 2? If not, you probably should not proceed.

(d) If the data pass the tests in (b) and (c), then perform the one-way ANOVA. Write your conclusions.

Some details of ANOVA

Now we will give the actual recipe for the ANOVA F statistic. We have SRSs from each of I populations. Subscripts from 1 to I tell us which sample a statistic refers to:

Population	Sample size	Sample mean	Sample std. dev.
1	n_1	$\bar{x}_1$	s_1
2	n_2	$\bar{x}_2$	s_2
⋮	⋮	⋮	⋮
I	n_I	$\bar{x}_I$	s_I

You can find the F statistic from just the sample sizes n_i, the sample means $\bar{x}_i$, and the sample standard deviations s_i. You don't need to go back to the individual observations.

The ANOVA F statistic has the form

$$F = \frac{\text{variation among the sample means}}{\text{variation among individuals}}$$

mean squares

The measures of variation in the numerator and denominator of F are called ***mean squares.*** A mean square is a more general form of a sample variance. An ordinary sample variance s^2 is an average (or mean) of the squared deviations of observations from their mean, so it qualifies as a "mean square."

The numerator of F is a mean square that measures variation among the I sample means $\bar{x}_1, \bar{x}_2, \ldots, \bar{x}_I$. Call the overall mean response, the mean of all N observations together, $\bar{x}$. You can find $\bar{x}$ from the I sample means by

$$\bar{x} = \frac{n_1\bar{x}_1 + n_2\bar{x}_2 + \cdots + n_I\bar{x}_I}{N}$$

The sum of each mean multiplied by the number of observations it represents is the sum of all the individual observations. Dividing this sum by N, the total number of observations, gives the overall mean $\bar{x}$. The numerator mean square in F is an average of the I squared deviations of the means of the samples from $\bar{x}$. It is called the ***mean square for groups,*** abbreviated as MSG:

MSG

$$\text{MSG} = \frac{n_1(\bar{x}_1 - \bar{x})^2 + n_2(\bar{x}_2 - \bar{x})^2 + \cdots + n_I(\bar{x}_I - \bar{x})^2}{I - 1}$$

Each squared deviation is weighted by n_i, the number of observations it represents.

The mean square in the denominator of F measures variation among individual observations in the same sample. For any one sample, the sample variance s_i^2 does this job. For all I samples together, we use an average of the individual sample variances. It is again a weighted average in which each s_i^2 is weighted by one fewer than the number of observations it represents, $n_i - 1$. Another way to put this is that each s_i^2 is weighted by its degrees of freedom $n_i - 1$. The resulting mean square is called the ***mean square for error,*** MSE: *MSE*

$$\text{MSE} = \frac{(n_1 - 1)s_1^2 + (n_2 - 1)s_2^2 + \cdots + (n_I - 1)s_I^2}{N - I}$$

Here is a summary of the ANOVA test.

THE ANOVA *F* TEST

Draw an independent SRS from each of I populations. The ith population has the $N(\mu_i, \sigma)$ distribution, where σ is the common standard deviation in all the populations. The ith sample has size n_i, sample mean $\bar{x}_i$, and sample standard deviation s_i.

The **ANOVA *F* statistic** tests the null hypothesis that all I populations have the same mean:

$$H_0: \mu_1 = \mu_2 = \cdots = \mu_I$$
$$H_a: \text{not all of the } \mu_i \text{ are equal}$$

The statistic is

$$F = \frac{\text{MSG}}{\text{MSE}}$$

The **mean squares** that make up F are

$$\text{MSG} = \frac{n_1(\bar{x}_1 - \bar{x})^2 + n_2(\bar{x}_2 - \bar{x})^2 + \cdots + n_I(\bar{x}_I - \bar{x})^2}{I - 1}$$

and

$$\text{MSE} = \frac{(n_1 - 1)s_1^2 + (n_2 - 1)s_2^2 + \cdots + (n_I - 1)s_I^2}{N - I}$$

When H_0 is true, F has the **F distribution** with $I - 1$ and $N - I$ degrees of freedom.

The denominators in the recipes for MSG and MSE are the two degrees of freedom $I - 1$ and $N - I$ of the F test. The numerators are called *sums of squares*, from their algebraic form. It is usual to present the results of ANOVA in an ***ANOVA table*** like that in the Minitab output.

ANOVA *table*

Source	Degrees of freedom	Sum of squares	Mean square	F
Groups	$I - 1$	$\Sigma_{\text{groups}} n_i(\bar{x}_i - \bar{x})^2$	SSG/DFG	MSG/MSE
Error	$N - I$	$\Sigma_{\text{groups}} (n_i - 1)s_i^2$	SSE/DFE	
Total	$N - 1$	$\Sigma_{\text{obs}} (x_{ij} - \bar{x})^2$	SST/DFT	

The table has columns for degrees of freedom (DF), sums of squares (SS), and mean squares (MS). Check that each MS entry in Figure 15.9 (page 838), for example, is the sum of squares SS divided by the degrees of freedom DF in the same row. The F statistic in the "F" column is MSG/MSE. The rows are labeled by sources of variation. In this output, variation among groups is labeled "COLOR." Other statistical software calls this line "Treatments" or "Groups." Variation among observations in the same group is called "ERROR" by most software. This doesn't mean a mistake has been made. It's a traditional term for chance variation.

Because MSE is an average of the individual sample variances, it is also called the *pooled sample variance*, written as s_p^2. When all I populations have the same population variance σ^2 (ANOVA assumes that they do), s_p^2 estimates the common variance σ^2. The square root of MSE is the ***pooled standard deviation*** s_p. It estimates the common standard deviation σ of observations in each group. Minitab, like most ANOVA programs, gives the value of s_p as well as MSE. It is the "Pooled StDev" value in Figure 15.9 (page 838).

pooled standard deviation

The pooled standard deviation s_p is a better estimator of the common σ than any individual sample standard deviation s_i because it combines (pools) the information in all I samples. We can get a confidence interval for any of the means μ_i from the usual form

$$\text{estimate} \pm t^* \text{SE}_{\text{estimate}}$$

using s_p to estimate σ. The confidence interval for μ_i is

$$\bar{x}_i \pm t^* \frac{s_p}{\sqrt{n_i}}$$

Use the critical value t^* from the t distribution with $N - I$ degrees of freedom, because s_p has $N - I$ degrees of freedom. These are the confidence intervals that appear in the Minitab ANOVA output.

EXAMPLE 15.8 ANOVA CALCULATIONS

We can do the ANOVA test comparing board colors in Example 15.7 using only the sample sizes, sample means, and sample standard deviations. Minitab gives these in Figure 15.9, but it is easy to find them with a calculator.

The overall mean of the 24 counts is

$$\begin{aligned}\bar{x} &= \frac{n_1\bar{x}_1 + n_2\bar{x}_2 + \cdots + n_I\bar{x}_I}{N} \\ &= \frac{(6)(14.833) + (6)(31.167) + (6)(16.167) + (6)(47.167)}{24} \\ &= \frac{656}{24} = 27.333\end{aligned}$$

The mean square for groups is

$$\begin{aligned}\text{MSG} &= \frac{n_1(\bar{x}_1 - \bar{x})^2 + n_2(\bar{x}_2 - \bar{x})^2 + \cdots + n_I(\bar{x}_I - \bar{x})^2}{I-1} \\ &= \frac{1}{4-1}[(6)(14.833 - 27.333)^2 + (6)(31.167 - 27.333)^2 \\ &\quad + (6)(16.167 - 27.333)^2 + (6)(47.167 - 27.333)^2] \\ &= \frac{4134.100}{3} = 1378.033\end{aligned}$$

The mean square for error is

$$\begin{aligned}\text{MSE} &= \frac{(n_1 - 1)s_1^2 + (n_2 - 1)s_2^2 + \cdots + (n_I - 1)s_I^2}{N - I} \\ &= \frac{(5)(5.345^2) + (5)(6.306^2) + (5)(3.764^2) + (5)(6.795^2)}{24 - 4} \\ &= \frac{643.372}{20} = 32.169\end{aligned}$$

Finally, the ANOVA test statistic is

$$F = \frac{\text{MSG}}{\text{MSE}} = \frac{1378.033}{32.169} = 42.84$$

Our work agrees with the computer output in Figure 15.9 (page 838) and the calculator output (up to roundoff) in the Technology Toolbox on p. 839. We don't recommend doing these calculations, because tedium and roundoff errors cause frequent mistakes.

The pooled estimate of the standard deviation σ in any group is

$$s_p = \sqrt{\text{MSE}} = \sqrt{32.169} = 5.672$$

A 95% confidence interval for the mean count of insects trapped by yellow boards, using s_p and 20 degrees of freedom, is

$$\bar{x}_4 \pm t^* \frac{s_p}{\sqrt{n_4}} = 47.167 \pm 2.086 \frac{5.672}{\sqrt{6}}$$
$$= 47.167 \pm 4.830$$
$$= 42.34 \text{ to } 52.00$$

This confidence interval appears in the graph in the Minitab ANOVA output in Figure 15.9.

EXERCISES

15.18 WEIGHTS OF NEWLY HATCHED PYTHONS A study of the effect of nest temperature on the development of water pythons separated python eggs at random into nests at three temperatures: cold, neutral, and hot. Exercise 13.35 (page 772) shows that the proportions of eggs that hatched at each temperature did not differ significantly. Now we will examine the little pythons. In all, 16 eggs hatched at the cold temperature, 38 at the neutral temperature, and 75 at the hot temperature. The report of the study summarizes the data in the common form "mean ± standard error" as follows:[6]

Temperature	n	Weight (grams) at hatching	Propensity to strike
Cold	16	28.89 ± 8.08	6.40 ± 5.67
Neutral	38	32.93 ± 5.61	5.82 ± 4.24
Hot	75	32.27 ± 4.10	4.30 ± 2.70

(a) We will compare the mean weights at hatching. Recall that the standard error of the mean is $s/\sqrt{n}$. Find the standard deviations of the weights in the three groups and verify that they satisfy our rule of thumb for using ANOVA.

(b) Starting from the sample sizes n_i, the means $\bar{x}_i$, and the standard deviations s_i, carry out an ANOVA. That is, find MSG, MSE, and the F statistic, and use Table D to approximate the P-value. Is there evidence that nest temperature affects the mean weight of newly hatched pythons?

15.19 PYTHON STRIKES The data in the previous exercise also describe the "propensity to strike" of the hatched pythons at 30 days of age. This is the number of taps on the head with a small brush until the python launches a strike. (Don't try this with adult pythons.) The data are again summarized in the form "sample mean ± standard error of the mean." Follow the outline in (a) and (b) of the previous exercise for propensity to strike. Does nest temperature appear to influence propensity to strike?

15.20 HOW MUCH CORN SHOULD I PLANT? Return to the data in Exercise 15.10 (page 829) on corn yields for different planting rates.

(a) Starting from the sample means and standard deviations for the five groups (Figure 15.5), calculate MSE, the overall mean yield $\bar{x}$, and MSG. Use the computer output in Figure 15.5 to check your work.

(b) Give a 90% confidence interval for the mean yield of corn planted at 20,000 plants per acre. Use the pooled standard deviation s_p to estimate σ in the standard error.

SUMMARY

One-way analysis of variance (ANOVA) compares the means of several populations. The **ANOVA *F* test** tests the overall H_0 that all the populations have the same mean. If the *F* test shows significant differences, examine the data to see where the differences lie and whether they are large enough to be important.

ANOVA assumes that we have an **independent SRS** from each population; that each population has a **normal distribution;** and that all populations have the **same standard deviation.**

In practice, ANOVA is relatively **robust** when the populations are nonnormal, especially when the samples are large. Before doing the *F* test, check the observations in each sample for outliers or strong skewness. Also verify that the largest sample standard deviation is no more than twice as large as the smallest standard deviation.

When the null hypothesis is true, the **ANOVA *F* statistic** for comparing I means from a total of N observations in all samples combined has the *F* distribution with $I - 1$ and $N - I$ degrees of freedom.

ANOVA calculations are reported in an **ANOVA table** that gives sums of squares, mean squares, and degrees of freedom for variation among groups and for variation within groups. In practice, we use software to do the calculations.

SECTION 15.2 EXERCISES

15.21 GO FISH! Table 15.3 gives the results of a study of fish caught in a lake in Finland. We are willing to regard the fish caught by the researchers a random sample of fish in this lake. The weight of commercial fish is of particular interest. Is there evidence that the mean weights of all bream, perch, and roach found in the lake are different? Before doing ANOVA, we must examine the data.

(a) Display the distribution of weights for each species of fish with side-by-side stemplots. Are there outliers or strong skewness in any of the distributions?

(b) Find the five-number summary for each distribution. What do the data appear to show about the weights of these species?

15.22 MORE FISHING Now we proceed to the ANOVA for the data in Table 15.3. The heaviest perch had six roach in its stomach when caught. This fish may be an outlier and its condition is unusual, so remove this observation from the data. Figure 15.10 gives the Minitab ANOVA output for the data with the outlier removed.

TABLE 15.3 Weights (grams) of three fish species

Bream			Perch				Roach	
13.4	13.9	14.1	16.0	14.5	15.7	17.6	14.0	13.6
13.8	15.0	14.9	13.6	15.0	14.8	17.6	13.9	15.4
15.1	13.8	15.5	15.2	15.0	17.9	15.9	13.7	14.0
13.3	13.5	14.3	15.3	15.0	14.6	16.2	14.3	15.4
15.1	13.3	14.3	15.9	17.0	15.0	18.1	16.1	15.6
14.2	13.7	14.9	17.3	16.3	15.0	14.5	14.7	15.3
15.3	14.8	14.7	16.1	15.1	15.8	17.8	14.7	
13.4	14.1		15.1	15.1	14.3	16.8	13.9	
13.8	13.7		14.6	15.0	15.4	17.0	15.2	
13.7	13.3		13.2	14.8	15.1	17.6	14.6	
14.1	15.1		15.8	14.9	17.7	15.6	15.1	
13.3	13.8		14.7	15.0	17.7	15.4	13.3	
12.0	14.8		16.3	15.9	17.5	16.1	15.2	
13.6	15.0		15.5	13.9	20.9	16.3	14.1	

Source: The data are part of a larger data set in the *Journal of Statistics Education* archive, accessible via the Internet. The original source is Pekka Brofeldt, "Bidrag till kaennedom on fiskbestondet I vaara sjoear. Laengelmaevesi," in T. H. Jaervi, *Finlands Fiskeriet*, Band 4, *Meddelanden utgivna av fiskerifo-ereningen I Finland*, Helsinki, 1917. The data were contributed to the archive (with information in English) by Juha Puranen of the University of Helsinki.

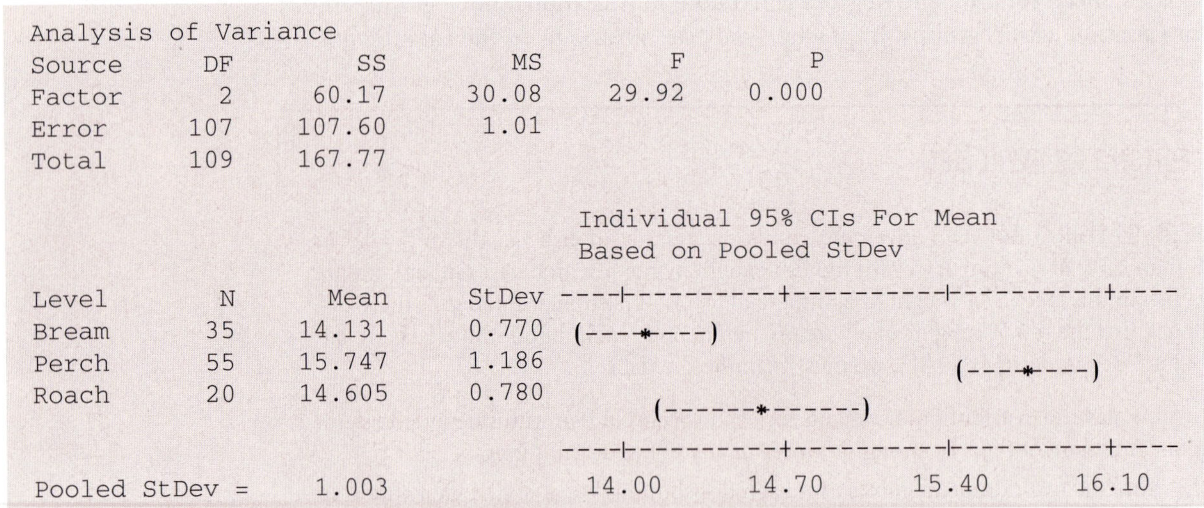

```
Analysis of Variance
Source     DF       SS       MS       F       P
Factor      2    60.17    30.08   29.92   0.000
Error     107   107.60     1.01
Total     109   167.77

                               Individual 95% CIs For Mean
                               Based on Pooled StDev
Level       N     Mean    StDev  ---+---------+---------+---------+---
Bream      35   14.131    0.770  (---*---)
Perch      55   15.747    1.186                      (---*---)
Roach      20   14.605    0.780       (------*------)
                                 ---+---------+---------+---------+---
Pooled StDev =   1.003             14.00     14.70     15.40     16.10
```

FIGURE 15.10 Minitab output for the data in Table 15.3 of weights of three species of fish.

(a) What null hypothesis does the ANOVA F statistic test? State this hypothesis both in words and in symbols.

(b) What is the value of the F statistic? What is its P-value?

(c) Based on your work in this and the previous exercise, what do you conclude about the weights of these species of fish?

15.23 Exercises 15.21 and 15.22 compare the weights of three species of fish.

(a) What are I, the n_i, and N for these data? (Remember that the heaviest perch was removed before doing ANOVA.) Identify these quantities in words, and give their numeric values.

(b) Find the degrees of freedom for the ANOVA F statistic. Check your work against the computer output in Figure 15.10.

(c) For these data, $F = 29.92$. What does Table D tell you about the P-value of this statistic?

15.24 Do the sample standard deviations for the fish weights in Table 15.3 allow use of ANOVA to compare the mean weights? (Use the computer output in Figure 15.10.)

15.25 ANOVA CALCULATIONS FOR FISH DATA

(a) Starting from the sample standard deviations in Figure 15.10, calculate MSE and the pooled standard deviation s_p. Use the computer output to check your work.

(b) Give a 95% confidence interval for the mean weight of perch, using the pooled standard deviation s_p. A graph of this interval appears in the computer output.

(c) Starting from the sample means in the computer output, find the overall mean weight $\bar{x}$. Then find MSG.

(d) Finally, combine MSG with MSE to find F. Use the calculator command Fcdf to find the P-value.

CHAPTER REVIEW

The F test can be used to perform inference for comparing the spread of two normal populations, but there are serious cautions for basic users of statistics. The procedure is extremely sensitive to nonnormal populations, and interpreting significant F-values is therefore difficult. Consequently, inference about population standard deviations at this level is not recommended.

Advanced statistical inference often concerns relationships among several parameters. The second section in this chapter introduces the ANOVA F test for one such relationship: equality of the means of any number of populations. The alternative to this hypothesis is "many-sided," because it allows any relationship other than "all equal." The ANOVA F test is an overall test that tells us whether the data give good reason to reject the hypothesis that all the population means are equal. You should always accompany the test by data analysis to see what kind of inequality is present. Plotting the data in all groups side-by-side is particularly helpful.

ANOVA requires that the data satisfy conditions similar to those for t procedures. The most important assumption concerns the design of the data production. Inference is most trustworthy when we have random samples or responses to the treatments in a randomized comparative experiment. ANOVA assumes that the distribution of responses in each population has a normal distribution. Fortunately, the ANOVA F test shares the robustness of the t procedures, especially when the sample sizes are not very small. Do beware of outliers, which can greatly influence the mean responses and the F statistic.

ANOVA also requires a new condition: the populations must all have the same standard deviation. Fortunately, ANOVA is not highly sensitive to unequal standard deviations, especially when the samples are similar in size. As a rule of thumb, you can use ANOVA when the largest sample standard deviation is no more than twice as large as the smallest.

After studying this chapter, you should be able to do the following.

A. RECOGNITION

1. Recognize when testing the equality of several means is helpful in understanding data.

2. Recognize that the statistical significance of differences among sample means depends on the sizes of the samples and on how much variation there is within the samples.

3. Recognize when you can safely use ANOVA to compare means. Check the data production, the presence of outliers, and the sample standard deviations for the groups you want to compare.

B. INTERPRETING ANOVA

1. Explain what null hypothesis F tests in a specific setting.

2. Locate the F statistic and its P-value on the output of a computer analysis of variance program.

3. Find the degrees of freedom for the F statistic from the number and sizes of the samples. Use Table D of the F distributions to approximate the P-value when software does not give it.

4. If the test is significant, use graphs and descriptive statistics to see what differences among the means are most important.

CHAPTER 15 REVIEW EXERCISES

15.26 COMPETITIVE ROWERS' WEIGHTS Exercise 11.48 (page 665) records the results of comparing weights of skilled and novice female competitive rowers. Is there significant evidence of inequality between the standard deviations of the two populations? Follow the Inference Toolbox

15.27 SOCIAL INSIGHT AMONG MEN AND WOMEN Do the data in Exercise 11.39 (page 657) provide evidence of different standard deviations for Chapin test scores in the population of female and male college liberal arts majors?

(a) State the hypotheses and carry out the test. Software can assess significance exactly, but inspection of the proper table is enough to draw a conclusion.

(b) Do the large sample sizes allow us to ignore the assumption that the population distributions are normal?

15.28 BUILDING AN ANOVA TABLE In each of the following situations, we want to compare the mean response in several populations. For each setting, identify the populations and the response variable. Then give I, the n_i, and N. Finally, give the degrees of freedom of the ANOVA F test.

(a) A study of the effects of smoking classifies subjects as nonsmokers, moderate smokers, or heavy smokers. The investigators interview a sample of 200 people in each group. Among the questions is "How many hours do you sleep on a typical night?"

(b) The strength of concrete depends on the mixture of sand, gravel, and cement used to prepare it. A study compares five different mixtures. Workers prepare six batches of each mixture and measure the strength of the concrete made from each batch.

(c) Which of four methods of teaching American Sign Language is most effective? Assign 10 of the 42 students in a class at random to each of three methods. Teach the remaining 12 students by the fourth method. Record the students' scores on a standard test of sign language after a semester's study.

15.29 CAN YOU HEAR THESE WORDS? To test whether a hearing aid is right for a patient, audiologists play a tape on which words are pronounced at low volume. The patient tries to repeat the words. There are several different lists of words that are supposed to be equally difficult. Are the lists equally difficult when there is background noise? To find out, an experimenter had subjects with normal hearing listen to four lists with a noisy background. The response variable was the percent of the 50 words in a list that the subject repeated correctly. The data set contains 96 responses.[7]

(a) Here are two study designs that could produce these data:

Design A: The experimenter assigns 96 subjects to 4 groups at random. Each group of 24 subjects listens to one of the lists. All individuals listen and respond separately.

Design B: The experimenter has 24 subjects. Each subject listens to all four lists in random order. All individuals listen and respond separately.

Does Design A allow use of one-way ANOVA to compare the lists? Does Design B allow use of one-way ANOVA to compare the lists? Briefly explain your answers.

(b) Figure 15.11 displays Minitab output for one-way ANOVA. The response variable is "Percent," and "List" identifies the four lists of words. Based on this analysis, is there good reason to think that the four lists are not all equally difficult? Write a brief summary of the study findings.

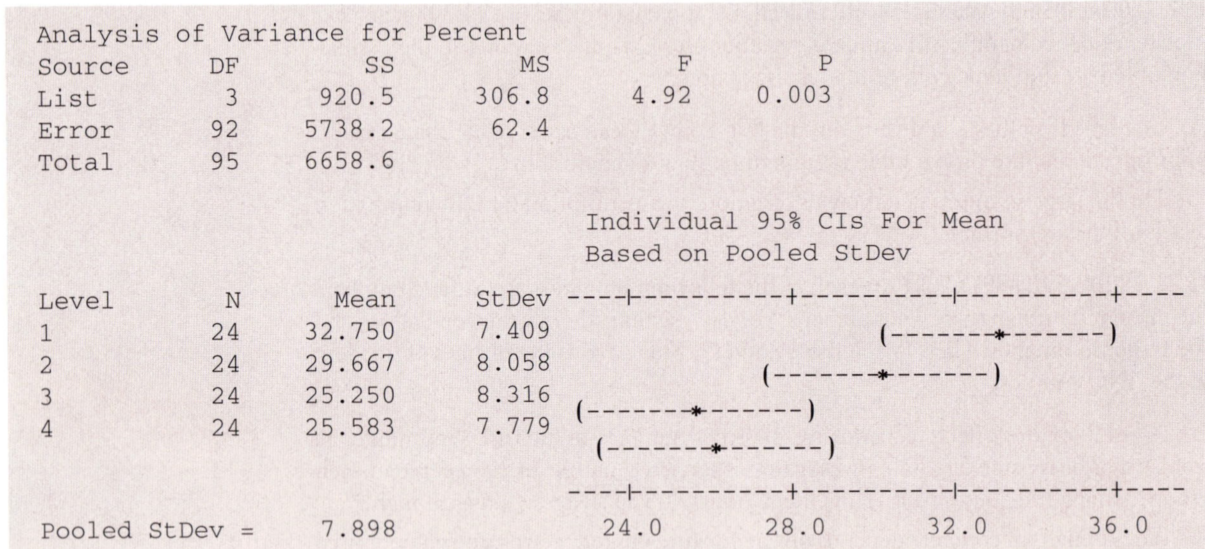

```
Analysis of Variance for Percent
Source      DF       SS        MS       F       P
List         3    920.5     306.8    4.92   0.003
Error       92   5738.2      62.4
Total       95   6658.6
                                   Individual 95% CIs For Mean
                                   Based on Pooled StDev
Level        N     Mean     StDev ---+---------+---------+---------+---
1           24   32.750     7.409                     (-------*------)
2           24   29.667     8.058               (-------*------)
3           24   25.250     8.316  (-------*------)
4           24   25.583     7.779   (-------*------)
                                  ---+---------+---------+---------+---
Pooled StDev =    7.898             24.0      28.0      32.0      36.0
```

FIGURE 15.11 Minitab ANOVA output for comparing the percents heard correctly in four lists of words.

15.30 LOGGING IN THE RAINFOREST "Conservationists have despaired over destruction of tropical rainforest by logging, clearing, and burning." These words begin a report on a statistical study of the effects of logging in Borneo. The study compared forest plots that had never been logged with similar plots nearby that had been logged 1 year earlier and 8 years earlier. Although the study was not an experiment, the authors explain why we can consider the plots to be randomly selected. Table 15.4 contains the data.

TABLE 15.4 The effects of logging in a tropical rainforest

Never logged		Logged 1 year ago		Logged 8 years ago	
Trees per plot	Species per plot	Trees per plot	Species per plot	Trees per plot	Species per plot
27	22	12	11	18	17
22	18	12	11	4	4
29	22	15	14	22	18
21	20	9	7	15	14
19	15	20	18	18	18
33	21	18	15	19	15
16	13	17	15	22	15
20	13	14	12	12	10
24	19	14	13	12	12
27	13	2	2		
28	19	17	15		
19	15	19	8		

Source: Data provided by Charles Cannon, Duke University. The study report is C. H. Cannon, D. R. Peart, and M. Leighton, "Tree species diversity in commercially logged Bornean rainforest," *Science*, 281 (1998), pp. 1366–1367.

(a) Use side-by-side stemplots to compare the distributions of number of trees per plot for the three groups of plots. Are there features that might prevent use of ANOVA?

(b) Figure 15.12 displays Minitab output for one-way ANOVA on the trees per plot data. Do the standard deviations satisfy our rule of thumb for safe use of ANOVA? What do the means suggest about the effect of logging on the number of trees per plot? Report the *F* statistic and its *P*-value and state your overall conclusion.

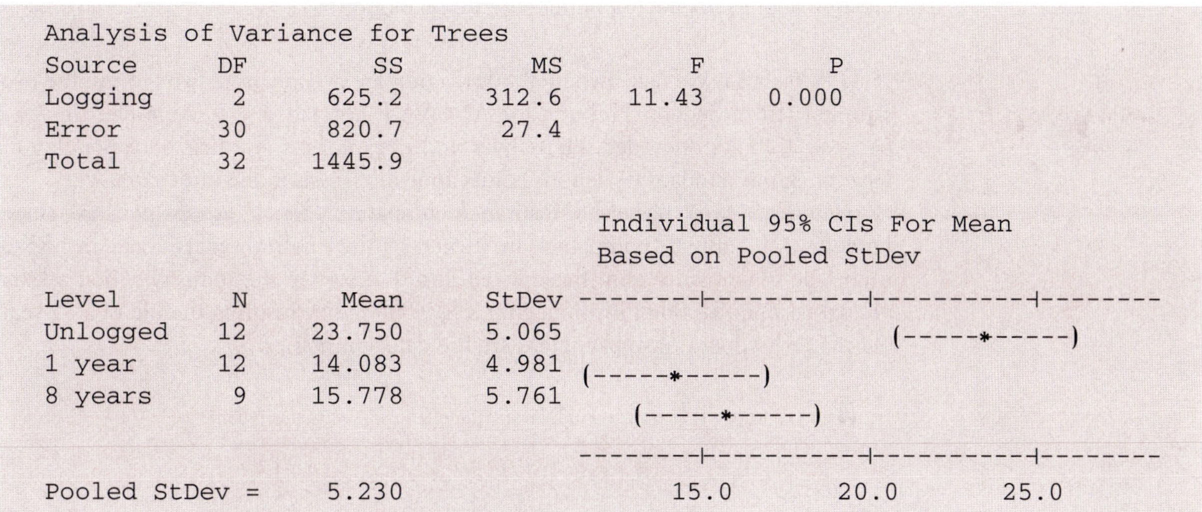

```
Analysis of Variance for Trees
Source     DF       SS       MS        F        P
Logging     2    625.2    312.6    11.43    0.000
Error      30    820.7     27.4
Total      32   1445.9
                                    Individual 95% CIs For Mean
                                    Based on Pooled StDev
Level       N     Mean    StDev  -------+----------+----------+-------
Unlogged   12   23.750    5.065                        (-----*-----)
1 year     12   14.083    4.981  (-----*-----)
8 years     9   15.778    5.761     (-----*-----)
                                 -------+----------+----------+-------
Pooled StDev =   5.230                15.0       20.0       25.0
```

FIGURE 15.12 Minitab ANOVA output for comparing the number of trees per plot in unlogged forest plots, plots logged 1 year ago, and plots logged 8 years ago.

15.31 LOGGING IN THE RAINFOREST, II Table 15.4 gives data on the number of tree species per forest plot as well as on the number of individual trees. In the previous exercise, you examined the effect of logging on the number of trees. Use software or your calculator to analyze the effect of logging on the number of species.

(a) Make a table of the group means and standard deviations. Do the standard deviations satisfy our rule of thumb for safe use of ANOVA? What do the means suggest about the effect of logging on the number of species?

(b) Carry out the ANOVA. Report the *F* statistic and its *P*-value and state your conclusion.

15.32 NEMATODES AND TOMATO PLANTS How do nematodes (microscopic worms) affect plant growth? A botanist prepares 16 identical planting pots and then introduces different numbers of nematodes into the pots. He transplants a tomato seedling into each plot. Here are data on the increase in height of the seedlings (in centimeters) 16 days after planting:[8]

Nematodes	Seedling growth			
0	10.8	9.1	13.5	9.2
1,000	11.1	11.1	8.2	11.3
5,000	5.4	4.6	7.4	5.0
10,000	5.8	5.3	3.2	7.5

(a) Make a table of means and standard deviations for the four treatments. Make side-by-side stemplots to compare the treatments. What do the data appear to show about the effect of nematodes on growth?

(b) State H_0 and H_a for the ANOVA test for these data, and explain in words what ANOVA tests in this setting.

(c) Use software or your calculator to carry out the ANOVA. Report your overall conclusions about the effect of nematodes on plant growth.

15.33 *F* VERSUS *t* We have two methods to compare the means of two groups: the two-sample t test of Section 11.2 and the ANOVA F test with $I = 2$. We prefer the t test because it allows one-sided alternatives and does not assume that both populations have the same standard deviation. Let us apply both tests to the same data.

There are two types of life insurance companies. "Stock" companies have shareholders, and "mutual" companies are owned by their policyholders. Take an SRS of each type of company from those listed in a directory of the industry. Then ask the annual cost per $1000 of insurance for a $50,000 policy insuring the life of a 35-year-old man who does not smoke. Here are the data summaries:[9]

	Stock companies	Mutual companies
n_i	13	17
$\bar{x}_i$	\$2.31	\$2.37
s_i	\$0.38	\$0.58

(a) Calculate the two-sample t statistic for testing H_0: $\mu_1 = \mu_2$ against the two-sided alternative. Use the conservative method to find the P-value.

(b) Calculate MSG, MSE, and the ANOVA F statistic for the same hypotheses. What is the P-value of F?

(c) How close are the two P-values? (The square root of the F statistic is a t statistic with $N - I = n_1 + n_2 - 2$ degrees of freedom. This is the "pooled two-sample t" mentioned on page 666. So F for $I = 2$ is exactly equivalent to a t statistic, but it is a slightly different t from the one we use.)

15.34 DO IT THE HARD WAY Carry out the ANOVA calculations (MSG, MSE, and F) required for part (c) of Exercise 15.32. Find the degrees of freedom for F and report its P-value as closely as Table D allows.

15.35 TEACHING READING A study of reading comprehension in children compared three methods of instruction. As is common in such studies, several pretest variables were measured before any instruction was given. One purpose of the pretest was to see if the three groups of children were similar in their comprehension skills. One of the pretest variables was an "intruded sentences" measure, which measures one type of reading comprehension skill. The data for the 22 subjects in each group are given in Table 15.5. The three methods of instruction are called basal, DRTA, and strategies. We use Basal, DRTA, and Strat as values for the categorical variable indicating which method each student received.

TABLE 15.5 Pretest reading scores

Group	Subject	Score	Group	Subject	Score	Group	Subject	Score
Basal	1	4	DRTA	23	7	Strat	45	11
Basal	2	6	DRTA	24	7	Strat	46	7
Basal	3	9	DRTA	25	12	Strat	47	4
Basal	4	12	DRTA	26	10	Strat	48	7
Basal	5	16	DRTA	27	16	Strat	49	7
Basal	6	15	DRTA	28	15	Strat	50	6
Basal	7	14	DRTA	29	9	Strat	51	11
Basal	8	12	DRTA	30	8	Strat	52	14
Basal	9	12	DRTA	31	13	Strat	53	13
Basal	10	8	DRTA	32	12	Strat	54	9
Basal	11	13	DRTA	33	7	Strat	55	12
Basal	12	9	DRTA	34	6	Strat	56	13
Basal	13	12	DRTA	35	8	Strat	57	4
Basal	14	12	DRTA	36	9	Strat	58	13
Basal	15	12	DRTA	37	9	Strat	59	6
Basal	16	10	DRTA	38	8	Strat	60	12
Basal	17	8	DRTA	39	9	Strat	61	6
Basal	18	12	DRTA	40	13	Strat	62	11
Basal	19	11	DRTA	41	10	Strat	63	14
Basal	20	8	DRTA	42	8	Strat	64	8
Basal	21	7	DRTA	43	8	Strat	65	5
Basal	22	9	DRTA	44	10	Strat	66	8

Source: Data from a study conducted by Jim Baumann and Leah Jones of the Purdue University School of Education.

(a) Use an appropriate statistical technique to compare the pretest scores for the three groups.

(b) The response variable is a measure of reading comprehension called COMP that was determined by a test taken after the instruction was completed. Figures 15.13, 15.14, and 15.15 provide information about the COMP scores in the three treatment groups. What conclusions can you draw? Give appropriate statistical evidence to support your conclusions.

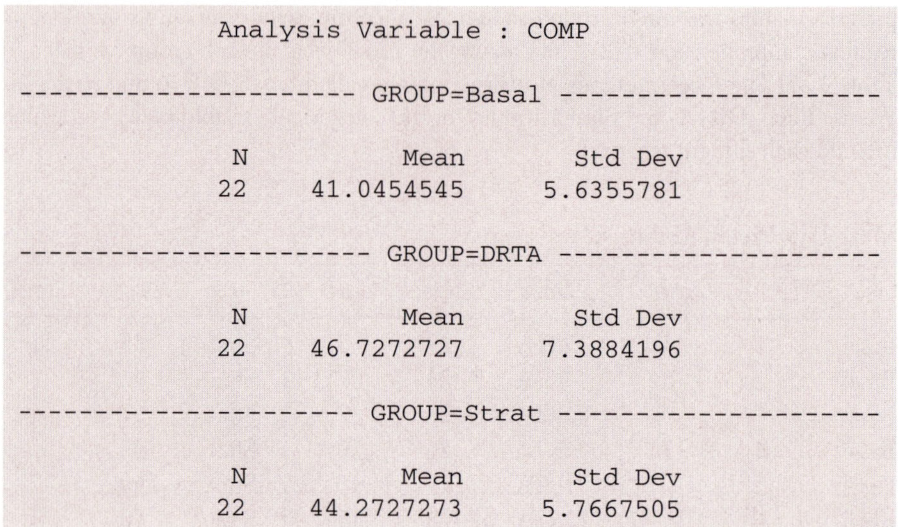

```
Analysis Variable : COMP

---------------------- GROUP=Basal ---------------------

        N            Mean         Std Dev
       22      41.0454545       5.6355781

---------------------- GROUP=DRTA ----------------------

        N            Mean         Std Dev
       22      46.7272727       7.3884196

---------------------- GROUP=Strat ---------------------

        N            Mean         Std Dev
       22      44.2727273       5.7667505
```

FIGURE 15.13 Summary statistics for the comprehension scores in the three groups of the reading comprehension study.

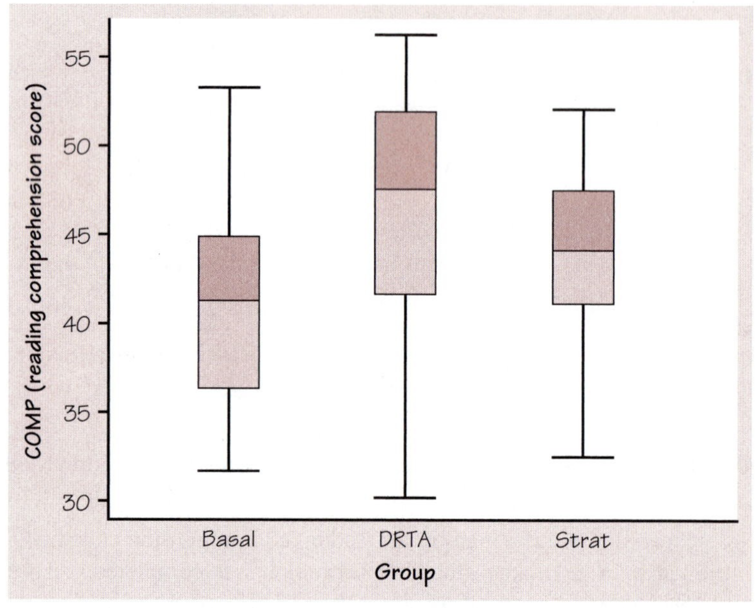

FIGURE 15.14 Side-by-side boxplots of the comprehension scores for the reading comprehension study.

```
                General Linear Models Procedure

Dependent Variable: COMP

                                 Sum of          Mean
Source                  DF      Squares        Square  F Value     Pr > F
Model                    2    357.30303     178.65152     4.48     0.0152
Error                   63   2511.68182      39.86797
Corrected Total         65   2868.98485

              R-Square         C.V.         Root MSE            COMP Mean
              0.124540     14.34531           6.3141               44.015

Contrast                DF  Contrast SS  Mean Square  F Value     Pr > F
B vs D and S             1    291.03030    291.03030     7.30     0.0088
D vs S                   1     66.27273     66.27273     1.65     0.2020
```

FIGURE 15.15 Analysis of variance output for COMP in the reading comprehension study.

NOTES AND DATA SOURCES

1. The problem of comparing spreads is difficult even with advanced methods. Common distribution-free procedures do not offer a satisfactory alternative to the F test, because they are sensitive to unequal shapes when comparing two distributions. A good introduction to the available methods is W. J. Conover, M. E. Johnson, and M. M. Johnson, "A comparative study of tests for homogeneity of variances, with applications to outer continental shelf bidding data," *Technometrics*, 23 (1981), pp. 351–361. Modern resampling procedures often work well. See Dennis D. Boos and Colin Brownie, "Bootstrap methods for testing homogeneity of variances," *Technometrics*, 31 (1989), pp. 69–82.
2. Simplified from W. L. Colville and D. P. McGill, "Effect of rate and method of planting on several plant characters and yield of irrigated corn," *Agronomy Journal*, 54 (1962), pp. 235–238.
3. Modified from M. C. Wilson and R. E. Shade, "Relative attractiveness of various luminescent colors to the cereal leaf beetle and the meadow spittlebug," *Journal of Economic Entomology*, 60 (1967), pp. 578–580.
4. Sanders Korenman and David Neumark, "Does marriage really make men more productive?" *Journal of Human Resources*, 26 (1991), pp. 282–307.
5. Patricia F. Campbell and George P. McCabe, "Predicting the success of freshmen in a computer science major," *Communications of the ACM*, 27 (1984), pp. 1108–1113.

6. R. Shine, T. R. L. Madsen, M. J. Elphick, and P. S. Harlow, "The influence of nest temperatures and maternal brooding on hatchling phenotypes in water pythons," *Ecology*, 78 (1997), pp. 1713–1721.

7. The data and the full story can be found in the Data and Story Library at http://lib.stat.cmu.edu/. The original study is by Faith Loven, "A study of interlist equivalency of the CID W-22 word list presented in quiet and in noise," M.S. thesis, University of Iowa, 1981.

8. Data provided by Matthew Moore.

9. Mark Kroll, Peter Wright, and Pochera Theerathorn, "Whose interests do hired managers pursue? An examination of select mutual and stock life insurers," *Journal of Business Research*, 26 (1993), pp. 133–148.

SOLUTIONS TO ODD NUMBERED EXERCISES

Chapter 1

1.1 (a) Cars. (b) Vehicle type and transmission type are categorical; number of cylinders, city mpg, and highway mpg are quantitative.

1.3 A few possibilities are: length (no. of pages), weight (ounces), thickness (cm).

1.5 (b) No. These percentages cannot be combined to form a whole (sum exceeds 100%).

1.7 Shape: skewed right; Center: 1; Spread: 0 to 39; Outliers: 26, 28, 39.

1.9 (a) Stems: thousands of dollars; Leaves: hundreds of dollars; data are rounded to nearest \$100. (b) Shape: skewed right; Center: about \$4500; Spread: about \$1300 to \$19,300; Outliers: maybe \$18,200 and \$19,300.

1.11 (a) The stemplot is skewed right with peak on the "2" stem. (b) This stemplot shows more distinct clusters and gaps. (c) Shape: skewed right; Center: about \$28; Spread: about \$3 to \$93. Over half the shoppers spent \$28 or less, but a few spent over \$80.

1.13 The histogram has a fairly indistinct shape and is centered at 35. There are no apparent outliers. The stemplot has the advantage of showing actual data values.

1.15 (a) The calculator's zoom feature may choose an unusual x-scale. (b) Shape: roughly symmetric; Center: 40 inches; Spread: 33 to 48 inches; might be useful in designing uniforms.

1.17 (a) Shape: skewed right; Center: 3 days; Spread: 0 to 10 days; Outliers: none. (b) 23.

1.19 (a) Answers will vary depending on choice of class intervals. (b) 90; 12.4%, meaning that 40% of states have 12.4% or less of their populations aged 65 or older.

1.21 (a) The timeplot shows an increasing trend in the cancer death rate from 1945 to 1995. (b) No; our ability to diagnose cancer has improved and elderly people make up a much higher proportion of the population now.

1.23 Categorical: Gender, vote; Quantitative: Age, household income.

1.25 You can make a bar graph or a pie chart. You will need an "other methods" category for the pie chart.

1.27 (a) Shape: skewed right; Center: 4 letters; Spread: 1 to 15 letters. (b) There are more 2-, 3-, and 4-letter words in Shakespeare's plays and more very long words in *Popular Science* articles.

1.29 (a) Roughly symmetric; no. (b) Answers will vary. (c) 8.42 min.; about 9 min. (d) About the 28th percentile.

1.31 (a) 85. (b) 79.33; the mean is not resistant to outliers. (c) A histogram.

1.33 \$30 million; No.

1.35 \$675,00 is the mean and \$330,00 is the median. A few extremely high salaries will pull the mean upward.

1.37 (a) About the same since the graph is symmetric. (b) min = 42, $Q_1 = 51$, $M = 55$, $Q_3 = 58$, max = 69 (c) 7 years (d) The boxplot is roughly symmetric. (e) Yes; Ronald Reagan.

1.39 (a) \$1735 (b) The boxplot is skewed right. The shoppers who spent \$85.76, \$86.37, and \$93.34 are outliers.

1.41 (a) $s = 15.609$ (b) $\bar{x} = 22.2$, $s = 10.244$; No.

1.43 (a) All four numbers the same. (b) 0, 0, 10, 10 (c) For (a).

1.45 (a) \$1,000; \$1,000 (b) No. (c) No.

1.47 **Calories:** Boxplots for meat and poultry hot dogs are very similar. Calorie content of poultry hot dogs is noticeably lower. 5-number summaries: Beef—111, 140, 152.5, 178.5, 190; Meat—107, 138.5, 153, 180.5, 195; Poultry—86, 100.5, 129, 143.5, 170.
Sodium: Boxplots of sodium content are skewed right for all three types. Meat and poultry hot dogs have higher sodium content than beef hot dogs. 5-number summaries: Beef—253, 320.5, 380.5, 478, 645; Meat—144, 379, 405, 501, 545; Poultry—357, 379, 430, 535, 588. The meat hot dog with 144 mg of sodium is an outlier.

1.49 (a) Use % for the vertical scales. (b) Both distributions are skewed right. The center of the male salary histogram is clearly higher than the female graph. (c) Roundoff error.

1.51 There are two distinct clusters of hot dogs: those with 135 to 153 calories and those with 172 to 195 calories.

1.53 Sales of full-length cassettes decrease steadily by about 6 million per year while sales of full-length CDs increase by about the same amount.

1.55 (a) $a = 0$, $b = 746$. (b) $a = 32$, $b = 9/5$. (c) $a = 10$, $b = 1$.

1.57 (a) Normal—272, 337, 358, 400.5, 462; New—318, 383.5, 406.5, 428.5, 477. Chicks fed new corn gain more weight than chicks fed normal corn. (b) Normal: $\bar{x} = 366.3$, $s = 50.805$; New: $\bar{x} = 402.95$, $s = 42.729$; about 36.6 grams. (c) Normal: $\bar{x} = 12.9$, $s = 1.792$; New: $\bar{x} = 14.2$, $s = 1.507$.

1.59 Number of employees, value of company stock, total salaries.

1.61 (a) There were as many deaths due to accidents in this age group as for the other six leading causes combined. (b) Total number of deaths due to other causes in this age group.

1.63 (a) The stemplot is roughly symmetric with no apparent outliers. (b) 50.4 %. (c) $Q_3 = 57.4\%$; The 1956, 1964, 1972, and 1984 elections.

1.65 (a) The distribution is roughly symmetric, excluding the two high outliers. (b) $M =$ 19 mpg, $IQR = 2$ mpg. (c) The midsize car distribution is skewed left. Aside from the two outliers, nearly all SUVs get lower highway gas mileage than midsize cars.

1.67 (a) −34.04%, −2.95%, 3.47%, 8.45%, 58.68%. (b) Roughly symmetric, unimodal with high and low outliers. (c) \$1586.78; \$659.57. (d) $IQR = 11.401$; Yes. It appears that the software uses this criterion.

1.69 Median speed.

1.71 The stemplot shows a bimodal distribution. $M = 37\%$. Since there are two distinct clusters of states—those where few students take the SAT and those where the majority take the SAT—the median gives a poor summary.

1.73 (a) Northeastern: 20.8, 21.2, 20, 17.8, 23.3, 25.2, 25.3, 28, 19.2; Southern: 33.1, 22.5, 26.9, 25.6, 29.1, 35.4, 21.6, 35.7, 30, 31.7, 32.9, 24.8, 34. (b) Side-by-side boxplots show that a higher percentage of adults did not graduate in the southern states than in the northeastern states. Since both distributions are skewed, use medians and IQRs to summarize. Northeastern: $M = 21.2\%$, $IQR = 5.65\%$; Southern: $M = 30$, $IQR = 8.35\%$.

Chapter 2

2.1 Answers will vary.

2.3 (b) 0.2. (c) 0.6. (d) 0.35 (e) By (d), the area between 0 and 0.2 is 0.35. The area between 0.4 and 0.8 is 0.4. Thus the "equal areas" point must lie between 0.2 and 0.4.

2.5 The uniform distribution. Each of the 6 bars should have a height of 20.

2.7 (a) 2.5% (this is 2 standard deviations above the mean). (b) 69 ± 5; that is, 64 to 74 inches. (c) 16%. (d) 84%.

2.9 (a) 50. (b) 2.5. (c) 84. (d) 99.85.

2.11 Approximately 0.2 (for the tall one) and 0.5.

2.13 (a) 266 ± 32, or 234 to 298 days. (b) Less than 234 days. (c) More than 298 days.

2.15 (b) 68%: (58.3, 67.9). 95%: (53.5, 72.7). 99.7%: (48.7, 77.5).

2.17 (a) Outcomes around 25 are most likely. (d) The distribution should be roughly symmetric, with center at about 25, single peaked at the center, standard deviation about 3.5, and few or no outliers. The normal density curve should fit this histogram well.

2.19 Eleanor: $z = \frac{680 - 500}{100} = 1.8$. Gerald: $z = \frac{27 - 18}{6} = 1.5$. Eleanor's score is higher.

2.21 (a) 0.9978. (b) 1 – 0.9978 = 0.0022. (c) 1 – 0.0485 = 0.9515. (d) 0.9978 – 0.0485 = 0.9493

2.23 (a) $z = 3/2.5 = 1.20$; 1 – 0.8849 = 0.1151; about 11.5%. (b) About 88.5%. (c) 72.2 inches.

2.25 (a) $x > 0.40$ corresponds to $z > \frac{0.40 - 0.37}{0.04} = 0.75$; Table A gives 1 – 0.7734 = 0.2266. (b) $0.40 < x < 0.50$ corresponds to $0.75 < z < \frac{0.50 - 0.37}{0.04} = 3.25$; this proportion is 0.9994 – 0.7734 = 0.2260. (c) The inequalities $x > 0.40$ and $0.40 < x < 0.50$ correspond (respectively) to $z > \frac{0.40 - 0.41}{0.02} = -0.5$ and $-0.5 < z < \frac{0.50 - 0.41}{0.02} = 4.5$. For the first one of these, Table A gives proportion 1 – 0.3085 = 0.6915; the second is essentially 0.6915 as well.

2.27 (a) The distribution is moderately symmetric with two gaps, one low outlier (9.4) and one high outlier (22.8). (b) The median is 15.75 and the mean is 15.59. That these two measures are nearly the same suggests that the distribution is fairly symmetric. (c) A stemplot, a dotplot, and a normal probability plot all show that the lengths are approximately normal. About 68% of the observations are within 1σ of the mean; about 97.7% are within 2σ, and 100% are within 3σ. (d) The normal probability plot shows an approximately straight pattern of points and the high and low outliers. (e) Yes, all of the above observations point to approximate normality of the data.

2.29 (a) $z = 0.52$. (b) $z = -1.04$. (c) $z = 0.84$. (d) $= -1.28$.

2.31 (a) About 5.2%. (b) About 55%. (c) Approximately 279 days or longer.

2.33 (a) At about ±0.675. (b) For any normal distribution, the quartiles are ±0.675 standard deviations from the mean; for human pregnancies, the quartiles are 266 ± 10.8, or 255.2 and 276.8.

2.35 (a) The mean is 1.442 and the median is 1.45. If the distribution were approximately normal, then these two numbers would be almost equal, and they are. (b) Approximately 66% of the data lie within 1 standard deviation of the mean; 100% of the data lie within 2σ; and 100% of the data lie within 3σ. (c) The data appear to be approximately normal. The histogram is bell-shaped with a single peak in the middle. The boxplot is symmetric. The normal probability plot is approximately linear, suggesting that the distribution is approximately normal.

2.37 The mean and the standard deviation of the standarized scores are $8.37(10^{-15})$, which is another name for 0, and 1, respectively.

2.39 (b) $M = 0.5$, $Q_1 \cong 0.3$, $Q_3 \cong 0.7$. (c) 25.2%. (d) 49.6%.

2.41 (a) 0.8997 (b) 0.6628 (c) 0.5625 (d) 0.6628.

2.43 The z-value corresponding to an area of 0.25 is $z = -0.6745$. Then $Q_1 = 150$. Similarly, $z = 170$ for the median, and 190 for Q_3.

2.45 13.

2.47 Soldiers whose head circumference is outside the range 22.8 ± 1.81, approximately, less than 21 inches or greater than 24.6 inches.

2.49 (a) The distribution is slightly skewed left with no obvious outliers (a boxplot shows that 10.17 is an outlier). (b) The remaining data are approximately normally distributed; the mean is slightly less than the median.

2.51 (a) Use the same window settings as in the Technology Toolbox. The command shadeNorm(110,1E99,100,15) produces an area of 0.2525. (b) shadeNorm(–1E99,85,100,15) = 0.1587. (c) shadeNorm(70,130,100,15) = 0.9545. Also 1–normalcdf(–1E99,70,100,15) = 0.9545.

2.53 No. Within 4 standard deviations is 0.999937 area. Going out to 5 standard deviations gives area = 0.999999, which rounds to 1 for 4 decimal place accuracy.

Chapter 3

3.1 (a) Explanatory: time spent studying; response: grade. (b) Explore the relationship. (c) Explanatory: rainfall; response: crop yield. (d) Explore the relationship. (e) Explanatory: father's class; response: son's class.

3.3 Sex is explanatory, and political preference in the last election is the response. Both are categorical.

3.5 The variables are: SAT math score; SAT verbal score. There is no explanatory/response relationship. Both variables are quantitative.

3.7 (a) Number of jet skis in use. (b) There is a strong straight line relationship between number of jet skis in use and accidents.

3.9 (a) Positive association. (b) Linear. (c) A fairly strong relationship, allowing for reasonably accurate prediction. With 716,000 boat registrations, expect about 50 manatee deaths/year.

3.11 (a) Speed is explanatory. (b) The relationship is curved—low in the middle, high at the extremes. Since low "mileage" is actually *good* (it means that we use less fuel to travel 100 km), this makes sense: Moderate speeds yield the best performance. Note that 60 km/hr is about 37 mph. (c) Above-average (that is, bad) values of "fuel efficiency" are found with both low and high values of "speed." (d) The relationship is very strong; there is little scatter around the curve, and it is very useful for prediction.

3.13 (b) The relationship is clearly curved. (d) The relationship is quite strong.

3.15 (a) A positive association between IQ and GPA would mean that students with higher IQs tend to have higher GPAs, and those with lower IQs generally have lower GPAs. The plot does show a positive association. (b) The relationship is positive, roughly linear, and moderately strong (except for three outliers). (c) The lowest point on the plot is for a student with an IQ of about 103 and a GPA of about 0.5.

3.17 (a) New York's median household income is about $32,800, and the mean per capita income is about $27,500. (b) The association should be positive since the more money households have, the more money we expect individuals to have. Since the money in a household must be divided among those in the household, we expect household income to be higher than personal income. (c) Income distributions tend to be right-skewed, which would raise the mean per capita income above the median. In the District of Columbia, this skewness (perhaps combined with small household sizes) overcomes the effect we described in (b). (d) Alaska's median household income is about $47,900. (e) Ignoring the outliers, the relationship is strong, positive, and moderately linear.

3.19 Either variable could go on the vertical axis. The plot shows a strong positive linear relationship, with no outliers; there appears to be only one species represented.

3.21 (a) Planting rate is explanatory. (b) See (d). (c) The pattern is curved—high in the middle, and lower on the ends. Not linear, and there is neither positive nor negative association. (d) 20,000 plants per acre seems to give the highest average yield.

3.23 (a) The means are (in the order given) 47.167, 15.667, 31.5, and 14.833. (b) Yellow seems to be the most attractive, and green is second. White and blue are poor attractors. (c) Positive or negative association make no sense here because color is a categorical variable (what is an "above-average" color?).

3.25 (a) The correlation in Figure 2.5 is positive but not near 1; the plot clearly shows a positive association, but with quite a bit of scatter. (b) The correlation in Figure 2.6 is closer to 1 since the spread is considerably less in this scatterplot. (c) The outliers in Figure 2.5 weaken the relationship, so dropping them would increase r. The one outlier in Figure 2.6 strengthens that relationship (since the relative scatter about the diagonal line is less when it is present), so the correlation would drop with it removed.

3.27 (a) The plot shows a strong positive linear relationship, with little scatter, so we expect that r is close to 1. (b) r would not change—it is computed from unitless standardized values.

3.29 (a) $r \cong -0.746$, consistent with a moderate negative association. (b) r does not change.

3.31 (a) $r = 0.82450$. This agrees with the positive association observed in the plot; it is not too close to 1 because of the outliers. (b) It has no effect on the correlation. If every guess had been 100 calories higher—or 1000, or 1 million—the correlation would have been exactly the same, since the standardized values would be unchanged. (c) The revised correlation is $r = 0.98374$. The correlation got closer to 1 because without the outliers, the relationship is much stronger.

3.33 $r = 0.25310$ for both sets of data. The points have been transformed, but the distances between corresponding points and the strength of association have not changed.

3.35 (a) Small-cap stocks. (b) A negative correlation.

3.37 (a) Gender has a nominal scale. (b) $r = 1.09$ is impossible. (c) Correlation has no units.

3.39 Answers will vary.

3.41 (b) The least squares equation is KILLED = −41.43 + 0.125 (REGISTRATIONS). (c) The predicted number of manatees killed is about 48. (d) The measures appear to be succeeding because in the last three years, three of the four points are below the regression line. The mean manatee deaths for the three years is 42. Our prediction of about 48 was too high.

3.43 (b) We predict $y \cong 147.4$ bpm, about 4.6 bpm lower than the actual value. (c) With y = time and x = pulse, $y = 43.10 - 0.0574x$, so the predicted time is

34.38 minutes---only 0.08 minutes (4.8 seconds) too high. **(d)** The results depend on which variable is viewed as explanatory.

3.45 **(a)** Stumps is explanatory; the plot shows a positive linear association. **(b)** The regression line is $\hat{y} = -1.286 + 11.89x$. **(c)** $r^2 \cong 83.9\%$.

3.47 **(b)** Let y be "guessed calories" and x be actual calories. Using all points: $\hat{y} = 58.59 + 1.3036x$ (and $r^2 = 0.68$). Excluding spaghetti and snack cake: $\hat{y}^* = 43.88 + 1.14721x$ (and $r^2 = 0.968$). **(c)** The two removed points could be called influential, in that when they are included, the regression line passes above every *other* point; after removing them, the new regression line passes through the "middle" of the remaining points.

3.49 **(b)** The equation is METF = 201.1616 + 24.026 (MASSF). The correlation is $r = 0.876$. Lean body mass explains 76.82% of the variation in female metabolic rate. **(c)** The residual plot shows no clear pattern, so the least squares line is an adequate model for the data. Lean body mass is on the vertical axis, and metabolic rate METF is on the horizontal axis. **(d)** The residual plot of residuals versus predicted value looks exactly like the previous residual plot of residual versus lean body mass (x).

3.51 **(a)** Weight $y = 100 + 40x$ g; the slope is 40 g/week. **(c)** When $x = 104$, $y = 4260$ grams, or about 9.4 pounds—a rather frightening prospect. The regression line is reliable only for young rats; like humans, rats do not grow at a constant rate throughout their entire lives.

3.53 **(a)** Since both variables are measured in dollars, the same scale is used on both axes. The plot shows (perhaps) a weak positive association, with one outlier. **(b)** The correlation is $r \cong 0.1426$; about $r^2 \cong 2\%$ of soda price variation is explained by a linear relationship with hot dog price. **(c)** The regression line is $\hat{y} = 1.9057 + 0.0619x$. The slope is near zero because the relationship is weak; regardless of the hot dog price x, our predicted soda price is near the mean soda price (about \$2.05). **(d)** The outlier is the point for the Cardinals, with an extremely expensive hot dog and a soda priced just below average. Without that point, the regression equation is $\hat{y} = 1.7613 + 0.1303x$ (the dashed line). Some might call this point influential since the line is slightly different without it (and r^2 increases to 5%); however, there is not much difference between the two lines over the range of the remaining hot dog prices, so whether we should call this point influential is a matter of opinion.

3.55 **(b)** $\hat{y} = 71.950 + 0.38333x$ cm. **(c)** When $x = 40$, $\hat{y} \cong 87.2832$ cm; when $x = 60$, $\hat{y} \cong 94.9498$ cm. **(d)** A change of 6 cm in 12 months is 0.5 cm/month ($0.5 = \frac{6}{60-48}$). Sarah is growing at about 0.38 cm/month—more slowly than normal.

3.57 **(a)** $b = 0.16$, $a = 30.2$. **(b)** $\hat{y} = 78.2$. **(c)** Only $r^2 = 36\%$ of the variability in y is accounted for by the regression.

3.59 **(a)** U.S.: –26.4%, 5.1%, 18.2%, 30.5%, and 37.6%. Overseas: –23.4%, 2.1%, 11.2%, 29.6%, and 69.4%. **(b)** The three middle numbers of the U.S. five-number summary are higher, but the minimum and maximum overseas returns are higher. **(c)** Overseas stocks are more volatile: The boxplot is more widely spread, and the low U.S. return appears to be an outlier, but the low overseas return is not.

3.61 **(b)** No: The pattern is not linear. **(c)** The sum is 0.01. The first two and last four residuals are negative, and those in the middle are positive.

3.63 **(a)** r is negative because the association is negative. The straight-line relationship explains about $r^2 \cong 71.1\%$ of the variation in death rates. **(b)** $\hat{y} \cong 168.8$ deaths per 100,000 people. **(c)** $b = rs_y/s_x$, and s_y and s_x are both positive.

3.65 (a) To three decimal places, the correlations are all approximately 0.816 (for Set D, r actually rounds to 0.817), and the regression lines are all approximately $\hat{y} = 3.000 + 0.500x$. For all four sets, we predict $\hat{y} \cong 8$ when $x = 10$. (c) For Set A, the use of the regression line seems to be reasonable; the data do seem to have a moderate linear association (albeit with a fair amount of scatter).

3.67 It is useful to compare the means for each nematode level: 10.65, 10.43, 5.60, and 5.45 cm. There is little difference in the growth when comparing 0 and 1000 nematodes, or 5000 and 10,000 nematodes—but the growth drops substantially between 1000 and 5000 nematodes.

3.69 The slope is 0.54; $\hat{y} = 33.67 + 0.54x$. We predict the husband's height will be about 69.85 inches.

3.71 (a) Put speed on the horizontal axis. (b) There is a very strong positive linear relationship; $r = 0.9990$. (c) $\hat{y} = 1.76608 + 0.080284x$. (d) $r^2 = 0.998$, so nearly all (99.8%) the variation in steps per second is explained by the linear relationship. (e) The regression line would be different; the line in (c) minimizes the sum of the squared *vertical* distances on the graph, while this new regression would minimize the squared *horizontal* distances. r^2 would remain the same, however.

3.73 (b) There is a moderately strong positive linear association. There are no really extreme observations, though Bank 9 did rather well. Franklin does not look out of place. (c) $\hat{y} = 7.573 + 4.9872x$. (d) Franklin's predicted income was $\hat{y}$ = 26.5 million dollars—almost twice the actual income. The residual is –12.7.

3.75 (a) The men's times and the women's times have been steadily decreasing for the last 100 years. A straight line appears to be a reasonable model for the men's times; the correlation for the men is $r = 0.983$. A straight line also appears to be a reasonable model for the women's times; the correlation for the women is $r = -0.9687$. In order to plot both scatterplots on the same axes, we deleted the first two men's times (1905 and 1915). The resulting plot suggests that while the women's scores were decreasing faster than the men's times from 1925 through 1965, the times for both sexes from 1975 have tended to flatten out. In fact, for the period 1925 to 1995, the best model may not be a straight line, but rather a curve, such as an exponential function. In any event, it is not clear, as Whitt and Ward suggest, that women will soon outrun men.

The two regression lines intersect at the point (2002.09, 96.956).

3.77 A scatterplot of fatalities versus number of jet skis in use shows a moderately strong, positive, linear association. The correlation, $r = 0.9382$ indicates that the strength of the association is strong. $R^2 = 0.868$, and the interpretation is that the number of jet skis in use explains about 87% of the variation in number of jet ski fatalities. The equation of the LS line is FATALITIES = 8.003 + 0.000066 (IN USE). The residual plot shows no clear pattern, so we conclude that the LS line is an adequate model for these data.

Chapter 4

4.1 (a) $y = 2.54x$ is monotonic increasing. (b) $y = (5/6)x$ is monotonic increasing. (c) circumference = π(diameter) is monotonic increasing. (d) SquaredError = $(\text{time} - 5)^2$ is not monotonic.

4.3 Strength = $c(\text{weight})^{2/3}$, where c is a constant.

4.5 Let y = average heart rate. Kleiber's Law says that energy consumed is proportional to the three-fourths power of body weight, or $c_1x^{3/4}$. But energy consumed is also proportional to the product of volume of blood pumped by the heart and the heart rate. Equating these quantities, and using the fact that volume of blood pumped by the heart is proportional to body weight, gives $c_1x^{3/4} = c_2x(y)$. Solving for y, we obtain $y = c_1x^{3/4} / c_2x = cx^{-1/4}$.

4.7 (a) A scatterplot of the data shows the characteristic exponential growth shape, and a plot of the ratios y_{n-1}/y_n shows some variability initially, but settles down about the line $y = \bar{x} = 2.7$. (b) Plotting log y versus x shows a straight line pattern with $r^2 = 0.995$. Regressing (log y on x) and then inspecting the residual plot for the transformed data shows no clear pattern, so a straight line is an appropriate model for the reexpressed data. We conclude that the number of transistors on a chip is growing exponentially.

4.9 16 and 1,048,576.

4.11 (b) The transformed data appear to be linear from 1790 to about 1880, and then linear again, but with a smaller slope. This reflects a lower exponential growth rate. (c) We will use data from 1920 to 2000 since the pattern over that range appears very linear. Regressing log POP on YEAR yields the equation log $\widehat{\text{POP}} = -8.2761 + 0.005366$ YEAR and correlation for the transformed data of 0.998. The residual plot shows random scatter, so the original data is exponential. The predicted population in the year 2010 is 323,531,803. (d) $r^2 = 0.995$ for the transformed data. (e) The exponential model for these data is quite good.

4.13 A power model with power 0.2182 appears to fit the data well. The model, Lifespan = (10^0.7617)(WT^0.2182), predicts a lifespan of 17.06 years for humans. Oh well.

4.15 (a) As height increases, weight increases. Since weight is a 3-dimensional attribute and height is 1-dimensional, weight should be proportional to the cube of the height. One might speculate a model of the form WEIGHT = $a \times$ HEIGHT^b, where a and b are constants. This is a power function. (b) Height is the explanatory variable, and weight is the response variable. (c) Calculate the logarithms of the heights and the logarithms of the weights. Plot log(WT) vs. log(HT). The plot appears to be very linear, so least squares regression is performed on the transformed data. The correlation is 0.99996. The regression line fits the transformed data extremely well. (d) A residual plot for the transformed data shows no discernable pattern. The line clearly fits the transformed data well. (e) An inverse transformation yields a power equation for the original overweight data: y = (10^–1.3927)(x^2.0037). The predicted severely overweight value for a 5'10" adult (70 inches tall) would be y = (10^–1.3927)(70^2.0037) = 201.5 pounds. The predicted severely overweight value for a 7 foot adult (84 inches tall) would be y = (10^–1.3927)(84^2.0037) = 290.4 pounds.

4.17 The values of the bond at the end of 1 through 10 years are: \$537.50, \$577.81, \$621.15, \$667.73, \$717.81, \$771.65, \$829.52, \$891.74, \$958.62, and \$1,030.52.

4.19 (a) The association is positive; as the length of the fish increases, its weight increases. The association is strong, and the pattern is clearly curved. (b) Considering that length is one-dimensional, and weight is 3-dimensional, we expect the weight of the fish to change proportionate to the cube of the length. A plot of log(WEIGHT) on log(LENGTH) shows a very linear pattern. The correlation of log(WEIGHT) and log(LENGTH) is 0.994 ($r^2 = 0.9886$).

4.21 (b) The first phase 0 to 6 hours when the mean colony size actually decreases. In the second phase, from 6 to 24 hours, the mean colony size increases exponentially. At 36 hours, mean growth is in the third phase, where growth is still increasing but at less than the previous exponential rate of growth. (c) The least squares line for the transformed data is log(mean size) = –0.55942 + 0.0851(hour). When hour = 10, this equation predicts log(mean size) = 0.25647, so mean size = 10^0.25647 = 1.805.

4.23 The correlation between log(mean size) and hour is $r = 0.99147$. The correlation between the log(individual sizes) and hour is 0.98456. We would expect the second correlation to be smaller since there is more variability with the individual observations.

4.25 The association is negative, nonlinear, and strong. Species with seeds that weigh the least produce the largest count of seeds. Species with the heaviest seeds produce the smallest number of seeds. (b) A plot of log(COUNT) on log(WEIGHT) shows a moderately strong, linear pattern with correlation $r = -0.93$ and $r^2 = 0.86$. The transformed data show a negative association.

4.27 (a) $\hat{y} = 1166.93 - 0.58679x$. (b) The farm population decreased about 590 thousand (0.59 million) people per year. The regression line explains 97.7% of the variation. (c) –782,100—but a population must be greater than or equal to 0.

4.29 The correlation would be smaller because there is much more variation among the individual data points.

4.31 The correlation would be lower; individual stock performances will be more variable, weakening the relationship.

4.33 Seriousness of the fire is a lurking variable: more serious fires require more attention. It would be more accurate to say that a large fire "causes" more firefighters to be sent, rather than vice versa.

4.35 (a) The straight-line relationship explains about $r^2 \cong 94.1\%$ of the variation in either variable. (b) With individual data, there is more variation, so r would be much smaller.

4.37 Explanatory: Whether or not a student has studied a foreign language; response: Score on the test. Lurking variable: English skills before taking (or not taking) the foreign language.

4.39 A = exposure to industrial chemicals, B = number of miscarriages, and C = number of hours spent standing up at work. Variable C is a confounding variable.

4.41 It could be that children with lower intelligence watch many hours of television. Or it could be that children from lower socio-economic households where the parent(s) are less likely to limit television viewing and are unable to help their children with their schoolwork because the parents themselves lack education. A = number of hours spent watching television, B = grade point average, and C = socio-economic level. Variables A and B change in common response to C.

4.43 A = takes SAT preparation course, B = SAT score, and C = knowledge gained as a result of taking the SAT exam previously. The increase in SAT is hopelessly confounded with variable C.

4.45 A = sales of television sets, and B = number of obese adolescents. Variable C could be the economy: as the economy has grown over the last 20 years, more families can afford television sets (even multiple TV sets), and TV viewing has increased as a result, and children have had less physical work to help make ends meet.

4.47 Higher income can cause better health: higher income means more money to pay for medical care, drugs and better nutrition, which in turn results in better health. Better health can cause higher income: if workers enjoy better health, they are better able to get and hold a job, which can increase their income.

4.49 (a) Yes. The relationship is not very strong. (b) In hospitals that treat very few heart attack cases, the mortality is quite high. One can almost see an exponentially decreasing pattern of points in the plot. The nonlinearity weakens the conclusion that heart attack patients should avoid hospitals that treat few heart attacks. The mortality rate is extremely variable for those hospitals that treat few heart attacks.

4.51 21.6%, 46.5%, and 32.0% (total is 100.1% due to rounding).

4.53 (a) Add all six numbers in the table: 5375 students. (b) $\frac{400+416+188}{5375}=\frac{1004}{5375}=18.7\%$. (c) Both parents smoke: 1780 (33.1%); One parent smokes: 2239 (41.7%); Neither parent smokes: 1356 (25.2%).

4.55 (a) 6014; 1.26%. (b) Blood pressure is explanatory. (c) Yes: among those with low blood pressure 0.785% died; the death rate in the high blood pressure group was 1.65%—about twice as high as the other group.

4.57 Among 35- to 54-year-olds: 11.2% never finished high school, 32.5% finished high school, 27.8% had some college, and 28.5% completed college. This is more like the 25–34 age group than the 55-and-over group.

4.59 (a) Use column percentages (e.g., $\frac{68}{225}\cong 30.2\%$ of females are in administration, etc.). See table below. The biggest difference between women and men is in Administration: a higher percentage of women chose this major. Meanwhile, a greater proportion of men chose other fields, especially Finance. (b) There were 386 responses; $\frac{336}{722}\cong 46.5\%$did not respond.

	Female	Male
Accting.	30.2%	34.8%
Admin.	40.4%	24.8%
Econ.	2.2%	3.7%
Fin.	27.1%	36.6%

4.61 (a) White defendant: 19 yes, 141 no. Black defendant: 17 yes, 149 no. (b) Overall death penalty: 11.9% of white defendants, 10.2% of black defendants. For white victims, 12.6% and 17.5%; for black victims, 0% and 5.8%. (c) The death penalty is more likely when the victim was white (14%) rather than black (5.4%). Because most convicted killers are of the same race as their victims, whites are more often sentenced to death.

4.63 (a) 41.1%. (b) 32.5%.

4.65 Older students: 7.1%, 43.2%, 13.5%, and 36.2%. For all students: 13.9%, 24.6%, 43.3%, and 18.1%.

4.67 89.3% of homicides were committed with handguns, 8.2% with long guns, and 2.5% unknown. Among suicides, 70.9% used handguns, 26.3% long guns, and 2.9% unknown. Long guns are more often used in suicides.

4.69 (a) 7.2%. (b) 10.1% of the restrained children were injured, compared to 15.4% of unrestrained children.

4.71 Examples will vary, of course; here is one very simplistic possibility (the two-way table is at the right; the three-way table is below). The key is to be sure that the three-way table has a lower percentage of overweight people among the smokers than among the nonsmokers.

	Early Death	
	Yes	No
Overweight	4	6
Not overweight	5	5

Smoker	Early Death	
	Yes	No
Overweight	1	0
Not overweight	4	2

Nonsmoker	Early Death	
	Yes	No
Overweight	3	6
Not overweight	1	3

4.73 (a) The scatterplot of period vs. length of pendulum suggests a power function model. The plot of log (period) versus log (length) appears linear. There is a positive association; as length increases, the period increases. (b) Least-squares regression is performed on the transformed data. The correlation between the transformed variables is extremely strong. The LSRL fits the transformed data like a glove. The power function model takes the form $\hat{y} = (10^{.0415})(x^{.5043})$. (c) The equation for the model says that the period of a pendulum is proportional to the square root of its length.

4.75 (a) A power model would be slightly better for men's times, based only on an $r = -0.9843$ value (versus $r = -0.9837$ for an exponential model). (b) The power model would also be slightly better for the women's times based on an $r = -0.9747$ value (versus $r = -0.9737$ for an exponential model). (c) Neither curve will eventually reach zero in the next million years, but the power curves for our power function models will intersect in 2033, when the common record will be predicted to be 96.13 seconds. Extrapolation is dangerous, of course, and whether women will overtake men is open to considerable speculation.

4.77 These data call for an exponential decay model. Plotting log(count) versus time produced a very linear pattern. Regressing the reexpressed data gives the LS equation log(count) = 2.5941 – .0949(minutes) and correlation –0.9942. A residual plot from the regression shows no clear pattern, so an exponential equation will fit the original data well.

4.79 The lurking variable is temperature or season. More flu cases occur in winter, and less ice cream is sold in winter. This is an example of common response.

4.81 (a) 19,039; roundoff error. (b) 25.1%, 54.7%, 10.0%, and 10.2%. (c) 18–24: 84.7%, 14.1%, 0.1%, 1.1%; 40–64: 8.4%, 70.1%, 5.5%, 16.0%. Among the younger women, more than 4 out of 5 have not yet married, and those who are married have had little time to become widowed or divorced. Most of the older group is or has been married—only about 8.4% are still single. (d) 58.1% of single women are 18–24, 26.7% are 25–39, 12.8% are 40–64, and 2.4% are 65 or older.

4.83 76.1% of the smokers stayed alive for 20 years; 68.6% of the nonsmokers did so. (b) For the youngest group, 96.2% of the nonsmokers and 93.4% of the smokers were alive 20 years later. For the middle group, 73.9% of the nonsmokers and 68.2% of the smokers were alive 20 years later. For the oldest group, 14.5% of the nonsmokers and 14.3% of the smokers were alive 20 years later. So the results are reversed when the data for the three age groups are combined. (c) 45.9% of the youngest group, 55.2% of the middle-aged group, but only 20.2% of the oldest group were smokers.

Chapter 5

5.1 The (desired) population is employed adult women; the sample is the 48 club members who returned the survey.

5.3 This an experiment: A treatment is imposed. The explanatory variable is the teaching method (computer assisted or standard), and the response variable is the increase in reading ability based on the pre- and posttests.

5.5 Observational. The researcher did not attempt to change the amount that people drank. The explanatory variable is alcohol consumption. The response variable is survival after 4 years.

5.7 Only persons with a strong opinion on the subject—strong enough that they are willing to spend the time, and at least 50 cents—will respond to this advertisement.

5.9 Label from 001 to 440; select 400, 077, 172, 417, 350, 131, 211, 273, 208, and 074.

5.11 Assign 01 to 30 to the students (in alphabetical order). The exact selection will depend on the starting line chosen in Table B; starting on line 123 gives 08-Ghosh, 15-Jones, 07-Fisher, and 27-Shaw. Assigning 0–9 to the faculty members gives (from line 109) 3-Gupta and 6-Moore. (We could also number faculty from 01 to 10, but this requires looking up 2-digit numbers.)

5.13 (a) Households without telephones, or with unlisted numbers. Such households would likely be made up of poor individuals (who cannot afford a phone), those who choose not to have phones, and those who do not wish to have their phone number published. (b) Those with unlisted numbers would be included in the sampling frame when a random-digit dialer is used.

5.15 The first wording would pull respondents towards a tax cut because the second wording mentions several popular alternative uses for tax money.

5.17 (a) 13147 + 15182 + 1448 = 29777. (b) There's nothing to prevent a person from answering several times. Also, the respondents were only those who went to that Web site and took the time to respond. We cannot define "non response" in this situation. (c) The results are slanted towards the opinions of men, who might be less likely to feel that female athletes should earn as much as men.

5.19 (a) The adults in the country. (b) All the wood sent by the supplier. (c) All households in the U.S.

5.21 Number the bottles across the rows from 01 to 25, then select 12–B0986, 04–A1101, and 11–A2220. (If numbering is done down columns instead, the sample will be A1117, B1102, and A1098.)

5.23 The blocks are already marked; select three-digit numbers and ignore those that do not appear on the map. This gives 214, 313, 409, 306, and 511.

5.25 It is *not* an SRS, because some samples of size 250 have no chance of being selected (e.g., a sample containing 250 women).

5.27 A smaller sample gives less information about the population. "Men" constituted only about one-third of our sample, so we know less about that group than we know about all adults.

5.29 Answers will vary, of course. One possible approach: Obtain a list of schools, stratified by size or location (rural, suburban, urban). Choose SRSs (not necessarily all the same size) of schools from each strata. Then choose SRSs (again, not necessarily the same size) of students from the selected schools.

5.31 Units are the individual trees. Factor is the amount of light. Treatments are full light and reduced light. Response variable is the weight of the trees.

5.33 The units are the individuals who were called. One factor is what information is offered. Treatments are (1) giving name, (2) identifying university, (3) both of these. Second factor is offering to send a copy of the results. The treatments are either offering or not offering. The response is whether the interview was completed.

5.35 (a) This is an experiment, since the teacher imposes treatments (instruction method). (b) The explanatory variable is the method used (computer software or standard curriculum), and the response is the change in reading ability.

5.37 (a) In a serious case, when the patient has little chance of surviving, a doctor might choose not to recommend surgery; it might be seen as an unnecessary measure, bringing expense and a hospital stay with little benefit to the patient. (b)

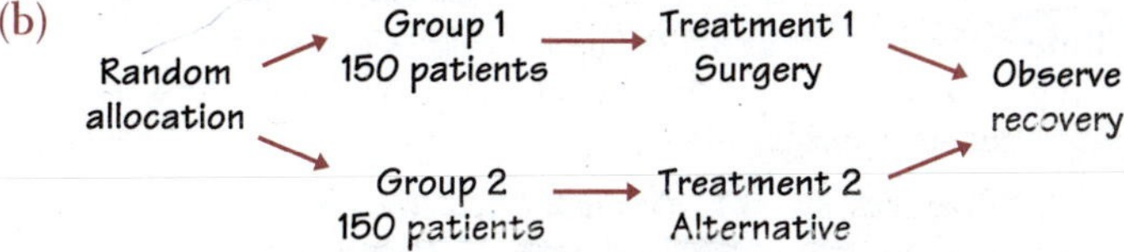

5.39 (a) Randomly select 20 women for Group 1, which will see the "childcare" version of Company B's brochure, and assign the other 20 women to Group 2 (the "no childcare" group). Allow all women to examine the appropriate brochures, and observe which company they choose. Compare the number from Group 1 who choose Company B with the corresponding number from Group 2. (b) Numbering from 01 to 40, Group 1 is 05-Cansico, 32-Roberts, 19-Hwang, 04-Brown, 25-Lippman, 29-Ng, 20-Iselin, 16-Gupta, 37-Turing, 39-Williams, 31-Rivera, 18-Howard, 07-Cortez, 13-Garcia, 33-Rosen, 02-Adamson, 36-Travers, 23-Kim, 27-McNeill, and 35-Thompson.

5.41 The second design is an experiment—a treatment is imposed on the subjects. The first is a study; it may be confounded by the types of men in each group. In spite of the researcher's attempt to match "similar" men from each group, those in the first group (who exercise) could be somehow different from men in the non-exercising group.

5.43 Because the experimenter knew which subjects had learned the meditation techniques, he (or she) may have had some expectations about the outcome of the experiment: if the experimenter believed that meditation was beneficial, he may subconsciously rate that group as being less anxious.

5.45 (a) Ordered by increasing weight, the five blocks are (1) Williams-22, Deng-24, Hernandez-25, and Moses-25; (2) Santiago-27, Kendall-28, Mann-28, and Smith-29; (3) Brunk-30, Obrach-30, Rodriguez-30, and Loren-32; (4) Jackson-33, Stall-33, Brown-34, and Cruz-34; (5) Birnbaum-35, Tran-35, Nevesky-39, and Wilansky-42. (b) The exact randomization will vary with the starting line in Table B. Different methods are possible; perhaps the simplest is to number from 1 to 4 within each block, then assign the members of block 1 to a weight-loss treatment, then assign block 2, etc. For example, starting on line 133, we assign 4-Moses to treatment A, 1-Williams to B, and 3-Hernandez to C (so that 2-Deng gets treatment D), then carry on for block 2, etc. (either continuing on the same line, or starting over somewhere else).

5.47 (a) Assign 10 subjects to Group 1 (the 70° group) and the other 10 to Group 2 (which will perform the task in the 90° condition). Record the number of correct insertions in each group. (b) All subjects will perform the task twice—once in each temperature condition. Randomly choose which temperature each subject works in first, either by flipping a coin, or by placing 10 subjects in Group 1 (70°, then 90°) and the other 10 in Group 2.

5.49 "Randomized" means that patients were randomly assigned either St. John's wort or the placebo. "Placebo controlled" means that we will compare the results for the group using St. John's wort to the group that received the placebo.
(b)

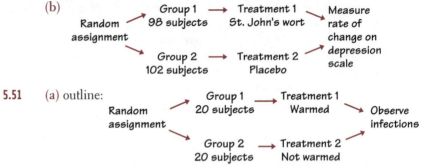

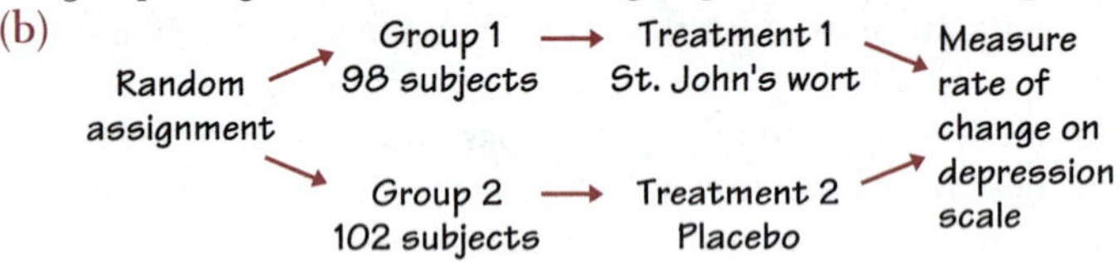

5.51 (a) outline:

Random assignment → Group 1, 20 subjects → Treatment 1, Warmed → Observe infections
Random assignment → Group 2, 20 subjects → Treatment 2, Not warmed → Observe infections

(b) Number the subjects from 01 to 40. Divide digits into groups of 2. Omit groups that are over 40. First 20 pairs will be Group 1. The rest will be Group 2.

5.53 (a) Randomly assign 20 men to each of two groups. Record each subject's blood pressure, then apply the treatments: a calcium supplement for Group 1, and a placebo for Group 2. After sufficient time has passed, measure blood pressure again and observe any change. (b) Number from 01 to 40 down the columns. Group 1 is 18-Howard, 20-Imrani, 26-Maldonado, 35-Tompkins, 39-Willis, 16-Guillen, 04-Bikalis, 21-James, 19-Hruska, 37-Tullock, 29-O'Brian, 07-Cranston, 34-Solomon, 22-Kaplan, 10-Durr, 25-Liang, 13-Fratianna, 38-Underwood, 15-Green, and 05-Chen.

5.55 Responding to a placebo does not imply that the complaint was not "real"—38% of the placebo group in the gastric freezing experiment improved, and those patients really had ulcers. The placebo effect is a *psychological* response, but it may make an actual *physical* improvement in the patient's health.

5.57 Three possible treatments are (1) fine, (2) jail time, and (3) attending counseling classes. The response variable would be the rate at which people in the three groups are re-arrested.

5.59 (a) Flip the coin twice. Let HH ↔ failure, and let the other three outcomes, HT,TH,TT ↔ success. (b) Let 1,2,3 ↔ success, and let 4 ↔ failure. If 5 or 6 come up, ignore them and roll again. (c) Peel off two consecutive digits from the table; let 01 through 75 ↔ success, and let 76 through 99 and 00 ↔ failure. (d) Let diamond, spade, club ↔ success, and let heart ↔ failure.

5.61 (a) Obtain an alphabetized list of the student body, and assign consecutive numbers to the students on the list. Use a random process (table or random digit generator) to select 10 students from this list. (b) Let the two-digit groups 00 to 83 represent a "Yes" to the question of whether or not to abolish evening exams and the groups 84 to 99 represent a "No." (c) Starting at line 129 in Table B ("Yes" in boldface) and reading across rows:

Repetiton 1: **36**, **75**, 95, 89, 84, **68**, **28**, **82**, **29**, **13** # "Yes": 7
Repetiton 2: **18**, **63**, 85, **43**, **03**, **00**, **79**, **50**, 87, **27** # "Yes": 8
Repetiton 3: **69**, **05**, **16**, **48**, **17**, 87, **17**, **40**, 95, **17** # "Yes": 8
Repetiton 4: 84, **53**, **40**, **64**, 89, 87, **20**, **19**, **72**, **45** # "Yes": 7
Repetiton 5: **05**, **00**, **71**, **66**, **32**, **81**, **19**, **41**, **48**, **73** # "Yes": 10

(Theoretically, we should achieve 10 "Yes" results approximately 10.7% of the time.)

5.63 The choice of digits in these simulations may of course vary from that made here. In (a)–(c), a single digit simulates the response; for (d), two digits simulate the response of a single voter.

(a) Odd digits — voter would vote Democratic
Even digits — voter would vote Republican
(b) 0, 1, 2, 3, 4, 5 — Democratic
6, 7, 8, 9 — Republican
(c) 0, 1, 2, 3 — Democratic
4, 5, 6, 7 — Republican
8, 9 — Undecided
(d) 00, 01, ..., 52 — Democratic
53, 54, ..., 99 — Republican

5.65 Let 1 = girl and 0 = boy. The command `randInt (0, 1)` produces a 0 to 1 with equal likelihood. Continue to press ENTER. In 50 repetitions, we get a girl 47 times, and all 4 boys three times. Our simulation produced a girl 94% of the time, vs. a theoretical probability of 0.938.

5.67 Let 01 to 15 ↔ break a racquet, and let 16 to 99 and 0 ↔ not breaking a racquet. Starting with line 141 in the random digit table, we peel two digits off at a time and record the results: 96 76 73 59 64 23 82 29 60 12. In the first repetition, Brian played 10 matches until he broke a racquet. Additional repetitions produced these results: 3, 11, 6, 37, 5, 3, 4, 11, 1. The average for these 10 repetitions is 9.1 We will learn in Chapter 8 that the expected number of matches until a break is about 6.67. More repetitions should improve our estimate.

5.69 For five sets of 30 repetitions, we observed 5, 3, 3, 8, and 4 numbers that were multiples of 5. The mean number of multiples of 5 in 30 repetitions was 3.6, so 3.6/30 = 12% is our estimate for the proportion of times a person wins the game.

5.71 (a) Let 000 to 999 ↔ at bats, 000 to 319 ↔ hits, and 320 to 999 ↔ no hits. (b) We entered 1 → c ENTER to set a counter. Then enter `randInt (0, 999, 20)` → L_1 : `sum (`$L_1 \geq$ `0 and` $L_1 \leq$ `319)` → L_2 `(C) : C + 1` → `C` and press ENTER repeatedly. The count (number of the repetition) is displayed on the screen to help you see when to stop. The results for the 20 repetitions are stored in list L_2. We obtained the following frequencies:

Number of hits in 20 at-bats:	4	5	6	7	8	9
Frequency:	3	5	4	3	2	3

(c) The mean number of hits in 20 at bats was $\bar{x} = 6.25$. And $6.25/20 = .3125$, compared with the player's batting average of .320. Notice that even though there was considerable variability in the 20 repetitions, ranging from a low of 3 hits to a high of 9 hits, the results of our simulation were close to the player's batting average.

5.75 (a) Explanatory: treatment method; response: survival times. (b) No treatment is actively imposed; the women (or their doctors) chose which treatment to use. (c) Doctors may make the decision of which treatment to recommend based in part on how advanced the case is—for example, some might be more likely to recommend the older treatment for advanced cases, in which case the chance of recovery is lower.

5.77 Divide the players into two groups—one to receive oxygen and the other without oxygen during the rest period. Match the players so each player receiving oxygen has a corresponding player of similar speed who does not receive oxygen.

5.79 (a) The chicks are the experimental units; weight gain is the response variable. (b) There are two factors: corn variety (2 levels) and percent of protein (3 levels). This makes 6 treatments, so 60 chicks are required.

5.81 The factors are whether or not the letter has a ZIP code (2 levels: yes or no), and the time of day the letter is mailed. The number of levels for the second factor may vary.

To deal with lurking variables, all letters should be the same size and should be sent to the same city, and the day on which a letter is sent should be randomly selected. Because most post offices have shorter hours on Saturdays, one may wish to give that day some sort of "special treatment" (it might even be a good idea to have the day of the week be a *third* factor in this experiment).

5.83 (a) Below. The two extra patients can be randomly assigned to two of the three groups. (b) No one involved in administering treatments or assessing their effectiveness knew which subjects were in which group. (c) The pain scores in Group A were so much lower than the scores in Groups B and C that they would not often happen by chance if NSAIDs were not effective. We can conclude that NSAIDs provide real pain relief.

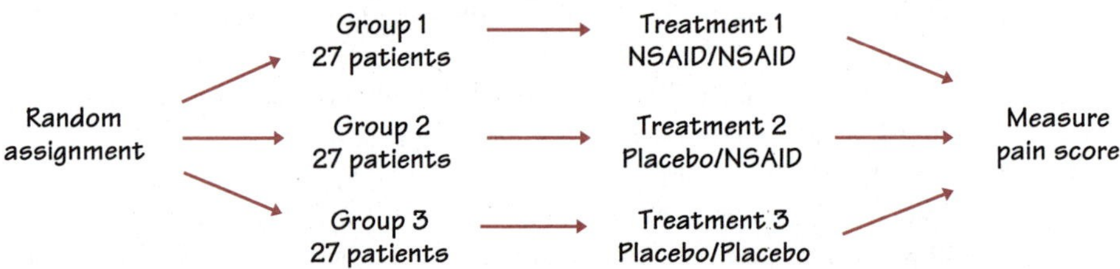

5.85 (a) Let 01 to 05 represent demand for 0 cheesecakes. Let 06 to 20 represent demand for 1 cheesecake. Let 21 to 45 represent demand for 2 cheesecakes. Let 46 to 70 represent demand for 3 cheesecakes. Let 71 to 90 represent demand for 4 cheesecakes and let 91 to 99 and 00 represent demand for 5 cheesecakes. (b) The baker should make 2 cheesecakes each day to maximize his profits.

5.87 (a) A single digit simulates one try, with 0 or 1 a pass and 2 to 9 a failure. Three independent tries are simulated by three successive random digits. (b) With the convention of (a), 50 tries beginning in line 120 gives 25 successes, so the probability of success is estimated as 25/50 = 1/2. [In doing the simulation, remember that you can end a repetition after 1 or 2 tries if the student passes, so that some repetitions do not use three digits. Though this is a proper simulation of the student's behavior, the probability of at least one pass is the same if three digits are examined in every repetition. The true probability is $1 - (0.8)^3 = 0.488$, so this particular simulation was quite accurate.] (c) No—learning usually occurs in taking an exam, so the probability of passing probably increases on each trial.

Chapter 6

6.1 Long trials of this experiment often approach 40% heads. One theory attributes this surprising result to a "bottle-cap effect" due to an unequal rim on the penny. We don't know. But a teaching assistant claims to have spent a profitable evening at a party betting on spinning coins after learning of the effect.

6.3 (a) The results have been stored in a list. (b) Shaq hit 52% of his shots. (c) The longest sequence of misses was 6 and the longest sequence of hits was 9.

6.5 There are 21 0s among the first 200 digits; the proportion is $\frac{21}{200} = 0.105$.

6.7 Obviously, results will vary with the type of thumbtack used. If you try this experiment, note that although it is commonly done when flipping coins, we do not recommend throwing the tack in the air, catching it, and slapping it down on the back of your other hand.

6.9 The study looked at regular season games, which included games against poorer teams, and it is reasonable to believe that the 63% figure is inflated because of these weaker opponents. In the World Series, the two teams will (presumably) be nearly the best, and home game wins will not be so easy.

6.11 (a) S = {germinates, fails to grow}. (b) If measured in weeks, for example, S = {0, 1, 2, ...}. (c) S = {A, B, C, D, F}. (d) Using Y for "yes (shot made)" and N for "no (shot missed)," S = {YYYY, NNNN, YYYN, NNNY, YYNY, NNYN, YNYY, NYNN, NYYY, YNNN, YYNN, NNYY, YNYN, NYNY, YNNY, NYYN}. (There are 16 items in the sample space.) (e) S = {0, 1, 2, 3, 4}.

6.13 S = {all numbers between_ and_}. The numbers in the blanks may vary. Table 1.10 has values from 86 to 195 cal; the range of values in S should include at *least* those numbers. Some students may play it safe and say "all numbers greater than 0."

6.15 (a) 10,000 (b) 5,040 (c) 11,110.

6.17 (b) 18 (c) There are 4 ways to get a sum of 5, and 5 ways to get a sum of 8. (d) Answers will vary but might include:
- The "number of ways" increases until "sum = 7" and then decreases.
- The "number of ways" is symmetrical about "sum = 7."
- Odd sums occur an even number of ways and even sums occur an odd number of ways.

6.19 (a) The given probabilities have sum 0.96, so P(type AB) = 0.04. (b) P(type O or B) = 0.49 + 0.20 = 0.69.

6.21 P (either CV disease or cancer) = 0.45 + 0.22 = 0.67; P (other cause) = 0.33.

6.23 (a) The sum is 1, as we expect since all possible outcomes are listed. (b) 1 – 0.41 = 0.59. (c) 0.41 + 0.23 = 0.64. (d) (0.41)(0.41) = 0.1681.

6.25 (a) Let x = number of spots. Then $P(x = 1) = P(x = 2) = P(x = 3) = P(x = 4) = 0.25$. Since all 4 faces have the same shape and the same area, it is reasonable to assume that the probability of a face being down is the same as for any other face. Since the sum of the probabilities must be one, the probability of each should be 0.25. (b) Outcomes (1,1 1,2 1,3 1,4 2,1 2,2 2,3 2,4 3,1 3,2 3,3 3,4 4,1 4,2 4,3 4,4) The probability of any pair is 1/16 = 0.0625. $P(\text{Sum} = 5) = P(1,4) + P(2,3) + P(3,2) + P(4,1) = (0.0625)4 = 0.25$.

6.27 Fight one big battle: His probability of winning is 0.6, compared to $0.8^3 = 0.512$. (Or he could choose to try for a negotiated peace.)

6.29 No: It is unlikely that these events are independent. In particular, it is reasonable to expect that college graduates are less likely to be laborers or operators.

6.31 An individual light remains lit for 3 years with probability 1 – 0.02; the whole string remains lit with probability $(1 - 0.02)^{20} = (0.98)^{20} \cong 0.6676$.

6.33 (a) $P(\text{Call does not reach a person}) = 0.8$; $P(\text{None of 5 calls reaches a person}) = (0.8)^5 = 0.32768$ (b) $P(\text{None of 5 calls to NYC reach a person}) = (0.92)^5 = 0.6591$.

6.35 (a) Legitimate. (b) Not legitimate, because probabilities sum to more than 1. (c) Not legitimate, because probabilities sum to less than 1.

6.37 (a) The sum of all 8 probabilities equals 1 and all probabilities are $0 \le p \le 1$. (b) $P(A) = 0.000 + 0.003 + 0.060 + 0.062 = 0.125$ (c) The chosen person is not white. $P(B^c) = 1 - P(B) = 1 - (0.060 + 0.691) = 1 - 0.751 = 0.249$ (d) $P(A^c \cap B) = 0.691$.

6.39 (a) $P(\text{undergraduate and score} \ge 600) = (0.40)(0.50) = 0.20$; $P(\text{graduate and score} \ge 600) = (0.60)(0.70) = 0.42$ (b) $P(\text{score} \ge 600) = 0.20 + 0.42 = 0.62$.

6.41 (a) $P(\text{under 65}) = 0.321 + 0.124 = 0.445$. $P(\text{65 or older}) = 1 - 0.445 = 0.555$. (b) $P(\text{tests done}) = 0.321 + 0.365 = 0.686$. $P(\text{tests not done}) = 1 - 0.686 = 0.314$. (c) $P(\text{A and B}) = 0.365$; $P(A)P(B) = (0.555)(0.686) = 0.3807$. A and B are not independent; tests were done less frequently on older patients than if these events were independent.

6.43 Look at the first five rolls in each sequence. All have one G and four Rs, so those probabilities are the same. In the first sequence, you win regardless of the sixth roll; for the second you win if the sixth roll is G, and for the third sequence, you win if it is R. The respective probabilities are $\left(\frac{2}{6}\right)^4\left(\frac{4}{6}\right) = \frac{2}{243} \cong 0.00823$, $\left(\frac{2}{6}\right)^4\left(\frac{4}{6}\right)^2 = \frac{4}{729} \cong 0.00549$, and $\left(\frac{2}{6}\right)^5\left(\frac{4}{6}\right) = \frac{2}{729} \cong 0.00274$.

6.45 (a) If independent, then $P(\text{A and B}) = P(A) \times P(B)$. Since A and B are nonempty, then $P(A) \ge 0$, $P(B) \ge 0$ and $P(A) \times P(B) > 0$. Therefore, $P(\text{A and B}) > 0$. So $A \cap B$ is not empty. (b) If A and B are disjoint, then $P(\text{A and B}) = 0$. But this cannot be true if A and B are independent. So A and B cannot be independent. (c) Example: A bag contains 3 red balls and 2 green balls. A ball is to be drawn from the bag, its color is noted and the ball is set aside. Then a second ball is drawn and its color is noted. Event A is that the first ball is red. Event B is that the second ball is red. Events A and B are not disjoint because both balls can be red. However, events A and B are not independent because whether the first ball is red or not alters the probability of the second ball being red.

6.47 (a) {A and B} represents both prosperous and educated. $P(\text{A and B}) = 0.080$. (b) {A and B^c} represents prosperous but not educated. $P(\text{A and } B^c) = 0.054$.

(c) $\{A^c$ and $B\}$ represents not prosperous but educated. $P(A^c$ and $B) = 0.174$.
(d) $\{A^c$ and $B^c\}$ represents neither prosperous nor educated. $P(A^c$ and $B^c) = 0.692$.

6.49 $P(A) \times P(B) = (0.6)(0.5) = 0.30$. Since this equals the stated probability for $P(A$ and $B)$, events A and B are independent.

6.51 In constructing the Venn diagram, start with the numbers given for "only tea" and "all three," then determine other values. For example, P(coffee and cola, but not tea) = P(coffee and cola) – P(all three). (a) 15% drink only cola. (b) 20% drink none of these.

6.53 (b) 0.3 (c) 0.3.

6.55 (a) 7842/59920 = 0.13087 (b) married, age 18 to 29 (c) 0.13087 is the proportion of women who are age 18 to 29 among those women who are married.

6.57 0.32.

6.59 First, concentrate on (say) spades. The probability that the first card dealt is one of those five cards (A♠, K♠, Q♠, J♠, or 10♠) is 5/52. The conditional probability that the second is one of those cards, given that the first was, is 4/51. Continuing like this, we get 3/50, 2/49, and finally 1/48; the product of these five probabilities gives P(royal flush in spades) ≅ 0.00000038477. Multiplying by four gives P(royal flush) ≅ 0.000001539.

6.61 Let G = {student likes Gospel} and C = {student likes country}. (a) $P(G \mid C) = P(G \text{ and } C)/P(C) = 0.1/0.4 = 0.25$. (b) $P(G \mid \text{not } C) = P(G \text{ and not } C) / P(\text{not } C) = 0.2/0.6 = \frac{1}{3} \cong 0.33$.

6.63 I = infection occurs, F = procedure fails. The percent of procedures that succeed and are free from infection is 0.84. $P(F^c \text{ and } I^c) = P(F^c) \times P(I^c \mid F^c) = (0.86)(0.84/0.86) = 0.84$.

6.65 (a) $P(antibody|test\ pos) = \frac{0.0009985}{0.0009985 + 0.005994} = 0.1428$.
(b) $P(antibody|test\ pos) = \frac{0.09985}{0.09985 + 0.0054} = 0.9487$.
(c) A positive result does not always indicate that the antibody is present. How common a factor is in the population can impact on the test probabilities.

6.67 (a) $\frac{856}{1,626} \cong 0.5264$. (b) $\frac{30}{74} = 0.4054$. (c) No: If they were independent, the answers to (a) and (b) would be the same.

6.69 (a) $\frac{770}{1,626} \cong 0.4736$. (b) $\frac{529}{770} = 0.6870$. (d) Using multiplication rule: P(male and bachelor's degree) = P(male)P(bachelor's degree | male) = (0.4736)(0.6870) = 0.3254. (Answers will vary with how much previous answers had been rounded.) Directly: $\frac{529}{1626} \cong 0.3253$. (Note that the difference between these answers is inconsequential, since the numbers in the table are rounded.)

6.71 Percent of "A" students involved in an accident is

$$P(C|A) = \frac{P(C \text{ and } A)}{p(A)} = \frac{0.01}{0.10} = 0.10$$

Or, from the tree diagram:

$$P(C|A) = \frac{0.01}{0.10 + 0.09} = 0.10$$

6.73 P(correct) = P(knows answer) + P(doesn't know, but guesses correctly) = 0.75 + (0.25)(0.20) = 0.8.

6.75 P (knows the answer | gives the correct answer) = $\frac{0.75}{0.80} = \frac{15}{16} = 0.9375$.

6.77 (a) 0.1; 0.9. (b) 9999 switches remain; 999 are bad; 999/9999 ≅ 0.09991. (c) 9999 switches remain; 1000 are bad; 1000/9999 ≅ 0.10001.

6.79 (a) P(*Type* AB) = 1 – (0.45 + 0.40 + 0.11) = 0.04. (b) P(*Type* B *or Type* O) = 0.11 + 0.45 = 0.56. (c) Assuming that the blood types for husband and wife are independent P(*Type* B *and Type* A) = (0.11)(0.40) = 0.044. (d) P(*Type* B *and Type* A) + P(*Type* B *and Type* A) = (0.11)(0.40) + (0.40)(0.11) = 0.088. (e) P(*Husband Type* O *or Wife Type* O) = P(*Husband Type* O) + P(*Wife Type* O) – P (*Husband and wife Type* O) = 0.45 + 0.45 – 0.2025 = 0.6975.

6.81 (a) To find P(A or C), we would need to know P(A and C). (b) To find P(A and C), we would need to know P(A or C).

6.83 (a) $P(\text{firearm}) = \frac{18,940}{31,510} = 0.6011$.
(b) $P(\text{firearm} \mid \text{female}) = \frac{2,559}{6,095} = 0.4199$.
(c) $P(\text{female and fireaem}) = \frac{2,559}{31,510} = 0.0812$.
(d) $P(\text{firearm} \mid \text{male}) = \frac{16,381}{25,415} = 0.6445$.
(e) $P(\text{male} \mid \text{firearm}) = \frac{16,381}{18,940} = 0.8649$

6.85 P(B | A) = P(both tosses have the same outcome | H on first toss) = 1/2 = 0.5. P(B) = P(both tosses have same outcome) = 2/4 = 0.5. Since P(B | A) = P(B), events A and B are independent.

6.87 The response will be "no" with probability 0.35 = (0.5)(0.7). If the probability of plagiarism were 0.2, then P(student answers "no") = 0.4 = (0.5)(0.8). If 39% of students surveyed answered "no," then we estimate that 2 × 39% = 78% have *not* plagiarized, so about 22% have plagiarized.

Chapter 7

7.1 $\frac{2}{6} = \frac{1}{3}$.

7.3 (a) 1% (b) All probabilities are between 0 and 1; the probabilities add to 1. (c) $P(X \le 3) = 0.48 + 0.038 + 0.08 = 1 - 0.01 - 0.05 = 0.94$. (d) $P(X < 3) = 0.48 + 0.38 = 0.86$. (e) Write either $X \ge 4$ or $X > 3$. The probability is 0.06. (f) Read two random digits from Table B. Here is the correspondence: 01 to 48 ↔ Class 1, 49 to 86 ↔ Class 2, 87 to 94 ↔ Class 3, 95 to 99 ↔ Class 4, and 00 ↔ Class 5. Repeatedly generate 2 digit random numbers. The proportion of numbers in the range 01 to 94 will be an estimate of the required probability.

7.5 (a) $P(x \ge 5) = 0.868$ (b) $P(x > 5)$ means "the probability that the chosen house has more than five rooms." $P(x > 5) = 0.658$. (c) Since the probability is for a specific value, including, or not including, that value changes the probability. For discrete random variables, $P(x < a)$ does not generally equal $P(x \le a)$.

7.7 (a) $P(X \le 0.49) = 0.49$. (b) $P(X \ge 0.27) = 0.73$. (c) $P(0.27 \le X \le 1.27) = P(0.27 \le X \le 1) = 0.73$. (d) $P(0.1 \le X \le 0.2 \text{ or } 0.8 \le X \le 0.9) = 0.1 + 0.1 = 0.2$. (e) $P(\text{not } [0.3 \le X \le 0.8]) = 1 - 0.5 = 0.5$. (f) $P(X = 0.5) = 0$.

7.9 (TI-83) For a *sample* simulation of 400 observations from the N(0.3, 0.023) observations: there are 4 observations less than 0.25, so the relative frequency is 4/400 = 0.01. The actual probability that $p < 0.25$ is 0.0149.

7.11 (a) $P(Y > 1) = 1 - P(Y = 1) = 1 - 0.25 = 0.75$. (b) $P(2 < Y \le 4) = P(Y = 3 \text{ or } Y = 4) = 0.17 + 0.15 = 0.32$ (c) $P(Y \uparrow 2) = 1 - P(Y = 2) = 1 - 0.32 = 0.68$.

7.13 (a) The 36 possible pairs of "up faces" are:

(1, 1)	(1, 2)	(1, 3)	(1, 4)	(1, 5)	(1, 6)
(2, 1)	(2, 2)	(2, 3)	(2, 4)	(2, 5)	(2, 6)
(3, 1)	(3, 2)	(3, 3)	(3, 4)	(3, 5)	(3, 6)
(4, 1)	(4, 2)	(4, 3)	(4, 4)	(4, 5)	(4, 6)
(5, 1)	(5, 2)	(5, 3)	(5, 4)	(5, 5)	(5, 6)
(6, 1)	(6, 2)	(6, 3)	(6, 4)	(6, 5)	(6, 6)

(b) Each pair must have probability 1/36. **(c)** Let X = sum of up faces. Then $P(X = 2) = 1/36$, $P(X = 3) = 2/36$, $P(X = 4) = 3/36$, $P(X = 5) = 4/36$, $P(X = 6) = 5/36$, $P(X = 7) = 6/36$, $P(X = 8) = 5/36$, $P(X = 9) = 4/36$, $P(X = 10) = 3/36$, $P(X = 11) = 2/36$, $P(X = 12) = 1/36$. **(d)** $P(7 \text{ or } 11) = 6/36 + 2/36 = 8/36$ or 2/9. **(e)** P (any sum other than 7) $= 1 - P(7) = 1 - 6/36 = 30/36 = 5/6$ by the complement rule.

7.15 **(a)** 75.2%. **(b)** All probabilities are between 0 and 1; the probabilities add to 1. **(c)** $P(X \geq 6) = 1 - 0.010 - 0.007 = 0.983$. **(d)** $P(X > 6) = 1 - 0.010 - 0.007 - 0.007 = 0.976$. **(e)** Either $X \geq 9$ or $X > 8$. The probability is $0.068 + 0.070 + 0.041 + 0.752 = 0.931$.

7.17 **(a)** The height should be 1/2, since the area under the curve must be 1. **(b)** $P(y \leq 1) = 1/2$. **(c)** $P(0.5 < y < 1.3) = 0.4$. **(d)** $P(y \geq 0.8) = 0.6$.

7.19 The resulting histogram should *approximately* resemble the triangular density curve of Figure 7.8, with any deviations or irregularities depending upon the specific random numbers generated.

7.21 In this case, we will simulate 500 observations from the N(0.15, 0.0092) distribution. Scrolling through the 500 simulated observations, we can determine the relative frequency of observations that are at least 0.16 by using the complement rule. For a sample simulation, there are 435 observations less than 0.16, thus the desired relative frequency is $1 - 435/500 = 65/500 = 0.13$. The actual probability that $p \geq 0.16$ is 0.1385. 500 observations yield a reasonably close approximation.

7.23 $\mu_{\text{owned}} = 1(0.003) + 2(0.002) + 3(0.023) + \cdots + 10(0.035) = 6.284$

$\mu_{\text{rented}} = 1(0.008) + 2(0.027) + 3(0.287) + \cdots + 10(0.003) = 4.187$

In 7.4, the distribution of owned units is approximately normal and peaks at 6. The distribution of rented units is right skewed and peaks at 4. The owned units should have a higher mean, which it does.

7.25 **(a)** The payoff is either \$0 or \$3; $P(x = 0) = 0.75$, $P(x = 3) = 0.25$. **(b)** For each \$1 bet, $\mu_x = (\$0)(0.75) + (\$3)(0.25) = \$0.75$. **(c)** The casino makes 25 cents for every dollar bet (in the long run).

7.27 $\mu_{\text{hsld}} = 1(0.25) + 2(0.32) + 3(0.17) + \cdots + 7(0.01) = 2.6$

$\mu_{\text{fam}} = 1(0) + 2(0.42) + 3(0.23) + \cdots + 7(0.02) = 3.14$

$\sigma^2_{\text{hsld}} = (1 - 2.6)^2\,(0.35) + (2 - 2.6)^2\,(0.32) + \cdots + (7 - 2.6)^2\,(0.01) = 2.02$

$\sigma_{\text{hsld}} = 1.421$

$\sigma^2_{\text{fam}} = (1 - 3.14)^2\,(0) + (2 - 3.14)^2\,(0.42) + \cdots + (7 - 3.14)^2\,(0.02) = 1.564$

$\sigma_{\text{fam}} = 1.249$

The distributions are fairly similar except for the fact that the family distribution cannot have a value of 1. Both are skewed right, but the lack of a lower value (1) yields a somewhat higher mean for the family distribution. The standard deviations are also somewhat similar and, again, the smaller standard deviation for the family distribution is consistent with the fact that a family must consist of at least 2 persons.

7.29 **(a)** $\mu_x = (0)(0.03) + (1)(0.16) + (2)(0.30) + (3)(0.23) + (4)(0.17) + (5)(0.11) = 2.68$. $\sigma^2_x = (0 - 2.68)^2\,(0.03) + (1 - 2.68)^2\,(0.16) + (2 - 2.68)^2\,(0.30) +$

$(3 - 2.68)^2\,(0.23) + (4 - 2.68)^2\,(0.17) + (5 - 2.68)^2\,(0.11) = 1.7176$, and $\sigma_x = \sqrt{1.7176} = 1.3106$. (b) Answers will vary.

7.31 Below is the probability distribution for L, the length of the longest run of heads or tails. P (You win) $= P(\text{run of 1 or 2}) = \frac{89}{512} \cong 0.1738$, so the expected outcome is $\mu = (\$2)(0.1738) + (-\$1)(0.8262) \cong \$0.4785$. On the average, you will lose about 48 cents each time you play. (Simulated results should be close to these exact results; how close depends on how many trials are used.)

Value of L:	1	2	3	4	5	6	7	8	9	10
Probability:	$\frac{1}{512}$	$\frac{88}{512}$	$\frac{185}{512}$	$\frac{127}{512}$	$\frac{63}{512}$	$\frac{28}{512}$	$\frac{12}{512}$	$\frac{5}{512}$	$\frac{2}{512}$	$\frac{1}{512}$

7.33 No: Assuming all "at-bats" are independent of each other, the 35% figure only applies to the "long run" of the season, not to "short runs."

7.35 (a) Do not assume independence. The points awarded from three cards depend on the first two cards (Event X). (b) Assume independence. There should not be any way the result of the first roll (Event X) will influence the probability for the second roll (Event Y).

7.37 (a) 11 + 20 = 31. (b) No. (c) The mean time will not be changed, since independence is not required in order to add means. If the two variables are dependent, the standard deviation will be altered, but the mean of the total will still remain the same.

7.39 Since the 5-minute process between the reactions is a constant, it will not affect the standard deviation of the complete procedure.

$$\sigma^2_{X+Y} = \sigma^2_X + \sigma^2_Y = 2^2 + 1^2 = 5$$

$$\sigma_{X+Y} = \sqrt{5} = 2.236$$

7.41 $F - L$ is $N\left(0, 2\sqrt{2}\right)$ so $P(|F - L| > 5) = P(|Z| > 1.7678) = 0.0771$.

7.43 The law of large numbers states that the actual outcome approaches the distribution as more trials are conducted. For example, the proportion of 1's in a sample of 10 numbers could vary greatly. But the proportion in a very large sample, say several million trials, should be near Benford's value of 0.30.

7.45 (a) Randomly selected students would presumably be unrelated. (b) $\mu_{f-m} = \mu_f - \mu_m = 120 - 105 = 15$. $\sigma^2_{f-m} = \sigma^2_f + \sigma^2_m = 28^2 + 35^2 = 2009$, so $\sigma_{f-m} = 44.82$. (c) Knowing only the mean and standard deviation, we cannot find that probability (unless we assume that the distribution is normal). Many different distributions can have the same mean and standard deviation.

7.47 Read two-digit random numbers. Establish the correspondence 01 to 10 $\leftrightarrow$ 540°, 11 to 35 $\leftrightarrow$ 545°, 36 to 65 $\leftrightarrow$ 550°, 66 to 90 $\leftrightarrow$ 555°, and 91 to 99, 00 $\leftrightarrow$ 560°. Repeat many times, and record the corresponding temperatures. Average the temperatures to approximate μ; find the standard deviations of the temperatures to approximate σ.

7.49 (a) Yes. This is always true; it does not depend on independence. (b) No. It is not reasonable to believe that X and Y are independent.

7.51 $\mu_{0.8m+0.2j} = 0.8\mu_m + 0.2\mu_j = 0.8(0.0114) + 0.2(0.0159)$
$= 0.0123 = 1.23\% > 1.14\%$
$\sigma^2_{0.8m+0.2j} = 0.8^2\sigma^2_m + 0.2^2\sigma^2_j = 0.8^2(0.0464)^2 + 0.2^2(0.0675)^2$
$= 0.0015601$

$\sigma_{0.8m+0.2j} = 0.0395 = 3.95\% < 6.75\%$

7.53 $\mu_R = 0.6(0.0114) + 0.2(0.0016) + 0.2(0.0159) = 0.01034$ (or 1.034%)
$\sigma_R^2 = 0.6^2(0.0464)^2 + 0.2^2(0.0361)^2 + 0.2^2(0.0675)^2 + 2(0.6)(0.2)(0.19) \times (0.0464)(0.0361) + 2(0.6)(0.2)(0.54)(0.0464)(0.0675) + 2(0.2)(0.2)(-0.17)(0.0361)(0.0675) = 0.0014586$
$\sigma_R = 0.0382$ or 3.82%.

7.55 The missing probability is 0.99058 (so that the sum is 1). This gives mean earnings $\mu_x = \$303.3525$.

7.57 $\sigma_X^2 = 94{,}236{,}826.64$, so that $\sigma_X \cong 9707.57$.

7.59 $X - Y$ is $N(0, 0.4243)$, so $P(|X - Y| \geq 0.8) = 0.0593$.

7.61 (a) $\mu_{Y-X} = \mu_Y - \mu_X = 2.001 - 2.000 = 0.001$ g. $\sigma_{Y-X}^2 = \sigma_Y^2 + \sigma_X^2 = 0.002^2 + 0.001^2 = 0.000005$, so $\sigma_{Y-X} = 0.002236$ g. (b) $\mu_Z = \frac{1}{2}\mu_X + \frac{1}{2}\mu_Y = 2.0005$ g. $\sigma_Z^2 = \frac{1}{4}\sigma_X^2 + \frac{1}{4}\sigma_Y^2 = 0.00000125$, so $\sigma_Z = 0.001118$ g. Z is slightly more variable than Y, since $\sigma_Y < \sigma_Z$.

7.63 (a) A single random digit simulates each toss, with (say) odd = heads and even = tails. The first round is two digits, with two odds a win; if you don't win, look at two more digits, again with two odds a win. (b) The probability of winning is $\frac{1}{4} + \left(\frac{3}{4}\right)\left(\frac{1}{4}\right) = \frac{7}{16}$, so the expected value is $(\$1)\frac{7}{16} + (-\$1)\frac{9}{16} = -\frac{2}{16} = -\$0.125$.

7.65 We know that $\sigma_{x+y}^2 = \sigma_x^2 + 2\rho\sigma_x\sigma_y + \sigma_y^2$. If $\rho = 1$, then $\sigma_{x+y}^2 = \sigma_x^2 + 2\sigma_x\sigma_y + \sigma_y^2 = (\sigma_x + \sigma_y)^2$. Taking the square root of both sides, $\sigma_{x+y} = \sigma_x + \sigma_y$.

7.67 $\mu_{a+bX} = a + b\mu_X = a + b(1400) = 0$
$\sigma_{a+bX}^2 = b^2\sigma_X^2 = b^2 20^2 = 1$
The system of equations $a + 1400b = 0$, and $400b^2 = 1$ yields the solution $a = -70$ and $b = 0.05$.

Chapter 8

8.1 (a) No: There is no fixed n (i.e., there is no definite upper limit to the number of defects). (b) Yes: It is reasonable to believe that all responses are independent (ignoring any "peer pressure"), and all have the same probability of saying "yes" since they are randomly chosen from the population. Also, a "large city" will have a population over 1000 (10 times as big as the sample). (c) Yes: In a "Pick 3" game, Joe's chance of winning the lottery is the same every week, so assuming that a year consists of 52 weeks (observations), this would be binomial.

8.3 (a) 0.2637. (b) The binomial probabilities for $x = 0, ..., 5$ are: 0.2373, 0.3955, 0.2637, 0.0879, 0.0146, 0.0010. (e) The cumulative probabilities for $x = 0, ..., 5$ are: 0.2373, 0.6328, 0.8965, 0.9844, 0.9990, 1. Compare with Corinne's cdf histogram; the bars in this histogram get taller, sooner. Both peak at 1 on the extreme right.

8.5 (a) $P(X \geq 1) = 1 - P(X = 0) = 1 - 0.0563 = 0.9437$. (b) $P(X = 0) = P$(none correct on the first) $\times$ P(none correct on the second) $\times$ P(none correct on the third) = (0.6667)(0.75)(0.8) = 0.4. Then P(at least one is correct) = 1 − 0.4 = 0.6.

8.7 $P(X = 11)$ = binompdf(20,0.8,11) = 0.007.

8.9 $P(X = 3) = (10)(0.25)^3(0.75)^2 = 0.0879$.

8.11 $P(X \geq 1) = 1 - P(X = 0) = 1 - \binom{5}{0}(0.25)^0(0.75)^5 = 0.7627$.

8.13 (a) $\binom{15}{3}(0.3)^3(0.7)^{12} = 0.17004$. (b) $\binom{5}{0}(0.3)^3(0.7)^{15} = 0.00475$.

8.15 (a) $np = 1500 \geq 10$, and $n(1-p) = 1000 \geq 10$. (b) 1 – binomcdf(2500, 0.6, 1519) = 0.213139. (c) $P(X \leq 1468) = 0.0994$, correct to four decimal places. For comparison, a normal approximation gives $P(X \leq 1468) = P(Z \leq -1.3064) = 0.0957$.

8.17 (a) $\mu = 16$ (if $p = 0.8$). (b) $\sigma = \sqrt{3.2} = 1.78885$. (c) if $p = 0.9$, then $\sigma = \sqrt{1.8} = 1.34164$. If $p = 0.99$, then $\sigma = \sqrt{0.198} = 0.44497$. As p gets close to 1, σ gets closer to 0.

8.19 (a) X = number of U.S. adults who approve of the president's response to the World Trade Center terrorist attacks in September 2001. X is $B(400, 0.92)$ because X satisfies the four requirements for a binomial setting (check this!). (b) $P(X \leq 358)$ = binomcdf(400,0.92,358) = 0.0441. (c) $E(X) = \mu_X = np = 400(0.92) = 368$. $\sigma_x = \sqrt{(np(1-p))} = \sqrt{((368)(0.08))} = 5.4259$. (d) $P(X \leq 358) = P(Z < -1.843) = 0.0327$. This is not a very satisfactory result because $p = 0.92$ is relatively far from 0.5 and the sample size of 400 for a poll is relatively small.

8.21 X is $B(12, 0.75)$. $P(X = 0) = 0.0000$, $(X = 1) = 0.0000$, $P(X = 2) = 0.0000$, $P(X = 3) = 0.0004$, $P(X = 4) = 0.0024$, $P(X = 5) = 0.0115$, $P(X = 6) = 0.0401$, $(X = 7) = 0.1032$, $P(X = 8) = 0.1936$, $P(X = 9) = 0.2581$, $P(X = 10) = 0.2330$, $P(X = 11) = 0.1267$, $P(X = 12) = 0.0317$. The command "cumSum" produces a cumulative sum as it adds the individual probabilities in list L_2/list2. The command binomcdf(12,0.75,L_1) gives the same result, that is, lists L_3/list3 and L_4/list4 are the same.

8.23 Let 0,1,2 ↔ Hispanic and let 3 to 9 ↔ non-Hispanic. Use random digit table. Or, using the calculator, repeat the command 30 times: `randBin(1,.3,15)`→ `L₁:sum(L₁)` → `L₂(1)`. In our simulation, the relative frequency of 3 or fewer Hispanics was 7/30 = 0.233.

8.25 (a) Let 0 ↔ never married, let 1,2,3, ↔ married, and use Table B. Or, using the calculator, repeat the command `randBin(1,.25,10)`→ `L₁:sum(L₁)`. Our relative frequency of 2 or fewer never married was 19/30 = $0.6\bar{3}$.

8.27 The count of 0s among n random digits has a binomial distribution with $p = 0.1$. (a) P(at least one 0) = 1 – P(no 0) = $1 - (0.9)^5 = 0.40951$. (b) $\mu = (40)(0.1) = 4$.

8.29 (a) $n = 5$ and $p = 0.65$. (b) X takes values from 0 to 5. (c) $P(X = 0) = 0.0053$, $P(X = 1) = 0.0488$, $P(X = 2) = 0.1816$, $P(X = 3) = 0.3364$, $P(X = 4) = 0.3124$, $P(X = 5) = 0.11603$. (d) $\mu = 3.25$, $\sigma = 1.067$.

8.31 $\mu = (200)(0.4) = 80$ and $\sigma = \sqrt{(200)(0.4)(0.6)} \cong 6.9282$ households, so $P(X \geq 100) \cong P(Z > 2.89) \cong 0.0019$. Regardless of how we compute the probability, this is strong evidence that the local percentage of households concerned about nutrition is higher than 40%.

8.33 (a) The normal approximation can be used, since the second rule of thumb is *just* satisfied: $n(1-p) = 10$. With $\mu = 90$ and $\sigma = \sqrt{9} = 3$ orders, we compute $P(X \leq 86) = P(Z < -1.33) = 0.0918$. (b) Even when the claim is correct, there will be some variation in sample proportions. In particular, in about 9% of samples we can expect to observe 86 or fewer orders shipped on time.

8.35 In this case, $n = 20$ and the probability that a randomly selected basketball player graduates is $p = 0.8$. We will estimate $P(X \geq 10)$ by simulating 30 observations of X = number graduated and computing the relative frequency of observations that are 10 or greater. The sequence of TI-83 commands is as follows: `randBin(1,.8,20)`→ `L₁:sum(L₁)`→ `L₂(1)`, where 1's represent players who graduated. (Press ENTER sufficient times to obtain 30 numbers. The actual value of $P(X \geq 10)$ is `1-binomcdf(20,.8,9)` = 0.9994 ≈ 1.

8.37 (a) Geometric setting; success = tail, failure = head; trial = flip of coin; $p = 1/2$. (b) Not a geometric setting. You are not counting the number of trials before the first success is obtained. (c) Geometric setting; success = jack, failure = any other card; trial = drawing of a card; $p = 4/52 = 1/13$. (Trials are independent because the card is replaced each time.) (d) Geometric setting, success = match all 6 numbers, failure = do not match all 6 numbers; trial = drawing on a particular day; the probability of success ($p = 0.0000102$) is the same for each trial; and trials are independent because the setting of a drawing is always the same and the results on different drawings do not influence each other. (e) Not a geometric setting. The trials (draws) are not independent because you are drawing without replacement. Also, you are interested in getting 3 successes, rather than just the first success.

8.39 (a) X = number of drives tested in order to find the first defective. Success = defective drive. This is a geometric setting because the trials (tests) on successive drives are independent, $p = .03$ on each trial, and X is counting the number of trials required to achieve the first success. (b) $P(X = 5) = (1 - .03)^{5-1}(.03) = (.97)^4(.03) = .0266$. (c) $P(X = 1) = 0.03$, $P(X = 2) = 0.0291$, $P(X = 3) = 0.0282$, $P(X = 4) = 0.0274$.

8.41 (a) X = number of flips until a head appears. (b) $P(X = 1) = 0.5$, $P(X = 2) = 0.25$, $P(X = 3) = 0.125$, $P(X = 4) = 0.0625$, $P(X = 5) = 0.03125$, etc. (c) $P(X \le 1) = 0.5$, $P(X \le 2) = 0.75$, $P(X \le 3) = 0.875$, $P(X \le 4) = 0.9375$, $P(X \le 5) = 0.96875$, etc.

8.43 (b) $P(X > 10) = (1 - 1/6)^{10} = (5/6)^{10} = 0.1615$. (c) The smallest positive integer k for which $P(X \le k) > .99$ is $k = 26$.

8.45 (a) That the shots are independent, and that the probability of success is the same for each shot. (b) 0.0655. (c) 0.738.

8.47 (a) Geometric setting; X = number of marbles you must draw to find the first red marble. We choose geometric in this case because the number of trials (draws) is the variable quantity. (b) $p = 20/35 = 4/7$ in this case, so $P(X = 2) = (1 - 4/7)^{2-1} \times (4/7) = (3/7)(4/7) = 12/49 = 0.2449$. $P(X \le 2) = 4/7 + (3/7)(4/7) = 4/7 + 12/49 = 40/49 = 0.8163$. $P(X > 2) = (1 - 4/7)^2 = (3/7)^2 = 9/49 = 0.1837$ (c) Use the TI-83 commands `seq(X, X, 1, 20)` $\rightarrow$ `L`$_1$, `(geometpdf (4/7, L`$_1$`)` $\rightarrow$ `L`$_2$, `cumSum (L`$_2$`)` $\rightarrow$ `L`$_3$, or `(geometcdf(4/7, L`$_1$`)` $\rightarrow$ `L`$_3$`)`.

8.49 (a) Success = getting a correct answer. X = number of questions Carla must answer in order to get the first correct answer. $p = 1/5 = 0.2$ (all 5 choices equally likely to be selected) (b) $P(X = 5) = (1 - 1/5)^{5-1}(1/5) = (4/5)^4(1/5) = 0.082$. (c) $P(X > 4) = (1 - 1/5)^4 = (4/5)^4 = 0.4096$. (d) $P(X = 1) = 0.2$, $P(X = 2) = 0.16$, $P(X = 3) = 0.128$, $P(X = 4) = 0.1024$, $P(X = 5) = 0.082$, ... (e) $\mu_x = 1/(1/5) = 5$.

8.51 (a) Letting G = girl and B = boy, the outcomes are: {G, BG, BBG, BBBG, BBBB}. Success = having a girl. (b) X = number of boys can take values of 0, 1, 2, 3, or 4. The probabilities are calculated by using the multiplication rule for independent events: $P(X = 0) = 1/2$, $P(X = 1) = (1/2)(1/2) = 1/4$, $P(X = 2) = (1/2)(1/2)(1/2) = 1/8$, $P(X = 3) = (1/2)(1/2)(1/2)(1/2) = 1/16$, $P(X = 4) = (1/2)(1/2)(1/2)(1/2) = 1/16$. Note that $\Sigma P(X) = 1$. (c) Let Y = number of children produced until first girl is seen. Then Y is a geometric variable for Y = 1 up to Y = 4, but then "stops" because the couple plans to stop at 4 children if it does not see a girl by that time. By the multiplication rule, $P(Y = 1) = 1/2$, $P(Y = 2) = 1/4$, $P(Y = 3) = 1/8$, $P(Y = 4) = 1/16$. Note that the event Y = 4 can only include the outcome BBBG. BBBB must be discarded. The probability distribution would begin

$P(Y = 1) = 1/2$, $P(Y = 2) = 1/4$, $P(Y = 3) = 1/8$, $P(Y = 4) = 1/16$. But note that this is not a valid probability model since $\Sigma P(Y) < 1$. The difficulty lies in the way Y was defined. It does not include the possible outcome BBBB. (d) Let Z = number of children per family. Then $P(Z = 1) = 1/2$, $P(Z = 2) = 1/4$, $P(Z = 3) = 1/8$, $P(Z = 4) = 1/16$ and $\mu_z = \Sigma Z \times P(Z) = (1)(1/2) + (2)(1/4) + (3)(1/8) + (4)(1/8) = 1/2 + 1/2 + 3/8 + 1/2 = 1.875$ (e) $P(Z > 1.875) = P(2) + P(3) + P(4) = 0.5$. (f) The only way in which a girl cannot be obtained is BBBB, which has probability 1/16. Thus the probability of having a girl, by the complement rule, is $1 - 1/16 = 15/16 = 0.938$.

8.53 We will approximate the expected number of children, μ, by making the mean $\bar{x}$ of 25 randomly generated observations of X. We create a suitable string of random digits (say of length 100) by using the command `randInt(0,9,100)` $\rightarrow L_1$ Now we scroll down the list L_1. Let the digits 0 to 4 represent a boy and 5 to 9 represent a girl. We read digits in the string until we get a "5 to 9" (girl) or until four "0 to 4"s (boys) are read, whichever comes first. In each case, we record X = the number of digits in the string = the number of children. We continue until 25 X-values have been recorded. Our sample string L_1 yields $\bar{x} = 45/25 = 1.8$, compared with the known mean $\mu = 1.875$.

8.55 (a) No: It is not reasonable to assume that the opinions of a husband and wife are independent. (b) The population is three times the size of the sample. The population should be at least 10 times larger.

8.57 (a) Symmetric; the shape depends on the value of the probability of success. (c) 0.0078125.

8.59 (a) $p = 0.2$. (b) There are 16 outcomes; here are 4 of the outcomes with their probabilities: SSSS with probability $(0.2)^4 = 0.0016$; FSSS with probability $(0.8)(0.2)(0.2)(0.2) = 0.0064$; FFSS with probability $(0.8)(0.8)(0.2)(0.2) = 0.0256$; and FFFS with probability 0.1024. (d) SSFF, SFSF, SFFS, FSSF, FSFS, FFFS. (e) Each of the outcomes has probability $(0.2)^2(0.8)^2 = 0.0256$, so $P(X = 2) = 6(0.0256) = 0.1536$. The probabilities of these six outcomes are the same because the trials are independent and multiplication is commutative.

8.61 (a) $P(X = 10) = 0.1652$. (b) $P(10 < X < 15) = 0.2708$ or $P(10 \leq X \leq 15) = 0.4423$, depending on your interpretation of "between" being exclusive or inclusive. (c) $P(X > 15) = 1 - P(X \leq 15) = 1 - 0.9980 = 0.0020$. (d) $P(X < 8) = P(X \leq 7) = 0.2241$.

8.63 X is geometric with $p = 0.325$. $1 - p = 0.675$. (a) $P(X = 1) = 0.325$. (b) $P(X \leq 3) = 0.325 + (0.675)(0.325) + (0.675)^2 (0.325) = 0.69245$. (c) $P(X > 4) = (0.675^4) = 0.208$. (d) The expected number of at-bats until Roberto gets his first hit is $\mu = 1/p = 1/0.325 = 3.08$. (e) To do this on the TI-83 use the commands `seq(X, X, 1, 10)`$\rightarrow L_1$, `geometpdf(.325, L1)`$\rightarrow L_2$, and `geometcdf(.325, L1)`$\rightarrow L_3$.

8.65 If X is $B(n, p)$ then $P(X \geq 1) = 1 - P(X = 0) = 1 - p^0(1 - p)^{n-0} = 1 - (1 - p)^n$.

Chapter 9

9.1 2.5003 is a parameter; 2.5009 is a statistic.

9.3 48% is a statistic; 52% is a parameter.

9.5 Answers will vary.

9.7 (a) The scores will vary depending on the starting row. Note that the smallest possible mean is 61.75 (from the sample 58, 62, 62, 65) and the largest is 77.25

(from 73, 74, 80, 82). Answers to (b) and (c) will vary. (d) The sampling distribution is slightly right skewed. The values of $\bar{x}$ and their frequencies are: 60 (2), 61.5 (1), 62 (2), 63.5 (2), 64 (2), 65 (1), 65.5 (2), 66 (1), 67 (2), 67.5 (2), 68 (2), 68.5 (1), 69 (3), 69.5 (2), 70 (2), 71 (2), 72 (2), 72.5 (2), 73 (2), 73.5 (2), 74 (1), 76 (1), 76.5 (1), 77 (2), 77.5 (1), 78 (1), 81 (1). (e) Shape: both sampling distributions are right skewed, but the one for samples of size 2 will be more skewed; Center: $\mu_{\bar{x}} = 69.4$ for both sampling distributions; Spread: the variability is greater for samples of size 2.

9.9 (b) 141.847 days. (c) Means will vary. (d) It would be unlikely (though not impossible) for all five $\bar{x}$ values to fall on the same side of μ. (e) The mean of the sampling distribution should be μ. (f) Answers will vary.

9.11 (a) Since the smallest number of total tax returns (i.e., the smallest population) is still more than 10 times the sample size, the variability will be (approximately) the same for all states. (b) Yes, it will change—the sample taken from Wyoming will be about the same size, but the sample in, e.g., California will be considerably larger, and therefore the variability will decrease.

9.13 (a) Statistic. (b) $\hat{p} = 0.045$.

9.15 (a) The histogram consists of bars of height 1/6 for values 1 through 6. $\mu = 3.5$ and $\sigma = 1.708$. (b) Rolling the die twice. (c) There are 36 possible samples of size 2. Organize the values of $\bar{x}$ in a chart similar to Figure 9.3. (d) Shape: symmetric, but not uniform; center: $\mu_{\bar{x}} = 3.5$; spread $\sigma_{\bar{x}} = \sigma/\sqrt{2} = 1.208$.

9.17 Assuming that the poll's sample size was less than 780,000—10% of the population of New Jersey—the variability would be practically the same for either population. (The sample size for this poll would have been considerably less than 780,000.)

9.19 (a) Mean: $\mu_{\hat{p}} = p = 0.7$; standard deviation: $\sigma_{\hat{p}} = \sqrt{(0.7)(0.3)/1012} = 0.0144$. (b) There are certainly more than 10(1012) = 10120 U.S. adults. (c) np = 1012(0.7) = 708.4 and $n(1 - p)$ = 1012(0.3) = 303.6. (d) 0.0186—this is a fairly unusual result if 70% of the population actually drink the cereal milk. (e) 4048. (f) Answers will vary.

9.21 For $n = 300$: $\sigma = 0.02828$ and $P = 0.7108$. For $n = 1200$: $\sigma = 0.01414$ and $P = 0.9660$. For $n = 4800$: $\sigma = 0.00707$ and $P = 1$ (approximately). Larger sample sizes give more accurate results (the sample proportions are more likely to be close to the true proportion.

9.23 (a) 0.86. (b) We use the normal approximation (Rule of Thumb 2 is *just* satisfied—$n(1 - p) = 10$). The standard deviation is 0.03, and $P(\hat{p} \le 0.86) = P(z \le -1.33) = 0.0918$. (*Note:* The exact probability is 0.1239.) (c) The amswers are the same.

9.25 (a) $\mu_{\hat{p}} = p = 0.15$, $\sigma_{\hat{p}} = \sqrt{(0.15)(0.85)/1540} = 0.0091$. (b) The population (U.S. adults) is considerably larger than 10 times the sample size (1540). (c) np = 231, $n(1 - p)$ = 1309—both are much bigger than 10. (d) $P(0.13 < \hat{p} < 0.17) = P(-2.198 < z < 2.198) = 0.9722$. (e) To achieve $\sigma = 0.0045$, we need a sample nine times as large: 13860.

9.27 Using normal approximation, we obtain 0.207. The exact answer is 0.213.

9.29 (a) $P(\hat{p} \le 0.70) = P(z \le -1.155) = 0.1241$. (b) $P(\hat{p} \le 0.70) = P(z \le -1.826) = 0.0339$. (c) The test must contain 1600 questions. (d) The answer is the same for Laura.

9.31 (a) $\mu_{\bar{x}} = \mu = -3.5\%$ and $\sigma_{\bar{x}} = \sigma/\sqrt{n} = 26\%/\sqrt{5} = 11.628\%$. (b) 0.3718. (c) 0.232. (d) 0.618.

9.33 (a) $\sigma_{\bar{x}} = 5.774$ mg. (b) $n = 12$. The average of several measurements is more likely than a single measurement to be close to the mean.

9.35 $\bar{x}$ has approximately a N(1.6, 0.0849) distribution; the probability is $P(z > 4.71)$—essentially 0.

9.37 (a) N(123, 0.04619). (b) $P(z > 21.65)$—essentially 0.

9.39 (a) $P(X > 295) = P(Z < -1) = 0.8413$. (b) $P(\bar{x} < 295) = P(z < -2.4495) = 0.0072$.

9.41 (a) N(2.2, 0.1941). (b) $P(z < -1.0304) = 0.1515$. (c) $P(\bar{x} < 100/52) = P(z < -1.4267) = 0.0768$.

9.43 (a) $p = 0.68$ is a parameter; $\hat{p} = 0.73$ is a statistic. (b) $\mu_{\hat{p}} = p = 0.68$, $\sigma_{\hat{p}} = \sqrt{0.68(0.32)/150} = 0.038$. (c) 0.0947.

9.45 2(0.139) = 0.278.

9.47 Although the probability of having to pay for a total loss for one or more of the 12 policies is very small, if this were to happen, it would be financially disastrous. On the other hand, for thousands of policies, the law of large numbers says that the average claim on many policies will be close to the mean, so the insurance company can be assured that the premiums they collect will (almost certainly) cover the claims.

9.49 (a) 0.3694. (b) $\mu_{\bar{x}} = 100$, $\sigma_{\bar{x}} = 1.936$. (c) 0.0049. (d) Part (a) would change. Parts (b) and (c) would not.

9.51 1.436 g/mi.

9.53 0.9537.

Chapter 10

10.1 (a) 44% to 50%. (b) We do not have information about the whole population; we only know about a small sample. We expect our sample to give us a good estimate of the population value, but it will not be exactly correct. (c) The procedure used gives an estimate within 3 percentage points of the true value in 95% of all samples.

10.3 No. The interval 267.8 to 276.2 was constructed using a method that captures the true mean NAEP score in 95% of all possible samples

10.5 0.830 to 0.851 grams per liter.

10.7 (a) The distribution is slightly skewed to the right. (b) 224.002 ± 0.029, or 223.973 to 224.031.

10.9 (a) 121.27 to 126.33. (b) Smaller. Increasing sample size reduces variability. (c) Narrower.

10.11 7.908.

10.13 69 (68.16).

10.15 (a) Yes. (b) Since the numbers are based on a voluntary response, rather than an SRS, the methods of this section cannot be used—the interval does not apply to the whole population.

10.17 (a) We can be 99% confident that between 63% and 69% of all adults favor such an amendment. (b) The survey excludes people without telephones (a large percentage of whom would be poor), so this group would be underrepresented. Also, Alaska and Hawaii are not included in the sample.

10.19 The method yields a result that is within 3 percentage points of the population value in 95% of all possible samples.

10.21 Larger sample size yields a smaller margin of error.

10.23 174 (173.19).

10.25 **(a)** Sampling error, question wording, question order. **(b)** Only sampling error.

10.27 **(a)** N(115, 6). **(b)** The actual result lies out toward the right tail of the curve, while 118.6 is fairly close to the middle. Assuming H_0 is true, observing a value like 118.6 would not be surprising, but 125.7 is less likely, and therefore provides evidence against H_0.

10.29 H_0: $\mu = 5$ mm; H_a: $\mu \uparrow 5$ mm.

10.31 H_0: $\mu = 50$; H_a: $\mu < 50$

10.33 **(a)** 0.2743; 0.0373. **(b)** A *P*-value of 0.2743 is not significant at either level. The *P*-value 0.0373 is significant at the $\alpha = 0.05$ level, but not at the $\alpha = 0.01$ level.

10.35 **(a)** $\bar{x} = 398$. **(b)** Because the population distribution is normal. **(c)** 0.0105. **(d)** It is significant at $\alpha = 0.05$, but not at $\alpha = 0.01$. This is pretty convincing evidence against H_0.

10.37 A difference in earnings as large as we observed in our sample would rarely occur if there was no difference in the average earnings of men and women. Meanwhile, the average earnings of blacks and whites in our sample were so close together that such results could easily arise when mean black and white incomes were equal.

10.39 $z = -0.7893$, $P = 0.2150$; this is reasonable variation when the null hypothesis is true, so we fail to reject H_0.

10.41 Use the command `rand(100)→L1` (`tistat.rand83(100)→list1` on the TI-89) to generate 100 random numbers in the interval (0, 1) and store them in L_1/list1. Then calculate 1-variable statistics.

10.43 **(a)** 99.86 to 108.41. **(b)** Because 105 falls in this 90% confidence interval, we cannot reject H_0: $\mu = 105$ in favor of H_a: $\mu \uparrow 105$.

10.45 **(a)** N(0, 0.113). **(b)** $\bar{x} = 0.09$ is very close to the hypothesized value of $\mu = 0$. $\bar{x} = 0.27$ lies far out in the right tail of the distribution, providing evidence against H_0. **(c)** 0.4274. **(d)** 0.0173.

10.47 H_0: $\mu = 18$ vs. H_a: $\mu < 18$.

10.49 **(a)** No. **(b)** No.

10.51 Significance at the 1% level means that the P-value for the test is less than 0.01. So it must also be less than 0.05. A result that is significant at the 5% level, may or may not be significant at the 1% level.

10.53 **(a)** $z > 1.645$. **(b)** $z < -1.96$ or $z > 1.96$. **(c)** In part **(a)**, we are performing a one-sided test. Values of z in the upper 5% of the standard normal curve provide evidence against H_0. Part **(b)** requires a two-sided test. Values of z in the lower 2.5% or the upper 2.5% provide evidence against H_0.

10.55 Yes. $z = 2.46$; $P = 0.007$.

10.57 **(a)** $z = 1.64$; not significant at 5% level ($P = 0.0505$). **(b)** $z = 1.65$; significant at 5% level ($P = 0.0495$).

10.59 **(a)** 452.24 to 503.76. **(b)** 469.85 to 486.15. **(c)** 475.42 to 480.58.

10.61 **(a)** No—in a sample of size 500, we expect to see about 5 people who have a "P-value" of 0.01 or less. These four *might* have ESP, or they may simply be among the "lucky" ones we expect to see. **(b)** The researcher should repeat the procedure on these four to see if they perform well again.

10.63 We might conclude that customers prefer design A, but perhaps not "strongly." Because the sample size is so large, this statistically significant difference may not be of any practical importance.

10.65 (a) The small *P*-values tell us that these differences in speed are very unlikely to be due to chance variation. (b) If there was not much variation in the speeds of cars with a radar detector or in those cars without a radar detector, a small difference in mean speeds could be statistically significant.

10.67 (a) Reject H_0 if $z < -2.326$. (b) 0.01 (the significance level). (c) We fail to reject H_0 if $\bar{x} \geq 270.185$, so when $\mu = 270$, $P(\text{Type II error}) = P(\bar{x} \geq 270.185) = P\left(\frac{\bar{x}-270}{60/\sqrt{840}} \geq \frac{270.185-270}{60/\sqrt{840}}\right) = 0.4644$.

10.69 (a) Type I error: You conclude that $\mu > \$45,000$ and decide (incorrectly) to open the restaurant; Type II error: You conclude that the mean income in the area is too low to open the restaurant when, actually, $\mu > \$45,000$. (b) The Type I error will probably cause your restaurant to fail. A Type II error represents a missed opportunity to make money. (c) H_0: $\mu = \$45,000$ versus H_a: $\mu > \$45,000$. (d) The choice depends on your willingness to make a Type I error. (e) For $\alpha = 0.01$, $\bar{x} \geq \$46,644.98$; For $\alpha = 0.05$, $\bar{x} \geq \$46,163.09$; For $\alpha = 0.10$, $\bar{x} \geq \$45,906.20$. (f) For $\alpha = 0.01$, the probability of a Type II error is 0.308; For $\alpha = 0.05$, $P(\text{Type II error}) = 0.118$; For $\alpha = 0.10$, $P(\text{Type II error}) = 0.061$.

10.71 $z \geq 2.326$ is equivalent to $\bar{x} \geq 460.4$, so $P(\text{reject } H_0 \text{ when } \mu = 460) = P(\bar{x} \geq 460.4 \text{ when } \mu = 460) = P(z \geq 0.089) = 0.4644$. This test is not very sensitive to a 10-point increase in the mean score.

10.73 (a) 0.5086. (b) 0.9543.

10.75 (a) We reject H_0 if $\bar{x} \geq 131.46$ or $\bar{x} \leq 124.54$. Power: 0.9246. (b) Power: 0.9246 (same as (a)). Over 90% of the time this test will detect a difference of 6 (in either the positive or the negative direction). (c) The power would be higher—it is easier to detect greater differences than smaller ones.

10.77 (a) $z = -4.99$; we reject the null hypothesis and conclude (at the 1% significance level) that the concentration is not as claimed. (b) 0.8935.

10.79 The plot is reasonably symmetric for such a small sample. (b) 26.06 to 34.74. (c) H_0: $\mu = 25$ vs. H_a: $\mu > 25$; $z = 2.44$; *P*-value = 0.007. This is strong evidence against H_0.

10.81 (a) H_0: $\mu = 32$ vs. H_a: $\mu > 32$. (b) $z = 1.864$; *P*-value is 0.0312. This is strong evidence against H_0—observations this extreme would only occur in about 3 out of 100 samples if H_0 were true.

10.83 No—"$P = 0.03$" *does* mean that the null hypothesis is unlikely, but only in the sense that the evidence (from the sample) would not occur very often if H_0 were true. *P* is a probability associated with the sample, not the null hypothesis; H_0 is either true or it isn't.

10.85 (a) The difference observed in this study would occur in less than 1% of all samples if the two populations actually have the same proportion. (b) The interval is constructed using a method that is correct (i.e., contains the actual proportion) 95% of the time. (c) No—treatments were not randomly assigned, but instead were chosen by the mothers. Mothers who choose to attend a job training program may be more inclined to get themselves out of welfare.

10.87 Yes—$z = 2.40$ and $P = 0.016$. This conclusion is still correct if the population distribution is nonnormal since the sample size is greater than 30.

10.89 (a) H_0: $\mu = 300$ pounds versus H_a: $\mu < 300$ pounds. (b) Type I error: concluding that the mean breaking strength of the company's chairs is less than 300 pounds when, in fact, $\mu = 300$ is true. Type II error: concluding that the mean breaking strength of the chairs is 300 pounds when it is actually lower than 300 pounds. The Type II error could lead to more serious problems. (c) $\bar{x} \leq 295.495$ pounds. (d) 0.0224.

Chapter 11

11.1 (a) 1.790. (b) 0.0173.

11.3 (a) 2.145. (b) 0.688.

11.5 (a) The $t(2)$ curve is a bit shorter at the peak and slightly higher in the tails. (b) The $t(9)$ curve has moved toward coincidence with the standard normal curve. (c) The $t(30)$ curve cannot be distinguished from the standard normal curve. As the degrees of freedom increase, the $t(\text{df})$ curve approaches the standard normal density graph.

11.7 (a) 14. (b) 1.82 is between 1.761 ($P = 0.05$) and 2.145 ($P = 0.025$). (c) The P-value is between 0.025 and 0.05 (in fact, $P = 0.0451$). (d) $t = 1.82$ is significant at $\alpha = 0.05$ but not at $\alpha = 0.01$.

11.9 (a) The data need to be an SRS from the population of interest and the population distribution must be approximately normal. We are only told that the data come from a random sample. It is difficult to check the normality of the population with only 8 observations. A stemplot shows that there are no outliers. (b) 16.5 to 28.5 mg/100 g. (c) Yes. $t = -6.883$, $P = 0.00023$.

11.11 A test of H_0: $\mu = 1200$ mg vs. H_a: $\mu < 1200$ mg yields $t = -3.953$ and $P = 0.000167$. We reject H_0 and conclude that participating subjects did have lower calcium intakes than recommended.

11.13 (a) H_0: $\mu = 0$ vs. H_a: $\mu > 0$, where μ is the mean improvement in score (posttest – pretest). (b) The differences are slightly left-skewed, with no outliers; the t-test should be reliable. (c) $\bar{x} = 1.450$; so $t = 2.02$. With df = 19, $0.025 < P < 0.05$. This result is significant at 5% but not at 1%. We have some evidence that scores improve, but it is not overwhelming. (d) 0.211 to 2.689 points.

11.15 (a) The scores are slightly left-skewed. (b) $\bar{x} = 3.618$, $s = 3.055$, $SE_{\bar{x}} = 0.524$. (c) 2.551 to 4.684.

11.17 (a) 1.54 to 1.80. (b) We are told the distribution is symmetric; because the scores range from 1 to 5, there is a limit to how much skewness there might be. In this situation, the condition that the 17 Mexicans are an SRS from the population is the most crucial.

11.19 The distribution is slightly skewed, but there are no outliers. We do not know whether the sample of 16 crankshafts is an SRS from the population. A test of H_0: $\mu = 224$ vs. H_a: $\mu \uparrow 224$ yields $t = 0.1254$ and $P = 0.9019$, so we have very little evidence against H_0.

11.21 (a) A test of H_0: $\mu = 0$ vs. H_a: $\mu > 0$ yields $t = 43.47$. The P-value is basically 0, so we can reject H_0 and conclude that the new policy would increase credit card usage. (b) \$312.14 to \$351.86. (c) The sample size is very large, and we are told that we have an SRS. This means that outliers are the only potential snag, and there are none. (d) Make the offer to an SRS of 200 customers, and choose another SRS of 200 as a control group. Compare the increase for the two groups.

11.23 (a) The power is 0.5287. (Reject H_0 if $t > 1.833$, i.e., if $\bar{x} > 0.4811$.) (b) The power is 0.9034. (Reject H_0 if $t > 1.711$, i.e., if $\bar{x} > 0.2840$.) (c) Type I error:

you conclude that variety A has a higher mean yield when there is actually no difference. Type II error: you conclude that there is no significant difference in the mean yield of the two varieties, but variety A actually has a higher mean yield.

11.25 $t > 3.174$ or $t < -3.174$.

11.27 22.964 to 27.878.

11.29 (a) A plot of the differences (Factory-Haiti) is somewhat left-skewed. There are no outliers. We are not told whether the marked bags are an SRS from the population. A test of H_0: $\mu = 0$ vs. H_a: $\mu > 0$ yields $t = 4.959$ and $P = 1.87 \times 10^{-5}$. We reject H_0 and conclude that the vitamin C content in the WSB did decrease. (b) 3.123 to 7.544 mg/100 g. (c) A plot of the vitamin C content in bags before leaving the factory is roughly symmetric. There are no outliers. A test of H_0: $\mu = 40$ vs. H_a: $\mu \uparrow 40$ yields $t = 3.091$ and $P = 0.0047$. This provides strong evidence that the mean vitamin C content in the bags is not 40 mg/100 g. We must be a bit careful with our conclusion since we do not know whether the sample is an SRS.

11.31 (a) Randomly assign 12 (or 13) into a group which will use the right-hand knob first; the rest should use the left-hand knob first. Alternatively, for each student, randomly select which knob he or she should use first. (b) μ is the mean difference between right-handed times and left-handed times; the null hypothesis is H_0: $\mu = 0$ (no difference). The alternative is H_a: $\mu_{R-L} < 0$. (c) A plot of the differences shows no outliers or strong skewness. $\bar{x} = -13.32$ (or + 13.22), $t = \pm 2.9037$, and $P = 0.0039$. We reject H_0 in favor of H_a.

11.33 (b) A test of H_0: $\mu = 105$ vs. H_a: $\mu \uparrow 105$ yields $t = -0.3195$, $P = 0.7554$. We do not reject the null hypothesis. The mean detector reading could be 105.

11.35 (a) 2.080. (b) Reject H_0 if $|t| \geq 2.080$, i.e., if $|\bar{x}| \geq 0.133$. (c) $P(|\bar{x}| \geq 0.133) = P(\bar{x} \leq -0.133$ or $\bar{x} \geq 0.133) = P(z \leq -5.207$ or $z \geq -1.047) = 0.853$.

11.37 (a) Two samples. (b) Matched pairs.

11.39 Both sample sizes are quite large, so we do not need to worry about the normality of the corresponding populations. A test of H_0: $\mu_1 = \mu_2$ vs. H_a: $\mu_1 \uparrow \mu_2$ yields $t = 0.654$, $P = 0.514$. These data give no evidence of a male/female difference in mean social insight score.

11.41 (a) The study was a randomized comparative experiment. The large sample sizes will help ensure the accuracy of the t-procedures. A test of H_0: $\mu_N = \mu_H$ vs. H_a: $\mu_N < \mu_H$ yields $t = -3.285$, $P = 6.217 \times 10^{-4}$. These data provide strong evidence that the use of warming blankets reduces the length of a patient's hospital stay. (b) -4.162 to -1.038; the use of warming blankets reduces a patient's hospital stay by one to four days.

11.43 H_0: $\mu_1 = \mu_2$ vs. H_a: $\mu_1 > \mu_2$, where μ_1 and μ_2 are the mean number of beetles on untreated (control) plots and malathion-treated plots, respectively. $t = 5.809$, which yields $P < 0.0001$ for a $t(12)$ distribution—this is significant at the 1% level.

11.45 No answer required.

11.47 (a) H_0: $\mu_{\text{skilled}} = \mu_{\text{novice}}$ vs. H_a: $\mu_s > \mu_n$. (b) The t statistic we want is the "unequal" value: $t = 3.1583$; its P-value is 0.0052. This is strong evidence against H_0. (c) Using $t^* = 1.895$ from a $t(7)$ distribution: 0.4691 to 1.876. Using $t^* = 1.8162$ from a $t(9.8)$ distribution (calculator): 0.4982 to 1.8474.

11.49 (a) With the small sample sizes, and with no way to assess the normality of the population distributions, the t-procedures will be only approximately accurate. A test of H_0: $\mu_N = \mu_C$ vs. H_a: $\mu_N \uparrow \mu_C$ yields $t = 2.003$ and $P = 0.0650$. There is slight evidence of a difference in mean number of trials, but the result is not significant at the $\alpha = 0.05$ level. (b) –4.698 to 77.298 for df = 9; 4.368 to 68.232. The first interval (more conservative) contains 0, but the second one does not.

11.51 (a) IDX: $n = 10$, $\bar{x} = 116$, $s = 17.709$; Untreated: $n = 10$, $\bar{x} = 88.5$, $s = 6.008$. (b) df = 9.

11.53 (a) Stemplots show little skewness, but one moderate outlier (85) for the control group. Nonetheless, the *t*-procedures should be fairly reliable since the total sample size is 44. (b) A test of H_0: $\mu_t = \mu_c$ vs. H_a: $\mu_t > \mu_c$ yields $t = 2.311$. Using $t(20)$ and $t(37.9)$ distributions, P equals 0.0158 and 0.0132, respectively; reject H_0. (c) Randomization was not possible, because existing classes were used—the researcher could not shuffle the students.

11.55 (a) Both distributions are clearly right-skewed. However, there are no outliers in either group. We should be safe applying the *t*-procedures. (b) –442.4 to 368.39 for (Reference-Generic).

11.57 (a) Since the sample sizes are quite large, we should be safe applying the *t*-procedures. We know that the data come from a random sample, but not necessarily an SRS. A test of H_0: $\mu_c = \mu_u$ vs. H_a: $\mu_c > \mu_u$ yields $t = 2.646$ and $P = 0.004$. We reject H_0 and conclude that coached students gained more on average than did uncoached students. (b) 0.184 to 15.816 points. (c) A difference this small may not be practically important.

11.59 E.g.: The difference between average female (55.5) and average male (57.9) self concept scores was so small that it can be attributed to chance variation in the samples ($t = -0.83$, df = 62.8, $P = 0.4110$). In other words, based on this sample, we have no evidence that mean self concept scores differ by gender.

11.61 (a) H_0: $\mu_A = \mu_B$ vs. H_a: $\mu_A \uparrow \mu_B$; $t = (\bar{x}_A - \bar{x}_B)\Big/\sqrt{S_A^2/n_A + S_B^2/n_B}$. (b) For a $t(349)$ distribution, $t^* = 1.967$; using a $t(100)$ distribution, take $t^* = 1.984$. (c) We reject H_0 when $|\bar{x}_A - \bar{x}_B| \geq 59.48$ (using $t^* = 1.967$). To find the power against $|\mu_A - \mu_B| = 100$, we choose *either* $\mu_A - \mu_B = 100$ or $\mu_A - \mu_B = -100$ (the probability is the same either way). Taking the former, we compute: $P[(\bar{x}_A - \bar{x}_B) \leq -59.48 \text{ or } (\bar{x}_A - \bar{x}_B) \geq 59.48] = P(z \leq -5.274 \text{ or } z \geq -1.340) = 0.9099$. Repeating these computations with $t^* = 1.984$ gives power 0.9071.

11.63 (a) The distribution is skewed left with one high outlier and one low value that is not an outlier. The mean and median are similar: $\bar{x} = 15.586$ and $M = 15.75$. $s = 2.550$. (b) 14.811 to 16.362 feet; Yes—since the interval does not contain 20. (c) How the data were obtained.

11.65 (a) Plots of the data reveal rough symmetry for the air-filled football and strong left-skewness for the helium-filled football. There were also four unusually short (outlier) kicks with the helium-filled ball. A two-sample *t*-test of H_0: $\mu_A = \mu_H$ vs. H_a: $\mu_A < \mu_H$ yields $t = -0.37$; $P = 0.356$. We fail to reject H_0; there is no significant difference in the lengths of kicks for the two balls. (b) Without the outliers, $t = -1.931$ and $P = 0.0287$. Now we might conclude that there is a significant difference in the mean distance traveled by air-filled and helium-filled footballs. (c) A time plot for each of the two balls shows an increasing trend, but there are also occasional short kicks that deviate from the trend. You could compare the lengths of the first 20 kicks and the remaining 19 kicks with each ball. For the air-filled ball, side-by-side boxplots suggest that the kicker may have improved. A two sample t-test yields $t = -1.736$ and $P = 0.046$.

11.67 (a) Using a $t(1361)$ distribution, you get \$1016.56 to \$1069.44; using a $t(2669.1)$ distribution, you get almost the same interval: \$1016.58 to \$1069.42. (b) Skewness will have little effect because the sample sizes are very large.

11.69 A 95% confidence interval for $\mu_C - \mu_F$ is –19.87 to –2.401. So maximum heart rate is between 2 and 20 beats per minute higher in the presence of a friend than alone. The 95% CI for $\mu_C - \mu_P$ is 18.203 to 37.797. Having a pet present seems to reduce maximum heart rate by 18 to 38 beats/minute. The 95% CI for $\mu_F - \mu_P$ is 28.464 to 49.803. On average, subjects' maximum heart rates will be between 28 and 50 beats/minute lower with a pet present than with a friend present.

11.71 The stemplot looks fairly symmetric except for the low outlier, 4.88. Our estimate is the mean, $\bar{x} = 5.4479$. The standard error of the mean is 0.0410; the margin of error depends on the confidence level chosen. For 90%: 0.0698; For 95%: 0.0840; For 99%: 0.1134.

11.73 (a) Plots show that both distributions are skewed right. There is one high outlier in the Active group. We do not know if the groups were assigned randomly. A test of H_0: $\mu_A = \mu_P$ vs. H_a: $\mu_A > \mu_P$ yields $t = 4.282$ and $P = 5.83 \times 10^{-5}$. We reject H_0 and conclude that active learning is superior. (b) 22.209 to 26.624 Blissymbols.

Chapter 12

12.1 (a) Population: The 175 residents of Tonya's dorm; p is the proportion who like the food. (b) $\hat{p} = 0.28$.

12.3 (a) The population is the 15,000 alumni, and p is the proportion who support the president's decision. (b) $\hat{p} = 0.38$.

12.5 (a) No—np_0 and $n(1 - p_0)$ are less than 10 (they both equal 5). (b) No—the expected number of failures is less than 10 ($n(1 - p_0) = 2$). (c) Yes—we have an SRS, the population is more than 10 times as large as the sample, and $np_0 = n(1 - p_0) = 10$.

12.7 (a) Have a TV in their room: 0.632 to 0.689; would choose Fox: 0.157 to 0.204. We are treating the sample as an SRS. The population of 13–17-year-olds is larger than 10(1048) = 10480. For both intervals, $n\hat{p}$ and $n(1 - \hat{p})$ are at least 10. (b) Both intervals have margins of error less than 3%. (c) Since $np_0 = 692(0.5)$ and $n(1 - p_0) = 692(0.5)$ are at least 10, we can perform a test of H_0: $p = 0.5$ vs. H_a: $p > 0.5$. Since $z = 10.379$ and $P = 1.577 \times 10^{-25}$, we conclude that more than half of all teenagers have a television in their room.

12.9 (a) $\hat{p} = 0.179$, $SE_{\hat{p}} = 0.0418$. (b) Checking conditions, $n\hat{p} = 84(15/84) = 15$ and $n(1 - \hat{p}) = 84(69/84) = 69$ are both at least 10. Provided that there are at least 10(84) = 840 applicants in the population of interest, we are safe constructing the confidence interval. The interval is 0.109 to 0.247. Between about 11% and 25% of applicants lie.

12.11 (a) $n = 1052$ (1051.74). (b) 1068 (1067.11); they differ by 16 people.

12.13 (a) We do not know that the examined records came from an SRS, so we must be cautious in drawing emphatic conclusions. Both $n\hat{p}$ and $n(1 - \hat{p})$ are at least 10. The interval is 0.295 to 0.339. (b) No—we only used available data. Only controlled experiments can help us infer causation.

12.15 (a) Checking conditions: we are treating the Gallup sample as an SRS; the population of U.S. adults is much larger than 10(1785); $n\hat{p} = 1785\ (750/1785)$ and $n(1 - \hat{p}) = 1785(1035/1785)$ are at least 10. We are 99% confident that the true proportion of U.S. adults who attended church or synagogue last week is between 0.39 and 0.45. (b) Since $np_0 = 1785(0.5)$ and $n(1 - p_0) = 1785(0.5)$ are both 10 or greater, we can perform a test of H_0: $p = 0.5$ vs. H_a: $p < 0.5$. $z = -6.75$ and P is practically 0. This provides strong evidence that less than half of all adults attended church or synagogue. (c) 16576.56—round up to 16577. The use of $p^* = 0.5$ is reasonable because our confidence interval shows that the actual p is in the range 0.3 to 0.7.

12.17 (a) H_0: $p = 0.5$ vs. H_a: $p > 0.5$, $z = 1.697$, $P = 0.0448$—reject H_0 at the 5% level. (b) 0.5071 to 0.7329. (c) The coffee should be presented in random order—some should get the instant coffee first, and others the fresh-brewed first.

12.19 (a) A test of H_0: $p = 0.533$ vs. H_a: $p > 0.533$ yields $z = 1.67$, $P = 0.047$. There is considerable evidence that Shaq has improved his free throw shooting. (b) Type I error: Conclude that Shaq's shooting has improved when it has not; Type II error: Decide that Shaq is shooting no better when he has actually improved. (c) Power = 0.2057. (d) P(Type I error) = 0.05; P(Type II error) = 1 – Power = 0.7943.

12.21 Tonya: $\hat{p} = 0.28$, $z = -0.752$, $P = 0.2261$; Frank: $\hat{p} = 0.28$, $z = -1.998$, $P = 0.0233$; Sarah: $\hat{p} = 0.28$, $z = -2.378$, $P = 0.0088$. Although Tonya, Frank, and Sarah all recorded the same sample proportion, $\hat{p} = 0.28$, the P-values were all quite different. Conclude: for a given sample proportion, the larger the sample size, the smaller the P-value.

12.23 –0.0518 to 0.3573. The population of mice are certainly more than 10 times as large as the samples, and the counts of successes and failures are more than 5 in both samples.

12.25 A test of H_0: $p_A = p_E$ vs. H_a: $p_A > p_E$ yields $z = 0.98$, $P = 0.1635$. We cannot conclude that the death rates are different.

12.27 Home computer: H_0: $p_1 = p_2$ vs. H_a: $p_1 > p_2$; $z = 1.01$; $P = 0.3124$. Access at work: same hypotheses; $z = 3.90$; $P < 0.0004$. There is no evidence of a difference in home computers, but have very strong evidence of a difference at work.

12.29 (a) H_0: $p_1 = p_2$ vs. H_a: $p_1 > p_2$; the populations are much larger than the samples, and 17 (the smallest count) is greater than 5. (b) $\hat{p} = 0.0632$, $z = 2.926$, and $P = 0.0017$—the difference is statistically significant. (c) Neither the subjects nor the researchers who had contact with them knew which subjects were getting which drugs—if anyone had known, this might confound the results by letting their expectations or biases affect the results.

12.31 H_0: $p_1 = p_2$ vs. H_a: $p_1 \uparrow p_2$; $z = 0.024$, $P = 0.9805$—insufficient evidence to reject H_0.

12.33 Fatal heart attacks: $z = -2.669$, $P = 0.008$—the difference between the groups is significant; Nonfatal heart attacks: $z = -4.578$, $P = 4.7 \times 10^{-6}$—the difference is significant; Strokes: $z = 1.433$, $P = 0.152$—insufficient evidence of a difference.

12.35 (a) 0.431 to 0.449. (b) 0.102 to 0.138. (c) The z-procedures require that individual occurrences of speeding be independent.

12.37 (a) 0.091 to 0.206, but this interval may be less accurate due to nonresponse. (b) Using $p^* = 0.149$, $n = 304$. (c) Food-and-drink businesses may not be typical of all small businesses.

12.39 (a) $\hat{p}_W = 0.097$, $\hat{p}_M = 0.142$, $z = 0.88$, $P = 0.378$. (b) $z = 2.12$, $P = 0.034$. (c) For part (a), –0.156 to 0.106; for part (b), 0.001 to 0.049. Larger samples decrease the width of the confidence interval.

12.41 (a) 0.247 to 0.336; yes. (b) No. $t = -0.866$, $P = 0.387$.

12.43 0.119 to 0.151.

12.45 (a) $p_0 = 0.791$. (b) $\hat{p} = 0.390$, $z = -29.115$, $P \oplus 0$. This is striking evidence of underrepresentation by Mexican Americans on juries. (c) $z = -29.195$, $P \oplus 0$. The answers are the same.

Chapter 13

13.1 (a) (i) $0.20 < P < 0.25$; (ii) $P = 0.235$. (b) (i) $0.02 < P < 0.025$; (ii) $P = 0.0204$. (c) (i) $P > 0.25$; (ii) $P = 0.317$.

13.3 $\chi^2 = 5.429$, df = 2, $P = 0.066$—there is insufficient evidence to reject the hypothesized genetic model.

13.5 (a) Use a χ^2 goodness-of-fit test. (b) Use a one-proportion z test. (c) You can construct the interval; however, your ability to generalize may be limited by the fact that your sample of bags is not an SRS. M&M's may be packaged by weight rather than count.

13.7 (a) H_0: $P_0 = P_1 = P_2 = \cdots = P_9 = 0.1$ vs. H_a: at least one of the P_i's $\uparrow 0.1$. (b) Using `randInt(0,9,200)→L₄`, we obtained these counts for digits 0 to 9: 19, 17, 23, 22, 19, 20, 25, 12, 27, 16. (e) $\chi^2 = 8.9$, df = 9, P-value = 0.447. There is no evidence that the sample data were generated from a distribution that is different from the uniform distribution.

13.9 (a) H_0: The distribution of heads and tails from spinning a 1982 penny shows equally likely outcomes vs. H_a: Heads and tails are not equally likely; df = 1; $\chi^2 = 9.68$; P-value = 0.0019. We reject H_0 and conclude that spinning a 1982 penny does not produce equally likely results. (b) Let p = the proportion of all spins that result in tails. H_0: $p = 0.5$ vs. H_a: $p \uparrow 0.5$; $z = 3.111$; P-value = 0.0019. (c) The results are identical.

13.11 H_0: Trix flavors are uniformly distributed vs. H_a: The flavors are not uniformly distributed; df = 5 – 1 = 4; $\chi^2 = 47.57$; P-value $\oplus 0$. Reject H_0 and conclude that either the Trix flavors are not uniformly distributed, or our box of Trix is not a *random* sample.

13.13 H_0: The wheel is balanced (the four outcomes are uniformly distributed) vs. H_a: The wheel is not balanced; df = 3; $\chi^2 = 24$. The P-value is 2.5×10^{-5}. Reject H_0 and conclude that the wheel is not balanced. Since Part IV: Win nothing shows the greatest deviation from the expected result, there may be reason to suspect that the carnival game operator may have tampered with the wheel to make it harder to win.

13.15 (a) $r = 4$, $c = 2$. (b) Females: HSC–HM: $\hat{p} = 0.209$, HSC–LM: $\hat{p} = 0.104$, LSC–HM: $\hat{p} = 0.313$, LSC–LM: $\hat{p} = 0.373$; Males: HSC–HM: $\hat{p} = 0.463$, HSC–LM: $\hat{p} = 0.269$, LSC–HM: $\hat{p} = 0.075$; LSC–LM: $\hat{p} = 0.194$.
(c)

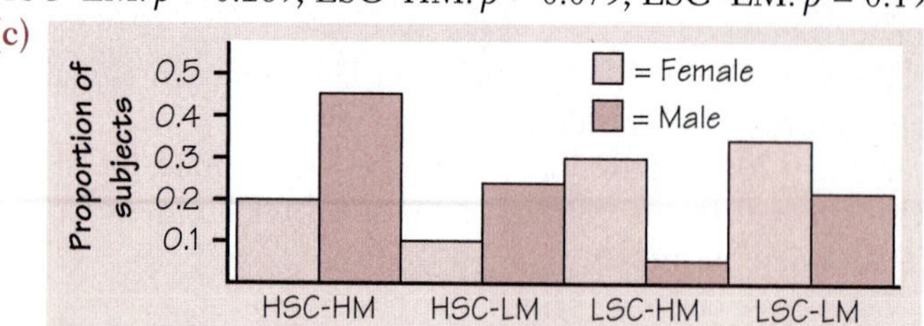

(d) HSC–LM: 12.5 for both genders; LSC–HM: 13.0 for both genders; LSC–LM: 19 for both genders. (e) There are fairly large differences between the observed and expected counts in every cell. The bar graph shows the discrepancy between male and female sports goals quite clearly.

13.17 (a) $\chi^2 = 3.211 + 3.211 + 2.42 + 2.42 + 4.923 + 4.923 + 1.895 + 1.895 = 24.898$. (b) $P < 0.0005$—reject H_0. We conclude that male and female distributions of sports goals differ significantly. (c) The two low social comparison–high mastery cells contribute most. We see that males are more likely to be motivated by social comparison goals and females are most likely to be motivated by mastery goals.

13.19 (a) 2×3. (b) 0.55, 0.747, 0.375—Some (but not too much) time spent in extracurricular activities seems to be beneficial. (d) H_0: There is no association between amount of time spent on extracurricular activities and grades earned in the course vs. H_a: There is an association. (e) Expected counts—C or better: 13.78, 62.71, 5.51; D or F: 6.22, 28.29, 2.49. (f) The first and last column have lower numbers than we expect in the "passing" row (and higher numbers in the "failing" row), while the middle column has this reversed—more passed than we would expect if all proportions were equal.

13.21 (a) Missing numbers: A, B, C, row—62.71, 5.51; D or F row—6.22; Chi-Sq = entries—0.447, 0.991. (b) df = 2; P-value = 0.0313. (c) The highest contribution comes from row 2, column 3 ("> 12 hours of extracurricular activities, D or F in the course"). (d) No—this study demonstrates association, not causation. Certain types of students may tend to spend a moderate amount of time in extracurricular activities and also work hard on their classes—one does not necessarily cause the other.

13.23 A test of H_0: there is no association between sleeping pattern and access to a car versus H_a: there is an association yields $\chi^2 = 5.915$ and $P = 0.052$. There is borderline evidence of an association between car access and sleeping pattern.

13.25 (a) Cardiac event: 3, 7, 12; No cardiac event: 30, 27, 28. (b) 0.9091, 0.7941, 0.7000. (c) 6.79, 6.99, 8.22; 26.21, 27.01, 31.78. All are greater than 5. (d) $P = 0.089$; we don't have significant evidence of a difference in success rates.

13.27 H_0: all proportions are equal versus H_a: some proportions are different. Phone interviews: 168 yes, 632 no; one-on-one: 200 and 600; Anonymous: 224 and 576. $\chi^2 = 10.619$ with 2 df, and $P = 0.0049$—good evidence against H_0, so we conclude that contact method makes a difference in response.

13.29 (a) H_0: $p_1 = p_2$, where p_1 and p_2 are the proportions of women customers in each city. $\hat{p}_1 = 0.8423$, $\hat{p}_2 = 0.6881$, $z = 3.9159$, and $P = 0.00009$. (b) $\chi^2 = 15.334$, which equals z^2. With 1 df, Table E tells us that $P < 0.0005$; a statistical calculator gives $P = 0.00009$. (c) 0.0774 to 0.2311.

13.31 The observed frequencies, marginal percents, and expected numbers of scores for the sample: "5"—167, 31.2, 81.855; "4"—158, 29.5, 117.7; "3"— 101, 18.9, 132.68; "2"—79, 14.8, 105.93; "1"—30, 5.6, 96.835. H_0: The distribution of scores in this sample is the same distribution of scores for all students who took this inaugural exam vs. H_a: The distribution of scores in this sample is different from the national results. df = 4; $\chi^2 = 162.9$; P-value ≈ 0. Reject H_0 and conclude that the distribution of

AP Statistics exam scores in this sample is different from the national distribution.

13.33 (a) H_0: $p_1 = p_2$ vs. H_a: $p_1 < p_2$. The z test must be used because the chi-square procedure will not work for a one-sided alternative. (b) $z = -2.8545$ and $P = 0.0022$. Reject H_0; there is a strong evidence in favor of H_a.

13.35 (a) Cold: Hatched = 16, Not = 11; Neutral: Hatched = 38, Not = 18; Hot: Hatched = 75, Not = 29. (b) Cold: 59.3%, Neutral: 67.9%, Hot: 72.1%. A lower percentage hatched in cold water, but the majority of eggs still hatched. (c) A test of H_0: $P_C = P_N = P_H$ vs. H_a: at least two of the proportions are not equal yields $\chi^2 = 1.703$ and $P = 0.427$. There is insufficient evidence to reject the null hypothesis that the proportions are equal.

13.37 (a) No. No treatment was imposed. (b) Among those who did not own a pet, 71.8% survived, while 94.3% of pet owners survived. Overall, 84.8% of the patients survived. (c) H_0: there is no relationship between patient status and pet ownership vs. H_a: there is a relationship between survival and pet ownership; $\chi^2 = 8.851$, df = 1, so $0.0025 < P < 0.005$. (d) Provided we believe there are no confounding or lurking variables, we reject H_0 and conclude that owning a pet improves survival. (e) A test of H_0: $p_1 = p_2$ vs. H_a: $p_1 > p_2$ yields $z = -2.975$ and $P = 0.0015$. The P-value for this test is half that of the χ^2-test.

13.39 (a) Yes, the evidence is very strong that a higher proportion of men died ($\chi^2 = 332.205$, df = 1). Possibly many sacrificed themselves out of a sense of chivalry (like Leonardo di Caprio). (b) For women, $\chi^2 = 103.767$, df = 2—a very significant difference. (c) For men, $\chi^2 = 34.621$, df = 2—another very significant difference (though not so strong as the women's value). Men with the highest status had the highest proportion surviving (over one-third). The proportion for low-status men was only about half as big, while middle-class men fared worst (only 12.8% survived).

Chapter 14

14.1 (a) $r = 0.994$, $\hat{y} = -3.660 + 1.1969x$. (b) β (estimated by 1.1969) represents how much we can expect the humerus length to increase with a 1-cm increase in femur length. The estimate of α is -3.660. (c) The residuals are -0.8226, -0.3668, 3.0425, -0.9420, and -0.9110. $s \cong 1.982$.

14.3 (a) HEIGHT = 11.547 + 0.84042 ARMSPAN. (b) The LS line is an appropriate model for the data because the residual plot shows no obvious pattern. (c) $a = 11.547$ estimates the true intercept, α. $b = 0.84042$ estimates the true slope, β. $s = 1.6128$ estimates σ.

14.5 Answers will vary.

14.7 (a) H_0: $\beta = 0$ (There is no association between number of jet skis in use and number of fatalities.) H_a: $\beta > 0$ (There is a positive association between jet skis in use and number of fatalities.) (b) The conditions are satisfied except for having independent observations. We will proceed with caution. (c) LinRegTTest (TI-83) reports that $t = 7.26$ with df = 8. The P-value is 0.000. With the earlier caveat, there is sufficient evidence to reject H_0 and conclude that there is an association between year and number of fatalities. As number of jet skis in use increases, the number of fatalities increases. (d) The confidence interval takes the form $b \pm t^*\text{SE}_b$. With $t^* = 2.8214$, and $\text{SE}_b = 0.00000913$, the 98% confidence interval is approximately (0.00004024, 0.00009176).

14.9 (a) Regression of fuel consumption on speed gives $b = -0.01466$, $\text{SE}_b = 0.02334$, and $t = -0.63$. With df = 13, we see that $P > 2(0.25) = 0.50$ (software reports 0.541), so we have no evidence to suggest a straight-line relationship. While the relationship between these two variables is very strong, it is definitely not linear.

14.11 (a) The plot shows a strong positive linear relationship. (b) β (the slope) is this rate; the estimate is listed as the coefficient of "year": 9.31868. (c) 11 df; $t^* = 2.201$; $9.31868 \pm (2.201)(0.3099) = 8.6366$ to 10.0008.

14.13 (a) The major difficulty is that the observations are not independent. The number of powerboat registrations for any year is related to the number of registrations the previous year. The other conditions can be assumed to be satisfied. (b) The confidence interval is (41.43, 49.59). The prediction interval is (33.35, 57.55). The confidence interval is more precise (i.e., narrower) since it is based on the mean of the observations, and the prediction interval is calculated for a single observation.

14.15 A scatterplot shows a positive association; $\hat{y} = 113.2 + 26.88x$. The confidence interval is 20.29 to 33.47 cal/kg; for each additional kilogram of mass, metabolic rate increases by about 20 to 33 calories. A stemplot of the residuals suggests that the distribution is right-skewed, and the largest residual may be an outlier.

14.17 (a) Stemplots and boxplots of the data show that both armspan and height are approximately normally distributed, with height slightly skewed right. It is reasonable to assume that the data are independent observations from normal populations; that for given armspan, heights would be approximately normally distributed, and that the standard deviation σ of heights is the same for all values of armspan. (b) 95% of the time, the prediction interval corresponding to armspan = 75 inches will capture the true height. (c) The 95% confidence interval corresponding to armspan = 75 inches predicts the mean height for all those individuals with armspan = 75 inches. The prediction interval establishes a range for the prediction of one student with armspan = 75, while a confidence interval establishes a range of heights for the mean height of all students with armspan = 75. Since averages have less variation than individual observations, the confidence interval will be shorter.

14.19 (a) Stumps (the explanatory variable) should be on the horizontal axis; the plot shows a positive linear association. (b) The regression line is $\hat{y} = -1.286 + 11.894x$. Regression on stump counts explains 83.9% of the variation in the number of beetle larvae. (c) Our hypotheses are H_0: $\beta = 0$ versus H_a: $\beta \uparrow 0$, and the test statistic is $t = 10.47$ (df = 21). The output shows $P = 0.000$, so we know that $P < 0.0005$ (as we can confirm from Table C); we have strong evidence that beaver stump counts help explain beetle larvae counts.

14.21 (a) U.S. returns (the explanatory variable) should be on the horizontal axis. Since both variables are measured in the same units, the same scale is used on both axes. (b) $t = b/\text{SE}_b = 0.6181/0.2369 \cong 2.609$. df = 25; since $2.485 < t < 2.787$, we know that $0.01 < P < 0.02$. Thus, we have fairly strong evidennce for a linear relationship—that is, that the slope is nonzero. (c) When $x = 15\%$,

$\hat{y} = 5.683\% + 0.6181x = 14.95\%$. For estimating an individual y-value, we use the prediction interval: –19.65% to 49.56%. **(d)** The width of the prediction interval is one indication that this prediction is not practically useful; another indication is that knowing the U.S. return accounts for only about $r^2 = 21.4\%$ of the variation in overseas returns.

14.23 **(a)** The linear relationship explains about 20.9% of the observed variation in corn yield. **(b)** A linear regressison t test of H_0: $\beta = 0$ versus H_a: $\beta < 0$ gives a t statistic of $t = -1.923$ with df = 14. The P-value is .0375. There is sufficient evidence to reject H_0 and conclude that corn yield decreases as the weeds per acre increases. **(c)** The small number of observations for each value of the explanatory variable (weeds/meter), the large variability in those observations, and the small value of r^2 will make prediction with this model somewhat imprecise. The LS equation is YIELD = 166.5 – 1.099 WEED/M. When x = 6, $\hat{y}$ = 159.9 bu/acre. The 95% confidence interval is (154.4, 165.3) from Minitab output.

14.25 **(a)** Use the prediction interval: 135.79 to 159.01 bpm. **(b)** Use df = 21 and t^* = 1.721.

14.27 **(a)** The plot shows a fairly strong curved pattern (weight increases with length). Two fish stray from the curve, but would not particularly be considered outliers. **(b)** Weight should be roughly proportional to volume; when all dimensions change by a factor of x, the volume increases by a factor of x^3. **(c)** This plot shows a strong, positive, linear association, with no particular outliers. **(d)** The correlations reflect the increased linearity of the second plot: With weight, $r^2 = 0.9207$; with weight$^{1/3}$, $r^2 = 0.9851$. **(e)** $\hat{y} = -0.3283 + 0.2330x$; $\hat{y} = 5.9623$ when x = 27 cm; 5.886 to 6.039 g$^{1/3}$. **(f)** The stemplot shows no gross violations of the assumptions, except for the high outlier for fish #143. The scatterplot suggests that variability in weight may be greater for larger lengths. Dropping fish #143 changes the regression line only slightly, and seems to alleviate both these problems (to some degree, at least).

Chapter 15

15.1 **(a)** $F^* = 3.68$. **(b)** Not significant at either 10% or 5% (in fact, $P = 0.1166$).

15.3 $F = 10.57$; P-value (for the two-sided alternative) is between 0.02 and 0.05 ($P = 0.0217$)—so this is significant at 5% but not at 1%.

15.5 H_0: $\sigma_1 = \sigma_2$; H_a: $\sigma_1 \uparrow \sigma_2$; $F = 8.68$; $0.002 < P < 0.02$. This is significant evidence that $\sigma_1 \uparrow \sigma_2$.

15.7 $F = 1.5443$; the P-value is 0.3725, so the difference is not statistically significant.

15.9 **(a)** The stemplots show no extreme outliers or skewness. **(b)** The means suggest that a dog reduces heart rate, but being with a friend appears to raise it. **(c)** $F = 14.08$; $P < 0.0005$. H_0: $\mu_P = \mu_F = \mu_C$; H_a: at least one mean is different. It appears that the mean heart rate is lowest when a pet is present, and highest when a friend is present.

15.11 **(a)** $I = 3$ (number of populations); $n_1 = n_2 = n_3 = 15$ (sample sizes from each population); $N = 45$ (total sample size). (b) $I - 1 = 2$ and $N - 1 = 42$. **(c)** Since $F > 9.22$, $P < 0.001$.

15.13 (a) Populations: tomato varieties; response: yield. $I = 4$; $n_1 = \cdots = n_4 = 10$; N = 40; 3 and 36 degrees of freedom. (b) Populations: consumers (responding to different package designs); response: attractiveness rating. $I = 6$; $n_1 = \cdots = n_6 = 120$; N = 720; 5 and 714 degrees of freedom. (c) Population: dieters (under different diet programs); response: weight change after six momths. $I = 3$; $n_1 = n_2 = 10$, $n_3 = 12$; N = 32; 2 and 29 degrees of freedom.

15.15 (a) Single men earn considerably less than the other group; widowed and married men earn the most. (b) Yes: 8119/5731 = 1.42. (c) 3 and 8231. (d) With large sample sizes, even *small* differences would be significant. (e) No: Age is the lurking variable.

15.17 No answer required.

15.19 (a) 32.32, 34.58, 35.51; 35.51/32.32 = 1.10. (b) MSG = 96.41, MSE = 1216, $F = 0.08$; df 2 and 126; $P > 0.100$. This is not enough evidence that nest temperature affects mean weight.

15.21 (a) The distribution for perch is slightly higher than the other two, and has an extreme high outlier (20.9); the bream distribution has a mild low outlier (12.0). Otherwise there is no strong skewness. (b) Bream: 12.0, 13.6, 14.1, 14.9, 15.5; Perch: 13.2, 15.0, 15.55, 16.675, 20.9; Roach: 13.3, 13.925, 14.65, 15.275, 16.1. The most important difference seems to be that perch are larger than the other two fish. It also appears that (typically) bream *may* be slightly smaller than roach.

15.23 (a) I, the number of populations, is 3; the sample sizes from the 3 populations are $n_1 = 35$, $n_2 = 55$ (after discarding the outlier), and $n_3 = 20$; the total sample size is N = 110. (b) Numerator ("factor"): $I - 1 = 2$, denominator ("error"): N – 1 = 107. (c) Since $F > 7.41$, the largest critical value for an $F(2,100)$ distribution in Table D, we conclude that $P < 0.001$.

15.25 (a) MSE =1/107 $[(34)(0.770)^2 + (54)(1.186)^2 + (19)(0.780)^2] = 107.67/107 = 1.0063$; $s_p = \sqrt{\text{MSE}} = 1.003$. (b) Use $t^* = 1.984$ from a $t(100)$ distribution (since $t(107)$ is not available). $15.747 \pm t^* . s_p / \sqrt{55} = 15.479$ to 16.015. Using software, we find that for a $t(107)$ distribution, $t^* = 1.982$; this rounds to the same interval.

15.27 (a) H_0: $\sigma_m = \sigma_w$ vs. H_a: $\sigma_m \uparrow \sigma_w$; $F = 1.16$, which gives $P = 0.3754$ (looking up values in Table D allows us to determine that $P > 0.2$). (b) No—F procedures are not robust against nonnormality, even with large samples.

15.29 (a) Only Design A allows use of one-way ANOVA. This procedure requires that the data come from independent random samples. In Design B, the data are paired within subjects. (b) The ANOVA F-test yields a P-value of 0.003. There is strong evidence that the mean percent heard correctly is not equal for the four lists. The 95% confidence intervals suggest that list 1 was easier to remember than list 3 or 4.

15.31 (a) Unlogged: $\bar{x} = 17.5$, $s = 3.529$; 1 year: $\bar{x} = 11.75$, $s = 4.372$; 8 years: $\bar{x} = 13.67$, $s = 4.5$. The rule of thumb concerning standard deviation is satisfied. It appears that the logged plots have noticeably fewer tree species than the unlogged plots. (b) $F = 6.02$, $P = 0.006$. There is considerable evidence that logging has reduced the number of tree species.

15.33 (a) $t = -0.323$, $P > 0.50$ ($P = 0.749$). (b) MSG = 0.02652, MSE = 0.237, $F = 0.112$, $P = 0.749$. (c) The P-values are the same.

15.35 (a) Normal probability plots for the three groups show that the data look reasonably normal. Because the largest standard deviation (3.34) is less than twice the smallest ($2 \times 2.69 = 5.38$), our rule of thumb tells us that we need not be concerned about violating the condition that the three populations have the same standard deviation. H_0: $\mu_1 = \mu_2 = \mu_3$; H_a: one of the means is different. With df 2 and 63, $F = 1.13$ and $P = 0.3288$. We have no evidence to reject the null hypothesis that the three populations have equal means. (b) The ANOVA null hypothesis is H_0: $\mu_B = \mu_D = \mu_S$ where the subscripts correspond to the group labels Basal, DRTA, and Strat. Figure 15.15 shows that $F = 4.48$ with degrees of freedom 2 and 63. The P-value is 0.0152. We have good evidence against H_0.

index

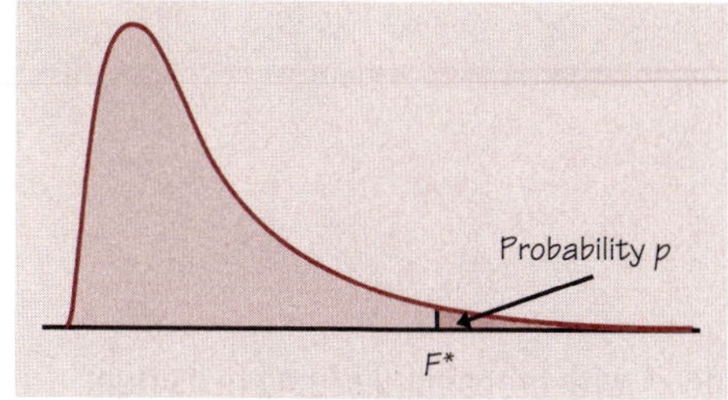

Table entry for p is the critical value F^* with probability p lying to its right.

TABLE D ***F* distribution critical values**

		Degree of freedom in the numerator							
Degrees of freedom in the denominator	p	1	2	3	4	5	6	7	8
	.100	39.86	49.50	53.59	55.83	57.24	58.20	58.91	59.44
	.050	161.45	199.50	215.71	224.58	230.16	233.99	236.77	238.88
1	.025	647.79	799.50	864.16	899.58	921.85	937.11	948.22	956.66
	.010	4052.2	4999.5	5403.4	5624.6	5763.6	5859	5928.4	5981.1
	.001	405284	500000	540379	562500	576405	585937	592873	598144
	.100	8.53	9.00	9.16	9.24	9.29	9.33	9.35	9.37
	.050	18.51	19.00	19.16	19.25	19.30	19.33	19.35	19.37
2	.025	38.51	39.00	39.17	39.25	39.30	39.33	39.36	39.37
	.010	98.50	99.00	99.17	99.25	99.30	99.33	99.36	99.37
	.001	998.50	999.00	999.17	999.25	999.30	999.33	999.36	999.37
	.100	5.54	5.46	5.39	5.34	5.31	5.28	5.27	5.25
	.050	10.13	9.55	9.28	9.12	9.01	8.94	8.89	8.85
3	.025	17.44	16.04	15.44	15.10	14.88	14.73	14.62	14.54
	.010	34.12	30.82	29.46	28.71	28.24	27.91	27.67	27.49
	.001	167.03	148.50	141.11	137.10	134.58	132.85	131.58	130.62
	.100	4.54	4.32	4.19	4.11	4.05	4.01	3.98	3.95
	.050	7.71	6.94	6.59	6.39	6.26	6.16	6.09	6.04
4	.025	12.22	10.65	9.98	9.60	9.36	9.20	9.07	8.98
	.010	21.20	18.00	16.69	15.98	15.52	15.21	14.98	14.80
	.001	74.14	61.25	56.18	53.44	51.71	50.53	49.66	49.00
	.100	4.06	3.78	3.62	3.52	3.45	3.40	3.37	3.34
	.050	6.61	5.79	5.41	5.19	5.05	4.95	4.88	4.82
5	.025	10.01	8.43	7.76	7.39	7.15	6.98	6.85	6.76
	.010	16.26	13.27	12.06	11.39	10.97	10.67	10.46	10.29
	.001	47.18	37.12	33.20	31.09	29.75	28.83	28.16	27.65
	.100	3.78	3.46	3.29	3.18	3.11	3.05	3.01	2.98
	.050	5.99	5.14	4.76	4.53	4.39	4.28	4.21	4.15
6	.025	8.81	7.26	6.60	6.23	5.99	5.82	5.70	5.60
	.010	13.75	10.92	9.78	9.15	8.75	8.47	8.26	8.10
	.001	35.51	27.00	23.70	21.92	20.80	20.03	19.46	19.03
	.100	3.59	3.26	3.07	2.96	2.88	2.83	2.78	2.75
	.050	5.59	4.74	4.35	4.12	3.97	3.87	3.79	3.73
7	.025	8.07	6.54	5.89	5.52	5.29	5.12	4.99	4.90
	.010	12.25	9.55	8.45	7.85	7.46	7.19	6.99	6.84
	.001	29.25	21.69	18.77	17.20	16.21	15.52	15.02	14.63
	.100	3.46	3.11	2.92	2.81	2.73	2.67	2.62	2.59
	.050	5.32	4.46	4.07	3.84	3.69	3.58	3.50	3.44
8	.025	7.57	6.06	5.42	5.05	4.82	4.65	4.53	4.43
	.010	11.26	8.65	7.59	7.01	6.63	6.37	6.18	6.03
	.001	25.41	18.49	15.83	14.39	13.48	12.86	12.40	12.05

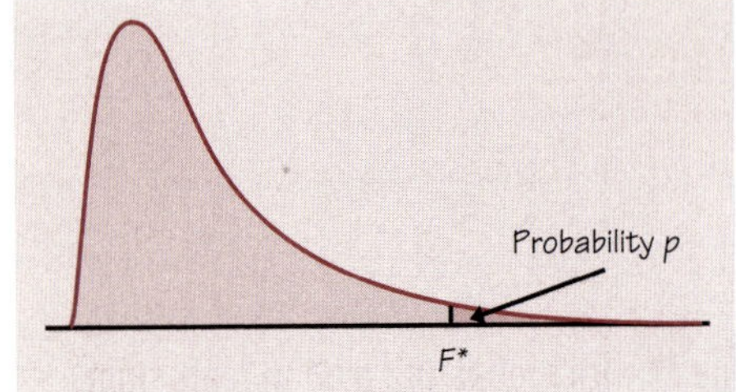

Table entry for p is the critical value F^* with probability p lying to its right.

TABLE D ***F*** **distribution critical values (*continued*)**

Degrees of freedom in the denominator	p	Degrees of freedom in the numerator: 9	10	15	20	30	60	120	1000
1	.100	59.86	60.19	61.22	61.74	62.26	62.79	63.06	63.30
	.050	240.54	241.88	245.95	248.01	250.10	252.20	253.25	254.19
	.025	963.28	968.63	984.87	993.10	1001.4	1009.8	1014	1017.7
	.010	6022.5	6055.8	6157.3	6208.7	6260.6	6313	6339.4	6362.7
	.001	602284	605621	615764	620908	626099	631337	633972	636301
2	.100	9.38	9.39	9.42	9.44	9.46	9.47	9.48	9.49
	.050	19.38	19.40	19.43	19.45	19.46	19.48	19.49	19.49
	.025	39.39	39.40	39.43	39.45	39.46	39.48	39.49	39.50
	.010	99.39	99.40	99.43	99.45	99.47	99.48	99.49	99.50
	.001	999.39	999.40	999.43	999.45	999.47	999.48	999.49	999.50
3	.100	5.24	5.23	5.20	5.18	5.17	5.15	5.14	5.13
	.050	8.81	8.79	8.70	8.66	8.62	8.57	8.55	8.53
	.025	14.47	14.42	14.25	14.17	14.08	13.99	13.95	13.91
	.010	27.35	27.23	26.87	26.69	26.50	26.32	26.22	26.14
	.001	129.86	129.25	127.37	126.42	125.45	124.47	123.97	123.53
4	.100	3.94	3.92	3.87	3.84	3.82	3.79	3.78	3.76
	.050	6.00	5.96	5.86	5.80	5.75	5.69	5.66	5.63
	.025	8.90	8.84	8.66	8.56	8.46	8.36	8.31	8.26
	.010	14.66	14.55	14.20	14.02	13.84	13.65	13.56	13.47
	.001	48.47	48.05	46.76	46.10	45.43	44.75	44.40	44.09
5	.100	3.32	3.30	3.24	3.21	3.17	3.14	3.12	3.11
	.050	4.77	4.74	4.62	4.56	4.50	4.43	4.40	4.37
	.025	6.68	6.62	6.43	6.33	6.23	6.12	6.07	6.02
	.010	10.16	10.05	9.72	9.55	9.38	9.20	9.11	9.03
	.001	27.24	26.92	25.91	25.39	24.87	24.33	24.06	23.82
6	.100	2.96	2.94	2.87	2.84	2.80	2.76	2.74	2.72
	.050	4.10	4.06	3.94	3.87	3.81	3.74	3.70	3.67
	.025	5.52	5.46	5.27	5.17	5.07	4.96	4.90	4.86
	.010	7.98	7.87	7.56	7.40	7.23	7.06	6.97	6.89
	.001	18.69	18.41	17.56	17.12	16.67	16.21	15.98	15.77
7	.100	2.72	2.70	2.63	2.59	2.56	2.51	2.49	2.47
	.050	3.68	3.64	3.51	3.44	3.38	3.30	3.27	3.23
	.025	4.82	4.76	4.57	4.47	4.36	4.25	4.20	4.15
	.010	6.72	6.62	6.31	6.16	5.99	5.82	5.74	5.66
	.001	14.33	14.08	13.32	12.93	12.53	12.12	11.91	11.72
8	.100	2.56	2.54	2.46	2.42	2.38	2.34	2.32	2.30
	.050	3.39	3.35	3.22	3.15	3.08	3.01	2.97	2.93
	.025	4.36	4.30	4.10	4.00	3.89	3.78	3.73	3.68
	.010	5.91	5.81	5.52	5.36	5.20	5.03	4.95	4.87
	.001	11.77	11.54	10.84	10.48	10.11	9.73	9.53	9.36

TABLE D *F* **distribution critical values (*continued*)**

Degrees of freedom in the denominator	*p*	Degrees of freedom in the numerator: 1	2	3	4	5	6	7	8
9	.100	3.36	3.01	2.81	2.69	2.61	2.55	2.51	2.47
	.050	5.12	4.26	3.86	3.63	3.48	3.37	3.29	3.23
	.025	7.21	5.71	5.08	4.72	4.48	4.32	4.20	4.10
	.010	10.56	8.02	6.99	6.42	6.06	5.80	5.61	5.47
	.001	22.86	16.39	13.90	12.56	11.71	11.13	10.70	10.37
10	.100	3.29	2.92	2.73	2.61	2.52	2.46	2.41	2.38
	.050	4.96	4.10	3.71	3.48	3.33	3.22	3.14	3.07
	.025	6.94	5.46	4.83	4.47	4.24	4.07	3.95	3.85
	.010	10.04	7.56	6.55	5.99	5.64	5.39	5.20	5.06
	.001	21.04	14.91	12.55	11.28	10.48	9.93	9.52	9.20
12	.100	3.18	2.81	2.61	2.48	2.39	2.33	2.28	2.24
	.050	4.75	3.89	3.49	3.26	3.11	3.00	2.91	2.85
	.025	6.55	5.10	4.47	4.12	3.89	3.73	3.61	3.51
	.010	9.33	6.93	5.95	5.41	5.06	4.82	4.64	4.50
	.001	18.64	12.97	10.80	9.63	8.89	8.38	8.00	7.71
15	.100	3.07	2.70	2.49	2.36	2.27	2.21	2.16	2.12
	.050	4.54	3.68	3.29	3.06	2.90	2.79	2.71	2.64
	.025	6.20	4.77	4.15	3.80	3.58	3.41	3.29	3.20
	.010	8.68	6.36	5.42	4.89	4.56	4.32	4.14	4.00
	.001	16.59	11.34	9.34	8.25	7.57	7.09	6.74	6.47
20	.100	2.97	2.59	2.38	2.25	2.16	2.09	2.04	2.00
	.050	4.35	3.49	3.10	2.87	2.71	2.60	2.51	2.45
	.025	5.87	4.46	3.86	3.51	3.29	3.13	3.01	2.91
	.010	8.10	5.85	4.94	4.43	4.10	3.87	3.70	3.56
	.001	14.82	9.95	8.10	7.10	6.46	6.02	5.69	5.44
25	.100	2.92	2.53	2.32	2.18	2.09	2.02	1.97	1.93
	.050	4.24	3.39	2.99	2.76	2.60	2.49	2.40	2.34
	.025	5.69	4.29	3.69	3.35	3.13	2.97	2.85	2.75
	.010	7.77	5.57	4.68	4.18	3.85	3.63	3.46	3.32
	.001	13.88	9.22	7.45	6.49	5.89	5.46	5.15	4.91
50	.100	2.81	2.41	2.20	2.06	1.97	1.90	1.84	1.80
	.050	4.03	3.18	2.79	2.56	2.40	2.29	2.20	2.13
	.025	5.34	3.97	3.39	3.05	2.83	2.67	2.55	2.46
	.010	7.17	5.06	4.20	3.72	3.41	3.19	3.02	2.89
	.001	12.22	7.96	6.34	5.46	4.90	4.51	4.22	4.00
100	.100	2.76	2.36	2.14	2.00	1.91	1.83	1.78	1.73
	.050	3.94	3.09	2.70	2.46	2.31	2.19	2.10	2.03
	.025	5.18	3.83	3.25	2.92	2.70	2.54	2.42	2.32
	.010	6.90	4.82	3.98	3.51	3.21	2.99	2.82	2.69
	.001	11.50	7.41	5.86	5.02	4.48	4.11	3.83	3.61
200	.100	2.73	2.33	2.11	1.97	1.88	1.80	1.75	1.70
	.050	3.89	3.04	2.65	2.42	2.26	2.14	2.06	1.98
	.025	5.10	3.76	3.18	2.85	2.63	2.47	2.35	2.26
	.010	6.76	4.71	3.88	3.41	3.11	2.89	2.73	2.60
	.001	11.15	7.15	5.63	4.81	4.29	3.92	3.65	3.43
1000	.100	2.71	2.31	2.09	1.95	1.85	1.78	1.72	1.68
	.050	3.85	3.00	2.61	2.38	2.22	2.11	2.02	1.95
	.025	5.04	3.70	3.13	2.80	2.58	2.42	2.30	2.20
	.010	6.66	4.63	3.80	3.34	3.04	2.82	2.66	2.53
	.001	10.89	6.96	5.46	4.65	4.14	3.78	3.51	3.30

TABLE D *F* **distribution critical values (*continued*)**

Degrees of freedom in the denominator	*p*	9	10	15	20	30	60	120	1000
		Degrees of freedom in the numerator							
9	.100	2.44	2.42	2.34	2.30	2.25	2.21	2.18	2.16
	.050	3.18	3.14	3.01	2.94	2.86	2.79	2.75	2.71
	.025	4.03	3.96	3.77	3.67	3.56	3.45	3.39	3.34
	.010	5.35	5.26	4.96	4.81	4.65	4.48	4.40	4.32
	.001	10.11	9.89	9.24	8.90	8.55	8.19	8.00	7.84
10	.100	2.35	2.32	2.24	2.20	2.16	2.11	2.08	2.06
	.050	3.02	2.98	2.85	2.77	2.70	2.62	2.58	2.54
	.025	3.78	3.72	3.52	3.42	3.31	3.20	3.14	3.09
	.010	4.94	4.85	4.56	4.41	4.25	4.08	4.00	3.92
	.001	8.96	8.75	8.13	7.80	7.47	7.12	6.94	6.78
12	.100	2.21	2.19	2.10	2.06	2.01	1.96	1.93	1.91
	.050	2.80	2.75	2.62	2.54	2.47	2.38	2.34	2.30
	.025	3.44	3.37	3.18	3.07	2.96	2.85	2.79	2.73
	.010	4.39	4.30	4.01	3.86	3.70	3.54	3.45	3.37
	.001	7.48	7.29	6.71	6.40	6.09	5.76	5.59	5.44
15	.100	2.09	2.06	1.97	1.92	1.87	1.82	1.79	1.76
	.050	2.59	2.54	2.40	2.33	2.25	2.16	2.11	2.07
	.025	3.12	3.06	2.86	2.76	2.64	2.52	2.46	2.40
	.010	3.89	3.80	3.52	3.37	3.21	3.05	2.96	2.88
	.001	6.26	6.08	5.54	5.25	4.95	4.64	4.47	4.33
20	.100	1.96	1.94	1.84	1.79	1.74	1.68	1.64	1.61
	.050	2.39	2.35	2.20	2.12	2.04	1.95	1.90	1.85
	.025	2.84	2.77	2.57	2.46	2.35	2.22	2.16	2.09
	.010	3.46	3.37	3.09	2.94	2.78	2.61	2.52	2.43
	.001	5.24	5.08	4.56	4.29	4.00	3.70	3.54	3.40
25	.100	1.89	1.87	1.77	1.72	1.66	1.59	1.56	1.52
	.050	2.28	2.2	2.09	2.01	1.92	1.82	1.77	1.72
	.025	2.68	2.61	2.41	2.30	2.18	2.05	1.98	1.91
	.010	3.22	3.13	2.85	2.70	2.54	2.36	2.27	2.18
	.001	4.71	4.56	4.06	3.79	3.52	3.22	3.06	2.91
50	.100	1.76	1.73	1.63	1.57	1.50	1.42	1.38	1.33
	.050	2.07	2.03	1.87	1.78	1.69	1.58	1.51	1.45
	.025	2.38	2.32	2.11	1.99	1.87	1.72	1.64	1.56
	.010	2.78	2.70	2.42	2.27	2.10	1.91	1.80	1.70
	.001	3.82	3.67	3.20	2.95	2.68	2.38	2.21	2.05
100	.100	1.69	1.66	1.56	1.49	1.42	1.34	1.28	1.22
	.050	1.97	1.93	1.77	1.68	1.57	1.45	1.38	1.30
	.025	2.24	2.18	1.97	1.85	1.71	1.56	1.46	1.36
	.010	2.59	2.50	2.22	2.07	1.89	1.69	1.57	1.45
	.001	3.44	3.30	2.84	2.59	2.32	2.01	1.83	1.64
200	.100	1.66	1.63	1.52	1.46	1.38	1.29	1.23	1.16
	.050	1.93	1.88	1.72	1.62	1.52	1.39	1.30	1.21
	.025	2.18	2.11	1.90	1.78	1.64	1.47	1.37	1.25
	.010	2.50	2.41	2.13	1.97	1.79	1.58	1.45	1.30
	.001	3.26	3.12	2.67	2.42	2.15	1.83	1.64	1.43
1000	.100	1.64	1.61	1.49	1.43	1.35	1.25	1.18	1.08
	.050	1.89	1.84	1.68	1.58	1.47	1.33	1.24	1.11
	.025	2.13	2.06	1.85	1.72	1.58	1.41	1.29	1.13
	.010	2.43	2.34	2.06	1.90	1.72	1.50	1.35	1.16
	.001	13.13	2.99	2.54	2.30	2.02	1.69	1.49	1.22

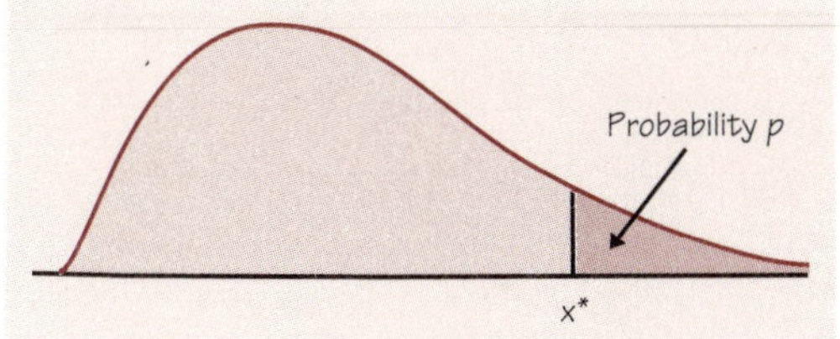

Table entry for p is the critical value x^* with probability p lying to its right.

TABLE E Chi-square distribution critical values

						p						
df	.25	.20	.15	.10	.05	.025	.02	.01	.005	.0025	.001	.0005
1	1.32	1.64	2.07	2.71	3.84	5.02	5.41	6.63	7.88	9.14	10.83	12.12
2	2.77	3.22	3.79	4.61	5.99	7.38	7.82	9.21	10.60	11.98	13.82	15.20
3	4.11	4.64	5.32	6.25	7.81	9.35	9.84	11.34	12.84	14.32	16.27	17.73
4	5.39	5.99	6.74	7.78	9.49	11.14	11.67	13.28	14.86	16.42	18.47	20.00
5	6.63	7.29	8.12	9.24	11.07	12.83	13.39	15.09	16.75	18.39	20.51	22.11
6	7.84	8.56	9.45	10.64	12.59	14.45	15.03	16.81	18.55	20.25	22.46	24.10
7	9.04	9.80	10.75	12.02	14.07	16.01	16.62	18.48	20.28	22.04	24.32	26.02
8	10.22	11.03	12.03	13.36	15.51	17.53	18.17	20.09	21.95	23.77	26.12	27.87
9	11.39	12.24	13.29	14.68	16.92	19.02	19.68	21.67	23.59	25.46	27.88	29.67
10	12.55	13.44	14.53	15.99	18.31	20.48	21.16	23.21	25.19	27.11	29.59	31.42
11	13.70	14.63	15.77	17.28	19.68	21.92	22.62	24.72	26.76	28.73	31.26	33.14
12	14.85	15.81	16.99	18.55	21.03	23.34	24.05	26.22	28.30	30.32	32.91	34.82
13	15.98	16.98	18.20	19.81	22.36	24.74	25.47	27.69	29.82	31.88	34.53	36.48
14	17.12	18.15	19.41	21.06	23.68	26.12	26.87	29.14	31.32	33.43	36.12	38.11
15	18.25	19.31	20.60	22.31	25.00	27.49	28.26	30.58	32.80	34.95	37.70	39.72
16	19.37	20.47	21.79	23.54	26.30	28.85	29.63	32.00	34.27	36.46	39.25	41.31
17	20.49	21.61	22.98	24.77	27.59	30.19	31.00	33.41	35.72	37.95	40.79	42.88
18	21.60	22.76	24.16	25.99	28.87	31.53	32.35	34.81	37.16	39.42	42.31	44.43
19	22.72	23.90	25.33	27.20	30.14	32.85	33.69	36.19	38.58	40.88	43.82	45.97
20	23.83	25.04	26.50	28.41	31.41	34.17	35.02	37.57	40.00	42.34	45.31	47.50
21	24.93	26.17	27.66	29.62	32.67	35.48	36.34	38.93	41.40	43.78	46.80	49.01
22	26.04	27.30	28.82	30.81	33.92	36.78	37.66	40.29	42.80	45.20	48.27	50.51
23	27.14	28.43	29.98	32.01	35.17	38.08	38.97	41.64	44.18	46.62	49.73	52.00
24	28.24	29.55	31.13	33.20	36.42	39.36	40.27	42.98	45.56	48.03	51.18	53.48
25	29.34	30.68	32.28	34.38	37.65	40.65	41.57	44.31	46.93	49.44	52.62	54.95
26	30.43	31.79	33.43	35.56	38.89	41.92	42.86	45.64	48.29	50.83	54.05	56.41
27	31.53	32.91	34.57	36.74	40.11	43.19	44.14	46.96	49.64	52.22	55.48	57.86
28	32.62	34.03	35.71	37.92	41.34	44.46	45.42	48.28	50.99	53.59	56.89	59.30
29	33.71	35.14	36.85	39.09	42.56	45.72	46.69	49.59	52.34	54.97	58.30	60.73
30	34.80	36.25	37.99	40.26	43.77	46.98	47.96	50.89	53.67	56.33	59.70	62.16
40	45.62	47.27	49.24	51.81	55.76	59.34	60.44	63.69	66.77	69.70	73.40	76.09
50	56.33	58.16	60.35	63.17	67.50	71.42	72.61	76.15	79.49	82.66	86.66	89.56
60	66.98	68.97	71.34	74.40	79.08	83.30	84.58	88.38	91.95	95.34	99.61	102.7
80	88.13	90.41	93.11	96.58	101.9	106.6	108.1	112.3	116.3	120.1	124.8	128.3
100	109.1	111.7	114.7	118.5	124.3	129.6	131.1	135.8	140.2	144.3	149.4	153.2

TABLE B Random digits

Line								
101	19223	95034	05756	28713	96409	12531	42544	82853
102	73676	47150	99400	01927	27754	42648	82425	36290
103	45467	71709	77558	00095	32863	29485	82226	90056
104	52711	38889	93074	60227	40011	85848	48767	52573
105	95592	94007	69971	91481	60779	53791	17297	59335
106	68417	35013	15529	72765	85089	57067	50211	47487
107	82739	57890	20807	47511	81676	55300	94383	14893
108	60940	72024	17868	24943	61790	90656	87964	18883
109	36009	19365	15412	39638	85453	46816	83485	41979
110	38448	48789	18338	24697	39364	42006	76688	08708
111	81486	69487	60513	09297	00412	71238	27649	39950
112	59636	88804	04634	71197	19352	73089	84898	45785
113	62568	70206	40325	03699	71080	22553	11486	11776
114	45149	32992	75730	66280	03819	56202	02938	70915
115	61041	77684	94322	24709	73698	14526	31893	32592
116	14459	26056	31424	80371	65103	62253	50490	61181
117	38167	98532	62183	70632	23417	26185	41448	75532
118	73190	32533	04470	29669	84407	90785	65956	86382
119	95857	07118	87664	92099	58806	66979	98624	84826
120	35476	55972	39421	65850	04266	35435	43742	11937
121	71487	09984	29077	14863	61683	47052	62224	51025
122	13873	81598	95052	90908	73592	75186	87136	95761
123	54580	81507	27102	56027	55892	33063	41842	81868
124	71035	09001	43367	49497	72719	96758	27611	91596
125	96746	12149	37823	71868	18442	35119	62103	39244
126	96927	19931	36809	74192	77567	88741	48409	41903
127	43909	99477	25330	64359	40085	16925	85117	36071
128	15689	14227	06565	14374	13352	49367	81982	87209
129	36759	58984	68288	22913	18638	54303	00795	08727
130	69051	64817	87174	09517	84534	06489	87201	97245
131	05007	16632	81194	14873	04197	85576	45195	96565
132	68732	55259	84292	08796	43165	93739	31685	97150
133	45740	41807	65561	33302	07051	93623	18132	09547
134	27816	78416	18329	21337	35213	37741	04312	68508
135	66925	55658	39100	78458	11206	19876	87151	31260
136	08421	44753	77377	28744	75592	08563	79140	92454
137	53645	66812	61421	47836	12609	15373	98481	14592
138	66831	68908	40772	21558	47781	33586	79177	06928
139	55588	99404	70708	41098	43563	56934	48394	51719
140	12975	13258	13048	45144	72321	81940	00360	02428
141	96767	35964	23822	96012	94591	65194	50842	53372
142	72829	50232	97892	63408	77919	44575	24870	04178
143	88565	42628	17797	49376	61762	16953	88604	12724
144	62964	88145	83083	69453	46109	59505	69680	00900
145	19687	12633	57857	95806	09931	02150	43163	58636
146	37609	59057	66967	83401	60705	02384	90597	93600
147	54973	86278	88737	74351	47500	84552	19909	67181
148	00694	05977	19664	65441	20903	62371	22725	53340
149	71546	05233	53946	68743	72460	27601	45403	88692
150	07511	88915	41267	16853	84569	79367	32337	03316